Handbook of Invasive Plant-Parasitic Nematodes

Handbook of Invasive Plant-Parasitic Nematodes

Ziaul Haque and Mujeebur Rahman Khan

CABI

CABI is a trading name of CAB International

CABI
Nosworthy Way
Wallingford
Oxfordshire OX10 8DE
UK

Tel: +44 (0)1491 832111
Fax: +44 (0)1491 833508
E-mail: info@cabi.org
Website: www.cabi.org

CABI
WeWork
One Lincoln St
24th Floor
Boston, MA 02111
USA

Tel: +1 (617)682-9015
E-mail: cabi-nao@cabi.org

A catalogue record for this book is available from the British Library, London, UK.

References to Internet websites (URLs) were accurate at the time of writing.

ISBN-13: 9781789247367 (hardback)
9781789247374 (ePDF)
9781789247381 (ePub)

DOI: 10.1079/9781789247367.0000

Commissioning Editor: Rebecca Stubbs
Editorial Assistant: Lauren Davies
Production Editor: James Bishop

Typeset by SPi, Pondicherry, India
Printed and bound in the UK by Severn, Gloucester.

On the Centenary Year of Aligarh Muslim University
Dedicated to the Founder

Sir Syed Ahmad Khan

A great visionary, statesman, reformer and the founder of Aligarh Muslim University.

Contents

Foreword

Nematodes are the love of my life. They have been fascinating organisms of study for me since 1975 when I was introduced to them as a graduate student while working on the International *Meloidogyne* Project at North Carolina State University. The more I have learned about them, the more impressed I have become. Unfortunately, whenever I read a book about life on this planet, I am usually disappointed because the author(s) minimize or completely ignore Earth's most common animal, the nematodes.

Fortunately, in this book Dr. Zia-ul Haque and Prof. M. R. Khan have given plant-parasitic nematodes the recognition that they deserve by focusing attention on all of the major species in the world. In an alphabet soup from *Anguina* to *Zygotylenchus*, the authors have described details of the taxonomy, morphological and molecular identification, movement and means of spread, distribution, host range, biology and life cycle, economic importance, tactics for management, and pest risk analysis. As a plant protectionist with more than nine years of research experience Dr. Haque has studied plant pathology and nematology by integrating basic and applied science to the biological control of nematodes and plant diseases. Likewise, Dr. Khan has had a long career researching nematodes and diseases of various agricultural plants. Both of them have published numerous research papers, books, and review articles on nematodes and plant diseases.

In this book, the authors have listed the invasiveness of each species according to numerous regulatory agencies from various regions around the world. This book is a valuable resource for plant pathologists, nematologists, plant health inspectors, agronomists, horticulturalists and other agriculturalists that work with the plant-parasitic nematodes. It is a comprehensive compilation of a plethora of information about this truly fascinating group of organisms.

Jonathan D. Eisenback, Ph.D.
Professor of Nematology
Virginia Tech
Blacksburg, VA
March 12, 2021

Preface

Plant-parasitic nematodes are potential pests of agricultural crops and cause quantitative as well as qualitative loss to crop produce. The worldwide distribution of plant-parasitic nematodes poses a high risk of spreading from one country to another, with the movement of plants and planting materials. In view of worldwide trade in seeds and other planting materials, it becomes mandatory to assess the status of invasive nematodes and their quarantine importance, especially in the country of destination. Collective information on the invasive nematodes and their looming threats is not available, with the exception of a few sporadic pieces of literature. This volume compiles and updates information on invasive nematodes and their looming threat in different countries. The present book offers a global perspective on invasive nematodes by presenting 17 chapters with information on more than 100 nematodes and their potential threat in different countries. Each nematode entry includes information on: authentic identification; geographical distribution; risk of introduction; host ranges; symptoms; biology and ecology; planting material liable to carry the nematode(s) and its vector, if any; chance of establishment; likely impact; phytosanitary measures; and a detailed account of diagnosis procedures, such as sampling, isolation/detection and identification with morphological and molecular characterization. The aim of the book is to provide basic and advanced knowledge on invasive nematodes with a global perspective, and it targets practitioners, professionals, scientists, researchers, students and government officials working on plant quarantine and biosecurity with regard to plant-parasitic nematodes.

Since the major portion of this book was completed at home during the first and second wave of COVID-19 pandemic, we would like to express our deep gratitude and indebtedness to our wives, children and other family members for the neglect and loss they endured during the preparation of this book and the love and support they showered upon us. We particularly thank the CABI Commissioning Editor Rebecca Stubbs, Managing Editor Ali Thompson, Editorial Assistant Lauren Davies, production editor James Bishop and copy editor Val Porter for their valuable assistance on English improvement and other editorial suggestions. We are highly grateful to the various journals and publishers for granting permission to reproduce illustrations and to several nematologists for their help in one way or other. We acknowledge with thanks Prof. Tariq Mansoor, Vice Chancellor, Aligarh Muslim University, for providing financial assistance to pay copyright charges for some figures. We are also grateful to Prof. Jonathan Eisenback of the School of Plant and Environmental Sciences, Virginia Polytechnic and State University, for contributing the valuable Foreword to this book

15 August, 2021

Ziaul Haque
Mujeebur Rahman Khan

List of Abbreviations

General abbreviations and symbols

1-DE = one-dimensional electrophoresis
1-DGE = one-dimensional gel electrophoresis
2-D PAGE = two-dimensional polyacrylamide gel electrophoresis
2-DE = two-dimensional electrophoresis
2-DGE = two-dimensional gel electrophoresis
3′ = hydroxyl-terminus of DNA molecule
5′ = phosphate-terminus of DNA molecule
ACRKN = Asian citrus root-knot nematode
AFLP = amplified fragment length polymorphism
ALPP = area of low pest prevalence
ArMV = Arabis mosaic virus
BCA = biological control agent
bp = base pair
CCN = cereal cyst nematode
cDNA = complementary DNA
COI, COX1 = cytochrome oxidase subunit I
CRLV = cherry rasp leaf virus
CT value = cycle threshold value
cv. = cultivar
Da = Dalton
DBCP = 1,2-dibromo-3-chloropropane
DD = 1,2-dichloropropane, 1, 3-dichloropropene
DNA = deoxyribose nucleic acid
dNTP = deoxyribonucleotide tri-phosphate (mix of A, C, G + T)
DOG = dorsal oesophageal gland
DP = diagnostic protocol
dsDNA = double-stranded deoxyribose nucleic acid
EDTA = ethylenediaminetetraacetic acid
ELISA = enzyme-linked immunosorbent assay
EM = electron microscopy
FA = formal-acetic
FAA = formal aceto-alcohol
fam. = family
FP = formal-propionic
GFLV = grapevine fanleaf virus
GMO = genetically modified organisms
HCl = hydrochloric acid
h = hour(s)
HCN = hop cyst nematode
hsp = heat shock protein
HWT = hot-water treatment
IEF = isoelectric focusing
IGS = intergenic spacer
ISPM = International Standards for Phytosanitary Measures
ITS = internal transcribed spacers
J_1 = first-stage juvenile
J_2 = second-stage juvenile
J_3 = third-stage juvenile
J_4 = fourth-stage juvenile
KCl = potassium chloride
kV = kilovolt
LAMP = loop-mediated isothermal amplification
LM = light microscope
LMO = living modified organisms
LNA = locked nucleic acid
LSU = large subunit gene
M = molar
mAbs = monoclonal antibodies
MALDI-TOF = matrix-assisted laser desorption/ionization–time of flight
MCCN = Mediterranean cereal cyst nematode
mdh = malate dehydrogenase
MEGA = molecular evolutionary genetic analysis
$MgCl_2$ = magnesium chloride
min = minute
ML = maximum likelihood
mM = millimolar
MP = maximum parsimony
MS = mass spectrometry
mtCOI = mitochondrial cytochrome c oxidase subunit I
mtDNA = mitochondrial DNA
NaCl = sodium chloride
NCHDT = North Carolina Host Differential Test
NEPO = Nematode-transmitted Polyhedral (ringspot viruses)
No., no. = number
PAB = propagule pressure; abiotic factors; biotic factors
pAbs = polyclonal antibodies
PAGE = polyacrylamide gel electrophoresis
PCN = potato cyst nematode
PCR = polymerase chain reaction

PCR-SCC	=	PCR-single-strand conformational polymorphism
PD	=	population density
PEBV	=	pea early browning virus
PepRSV	=	pepper ringspot virus
PEQ	=	post-entry quarantine
PFA	=	pest-free area
PPN	=	plant-parasitic nematode
PQ	=	Plant Quarantine (regulations)
PRA	=	pest risk analysis
PRMV	=	peach rosette mosaic nepovirus
PRN	=	potato rot nematode
PT	=	phytosanitary treatment
PWN	=	pine wilt/wood nematode
qPCR	=	quantitative polymerase chain reaction
RAPD	=	random amplification of polymorphic DNA
rDNA	=	ribosomal DNA
RFLP	=	restriction fragment length polymorphism
RKN	=	root-knot nematode
RLN	=	root-lesion nematode
RNA	=	ribonucleic acid
RNQP	=	regulated non-quarantine pest
RPA	=	recombinant polymerase amplification
RSN	=	rice stem nematode
rpm	=	revolutions per minute
RRN	=	red ring nematode
rRNA	=	ribosomal ribonucleic acid
RRSV	=	raspberry ringspot virus
RSN	=	rice stem nematode
s	=	second
satDNA	=	satellite DNA
SBN	=	stem and bulb nematode
SCAR	=	sequence-characterized amplified region
SCN	=	soybean cyst nematode
SCN	=	sugar beet cyst nematode
SDNA	=	satellite DNA
SDS	=	Sequence Detection Software
SDS	=	sodium dodecyl sulfate
SDS-CGE	=	sodium dodecyl sulfate–capillary gel electrophoresis
SEM	=	scanning electron microscope
sensu	=	according to
sensu lato	=	in broad sense
sensu stricto	=	in the strict sense
SLRSV	=	strawberry latent ringspot virus
SLV	=	spoon leaf virus
sp kn	=	spear (stylet) knob
sp. (pl. spp.)	=	species
SPS	=	sanitary and phytosanitary
SSCP	=	single-strand conformational polymorphism
SSR	=	single sequence repeat
SSU	=	small subunit (gene)
subfam.	=	subfamily
subsp., ssp	=	subspecies
TAE buffer	=	tris-acetic acid–EDTA electrophoresis buffer
TAE	=	tris-acetate
TAF	=	triethanol amine formalin
Taq	=	*Thermus aquaticus* DNA polymerase
TBRV	=	tomato black ring virus
TCN	=	tobacco cyst nematode
TEM	=	transmission electron microscope
TFA	=	trifluoroacetic acid
TOBRA	=	tobacco rattle virus
TobRSV	=	tobacco ringspot nepovirus
TomRSV	=	tomato ringspot nepovirus
tris-HCL	=	tris-hydrochloric acid
TRV	=	tobacco rattle virus
UV	=	ultraviolet
V	=	volt
VAM	=	vesicular–arbuscular mycorrhizal
WLB	=	worm lysis buffer
μ	=	micro
× *g*	=	unit for relative centrifugal force

Abbreviations used for nematode morphometrics (measurements)

de Man's ratios

L = total body length of nematode

$$a = \frac{\text{body length}}{\text{maximum body width}}$$

$$b = \frac{\text{body length}}{\text{oesophageal length}}$$

$$b' = \frac{\text{body length}}{\text{distance from head end to posterior end of oesophageal gland}}$$

$$c = \frac{\text{body length}}{\text{tail length}}$$

$$c' = \frac{\text{tail length}}{\text{body width at anus}}$$

$$V = \frac{\text{distance from head end to vulva}}{\text{body length}} \times 100$$

$$T = \frac{\text{distance from the cloacal aperture to anterior end of testis}}{\text{body length}} \times 100$$

h = length of hyaline part of tail.
m = length of conus as percentage of total stylet length
O = distance between stylet base and orifice of dorsal oesophageal gland as percentage of stylet length
Ost = length of odontostylet
Oph = length of odontophore
R = total number of body annules
Roes = number of annules in oesophageal region
Rex = number of annules between anterior end of body and excretory pore
RV = number of annules between posterior end of body and vulva

RVan = number of annules between vulva and anus
Ran = number of annules on tail
st = length of stylet
VL/VB = distance between vulva and posterior end of body divided by body width at vulva

Abbreviations used for nematode morphological characters

ac	=	anterior cephalid
an	=	anus
ap sp	=	anterior protractor of spicule
bur	=	bursa
da	=	dilator of anus
dn	=	dorsal oesophageal gland nucleus
dv	=	dilator of vulva
exp	=	excretory pore
gub	=	gubernaculum
hd	=	hemizonid
lat fd	=	lateral field
mob	=	median oesophageal bulb
nr	=	nerve ring
odg	=	orifice of dorsal gland nucleus
oij	=	oesophageal-intestine junction
ov	=	ovary
ovd	=	oviduct
pc	=	posterior cephalid
pg	=	protractor of gubernaculum
pp sp	=	posterior protractor of spicule
r sp	=	retractor of spicule
scu	=	scutellum
spg	=	spear (stylet) guide
spm	=	spermatheca
ss	=	sensilla sac
svn	=	subventral gland nuclei
trc	=	tricolumella
ut c	=	uterine columella

International organizations and agencies

APHIS	=	Animal and Plant Health Inspection Service
APPPC	=	Asia and Pacific Plant Protection Commission
CABI	=	CAB (formerly Commonwealth Agricultural Bureaux) International
CAC	=	*Codex Alimentarius* Commission
CAHFSA	=	Caribbean Agricultural Health and Food Safety Agency
CAN	=	Comunidad Andina
CBD	=	Convention on Biological Diversity
CISEH	=	Center for Invasive Species and Ecosystem Health
COSAVE	=	Comité de Sanidad Vegetal del Cono Sur
CPC	=	Crop Protection Compendium
CPPC	=	Caribbean Plant Protection Commission [abolished in 2014]
EPPO	=	European and Mediterranean Plant Protection Organization
EU	=	European Union
FAO	=	Food and Agriculture Organization of the United Nations
GISD	=	Global Invasive Species Database
GISP	=	Global Invasive Species Programme
I3N	=	IABIN Invasive Information Network
IAPSC	=	Inter-African Phytosanitary Council
IPPC	=	International Plant Protection Convention
ISC	=	Invasive Species Compendium
IUCN	=	International Union for Conservation of Nature and Natural Resources
ISSG	=	Invasive Species Specialist Group
NAPPO	=	North American Plant Protection Organization
NEBN-IAS	=	North European and Baltic Network on Invasive Alien Species
NEPPO	=	Near East Plant Protection Organization
NPPO	=	National Plant Protection Organization
OIRSA	=	Organismo Internacional Regional de Sanidad Agropecuaria
PCIT	=	Phytosanitary Certificate Issuance and Tracking System
PPPO	=	Pacific Plant Protection Organization
RPPO	=	Regional Plant Protection Organization
SCOPE	=	Scientific Committee on Problems of the Environment
UNEP	=	United Nations Environment Programme
USDA	=	United State Department of Agriculture
WTO	=	World Trade Organization

Authors Biography

Dr. Ziaul Haque

Dr. Haque is an Assistant Professor in the Department of Plant Protection, Faculty of Agricultural Science, Aligarh Muslim University, India. He has a PhD in Plant Pathology & Nematology, and over a decade of research and teaching experience in plant protection. He has contributed over 30 research articles in peer-reviewed journals and also published five books. He is member of six scientific bodies/organizations, reviewer to some journals and recipient of Maulana Azad National Fellowship, Sir Syed Ahmad Khan Memorial Award and commendation of Prof. M.J. Narsimhan Merit Award. His research focuses on the management of plant diseases caused by fungi and nematodes with integrating basic and applied science.

Dr. Mujeebur Rahman Khan

Dr. Khan is a Professor of Nematology in the Department of Plant Protection, Aligarh Muslim University, India, and a fellow of the National Academy of Agricultural Sciences. He obtained his Ph.D. (Plant Nematology) in 1988, Post doc. at the North Carolina State University and California Department of Food and Agriculture, U.S.A. in 1988-89 and also trained at the Commonwealth Institute of Parasitology, U.K. Dr. Khan has vast experience of teaching and research in plant protection, and has over 300 research publications and review articles, 14 books, 2 patents and 8 research projects to his credit. He has conferred with some of the most prestigious awards for research in Plant Pathology & Nematology in India. His major area of research is biocontrol of plant diseases; entomopathogenic nematodes, impact of climate change on plant diseases, and nanotechnology in plant disease management.

Introduction

I.1 Plant-parasitic Nematodes

Nematodes are vermiform or thread-like animals and belong to the most diverse group of pseudocoelomates. The nematode body is thin, flexible, generally elongated (0.3–11 mm long), cylindrical to fusiform and tapering toward both ends. By and large, nematodes are the most abundant, ubiquitous, multicellular organisms that are found in all types of environments, be it freshwater, marine, or terrestrial environments. Primarily, nematodes are aquatic animals thriving best in water but they have adapted to terrestrial habitats and constitute 80–90% of all the multicellular animals (Khan, 2008). Nematodes parasitize a large number of organisms, including plants, fungi, animals and even human beings, and the rest are free-living, surviving on various substrates. Fortunately, only a fraction of this number possesses the ability to parasitize plants. The plant-parasitic nematodes (PPNs) are of greater importance, as they reduce the quantity and quality of crop produce (Khan, 2016). They are one of the important groups of pathogens that infest and affect crop production. Nematodes generally attack roots by feeding on the epidermal cells, root hairs, cortical and/or stealer cells (Decraemer and Hunt, 2013). They usually cause a severe reduction in plant growth by impairing the absorption capacity of the root system (Khan *et al*., 2009). Nematode genera such as *Tylenchorhynchus*, *Rotylenchus*, *Helicotylenchus*, *Belonolaimus*, *Hoplolaimus*, *Trichodorus*, *Longidorus*, *Xiphinema*, etc., are ectoparasites and feed on the root surface (Siddiqi, 2000). Some nematodes, such as reniform nematodes (*Rotylenchulus reniformis*) and citrus nematodes (*Tylenchulus semipenetrans*), are semi-endoparasites as they partially enter the host tissue. Many nematodes fully enter the host root and are called endoparasites, such as root-knot nematodes (*Meloidogyne* spp.), cyst-forming nematodes (*Heterodera* and *Globodera* spp.), root-lesion nematodes (*Pratylenchus* spp.) and burrowing nematodes (*Radopholus similis*). A few nematodes, such as *Anguina*, *Aphelenchoides* and *Ditylenchus*, which attack stem, leaves, flowers, seeds, etc., are called foliar nematodes (Khan, 2008).

Nematodes generally cause hidden damage to the plant and the most common effect is the debilitation of the plant without any discernible symptoms (Sasser, 1989). In fields, symptoms of nematode infestation generally appear as patches of irregularly distributed plants showing stunted growth and sparse dull green or pale yellow foliage (Luc *et al*., 2005). However, in heavily infested fields discernible and characteristic symptoms appear on roots or shoots. Specific symptoms develop more frequently on roots, because most nematodes are root feeders, but also appear on the foliage depending on the nematode species. Specific symptoms are root-knots (*Meloidogyne*, *Nacobbus*, *Xiphinema*, *Longidorus* spp.), root lesions (*Pratylenchus*, *Hirschmanniella* spp.), root-rot (*Radopholus* spp.), root pruning (*Trichodorus*, *Hemicycliophora* spp.), cessation of root growth (*Tylenchorhynchus*, *Criconema* spp.), etc. (Khan, 2008). Some nematodes cause recognizable symptoms on above-ground parts, such as seed and leaf galls (*Anguina tritici*), leaf necrosis (*Aphelenchoides ritzemabosi*), white leaf tip (*Aphelenchoides besseyi*), flower bud deformation (*Ditylenchus dipsaci*, *Aphelenchoides fragariae*) and wilt (*Bursaphelenchus xylophilus*). In addition to direct damage, nematodes often aid or exaggerate diseases caused by bacteria, fungi and viruses or may break the resistance of cultivars to pathogens (Khan, 1993). Wilt-resistant crop cultivars may become susceptible to the wilt-inducing fungi (Francl and Wheeler, 1993) and bacteria (Sitaramaiah and Pathak, 1993) in the presence of plant-parasitic nematodes. Plant nematodes may also act as vectors for bacteria, fungi and viruses. The hairy root of roses caused by *Agrobacterium rhizogenes* is of minor importance, but in the presence of *Pratylenchus vulnus* the disease becomes severe (Sitaramaiah and Pathak, 1993). *Anguina tritici* carries *Clavibacter tritici* and *Dilophospora alopecuri* to the shoot meristems of wheat (Khan and Dasgupta, 1993). Ringspot viruses (NEPO viruses), e.g. tobacco ringspot virus, are transmitted by *Xiphinema* and *Longidorus* spp. *Trichodorus* and *Paratrichodorus* spp. act as vectors for tobacco rattle (TOBRA) virus, pepper ringspot virus and pea early browning viruses (Taylor and Brown, 1997).

I.2 Agricultural Importance of Plant-parasitic Nematodes

Plant-parasitic nematodes (PPNs) are potential pests of agricultural crops and their economic importance comes next to fungi, bacteria and viruses. PPNs attack all kinds of agricultural crops belonging to cereals, pulses, vegetables, ornamentals, forage, fibre, plantation, medicinal, spices, aromatic plants and forest trees. Depending on the species, the

DOI: 10.1079/9781789247367.0001

nematode infects roots, stems, bark, leaves, crowns, inflorescence, flowers, seeds and fruits. The crop damage and subsequent yield loss depend on the plant species or cultivar, nematode species, level of soil infestation and the prevailing environmental conditions (Khan, 2016). The nematode infestation results in the suppression of plant growth and yield, both quantitatively and qualitatively. Annual and seasonal plants are relatively more susceptible to nematode infestation because of their soft and succulent roots and other parts, and because their cultivation under high soil moisture conditions generally favours nematode infestation. This is not invariably true, as nematodes such as *Pratylenchus coffeae*, *Meloidogyne* spp., *Bursaphelenchus cocophilus* and *B. xylophilus* also cause severe damage to perennials, palm trees and conifers (Khan, 2020). Globally PPNs cause 8.8–14.6% yield loss to different crops annually (Nicol *et al.*, 2011). The monetary value of food loss due to nematode attack on crops has been estimated at over US$100 billion per year globally (Sikora *et al.*, 2018). Yield losses vary greatly, depending on the inoculum level and host species (Khan *et al.*, 2009). In tropical and subtropical climates, crop production losses attributable to PPN have been estimated at 14.6% (Nicol *et al.*, 2011). In a specific region or crop, the loss may be much greater. A severe infection may result in as much as 80–90% yield decline in an individual field and sometimes the entire crop fails to give yield of economic value (Table 1). In the USA alone, annual monetary loss due to nematodes is more than US$6 billion (Nicol *et al.*, 2011). In the UK in 2010, DEFRA estimated that the cyst nematodes *Globodera rostochiensis* and *G. pallida* were responsible for crop losses worth about US$70 million, or 9% of UK production per annum. Crop losses due to PPNs are greater in developing countries than in developed countries. In India, about 10–20% of crop losses are due to nematode infestations (Khan, 2008). All agricultural crops are susceptible to nematode attack, and the literature reveals knowledge of numerous PPN species that cause economically significant yield losses to important agricultural crops. The aetiology and damage potential of only 5% of known PPNs have been properly studied (Table 1) and the majority of other known species are yet to be explored for their economic significance.

I.3 Nematode Diversity and Global Agricultural Trade

Despite invariant body plan, nematodes show astounding biodiversity and inhabit a wide range of habitats and ecosystems (Yeates *et al.*, 2009). They are mostly found in oceans, lakes and the soil of all geographical regions, from snowy mountains to deserts (Khan, 2008). In soil, they are most common of all the soil fauna, numbering 1.8–120 million per square metre of the topsoil (Khan and Jairajpuri, 2010a). There are approximately 25,000 described nematode species (Hodda, 2011), but total nematode species present on Earth is estimated to be a million or more (Kiontke and Fitch, 2014). The nematodes have an obvious difference in feeding behaviour, which includes bacterial and fungal feeders, free-living forms, and plant and animal parasites (Nickle, 1992). Of all the nematodes documented, most of the information available is on the parasitic form, because of their significance as agricultural pests (Neher, 2010) or as a parasite of vertebrates, including humans (Hoberg, 2010), or as biocontrol agents (Dillon *et al.*, 2008), even though these groups represent less than half of the recognized nematode species (Hodda *et al.*, 2009). Although over 4200 species of plant-parasitic nematodes have been identified (Perry and Moens, 2013), new species are continually emerging while others, previously viewed as benign or non-damaging, are becoming pests with changing cropping patterns (Nicol, 2002). On a global scale, the distribution of nematode species varies greatly. Some are cosmopolitan, such as certain *Meloidogyne* species, while others are geographically restricted, e.g. *Nacobbus* spp.; some are highly host-specific, such as *Heterodera carotae* which mainly attacks carrots (Greco, 1986), while others are extremely polyphagous, e.g. *Pratylenchus penetrans*. Some crops may have very few nematode pests, while others have a particularly wide range of genera and species associated with them, leading to difficulties with nematode control strategies (Khan and Jairajpuri, 2010b). Based on the criteria and available information on species characteristics of a plant-parasitic nematode, its association with economically important host crops and capability to transmit viruses or incite a disease complex with other pathogens, approximately 250 plant-parasitic nematode species from 43 genera are considered to be of potential phytosanitary importance (Singh *et al.*, 2013b). However, out of these 250, only 126 nematode species from 33 genera are currently listed as regulated or quarantine pests in different countries.

The world has truly become a global village regarding communication and transport. The globalization of trade and disappearance of tariff barriers in trade have opened floodgates to alien species to move around the world at a faster pace. The later part of the World Trade Organization (WTO) regime is witnessing the mass invasion of pests into exotic areas, causing devastation and imbalance in the economy of many nations. Developing and underdeveloped countries are the most affected in the absence of preparedness to combat the pest invasion. The underestimation of nematode diversity and uninvestigated invasive nematode species pose a very high risk of their spread from one country to another, with the movement of plants and planting materials. In view of the large quantities and the wide variety of plants and planting materials being imported and exported, it has become necessary to assess the status of invasive nematode species and their implications in the context of a nation's agricultural biosecurity. Reviewing the species characteristics and procedures which lead to invasions by PPN may offer insights into how and why biological invasions by nematodes occur and will contribute towards strengthening measures to prevent future nematode invasions.

I.4 Invasive Plant-parasitic Nematodes

The Convention on Biological Diversity (CBD) defines an alien species as *'a species, subspecies or lower taxon, introduced outside its natural past or present distribution; includes any part, gametes, seeds, eggs, or propagules of such species that might*

Table 1. Crop losses caused by plant-parasitic nematodes to different agricultural crops (adapted from Khan and Jairajpuri, 2010a,b and 2012).

Crops	Important nematodes	Yield loss (%)
Cereal crops		
Wheat	*Anguina tritici*, *Ditylenchus dipsaci*, *Heterodera avenae*, *H. filipjevi*, *H. ustinovi*, *Meloidogyne graminicola*, *M. triticoryzae*, *M. javanica*, *M. oryzae*, *M. arenaria*, *Pratylenchus* spp.	7–19
Rice	*Aphelenchoides besseyi*, *Ditylenchus angustus*, *Hirschmanniella oryzae*, *Heterodera oryzicola*, *Hoplolaimus* spp., *M. arenaria*, *M. graminicola*, *M. incognita*, *M. javanica*, *M. oryzae*, *M. triticoryzae*, *Pratylenchus penetrans*	10–22
Maize	*Belonolaimus* spp., *Criconemoides* spp., *Ditylenchus* spp., *Helicotylenchus* spp., *Heterodera zeae*, *H. avenae*, *Hoplolaimus* spp., *Longidorus* spp., *Meloidogyne* spp., *Pratylenchus* spp., *Tylenchorhynchus* spp., *Trichodorus* spp., *Xiphinema americanum*	6–10
Barley and oats	*Anguina tritici*, *Ditylenchus dipsaci*, *Heterodera avenae*, *H. filipjevi*, *H. ustinovi*, *Meloidogyne arenaria*, *M. graminicola*, *M. javanica*, *M. oryzae*, *M. triticoryzae*, *Pratylenchus* spp.	17–77
Sorghum and pearl millets	*Belonolaimus longicaudatus*, *Criconemoides ornate*, *Helicotylenchus* spp., *Heterodera zeae*, *H. sorghi*, *Hoplolaimus indicus*, *Meloidogyne incognita*, *M. javanica*, *M. africanus*, *M. arenaria*, *M. hapla*, *M. graminicola*, *M. triticoryzae*, *Rotylenchulus reniformis*, *Tylenchorhynchus* spp., *Pratylenchus* spp., *Nanidorus minor* and *Trichodorus* spp.	5–15
Pulses		
Chickpea,lentil, mungbean and pigeonpea	*Helicotylenchus* spp., *Heterodera cajani*, *H. ciceri*, *H. glycines*, *Meloidogyne arenaria*, *M. hapla*, *M. incognita*, *M. javanica*, *Pratylenchus thornei*, *Radopholus similis*, *Rotylenchulus reniformis*, *Scutellonema bradys*, *Xiphinema* spp.	6–23
Pea	*Ditylenchus dipsaci*, *Heterodera cajani*, *H. glycines*, *Meloidogyne incognita*, *M. javanica*, *Pratylenchus thornei*, *Rotylenchulus reniformis*	12–16
Oilseed crops		
Soybean	*Helicotylenchus dihystera*, *H. pseudorobustus*, *Heterodera glycines*, *Hoplolaimus galeatus*, *H. seinhorsti*, *Longidorus* spp., *Macroposthonia xenoplax*, *Meloidogyne arenaria*, *M. hapla*, *M. incognita*, *M. javanica*, *Nanidorus minor*, *Paratrichodorus porosus*, *Pratylenchus* spp., *Quinisulcius acutus*, *Radopholus similis*, *Rotylenchulus parvus*, *R. reniformis*, *Scutellonema brachyurus*, *Tylenchorhynchus* spp., *Xiphinema americanum*	9–27
Ground nut (peanut)	*Pratylenchus brachyurus*, *Aphelenchoides arachidis*, *Belonolaimus gracilis*, *Belonolaimus longicaudatus*, *Criconemoides* spp., *Ditylenchus africanus*, *D. destructor*, *Helicotylenchus* spp. *Hoplolaimus* spp., *Heterodera schachtii*, *Meloidogyne arenaria*, *M. hapla*, *M. javanica*, *M. incognita*, *Nanidorus minor*, *Pratylenchus* spp., *Radopholus similis*, *Rotylenchulus* spp., *Scutellonema* spp., *Trichodorus* spp., *Tylenchorhynchus brevilineatus*, *Tylenchorhynchus* sp., *Xiphinema americanum*, *X. diversicaudatum*	7–17
Vegetable crops		
Potato	*Belonolaimus longicaudatus*, *Ditylenchus dipsaci*, *D. destructor*, *Globodera rostochiensis*, *G. pallida*, *Meloidogyne* spp., *Pratylenchus* spp., *R. similis*, *Trichodorus* spp., *Paratrichodorus* spp.	Up to 80
Tomato	*Aphelenchoides fragariae*, *A. ritzemabosi*, *Belonolaimus longicaudatus*, *Ditylenchus dipsaci*, *Globodera pallida*, *G. rostochiensis*, *G. tabacum*, *Helicotylenchus* spp., *Hemicycliophora arenaria*, *Heterodera* spp., *Hirschmanniella* spp., *Hoplolaimus galeatus*, *H. indicus*, *H. seinhorsti*, *Longidorus* spp., *Meloidogyne* spp., *Nanidorus minor*, *Pratylenchus* spp., *Radopholus similis*, *Rotylenchulus parvus*, *R. reniformis*, *Rotylenchus robustus*, *Scutellonema brachyurus*, *Trichodorus* spp., *Tylenchorhynchus* spp., *Xiphinema* spp., *Zygotylenchus* sp.	9–29 (up to 60)
Eggplant	*Belonolaimus longicaudatus*, *Helicotylenchus indicus*, *Meloidogyne* spp., *Pratylenchus* spp., *Rotylenchulus reniformis*	7–18

Continued

Table 1. Continued.

Crops	Important nematodes	Yield loss (%)
Okra	*Belonolaimus longicaudatus*, *Helicotylenchus dihystera*, *Helicotylenchus multicinctus*, *Hirschmanniella imamuri*, *Hirschmanniella oryzae*, *Hoplolaimus indicus*, *Meloidogyne arenaria*, *Meloidogyne incognita*, *Meloidogyne javanica*, *Nanidorus minor*, *Paratrichodorus porosus*, *Pratylenchus* spp., *Radopholus similis*, *Rotylenchulus reniformis*, *Scutellonema* sp.	10–37 (up to 81)
Chilli, bell pepper	*Helicotylenchus* spp., *Meloidogyne* spp., *Nanidorus minor*, *Paratylenchus* spp., *Radopholus similis*, *Scutellonema brachyurus*, *Trichodorus* spp., *Tylenchorhynchus* sp. *Xiphinema americanum*, *X. index*	10–25 (up to 52)
Onion, garlic	*Aphelenchoides besseyi*, *Aphelenchoides ritzemabosi*, *Belonolaimus longicaudatus*, *Ditylenchus dipsaci*, *Meloidogyne* spp., *Trichodorus* sp., *Xiphinema* sp.	5–16 (up to 39)
Cabbage, cauliflower	*Belonolaimus* spp., *Dolichodorus*, *Heterodera schachtii*, *H. cruciferae*, *Meloidogyne* spp., *Pratylenchus pratensis*, *Tylenchorhynchus brassicae*, *Paratylenchus* sp.	5–21
Cucurbits	*Belonolaimus gracilis*, *Criconemoides* sp., *Ditylenchus dipsaci*, *Helicotylenchus dihystera*, *Helicotylenchus microlobus*, *Heterodera* sp., *Heterodera trifolii*, *Hirschmanniella* sp., *Hoplolaimus columbus*, *Longidorus* sp., *Meloidogyne* spp., *Nacobbus aberrans*, *Nanidorus minor*, *Paralongidorus maximus*, *Pratylenchus* spp., *Radopholus similis*, *Rotylenchulus reniformis*, *Trichodorus teres*, *Tylenchorhynchus* sp., *Xiphinema diversicaudatum*	12–35
Carrot and radish	*Ditylenchus destructor*, *D. dipsaci*, *Helicotylenchus dihystera*, *H. multicinctus*, *Heterodera carotae*, *Longidorus* sp., *Meloidogyne* spp., *Nacobbus aberrans*, *Nanidorus minor*, *Paralongidorus maximus*, *Pratylenchus neglectus*, *Pratylenchus penetrans*, *Radopholus similis*, *Rotylenchulus reniformis*, *Trichodorus* sp., *Tylenchorhynchus* sp., *Xiphinema americanum*	10–20 (up to 60)
Leafy vegetables	*Belonolaimus gracilis*, *Criconemoides* sp., *Ditylenchus dipsaci*, *Helicotylenchus* spp., *Heterodera* spp., *Hirschmanniella* sp., *Hoplolaimus* spp., *Longidorus* sp., *Meloidogyne* spp., *Nanidorus minor*, *Paralongidorus maximus*, *Pratylenchus* spp., *Radopholus similis*, *Rotylenchulus reniformis*, *Trichodorus* spp., *Tylenchorhynchus* sp., *Xiphinema* spp.	9–20
Vegetable legumes	*Helicotylenchus* spp., *Hemicriconemoides* sp., *Heterodera* spp., *Hirschmanniella* sp., *Hoplolaimus* spp., *Longidorus* spp., *Meloidogyne* spp., *Nanidorus minor*, *Rotylenchulus reniformis*, *Paratrichodorus* sp., *Pratylenchus* spp., *Radopholus* sp., *Scutellonema* sp., *Tylenchorhynchus* spp., *Xiphinema* spp.	5–35
Fruit crops		
Banana	*Criconemoides* spp., *Helicotylenchus dihystera*, *H. multicinctus*, *H. pseudorobustus*, *H. varicaudatus*, *Helicotylenchus* sp., *Hemicriconemoides cocophilus*, *Heterodera oryzae*,*Hoplolaimus pararobustus*, *Hoplolaimus* sp., *Longidorus* sp., *Meloidogyne* spp., *Nanidorus minor*, *Paratrichodorus* sp., *Paratylenchus* sp., *Pratylenchus goodeyi*, *P. coffeae*, *P. vulnus*, *Radopholus similis*, *Rotylenchulus reniformis*, *Scutellonema cavenessi*, *Trichodorus* sp., *Trophotylenchulus* sp., *Tylenchorhynchus* sp., *Xiphinema* spp.	7–70
Citrus	*Belonolaimus longicaudatus*, *Criconemoides* spp., *Helicotylenchus dihystera*, *H. multicinctus*, *H. pseudorobustus*, *Longidorus* sp., *Meloidogyne* spp., *Nanidorus minor*, *Paratrichodorus* sp., *Pratylenchus* spp., *Radopholus similis*, *R. citri*, *Tylenchulus semipenetrans*, *Rotylenchulus reniformis*, *Scutellonema brachyurus*, *Tylenchorhynchus* sp., *Trichodorus* sp., *Xiphinema* spp.	10–80
Pineapple	*Criconemoides* sp., *Helicotylenchus dihystera*, *Helicotylenchus* sp., *Heterodera schachtii*, *Hirschmanniella* sp., *Meloidogyne acronea*. *M. hapla*, *M. incognita*, *M. javanica* *Pratylenchus brachyurus*, *Radopholus similis*, *Rotylenchulus reniformis*, *Trichodorus* sp., *Tylenchorhynchus* sp., *Xiphinema* sp.	21–60
Temperate fruits	*Belonolaimus longicaudatus*, *Meloidogyne hapla*, *Pratylenchus* spp., *Longidorus* spp., *Trichodorus* spp., *Xiphinema* spp.	5–15

Continued

Table 1. Continued.

Crops	Important nematodes	Yield loss (%)
Ornamental crops		
Chrysanthemum	*Aphelenchoides ritzemabosi*, *Heterodera mothi*, *Meloidogyne arenaria*, *M. incognita*, *M. javanica*, *Rotylenchulus reniformis*, *Pratylenchus coffeae*, *P. chrysanthus*, *Tylenchorhynchus brevilineatus*	8–42
Rose	*Helicotylenchus nannus*, *Hemicycliophora typica, Hoplolaimus galeatus*, *Meloidogyne* spp., *Pratylenchus* spp., *Tylenchorhynchus dubius*, *Xiphinema diversicaudatum*	5–17 (up to 72)
Tuberose	*Aphelenchoides ritzemabosi*, *A. besseyi*, *Meloidogyne arenaria*, *M. hapla*, *M. incognita*, *M. javanica*, *Rotylenchulus reniformis*	10–21 (up to 65)
Gladiolus	*Belonolaimus* sp., *Ditylenchus dipsaci*, *Hemicycliophora* sp., *Meloidogyne* spp., *Pratylenchus* spp., *Rotylenchus* sp., *Trichodorus*, *Xiphinema americanum*	10–17
Palms	*Bursaphelenchus* (=*Rahdinaphelenchus*) *cocophilus*, *Hoplolaimus* sp., *Criconemoides* sp., *Helicotylenchus* sp., *Hemicriconemoides cocophilus*, *Hirschmanniella* sp., *Hemicycliophora* sp., *Longidorus* sp., *Pratylenchus* sp., *Radopholus similis*, *Scutellonema bradys*, *Tylenchorhynchus* sp., *Xiphinema* spp.	8–14 (up to100)
Turf	*Anguina tritici*, *A. agrostis*. *Belonolaimus* spp., *Criconemoides* sp., *Helicotylenchus* spp., *Hemicycliophora* sp., *Hoplolaimus* spp., *Meloidogyne* spp., *Paratrichodorus* spp., *Peltamigratus* sp., *Trichodorus*, *Tylenchorhynchus* spp.	3–17 (up to100)
Medicinal, aromatic and spices		
Medicinal plants	*Helicotylenchus* spp., *Hemicriconemoides* sp., *Heterodera* spp., *Hirschmanniella* sp., *Hoplolaimus* spp., *Longidorus* spp., *Meloidogyne* spp., *Nanidorus minor*, *Rotylenchulus reniformis*, *Paratrichodorus* sp., *Pratylenchus* spp., *Radopholus* sp., *Scutellonema* sp., *Tylenchorhynchus* spp., *Xiphinema* spp.	25–30 (up to 70)
Spices	*Helicotylenchus* spp., *Hirschmanniella* sp., *Hoplolaimus* spp., *Longidorus* spp., *Meloidogyne* spp., *Paratrichodorus* sp., *Pratylenchus* spp., *Radopholus similis*, *Scutellonema* sp., *Tylenchorhynchus* spp., *Xiphinema* spp.	10–24
Aromatic plants	*Helicotylenchus* spp., *Hirschmanniella* sp., *Hoplolaimus* spp., *Longidorus* spp., *Meloidogyne* spp., *Paratrichodorus* sp., *Pratylenchus* spp., *Radopholus similis*, *Scutellonema* sp., *Tylenchorhynchus* spp., *Xiphinema* spp.	5–12
Ginger	*Meloidogyne* spp., *Radopholus similis*, *Pratylenchus coffeae*, *P. brachyurus*, *P. zeae*	40–46
Fibre crops		
Cotton	*Belonolaimus longicaudatus*, *Helicotylenchus* spp., *Hoplolaimus columbus*, *H. galeatus*, *Meloidogyne incognita*, *M. javanica*, *M. arenaria*, *Nanidorus minor*, *Pratylenchus* spp., *Rotylenchulus reniformis*, *Scutellonema* spp., *Trichodorus* spp., *Tylenchorhynchus* spp., *Xiphinema* spp.	7–28
Jute	*Helicotylenchus dihystera*, *Hemicriconemoides mangiferae*, *Hoplolaimus indicus*, *Hirschmanniella oryzae*, *Longidorus macrampohis*, *Meloidogyne incognita*, *M. javanica*, *Tylenchorhynchus annulatus*, *T. mashoodi*, *Rotylenchulus reniformis*, *Xiphinema* spp.	5–20
Plantation crops		
Tobacco	*Aphelenchoides fragariae*, *A. ritzemabosi*, *Belonolaimus longicaudatus*, *Ditylenchus dipsaci*, *Globodera pallida*, *G. rostochiensis*, *G. tabacum*, *Helicotylenchus* spp., *Hemicycliophora arenaria*, *Heterodera* spp., *Hirschmanniella* spp., *Hoplolaimus galeatus*, *H. indicus*, *H. seinhorsti*, *Longidorus* spp., *Meloidogyne* spp., *Nanidorus minor*, *Paratylenchus* spp., *Radopholus similis*, *Rotylenchulus parvus*, *R. reniformis*, *Rotylenchus robustus*, *Scutellonema brachyurus*, *Trichodorus* spp., *Tylenchorhynchus* spp., *Xiphinema* spp., *Zygotylenchus* sp.	5–35 (up to 60)
Sugarcane	*Belonolaimus* spp., *Criconemoides* spp., *Ditylenchus* spp., *Helicotylenchus* spp., *Heterodera* sp., *H. sacchari*, *Hoplolaimus* spp., *Longidorus* spp., *Meloidogyne* spp., *Pratylenchus zeae*, *Tylenchorhynchus* spp., *Trichodorus* spp., *Xiphinema americanum*	3–16

Continued

Table 1. Continued.

Crops	Important nematodes	Yield loss (%)
Tea	*Criconemoides* sp., *Helicotylenchus* sp., *Hemicriconemoides* sp., *Meloidogyne arenaria*, *M. brevicauda*, *M. exigua*, *M. hapla*, *Merlinius camelliae*, *Pratylenchus brachyurus*, *P. pratensis*, *P. loosi*, *Radopholus similis*, *Tylenchulus* sp.	4–10
Coffee	*Criconemoides goodeyi*, *Ditylenchus procerus*, *Helicotylenchus dihystera*, *H. erythrinae*, *Helicotylenchus pseudorobustus*, *Hoplolaimus pararobustus*, *Longidorus laevicapitatus*, *Meloidogyne africana*, *M. coffeicola*, *M. decalineata*, *M. exigua*, *M. hapla*, *M. incognita*, *M. javanica*, *Paratylenchus besoekianus*, *Pratylenchus coffeae*, *P. pratensis*, *Pratylenchus* sp., *Radopholus similis*, *Rotylenchulus reniformis*, *Rotylenchulus* sp., *Trichodorus* sp., *Trophotylenchulus obscurus*, *Trophotylenchulus* sp., *Tylenchorhynchus* sp., *Xiphinema* spp.	3–12 (up to 50)
Coconut	*Bursaphelenchus* (=*Rahdinaphelenchus*) *cocophilus*, *Hoplolaimus* sp., *Criconemoides* sp., *Helicotylenchus* sp., *Hemicriconemoides cocophilus*, *Hirschmanniella* sp., *Hemicycliophora* sp., *Longidorus* sp., *Pratylenchus* sp., *Radopholus similis*, *Scutellonema bradys*, *Tylenchorhynchus* sp., *Xiphinema* spp.	10–18 (up to 100)
Forest trees		
Pine	*Belonolaimus longicaudatus*, *Bursaphelenchus* sp., *Bursaphelenchus xylophilus*, *Ditylenchus dipsaci*, *Helicotylenchus dihystera*, *Helicotylenchus pseudorobustus*, *Hoplolaimus galeatus*, *Hoplolaimus indicus*, *Hoplolaimus* sp., *Longidorus* sp., *Macroposthonia xenoplax*, *Meloidogyne arenaria*, *M. incognita*, *Nanidorus minor* *Paralongidorus* sp., *Paratrichodorus porosus*, *P. renifer*, *Paratrichodorus* sp., *Pratylenchus* spp., *Radopholus similis*, *Rotylenchus* sp., *Scutellonema* sp., *Tylenchorhynchus* spp., *Xiphinema* spp.	5–15 (up to 100)
Bamboo	*Longidorus* spp., *Meloidogyne* spp., *Pratylenchus* spp., *Trichodorus* spp., *Xiphinema* spp.	2–15
Acacia, sal, teak	*Helicotylenchus* spp., *Hemicriconemoides* sp., *Heterodera* sp., *Hoplolaimus* spp., *Longidorus* sp., *Meloidogyne* spp., *Paratrichodorus* sp., *Pratylenchus* spp., *Trichodorus* sp., *Tylenchorhynchus*, *Xiphinema* spp.	3–10
Forage crops		
Forage plants	*Anguina* spp., *Ditylenchus* spp., *Helicotylenchus* spp., *Heterodera* sp., *Hoplolaimus* spp., *Longidorus* sp., *Meloidogyne* spp., *Pratylenchus* spp., *Trichodorus* sp., *Tylenchorhynchus*, *Xiphinema* spp.	8–59
Mushrooms		
Mushrooms	*Aphenchoides composticola*, *A. neocomposticola*, *A. agarici*, *A. swarupi*, *A. asterocaudatus*, *Aphelenchus avenae*, *A. radicicolus*, *Ditylenchus myceliophagus*	5–23

survive and subsequently reproduce' and further prescribes that an invasive alien species *is 'an alien species whose introduction and/or spread threaten biological diversity'* (CBD, 2002). The International Plant Protection Convention (IPPC) (see Chapter 16, section 111), on the other hand, defines a [plant] pest as *'any species, strain or biotype of plant, animal or pathogenic agent injurious to plants or plant products'* and a quarantine pest as *'a pest of potential economic importance to the area endangered thereby and not yet present there, or present but not widely distributed and being officially controlled'* (FAO, 1997). The IPPC definition of a quarantine pest covers much, but not all, of what is considered as an invasive alien species under the CBD. Both definitions cover any organism that is injurious to plants and that has an environmental impact (threatens biological diversity). Further, both definitions prescribe that the environmental impact results from the organism's introduction and/or spread, in different words. It can be argued that most quarantine pests are invasive alien species and that those invasive alien species that are directly or indirectly injurious to plants are quarantine pests (IPPC Secretariat, 2005). The relationship between invasive alien species and quarantine pests concerning plant-parasitic nematodes overlaps the mandates of invasiveness; hence in this volume we use the term invasive species (nematode) rather than invasive alien species or/and quarantine pests (nematode). Based on the core of all definitions, an invasive plant-parasitic nematode can be defined as *'a plant-parasitic nematode species that threatens livelihoods and the environment worldwide'*. Hence, the establishment and spread of economically potential plant-parasitic nematode species to areas outside their native range and their subsequent negative impact on plants or plant products and biological diversity can justify their being considered as invasive plant-parasitic nematodes species.

The impact of invasive species is immense, insidious and often irreversible. They are considered as the second largest threat to native biodiversity next to habitat destruction. Botanists, ecologists, microbiologists, zoologists and conservationists have defined invasive species in different ways (Singh *et al.*, 2013a). However, the definitions of invasive species frequently do not specify the rates of spread or the impacts necessary to be categorized as invasive; thus, depending on the organism, there are differences in grouping them as invasive species. For example, the spread rates of plants to be grouped as invasive can be very different from that of invertebrates, which have different mechanisms and rates of dispersion (Richardson *et al.*, 2000). Besides, the impact of some invasive species, such as rabbits and weeds, are more easily measured than those of invasive microorganisms like nematodes and bacteria, whose impacts, although important, are difficult to measure (Simberloff, 2011). Therefore, taxa that are easily observed, calculated and reported are better exemplified in the literature on invasive species, while invasive microorganisms are not well represented. Plant nematodes are mostly microscopic and their invasiveness is poorly studied among microorganisms and the risks they pose are not properly understood. Substantial academic and research efforts are required for systemic documentation of invasive nematode species. Further, no comprehensive book on invasive nematode species is available. This volume is the first book of its type which presents relevant information on suspected invasive plant-parasitic nematode species, based on the criteria discussed below.

1.5 Assessing Criteria for a Nematode as an Invasive Species

The common assessment criteria for the invasiveness of a nematode species are based on the pest risk analysis framework (ISPM 2, 2016), pest risk analysis for quarantine pest (ISPM 11, 2017) and pest risk analysis for regulated non-quarantine pests (ISPM 21, 2004) IPPC of the Food and Agriculture Organization (FAO). International Standards for Phytosanitary Measures (ISPMs) are standards adopted by the Commission on Phytosanitary Measures (CPM), which is the governing body of the IPPC.

Details of ISPMs 2, 11 and 21 are discussed in Chapter 16 under section 115.1. The other invasion characteristics that may also help in classifying invasive nematodes are based on the 'PAB' framework proposed by Catford *et al.* (2009), where: P = propagule pressure, based on size and frequency of introductions; A = abiotic factors, incorporating ecosystem invasibility based on physical conditions; and B = biotic factors, including the characteristics of invading species, recipient communities and their interactions. Using these criteria, important characteristics of invasive nematode species are described below.

1.5.1 Adaptability to soil edaphic factors

The distribution of PPNs is influenced by edaphic factors of soil such as soil type, organic matter composition, soil pH, soil moisture and host availability (Robertson and Freckman, 1995; Khan, 2008). An invasive nematode species must be able to tolerate a wide range of soil conditions, weather parameters and altitude.

1.5.2 Distribution (native or non-native origin)

Most of the nematode species have restricted distributions (Powers *et al.*, 2009; Monroy *et al.*, 2012). Determining the native range of PPNs remains a major challenge, because the origins of most species are still controversial (Trudgill, 1995). Improper information on the native origin of PPN species presents a major problem in defining their status as alien or native species. Further, knowledge of prior and current community composition is also important to determine the native or invasive status of a species (Litchman, 2010). However, substantial efforts have been made to recognize the origin of economically important and quarantine nematode species.

1.5.3 Survival adaptation

PPNs have varied adaptation mechanisms that enable them to survive under adverse conditions. These mechanisms include cryptobiosis, anhydrobiosis, cyst formation, egg mass production and sex reversal. For example, *Ditylenchus* spp. show cryptobiosis and form 'nemawool' by loosening body water and coiling/clumping, which enables them to survive in anhydrobiotic and cryptic conditions (Khan, 2008). Similarly, some PPNs, such as *Globodera* and *Heterodera* species, form cysts, which are dead females and act as a protective layer surrounding hundreds of viable but quiescent eggs (Turner and Subbotin, 2013). Seed gall nematodes can survive under anhydrobiotic conditions for 32 years as 'dauer' (enduring) larvae (Limber, 1973). These mechanisms assist the nematode species and increase their chances of establishment and persistence in new areas.

1.5.4 Wide host range

The PPN species with a wide host range survive better than the ones with a narrow host range. The majority of the invasive species are reported to be polyphagous. For example, the cosmopolitan root-knot nematode species *Meloidogyne incognita*, *M. javanica* and *M. arenaria* are known to attack a wide range of hosts and also reproduce on alternative hosts such as weeds (Karssen *et al.*, 2013), thus empowering them to survive in the absence of a main host and favouring their establishment and spread (Singh *et al.*, 2010).

1.5.5 Short life, high fecundity and multivoltine nature

A short life cycle, high reproductive ability and multiple generations in a year or cropping season are common traits of invasive species (*Aphelenchoides*, *Bursaphelenchus*) and most PPNs have these characteristics, with some variation (Litchman, 2010). The majority of the invasive nematode species (*Anguina*, *Ditylenchus*, *Globodera*) have very high fecundity and produce a large number of eggs/juveniles even without mating (*Meloidogyne*, *Heterodera*). They tend to have multiple generations per year in the absence of competitors and natural enemies. Some of the univoltine nematodes tend to have more than one generation in introduced areas if the weather parameters are congenial (Singh *et al.*, 2013a).

I.5.6 Dispersal rate and pathways of invasion

PPNs cannot move long distances by their own means. However, they have high dispersal rate due to their smaller size, light weight and pathways. Many invasion pathways have been reported for national and international movement of PPNs over long distances and their introduction into exotic areas. Human-assisted pathways appear most imperative for nematodes to spread both locally and internationally along with the trading of produce, seeds, movement of people and machinery (Singh *et al.*, 2013a). Infected planting and propagative materials, seeds or ornamental plants and pot gardens are the most common means of PPN spread. A wide range of animate (plant parts or animals) and inanimate contaminants (footwear, soiled clothing, machinery with soil, construction or building materials) can also act as a common pathway for the spread of PPNs (Singh *et al.*, 2013b). Other pathways include: movement of goods (species translocated in containers, planting media, untreated wood packaging, some food products); movement of people (by air, road, rail and sea transport); postal and courier services (including biological material purchased via the internet); aquarium trade; and spread/movement (including transboundary movement) of wind, water, wildlife and livestock.

I.5.7 Persistence and competition

The persistence of an introduced nematode species is influenced by its ability to adapt to the new environment (Singh *et al.*, 2013a). The invading PPN must either coexist or outcompete and persist with indigenous nematodes, fungi, bacteria and other fauna in the soil. The invasive species must have persistent and/or aggressive colonizing characteristics. An introduced species might become invasive if it can outcompete native species for resources, such as physical space, nutrients, light, food or water. Many nematode species fall within this category and are known to colonize aggressively and replace native pests in the absence of natural enemies (Singh *et al.*, 2013a).

Based on the above examples, PPNs have a clear potential to become invasive in non-native regions; however, which species become invasive and where, needs to be addressed properly to better understand the processes of invasion and the biosecurity risks (Singh *et al.*, 2013a).

I.6 Invasiveness Rating

The following invasiveness ratings (scores) are proposed for categorizing low, medium and highly invasive nematodes. The ratings are based on pest rating profiles (USDA-PCIT, 2017), which include: nematode adaptability to a wide range of climate, soil types, weather and altitude and host interaction; ability to live off a wide host range; dispersal rate and multiple pathways, short life, high fecundity and multiple generations in a year or crop season; economic impact; environmental impact, adaptation for survival, persistence and highly competitive nature; geographical distribution; and status as native or alien.

I.6.1 Climate/host interaction rating (score)

Low (1) – Nematode is not likely to establish in non-native areas.
Medium (2) – Nematode may be able to establish in a larger but limited area.
High (3) – Nematode likely to establish a widespread distribution.

I.6.2 Host range rating (score)

Low (1) – Nematode has a very limited host range.
Medium (2) – Nematode has a moderate host range.
High (3) – Nematode has a wide host range.

I.6.3 Nematode dispersal potential rating (score)

Low (1) – Nematode does not have high reproductive or dispersal potential.
Medium (2) – Nematode has either high reproductive or dispersal potential.
High (3) – Nematode has both high reproduction and dispersal potential.

I.6.4 Economic impact rating (score)

The economic impact of the nematode will be evaluated using the criteria A, B, C, D, E, F as follows.

A. The nematode could lower crop yield.
B. The nematode could lower crop value (includes increasing crop production costs).
C. The nematode could trigger the loss of markets (includes quarantines).
D. The nematode could negatively change normal cultural practices.
E. The nematode can vector, or is vectored by, another pestiferous organism.

Table 2. Characteristics of invasive plant-parasitic nematodes (adapted from Singh *et al.*, 2013a; CABI Invasive Species Compendium, 2020).

Highly Invasive Plant-Parasitic Nematodes

S. No.	Species	Common name	Distribution	Host range	Main host(s)	Feeding habit	Mode of reproduction	Duration of lifecycle	Survival adaptations	Main mode of dispersal
1.	*Anguina tritici*	Wheat seed gall nematode	Worldwide	Monophagous	Wheat, rye, barley	Migratory endoparasite or ectoparasite	Amphimictic	113 days	Anhydrobiosis	Seeds
2.	*Aphelenchoides arachidis*	Groundnut testa nematode	Africa (Egypt and Nigeria)	Monophagous	Groundnut	Migratory endoparasite or ectoparasite	–	–	Anhydrobiosis	Seeds
3.	*Aphelenchoides besseyi*	Rice leaf nematode	Worldwide	Polyphagous	Rice, strawberry, tuberose, bean	Migratory ectoparasite or endoparasite	Amphimictic and Parthenogenesis	10–12 days at 30°C	Anhydrobiosis	Seeds
4.	*Aphelenchoides fragariae*	Strawberry crimp nematode	Asia, Europe, Oceania, North America, South America	Polyphagous	Strawberry, aster, begonia	Migratory ectoparasite or endoparasite	Amphimictic	10–11 days at 18°C	Anhydrobiosis	Bulbs, cut flowers and propagative materials
5.	*Aphelenchoides ritzemabosi*	Chrysanthemum foliar nematode	Worldwide	Polyphagous	Strawberry, Chrysanthemum, dahlia	Migratory ectoparasite or endoparasite	Bisexual	10–13 days	Anhydrobiosis	Bulbs, cut flowers and propagative material
6.	*Belonolaimus longicaudatus*	Sting nematode	Asia, North America	Polyphagous	Cotton, turf, citrus, carrots, maize	Migratory ectoparasite	Amphimictic	28 days at 30°C	–	Propagative material, turf, soil contaminant
7.	*Bursaphelenchus cocophilus*	Red ring nematode	North America, South America	Monophagous	Coconut, palm trees	Migratory endoparasite	–	9–10 days	Dauer stage juveniles	Dispersal by weevil vector, wood materials
8.	*Bursaphelenchus xylophilus*	Pine wilt nematode	Asia, Africa, Europe, North America	Monophagous	*Pinus* spp.	Migratory endoparasite	Amphimictic	12 days at 15°C	Dauer stage juveniles	Dispersal by beetle vector, wood chips
9.	*Ditylenchus africanus*	Peanut pod nematode	Africa	Polyphagous	Groundnut	Migratory ectoparasite and endoparasite	–	6–7 days at 28°C	Anhydrobiosis	Groundnut seeds
10.	*Ditylenchus angustus*	Rice stem nematode	Asia	Monophagous	Rice	Migratory ectoparasite and endoparasite	–	10–20 days at 30°C	Anhydrobiosis	Seeds, infected planting materials
11.	*Ditylenchus destructor*	Potato tuber nematode	Worldwide	Polyphagous	Potato, bulbs, ornamentals	Migratory endoparasite	Amphimictic	6–7 days at 28°C	Anhydrobiosis, Nemawool by coiling and clumping	Seeds and propagative materials
12.	*Ditylenchus dipsaci*	Stem and bulb nematode	Worldwide	Polyphagous	Onions, garlic, cereals, legumes, ornamentals	Migratory endoparasite	Amphimictic	20 days at 15°C	Anhydrobiosis, Nemawool by coiling and clumping	Seeds and propagative materials
13.	*Globodera pallida*	White cyst nematode	Worldwide	Monophagous	Potato, tomato, eggplant	Sedentary endoparasite	Amphimictic	45 days	Cysts	Propagative material
14.	*Globodera rostochiensis*	Golden cyst nematode	Worldwide	Monophagous	Potato, tomato, eggplant	Sedentary endoparasite	Amphimictic	45 days	Cysts	Propagative material
15.	*Globodera tabacum*	Tobacco cyst nematode	Asia, Africa, Europe, North America, South America	Monophagous	Tobacco, tomato	Sedentary endoparasite	Amphimictic	32–36 days at 26–30°C, 72 days at 20°C	Cysts	Propagative material, soil contaminants

Continued

Table 2. Continued.

Highly Invasive Plant-Parasitic Nematodes

S. No.	Species	Common name	Distribution	Host range	Main host(s)	Feeding habit	Mode of reproduction	Duration of lifecycle	Survival adaptations	Main mode of dispersal
16.	*Heterodera avenae*	Cereal cyst nematode	Worldwide	Polyphagous	Wheat, barley, oat, rye	Sedentary endoparasite	Amphimictic	5–9 week	Cysts	Propagative material, soil contaminant
17.	*Heterodera cajani*	Pigeon pea cyst nematode	India, Pakistan, Egypt	Polyphagous	Peas, beans	Sedentary endoparasite	Amphimictic	16 days at 29°C,45–80 days at 10–25°C	Cysts	Contaminated soil, cyst dispersal
18.	*Heterodera glycines*	Soybean cyst nematode	Asia, Africa, Europe, North America, South America	Polyphagous	Soybean	Sedentary endoparasite	Amphimictic	22–35 days	Cysts	Soil contaminants and propagative materials
19.	*Heterodera sacchari*	Sugarcane cyst nematode	Asia, Africa, North America (Jamaica)	Polyphagous	Sugarcane, rice	Sedentary endoparasite	Mitotic parthenogenesis	30–40 days	Cysts	Propagative material, soil contaminant
20.	*Heterodera schachtii*	Sugarbeet cyst nematode	Worldwide	Polyphagous	Sugarbeet	Sedentary endoparasite	Amphimictic	22–35 days	Cysts	Propagative material, soil contaminant
21.	*Heterodera zeae*	Corn cyst nematode	Asia, Egypt, USA, Greece, Portugal	Polyphagous	Corn, sorghum	Sedentary endoparasite	Amphimictic	–	Cysts	Propagative material, soil contaminant
22.	*Longidorus elongatus*	Needle nematode	Africa (South Africa), Asia, North America, Oceania (New Zealand)	Polyphagous	Beetroot, carrot, strawberry, peppermint, rye grass	Ectoparasite	Parthenogenesis	9 weeks at 30°C	–	Propagative material, soil contaminant
23.	*Meloidogyne arenaria*	Peanut root-knot nematode	Worldwide	Polyphagous	Many hosts	Sedentary endoparasite	Mitotic parthenogenesis	21 days	Egg mass, gelatinous matrix protects eggs	Planting material, soil contaminant
24.	*Meloidogyne chitwoodi*	Columbia root-knot nematode	Worldwide	Polyphagous	Potato, carrots, lucerne	Sedentary endoparasite	Meiotic parthenogenesis	21–28 days	Gelatinous egg mass	Propagative material, soil contaminant
25.	*Meloidogyne enterolobii*	Pacara earpod tree root-knot nematode	Worldwide	Polyphagous	Bean, coffee, cucumber, guava, watermelon, tomato etc.	Sedentary endoparasite	Mitotic parthenogenesis	24 days	Egg mass gelatinous matrix protects eggs	Planting material, soil contaminant
26.	*Meloidogyne ethiopica*	Root-knot nematode	Africa, Europe, South America	Polyphagous	Cabbage, bell pepper, cucurbits	Sedentary endoparasite	Parthenogenetic	–	Gelatinous egg mass	Propagative material, soil contaminant
27.	*Meloidogyne exigua*	Coffee root-knot nematode	Europe, North America, South America	Polyphagous	Coffee, capsicum, onion, citrus	Sedentary endoparasite	Facultative parthenogenesis	–	Gelatinous egg mass	Propagative material, soil contaminant
28.	*Meloidogyne fallax*	False Columbia root-knot nematode	Europe, Oceania	Polyphagous	Carrots, potato, cucurbits	Sedentary endoparasite	Meiotic parthenogenesis	–	Gelatinous egg mass	Propagative material, soil contaminant
29.	*Meloidogyne graminicola*	Rice root-knot nematode	Asia, Africa, North America, South America	Polyphagous	Rice, wheat	Sedentary endoparasite	Meiotic parthenogenesis	19–27 days at 22–29°C	Gelatinous egg mass	Propagative material, soil contaminant
30.	*Meloidogyne hapla*	Northern root-knot nematode	Worldwide	Polyphagous	Carrots, lettuce, peanuts, sugarbeet, soybean, tomato, potato	Sedentary endoparasite	Race A-Meiotic parthenogenesis, Race B-Mitotic parthenogenesis	35 days	Survives at 0°C, Egg mass gelatinous matrix protects eggs	Planting material, soil contaminant

31.	*Meloidogyne incognita*	Southern root-knot nematode	Worldwide	Polyphagous	More than 500 plant hosts	Sedentary endoparasite	Mitotic parthenogenesis	20–25 days	Gelatinous matrix protects eggs	Planting material, soil contaminant
32.	*Meloidogyne javanica*	Sugarcane root-knot nematode	Worldwide	Polyphagous	Many hosts	Sedentary endoparasite	Mitotic parthenogenesis	27 days at 28–35°C	Gelatinous egg mass	Propagative material, soil contaminant
33.	*Nacobbus aberrans*	False root-knot nematode	North America, South America	Polyphagous	Winter pea, batata, beetroot, broccoli, etc.	Migratory and sedentary endoparasite	Parthenogenetic	–	Egg mass gelatinous matrix protects eggs	Propagative material, soil contaminant
34.	*Nanidorus minor* (=*Paratrichodorus minor*)	Stubby root nematode	Worldwide	Polyphagous	Many hosts	Migratory ectoparasite	Parthenogenetic	–	–	Propagative material, soil contaminant
35.	*Pratylenchus brachyurus*	Lesion nematode	Worldwide	Polyphagous	Maple, pineapple, groundnut, pigeon pea, tea	Migratory endoparasite	Amphimictic	3 and 4 weeks at 30°C	–	Propagative material, soil contaminant
36.	*Pratylenchus coffeae*	Root-lesion nematode	Worldwide	Polyphagous	Banana, yam, coffee, citrus, ginger	Migratory endoparasite	Amphimictic	27 days at 25–30°C	–	Planting material, yam tubers, soil contaminants
37.	*Pratylenchus goodeyi*	Banana lesion nematode	Africa, Europe, Oceania	Polyphagous	Banana	Migratory endoparasite	Amphimictic	24–35 days	–	Planting material (suckers), soil contaminant
38.	*Pratylenchus loosi*	Root-lesion nematode	Africa (Senegal), Asia, Europe (Bulgaria), North America, Oceania, South America	Polyphagous	Tea	Migratory endoparasite	Amphimictic	45–48 days	–	Propagative materials, soil contaminant
39.	*Pratylenchus neglectus*	Root-lesion nematode	Worldwide	Polyphagous	Alfalfa, barley, chick pea, oat, maize, wheat	Migratory endoparasite	Mitotic parthenogenesis	30–45 days	–	Propagative materials, soil contaminant
40.	*Pratylenchus penetrans*	Root-lesion nematode	Worldwide	Polyphagous	Field, horticultural, fruit crops	Migratory endoparasite	Amphimictic	30 days	–	Propagative materials, soil contaminant
41.	*Pratylenchus thornei*	Root-lesion nematode	Worldwide	Polyphagous	Chickpea, wheat	Migratory endoparasite	Amphimictic	6–8 weeks	–	Propagative materials, soil contaminant
42.	*Pratylenchus vulnus*	Walnut root lesion nematode	Worldwide	Polyphagous	Tara vine, kiwifruit, Japanese barberry, cabbage	Migratory endoparasite	Amphimictic	3 and 4 weeks at 30°C	–	Propagative material, soil contaminant
43.	*Pratylenchus zeae*	Root-lesion nematode	Worldwide	Polyphagous	Rice, maize, sorghum and sugarcane	Migratory endoparasite	Parthenogenesis	21–30 days	–	Propagative materials, soil contaminant
44.	*Radopholus similis*	Burrowing nematode	Worldwide	Polyphagous	Banana, coffee, pineapple, pepper, ginger, anthuriums, citrus	Migratory endoparasite	Hermaphroditic or Amphigony	20–25 days at 24–32°C	Males survive longer than females under stress conditions	Planting material (banana suckers), soil contaminant
45.	*Rotylenchulus reniformis*	Reniform nematode	Worldwide	Polyphagous	Banana, cotton, pineapple, vegetables	Semi–endo parasite	Amphimictic, occasionally parthenogenetic	17–23 days	–	Propagative plant materials, soil contaminants
46.	*Scutellonema brachyurus*	Carolina spiral nematode	Worldwide	Polyphagous	Asparagus, soybean, cotton, banana, tobacco, peach	Migratory endoparasite	Parthenogenesis	–	Undergo anhydrobiosis	Planting material, yams tubers, soil contaminant
47.	*Scutellonema bradys*	Yam nematode	Asia, Africa, North America, South America	Polyphagous	Yams, cowpeas	Migratory endoparasite	Amphimictic	–	Undergo anhydrobiosis	Planting material, yams tubers, soil contaminant

Continued

Table 2. Continued.

S. No.	Species	Common name	Distribution	Host range	Main host(s)	Feeding habit	Mode of reproduction	Duration of lifecycle	Survival adaptations	Main mode of dispersal
Highly Invasive Plant-Parasitic Nematodes										
48.	*Tylenchulus semipenetrans*	Citrus root nematode	Worldwide	Polyphagous	Citrus, olives, grapevine	Semi-endoparasite	Amphimictic or meiotic parthenogenesis	6–8 weeks at 24–26 ºC	Gelatinous matrix protects eggs	Planting material, soil contaminant
49.	*Xiphinema americanum*	American dagger nematode	Worldwide	Polyphagous	Field, horticultural, grassland and forest trees	Ectoparasite	Amphimictic	1 year	–	Infested plants, contaminated soil and machinery
50.	*Xiphinema diversicaudatum*	Dagger nematode	Worldwide	Polyphagous	Grapevine, strawberry, plum, peach, roses	Migratory ectoparasite	Amphimictic	2 years to develop from egg to adult	–	Planting material, soil Contaminant
51.	*Xiphinema index*	Fan–leaf virus nematode	Worldwide	Polyphagous	Fig, grapevine	Migratory ectoparasite	Amphimictic	7–9 months at 20–23°C	–	Planting material, soil contaminant
52.	*Xiphinema rivesi*	Dagger nematode	Europe, North America, South America	Polyphagous	Apple, grapevine	Ectoparasite	Amphimictic	1 year	–	Infested soil, plants, and contaminated machinery
53.	*Zygotylenchus guevarai*	Root lesion nematode	Asia, Europe, North America	Polyphagous	Ornamental trees, alfalfa, almond, broad bean	Migratory endoparasite	Amphimictic	45 day	–	Propagative materials, soil contaminants
Medium Invasive Plant-Parasitic Nematodes										
54.	*Achlysiella williamsi*	Lesion nematode	Mauritius, Papua New Guinea, Pacific region	Monophagous	Sugarcane	Sedentary endoparasite	–	–	Gelatinous egg sac	Propagative materials, soil contaminants
55.	*Anguina agrostis*	Bentgrass nematode	Europe, Asia, Africa, North America, Oceania	Polyphagous	Poaceae plants	Migratory endo/ ectoparasite	Amphimictic	3–4 weeks.	Anhydrobiosis	Seeds
56.	*Anguina funesta*	Seed gall nematode	Australia, USA	Polyphagous	Rye grass	Migratory endo/ ectoparasite	–	–	Anhydrobiosis	Seeds
57.	*Aphasmatylenchus straturatus*	–	Africa (Burkina Faso)	Polyphagous	Leguminous plants	Migratory endo/ ectoparasitic	–	–	Undergo anhydrobiosis	Propagative material, soil contaminant
58.	*Cactodera cacti*	Cactus cyst nematode	Worldwide	Monophagous	Cactus	Sedentary endoparasite	Amphimictic	29 to 34 days 18 to 26°C	Cysts	Propagative material, soil contaminant
59.	*Criconemoides* spp.	Ring nematode	Worldwide	Polyphagous	Citrus, coffee, groundnut, rice, grapes, peach, pineapple, Brassica spp.	Migratory ectoparasite	Parthenogenesis	25–34 days	–	Propagative material, soil contaminant
60.	*Hemicriconemoides mangiferae*	Sheathoid nematode	Asia, Africa, Europe, North America, Oceania, South America	Polyphagous	Citrus, mango, litchi	Migratory ectoparasite, occasionally endoparasite	Parthenogenesis	–	–	Propagative material, soil contaminant
61.	*Helicotylenchus dihystera*	Spiral nematode	Worldwide	Polyphagous	Beetroot, okra, celery, fruit tree, pineapple, chili	Ecto- or semi-endoparasite	Amphimictic	–	–	Propagative material, soil contaminant
62.	*Helicotylenchus multicinctus*	Banana spiral nematode	Worldwide	Polyphagous	Banana	Migratory ectoparasite	Amphimictic	39 days	–	Propagative materials, banana suckers

63.	*Helicotylenchus pseudorobustus*	Spiral nematode	Worldwide	Polyphagous	Oats, barley, rye, wheat, maize	Ecto- or semi-endoparasite	Amphimictic	–	–	Propagative material, soil contaminant
64.	*Hemicycliophora arenaria*	Sheath nematode	Australia, USA, Brazil	Polyphagous	Citrus	Ectoparasite	Parthenogenetic	–	–	Propagative plant material and soil contaminants
65.	*Heterodera carotae*	Carrot cyst nematode	Europe, Asia, South America	Polyphagous	Carrot	Sedentary endoparasite	Amphimictic	–	Cysts	Soil contaminant
66.	*Heterodera ciceri*	Chickpea cyst nematode	Mediterranean Europe	Monophagous	Chickpeas, lentil	Sedentary endoparasite	Amphimictic	38 days	Cysts	Propagative material, soil contaminants
67.	*Heterodera goettingiana*	Pea cyst eelworm	Asia, Africa, Europe, North America	Monophagous	Pea, broad bean	Sedentary endoparasite	Amphimictic	40–55 days	Cysts	Soil contaminants
68.	*Heterodera humuli*	Hops cyst nematode	Africa, Europe, North America, Oceania	Polyphagous	Hemp and hop	Sedentary endoparasite	Amphimictic	40 days	Cysts	Propagative material, soil contaminant
69.	*Heterodera latipons*	Mediterranean cereal cyst nematode	Europe, Africa, Asia	Polyphagous	Barley and wheat	Sedentary endoparasite	Amphimictic		Cysts	Propagative material, soil contaminant
70.	*Hirschmanniella diversa*	Lotus root nematode	China, Japan, USA, Vietnam	Monophagous	Lotus	Migratory endoparasite	–	–	–	Propagative plant material and soil contaminants
71.	*Hirschmanniella miticausa*	Taro nematode	Oceania	Monophagous	Taro, Colocasia esculenta	Migratory endoparasite	–	–	–	Planting material, soil contaminant
72.	*Hirschmanniella oryzae*	Rice root nematode	Asia, Africa, Europe (France), USA, South America	Polyphagous	Rice	Migratory endoparasite	–	–	–	Propagative plant material and soil contaminants
73.	*Hirschmanniella spinicaudata*	Rice root nematode	Asia, Africa, North America, South America	Polyphagous	Rice	Migratory endoparasite	–	–	–	Propagative plant material and soil contaminants
74.	*Hoplolaimus columbus*	Columbia lance nematode	India, Pakistan, Egypt, USA, Trinidad and Tobago	Polyphagous	Cotton, maize, soybean, wheat, turfgrass	Migratory ecto- and endoparasite	Parthenogenesis	45–49 days	–	Propagative plant material and soil contaminants
75.	*Hoplolaimus galeatus*	Lance nematode	Africa (Egypt, Tanzania), Asia (India), North America, South America	Polyphagous	Turf grass, cotton, pine, oak, wheat, corn, beans	Migratory ecto- and endoparasite	Parthenogenesis	–	–	Propagative plant material and soil contaminants
76.	*Hoplolaimus indicus*	Lance nematode	Bangladesh, Canada, China, India, Iran Pakistan	Polyphagous	Rice, vegetable, fruit, ornamental crops	Migratory endoparasite	Parthenogenesis	–	–	Propagative plant material and soil contaminants
77.	*Hoplolaimus pararobustus*	Lance nematode	Africa, Asia, Central America	Polyphagous	Banana, cotton, citrus	Migratory endoparasite	–	–	–	Propagative material, soil contaminant
78.	*Hoplolaimus seinhorsti*	Lance nematode	Asia, Africa North America, Oceania, South America	Polyphagous	Barley, groundnut, nana, pineapple, rice, tomato	Migratory endoparasite	–	–	–	Propagative material, soil contaminant
79.	*Longidorus attenuates*	Needle nematode	Europe	Polyphagous	Barley, cabbage, clover	Migratory ectoparasite	–	–	–	Propagative plant material and soil contaminants
80.	*Longidorus macrosoma*	Needle nematode	Europe	Polyphagous	Apple, Oak, roseberry	Migratory ectoparasite	–	–	–	Propagative plant material and soil contaminants

Continued

Table 2. Continued.

Medium Invasive Plant-Parasitic Nematodes

S. No.	Species	Common name	Distribution	Host range	Main host(s)	Feeding habit	Mode of reproduction	Duration of lifecycle	Survival adaptations	Main mode of dispersal
81.	*Meloidogyne acronea*	African cotton root nematode	Africa	Polyphagous	Cotton	Sedentary endoparasite	Amphimictic	19 days at 33°C	Gelatinous egg mass	Propagative material, soil contaminant
82.	*Meloidogyne artiellia*	British root-knot nematode	Europe, Asia	Polyphagous	Barley, sorghum and wheat	Sedentary endoparasite	Parthenogenetic	–	Egg mass gelatinous matrix protects eggs	Propagative material, soil contaminant
83.	*Meloidogyne brevicauda*	Indian root-knot nematode	India, Sri Lanka, Azerbaijan	Monophagous	Tea, saffron	Sedentary endoparasite	Parthenogenetic	–	Egg mass gelatinous matrix protects eggs	Propagative material, soil contaminant
84.	*Meloidogyne citri*, *M. donghaiensis*, *M. fujianensis*, *M. indica*, *M. jianyangensis*, *M. kongi* and *M. mingnanica*	Asian citrus root-knot nematode	Asia (China, India)	Polyphagous	Citrus	Sedentary endoparasite	–	–	Gelatinous egg mass	Propagative material, soil contaminant
85.	*Meloidogyne coffeicola*	Coffee root-knot nematode	Brazil, East Africa, USA	Polyphagous	Coffee	Sedentary endo or semi-endoparasite	–	–	Gelatinous egg mass	Propagative material, soil contaminant
86.	*Meloidogyne decalineata*	African coffee root-knot nematode	Africa	Polyphagous	Coffee	Sedentary endoparasite	Mitotic parthenogenesis	–	Gelatinous egg mass	Propagative material, soil contaminant
87.	*Meloidogyne mali*	Root-knot nematode	France, Italy, Japan, the Netherland, USA	Polyphagous	Apple, Dutch elm	Sedentary endoparasite	Amphimictic	–	Gelatinous egg mass	Propagative material, soil contaminant
88.	*Meloidogyne naasi*	Root-knot nematode	Asia, Europe, North America, South America	Polyphagous	Grasses	Sedentary endoparasite	Meiotic parthenogenesis	–	Gelatinous egg mass	Propagative material, soil contaminant
89.	*Meloidogyne paranaensis*	Coffee root-knot nematode	Brazil	Polyphagous	Coffee, tobacco, tomato and watermelon	Sedentary endoparasite	Parthenogenetic	–	Gelatinous egg mass	Propagative material, soil contaminant
90.	*Paralongidorus maximus*	Needle nematode	Africa (South Africa), Europe	Polyphagous	Field crops, ornamentals, and forest trees	Migratory ectoparasite	Amphimictic	Up to 2 years	Anaerobic condition	Infested plants, soil contaminant
91.	*Paratrichodorus porosus*	Stubby root nematode	Worldwide	Polyphagous	Many hosts	Migratory ectoparasite	Parthenogenetic	–	–	Propagative material, soil contaminant
92.	*Pratylenchus crenatus*	Root-lesion nematode	Worldwide	Polyphagous	Cereals and grasses	Migratory endoparasite	Amphimictic	–	–	Propagative material, soil contaminant
93.	*Punctodera chalcoensis*	Mexican corn cyst nematode	Mexico	Monophagous	Corn	Sedentary endoparasite	–	30–50 days	Cysts	Propagative material, soil contaminant
94.	*Punctodera punctata*	Grass cyst nematode	Europe, Argentina, Chile, Israel, Japan, Morocco, North America	Polyphagous	Poaceae plants	Sedentary endoparasite	Amphimictic	–	Cysts	Propagative material, soil contaminant
95.	*Pratylenchus scribneri*	Root-lesion nematode	Worldwide	Polyphagous	Potato, corn and soybean	Migratory endoparasite	Amphimictic	–	–	Propagative materials, soil contaminant
96.	*Radopholus citri*	Citrus burrowing nematode	Asia (Indonesia)	Monophagous	Citrus	Migratory endoparasite	–	–	–	Planting material, soil contaminant

97. *Rotylenchulus macrodoratus*	Mediterranean reniform nematode	Mediterranean region, South Africa	Polyphagous	Many fruit and ornamental trees	Semi-endo parasite	Both amphimictic and parthenogenetic	–	–	Propagative plant materials, soil contaminants
98. *Rotylenchulus parvus*	Reniform nematode	Africa, Asia, Europe (Cyprus), Oceania	Polyphagous	Bean, cowpea, groundnut, maize, potato	Semi-endo parasite	Parthenogenetic	27–36 days	–	Propagative plant materials, soil contaminants
99. *Trichodorus viruliferus*	Stubby root nematode	Europe and USA	Polyphagous	Apple, barley, maize, pea, potato, rye, wheat	Migratory ectoparasite	–	45 days at 15–20 °C	–	Propagative material, soil contaminant
100. *Tylenchorhynchus claytoni*	Stunt nematode	Worldwide	Polyphagous	Azaleas, corn, pea, pine, peach, sugarcane, tobacco and grasses	Migratory ectoparasite	Amphimictic	31–38 days	–	Propagative material, soil contaminant
101. *Xiphinema ifacolum*	Dagger nematode	South America, Africa, Asia	Polyphagous	Rice	Ectoparasite	–	–	–	Infested soil, plants, and contaminated machinery
102. *Xiphinema italiae*	Dagger nematode	Mediterranean Europe, Africa	Polyphagous	Grapevine	Ectoparasite	–	–	–	Infested soil, plants, and contaminated machinery
Low Invasive Plant-parasitic Nematodes									
103. *Helicotylenchus variocaudatus*	Spiral nematode	Africa (Côte d'Ivoire, São Tomé and Principe island)	Polyphagous	Banana, black pepper	Ectoparasite	–	–	–	Propagative material, soil contaminant
104. *Heterodera elachista*	Rice cyst nematode	Asia	Monophagous	Rice, maize	Sedentary endoparasite	Amphimictic	–	Cysts	Propagative material, soil contaminant
105. *Heterodera oryzae*	Rice cyst nematode	Africa, Asia	Monophagous	Rice, maize	Sedentary endoparasite	Amphimictic	30 days	Cysts	Propagative plant material and soil contaminants
106. *Heterodera oryzicola*	Paddy cyst nematode	India	Monophagous	Rice	Sedentary endoparasite	Amphimictic	–	Cysts	Propagative plant material and soil contaminants
107. *Hirschmanniella belli*	Rice root nematode	China, USA	Polyphagous	Rice, sorghum, Sudan grass	Migratory endoparasite	–	–	–	Propagative plant material and soil contaminants
108. *Longidorus diadecturus*	Needle nematode	USA.	Polyphagous	Peach trees and grape vine	Migratory ectoparasite	–	–	–	Propagative material, soil contaminant
109. *Longidorus leptocephalus*	Needle nematode	Europe	Polyphagous	Cereals, carrot, potatoes, rose, sugarbeet, tulip, turnip, etc.	Migratory ectoparasite	–	–	–	Propagative material, soil contaminant
110. *Meloidogyne arabicida*	Coffee root-knot nematode	Costa Rica	Monophagous	Coffee	Sedentary endoparasite	Parthenogenetic	–	Gelatinous egg mass	Propagative material, soil contaminant
111. *Meloidogyne ardenensis*	Root-knot nematode	Europe	Polyphagous	Ligustrum, liverleaf, oak, periwinkle	Sedentary endoparasite	Parthenogenetic	–	Gelatinous egg mass	Propagative material, soil contaminant
112. *Meloidogyne marylandi*	Root-knot nematode	Costa Rica, Israel, Japan, USA	Polyphagous	Turfgrass	Sedentary endoparasite	–	–	Gelatinous egg mass	Propagative material, soil contaminant
113. *Meloidogyne kikuyensis*	Root-knot nematode	Africa	Polyphagous	Coffee, cowpea, kikuyu grass and sugarcane	Sedentary endoparasite	–	–	Gelatinous egg mass	Propagative material, soil contaminant
114. *Meloidogyne salasi*	Root-knot nematode	Costa Rica and Panama	Polyphagous	Rice	Sedentary endoparasite	Parthenogenetic	–	Gelatinous egg mass	Propagative material, soil contaminant

Continued

Table 2. Continued.

Low Invasive Plant-parasitic Nematodes

S. No.	Species	Common name	Distribution	Host range	Main host(s)	Feeding habit	Mode of reproduction	Duration of lifecycle	Survival adaptations	Main mode of dispersal
115.	*Quinisulcius acutus*	Stylet-stunt nematode	Asia, North, America	Polyphagous	Soyabean, lucerne, tobacco, bean, sorghum	Migratory ectoparasite	–	–	–	Propagative material, soil contaminant
116.	*Paralongidorus australis*	Needle nematode	Australia	Monophagous	Rice	Migratory ectoparasite	Amphimictic	Up to 2 years	Anaerobic condition	Infested plants, soil contaminant
117.	*Pratylenchus alleni*	Root-lesion nematode	Argentina, Canada, India, Russia, Turkey USA	Polyphagous	Soybean, vetch, chickpea, castor bean, cotton, wheat, corn, potato, tomato	Migratory endoparasite	Amphimictic	45–65 days	–	Propagative material, soil contaminant
118.	*Pratylenchus fallax*	Root-lesion nematode	Europe, Canada, Japan	Polyphagous	Alfalfa, barley, corn, oat, wheat, sugarbeet	Migratory endoparasite	Amphimictic	–	–	Propagative material, soil contaminant
119.	*Trophotylenchulus piperis*	–	India	Monophagous	Black pepper	Semi-endoparasite		55 days	Gelatinous matrix protects eggs	Propagative material, soil contaminant
120.	*Tylenchorhynchus annulatus*	Stunt nematode	Worldwide	Polyphagous	Rice, sugarcane and other grasses	Migratory ectoparasite	–	–	–	Propagative material, soil contaminant

F. The nematode is injurious or poisonous to agriculturally important animals.

Low (1) – Nematode causes 0 or 1 of these impacts.
Medium (2) – Nematode causes 2 of these impacts.
High (3) – Nematode causes 3 or more of these impacts.

I.6.5 Environmental impact rating (score)

The environmental impact of the nematode will be evaluated using the criteria A, B, C, D, E as follows.

A. The nematode could have a significant environmental impact such as lowering biodiversity, disrupting natural communities, or changing ecosystem processes.
B. The nematode could directly affect threatened or endangered species.
C. The nematode could impact on threatened or endangered species by disrupting critical habitats.
D. The nematode could trigger additional official or private treatment programmes.
E. The nematode significantly impacts cultural practices, home/urban gardening or ornamental plantings.

Low (1) – Nematode causes none of the above to occur.
Medium (2) – Nematode causes 2 of these impacts.
High (3) – Nematode causes more than two of the above to occur.

I.6.6 Post-entry distribution rating (score)

Not known (0) – Post-entry distribution of nematode in the non-native area is not known.
Low (–1) – Nematode has a localized distribution, or is established in one suitable climate/host area (region).
Medium (–2) – Nematode is widespread but not fully established in the endangered area; or nematode established in two contiguous suitable climate/host areas.
High (–3) – Nematode is widespread and has fully established in the endangered area; or nematode is reported in more than two contiguous or non-contiguous suitable climate/host areas.

I.6.7 Final rating (score)

After adding up the total score (15) minus (–) the post-entry distribution, the final rating is:

Low invasive nematode = 5–8 points
Medium invasive nematode = 9–11 points
Highly invasive nematode = 12–15 points

Using the above characteristics and rating criteria, we have examined over 120 suspected plant-parasitic nematodes for their invasiveness and categorized them as high, medium and low invasive species in this volume (Table 2). Besides other sources, the baseline data for the book are collected from the CABI invasive species compendium (CABI-ISC, 2020), Center for Invasive Species and Ecosystem Health (CISEH, 2019), Global Invasive Species Database (GISD, 2019), Global Invasive Species Programme (GISP, 2019), North American Plant Protection Organization (NAPPO, 2019), European and Mediterranean Plant Protection Organization (EPPO, 2019), North European and Baltic Network on Invasive Alien Species (NEBN-IAS, 2019) and IABIN Invasive Information Network (I3N, 2019).

However, uncertainty about many of the traits related to invasiveness discussed above, for many PPN species, makes it very tough to examine their invasiveness accurately. When information is insufficient, then non-invasiveness is assumed under many of the codes that govern pest risk analysis (ISPM 11, 2017; ISPM 2, 2016). Hence, PPNs may be wrongly categorized as non-invasive purely on the absence of information rather than characteristics or evidence suggesting non-invasiveness (Singh *et al.*, 2013a). To mitigate this issue, we also include some probable nematodes species and categorize them as low invasive species (Table 2). This would offer a chance of considering the PPN species that may pose a risk of becoming invasive species but which otherwise would have been neglected mainly because their impacts are poorly recognized.

1

Anguinidae: *Anguina* and *Ditylenchus* Species

Anguina Spp.

The genus *Anguina* covers the first plant-parasitic nematode to be recorded, *Anguina tritici*, discovered by Needham in 1743 from wheat seed gall. The gall-forming nematodes, *Anguina* spp., are obligate parasites and attack aerial plants parts of cereals and forage grasses. More than 40 nominal species of *Anguina* have been described (Subbotin and Riley, 2012) and important species include:

Anguina agropyri Kirjanova, 1955
A. agropyronifloris Norton, 1965
A. agrostis (Steinbuch, 1799) Filipjev, 1936
A. amsinckiae (Steiner and Scott, 1935) Thorne, 1961
A. australis Steiner, 1940
A. balsamophila (Thorne, 1926) Filipjev, 1936
A. caricis Solovyeva and Krall, 1982
A. cecidoplastes (Goodey, 1934) Filipjev, 1936
A. funesta Price, Fisher and Kerr, 1979
A. graminis (Hardy, 1850) Filipjev, 1936
A. microlaenae (Fawcett, 1938) Steiner, 1940
A. pacificae Cid del Prado Vera and Maggenti, 1984 (Ferris, 2013).
A. tritici (Steinbuch, 1799) Filipjev, 1936

Three *Anguina* species, *A. tritici*, *A. agrostis* and *A. funesta*, are considered as invasive nematode species because of their economic importance and quarantine significance in various countries (Chizhov and Subbotin, 1990; Krall, 1991). Other species are of lower importance in a limited geographical range.

Systematic position of *Anguina* species

	Classical (after Siddiqi, 2000)	Molecular (after Decraemer and Hunt, 2013)
Phylum	Nematoda	Nematoda
Class	Secernentea	Chromadorea
Subclass	Tylenchia	Chromadoria
Order	Tylenchida	Rhabditida
Suborder	Tylenchina	Tylenchina
Infra-order	Anguinata	Tylenchomorpha
Superfamily	Anguinoidea	Sphaerularioidea
Family	Anguinidae	Anguinidae
Subfamily	Anguininae	Anguininae
Genus	*Anguina*	*Anguina*

1 *Anguina tritici*

1.1 Common name

Seed gall nematode; bunted wheat; ear cockle

1.2 Scientific name

Anguina tritici (Steinbuch, 1799) Filipjev, 1936

1.3 Synonyms

Anguillula scandens Schneider, 1866
Anguillula tritici (Steinbuch) Grube, 1849
Anguillulina scandens (Schneider) Goodey, 1932
Anguillulina tritici (Steinbuch) Gervais and Van Beneden, 1859
Rhabditis tritici (Steinbuch) Dujardin, 1845
Tylenchus tritici (Steinbuch) Bastian, 1865
Tylenchus scandens (Schneider) Cobb, 1890
Vibrio tritici Steinbuch, 1799

1.4 EPPO code

ANGUTR

1.5 Diagnosis

To facilitate the identification of *Anguina tritici*, morphological, biochemical and molecular diagnosis approaches are presented.

1.5.1 Morphological diagnosis

Important morphological descriptions of *A. tritici* after Southey (1972) and Krall (1991) are summarized below. For measurements, see Table 3.

Mature female: Body obese, spirally coiled ventrally. Lip region low and flattened, slightly offset, cephalic framework weak. Cuticle very finely annulated. Procorpus often swollen by gland

DOI: 10.1079/9781789247367.0002

Table 3. Morphometric data for *Anguina tritici*, *A. agrostis* and *A. funesta* (Adapted from ISPM-27, DP 18: *Anguina* spp.).

Morphometric characters		Range, numerous populations		
		Anguina tritici	*Anguina agrostis*	*Anguina funesta*
Mature females	Length (mm)	3.0–5.2	1.3–2.7	1.65–2.44
	Stylet (µm)	8–11	8–12	7–10
	a (nematode body length/greatest width)	13–30	13.8–25.4	16.8–20.1
	b (nematode body length/pharynx length valve)	9.8–25.0	8.0–28.7	9.3–34.0
	c (body length/tail length)	24–63	25.2–44.0	18.1–41.2
	V% (distance from the anterior end to the vulva / nematode body length)	70–95	87–92	86.9–94.0
Males	Length (mm)	1.9–2.5	1.05–1.68	0.78–1.52
	Stylet (µm)	8–11	10–12	7–10
	a	21–30	23–38	20.3–30.9
	b	6.3–13.0	6–9	6.3–9.5
	c	17–28	20.0–28.4	16.1–24.9
	Spicules (µm)	35–40	25–40	16–28
	Gubernaculum (µm)	9–10	10–14	9–14
Second-stage juveniles	Length (mm)	0.75–0.95	0.55–1.25	0.81–0.87
	Stylet (µm)	10	10	7–10
	a	47–59	44–71	48–53
	b	4.0–6.3	3.2–6.1	4.2–4.6
	c	23–28	11.7–20	12.3–15.1

secretions but constricted at junction of metacorpus. Isthmus sometimes posteriorly swollen, offset from pharyngeal glands by a deep constriction. Pharyngeal glands not overlapping, or with very slight overlap of intestine. Excretory pore near junction of pharynx and intestine or slightly more posterior. Vulval lips protruding, orifice of small glands visible on vulval lips anterior and posterior to vulva. Ovary with two or more flexures, often reaching to pharyngeal region with oogonia in multiple rows arranged around a rachis. Spermatheca pyriform, separated from oviduct by a sphincter. Postvulval uterine sac present (Fig. 1.1). Tail conoid, tapering to obtuse or rounded tip, not mucronate.

MALE: Slenderer than female. Habitus upon heat relaxation curved either ventrally or dorsally. Testis with one or two flexures. Spicules stout, arcuate, with two ventral ridges running from tip to widest part (Fig. 1.1). Capitulum with distinct ventral folding at anterior. Gubernaculum simple, trough-like. Bursa leptoderan. Tail conoid, tip rounded or obtuse.

SECOND-STAGE JUVENILE: Body slender, not spirally coiled. Tail conoid, pointed (Fig. 1.1).

1.5.2 Biochemical and cytogenetic diagnosis

Biochemical and cytogenetic techniques are used to discover the degree of genetic similarity in a taxonomic group. Such data must be obtained for several species to compare with both in-group and out-group members. A comparative study of such data provides valuable information about characters and features that are the result of common genetic material, as well as those that are unique for a particular member of the group. The uniqueness determines the identity of the taxon (Siddiqi, 2000). Biochemical approaches such as the separation of proteins (general proteins and isozymes) by one-dimensional gel electrophoresis, isoelectric focusing, two-dimensional gel electrophoresis and sodium dodecyl sulphate–capillary gel electrophoresis are used for the biochemical diagnosis of plant-parasitic nematodes (Abrantes *et al.*, 2004). Serological techniques have also been found effective in the identification and quantification of nematodes, monoclonal antibodies being a more useful immunological tool than polyclonal antibodies (Abrantes *et al.*, 2004). Riley *et al.* (1988) characterized *A. agrostis*, *A. funesta* and *A. tritici* based on isozymes, using one-dimensional native gel electrophoresis.

1.5.3 Molecular diagnosis

Molecular diagnosis of *Anguina* spp. is based on polymerase chain reaction (PCR)–restriction fragment length polymorphism (RFLP) (Powers *et al.*, 2001), real-time PCR (Ma *et al.*, 2011; Li *et al.*, 2015) or sequencing of the internal transcribed spacer (ITS) region of ribosomal (r) RNA (Subbotin *et al.*, 2004). The choice of test depends on whether identification requires confirmation of both the presence and the absence of particular species, and on the availability of species standards for controls. The method described by Ma *et al.* (2011) is limited to the positive identification of *A. agrostis*, while the other methods can simultaneously distinguish multiple species within the same test. Powers *et al.* (2001) were the first to sequence the ITS1 region of *Anguina* spp. They combined PCR with analysis of RFLP and simultaneously distinguished a range of *Anguina* species from each other (Table 4). Subbotin *et al.* (2004) subsequently sequenced 58 populations of *Anguina*, *Ditylenchus*, *Heteroanguina* and *Mesoanguina* for phylogenetic analysis. At present there are 71 sequence accessions of rRNA fragments of *Anguina* spp., collected from different localities and host plants, in the United States National Center for Biotechnology Information

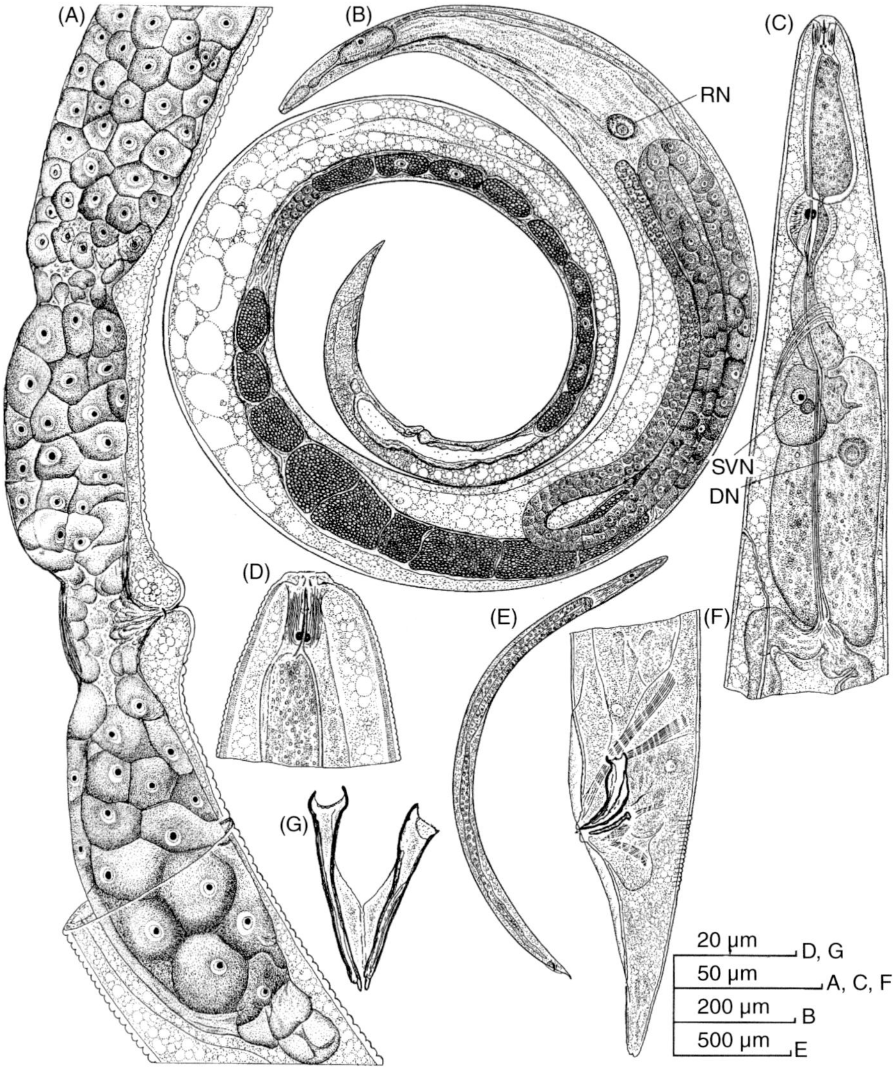

Fig. 1.1. Line drawing of *Anguina tritici from wheat grain*. Vulval region with uterus and postvulval uterine sac (A); female (B); oesophageal region of male (C); head end of female (D); male (E); tail end of male (F); spicule (G). DN, nucleus of dorsal oesophageal gland; SVN, nuclei of subventral glands; RN, nucleus of renette cell. [Reproduced from M.R. Siddiqi, 2000; courtesy CAB International]

Table 4. Restriction fragment sizes for *Anguina* species and associated restriction fragment length polymorphism (RFLP) patterns (after Powers *et al.*, 2001).

Species	*AluI*	#	*BsrI*	#	*EcoRI*	#	*HaeIII*	#	*HhaI*	#	*HinfI*	#	*TaqI*	#
Anguina agrostis	548	A	295, 238, 15	B	299, 249	A	548	A	548	A	448, 100	B	355, 135, 58	A
Anguina funesta	548	A	295, 238, 15	B	299, 249	A	548	A	548	A	448, 100	B	490, 58	B
Anguina tritici	277, 274	B	550	C	550	B	550	A	462, 88	B	550	E	492, 58	B

Code for the RFLP profile for each restriction enzyme.

(NCBI) public database. Details of some molecular methods for diagnosis of *Anguina* spp. are described below.

Li *et al.* (2015) designed a TaqMan real-time PCR to identify *A. tritici*, *A. agrostis*, *A. funesta* and *A. pacificae*. This test included forward and reverse genus-specific primers combined with a fluorescent probe. These primers and probe set were designed to serve as an internal control for confirming the presence of *Anguina* spp. as well as the integrity of the PCR components and user performance. The test also included primers and probe sets specifically designed for the detection of each of the target species mentioned above and was intended for identification of single juveniles. The sensitivity of the test was demonstrated through construction of standard curves from reactions using serially diluted nematode DNA and the test was able to detect as little as 1.25 copies of the ITS rDNA. The specificity of each primer and probe set was demonstrated in single-plex and duplex reactions (i.e. with the species-specific and genus-specific primers and probe sets) tested against all of the target species as well as several non-target nematodes, including *Meloidogyne* spp., *Pratylenchus* spp. and *Ditylenchus* spp. For the DNA extraction procedure, see Chapter 15, section 110.1. The ITS rRNA genus-and species-specific TaqMan primers and probes are:

A. tritici (ATfpr primers-probe set)

ATf (forward): 5′-GTTGCCTACGGCCGT-3′
ATr (reverse): 5′-ATGTAATCGATGTGGTACAGCCAT-3′
ATp (probe): 5′-6-FAM/ATCATGTCTTGGCTAGTG TAGACGTATCTG/BHQ-1-3′

A. agrostis (AAfpr primers-probe set)

AAf (forward): 5′-CGGTTGTTTACGGCCGT-3′
AAr (reverse): 5′-ATGTAGTCGGTGTGAAAACAGCCAT-3′
AAp (probe): 5′-6-FAM/ATCATGTCTTGGCTATTGTAGA CGTATCTG/BHQ-1-3′

A. funesta (AFfpr primers-probe set)

AFf (forward): 5′-GGTTGCTTACGGCCC-3′
AFr (reverse): 5′-GTGTAATCGATGTGATACAGCCCC-3′
AFp (probe): 5′-6-FAM/ATCATGTCTTGGCTATTAT AGACGTATCTG/BHQ-1-3′

A. pacificae (APfpr primers-probe set)

APf (forward): 5′-ACCGGTTGAATATTGGCTGT-3′
APr (reverse): 5′-ATGTAATCGATGTGAAACAGCCGT-3′
APp (probe): 5′-6-FAM/ATCATGTCTTGGAAAGTTTA GACGTATCTG/BHQ-1-3′

Anguina spp. (ASfpr primers-probe set)

ASf (forward): 5′-GTCTTATCGGTGGATCACTCGG-3′
ASr (reverse): 5′-TGCAGTTCACACCATATATCGCAG-3′
ASp (probe): 5′-TET/TCATAGATCGATGAAGAACG CAGCCA/BHQ-2-3′

The PCR amplification was performed in a real-time PCR using the following reagents and cycling parameters.

Reagent	–	Final concentration
Polymerase chain reaction (PCR)-grade water	–	final reaction volume of 25 µl
PCR buffer (including $MgCl_2$)	–	1× $MgCl_2$ 2.60 mM
dNTPs	–	0.24 mM
Species-specific primer (forward)	–	240 nM
Species-specific primer (reverse)	–	240 nM
Species-specific probe	–	120 nM
ASf internal control primer (forward)	–	160 nM
ASr internal control primer (reverse)	–	160 nM
ASp internal control probe	–	120 nM
DNA polymerase	–	1.0 U
DNA (volume)	–	1 µml
Cycling parameters		
Initial denaturation	–	95°C for 20 s
Number of cycles	–	40
Denaturation	–	95°C for 1 s
Optics	–	OFF
Annealing	–	60°C for 40 s
Optics	–	ON
Ramp	–	5°C per s
Expected amplicons size	–	74–85 base pairs

Ma *et al.* (2011) designed a species-specific TaqMan real-time PCR to identify juveniles of *A. agrostis*. It was evaluated against the target species *A. agrostis* as well as non-target species *A. tritici*, *A. wevelli* and *Ditylenchus destructor*. The ITS rRNA species-specific primers are:

PF: 5′-GTTTGCCTACCGGTTGTTTACG-3′
PR: 5′-CCACATGCAGTCGGTGTGAA-3′
TaqMan Pb: 5′-FAM-TCATGTCTTGGCTATTGTAGACG TATCTGA-TAMRA-3′

The amplification was performed in a real-time PCR using the Light Cycler according to the following cycling parameters.

Reagent	–	Final concentration
Polymerase chain reaction (PCR)-grade water	–	final reaction volume of 10 µl
PCR buffer	–	1× $MgCl_2$ 1.25 mM
dNTPs	–	0.2 mM
Primer PF (forward)	–	0.4 µM
Primer PR (reverse)	–	0.4 µM
Probe Pb	–	0.02 µM
DNA polymerase	–	0.5 U
DNA (volume)	–	1 µl
Cycling parameters		
Initial denaturation	–	94°C for 3 min
Number of cycles	–	45
Denaturation	–	94°C for 10 s
Annealing	–	60°C for 30 s
Expected amplicons size	–	88 base pairs

Detailed methodology and protocols for molecular diagnosis including DNA isolation, sequence analysis, construction of phylogenetic tree etc. are described in Chapter 15, section 110.1.

1.6 Geographical distribution

The seed gall nematode, *A. tritici* is worldwide in distribution and has been reported from the majority of wheat-growing

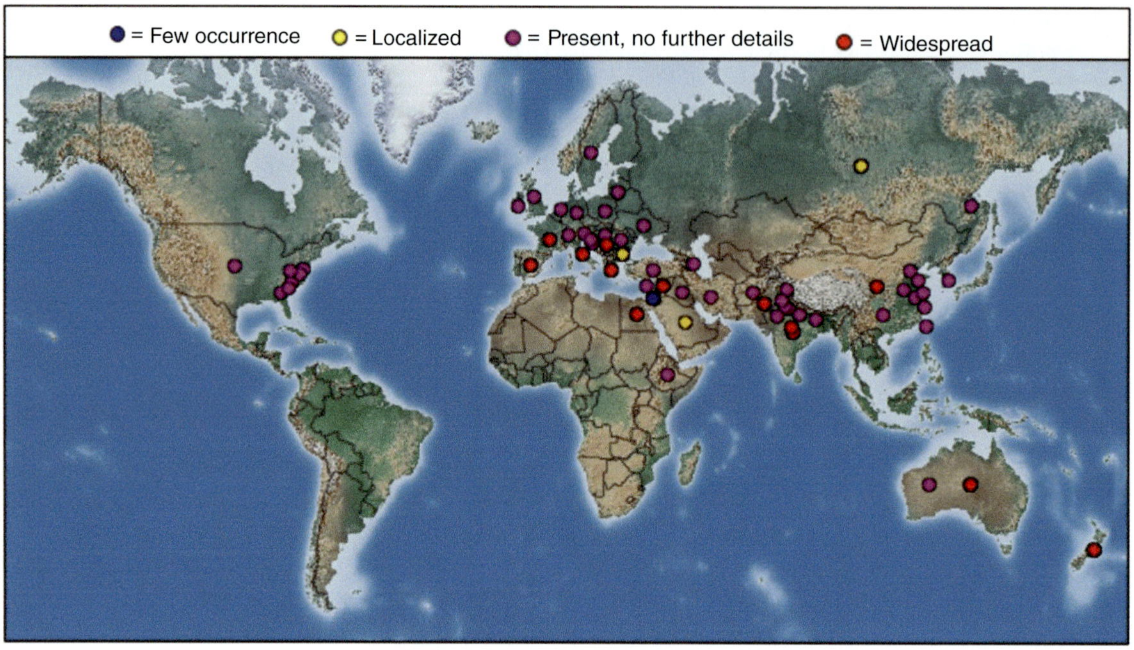

Fig. 1.2. Distribution map of *Anguina tritici.* [Adapted from CABI-ISC, 2020]

areas of the world (Fig. 1.2). At present, this nematode has almost been eradicated from all advanced European countries. *A. tritici* was present in several states of the USA, but was eliminated in 1975 and currently is not reported in the USA (Randhawa and Chitambar, 2017). However, it is still considered an important problem in several underdeveloped and developing countries like India, Pakistan, Ethiopia, Romania, Iraq, Syria and Yugoslavia (Nandal *et al.*, 2010) where the farmers have been using their old stock of seed for a long time. There are indications that *A. tritici* is becoming rare or extinct in countries or regions where it had been reported previously. Therefore, past records should be treated with caution. *A. tritici* has been known to occur in the following countries (CABI-ISC, 2020).

Africa: Egypt, Ethiopia.
Asia: Afghanistan, Azerbaijan, China, India, Iran, Iraq, Israel, Pakistan, Saudi Arabia, South Korea, Syria, Taiwan, Turkey.
Europe: Austria, Bulgaria, Croatia, Cyprus, Denmark, France, Germany, Greece, Hungary, Ireland, Italy, Lithuania, Netherlands, Poland, Romania, Russia, Serbia, Serbia and Montenegro, Slovenia, Spain, Sweden, Switzerland, Ukraine, United Kingdom.
Oceania: Australia, New Zealand.

1.7 Hosts

Wheat (*Triticum aestivum*) is considered the major host of this nematode, but some other crops like triticale (× *Tritico secale*), barley (*Hordeum vulgare*), oats (*Avena sativa*), rye (*Secale orientalis*), emmer (*Triticum dicoccum*), spelt (*Triticum spelta*) and littleseed canarygrass (*Phalaris minor*) have also been reported to be attacked by this nematode (Nandal *et al.*, 2010).

1.8 Symptoms

1.8.1 Ear cockle

The seed gall nematode causes typical above-ground symptoms at both the vegetative and the reproductive phases of the plant growth. Initially, the diseased plants show basal swelling of the stem (Nandal *et al.*, 2010), crinkling, curling, twisting and rolling of the leaves, profuse tillering and stunted growth (Fig. 1.3A). The symptoms on leaves are more conspicuous at the early stage of plant growth and occur due to ectoparasitic feeding (Khan and Athar, 1996). The symptoms on vegetative parts normally disappear at the later stage (Lal and Lal, 2006). During the reproductive phase of plant growth, the nematode larvae enter into the developing grain and feed endoparasitically (Clark *et al.*, 1991). Due to this feeding, diseased ear heads are formed (Fig. 1.3B), which are generally smaller and broader than the normal ones, with or without awns (Nandal *et al.*, 2010). In such ear heads, instead of formation of grains, cockles (seed galls) are formed which are initially green and turn brown to black at maturity and hard to the touch (Khan and Athar, 1998; Fig. 1.3C). Some or all grains are transformed into galls and filled with numerous infective juvenile nematodes (Subbotin and Riley, 2012). Green immature cockles are soft in consistency and contain all the stages of the nematode, whereas black mature cockles are hard in texture and contain a

Fig. 1.3. Symptoms of *Anguina tritici* (A–F), *A. agrostis* (G–I) and *A. funesta* (J–L). Infected wheat plant (A,B); infected and healthy ear heads (C); healthy seeds and galls (D); wheat ear heads showing yellow ear rot or *tundu* disease (E); emergence of second-stage juvenile from infested seed (F); (right) healthy and (centre, left) infested bentgrass with *A. agrostis* (G); an elongated, black seed gall and sliced SEM of gall containing anhydrobiotic juveniles (H); juveniles and eggs from seed gall of the grass *Holcus mollis* (I); (left) healthy seed of *Lolium rigidum*, (centre) galled and (right) gall colonized by the bacterium *Rathayibacter toxicus* (J,K); gumming disease of ryegrass due to *Rathayibacter toxicus* (L). [Courtesy A,E: NASI, India; B: Author's own photo; C: A.L. Taylor, Nemapix picture set, Bugwood.org; D: © 2011, Ian Riley, Adelaide, Australia; F: M. McClure, University of Arizona, Bugwood.org; G: Malcolm Storey, www.bioimages.org.uk; H: J. Eisenback; I: J. Bridge/CABI BioScience; J,L: J. Allen; K: © 2006, Ian Riley, Adelaide, Australia]

large number of second-stage juveniles (J_2) of *A. tritici* in the anhydrobiotic state (dormant/quiescent, Fig. 1.3F) which may survive for as long as 38 years (Ferris, 2013). Sometimes cockles (galls) are also formed on non-floral parts like leaves, glume, lemma, palea, awns, etc.

1.8.2 Yellow ear rot (tundu disease)

Tundu disease is caused by combined infection by the nematode *A. tritici* and the bacterium *Clavibacter tritici* (= *Rathayibacter tritici*). The vector role of the nematode is essential where

A. tritici carries the bacterial cells to the shoot meristem which is well protected by the compact leaf scales, and causes yellow ear rot disease (Khan, 1993). The initial symptoms of the disease are similar to those of ear cockle disease. However, at the later stage when low temperature, high humidity and cloudy weather prevail, a bright yellow slime or gum-like substance (Fig. 1.3E) is profusely exuded by the bacterium on abortive ear heads, leaves and stems. It becomes hard and brittle on drying (Nandal *et al.*, 2010). In this disease, normally no grains are produced, due to the destruction of floral organs by the bacterium. In partially diseased ear heads, some wrinkled shaped grains of poor quality are produced (Paruthi *et al.*, 1987). Normally in such ear heads, cockles are not formed. Sahebani *et al.* (2007) reported that attachment of a large number of bacterial cells of *C. tritici* (= *R. tritici*) on the nematode body induces weakness and mortality to nematodes. Freshly harvested infected wheat cockles containing the bacterium are also toxic to cattle and sheep (Anwar *et al.*, 2001).

1.9 Biology and life cycle

The life cycle of *A. tritici* synchronizes with the development of wheat plants and one generation completes in one season on wheat (Nandal *et al.*, 2010). The cockled seeds which contain quiescent larvae in the anhydrobiotic state are the sole source of nematode inoculum (Subbotin and Riley, 2012). The nematode reaches the field whenever nematode-infested cockles are sown along with healthy wheat seeds. The cockles, when coming into contact with soil moisture, absorb water and swell. This terminates the quiescent state and facilitates the migration of the larvae. The activation of dormant larvae and their complete release into the soil may occur within 10–15 days, depending upon temperature, moisture content and soil type (Paruthi *et al.*, 1990). These larvae move here and there in the soil to make their way to the growing points of young wheat plants. A layer of moisture is essentially required for the movement of larvae on the plant surface (Khan *et al.*, 1998). The second-stage larvae, during the vegetative growth of plant phase, remain ectoparasitic under the well-protected innermost layer of compact leaf scales and bud scales at the growing point (Paruthi *et al.*, 2009). At the time of initiation of floral primordia, the larvae invade the floral tissues and feed endoparasitically. This stimulates the development of galls. During the endoparasitic feeding, the larvae become enclosed in the cavities formed by overgrowth of floral initials which eventually are turned into cockles or galls. The larvae undergo three successive moults very quickly within 3–5 days of the invasion of floral primordia and develop into adult males and females in the ratio of 1:2 (Midha and Swarup, 1972). Reproduction is amphimictic and egg-laying starts within 6–12 days, with each female laying upto 30,000 or more eggs (Subbotin and Riley, 2012). First moulting takes place inside the egg and hatching occurs quickly, giving rise to second-stage larvae (McClure, 1988). This stage coincides with the drying and maturing of the wheat plants. Under the dry conditions the larvae lose body water and enter into a quiescent state of anhydrobiosis. The nematode completes one generation from J_2 to J_2 in a growing season (Nandal *et al.*, 2010).

1.10 Movement and means of dispersal

Trading of wheat grains infested with seed galls/cockles or sowing of cockles along with healthy seeds in fields are the main means of dispersal of *A. tritici* (Nandal *et al.*, 2010). Other means of spread include infected crop straw, flooding and rainfall, natural migration (maximum 20–30 cm), and sheep, cow or bird manure. *A. tritici* is also spread by wind and by animal feet, farm machinery and implements (Subbotin and Riley, 2012).

1.10.1 Plant parts liable to carry the nematode in trade/transport

Plant parts, including roots, flowers, inflorescences, cones, calyx, leaves, seedlings, micro-propagated plants, stems (above ground), shoots, tillers and true seeds (including grain), are liable to carry all stages (eggs, juveniles and adults) of the nematode.

1.10.2 Plant parts not known to carry the nematode in trade/transport

Bark and wood.

1.11 Economic importance

Wheat is the third-largest cereal crop, cultivated throughout the world after maize and rice (FAOSTAT, 2019). Ear cockle nematodes cause quantitative and qualitative losses in wheat, through either the formation of galls (Nandal *et al.*, 2010; Subbotin and Riley, 2012) or the occurrence of yellow ear-rot disease (Ferris, 2013). Wheat produced from partially attacked ear heads or yellow ear-rot disease has less market value (Anwar *et al.*, 2001). The average incidence of the diseases and loss in yield of wheat caused by this nematode from India has been reported as 1–3% but the losses on individual farms may be up to 30–50% (Nandal *et al.*, 2010). Khan and Athar (1996) reported that yield reduction in wheat is dependent on the inoculum level of the nematode; 15,000, 20,000 and 30,000 J_2/plant caused 6–18%, 12–28% and 24–43% yield decline to wheat varieties, respectively.

1.12 Pest risk analysis and looming threat from introduction

The wheat seed gall nematode *A. tritici* is the oldest-recorded plant-parasitic nematode, reported in 1743 by Needham. In the early 20th century, this nematode was considered as one of the important problems of wheat throughout the world. Due to the adoption of seed-cleaning techniques, its occurrence has become rarest in European countries but it is still considered an important problem in some underdeveloped and developing countries, including India. In the countries where the nematode is prevalent, it is considered as an important pest of wheat. The nematode infection in wheat inflorescence results in ear-cockle disease. The nematode in association with a bacterium, *Clavibacter tritici* (= *Rathayibacter tritici*), causes yellow ear-rot or *tundu* disease.

A. tritici is considered as a highly invasive nematode species due to its global distribution, the crop damage potential, seed-borne nature and long survival strategy. Infested grains are the principal means of international transport of the nematode, which can be very easily carried by wheat seed being moved as a food commodity. *A. tritici* has been intercepted on several occasions in consignments of wheat as a food commodity. The likelihood of introduction or spread of *A. tritici* into new areas is minimal, largely due to effective ongoing modern seed grading and other practices, but the potential economic damage by this pest should not be underestimated. As stated earlier, wheat seeds infested with low populations of the nematode may not show typical symptoms of seed galling and pose the probability of accidental introductions. Therefore, the likely risk posed to the wheat crop and the looming threat from its introduction are evaluated below.

1.12.1 Climate/host interaction

Wheat, the main host of *A. tritici*, is the third-largest cereal crop of the world after maize and rice. More of the Earth's surface is covered by wheat than by any other food crop. Wheat is a hardy crop that can grow in a wide range of climatic conditions and that permits large-scale cultivation and long-term storage. If *A. tritici* is introduced into non-native areas, it is likely to establish wherever wheat is grown.

1.12.2 Host range

Primary hosts of *A. tritici* include wheat, rye, spelt and emmer wheat. Oats, barley, durum wheat and rivet wheat are considered minor hosts.

1.12.3 Pest dispersal potential

Reproduction of *A. tritici* is amphimictic. Mating occurs and a single female produces up to 30,000 eggs over several weeks. These eggs hatch and produce second-stage juvenile (J_2s), which remain within the galls as the survival stage perpetuates as the invasive stage in the following years. Dry galls are harvested along with the healthy grains, and each gall may contain thousands of J_2 (Subbotin and Riley, 2012). *A. tritici* produces one generation per year (Ferris, 2013). It is likely to enter new areas through the transport of infected host plants or seed material, since wheat is an important commodity for import and export and its international trade is allowed in many countries, some of which are known to have *A. tritici* (USDA, 2015). Because of its high reproduction and dispersal potential, *A. tritici* receives a high rating in this category.

1.12.4 Economic impact

Seed gall nematodes can cause severe crop losses to wheat (70%) and rye (35–65%) in underdeveloped and developing countries due to poor agricultural practices, monoculture and use of seeds from the previous season or non-certified seeds (Nandal *et al.*, 2010). Growth in global wheat consumption has been facilitated by imports, particularly into developing countries, including many tropical non-wheat-producing countries and those where increasing land and water constraints hinder wheat cultivation (FAOSTAT, 2019). Introduction of seed gall nematodes into developed countries may cause insignificant damage to the wheat and rye, due to the use of modern agricultural production systems in these countries. However, the introduction can severely hamper the export of wheat grains to international markets, due to the presence of the nematode in production areas. The interaction of this nematode and *C. tritici* results in *tundu* or spike blight or yellow ear-rot disease (Ferris, 2013). *Dilophospora alopecuri* in association with *A. tritici* could result in dilophosphorosis disease of wheat where the spike is covered by a black sticky mass. Since the nematode directly affects the crop produce (grain), it could lower crop yield as well as crop value (including increasing crop production costs) and could trigger the loss of markets (including quarantines). Further, the nematode can vector another pestiferous organism and can be injurious or poisonous to agriculturally important animals.

1.12.5 Environmental impact

A. tritici infestation is likely to cause ear cockle disease in all wheat-growing areas and could trigger additional official or private treatment programmes. To prevent ear cockle and the spread of this nematode species, growers would require seed cleaning and crop rotation to eliminate the pest. Hot-water treatments would be needed to eradicate *A. tritici* from seed lots or mechanical separation of nematode galls from the healthy grains using sieves or brine solution (PKB, 2016). Additionally, infestations of *A. tritici* could significantly impact irrigation, ornamental plantings and other cultural practices.

1.12.6 Global trade and nematode interceptions

The nematode has been found in shipments of wheat seed being traded as a food commodity. Shipments from infested countries are assumed to be of serious concern. Introduction of the nematode species into new areas can likely occur in seeds that pass inspection, because they may be infested with low numbers of nematode inoculum and not transformed into typical galls. The nematode has an invasive nature and alone is responsible for causing ear cockle disease, but in association with the bacterium *Clavibacter tritici* it causes yellow ear rot or *tundu* disease. Both these diseases, being seed-borne in nature, can be easily detected and controlled. The dispersal of the nematode from one country to another is through the transportation of infected seeds, straw from an infected crop, or manure of cows, sheep, sparrows, pigeons, etc. fed on the cockle grains.

1.13 Invasiveness rating

Considering the consequences of introduction and pest risk analysis of *A. tritici*, the following invasiveness rating has been assigned.

Climate/host interaction	–	High (3)
Host range	–	Medium (2)
Dispersal potential	–	High (3)
Economic impact	–	High (3)
Environmental impact	–	High (3)
Post-entry distribution	–	Medium (–2)
Final rating (score)	–	High (12)

Based on the above rating, *A. tritici* receives a score of 12 out of 15 and is rated as HIGHLY INVASIVE NEMATODE.

1.14 Region-wise status of invasiveness

A. tritici is in the 'Harmful Organisms' list for Argentina, Brazil, Chile, Colombia, Ecuador, Egypt, Guatemala, Indonesia, Israel, Madagascar, Namibia, Nepal, New Zealand, Paraguay, Peru, South Africa, Taiwan, Thailand, Timor-Leste, and Uruguay (USDA-PCIT, 2017). Based on the worldwide geographical distribution of *A. tritici* and likely introduction in the major wheat-producing countries where it has so far been absent, the region-wise status of invasiveness is evaluated (for Regional Plant Protection Organizations, see Chapter 16, section 112). Region-wise top wheat-producing countries where *A. tritici* attains invasive status are:

COSAVE: Argentina, Brazil
EPPO: Belarus, Belgium, Czech Republic, Kazakhstan, Morocco, Turkmenistan, Uzbekistan
IAPSC: Algeria
NAPPO: Canada, Mexico, USA.

1.15 Management measures

1.15.1 Cultural methods

Since both the pathogens (*A. tritici* and *C. tritici*) are seed-borne in nature, use of cockle-free seeds is the best approach to control the disease (Nandal *et al.*, 2010). Seeds can be cleaned and cockles can be separated from healthy seeds using mechanical and physical methods, or healthy seeds can be procured from seed agencies. Sukul (1970) reported decreased incidence of disease caused by *A. tritici* by growing wheat along with *Polygonium hydropiper*. Inter-culturing of mustard (*Brassica campestris*) and rapeseed (*B. juncea*) with wheat also showed a remarkable decrease in the diseases caused by *A. tritici* and improved wheat yield (Verma, 1991). Singh *et al.* (2009) studied the effect of sowing dates on the incidence of ear cockle disease of wheat and reported that delay in sowing by 35 days from November onwards increased the incidence of disease in most of the cultivars of wheat.

1.15.2 Mechanical methods of separation of cockles

A method of separation that is common in certain parts of India and China is to use a winnowing basket in an open area or in front of a fan. The contaminated seeds (cockles), being lighter in weight, fall away from the heap of healthy grains and can be separated. The efficiency of removal of galls by this method depends upon the velocity of wind at the time of winnowing. An alternative method is screening infested grains by putting them on a sieve (3 mm size) and rotating sidewise for some time. The galls, being smaller than the healthy seeds, fall down while the healthy seeds are retained on the sieve. Again, complete removal of galls is not achieved in this method, because some large-sized cockles are retained on the sieve along with the healthy grains (Paruthi *et al.*, 2009).

1.15.3 Physical methods of separation of cockles

WATER FLOTATION This method is based on putting the nematode-infested wheat seeds into a drum containing water and stirring it well with a wooden stick or by hand for some time. The galls being lighter in weight than the healthy grains start floating on the surface of the water and can then be removed by ordinary sieves. The cockles should be removed as early as possible, because they start sinking along with the seed after absorbing water.

SALT SEDIMENTATION Another method of removing the cockles from the grains is putting the seed in different concentrations of salt solution. Addition of salt increases the range and duration of floating by the cockles, hence they are efficiently removed from healthy seeds. Complete removal of cockles from the grains can be achieved at or above 10% concentration of salt. This method is more efficient than the water flotation method but is less often adopted because the treated seeds need to be washed in water several times before sowing, to avoid the injurious effect of salt on germination of seeds (Paruthi and Bhatti, 1992).

HOT-WATER TREATMENT The hot-water method is based on the principle that animal tissue is slightly more sensitive to temperature (45–55°C) than the plant tissue. Hence immersing the nematode-infested wheat seeds in hot water at 54–56°C for 10–12 min or 50°C for 30 min or 30°C for 4 h effectively killed *A. tritici* larvae inside the galls (Nandal *et al.*, 2010). Although this method has been found very effective, it could not be used in farmers' fields due to the involvement of several technical difficulties in maintaining the temperature of the water, and duration of exposure.

MICROWAVE TREATMENT A few researchers have explored the feasibility of using the energy of microwaves to kill J_2 of *A. tritici* inside the gall. Khan and Hayat (1999) exposed water-soaked and unsoaked wheat seeds and galls to microwaves at 2450 MHz/sec for 4–30 s. The exposure for 14 s onwards induced 40–76% and 20–48% mortality in the juveniles from soaked and unsoaked seeds, respectively.

RESISTANT VARIETIES The use of resistant varieties is one of the most economical and feasible methods of controlling the disease. A large number of varieties and lines have been screened by various workers without much success (Khan and Athar, 1998). However, some lines and varieties, such as Saberbeg (Saleh and Fattah, 1990) and AUS 15854 (Baheti and Verma, 2001), have been reported to be resistant against this nematode.

1.15.4 Chemical control

Limited information is available on chemical control of *A. tritici*. Spraying of ethyl parathion at 0.05% provided complete control of *A. tritici* on wheat (Singh and Prasad, 1972). Foliar application of monocrotophos on 1-month-old wheat plants at 1 kg/ha led to the recovery of plants from nematode attack and improved the yield of commonly grown varieties of wheat by 46% (Khan *et al.*, 1998). Soil application of carbofuran at 1 kg a.i./ha after 50 days of sowing resulted in 18.3% disease incidence caused by *A. tritici* in comparison with 46.7% in an untreated check (Paruthi *et al.*, 2009).

1.15.5 Phytosanitary measures

Ear cockles are the only source for the perpetuation of the disease and their removal from contaminated seed lots can completely eradicate the disease (Luc *et al.*, 1990). *A. tritici* has been eliminated, or reduced to a minimal number of infestations, in Europe and the USA by seed cleaning, crop rotation and fallow (Brown, 1987). Selective quarantine should be imposed in regions of the world where the disease is not under control. For further details on International Sanitary and Phytosanitary Measures (ISPMs), see Chapter 16 (International Regulations), section 115.

2 *Anguina agrostis*

2.1 Common name

Bentgrass nematode

2.2 Scientific name

Anguina agrostis (Steinbuch, 1799) Filipjev, 1936

2.3 Synonyms

Anguillula agrostidis (Steinbuch, 1799) Warming, 1877
Anguillula agrostis (Steinbuch, 1799) Ehrenberg, 1838
Anguillula phalaridis (Steinbuch, 1799) Ehrenberg, 1838
Anguillulina agrostis (Steinbuch, 1799) Goodey, 1932
Anguillulina phalaridis (Steinbuch, 1799) Goodey, 1932
Anguina phalaridis (Steinbuch, 1779) Chizhov, 1980
Anguina poophila Kiryanova, 1952
Anguina lolii Price, 1973
Tylenchus agrostidis Bastian, 1865
Tylenchus agrostis (Steinbuch, 1799) Goodey, 1930
Tylenchus phalaridis (Steinbuch, 1799) Orley, 1880
Tylenchus phlei Horn, 1888
Vibrio phalaridis Steinbuch, 1799
Vibrio agrostis Steinbuch, 1799

2.4 EPPO code

ANGUAG

2.5 Diagnosis

2.5.1 Morphological diagnosis

Important morphological descriptions of *Anguina agrostis* after Southey (1973) and Krall (1991) are summarized below. For measurements, see Table 3.

MATURE FEMALE: Based on specimens from the type host. Body obese, C-shaped to spirally coiled ventrally. Lip region low and flattened, offset by a fine constriction. Cuticle marked by fine annulations. Lateral fields not discernible on fully developed adult females, six incisures visible on immature specimens. Neither procorpus nor isthmus exhibiting marked swellings, the former slightly contracted at junction with metacorpus. Isthmus occasionally folded in mature specimens by forward pressure of gonad. Pharyngeal glands not overlapping, or with very slight overlap of intestine. Vulval lips prominent. Ovary usually with two flexures, often reaching to pharyngeal region with oogonia in multiple rows arranged around a rachis. Spermatheca pyriform, separated from oviduct by a constriction. Postvulval uterine sac present, 36–63% of vulva–anus distance. Tail conoid with acute terminus (Fig. 1.4A).
MALE: Smaller and slenderer than female. Habitus upon heat relaxation curved ventral to almost straight. Lateral field difficult to discern on mature specimens, reported to have six incisures. Testis usually reflexed once, spermatocytes arranged about a rachis. Spicules slenderer in build than those of *A. tritici*, capitulum showing little or no ventral folding at anterior. Two ventral ridges running from tip of each spicule to widest part, before converging and joining capitulum. Gubernaculum simple, bursa leptoderan, ending just short of acute or finely mucronate tail tip (Fig. 1.4B).
SECOND-STAGE JUVENILE: Body slender, not spirally coiled; tail conoid, pointed.

2.5.2 Biochemical and molecular diagnosis

Morphologically *A. agrostis* can be easily differentiated from *A. tritici* but not from *A. funesta*. To distinguish *A. agrostis* from *A. funesta*, biochemical diagnosis using isozymes (Riley *et al.*, 1988) and molecular analyses (Powers *et al.*, 2001; Li *et al.*, 2015) are more reliable. For details, see section 1.5.3.

2.6 Geographical distribution

A. agrostis is worldwide in distribution and has been reported from different countries on bentgrass, especially in the Pacific

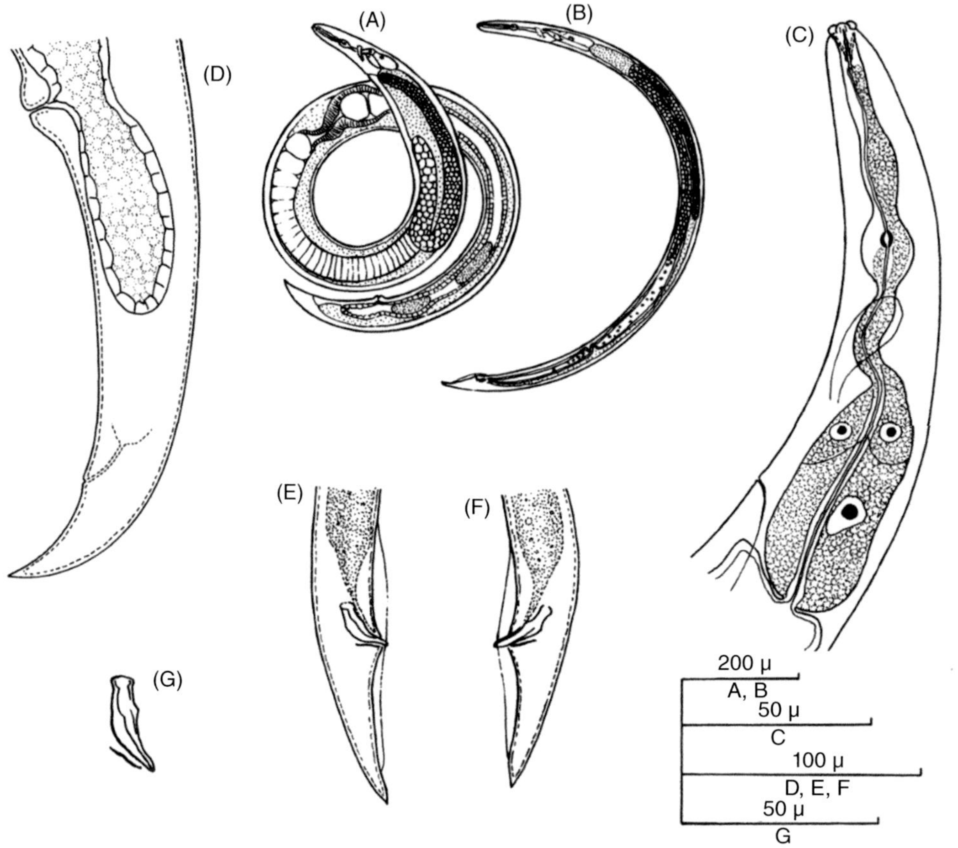

Fig. 1.4. Line drawing of *Anguina agrostis*. Adult female (A); adult male (B); head and oesophageal region of female (C); tail of female (D); bursa (E,F); spicule (G). [Reproduced from Southey, 1973, CIH Descriptions of Plant-parasitic Nematodes, Set 2, No. 20; courtesy CAB International]

Northwest USA and New Zealand. The nematode is known to occur in the following countries (CABI-CPC, 2010).

Africa: South Africa
Asia: China, Georgia, Kyrgyzstan
Europe: Czech Republic, Estonia, Finland, Germany, Ireland, Netherlands, Norway, Poland, Russian Federation, Slovakia, Sweden, Ukraine, United Kingdom
North America: Canada, USA
Oceania: Australia, New Zealand

2.7 Host plants

A. agrostis is considered to be a species complex with pathotypes differing in host range (Krall, 1991; Brzeski, 1998). Subbotin *et al.* (2004) supported the concept of narrow specialization of seed gall nematodes, concluding that *A. agrostis* occurs in only one host, *Agrostis capillaris*, and that other *Agrostis* species are hosts for two further undescribed species of *Anguina*. The type host of *A. agrostis* is *Agrostis tenuis*. In addition to various bentgrass species like *Agrostis canina* (velvet bentgrass), *Agrostis capillaris* (brown bentgrass), *Agrostis exarata* and *Agrostis stolonifera* var. *palustris* (bentgrass). *Anguina agrostis sensu lato* has been reported from other grass genera, including *Bromus erectus* (upright brome grass), *Dactylis glomerata* (orchard grass), *Festuca nigrescens* (Chewings fescue), *Festuca ovina* (sheep's fescue), *Lolium rigidum* (rigid ryegrass), *Phleum boehmeri*, *Phleum phleoides*, *Phleum pratense* (timothy), *Poa alpina*, *Poa annua* (annual meadowgrass), *Poa pratensis* (smooth-stalked meadowgrass) and *Trisetum flavescens* (yellow oatgrass).

2.8 Symptoms

A. agrostis causes no obvious symptoms on grass shoots except that of the inflorescence, which is so distorted that the parasitized plants were once described as a new plant species, *Agrostis sylvaticus*. In grasses, seed galls are difficult to detect as they are covered by lemmas and paleae (Fig. 1.3G). A small scarifier can be used to remove lemmas and paleae without damage to seeds or galls. This allows visual identification of galls (Alderman *et al.*, 2003). Galled flowers have glumes of two or three times the normal length, lemmas five to eight times the normal length, projecting beyond the glumes as a sharp point, and paleae developing to about four times the normal length. Lodicules, stamens and sometimes other flower parts are suppressed in such parasitized flowers (Goodey, 1932). Galls are at first greenish and later become dark purple-brown. Seed galls containing the nematodes are dark brown (Fig. 1.3H). They may look similar to normal seeds but are less heavy and hence can be separated mechanically from them. Adult nematodes (one to three females, one to three males) are found within these galls. Later, the females lay eggs and die, the eggs hatch and the second-stage juveniles fill the gall, which then turns into a seed gall.

2.9 Biology and life cycle

Reproduction is amphimictic and females and males mate within the gall. The female lays as many as 1000 eggs. Shortly after oviposition, the females and males die and disintegrate. The eggs are embryonated and the first-stage juveniles moult within the egg. The second-stage juveniles hatch and aggregate and undergo quiescence as the gall dries off (Bird and Stynes, 1981). The optimum temperature for egg development, from laying to hatching, is 18–20°C. The minimum time recorded for embryogenesis to hatching is 9–10 days. Embryogenesis is inhibited at temperatures of 27°C and above and hatching at temperatures greater than 23°C. Second-stage juveniles hatch and invade young grasses early in May (Europe) or at the end of August (Australia). The juveniles migrate to the inflorescence and incite seed galls. Only one generation completes in a year (Bird and Stynes, 1981).

In dry galls, the second-stage juveniles remain viable for up to 10 years. During rains, galls are softened and the juveniles escape (Southey, 1974). When the galls are soaked in water, the mucoprotein in which the juveniles are embedded swells, causing the walls of the gall to rupture, leading to expulsion of the juvenile mass. The second-stage juveniles feed ectoparasitically on the growing points of the shoots of their host grasses until the plant begins to form its inflorescence. The larvae enter young flower tissues, invade developing ovules and quickly moult three times to become adults. The ovary of each flower is thus replaced by a gall which contains one to three adult females and about as many males, although sometimes only a single individual is found in a gall. In the Pacific Northwest USA, the life cycle is completed in 3–4 weeks (Bird and Stynes, 1981).

2.10 Interactions with other pathogens

The nematodes sometimes introduce the bacterium *Corynebacterium rathayi* into the galls induced by *A. agrostis* in annual ryegrass (*Lolium rigidum*), which subsequently become toxic to grazing animals (Bird and Riddle, 1984). The bacterium *C. rathayi* is carried on the cuticle surface of the nematodes. Dauer juveniles of *A. agrostis* are not attracted to *C. rathayi* (from bacterial galls), but on contact, the bacteria attach to the nematode cuticle surface (Bird and Riddle, 1984). Shaw and Muth (1949) reported that in Oregon, USA, cattle, horses and sheep developed symptoms of poisoning such as nervous disorders, muscle trembling and lack of coordination, sometimes resulting in death, when they had fed on *Festuca nigrescens* infested with *A. agrostis* galls.

2.11 Economic impact

Bentgrass represents the most widely used grass for golf-course putting greens, gardens, etc. *A. agrostis* is considered to be a serious and potentially important nematode pest of bentgrass, especially in the Pacific Northwest USA and New Zealand (Bird, 1985). Samples from Oregon, USA, during 1996–2000 recovered 35–50% of orchard grass seeds containing *Anguina* galls and it caused 75% reduction in grass seed production. Fresh galls of *A. agrostis* from Chewing fescue (*Festuca rubra*) are poisonous to cattle and sheep (Apt *et al*., 1960).

Galls induced by this nematode in annual ryegrass *Lolium rigidum* in Western Australia replace the seeds and the nematode sometimes introduces *Rathayibacter rathayi* into the galls (Riley and McKay, 1990). These subsequently become toxic to grazing animals. Seed galls containing the nematodes are dark brown and those colonized by *R. rathayi* are yellow. Walls of the former are twice as thick as those of the latter. Both types of gall are originally initiated by the nematode (Bird *et al*., 1980).

2.12 Movement and means of dispersal

The primary means of dispersal is by seeds infested with seed galls/cockles. Other means of spread include the following.

2.12.1 Plant parts liable to carry the nematode in trade/transport

Plant parts including roots, flowers, inflorescences, cones, calyx, leaves, seedlings, micro-propagated plants, stems (above ground), shoots, tillers and true seeds (including grain) are liable to carry all stages (eggs, juveniles and adult) of the nematode.

2.12.2 Plant parts not known to carry the nematode in trade/transport

Bark and wood.

2.13 Pest risk analysis and looming threat from introduction

A. agrostis is considered as a serious and potentially important nematode pest of bentgrass (Southey, 1973). Due to its global distribution, wide host range and crop damage potential, the likelihood of introduction or spread of *A. agrostis* into new areas puts golf-course greens and gardens at risk. The chance of introduction is minimal, largely due to dispersal potential, but the economic damage potential of this pest should not be underestimated.

2.14 Invasiveness rating

Considering the consequences of introduction and pest risk analysis of *A. agrostis*, the following invasiveness rating has been assigned.

Climate/host interaction	–	Medium (2)
Hosts range	–	High (3)
Dispersal potential	–	Low (1)
Economic impact	–	High (3)
Environmental impact	–	Medium (2)
Post-entry distribution	–	Low (–1)
Final rating (score)	–	Medium (10)

Based on the above invasiveness rating, *A. agrostis* receives a score of 10 out of 15 and is rated as MEDIUM INVASIVE NEMATODE.

2.15 Region-wise status of invasiveness

APPPC: Afghanistan, Armenia, Azerbaijan, India, Iran, Iraq, Israel, Jordan, Kuwait, Lebanon, Pakistan, Saudi Arabia, Syria.
COSAVE: Argentina, Chile.
EPPO: Albania, Bulgaria, France, Greece, Italy, Morocco, Portugal, Spain, Switzerland, Turkmenistan, Uzbekistan, Yugoslavia (former).
IAPSC: Algeria, Egypt, Libya, Tunisia.

2.16 Management measures

2.16.1 Mechanical control

In infested areas, all harvesting machinery, including cutting and threshing machines, tractors, trucks and other equipment, including seed bags, clothing and boots of farm workers, should be cleaned to remove all trash and chaff before moving to clean areas. Mechanical separation to remove lighter infested galls removes most but not all galls from grass seed. Seed harvesting procedures can also minimize the disease, as indicated by the reduction of seed head galls in a *Trisetum flavescens* field from 8.2% to 0.11% by cutting for fodder in one year and taking the first seed harvest in the following year (Wagner and Domling, 1974).

2.16.2 Hot-water treatment

Hot-water treatment of infested seeds at 52°C for 15 min, after a 2 h pre-soak in tepid water, controls nematodes and reduces germination by only 5%. Hot-water treatment is effective only when the galls are 'naked', not covered by the inner paleae of the inflorescence (Christie, 1959).

2.16.3 Chemical control

Satisfactory control without serious effect on seed germination was obtained by fumigation of *Agrostis tenuis* seeds with methyl bromide (Hague, 1963).

2.16.4 Phytosanitary measures

See Chapter 16, section 115.

3 *Anguina funesta*

3.1 Common name

Seed gall nematode

3.2 Scientific name

Anguina funesta Price, Fisher and Kerr, 1979

3.3 Synonym

Anguina lolii Bird and Stynes, 1977

3.4 EPPO code

ANGUFU

3.5 Diagnosis

3.5.1 Morphological diagnosis

Important morphological descriptions of *Anguina funesta* after Price *et al.* (1979) are summarized below. For morphometric measurements, see Table 3.

FEMALE: Young females are fully motile, but older females, in which gross expansion of the ovary has occurred, are strongly ventrally curved and capable of only weak movements of head and tail. Habitus following heat relaxation ventrally curved, forming a complete circle, with head and tail overlapping (Fig 1.5B). Lips slightly offset and rounded in front, cephalic framework lightly cuticularized. Stylet length 7–10 µm with conus and shaft of roughly equal length and with well-developed knobs. Pharynx 64–178 µm long with a wide procorpus opening to a muscular metacorpus, ovate to spheroid, 17–25 µm long. Pharyngeal–intestinal junction obscured by large dorsal gland, 45–72 µm in length, ovate to spatulate with a prominent nucleus. Pharyngeal glands slightly overlapping the intestine. Hemizonid between base of metacorpus and anterior end of dorsal pharyngeal gland, at 80–100 µm from anterior. Excretory pore located more posteriorly, 105–155 µm from anterior. Lateral field difficult to discern. Vulva with prominent lips. Anterior ovary with

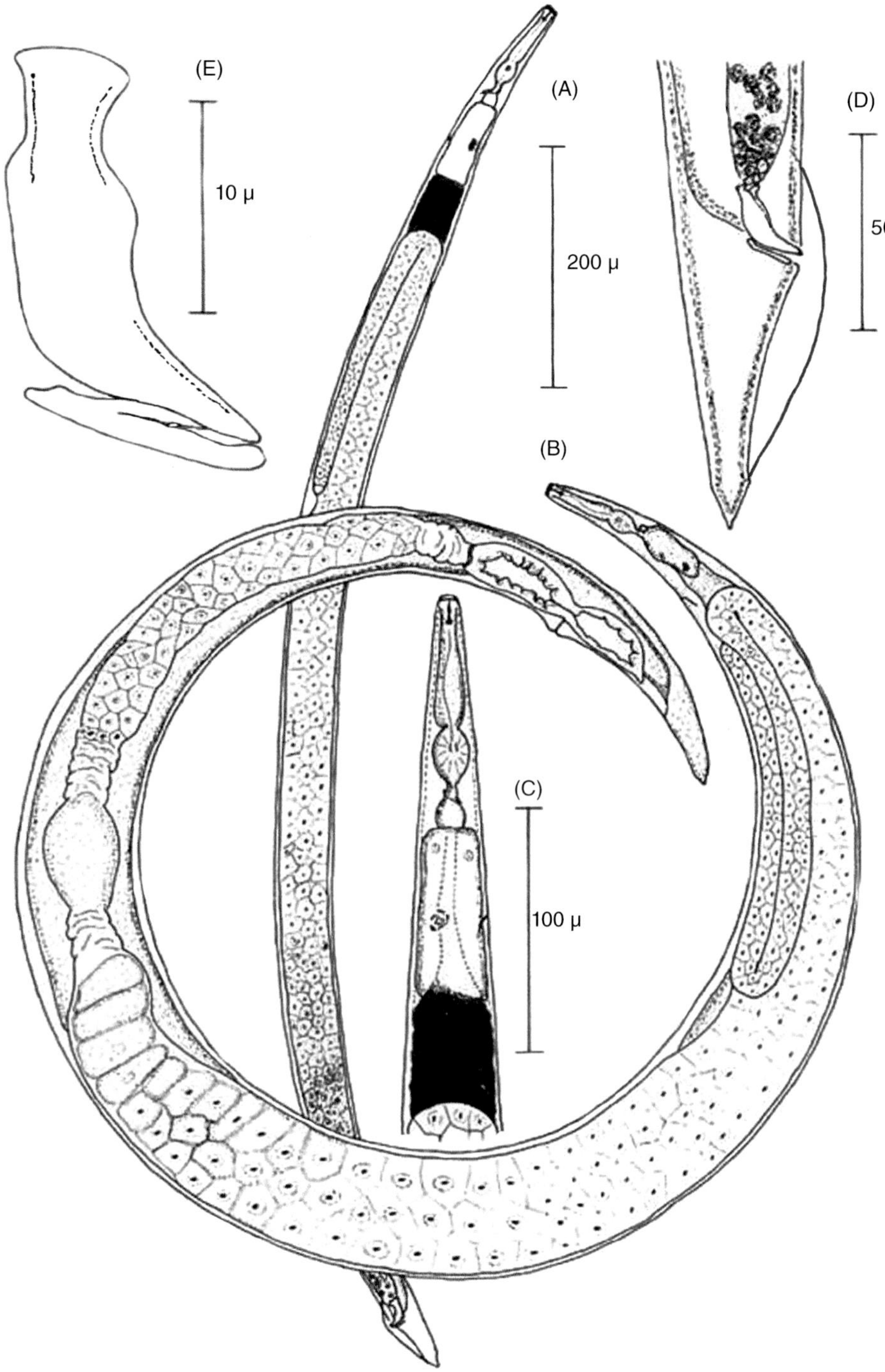

Fig. 1.5. Line drawing of *Anguina funesta*. Adult male (A); adult female (B); male anterior (C); male tail (D); and spicule and gubernaculum (E). [Price *et al.*, 1979; courtesy Nematologica]

one or two flexures. Rarely, gonad not reflexed but rather extended anteriorly to base of procorpus. Oocytes arranged in three or four rows about a rachis, except near base of ovary where these increase to five or six rows. Spermatheca ovate or elongated, long and 25–40 μm wide. Crustaformeria long and slender, made up of more than four columns of cells and up to 350 μm long, separated from spermatheca and uterus by short constrictions, 10–15 μm long. Crustaformeria often containing sperm cells and up to eight eggs. Uterus thick-walled, 70–100 μm long. Postvulval uterine sac approximately same length as uterus, 62–112 μm. Tail 48–112 μm long, vulva–anus distance 86–73 μm. Body width at anus approximately half that at vulva. Tail occasionally bluntly rounded, more usually conically pointed, sometimes with mucronate tip (Fig. 1.5B).

Male: Shorter and thinner than female (Fig. 1.5A). Habitus upon heat relaxation straight or slightly curved. Lip and pharyngeal regions of male similar to that of female, although dorsal pharyngeal gland larger, almost rectangular, 50–78 μm long (Fig. 1.5C). Hemizonid and excretory pore 71–90 μm and 102–147 μm from anterior, respectively. Lateral field with five or six incisures, broken at intervals and occupying one-quarter to one-third of body width. Testis nearly always reflexed once. Testis with spermatocytes in multiple rows about a rachis. Spicules paired, non-fused and arcuate, each 16–28 μm long with characteristic bulges on manubrium and where manubrium and shaft join. Gubernaculum slim and trough-like (Fig. 1.5D). Bursa leptoderan, extending almost to tail tip, 44–114 μm in length. Tail 43–72 μm long with terminus conically pointed (Fig. 1.5C). Body width at cloaca 17–43 μm.

3.5.2 *Biochemical and molecular diagnosis*

Morphologically *A. agrostis* and *A. funesta* are similar and *A. funesta* was considered a junior synonym of *A. agrostis*. However, despite their morphological affinities, these two species can be differentiated by biochemical diagnosis using isozymes (Riley *et al.*, 1988) and molecular analyses (Powers *et al.*, 2001; Li *et al.*, 2015). For details, see section 1.5.3.

3.6 Geographical distribution

A. funesta was reported from Australia (Price *et al.*, 1979) and has been detected in Oregon, USA (Meng *et al.*, 2012).

3.7 Host plant

The principal host of *A. funesta* is annual ryegrass *Lolium rigidum*.

3.8 Symptoms

The nematodes congregate near the apical meristems until ovary initiation, then stimulate ovary primordia to develop into galls. Occasionally, galls are produced in stamen primordia or, in very heavily infested plants, on glumes or rachis (McCay and Ophel, 1993). *A. funesta* is recorded as a vector of *Rathayibacter toxicus*, which causes the disease annual ryegrass toxicity when consumed by livestock. Rangeland infested by the nematode and bacterium is unusable for grazing. Symptoms of infestation are shown in Fig. 1.3J-L.

3.9 Biology and life cycle

The life cycle of *A. funesta* is similar to that of *A. agrostis*. The second-stage juveniles (J_2) emerge from the seed galls in the soil and crawl on the newly germinated ryegrass seedlings. The J_2 become located between young leaves and penetrate flower buds at the time of flower bud initiation, where they stimulate gall formation in place of normal grains. Juvenile development is completed in the galls. Newly formed females deposit eggs, which hatch, producing J_2, which remain, encased in the galls, and perpetuate the plant infection in the following years (Price *et al.*, 1979). Dried galls are harvested with seeds. During dry summers, *A. funesta* survives in the soil within these seed galls as anhydrobiotic second-stage juveniles. During winter, the nematodes are released from decaying galls and, via water droplets in moist soil, they invade new host seedlings, where they feed upon the young leaves. The nematodes congregate near the apical meristems until ovary initiation, then invade and stimulate ovary primordia to develop into galls. Occasionally, galls are produced in stamen primordia or, in very heavily infested plants, on glumes or rachis (McCay and Ophel, 1993).

3.10 Interactions with other pathogens

A. funesta is associated with a yellow slime bacterium, *Rathayibacter toxicus* (formerly *Clavibacter toxicus* and *Corynebacterium toxicus*). The J_2 can acquire bacterial cells from the soil or from infected plants. The bacterium is introduced in the plant tissue during seed gall formation. Such galls are fatally toxic to grazing livestock.

3.11 Economic impact

A. funesta is recorded as a vector of *R. toxicus*, which causes the disease annual ryegrass toxicity when consumed by livestock. Rangeland infested by the nematode and bacterium is unusable for grazing. Annual ryegrass toxicity is responsible for severe losses in the livestock industry in Australia (Price *et al.*, 1979).

3.12 Movement and means of dispersal

See section 2.12.

3.13 Pest risk analysis and looming threat from introduction

The nematode is a serious damaging pest for the livestock industry, pastures, rangeland and land for recreational uses. It is also of major regulatory significance. A complete pest risk assessment should be prepared for this nematode. The accidental introduction of this nematode would have devastating repercussions on grain exports to domestic and international markets because of the quarantines imposed against this pest by many countries.

3.14 Invasiveness rating

Considering the consequences of introduction and pest risk analysis of *A. funesta*, the following invasiveness rating has been assigned.

Climate/host interaction	–	Medium (2)
Hosts range	–	Low (1)
Dispersal potential	–	Low (1)
Economic impact	–	High (3)
Environmental impact	–	Medium (2)
Post-entry distribution	–	Not known (0)
Final rating (score)	–	Medium (9)

Based on the above invasiveness rating, *A. funesta* receives a score of 9 out of 15 and is rated as LOW INVASIVE NEMATODE.

3.15 Region-wise status of invasiveness

APPPC: Afghanistan, Armenia, Azerbaijan, Georgia, India, Iran, Iraq, Israel, Jordan, Kuwait, Lebanon, Pakistan, Saudi Arabia, Syria.
COSAVE: Argentina, Chile.
EPPO: Albania, Bulgaria, France, Greece, Italy, Morocco, Portugal, Russian Federation, Spain, Switzerland, Turkey, Turkmenistan, Ukraine, Uzbekistan, Yugoslavia (former).
IAPSC: Algeria, Egypt, Libya, South Africa, Tunisia.
NAPPO: Mexico.

3.16 Management measures

Management measures, including phytosanitary methods, for *A. funesta* are similar to those for *A. agrostis*. For details, see section 2.16.

Ditylenchus Spp.

The genus *Ditylenchus* (Filipjev, 1936) contains a large number of species which are mostly myceliophagous. A few species are phytophagous, but are of great importance as they attack more than 1200 species of cultivated and wild plants (Sturhan and Brzeski, 1991). Over 60 species of *Ditylenchus* have been described (Siddiqi, 2000; Jeszke *et al.*, 2013) and important plant-parasitic species are:

Ditylenchus africanus Wendt, Swart, Vrain and Webster, 1995
D. angustus (Butler, 1913) Filipjev, 1936
D. destructor Thorne, 1945
D. dipsaci (Kuhn, 1857) Filipjev, 1936
D. humli Skarbilovich, 1972
D. gigas Vovlas *et al.*, 2011
D. medicaginis Wasilewska, 1965
D. melongena, Bhatnagar and Kadyan, 1969
D. procerus (Bally and Reydon, 1930) Filipjev, 1936
D. sorghi, Verma, 1966
D. weischeri Chizhov *et al.*, 2010

Among these, *D. africanus*, *D. angustus*, *D. destructor*, *D. gigas* and *D. dipsaci* are considered as highly invasive nematode species because of their economic importance and quarantine significance in various countries (Sturhan and Brzeski, 1991; Lal and Lal, 2006; Bora and Rahman, 2010; EPPO, 2013). These *Ditylenchus* species attack more than 450 higher plant species (Siddiqi, 2000; Chizhov *et al.*, 2010; Vovlas *et al.*, 2011). Other species are of lesser importance, having limited geographical distribution.

Systematic position of *Ditylenchus* species

	Classical (after Siddiqi, 2000)	Molecular (after Decraemer and Hunt, 2013)
Phylum	Nematoda	Nematoda
Class	Secernentea	Chromadorea
Subclass	Tylenchia	Chromadoria
Order	Tylenchida	Rhabditida
Suborder	Tylenchina	Tylenchina
Infra-order	Anguinata	Tylenchomorpha
Superfamily	Anguinoidea	Sphaerularioidea
Family	Anguinidae	Anguinidae
Subfamily	Anguininae	Anguininae
Genus	*Ditylenchus*	*Ditylenchus*

4 *Ditylenchus africanus*

4.1 Common name

Peanut or groundnut pod nematode

4.2 Scientific name

Ditylenchus africanus Wendt, Swart, Vrain and Webster, 1995

4.3 EPPO code

DITYAF

4.4 Diagnosis

The confounding species of *Ditylenchus* cannot be distinguished categorically from the target species based only on morphological characters. When indispensable, the distinction should be supported by molecular and biochemical identification of the species. The morphological, biochemical and molecular characters of *Ditylenchus africanus* are presented below.

4.4.1 Morphological diagnosis

Detailed morphological descriptions of *D. africanus* were given by De Waele *et al.* (1989) and Wendt *et al.* (1995) and are summarized below. For morphometric measurements, see Table 5.

Table 5. Morphometric data for *Ditylenchus africanus*, *D. angustus*, *D. destructor*, *D. dipsaci* and *D. gigas* (adapted from ISPM 27, DP 8).

Morphometric characters		Range, numerous populations				
		D. africanus (after Wendt *et al.*, 1995)	*D. angustus* (after Seshadri and Dasgupta, 1975)	*D. destructor* (after Goodey, 1952; Hooper, 1973a)	*D. dipsaci* (after Hooper, 1972)	*D. gigas* (after Vovlas *et al.*, 2011)
Females	Length (mm)	0.7–1.1	0.8–1.20	0.8–1.9	1.0–1.7	1.6–2.2
	Stylet (µm)	8–10	10–11	10–14	10–12	10.5–13.0
	a	–	50–62	32 (18–49)	62 ± 5.6	
	b	–	6–9	7 (4–12)	15 ± 1.4	
	c	8.8–16.9	18–24	17 (9–30)	11–20	15.7–27.6
	V%		78–80	80 (73–90)	80 ± 1.5	
	Post uterine sac/vulva–anus length (%)	37–85		53–90	40–70	About 50
	Lateral lines	6–15	4	6	4	4
	Tail terminus	Rounded	Conoid	Rounded	Pointed	Pointed to finely rounded
	Posterior bulb	Short, dorsally overlapping	Slightly overlapping	Short, dorsally overlapping	Not overlapping	Slightly overlapping
Males	Length (mm)		0.7–1.18	0.96 (0.76–1.35)	1.3 mm ± 0.017	
	Stylet (µm)		10	10–12		
	a		40–55	35 (24–50)	63 ± 11.3	
	b		6–8	7 (4–11)	15 ± 1.7	
	c		19–26	14 (11–21)	14 ± 2.1	
	Tail		60–73	65 (40–84)	72	
	Bursa length (as % of tail length)	48–66	–	50–70	40–70	72–76
	Spicules (µm)	17–21	16–21	24–27	23–28	23.5–28
Host	Host preference	Groundnuts and fungi	Rice	Higher plants and mycelia of fungi	Higher plants and fungi	Higher plants

Female: Flattened head, about 1.3 µm high and 6.4–7.3 µm wide, narrower than rest of body, but not offset. Labial area with pore-like stomal opening encircled by six outer labial papillae and two large medial lips, each with single pair of cephalic papillae. Outline of labial region with hexagonal head. Amphidial aperture oval, directed towards stoma opening. First annule of head discontinuous, instigated by position of amphidial apertures. Four lip annuli present in lip region, apart from labial disc. Stylet delicate with distinct knobs, detached, sloping backwards. Shaft long, around 60% of stylet length. Median bulb with crescentic valves and overlapping the intestine (Fig. 1.6A). Postvulval uterine sac long (50–143 µm), comprising about 8% of total body length or 37–85% of vulva–anus distance and equal to 1.5–3.7 times vulval body diameter. Tail conoid elongate, tapering in posterior one-third to finely rounded terminus (Fig. 1.6C).

Male: Male similar to female (Fig. 1.6B). Bursa long (33–60 µm), leptoderan, cover 48–66% of tail length. Spicule ventrally arcuate, cephalated slightly (Fig. 1.6D).

4.4.2 Biochemical diagnosis

One-dimensional gel electrophoresis was the first biochemical method to be applied in nematology, to distinguish *Ditylenchus* from *Panagrellus* (Benton and Myers, 1966) and four *Meloidogyne* species (Dickson *et al.*, 1970). Immunological approaches using polyclonal antibodies (pAbs) which identify multiple antigenic determinants in nematodes have been applied to the nematode diagnosis, including *Ditylenchus* spp. (Webster and Hooper, 1968; Abrantes *et al.*, 2004).

4.4.3 Molecular diagnosis

Molecular diagnosis using DNA-based approaches, such as microsatellites or single sequence repeats (SSRs), satellite DNA (satDNA), sequence characterized amplified regions (SCARs), PCR–single-strand conformational polymorphism (PCR–SSC) and DNA microarrays have been used to characterize different *Ditylenchus* spp. (Zouhar *et al.*, 2002; Abrantes *et al.*, 2004; Kerkoud *et al.*, 2007; Subbotin *et al.*, 2011b). Wendt *et al.* (1993 and 1995) used restriction fragment length polymorphisms (RFLPs) of ITS regions to separate *D. africanus*, *D. destructor*, *D. dipsaci*, *D. gigas* and *D. myceliophagus* by four restriction enzymes. Numerous adults or juveniles were used for DNA extraction using the method described by Webster *et al.* (1990; see Chapter 15, section 110.2.1). The ITS rRNA universal primers (Vrain *et al.*, 1992) used in this test are:

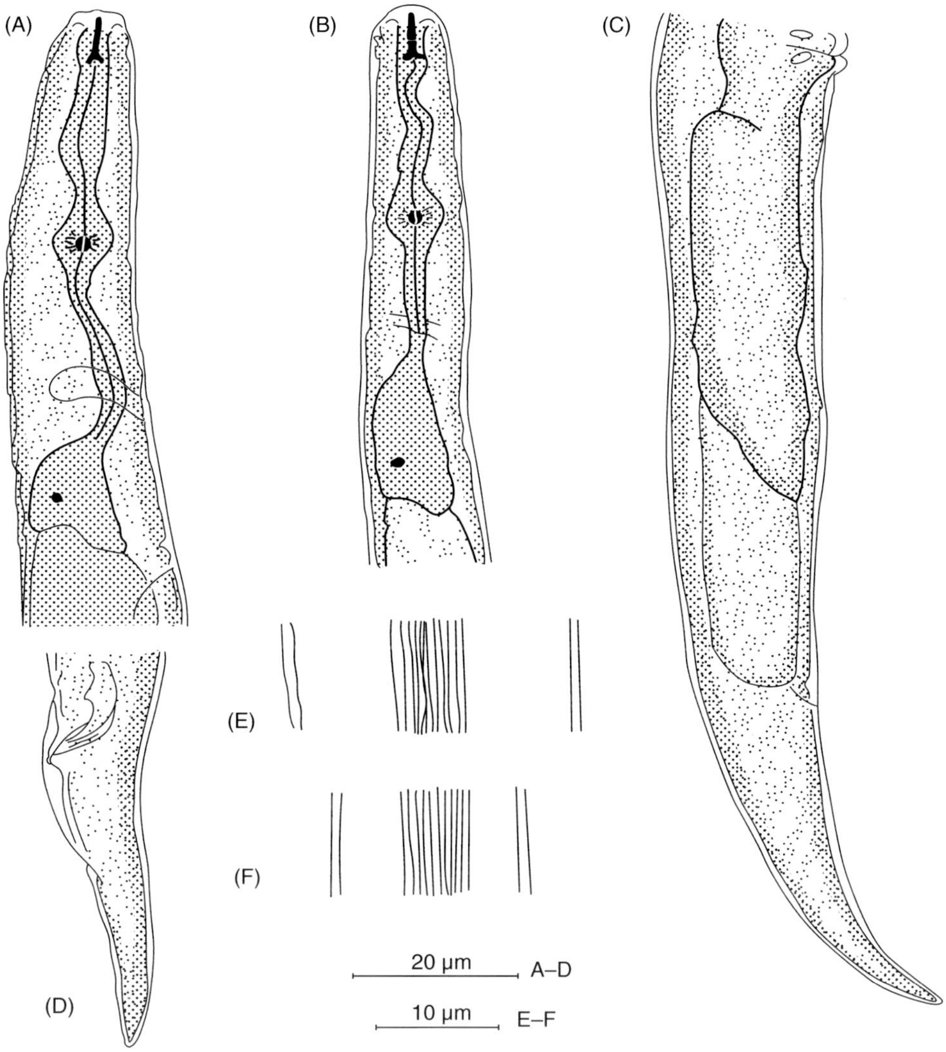

Fig. 1.6. Line drawing of *Ditylenchus africanus*. Anterior part of female body (A); anterior part of male body (B); posterior part of female body (C); male tail (D); female lateral field (E); male lateral field (F). [De Waele *et al.*, 1989; courtesy *Journal of Nematology*]

18S: 5′-TTG ATT ACG TCC CTG CCC TTT-3′
26S: 5′-TTT CAC TCG CCG TTA CTA AGG-3′

The PCR cycling parameters comprise the first cycle of 1.5 min at 96°C, 30 s at 50°C and at 72°C for 4 min; 40 cycles at 96°C for 45 s, at 50°C for 30 s and at 72°C for 4 min; and a final cycle at 96°C for 45 s, at 50°C for 30 s and at 72°C for 10 min. The 2–5 µl amplified DNA is analysed by 0.1% agarose gel electrophoresis. The amplicon size is 1000 base pairs (bp) for *D. africanus*, 900 bp for *D. dipsaci*, *D. gigas* and *D. myceliophagus* and 1200 bp for *D. destructor*. The approximate lengths (bp) of RFLP fragments of the ITS-rRNA generated by four restriction enzymes are presented in Table 6.

4.5 Geographical distribution

The peanut or groundnut pod nematode has limited distribution and has only been reported from South Africa (Wendt *et al.*, 1995) and Mozambique (CABI-ISC, 2020). In China, the reports of *D. destructor* from American ginseng (Zhang *et al.*, 2009) and *D. arachis* from groundnut (Zhang *et al.*, 2014) are possibly junior synonyms of *D. africanus* and both nematodes need to be further studied and validated.

4.6 Hosts

Groundnut (*Arachis hypogea*) is the principal host (Dickson and De Waele, 2005). Other agronomic crops and weeds are listed in Table 7.

4.7 Symptoms

On the below-ground parts of groundnuts, *D. africanus* does not cause any visible lesions. However, on roots, the infestation

Table 6. Approximate length (base pair) of RFLP fragments of the ITS-rRNA for *Ditylenchus* species generated by four restriction enzymes (adapted from ISPM-27, DP 8).

Species	Unrestricted PCR product	*Hae*III	*Hpa*II	*Hin*fI	*Rsa*I
D. africanus	1000	650, 540	950	450, 340, 150, 130, 100	690, 450
D. destructor	1200	450, 170	1000	780, 180	600, 250, 170
D. dipsaci	900	900	320, 200, 180	440, 350, 150	450, 250, 140
D. gigas	900	800, 200	600, 200	350, 150	490, 450
D. myceliophagus	900	450, 200	900	630, 310	900

Source: Wendt *et al.* (1993, 1995).

Table 7. Hosts of *Ditylenchus africanus* (adapted from CABI-ISC, 2020).

Main hosts
Peanut/groundnut (*Arachis hypogea*)
Other hosts
Glycine max (soybean)
Gossypium hirsutum (Bourbon cotton)
Helianthus annuus (sunflower)
Lupinus albus (white lupin)
Medicago sativa (lucerne)
Nicotiana tabacum (tobacco)
Phaseolus vulgaris (common bean)
Pisum sativum (pea)
Solanum tuberosum (potato)
Sorghum bicolor (sorghum)
Triticum aestivum (wheat)
Vigna unguiculata (cowpea)
Zea mays (maize)
Wild hosts
Chenopodium album (fat hen)
Datura stramonium (jimson weed)
Eleusine indica (goosegrass)
Tagetes minuta (stinking Roger)
Xanthium strumarium (common cocklebur)

affects appearance and sometimes reduces weight. The germination of the seeds is also reduced (Dickson and De Waele, 2005). Infected hulls exhibit greyish-black to brown necrotic tissue mainly at the point of connection with the peg, and broad bands are formed along the longitudinal veins that extend in the exocarp just beneath the pod surface (Dickson and De Waele, 2005). Infected seeds are usually shrivelled with dark brown to black micropyles and yellow to dark flaccid testae with dark vascular strands (Fig. 1.7A-C). The embryos may also become darkly discoloured (Jones and De Waele, 1990). Around 40–60% of the pods and seeds are destroyed in a heavily infested field (Dickson and De Waele, 2005).

4.8 Biology and life cycle

The biology of *D. africanus* coincides with the growth of the groundnut plant (Dickson and De Waele, 2005). The nematode usually survives in very low numbers in the soil on fungi, groundnut roots, alternative hosts and weeds, until the groundnut pegs appear in the soil (about 60 days after sowing). The nematode penetrates the young pod at its connection point with the peg, entering the exocarp and moving either longitudinally in the cell layer towards the pod beak-end and producing the pod discoloration, or through the mesocarp into the endocarp of the hull-shell (Venter *et al.*, 1995). It then moves to the seed micropyle, from where it invades the seed testa and embryo and produces symptoms on the seeds. The nematode has not been recorded within the cotyledons of the seed (Jones and De Waele, 1990). The nematode multiplies within the pod and seed, and the life cycle from egg to adult is completed in 8 days at 25°C (De Waele and Wilken, 1990). At 28–30°C, egg hatching is initiated at around 3 days and by the 6th day 90% of eggs hatch (Dickson and De Waele, 2005). The nematode can enter into a state of anhydrobiosis, with around one-third of anhydrobiotic nematodes becoming active after rehydration to invade hulls and seeds of the newly planted crop (Dickson and De Waele, 2005). From 17 to 21 weeks after sowing (varying among cultivars), as the plant matures the relative number of eggs and anhydrobiotes in the pod and seed tissues increases (Venter *et al.*, 1995).

The infective stage is unknown and the nematode can survive for at least 28–32 weeks in planting seeds, which may be symptomless, and in the field soil in the absence of the host plant, or in groundnut hulls buried in the soil (Basson *et al.*, 1993). This period is long enough to survive the dry winter season in South Africa. With the first spring rains, the eggs hatch and anhydrobiotes rehydrate. Basson *et al.* (1993) showed that rehydrated soil populations of the nematode can invade the next groundnut crop and cause damage. Although comparatively few *D. africanus* survive in whole stored seed at 10°C, the surviving nematode population is also able to reproduce to levels capable of causing damage to the next crop (Basson *et al.*, 1993).

4.9 Economic importance

D. africanus was first reported from south-western Transvaal province in South Africa in 1987 (De Waele *et al.*, 1989). In South Africa, it was found in all the major groundnut production areas on about 75% of all fields (De Waele *et al.*, 1989). Nematode infection causes serious qualitative damage and results in 65% of under-graded groundnut seeds in South Africa. In the greenhouse, the fresh weight of harvested seed was suppressed by 20–50% (Dickson and De Waele, 2005). A 1987 survey showed that 73% of the samples from 877 farms of the major groundnut production areas were infected with *D. africanus* (De Waele *et al.*,

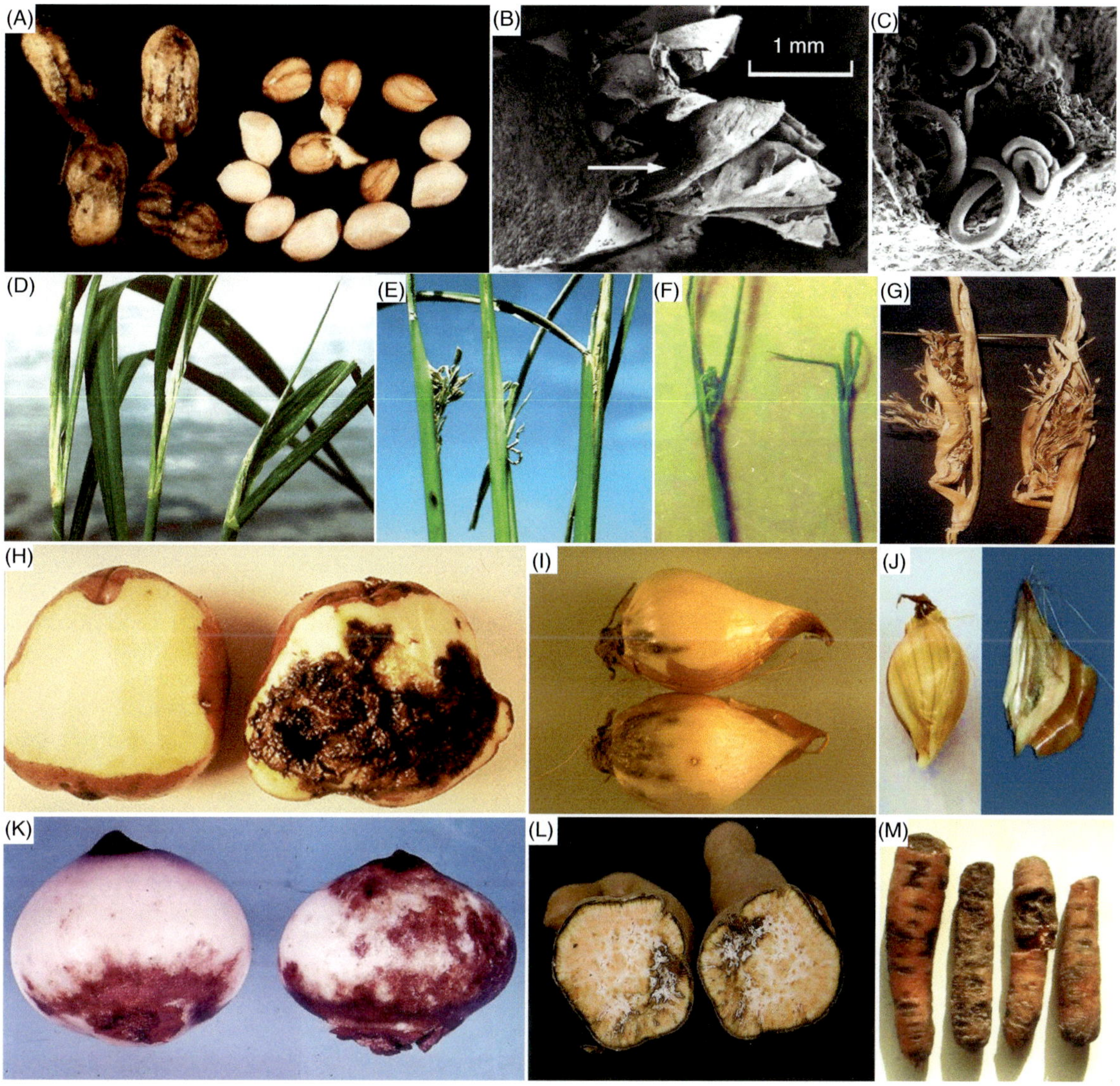

Fig. 1.7. Symptoms of *Ditylenchus africanus* on groundnut (A–C), *D. angustus* on rice (D–G) and *D. destructor* on different hosts (H–M). Nematodes on pods and seed of groundnut (A); SEM overview of *D. africanus* on the embryo of groundnut seed (B); close-up SEM (C); initial symptoms of *ufra* disease – white patches on the leaf bases (D); *ufra* symptom – partial emergence and distorted panicles due to nematode infection (E); twisting and loop formation (F); swollen *ufra* (G); infected potato tuber (H); infected iris rhizome (I); healthy and infected iris rhizome (J); healthy and infected tulip bulbs (K); infected sweet potato (L); infected carrots (M). [Photo courtesy A-C: ARC-Grain Crops Institute, South Africa; D,E: J. Bridge, CABI BioScience; F,G: NASI, India; H,M: Ephytia, INRAe; I: M.V. McKenry and P.A. Roberts, University of California; J: USDA; K: PPO Wageningen UR; L: : © 2007, Ian Riley, Adelaide, Australia]

1989). The average number of nematodes per seed was 160, with up to 3338 per seed in the south-western Transvaal area. Further, infestations of up to 97,000 nematodes per pod are not uncommon for groundnut grown in the field (Dickson and De Waele, 2005). The nematode can multiply to heavy infections (100,000 nematodes per pod), causing 100% losses to groundnut crops in some fields (Venter, 1994). In heavy infestations, the final nematode density exceeds 700 nematodes per seed and the seed mass may be reduced by up to 50%. Lower germination of seeds with 25% decrease in the number of harvestable seeds has been recorded (Venter *et al.*, 1991).

4.10 Similarities to other species/conditions

Morphologically *D. africanus* is very similar to *D. myceliophagous*, *D. destructor*, *D. dipsaci and D. gigas* (see Table 5). Earlier, *D. africanus* was reported as *D. destructor* on groundnut/peanut in South Africa, later separated as *D. africanus* (Wendt *et al.*, 1995). The histopathology of *D. africanus* resembles that of *Aphelenchoides arachidis* on groundnut. In South Africa, groundnut producers often confuse *D. africanus* symptoms with those of the fungus *Chalara elegans*. However, symptoms can be differentiated on the pattern of development of pod discoloration. The *D. africanus*-infected seeds of pods have yellow to brown flaccid testae, frequently with darkened veins, and discoloured embryos, while *C. elegans*-infected seeds are not discoloured (Dickson and De Waele, 2005).

4.11 Movement and means of dispersal

The main means of dispersal is by seeds containing nematodes in trade and by the sowing of infected seeds in fields. Other means of spread include infected crop straw, flooding and rainfall, natural migration (maximum 20–30 cm), and sheep, cattle or bird manure.

4.11.1 Plant parts liable to carry the nematode in trade/transport

Plant parts including roots, flowers, inflorescences, cones, calyx, leaves, seedlings, micro-propagated plants, stems (above ground), shoots, tillers and true seeds, and fruits (including pods) are liable to carry all stages (eggs, juveniles and adults) of the nematode.

4.11.2 Plant parts not known to carry the nematode in trade/transport

Bark and wood.

4.12 Pest risk analysis and looming threat from introduction

D. africanus is a damaging pest in South Africa but is not present in any quarantine regions within the continent of Africa. This nematode is also reported from Mozambique. However, *D. africanus* can become a major economic pest if introduced in other major groundnut-producing countries, especially China, India, the USA and Sudan. Due to its limited distribution, the chances of introduction into other groundnut-producing countries are very limited. However, a complete pest risk assessment should be prepared for this nematode. Its accidental introduction would have devastating repercussions on groundnut exports to domestic and international markets; hence, strict quarantine regulation and extreme cautions should be taken in importing groundnut pods or seeds from Africa.

4.13 Invasiveness rating

Considering the consequences of introduction and pest risk analysis, the following invasiveness rating has been assigned to *D. africanus*.

Climate/host interaction	–	Medium (2)
Hosts range	–	High (3)
Dispersal potential	–	High (3)
Economic impact	–	High (3)
Environmental impact	–	Medium (2)
Post-entry distribution	–	Low (–1)
Final rating (score)	–	High (12)

Based on the above invasiveness rating, *D. africanus* receives a score of 12 out of 15 and is rated as HIGHLY INVASIVE NEMATODE.

4.14 Region-wise status of invasiveness

Region-wise, groundnut-producing countries where *D. africanus* is absent and attains invasive status are:

APPPC: Bangladesh, Cambodia, China, India, Indonesia, Iraq, Israel, Japan, Jordan, Kazakhstan, Kyrgyzstan, Lebanon, Malaysia, Myanmar, Pakistan, Philippines, Saudi Arabia, South Korea, Sri Lanka, Syria, Tajikistan, Thailand, Vietnam
CAHFSA: Barbados, Belize, Cuba, Dominican Republic, Haiti, Jamaica, Saint Kitts & Nevis
CAN: Bolivia, Colombia, Ecuador, Peru
COSAVE: Argentina, Brazil, Guyana, Paraguay, Suriname, Uruguay
EPPO: Bulgaria, Cyprus, Greece, Hungary, Italy, Morocco, Portugal, Spain, Turkey, Uzbekistan
IAPSC: Algeria, Angola, Benin, Botswana, Burkina Faso, Burundi, Cameroon, Cape Verde, Central African Republic, Chad, Comoros, Congo, Côte d'Ivoire, Egypt, Eritrea, Ethiopia, Gabon, Gambia, Ghana, Guinea, Guinea-Bissau, Kenya, Liberia, Madagascar, Malawi, Mali, Mauritania, Mauritius, Namibia, Niger, Nigeria, Réunion, Rwanda, Senegal, Sierra Leone, Somalia, Sudan, Swaziland, Togo, Uganda, Zambia, Zimbabwe
NAPPO: Mexico, USA
OIRSA: Costa Rica, El Salvador, Guatemala, Honduras, Nicaragua
PPPO: Australia, Fiji, Papua New Guinea, Tonga, Vanuatu

4.15 Management measures

4.15.1 Prevention and control

The spread of the pest can be stopped through the elimination of infection sources by burning off the infested groundnut (Dickson and De Waele, 2005). Timely harvesting also helps in the drying of the hulls and seeds before eggs are laid, and such hull stubble and seed for planting contain lower numbers of anhydrobiotic nematodes (Venter, 1994). Recently, the groundnut breeding line PC254K1 has been identified as resistant to *D. africanus* and has been included in the groundnut breeding programme (Steenkamp *et al.*, 2016).

4.15.2 Phytosanitary measures

See Chapter 16, section 115.

5 *Ditylenchus angustus*

5.1 Common name

Rice stem nematode; *ufra* disease

5.2 Scientific name

Ditylenchus angustus (Butler, 1913) Filipjev, 1936

5.3 Synonyms

Anguillulina angusta (Butler, 1913) Goodey, 1932
Tylenchus angustus Butler, 1913

5.4 EPPO code

DITYAN

5.5 Diagnosis

Morphological, biochemical and molecular diagnosis approaches of *Ditylenchus angustus* are presented below.

5.5.1 Morphological diagnosis

Detailed morphological descriptions of *D. angustus* were given by Seshadri and Dasgupta (1975). Additional morphological detail was provided by Mian and Latif (1995). For measurements, see Table 5.

FEMALE: Slender body, nearly straight to ventrally arcuate when heat relaxed. Cuticle with fine annules, transverse striations about 1 μm wide at mid-body (Fig. 1.8G). Lip region low, flattened, not distinctly offset, unstriated, wider than high at lip base. Cephalic framework hexaradiate, (Fig. 1.8D) slightly sclerotized, *en face* view shows six lips of roughly equal size (Fig. 1.8C). Lateral fields with four incisures, outer incisures different than inner ones, one-fourth of body width or marginally less, extending almost to tail tip. Deirids present, nearly posterior to the level of excretory pore. Stylet moderately developed, conus reduced, around 45% of total stylet length, knobs small about 2 μm across but distinct, frequently with posteriorly sloping anterior surfaces, rather amalgamated with one another. Procorpus cylindrical, narrowing at median oesophageal bulb junction, 3–3.6 times as long as body width in that region. Median oesophageal bulb elliptical, with distinct valvular plate anterior to centre. Isthmus cylindrical, narrow, 1.5–1.9 times as long as procorpus; posterior oesophageal bulb usually clavate, ventrally overlapping the intestine, with three distinct gland nuclei. Cardia absent. Nerve ring conspicuous, behind median oesophageal bulb. Excretory pore slightly anterior to starting of posterior oesophageal bulb. Hemizonid present, anterior to excretory pore. Vulva with a transverse slit, vaginal tube slightly oblique, attaining more than halfway across body. Spermatheca extremely elongated, filled with large rounded sperms. Genital tract single anteriorly outstretched ovary, oocytes in single row, rarely in double rows. Post-uterine sac present, collapsed, without sperm, 2.5 times as long as vulval body width, extending about half to two-thirds distance to anus. Tail conoid, around five times anal body width in length, narrowing to sharply pointed terminus looking like a mucro (Fig. 1.8F). Phasmids pore-like, difficult to see, close behind mid-part of tail.

MALE: Morphologically similar to females. Body approximately straight to slightly curved ventrally when heat fixed. Bursa present, but narrow in some specimens, starting opposite the proximal end of spicules and extending nearly to tail tip. Spicules simple, ventrally curved. Gubernaculum simple and short (Fig. 1.8I).

JUVENILE: Morphologically similar to adults. Oesophagus proportionally shorter than adult's.

5.5.2 Biochemical diagnosis

Ibrahim *et al.* (1994b) used esterase and protein patterns to differentiate *D. angustus* from two new species of *Aphelenchoides* on rice. Each nematode species exhibited distinct protein bands and a characteristic esterase pattern. *D. angustus* showed two moderately strong protein bands at Rm = 0.63 and 0.65.

5.5.3 Molecular diagnosis

PCR–RFLP involving the amplification of the ITS region of the ribosomal DNA and restriction enzyme digestion has been used to differentiate *D. angustus* from an *Aphelenchoides* species and an undescribed *Aphelenchoides* population (Ibrahim *et al.*, 1994a). For details of molecular characterization of *Ditylenchus* species, see section 6.5.3 below.

5.6 Geographical distribution

The rice stem nematode (RSN) has been reported to occur in South-East Asia (Bora and Rahman, 2010). Since the report by Butler (1913a) on occurrence of the nematode in Noakhali and Tippera of East Bengal (now Bangladesh), this nematode has been recorded in many other countries, including Bangladesh, India, Indonesia, Malaysia, Myanmar, Philippines and Thailand (CABI-ISC, 2020).

5.7 Hosts

Almost all the cultivated deep-water rice cultivars, including some of the transplanted rice varieties, are good hosts of *D. angustus* (Bora and Rahman, 2010). Thirteen species of *Oryza*, namely, *O. perennis*, *O. sativa* var. *fatua*, *O glaberima*, *O. cubensis*, *O. officinalis*, *O. meyaiana*, *O. latifolia*, *O. eichingeri*, *O. alta*, *O. minuta*, *O. nivera*, *O. rufipogon* and *O. spontanese* (Bridge *et al.*, 2005a; Peng *et al.*, 2018) are reported as host crops of *D. angustus*. Apart from *Oryza* spp., duckweed *Hygroryza aristata*, swamp rice grass *Leersia hexandra* (Miah and Bakr, 1977), *Paspalum scorbiculatum* (Pathak, 1992), *Sacciolepsis interrupta* and *Echinochola colona* (Cuc,1982) are recorded as alternative hosts of *D. angustus*. Four different fungi, namely *Curvularia oryzae*, *Epicoccum purpurascens*, *Pestalotia oryzae* and *Rhizoctonia solani*, can also harbour this nematode.

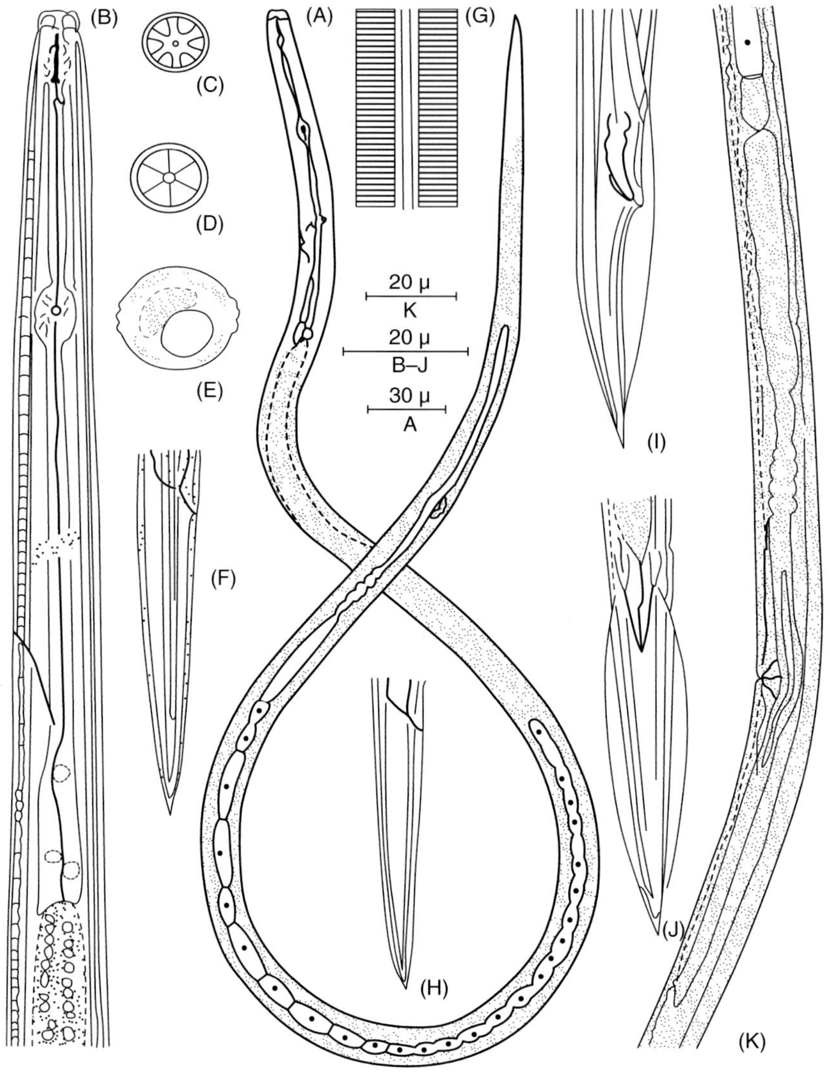

Fig. 1.8. Line drawing of *Ditylenchus angustus*. Whole female body (A); oesophageal region of a female (B); *en face* view (C); cephalic framework (D); cross-section of a female at mid-body (E); female tail (F); lateral field at mid-body (G); juvenile tail (H); lateral view of a male tail (I); ventral view of a male tail (J); vulva region of an adult female (K). [Source: Nemaplex, http://nemaplex.ucdavis.edu/, H. Ferris]

5.8 Symptoms

RSNs cause characteristic symptoms of *ufra* which vary considerably with the degree and time of infestation. The infection starts from the early seedling stage but it is difficult to recognize the nematode attack at this stage. In seedling stages, the first visible symptom appears as mosaic-like discoloration arranged in a splash pattern, or chlorosis or white streaks on young leaves and sheaths; becoming more evident with time. The entire leaf may become twisted or severely malformed. In some cases, the basal portion of the young leaf becomes wrinkled, following whitish-green discoloration (Bora and Rahman, 2010). Gradually the chlorotic portion becomes brown to dark brown with the advancement of the disease. In severe cases, the leaf margins may crinkle and the leaf tip is twisted. Often, the central leaf may emerge from the side of the stem by forming a loop or the central leaf may be curly, and later it becomes dried (Ou, 1985). Sometimes several branches are produced from the infested node which make the plant bushy. Sometimes the symptoms are masked and the plant looks healthy. Before the heading stage, plants remain symptomless. Most typical and conspicuous symptoms, which again vary considerably, are seen at the heading stage of the crop.

Butler (1913b) classified the disease into two types: *thor ufra* or swollen *ufra*; and *pucca ufra* or ripe *ufra*. Swollen *ufra* develops when a large number of nematodes are present which damage the panicle at an early stage; the panicle coils and remains enclosed within the leaf sheath (Fig. 1.7D–F). Ripe *ufra* represents the secondary or tertiary infestation where the nematode population is comparatively less, which damages the panicle at a later stage; the panicle emerges but is distorted, bears sterile or empty grains at the base and produces normal grains only near the tip (Fig. 1.7G). Cox (1980) reclassified the symptoms into three types: *ufra* I, where the panicle remains completely enclosed within the leaf sheath (equivalent to swollen *ufra*); *ufra* II, where the panicle partially emerges and is weak, with unfilled grains (Fig. 1.7E); and *ufra* III, where the panicle emerges normally with partly filled grains near the tip (equivalent to ripe *ufra*). *Ufra* symptoms generally appear in patches in the field and expand with the advancement of crop age. Severely infected plants with *ufra* I symptoms are more often noticed at the centre of the patch than at the edges. In fields with severe *ufra* I and *ufra* II symptoms, yield may reduce to zero; whereas with *ufra* III, fields may produce some yield.

5.9 Biology and life cycle

The rice stem nematode *D. angustus* is an obligate ectoparasite and stubble-borne in nature. Infested plant residues are believed to be the primary source of inoculum. After harvesting of the infested crop, the nematode survives on wild rice, ratoons, weed grass and left-over stubbles (Miah and Bakr, 1977). In the infested rice stubbles, the fourth-stage larvae remain dormant, forming a coil, particularly inside peduncles, under upper leaf sheaths and within the glumes of the lower grains of the panicle. The number of nematodes in single infested stubbles varies from 1 to 30,000 (Rathaiah, 1988) and in seeds varies from 5.3 to 2400 (Ibrahim and Perry, 1993). Occurrence of this nematode in soil and seed has been reported (Cuc and Giang, 1982), but this nematode is not soil-borne (Pathak, 1992). In seed, the larvae are located at the overlapping region of the rice husk. The infestation starts at seedling stages of a few days old. Under favourable environmental conditions, with temperatures at 28–30°C and humidity above 85% (McGeachie and Rahman, 1983), the active nematodes climb up on the surface of newly germinated rice seedlings and feed ectoparasitically on the epidermal cells of the tender growing tissues and developing inflorescence (Ou, 1985). Because of the compact nature of the innermost leaf that ensheaths the shoot meristem, the nematode is not able to invade the apical meristem. It is probably for this reason that the infested plants are not killed, but grow and produce flower and seeds. The nematode does not produce any toxic substances while feeding but continues to suck sap through its small stylet.

The fourth-stage larvae, usually under the compact leaves, moult to become adults. The reproduction of the nematode is amphimictic and takes place on the host plant during the months of May–June and November, but the number of generations in a single season is not certain and quantitative estimation of eggs laid per female is difficult to ascertain (Perry, 1995). Three generations in a cropping season and 50–100 eggs per female was reported by Butler (1913a), and this was confirmed by the observation of three population peaks in a cropping season by Cox and Rahman (1979). In artificial conditions, the developmental cycle of *D. angustus* takes 15 days from second-stage juvenile (J_2) to become adult, 21 days from J_2 to egg and 24 days to complete its life cycle (Cuc, 1989). However, Plowright and Gill (1994) mentioned that the period of the life cycle of *D. angustus* is 10–20 days at 30°C. When crops proceed towards maturity, nematodes are found in large numbers at the top of the stem, in the panicle and in the seeds. At near maturity of the crop, predominantly fourth-stage larvae become inactive and undergo quiescence, each being tightly coiled circularly with the head at the centre. This state is commonly known as eelwool. When placed in water, larvae in the eelwool state uncoil and become active with vigorous wrinkling. The larvae become active more quickly at 31°C. The larvae can move on a solid surface at an atmospheric humidity of 85% or more.

D. angustus can survive in the quiescent stage in grains (under glumes or husk), in dry paddy stubbles, in the infested field, or on the surface of the soil until the next cropping season (Bora and Rahman, 2010). It also survives in the active stage in ratoons, weeds and wild rice after the harvest of paddy (Pathak, 1992). In dry paddy stubbles, the nematode can survive desiccation for at least 15 months (Catling *et al.*, 1979). However, after harvest, the population in stubbles reduces drastically, leaving no live nematodes 4–5 months after harvest (Kinh, 1981). There are no specific ‘survival stages’ of *D. angustus* and in dry diseased stubbles three stages of *D. angustus* are found: J_3, J_4 and adult, with J_4 being predominant (Bora and Rahman, 2010).

5.10 Economic importance

Although this disease is most devastating to deep-water rice, it can also attack transplanted winter rice and summer rice (Chakraborti *et al.*, 1985). The infestation is greater if the nematode inoculums are present in the field at the time of sowing (Rahman and Evans, 1987). Field experiments have revealed that a level of 10% infested seedlings at the time of transplanting is sufficient to cause significant yield loss (Mondal *et al.*, 1989). Loss in grain yield depends on the percentage of *ufra* infestation. If there are more than 40% *ufra* II symptoms in a field, the yield loss may reach 100% (Cox and Rahman, 1980). In India, the yield reduction may be 20–90% (Rao *et al.*, 1986a). Sometimes the disease is so severe that farmers cut the crop for cattle feed (Bora and Rahman, 2010). Miah and Bakr (1977) recorded that *ufra* disease affected only 2% of deep-water rice in Bangladesh, resulting in an average yield loss of 50% and even 100% in individual farms. Similarly, Mondal and Miah (1987b) reported 60–70% yield reduction in rice in low-lying areas in Bangladesh and 10–100% in transplanted rice (Rahman *et al.*, 1994).

5.11 Similarities to other species/conditions

D. angustus is very similar to other *Ditylenchus* species, but is the only pest of rice. Morphologically it appears similar to *Aphelenchoides besseyi* under low microscopic resolution, but can be distinguished by head shape and type of oesophageal bulb. Males are more simply differentiated based on tail shape, spicule shape and presence or abundance of bursa.

5.12 Movement and means of dispersal

Dissemination of *D. angustus* from one field to another or from one plant to another takes place through plant residues, water (flood or irrigation water), rain splash and stem and leaf contact under conditions of high humidity (Rahman and Evans, 1987). Chances of transmission of this nematode through seed are minimal (Seshadri and Dasgupta, 1975). However, Prasad and Varaprasad (2001) showed that *D. angustus* is seed-borne and there is a possibility of dispersal of this nematode through seeds from infested field to other rice belts, located away from the RSN endemic zone. Considerable numbers of *D. angustus* have also been reported from filled grains containing more than 12% moisture (Ibrahim and Perry, 1993). Mondal and Miah (1987a) reported that tidal water plays an important role in the spread of this nematode.

5.12.1 Plant parts liable to carry the nematode in trade/transport

Plant parts including roots, flowers, inflorescences, cones, calyx, panicle (including ear heads), leaves, seedlings, micropropagated plants, stems (above ground), shoots, tillers and true seeds (including grain) are liable to carry all stages (eggs, juveniles and adults) of the nematode.

5.12.2 Plant parts not known to carry the nematode in trade/transport

Bark and wood.

5.13 Pest risk analysis and looming threat from introduction

D. angustus is solely a pest of rice, mainly in different types of flooded rice from deep-water to lowland irrigated. The nematode is confined to rice cultivation in India, Bangladesh, Myanmar and Thailand, which are major rice-growing countries of the world. There are also discrete reports of its occurrence in Egypt, Indonesia, Iran, Madagascar, Malaysia, Pakistan, Philippines and Sudan but there has not been any evidence about the economic damage by *D. angustus* in these countries. At a global level, *D. angustus* draws significant attention as a quarantine pest of rice in the countries where it is absent or has limited distribution or interception. Samplings during quarantine inspection have shown large numbers of seed lots to be infected with *D. angustus* traded from the nematode's endemic zone. The accidental introduction of *D. angustus* would have devastating repercussions on rice export to domestic and international markets.

5.14 Invasiveness rating

Considering the consequences of introduction and pest risk analysis, the following invasiveness rating for *D. angustus* has been assigned.

Climate/host interaction	–	High (3)
Hosts range	–	High (3)
Dispersal potential	–	High (3)
Economic impact	–	High (3)
Environmental impact	–	Medium (2)
Post-entry distribution	–	Low (–1)
Final rating (score)	–	High (13)

Based on the above invasiveness rating, *D. angustus* receives a score of 13 out of 15 and is rated as HIGHLY INVASIVE NEMATODE.

5.15 Region-wise status of invasiveness

Region-wise, top rice-producing countries where *D. angustus* attains invasive status are:

APPPC: Afghanistan, Cambodia, China, Iran, Japan, Nepal, North Korea, Pakistan, South Korea, Sri Lanka
CAHFSA: Lao People's Democratic Republic
CAN: Colombia, Ecuador, Peru, Venezuela
COSAVE: Argentina, Brazil, Uruguay
EPPO: Italy, Russian Federation, Spain, Turkey
IAPSC: Côte d'Ivoire, Egypt, Guinea, Mali, Madagascar, Nigeria, Sierra Leone, Senegal, Tanzania
NAPPO: USA
PPPO: Australia, Melanesia, Micronesia, New Zealand

5.16 Management measures

5.16.1 Seed treatment

Seed treatment is quite an effective and economically feasible management practice for *D. angustus*. Seed treatments include soaking of rice seeds in thiabendazole (0.1%) or ethylthiocyanoacetate (0.13%) or phosphamidon (0.1%) for 24 h (Vuong and Rodriguez, 1972). Dipping of seedling roots in carbofuran and applications of carbofuran during transplanting and every 20 days thereafter will effectively control the nematode infestation in rice (Chakraborti, 2000).

5.16.2 Cultural methods

Burning of diseased stubbles, followed by ploughing, reduces the incidence of *ufra* disease and is suggested as an effective control measure (Ou, 1985), but burning may create a problem of air pollution which has been observed in some states of India in recent years. Rotation of deep-water rice with a non-host crop such as jute (*Corchorus olitorius*) and mustard (*Brassica* sp.) reduces the *ufra* disease in the next crop (Butler, 1919). In areas with good irrigation facilities, cultivation of autumn rice (*boro* rice) followed by summer rice (*ahu* rice) will control the *ufra*, as the nematode will be deprived of the susceptible host during its active period. However, this practice may not be feasible in most of the deep-water rice-growing areas, where floodwater rises to a high level that does not allow the cultivation of *ahu* rice.

5.16.3 Host resistance

A good number of rice cultivars have expressed tolerance/ resistance against *D. angustus*, including Bazail-65, Jalamagna, Basudeu, AR-9(C), IR 13437-20-4E-PI, IR 17643-4, Lakhi, Karkati, BR 308-3-3-2, Rayada 16-011, Rayda 16-013, Rayada 16-05, Rayada 16-06, Rayada 16-07, Ba Tuc, etc. (Bora and Rahman, 2010).

5.16.4 Chemical control

Many chemical pesticides have been known to be useful for controlling *ufra* disease. Soil application of carbofuran at the time of transplanting, at 0.67 kg a.i./ha, followed by benomyl spray at 2.5 kg a.i./ha effectively suppressed *ufra* infestation (Rahman, 1993). Root-dip treatment of nursery stock with either carbofuran 3%, Miran 3% at 7.5–10% or Tecto 40FL at 2.5–10% for 12 h resulted in a healthy and *ufra*-free crop (Mondal and Miah, 1987b). Seed dressing with carbosulfan 25 STD plus soil application of carbofuran 3G at 3 kg a.i./ha plus foliar spray with carbosulfan 40 EC at 40, 120 days after sowing the seeds and before panicle initiation stage satisfactorily controlled *ufra* disease in deep-water rice (Das, 2004).

5.16.5 Phytosanitary measures

See Chapter 16, section 115.

6 *Ditylenchus destructor*

6.1 Common names

Potato tuber nematode; potato rot nematode

6.2 Scientific name

Ditylenchus destructor Thorne, 1945

6.3 EPPO code

DITYDE

6.4 Diagnosis

Morphological, biochemical and molecular diagnosis approaches of *Ditylenchus destructor* are presented below.

6.4.1 Morphological diagnosis

Morphological descriptions of *D. destructor* after Sturhan and Brzeski (1991) and Brzeski (1998) are summarized below. Substantial morphometric variation occurs in adults according to their host and age. For morphometric measurements, see Table 5.

FEMALE: Minute, around 800–1400 μm long, 23–47 μm wide and ventrally arcuate slightly. Lateral field with six incisures (Fig. 1.9F), narrowed to two on neck and tail areas. Cuticle with fine annulation including on head. About four head annules distinguished by scanning electron microscopy (Wendt *et al.*, 1995). Head often narrower than adjoining body. Stylet long, 10–12 μm in length, sometimes as long as 14 μm. Stylet cone comprising 45–50% of stylet length, distinct knobs, rounded and sloping backwards. Muscular median bulb, with thickenings of lumen valve and around 3 μm long. Posterior oesophageal bulb overlaps intestine dorsally for a short distance (Fig. 1.9D), though specimens with offset glandular bulb are also rarely seen. Excretory pore behind oesophageal glands. Postvulval sac present spreading about three-quarters of vulva–anus distance (Fig. 1.9D). Eggs twice as long as wide (Andrássy, 2007). Lips of vulva elevated and thick (Fig. 1.9E). Genital tract single with ovary outstretched anteriorly, sometimes touching oesophageal region. Postvulval uterine sac 40–98% of vulva–anus distance, but not functioning as a spermatheca (Fig. 1.9D).
MALE: Similar to female in general morphological appearance (Fig. 1.9A). Bursa covers 50–90% of the tail length. Spicules long, 24–27 μm (Fig. 1.9C). Testis outstretched reaching base of oesophagus. Tail conical like female, usually ventrally curved, three to five anal body widths long with rounded terminus.
JUVENILE: Four juvenile stages present, first stage occurring within the egg, almost twice as long as wide. Similar to female in general morphology but genital structures absent.

6.4.2 Biochemical diagnosis

One-dimensional SDS gel electrophoresis of proteins has been used to distinguish species and races of *Ditylenchus*, including *D. destructor* (Evans, 1971), *D. dipsaci* (Tenente and Evans, 1997) and *D. myceliophagus* (Ibrahim *et al.*, 1994b).

6.4.3 Molecular diagnosis

Molecular diagnosis of *D. destructor* based on PCR–RFLP or sequencing of the ITS region of the rRNA gene has been given by different workers. Wendt *et al.* (1993, 1995) used PCR–RFLPs of ITS regions to separate *D. destructor*, *D. africanus*, *D. dipsaci* and *D. myceliophagus* by four restriction enzymes; see section 4.4.3, above, and Table 6. Several populations of *D. destructor* from sweet potato have been characterized by RFLP profiles and showed differences in their RFLP profiles (Ji *et al.*, 2006; Subbotin *et al.*, 2011b). More than 50 ITS1 region sequence accessions of rRNA fragments acquired from *D. destructor* collected from different geographical regions and host plants are currently available in the GenBank database (Powers *et al.*, 2001).

6.5 Geographical distribution

The potato rot nematode (PRN) is a near-cosmopolitan species and most common in temperate regions (EPPO, 2013). It has

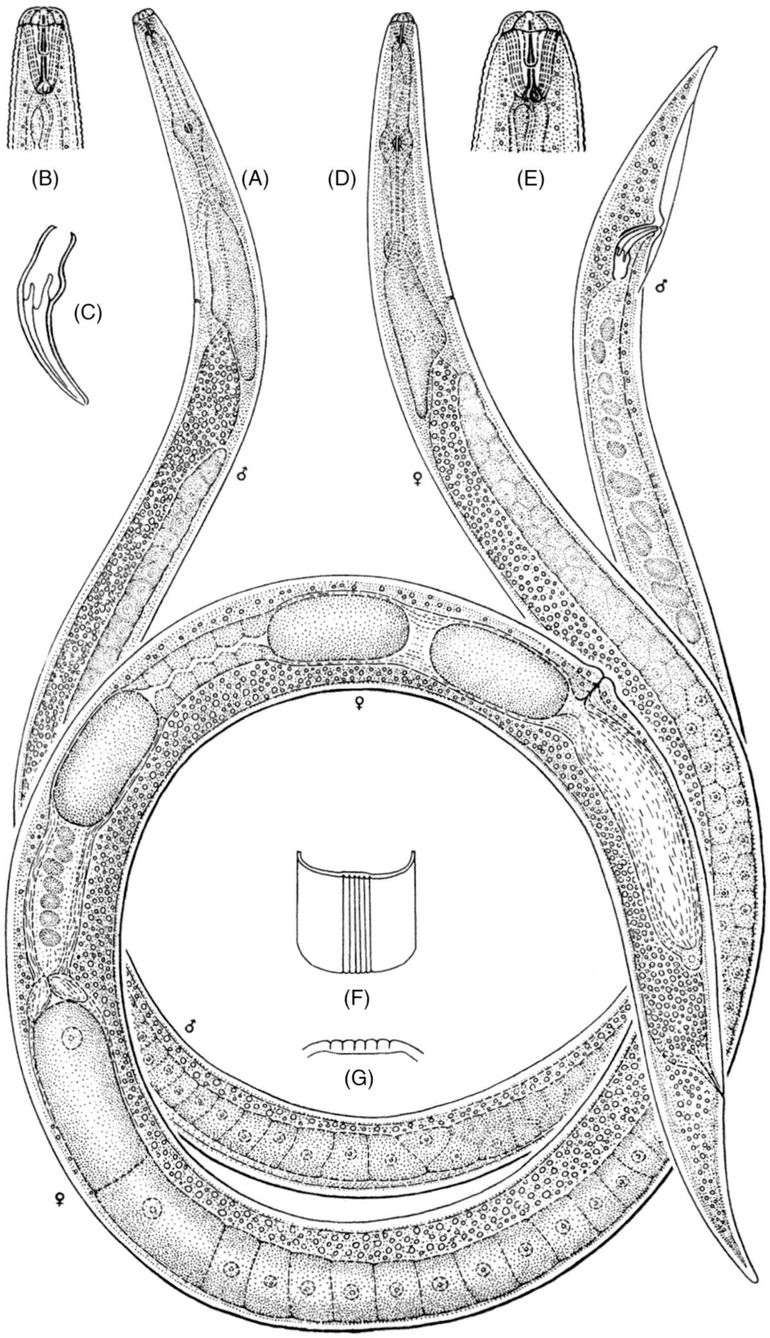

Fig. 1.9. Line drawing of *Ditylenchus destructor*. Male (A); head of male (B); spicule (C); female (D); head of adult female (E); section of cuticle at mid-body showing 6 incisures of lateral field (F); cross-section of lateral field (G). [Reproduced from CIH description, Set 2, No. 21 (after Thorne, 1945); courtesy Helminthology Society of Washington].

been reported to occur worldwide (Fig. 1.10). *D. destructor* is known to occur in the following countries (CABI-ISC, 2020).

Africa: South Africa
Asia: Azerbaijan, China, Iran, Japan, Kazakhstan, Kyrgyzstan, Pakistan, Saudi Arabia, South Korea, Tajikistan, Turkey, Uzbekistan.
Europe: Albania, Austria, Belarus, Belgium, Bulgaria, Czech Republic, Estonia, France, Germany, Greece, Hungary, Ireland, Italy, Jersey, Latvia, Lithuania, Luxembourg, Moldova, Netherlands, Norway, Poland, Romania, Russia, Slovakia, Slovenia, Sweden, Switzerland, Ukraine, United Kingdom.
North America: Canada, Mexico, USA.
Oceania: New Zealand.
South America: Ecuador, Peru.

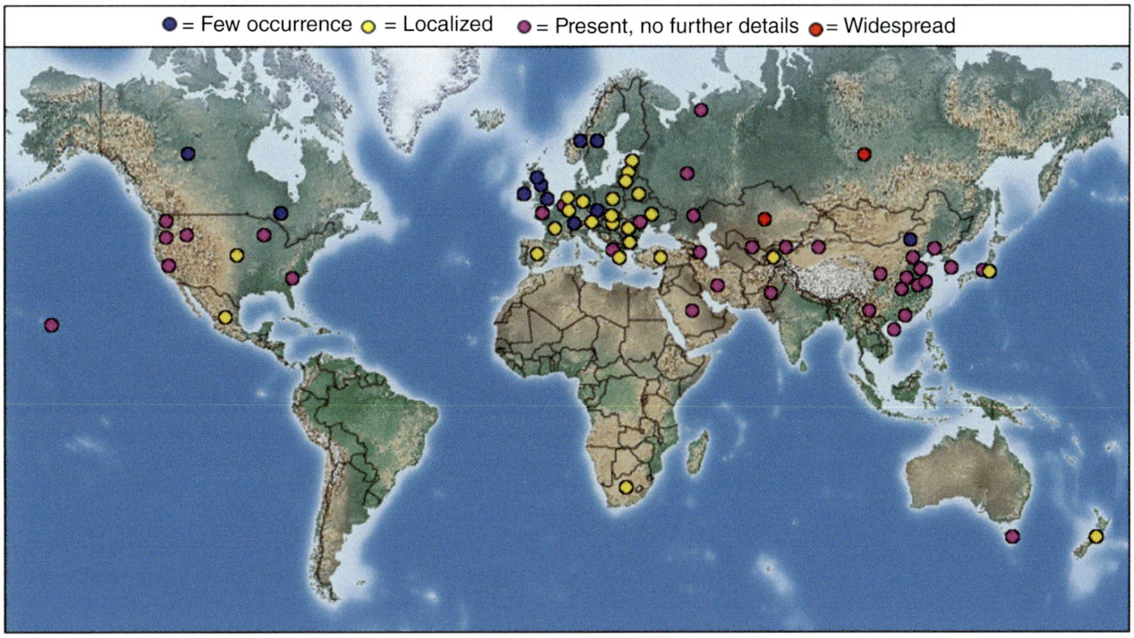

Fig. 1.10. Distribution map of *Ditylenchus destructor*. [Adapted from CABI-ISC, 2020]

6.6 Hosts

PRN attacks exclusively the below-ground parts of plants (tubers, rhizomes and stem-like subterranean parts). Potato is the principal host, but the nematode has an extensive host range, comprising more than 90 plant species, which include crop plants, ornamental plants and weeds (Sturhan and Brzeski, 1991). Main, and wild hosts of *D. destructor* are listed in Table 8.

6.7 Symptoms

PRN attacks mainly below-ground plant parts (tubers and stolons of potato, rhizomes of mint, and roots of hop and lilac), instigating discoloration and rotting of plant tissue. The above-ground parts are sometimes also infected, showing dwarfing, thickening and branching of the stem and curling, dwarfing and discoloration of the leaves (Sturhan and Brzeski, 1991). Symptoms on different crops are described below.

6.7.1 Potato

No noticeable symptoms appear during the growth period. The nematodes enter potato tubers typically through the stolons. The maximum numbers of nematodes are located at the edge of the undamaged and browning parts; here, the mass of small nematodes is visible even with a simple magnifying glass. The initial symptoms of *D. destructor* infection are small, white, chalky or light-coloured spots that can be seen just below the skin of the tuber (Brodie, 1998). These spots later become larger and slowly darker (through grey to dark brown and black) and develop a spongy texture (Fig. 1.7H). This is mainly a result of the secondary attack by fungi, bacteria and other saprophytic nematodes (Brodie, 1998). On severely affected tubers there are typically vaguely sunken areas with wrinkled, cracked, papery skin. The skin is not damaged but becomes thin and cracked as core infected tissues become dry and shrink (Brodie, 1998). Finally, the whole tuber may be mummified. Such damaged tubers drift in water (Fig. 1.7H). In contrast, the skin of potato infested with *D. dipsaci* is generally not cracked. The *D. destructor* nematodes continue to multiply inside the tubers even after harvest and may build up to huge numbers. Symptoms are more visible after storage, and secondary infections by fungi, bacteria and free-living nematodes occur.

6.7.2 Ornamental plants (iris, tulip, dahlia, etc.)

Infestation on iris results in greyish linear marks that expand upwards from the basal plate on the outer fleshy scales (Haglund, 1983). As nematode infestation intensifies, the destruction spreads over and reaches across the tissue of the bulb (Fig. 1.7I-K). This leads to a secondary dry, fibrous rotting which results in the breakdown of the bulb. Ring-like brown spots are noticeable in the cross-section of infested iris and tulip bulbs (Southey, 1993). Infested dahlia tubers develop symptoms similar to those in potato (Hooper, 1973a). Yellowing and dieback of the leaves are secondary symptoms caused by injury to the bulb and subsequent cessation of root functioning. *D. destructor* infestation of ornamental *Liatris spicata* corms ('Gayflower', 'Blazing Star' or 'Button Snakeroot') in cold storage in South Africa caused a blackish rot with living nematodes at different stages in the tissue adjoining the decaying areas (Van der Vegte and Daiber, 1983).

Table 8. List of main and wild host plants of *Ditylenchus destructor* (adapted from CABI-ISC, 2020).

Main hosts
Allium cepa (onion)
Allium sativum (garlic)
Arachis hypogaea (groundnut)
Beta vulgaris (beetroot)
Beta vulgaris var. *saccharifera* (sugar beet)
Camellia sinensis (tea)
Capsicum annuum (bell pepper)
Chrysanthemum morifolium
Citrus sinensis (navel orange)
Cucumis sativus (cucumber)
Cucurbita moschata (pumpkin)
Dahlia hybrids
Daucus carota (carrot)
Fragaria ananassa (strawberry)
Gladiolus hybrids (sword lily)
Glycine max (soybean)
Humulus lupulus (hop)
Ipomoea batatas (sweet potato)
Iris (irises)
Mentha (mints)
Panax ginseng (Asiatic ginseng)
Panax quinquefolius (American ginseng)
Solanum lycopersicum (tomato)
Solanum melongena (aubergine)
Solanum tuberosum (potato)
Trifolium (clovers)
Triticum aestivum (wheat)
Tulipa (tulip)
Zea mays (maize)
Wild hosts
Chenopodium album (fat hen)
Cyperus rotundus (purple nutsedge)
Datura stramonium (jimsonweed)
Eleusine indica (goosegrass)
Elymus repens (quackgrass)
Fumaria officinalis (common fumitory)
Solanum (nightshade)
Solanum nigrum (black nightshade)
Sonchus arvensis (perennial sowthistle)
Tagetes minuta (stinking Roger)
Taraxacum officinale complex (dandelion)
Xanthium strumarium (cocklebur)

6.7.3 Carrot

Infestation results in diagonal cracks in the skin of the carrot with white stripes in the cortical tissue (Fig. 1.7M). Secondary infections by fungi and bacteria in these areas may also occur, resulting in decay of the tissue. The damage is simply seen in a cross-section of the carrot. The nematode continues its damaging activity during storage and the carrots become unacceptable for consumption (Stoeen, 1977).

6.7.4 Beetroot and sweet potato

In sugar beet, infestation causes dark, necrotic lesions on rhizomes and roots. The symptoms are similar to crown canker (Dallimore and Thorne, 1951). On sweet potato, symptoms develop similar to potato. The damage is conspicuous in a cross-section of the tuber (Fig. 1.7L). In addition to yield loss, sugar content decreases.

6.8 Biology and life cycle

D. destructor is a migratory endoparasite of roots and below-ground modified plant parts. The nematode seldom attacks the aerial parts of plants. It enters potato tubers through the lenticels, and then starts to multiply quickly and colonize the whole tuber. The life cycle is completed inside potato tubers, where nematodes feed on starch grains. They live inside the living tissue, where they aggregate rapidly and the fecund females produce up to 250 eggs individually. Two days after being laid, eggs hatch at 28°C, with an average interval of 4.4 days between egg-laying and hatch, and development from egg to adult completes in 6–7 days. The nematodes can survive and develop within harvested tubers as well as in stored tubers during the winter. Soil plays only a secondary role in the transfer of this nematode (Hooper, 1973a). The PRN attacks carrots at the base of the lateral roots, and tissue breaks down the cortex. The damaged tissue becomes discoloured; subsequently, external lesions appear which serve as infection sites for other secondary pathogens (Stoeen, 1977). It was also found that PRNs invade stems, buds and leaves of *Cimicifuga racemosa* (Planer, 1972) and roots of ginseng in Korea (Young and Seung, 1995). Stem infestations are unusual but have also been reported on potato haulm (Goodey, 1951) and *Vicia sativa* (Duggan and Moore, 1962).

D. destructor can survive on other host plants and about 70 crops and weeds have been recorded as hosts. Unlike the closely related *D. dipsaci*, it does not form 'eelworm wool'. *D. destructor* is incapable of withstanding excessive desiccation, and for this reason is generally important only in cool and moist soils. Without a resistant resting stage, this species hibernates in the soil as adults or larvae and may even reproduce by feeding on alternative weed hosts like *Mentha arvensis*, *Sonchus arvensis* and fungal mycelia. It may also possibly overwinter as eggs and hatch in the spring and larvae are instantly able to parasitize hosts.

6.9 Economic importance

D. destructor is accountable for severe losses in potato and hops production (EPPO, 2013). The intensity of infestation of potato tubers by *D. destructor* on Estonian farms ranged from 2 to 9% and up to 80–90% of tubers from various fields became infected during storage (Kikas, 1969). *D. destructor* was frequently recorded on seed potatoes from Central Asia and it is prevalent on potatoes in Kazakhstan (German, 1972) and Azerbaijan (Ismailov, 1976) and causes substantial yield losses. In Uzbekistan, PRN represented 84.7% of the total nematodes found on potato tubers (Adylova and Vasilevskii, 1983). Severe infestations of potatoes have been recorded in the Samarkand, Tashkent and Fergana regions of Uzbekistan (Usmanova, 1972). Sturhan and Brzeski (1991) reported that *D. destructor* is also able to destroy the hyphae of cultivated mushroom (*Agaricus hortensis*).

6.10 Interaction with other pathogens

Rojankovski and Ciurea (1986) found 55 species of fungi and bacteria associated with *D. destructor* in potato tubers, with *Fusarium* spp. being the most common. *Rhizoctonia solani* infections of potato tubers were maximum in pots to which the largest number of *D. destructor* was added (Janowicz and Mazurkiewicz, 1982). The damage to potato tubers was greater when both the dry rot fungus (*Fusarium solani* var. *coeruleum*, *F. culmorum* and *F. oxysporum*) and *D. destructor* were present concomitantly (Janowicz, 1984).

6.11 Similarities to other species

D. destructor is similar to *D. dipsaci* but is distinguished by the lateral field demonstrating six incisures (Fig. 1.9F), the longer postvulval sac and the finely rounded tail terminus (Fig. 1.9D). Morphologically *D. destructor* differs from *D. africanus* mainly in its stylet length, which may overlap slightly, and the spicule length, which implies that males must be present in the population. The spiculum shape of *D. dipsaci* differs from *D. destructor* in having a ventral tumulus in the calomus area (Karssen and Willemsen, 2010). When observed in the lateral view, the spicule is less arched in *D. dipsaci* than in *D. destructor* (Fig. 1.9C).

6.12 Movement and means of dispersal

The nematode can be easily disseminated through the seed, tuber, bulbs and other plant parts.

6.12.1 Plant parts liable to carry the nematode in trade/transport

Roots, bulbs, tubers, corms, rhizomes, fruits (including pods), flowers, inflorescences, cones, calyx, leaves, stems (above ground), shoots, true seeds, seedlings/micro-propagated plants.

6.12.2 Plant parts not known to carry the nematode in trade/transport

Bark, wood and trunks.

6.13 Pest risk analysis and looming threat from introduction

D. destructor has a very wide host range and attacks several important food crops and ornamental plants in most of the temperate areas of the world (CABI-ISC, 2020). *D. destructor* is of quarantine significance for many countries, particularly in APPPC and COSAVE regions, and at present it is targeted in regulatory programmes worldwide. However, its dissemination varies between regions, countries and localities. Seed potatoes and bulbous ornamentals transmit the infection and disperse the disease, and hence must be subject to proper phytosanitary regulations.

6.14 Invasiveness rating

Considering the consequences of introduction and pest risk analysis, the following invasiveness rating has been assigned to *D. destructor*.

Climate/host interaction	–	High (3)
Hosts range	–	High (3)
Dispersal potential	–	High (3)
Economic impact	–	High (3)
Environmental impact	–	Medium (2)
Post-entry distribution	–	Medium (–2)
Final rating (score)	–	High (12)

Based on the above invasiveness rating, *D. destructor* receives a score of 12 out of 15 and is rated as HIGHLY INVASIVE NEMATODE.

6.15 Region-wise status of invasiveness

At the international level, *D. destructor* draws significant quarantine attention and has attained invasive status in the following countries, where it is absent or has very limited distribution.

APPPC: Bangladesh, Cambodia, India, Indonesia, Lao, Malaysia, Myanmar, Nepal, Philippines, Sri Lanka, Thailand, Tonga, Vietnam
CAHFSA: Anguilla, Bahamas, Barbados, Bermuda, Cayman Islands, Colombia, Cuba, Dominica, Dominican Republic, French Guiana, Grenada, Guadeloupe, Guyana, Haiti, Honduras, Jamaica, Martinique, Netherlands Antilles, Nicaragua, Panama, Puerto Rico, Saint Lucia, Saint Kitts & Nevis, Suriname, Trinidad and Tobago, Turks and Caicos Islands, Virgin Islands (British), Virgin Islands (US).
CAN: Bolivia, Venezuela
COSAVE: Argentina, Chile, Paraguay, Uruguay
EPPO: Croatia, Cyprus, Finland, Georgia, Guernsey, Jordan, Macedonia, Serbia, Tunisia
IAPSC: Angola, Benin, Botswana, Burkina Faso, Burundi, Cameroon, Cape Verde, Central African Republic, Chad, Comoros, Congo (Republic of), Congo (Democratic Republic of), Côte d'Ivoire, Djibouti, Egypt, Equatorial Guinea, Eritrea, Ethiopia, Gabon, Gambia, Ghana, Guinea, Guinea-Bissau, Kenya, Lesotho, Liberia, Madagascar, Malawi, Mali, Mauritania, Mauritius, Mozambique, Niger, Nigeria, Rwanda, São Tomé & Principe, Senegal, Seychelles,

Sierra Leone, Somalia, Sudan, Swaziland, Tanzania, Togo, Uganda, Zaire, Zambia, Zimbabwe
NEPPO: Algeria, Jordan, Libya, Malta, Morocco, Syria
OIRSA: Costa Rica, El Salvador, Guatemala, Honduras
PPPO: American Samoa, Australia, Cook Islands, Fiji, French Polynesia, Guam, Kiribati, Marshall Islands, Micronesia, Nauru, New Caledonia, Niue, Northern Mariana Islands, Palau, Pitcairn, Samoa, Solomon Islands, Tokelau, Tonga, Tuvalu, Vanuatu, Wallis and Futuna Islands

6.16 Management measures

6.16.1 Prevention and control

The spread of *D. destructor* can be prevented through the removal of infection sources by fumigation and the application of strict quarantine restricting the movement of infected tubers. Other phytosanitary methods for controlling *D. destructor* include the following.

TREATMENT OF POTATO TUBERS Soaking potatoes in nematicides before planting is effective against *D. destructor*. In Azerbaijan, sorting out of 'clean' seed potatoes, or treatment with dimethoate and carbathion (metam), were both effective against *D. destructor*. Dipping potato tubers in thionazin controlled PRN and enhanced the yield (Wilski, 1972).

TREATMENT OF BULBS Application of diazinon as bulb treatment decreased *D. destructor* populations by 37–48% (Rasinya, 1972). Immersion of iris bulbs in hot water containing 0.5% formaldehyde at 43.5°C for 3 h can control PRN infestations. The bulbs may be soaked for 2.5 h in a solution of thionazin plus formaldehyde. In another study, hot-water treatment of iris bulbs (cv. 'Wedgewood') for 3 h at 43.6°C after warm storage for 7 days at 30°C effectively controlled *D. destructor* (ISEHS, 1974). In Japan, infestation in iris bulbs can be controlled by immersion in water containing formaldehyde at 43.5°C for 2–3 h, but some varieties may be damaged during this treatment. In garlic bulbs, *D. destructor* was managed by drying at 34–36°C for 12–17 days (Fujimura *et al.*, 1989).

6.16.2 Phytosanitary measures

See Chapter 16, section 115.

7 *Ditylenchus dipsaci*

7.1 Common name

Stem and bulb nematode

7.2 Scientific name

Ditylenchus dipsaci (Kuhn, 1857) Filiopjev, 1936

7.3 Synonyms

Anguillula devastatrix Kühn, 1869
Anguillula dipsaci Kühn, 1857
Anguillula secalis Nitschke, 1868
Anguillulina dipsaci (Kühn, 1857) Gervais and Van, 1859
Anguillulina dipsaci var. *communis* Steiner and Scott, 1935
Ditylenchus allocotus (Steiner, 1934) Filip'ev and Sch. Stek., 1
Ditylenchus amsinckiae (Steiner & Scott, 1935) Filipev and Sch.
Ditylenchus dipsaci var. *tobaensis Schneider, 1937*
Ditylenchus fragariae Kir'yanova, 1951
Ditylenchus sonchophila Kir'yanova, 1958
Ditylenchus trifolii Skarbilivich, 1958
Tylenchus allii Beijerinck, 1883
Tylenchus devastator
Tylenchus devastatrix (Kühn) Oerley
Tylenchus dipsaci (Kühn, 1857) Bastian, 1865
Tylenchus havensteinii Kühn, 1881
Tylenchus hyacinthi Prillieux, 1881
Tylenchus putrefaciens Kühn, 1879

7.4 EPPO code

DITYDI

7.5 Diagnosis

Morphological, biochemical, karyological and molecular studies of different populations and races of *Ditylenchus dipsaci sensu lato* (*s.l.*) revealed it as a complex species composed of a large number of biological races mainly differing in host preference (Jeszke *et al.*, 2013). Subsequently, a total of 13 nominal species have been synonymized with *D. dipsaci* and up to 30 biological races have been distinguished, mainly differentiated by host range and usually named after their primary host plant. Further, Jeszke *et al.* (2013) grouped this complex into two, the first comprising diploid (n =12) populations categorized by their 'normal' size and named *D. dipsaci sensu stricto* (*s.s.*). This group contains most of the populations documented so far. The second group is polyploidal (n = 18–28) and currently includes: the 'giant race' of *D. dipsaci* (*Ditylenchus gigas* Vovlas *et al.*, 2011) parasitizing broad bean (*Vicia faba*); *D. weischeri* Chizhov *et al.*, 2010, parasitizing creeping thistle (*Cirsium arvense*); and three undescribed *Ditylenchus* spp. called D, E and F. D is associated with plant species of the Fabaceae, E is associated with plant species of the Asteraceae and F is associated with the plant species of Plantaginaceae (Jeszke *et al.*, 2013). Of all these *Ditylenchus* species, only *D. dipsaci s.s.* and its morphologically bigger variant *D. gigas* are plant parasites of higher plants and of economic importance. The following diagnosis protocols facilitate the precise identification of *D. dipsaci s.s* using morphological, biochemical, karyological and molecular diagnosis approaches.

7.5.1 Morphological diagnosis

Detailed morphological descriptions of *D. dipsaci* are given after Hooper (1972), Sturhan and Brzeski (1991), Wendt *et al.* (1995) and Brzeski (1998). For morphometric measurements, see Table 5.

FEMALE: Body almost straight after heat relaxed. Lateral field with four incisures (Fig. 1.11F). Head continuous with adjoining body (Fig. 1.11B). Stylet long, 10–12 µm. Stylet cone around half of stylet length, knobs well developed and rounded. Phasmid-like structures located dorsal to the lateral fields. Lip region unstriated, low, somewhat flattened, hardly offset from body. Median bulb muscular, 4–5 µm long with thickenings of lumen walls (Fig. 1.11A). Basal bulb overlapping intestine for a few micrometres or occasionally offset. Excretory pore behind posterior part of glandular bulb or isthmus. Postvulval uterine sac covering about half to marginally more of vulva–anus distance. Tail conical with a pointed tip (Fig. 1.11E).

MALE: Anterior region similar to female (Fig. 1.11G). Stylet 10–12 µm long. Bursa envelops tail about three-quarters (Fig. 1.11D, H). Spicules 23–28 µm long. Gubernaculum short and simple.

7.5.2 Biochemical and karyological diagnosis

Different workers used one-dimensional SDS gel electrophoresis of isozymes and proteins to distinguish species and races of *D. dipsaci* (Hussey and Krusberg, 1971). *D. dipsaci* was also categorized based on enzymes, using isoelectric focusing (Ibrahim, 1991). Two-dimensional gel electrophoresis (2-DGE) is a significantly more informative demonstration of proteins in a nematode extract (Abrantes *et al.*, 2004) and has been used to

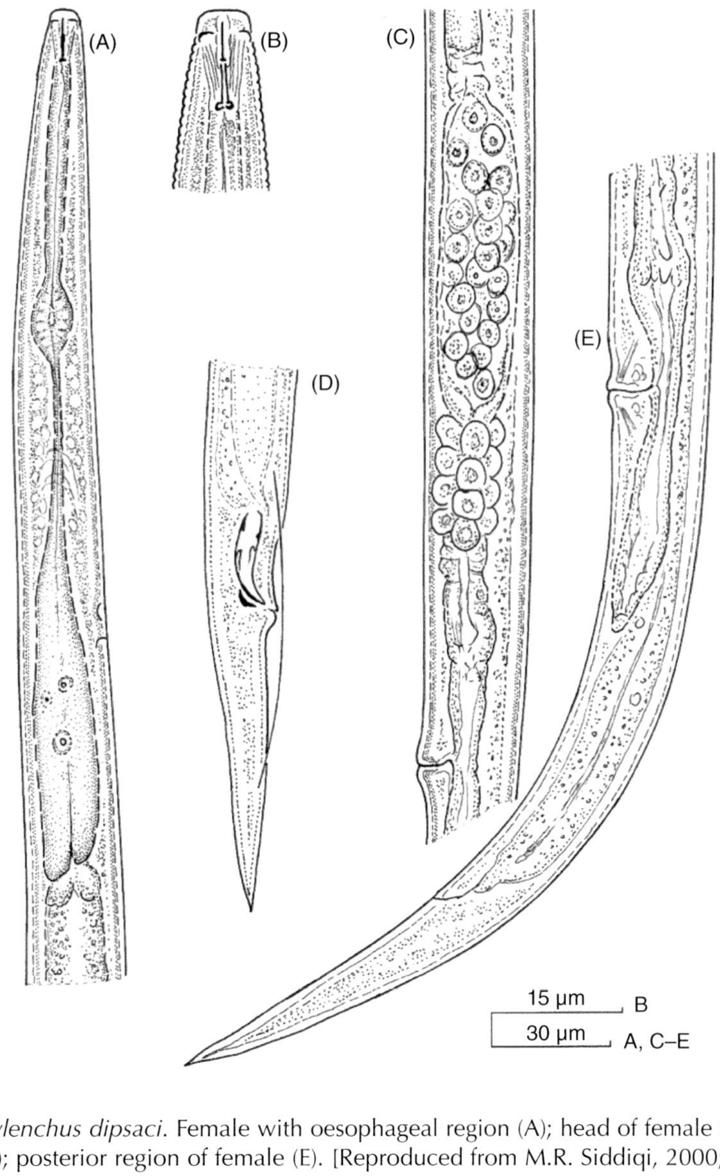

Fig. 1.11. Line drawing of *Ditylenchus dipsaci*. Female with oesophageal region (A); head of female (B); female reproductive system (C); tail end of male (D); posterior region of female (E). [Reproduced from M.R. Siddiqi, 2000; courtesy CAB International].

examine inter- and intraspecific variations in proteins of *D. dipsaci* (Bossis *et al.*, 1998). Southern hybridization (Wendt *et al.*, 1993) and electrophoresis (Palazova and Baicheva, 2002) have also been used to examine the races concept within *D. dipsaci* and genetic diversity amongst *Ditylenchus* species.

Barabashova (1978) distinguished the *D. dipsaci* complex in two groups based on karyological analyses. The first forms a fairly homologous group that includes the parasites of cultivated plants, with chromosome number $n = 12$. The majority of the races in the first group cross-breed successfully. The second forms a heterogeneous group that includes parasites of wild plants and these differ in both karyotype ($n = 18$ to 28) and morphology. The 'giant race', parasitic in *Vicia faba* var. *equina*, where $n = 27$, is also considered in the second group. At least partial incompatibility is perceived among races in the second group. Genetically, the two groups are incompatible.

7.5.3 Molecular diagnosis

The literature reveals that numerous molecular methods have been developed and used for *D. dipsaci* diagnosis. Most of the molecular approaches are PCR or PCR–RFLP based, and for identification and detection of population difference by sequence analysis (Leal-Bertioli *et al.*, 2000; Esquibet *et al.*, 2003; Subbotin *et al.*, 2005a; Zouhar *et al.*, 2007). Madani *et al.* (2015) designed species-specific primer sets based on the nucleotide sequence of the heat shock protein (hsp90) gene for *D. dipsaci*. Some of these techniques are described below.

18S AND ITS1-SPECIFIC PCR TEST Subbotin *et al.* (2005a) developed a species-specific identification for *D. dipsaci s.s.* (normal race only; the giant race, later called *D. gigas*, not included). The test was evaluated against *D. destructor* (one population), *D. dipsaci* (18 populations from various hosts and locations) and *Ditylenchus* sp. (12 populations from different hosts and locations). The *D. dipsaci*-specific primers are:

rDNA2: 5′-TTT CAC TCG CCG TTA CTA AGG-3′ (Vrain *et al.*, 1992)
DitNF1: 5′-TTA TGA CAA ATT CAT GGC GG-3′

The amplification is performed using the following cycling parameters.

Reagent	–	Final concentration
PCR mixture	–	25 µl (1× from 10× PCR buffer)
$MgCl_2$	–	15 mM
dNTPs	–	0.5 mM
Primer rDNA2	–	60 nM
Primer DitNF1	–	60 nM
Taq DNA polymerase	–	1 U
Cycling parameters		
Initial denaturation	–	94°C for 4 min
Number of cycles	–	35
Denaturation	–	94°C for 15s
Annealing	–	57°C for 30s
Extension	–	72°C for 30s
Final elongation	–	72°C for 10 min

The PCR products are analysed by agarose gel electrophoresis. The amplicon size is almost 263 bp for *D. dipsaci s.s.* (normal race). No amplification is observed with non-target species.

5.8S RDNA-SPECIFIC PCR TEST Marek *et al.* (2005) developed a species-specific test for *D. dipsaci*. It was evaluated against three European populations of *D. dipsaci* obtained from different hosts and some non-target genus populations such as *Globodera pallida*, *Bursaphelenchus xylophilus* and *Rhabditis* spp. Two specific primer sets were designed for *D. dipsaci* identification, but the most sensitive (10 pg of target DNA detected) primers are:

PF1: 5′-AAC GGC TCT GTT GGC TTC TAT-3′
PR1: 5′-ATT TAC GAC CCT GAG CCA GAT-3′

The PCR amplification is performed using the following reagents and cycling parameters.

Reagent	–	Final concentration
PCR mixture	–	25 µl (1× Taq buffer)
$MgCl_2$	–	1.5 mM
dNTPs	–	200 µM
Primer PF1	–	10 pmol
Primer PR1	–	10 pmol
Taq DNA polymerase	–	1.5 U
Cycling parameters		
Initial denaturation	–	94°C for 3 min
Number of cycles	–	30
Denaturation	–	94°C for 2 min
Annealing	–	62°C for 30 s
Extension	–	72°C for 2 min
Final elongation	–	72°C for 10 min

The PCR products are analysed by agarose gel electrophoresis. The amplicon size with this primer set is around 327 bp for *D. dipsaci*.

5.8S RDNA AND ITS-SPECIFIC PCR TEST FOR *D. DIPSACI* Kerkoud *et al.* (2007) developed 5.8S rDNA and ITS-specific PCR test for *D. dipsaci* and it was evaluated against *D. dipsaci*, *D. africanus*, *D. destructor*, *D. myceliophagus*, *Ditylenchus* sp. (now described as *D. gigas*) and *Aphelenchoides ritzemabosi*. Two sets of specific primers are used, one for *D. dipsaci* alone and one for *D. gigas* and *D. dipsaci* combined. The primer sets allow separation of *D. dipsaci* from *D. gigas*.

The first primer set is:

DdpS1: 5′-TGG CTG CGT TGA AGA GAA CT-3′
rDNA2: 5′-TTT CAC TCG CCG TTA CTA AGG-3′ (Vrain *et al.*, 1992)

The second primer set is:

DdpS2: 5′-CGA TCA ACC AAA ACA CTA GGA ATT-3′
rDNA2: 5′-TTT CAC TCG CCG TTA CTA AGG-3′ (Vrain *et al.*, 1992)

The amplification is performed using the following reagents and cycling parameters.

Reagent	–	Final concentration
PCR mixture	–	20 µl
PCR buffer	–	1.5 mM
$MgCl_2$	–	5 mM
dNTPs	–	200 µM
First primer DdpS1	–	0.5 µM
First primer rDNA2	–	0.5 µM
Second primer DdpS1	–	1 µM
Second primer rDNA2	–	1 µM
Taq DNA polymerase	–	1 U
Cycling parameters		
Initial denaturation	–	94°C for 1 min
Number of cycles	–	40
Denaturation	–	94°C for 30 s
Annealing	–	60°C for 45 s
Extension	–	72°C for 45 s

The PCR products are analysed by agarose gel electrophoresis. The amplicon size for *D. dipsaci* alone is approximately 517 bp. No amplification is recorded with non-target species, including *D. gigas*. When both primer sets are used, the amplicon size for *D. dipsaci* and *D. gigas* is approximately 707 bp.

SCAR PCR TEST Esquibet *et al.* (2003) developed a species-specific test for *D. dipsaci* using sequence characterized amplified region (SCAR) PCR test and it was designed to differentiate normal and giant races of *D. dipsaci*. The test was evaluated against *D. dipsaci* normal race (11 populations from different hosts and locations), *D. dipsaci* giant race (*D. gigas*, 11 populations isolated from *V. faba* from different locations) and *D. myceliophagus* (one population). The *D. dipsaci*-specific primers used are as follows.

Normal race:

H05: 5′-TCA AGG TAA TCT TTT TCC CCA CT-3′
H06: 5′-CAACTG CTA ATG CGT GCT CT-3′

Giant race:

D09: 5′-CAA AGT GTT TGA TCG ACT GGA-3′
D10: 5′-CAT CCC AAA ACA AAG AAA GG-3′

The amplification is performed using the following reagents and cycling parameters.

Reagent	–	Final concentration
PCR mixture	–	10 µl
$MgCl_2$	–	1.5 mM
dNTPs	–	250 µM
Primer H05-H06 or D09-D10	–	690 nM (for duplex PCR)
Primer H05-H06 or D09-D10	–	500 nM (for multiplex PCR)
Taq DNA polymerase	–	0.5 U
Cycling parameters		
Initial denaturation	–	94°C for 3 min
Number of cycles	–	30
Denaturation	–	94°C for 1 min
Annealing	–	59°C for 1 min
Extension	–	72°C for 1 min

The PCR products are analysed by agarose gel electrophoresis. The amplicon size is approximately 242 bp for *D. dipsaci* (normal race) and 198 bp for *D. dipsaci* (giant race). When both primer sets are used, no amplification is recorded with non-target race and non-target species (Esquibet *et al.*, 2003).

Another SCAR PCR test was designed by Zouhar *et al.* (2007) using a species-specific primer for *D. dipsaci*. It was evaluated against ten European populations of *D. dipsaci* from different hosts. The two following specific primer sets have been developed.

First primer:

DIT_2 forward: 5′-GCA ATG CAC AGG TGG ATA AAG-3′
DIT_2 reverse: 5′-CTG TCT GTG ATT TCA CGG TAG AC-3′

Second primer:

DIT_5 forward: 5′-GAA AAC CAA AGA GGC CGT AAC-3′
DIT_5 reverse: 5′-ACC TGA TTC TGT ACG GTG CAA-3′

The PCR is performed with the following reagents and cycling parameters.

Reagent	–	Final concentration
PCR mixture	–	25 µl
$MgCl_2$	–	1.5 mM
dNTPs	–	200 µN
Forward primer DIT_2 or DIT_5	–	10 pmol
Reverse primer DIT_2 or DIT_5	–	10 pmol
Taq DNA polymerase	–	1.5 U
DNA template	–	50 ng
Cycling parameters		
Initial denaturation	–	94°C for 3 min
Number of cycles	–	30
Denaturation	–	94°C for 1 min
Annealing	–	60°C for 1 min
Extension	–	72°C for 1 min
Final elongation	–	72°C for 10 min

The PCR products are analysed by agarose gel electrophoresis. The amplicon with the first primer set is approximately 325 bp for *D. dipsaci*. The amplicon with the second primer set is approximately 245 bp for *D. dipsaci*.

7.6 Geographical distribution

The stem and bulb nematode (SBN) has been reported to occur all over the world, especially in temperate regions (Fig. 1.12). There are several reports about the presence of this nematode in India (Sturhan and Brzeski, 1991; Chakraborti, 2001), although other workers noted its absence (EPPO, 2009b; CABI-ISC, 2020). *D. dipsaci* is known to occur in the following countries (EPPO, 2013; CABI-ISC, 2020).

Africa: Algeria, Kenya, Morocco, Réunion, South Africa, Tunisia.

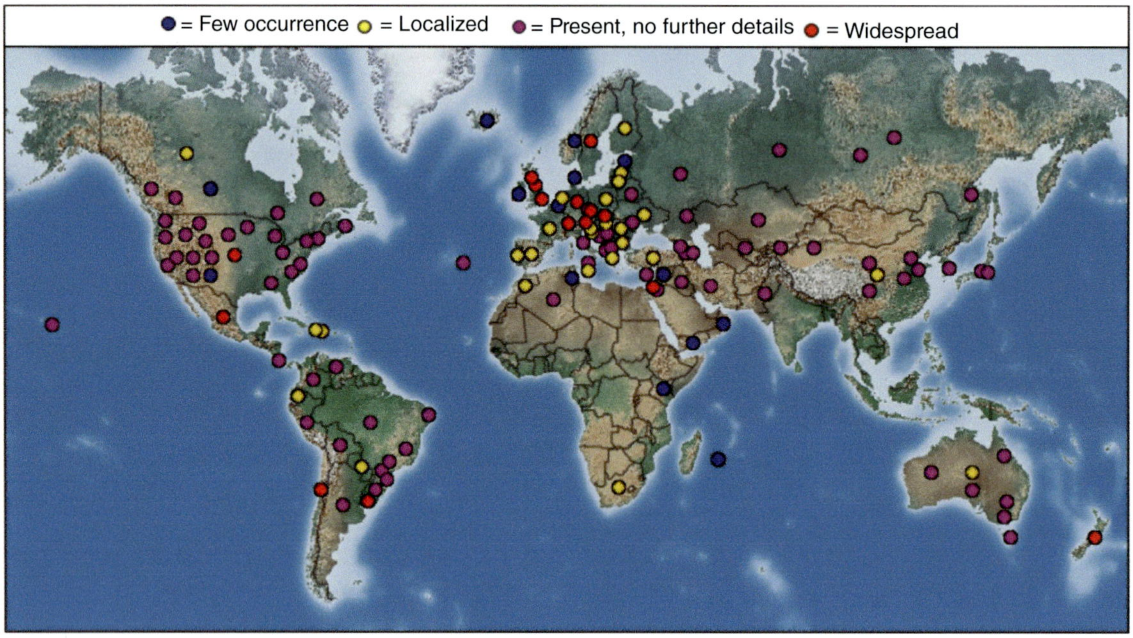

Fig. 1.12. Distribution map of *Ditylenchus dipsaci*. [Adapted from CABI-ISC, 2020]

Asia: Armenia, Azerbaijan, China, Georgia, Iran, Iraq, Israel, Japan, Jordan, Kazakhstan, Kyrgyzstan, Oman, Pakistan, South Korea, Syria, Taiwan, Turkey, Uzbekistan, Yemen.

Europe: Albania, Austria, Belarus, Belgium, Bosnia–Herzegovina, Bulgaria, Croatia, Cyprus, Czech Republic, Denmark, Estonia, Finland, France, Germany, Greece, Hungary, Iceland, Ireland, Italy, Latvia, Lithuania, Macedonia, Malta, Moldova, Netherlands, Norway, Poland, Portugal, Romania, Russian, Serbia, Slovakia, Slovenia, Spain, Sweden, Switzerland, Ukraine, United Kingdom.

North America: Canada, Costa Rica, Dominican Republic, Haiti, Mexico, USA.

Oceania: Australia, New Zealand.

South America: Argentina, Bolivia, Brazil, Chile, Colombia, Ecuador, Paraguay, Peru, Uruguay, Venezuela.

7.7 Hosts

SBNs parasitizes more than 1200 species of cultivated and wild plants (Jeszke *et al*., 2013). Among these over 40 angiosperm families are known to be hosts of *D. dipsaci*. According to Sturhan and Brzeski (1991), the primary hosts of *D. dipsaci* from different families are:

Amaryllidaceae: *Narcissus* spp., *Allium cepa*, *A. sativum*, *Tulipa* spp.
Brassicasae (=Cruciferae): *Brassica campestris*
Poaceae (=Gramineae): *Avena sativa*, *Secale cereale*, *Zea mays*, *Triticum aestivum*
Fabaceae (=Leguminosae): *Medigo sativa*, *Vicia* spp., *Pisum sativum*, *Trifolium* spp.
Solanaceae: *Solanum tuberosum*, *Nicotiana* spp.

Several weeds and grasses are also recorded as hosts for *D. dipsaci* and may play an important role in its survival in the absence of cultivated plants. Many of the biological races of *D. dipsaci* have limited host ranges (EPPO, 2013). Some important host plants of *D. dipsaci* are presented in Table 9.

7.8 Symptoms

D. dipsaci is one of the most devastating nematode pests on a wide range of plants. Specific symptoms in the principal hosts are as follows.

7.8.1 Poaceae (=Gramineae)

OATS, RYE AND WHEAT Infested leaves are distorted, stems swollen with abnormal number of tillers. Overall, the plant is stunted, bushy and short. In rye (*Secale cereal*), *D. dipsaci* occurs mostly in light soils with poor humus and naturally in areas where rye is regularly cultivated. The first symptom can be detected in late autumn, but they are most noticeable in spring. In oats (*Avena sativa*), several patches of plants with stunted growth in the field indicate damage by the nematode. As infestation intensifies, plant growth ceases and the damage is visible as chlorosis of the crop (McDonald and Nicol, 2005). Wheat plants attacked by *D. dipsaci* show symptoms similar to those in rye and oats (Rivoal and Cook, 1993).

MAIZE Maize is a poor host of *D. dipsaci* but young plants may be killed. Invasion by *D. dipsaci* on young stem tissues produces necrosis, mainly in young plants, and the plants eventually die (Rivoal and Cook, 1993). Infected plants show crispy and twisted leaves resembling a corkscrew. The internodes are also shortened, and the bottom of the stem becomes hollow, causing the maize plant to fall over before harvest.

Table 9. List of host plants of *Ditylenchus dipsaci* (adapted from CABI-ISC, 2020).

Main host
Allium cepa (onion)
Allium porrum (leek)
Allium sativum (garlic)
Avena sativa (oats)
Begonia
Beta vulgaris (sugar beet)
Cannabis sativa (hemp)
Fragaria ananassa (strawberry)
Gladiolus hybrids (sword lily)
Hyacinthus orientalis (hyacinth)
Medicago sativa (lucerne)
Narcissus (daffodil)
Narcissus pseudonarcissus (wild lent lily)
Nicotiana tabacum (tobacco)
Phaseolus (beans)
Phlox drummondii (annual phlox)
Phlox paniculata (perennial phlox)
Pisum sativum (pea)
Polyphagus (polyphagous)
Secale cereale (rye)
Solanum tuberosum (potato)
Trifolium pratense (purple clover)
Trifolium repens (white clover)
Tulipa (tulip)
Vicia faba (broad bean)
Zea mays (maize)
Other hosts
Apium graveolens (celery)
Allium cepa var. *aggregatum*
Brassica napus var. *napus* (rape)
Carduus acanthoides
Crocus sativus (saffron)
Cucurbitaceae (cucurbits)
Dianthus caryophyllus
Helianthus annuus (sunflower)
Hydrangea (hydrangeas)
Ipomoea batatas (sweet potato)
Lens culinaris subsp. *culinaris* (lentil)
Onobrychis viciifolia (sainfoin)
Phaseolus coccineus (runner bean)
Pimpinella anisum (aniseed)
Petroselinum crispum (parsley)
Triticum (wheat)
Wild host
Astrantia
Avena sterilis (winter wild oat)
Bergenia
Brassica rapa subsp. *rapa* (turnip)
Chenopodium murale
Cirsium arvense (creeping thistle)
Convolvulus arvensis (bindweed)
Hieracium pilosella
Lamium album
Lamium amplexicaule (henbit deadnettle)
Lamium purpureum (purple deadnettle)
Myriophyllum verticillatum
Nerine sarniensis (Guernsey lily)
Stellaria media (common chickweed)
Raphanus raphanistrum (wild radish)
Ranunculus arvensis (corn buttercup)
Taraxacum officinale complex

7.8.2 Amaryllidaceae

NARCISSUS SPP. On above-ground parts, symptoms are the presence of chlorotic leaves with blister-like swellings (spickels). Swellings are conspicuous before flowering when leaves are actively growing. In low infestations, the swellings can be more easily felt between finger and thumb than seen. On bulbs, concentric brown rings can be seen when they are cut transversely (Fig. 1.13A-C). The necrosis can also be seen when bulbs are cut longitudinally. The necrosis starts at the neck and spreads downwards. Nematode infection can be noticed in dry bulbs with nominal bulb damage by cutting just below the neck. In the early stages of infestation, careful inspection reveals shining, spongy areas where cells have been detached. This is promptly followed by brown necrosis (Southey, 1993). Other Amarylliaceae plants usually show similar symptoms as observed in *Narcissus* spp.; for example, *Galanthus* spp. and *Nerine* spp. also express swellings on their foliage and concentric, brown necrotic rings in the bulbs.

ALLIUM SPP. (ONION, GARLIC, LEEK AND SHALLOT) In most *Allium* spp. the characteristic symptoms of SBN infestation appear on the leaves and bulbs, which become deformed (Fig. 1.13D,E,J). The base of young plants becomes swollen and leaves become twisted. Older infected bulbs exhibit bloat (swelling) of scales with open cracks frequently occurring at the root disc of the bulbs (Potter and Olthof, 1993). In onion (*A. cepa*), nematode feeding in leaves causes maceration and dissolution of cells, resulting in blisters and frosted appearance of the plants (Ferris and Ferris, 1998). In storage, the infested bulbs are predisposed to rot readily (Duncan and Moens, 2013). The inner scales of the bulb are generally more severely attacked by the nematode than the outer scales. As the season progresses the bulbs become soft and, when cut open, show browning of the scales in concentric circles. Above-ground symptoms are stunting and yellowing, while symptoms in the bulbs are necrosis, under-development and distortion. In garlic, in contrast to onion, *D. dipsaci* does not prompt deformation of leaves or swelling but does cause leaf yellowing and death of the plants (Netscher and Sikora, 1990). No visible symptoms of nematode infestation are observed on infested seeds of *Allium* spp.

7.8.3 Liliaceae

TULIP On tulips, the symptoms of *D. dipsaci* attack appear on both growing plants and bulbs, and they are quite different from those in *Narcissus* spp. In the field, infestation is best noticed at flowering. The first sign of nematode infestation is a pale or purplish lesion on one side of the stem immediately below the flower, which bends in the direction of the lesion (Southey, 1993). The lesion size increases and the epidermis

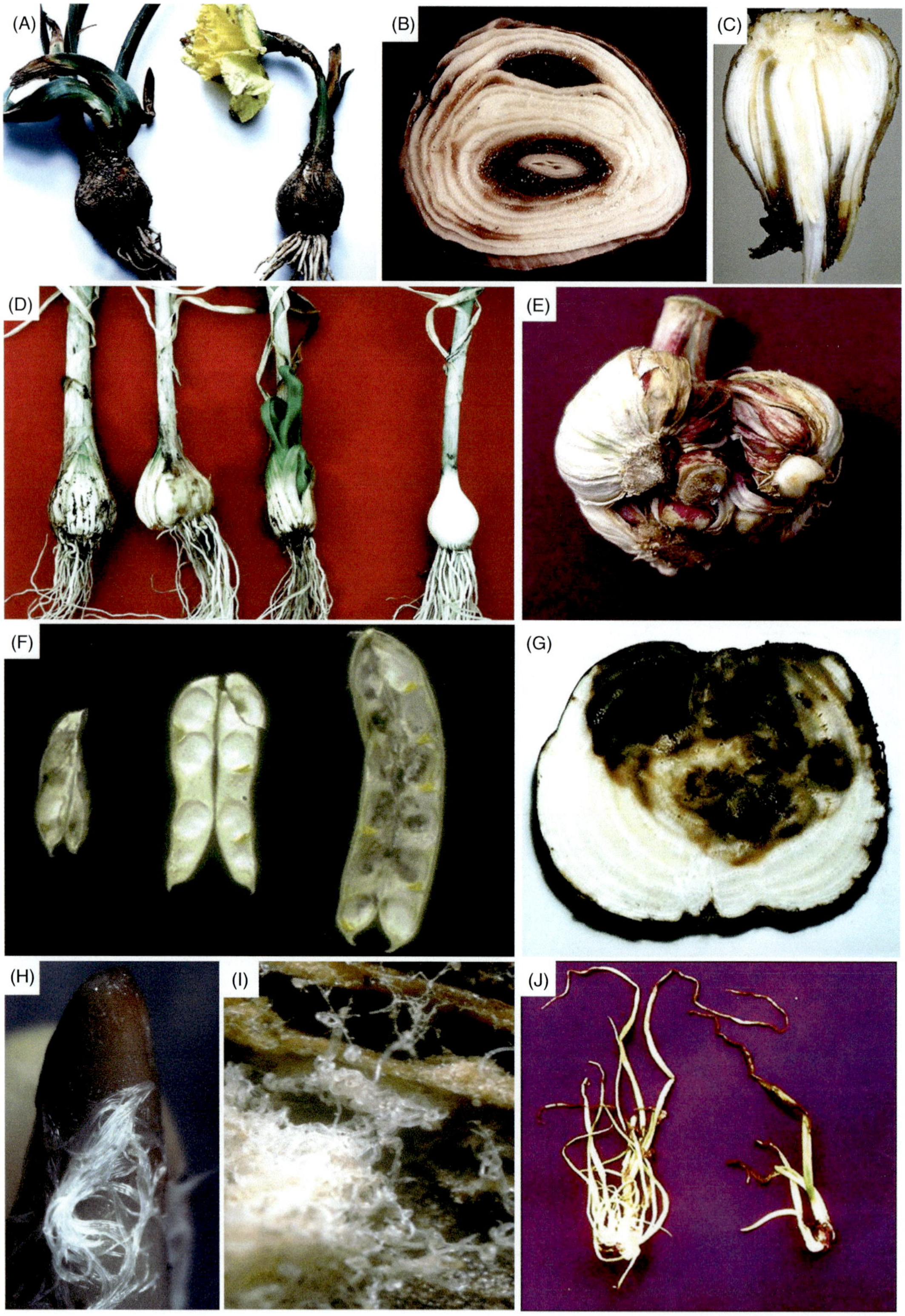

Fig. 1.13. Symptoms of *Ditylenchus dipsaci* on different plants. On narcissus bulbs (A); cross-section of narcissus (B,C); on garlic bulb (D,E); bean (F); cross-section of sugar beet (G); nematode wool (H); nematode in narcissus bulb tissue (I); young onion plants (J). [Courtesy A,D,E,H: G. Caubel, Nemapix, (1999); B: C.W. Laughlin, Nemapix, (2002); C,I: Central Science Laboratory, Harpenden, British Crown, Bugwood.org; F: Dr Augustin/Landesanstalt fur Pflanzenbau und Pflanzenschutz, Mainz, Bugwood.org; G: C. Hogger, Nemapix, (1999); J: E. Hennig, State Plant Health and Seed Inspection Service, Torun, Poland]

splits, revealing typical loose tissue beneath. The damage spreads downwards and often upwards on to the petals. In a severe infestation, the lesions extend down stems from leaf axils and growth may become retarded. On below-ground parts, the infestations start at the base of new bulbs, which arise as lateral offset buds from the base of the previous stems. The infection can be seen and detected on the removal of the outer brown scales as grey or brown soft patches on the outer fleshy scales. Infected tulip bulbs do not show brown rings as in narcissus and hyacinth (Southey, 1993).

7.8.4 Fabaceae (=Leguminosae)

FABA BEANS, VETCH, CHICKPEA, PEA AND LENTIL On leguminous crops, *D. dipsaci* induces swelling and distortion of stem tissue or lesions which turn reddish-brown and then black, depending on environmental factors and cultivar. Newly developed pods take on a dark brown appearance (Fig.1.13F). The lesions wrap the stem and increase in length, often progressing to the edge of an internode. Leaf and petiole necrosis are also frequent under heavy infestations but can be confused with symptoms induced by fungal pathogens. Infected seeds are darker, distorted, smaller in size and may have speckle-like spots on the surface. Heavy infestations regularly kill the main shoots, stimulating secondary tiller formation. The symptoms with greater severity are generally caused by the 'giant race' (Sikora and Greco, 1990). On faba bean (*Vicia faba*), *D. dipsaci* (now *D. gigas*) causes swelling or necrosis of the tissue. Infested stems of lentil and vetch (*Vicia* spp.) are also swollen and display shortened internodes. *D. dipsaci* induces confined necrosis on pea and complete necrosis of the stem on vetch (Caubel *et al*., 1998).

LUCERNE *D. dipsaci* is one of the most important nematode pests of lucerne (*Medicago sativa*). The infestation is more common in heavier soils and during high rainfall or in sprinkler-based irrigated areas. 'White flagging' related to loss of leaf chlorophyll is often a feature of infested crops under conditions of low moisture (Griffin, 1985). Infested fields exhibit irregular areas of sparse growth. Typical symptoms of nematode attack include basal swelling, dwarfing and twisting of stalks and leaves, shortening of internodes, and the formation of many axillary buds, producing an abnormal number of tillers that give the plant a bushy appearance (McDonald and Nicol, 2005). Infested plants occasionally do not grow high enough for hay (Ferris and Ferris, 1998) and they often fail to produce flower spikes (McDonald and Nicol, 2005). No symptoms of nematode infestation are observed in *Medicago* seeds.

CLOVER In general, the symptoms on clover are very similar to those expressed by *M. sativa*, except in red and white clovers. The nematode invades red clover in cool, rainy climates. Large, patchy areas of diseased plants appear in the field, with more diseased plants towards the inner side of the area. Central plants often show wilting (Cook and Yeates, 1993). The infested plants have swollen bases like bulbs and shrivelled papery leaves with conspicuously thick veins. Flower initiations are swollen like galls, and a single flower gall may comprise 5000 nematodes (Courtney, 1962). On white clover, the infected stems are swollen and short, buds are tufty, and the diseased parts become brown in summer or autumn. Leaves are slimmer than usual, and their petioles are shorter and thicker. Infested flower buds are swollen at their bases (Andrássy and Farkas, 1988).

7.8.5 Solanaceae

POTATO On potato, *D. dipsaci* produces a funnel-shaped rot, which expands further into the tuber, hence differing from the superficial rot produced by *D. destructor*. Leaves and stems are also invaded by *D. dipsaci* and this results in the characteristic stunting of the plant, supplemented by severe alteration of stems and petioles (Evans *et al*., 1993).

TOBACCO On tobacco, *D. dipsaci* enters the leaves and stems of seedlings during rainy weather and produces small, yellow galls (swellings) that may reach 40 cm or more above the soil. As disease progresses, the number of galls increases and plant tissue starts to die prematurely. The lower leaves may fall off and upper leaves may turn chlorotic. Galls ultimately rot and further growth is stopped in infected plants. In heavy soils, during cool and damp weather, the infested stems break down and the plants dislodge (Johnson, 1998).

7.8.6 Brassicaceae (=Cruciferae)

In mature *B. campestris*, severe crown rot may develop after infestation with *D. dipsaci*.

7.8.7 Other hosts

HYACINTHS The symptoms produced on hyacinth bulbs are similar to those on *Narcissus* spp., but prominent swellings are not typically seen on the plant leaves. The leaves may produce yellow streaks, distortion and frequently slight swelling (Southey, 1993).

STRAWBERRY *D. dipsaci* is the primary species of *Ditylenchus* that is regarded as an important pest of strawberry (Brown *et al*., 1993). The symptoms appear as small, twisted leaves with short, thick and distorted petioles.

BEETROOT Infestation by *D. dipsaci* results in the death of the growing point of seedlings and leads to the formation of multiple crowns. The cotyledons and leaves may become distorted, swollen and twisted. Galls may also develop on the leaves or petioles, usually on older plants. In severe conditions, feeding by the nematode on the crown may result in crown canker, crown rot or collar rot. The first noticeable symptom appears as raised, greyish pustules, generally among the leaf scars, followed by rotting outwards and downwards, expanding across the shoulder of the plant. Rotten crowns are easily detached when pulled (Cooke, 1993). The cross-section cutting of sugar beet root shows necrotic infection (Fig. 1.13G).

CARROT On carrot, the main infestation occurs on the root and stem of the plant around 1 inch (2.5 cm) below and above the ground level (Cooke, 1993). The symptoms may appear as overlapped leaves and discoloration of the main root head below-ground. Serious damage causes the death of leaves and crown rot, particularly in autumn.

PHLOX AND OTHER ORNAMENTAL PLANTS On phlox, *D. dipsaci* causes typical thickening of shoots, brittleness of stems and shortening of internodes, which tend to split. Crinkling and reduction of the laminae of upper leaves and reducing of the uppermost leaf to attenuated filaments are characteristic and unique symptoms on phlox (Southey, 1993).

Other ornamental plants, such as *Anemone*, *Calceolaria*, *Cheiranthus*, *Gypsophila*, *Helenium*, *Heuchera*, *Lychnis*, *Lysimachia*, *Penstemon* and *Primula* spp., have been recorded as hosts of *D. dipsaci*, with symptoms including abnormal growth, swelling, rotting and failure to produce flowers (Roberts *et al.*, 1981). Woody plants are tougher to attack than succulent plants, but *Hydrangea* may be invaded by SBN, causing alteration of non-woody shoots and swelling of petioles and main veins. Pronounced crinkling of leaf laminae also occurs. Another woody plant, *Yucca smaliana*, shows distortion and blister-like swellings on leaves (ISPM 27, DP-8, 2015).

7.9 Biology and life cycle

The life cycle is similar to that of other *Ditylenchus* species. Reproduction is by amphimixis and numerous generations can occur in a single growing season. Population growth can be very rapid, as the females have very high fecundity rate and may lay 200–500 eggs, from which the J_2 hatch within 2 days and develop into females within 4–5 days. Females live more than 10 weeks and the life cycle requires 19–23 days at 15°C (Duncan and Moens, 2013). Within onion and garlic tissue, the life cycle completes from egg to egg in approximately 21 days at 15°C. The best temperature range for egg development is 15–22°C (Becker and Westerdahl, 2018). The pre-adults or fourth-stage larvae can survive in freezing or extremely dry conditions in an anhydrous state as 'nematode wool' (Fig. 1.13H) for quite long periods, sometimes 20 years or more in plant tissue, leaves, stems, bulbs, seeds or in soil (Agrios, 2005). The rapid population growth of this nematode can result in severe crop damage even when the initial population density is low (Duncan and Moens, 2013).

7.10 Economic importance

D. dipsaci is one of the highly devastating plant-parasitic nematodes, mainly in temperate regions. The nematode can cause complete failure of host crops such as garlic, onions, cereals, legumes, strawberries and ornamental plants and affect the production of flower and bulbs. Sturhan and Brzeski (1991) observed that crop losses of 60–80% are not uncommon in severe infestations. In onion, up to 60% of seedlings died before attaining the planting stage in Italy. Losses of 50% were recorded on garlic in Italy and more than 90% in France and Poland.

7.11 Similarities to other species/conditions

In the first instance, *D. dipsaci* looks similar to other *Ditylenchus* species (*D. destructor*, *D. gigas*, *D. myceliophagus*, etc.). However, morphologically the number of lateral incisures (four), relatively long stylet, length of the postvulval sac and the pointed tail terminus are the differentiating characters of this species (Andrássy, 2007). *D. dipsaci* can be differentiated from *D. gigas* by the shorter body of females and the longer vulva–anus distance (Vovlas *et al.*, 2011). In the lateral view, the spicule of *D. dipsaci* is more arched than in *D. destructor*. Several plants, such as lucerne, beets and clover, are infested by *D. dipsaci* and *D. destructor*, but the two species seldom occur together in the same plant (Andrássy and Farkas, 1988).

7.12 Movement and means of dispersal

Dispersal of SBN is primarily through infested plant material such as leaves, stems, bulbs and seeds, and infested soil. Contaminated agricultural tools and equipment, irrigation and splash water also help in dispersal and spread of *D. dipsaci* (Chitambar, 2018). Dissemination from one field to another or from one plant to another takes place through plant residues, water (flood or irrigation water), rain splash and stem and leaf contact. *D. dipsaci* lives as an endoparasite in stems, leaves, flowers, bulbs, tubers and rhizomes. The seed-borne nature of this nematode has also been reported by Sousa *et al.* (2003) and Sikora *et al.* (2005) in *V. faba*, *Medicago sativa* (alfalfa), *Allium cepa* (onion), *Trifolium* spp. (clovers), *Dipsacus* spp. (teasel) and *Cucumis melo* (melon). There is a possibility of dispersal of this nematode to new host areas through seeds from infested fields.

7.12.1 Plant parts liable to carry the nematode in trade/transport

Plant parts including roots, flowers, inflorescences, rhizomes, tubers, bulbs, cones, calyx, panicle (including ear heads), leaves, seedlings, micro-propagated plants, stems (above ground), shoots, tillers and true seeds are liable to carry all stages (eggs, juveniles and adult) of the nematode.

7.12.2 Plant parts not known to carry the nematode in trade/transport

Bark and wood.

7.13 Pest risk analysis and looming threat from introduction

At present, *D. dipsaci* is distributed throughout the world, but the distribution of different races is sporadic. *D. dipsaci* is categorized as a major quarantine nematode in many countries and in some countries official control measures have been implemented to limit the spread of this nematode. However, due to its wide host range, economic significance and seed-borne nature, imports of planting materials, stems, leaves, cut flowers, bulbs, tubers, rhizomes and seeds of host plants from countries where this nematode occurs should be restricted.

7.14 Invasiveness rating

Considering the consequences of introduction and pest risk analysis, the following invasiveness rating has been assigned to *D. dipsaci*.

Climate/host interaction	–	High (3)
Hosts range	–	High (3)
Dispersal potential	–	High (3)
Economic impact	–	High (3)
Environmental impact	–	Medium (2)
Post-entry distribution	–	Low (–1)
Final rating (score)	–	High (13)

Based on the above invasiveness rating, *D. dipsaci* receives a score of 13 out of 15 and is rated as HIGHLY INVASIVE NEMATODE.

7.15 Region-wise status of invasiveness

Currently, *D. dipsaci* is on the 'Harmful Organisms' lists for many countries (USDA-PCIT, 2018). Region-wise countries where *D. dipsaci* is absent and attain invasive status are;

APPPC: Bangladesh, India, Indonesia, Sri Lanka, Taiwan, Thailand, Timor-Leste, United Arab Emirates, Vietnam
CAHFSA: Antigua and Barbuda, French Polynesia, Georgia, Grenada, Guatemala, Honduras, Nicaragua, San Marino
CAN: Cuba, El Salvador, Panama
COSAVE: Costa Rica
EPPO: Lebanon, Holy See (Vatican City State)
IAPSC: Egypt, Madagascar, Monaco, Namibia, Tanzania United Republic, South Africa, Uganda
PPPO: Cook Islands, New Caledonia

7.16 Management measures

7.16.1 Prevention and control

Preventing the nematode spread is an essential measure to save the crop from damage by *D. dipsaci*. Growers are strongly recommended to use certified nematode-free seeds and planting material. Hot-water treatments with different temperature–time regimes, depending on type and state of infected planting material and seed, are practically efficient for controlling *D. dipsaci* (Gratwick and Southey, 1972). Hot-water treatments of narcissus bulbs that may effectively control the nematode comprise either storage for 1–2 weeks at 25–30°C, followed by dipping in water for 24 h and hot-water treatment at 45°C for 4 h, or storage for 1–2 weeks at 25–30°C, followed by hot-water treatment at 47°C for 4 h (Hooper, 1991). Commercial cleaning of the seed along with associated plant debris has proved to be useful in removing nematode contamination.

7.16.2 Phytosanitary measures

See Chapter 16, section 115.

2

Aphelenchidae: *Aphelenchoides* and *Bursaphelenchus* Species

Aphelenchoides Spp.

More than 180 species of *Aphelenchoides* (and 19 of uncertain status) have been recorded (Sánchez-Monge *et al.*, 2015). The majority of *Aphelenchoides* species are myceliophagous. Some species, besides feeding and reproducing on fungi also parasitize higher plants (Hunt, 1993). Important plant-parasitic species of *Aphelenchoides* include:

Aphelenchoides aligarhiensis Siddiqi, Husain and Khan, 1967
A. arachidis Bos, 1977
A. besseyi Christie, 1894
A. blastophthorus Frankalin 1952
A. bicaudatus (Imamura, 1931) Filipjev and Schuurmans Stekhoven, 1941
A. brassicae Edward and Mishra, 1969
A. fragariae Ritzema Bos, 1891
A. graminis Baranovskaya and Haque, 1968
A. ritzemabosi Schwartz, 1911
A. rosei Dmitrenko, 1966
A. sacchari Hooper, 1958
A. subtenuis (Cobb, 1926) Steiner Buhrer, 1932
A. tagetae Steiner, 1941

Among these, four species, namely, *A. arachidis*, *A. besseyi*, *A. fragariae* and *A. ritzemabosi*, are considered as invasive nematode species because of their economic importance and quarantine significance in several countries (Chizhov and Subbotin, 1990; Krall, 1991; Duncan and Moens, 2013). Other species are of lower importance and have limited information on geographical distribution and host range.

Systematic position of *Aphelenchoides* species

	Classical (after Hunt, 1993)	Molecular (after Decraemer and Hunt, 2013)
Phylum	Nematoda	Nematoda
Class	Secernentea	Chromadorea
Subclass	Diplogasteria	Chromadoria
Order	Aphelenchida	Rhabditida
Suborder	Aphelenchina	Tylenchina
Infra-order	–	Tylenchomorpha
Superfamily	Aphelenchoidea	Aphelenchoidea
Family	Aphelenchoididae	Aphelenchoididae
Subfamily	Aphelenchoidinae	Aphelenchoidinae
Genus	*Aphelenchoides*	*Aphelenchoides*

8 *Aphelenchoides arachidis*

8.1 Common name

Groundnut testa nematode

8.2 Scientific name

Aphelenchoides arachidis Bos, 1977

8.3 EPPO code

APLOAR

8.4 Diagnosis

8.4.1 Morphological diagnosis

Aphelenchoides arachidis was first described by Bos (1977a) in Nigeria. Important morphological characters are summarized below. For morphometric measurements, see Table 10.

Female: Body slender with fine annules 0.8 µm apart at mid-body. Lateral field 1.8–2.4 µm wide at mid-body and marked by two incisures. Lip region more or less flattened anteriorly, 2.5–3 µm high and 6–8.5 µm wide, weakly sclerotized. Stylet long with distinct basal knobs. Nerve ring one body width or less behind the median bulb, excretory pore about a further

DOI: 10.1079/9781789247367.0003

Table 10. Morphometric data for *Aphelenchoides arachidis*, *A. besseyi*, *A. fragariae* and *A. ritzembosi* (adapted from ISPM-27, DP 17).

Morphometric characters		Range, numerous populations			
		A. arachidis (Bos, 1977a; Bridge and Hunt, 1985)	*A. besseyi* (Fortuner, 1970; Franklin and Siddiqi, 1972; Andrássy, 2007)	*A. fragariae* (Allen, 1952; Siddiqi, 1975b; Shahina, 1996)	*A. ritzemabosi* (Allen, 1952; Siddiqi, 1974c; Andrássy, 2007)
Females	Length (mm)	0.510–1.000	0.57–0.84	0.45–0.80	0.77–1.20
	Stylet (µm)	11–12	10.0–12.5	10–11	12
	a	39–50	39–53	36–63	40–45
	b	11–18	9.2–13.1	8–15	10–13
	b'	5–8	4.06–5.77	–	
	c	25–42	13.8–20.4	12–20	18–24
	c'	2–3	3.5–5.0	4.9	4–5
	V	67–74%	68.7–73.6%	64–71%.	66–75%
	Lateral lines	2	4	2	4
	Tail type	Subcylindroid	Conoid	Elongate conoid	Elongate conoid
	Terminal mucro shape	Single central spine	Star	Single central spine	Peg with 2–4 minute processes
Males	Length (mm)	0.560–1.040	0.53–0.61	0.48–65	0.70–0.93
	Stylet (µm)	10–11	10.0–12.5		11–12
	a	37–60	40.7–46.9	46–63	31–50
	b	10–18	8.87–10.70	9–11	10–14
	b'	5–9	3.5--4.91	–	–
	c	20–39	16–20	16–19	16–30
	c'	2–3	3.5–5.0	4.9	–
	Tail	40–84	28–52	38–42	47
	Spicules (µm)	15–25	18–21	14–17	20–22

body width posterior. Oesophageal glands in a dorsal lobe overlapping intestine (Fig. 2.1A-B). Vulval aperture a transverse slit with slightly swollen lips. Genital tract monoprodelphic outstretched but may show flexures when well developed. Developing oocytes in single file. Postvulval uterine sac extending about half of the vulva–anus distance; often containing sperm. Tail subcylindroid with bluntly rounded tip and terminal mucro (Fig. 2.1C-D).

Male: Testis outstretched, with one or more flexures. Three pairs of caudal papillae, adanal, slightly posterior to middle of tail, just before tail tip (Fig. 2.1H-J). Spicules rose-thorn shaped with dorsal limb 15–25 µm long. Tail conoid with finely pointed, narrowly rounded or irregular terminus. Small terminal mucro sometimes present.

8.5.2 Biochemical diagnosis

Biochemical approaches such as one-dimensional SDS gel electrophoresis of proteins have been used to differentiate species and races of various groups of phytonematodes, including *Aphelenchoides* species (Eriksson and Granberg, 1969). Ibrahim *et al.* (1994b) also characterized *Aphelenchoides* species (*A. arachidis*, *A. besseyi*, *A. bicaudatus*, *A. fragariae*, *A. hamatus*) based on esterase and protein, by one-dimensional SDS gel electrophoresis.

8.5.3 Molecular diagnosis

Several plant-parasitic nematodes are characterized by PCR–RFLP of ITS regions, including *A. arachidis* (Ibrahim *et al.*, 1994a). Some PCR-based molecular techniques for the diagnosis of *Aphelenchoides* species have been used by McCuiston *et al.* (2007) and Rybarczyk-Mydłowska *et al.* (2012). Rybarczyk-Mydłowska *et al.* (2012) developed a small subunit (SSU) ribosomal (r)DNA-based species-specific PCR for four *Aphelenchoides* species, i.e. *A. besseyi*, *A. fragariae*, *A. ritzemabosi* and *A. subtenuis*. The species-specific primers were intended to use the full-length SSU rDNA sequences of these four foliar nematodes. They were used for real-time PCR to rapidly identify one or more *Aphelenchoides* species isolated from soil and plant material. These PCR assay techniques can be modified and will be able to detect the presence of *A. arachidis* within the infected plant. For details, see section 9.5.3, below.

8.6 Geographical distribution

A. arachidis is widely distributed throughout Nigeria (Bos, 1977a; Bridge *et al.*, 1977; Khan and Misari, 1992). The nematode has also been recorded from Egypt (Montasser *et al.*, 2008) and South Africa (Lesufi *et al.*, 2015).

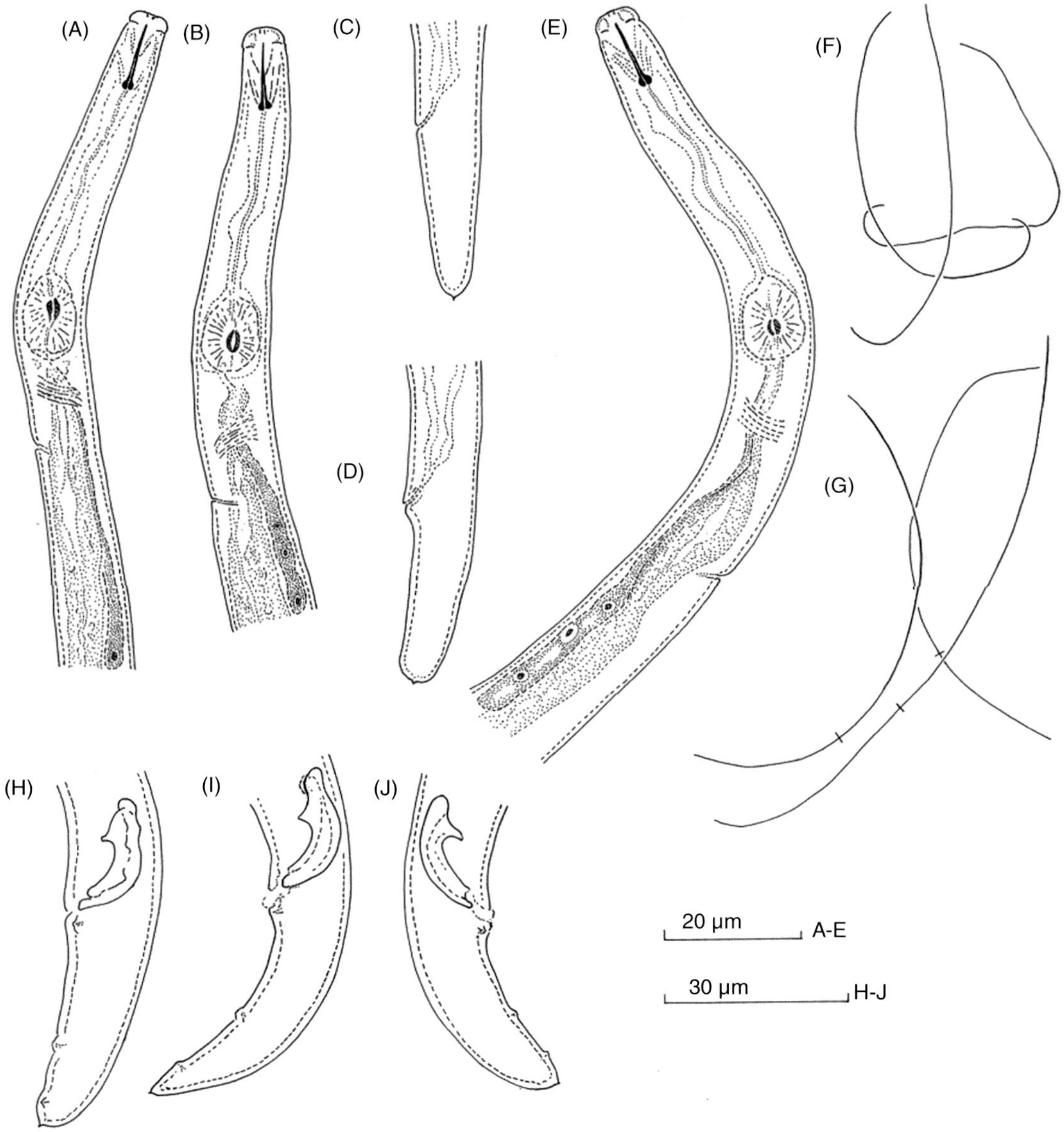

Fig. 2.1. Line drawing of *Aphelenchoides arachidis*. Anterior region of different females (A–B); tails of different females (C–D); anterior region of male (E); habitus of different males (F); habitus of different females (G); tails of different males (H–J). [Lesufi *et al.*, 2015; courtesy Nematology]

8.7 Host plant

The principal host of *A. arachidis* is *Arachis hypogea* (groundnut). Other hosts include maize (*Zea mays*), rice (*Oryza sativa*), pearl millet (*Pennisetum glaucum*), sugarcane (*Saccharum officinarum*) and sorghum (*Sorghum bicolor*) (Bos, 1977b).

8.8 Symptoms

A. arachidis infects pods, testae, roots and hypocotyls (Dickson and De Waele, 2005). In groundnut seeds, the nematode is only found in the tissues of the testa, usually in the sub-epidermal parenchymatous layer, and near the tracheids (Fig. 2.2A-B). These seed coats become discoloured when large numbers of nematodes (> 2000/testa) are present. Severely infested seeds, just after removal from fresh, mature pods, have translucent testae, are a lighter brown and have darker vascular strands within the testae (Dickson and De Waele, 2005). When dried it appears to be abnormal moisture loss and the seeds become shrivelled and a darker brown than normal seeds (Bridge *et al.*, 1977). Infested testae are thicker and more irregular than normal testae. The epidermal layer is condensed in infested testae and the basal tissues comprising the aleurone layer are disorganized. Many larvae can be seen in pod shells of groundnuts without any obvious symptoms.

Fig. 2.2. *Aphelenchoides arachidis* in a groundnut seed testa (A); SEM of shell showing nematode presence of *A. arachidis* (B); *Aphelenchoides besseyi* symptom on rice crop showing characteristic 'white tip' (C,D); strawberry plants infected by *A. besseyi* (E,F). [Courtesy A,D: J. Bridge/CABI; B: Lesufi *et al.*, 2015, Nematology; C: William M. Brown Jr., Bugwood.org (inset photo by D. Groth, Bugwood.org); E: Janet S. McCulloch, Department of Primary Industries and Fisheries, Queensland, Bugwood.org; F: Ministry of Agriculture and Regional Development, Bugwood.org]

8.9 Biology and life cycle

A. arachidis is a facultative endoparasite of groundnut and feeds on seed testa, pod shell, roots and hypocotyl of groundnuts (Dickson and De Waele, 2005). It is also capable of feeding ectoparasitically on the roots and also on fungi. The nematodes are found in the parenchymatous tissues, testa, root cortex and hypocotyl of the groundnut, but not in the vascular bundles or central stele. The nematodes are not found in the cotyledons or embryos of the seeds. In the hypocotyl, they occur intercellularly. In testae, nematodes are found mostly in the sub-epidermal parenchymatous layer where walls are broken and cells enlarged, and around the tracheids. Development of all stages of *A. arachidis* occurs within the pod shells and in testae of young seeds, but at the end of the growing season it can be found in heavily infested mature seeds. Testae mainly contain the juvenile stages, with few adults (Dickson and De Waele, 2005).

Invasion of developing pods occurs 10 days after the groundnut fruiting pegs penetrate the soil. The population does not increase and only a few nematodes are found in pods until 30 days after pod development. Thereafter nematode numbers increase rapidly in infested pods, mostly found at about 60 days (Bridge and Hunt, 1985). The nematode occurs within the tissues of the roots, pods, testae and hypocotyls, but not within the embryos, cotyledons or other plant parts. Higher numbers of *A. arachidis* are found in pods than in testae; freshly harvested whole pods may contain > 113,000 active nematodes. In testae of young seeds, the numbers exceed 35,000 nematodes per testa. Generally, testae do not show any external symptoms and contain mostly adults and eggs, often arranged along the vascular elements of the seed coats (Dickson and De Waele, 2005). In mature seeds, at the end of the growing season, heavily infested testae comprise mainly juvenile stages, with few adults. In pods, the nematodes are confined to the inner tissues of the shell (Bos, 1977a).

A. arachidis can survive by desiccation in partly dried stored groundnut pods for 12 months (Dickson and De Waele, 2005). However, numbers rapidly decline and, generally, only 2–3% of nematodes remain alive after 4 months of storage. All stages of juveniles can survive, with no specific stage predominating, but occasionally adults are found alive in either testae or partly dried pods in storage. No active nematodes were isolated from infested pods sun-dried in the field in the very dry conditions of northern Nigeria before storage (Bridge *et al.*, 1977).

8.10 Similarities to other nematodes

The damage symptoms caused by *A. arachidis* on groundnuts are very similar to those caused by another nematode, *Ditylenchus africanus* (Wendt *et al.*, 1995). The nematodes are not easy to distinguish but differ in the shape of the median bulb (more elongated in *D. africanus*) and the type of spicule in males (rose-thorn shaped in *A. arachidis*).

8.11 Economic importance

A. arachidis has only been recorded as a seed-borne pest of groundnuts in Nigeria (Dickson and De Waele, 2005). In northern Nigeria, the incidence of seed infestation in all groundnut varieties examined showing symptoms of discoloration was 99–100% (Bridge *et al.*, 1977). In one sample more than 25% of the seeds were found infested with the nematode (McDonald *et al.*, 1979). Surveys have shown the nematode to be widely distributed in the main groundnut-growing areas of Nigeria and occurring in over 85% of groundnuts sampled (Khan and Misari, 1992). Both qualitative and quantitative losses occur due to *A. arachidis*. It reduces the marketable value of confectionery groundnuts because it causes shrivelled and discoloured seeds and the seed weights are also reduced by as much as 34% (Bridge *et al.*, 1977). It does not reduce yields of freshly harvested produce but causes a reduction in the weight of stored seeds. Nematode infection does not suppress yield directly but causes serious qualitative damage and also predisposes seeds to fungal infection. The infected seeds are rendered under-graded and unmarketable (Minton and Baujard, 1990).

8.12 Movement and means of dispersal

The nematode is spread by infected groundnut shell and seeds. It can survive desiccation in partly dried stored pods and seeds of groundnut but few nematodes remain alive after storage. All stages of nematode can be revived from stored seeds. Although seed transmission is possible, rapid drying of the pods after harvesting kills the nematodes in the testae under dry conditions. Where conditions are more humid, sun-drying of pods is less effective and the nematodes may survive and be dispersed in the seed (Bridge *et al.*, 1977).

8.12.1 Plant parts liable to carry the nematode in trade/transport

Roots, fruits (including pods), true seeds, seedlings, micropropagated plants.

8.12.2 Plant parts not known to carry the nematode in trade/transport

Bark, wood, leaves, stems (above ground), shoots, trunks, branches.

8.12.3 Transport pathways for long-distance movement

Containers and packing, groundnut straw, soil and gravel from groundnut fields.

8.13 Pest risk analysis and the looming threat from the introduction

Groundnut is one of the major oilseeds and food legume crops; it was originated in South America from southern Bolivia to north-western Argentina (CABI-ISC, 2020). Groundnuts are now grown in most subtropical, tropical and temperate countries. *A. arachidis* is a major nematode problem in Nigeria, occurring in over 85% of groundnuts sampled (Khan and Misari, 1992). Currently, *A. arachidis* has been reported only from Nigeria, Egypt and South Africa, but this nematode can become a major economic pest if introduced into other major groundnut-producing countries, especially China (annual production 15.3 million tonnes (Mt)), India (7.5 Mt), USA (1.9 Mt) and Sudan (1.2 Mt) (FAOSTAT, 2019). Because of its limited distribution (Nigeria, Egypt and South Africa), the chances of introduction of this nematode into other groundnut-producing areas are very limited. However, a complete pest risk analysis should be prepared for this nematode. An accidental introduction of this nematode would have devastating consequences on groundnut export to international as well as domestic markets. Hence, strict quarantine regulations and extreme caution should be taken in importing groundnuts whole or as seeds from host countries.

8.14 Invasiveness rating

Considering the consequences of introduction and pest risk analysis, the following invasiveness rating has been assigned to *A. arachidis*.

Climate/host interaction	–	Medium (2)
Hosts range	–	High (3)
Dispersal potential	–	High (3)
Economic impact	–	High (3)
Environmental impact	–	Medium (2)
Post-entry distribution	–	Low (–1)
Final rating (score)	–	High (12)

Based on the above invasiveness rating, *A. arachidis* receives a score of 12 out of 15 and is rated as HIGHLY INVASIVE NEMATODE.

8.15 Region-wise status of invasiveness

Based on the probable introduction of *A. arachidis*, a high invasiveness rating has been given to the following groundnut-producing countries (FAOSTAT, 2019), where this nematode has not been detected.

APPPC: Bangladesh, Cambodia, China, India, Indonesia, Iraq, Israel, Japan, Jordan, Kazakhstan, Kyrgyzstan, Lebanon, Malaysia, Myanmar, Pakistan, Philippines, Saudi Arabia, South Korea, Sri Lanka, Syria, Tajikistan, Thailand, Vietnam
CAHFSA: Barbados, Belize, Cuba, Dominican Republic, Haiti, Jamaica, Saint Kitts & Nevis
CAN: Bolivia, Colombia, Ecuador, Peru
COSAVE: Argentina, Brazil, Guyana, Paraguay, Suriname, Uruguay
EPPO: Bulgaria, Cyprus, Greece, Hungary, Italy, Morocco, Portugal, Spain, Turkey, Uzbekistan
IAPSC: Algeria, Angola, Benin, Botswana, Burkina Faso, Burundi, Cameroon, Cape Verde, Central African Republic, Chad, Comoros, Congo, Côte d'Ivoire, Eritrea, Ethiopia, Gabon, Gambia, Ghana, Guinea, Guinea-Bissau, Kenya, Liberia, Madagascar, Malawi, Mali, Mauritania, Mauritius, Mozambique, Namibia, Niger, Réunion, Rwanda, Senegal, Sierra Leone, Somalia, Sudan, Swaziland, Togo, Uganda, Zambia, Zimbabwe
NAPPO: Mexico, USA
OIRSA: Costa Rica, El Salvador, Guatemala, Honduras, Nicaragua
PPPO: Australia, Fiji, Papua New Guinea, Tonga, Vanuatu

8.16 Management measures

8.16.1 Cultural method

Rapid sun-drying of the pods after harvesting is the agricultural practice used in northern Nigeria, where relative humidity can drop to as low as 4%. This kills the nematodes in the testae (Bridge *et al.*, 1977). Severely infested seeds are detectable and can be sorted out before planting, but seeds with only a few nematodes are non-detectable from healthy seeds and, if planted, could spread *A. arachidis*.

8.16.2 Seed treatment

Hot-water treatment is quite effective in killing all stages of nematodes within seed coats and requires only a simple heating device. Seeds are immersed for 5 min in 4 times their volume of water which has been pre-heated to 60°C and removed from the heat. The hot-water treatment does not affect seed germination or yield (Bridge *et al.*, 1977). The hot-water treatment may be valuable to control nematodes in groundnut-growing areas where sun-drying is less effective.

8.16.3 Phytosanitary measures

See Chapter 16, section 115.

9 *Aphelenchoides besseyi*

9.1 Common names

White tip nematode; leaf and bud nematode

9.2 Scientific name

Aphelenchoides besseyi Christie, 1942

9.3 Synonyms

Aphelenchoides oryzae Yokoo, 1948
Asteroaphelenchoides (Christie) Drozdovsky, 1967

9.4 EPPO code

APLOBE

9.5 Diagnosis

9.5.1 Morphological diagnosis

The species has been described by several workers but a comprehensive description of *A. besseyi* was given by Franklin and Siddiqi (1972) as below. For measurements, see Table 10.

Female: Slender body, straight to slightly ventrally arcuate when relaxed. Body annules fine, indistinct, around 0.9 µm wide near mid-body (Fig. 2.3A). Cephalic region unstriated, rounded, slightly offset and at lip base wider than body. Labial framework hexaradiate, lightly sclerotized. Lateral fields with four incisures, about one-fourth as wide as body. Anterior part of stylet about 45% of total stylet length sharply pointed, while posterior part with slight basal swellings 1.75 µm across. Median bulb oval shape, with distinct valvular apparatus slightly behind its centre. Oesophageal glands covering dorsally and subdorsally for four to eight body widths overlapping intestine. Nerve ring behind median bulb, about one body width. Excretory pore generally near anterior edge of nerve ring (in neotype, anterior to nerve ring). Hemizonid 11–15 µm behind excretory pore. Vulva transverse, with lightly raised lips. Spermatheca elongate oval, usually packed with sperm and up to eight times as long as wide when fully distended. Relatively short ovary and not extending to oesophageal glands, with oocytes in two to four rows. Post-uterine sac inconspicuous, narrow, not containing sperms, 2.5–3.5 times anal body width long but less than one-third of distance from vulva to anus. Tail conoid 3.5–5 times anal body widths long, with terminus bearing a mucro of diverse shape with three to four pointed processes.

Male: Morphologically similar to females. Posterior end of body curved by around 180° when specimens relaxed. Cephalic region, stylet and oesophagus as designated for female. Tail conoid, with terminal mucro with two to four pointed processes. First pair of ventro-submedian papillae adanal, second somewhat behind middle of tail and third sub-terminal. Spicules typically similar to other *Aphelenchoides* spp. except proximal end lacks apex and has only modified developed rostrum. Testis outstretched, single.

9.5.2 Biochemical diagnosis

A. besseyi has been characterized on the basis of esterase and protein pattern using one-dimensional SDS gel electrophoresis of proteins (Ibrahim *et al.*, 1994b).

9.5.3 Molecular diagnosis

A. besseyi can be characterized using the PCR–RFLP of ITS regions (Ibrahim *et al.*, 1994a; Iwahori *et al.*, 1998). Rybarczyk-Mydłowska *et al.* (2012) developed a small subunit (SSU) ribosomal (r)DNA-based species-specific PCR for *A. besseyi*. The primers were designed for full-length SSU rDNA sequences of four foliar nematode species, including *A. besseyi*. They used real-time PCR for quick identification of one or more *Aphelenchoides* species isolated from plant material and soil. The primer pair designed with locked nucleic acid (LNA) (# below) and recommended for the diagnosis of *A. besseyi* is:

1770: 5′-GCG GGA TTC GTG GTT C#T-3′
1772: 5′-CGA CAT GCC GAA ACA GAG-3′

Each real-time PCR is performed according to the following described conditions.

Reagent	–	Final concentration
PCR-grade water	–	25 µl
PCR buffer Absolute qPCR mix	–	1×
Specific forward primer	–	0.2 µM
Specific reverse primer	–	0.2 µM
DNA (volume)	–	3 µl
Cycling parameters		
Initial denaturation	–	95°C for 15 min
Number of cycles	–	60
Denaturation	–	95°C for 30 s
Annealing	–	63°C for 60 s
Elongation	–	72°C for 30 s

The PCR products are analysed by agarose gel electrophoresis. The amplicon is approximately 325 bp for *A. besseyi*.

9.6 Geographical distribution

A. besseyi is worldwide in distribution (Fig. 2.4) and has been recorded from the areas and regions where rice, strawberry and chrysanthemum are cultivated. Region-wise, the nematode is present in the following countries (CABI-ISC, 2020).

Africa: Benin, Burkina Faso, Burundi, Cameroon, Central African Republic, Chad, Comoros, Congo Democratic Republic, Côte d'Ivoire, Egypt, Gabon, Gambia, Ghana, Guinea, Ivory Coast, Kenya, Madagascar, Malawi, Mali, Nigeria, Senegal, Sierra Leone, South Africa, Tanzania, Togo, Uganda, Zambia, Zimbabwe.
Asia: Afghanistan, Azerbaijan, Bangladesh, Cambodia, China, Georgia, India, Indonesia, Iran, Japan, Kyrgyzstan, Laos, Malaysia, Myanmar, Nepal, North Korea, Pakistan, Philippines, Singapore, South Korea, Sri Lanka, Taiwan, Tajikistan, Thailand, Turkey, Uzbekistan, Vietnam.
Europe: Bulgaria, Hungary, Italy, Netherlands, Romania, Russia, Ukraine.
North America: Cuba, Dominica, Dominican Republic, El Salvador, Guadeloupe, Mexico Panama, USA.
Oceania: Australia, Cook Islands, Fiji, Papua New Guinea.
South America: Brazil, Ecuador.

9.7 Host plant

Rice (*Oryza sativa*) is a major host of *A. besseyi*, which causes white tip disease. Besides rice, *A. besseyi* attacks over 35 genera of higher plants (Bridge *et al.*, 2005a; Peng *et al.*, 2018) (Table 11). The nematode infests *Fragaria* spp. and is the cause of summer crimp/dwarf disease in Australia, USA and more recently in Europe. The nematode has been identified as the causal agent of black spot disease on *Phaseolus vulgaris* (bean) (Chaves *et al.*, 2013). Other hosts include food crops (*Dioscorea*, *Ipomea* and *Zea mays*), ornamentals (*Begonia* and *Chrysanthemum*), vegetables (*Allium*) and grasses (*Panicum*, *Pennisetum* and *Setaria*) (Bridge *et al.*, 2005a; Peng *et al.* 2018). Different plant species attacked by *A. besseyi* are listed in Table 11.

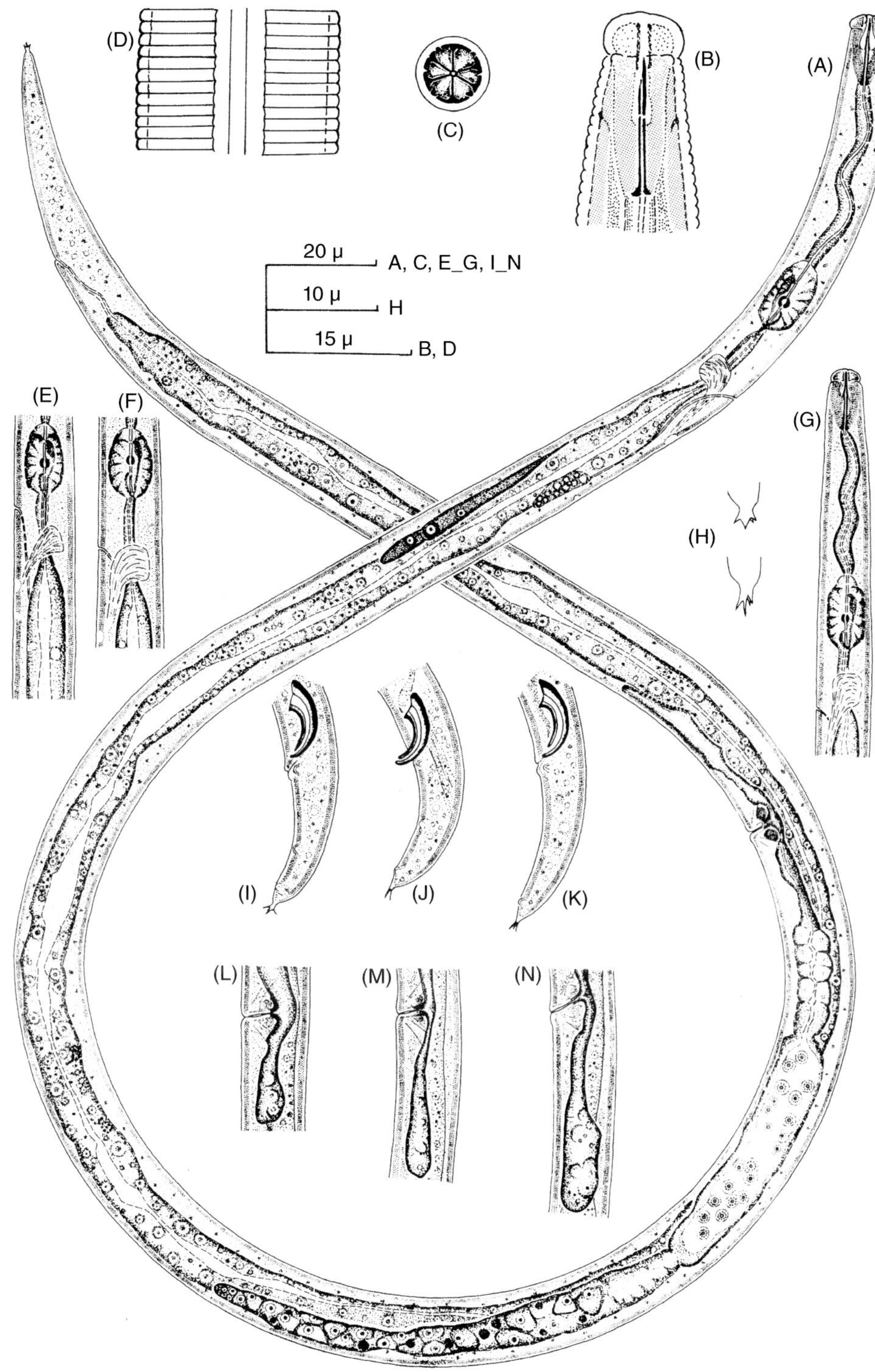

Fig. 2.3. Line drawing of *Aphelenchoides besseyi.* Female (A); female head end (B); female *en face* view (C); lateral field (D); variation in female median oesophageal bulb and position of excretory pore with respect to nerve ring (E,F); male anterior end (G); female tail termini showing variation in shape of mucro (H); male tail ends (I,J,K); variation in postvulval sac (L,M,N). [Reproduced from D.J. Hunt, 1993; B,D after Franklin and Siddiqi (1972); the rest after Fortuner (1970), courtesy CAB International]

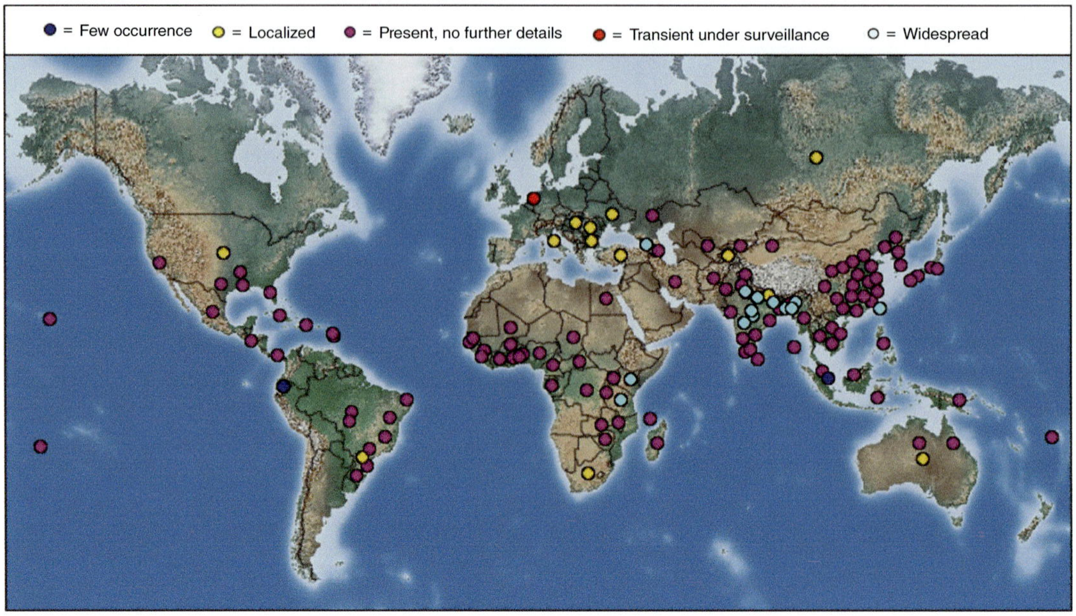

Fig. 2.4. Distribution map of *Aphelenchoides besseyi*. [Adapted from CABI-ISC, 2020]

Table 11. Hosts of *Aphelenchoides besseyi* (adapted from Khan, 2010b).

Hosts	Common name
Allium cepa	Onion
Allium sativum	Garlic
Armoracia rusticana	Horseradish
Avena strigosa	Black oat
Avena sativa	Oats
Brachiaria dictyoneura	Brachiaria
Boehmeria nivea	Ramie
Callistephus chinensis	Callistephus
Calendula officinalis	Calendula
Capsicum annum var. *longum*	Chilli
Cattleya sp.	Orchid
Chrysanthemum maximum	Chrysanthemum
Chrysanthemum morifolium	Chrysanthemum
Chloranthus spicatus	Chullan
Coleus blumet	Coleus
Colocasia esculenta	Taro
Cyperus iria	Motha
Cyperus sp.	Motha
Dahlia variabilis	Dahlia
Dianthus caryophyllus,	Dianthus
Digitaria adscendens	Summergrass
Digitaria sanguinalis	Crab grass
Dioscorea trifida	Yam
Echinochloa spp.	Shyama grass
Erechtites praealta	Erechtites
Euchlaena mexicana Schrad	Teosinte
Fern	Fern
Ficus elastica var. *Decora*	Rubber plant
Fraxinus americana	American ash
Fragaria chiloensis var. *ananassa*	Strawberry
Gerbera spp., *G. jamesonii*,	Gerbera
Glycine max	Soybean
Hibiscus bracherridgii	Hibiscus
Hydrangea macrophylla	Hydrangea
Impatiens balsamina	Balsam
Imperata cylindrica	Imperata
Ipomoea batatas	Sweet poptato
Jasminum volubile	Jasmine
Lactuca sativa	Lettuce
Agaricus brunnescens	Mushroom
Phalaenopsis sp.	Dendrobium
Panicum bisulcatum	–
Panicum crusgalli var. *frumentaceum*	–
Panicum sanguinale	Crab grass
Panicum miliaceum	Proso millet
Pennistum typhoides	Pearl millet
Phaseolus vulgaris	Bean
Pluchea odorata	Pluchea
Polianthes tuberosa	Tuberose
Pycreus polystachyus	Pycreus
Saccharum officinarum	Sugarcane
Saintapaulia ionantha	Saintapaulia
Setaria italica Beauv.	Italian millet
Setaria viridis	Foxtail millet
Sporobolus poirettii	Sporobolus
Stylosanthes hamata	Stylo
Tagetes sp.	Marigold
Tithonia diversifolia	–
Torenia fournieri	–
Vanda sp.	–
Zea mays	Maize
Zinnia elegans	Zinnia

9.8 Mycophagous nature

In addition to the above host plants, *A. besseyi* naturally feeds on many saprophytic and plant-pathogenic fungi such as *Alternaria* sp., *A. oleracea*, *A. tenuis*, *A. citri*, *Curvularia* sp., *Curvularia geniculata*, *Fusarium* sp., *F. oxysporum*, *F. solani*, *Bipolaris sorokinianum* (= *Helminthosporium sativum*), *Nigrospora* sp., *Sclerotium oryzae*, *Botrytis cineraria*, *Sclerospora* sp. and edible fungi, namely *Agaricus* spp. Some other fungi, like *Chaetomium convolutum*, *Epicoccum oryzae*, *Trichoconiella padwickii* and *Stachybotrys* sp., are also good hosts of *A. besseyi* (Gokte and Mathur, 1996) and have been used for successful culturing of the nematode.

9.9 Pathotypes/physiological races

Pathotypes or physiological races are known to occur in *A. besseyi*. At least two races are reported: one that causes 'white tip' disease in rice (*Oryza sativa*) and the other that induces 'summer dwarf' in strawberry, *Fragaria* sp. (Riggs, 1991). The population of *A. besseyi* infecting rice also causes foliar disease (Holtzmann, 1968) and floral malady in tuberose in India (Khan, 2010a,b). Cross-inoculation studies have further proved that *A. besseyi* populations causing white tip disease in rice are also causative of floral malady in tuberose (Khan, 2004). It is possible that more races in *A. besseyi* populations exist, because the species infects various other host plants and the parasitism varies considerably with hosts like *Ficus elastica* (Marlatt, 1970), *Hibiscus brachenridgii* (Raabe and Holtzmann, 1965), tuberose (Trujillo, 1968), etc. The mode of parasitism of the nematode also varies, sometimes as an ectoparasite (on *Vanda* sp.) or as an endoparasite (on *Saintapaulia* sp.) (Allen, 1952).

9.10 Symptoms

A. besseyi may inflict damage without producing symptoms on susceptible plants. At early growth stages, infection of *A. besseyi* manifests as chlorotic tip of newly emerged leaves. Later the leaf tip, 2–5 cm from the tip, turns bronze (Fig. 2.2C-D) with crinkled margins and then there is browning followed by death and shredding of the leaf tip. Chlorotic patches are also observed in the interveinal area of leaf sheath and leaf blade and sometimes the sheath appears abnormally dark green with greater chlorophyll content as compared with the healthy one. The flag leaf characteristically becomes shorter and often twisted and curled, with the leaf margin corrugated and distorted at booting stage. Initiation of panicle and development is arrested and the flag leaf emerges only in a shortened and distorted form. Affected panicles are shorter and often swollen at the tips, bearing fewer grains. Severely affected panicles produce empty grains, giving the appearance of chaffiness with twisted husks. There may be abnormal elongation of glumes in some of the spikelets. *A. besseyi* is a seed-borne ectoparasite infesting rice mainly during the three-leaf stage of the plant. The nematode can survive for over 1 year inside rice seeds and 53 days in water at less than 10°C (Qiu *et al.*, 1991). It is mainly located near the embryonic region of rice seeds, mostly between lemma and palea, and the fourth-stage juvenile is the infective stage. The infected abnormal rice crop will yield small grains and the plants have erect and shortened panicles with smaller (in size and number) distorted grains that are brown to black; sometimes the flag leaf remains normal (Lin *et al.*, 2005).

On *Fragaria* spp., *A. besseyi* is the causal agent of 'summer dwarf' disease (Perry and Moens, 2006). Symptoms include leaf crinkling and distortion, and dwarfing of the plant with an associated reduction in flowering (Fig. 2.2E-F). Symptoms may be similar to and therefore confused with those caused by other *Aphelenchoides* species, emphasizing the importance of correct identification. In *O. sativa* and *Fragaria* spp., *A. besseyi* feeds ectoparasitically, but the nematode may also be endoparasitic, as in *Ficus elastica* and *Polianthes tuberosa*, in which it causes leaf drop and leaf lesions, respectively. On *Capsicum annum* var. *longum* the infestation appears to result in rotting of the pods and premature pod drop, similar to some fungal diseases (Hockland and Eng, 1997). In the grass *Sporobolus poirettii*, the nematode stimulates growth, resulting in increased flowering.

In addition to direct effects, this nematode also acts as a wounding agent for the establishment of fungi and bacteria in the grain, thereby enhancing grain discoloration. *A. besseyi* plays an important role in the development of rice disease complex, spreading and establishing several pathogenic fungi such as *Acrocylindrium* (*Sarocladium*) *oryzae*, *Curvularia* spp., *Fusarium* spp., etc. (Rao and Prakash, 2002).

9.11 Biology and life cycle

A. besseyi is a facultative ectoparasite of the leaves and young tissues of rice. It is a seed-borne nematode surviving between the lemma and palea and the base of the glume. Infected seeds are the main source of inoculum, although the nematode can also survive on crop debris for a shorter period. Upon sowing of the infected seed, the predominantly adult females in quiescent state will revive when coming in contact with soil moisture. Later, the nematodes slowly migrate to the growing points of rice seedlings and feed on the leaf tips ectoparasitically (Bridge *et al.*, 2005a; Peng *et al.* 2018). As the plant grows, the nematodes are carried upwards and oviposit in the leaf axils and leaf sheaths. Eggs hatch soon giving rise to J_2 when moving to the apical portion and feed ectoparasitically. Under high humidity and rainy conditions, the nematode migrates on the plant surface. Ultimately, it settles on the apical portion, infects the young panicles and moves inside the hulls. At least five to six nematodes per grain are required to produce symptoms on rice grains. *A. besseyi* increases in number at late tillering or when the rice reaches a reproductive phase and it enters into the spikelets at the time of anthesis and feeds on reproductive parts (Khan, 2010b).

Although the mode of reproduction is usually amphimictic, parthenogenetic reproduction is also known (Gokte *et al.*, 2001). In the amphimictic type of reproduction, the number of males of *A. besseyi* is most abundant at the flower fertilization stage (Tsay *et al.*, 1998). Many nematodes (predominantly adults) enter into a cryptic state towards the end of the growing season and survive under the rice hull (Taylor, 1969). The immature stages (3rd and 4th) of the nematode are concealed between the lemma and palea of the rice grain in a quiescent state. The length of the life cycle depends much on ecological conditions, taking 29 days at 14.7°C, 9 days at 20.6°C, 6 days at 25°C and 3 days at 31.8°C (Tikhonova, 1966). *A. besseyi* usually takes 8–12 days to complete the life cycle at 30°C and no development occurs at temperatures below 13°C (Sudakova, 1968). The number of generations per season varies depending upon environmental conditions and hosts. The minimum temperature for nematode activity is 4°C and the thermal death point is 49°C for 10 min. In hot water at 41–44°C the nematode dies within 1 h (Qiu *et al.*, 1991). It prevails between temperatures of 13–42°C and survives for a much shorter period at 10°C (Peng *et al.*, 2018). The nematode is primarily seed-borne and can survive in a quiescent state (anhydrobiosis) for 8 months to 3 years (Bridge *et al.*, 2005a) on stored grains, depending upon the moisture content, but survival is much shorter in soil among the crop under field conditions (Giudici *et al.*, 2004). The nematode could survive in the straw, wild rice and weeds like *Cyperus* spp. and *Echinochloa* spp. (Giudici *et al.*, 2004) and anabiotically on rice husks or on the plant debris after harvest (Peng *et al.*, 2018).

9.12 Economic importance

The white tip nematode, *A. besseyi*, has assumed the status of a major pest in the areas and regions where rice is cultivated under deep-water condition (Davide, 1988). It makes a significant contribution to the estimated US$16 billion worth of damage caused by nematodes to rice crop (Lilley *et al.*, 2011). This nematode is widespread in many countries and its economic importance varies with region and also depends on cultural practices, variety of rice grown and prevailing weather conditions. It may cause damage to the rice without manifestation of white tip symptoms and inflicts substantial yield loss. Yield losses due to *A. besseyi* have been recorded up to 40% in India (Khan, 2010a), 20–60% in Thailand (Taylor, 1969) and 10–30% in China (Lin *et al.*, 2005). *A. besseyi* is also a problem in Bangladesh, where more than 50% of fields are infested with *A. besseyi* and the weight of the panicle is reduced to one-third at an infestation level of 112–650 nematodes/100 seeds (Rahman and Taylor, 1983). In upland rice, the white tip nematode may cause losses of up to 50% (Silva, 1992). In India, *A. besseyi* has been reported to inflict yield losses in rice of 10–15% (Thakar *et al.*, 1987) and 45–50% (Tsay *et al.*, 1998). Overall, white tip of rice is prevalent all over India but appears to be most severe in southern and eastern states, causing up to 20% yield losses (Prasad *et al.*, 1986).

9.13 Movement and means of dispersal

The nematode is primarily seed-borne and can survive in an anhydrobiotic state (quiescent) for 8 months to 3 years on stored grain. *A. besseyi* is widely disseminated through the seed and can also survive in the straw and rice husk. The nematode is myceliophagous and feeds on the mycelium of fungi such as *Curvularia*, *Fusarium* and *Agaricus* in the harvested straw.

9.13.1 Plant parts liable to carry the nematode in trade/transport

Roots, bulbs, tubers, corms, rhizomes, fruits (including pods), flowers, inflorescences, cones, calyx, leaves, stems (above ground)/shoots, true seeds, seedlings, micro-propagated plants.

9.13.2 Plant parts not known to carry the nematode in trade/transport

Bark, wood, trunks, branches.

9.13.3 Transport pathways for long-distance movement

Containers and packing, rice straw/husk packing materials, soil and gravel from rice fields.

9.14 Pest risk analysis and looming threat from introduction

A. besseyi is responsible for inducing 'white tip' disease in rice, 'summer crimp or dwarf' in strawberry, 'greasy leaf blotch' symptom on the leaves of dendrobium Lady Fay (*Phalaenopsis* sp.) and 'floral malady' in tuberose. It also attacks several other important crops, including onion, oats, Chinese cabbage, soybean, sweet potato, yam, sugarcane, maize, etc. It inflicts damage both ecto- and endoparasitically in crops and remains in a quiescent state in the affected parts for up to 8 years. In addition to parasitizing higher plants, the species can multiply and survive by feeding on fungi. *A. besseyi* is widely distributed and occurs in most of the rice-growing areas of the world (Khan, 2010b; CABI-ISC, 2020). However, its dissemination varies between regions, countries and localities. At the international level, *A. besseyi* draws significant attention as a quarantine pest of rice in the countries where it is absent or has limited distribution. Samplings during quarantine inspection have shown large numbers of seed lots to be infected with *A. besseyi* throughout the main rice-producing areas of the world. The accidental introduction of this nematode would have devastating repercussions on rice exports to domestic and international markets.

9.15 Invasiveness rating

Considering the consequences of introduction and pest risk analysis, the following invasiveness rating has been assigned to *A. besseyi*.

Climate/host interaction	–	High (3)
Hosts range	–	High (3)
Dispersal potential	–	High (3)
Economic impact	–	High (3)
Environmental impact	–	High (3)
Post-entry distribution	–	Medium (–2)
Final rating (score)	–	High (13)

Based on the above invasiveness rating, *A. besseyi* receives a score of 13 out of 15 and is rated as HIGHLY INVASIVE NEMATODE.

9.16 Region-wise status of invasiveness

Based on the worldwide geographical distribution of *A. besseyi* and likely introduction in the major rice-producing countries where it is absent or intercepted only, the region-wise status of invasiveness of *A. besseyi* is as under:

APPPC: Bhutan, Brunei Darussalam, Iraq, Israel, Kazakhstan, Saudi Arabia, Syria, Turkmenistan
CAHFSA: Belize, Haiti, Jamaica, Puerto Rico, Trinidad and Tobago
CAN: Bolivia, Colombia, Peru
COSAVE: Chile, Guyana, Paraguay, Suriname, Uruguay
EPPO: Albania, Greece, Morocco, Romania, Spain
IAPSC: Algeria, Angola, Côte d'Ivoire, Ethiopia, Guinea-Bissau, Liberia, Mauritania, Mauritius, Mozambique, Niger, Réunion, Rwanda, Somalia, Sudan, Swaziland
OIRSA: Guatemala, Honduras, Nicaragua
PPPO: Micronesia, New Zealand, Solomon Islands

9.17 Management measures

Important methods for the management of white tip disease of rice are as follows.

9.17.1 Cultural practices

The nematode can be easily eradicated through the adoption of seed-cleaning technique. The primary source of infestation is infected seeds; disinfestation of seeds may provide complete elimination of white tip disease (Todd and Atkins, 1958). Early planting during cooler conditions in the USA and Japan eliminates or reduces the infection (Bridge *et al.*, 2005a). Crop rotation with non-host plants such as beans, vegetables, etc. may effectively reduce the field population of *A. besseyi* (Peng *et al.*, 2018).

9.17.2 Resistant or tolerant cultivars

Growing of tolerant or resistant cultivars reduces *A. besseyi* populations in rice and thus prevents yield losses. However, a major limitation is the availability and acceptability of tolerant/resistant cultivars to growers. There are many reports on resistance to *A. besseyi*. Cultivars such as Bluebonnet, Bluebonnet 50 and Starbonnet in Brazil (Papova *et al.*, 1994) and IR 841, IAC 435, IAC 120 and IAC 899 in India are immune or tolerant to the nematode (Silveira *et al.*, 1989).

9.17.3 Seed treatment

The most effective means of controlling *A. besseyi* is seed disinfection with hot-water treatment (HWT) but it involves some risk of injury to germination. Thermal wet treatment with hot water at 55–60°C for 15 min (Gergon and Prot, 1993) and at 53–54°C for 15 min (Tacconi *et al.*, 1999) is a highly effective disinfestation treatment against *A. besseyi* without any effect on seed germination or viability. Pre-soaking of seeds for 12–15 h followed by HWT at 52–53°C for 10 min is more effective (Nandakumar *et. al.*, 1975). Sun-drying rice on a concrete floor for 6 h for four consecutive days may also reduce nematode infestation levels in grain (Kuriyan, 1995).

9.17.4 Chemical control

Chemical treatments of seed with carbofuran, sulfone or methomyl at 0.1% WP are most promising for controlling *A. besseyi* but germination may be impaired by 11–29% (Kuriyan, 1995). Benomyl as a seed treatment and two sprayings, at the nematode infestation stage, controlled white tip disease in rice (Gergon and Prot, 1993). Seed treatment with carbofuran, thiobendazole, benomyl or fenitrothion was effective for reducing the nematode to very low levels (Tenente and Manso, 1994). The application of cartap and 16% prochloraz cartap as seed treatment through immersion for 48–60 h before sowing satisfactorily controlled white tip nematode disease in rice (Wang *et al.*, 2006). Oxamyl and fenamiphos controlled the nematode most effectively (Lin *et al.*, 1992).

9.17.5 Biological control

Bio-management of the white tip nematode has not evoked much interest. The use of *Pseudomonas fluorescens* as a seed treatment (at 10 g/kg seed) followed by three foliar sprayings with the same at 1 kg/ha effectively controlled *A. besseyi* in rice (Ramakrishnan and Rajendran, 2002).

9.17.6 Phytosanitary measures

See Chapter 16, section 115.

10 *Aphelenchoides fragariae*

10.1 Common names

Strawberry crimp nematode; fern nematode; strawberry spring dwarf nematode

10.2 Scientific name:

Aphelenchoides fragariae (Ritzema – Bos, 1891) Christie, 1932

10.3 Synonyms:

Aphelenchoides olesistus (Ritzema Bos, 1893) Steiner, 1932
Aphelenchoides olesistus var. *longicollis* (Schwartz, 1911) Goodey, 1933
Aphelenchoides pseudolesistus (Goodey, 1928) Goodey, 1933
Aphelenchus fragariae Ritzema Bos, 1891
Aphelenchus olesistus Ritzema Bos, 1893
Aphelenchus olesistus var. *longicollis* Schwartz, 1911
Aphelenchus pseudolesistus Goodey, 1928

10.4 EPPO code

APLOFR

10.5 Diagnosis

10.5.1 Morphological diagnosis

Allen (1952) and Siddiqi (1975b) provided a comprehensive description of *Aphelenchoides fragariae*. These morphological descriptions were modified by Hunt (1993). For morphometric measurements, see Table 10.

Female: Body slender, straight to ventrally arcuate when relaxed. Cuticle finely annulated, lateral field with two incisures. Cephalic region almost continuous with body, appears smooth under microscope, and four to five annuli visible by scanning electron microscopy (Khan *et al.*, 2008). Stylet slender with conus and shaft nearly equal in length; basal knobs minute but distinct. Pharynx characteristic of the genus, metacorpus oval and highly muscular with central valve plates, pharyngeal gland lobe with dorsally overlapping the intestine, two to four body widths long. Nerve ring encircling isthmus near its base and about one body width behind metacorpus. Excretory pore level with or close behind nerve ring. Ovary monoprodelphic, outstretched, with oocytes in single row, never reaching pharynx. Postvulval uterine sac long, extending over half distance between vulva and anus. Tail conoid, elongate with single simple spike or minute mucro at tail tip (Fig. 2.5).
Male: Abundant in number and morphologically similar to female (Fig. 2.5). Tail arcuate through 45–90 degrees when relaxed but not sharply curved like a hook, with simple terminal spine. Three pairs of caudal papillae. Spicules rose-thorn shaped with slightly developed apex and rostrum, dorsal limb 10–19 µm long.

10.5.2 Biochemical diagnosis

One-dimensional SDS gel electrophoresis protein patterns were used to distinguish species and races of *A. fragariae* (Platzer, 1981). Ibrahim *et al.* (1994b) also characterized *A. fragariae* based on esterase and protein, using one-dimensional SDS gel electrophoresis.

10.5.3 Molecular diagnosis

Several molecular diagnostic tests for *Aphelenchoides* species have been established and are now in use (Chizov *et al.*, 2006; McCuiston *et al.*, 2007; Rybarczyk-Mydłowska *et al.*, 2012). Different foliar nematodes are characterized by PCR–RFLP of ITS regions, including *A. fragariae* (Ibrahim *et al.*, 1994a; Iwahori *et al.*, 1998). Real-time PCR with species-specific primers can also be used for diagnosis of *Aphelenchoides* species directly from plant material (Rybarczyk-Mydłowska *et al.*, 2012). The species-specific primers for *A. fragariae* (located in the SSU rDNA sequence) used in this PCR are:

1469: 5′-CTT ATC GCA CGA CTT TAC G-3′
1472: 5′-TCA AAG TAA TCC GCA TCC AAT-3′

Each real-time PCR is performed according to the following protocols.

Reagent	–	Final concentration
PCR-grade water	–	25 µl
PCR buffer Absolute qPCR mix	–	1×
Specific forward primer	–	0.2 µM
Specific reverse primer	–	0.2 µM
DNA (volume)	–	3 µl
Cycling parameters		
Initial denaturation	–	95°C for 15 min
Number of cycles	–	60
Denaturation	–	95°C for 30 s
Annealing	–	63°C for 60 s
Elongation	–	72°C for 30 s

The PCR products are analysed by agarose gel electrophoresis. The amplicon size is approximately 470 bp for *A. fragariae*. McCuiston *et al.* (2007) developed internal transcribed spacer (ITS)1-PCR for early detection and identification of *A. fragariae* directly from host plant material using species-specific primers (for detailed procedure of DNA extraction, see Chapter 15, section 110.1). The primers (within the rDNA-ITS1 region) used in this PCR are:

AFragF1 (forward) 5′-GCA AGT GCT ATG CGA TCT TCT-3′
AFragR1 (reverse) 5′-GCC ACA TCG GGT CAT TAT TT-3′

Each real-time PCR is performed according to the following protocols.

Reagent		Final concentration
PCR-grade water	–	25 µl
PCR buffer (20 mM Tris-HCl (pH 8.4), 50 mM KCl)	–	1×
$MgCl_2$	–	1 mM
dNTPs	–	0.2 mM
Primer AFragF1 (forward)	–	0.4 µM
Primer AFragR1 (reverse)	–	0.4 µM
DNA polymerase (GoTaq Flexi DNA Polymerase1	–	1.25 U
DNA (volume)	–	2 µl
Cycling parameters		
Initial denaturation	–	94°C for 2 min
Number of cycles	–	40
Denaturation	–	94°C for 1 min
Annealing	–	53°C for 40 s
Elongation	–	72°C for 1 min
Final elongation	–	72°C for 10 min

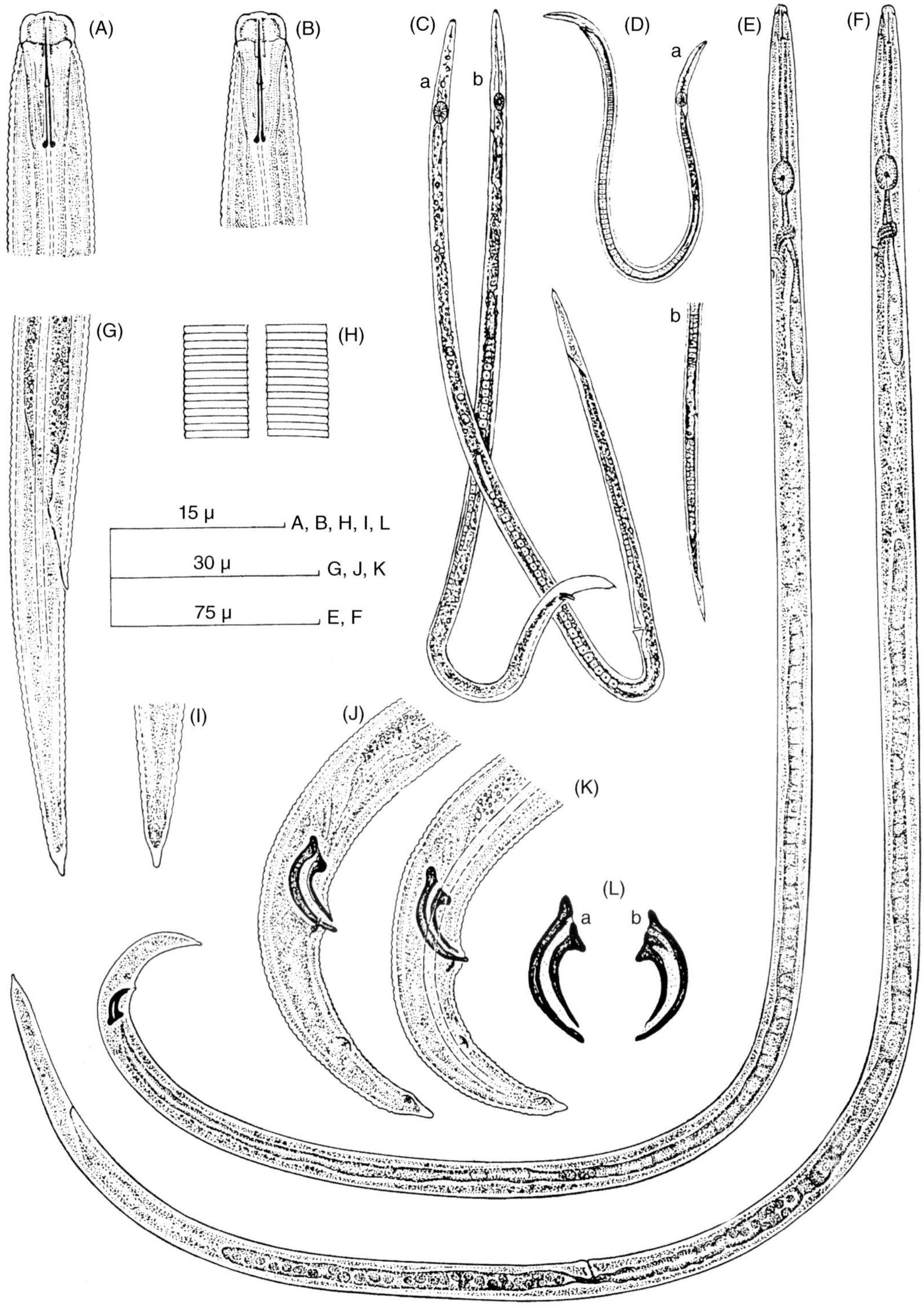

Fig. 2.5. Line drawing of *Aphelenchoides fragariae*. Female anterior or lip region (A, N); male anterior or lip region (B); a-female and b-male (C); a- male and b- posterior of female (D); male (E); female (F); female tail (G); lateral field (H,O); female tail tip (I); male tails (J,K); spicules (L). [Reproduced from Hunt, 1993; courtesy CAB International]

The PCR products are analysed by agarose gel electrophoresis. The expected amplicon size is approximately 169 bp for *A. fragariae*. These species-specific primers amplify the DNA only from *A. fragariae*, not from other plant-parasitic nematode species such as *A. besseyi*, *A. ritzemabosi*, *Caenorhabditis elegans*, *D. dipsaci*, *Heterodera schachtii*, *Meloidogyne incognita*, *Pratylenchus penetrans*, etc. The PCR is sensitive enough in detecting a single nematode in a background of plant material extract. The test has precisely detected *A. fragariae* in over 100 naturally infected samples, including 50 ornamental plant species.

10.6 Geographical distribution

A. fragariae is geographically concentrated in the temperate regions of the world (Fig. 2.6). However, it has been recorded from almost all areas and regions where ornamental crops are cultivated. Region-wise, *A. fragariae* is known to occur in the following countries (CABI-ISC, 2020).

Asia: China, India, Israel, Japan, Kyrgyzstan, South Korea, Turkey.
Europe: Belgium, Bulgaria, Denmark, Estonia, France, Germany, Hungary, Ireland, Italy, Latvia, Moldova, Netherlands, Norway, Poland, Portugal, Russia, Slovakia, Spain, Sweden, Switzerland, Ukraine, United Kingdom.
North America: Canada, Cuba, Mexico, USA.
Oceania: Australia, New Zealand, Papua New Guinea.
South America: Brazil.

10.7 Host plants

Over 250 plants from 47 families are recorded as hosts of *A. fragariae* (Sturhan, 1962). Some earlier records may also refer to *A. ritzemabosi* which occurs sympatrically in about 28 hosts, including strawberry, aster and begonia (Siddiqi, 1975b). Hosts mostly include ferns and members of Liliaceae, Primulaceae and Ranunculaceae, compared with *A. ritzemabosi* which mainly parasitizes members of Compositae. It has been recorded on 27 plant species in California, USA (Siddiqi *et al.*, 1973). Different plant species attacked by *A. fragariae* are listed in Table 12.

10.8 Symptoms

The common symptoms of *A. fragariae* damage on plants are chlorosis, necrosis, distortion, deformation and dwarfing of the leaves, stems, flowers or bulbs, leaf tattering and defoliation. The symptoms are frequently confused with those of powdery mildew. The symptoms typically appear as vein encircled lesions or blotches that start as lightly chlorotic lesions and later turn brown to black or necrotic and dry (Fig. 2.7). Symptom appearance may be extremely variable as a result of the different characteristics of host plant species and the influence of environmental conditions. Infested plants occasionally do not exhibit symptoms until the plant is severely infested with the nematodes (Knight *et al.*, 2002; Khan *et al.*, 2008). The shape and pattern of the blotches are closely related to the venation pattern of the leaf, the infested leaves appearing pale green to tan in colour or showing dark brown mosaic spots or angular necrotic lesions (Kohl, 2011) (Fig. 2.7A-D).

The nematode normally attacks above-ground parts of plants and may be endo- or ectoparasitic. *A. fragariae* infests buds or the crown, causing buds to decay, flowers to shrivel, and leaves, petioles and stems expanding from the infested buds to become misshapen, crinkled and stunted and to develop brown scars (Southey, 1993).

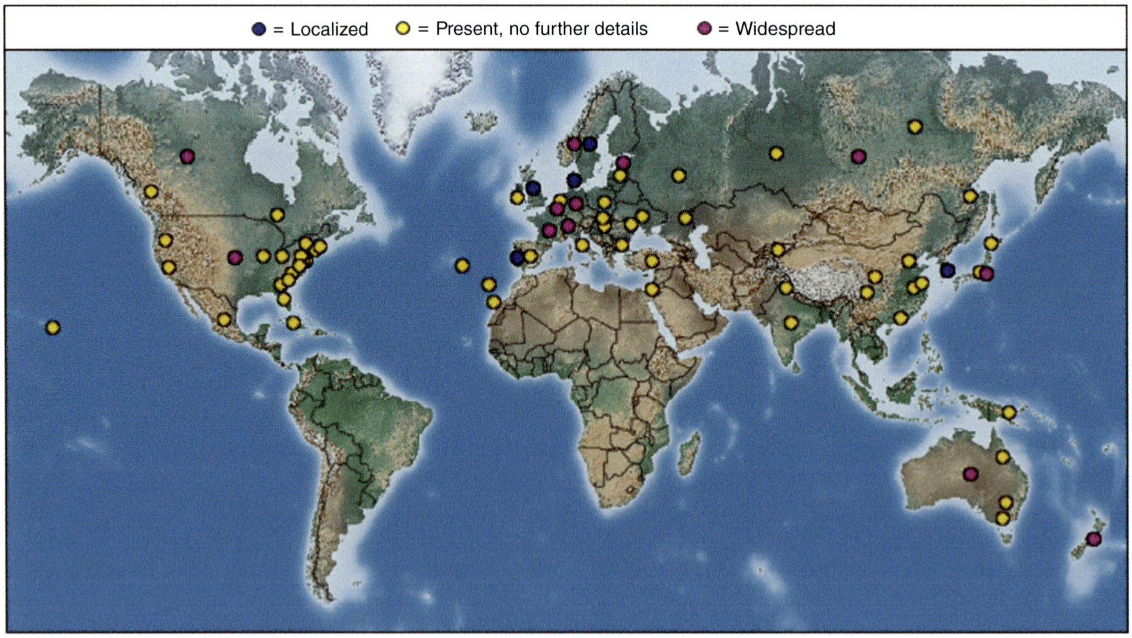

Fig. 2.6. Distribution map of *Aphelenchoides fragariae*. [Adapted from CABI-ISC, 2020]

Table 12. Hosts of *Aphelenchoides fragariae* (adapted from CABI-ISC, 2020).

Main hosts
Anthurium andreanum
Azaleas
Barleria cristata (Philippine violet)
Begonia spp.
Chloranthus spicatus
Fragaria ananassa (strawberry)
Hibiscus rosa-sinensis (China-rose)
Hosta spp.
Hydrangea macrophylla (French hydrangea)
Lilium spp.
Maranta leuconeura (banded arrowroot)
Peony
Pimpinella diversifolia
Primula sp. (primrose)
Psychotria nervosa
Rhododendron simsii (Sim's azalea)
Ruscus hypophyllum
Saintpaulia ionantha (African violet)
Saxifraga (saxifrage)
Tolmiea menziesii (pick-a-back plant)
Veronica arvensis (corn speedwell)
Viola odorata (English violet)
Wulfenia carinthiaca
Other hosts
Allium cepa (onion)
Allium sativum (garlic)
Anigozanthos sp.
Avena sativa (oats)
Cornus canadensis (creeping dogwood)
Eriobotrya japonica (loquat)
Ficus carica (common fig)
Ficus macrophylla (Moreton Bay fig)
Helianthus tuberosus (Jerusalem artichoke)
Ipomoea batatas (sweet potato)
Lamium maculatum (spotted deadnettle)
Osmunda regalis
Prunus persica (peach)
Habitat/association
Asplenium nidus (bird's nest fern)
Capsella bursa-pastoris (shepherd's purse)
Chenopodium album (fat hen)
Cobotium chamissoi
Erigeron annuus (annual fleabane)
Ficus elastica (rubber plant)
Polygonum blumei (tufted knotweed (USA))
Solanum nigrum (black nightshade)
Stellaria media (common chickweed)

On strawberry (*Fragaria* spp.), the early symptoms of infestation are stunted plant growth and distortion of buds, leaves and flowers; infested plants show malformations including twisting and puckering of leaves, discoloured areas with hard and rough surfaces, undersized leaves with crinkled edges, tight aggregation or death of crowns, reddened and stunted petioles, and flower stalks with aborted or partly aborted flowers (Fig. 2.7A). Heavily infested plants do not produce fruit (Siddiqi, 1975b).

On *Zinnia* sp., leaf blotch symptoms appear as long, narrow necrotic patches bounded by longer veins; in severe cases, the entire leaf dries and dies (Fig. 2.7C). The leaf blotch symptoms on ferns appear as narrow linear patches perpendicular to the midrib of the frond, corresponding to closely spaced lateral veins, as chevron-like stripes (Cobon and O'Neill, 2011; Zhen *et al*., 2012).

On *Tolmiea menziesii* (Fig. 2.7B), African violet (Fig. 2.7D), *Cyclamen* spp. and *Andrographis paniculata*, infested leaves show water-soaked irregular patches that later turn brown (dan Supriadi, 2008). In general, the blotches form more or less angular chlorotic areas in ternate or palmate leaves with reticulate venation or with main veins radiating from the petiole–lamina junction, while infected thicker and succulent leaves initially show water-soaked irregular patches that subsequently become necrotic without defined margins; ultimately, the entire leaf dies (Richardson and Grewal, 1993). In *Begonia* spp., the nematode feeds on, and destroys, mesophyll cells of the leaves and may cause reddening along the veins, causing the entire leaf blade to appear red; severe necrosis may result in the presence of *Xanthomonas begoniae* (Riedel and Larsen, 1974).

10.9 Biology and ecology

A. fragariae is an obligate parasite of above-ground plant parts and may feed ecto- or endoparasitically. On strawberry, it is ectoparasitic on folded crown and runner buds and its feeding causes small, dry brown areas delimited by the midrib and major veins. The nematodes can also be found feeding endoparasitically on leaf tissues and in fruit pulp (Tacconi, 1972). In violets, it feeds ectoparasitically in the unopened leaf and flower buds and has been found within the ovary. It is endoparasitic in leaves of ferns, begonias, peonies, etc., and causes leaf-blotch symptoms. The nematode enters the leaf through the stomatal opening but always requires a thin film of water (Klingler, 1970). It also penetrates the epidermis of the undersurface leaves (Strümpel, 1967).

A. fragariae is an amphimictic species (bisexual) with $n = 2$ (Cayrol and Dalmasso, 1975). The life cycle is completed in 10–11 days at 18°C in the leaves of Lorraine begonia. Egg hatching takes place in 4 days and the juveniles require 6–7 days to mature. Approximately 32 eggs are laid by a single female (Strümpel, 1967). The populations of *A. fragariae* are at a peak from March to May and November to January but are influenced by temperature and moisture (Szczygiel and Hasior, 1972). The nematode cannot survive in soil without any host for more than 90 days (Szczygiel and Hasior, 1971). The nematode population per number of hearts did not change when strawberry plants infested with *A. fragariae* and *A. ritzemabosi* were stored at temperatures of 14–15°C or in an unheated glasshouse in winter. However, at 20°C, the population increased many times. In cold-storage, it performed well in plant tissues at –2 to –1°C (Tacconi, 1972). At –20°C only a few individuals of *A. fragariae* can survive, while under dry conditions

Fig. 2.7. Symptoms caused by *Aphelenchoides fragariae* (A–D) and *A. ritzemabosi* (E–H) on different plants. Strawberry (A); *Tolmiea menziesii* (B); common zinnia (C); African violet (D); aster (E); tobacco (F-a & b); chrysanthemum (G,H). [Photo courtesy A: Cobon and O'Neill, 2011; B,G: Penn State Department of Plant Pathology & Environmental Microbiology Archives, Penn State University, Bugwood.org; C: Department of Plant Pathology, North Carolina State University, Bugwood.org; D: Bonsak Hammeraas, NIBIO - The Norwegian Institute of Bioeconomy Research, Bugwood.org; E: Source Nemaplex, H. Ferris; F: D. Blancard (INRA), http://ephytia.inra.fr/en/D/2264; H: J. Bridge, CABI BioScience]

A. fragariae survived in damaged lily leaves for almost 2 years (Yamada and Takakura, 1987).

10.10 Interaction with other organisms

On strawberry, interaction with the bacterium *Rhodococcus fascians* (=*Corynebacterium fascians*) resulted in cauliflower disease (Crosse and Pitcher, 1952). The bacterium alone causes secondary crowns and galls (Sitaramaiah and Pathak, 1993). In Russia, *A. fragariae* and virulent strains of *R. fascians* are responsible for cauliflower symptoms and less virulent strains cause alamination of leaves or red plants (Drozdovski *et al.*, 1971). On Rieger begonias, in the presence of *A. fragariae*, the bacterial leaf spot disease (*Xanthomonas begonia*) was more severe and developed more speedily than when bacteria alone were present (Riedel and Larsen, 1974).

10.11 Economic importance

A. fragariae decreased yield of strawberry by up to 60% in Ireland (Duggan, 1969). In France, *A. fragariae* has resulted in a decline in strawberry production (Clerjeau *et al.*, 1983). The crown weight of strawberry cv. 'Senga Sengana' was reduced by 41% by *A. fragariae*. Fruit yield in the first year was also reduced, largely due to declines in fruit number, by 54%. The number of runners was reduced by 11–15% by *A. fragariae* (Bohmer, 1981). The susceptible strawberry cultivars 'Macherauchs frühernte' and 'Cambridge favourite' exhibited 65% and 82% reduction in yield, respectively, after 2 years of infestation in Poland (Szczygiel, 1966). Foliar blight of anthurium (*Anthurium andraeanum*) in Hawaii, which is often lethal in young plants, was found to be caused by *A. fragariae*. The nematode also attacks and destroys anthurium seeds (Hunter *et al.*, 1974).

10.12 Similarities to other species/conditions

A. fragariae may occur sympatrically with *A. ritzemabosi* on 28 hosts, including strawberry, aster and begonia, but only on one species of fern (*Struthiopteris orientalis*) (Siddiqi, 1975b). Most of the *A. fragariae* hosts belong to ferns, Liliaceae, Primulaceae and Ranunculaceae, whereas those of *A. ritzemabosi* occur mostly in Compositae (Siddiqi, 1975b). These two species can be differentiated based on their morphological characters (Cayrol and Dalmasso, 1975)

10.13 Movement and means of dispersal

The nematode can be easily disseminated through infected plant parts. For more details, see section 9.13, above.

10.14 Pest risk analysis and looming threat from introduction

A. fragariae is one of a group of nematodes that are presently targeted in regulatory programmes in different countries worldwide (EPPO, 2017b). In Russia, it is one of the major nematode species designated for quarantine measures (Chizov *et al.*, 2006). *A. fragariae* was intercepted on strawberry seedlings imported to Tianjin, China, from the USA during quarantine checks (Zhang and Wang, 1989). *A. fragariae* is widely distributed and occurs in most of the temperate areas of the world (CABI-ISC, 2020). However, its dissemination varies between regions, countries and localities. The accidental introduction of this nematode would have devastating impacts on the export of ornamental crops such as cut flowers to domestic and international markets.

10.15 Invasiveness rating

Considering the consequences of introduction and pest risk analysis, the following invasiveness rating has been assigned to *A. fragariae*.

Climate/host interaction	–	High (3)
Hosts range	–	High (3)
Dispersal potential	–	High (3)
Economic impact	–	High (3)
Environmental impact	–	High (3)
Post-entry distribution	–	Medium (–2)
Final rating (score)	–	High (13)

Based on the above invasiveness rating, *A. fragariae* receives a score of 13 out of 15 and is rated as HIGHLY INVASIVE NEMATODE.

10.16 Region-wise status of invasiveness

A. fragariae has a very wide host range and attacks several important ornamental and food crops. It is widely distributed and occurs in most of the temperate areas of the world (CABI-ISC, 2020). At the international level, *A. fragariae* attains invasive status in the following countries, where it is absent or has very limited distribution.

APPPC: Bangladesh, Cambodia, Indonesia, Lao, Malaysia, Myanmar, Nepal, North Korea, Pakistan, Philippines, Sri Lanka, Thailand, Tonga, Vietnam
CAHFSA: Anguilla, Bahamas, Barbados, Bermuda, Cayman Islands, Colombia, Dominica, Dominican Republic, French Guiana, Grenada, Guadeloupe, Guyana, Haiti, Honduras, Jamaica, Martinique, Netherlands Antilles, Nicaragua, Panama, Puerto Rico, Saint Lucia, Saint Kitts & Nevis, Suriname, Trinidad and Tobago, Turks and Caicos Islands, Virgin Islands (British), Virgin Islands (US)
CAN: Bolivia, Ecuador, Venezuela
COSAVE: Argentina, Chile, Paraguay, Peru, Uruguay
EPPO: Albania, Austria, Belarus, Croatia, Cyprus, Czech Republic, Finland, Georgia, Greece, Guernsey, Jersey, Jordan, Lithuania, Luxembourg, Macedonia, Romania, Serbia, Slovenia, Tunisia, Uzbekistan
IAPSC: Angola, Benin, Botswana, Burkina Faso, Burundi, Cameroon, Cape Verde, Central African Republic, Chad, Comoros, Congo (Democratic Republic of), Congo (Republic of), Côted'Ivoire, Djibouti, Egypt, Equatorial Guinea, Eritrea, Ethiopia, Gabon, Gambia, Ghana, Guinea, Guinea-Bissau, Kenya, Lesotho, Liberia, Madagascar, Malawi, Mali, Mauritania, Mauritius, Mozambique, Niger, Nigeria, Rwanda, Sao Tome & Principe, Senegal, Seychelles, Sierra Leone, Somalia, Sudan, Swaziland, Tanzania, Togo, Uganda, Zaire, Zambia, Zimbabwe
NEPPO: Algeria, Jordan, Libya, Malta, Morocco, Syria
OIRSA: Costa Rica, El Salvador, Guatemala, Honduras
PPPO: American Samoa, Cook Islands, Fiji, French Polynesia, Guam, Kiribati, Marshall Islands, Micronesia, Nauru, New Caledonia, Niue, Northern Mariana Islands, Palau, Pitcairn, Samoa, Solomon Islands, Tokelau, Tonga, Tuvalu, Vanuatu, Wallis and Futuna Islands

10.17 Management measures

The diseases of *A. fragariae* can be managed or avoided by different methods, as discussed below.

10.17.1 Chemical control

Application of aldicarb in strawberry runner beds effectively controlled *A. fragariae* and *A. ritzemabosi* in strawberry plants grown for production of runner plants in September or October in southern Germany (Blank, 1985). Dipping fresh strawberry plants in thionazin before freeze preservation virtually eradicated *A. fragariae* infection (Tacconi *et al.*, 1982). Strawberry cv. 'Redgauntlet' and 'Tioga' infested with *A. fragariae* soaked in a solution of fenamiphos produced higher yields, crown weights and fruit numbers per plant than treated plants and untreated infested plants in New Zealand (Soteros, 1985). Two foliar applications of oxamyl, with a 2-week interval, gave excellent control of *A. fragariae* in two begonia cultivars and completely reduced symptom development (Hansen *et al.*, 1972).

10.17.2 Physical control

Hot-water treatment (HWT) of infested aerial plant parts has long been used for controlling *A. fragariae* and *A. ritzemabosi*. HWT of strawberry runners at 47°C for 15 min and at 46°C for 10 min followed by a cold plunge has been recommended (Strümpel, 1967). HWT of strawberry runner initials at 45°C and 50°C for 10 and 15 min, respectively, controlled *A. fragariae*, *A. ritzemabosi* and *Ditylenchus dipsaci*. HWT of nematode-infested lily bulbs at 45°C for 20–30 min, at 41°C for 1 h or at 36°C for 6 h was quite effective. On Lorraine begonia, HWT at 46°C for 10 min effectively controlled *A. fragariae* and *A. ritzemabosi*.

10.17.3 Cultural control

Cultural methods include thorough and constant roguing of plants displaying signs of nematode infestation, burning of all infected plant materials, propagating only from healthy stocks and in clean containers and soil, and avoiding contacts between plants and excessive surface moisture of the leaves.

10.17.4 Phytosanitary measures

See Chapter 16, section 115.

11 *Aphelenchoides ritzemabosi*

11.1 Common names

Chrysanthemum foliar nematode; leaf and bud nematode

11.2 Scientific name

Aphelenchoides ritzemabosi (Schwartz, 1911) Steiner and Buhrer, 1932

11.3 Synonyms

Aphelenchoides ribes (Taylor, 1917) Goodey, 1933
Aphelenchus phyllophagus Stewart, 1921
Aphelenchus ribes (Taylor, 1917) Goodey, 1923
Aphelenchus ritzema-bosi Schwartz, 1911
Pathoaphelenchus ritzemabosi (Schwartz, 1911) Steiner, 1932
Pseudaphelenchoides ritzemabosi (Schwartz, 1911) Drozdovski, 1967
Tylenchus ribes Taylor, 1917

11.4 EPPO code

APLORI

11.5 Diagnosis

11.5.1 Morphological diagnosis

The morphological descriptions of *A. ritzemabosi* are after Siddiqi (1974c). For morphometric measurements, see Table 10.

Female: Body slender with distinct annules 0.9–1.0 μm wide. Lateral fields with four incisures, one-sixth to one-fifth as wide as body. Lip region hemispherical, set off by a constriction, slightly wider than adjacent body (Fig. 2.8), under a light microscope no annulations visible. Cephalic framework hexaradiate, weakly sclerotized. Stylet about 12 μm long, with distinct basal knobs and sharply pointed anterior. Procorpus slender. Large somewhat oval metacorpus, highly muscular, with prominent internal cuticular thickening and orifices of dorsal and subventral pharyngeal glands. In neotype, nerve ring 1.5 body widths behind bulb. Excretory pore 0.5–2 body widths posterior to nerve ring. Three pharyngeal glands forming a lobe extending about four body widths over intestine dorsally. Pharyngo-intestinal junction about 8 μm behind metacorpus, indistinct and valve not discernible. Intestine with small spherical granules and a distinct lumen throughout. Vulva protruding slightly, with transverse slit. Postvulval uterine sac extending for more than half the distance between vulva and anus, usually containing sperm. Ovary single outstretched anteriorly, oocytes in multiple rows. Tail conoid elongate, bearing a terminal peg which has two to four minute processes pointing posteriorly with paintbrush-like appearance (Fig. 2.8).
Male: Males are common. Posterior end of body usually curved through 180 degrees upon relaxation. Morphology of lip region, stylet and pharynx similar to that in female. Testis single monoarchic, outstretched. Three pairs of ventro-submedian caudal papillae: first pair adanal, second midway on tail, third near tail end. Spicules well curved, rose-thorn shaped, lacking a dorsal or ventral process at the proximal end, dorsal limb 20–22 μm long. Tail peg with two to four processes, of different shapes (Fig. 2.8).

11.5.2 Biochemical diagnosis

Eriksson and Granberg (1969) used one-dimensional SDS gel electrophoresis of proteins to differentiate species and races of *A. fragariae*. Ibrahim *et al.* (1994b) also characterized *A. ritzemabosi* based on esterase and protein, using one-dimensional SDS gel electrophoresis.

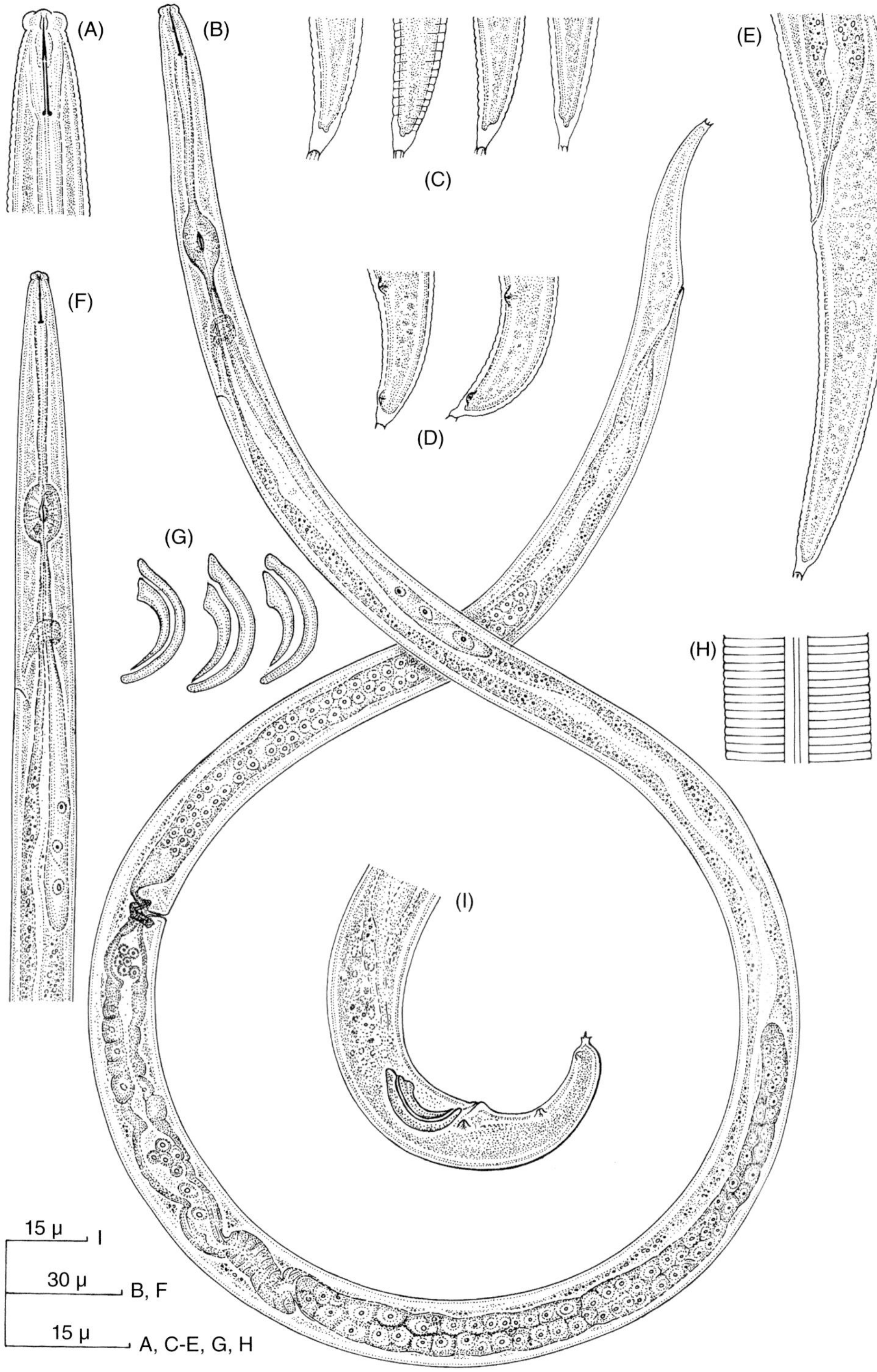

Fig. 2.8. Line drawing of *Aphelenchoides ritzemabosi*. Female head end (A); female (B); female tail ends (C); male tail ends (D); female tail (E); female pharyngeal region (F); spicules (G); lateral field (H); male tail region (I). [Reproduced from Hunt, 1993; courtesy M.R. Siddiqi/CABI]

11.5.3 Molecular diagnosis

Rybarczyk-Mydłowska *et al.* (2012) developed small subunit (SSU) ribosomal (r)DNA-based species-specific primers for diagnosis of *A. ritzemabosi*. The species-specific primers were used for real-time PCR to rapidly identify one or more foliar nematode species isolated from plant material and soil. The species-specific primers for *A. ritzemabosi* (designed with LNA, labelled by # below) are:

1496: 5′-CGC TGG TGG GTT TCG A-3′
1499: 5′-CCC GCT AAG AAA TGA TCA C#C-3′

Each real-time PCR is performed according to the following described conditions.

Reagent	–	Final concentration
PCR-grade water	–	25 µl
PCR buffer Absolute qPCR mix	–	1×
Specific forward primer	–	0.2 µM
Specific reverse primer	–	0.2 µM
DNA (volume)	–	3 µl
Cycling parameters		
Initial denaturation	–	95°C for 15 min
Number of cycles	–	60
Denaturation	–	95°C for 30 s
Annealing	–	63°C for 60 s
Elongation	–	72°C for 30 s

The PCR products are analysed by agarose gel electrophoresis. The amplicon size is approximately 470 bp for *A. ritzemabosi*.

11.6 Geographical distribution

A. ritzemabosi is geographically concentrated in the temperate regions of the world (Fig. 2.9). However, it has been recorded from almost all areas or regions where ornamental crops are cultivated. Region-wise, *A. ritzemabosi* is known to occur in the following countries (CABI-ISC, 2020).

Africa: Mauritius, South Africa.
Asia: China, India, Japan, Iran, Japan, Kazakhstan, South Korea, Uzbekistan.
Europe: Bulgaria, Croatia, Germany, Hungary, Ireland, Italy, Latvia, Netherlands, Poland, Portugal, Russia, Serbia, Serbia and Montenegro, Slovenia, Spain, Switzerland, Ukraine, United Kingdom.
North America: Cuba, Mexico, USA.
Oceania: Fiji, New Zealand.
South America: Brazil, Chile, Venezuela.

11.7 Host plant

A. ritzemabosi is a polyphagous nematode and parasitizes a large number of cultivated and wild plants. Major, minor and wild host plants of *A. ritzemabosi* are listed in Table 13.

11.8 Symptoms

On *Chrysanthemum* spp., infested leaves show characteristic angular blotches surrounded by the principal veins. The discoloration advances from translucent yellowish and brownish-green to dark brown (Fig. 2.7G-H). At a later stage leaves shrivel and die, hanging down, extending to the top of the plant (Johnson, 1998). Although some stems of a given plant may bear dead leaves, other stems may be symptomless. The nematodes also invade and feed within the buds, sometimes killing the growing point and preventing flowering, or producing malformed leaves with surface irregularities and rough brown scars (Farkas *et al.*, 1985).

On strawberry (*Fragaria* spp.), the most noticeable damage by *A. ritzemabosi* appears on newly formed leaves, which become distorted and puckered and sometimes show rough, greyish feeding areas near the base of the central veins. The

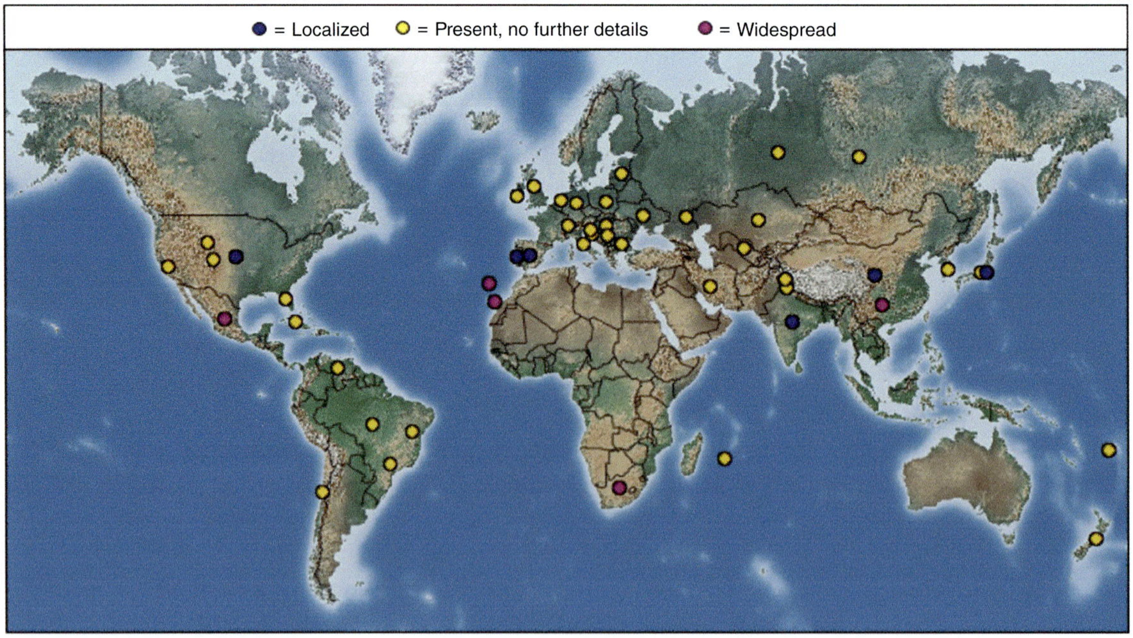

Fig. 2.9. Distribution map of *Aphelenchoides ritzemabosi*. [Adapted from CABI-ISC, 2020]

Table 13. Hosts of *Aphelenchoides ritzemabosi* (adapted from CABI-CPC, 2010).

Main hosts
Chrysanthemum morifolium (florist's chrysanthemum)
Dahlia (dahlia)
Fragaria (strawberry)
Fragarian ananassa (strawberry)
Minor hosts
Asplenium nidus (bird's nest fern)
Bergenia (elephant-leaved saxifrage)
Calceolaria (pouch flower)
Callistephus chinensis (China aster)
Campanula (campanulas)
Campanula pyramidalis (chimney bellflower)
Consolida ambigua (rocket larkspur)
Crassula coccinea
Geranium (cranesbill)
Helianthus annuus (sunflower)
Helleborus niger (Christmas rose)
Hibiscus rosa-sinensis (China-rose)
Lavandula angustifolia (lavender)
Limonium sinuatum (sea pink)
Lotus corniculatus (bird's-foot trefoil)
Lupinus (lupins)
Lycopersicon esculentum (tomato)
Medicago sativa (lucerne)
Mimulus (monkey-flower)
Nicotiana tabacum (tobacco)
Ocimum basilicum (basil)
Passiflora edulis (passionfruit)
Peperomia
Phaseolus vulgaris (common bean)
Rheum hybridum (rhubarb)
Rhododendron (azalea)
Ribes nigrum (blackcurrant)
Ribes uva-crispa (gooseberry)
Rudbeckia (coneflower)
Saintpaulia (African violet)
Saintpaulia ionantha (African violet)
Salvia splendens (scarlet sage)
Sinningia speciosa (gloxinia)
Solanum nigrum (black nightshade)
Verbena (vervain)
Viburnum
Viola (violet)
Zinnia elegans (zinnia)
Wild hosts
Buddleia (butterflybush)
Echium
Eleusine coracana (finger millet)
Ranunculus (buttercup)
Rumex (dock)
Sambucus (elderberry)
Senecio (groundsel)
Sonchus (sowthistle)
Stellaria media (common chickweed)
Veronica (speedwell)

cauliflower disease of strawberry, resulting from the continuous production of axillary buds on infected crowns, was experimentally induced in runners of *Fragaria* spp. through simultaneous inoculation of *A. ritzemabosi* and *R. fascians* by Crosse and Pitcher (1952).

On tobacco (*Nicotiana* spp.), *A. ritzemabosi* causes polygonal blotches that are bound by veins on the leaves of nematode-infested plants (Shepherd and Barker, 1990) (Fig. 2.7F). *A. ritzemabosi* also infests many herbaceous plants and most plants show the typical interveinal leaf blotches and distortions of the upper leaves resulting from bud infestation. *A. ritzemabosi* is also associated with the death of lower leaves and buds and malformed growth of shoots in *Lavandula* spp. Woody plants that are attacked by *A. ritzemabosi* include *Buddleia*, in which the nematode causes death of buds and leaf distortions. Infested *Viola* spp. are stunted and affected leaves curl downwards, wither and die, while the undersides of leaves show typical water-soaked blotches (Thomas, 1968; Southey, 1993). On *Crassula coccinia*, stunting and shoot blindness occur (Atkinson, 1964). The combined infestation of *A. ritzemabosi* with *Ditylenchus dipsaci* in stem tissues shows as discoloration caused by feeding of the nematode after cell separation by *D. dipsaci* (Southey, 1993).

11.9 Biology and ecology

Similar to *A. fragariae*, *A. ritzemabosi* is also an obligate plant parasite and inhabits leaves, buds, growing points and outer layers of stem on above-ground parts. In soil, it neither completes its life cycle nor survives during the winter. It can feed both endoparasitically and ectoparasitcally. On mesophyll cells of leaves it feeds endoparsitically, while on buds and growing points it feeds ectoparasitically (Siddiqi, 1974c). On chrysanthemum, infestation through the soil, dead leaves or weed hosts progresses from the base of the plant upwards under moist conditions. The nematodes move through the water film over plants, not within the stem tissue, to reach the leaves and buds. Rain splashes and leaf contacts contribute to reinfestation and spread. The nematode enters the leaves through stomata, feeds on the parenchymatous tissue of the mesophyll and destroys the cells, resulting in leaf spots or blotches, which can be easily seen on the under-surface. The nematodes migrate out from the necrotic tissue through the stomata and move in the film of water on the surface to infect terminal flower buds, which produce deformed and undersized flowers. Approximately 15,000 nematodes can be found in one chrysanthemum leaf, and over 300 individuals per 14 g of infested aster seeds have been recovered (Southey, 1993).

A. ritzemabosi is amphimictic and typically multiplies bisexually, not by parthenogenesis. Fertilized females go on reproducing for 6 months without further fertilization. A female lays around 25–30 eggs in compact leaves in chrysanthemum. Eggs hatch in 3–4 days and the juveniles mature in 9–10 days. The life cycle is completed in 10–13 days (Wallace, 1961). In susceptible chrysanthemum varieties, the female remains in the leaf at one place, feeding on adjoining cells, and lays many eggs within the leaves which hatch rapidly. However, in resistant varieties, the female moves around rapidly through the tissues, feeding on cells, but lays only a few eggs and the emerging juveniles fail to reach maturity (Wallace, 1961).

A. ritzemabosi possesses multiple contraction ability, i.e. coiling coupled with transverse and longitudinal folding of the cuticle, and survives unfavourable conditions through anhydrobiosis. *A. ritzemabosi* overwinters in dormant buds and growing points of chrysanthemum stools. Hence, mainly stools, rather than the soil, serve as the source of infestation (Wallace, 1961). The nematode is also seed-borne and carried through the seeds of aster (Brown, 1956). *A. ritzemabosi* also has the ability to reproduce on fungal mycelium (Hooper and Cowland, 1986).

11.10 Economic importance

A. ritzemabosi is a major and widespread pest of chrysanthemum in all of Europe, North America, South Africa and New Zealand. It is also a serious pest of strawberry in Ireland, where up to 60% yield reduction has been recorded due to *A. ritzemabosi* and *A. fragariae* infestations (Duggan, 1969). The crown weight of strawberry cv. 'Senga Sengana' was reduced by 51% by *A. ritzemabosi* and 41% by *A. fragariae*. Consequently, the fruit yield in the first year was reduced, largely due to a decline in fruit number, by 65% and 54%, respectively, by these two species. The number of runners was reduced by 25–30% by *A. ritzemabosi*, but only 11–15% by *A. fragariae*. Population density was directly related to the intensity of damage to the plant crowns and yield reduction in winter and spring, but reduced runner production was due to the summer population density (Bohmer, 1981). *A. ritzemabosi* infestation destroyed 45% of chrysanthemum plants on a holding, and for the most susceptible varieties the figure was as high as 92% in Poland (Karnkowski, 1999).

11.11 Similarities to other species/conditions

See section 10.12, above.

11.12 Movement and means of dispersal

See section 9.13, above.

11.13 Pest risk analysis and looming threat from introduction

A. ritzemabosi is one of a group of nematodes that is being targeted in regulatory programmes in different countries (CABI-ISC, 2020). *A. ritzemabosi* has a wide host range, including chrysanthemum and strawberry. Globally this nematode has not been reported in many countries. Detection and any new report of this pest in the areas where it is absent necessitate immediate regulatory action to protect the area against the introduction of this invasive nematode.

11.14 Invasiveness rating

Considering the consequences of introduction and pest risk analysis, the following invasiveness rating has been assigned to *A. ritzemabosi*.

Climate/host interaction	–	High (3)
Hosts range	–	High (3)
Dispersal potential	–	High (3)
Economic impact	–	High (3)
Environmental impact	–	High (3)
Post-entry distribution	–	Medium (–2)
Final rating (score)	–	High (13)

Based on the above invasiveness rating, *A. ritzemabosi* receives a score of 13 out of 15 and is rated as HIGHLY INVASIVE NEMATODE.

11.15 Region-wise status of invasiveness

A. ritzemabosi is geographically concentrated in temperate regions. However, it has been absent in many countries of the world where ornamental crops are cultivated. Region-wise, the status of invasiveness is as follows.

APPPC: Bangladesh, Cambodia, Indonesia, Iraq, Israel, Jordan, Kyrgyzstan, Lebanon, Malaysia, Myanmar, Pakistan, Philippines, Saudi Arabia, Sri Lanka, Syria, Tajikistan, Thailand, Vietnam
CAHFSA: Barbados, Belize, Cuba, Dominican Republic, Haiti, Jamaica, Saint Kitts & Nevis
CAN: Bolivia, Colombia, Ecuador, Peru
COSAVE: Argentina, Guyana, Paraguay, Suriname, Uruguay
EPPO: Cyprus, Greece, Morocco, Turkey
IAPSC: Algeria, Angola, Benin, Botswana, Burkina Faso, Burundi, Cameroon, Cape Verde, Central African Republic, Chad, Comoros, Congo, Côte d'Ivoire, Egypt, Eritrea, Ethiopia, Gabon, Gambia, Ghana, Guinea, Guinea-Bissau, Kenya, Liberia, Madagascar, Malawi, Mali, Mauritania, Mozambique, Namibia, Niger, Réunion, Rwanda, Senegal, Sierra Leone, Somalia, Sudan, Swaziland, Togo, Uganda, Zambia, Zimbabwe
NEPPO: Malta, Tunisia
OIRSA: Costa Rica, El Salvador, Guatemala, Honduras, Nicaragua
PPPO: Australia, Papua New Guinea, Tonga, Vanuatu

11.16 Management measures

For management measures, see section 10.17, above.

Bursaphelenchus Spp.

Over 110 species have been described within the genus *Bursaphelenchus*, with *Bursaphelenchus piniperdae* as type species (Mota and Vieira, 2004; Ryss *et al.*, 2005; Futai, 2013). Another closely related genus to *Bursaphelenchus* is monotypic *Rhadinaphelenchus* (Hunt, 1993). *Bursaphelenchus* species are split into 13 typological groups, namely *abietinus*, *africanus*, *cocophilus*, *eggersi*, *eremus*, *fungivorus*, *hofmanni*, *kevini*, *leoni*, *okinawaensis*, *sexdentati*, *sinensis* and *xylophilus*, and one unassigned group (Braasch *et al.*, 2009; Ye and Giblin-Davis, 2013). Some important *Bursaphelenchus* species are:

Bursaphelenchus baujardi Walia, Negi, Bajaj and Kalia, 2003
B. chitwoodi Ruhm, 1956
B. cocophilus (Cobb, 1919) Baujard, 1989, syn. *Rhadinaphelenchus cocophilus* (Cobb, 1919) Goodey, 1960
B. conicaudatus Kanzaki, Tsuda and Futai, 2000
B. doui Braasch, Gu, Burgermeister and Zhang, 2004
B. elytrus Massey, 1971
B. firmae Kanzaki, Maehara, Aikawa and Matsumato, 2012
B. fraudulentus Rühm, 1956 (Goodey, 1960)
B. fungivorus Franklin and Hooper, 1962
B. gillanii Schönfeld, Braasch, Riedel and Gu, 2013
B. koreanus Gu, Wang and Chen, 2013
B. leoni Baujard, 1980
B. luxuriosae Kanzaki and Futai, 2003
B. macromucronatus Gu, Zheng, Braasch and Burgermeister, 2008
B. mucronatus kolymensis, Braasch, Gu and Burgermeister, 2011
B. mucronatus mucronatus (Mamiya and Enda, 1979) Braasch *et al.* 2011
B. paraluxuriosae Gu, Wang and Braasch, 2012
B. platzeri Gablin-Davis *et al.*, 2006
B. populi Tomalak and Filipiak, 2010
B. silvestris (Lieutier and Laumond, 1978) Baujard, 1980
B. singaporensis Gu, Zhang, Braasch and Burgermeister, 2005
B. wili (Massey, 1974) Baujard, 1989
B. xylophilus (Steiner and Buhrer, 1934), Nickle, 1970

Among these, four species – *B. cocophilus*, *B. xylophilus*, *B. mucronatus kolymensis* (previously *B. mucronatus* European type) and *B. mucronatus mucronatus* (previously *B. mucronatus* East Asian type) – are considered as invasive nematode species, because of their economic importance and quarantine significance in various countries (Mota and Vieira, 2004; Ryss *et al.*, 2005; Hunt, 2008; Zhao *et al.*, 2008; Braasch *et al.*, 2009; Duncan and Moens, 2013). However, *B. mucronatus* is solely a fungal feeder (Jones *et al.*, 2013), while *B. xylophilus* is transmitted by a beetle vector (*Monochamus* spp.).

Systematic position of *Bursaphelenchus* species

	Classical (after Baujard, 1989)	Molecular (after Decraemer and Hunt, 2013)
Phylum	Nematoda	Nematoda
Class	Secernentea	Chromadorea
Subclass	Diplogasteria	Chromadoria
Order	Aphelenchida	Rhabditida
Suborder	Aphelenchina	Tylenchina
Infra-order	–	Tylenchomorpha
Superfamily	Aphelenchoidea	Aphelenchoidea
Family	Aphelenchoididae	Aphelenchoididae
Subfamily	Parasitaphelenchinae	Parasitaphelenchinae
Genus	Bursaphelenchus	Bursaphelenchus

12 *Bursaphelenchus cocophilus* (= *Rhadinaphelenchus cocophilus*)

12.1 Common name

Red ring nematode

12.2 Scientific name

Bursaphelenchus cocophilus (Cobb, 1919) Baujard, 1989

12.3 Synonyms

Aphelenchoides cocophilus (Cobb, 1919) Goodey, 1933
Aphelenchus cocophilus Cobb, 1919
Chitinoaphelenchus cocophilus (Cobb, 1919) Chitwood in Corbett, 1959
Rhadinaphelenchus cocophilus (Cobb, 1919) Goodey, 1960

12.4 EPPO code

RHAACO

12.5 Diagnosis

Bursaphelenchus cocophilus is also known as *Rhadinaphelenchus cocophilus* (Hunt, 1993). Giblin-Davis *et al.* (1989) presented morphological similarities between genus *Bursaphelenchus* and genus *Rhadinaphelenchus*, while Hunt (1993) retained *Rhadinaphelenchus* as a monotypic genus. However, molecular characterization and real-time PCR assay based on DNA sequences of the ribosomal DNA small subunit (rDNA SSU), large subunit (LSU) D2–D3, internal transcribed spacer (ITS) and mitochondrial DNA (mtDNA) cytochrome oxidase subunit (COI) of 97 species of *Bursaphelenchus*, including *B. cocophilus*, support its attachment to the genus *Bursaphelenchus* (Ye and Giblin-Davis, 2013) and now the majority of nematologists refer to this species as *B. cocophilus* (Duncan and Moens, 2013; Griffith *et al.*, 2018).

12.5.1 Morphological diagnosis

Important morphological characters of *Bursaphelenchus cocophilus* are based on descriptions after Brathwaite and Siddiqi (1975) and Hunt (1993) and are summarized below. For measurements, see Table 14.

FEMALE: Body very slender and medium size to long, straight to nearly ventrally arcuate when heat relaxed (Fig. 2.10F). Cuticle finely annulated with thin and fine transverse striae. Lateral fields marked by four incisures and a faint median line resembling a fifth incisure, comprising one-fourth of body

Table 14. Morphometric measurements of *Bursaphelenchus cocophilus* by different authors.

Morphometric characters		Range, numerous populations			
		Cobb (1919)	Lordello and Zamith (1954)	Goodey (1960)	Thorne (1961)
Females	Length (mm)	1.0	0.83–1.11	0.97–1.18 (1.05)	1.2
	Stylet (μm)	–	15.3	11–13	–
	a	71	60–96	78–96 (87)	69
	b	7.7	11–14	8.7	8.4
	c	12.5	9.4–10.7	9.5–13.2 (11.6)	12
	V%	68	66–69	64–68 (66)	65
Males	Length (mm)	1.0	0.82–1.42	0.84–1.16 (1.02)	1.0
	Stylet (μm)	–	10.7–13.6	–	–
	a	100	92–143	100–179 (120)	65
	b	7.7	11–19	6.5	7.8
	c	20	22–47	24–35 (28)	19
	Spicules (μm)	–	–	9–11	–
	Tail (mm)	45	–	50–68 (59)	42

width. Cephalic region high, smooth, slightly narrower than body and anteriorly flattened with somewhat straight sides (Fig. 2.10B). Head framework protruding, sclerotized. Stylet 11–13 μm long, slender with basal knobs. Procorpus elongate, cylindrical leading to the rounded rectangular median bulb, usually about twice as long as wide, with noticeable valve-plates slightly posterior to centre (Fig. 2.10L). Dorsal oesophageal gland orifice between anterior margin of bulb and valve plates. Oesophageal glands usually obscure, dorsally overlapping intestine. Nerve ring surrounding isthmus, one to two body widths behind bulb. Excretory pore slightly behind nerve ring and anterior to hemizonid. Intestine with indistinct lumen and small granules. Vulva post-median, slit-like, appearing as open C in ventral view (Fig. 2.10U). Dorsal lip strongly developed and forms a flap, slightly overhung sclerotized posterior lip. Vagina thick-walled and slightly curved. Genital tract monoprodelphic, outstretched, and anterior gonad well developed. Developing oocytes in a single row. Post-uterine sac very long, extending to about 75% of vulva–anus distance and often with spermatozoa. Rectum about 1.5 anal body widths long; anus distinct. Tail elongate-subcylindrical, narrow with a rounded terminus, 10–17 anal body widths long (Fig. 2.10K). Phasmids and deirids absent.

MALE: Body arcuate ventrally, further strongly curved in tail region (Fig. 2.10N,O). Head, stylet and oesophagus similar to female. Testis monoarchic, outstretched anteriorly, with spermatogonia in single row. Spicules small, paired, rose-thorn shaped. Distal tip notched with an elongated rounded terminus. Rostrum small rounded. Gubernaculum absent, but dorsal wall of spicular pouch thick to form an apophysis. Bursa terminal, prominent in dorsal or ventral view with finely striated margins, extending for 40–50% of tail length. Tail prominently curved, initially cylindrical and then tapering to become conoid with pointed terminus (Fig. 2.10N). Seven caudal papillae present, two pairs ventro-submedian near base of bursa and one pair preanal about 50% of the spicule length anterior to cloaca.

JUVENILE: Dome-shaped high head, not offset (Fig. 2.10S). Tails of second- and third-stage juvenile conoid with mucronate tips (Fig. 2.10Q). Fourth-stage juvenile shows dimorphic tail, female juvenile has rounded tip whereas male has sharply pointed tip.

12.5.2 Biochemical diagnosis

Different biochemical methods have been developed for distinguishing *B. cocophilus* from other *Bursaphelenchus* species (Hotchkin and Giblin, 1984). Biochemical methods assisted the morphological identification of the nematode species but their value must be supported with molecular diagnosis (Li, 2008).

12.5.3 Molecular diagnosis

Many methods are available for the molecular identification of *Bursaphelenchus* species but limited information is available on *B. cocophilus*. DNA sequences on the rDNA SSU, LSU D2–D3, ITS and mtDNA COI using real-time-PCR were reported by Ye and Giblin-Davis (2013) for rapid and accurate identification of 57 species of *Bursaphelenchus*, including *B. cocophilus*. In this method, one to ten nematodes can be used for DNA extraction either directly from wood or from the vectors (see Chapter 15, section 110.1.3). Primers for partial rDNA-LSU D2/D3 are:

Forward primer D2a (5′ ACAAGTACCGTGAGGGAAAGT 3′)
Reverse primer D3b (5′ TGCGAAGGAACCAGCTACTA 3′)

Amplification is performed using 25 μl PCR with the following cycling parameters.

Initial denaturation	–	95°C for 5 min
Number of cycles	–	40
Denaturation	–	94°C for 30 s
Annealing	–	55°C for 45 s
Extension	–	72°C for 10 min

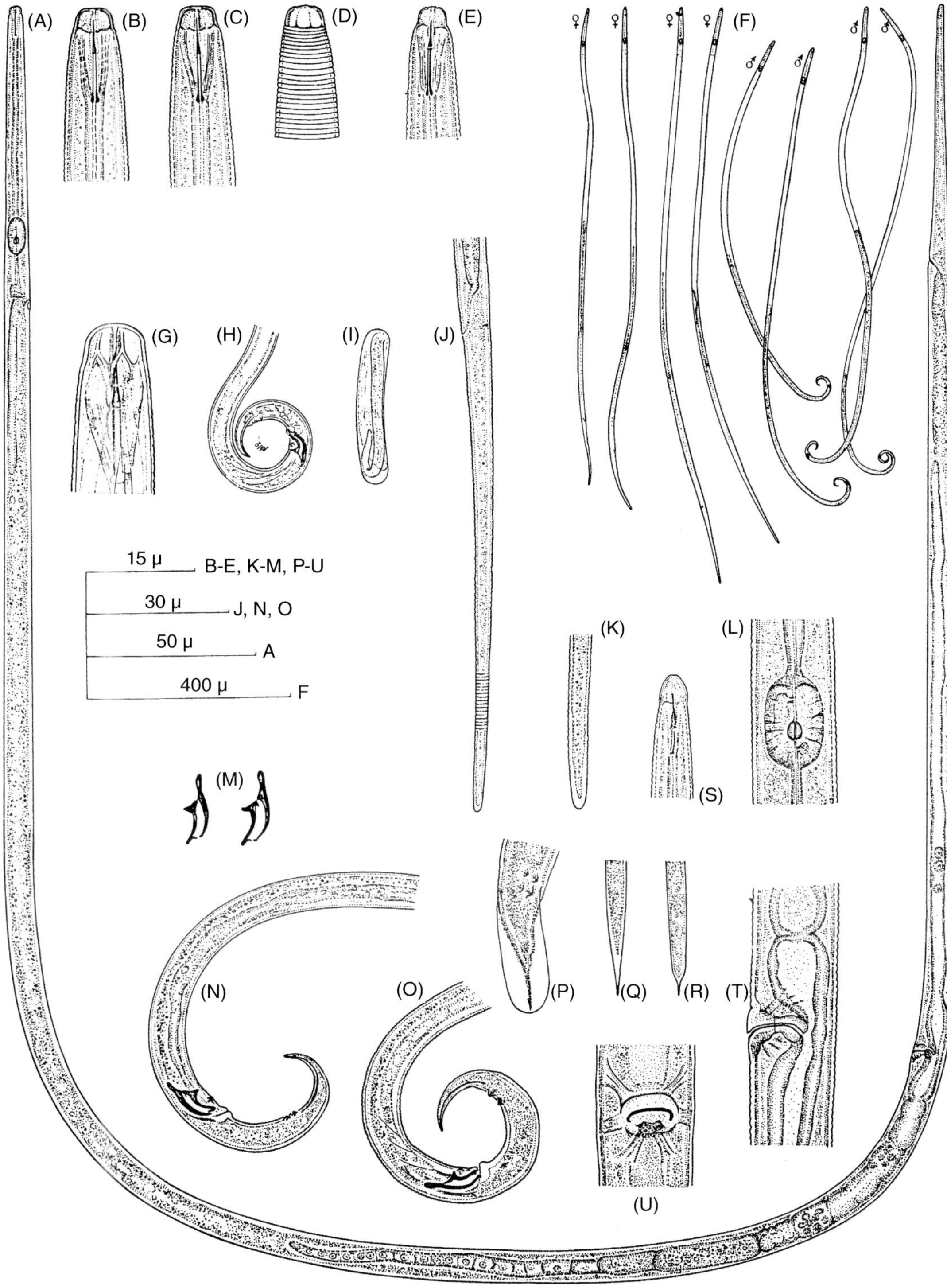

Fig. 2.10. Line drawing of *Bursaphelenchus cocophilus*. Adults (A,F); female head (B-D, G); male head end (E); male tail end (H,N,O); egg (I); female tail tip (J); female median oesophageal bulb (L); spicules (M); bursa dorsal view (P); juvenile tail tip (Q,R); juvenile head (S); vulval in lateral and ventral view, respectively (T,U). [Reproduced from Hunt, 1993; G-I after Cobb, 1919; rest after Brathwaite and Siddiqi, 1975; courtesy CAB International]

PCR products are purified and sequenced using DNA analyser. Then the molecular sequences are compared with other closely related nematode species available at the GenBank sequence database using the BLASTn homology search program.

12.6 Geographical distribution

The red ring nematode (RRN) was first described from Trinidad by Hart (1905). Currently, RRN has a limited distribution and has only been recorded from the West Indies and Latin American countries (Fig. 2.11). RRN occurs in Guatemala but does not occur outside the Western Hemisphere (Dean, 1979). Several reports of the presence of RRN in Bahamas, Barbados, Dominica, Dominican Republic, Haiti, Jamaica and Puerto Rico have been negated. The nematode is mainly distributed from Mexico to South America and in the Lower Antilles along with its primary vector, the palm weevil, *Rhynchophorus palmarum* (Sullivan, 2013). Worldwide distribution of *B. cocophilus* (CABI-ISC, 2020) is as follows.

North America: Belize, Costa Rica, El Salvador, Grenada, Guatemala, Honduras, Mexico, Nicaragua, Panama, Saint Vincent and the Grenadines, Trinidad and Tobago.

South America: Brazil, Colombia, Ecuador, French Guiana, Guyana, Peru, Suriname, Venezuela.

12.7 Hosts and vector species

Red ring is primarily a disease of coconut palm and most common in oil and date palm (Griffith *et al.*, 2005). However, more than 17 species from the family Palmae can act as host of this nematode (Chitamber, 2016). The major and minor hosts of *B. cocophilus* are as follows.

Major hosts: *Cocos nucifera* (coconut), *Elaeis guineensis* (African oil palm), *Phoenix dactylifera* (date-palm).

Minor hosts: *Acrocomia aculeate* (gru gru palm), *Attalea cohune* (cohune palm), *Mauritia flexuosa* (ita palm), *M. caribea* (cocorite), *Phoenix canariensis* (Canary Island palm), *Roystonea oleracea* (royal or cabbage palm), *Sabal palmetto* (sabal palmetto).

12.7.1 Vectors

The American palmetto weevil (*Rhynchophorus cruentatus*) and the palm weevil (*R. palmarum*) are the main vectors of RRN. The palm weevil does not transmit the RRN to other non-palm host species (Griffith *et al.*, 2005). Other palm-associated weevils, such as *Dynamis borassi* and *Metmasius hemipterus*, have also been noted as initiating red ring disease (Giblin-Davis, 2001). Esser and Meredith (1987) reported that out of 22.3% infected palms, 92% had been invaded by palm weevils and approximately 72% of those weevils had vectored the RRN. Similarly, over 90% of newly emerged adult weevils were infested with RRN juveniles, and over 47% of the weevils contained more than 1000 RRNs in Trinidad and Tobago. Around 32% of weevils were found carrying the nematode. These weevils were around 5 mm smaller than the healthy weevil not containing the nematode (Gerber and Giblin-Davis, 1990).

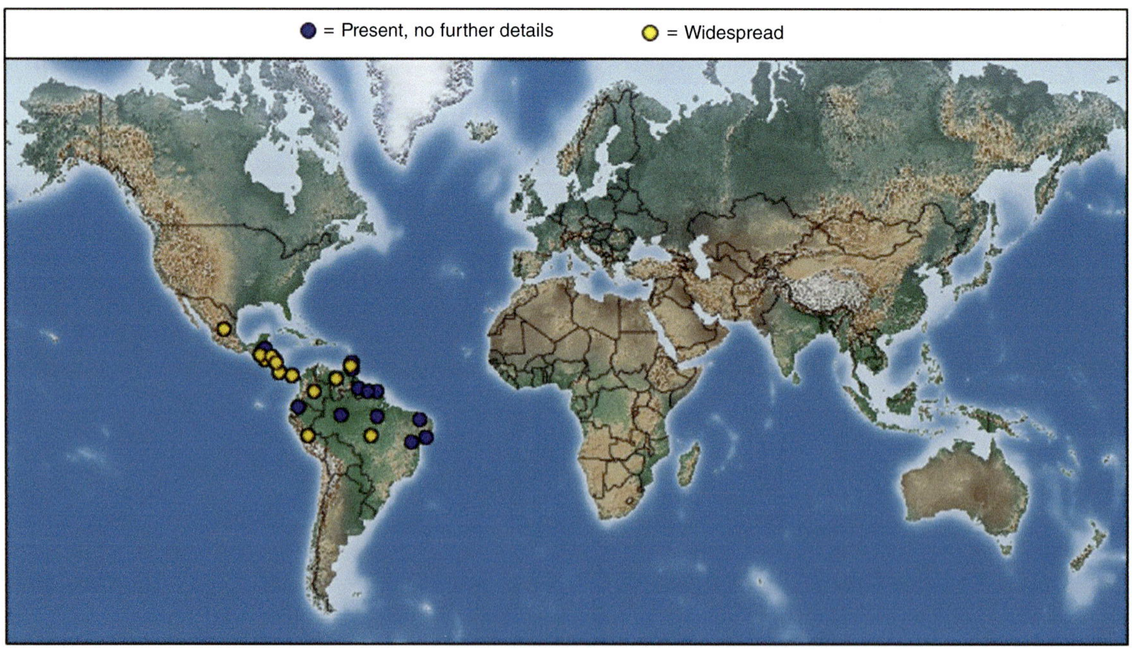

Fig. 2.11. Distribution map of *Bursaphelenchus cocophilus*. [Adapted from CABI-ISC, 2020]

12.8 Symptoms

Symptoms of RRN disease differ greatly with cultivar, palm species, age of palm and environmental factors (Chitamber, 2016). Young palms succumb more easily to RRN than older ones. The RRN mostly attacks trees that are 2.5–10 years old, with high incidence in those that are 5–7 years old (Griffith *et al.* 2005). The trees, once affected, never fully recover from RRN infestation. Both internal and external symptoms may appear on infected palms.

12.8.1 External symptoms

The disease is not recognizable externally in its early stage. The first noticeable external symptoms in infested coconut palms are chlorosis at the tips of the oldest leaves, spreading towards their bases and progressing to younger leaves (Fig. 2.12A). Later the plants become deformed and dwarfed, with yellow-bronzed leaves that turn deep reddish-brown. The chlorotic leaves ultimately die and will often break close to the petiole or remain drooping from the stem (Fig. 2.12B). Nuts are shed prematurely, concurrent with the leaf symptoms. In 4–6 weeks the crown topples and the trunk remains as a standing pole in the field, mainly due to indirect damage (Hunt, 1993).

12.8.2 Internal symptoms

The most characteristic symptom is internal red lesions that appear as two separate bands in longitudinal section and as a ring in the cross-section (Fig. 2.12C,E). A red ring, 2–4 cm wide and at a distance of 3–5 cm from the periphery, is visible in the cross-section of the susceptible palm within 2–3 weeks after infestation. In longitudinal section, the infested tissue may appear U-shaped with two united bands in the region of pole. The lesion colour may vary from red to light pink or creamy to dark brown (Chitamber, 2016).

12.9 Biology and life cycle

The biology and life cycle of *B. cocophilus* are closely related to the associated weevil vectors. It has been recognized that the nematode does not build up a large population in the soil, but red ring disease can be mainly initiated by *B. cocophilus* entering the plant via the root system. On the infested palm, the weevil larvae feed by burrowing through coconut stems and, during feeding, the larvae become infected with the *B. cocophilus*. The larvae moult, become adults and emerge from infected trees carrying the nematode to transmit to a new tree. *B. cocophilus* enters the haemocoel of weevil larvae through the gut, body cavity and parts of ovipositors (Griffith *et al.*, 2005). The survival of the nematode depends upon the third-stage juveniles, which persist through metamorphosis in the insect, apparently without moulting as dauer juveniles, and they disperse to healthy coconut palms during oviposition by the adult weevil, frequently at the leaf internodes or axils. The number of nematode juveniles associated with the weevil varies considerably and as many as 13,000 *B. cocophilus* juveniles per insect have been recorded (Griffith *et al.*, 2005). Natural cannibalism in the insect larvae also affects the number of emerging weevils. The nematode aggregates near the genital capsule of the adult weevil and when the adult weevils emerge from their cocoons in the infested palm, they apparently disperse the nematode to healthy or already infested and dying palms and complete their life cycle (Griffith *et al.* 2005).

Fig. 2.12. Symptoms of red ring disease. Infested coconut showing chlorotic leaves (A); infested palm (B); sectors of concentric red ring on coconut log (C, E); An adult palm weevil, *Rhynchophorus palmarum* (D). [Courtesy A,C: H. Ferris, source Nemaplex; B,E: Robin Giblin-Davis, University of Florida; D: Jennifer C. Giron Duque, University of Puerto Rico, Bugwood.org]

Naturally *B. cocophilus* only invades parenchymatous tissue in the roots, stems and leaves, but artificially it also infests nuts. At the start, the nematodes intercellularly invade newly infested tissue but later they can feed both intracellularly and intercellularly. Sometimes, lysigenous cavities are formed in the discoloured tissue and enormous numbers of nematodes are found in it. Such tissue can comprise as many as 10,000 nematodes per 1 g of tissue. *B. cocophilus* has never been found in the xylem vessels (Griffith *et al.* 2005). Artificially inoculated nematode into the mesocarp of coconuts completed the life cycle in 9–12 days (Blair, 1970). *B. cocophilus* can persist in coconut tissue for around 3 months, but it is very susceptible to desiccation (Griffith *et al.* 2005).

12.10 Similarities to other species

Morphologically and molecularly, this RRN is similar to other species of *Bursaphelenchus*. However, no other nematode pest causes the characteristic red ring disease as incited by *B. cocophilus* on palms.

12.11 Economic importance

The RRN, *B. cocophilus* seriously damages oil palms and inflicts major crop losses in coconut plantations, although the distribution area is limited. The percentage loss can vary from a few per cent to complete destruction of young coconuts palms. The RRN kills palm trees and there is no record of recovery of an infested palm. Most Latin American and Caribbean countries show crop losses ranging from less than 1% to more than 20% in 2–10-year-old palms. A 35% mortality of young coconut palms has been reported in Trinidad and 80% loss in a single plantation in Tobago (Esser, 1969). Within 10 years, 35% of oil palms died in Venezuela (Oostenbrink, 1963). In the Neotropics, the RRN causes up to 10–15% annual losses to coconut palm and African oil palms. In Grenada, 22.3% of coconut palms were found infected with RRN (Esser and Meredith, 1987). At present, losses of 35–80% in coconut and oil palm production are reported in the tropics (Chitamber, 2016).

12.12 Movement and means of dispersal

B. cocophilus mainly spreads through the adult palm weevil, especially *Rhynchophorus palmarum*, and the sugarcane weevil *Metamasius* spp. (Agrios, 2005). Other potential weevil vectors are *Metamasius hemipterus* and *Rhynchophorus cruentatus* (Giblin-Davis, 2001).

12.12.1 Pathway vectors

Palm weevil, containers, wood packaging material, land vehicles, soil, sand, gravel, etc.

12.12.2 Plant parts liable to carry the pest in trade/transport

Stems (above ground), leaves (needles), fruits (including nuts), roots, shoots, trunks, branches, bark, wood logs, wood chips, seedlings, micro-propagated plants and growing medium accompanying plants can all carry the eggs, juveniles and adults of the nematode.

12.13 Pest risk analysis and looming threat from introduction

B. cocophilus is a highly invasive species and has the potential to colonize and damage cultivated or natural palms in areas outside its native range (Chitamber, 2016). It poses potential phytosanitary risks on coconut and other palms in tropical countries, particularly where the palm weevil is known to be present. Introduction of this RRN would destroy landscape/ornamental palms, date fruit, palm oil industries and tourism as well as international and domestic trading of palm. Hence, quarantines against the nematode would be implemented in the countries where it is absent.

12.14 Invasiveness rating

Considering the consequences of introduction and pest risk analysis, the following invasiveness rating has been assigned to *B. cocophilus*.

Climate/host interaction	–	High (3)
Hosts range	–	Medium (2)
Dispersal potential	–	High (3)
Economic impact	–	High (3)
Environmental impact	–	High (3)
Post-entry distribution	–	Low (–1)
Final rating (score)	–	High (13)

Based on the above invasiveness rating, *B. cocophilus* receives a score of 13 out of 15 and is rated as HIGHLY INVASIVE NEMATODE.

12.15 Region-wise status of invasiveness

APPPC: Bangladesh, Cambodia, China, India, Indonesia, Iraq, Israel, Japan, Jordan, Kazakhstan, Kyrgyzstan, Lebanon, Malaysia, Myanmar, Philippines, South Korea, Sri Lanka, Syria, Tajikistan, Thailand, Vietnam, Uzbekistan
CAHFSA: Barbados, Cuba, Dominican Republic, Haiti, Jamaica, Saint Kitts & Nevis
CAN: Bolivia
COSAVE: Argentina, Chile, Paraguay, Uruguay
EPPO: Albania, Austria, Belarus, Belgium, Bulgaria, Croatia, Cyprus, Czech Republic, Denmark, Estonia, Finland, France, Georgia, Germany, Greece, Guernsey, Hungary, Ireland, Israel, Italy, Jersey, Jordan, Kazakhstan, Kyrgyzstan, Latvia, Lithuania, Luxembourg, Macedonia, Malta, Moldova, Morocco,

Netherlands, Norway, Poland, Portugal, Romania, Russia, Serbia, Slovakia, Slovenia, Spain, Sweden, Switzerland, Tunisia, Turkey, Ukraine, United Kingdom, Uzbekistan

IAPSC: Algeria, Angola, Benin, Botswana, Burkina Faso, Burundi, Cameroon, Cape Verde, Central African Republic, Chad, Comoros, Congo, Côte d'Ivoire, Egypt, Eritrea, Ethiopia, Gabon, Gambia, Ghana, Guinea, Guinea-Bissau, Kenya, Liberia, Madagascar, Malawi, Mali, Mauritania, Mauritius, Mozambique, Namibia, Niger, Réunion, Rwanda, Senegal, Sierra Leone, Somalia, South Africa, Sudan, Swaziland, Togo, Uganda, Zambia, Zimbabwe

NAPPO: Canada, USA

NEPPO: Egypt, Jordan, Libya, Malta, Sudan, Syria, Tunisia

PPPO: Australia, Fiji, New Zealand. Papua New Guinea, Tonga, Vanuatu

12.16 Management measures

12.16.1 Prevention

The best method for managing RRN is prevention through the early removal and destruction of infested palms. Reducing the weevil vector population and the sources for nematode inoculum is the main course of this strategy. As soon as RRN-induces small and chlorotic leaf symptoms become visible in infested palms, they should be demolished.

12.16.2 Chemical treatments

The leaf axils of diseased palms should be sprayed with 0.1% Lannate® (methomyl) insecticide suspension when early symptoms appear. Then the palm should be poisoned by adding one tablespoon of weedicide '100' in three holes (2 cm diameter and 10 cm deep) bored around the trunk of the tree at a level of 15 cm above the soil. At a later stage, the dead palms should be cut down and the piece and stumps sprayed with at least 4.5 l of 0.1% Lannate® suspension. Guard baskets made of 2 cm mesh wire are used to protect against frequent outbreaks of the disease. These baskets are filled with fresh infected tissues and sprayed with 0.1% Lannate® suspension. The palm weevils are attracted to the tissue in the basket. After 2 weeks the tissue in the basket is burnt. One guard basket is used per 0.4 ha of palms (Griffith and Koshy, 1990).

12.16.3 Trapping

Mass-trapping with baited traps using sugarcane or synthetic pheromone (Rhyncholure®; racemic 6-methyl-2-hepten-4-ol) reduces the numbers of palm weevil, affects their distribution patterns and ultimately helps to reduce the disease incidence. At the beginning of mass trapping, most *R. palmarum* can be captured in 'border' traps at the test site. A combination of perimeter and 'internal' traps is most efficient for mass trapping (Moura *et al*., 2000).

12.16.4 Biological control

The weevil vector is found to be parasitized by several species of Rhabditidae or Heterorhabditidae throughout Latin America. Since the vector insects can be parasitized with the above nematodes, selective pressure can be introduced against the vectors. Such measures are being employed in Trinidad with a species of Rhabditidae (Griffith and Koshy, 1990).

12.16.5 Phytosanitary measures

It is recommended that the introduction of *B. cocophilus* and its vectors, covering plants and wood of all palm for planting, should be prohibited from countries where the nematode occurs. In the case of the interception of *B. cocophilus* and its vector, after confirmation the diseased plants or planting material should be rejected and burned or returned to the exporter. In the case of wood packing material, ISPM 15 should apply. Sulfuryl fluoride fumigation treatment is also recommended for nematodes and the insect vector in debarked wood (ISPM 28, PT-23, 2017). See Chapter 16, section 115, for further details.

13 *Bursaphelenchus xylophilus*

13.1 Common name

Pine wilt nematode

13.2 Scientific name

Bursaphelenchus xylophilus (Steiner and Buhrer) Nickle, 1970

13.3 Synonyms

Aphelenchoides xylophilus Steiner and Buhrer, 1934
Bursaphelenchus lignicolus Mamiya and Kiyohara, 1972

13.4 EPPO code

BURSXY

13.5 Diagnosis

For an accurate diagnosis of *Bursaphelenchus* species, a combination of morphological and molecular analysis is recommended (Mota and Vieira, 2004). The morphological data assessed with phylogenetic studies based on a molecular approach are discussed below;

13.5.1 Morphological diagnosis

The general morphological characters of *B. xylophilus* are similar to those of other *Bursaphelenchus* species. Important morphological characters based on Nickle (1970) and Hunt (1993)

are summarized below. Measurements from different authors (Nickle *et al.*, 1981; Mamiya and Kiyohara, 1972; Mota *et al.*, 1999; Peñas *et al.*, 2008) and localities (USA, Japan, Portugal) are presented in Table 15.

FEMALE: Long and slender in overall body shape (Fig. 2.13A). Lip region high with six lips and offset by a constriction (Fig. 2.13F). Stylet well developed, usually with small and thick basal portion (Fig. 2.13F). Metacorpus well developed (Fig. 2.13F). Vulva conspicuous with a vulval flap overlapping anterior lip (Fig. 2.13G). Vulva posterior and typically at 70–80% of body length. Long post-uterine sac usually at three to six body widths in length. In most populations, tail sub-cylindrical with a broadly rounded tip (Fig. 2.13H). Sometimes tail tip may have nipple-like terminal extension or small mucro (Fig. 2.13I).

MALE: Long and slender, smaller than female (Fig. 2.13B). Strongly ventrally curved tail with sharply pointed terminal (Fig. 2.13C). Bursa present, small terminal at the tail tip and can be seen when nematode is in dorso-ventral position (Fig 2.13D). Spicules strongly arcuate, robust with large rounded apex and rostrum prominent and sharply pointed. Spicule end having disc-like projection (cucullus). Gubernaculum absent (Fig. 2.13C). Caudal papillae present as adanal pair a little anterior to the anus, two post-anal pairs a little anterior to the bursa origin and one single pre-anal median papillae.

13.5.2 Biochemical diagnosis

The protein analysis using SDS-PAGE has been applied for differentiating *Bursaphelenchus* species (Hotchkin and Giblin, 1984). Enzyme electrophoresis based on enzyme staining patterns analysed after isoelectric focusing (IEF) has also been used to differentiate *B. mucronatus* and *B. xylophilus* isolates (Hu and Yang, 1995). Immunological approaches have been designed using polyclonal antibodies that distinguish specific antigens of a few *B. xylophilus* isolates on western blots (Lawler and Harmey, 1993). Monoclonal antibodies have also been developed to identify different *B. xylophilus* isolates (Jiang *et al.*, 2006). Biochemical methods have assisted in the identification of *Bursaphelenchus* species, but their value is inadequate until supplied with differential gene expression by molecular diagnosis during the various life stages of a nematode (Li, 2008).

13.5.3 Molecular diagnosis

Molecular methods have been extensively used in taxonomic studies, analysis of genetic variation and analysis of phylogenetic relationships among *Bursaphelenchus* spp. These methods can be performed by conventional DNA-based approaches or by real-time PCR methods for the identification of *B. xylophilus*.

DNA-BASED APPROACHES DNA-based methods using genomic DNA, ribosomal DNA and mitochondrial DNA have been studied by different approaches for inter- and intraspecific identification of *Bursaphelenchus* spp. (Li, 2008). Total genomic DNA can be analysed for species differentiation by using DNA hybridization methods. Bolla *et al.* (1988) distinguished *B. mucronatus* and *B. xylophilus* isolates using restriction enzymes (RFLP) followed by hybridization of DNA probe. Other probes have been used in other similar analyses, including the *unc-22* gene of *Caenorhabditis elegans* (Abad *et al.*, 1991), ribosomal DNA (Webster *et al.*, 1990), satellite DNA (Tarés *et al.*, 1993) and homologous DNA probes (Harmey and Harmey, 1994) to distinguish the isolates of *Bursaphelenchus* spp. within the *xylophilus* group. Besides their complexity and expensiveness, these methods are of restricted effectiveness due to the need for large amounts of pure DNA.

Table 15. Morphometric measurements of *Bursaphelenchus xylophilus* by different authors and localities (adapted from ISPM 27, DP 10).

Morphometric characters		Measurements (mean and range in parenthesis) of some isolates: Nickle *et al.* (1981) USA (*n* = 5)	Mamiya and Kiyohara (1972) Japan (*n* = 40)	Mota *et al.* (1999) Portugal (*n* = 12)	Peñas *et al.* (2008) Portugal (*n* = 20)
Females	Length (mm)	0.52 (0.45–0.61)	0.81 (0.71–1.01)	1.05 (0.89–1.29)	0.58 (0.51–0.66)
	Stylet (µm)	12.8 (12.6–13.0)	15.9 (14–18)	12.3 (11–15)	11.2 (10.0–12.5)
	a	42.6 (37–48)	40.0 (33–46)	50.0 (41–58)	41.9 (32.8–50.6)
	b	9.6 (8.3–10.5)	10.3 (9.4–12.8)	13.8 (12.7–16.4)	10.1 (9.1–11.2)
	c	27.2 (23–31)	26.0 (23–32)	26.6 (22–32)	25.4 (20.2–29.0)
	V%	74.7 (73–78)	72.7 (67–78)	73.3 (70–76)	71.5 (70.1–72.9)
Males	Length (mm)	0.56 (0.52–0.60)	0.73 (0.59–0.82)	1.03 (0.80–1.30)	0.57 (0.45–0.69)
	Stylet (µm)	13.3 (12.6–13.8)	14.9 (14–17)	12.6 (11–16)	11.0 (10–14)
	a	40.8 (35–45)	42.3 (36–47)	49.4 (44–56)	46.0 (40.2–58.5)
	b	9.4 (8.4–10.5)	9.4 (7.6–11.3)	13.3 (11.1–14.9)	9.6 (8.2–10.7)
	c	24.4 (21–29)	26.4 (21–31)	28.0 (24–32)	21.6 (19.1–24.6)
	Spicules (µm)	21.2 (18.8–23.0)	27.0 (25–30)	24 (22–25)	19.3 (16.5–24.0)

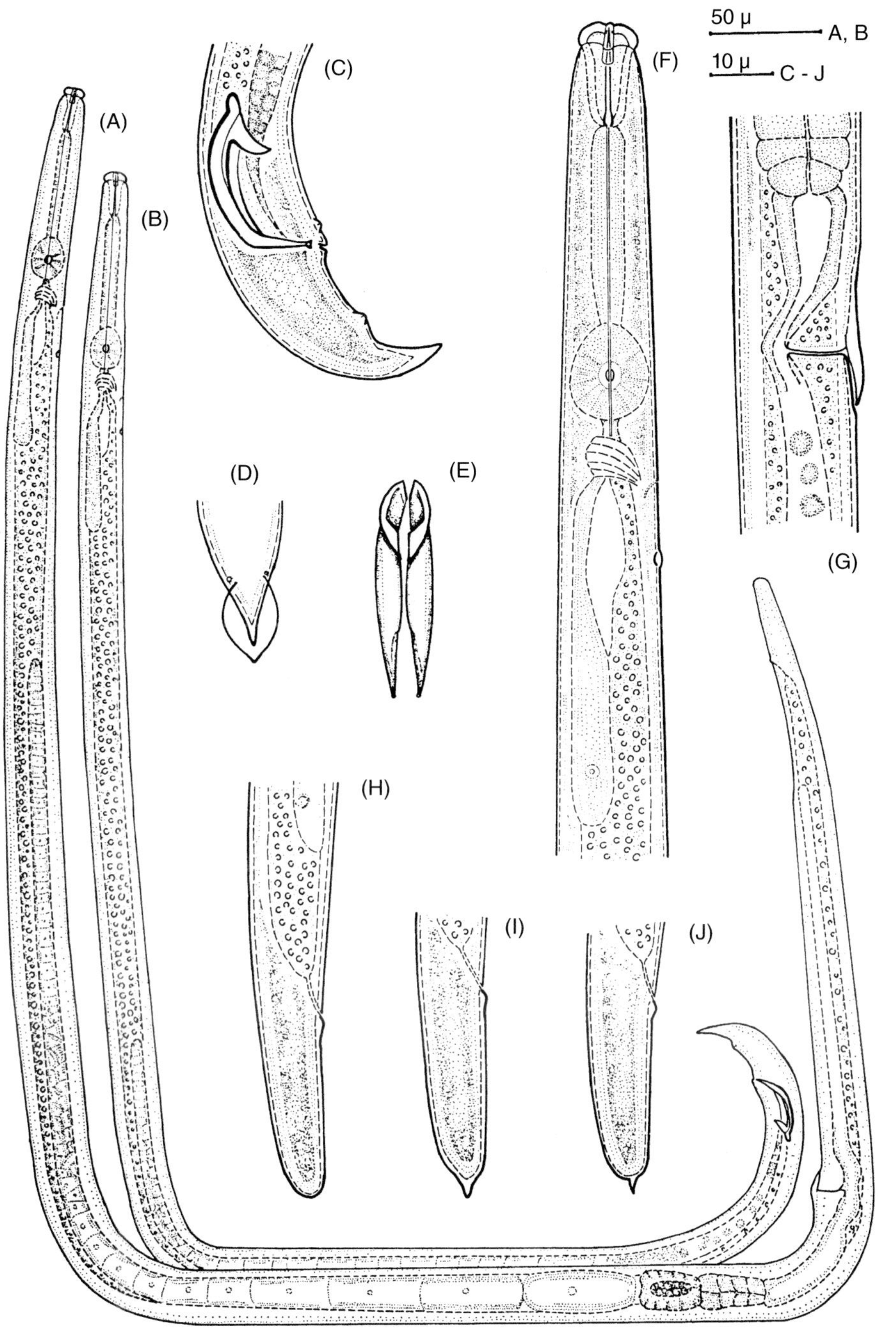

Fig. 2.13. Line drawing of *Bursaphelenchus xylophilus*. Female (A); male (B); male tail (C); ventral view of male tail, tip with bursa (D); ventral view of spicules (E); female, anterior portion (F); female vulva (G); female tail (H–J). [Mamiya and Kiyohara, 1972; courtesy Nematologica]

PCR-BASED METHODS Amplification and analysis of specific regions of DNA using PCR have proved to be an efficient method for the identification of inter- and intraspecific differences in the genus *Bursaphelenchus*. Zheng *et al.* (1998) used an arbitrary primer for random amplification of fragments of any part the genome and could generate polymorphic DNA patterns through RAPD-PCR. Other studies have used RAPD-PCR for analysis of intraspecific variation in *Bursaphelenchus* isolates from different countries such as China (Zhang *et al.*, 2006), Japan (Kusano *et al.*, 1999), Portugal (Vieira *et al.*, 2007), isolates of various geographical locations (Wang *et al.*, 2001; Zhang *et al.*, 2002) and native and non-indigenous isolates (Metge and Burgermeister, 2006).

Recent studies are more concentrated on the analysis of specific genomic regions, like ribosomal DNA (rDNA), mitochondrial DNA (mtDNA) and satellite DNA (sDNA) for phylogenetic analysis and diagnostic purposes. Use of external and internal transcribed (ITS) regions has also received significant attention for the identification of *Bursaphelenchus* species. Different workers used restriction fragment length polymorphisms of ITS regions (ITS-RFLP) for the diagnosis of *Bursaphelenchus* species (Liao *et al.*, 2001; Zheng *et al.*, 2003). Burgermeister *et al.* (2009) suggested the ITS-RFLP patterns created by digestion of ITS regions with five restriction endonuclease enzymes (*Rsa*I, *Hae*III, *Msp*I, *Hinf*I and *Alu*I) as a tool for identification of *Bursaphelenchus* species (Table 16) and also used to support the description of new species (Giblin-Davis *et al.*, 2005; Peñas *et al.*, 2006).

Precise detection of *B. xylophilus* has been attained by developing specific primer sets based on ITS sequences followed by conventional PCR (Leal *et al.*, 2005), or PCR-SSCP (Zhang *et al.*, 2001), nested PCR (Takeuchi *et al.*, 2005), duplex PCR (Zhao *et al.*, 2004) and real-time PCR assays (Leal *et al.*, 2007). A loop-mediated isothermal amplification (LAMP) approach was developed for direct detection and identification of *B. xylophilus* from wood by Kikuchi *et al.* (2009). A review of the literature revealed several PCR-based techniques using species-specific primers. Some useful PCR-based methods with *B. xylophilus* specific primers are:

PCR targeting ITS rDNA (Matsunaga and Togashi, 2004)

X-F: 5′-ACG ATG ATG CGA TTG GTG AC-3′
X-R: 5′-TAT TGG TCG CGG AAC AAA CC-3′

PCR targeting satellite DNA (Castagnone *et al.*, 2005)

J10-1: 5′-GGT GTC TAG TAT AAT ATC AGA G-3′
J10-2Rc: 5′-GTG AAT TAG TGA CGA CGG AGT G-3′

Real-time PCR targeting satellite DNA sequences (François *et al.*, 2007)

BsatF: 5′-TGA CGG AGT GAA TTG ACA AGA CA-3′
BSatRV: 5′-AAG CTG AAA CTT GCC ATG CTA AA-3′

Real-time PCR test targeting an hsp70 gene sequence (Leal *et al.*, 2007)

BxLNAF: 5′-TAA GAT GTc TTT tAc AGA TGc CAA G-3′
BxLNAR: 5′-GCc TGG ACG AcC TTG AAT-3′

Two-step real time-PCR targeting an hsp70 DNA sequence (Leal *et al.*, 2013):

Hsp23F1: 5′-ACC CAA GTT TGA GTT GTA TTG TTT-3′
Hsp19R2: 5′-ACG GTA ACA ACG GCA TCC T-3′
BxActF3: 5′-TCG TCA CCA ACT GGG ATG ATA-3′
BxActR3: 5′-CAC CAG TGG TAC GAC CG-3′

Loop-mediated isothermal amplification (Kikuchi *et al.*, 2009)

ITS (ID19) F3: 5′-GCA GAA ACG CCG ACT TGT-3′
ITS (ID19) B3: 5′-TCA TCC GAA CGT CCC TGA C-3′
ITS (ID19) FIP: 5′-CGC GGA ACA AAC CGC GTA AAA C-CG TTG TGA CAG TCG TCT CG-3′
ITS (ID19) BIP: 5′-AGA GGG CTT CGT GCT CGA TTG GCC GTT GAA ACA ACA TCACC-3′
ITS (ID19) LF: 5′-AGA TGG TGC CTA ACA TTG CG-3′

Burgermeister *et al.* (2009) developed a PCR-based ITS-RFLP method for distinguishing *B. xylophilus* from 43 other *Bursaphelenchus* species. Almost all new descriptions of *Bursaphelenchus* species published after 2009 contain the

Table 16. Restriction fragment length polymorphism (RFLP) patterns of *Bursaphelenchus* species (adapted from Burgermeister *et al.*, 2009).

Species	PCR product (base pairs)	Restriction fragments (base pairs) produced by restriction enzyme				
		*Rsa*I	*Hae*III	*Msp*I	*Hinf*I	*Alu*I
B. mucronatus East Asian type = *B. mucronatus mucronatus*	920	486	621	355	408	674
		412	299	302	232	246
		12		263	121	
					86	
					49	
					24	
B. mucronatus European type = *B. mucronatus kolymensis*	925	413	625	356	412	678
		263	195	303	232	247
		227	105	266	121	
		22			87	
					49	
					24	
B. xylophilus	925	483	728	562	263	433
		420	197	363	232	256
		22			142	142
					139	96
					125	
					24	

ITS-PCR RFLP patterns based on this method. An example of species distinction by ITS-RFLP restriction fragment patterns for *B. mucronatus mucronatus*, *B. mucronatus kolymensis* isolates and *B. xylophilus* is provided in Table 16.

In this technique, one to 30 nematodes were processed for DNA extraction (see Chapter 15, section 110.1) and ITS-PCR RFLP analysis is carried out by using PCR followed by RFLP on the PCR product. The rDNA containing the ITS1 and ITS2 was amplified by PCR using the following primer pair:

ITS1-forward (F): 5′-CGT AAC AAG GTA GCT GTA G-3′ (Ferris *et al.*, 1993)

ITS2-reverse (R): 5′-TTT CAC TCG CCG TTA CTA AGG-3′ (Vrain, 1993)

Each PCR is performed according to the following described conditions.

Reagent	–	Final concentration
PCR mixture	–	50 µl
ITS1-forward primer	–	0.6 µM
ITS2-reverse primer	–	0.6 µM
Taq DNA polymerase	–	2 U
Tris-HCl (pH 8.8)	–	10 mM
KCl,	–	50 mM
$MgCl_2$	–	2 mM
dNTPs	–	0.2 mM
DNA (template)	–	2 ng
Cycling parameters		
Initial denaturation	–	94°C for 2.5 min
Number of cycles	–	40
Denaturation	–	94°C for 1 min
Annealing	–	55°C for 1 min
Elongation	–	72°C for 5 min

The amplified DNAs were digested with 5 U restriction endonucleases *Alu*I, *Hae*III, *Hinf*I, *Msp*I and *Rsa*I. Then PCR product (5 µl) was analysed by gel electrophoresis and the size of the amplicon is 920–925 bp. Finally, *B. xylophilus* is identified based on the species-specific DNA restriction fragment patterns (Fig. 2.14). Numbers and sizes of DNA restriction fragments for at least the 57 *Bursaphelenchus* species have been described by Gu (2014). Two species, *B. hunanensis* and *B. lini*, are proposed to be regrouped and therefore no longer belong to the genus *Bursaphelenchus*.

13.6 Geographical distribution

13.6.1 Nematode

B. xylophilus is indigenous to North America, predominantly the USA (Table 17), and has possibly been introduced elsewhere by infected timber/wood products from there. It is supposed that the nematode originated in North America and was transported from there to the Kyushu island of Southern Japan by infested timber around the beginning of the 20th century (Mamiya, 1983). Most American native conifers are resistant to the nematode, while Japanese species are susceptible, which tends to support this view. *B. xylophilus* has spread from Japan to other Asian countries (Li *et al.*, 1983), but not to neighbouring Russia (Kulinich *et al.*, 1994). In Europe, *B. xylophilus* has been detected in imported timber in Finland, France, Norway and Sweden (Braasch, 2001). Despite the strict phytosanitary restrictions that have been endorsed by the EU, *B. xylophilus* was recorded for the first time in Portugal in 1999 on *Pinus pinaster*, which was killed by the nematode within a few months after infestation (Fonseca *et al.*, 2012). Later, in 2008, *B. xylophilus* was found in Spain (Abelleira *et al.*, 2011). The worldwide distribution of *B. xylophilus* (Fig. 2.15) and their country of origin are presented in Table 17.

13.6.2 Vector

B. xylophilus is mainly transmitted by pine sawyer beetle (*Monochamus* spp.) to the host trees. *Monochamus* spp. are recorded throughout the northern hemisphere, and many other species have been recognized as associates of *Bursaphelenchus*. It is presumed that the majority of the *Monochamus* species would have been capable of transmission of the nematode to a

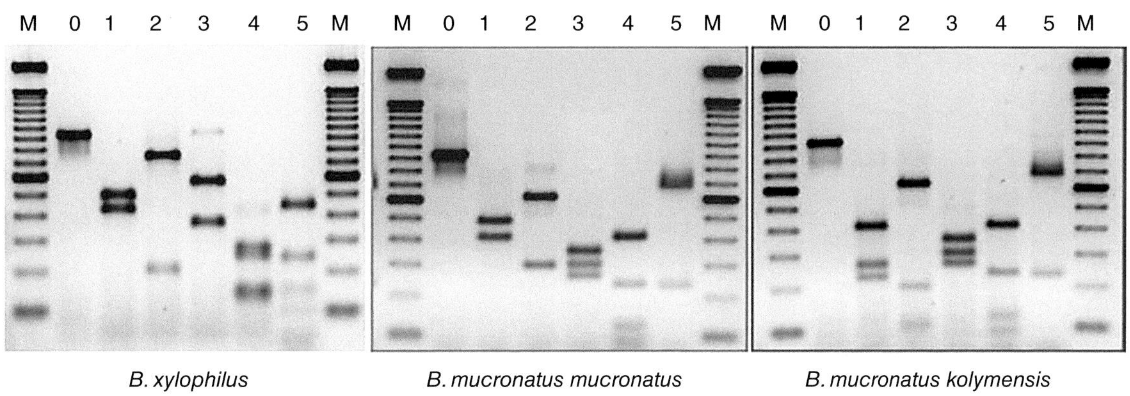

Fig. 2.14. Internal transcribed spacer (ITS)-restriction fragment length polymorphism (RFLP) patterns of *Bursaphelenchus xylophilus* (left), *B. mucronatus* (middle) and *B. mucronatus kolymensis* (right). Restriction fragments were obtained by digestion of the amplified ribosomal (r)DNA fragment (0) with *Rsa*I (1), *Hae*III (2), *Msp*I (3), *Hinf*I (4) and *Alu*I (5). M, DNA marker (100 base pair ladder). [Burgermeister *et al.*, 2009; courtesy Nematology].

lesser or greater extent. However, the main vector species known to transmit *B. xylophilus* and likely to carry the nematode were introduced from North America and Eastern Asia (Li *et al.*, 1983). Of the species reported as a vector of *B. xylophilus*, *Monochamus alternatus* (Fig. 2.16B,C) occurs in China, Japan, Laos, South Korea and Taiwan, and *M. nitens* and *M. saltuarius* in Northern China, Japan and Siberia. Other *Monochamus* species occur in Canada and the USA (Evans *et al.*, 1996).

Table 17. Worldwide distribution of *Bursaphelenchus xylophilus* and country of origin (adapted from CABI-ISC, 2020).

Country	Distribution	Origin
Asia		
China	Restricted distribution	Introduced
Hong Kong	Present, no further details	Introduced
Japan	Widespread	Introduced
South Korea	Restricted distribution	Introduced
Taiwan	Widespread	Introduced
Africa		
South Africa	Present, few occurrences	Introduced
Europe		
Portugal	Present, few occurrences	Introduced
Spain	Present, transient under eradication	Unknown
Sweden	Absent, intercepted only	Introduced
North America		
Canada	Widespread	Native
Mexico	Present, no further details	Native
USA	Widespread	Native

13.7 Hosts and vector species

Most species of *Pinus* can act as hosts, but the Asiatic pines *Pinus densiflora*, *P. thunbergii*, *P. luchuensis*, the European pines *P. nigra*, *P. sylvestris* and *P. pinaster*, and the American pines *P. lambertiana*, *P. radiata* and *P. echinata* are main hosts of pine wilt nematode (PWN). Other conifers can also act as hosts, but reports of their being damaged are occasional. In the USA, isolated cases of death of *Picea* and *Pseudotsuga* spp. due to PWN have been reported (Malek and Appleby, 1984). *Pinus koraiensis*, has been reported as a susceptible host to the nematode in Korea (Han *et al.*, 2008). *B. xylophilus* has also been detected on *P. pinaster*, *P. nigra* and *P. radiata* in Portugal and Spain (Inácio *et al.*, 2014; Zamora *et al.*, 2015). The expanded list of hosts of *B. xylophilus* is presented in Table 18.

Longhorn beetles of the genus *Monochamus* (Coleoptera: Cerambycidae) serve as a vector to transmit pinewood nematodes. These beetles are called sawyers because larvae, as they feed, make a loud chewing noise. The larvae are long white grubs, with no noticeable legs. The adult beetles emerge mostly in April and May; but they are active throughout the summer and even in warm spells during winter. Of these, *M. carolinensis* is the major vector in North America and *M. alternatus* is the major vector in Japan. Major and other less efficient vectors are listed in Table 18.

13.8 Symptoms

The first noticeable external symptom is chlorosis and wilting of the needles. A decrease of oleoresin production in response to wounding is also an indication of the presence of nematodes in the branch or entire tree (Khan, 2020). Transpiration from

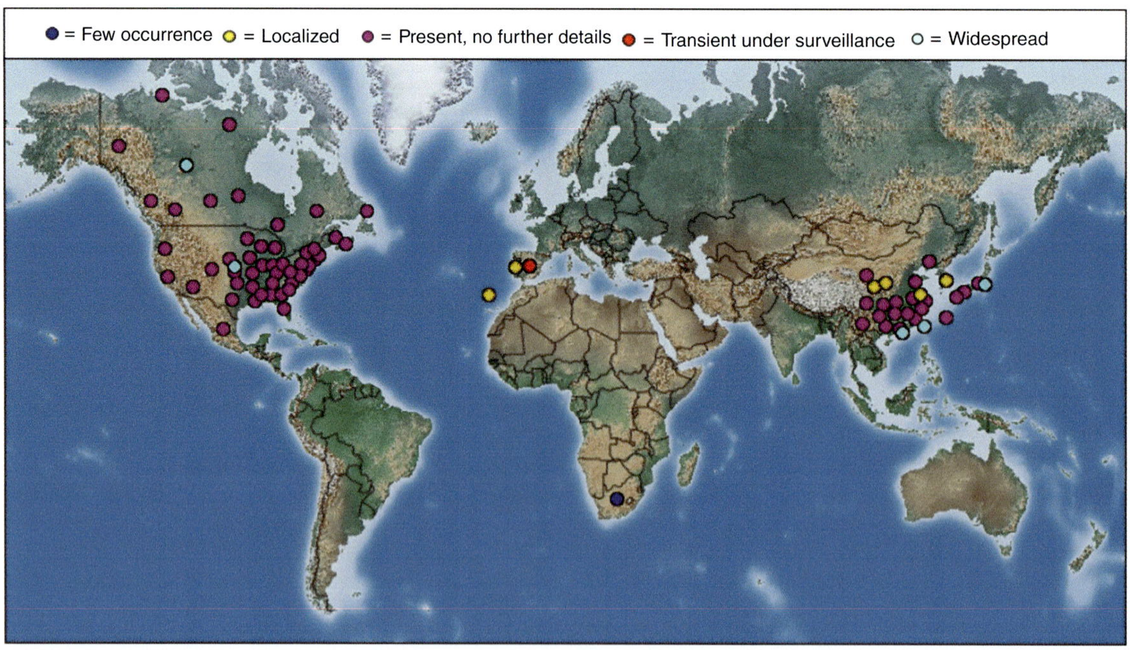

Fig. 2.15. Distribution map of *Bursaphelenchus xylophilus*. [Adapted from CABI-ISC, 2020]

Fig. 2.16. Symptoms of *Bursaphelenchus xylophilus*. Development of wilt symptoms on pine, *Pinus nigra* (A); beetle feeding on tree (B); an adult pine sawyer beetle, *Monochamus alternatus* (C); cobalt-blue discoloration on logs of mature pine tree suffering from pine wilt (D); damaged Japanese red pine, *Pinus densiflora* (E). [Courtesy A,D,E: USDA Forest Service - North Central Research Station, USDA Forest Service, Bugwood.org; B: Jijing Song and Juan Shi, Beijing Forestry University, Bugwood.org; C: Christopher Pierce, USDA APHIS PPQ, Bugwood.org].

Table 18. Important hosts of pine wilt nematode and its vector (adapted from Evans *et al.*, 1996).

Main hosts
Pinus banksiana (jackpine)
Pinus densiflora (Japanese umbrella pine)
Pinus lambertiana (big pine)
Pinus luchuensis (Luchu pine)
Pinus nigra (black pine)
Pinus radiata (radiata pine)
Pinus resinosa (red pine)
Pinus strobus (eastern white pine)
Pinus sylvestris (Scots pine)
Pinus taeda (loblolly pine)
Pinus thunbergii (Japanese black pine)
Other hosts
Abies amabilis (Pacific silver fir)
Abies balsamea (balsam fir)
Abies firma (momi fir)
Abies grandis (grand fir)
Abies sachalinensis (Sakhalin fir)
Cedrus atlantica (Atlas cedar)
Cedrus deodara (Himalayan cedar)
Chamaecyparisnootkatensis (Alaska cedar)
Larix decidua (common larch)
Larix kaempferi (Japanese larch)
Larix laricina (American larch)
Larix occidentalis (western larch)
Picea abies (common spruce)
Picea engelmannii (Engelmann spruce)
Picea glauca (white spruce)
Picea jezoensis (Yeddo spruce)
Picea mariana (black spruce)
Picea pungens (blue spruce)
Picea rubens (red spruce)
Picea sitchensis (Sitka spruce)
Pinus ayacahuite (Mexican white pine)
Pinus brutia (brutian pine)
Pinus bungeana (lace bark pine)
Pinus caribaea (Caribbean pine)
Pinus contorta (lodgepole pine)
Pinus halepensis (Aleppo pine)
Pinus hartwegii (Hartweg pine)
Pinus jeffreyi (Jeffrey pine)
Pinus koraiensis (fruit pine)
Pinus leiophylla(smooth-leaved pine)
Pinus massoniana (masson pine)
Pinus monticola (western white pine)
Pinus mugo (mountain pine)
Pinus palustris (longleaf pine)
Pinus pinaster (maritime pine)
Pinus pinea (stone pine)
Pinus ponderosa (Ponderosa pine)
Pinus pungens (Table Mountain pine)
Pinus strobiformis (southwestern white pine)
Pinus wallichiana (blue pine)
Pinus yunnanensis (Yunnan pine)
Pseudotsuga menziesii (Douglas-fir)
Major vectors
Monochamus alternatus
Monochamus carolinensis
Monochamus carolinensis
Less efficient vectors
Monochamus nitens
Monochamus saltuarius
Monochamus marmorator
Monochamus mutator (*M. clamator*)
Monochamus obtusus
Monochamus scutellatus
Monochamus titillator

the leaves reduces and later stops completely. As the disease progresses, the needles turn greyish green, then tan-coloured to brown, leading to the ultimate death of the pine tree (Mamiya, 1983). The wilting may first appear only on flag branches; later the whole tree may show the symptoms (Malek and Appleby, 1984) and within a few weeks to a few months it kills the tree (Fig. 2.16A,E). Susceptible pine species may die within 1–3 months after the first discernible symptoms, or longer for less susceptible tree species.

PWN symptoms may be easily confused with those of root rot caused by several bark beetles and the fungus *Fomes annosus*. To confirm the association of pinewood nematode as a causal agent, the nematode should be recovered from the dead or dying pine. It has been noticed that sawyer beetles generally infest trees that have been recently killed or trees that are under stress. Sawyer beetles mainly feed on the bark and phloem of twigs of susceptible trees (primary transmission) or when the female beetles oviposit the eggs in freshly cut timber or dying trees (secondary transmission). The larvae burrow into the wood and damage the value of the wood for lumber. When bark beetles bore into dying pines, blue-strain fungi infesting the beetles are able to enter the tree. These fungi rapidly colonize the wood of the dying tree, leaving behind a characteristic cobalt-blue discoloration (Fig. 2.16D). PWNs also feed on these fungi (Dwinell, 1997).

13.9 Biology and life cycle

This PWN is known to have two kinds of life cycles: phytophagous and mycophagous (Wingfield, 1983). In both life cycles, the nematodes are transmitted from one host to the next by the vector (*Monochamus* spp.). In the mycophagous type of life cycle, the fourth-stage juveniles (J_4) of *B. xylophilus* are transmitted by pine sawyer beetles from diseased trees to healthy trees during oviposition by the female vector (Fig. 2.17). While the vector feeds on a healthy tree, the dauers of PWNs (J_4) leave the beetle and enter the tree through the wounds created by the vector to lay its eggs. Inside the wood, the nematodes feed on the mycelial hyphae of fungi (frequently blue-strain fungi, *Ceratocystis* spp.) which are also transmitted to the wood by ovipositing beetles. The nematode larvae moult to adults immediately after entering the wood and start laying eggs.

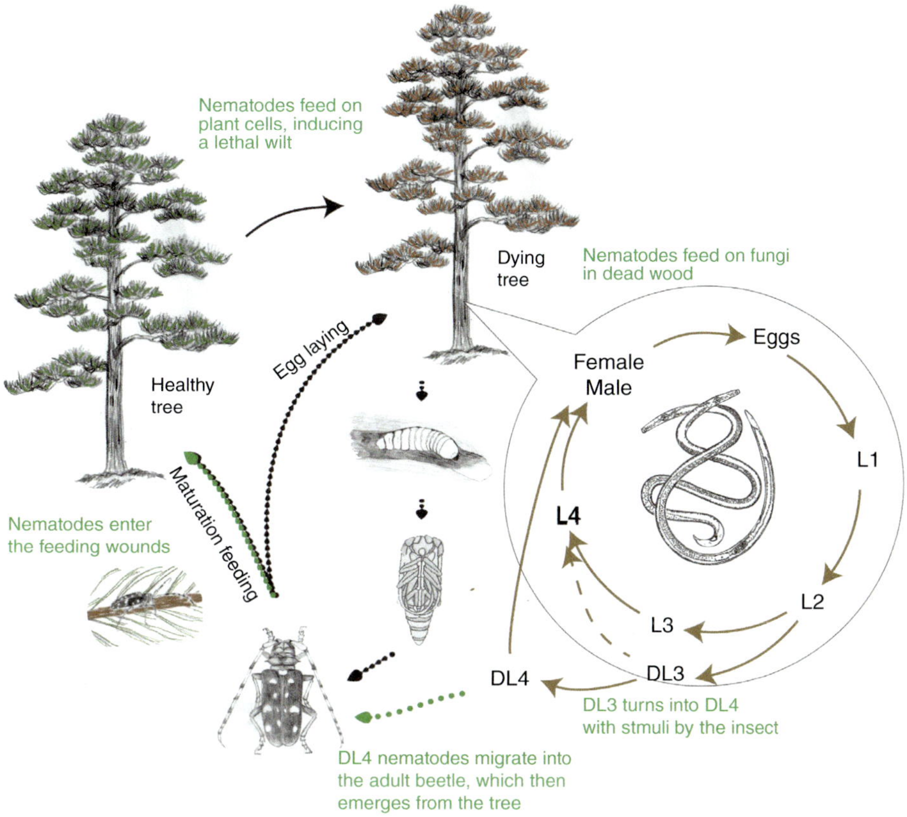

Fig. 2.17. Diagram summarizing the life cycle of the pine wilt nematode *Bursaphelenchus xylophilus*. L, Larval stage; DL, Dispersal larvae [Courtesy Kikuchi *et al.*, 2011; PLoS Pathogens]

Subsequently, rapid multiplication occurs and the males, females and fourth-stage juveniles develop in this propagative stage (Khan, 2020). The population ceases to multiply at a certain time after the initial invasion and the nematode population begins to decline. At this stage, a different type of third-stage juvenile is produced. This stage is called the pre-dispersal third-stage juvenile (Mamiya, 1983). Being capable of resisting adverse conditions, this stage is more correctly a survival stage. The population construction of the PWN changes in reaction to a decline in food availability, when the fungus has fully oppressed the wood. The 'dispersal' juveniles gather in the wood near the pupal chamber of the *Monochamus* vector, probably under the influence of substances emitting from the developing pupa. Near the time of beetle emergence, the nematodes moult into the special fourth-stage juveniles (dauer larvae). The fungal hyphae also develop nearby the pupation chambers and form long-necked perithecia protruding into the chamber. The nematodes accumulate at the tips of the perithecia and when the young adult beetle emerges, it encounters the perithecial necks, picking up the nematodes, which settle below the elytra and in the tracheae, in particular. The young adult beetle then flies from the infested wood, carrying nematodes (Mamiya, 2008). The mycophagous type of life cycle is considered to be the normal life cycle of *B. xylophilus* and is akin to that of most other *Bursaphelenchus* species that have phoretic relationships with the beetles (Wingfield, 1983). It is apparently the most common condition in North America.

However, in both North America and Asia, wherever the PWN comes into contact with susceptible or non-native *Pinus* species, the phytophagous type of life cycle prevails. In this life cycle, the nematodes are transmitted by immature adult beetles after emergence from their pupal chambers, during which they fly to feed on young *Pinus* shoots. The nematodes invade the shoots through the wounds created by the feeding vector. This form of transmission only occurs on susceptible species of *Pinus* probably because native *Pinus* species of the area are resistant. *B. xylophilus* occurs in the area to develop physical or biochemical barriers to prevent the direct invasion of healthy tissues. This may be the reason why, when the PWN is introduced into a non-native area, it encounters new host species of *Pinus*, some of which are extremely susceptible and the nematode follows the phytophagous life cycle (Fig. 2.17).

Once *B. xylophilus* infests the young *Pinus* shoots, the nematode multiplies rapidly in the resin canals, damaging their epithelial cells and clogging the tree's water transport system through resin canals. The tree shows the first symptoms of drying out within 3 weeks. Now the nematodes can freely move throughout the wilting tree. At this stage, intensified yellowing and wilting of the pine needles are seen, which prompts the adult insects on the trunks to mate. Wilted pines attract not only egg-laying pine sawyer beetles but also bark beetles. Bark beetles are not directly associated with the pine wilt disease cycle, but their activities ultimately help in nematode nutrition. The tree or branch dies 30–40 days after infection under hot and dry climatic conditions and may comprise millions of nematodes throughout the trunk, branches and roots (up to 1000 larvae/g wood). The rest of the phytophagous life cycle is parallel to that of the mycophagous type and the nematodes locate the pupae of the beetle vector just before their emergence (Mamiya, 2008).

Pine sawyer beetles are attracted to weakened trees, or recently cut logs, where they mate and lay eggs. The beetles oviposit the eggs only on trees or logs with bark attached. The beetle larvae hatch within a week and feed on the phloem. The larvae bore to form tunnels into the xylem (wood) and make an oval entrance hole with U-shaped galleries. Sawyer beetle larvae overwinter and then pupate within an enlarged portion of the hole's gallery. The PWN juveniles, present in the wood, invade the thoracic spiracles and tracheae of the beetle pupae in numbers as high as 289,000/larva (Mamiya, 2008). The adult beetle emerges from the tree through a round exit hole in the tree trunk (Khan, 2020).

13.10 Similarities to other species/groups

There are several groups among the *Bursaphelenchus* species. *B. xylophilus* belongs to the *xylophilus* group *sensu* (Braasch, 2001). Based on their morphological characters, currently at least 15 species or subspecies belong to the *xylophilus* group (Braasch and Schönfeld, 2015). Further, *B. xylophilus* can be distinguished into two populations or forms: round-tailed and mucronate-tailed (Gu *et al.*, 2011), both found in North America, and they are quite similar to *B. mucronatus kolymensis*. A detailed key to all *xylophilus* group species has been compiled by Braasch and Schönfeld (2015). Among the *xylophilus* group, the most widespread species are *B. mucronatus mucronatus* and *B. mucronatus kolymensis*.

13.11 Economic importance

B. xylophilus is a highly invasive nematode species. It can colonize and damage natural or managed pine forests in areas outside its native range and it presents a direct threat to the survival of the local endemic pines and other conifers (Kanetani *et al.*, 1999; Khan, 2020). The disease was first reported in 1913 in the Nagasaki region in Japan. The disease then spread northwards in Japan, causing very severe losses to conifers; by the end of the 1940s, over 1,000,000 m^3 of pinewood were being lost annually (Futai, 2013). A campaign was started for destroying the infected trees and it reduced this figure to below 500,000 m^3 per year. However, the loss curve turned sharply upwards from the 1970s because of industrialization, which has reduced the availability of manpower for the forests, and wood has been replaced by oil as a fuel. In consequence, infected trees again remain standing as reservoirs of the nematode and the loss exceeds 2,000,000 m^3 per year (CABI-CPC, 2010). Monetary losses are substantial with losses of 2,000,000 m^3 of timber under certain environmental conditions (Nose and Shiraishi, 2007), but the nematode can also cause devastation on an ecosystem scale, as all infected trees in an area can die when conditions are suitable for the disease (Jones *et al.*, 2013).

Losses from pine wilt disease can occur in some non-native pine plantations and landscape trees in the USA. The greatest economic effect of the nematode is in North America and has resulted from restrictions on untreated wood chips and timber by importing countries. Before the bans on softwood (from 1972 to 1989), Canada and the USA shipped almost 5 million tonnes of conifer chips to Scandinavia. The exact losses to the American forest industry are hard to confirm because of a variety of factors affecting international trade; however, one probable estimate accounts for a loss of US$100 million annually in green lumber exports to Europe (Zhao *et al.*, 2008).

13.12 Movement and means of dispersal

In nature, *B. xylophilus* is transferred from one tree to another tree through the activity of the adult *Monochamus* spp. (Khan, 2020). The beetle can transmit the nematode either to the shoots of living trees during feeding or to the trunks or larger branches, including wood waste or debris, or weakened or recently dead trees during oviposition by the females. Within the wood tissues, the nematodes can actively move and can leave one piece of wood to move into an adjacent piece. However, without their vectors, they are unable to move from one host tree to another. The adult vector beetles can fly vigorously and a peak of flight movement is reached about 5 days after emergence (Mota and Vieira, 2004).

Infested wood is the most feasible means of international transport of the nematode, and *B. xylophilus* has been intercepted on several occasions on wood chips, sawn wood and roundwood imported from Canada and the USA. The most likely pathway of the introduction of *B. xylophilus* is in raw and unprocessed wood when imported together with vector insects which may carry the nematodes to coniferous trees (Zhao *et al.*, 2008; Khan, 2020).

13.12.1 Pathway vectors

Containers, wood logs, wood packaging material, land vehicles, soil, sand, gravel, etc.

13.12.2 Plant parts liable to carry the pest in trade/transport

Stems (above ground), leaves (needles), shoots, trunks, branches, bark, pinewood logs, woods (chips, sawn, roundwood), seedlings and micro-propagated plants can carry the eggs, juveniles and adults of the nematode.

13.12.3 Plant parts not known to carry the pest in trade/transport

Fruits, true seeds, roots.

13.13 Pest risk analysis and looming threat from introduction

B. xylophilus is a highly invasive species and has the potential to colonize and damage natural or managed conifer forests in areas outside its natural range (Jones *et al*., 2013). The nematode is native to North America (Canada, USA) and is thought to have been carried to Japan at the beginning of the 20th century on timber exports (Nickle *et al*., 1981). In Japan, the nematode has been triggering massive mortality of native pine trees, *Pinus densiflora*, *P. luchuensis* and *P. thunbergii*. It has also spread to China, Korea, Portugal, Taiwan and recently Spain. The PWN and its vectors, *Monochamus* spp., are treated as quarantine pests in many countries. When introduced into new geographical areas, it shares the same ecological niche as that of the closely related indigenous species *B. mucronatus*. Reports of a new closely related species, *B. macromucronatus*, from wooden packing boxes from India (Gu *et al*., 2008) alerted the quarantine agencies for emergency preparedness, particularly to protect the Himalayan pines. Competition between native and introduced species may affect the spread of this invasive species. Vincent *et al*. (2008) investigated the possible interaction between these two nematode species transmitted by the same insect vector, *M. galloprovincialis*.

The nematode is very easily carried by wood moving in trade, either as a commodity or as wood packaging material, being transported with other commodities. Infested wood is the most likely means of movement of *B. xylophilus* in international transport, and the species has been intercepted on several occasions in wood chips, sawn wood and roundwood imported from Canada and the USA. The introduction of the nematodes within wood imports does not, in itself, represent a looming threat to the forests. A pathway needs to be found for *B. xylophilus* to come into contact with a native insect vector and this can only be achieved if the nematode first invades wood that contains juveniles or pupae of a probable vector (CAB-ISC, 2020).

13.13.1 Environmental impact

The PWN is a native of North America, but does not cause damage to native pines, rather it attacks non-native pines that include Austrian pine, Scotch pine, Japanese red pine and black pine in different countries (Khan, 2020). In Japan, forests of *P. densiflora* and *P. thunbergii* strongly affected by *B. xylophilus* are inclined to be replaced by deciduous oak forest (*Quercus serrata*) and evergreen oak forest (*Q. glauca*) and the pine forest on upper slopes only persists because the damage is less severe (Toyohara and Fujihara, 1998). Species diversity was also affected and initially reduced, but gradually recovered by oak forestation (Katagiri *et al*., 1995). In the southern island of Kyushu, pine was replaced by more tropical evergreen trees like *Mallotus japonicus* and *Rhus succedanea* (Inoue, 1995). Similarly, in Taiwan, *P. luchuensis* was replaced by the evergreen *M. paniculatus* (Ma *et al*., 2002).

13.13.2 Biodiversity impact

B. xylophilus poses a direct threat to the survival of the very local endemic *P. armandii* var. *amamiana* in southern Japan (Kanetani *et al*., 1999). The nematode can also cause devastation on an ecosystem scale, as all infected trees in an area can die when conditions are suitable for the disease (Jones *et al*., 2013)

13.13.3 Social impact

In Japan, destruction of pine forests initially affected local employment, but this was overtaken by post-war economic expansion involving a much-reduced reliance of local populations on forest products. In fact, this decline in forest use has worsened the losses due to PWN, as damaged forests tended to be abandoned.

13.13.4 Dispersal potential, global trade and nematode interceptions

B. xylophilus is one of the most important invasive nematodes worldwide and can be dispersed from one country to another, even from one continent to another. Shipments from infested countries are of serious concern. The nematode has been intercepted in shipments of conifer chips, sawn wood, roundwood, wood chips and wood packaging, in unseasoned lumber and in wood packing cases (Braasch *et al*., 2001; Gu *et al*., 2006). The beetle vectors, *Monochamus* spp., have been found in wood pallets, crates and dunnage. The nematode has been intercepted in imported timber in different countries of Europe such as Finland, France, Norway and Sweden. Surveys conducted in Germany, the Netherlands, Poland and the UK did not reveal nematode infestation in local forests. The nematode has been often detected by Chinese inspectors in imported wooden packing boxes from Japan, Canada, the USA and other infested areas (Zhang *et al*., 2008). Similarly, the nematode has been detected in Chinese wooden boxes when inspected in other countries. *Bursaphelenchus* larvae were extracted from 12 samples taken from imported wood packaging originating from seven countries (Li *et al*., 2009). Because of the close relationship between the nematode and its vector, shipments from areas where the nematode occurs can be assumed to be at risk of also carrying the nematode and/or its vector. More than 29 new species of *Bursaphelenchus* were described in Ryss *et al*. (2005). Thirteen of these were described from dead wood or dying trees of *Pinus* spp., nine from insect organs and seven were recovered from wood packaging intercepted during quarantine measures. Phylogenetic and biogeographical investigations of these species of *B. xylophilus* are therefore important to understand the origin, source and spread of this species worldwide.

13.14 Invasiveness rating

Considering the consequences of introduction and pest risk analysis, the following invasiveness rating has been assigned to *B. xylophilus*.

Climate/host interaction	–	High (3)
Hosts range	–	High (3)
Dispersal potential	–	High (3)
Economic impact	–	High (3)
Environmental impact	–	High (3)
Post-entry distribution	–	Low (–1)
Final rating (score)	–	High (14)

Based on the above invasiveness rating, *B. xylophilus* receives a score of 14 out of 15 and is rated as HIGHLY INVASIVE NEMATODE.

13.15 Region-wise status of invasiveness

B. xylophilus is geographically concentrated in North America and East Asia. However, it has been introduced and intercepted in many countries of the world. *B. xylophilus* has been given a very high invasiveness rating in all the countries where the nematode is still absent. Region-wise status of invasiveness is as follows.

APPPC: Bangladesh, Cambodia, India, Indonesia, Iraq, Israel, Japan, Jordan, Kazakhstan, Kyrgyzstan, Lebanon, Malaysia, Myanmar, Philippines, Sri Lanka, Syria, Tajikistan, Thailand, Uzbekistan, Vietnam

CAHFSA: Barbados, Belize, Cuba, Dominican Republic, Haiti, Jamaica, Saint Kitts & Nevis

CAN: Bolivia, Colombia, Ecuador, Peru, Venezuela

COSAVE: Argentina, Brazil, Chile, Guyana, Paraguay, Suriname, Uruguay

EPPO: Austria, Belarus, Belgium, Bulgaria, Croatia, Cyprus, Czech Republic, Denmark, Estonia, Finland, France, Georgia, Germany, Greece, Guernsey, Hungary, Ireland, Israel, Italy, Jersey, Jordan, Kazakhstan, Kyrgyzstan, Latvia, Lithuania, Luxembourg, Macedonia, Malta, Moldova, Morocco, Netherlands, Norway, Poland, Romania, Russia, Serbia, Slovakia, Slovenia, Sweden, Switzerland, Tunisia, Turkey, Ukraine, United Kingdom, Uzbekistan

IAPSC: Algeria, Angola, Benin, Botswana, Burkina Faso, Burundi, Cameroon, Cape Verde, Central African Republic, Chad, Comoros, Congo, Côte d'Ivoire, Egypt, Eritrea, Ethiopia, Gabon, Gambia, Ghana, Guinea, Guinea-Bissau, Kenya, Liberia, Madagascar, Malawi, Mali, Mauritania, Mauritius, Mozambique, Namibia, Niger, Réunion, Rwanda, Senegal, Sierra Leone, Somalia, Sudan, Swaziland, Togo, Uganda, Zambia, Zimbabwe

NEPPO: Egypt, Jordan, Libya, Malta, Sudan, Syria, Tunisia

OIRSA: El Salvador, Guatemala, Honduras, Nicaragua

PPPO: Australia, Fiji, New Zealand, Papua New Guinea, Tonga, Vanuatu

13.16 Management measures

It has been proved that it is almost impossible to control PWN once introduced into a tree. Therefore, control measures have concentrated on the integration of different management practices, including the following.

13.16.1 Prevention

Management of pine wilt disease is primarily focused on prevention. There is no remedy for PWN once a susceptible tree becomes infested. The most effective prevention approach is to avoid planting susceptible pine species, such as Scotch and Austrian pine, where the mean summer temperature is greater than 20°C. Some conifer species are rarely colonized by the pinewood nematode. These species include Douglas-fir (*Pseudotsuga menziesii*), redwood (*Sequoia sempervirens*), white fir (*Abies concolor*), western red cedar (*Thuja plicata*), eastern hemlock (*Tsuga canadensis*) and western hemlock (*T. heterophylla*). The starting point is restricting the import of infested logs or packing material made of coniferous wood from infested countries. Surveys of natural forest and introduced planted areas for disease symptoms should be made and, on its first record, containment of the disease should be made through sanitation. Dead pines can become beetle reservoirs, so they should be cut and burnt or debarked (Khan, 2020). Avoid saving wilt-killed pine as firewood, because beetles can continue to emerge from the logs. It is also very difficult to exterminate the nematode in infected wood. Insecticide treatments to control the beetles that carry the nematodes are impractical, because the beetle has a long period of activity.

13.16.2 Cultural practices

Management of *B. xylophilus* is mainly concentrated on cultural practices, including removal of dead or dying trees from the forest to avoid their use as a source of further infection, and control of the beetle vector by insecticidal treatment. In PWN hit areas, stripping off bark and removal of diseased tree may greatly reduce the population of insect vector (Khan, 2020). The main risk of infection in cut timber is in the period between felling and removal from the forest. Cut logs can be protected from oviposition by the beetle by insecticidal treatment. Such treatment is more effective in killing the insect larvae already present under the bark. Other means of reducing the risk of attack are to cover logs after felling, to leave trap logs exposed nearby and to ensure that the felling is conducted outside the flight period of the beetles (Dominik, 1981). In the long term, resistant pines should be selected. Christmas trees and nursery stock that are not watered during prolonged droughts can be protected from borers such as sawyers by spraying them with lindane.

13.16.3 Heat treatment

The only known effective treatment for wood already infected with *B. xylophilus* and its vectors appears to be heat treatment, in which all parts of the wood reach a temperature of 56°C for at least 30 min; commercial kiln practices normally achieve this. Fumigation with aluminium phosphine has been effective in eliminating the nematode from wood chips.

13.16.4 Chemical control

Application of nematicides in combination with insecticides is another option for the management of pine wilt disease. In localized areas, garden or parks, chemical treatment by injecting nematicides such as disulfoton, fensulphothion, thionazin, abamectin, emamectin benzoate etc. in tree trunk can effectively kill the PWN in an individual tree. In addition, the application of fenitrothion or carboxyl can substantially reduce the vector population leading to a decrease in the nematode spread to new trees (Khan, 2020).

13.16.5 Phytosanitary measures

The long-distance movement of the nematode and its vector occurs through infested wood and wood products. The primary tool in managing the threat of introduction of the nematode into new areas is that the importation of wood from nematode-infested countries should be avoided. If the import is indispensable, the wood should be inspected and fumigated or kiln-dried. On the first report or occurrence of PWN in a new area, the appropriate authority should immediately be informed and immediate phytosanitary measures should be taken to prevent further spread. The suggested monitoring and quarantine efforts need to be strengthened to keep the area(s) disease-free. Adequate measures should be enforced to prevent the introduction of PWN and infested insect vectors from infested countries and the strict implementation of quarantine strategies. Existing quarantine legislation on movement of wood or wood products that can carry pine wilt nematode are summarized below.

1. Most countries have already stated in their national regulations that wood packing material (such as veneer peeler cores, sawdust, woodwool and shavings and thin wood pieces) must be free from pinewood nematode.
2. Fumigation with aluminium phosphide is effective in eliminating the nematode from wood chips.
3. The European Plant Protection Organization (EPPO) has recommended that Europe should ban conifer products from countries that have the nematode unless the products have been kiln-dried.
4. Further packing material needs to be properly marked, as per international systems, after appropriate treatment; or otherwise, the consignment should be accompanied by a Phytosanitary Certificate in respect of the packing material with the endorsement of the treatment in the certificate (ISPM 15).
5. Sulfuryl fluoride fumigation treatment for nematodes and insect vectors in debarked wood should be practised (ISPM 23 PT-23, 2017).
6. In India, all packing materials are permitted only after appropriate treatment (heat kiln at 56°C for 30 min, or methyl bromide fumigation at 48 g/m^3 for 16 h at 21°C, or chemical impregnation: see Plant Quarantine Order 2003; details available at http://www.plantquarantine.org.in).

3

Belonolaimidae: *Belonolaimus* Species

Belonolaimus Spp.

The sting nematode, *Belonolaimus* spp., is one of the most destructive nematode pests of turfgrass and attacks a wide range of monocotyledonous plants and gymnosperms, including vegetables, legumes, cereals, etc. The genus *Belonolaimus* contain ten species (Siddiqi, 2000; Decraemer and Geraert, 2013) and some of economic importance are:

Belonolaimus euthychilus Rau 1963
B. gracilis, Steiner, 1949
B. longicaudatus Rau, 1958
B. martimus Rau, 1958
B. nortoni Rau, 1963
B. maluceroi Cid Del Prado Vera and Subbotin, 2012
B. longicaudatus is considered as one of the most important plant pathogens in Florida, USA (Dickson and De Waele, 2005). Other species are relatively uncommon compared with *B. longicaudatus*.

Systematic position of *Belonolaimus* species

	Classical (after Siddiqi, 2000)	Molecular (after Decraemer and Hunt, 2013)
Phylum	Nematoda	Nematoda
Class	Secernentea	Chromadorea
Subclass	Tylenchia	Chromadoria
Order	Tylenchida	Rhabditida
Suborder	Hoplolaimina	Tylenchina
Infra-order	Anguinata	Tylenchomorpha
Superfamily	Dolichdoroidea	Tylenchoidea
Family	Belonolimidae	Dolichodoridae
Subfamily	Belonoliminae	Belonoliminae
Genus	*Belonolaimus*	*Belonolaimus*

14 *Belonolaimus longicaudatus*

14.1 Common name

Sting nematode

14.2 Scientific name

Belonolaimus longicaudatus Rau, 1958

14.3 EPPO code

BELOLO

14.4 Diagnosis

14.4.1 Morphological diagnosis

Belonolaimus longicaudatus was first described by Rau (1958). Morphological descriptions and morphometric measurements of the nematode after Rau (1958, 1961) are as follows.

Female: Lip region hemispherical, divided by dorsal, ventral and lateral grooves into four main lobes, each lobe bearing six or more horizontal striations. Lip region usually offset from body by a profound constriction (Fig. 3.1A,E). Lateral fields marked by single incisure extending from lip base and going up to tail terminus. Stylet spear thin, flexible, with round knobs. Median bulb with prominent valve plates (Fig. 3.1A). Excretory pore posterior to median bulb and hemizonid just anterior to it. Oesophageal glands lobe-like and extending over anterior portion of intestine. Intestine extending nearly to tail terminus. Serpentine canals present (Fig. 3.1C). Vulva with a transverse slit, located in mid-region of body, and vulval lips not protruding (Fig. 3.1C). In lateral view, vagina with opposing pairs of cuticularized pieces. Genital tract amphididelphic and outstretched. Spermathecae present. Tail long, about five times anal body width and subcylindroid with rounded terminus. Phasmids inconspicuous.
Measurements: L = 2.0–3.0 mm; a = 55–75; b = 7.2–12.6; c = 13–21; c' = 3.5–6.0; V = 46–54; spear = 100–140 μm; tail length = 115–183 μm; phasmid to terminus = 79–129 μm; spear length divided by tail length = 0.67–1.14.
Male: Lip region more flattened at sides than in female (Fig. 3.1B). Morphological characters similar to female. Spicules arcuate with ventral flanges. Distal ends of spicules pointed with small apical notches. Gubernaculum well evolved. Bursa present and over terminal, enveloping tail (Fig. 3.1D).

DOI: 10.1079/9781789247367.0004

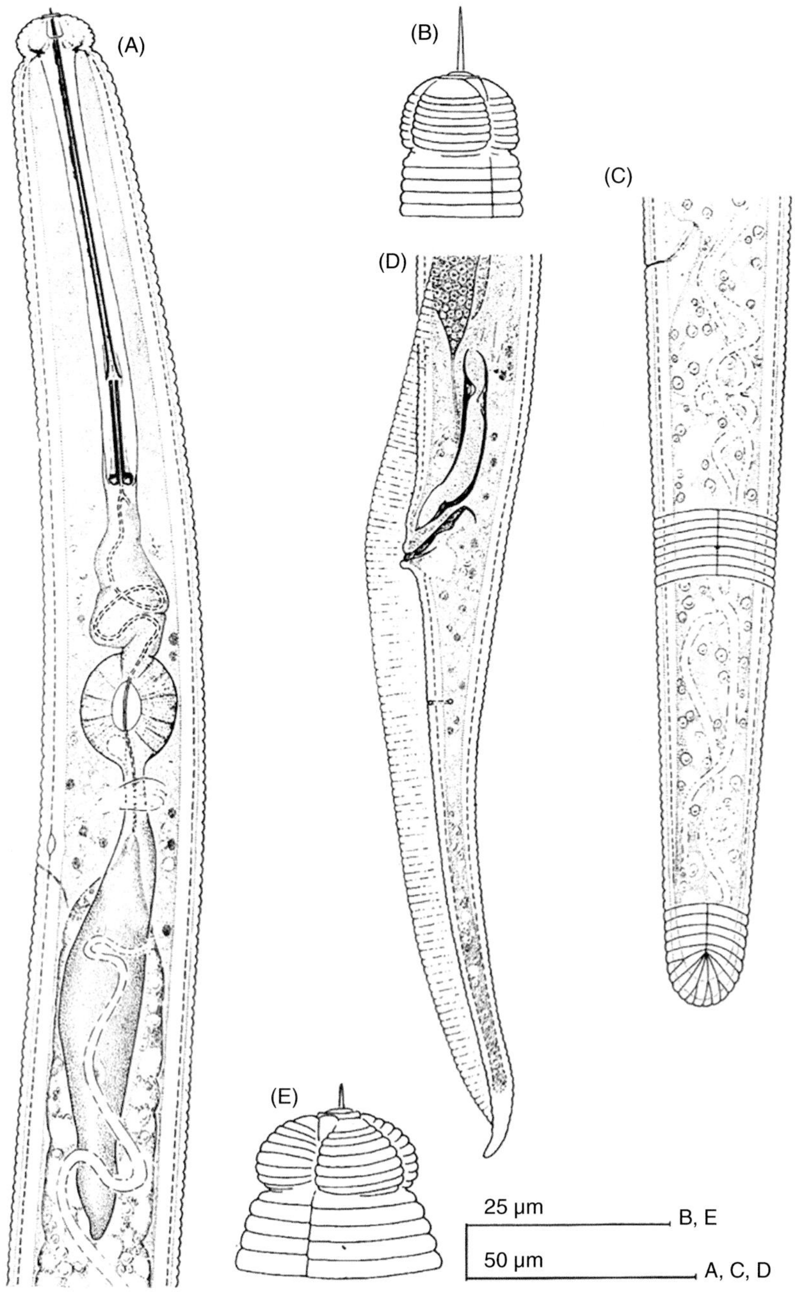

Fig. 3.1. Line drawing of *Belonolaimus longicaudatus*. Female oesophageal region (A); male head (B); female tail (C); male tail region (D); female head end (E). [Reproduced from M.R. Siddiqi, 2000; courtesy CAB International]

Measurements: L = 2.0–3.0 mm; a = 54–76; b = 6.3–8.1; c = 13–17; spear = 107–132 µm; tail length = 100–157 µm; phasmid to terminus = 74–129 µm; spear length divided by tail length = 0.76–1.19; spicule length along chord = 38–49 µm; gubernaculum = 15–18 µm.

14.4.2 Biochemical and cytogenetic diagnosis

Biochemical and cytogenetic techniques are used to discover the degree of genetic similarity in a taxonomic group (Siddiqi, 2000). Biochemical approaches such as the separation of proteins and isozymes by one-dimensional gel electrophoresis, isoelectric focusing, two-dimensional gel electrophoresis, and sodium dodecyl sulfate–capillary gel electrophoresis are used for the biochemical diagnosis of plant-parasitic nematodes (Abrantes *et al.*, 2004). Cytogenetic techniques are also used to determine chromosome numbers and successfully classified higher taxa and the identification of species. All the members of superfamily Dolichodoroidea, including *B. longicaudatus*, have $n = 8$ (Siddiqi, 2000).

14.4.3 Molecular diagnosis

Numerous plant-parasitic nematodes have been characterized by PCR–RFLP of internal transcribed spacer (ITS) regions, including *B. longicaudatus* (Abrantes *et al*., 2004). Cherry *et al.* (1997) compared six *Belonolaimus* isolates from different states of the USA by restriction endonuclease digestion of amplified first internal transcribed spacer region (ITS1) of the nuclear ribosomal genes. Approximately 10–14 individual nematodes from an isolate were processed with seven restriction enzymes for evaluation based on restriction sites inferred from the nucleotide sequence (see Chapter 15, section 110.2.1). Two PCR amplification primers were designed by comparative sequence alignments of various nematode species. The primers are:

rDNA2 (5'-TTGATTACGTCCCTGCCCTTT-3') described by Vrain *et al.* (1992)
rDNA1.58S (3'-GCCACCTAGTGAGCCGAGCA-5')

The PCR amplification is performed according to the following protocol.

Reagent		Final concentration
Buffer ($MgCl_2$)	–	1.5 mM
Specific forward primer	–	0.3 μM
Specific reverse primer	–	0.3 μM
DNA (volume)	–	50 μl
Cycling parameters		
Initial denaturation	–	94°C for 2 min
Number of cycles	–	40
Denaturation	–	94°C for 1 min
Annealing	–	57°C for 1 min
Extension	–	72°C for 2 min

The PCR products are analysed by agarose gel electrophoresis and the amplified product size from individuals of each isolate is approximately 700 bp. Other molecular analyses made by Gozel *et al.* (2006), Han *et al.* (2006) and Cid Del Prado Vera and Subbotin (2012) can also be used to differentiate between the populations of *Belonolaimus* spp.

14.5 Geographical Distribution

The sting nematode *B. longicaudatus* is primarily distributed in sandy soil and one of the most common plant-parasitic nematodes in the subtropical region of the USA (Sikora and Fernandez, 2005). Elsewhere it has limited geographical distribution. Continent-wise, the countries where *B. longicaudatus* has been reported are:

Asia: Pakistan, Saudi Arabia, Turkey.
North America: Bahamas, Bermuda, Costa Rica, Mexico, Puerto Rico, USA.

14.6 Hosts

B. longicaudatus has a very wide host range, including many grasses, crop plants and woody shrubs and trees (Smart and Nguyen, 1991; Sikora and Fernandez, 2005). The extended list of hosts of *B. longicaudatus* is presented in Table 19.

14.7 Symptoms

The above-ground symptoms of *B. longicaudatus* infestation include severe stunting, leaf yellowing, wilting in dry conditions and, in severe cases, death of the plant (Fig. 3.2A). On the roots, the nematode feeds ectoparasitically near the root tip and along the succulent root. In vegetables, nematode feeding induces stubby root and root lesion, which can extend to the root girdle (Sikora and Fernandez, 2005). In groundnut, a reduced root system (Fig. 3.2B) with stubby side branches and terminal galling appears. Dark lesions may appear on the outer root surface at the point of penetration. Yield and quality of groundnut may be severely hampered (Dickson and De Waele 2005).

On turfgrass, plants with sting nematode infestations exhibit nutrient and water deficiencies and these areas could be over-irrigated and fertilized in an attempt to get a response from the turf. Symptoms of damage on turfs include irregular patches and yellowing and drying of grass (Fig. 3.2C). Occasionally turf areas with heavy sting nematode infestations can be completely killed (Jonathan *et al.*, 2012).

14.8 Biology and life cycle

B. longicaudatus is a migratory ectoparasite and obligate parasite on plant roots (Sikora and Fernandez 2005). The nematode reproduces amphimictically and has three juvenile stages (J_2, J_3 and J_4) in the soil. The first moult occurs within the egg (Smart and Nguyen, 1991), resulting in the hatching of the second-stage juvenile. The nematode completes one generation in 28 days at 28–30°C (Sikora and Fernandez, 2005). Although *B. longicaudatus* is an ectoparasite, the remarkably long stylet spear permits the nematode to feed very deep in the root tissue, producing severe damage to the host plant.

14.9 Ecology

Ecological requirements such as soil texture, moisture and temperature have a high impact on the nematode's life cycle and reproduction. It thrives well in sandy soil, but finely textured soil inhibits movement and reproduction (Dickson and DeWaele, 2005). Its reproduction is favoured in soil with more than 80% sand and less than 10% clay (Robbins and Baker, 1974). The optimum temperature for reproduction is around 30°C, but the nematode remains active and can feed up to 35°C (Dickson and DeWaele, 2005). However, at 39.5°C or higher, the nematode is either killed or migrates downwards (Boyd and Perry, 1971). A higher population of the nematode is reported in the upper layer of soil (up to 20 cm) than in the deeper layer (Perez *et al.*, 2000).

14.10 Physiological races

Several studies have suggested the presence of physiological races of *B. longicaudatus* on different hosts (Dickson and DeWaele, 2005).

Table 19. Hosts of *Belonolaimus longicaudatus* (adapted from CABI-CPC, 2010).

Major hosts
Arachis hypogaea (groundnut)
Brassica spp.
Citrus sinensis (navel orange)
Cucumis melo (melon)
Cynodon dactylon (Bermuda grass)
Daucus carota (carrot)
Fragaria (strawberry)
Glycine max (soybean)
Gossypium hirsutum (Bourbon cotton)
Zea mays (maize)
Minor hosts
Abelmoschus esculentus (okra)
Acer (maples)
Acer pseudoplatanus (sycamore)
Agrostis stolonifera var. *palustris* (bentgrass)
Allium cepa (onion)
Apium graveolens (celery)
Baccharis halimifolia (groundsel-bush)
Beta vulgaris (beetroot)
Brassica oleracea (cabbages, cauliflowers)
Brassica oleracea var. botrytis (cauliflower)
Brassica oleracea var. capitata (cabbage)
Brassica rapa ssp. *oleifera* (turnip rape)
Capsicum annuum (bell pepper)
Casuarina equisetifolia (casuarina)
Chrysanthemum (daisy)
Citrullus lanatus (watermelon)
Cucumis sativus (cucumber)
Desmodium tortuosum (Florida beggarweed)
Digitaria decumbens (pangolagrass)
Digitaria sanguinalis (large crabgrass)
Diospyros kaki (persimmon)
Eremochloa ophiuroides (centipedegrass)
Festuca arundinacea (reed fescue)
Helianthus annuus (sunflower)
Hordeum vulgare (barley)
Ilex (holly)
Ipomoea batatas (sweet potato)
Lactuca sativa (lettuce)
Liquidambar styraciflua (sweet gum)
Lolium multiflorum (Italian ryegrass)
Lycopersicon esculentum (tomato)
Mentha spicata (spearmint)
Ocimum basilicum (basil)
Oxalis (wood sorrels)
Paspalum notatum (bahiagrass)
Pennisetum glaucum (pearl millet)
Pennisetum purpureum (elephant grass)
Pharbitis purpurea (tall morning glory)
Phaseolus vulgaris (common bean)
Pinus palustris (longleaf pine)
Pinus taeda (loblolly pine)
Pisum sativum (pea)
Pittosporum tobira (Japanese pittosporum)
Poaceae (grasses)
Rumex crispus (curled dock)
Saccharum officinarum (sugarcane)
Schinus terebinthifolius (Brazilian pepper tree)
Secale cereale (rye)
Sesbania exaltata (coffeebean)
Solanum melongena (aubergine)
Solanum tuberosum (potato)
Sorghum sudanense (Sudan grass)
Stenotaphrum secundatum (buffalograss)
Trifolium (clovers)
Triticum aestivum (wheat)
Turfgrasses
Ulmus parvifolia (lacebark elm)
Vaccinium corymbosum (blueberry)
Vigna unguiculata (cowpea)
Vitis rotundifolia (Muscadine grape)

14.11 Economic importance

B. longicaudatus causes serious crop losses to groundnut, soybean, *Phaseolus vulgaris*, beet, crucifers, celery, okra, onion, pea, pepper, potato, maize, forage crops and turf grasses (Perry and Rhoades, 1982a; Sikora and Fernandez, 2005). The nematode has been considered responsible for heavy losses in vegetables in Florida, USA (Sikora and Fernandez, 2005). Economic losses in groundnut have been reported from North Carolina (Dickson and De Waele, 2005). The nematode is economically important in amenity grassland such as golf courses (Perry and Rhoades, 1982a). In Florida, *B. longicaudatus* caused losses to sweetcorn in sandy soil (McDonald and Nicol, 2005). On cotton, the presence of the nematode also breaks resistance to Fusarium wilt (*Fusarium oxysporum* f. sp. *vasinfectum*), leading to high crop losses in the field (Perry and Rhoades, 1982a).

14.12 Economic threshold level

The economic threshold level varies from two to five larvae of *B. longicaudatus* per 130 cm^3 of soil (Dickson and De Waele, 2005).

14.13 Similarities to other species

B. longicaudatus is superficially similar to other species of the genus and also to members of the genus *Dolichodorus*. It can be best distinguished from other species by the presence of one incisure in the lateral field and also by the length of the spear (> 100 μm long) (Smart and Nguyen, 1991).

14.14 Pest risk analysis and looming threat

B. longicaudatus is a common parasite of vegetables, legumes, grasses and other plant crops, particularly in south-eastern USA. It is widespread throughout the Atlantic coastal plain from Virginia to Florida and attacks a wide range of fruit, vegetable

Fig. 3.2. Symptoms of *Belonolaimus longicaudatus*. Infested cabbage field (A); healthy and damaged peanuts plants (B); patches of yellowing and drying of *Zoysia* grass due to *Belonolaimus* infestation (C). [Photo courtesy A: J. Ole Becker, DAFF Archive, Bugwood.org; B: H. Ferris, source Nemaplex; D: NASI, India]

and fibre crops such as citrus, cotton, ornamentals and forage (Dickson and De Waele, 2005). It is also one of the most destructive nematode pests of turfgrass. Besides USA, higher populations have been reported from Mexico and Central America. Reports from Bermuda, Bahamas and Puerto Rico that deceptively refer to golf courses were due to imported infected turf from the USA (Perry and Rhoades, 1982a). Because of the wide host range, economic impact and limited distribution, *B. longicaudatus* is an important regulated/quarantine nematode in many countries.

14.15 Region-wise status of invasiveness

B. longicaudatus has limited geographical distribution and it has not been recorded from many countries of the world. Region-wise the status of invasiveness is as follows.

APPPC: Bangladesh, Cambodia, China, India, Indonesia, Iraq, Israel, Japan, Jordan, Kazakhstan, Kyrgyzstan, Lebanon, Malaysia, Myanmar, Philippines, South Korea, Sri Lanka, Syria, Tajikistan, Thailand, Vietnam, Uzbekistan

CAHFSA: Barbados, Belize, Cuba, Dominican Republic, Haiti, Jamaica, Saint Kitts & Nevis

CAN: Bolivia, Colombia, Ecuador, Peru, Venezuela

COSAVE: Argentina, Brazil, Chile, Guyana, Paraguay, Suriname, Uruguay

EPPO: Bulgaria, Cyprus, Germany, Greece, Hungary, Ireland, Italy, Latvia, Morocco, Netherlands, Poland, Portugal, Russia, Serbia and Montenegro, Spain, Switzerland, Ukraine, United Kingdom, Uzbekistan

IAPSC: Algeria, Angola, Benin, Botswana, Burkina Faso, Burundi, Cameroon, Cape Verde, Central African Republic, Chad, Comoros, Congo, Côte d'Ivoire, Egypt, Eritrea, Ethiopia, Gabon, Gambia, Ghana, Guinea, Guinea-Bissau, Kenya, Liberia, Madagascar, Malawi, Mali, Mauritania, Mauritius, Mozambique, Namibia, Niger, Réunion, Rwanda, Senegal, Sierra Leone, Somalia, South Africa, Sudan, Swaziland, Togo, Uganda, Zambia, Zimbabwe

NAPPO: Canada

NEPPO: Malta, Tunisia

OIRSA: El Salvador, Guatemala, Honduras, Nicaragua

PPPO: Australia, Fiji, New Zealand, Papua New Guinea, Tonga, Vanuatu.

14.16 Invasiveness rating

Considering the consequences of introduction and pest risk analysis, the following invasiveness rating has been assigned to *B. longicaudatus*.

Climate/host interaction	–	High (3)
Hosts range	–	High (3)
Dispersal potential	–	Medium (2)
Economic impact	–	High (3)
Environmental impact	–	High (3)
Post-entry distribution	–	Low (–1)
Final rating (score)	–	High (13)

Based on the above invasiveness rating, *B. longicaudatus* receives a score of 13 out of 15 (High) and is rated as HIGHLY INVASIVE NEMATODE.

14.17 Movement and means of dispersal

The nematode can be disseminated by any means that can transport infected sods of turf in trade, infected soil, farm implements that have soil attached, animals, water and transplants.

14.17.1 Plant parts liable to carry the nematode in trade/transport

Plant parts including roots, seedlings, infested seed (including pods) amd micro-propagated plants are liable to carry all stages (eggs, juveniles and adult) of the nematode.

14.17.2 Plant parts not known to carry the nematode in trade/transport

Bark, leaves, flowers, inflorescences, cones, calyx, fruits, true seeds, shoots and wood.

14.18 Management measures

14.18.1 Cultural methods

Incorporation of organic amendments that alter the soil biota has been shown to suppress the nematode (Overman *et al.*, 1987; Sikora and Fernandez, 2005). Crop rotation with non-host crops, soil solarization and summer fallowing with cover crops also reduce the nematode population and enhance yield (Perry and Rhoades, 1982a; Smart and Nguyen, 1991).

14.18.2 Chemical methods

Various chemical nematicides are highly effective, due to the porous nature of the sandy soils that favour the nematode, and have been widely used to control *B. longicaudatus* (Perry and Rhoades, 1982a). Good control was observed in onion and cabbage with a pre-plant application of fumigant and non-fumigant nematicides (Sikora and Fernandez, 2005).

14.18.3 Biological methods

Pasteuria penetrans inhibited populations of *B. longicaudatus* after 1 year of application (Giblin-Davis, 1990).

14.18.4 Phytosanitary measures

See Chapter 16, section 115.

4

Criconematidae: *Criconemoides* and *Hemicriconemoides* Species

Criconemoides Spp.

The ring nematode, *Criconemoides*, was first described by Hofmanner and Menzel in 1914 as *Criconema morgense*. The taxonomic status is still controversial and taxonomists have not reached a consensus about the validity and species composition of the genus *Criconemoides*. Decraemer and Hunt (2006) synonymized the genus *Criconemoides* with *Mesocriconema* de Man, 1880, while Siddiqi (2000) considered *Mesoscriconema* as a junior synonym of *Macroposthonia*. In this chapter, we refer to the genus as *Criconemoides* with *Criconemella*, *Mesoscriconema* and *Macroposthonia* as synonyms. Recent keys to the genus *Criconemoides* are given by various workers (Geraert, 2010; Eskandari *et al.*, 2010; Cordero *et al.*, 2012; Van Den Berg *et al.*, 2018). At present, the genus *Criconemoides* contains more than 42 valid species, with *Criconemoides morgensis* as its type species (Siddiqi, 2000; Van Den Berg *et al.*, 2018). Some of the *Criconemoides* species are:

Criconemoides brevistylus Singh and Khera, 1976
C. curvatus Raski, 1952
C. helicus Eroshenko and Nguent Vu Tkhan', 1981
C. informis (Micoletzky, 1922) Taylor, 1936
C. morgensis (Hofmanner and Menzel, 1914) Taylor, 1936
C. obtusicaudatus Heyns, 1962
C. onoensis Luc, 1959
C. onostris (Phukan and Sanwal, 1980) Hunt, Luc and Manzanilla-López, 2005
C. ornatus Rski, 1958
C. paronostris (Deswal and Bajaj, 1987) Hunt, Luc and Manzanilla-López, 2005
C. parvus Raski, 1952
C. sphaerocephalus Taylor, 1936
C. xenoplax Raski, 1952
C. zavadskii (Taulaganov, 1941) Raski, 1958

Among these, *C. curvatus* (syn. *Cricinemella curvata*; *Macroposthonia curvata*), *C. onoensis* (syn. *Cricinemella onoensis*; *Macroposthonia onoensis*), *C. ornatus* (*Cricinemella ornata*; *Macroposthonia ornata*), *C. sphaerocephalus* (syn. *Cricinemella sphaerocephalus*; *Macroposthonia sphaerocephalus*) and *C. xenoplax* (*Cricinemella xenoplax*; *Macroposthonia xenoplax*) are of economic significance to tropical and subtropical agriculture (Orton Williams, 1973a; Loof, 1974; Geraert, 2010).

Systematic position of *Criconemoides* species

	Classical (after Siddiqi, 2000)	Molecular (after Decraemer and Hunt, 2013)
Phylum	Nematoda	Nematoda
Class	Secernentea	Chromadorea
Order	Tylenchida	Rhabditida
Suborder	Criconematina	Criconematina
Infra-order	–	Tylenchomorpha
Superfamily	Criconematoidea	Criconematoidea
Family	Criconematidae	Criconematidae
Subfamily	Macroposthoniinae	Macroposthoniinae
Genus	*Criconemoides*	*Criconemoides*

15 *Criconemaoides* species

15.1 Common name

Ring nematode

15.2 Scientific name

Criconemoides Taylor, 1936

15.3 Synonyms

Criconemella De Grisse and Loof, 1965
Crossonemoides Eroshenko, 1981
Macroposthonia de Man, 1880
Madinema Khan, Chawla and Saha, 1976
Mesocriconema Andrassy, 1965
Neobakernema Ebsary, 1981
Seshadriella Darekar and Khan, 1981
XenoCriconemoides De Grisse and Loof, 1965

DOI: 10.1079/9781789247367.0005

15.4 EPPO code

CRIMSP (*Criconemella* sp.)

15.5 Diagnosis

Usually, pest risk analysis and invasiveness ratings are employed at the species level, not at the genus or group level. An exception has been made for *Criconemoides* mainly because, over the past several decades, the genus *Criconemoides* was seldom distinguished to species level by taxonomists (Chitambar *et al.*, 2018).

15.5.1 Morphological diagnosis

Criconemoides species are sexually dimorphic and the following morphological description of *Criconemoides* species is from Siddiqi (2000).

FEMALE: Body 0.2–1.0 mm long. Body annules 42–200; posterior end smooth or finely crenate (Fig. 4.1). Submedian lobes usually well developed, occasionally poorly developed, or even not present in some species. First annules may be reduced or sometimes divided into plates. Spear strong, rarely thin and flexible, exceptionally short with rounded basal knobs. Oesophagus criconematid, isthmus and basal bulb very short. Genital tract monoprodelphic without post-vulval uterine sac. Vulval lips closely appressed (closed vulva) to somewhat widely separated ('open' vulva); anterior lip region may be ornamented.

MALE: Vermiform with degenerated stylet and oesophagus. Head end rounded to conoid, lateral lines generally four but may be two to three; bursa distinct or, exceptionally, absent.

JUVENILES: Similar to the female. Annules smooth to crenate with no row of scales.

15.5.2 Molecular diagnosis

Criconemoides and its synonyms have been molecularly characterized by several works (Van Den Berg *et al.*, 2012; Powers *et al.*, 2014; Munawar *et al.*, 2018). Van Den Berg *et al.* (2012) characterized *Criconemoides* species using D2–D3 of 28S rRNA and ITS-rRNA. The DNA was extracted using the protocol described by Tanha Maafi *et al.* (2003) (see Chapter 15, section 110.1.1). The primers used in this study are:

For D2–D3 expansion segments of the 28S-rDNA (Subbotin *et al.*, 2007)

D2A forward primer (5′-ACAAGTACCGTGAGGGAA AGTTG-3′)
D3B reverse primer (5′-TCGGAAGGAACCAGCTACTA-3′),

For the ITS-rRNA region (Tanha Maafi *et al.*, 2003)

TW81 forward primer (5′-GTTTCCGTAGGTGAACC TGC-3′)
AB28 reverse primer (5′-ATATGCTTAAGTTCAGCGGGT-3′)

Fig. 4.1. Four females and one egg of *Criconemoides* sp. [Courtesy J.D. Eisenback/Nemapix Vol. 1]

The PCR amplifications are performed using the following reagents and cycling parameters.

Reagents		
PCR reaction	–	25 μl
Reaction buffer	–	5 μl
dNTPs	–	2.5 μl (200 μM)
Forward primer	–	2.5 μl (0.5 μM each)
Reverse primer	–	2.5 μl (0.5 μM each)
DNA polymerase	–	1 U
DNA template	–	4 ng
Cycling parameters		
For D2–D3 expansion segments of the 28S-rDNA (Subbotin *et al.*, 2007):		
Initial denaturation	–	94°C for 4 min
Number of cycles	–	30
Denaturation	–	94°C for 1 min
Annealing	–	55°C for 1.5 min
Extension	–	72°C for 2 min
Final extension	–	72°C for 10 min
For the ITS-rRNA region (Tanha Maafi *et al.*, 2003):		
Initial denaturation	–	94°C for 2 min
Number of cycles	–	30
Denaturation	–	94°C for 30 s
Annealing	–	59°C for 1 min
Extension	–	72°C for 1 min
Final extension	–	72°C for 10 min

The PCR products are analysed by agarose gel electrophoresis and the amplified product size is approximately 800 and 760 bp for the ITS rRNA gene and D2–D3 of 28S rRNA gene, respectively. Munawar *et al.* (2018) identified *Criconemoides* species from China using near full-length 18S region of rDNA and mitochondrial cytochrome c oxidase subunit I (COI) region gene with the following primers.

For amplification of the nearly full-length 18S rRNA (Olson *et al.*, 2017):

Forward 18S39F (5′-AAA GAT TAA GCC ATG CAT G-3′)
Reverse 18S977R (5′-TTT ACG GTT AGA ACT AGG GCG G-3′),
Forward 18S900F (5′-AAG ACG GAC TAC AGC GAA AG-3′)
Reverse 18S1713R (5′-TCA CCT ACA GCT ACC TTG TTA CG-3′)

For the amplification of COI (Powers *et al.*, 2014):

COI-F5-(5′-AATWTWGGTGTTGGAACTTC TTGAAC-3′)
COI-R9-(5′CTTAAAACATAATGRAAATGWGCWACWA CATAATAAGTATC-3′)

15.6 Geographical distribution

Criconemoides spp. are widely distributed in the tropical and subtropical regions of the world (Fig. 4.2). These nematodes are known to occur in the following countries (CABI-ISC, 2020).

Africa: Burkina Faso, Cameroon, Canary Islands (Spain), Côte d'Ivoire, Egypt, Ghana, Guinea, Kenya, Libya, Mauritius, Réunion, Senegal, Seychelles, South Africa, Sudan, Togo, Zimbabwe.

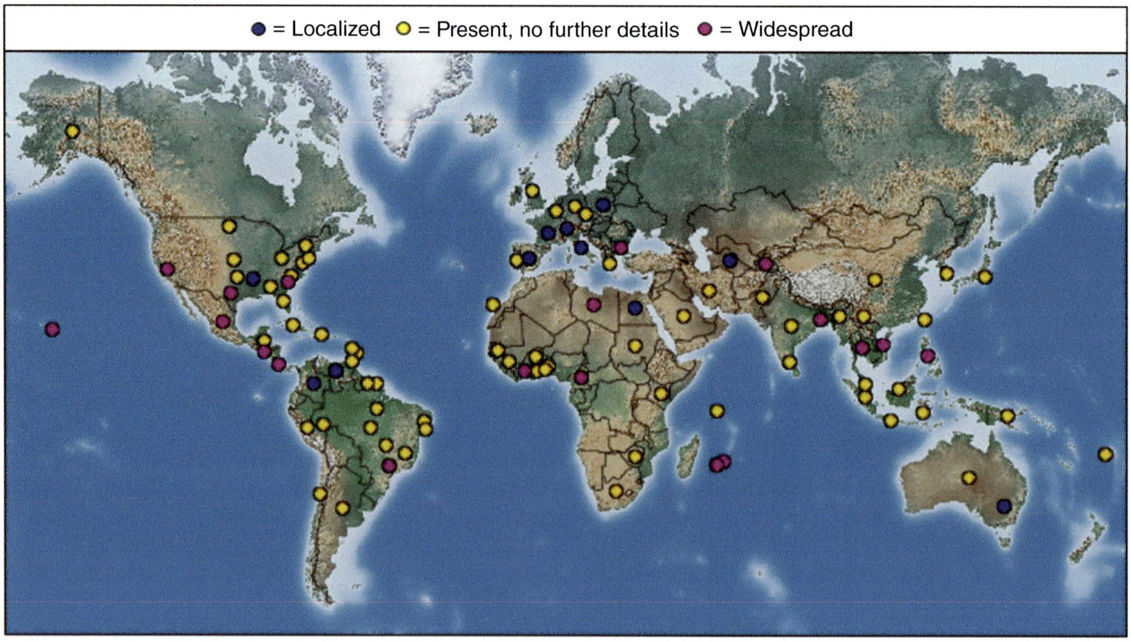

Fig. 4.2. Distribution map of *Criconemoides* species. [Adapted from CABI-ISC, 2020]

Asia: India, Indonesia, Japan, Malaysia, Pakistan, Philippines, South Korea, Taiwan, Tajikistan, Thailand, Turkmenistan, Vietnam.
Europe: Belgium, Bulgaria, Czechoslovakia (former), France, Germany, Greece, Italy, Poland, Portugal, Spain, Switzerland, United Kingdom.
North America: Barbados, Belize, Costa Rica, Cuba, Dominican Republic, El Salvador, Martinique, Mexico, Trinidad and Tobago, USA.
Oceania: Australia, Fiji, Papua New Guinea.
South America: Argentina, Brazil, Chile, Colombia, French Guiana, Peru, Suriname, Venezuela.

15.7 Hosts

Criconemoides spp. have a wide host range, including field crops, fruit trees, ornamentals, vegetables, nursery stock, shrubs, grasses, woody perennials and weeds. However, the host status of parasitized plants is not fully known (Siddiqi *et al.*, 1973). The list of main host plants is given in Table 20.

15.8 Symptoms

Ring nematodes do not enter the root and feed ectoparasitically on tips or along older roots. They feed deep into root tissues using a long stout stylet. The feeding causes damage to the root cortex and reduces small feeder roots. Sometimes these small roots become abnormal with tufted growth. The highest population densities can be found in deeper soil (around 18 inches (45 cm)) where roots are most abundant (Chitamber *et al.*, 2018). Ring nematodes parasitize a wide range of agriculturally important plants, including rice, groundnut, tobacco, cotton, oats, grapes, peach, carnation and maize. Symptoms on some of these plant species are specified below.

RICE *C. onoensis* does not produce any obvious symptoms in flooded rice fields, but in pot trials the nematode produces severe stunting and chlorosis. It mainly damages primary and secondary roots, which become stunted with lesions near hook-shaped root tips (Chinappen *et al.*, 1988).

GROUNDNUT *C. ornata* causes severe discoloration of roots, pods and pegs of groundnut plants, with small brown necrotic lesions in heavily infested micro-plots (Minton and Bell, 1969). These small necrotic lesions are apparent, but in large lesions the necrosis stretches deep into the tissues. Several lateral root primordia and new roots are killed, leading to abridged lateral roots. Infected plants are severely stunted with pronounced chlorosis and show characteristic 'peanut yellows'.

GRAPES Infestation of *C. xenoplax* may significantly reduce top growth and cause chlorotic foliage of potted *Vitis vinifera* in the glasshouse. Several young roots are damaged, resulting in local necrosis and destruction. The necrotic areas extend, leading to the breakdown of roots within a few days. The older lignified roots are also attacked and the entire root system is evidently suppressed (Klinger, 1975).

Table 20. List of main hosts affected by *Criconemoides* species (adapted from CABI-ISC, 2020).

Agrostis (bentgrasses)
Ananas comosus (pineapple)
Arachis hypogaea (groundnut)
Brassica spp.
Carya illinoinensis (pecan)
Cinnamomum
Citrus
Coffea (coffee)
Curcuma longa (turmeric)
Cyperaceae (sedges)
Dianthus caryophyllus (carnation)
Gossypium hirsutum (Bourbon cotton)
Ipomoea batatas (sweet potato)
Juglans regia (walnut)
Juniperus (junipers)
Litchi chinensis (lychee)
Mangifera indica (mango)
Manihot esculenta (cassava)
Medicago (medic)
Morus (mulberry tree)
Musa (banana)
Nicotiana tabacum (tobacco)
Oryza sativa (rice)
Pennisetum glaucum (pearl millet)
Piper nigrum (black pepper)
Poaceae (grasses)
Prunus dulcis (almond)
Psidium guajava (guava)
Rosa (roses)
Saccharum officinarum (sugarcane)
Solanum melongena (aubergine)
Solanum tuberosum (potato)
Sorghum bicolor (sorghum)
Syzygium aromaticum (clove)
Theobroma cacao (cocoa)
Trifolium (clovers)
Vitis vinifera (grapevine)
Zea mays (maize)

PEACHES *C. xenoplax* causes short-life disease syndrome of peach trees. The nematode primarily attacks the roots and, in association with bacterial canker caused by *Pseudomonas syringae* pv. *syringae* or cold injury, there is an abrupt collapse of new growth, just above the soil line, and consequently death of the above-ground part of the peach tree takes place (Nyczepir, 1989).

CARNATIONS In pot trials, inoculation of carnations (*Dianthus caryophyllus*) with *C. curvata* suppressed 33% of the root weight in 6 months and the root system visibly reduced within 37 days, leading to necrosis of the cortical parenchyma (Cuany and Ritter, 1977).

15.9 Biology and life cycle

Criconemoides is an obligate root ectoparasite and feeds on root tips or anywhere along the root. Like other nematodes, the developmental stage consists of four moults. The first moult occurs inside the egg, and second-stage juvenile (J_2) hatch from eggs. Feeding is essential for the development of all juvenile stages. *C. rustica* females lay up to 18 eggs and hatching takes place in 14–21 days. In a study, about 66% of eggs hatched at 13–32°C and 42% at 10°C, but there was no hatching at temperatures above 32.5°C in distilled water (Townshend and Davidson, 1989). The female completes its life cycle in 25–34 days and survives for over 60 days under laboratory conditions. Lighter and warmer soils favour the build-up of population densities of *Criconemoides* spp. *C. ornata* appears to be well adapted to groundnut fields in the southern USA. *C. xenoplax* is very sensitive to drying and the population decreases proportionately to soil moisture. The population also rapidly declines in the absence of hosts or in the presence of a poor host. *C. onoensis* survives well on common weeds of Cyperaceae and Poaceae (Hollis, 1977).

15.10 Economic importance

Generally, *Criconemoides* spp. are considered as mild pathogens. However, high populations cause significant damage. *C. ornata* is a major pest of groundnuts in the USA, causing up to 50% losses in pod yield (Minton and Bell, 1969). Rickard *et al.* (1977) reported 18.7% yield losses for each tenfold increase in the initial infestation. *C. xenoplax* is an important pest of orchard crops, and all *Prunus* accessions were found susceptible hosts to the nematode in the USA (Westcott *et al.*, 1994) and vineyards in Switzerland (Guntzell *et al.*, 1987). *C. onoensis* caused 63–72% reduction in paddy yield in Mauritius (Chinappen *et al.*, 1988) and up to 15% in Louisiana and Texas, USA (Hollis *et al.*, 1984). In West Bengal, India, both *C. xenoplax* and *C. ornata* are serious pests of pineapples (Rama and Dasgupta, 1987), while *C. ornata* is an important pest of brassicas, okra, sorghum and ragi (Acharya and Das, 1983). In Indonesia, *C. xenoplax* was considered to be one of the most significant pests of sugarcane (Harper, 1975). *Criconemoides* spp. are also limiting factors in vegetable production in tropical and subtropical regions (Netscher and Sikora, 1990).

15.11 Movement and means of dispersal

The main mode of long and short distance dispersal of *Criconemoides* species is through the artificial movement of infected planting stock, contaminated soil, farm tools and machinery. The ring nematodes can also spread from infested to non-infested sites through water run-off, irrigation and human activity.

15.11.1 Plant parts known to carry the pest in trade/transport

Bulbs, tubers, corms, rhizomes (sucker), growing medium accompanying plants, roots, seedlings and micro-propagated plants.

15.11.2 Plant parts not known to carry the pest in trade/transport

Bark, flowers, inflorescences, cones, calyx, fruits (including pods), leaves, stems (above ground), shoots, trunks, branches, wood and true seeds (including grain).

15.12 Pest risk analysis and looming threat from introduction

Criconemoides spp. are widely spread in the tropical and subtropical regions of the world and are fewer in temperate countries. Considering the wide host range and the potential damage to economically important crops, *Criconemoides* spp. pose a definite threat to the production of rice, groundnut and other horticultural crops. Introduction of these nematodes into new areas, particularly in temperate European countries, should be prevented. The nematode can easily spread through rootstocks for breeding purposes.

15.13 Invasiveness rating

Considering the consequences of introduction and probable economic damage caused by *Criconemoides* spp., the following invasiveness rating has been assigned.

Climate/host interaction	–	High (3)
Hosts range	–	High (3)
Dispersal potential	–	Medium (2)
Economic impact	–	Medium (2)
Environmental impact	–	Medium (2)
Post-entry distribution	–	Medium (–2)
Final rating (score)	–	Medium (10)

Based on the above invasiveness rating, *Criconemoides* spp. receives a score of 10 out of 15 and is rated as MEDIUM INVASIVE NEMATODE.

15.14 Region-wise status of invasiveness

Because of the wide host range and extensive distribution, *Criconemoides* spp. have been given a medium invasive rating for all the countries where this nematode is absent.

15.15 Management measures

15.15.1 Cultural control

Solarization using clear plastic mulch reduced *C. xenoplax* populations in horticultural soils by 30–100% (Barbercheck and Broembsen, 1986). In India, losses in the groundnut crop due to *C. ornata* were considerably reduced by rotations involving *Brassica rapa*, *Phaseolus vulgaris* and groundnut (Das *et al.*, 1988). In the USA, rotations of cotton, maize, soybean, sorghum or tobacco with groundnut also suppressed *Criconemoides* populations (Kinloch and Lutrick, 1975). Crop rotation with *Desmodium decumbens* for 18 months eradicated *Criconemoides* species in pineapples (Haroon and Smart, 1983). In Brazil, *C. ornata* populations on maize were significantly decreased by rotation with *Tagetes patula* or green manuring with *Crotalaria paulina* (Sharma and Scolari, 1984).

15.15.2 Chemical control

Nematicides are the most effective means to control *Criconemoides* and extensive work has been done on various crops such as rice (Chinappen *et al.*, 1988), groundnut (Phipps, 1987), maize (Phipps and Elliot, 1983), peaches (Johnson *et al.*, 1987), vines (Santo *et al.*, 1985), carnations (Cuany and Ritter, 1977) and bermudagrass on golf courses (Lucas, 1982). Application of nematicide (fenamiphos) against *C. onoensis* increased rice yields by 17% (Hollis, 1977).

15.15.3 Phytosanitary measures

See Chapter 16, section 115.

Hemicriconemoides spp.

Hemicriconemoides was proposed by Chitwood and Birchfield in 1957 as an intermediate genus between *Criconemoides* and *Hemicycliophora* (Siddiqi, 2000). These nematodes are called 'sheathoid' nematodes, as their body cuticle is covered by an outer additional layer or sheath with smooth annuli in adult stages. Dasgupta *et al.* (1968) revised the genus and gave a detailed account and validated the species with five new species. At present, the genus has over 52 valid species, with *Hemicriconemoides wessoni* as type species (Siddiqi, 2000; Geraert, 2010; Van Den Berg *et al.*, 2015). Some of the *Hemicriconemoides* species are:

Hemicriconemoides alexis Vovlas,1980
H. chitwoodi Esser, 1960,
H. californianus Pinochet & Raski, 1975
H. cocophillus (Loos, 1949) Chitwood and Birchfield, 1957
H. gaddi (Loos, 1949) Chitwood and Birchfield, 1957
H. kanayaensis Nakasono and Ichinohe, 1961
H. mangiferae Siddiqi, 1961
H. ortonwilliamsi Ye and Siddiqi, 1979
H. parasinensis Chen and Liu, 2003
H. pseudobrachyurus De Grisse, 1964
H. wessoni Chitwood and Birchfield, 1957

Hemicriconemoides spp. are migratory root ectoparasites and widespread in warmer areas of the tropics and subtropics. These sheathoid nematodes parasitize many agricultural and horticultural crops, including fruit trees. *H. mangiferae* attacks citrus (Crozzoli *et al.*, 1998), banana, date, guava, lychee, mango, pineapple, sapodilla (Siddiqi, 1977) and grapevine (Deimi and Mitkowski, 2010). *H. cocophillus* attacks small millets (Jain, 2009), rice (Sharma *et al.*, 1992a), sugarcane (Cadet and Albrecht,1992) and tuber crops (Ray *et al.*, 1992). *H. kanayaensis* attacks tea (Nakasono and Ichinohe, 1961), *H. wessoni* attacks sod grasses (Crow, 2013). Among these species, economic damage is documented for *H. mangiferae* (McSorley, 1992). *H. mangiferae* (syn. *H. abberans*, *H. birchfeildi*, *H. litchi* and *H. strictathecatus*) is one of the most commonly described and economically important species and is considered for invasiveness in this chapter.

Systematic position of *Hemicriconemoides* species

	Classical (after Siddiqi, 2000)	Molecular (after Decraemer and Hunt, 2013)
Phylum	Nematoda	Nematoda
Class	Secernentea	Chromadorea
Order	Tylenchida	Rhabditida
Suborder	Criconematina	Criconematina
Infra-order	–	Tylenchomorpha
Superfamily	Criconematoidea	Criconematoidea
Family	Criconematidae	Criconematidae
Subfamily	Hemicriconemoidinae	Hemicriconemoidinae
Genus	Hemicriconemoides	Hemicriconemoides

16 *Hemicriconemoides mangiferae*

16.1 Common name

Sheathoid nematode

16.2 Scientific name

Hemicriconemoides mangiferae Siddiqi, 1961

16.3 Synonyms

Hemicriconemoides abberans Phukan and Sanwal, 1983
Hemicriconemoides birchfieldi Edward *et al.*, 1965
Hemicriconemoides litchi Edward and Mishra, 1964
Hemicriconemoides strictathecatus Esser, 1960

16.4 EPPO code

HEMRMA

16.5 Diagnosis

Decraemer and Geraert (1996) synonymized *Hemicriconemoides abberans*, *H. birchfeildi*, *H. litchi* and *H. mangiferae* with *H. strictathecatus*. However, Siddiqi (2000) considered *H. mangiferae* and *H. strictathecatus* as distinct species. Morphological and molecular analysis by Van Den Berg *et al.* (2015) clarified the taxonomic status and confirmed that both belong to the same taxon.

16.5.1 Morphological diagnosis

The morphological and morphometric descriptions of *H. mangiferae* below are after Siddiqi (1961 and 1977).

FEMALE: Body tubular, slightly arcuate when heat fixed and tapering towards both ends. Body enclosed in a circular sheath (Fig. 4.3A). Sheath attached at head end, excretory pore, vulva, anus and occasionally at tail terminus. Sheath annuli smooth but sometimes with several longitudinal indentations 3-4 µm wide near mid-body. Lateral fields absent. Number of sheath or body annuli ranging from 111 to 152. Lip region flattened anteriorly, with two annuli and the anterior annulus angular and directed forward, while posterior annulus round and nearly continuous with body annuli. Labial framework sclerotized, hexaradiated, with labial disc inconspicuous or slightly offset and rounded on top. Stylet long, 53–90 µm with basal knobs directed anteriorly, anchor-shaped. Dorsal gland opening 5–7 µm behind stylet base. Oesophagus criconematoid type with massive median bulb, continuous with wide procorpus; isthmus short and broad, enveloped by nerve ring. Basal bulb short, slightly wider than isthmus and spherical at base. Excretory pore on 29–39 annuli from anterior end. Vulva wide transverse slit, 11 µm in length and on 12–15 annuli from tail end. Lateral vulval flaps or vulval sheath not present. Vagina directed inwards and forwards. Ovary single, outstretched anteriorly, occasionally with one flexure and sometimes reaching to oesophagus. Oocytes in single row except in ripening zone. Spermatheca offset, small, generally with rounded small sperms. Anus 6–11 annuli from tail end and 2–6 annuli behind vulva. Tail subcylindrical with hemispherical annulated terminus or conoid with terminal annulus smoothly rounded (Fig. 4.3I,H). Eggs 67–71 × 16–18 µm.
Measurements: L = 0.43–0.56 mm; a = 16–26; b = 4.2–5.7; c = 18–26; V = 90.7–94.7; stylet = 65–76µm; R = 117–133; Rex = 30–36; RV = 11–14; Rvan = 3–4; Ran = 7–10.
MALE: Largely common. Distinctly different from female. Body slender without cuticular sheath. Stylet and oesophagus degenerated (Fig. 4.3D). Annuli fine, around 1.5 µm wide at mid-body. Lateral fields with four lines and one-fourth of body width, inner ones more discrete than outer ones. Lip region raised, conoid-truncate, incessant with body, with three to six annuli and a labial disc (Fig. 4.3D). Testis outstretched, single. Spicules arcuate, long, slender, cephalated with pointed tips (Fig. 4.3E). Gubernaculum simple, trough-shaped. Bursa reduced with coarse crenate margins, terminating before tail terminus. Tail subcylindrical with rounded tip, almost as long as spicules, consisting of 17–23 ventral annuli. Anal lips protruding. Phasmids not observed.
Measurements: L = 0.35–0.38 mm; a = 25–29; c = 14.8–17.6; T = 35–42; spicules = 22–24µm; gubernaculum = 4.5–5.5µm.
JUVENILES: Head annuli resemble female's. Body cuticle consists of 12 longitudinal rows of pointed, scale-like spines in all juvenile stages except second stage; second-stage juvenile has spineless cuticle. Length of stylet enlarged greatly between fourth and adult stages. Third-stage juveniles can be differentiated from second-stage juveniles by larger ovary length. Sex differentiation happens in fourth stage.

16.5.2 Molecular diagnosis

Van Den Berg *et al.* (2015) developed species-specific primers to characterize 12 species of *Hemicriconemoides* from China, Greece, Japan, Myanmar, South Africa, Spain and the USA using ITS-rRNA and D2–D3 of 28S rRNA (for detail protocol, see Chapter 15, section 110.1.1). The PCR amplification was done using the following primers.

For ITS-rRNA region (Tanha Maafi *et al.*, 2003) (amplicon size 850–880 bp):

TW81 (5′-GTTTCCGTAGGTGAACCTGC-3′)
AB28 (5′-ATATGCTTAAGTTCAGCGGGT-3′)

For D2–D3 expansion segments of the 28s-rRNA (Subbotin *et al.*, 2007) (amplicon size 780 bp):

D2A forward primer (5′-ACAAGTACCGTGAGGGAAAGTTG-3′)
D3B reverse primer (5′-TCGGAAGGAACCAGCTACTA-3′)

For ITS-rRNA region with species-specific primer of *H. strictathecatus* (amplicon size 730 bp):

TW81 (5′-GTTTCCGTAGGTGAACCTGC-3′)
H_strict (5′-CAGTCGTCAGTGAACAAGTCA-3′)

For ITS-rRNA region with species-specific primer of *H. chitwoodi* (amplicon size 333 bp):

TW81 (5′-GTTTCCGTAGGTGAACCTGC-3′)
H_chitw (5′-CGCACCGCGTATCAGTGC-3′)

For ITS-rRNA region with species-specific primer of *H. californianus* (amplicon size 186 bp):

TW81 (5′-GTTTCCGTAGGTGAACCTGC-3′)
H_califor (5′-CTATTCCGAAAGGGGTGTTC-3′)

16.6 Geographical distribution

H. mangiferae is widely distributed in the warmer regions of tropical and subtropical parts of the world but not in Mediterranean Europe (Fig. 4.4). The nematode is known to occur in the following countries (CABI-ISC, 2020).

Africa: Côte d'Ivoire, Egypt, Ghana, Nigeria, South Africa, Sudan.
Asia: Bangladesh, Brunei, China, India, Indonesia, Iran, Israel, Oman, Pakistan, Philippines, Saudi Arabia, South Korea, Thailand, Vietnam.
North America: Costa Rica, Dominica, Grenada, Mexico, Saint Kitts & Nevis, Saint Lucia, Trinidad and Tobago, USA.
Oceania: American Samoa, Fiji, New Caledonia, Papua New Guinea, Samoa, Tonga.
South America: Brazil, Ecuador, Peru, Venezuela.

16.7 Hosts

H. mangiferae parasitizes a wide range of hosts, including agronomic crops, fruit trees, ornamentals, vegetables, shrubs,

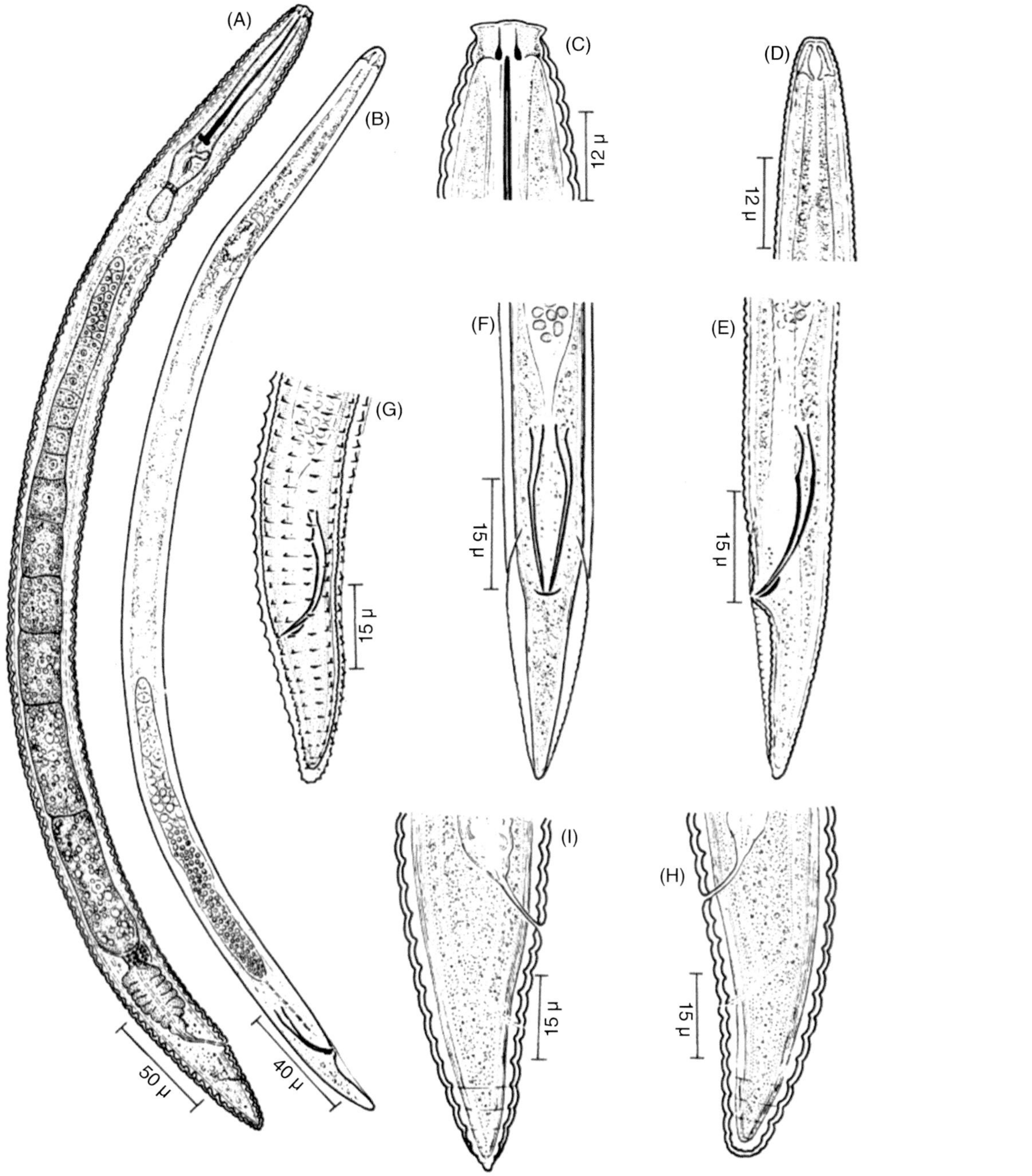

Fig. 4.3. Line drawing of *Hemicriconemoides mangiferae*. Entire female (A); entire male (B); female head end (C); male head (D); male tail end (E,F); ensheathed pre-adult male juvenile (G); female tail end (H,I). [Reproduced from Siddiqi, 1977. CIH Descriptions of Plant-parasitic Nematodes. Set 7, No. 99; courtesy Helminthological Society of Washington]

grasses, woody perennials and weeds. The extended list of hosts of *H. mangiferae* is given in Table 21.

16.8 Symptoms

Symptoms induced by sheathoid nematodes on susceptible plants include stunting, premature wilting, leaf chlorosis, root damage, cortical root tissue necrosis and other symptoms related to nutrient deficiencies (McSorley *et al.*, 1981). Lychee trees infected by *H. mangiferae* exhibit bare branches and twigs, leaf yellowing, leaf-tip burn, reduced flowering and extensive fruit drop that sometimes cause the death of the tree in South Africa. Replanted young trees are slow to become established in infested areas and also begin to show symptoms of decline after 7–8 years (Milne, 1982b). The nematode can

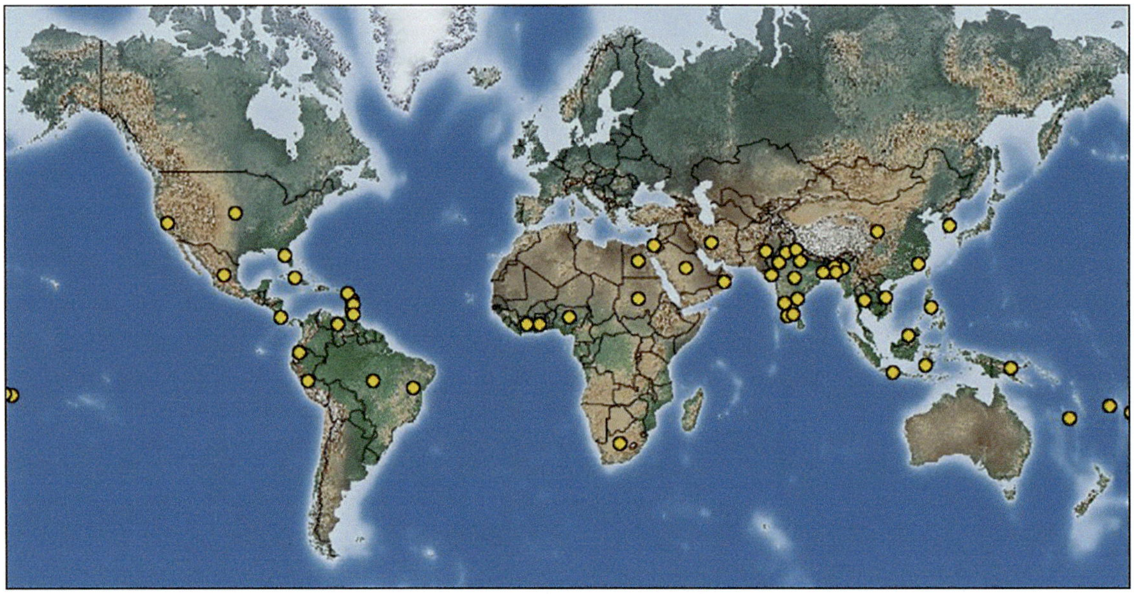

Fig. 4.4. Distribution map of *Hemicriconemoides mangiferae*. [Adapted from CABI-ISC, 2020]

Table 21. List of hosts affected by *Hemicriconemoides* species (adapted from CABI-ISC, 2020).

Main hosts
Citrus
Litchi chinensis (lychee)
Mangifera indica (mango)
Musa × *paradisiaca* (plantain)
Other hosts
Ananas comosus (pineapple)
Artocarpus heterophyllus (jackfruit)
Artocarpus integer (champedak)
Bambusa vulgaris (common bamboo)
Carica papaya (papaw)
Cocos nucifera (coconut)
Dimocarpus longan (longan tree)
Gossypium hirsutum (Bourbon cotton)
Grewia asiatica (phalsa)
Manihot esculenta (cassava)
Manilkara zapota (sapodilla)
Phoenix dactylifera (date-palm)
Piper (pepper)
Prunus armeniaca (apricot)
Psidium guajava (guava)
Saccharum officinarum (sugarcane)
Theobroma cacao (cocoa)
Vitis vinifera (grapevine)
Zea mays (maize)
Wild host
Alnus (alders)
Eriobotrya japonica (loquat)
Rosa (roses)
Salix (willows)
Umbellularia californica (California laurel)

penetrate deeply into the roots in lychee and mango and becomes endoparasitic. Feeding and movement within the roots result in the devastation of cell walls, leading to necrosis of affected tissues. The affected root cortex may slough off and may assume 'stubby root' symptoms (Milne *et al.*, 1971). Under high population densities, multiple invasions of the roots take place and result in the rupture of the endodermis and breakdown of the exodermis and cortical parenchyma (Milne *et al.*, 1975). The root damage adversely affects water and nutrient intake, leading to decline of the lychee tree (Milne *et al.*, 1971).

16.9 Biology and life cycle

H. mangiferae is usually an ectoparasite but it can penetrate deeply into the roots and in some hosts such as lychee and mango the nematodes may be found wholly inside the roots (Milne *et al.*, 1971). *H. mangiferae* populations are mainly encountered in the rhizosphere and concentrated at a depth of 23–46 cm and a horizontal distance of up to 46 cm (Saeed, 1974). Summer drought and high soil temperatures drastically reduced the field populations in mango orchards at Aligarh, North India (Khan *et al.*, 1971). A soil temperature of 19–22°C and moisture of 12–16% seemed favourable for nematodes in sapodilla roots (Saeed and Ghaffar, 1986).

16.10 Economic importance

H. mangiferae attacks citrus (Crozzoli *et al.*, 1998), banana, date, lychee, mango, pineapple (Siddiqi, 1977) and grapevine

(Deimi and Mitkowski, 2010) and causes significant yield loss to these plants. Approximately 40% of lychee trees are known to have been killed by *H. mangiferae* in association with *Xiphinema brevicolle* in South Africa (Milne *et al.*, 1971). In Florida, USA, *H. mangiferae* at more than 100 nematodes per 100 ml of soil significantly reduced the growth of mango trees (McSorley *et al.*, 1981).

16.11 Movement and means of dispersal

See section 15.11.

16.12 Pest risk analysis and looming threat from introduction

Hemicriconemoides mangiferae is widespread in warmer parts of the tropical and subtropical regions of the world and fewer in temperate countries. *H. mangiferae* may have evolved with mango (*Mangifera* spp.) in South-East Asia (Van Den Berg *et al.*, 2015). However, the presence of this sheathoid nematode in the USA indicates that it was most probably introduced through the import of a large number of grafted Indian mango varieties infested with the nematode into Florida from India in 1889, and then became established in the USA (Van Den Berg *et al.*, 2015). Considering this argument and the damage potential to a wide variety of economically important plants, this nematode poses a definite threat to the production of fruit trees and other horticultural crops in the host countries. The nematode can easily spread through rootstocks for breeding purpose and introduction of the nematode in new areas, particularly in the Mediterranean and European countries, should be prevented by strict legislative measures.

16.13 Invasiveness rating

Considering the consequence of the introduction and probable economic damage caused by *H. mangiferae*, the following invasiveness rating has been assigned.

Climate/host interaction	–	High (3)
Hosts range	–	High (3)
Dispersal potential	–	Medium (2)
Economic impact	–	Medium (2)
Environmental impact	–	Medium (2)
Post-entry distribution	–	Medium (–2)
Final rating (score)	–	Medium (10)

Based on the above invasiveness rating, *H. mangiferae* receives a score of 10 out of 15 and is rated as MEDIUM INVASIVE NEMATODE.

16.14 Region-wise status of invasiveness

Due to the wide host range and extensive distribution, *H. mangiferae* has been given a medium invasive rating for all the countries where it is absent.

16.15 Management measures

See section 15.15.

5

Hemicycliophoridae: *Hemicycliophora* Species

Hemicycliophora Spp.

The sheath nematode *Hemicycliophora* de Man, 1921 covers approximately 150 valid species, with *Hemicycliophora typica* as its type species (Siddiqi, 2000; Subbotin *et al.*, 2014). Keys to *Hemicycliophora* spp. are given by Chitambar and Subbotin (2014) and Van Den Berg *et al.* (2018). The important plant-parasitic nematodes from *Hemicycliophora* species are:

Hemicycliophora arenaria Raski, 1958
H. argiensis Khan and Nanjappa, 1972
H. astriatus Khan and Nanjappa, 1972
H. attapdii Rahman, Ahamd and Jairajpuri, 1996
H. chatami Yeates, 1978
H. chilensis Brzeski, 1974
H. labiate Colbran, 1971
H. loofi Maas, 1970
H. onubensis Van den Berg *et al.* 2018
H. oryzae De Waele and Van Den Berg, 1988
H. paracamelliae Maria *et al.*, 2018a
H. parvana Tarjan, 1952
H. penetrans Thorne, 1955
H. poranga Monteiro and Lordello, 1978
H. robbinsi Van den Berg *et al.* 2018
H. similis Thorne, 1955
H. subbotini Maria *et al.*, 2018b
H. thienemanni (Schneider, 1925) Loos, 1948
H. typica de Man, 1921
H. utkali Ray and Das, 1981
H. wyei Cordero López *et al.*, 2013

Among these, *H. arenaria* and *H. typica* are of economic importance and associated with plant growth suppression on different agricultural crops (Siddiqi, 2000; Chitambar and Subbotin, 2014; Hunt, 2018).

Systematic position of *Hemicycliophora* species

	Classical (after Siddiqi, 2000)	Molecular (after Decraemer and Hunt, 2013)
Phylum	Nematoda	Nematoda
Class	Secernentea	Chromadorea
Order	Tylenchida	Rhabditida
Suborder	Criconematina	Criconematina
Infra-order	–	Tylenchomorpha
Superfamily	Hemicycliophoroidea	Criconematoidea
Family	Hemicycliophoridae	Hemicycliophoridae
Subfamily	Hemicycliophorinae	Hemicycliophorinae
Genus	*Hemicycliophora*	*Hemicycliophora*

17 *Hemicycliophora arenaria*

17.1 Common name

Sheath nematode

17.2 Scientific name

Hemicycliophora arenaria Raski, 1958

17.3 EPPO code

HEMCAR

17.4 Diagnosis

Morphological and molecular diagnosis approaches are strongly recommended for correct species identification of *Hemicycliophora arenaria*.

17.4.1 Morphological diagnosis

Morphological and morphometric descriptions (after Franklin and Stone, 1974) are summarized below.

Female: Body with cuticular sheath; annules coarse, about 5 µm wide, numbering 172 in holotype (Fig. 5.1). Lateral field indistinct, lip region bluntly rounded with two annules. Spear slender, 86–100 µm long with well-developed basal knobs that slope posteriorly. Oesophagus criconematoid with amalgamated procorpus and median bulb (Fig. 5.1E). Anus on 11th annule from terminus in

DOI: 10.1079/9781789247367.0006

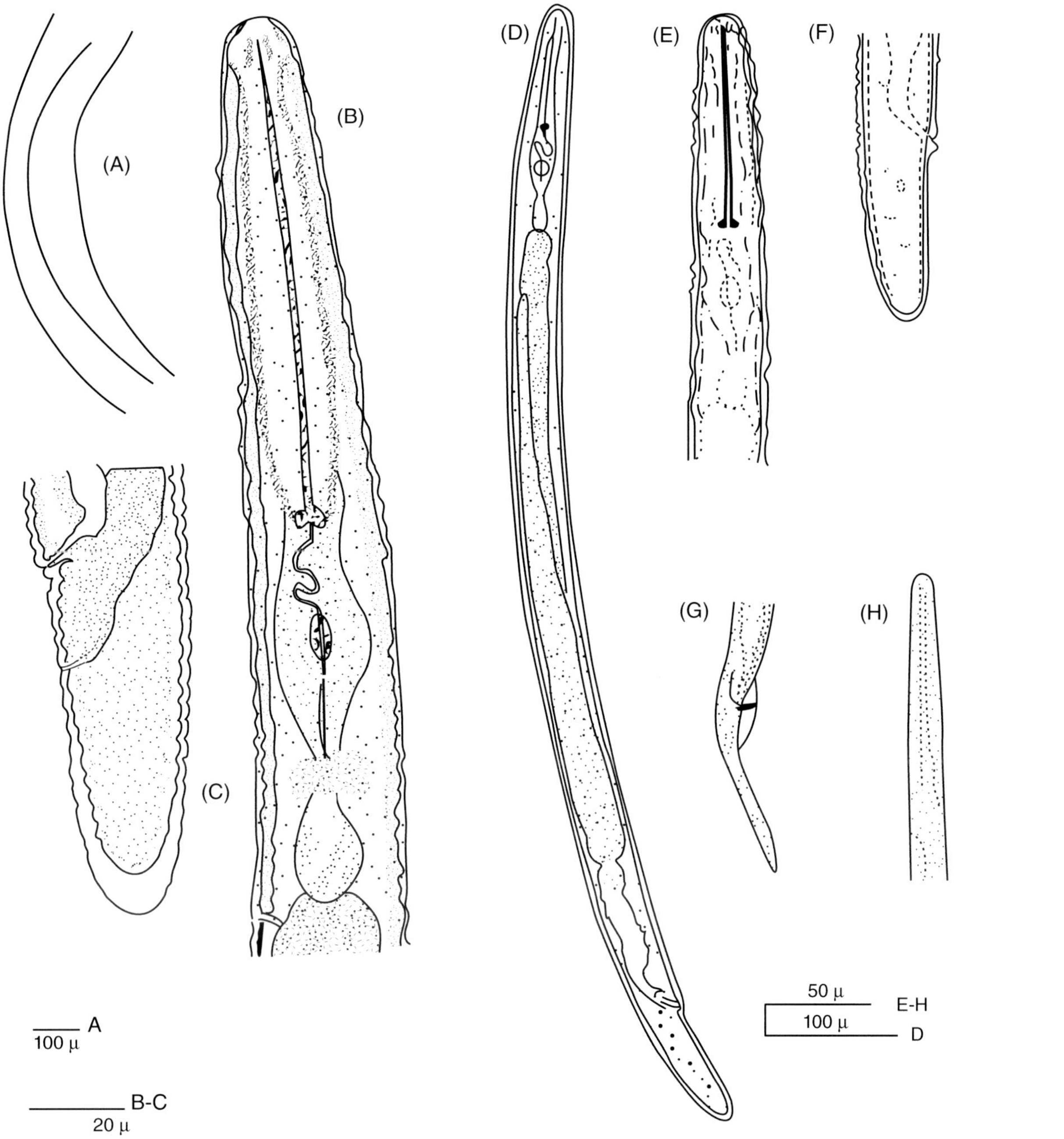

Fig. 5.1. Line drawing of *Hemicycliophora arenaria*. Shape when relaxed (A); anterior end of female (B); postvulval region (C); whole female (D); female oesophageal region (E); female posterior region (F); male tail (G); male anterior region (H). [Courtesy A-C: J. Nicol, University of Sydney; D: A.R. Stone; E-G: after Raski 1958; D-G adapted from CIH Descriptions set 3, CABI]

holotype; tail cylindroid with a bluntly rounded terminus. Vulva posterior and with slightly protruding lips. Genital tract monoprodelphic.

Measurements: L = 813 (770–885) µm; width = 36 (34–38) µm; tail 39 (36–42) µm; a = 23 (21–24); b = 5.3 (4.9–5.7); c = 20.0 (19.1–22.3); V = 91 (89–92); spear = 89 (87–91) µm; anterior end to excretory pore = 169 (160–176) µm.

MALE: Body slender, cylindroid, tapering at each extremity. Annules fine, less than 2 µm in width. Lateral field with four incisures. Lip region hemispheroid and lacking annules. Spear lacking and oesophagus degenerate. Spicules sickle-shaped, gubernaculum simple. Bursa adcloacal and about three times as long as width of cloacal body; tail gradually tapering to a finely rounded tip (Fig. 5.1G).

Measurements: L = 652 (610–706) µm; width = 23 (18–28) µm; tail = 47 (40–55); a = 296 (23.7–36.3); c = 14.1 (12.7–15.4); T = 22 (19–24); tip of testis to cloaca = 143 (125–1588) µm.

17.4.2 Molecular diagnosis

Van Den Berg *et al.* (2018) characterized 18 valid species of *Hemicycliophora*, two new species and 13 unidentified species from Canada, South Africa, Spain and the USA, based on molecular analysis of the partial *COI* mtDNA gene, D2–D3 of the 28S-rRNA the and ITS rRNA regions using standard PCR. The nucleic acid was extracted from single or numerous individuals of a species using worm lysis buffer (Tanha Maafi *et al.*, 2003) (see Chapter 15, section 110.1.1). The primers used in the study are:

For D2–D3 expansion segments of the 28S-rRNA (Subbotin *et al.*, 2007):

D2A forward primer (5′-ACA AGT ACC GTG AGG GAA AGT TG-3′)
D3B reverse primer (5′-TCG GAA GGA ACC AGC TAC TA-3′),

For ITS1-5.8S-ITS2 of rRNA gene (Curran *et al.*, 1994):

Forward primer TW81 (5′-GTT TCC GTA GGT GAA CCT GC-3′)
Reverse AB28 (5′-ATA TGC TTA AGT TCA GCG GGT-3′)

For partial *COI* mtDNA gene primers (Powers *et al.*, 2014):

Forward COIF5 (5′-AATWTWGGTGTTGGAACTTCT TGAAC-3′)
Reverse COIR9 (5′-CTTAAAACATAATGRAAATG WGCWACWACATAATAAGTATC-3′)

The PCR amplifications are performed using the procedure described by Tanha Maafi *et al.* (2003) (see Chapter 4, section 15.5.2). The PCR products are sequenced directly in both directions and the sequences are deposited to the NCBI GenBank under the accession numbers MG019814–MG019907. The study distinguished six valid species: *H. californica*, *H. gracilis*, *H. parvana*, *H. poranga*, *H. raskii* and *H. signata*.

17.5 Geographical distribution

H. arenaria has limited distribution and has been reported from Australia, the USA and Brazil (CABI-CPC, 2010). An earlier report from Argentina (Brugni and Chaves, 1994) has not been validated and it is considered as absent (EPPO, 2009b).

17.6 Hosts

H. arenaria is polyphagous and parasitizes a wide range of hosts, including citrus, tomato and cucurbits. The major and minor hosts are as follows (CABI-CPC, 2010).

MAJOR HOSTS: *Citrus aurantiifolia* (lime), *Citrus limonia* (mandarin lime), *Citrus reticulata* (mandarin), *Cucurbita moschata* (pumpkin), *Lycopersicon esculentum* (tomato).

MINOR HOSTS: *Apium graveolens* (celery), *Capsicum frutescens* (chilli), *Citrus limettioides* (Palestine sweet lime), *Citrus macrophylla* (alemow), *Citrus taiwanica* (nanshodaidai), *Libocedrus chilensis* (Chilean cedar), *Saccharum officinarum* (sugarcane), *Vigna unguiculata* (cowpea), *Vitis vinifera* (grapevine).

17.7 Symptoms

The nematodes feed near the root tips and cause galling of the affected area. Lateral roots are preferably attacked and become galled (Fig. 5.2). The overall effect is a stubby root system with galling at the root tip, reduced vigour and root function and associated effects on the above-ground plant parts. This nematode causes damage to citrus in nurseries (McElroy *et al.*, 1966) and has been brought under quarantine regulation (Van Gundy, 1959). The nematode feeds on root tips, leading to the formation of typical galls on the roots as a

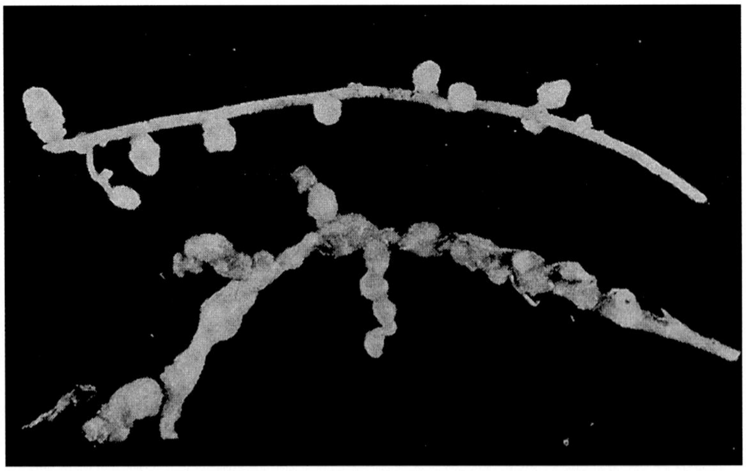

Fig. 5.2. Gall on rough lemon (*Citrus jambhiri*) induced by *Hemicycliophora*. [Courtesy H. Ferris, source Nemaplex; photo credit S.D. Van Gundy]

result of hypertrophy and hyperplasia (Fig. 5.2) (Colbran, 1963). Other species of *Hemicycliophora* are among the most widespread and abundant plant-parasitic nematodes associated with sugarcane in Peru (Carbonell, 1978), causing reduction in the growth of the cane and distortion to the roots. The roots acquire a cork-like consistency and the smaller roots collapse.

17.8 Biology and life cycle

H. arenaria is an obligate ectoparasite of plant roots and inhabits moist soils. It appears to be polyphagous to a certain degree, with citrus being the main associated host. It attacks both soft-rooted and woody plants. The second-stage juvenile hatches from the egg and there are two further juvenile stages before developing to adult males and females. Males have a degenerated spear and oesophagus and do not feed. They are not essential for reproduction. The nematodes are active from 6°C to 39ºC. At the type locality, the soil was 75% sand, 16% clay and 9% silt. In pot experiments, the nematodes multiplied best in soil containing 90% sand, 5% silt and 5% clay (Van Gundy and Rackham, 1961). Adequate soil aeration is important to the nematodes and they may be killed by prolonged irrigation cycles, a process that reduces soil aeration (Chitambar and Subbotin, 2014).

17.9 Similarities to other species

Most species of the genus *Hemicycliophora* are rather similar to one another and require expert examination for species-level identification. *H. arenaria* belongs to the small group of species having a cylindroid, bluntly rounded tail.

17.10 Economic importance

H. arenaria is known to damage agricultural crops in various countries (Chitambar and Subbotin, 2014). The major crop affected by this nematode is citrus in California, USA. A 10–20% reduction in the plant growth and yield of tomato and squash in field plots was reported by McElroy and Van Gundy (1968). Moura and Almeida (1982) regarded *H. arenaria* as being one of the nematodes responsible for the low yield of sugarcane in north-east Brazil.

17.11 Movement and means of dispersal

The likeliest mode of spread of *H. arenaria* is through the roots of infected citrus rootstocks from nurseries. Transmission via infected soil and farm implements or machinery is also possible.

17.11.1 Plant parts known to carry the pest in trade/transport

Bulbs, tubers, corms, rhizomes (sucker), growing medium accompanying plants, roots, seedlings and micro-propagated plants.

17.11.2 Plant parts not known to carry the pest in trade/transport

Bark, flowers, inflorescences, cones, calyx, fruits (including pods), leaves, stems (above ground), shoots, trunks, branches, wood and true seeds (including grain).

17.12 Pest risk analysis and looming threat from introduction

Because of the wide host range and economic damage to citrus in the USA, a complete pest risk analysis needs to be instigated for *H. arenaria*. However, due to its restricted distribution, preventive measures should be taken to avoid introduction into new host areas. The nematode can easily spread through citrus rootstocks for breeding purposes.

17.13 Invasiveness rating

Considering the consequences of the introduction and probable economic damage caused by *H. arenaria*, the following invasiveness rating has been assigned.

Climate/host interaction	–	Medium (2)
Hosts range	–	High (3)
Dispersal potential	–	Medium (2)
Economic impact	–	High (3)
Environmental impact	–	Low (1)
Post-entry distribution	–	Not known (0)
Final rating (score)	–	Medium (11)

Based on the above invasiveness rating, *H. arenaria* receives a score of 11 out of 15 (Medium) and is rated as MEDIUM INVASIVE NEMATODE.

17.14 Region-wise status of invasiveness

Because of the wide host range but limited distribution, *H. arenaria* has been given a medium invasive rating for the following countries where it is still absent or not recorded.

APPPC: Bangladesh, Cambodia, China, India, Indonesia, Iraq, Israel, Japan, Jordan, Kazakhstan, Kyrgyzstan, Lebanon, Malaysia, Myanmar, Pakistan, Philippines, South Korea, Sri Lanka, Syria, Tajikistan, Thailand, Vietnam, Uzbekistan

CAHFSA: Barbados, Belize, Cuba, Dominican Republic, Haiti, Jamaica, Saint Kitts & Nevis

CAN: Bolivia, Colombia, Ecuador, Peru, Venezuela

COSAVE: Argentina, Chile, Guyana, Paraguay, Suriname, Uruguay

EPPO: Austria, Belarus, Belgium, Bulgaria, Croatia, Cyprus, Czech Republic, Denmark, Estonia, Finland, France, Georgia, Germany, Greece, Guernsey, Hungary, Ireland,

Israel, Italy, Jersey, Jordan, Kazakhstan, Kyrgyzstan, Latvia, Lithuania, Luxembourg, Macedonia, Malta, Moldova, Morocco, Netherlands, Norway, Poland, Portugal, Romania, Russia, Serbia, Slovakia, Slovenia, Spain, Sweden, Switzerland, Tunisia, Turkey, Ukraine, United Kingdom, Uzbekistan

IAPSC: Algeria, Angola, Benin, Botswana, Burkina Faso, Burundi, Cameroon, Cape Verde, Central African Republic, Chad, Comoros, Congo, Côte d'Ivoire, Egypt, Eritrea, Ethiopia, Gabon, Gambia, Ghana, Guinea, Guinea-Bissau, Kenya, Liberia, Madagascar, Malawi, Mali, Mauritania, Mauritius, Mozambique, Namibia, Niger, Réunion, Rwanda, Senegal, Sierra Leone, Somalia, South Africa, Sudan, Swaziland, Togo, Uganda, Zambia, Zimbabwe

NAPPO: Canada, Mexico

NEPPO: Egypt, Jordan, Libya, Malta, Sudan, Syria, Tunisia

OIRSA: El Salvador, Guatemala, Honduras, Nicaragua

PPPO: Fiji, New Zealand, Papua New Guinea, Tonga, Vanuatu

17.15 Management measures

Pre-planting application of fumigant nematicides such as DBCP (1,2-dibromo-3-chloropropane) or DD mixture to basins around individual citrus trees is highly effective in controlling the nematode population (Van Gundy and McElroy, 1969). However, these fumigant nematicides have been superseded and effective control, or even complete eradication, can be achieved by using organophosphate nematicides as soil drenches. The spread of this nematode from the nursery can be checked by subjecting bare-rooted nursery seedlings to hot-water treatment at 46°C for 10 min. For phytosanitary measures, see Chapter 16, section 115.

6

Heterodoridae: *Cactodera*, *Globodera*, *Heterodera* and *Punctodera* Species

The family Heterodoridae consists of several cyst-forming species from eight genera, including four important ones: *Cactodera*, *Globodera*, *Heterodera* and *Punctodera* (Turner and Subbotin, 2013). The cyst nematodes are highly evolved and economically important plant-parasitic nematodes. The genera of this group stand out for their remarkably efficient parasitic adaptation and sexual dimorphism (Siddiqi, 2000). The juveniles and male remain vermiform, while the adult female is sedentary and becomes swollen and exposed on the root surface. Later the female body turns into a hard-walled protective cyst.

Cactodera Spp.

Adam (1932) described a cyst nematode for the first time, infesting cactus, and called it *Heterodera schachtii*. Filipjev and Schuurmans Stekhoven (1941) considered this cyst nematode to be a distinct species and re-described it as *Heterodera cacti*. The genus *Cactodera* was created by Krall and Krall (1978) within subfamily Heterodorinae (family Heterodoridae). The recent key of the genus is given by Subbotin *et al.* (2010). At present *Cactodera* includes the following 14 valid species.

Cactodera acnidae (Schuster and Brezina 1979) Wout, 1985
C. amaranthi (Stoyanov, 1972) Krall and Krall, 19788
C. aquatic (Krjanova, 1971) Krall and Krall, 1978
C. betulae (Hirschmann and Riggs, 1969) Krall and Krall, 1978
C. cacti Filipjev and Schuurmans Stekhoven, 1941
C. eremica Baldwin and Bell, 1985
C. estonica (Kirjanova and Krall, 1963) Krall and Krall, 1978
C. galinsogae Tovar, Prado, Nicol, Evans, Islas and Garza, 2003
C. milleri Graney and Bird, 1990
C. rosae Cid Del Prado and Miranda, 2008
C. salina Baldwin Mundo-Ocampo and McClure, 1997
C. thornei (Golden and Raski, 1977) Krall and Krall, 1978
C. torreyanae Cid Del Prado Vera and Subbotin, 2014
C. weissi (Steiner, 1949) Krall and Krall, 1978

Of these species, *C. amranthi* and *C. cacti* are considered economically important nematodes (Baldwin and Mundo-Ocampo, 1991; Sikora and Fernandez, 2005). *C. amranthi* has a limited host range and has been found attacking spinach in central Mexico (Sosa-Moss, 1986), *Amaranthus viridis* in Cuba (Stoyanov, 1971), *A. spinosus* and *A. retroflexus* in equatorial and hot tropical regions (Luc, 1986). *C. cacti* is usually found in association with succulent plants from the families Cactaceae and Euphorbiaceae.

Systematic position of *Cactodera* species

	Classical (after Siddiqi, 2000)	Molecular (after Decraemer and Hunt, 2013)
Phylum	Nematoda	Nematoda
Class	Secernentea	Chromadorea
Order	Tylenchida	Rhabditida
Suborder	Hoplolaimina	Tylenchina
Infra-order	–	Tylenchomorpha
Superfamily	Holpolaimoidea	Tylenchoidea
Family	Heteroderidae	Hoplolaimidae
Subfamily	Heteroderinae	Heteroderinae
Genus	*Cactodera*	*Cactodera*

18 *Cactodera cacti*

18.1 Common name

Cactus cyst nematode

18.2 Scientific name

Cactodera cacti Filipjev and Schuurmans Stekhoven, 1941

18.3 Synonym

Heterodera cacti Filipjev and Schuurmans Stekhoven, 1941

18.4 EPPO code

HETDCC

DOI: 10.1079/9781789247367.0007

18.5 Diagnosis

Morphological and molecular characters of *C. cacti* are as follows.

18.5.1 Morphological diagnosis

Important morphological characters, including morphometric measurements of *C. cacti*, are based on the descriptions from Baldwin and Mundo-Ocampo (1991) and Subbotin *et al.* (2010).

EGGS: Eggs always retained within cyst body and no egg sac is produced by the female (Cordero, 1989). The eggshell surface is covered by tiny punctations which resemble microvilli when viewed under scanning electron microscopy.

FEMALE: Emerges from root cortex, pearly-white in colour, turning yellow as it matures. Mature female usually small and spherical with low cone at posterior end of body, which is not always directly opposite the head (Fig. 6.1). Cuticle surface displays irregular transverse ridges and a distinct 'D' layer is present (Baldwin and Mundo-Ocampo, 1991).

Measurements: Length = 550 ± 5 µm, width = 450 ± 62 µm; stylet length = 29–35 µm; stylet base to dorsal oesophageal gland duct = 5.4 µm; vulval slit length = 9.5–15 µm; head tip to excretory pore = 124 ± 21 µm.

JUVENILE: Second-stage juvenile vermiform, cuticle regularly annulated, annules approximately 2 µm wide at mid-body (Fig. 6.1). Four incisures in lateral field, two outer ones crenate. Head four to five annules, lip region offset. *En face* view of lip pattern normally shows six distinct lips, dorso-ventrally elongated labial disc, four submedial lips and two greatly reduced lateral lips. Fusion often takes place between submedial and lateral lips (Graney and Bird, 1990). Framework of head skeleton strongly sclerotized but cephalids indistinct. Stylet robust and basal knobs slightly anteriorly faced. Dorsal oesophageal gland duct opening 4–5 µm from stylet base. Median oesophageal bulb well

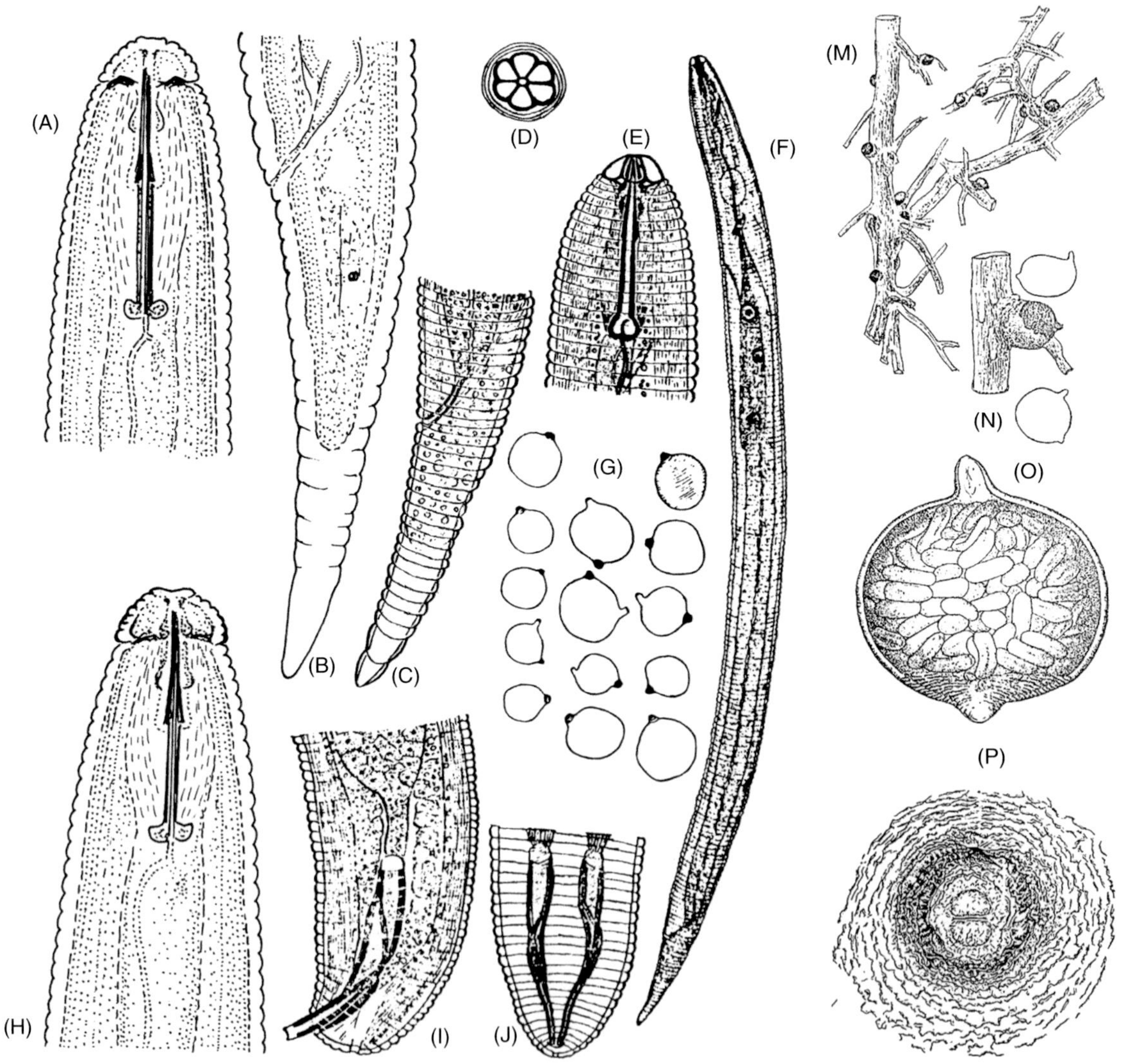

Fig. 6.1. Line drawing of *Cactodera cacti* second-stage juvenile (A–F), Cysts (G,M–P) and male (H–J). Head ends (A,E); tail end (B,C); *en face* view (D); head end (H); tail end, ventral view (I); female on roots of *Phylocactus* sp. (M); cysts (N,O); vulval cone top (P). [A–I adapted from CIH description, Set 7 No. 96; M–P adapted from M.R. Siddiqi, 2000, after Kirjanova and Terenteva, 1961].

developed, oesophageal glands overlap intestine ventrally and laterally and fill large proportion of anterior body cavity. Nerve ring encircles isthmus approximately 85–95 µm from head tip. Hemizonid located either adjacent to or 2 µm in front of excretory pore. Genital primordium located in mid-body region. Tail pointed, length 44–50 µm. Hyaline tail length 12–20 µm. Phasmids small, pore-like and located 12–18 µm behind anus.
Measurement: Length = 430–485 µm; a = 21–25, body width at excretory pore = 19 µm; stylet length = 24 µm; stylet base to dorsal oesophageal gland duct = 4 µm; head tip to median bulb valve = 66–70 µm; head tip to excretory pore = 101–108 µm; anus to tail terminus = 44 µm; tail width at anus = 10 µm; hyaline part of the tail = 18 µm.
CYSTS: Spherical to lemon-shaped with low protruding vulval cone (Fig. 6.1). Cuticle thin sac-like shape resembling dead female from which it originated. Cysts light brown to reddish-brown on maturation. Outer surface of cuticle covered in irregular transverse ridges, becoming characteristically striated near mid-body (a feature of genus *Cactodera* and unknown in other cyst nematodes except *Dolichodera* (Baldwin and Mundo Ocampo, 1991)). Subcrystalline layer not observed, but new cysts appear to have 'waxy' bloom (Southey, 1957). Egg mass not produced. Cysts abullate but have discrete vulval denticles. Vulval denticles tooth-like structures present on vulval cone wall at depth within cone of around 20 µm. In *C. cacti* they occur in small groups or singly, in other *Cactodera* species they may encircle the fenestra. Presence of vulval or cyst cone with circumfenestrate arrangement unique amongst cyst nematodes. No anal fenestration or pattern surrounding anus; anus distinct and located around 50 µm or 10% of cyst length from cone tip.
Measurements: Length excluding neck = 360–590 µm; maximum width = 340–580 µm; length of fenestra = 25–38 µm; length of vulval slit = 19–25 µm; vulva to anus = 70–75 µm; width of fenestra = 20–26 µm.
MALE: Body vermiform and twisted between 90° and 180° near tail, with four incisures in lateral field. Outer parts of lateral fields crenate with areolations seen under light microscopy (× 100 objective). Head small, dome-shaped, with six head annules. Head skeleton strongly sclerotized but cephalids indistinct. Stylet well developed, 26–28 µm in length, conus being half of this length. Stylet knob shape anteriorly concave. Dorsal oesophageal duct terminates approximately 5 µm from stylet knob base. Median bulb well developed. Oesophageal glands extend ventrally over intestine. Excretory pore located 120–150 µm from head tip. Hemizonid about 4 µm in length, found from three to seven body annules anterior to excretory pore. Spicules have bifid ends, typical of heteroderid males (Fig. 6.1). Gubernaculum simple. Testis singular, about half the body length. Phasmids inconspicuous. Head of spicule of *C. cacti* from population from Netherlands depicted as straight rather than curved (Graney and Bird, 1990).
Measurements: Length = 1000 µm; stylet length = 27 µm; dorsal oesophageal gland duct = 4 µm; tail length = 5 µm; spicule length = 33 µm; gubernaculum length = 10–15 µm; a = 33–40; head tip to excretory pore = 120–150 µm.

18.5.2 Molecular diagnosis

Tanha Maafi *et al.* (2003) used restriction fragment length polymorphism (RFLP) and sequences of internal transcribed spacer (ITS) rDNA of 45 populations of cyst-forming nematodes collected from different parts of Iran and identified as *Cactodera cacti*, *Heterodera avenae*, *H. elachista*, *H. filipjevi*, *H. humuli*, *H. fici*, *H. glycines*, *H. goettingiana*, *H. hordecalis*, *H. latipons*, *H. mothi*, *H. schachtii*, *H. trifolii* and *H. turcomanica*, and several unidentified species attributed to the genus *Heterodera*. For DNA extraction, one to four cysts are punctured on a glass slide under a dissecting microscope, releasing second-stage juveniles and eggs, and processed using worm lysis buffer (for details, see Chapter 15, section 110.1.1). The ITS-rDNA is amplified with the primers Forward TW81 (5′-GTTTCCGTAGGTGAACCTGC-3′) and Reverse AB28 (5′-ATATGCTTAAGTTCAGCGGGT-3′) (for details, see Chapter 4, section 15.5.2). The PCR product is purified and digested with one of the eight restriction enzymes (*Alu*I, *Ava*I, *Bsh1236*I, *BsuR*I, *Cfo*I, *Mva*I, *Pst*I, and *Rsa*I) to obtain RFLP profiles. The digested DNA is run on a 1.5% TBE buffered agarose gel, stained with ethidium bromide and photographed using Scientific Imaging System under UV light. The exact lengths of restriction fragments from the PCR products are obtained by virtual digestion of the sequences using WebCutter 2.0. Several clones of each species are isolated using blue/white selection and submitted to PCR with vector primers. PCR products from each sample or clone are sequenced using the TW81, AB28 primer and internal reverse 5.8SM5 primer (5′-GGCGCAATGTGCATTCGA-3′) as described by Zheng *et al.* (2000) with the BigDye Terminator Cycle Sequencing Ready Reaction Kit according to the manufacturer's instructions. The programme used for sequencing reactions is 30 s at 94°C, 30 s at 50°C and 3 min 30 s at 60°C, repeated for 25 cycles. The resulting products are purified and sequences are run on a 377 DNA Sequencer. The sequences obtained in this study were deposited in GenBank (AF498393 for *C. cacti*). The amplification of the ITS-rDNA region yielded 981 bp in *C. cacti*. Subbotin *et al.* (2011a) also used PCR–RFLP of ITS-rDNA gene sequence to characterize circumfenestrate cyst nematode, including *C. cacti*.

Cid Del Prado Vera and Subbotin (2014) used ITS1-5.8S-ITS2 of rRNA coupled with D2–D3 expansion segment to identify the new species of cyst nematode, *C. torreyanae*, and to compare similarity with other *Cactodera* species, including *C. cacti*. Skantar *et al.* (2019) used ITS-rDNA sequences to evaluate the intraspecific variation among all available *C. cacti* ITS sequences. The ITS-rDNA sequences from the Idaho population formed a strongly supported clade with 18 other sequences of *C. cacti*. Three 28S rDNA amplicons of 721 bp were obtained from three separate nematode juveniles and gave rise to identical sequences (MH478572–MH478574). MegablastN search of the NCBI NR database showed 99% identity to a single sequence of *C. cacti* (DQ328702) from Germany but also to several *C. estonica* sequences (KX230467–KX230473), indicating that 28S is unable to resolve the identity of the population.

18.6 Geographical distribution

C. cacti was first recorded and described from the Netherlands and later reported from Central America, the southern part of

North America and in the arid zone of north and central South America (Siddiqi, 2000). However, it is now considered to have originated in Mexico (Krall and Krall, 1978) and is distributed worldwide, with more concentration in European countries (Fig. 6.2). There are several records of major infestations in Russia, the USA and Central America (Sosa-Moss, 1987). It has also been reported from some Asian countries such as India (Kumar, 1964), Iran (Tanha Maafi *et al.*, 2003) and China (Duan *et al.*, 2012). Continent-wise, *C. cacti* is distributed in the following countries.

Africa: Algeria.
Asia: China, India, Iran, Israel, Japan, Singapore, South Korea, Uzbekistan, Vietnam.
North America: Canada, Cuba, Mexico, USA.
Europe: Austria, Belgium, Czechia, Denmark, France, Germany, Hungary, Italy, Malta, Netherlands, Poland, Romania, Russia, Spain, Sweden, Switzerland, United Kingdom, former Yugoslavia.
Oceania: Australia.
South America: Argentina, Brazil, Colombia.

18.7 Hosts

The cactus cyst nematode *C. cacti* has a very restricted host range, with most hosts being succulent plants belonging to the families Eurphorbiaceae and Cactaceae (Siddiqi, 2000). The nematode was first described from the cacti *Disocactus akkermanni* and *Cereus speciosus*. The first record of hosts from the USA came from Johnson (1968). Paneque and Sampedro (1986) conducted a host range test on other economically important plants in Cuba and found that sugarcane, kenaf, tomatoes, squash and maize are not a host of this nematode. It mainly attacks ornamental plants like *Echinopsis*, *Epiphyllum ackermannii*, *Erysimum allionii*, *Ferocactus*, *Opuntia microdasys* (angel's wings), *Opuntia stricta* (erect prickly pear) and *Schlumbergera truncata* (Christmas cactus). Chen *et al.* (2012) reported *C. cacti* infection on pitaya (dragon fruit) in Taiwan.

18.8 Symptoms

The symptoms include branching of roots and the production of more rootlets than normal. Necrosis found at feeding sites may spread and contribute to secondary root-rotting, allowing saprozoic nematodes and pathogenic fungi and bacteria into the plant tissues. Observations on the top growth of infected plants range from no adverse effects to top growth that is reddish-brown to yellow in colour, stunting, wilting, decreased flower production and reduced flowering period (O'Bannon and Esser, 1970). Death of plants may occur when population densities of *C. cacti* are high. Hosts which appear to be very susceptible, such as *Zygocactus truncatus* (= *Schlumbergera truncate*) and *Cereus baumanni*, may die after several years of heavy infestation by *C. cacti* (Narbaev, 1973).

18.9 Biology and life cycle

The biology and life cycle of *C. cacti* are very similar to other cyst nematodes. First-stage juveniles moult within the egg to become infective second-stage juveniles, which are normally folded five times within the egg (Golden and Raski, 1977). Once the host–parasite relationship is established, the life cycle takes 29–34 days to complete at 18–26°C under glasshouse conditions (Shmal'ko, 1959). Young females are white and are often found in the parenchyma of the root cortex, usually breaking through the cortical cells on to the root surface. By this time the females have become a golden yellow colour, before they tan to a light reddish-brown at maturation. The cyst contains 200–600 eggs (Graney and Bird, 1990). Males are usually present but not in large numbers. As the roots of succulent plants parasitized by *C. cacti* are themselves

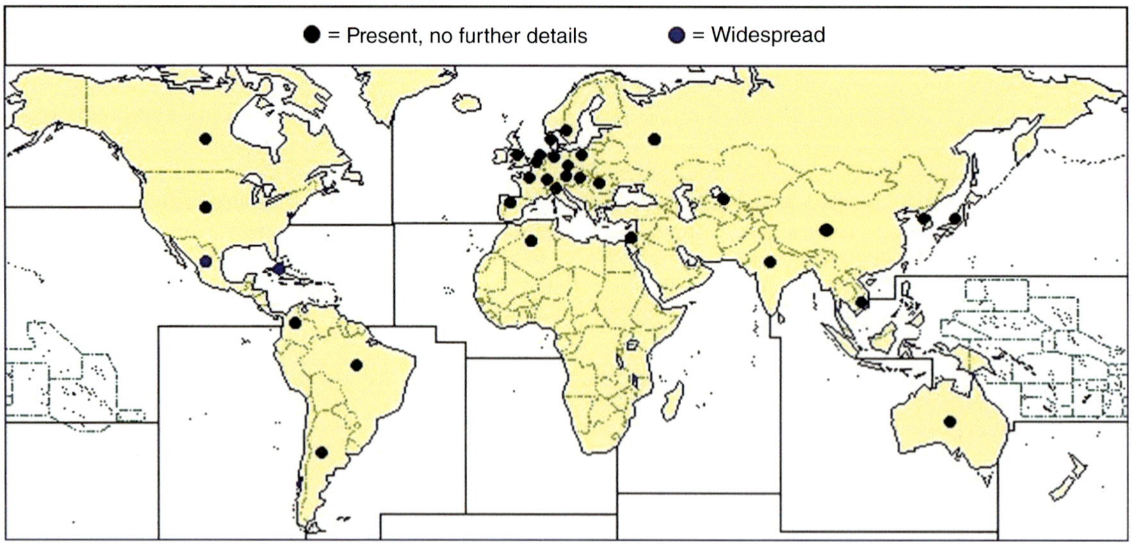

Fig. 6.2. Distribution map of *Cactodera cacti*. [Adapted from CABI-CPC, 2010]

long-lived, they are probably capable of supporting the nematode population for a considerable period (Baldwin and Mundo-Ocampo, 1991).

18.10 Similarities to other species

Cactodera forms a relatively small group of cyst nematodes, with 14 described species. *C. cacti* is most likely to be confused with *C. milleri* (Graney and Bird, 1990). However, the major host of *C. milleri* is the weed *Chenopodium album*. Among other species, *C. betulae* has an entirely different host range, exclusively from Betulaceae. The punctations are an important diagnostic character within the *Cactodera* group, as some species have smooth egg surfaces, such as *C. weissi* and *C. amaranthi* (Golden and Raski, 1977), while others, such as *C. cacti*, *C. milleri*, *C. eremica* and *C. thornei*, are punctated (Graney and Bird, 1990; Turner and Subbotin, 2013).

18.11 Economic importance

C. cacti is not significantly important with regard to economic loss. In Mexico, several states produce cacti for food, in the form of fruit and calodens, and *C. cacti* can reduce the profitability of these plants (Baldwin and Mundo-Ocampo, 1991). Recently it was found to infect dragon fruit in Taiwan (Chen *et al.*, 2012).

18.12 Movement and means of dispersal

The main means of dispersal to new areas is by passive movement of infested ornamental plants.

18.12.1 Plant parts liable to carry the pest in trade/transport

Roots, seedlings, micro-propagated plants and growing medium accompanying plants can carry the eggs, juveniles, adults and cysts of the nematode.

18.12.2 Plant parts not known to carry the pest in trade/transport

Stems (above ground), leaves (needles), shoots, trunks, branches, flowers, fruits, true seeds.

18.13 Pest risk analysis and looming threat from introduction

The cactus cyst nematode has been recorded mainly on ornamental plants of the families Cactaceae and Eurphorbiaceae, particularly when grown in greenhouses or protected cultivation. The dispersal of *C. cacti* from the native region to the Americas is beyond doubt associated with the international trade of ornamental cactus plants around the world (Turner and Subbotin, 2013). Infested plants are stunted and wilted, with reduced flower production. With high population densities of *C. cacti*, death of plants may also occur. The nematode is distributed throughout the world but, due to its relatively lower economic importance than many other cyst nematodes, it is not extensively studied or explored. Since it attacks ornamental plants that have global demand, quarantine measures need to be implemented in the areas where it has not yet been reported.

18.14 Invasiveness rating

Considering the consequences of introduction and pest risk analysis, *C. cacti* has been assigned the following invasiveness rating.

Climate/host interaction	–	Medium (2)
Hosts range	–	Medium (2)
Dispersal potential	–	High (3)
Economic impact	–	Medium (2)
Environmental impact	–	Medium (2)
Post-entry distribution	–	Not known (–1)
Final rating (score)	–	Medium (10)

Based on the above invasiveness rating, *C. cacti* receives a score of 10 out of 15 and is rated as MEDIUM INVASIVE NEMATODE.

18.15 Region-wise status of invasiveness

C. cacti is a regulated quarantine pest in many countries. Based on worldwide distribution and likely introduction in the countries where the nematode is still absent, *C. cacti* attains medium invasive status in the countries where the nematode is not known to occur.

18.16 Management measures

Most of the infested plants can be individually freed from nematodes either by taking stem cuttings and potting them into new, non-infested soil, or by cutting off the old root system and repotting into clean soil. When large numbers of potted plants require treatment, a hot-water method can be used effectively. This involves immersing the pot or root-ball in ten times its own volume of water heated to 43–45°C for 30 min (Stelter and Kühn, 1973). Effective control, even complete eradication, has been claimed when using organophosphate nematicides as soil drenches at 300, 600 or 1000 ppm (O'Bannon and Esser, 1970).

Since it has a wide host range, including ornamental crops, immediate regulatory measures should be taken if *C. cacti* is detected in any consignment, especially ornamental plants. It is a regulated quarantine nematode in many countries and an additional phytosanitary certificate is needed to import the planting material from the host country. Details of ISPMs are discussed in Chapter 16, section 115.

Globodera Spp.

Skarbilovich (1959) proposed the subgenus *Globodera* for *Heterodera* spp. having a globular cyst. Later she suggested the separate genus *Globodera*, with *G. rostochiensis* as its type species (Siddiqi, 2000). *Globodera* spp. parasitize plants from Solanaceae and Asteraceae. Potato cyst mematodes (PCNs) were introduced into Europe in the 1850s with potato collections used in breeding for late blight resistance following the Irish Famine (Evans and Stone, 1977). The first report of cyst-forming nematodes on potatoes was made by Kühn in 1881 from Rostoc, Germany. Wollenweber (1923) described this nematode as *Heterodera rostochiensis* and so it was distinguished from the beet cyst nematode, *H. schachtii*. At present, 16 *Globodera* species have been described by different authors, including two new species, *G. ellingtonae* and *G. capensis*:

Globodera archilleae (Golden and Klindic, 1973) Behrens, 1975
G. artemisiae (Eroshenko and Kazachenko, 1970) Behrens, 1975
G. capensis Knoetze, Swart and Lowwrens, 2013
G. chaubattia (Gupta and Edward, 1973) Wout, 1984
G. ellingtonae Handoo *et al.* 2012
G. hypolysi Ogawa, Oshima and Ichinohe, 1983
G. leptonepia (Cobb and Taylor, 1953) Skarbilovich, 1959
G. millefolii (Kirjanova and Krall, 1965) Behren, 1975
G. mirabilis (Kirjanova, 1971) Mulvey and Stone, 1976
G. pallida (Stone, 1973) Behren, 1975
G. pseudorostochiensis (Kirjanova, 1971) Mulvey and Stone, 1976
G. rostochiensis (Wollenweber 1923) Skarbilovich, 1959
G. tabaccum solanacearum (Miller and Gray, 1972) Behren, 1975
G. tabaccum tabaccum (Lownsbery and Lownsbery, 1954) Skarbilovich, 1959
G. tabaccum virginiae (Miller and Gray, 1968) Stone, 1973
G. zelandica Wout, 1984

For many years, *G. rostochiensis* and *G. pallida* were classified as a single species, *H. rostochiensis*, with the common names golden nematode or potato cyst nematode (PCN). Later the PCN was separated into two species, *H. rostochiensis* (with golden females) and *H. pallida* (with white or cream-coloured females) by Stone (1973). PCNs are a menace to potato growers in Europe and other parts of the world and one of the important quarantine pests in many countries (Turner and Subbotin, 2013).

Systematic position of *Globodera* species

	Classical (after Siddiqi, 2000)	Molecular (after Decraemer and Hunt, 2013)
Phylum	Nematoda	Nematoda
Class	Secernentea	Chromadorea
Order	Tylenchida	Rhabditida
Suborder	Hoplolaimina	Tylenchina
Infra-order	–	Tylenchomorpha
Superfamily	Holpolaimoidea	Tylenchoidea
Family	Heteroderidae	Hoplolaimidae
Subfamily	Heteroderinae	Heteroderinae
Genus	*Globodera*	*Globodera*

19 *Globodera pallida*

19.1 Common name

White potato cyst nematode

19.2 Scientific name

Globodera pallida (Stone, 1973) Behrens, 1975

19.3 Synonyms

Heterodera pallida Stone, 1973

19.4 EPPO code

HETDPA

19.5 Diagnosis

Morphological and molecular characters are described below.

19.5.1 Morphological diagnosis

The following morphological characters of *G. pallida* are based on descriptions from Baldwin and Mundo-Ocampo (1991) and Subbotin *et al.* (2010). For morphometric measurements (after Stone, 1973) see Table 22.

Eggs: Eggs retained within cyst, no egg sac formed. Eggshell surface smooth with no microvilli.
Juvenile: Juvenile folded four times within the egg, tail tapers to rounded point. Second-stage juvenile almost straight after heat fixation (Fig. 6.3A). Head rounded, strongly sclerotized, hexaradiated (Fig. 6.3D), offset and bears 4–6 annules. Cephalids located at body annules 2 and 3 and posteriorly at annules 6–8. Stylet long and robust, with anteriorly facing stylet knobs. Lateral field with four incisures along body length (Fig. 6.3G). Nerve ring girdles the oesophagus. Excretory pore about 110 µm from head. Body cavity extends halfway along tail, ending at anus. Hyaline region of tail about 20 µm long

Table 22. Morphometric data for *Globodera pallida*, *G. rostochiensis* and *G. tabacum*.

Morphometric characters		Range, numerous populations		
		G. pallida (after Stone, 1973)	*G. rostochiensis* (after Stone, 1973)	*G. tabacum* (after Mota and Eisenback, 1993c)
Females	Stylet (µm)	27.4 ± 1.1	23 ± 1	23 (18.5–24)
	Head width at base (µm)	5.2 ± 0.5	5.2 ± 0.7	310 (201–516)
	Stylet base to dorsal gland duct (µm)	5.4 ± 1.1	6 ± 1	5.5 (3.8–8.0)
	Head tip to median bulb valve (µm)	67.2 ± 18.7	73 ± 14.6	–
	Median bulb valve to excretory pore (µm)	71.2 ± 22	65 ± 2.0	–
	Head tip to excretory pore (µm)	139.7 ± 15.5	145 ± 17	–
	Mean diameter of vulval basin (µm)	24.8 ± 3.7	22 ± 2.8	–
	Length of vulval slit (µm)	11.5 ± 1.3	9.7 ± 2.0	9 (8-10)
	Anus to vulval basin (µm)	44.6 ± 10.9	60 ± 10	–
	Number of ridges on anal–vulval axis	12.5 ± 3.1	21 ± 3.0	5–10
Males	Length (mm)	1.20 ± 10	0.89–1.27	1.20 (0.7–1.3)
	Body width at excretory pore (µm)	28.4 ± 1.0	28 ± 1.7	34.7 ± 4.1
	Head width at base (µm)	12.3 ± 0.5	11.8 ± 0.6	11.0 ± 0.8
	Head length (µm)	6 ± 0.3	7.0 ± 0.3	5.7 ± 0.6
	Stylet length (µm)	27.5 ± 1.0	26 ± 1.0	26 (24–27)
	Stylet base to dorsal gland duct (µm)	3.0 ± 1.0	5.3 ± 1.0	4 (2.0-3.6)
	Head tip to median bulb (µm)	66 ± 7.1	98.5 ± 7.4	95.7 ± 7.0
	Median bulb to excretory pore (µm)	81.0 ± 11	74 ± 9.0	64.4 ± 4.3
	Head tip to excretory pore (µm);	176.4 ± 14.5	172 ± 12.0	160.1 ± 11.3
	Tail length (µm)	5.2 ± 1.4	5.4 ± 1.0	4.0 (2.0–3.6)
	Tail width at anus (µm)	13.5 ± 2.1	13.5 ± 0.4	–
	Spicule length (µm)	36.3 ± 4.1	35.0 ± 3.0	34 (26–34)
	Gubernaculum length (µm)	11.3 ± 1.6	10.3 ± 1.5	12
Cysts	Length, excluding neck (µm)	579 ± 70	445 ± 50	550 (337–740)
	Neck length (µm)	188 ± 20	104 ± 19	86 (10–100)
	Width (µm)	534 ± 66	382 ± 60	500 (232–645)
	Mean fenestral diameter (µm)	24.5 ± 5.0	19.0 ± 2.0	22 (15–28)
	Anus to fenestra distance (µm)	50 ± 13.4	66.5 ± 10.3	35
	Granek's ratio	2.2 ± 1.0	3.6 ± 0.8	1.5-2.0
Second stage juveniles	Length (µm)	486 ± 28	468 ± 100	500 (410–527)
	Body width at excretory pore (µm)	19.3 ± 0.9	18 µm ± 0.6	24 (22–24)
	Stylet length (µm);	23.0 ± 1.0	22 ± 0.7	22 µm (19–24
	Stylet base to dorsal gland duct (µm)	5.3 ± 0.9		5 µm (4.3–6.8)
	Head tip to median bulb valve (µm)	68.7 ± 2.7	69 ± 2.0	66.5 ± 3.6
	Head tip to excretory pore (µm)	108.6 ± 4.1	100 ± 2.0	102.6 ± 4.6
	Tail length (µm)	51.1 ± 2.8	44 ± 12	52 (46–59)
	Tail width at anus (µm)	12.1 ± 0.4	11.4 ± 0.6	13.8 ± 1.3
	Length of hyaline terminus (µm)	26.6 ± 4.1	26.5 ± 2.0	27 (23–31)
Eggs	Length (µm)	108.3 ± 2.0	101–104	90–100
	Width (µm)	43.2 ± 3.2	46–48	40–50
	L/W ratio	2.2–2.6	2.1–2.5	2.4 ± 0.2

(Fig. 6.3E). Hemizonid present just before excretory pore and about 2 annule-widths long. Hemizonion 5–6 annules opposite excretory pore. Genital primordium located 60% of body length from head tip.

MALE: Vermiform and generally an open C-shape upon heat fixation. Body annules regular along body. Lateral field with three bands, narrowing both anteriorly and posteriorly. Head offset, hexaradiate and heavily sclerotized, bears 6 or 7 annules. Anterior cephalids situated at head annules 2 and 4, and posteriorly at annules 6–9. Stylet strong with backward-sloping knobs. Median bulb rounded with crescentic strong valve. Oesophagus surrounded by nerve ring. Dorsal oesophageal gland nucleus the most obvious of the three gland nuclei, lobe extends nearly to excretory pore. Tail short-rounded, twisting

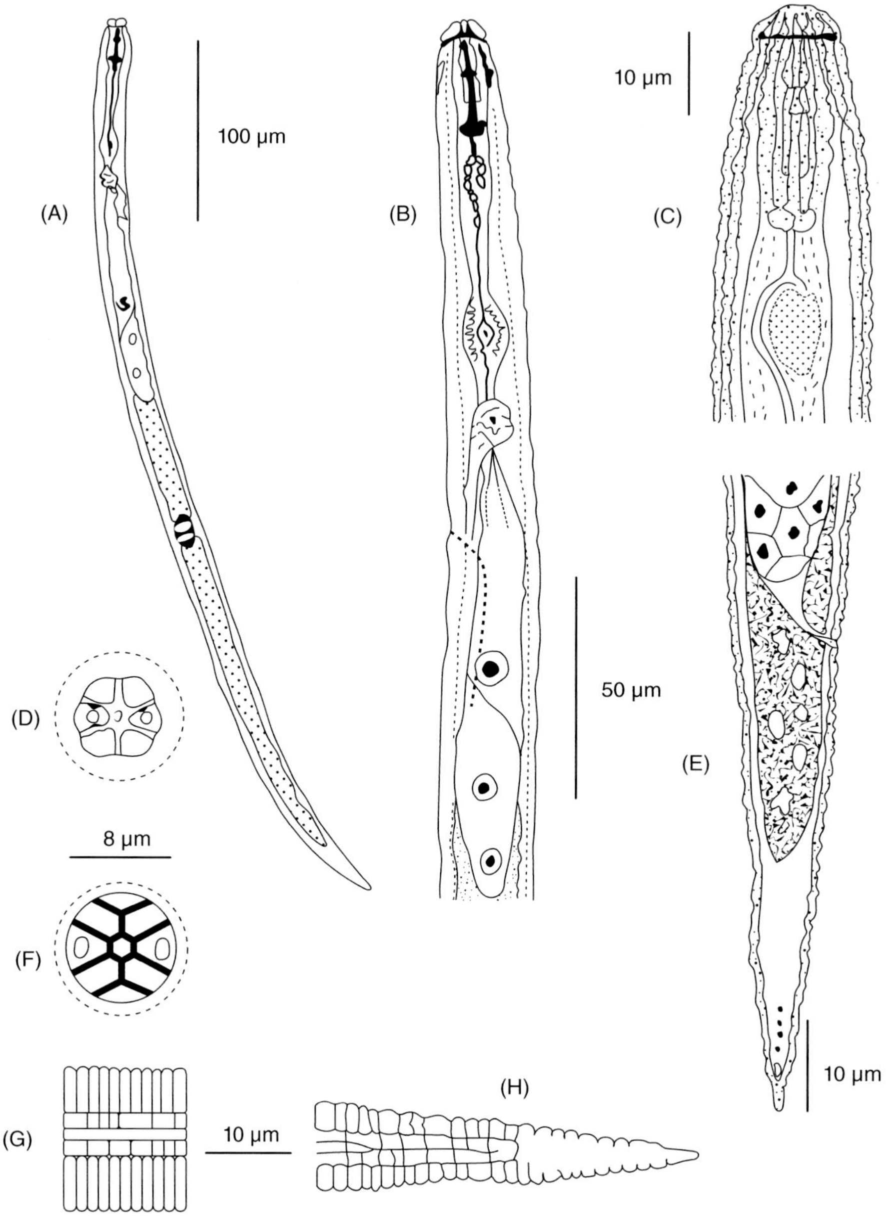

Fig. 6.3. Line drawing of *Globodera pallida* second-stage juvenile. Entire (A); anterior (B); head (C); head and face at level of lips (D); tail (E); head and face at level of base (F); lateral field mid-body region (G); lateral field tail (H). [Stone, 1973; courtesy Nematologica]

through 90–180°. Male genital tract with single testis extending for around 60% of body length. Spicules arcuate and with single points. Gubernaculum small and without ornamentation.

FEMALE: Nearly spherical body, protruding neck and head. Either cream or white. Head skeleton weak and hexaradiate. Stylet consists of equal proportion of conus and shaft with anteriorly facing knobs. Median bulb well developed with large crescentic pump. Oesophageal gland lobes frequently displaced into a forward position due to expansion of large paired ovaries in body cavity. Excretory pore at base of neck. Vulval slit located in vulval basin with round depression at opposite pole of body to neck. Vulval slit encircled by translucent thin cuticle which carries papillae.

CYST: Cuticle of dead female tans to deep brown colour and becomes tough, a hardened cyst. Dimensions of cyst nearly equal to those of female, while head is normally lost. Anus generally conspicuous. Fenestrae have usually been lost by mature cyst stage and only a hole remains. Vulval cone abullate, but rarely small rounded brown-pigmented bodies are found, called vulval bodies.

19.5.2 Biochemical diagnosis

Biochemical techniques to differentiate species, races or pathotypes of *G. pallida* and *G. rostochiensis* have been described by different workers (Den Nijs and Lock, 1990; Fleming and Powers, 1998; Abrantes *et al.*, 2004). Usually, these techniques are based on two-dimensional gel electrophoresis (2-DGE) using proteins to examine inter- and intraspecific differences of *Globodera*, *Heterodera*, *Ditylenchus* and *Meloidogyne* species (Cunha *et al.*, 2000; Fullaondo *et al.*, 2001). Immunological approaches using monoclonal antibodies (mAbs) and polyclonal antibodies (pAbs) are also tested to recognize multiple antigenic determinants in *Globodera* species (De Boer *et al.*, 1996). Fleming and Powers (1998) gave a detailed account of the biochemical diagnostic technique for potato cyst nematode.

19.5.3 Molecular diagnosis

Molecular characterization based on PCR-RFLP of ITS regions, microsatellites or single sequence repeats (SSRs) and random amplified polymorphic DNA (RAPD)-PCR has been exclusively studied for *G. pallida* (Subbotin *et al.*, 1999; Fleming *et al.*, 2000; Garcia *et al.*, 2009), *G. rostochiensis* (Radivojevic *et al.*, 2001; Subbotin *et al.*, 2010) and *G. tabacum* (Thiéry and Mugniéry, 1996; Fleming *et al.*, 2000; Subbotin *et al.*, 2010). PCR single-strand conformational polymorphism (PCR-SSCP) has also been designed for *G. pallida* and *G. rostochiensis* (Clapp *et al.*, 2000). More recently DNA barcoding, phylogeny and phylogeography from 148 populations of *Globodera* species from 23 countries have been analysed (Subbotin *et al.*, 2019b), which has provided comprehensive phylogenetic analyses of 455 ITS rRNA, 219 COI and 164 cytb gene sequences of 11 valid and two undescribed species of *Globodera*. The study revealed that *Globodera* displayed two main clades: (i) *Globodera* from South and North America parasitizing plants from Solanaceae; and (ii) *Globodera* from Africa, Europe, Asia and New Zealand parasitizing plants from Asteraceae and other families. Some useful molecular methods are described below.

MULTIPLEX PCR TEST Bulman and Marshall (1997) developed a protocol to differentiate *G. pallida* and *G. rostochiensis* populations from different pathotypes and geographical origins. The nucleic acid is directly extracted from the cysts. The test is designed for 18S rRNA gene of ITS5 region using the following primers (for detailed protocol, see Chapter 15, section 110.2.4).

Universal primer ITS5: 5′-GGAAGTAAAAGTCGTAACAAGG-3′
G. pallida-specific primer PITSp4: 5′-ACAACAGCAATCG TCGAG-3′
G. rostochiensis-specific primer PITSr3: 5′-AGCGCAGACATG CCGCAA-3′

The amplicon size of the PCR product of the reaction with the universal primer ITS5 and *G. pallida*-specific primer is 265 bp and with *G. rostochiensis*-specific primer it is 434 bp. As the primers are *Globodera*-specific, this test can only be used for the nematodes morphologically identified as *Globodera* spp.

REAL-TIME PCR FERA (2009) developed a TaqMan-realtime PCR method for *G. pallida* and *G. rostochiensis*. Over 100 populations of *G. pallida* from 26 countries, 30 populations of *G. rostochiensis* from 10 countries and some non-target populations of *G. tabacum*, *G. achillae/millefolii* were included in this protocol. The nucleic acids are extracted from either a single cyst or the anterior half of a single cyst or single juveniles of *Globodera* spp. The real-time PCR test targets ITSI genes with the following primer sets (for details, see Chapter 15, section 110.2.5).

Forward primer Glob 531F: 5′-TGT-AGG-CTG-CTA-YTC-CAT-GTY-GT-3′
Reverse primer Glob 601R: 5′-CCA-CGG-ACG-TAG-CACACA-AG-3′

Two locked nucleic acids (LNA) probes:

GP LNA: 5′-TGC-CGT-ACC-(C)(A)G-CGG-CAT-3′
GR LNA: 5′-GCC-GTA-CC(T)-(T)GC-GGC-AT-3′

The forward primer is located at base 521 of *G. pallida* and base 356 of *G. rostochiensis*. The amplicon length including primers is 71 bp. The sequences are submitted in GenBank with accession numbers *G. pallida* (AF016871) and *G. rostochiensis* (EF622531).

HIGH-THROUGHPUT DIAGNOSIS OF POTATO CYST NEMATODES (*GLOBODERA* SPP.) IN SOIL SAMPLES USING REAL-TIME PCR Reid *et al.* (2015) developed a high-throughput diagnosis protocol using ITS1 region of rDNA repeat unit. Besides *G. pallida* and *G. rostochiensis*, the primers were designed to take account of non-target other cyst species, including *Heterodera avenae* (two populations), *Puntodera punctipenis* (three populations) and *G. achillaea* (six populations). The DNA was extracted from the soil floats (cysts) (see Chapter 15, section 110.1.2). The amplification was performed using the following oligonucleotides.

Forward primer *G. pallida* and *G. rostochiensis*: 5′-CGTTTGTT GTTGACGGACAYA-3′
Reverse primer *G. pallida* and *G. rostochiensis*: 5′-GGCGCTGTC CRTACATTGTTG-3′

Two probes:

G. pallida 5′-CCGCTATGTTTGGGC-3′
G. rostochiensis 5′-CCGCTGTGTATKGGC-3′

The PCR product was sequenced and matched with available sequences from the NCBI database and submitted to GenBank with accession number FJ212165 for *G. pallida*.

QUALITATIVE DETECTION OF VIABILITY AND IDENTIFICATION OF PCNS (*GLOBODERA* SPP.) USING RNA-SPECIFIC REAL-TIME PCR Beniers *et al.* (2014) developed a protocol for the detection of viability and identification of *Globodera* spp. by real-time PCR. Total nucleic acids are extracted from a nematode sample containing from one to 50 *Globodera* cysts. The primer set is composed of:

Forward primer Fw_EF_Grp_mRNA: 5′-ACAAGATCGGA GGTATCG-3′)
Reverse primer Rv_EF_Gp_mRNA-1: 5′-GTGGTTCA TGATGATGA CCTG-3′

One probe is used for each species:

EF_Gpal_probe: 5′-Yakimo Yellow-CGAAGA{A}{T}GACCCGGC- BHQ1-3′
p_EF-Gros_2 LNA: 5′-6 FAM-CTCGAAGAG{C}GAC{C}CTG-BHQ1-3′

The amplification is performed using the following reagents and cycling parameters.

Reagent	– Final concentration
PCR-grade water	– 2.25 µl (per reaction)
One Step RT-PCR Buffer III (Takara)	– 12.5 µl
Primer Fw_EF_Grp_mRNA	– 0.3 µM
Primer Rv_EF_Gp_mRNA-1	– 0.3 µM
Probe EF_Gpal_probe	– 0.5 µM
Probe p_EF-Gros_2LNA	– 0.1 µM
Prime Script TM RT Enzyme Mix II (Takara)	– 0.5 µM
TaqTM HS (Takara)	– 2.5 U
ROXTM Dye II*	– 0.25 µl
Cycling parameters (42°C for 5 min)	
Initial denaturation	– 95°C for 10 s
Number of cycles	– 40
Denaturation	– 95°C for 10 s
Annealing	– 60°C for 1 min
Extension	– 60°C for 1 min

The method is quite sensitive and can detect a single viable juvenile or egg. The specificity is also very high and identifies *Globodera* spp., *G. pallida*, *G. rostochiensis* and no cross-reaction with other nematodes such as *H. betae*, *H. glycines*, *H. schachtii*, *H trifolii*, *G. tabacum* and *Cactodera cacti*.

QUANTITATIVE DNA DETECTION SYSTEMS FOR *GLOBODERA* SPP. Adams *et al.* (2009) developed this technique, which is able to isolate the DNA from float material (cyst) of both *G. pallida* and *G. rostochiensis*. The primer sets used in this technique are:

PCNTQPr1f: 5′-CACATGCCTCCGTTTGTTGT-3′
PCNTQPr1r: 5′-CGCTCAACGACGCACAGA-3′

The fluorescent probes are:

Pallida 5′ Fam-ACAGCAATCGTCGAGTCACCCATTG-Tamra3′
Rostochiensis 5′ Viv-CAGCAATCGTCGGCTCACCCATA-Tamra3′

The amplification is performed using the following reagents and cycling parameters.

Reagent	– Final concentration
TaqMan™ universal PCR master mix	– 1×
Primer PCNTQPr1f 5′	– 300 nM
Primer PCNTQPr1r	– 300 nM
Probe pallida	– 200 nM
Probe rostochiensis	– 100 nM
Cycling parameters (42°C for 5 min)	
Enzyme activation	– 50°C for 2 min
Gold activation	– 50°C for 2 min
Number of cycles	– 40
Denaturation	– 95°C for 10 min
Annealing	– 60°C for 1 min
Extension	– 60°C for 1 min

The rostochiensis primers and probe are specific for *G. rostochiensis* and will not pick up any *G. pallida*. Similarly, the pallida primer will not pick up *G. rostochiensis*.

19.6 Geographical distribution

The potato cyst nematode (PCN) (*G. pallida* and *G. rostochiensis*) is considered to have originated in the Andes mountains of South America and has been reported growing on *Solanum acaule* on the pre-Columbian agricultural terraces of Peru (Subbotin *et al.*, 2011a). PCNs were introduced into Europe in the 1850s with potato collections used in breeding for late blight resistance following the Irish Famine (Evans and Stone, 1977). From Europe, PCNs have spread to many parts of the world probably by the transport of potatoes and are found mostly in temperate regions, but also occur in coastal areas of the tropics (Evans and Brodie, 1980; EPPO, 2017c; CABI-ISC, 2020). At present the distribution covers temperate zones down to sea level and in the tropics at higher altitudes (Fig. 6.4). In these areas, distribution is linked with that of the potato crop. Currently, *G. pallida* is distributed in the following countries (CABI-ISC, 2020).

Africa: Algeria, Kenya, Tunisia.
Asia: India, Iran, Japan, Pakistan, Turkey.
Europe: Austria, Belgium, Bosnia and Herzegovina, Bulgaria, Croatia, Cyprus, Czech Republic, Denmark, Faroe Islands, Finland, France, Germany, Greece, Hungary, Iceland, Ireland, Italy, Luxemburg, Malta, Netherlands, Norway, Poland, Portugal, Romania, Slovenia, Spain, Sweden, Switzerland, United Kingdom.
North America: Canada, Costa Rica, Panama, USA.
Oceania: New Zealand.
South America: Argentina, Bolivia, Chile, Colombia, Ecuador, Falkland Islands, Peru, Venezuela.

19.7 Hosts

G. pallida is polyphagous and many solanaceous plants are its hosts. The only plant species outside Solanaceae, viz. *Antirrhinum majusi*, is a poor host of *G. pallida*. Major and minor hosts of *G. pallida* are listed below (CABI-ISC, 2020).

MAJOR HOSTS: *Lycopersicon esculentum* (tomato), *Solanum melongena* (aubergine), *Solanum tuberosum* (potato)
MINOR HOSTS: *Datura stramonium* (jimsonweed), *Lycopersicon pimpinellifolium* (currant tomato), *Solanum aviculare* (kangaroo apple), *Solanum gilo* (gilo), *Solanum indicum*, *Solanum marginatum*

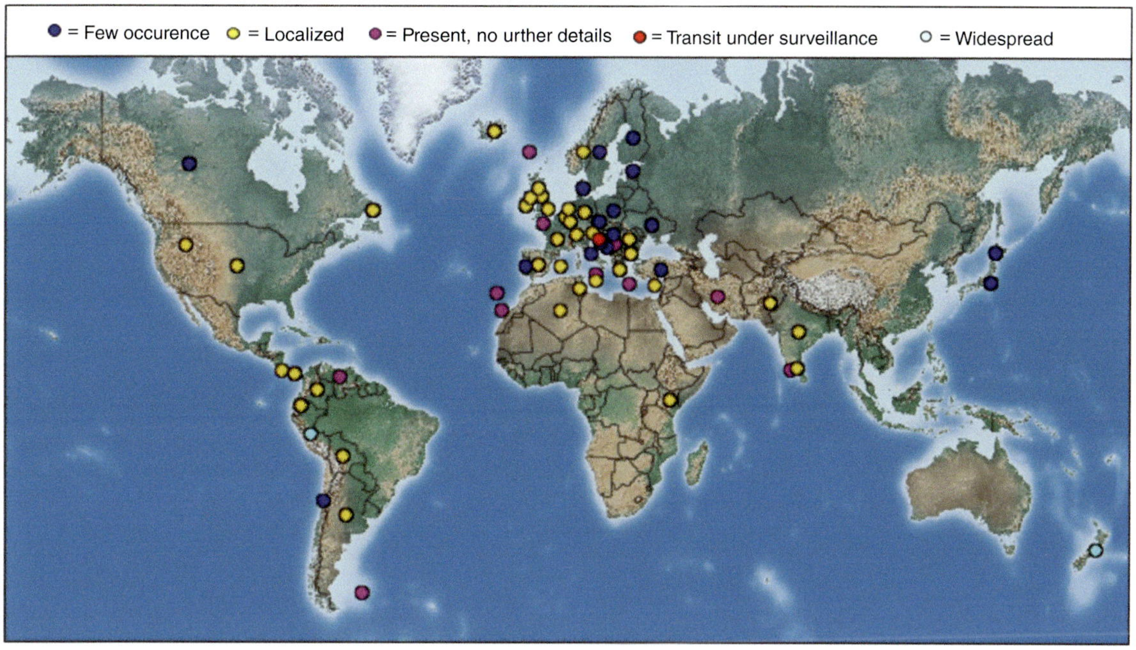

Fig. 6.4. Distribution map of *Globodera pallida*. [Adapted from CABI-ISC, 2020]

(white-edged nightshade), *Solanum mauritianum* (tree tobacco), *Solanum nigrum* (black nightshade), *Solanum quitoense* (narangillo), *Solanum sarrachoides* (green nightshade).

19.8 Symptoms

Generally, PCNs do not cause specific visible symptoms on above-ground parts of the plant and due to this the nematode often goes unnoticed. In severe condition, plants in patches show poor growth with yellowing and wilting or death of foliage (Fig. 6.5A) similar to that of many soil-borne roots pathogens. Plants show tufting of leaves on the top and the outer leaves show yellowing and subsequently die (Devrajan *et al*., 2012). The infested plants are also stunted and show day wilting during the hotter part of the day. Tuber production in terms of number and size is reduced. Roots are extensively branched, with soil stuck to them. In areas of severe infestation, roots are sometimes killed. The root system is smothered and secondary roots are produced at the colour region. Such plants can be easily pulled out (Marks and Brodie, 1998).

The above-ground symptoms described for PCN can have many other causes and cannot be taken as proof. Plants should, therefore, be pulled out for a visual verification for the presence of cysts and young females on the roots, or a soil sample should be taken for testing. Young females and cysts are just visible to the naked eye as small round white, yellow or brown pinheads on the roots of the plant (Fig. 6.5C,D). The cyst attached to the root can be observed by pulling plants from soil (Fig. 6.5F), but they can easily be lost at lifting. Soil testing is the best way to determine the presence of PCNs. The cysts can be seen by the naked eye but are best viewed with a hand lens. They are readily dislodged from roots, which serve as a source for future infestation.

Specific identification is just possible by examining the female colour at the suitable stage of development, either a change from white to yellow in *G. rostochiensis* or prolonged white (slightly cream but no yellow phase) in *G. pallida*. Thereafter, microscopical observation of juveniles, females or cysts is essential for precise identification (Subbotin *et al*. 2010).

19.9 Biology and life cycle

The active part of the life cycle starts when the second-stage juvenile (J_2) hatches from the egg. Hatching is stimulated by potato root diffusate (Rawsthorne and Brodie, 1986; Khan, 2008). After penetrating roots, the J_2 soon moults, but J_3 and J_4 stages moult after several weeks. Sex can be determined at this stage by nutritional status; those juveniles which induce sufficiently large giant cells tend to become females, whereas adult males can develop under conditions of nutrient stress (Trudgill, 1967). A final moult produces adult males and females. At this stage in the life cycle, the white or cream females gradually swell and develop into saccate shapes. As the size of the adult female enlarges, it ruptures through the root cortex so that the spherical body lies exposing the genitalia, with only the head and neck embedded in the root. The females secrete pheromones which attract males to mate with young adult females (Mugniery *et al*., 1992). Adult vermiform males leave the root and mate with as many available females as possible. Both sexes are capable of mating several times; fertilization is essential in *Globodera* spp. Males, once in the soil environment, survive for about 10 days (Evans, 1970; Marks and Brodie, 1998).

The adult female, initially white, becomes cream (*G. pallida*) or yellow (*G. rostochiensis*) at maturity. The female apparently dies, due to the obliteration of organs resulting from the massive

Fig. 6.5. Symptoms of *Globodera pallida* (A–F), *G. rostochiensis* (G–I) and *G. tabacum* (J–K) on host crops. Potato field infested with potato cyst nematode (A); healthy and infested potato plant (B); symptoms on potato root (C); cysts on potato tuber (D); close-up of cysts on the roots (E); damaged potato tuber (F); infested potato field (G); cysts on potato root in golden yellow phase (H); whole cysts (I); infested flue-cured tobacco field (J); brown cysts on tobacco root (K). [Photo courtesy: A,D: Florida Division of Plant Industry, Florida Department of Agriculture and Consumer Services, Bugwood.org; B: Christopher Hogger, Swiss Federal Research Station for Agroecology and Agriculture, Bugwood.org; C: NASI, India; E: SASA © Crown, https://www.flickr.com/photos/141760766@N08/28529191834/; F: Melanie Matthews, https://www.bhaf.org.uk/wp-content/uploads/cms/p/potato_cyst_nematode_damage.jpg; G: Bonsak Hammeraas, NIBIO - The Norwegian Institute of Bioeconomy Research, Bugwood.org; H: Pests and Diseases-Image Library Australia; I: NAK, NL; J: Janet A. Rowe, source CABI; K: R.J. Reynolds Tobacco Company, Bugwood.org]

formation of eggs (Khan, 2008). Upon death, the cuticle tans to form a brown protective cyst containing 200–500 embryonated eggs. Embryonic development occurs and the first moult occurs inside the egg; hence, each egg contains a dormant J_2. Cysts are dislocated from host-plant roots at harvest and are dispersed through the soil, where they may persist for 25 years or more (Turner and Subbotin, 2013). The encysted J_2 often remains dormant for many years and dormancy has been divided into quiescence and diapause. Quiescence is an arrest in development induced in response to unfavourable conditions; development is resumed soon after the return of favourable conditions, i.e. after the hatching is stimulated by potato root diffusates (Khan, 2008). However, some J_2 may enter into the state termed diapause, where development has been arrested and cannot be resumed until specific requirements have been satisfied, even if favourable conditions return. Sometimes, quiescence may follow after diapause has ended (Perry, 1998). The generation interval (egg to egg) of potato cyst nematode and the rate of increase of populations depend on the soil environment and the condition of the host plant. The life cycle takes around 6–10 weeks to complete (Turner and Subbotin, 2013). In India, at 15–21°C under laboratory conditions, the life cycle completed in 37–39 days in summer and 40–42 days in autumn, this being multiplied 7–13 and 6–11 times through each generation, respectively (Devrajan *et al.*, 2012). The nematodes generally complete one generation during a crop season. In cooler climates, hatching to the adult phase requires 38–45 days. The J_2 becomes active at 10°C; maximum root invasion takes place at 16°C. Females mate within 58 days of the invasion. The tolerance limit or measurable injury level is between 1.2 and 2.1 eggs/g of soil for both species (Dasgupta, 1998).

Even though PCNs are generally pests of temperate regions, or regions with a temperate climate (such as those high-altitude areas in tropical regions), they have also become adapted to various extents to some subtropical regions (Jones, 1961). Generally, the conditions that are optimum for potato growing are also optimum for cyst nematodes. Optimum conditions for juvenile survival and activity including movement are well-drained, aerated sandy, silty or peat soils, 50–75% water-holding capacity and a minimum temperature of 10°C. The eggs can remain viable in cysts for 20–30 years in soil under severe environmental stress. The cysts can withstand extreme cold (–15°C) and soil desiccation for long periods.

19.10 Host–parasite relationship

Exudations from the roots of a host plant stimulate hatching of viable eggs present in a cyst (Marks and Brodie, 1998). The J_2 is attracted to the growing root tip, migrates through the soil, enters the root near the growing point or lateral root and moves towards the vascular cylinder through intra- and intercellular migration, leaving a trail of ruptured cells (Evans and Stone, 1977). Thereafter, it becomes sedentary with its head towards the stele and the specific interaction induces enlargement of root cells and breakdown of cell walls to form a large, syncytial 'transfer cell' with dense, granular cytoplasm in the pericycle, primary phloem or endodermis. During feeding, the J_2 injects saliva into the cell, causing its enlargement and cell wall breakdown, leading to formation of three to 12 large transfer cells called syncytia (Jones and Northcote, 1972). The syncytia develop cell wall ingrowths adjacent to the conducting tissue, which increase the internal surface area, facilitating the passage of nutrients into the syncytia. The nematode continues to feed on the syncytium until its development is completed (Perry *et al.*, 2018).

19.11 Pathotypes

For *Globodera*, the term 'pathotype' is used by the International PCN Pathotype Scheme proposed by Kort *et al.* (1977) but is now considered too general. Many PCN populations cannot conclusively be allocated to the pathotypes described in this scheme. There are variances in virulence between the two PCN species, in particular within the populations of *G. pallida* (Hockland *et al.*, 2012). Any population showing signs of a new or unusual virulence should be verified as soon as possible. In practice, the virulence of populations can be tested on a set of cultivars used in each country. An EPPO Standard, PM 3/68 Testing of potato varieties to assess resistance to *G. rostochiensis* and *G. pallida*, was adopted in 2006.

19.12 Similarities to other species

G. pallida, a sibling species of *G. rostochiensis*, can be differentiated by the colour of the mature female. In most cases, the genus has yellow to golden females, whereas *G. pallida* have white or cream females, depending on the pathotype. *G. pallida*, *G. rostochiensis* and the *G. tabacum* complex share a wide host range mainly from solanaceous species, with the *G. tabacum* complex more frequently found on the spiny species whereas PCNs have a preference for tuber-forming species. *G. 'mexicana'* appears to have greatest affinity with *G. pallida* and these two species can interbreed (Mugniery, 1978).

19.13 Interaction with other organisms and disease complex

The PCN *G. pallida* and the fungus *Rhizoctonia solani* (black scurf/stem canker) cause great loss in concomitant situations (Back *et al.*, 2006). The nematode also interacts with *Verticillium dahliae* and enhances the severity of wilt disease (Storey and Evans, 1987).

19.14 Economic importance

Potato cyst nematodes are widely distributed in potato-growing temperate areas, affecting yield quantity and quality. Up to 80% yield losses are not uncommon in the hills of the tropics under high infestation and continuous potato culture. Under poor nematode management, crop losses induced by *G. pallida* range from 2% to 70% (Devrajan *et al.*, 2012). PCNs causes extensive damage, particularly in temperate areas. The nematodes are economically so important that US$10 million were spent on an eradication programme in the USA in 1958. At exceedingly high population levels, plants may die, but more frequently plants survive, though the yield is severely reduced. The actual tuber quality is not directly affected, but the size of the tubers may be reduced due to poor plant nutrition and root damage by the nematode. A few reliable estimates of yield losses caused by PCNs indicate that up to 9% of potato yields are lost annually due to PCNs (Evans and Brodie, 1980). Prasad (1992) reported up to 80% yield loss in the Nilgiri hills of Tamil Nadu, where both species of PCN are the major constraint in potato cultivation. The presence of PCNs in potato-producing areas for the international market is prohibited by many countries.

19.15 Movement and means of dispersal

The movement of PCNs to new localities is mainly by passive transport. PCNs have proved to be highly adaptable to new environments, being transported passively and often undetected around the globe in intimate association with potato tubers. The cysts can spread through the displacement of soil along with seed material for a very long distance. Movement of farm machinery and tools helps to carry the infested soil along with them. The cyst, being very light in weight, can be easily carried by wind over long distances. The main routes of spread are infested seed potatoes and movement of soil (e.g. on farm machinery) from infested land to other areas. Infested soil adhering to tubers and bags has been held responsible for the introduction of PCNs in many countries. Chitwood (1951) recorded 32,000 cysts from the tractor wheels in an infested area. Cysts can also be spread by wind and flood water, covering a long distance.

19.15.1 Plant parts liable to carry the nematode in trade/transport

Tuber (below-ground), roots, seeds (including seed tuber), seedlings, micro-propagated plants amd growing medium accompanying plants can carry the eggs, juveniles, adults and cysts of the nematode.

19.15.2 Plant parts not known to carry the nematode in trade/transport

Stems (above ground), leaves, shoots, trunks, branches, bark, flowers, inflorescences, cones, calyx, fruits (including pods) and wood.

19.16 Pest risk analysis and looming threat from introduction

Potato cyst nematodes are quarantine pests for EPPO, APPPC and NAPPO regions. Virtually all potato-cultivating areas are already contaminated with PCNs. These areas are generally closely monitored and domestic quarantine has been implemented in many places, including India. Until recently, PCNs were generally recognized as problematic only in temperate areas and phytosanitary clearances demanded in the increasing international trade. Now, the problem has well been recognized in other regions and there has been increasing research on PCNs in Asia, particularly the Indian subcontinent. Cysts containing viable eggs can survive in the soil for up to 20 years and can also be transported through seed tubers and by wind and floodwater. It may take 6–7 years from its introduction into a new potato area to reach a detectable level of the nematode population. Once the PCNs infest a new area, they are nearly impossible to eradicate and effective control options are quite expensive. Hence, it is important to keep potato production areas free from this invasive nematode.

19.17 Invasiveness rating

Considering the consequences of introduction and the economic damage potential of *G. pallida*, the following invasiveness rating has been assigned.

Climate/host interaction	–	High (3)
Hosts range	–	High (3)
Dispersal potential	–	High (3)
Economic impact	–	High (3)
Environmental impact	–	High (3)
Post-entry distribution	–	Medium (–2)
Final rating (score)	–	High (13)

Based on the above invasiveness rating, *G. pallida* receives a score of 13 out of 15 and is rated as HIGHLY INVASIVE NEMATODE.

19.18 Region-wise status of invasiveness

Region-wise invasiveness status is evaluated based on the worldwide geographical distribution of *G. pallida* and likely introduction in the major potato-producing countries where the PCN is absent to date. *G. pallida* may attain invasive status in the following major potato-producing countries.

APPPC: Afghanistan, Armenia, Australia, Azerbaijan, Bahrain, Bangladesh, Bhutan, China, Indonesia, Iraq, Kuwait, Myanmar, Nepal, Mongolia, Oman, Philippines, Tajikistan, Qatar, South Korea, Sri Lanka, Syria, Saudi Arabia, Thailand, Turkmenistan, United Arab Emirates, Vietnam, Yemen

OIRSA: Guatemala, Nicaragua, Norfolk Island

CAHFSA: Bahamas, Bermuda, Cuba, Dominica, Dominican Republic, Guadeloupe, Haiti, Honduras, Jamaica, Martinique, Saint Kitts & Nevis

COSAVE: Brazil, Paraguay, Uruguay

EPPO: Israel, Kazakhstan, Kyrgyzstan, Lebanon, Liechtenstein, Morocco, Moldova, Uzbekistan

IAPC: Benin, Burkina Faso, Burundi, Cameroon, Cape Verde, Central African Republic, Chad, Comoros, Congo, Côte d'Ivoire, Egypt, Eritrea, Ethiopia, Guinea, Lesotho, Madagascar, Malawi, Mali, Mauritania, Mauritius, Mozambique, Niger, Nigeria, Réunion, Rwanda, Senegal, Sierra Leone, Sudan, Swaziland, Uganda, Zambia, Zimbabwe

NAPPO: Mexico

NEPPO: French Polynesia, Jordan

PPPO: Belize, El Salvador, New Caledonia, Papua New Guinea

19.19 Management measures

19.19.1 Cultural practices

Large populations of nematodes normally build up where there is a continuous crop of potatoes. Nematode numbers can be decreased by not planting potatoes for several years in heavily infested fields, which is not always practical. It is possible to keep the PCN population under check with an adaptation of a short-term rotation with non-host crops and growing resistant potato cultivars. Potatoes may be grown once in 3 years in cases of low infestation and 4–5 years in cases of heavy infestation (Devrajan *et al.*, 2012). A 2-year rotation with non-host crops like French bean, pea and garlic could cause a decline in the nematode population by 80–90%. Potato–wheat or potato–strawberry rotation is recommended and also cultivation of a cyst nematode-resistant cultivar, like 'Kufri Swarna' in India. Inclusion of a resistant variety in a 4-year rotation programme can decrease the disease by 67–78%.

19.19.2 Chemical control

Soil treatment with nematicides has been an established practice in different countries for the control of cyst-forming nematodes of the genus *Globodera*, though it is expensive for the farming community. DBCP or EDB at 500 l/ha or mocap, carbofuran, fensulfothion or aldicarb at 30 kg/ha, two doses at intervals of 15–20 days is effective. Aldicarb at 2 kg a.i./ha as spot treatment was found to be an economical and effective method of management. Potato seed tuber treatments with several chemicals, including 10% formalin for 1 h or 9% calcium hypochlorite for 30 min, was also found effective (Dasgupta, 1998). Application of carbofuran 3G at 1.125 kg a.i./ha during planting was suggested by Subramaniyan *et al.* (1989) to manage the PCN.

19.19.3 Biological control

In the wake of acceptance of eco-friendly approaches, PCN management tactics may depend on the use of organic amendments (Kantharaju and Reddy, 2001), antagonistic plants (Franco *et al.*, 1999) and biocontrol agents (Mani *et al.*, 1998). Jatala *et al.* (1979) had observed the reduction of PCN on potatoes due to the application of *Paecilomyces lilacinus* (= *Purpureocillium lilacinum*). *P. lilacinus* and *Pseudomonas fluorescens* at 10 kg/ha were also found effective in reducing the PCN population and increasing potato yield (Seenivasan *et al.*, 2007). In addition to parasites, phosphate solubilizing microorganisms such as *Aspergillus niger*, *Penicillium digitatum*, *Bacillus subtilis* and *Pseudomonas fluorescens* possess great potential for use against PCN (Khan *et al.*, 2009). Cronin *et al.* (1997) reported that the antibiotic 2-4 diacetyl phloroglucinol, produced by *P. fluorescens*, was inhibitory to *G. rostochiensis*.

19.19.4 Integrated management

To manage PCN populations, an integrated approach is highly effective. Mustard intercrop, when combined with the application of carbofuran, reduced nematode infestation significantly (Devrajan and Balasubramanian, 2008). Integration of *P. fluorescens*, neem cake and mustard intercrop reduced populations of potato cyst nematodes and increased tuber yields (Devrajan *et al.*, 2004).

19.19.5 Phytosanitary measures

See Chapter 16, section 115.

20 *Globodera rostochiensis*

20.1 Common name

Yellow or golden potato cyst nematode

20.2 Scientific name

Globodera rostochiensis (Wollenweber, 1923) Behrens, 1975

20.3 Synonyms

Heterodera schachtii solani Zimmerman, 1927
Heterodera schachtii rostochiensis Wollenweber, 1923
Heterodera (*Globodera*) *rostochiensis* Wollenweber, 1923 (Skarbilovich, 1959)
Globodera rostochiensis (Wollenweber, 1923) Mulvey & Stone, 1976
Heterodera rostochiensis Wollenweber, 1923

20.4 EPPO code

HETDRO

20.5 Diagnosis

The morphological and molecular diagnosis approach for *Globodera rostochiensis* is described below.

20.5.1 Morphological diagnosis

Important morphological characters of *G. rostochiensis* from Baldwin and Mundo-Ocampo (1991) and Subbotin *et al.* (2010) are described below. For measurements, see Table 22.

EGGS: Eggs retained within cyst body, no egg sacs. Surface of eggshell smooth, no microvilli present.

FEMALE: Initially pure white, later turning golden yellow on maturation. When mature, around 500 μm in circumference without a cone. Cuticle occasionally has thin subcrystalline layer. Head bears one to two annules, neck region has plentiful tubercules. Head skeleton hexaradiate but weak. Stylet equally distributed in length between conus and shaft. Stylet knobs backwardly sloped. Median bulb well developed, large and circular (Fig. 6.6I); often oesophageal glands displaced due to

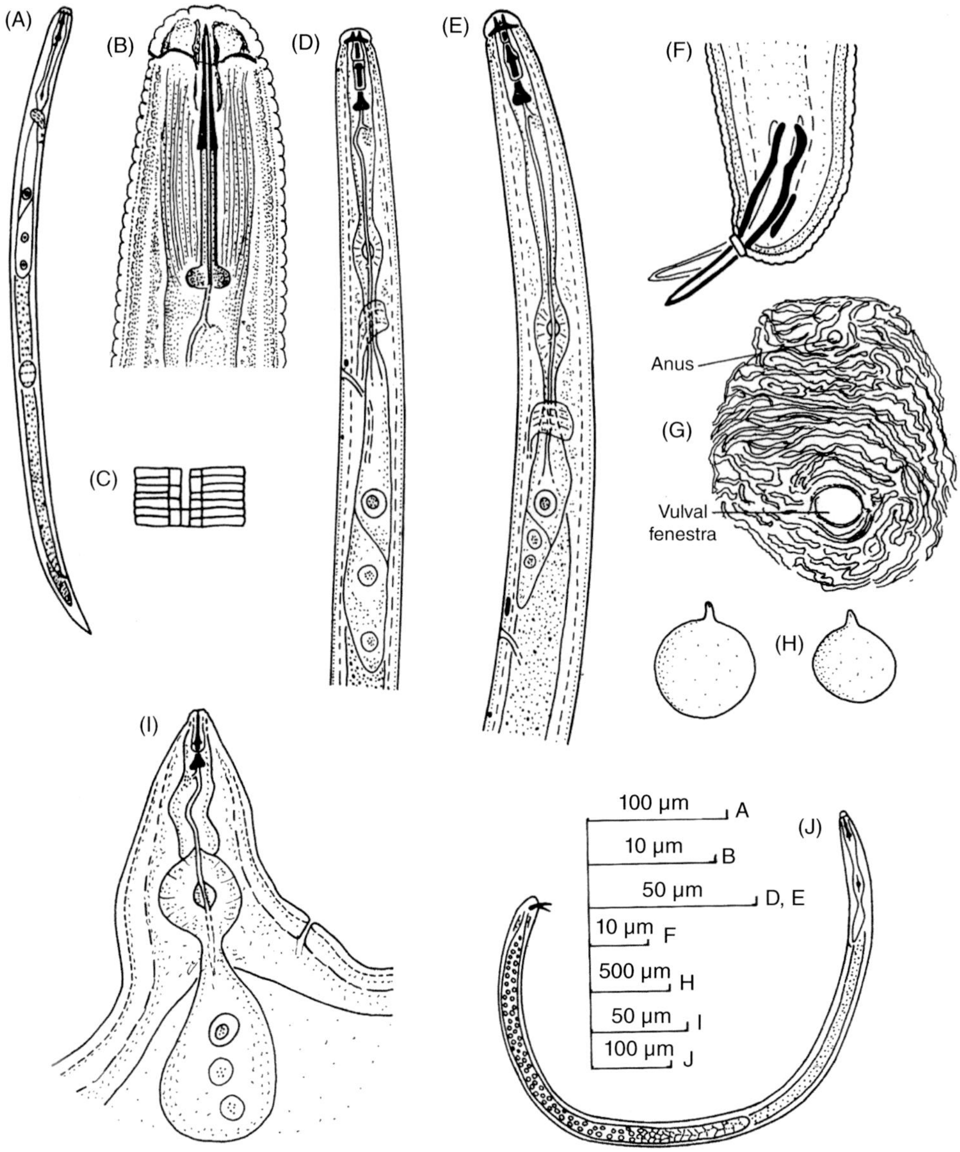

Fig. 6.6. Line drawing of *Globodera rostochiensis*. Entire second-stage juvenile, J_2 (A); head region of J_2 (B); lateral field of J_2 at mid-body (C); oesophageal region of J_2 (D); oesophageal region of male (E); tail of male (F); vulval region of female (G); entire cysts (H); head and neck of female (I); entire male (J). [Reproduced from M.R. Siddiqi, 2000; courtesy CAB International]

large paired ovaries. Excretory pore distinct at base of neck. Vulval slit in centre of vulval basin region, flanked on either side by papillae. Anus discrete, frequently seen at point in cuticle where end tapers to 'V' shape. Number of cuticular ridges lying in area between anus and edge of the fenestra aids in identification. Entire cuticle covered in small subsurface punctations.

MALE: Vermiform with short tail and without bursa (Fig. 6.6F). Body assumes curved shape with posterior region twisted at 90° angle to remainder of body after heat fixation (Fig. 6.6J). Lateral field with four incisures in mid-body (Fig. 6.6C) and three bands, which terminate on tail. Head rounded, hexaradiate, strongly sclerotized, offset and bears 6–7 annules. Cephalids situated at body annules 2–4 and 6–9, respectively. Stylet strong with backward sloping knobs. Median bulb well developed with large crescentic valve (Fig. 6.6E). Nerve ring positioned near oesophagus between median bulb and intestine. Hemizonid present with 2–3 annules anterior to excretory pore and two annules in length. Single monoarchic testis covers half body cavity. Spicules paired, arcuate, ending with single tips (Fig. 6.6F). Gubernaculum lies dorsal to spicules.

JUVENILE: Morphologically vermiform and similar to male. Head rounded and tail finely tapered (Fig. 6.6A). Hyaline portion of tail characterizes about two-thirds of its length. Lateral field with four incisures in mid-body region reducing to three at tail and anterior end. Head well developed, hexaradiate, slightly offset, bearing 4–6 annules (Fig. 6.6B). Cephalids situated at body annules 2–3 and 6–8, respectively. Stylet strong, conus about 45% of total length. Stylet knobs typically slope backwards. Median bulb well developed, oval with large central valve (Fig. 6.6D). Nerve ring girdles oesophagus between median valve and intestine. Hemizonion around one body annule in width. situated five body annules posterior from excretory pore. Hemizonid nearly two body annules in width. just anterior to excretory pore. Gonad primordium four-celled, situated at around 60% of body length.

20.5.2 Biochemical and molecular diagnosis

Biochemical and molecular characterizations of *G. rostochiensis* are similar to *G. pallida* (see section 19.5).

20.6 Geographical distribution

The yellow or golden potato cyst nematode *G. rostochiensis* is widely distributed and usually occurs in more areas than *G. pallida* (Fig. 6.7). In some countries, such as Armenia, Australia, Estonia, Georgia, Indonesia, Iran, Lebanon, Liechtenstein, Norfolk Island, Oman, Philippines, Russia, Rwanda, Serbia, Sierra Leone, Slovakia, South Africa, Sri Lanka, Tajikistan and Zimbabwe, only *G. rostochiensis* has been reported. The countries where *G. rostochiensis* has been reported are listed below (CABI-ISC, 2020).

Africa: Algeria, Kenya, Libya, Rwanda, Sierra Leone, South Africa, Tunisia.

Asia: Armenia, Georgia, India, Indonesia, Iran, Japan, Lebanon, Oman, Pakistan, Philippines, Sri Lanka, Tajikistan, Turkey.

Europe: Albania, Austria, Belarus, Belgium, Bosnia and Herzegovina, Bulgaria, Croatia, Cyprus, Czech Republic, Denmark, Estonia, Faroe Islands, Finland, France, Germany, Greece, Hungary, Iceland, Ireland, Italy, Latvia, Liechtenstein, Lithuania, Luxembourg, Malta, Netherlands, Norway, Poland, Portugal, Romania, Russia, Serbia, Slovakia, Slovenia, Spain, Sweden, Switzerland, Ukraine, United Kingdom.

North America: Canada, Costa Rica, Cuba, Mexico, Panama, USA.

Oceania: Australia, New Zealand, Norfolk Island.

South America: Bolivia, Chile, Colombia, Ecuador, Peru, Venezuela.

20.7 Hosts

Hosts are similar to those of *G. pallida* (see section 19.7, above).

20.8 Symptoms

Potato cyst nematodes do not cause specific symptoms of infestation on above-ground parts. The symptoms appear as stunted growth in patches (Fig. 6.5G) and are similar to those caused by *G. pallida*. The characteristic symptom of *G. rostochiensis* is the presence of yellow to golden females on the roots and tubers (Fig. 6.5H) which later turn into dark brown cysts (Fig. 6.5I). For details, see section 19.8.

NOTE: The biology, life cycle and ecology, economic importance, movement and means of dispersal, pest risk analysis and looming threat from introduction, and invasiveness rating of *G. rostochiensis* are similar to *G. pallida* (for details, see sections 19.9–19.17, above).

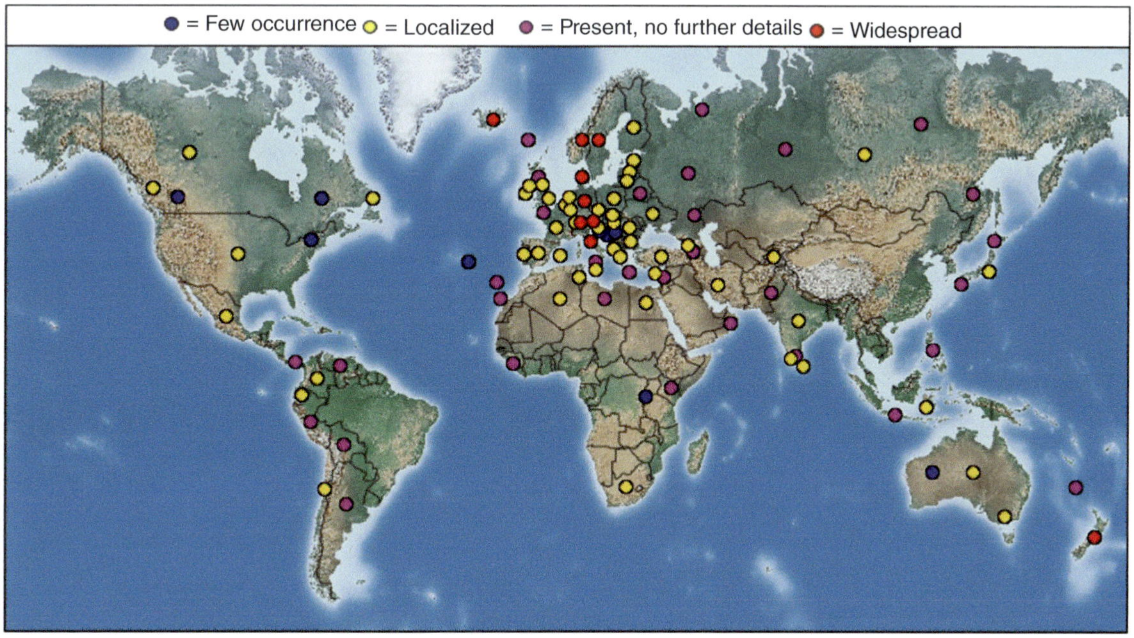

Fig. 6.7. Distribution map of *Globodera rostochiensis*. [Adapted from CABI-ISC, 2020]

20.9 Region-wise status of invasiveness

The worldwide distribution and introduction of *G. rostochiensis* outside its native region (South America) indicate that it is likely to be introduced into the major potato-producing countries where it is currently absent. *G. rostochiensis* attains highly invasive status in the following potato-producing countries where the nematode is still absent.

APPPC: Afghanistan, Azerbaijan, Bahrain, Bangladesh, Bhutan, China, Iraq, Kuwait, Myanmar, Nepal, Mongolia, Qatar, Syria, Saudi Arabia, South Korea, Thailand, Turkmenistan, United Arab Emirates, Vietnam, Yemen
OIRSA: Guatemala, Nicaragua
CAHFSA: Bahamas, Bermuda, Dominica, Dominican Republic, Guadeloupe, Haiti, Honduras, Jamaica, Martinique, Saint Kitts & Nevis
EPPO: Kazakhstan, Kyrgyzstan, Moldova, Uzbekistan
NEPPO: French Polynesia, Jordan
PPPO: Belize, El Salvador, New Caledonia, Papua New Guinea
IAPC: Benin, Burkina Faso, Burundi, Cameroon, Cape Verde, Central African Republic, Chad, Comoros, Congo, Côte d'Ivoire, Eritrea, Ethiopia, Guinea, Lesotho, Madagascar, Malawi, Mali, Mauritania, Mauritius, Mozambique, Niger, Nigeria, Réunion, Senegal, Sudan, Swaziland, Uganda, Zambia
COSAVE: Paraguay, Uruguay.

20.10 Management measures

Similar to *G. pallida* (see section 19.20, above).

21 *Globodera tabacum*

21.1 Common name

Tobacco cyst nematode

21.2 Scientific name

Globodera tabacum (Lownsbery and Lownsbery, 1954) Behrens, 1975

21.3 Synonym

Heterodera tabacum Lownsbery, 1954

21.4 EPPO code

HETDTA

21.5 Diagnosis

Stone (1983) suggested that *G. tabacum* should be considered a species complex, to include *G. solanacearum* (Miller and Gray, 1972) and *G. virginiae* (Miller and Gray, 1968) as subspecies. Baldwin and Mundo-Ocampo (1991), Mota and Eisenback (1993a,b) and Subbotin *et al.* (2010) also supported a species complex and considered *G. tabacum tabacum*, *G. tabacum solanacearum* and *G. tabacum virginiae* as subspecies of *G. tabacum*. The *G. tabacum* complex can be extremely difficult to diagnose and variation is common within each subspecies. Hence, both morphological and molecular diagnosis approaches are summarized below for correct species diagnosis of the *G. tabacum* complex.

21.5.1 Morphological diagnosis

Detailed morphological diagnosis of the *G. tabacum* complex is discussed by Mota and Eisenback (1993a,b) and Subbotin *et al.* (2010). For morphometric measurements, see Table 22.

EGGS: Egg sac not produced.
JUVENILE: Second stage vermiform, with three bands in lateral field. Head offset, with 3–4 head annules. Stylet three anchor-shaped, anteriorly facing knobs (Fig. 6.8A). Stylet conus and shaft equal in length. Oesophageal glands one-third of body length, overlapping intestine. Median bulb robust, rounded and approximately 70 μm from head tip. Excretory pore about 110 μm from head tip. Tail finely pointed, hyaline portion of tail usually half tail length (Fig. 6.8E).
FEMALE: Globose shape when first emerging from root (Fig. 6.8C). Head small. with four annules (Fig. 6.8G). Excretory pore at base of neck. Mid-body cuticle covered with zigzag markings, changing to parallel ridges in fenestral area. Vulva contained within area of thin-walled cuticle, having circumfenestrate fenestration. Vulval slit flanked either side by crescentic arrangement of wide and discrete tubercles. No underbridge. Anus distinct.
CYST: Globose in shape, lacks vulval cone (Fig. 6.8 D,H). Cuticle is strong and dark brown, eggs contained within it. A subcrystalline layer usually absent. No underbridge. No bullae.
MALE: Posterior region twisted through 90° to remainder of body. Four lateral lines at the mid-body, narrowing to two lateral lines at both ends. Lateral field extends around tail terminus. Head offset, with six head annules. Cephalic framework robust and heavily sclerotized (Fig. 6.8B). Stylet strong, three stylet knobs well developed and rounded, sloping posteriorly. Median bulb ellipsoidal and prominent. Nerve ring anterior to junction of oesophagus and intestine, encircles oesophagus. Oesophageal glands overlap intestine ventro-laterally. Tail short, no bursa (Fig. 6.8F). Phasmid near tail terminus. Spicules arcuate and single tipped.

21.5.2 Biochemical and molecular diagnosis

The taxonomy of the *G. tabacum* complex was originally based on biochemical, hybridization and morphometric data (Johnson and Chinheya, 2018). The biochemical characterization of *G. tabacum* is similar to *G. pallida* (see section 19.5).

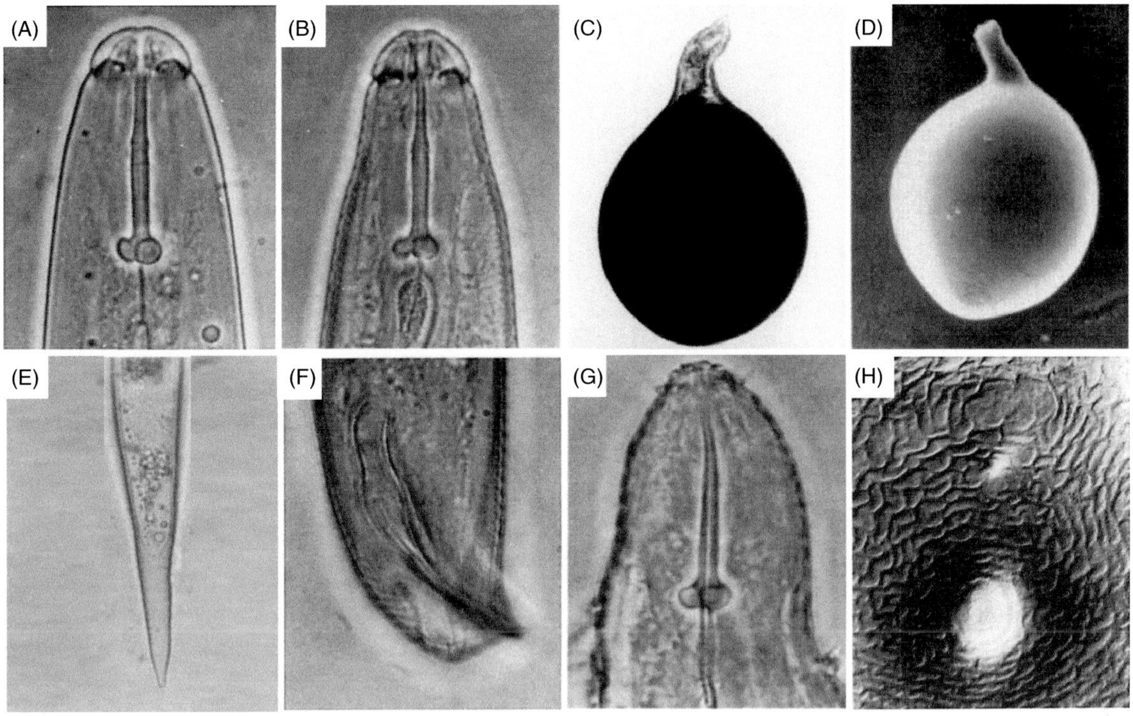

Fig. 6.8. Light microscopy images of *Globodera tabacum*. Anterior region of second stage juvenile (A); anterior region of male (B); female (C); cyst (D); tail region of second-stage juvenile (E); tail region of males (F); anterior region of females (G); terminal area of cyst (H). [Mota and Eisenback, 1993a,b; courtesy Journal of Nematology, The Society of Nematologists]

21.5.3 Molecular diagnosis

Techniques such as random amplified polymorphic DNA (RAPD), amplified fragment length polymorphism (AFLP), restriction fragment length polymorphism (RFLP) and real-time PCR have been used to confirm the identity of the *G. tabacum* complex (Thiéry and Mugniéry, 1996; Johnson, 1998; EPPO, 2017c). Thiéry and Mugniéry (1996) used ITS PCR-RFLP protocol and included five populations of the *G. tabacum* complex (*G. tabacum tabacum*, *G. tabacum virginae* and *G. tabacum solanacearum*), 11 populations of *G. pallida*, four populations of *G. rostochiensis*, one population of *G. mexicana* and *G. artemisiae* from different geographical areas. The primers set are (Vrain *et al.*, 1992):

18S 5′-TTG ATT ACG TCC CTG CCC TTT-3′
26S 5′-TTT CAC TCG CCG TTA CTA AGG-3′

The PCR amplification is performed with the following reagents and cycling parameters.

Reagent		Final concentration
Molecular-grade water	–	30 µl
Taq DNA polymerase buffer	–	5 µl
$MgCl_2$	–	2 mM
dNTPs	–	100 µM
Forward primer 18S	–	0.5 µM
Reverse primer 26S	–	0.5 µM
Reverse primer PITSr3	–	0.25 µM
Taq DNA polymerase	–	0.5 U
Cycling parameters		
Initial denaturation	–	94°C for 2 min
Number of cycles	–	30
Denaturation	–	94°C for 1 min
Annealing	–	60°C for 50 s
Extension	–	72°C for 1 min

The amplicon size of the PCR product is 1200 bp for *Globodera* spp. The size of restriction fragment length polymorphism (RFLP) reaction pattern of all the *G. tabacum* complex is presented in Table 23.

21.6 Geographical distribution

The tobacco cyst nematode (TCN) is a species complex (*G. tabacum tabacum*, *G. tabacum solanacearum* and *G. tabacum virginiae*) and is known to occur in North and Central America and southern Europe (Fig. 6.9). At present *G. tabacum* has been reported from the following countries (CABI-ISC, 2020).

Africa: Madagascar, Morocco.
Asia: China, Japan, Pakistan, South Korea.

Europe: Bulgaria, France, Greece, Italy, Slovenia, Spain, former USSR, former Yugoslavia.
North America: Canada, Mexico, USA.
South America: Argentina, Colombia.

21.7 Hosts

The *G. tabacum* complex parasitizes *Nicotiana tabacum* (tobacco) and some other solanaceous plants but not potato (Johnson *et al.*, 2005). Major and minor hosts of *G. tabacum* (CABI-ISC, 2020) are as follows.

MAJOR HOSTS: tomato (*Lycopersicon esculentum*), tobacco (*Nicotiana tabacum*) and wild tobacco (*Nicotiana rustica*).
MINOR HOSTS: black nightshade (*Solanum nigrum*), eggplant (Solanum melongena), gilo (*Solanum gilo*), Indian nightshade (*Solanum indicum*), narangillo (*Solanum quitoense*), potato (*Solanum tuberosum*) and tree tobacco (*Solanum mauritianum*).

Table 23. Sizes of RFLP fragments of *Globodera* spp. (adapted from Thiéry and Mugniéry, 1996).

Species	*Bsh* 1236I RFLP pattern
G. rostochiensis (European populations)	900, 190, 110
G. pallida (European populations)	500, 400, 350, 190, 110
G. 'mexicana'	500, 400, 190, 110
G. tabacum tabacum	445, 400, 190, 110
G. tabacum virginae	445, 400, 190, 110
G. tabacum solanacearum	445, 400, 190, 110

21.8 Symptoms

TCN can cause severe stunting of the above-ground parts of tobacco plants (Fig. 6.5J) that resembles root-knot and lesion nematode infestation (Johnson and Chinheya, 2018). On below-ground parts, TCN-infested plants have smaller root systems than normal. Females and cysts are attached to the root system. Cysts are initially pearly white to reddish-brown in colour and barely visible to the naked eye. However, 6–8 weeks after planting they turn dark brown and can normally be seen with the naked eye on infected plant roots (Fig. 6.5K).

21.9 Biology, life cycle and ecology

The life cycle and the biology of the *G. tabacum* complex are very similar to those of potato cyst nematodes (Miller and Gray 1972). The first-stage juvenile moults inside the egg and the second-stage juvenile hatches. *G. tabacum* could produce two generations at 26°C on eggplant, each taking around 36 days to complete the life cycle. However, at 20°C it requires 72 days to complete one generation and 32 days at 30°C. The maximum number of generations produced on eggplant in Italy is three (Miller and Gray 1972), whereas Barker and Lucas (1984) reported four to five generations from tobacco

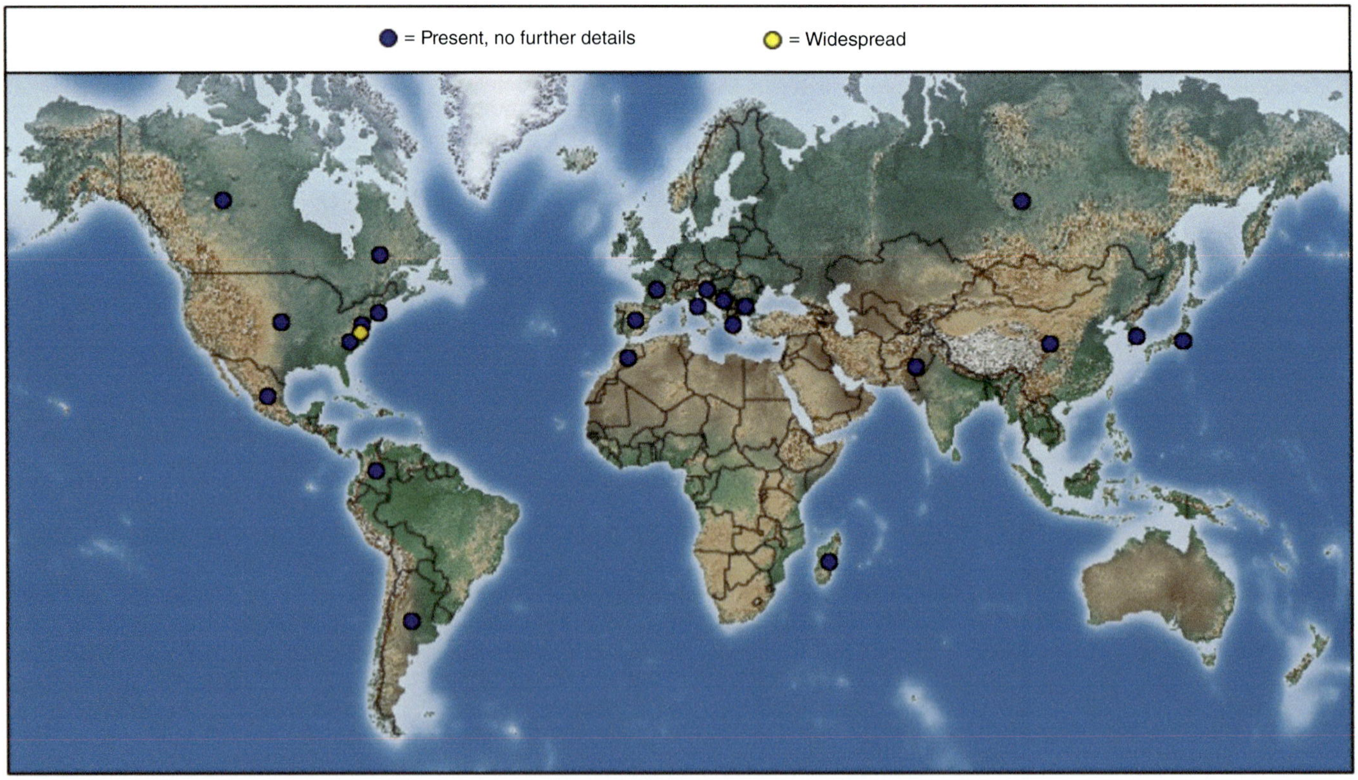

Fig. 6.9. Distribution map of *Globodera tabacum*. [Adapted from CABI-ISC, 2020]

fields in the USA. *G. t. virginiae* and *G. t. solanacerum* have temperature requirements and life cycles similar to those reported in Italy. TCN cysts may contain more than 100 eggs and the eggs within these cysts can survive for 11 years or more until stimulated to hatch by the temperature of host exudates (Johnson and Chinheya, 2018).

21.10 Interaction with other organisms

The TCN complex is often associated with bacterial wilt, black shank and Fusarium wilt disease complex and enhances the damage (Johanson *et al.*, 2005). *G. t. tabacum* increases the *Fusarium* infection in roots to a greater extent (LaMondia, 1995). TCNs may also suppress the reproduction of mycorrhizal fungi (Johnson, 1998).

21.11 Economic importance

The TCN complex has the potential to reduce crop yield considerably. Farmers in Virginia, USA, have recorded complete crop failure, but 15% average yield losses accounting for US$0.7 million generally occur in tobacco (Johnson, 1998). High populations in the early growing season can reduce yield by 27–50% (Johnson and Chinheya, 2018). In Connecticut, USA, large areas of the river valleys are infested with TCN, causing large yield losses and poor-quality produce. Nematicides were used extensively in that area (Miller, 1986). A population density of 100 juveniles/cm^3 may reduce shade tobacco leaf yield by less than 5%, while populations of 500–1000 juveniles/cm^3 can cause 45% decline in yield (Johnson, 1998).

21.12 Movement and means of dispersal

See section 19.15.

21.13 Pest risk analysis and looming threat from introduction

Introduction of the *G. tabacum* complex to non-infested areas where tobacco, tomato and other solanaceous crops are grown poses a serious threat to the production of these crops. Numerous cultivated and wild solanaceous plants are hosts of the TCN complex and will act as sources of its establishment. Cysts of *G. tabacum* contain viable eggs that can survive up to 11 years in the soil. It is nearly impossible to eradicate the TCN once it enters a new area. Hence, it is important to prevent its introduction into new areas.

21.14 Invasiveness rating

Considering the consequences of the introduction and the economic damage potential of *G. rostochiensis*, the following invasiveness rating has been assigned.

Climate/host interaction	–	High (3)
Hosts range	–	High (3)
Dispersal potential	–	High (3)
Economic impact	–	High (3)
Environmental impact	–	Low (1)
Post-entry distribution	–	Low (–1)
Final rating (score)	–	High (12)

Based on the above invasiveness rating, *G. tabacum* receives a score of 12 out of 15 and is rated as HIGHLY INVASIVE NEMATODE.

21.15 Region-wise status of invasiveness

The *G. tabacum* complex attains highly invasive status in the following host countries where the nematode is still absent.

APPPC: Bangladesh, Cambodia, India, Indonesia, Iraq, Israel, Kazakhstan, Kyrgyzstan, Lebanon, Malaysia, Myanmar, Philippines, Sri Lanka, Tajikistan, Thailand, Vietnam, Uzbekistan
CAHFSA: Barbados, Belize, Cuba, Dominican Republic, Haiti, Jamaica, Saint Kitts & Nevis
CAN: Bolivia, Ecuador, Peru, Venezuela
COSAVE: Brazil, Chile, Guyana, Paraguay, Suriname, Uruguay
EPPO: Austria, Belarus, Belgium, Croatia, Cyprus, Czech Republic, Denmark, Estonia, Finland, Georgia, Germany, Guernsey, Hungary, Ireland, Israel, Jersey, Jordan, Latvia, Lithuania, Luxembourg, Macedonia, Malta, Moldova, Netherlands, Norway, Poland, Portugal, Romania, Serbia, Slovakia, Sweden, Switzerland, Tunisia, Turkey, Ukraine, United Kingdom, Uzbekistan
IAPSC: Algeria, Angola, Benin, Botswana, Burkina Faso, Burundi, Cameroon, Cape Verde, Central African Republic, Chad, Comoros, Congo, Côte d'Ivoire, Eritrea, Ethiopia, Gabon, Gambia, Ghana, Guinea, Guinea-Bissau, Kenya, Liberia, Malawi, Mali, Mauritania, Mauritius, Mozambique, Namibia, Niger, Réunion, Rwanda, Senegal, Sierra Leone, Somalia, South Africa, Swaziland, Togo, Uganda, Zambia, Zimbabwe
NEPPO: Egypt, Jordan, Libya, Malta, Sudan, Syria, Tunisia
OIRSA: El Salvador, Guatemala, Honduras, Nicaragua
PPPO: Australia, Fiji, New Zealand, Papua New Guinea, Tonga, Vanuatu

21.16 Management measures

21.16.1 Prevention and control

The control of tobacco cyst nematodes is very difficult as the viable eggs in cysts can survive in the soil for many years. The most suitable means of management is preventive measures. Use of cyst-resistant cultivars is the most effective measure and further application of contact nematicides to reduce the nematode population densities. Since *G. tabacum* mainly attacks solanaceous crops, other alternative measures are crop rotation with non-host crops. Use of trap crops like tobacco, black nightshade (*Solanum nigrum*) and tomato is also very effective in reducing *G. tabacum* populations in both microplots and field plots over three seasons. Numbers of tobacco cyst nematodes were reduced by 80–96% when the host crop was destroyed after 3–6 weeks (LaMondia, 1996).

21.16.2 Phytosanitary measures

See Chapter 16, section 115.

Heterodera Spp.

Species of *Heterodera* Schmidt, 1871 are perhaps as important in parasitizing and as damaging to agricultural crops in temperate regions as root-knot nematodes (*Meloidogyne* spp.) are in the tropics. However, more than 30 species of *Heterodera* are known to adapt and occur in the tropics. At present more than 80 species of cyst nematodes have been described, with *Heterodera schachtii* Schmidt as the type species (Siddiqi, 2000; Turner and Subbotin, 2013; Perry *et al.*, 2018). Based on fundamental morphological characters and molecular analyses, the genus *Heterodera* is currently divided into seven groups of species (Subbotin *et al.*, 2010; Smiley *et al.*, 2017):

Afenestrata group (*H. koreana*, *H. orientalis*)
Avenae group (*H. arenaria*, *H. auklandica*, *H. australis*, *H. avenae*, *H. filipjevi*, *H. hordecalis*, *H. latipons*, *H. mani*, *H. pratensis*, *H. riparia*, *H. sturhani* and *H. ustinovi*)
Cyperi group (*H. bifenestra*, *H. cardiolata*, *H. cyperi*, *H. oryzicola*, *H. elachista*, *H. mothi*)
Goettingiana group (*H. carotae*, *H. circeae*, *H. cruciferae*, *H. goettingiana*, *H. mediterranea*, *H. persica*, *H. scuttelariae*, *H. urticae*)
Humuli group (*H. fici*, *H. humuli*, *H. litoralis*, *H. ripae*, *H. turcomanica*, *H. vallicola*, *H. zeae*)
Sacchari group (*H. goldeni*, *H. leuceilyma*, *H. sacchari*, *H. sinensis*, *H. sorghi*)
Schachtii group (*H. betae*, *H. cajani*, *H. ciceri*, *H. deverti*, *H. galeopsidis*, *H. glycines*, *H. lespedezae*, *H. medicaginis*, *H. mediterrenea*, *H. schachtii*, *H. sonchophyla*, *H. trifolii*).

Keys to *Heterodera* spp. are given by Subbotin *et al.* (2010) and Perry *et al.* (2018). The important plant-parasitic nematodes from *Heterodera* species are:

Heterodera auklandica Wout and Sturhan, 1995
H. australis Subbotin *et al.*, 2002
H. avenae Wollenweber, 1924
H. bifenestrata Cooper, 1956
H. cajani Koshy, 1967
H. carotae Jones, 1950
H. ciceri Vovlas *et al.* 1985
H. cruciferae Franklin, 1945
H. elachista Ohshima., 1974
H. fici Kirjanova 1954
H. filipjevi (Madzhidove, 1981) Steiner and Stelter, 1984
H. glycines Ichinohe, 1952
H. goettingiana Liebscher, 1892
H. graminis Stynes, 1971
H. hordecalis Anderson, 1975
H. humuli Filipjev, 1934
H. koreana (Vovlas *et al.*, 1992) Mundo-Ocampo *et al.*, 2008
H. latipons Franklin, 1969
H. medicaginis Kirjanova and Krall, 1978
H. mediterranea Vovlas *et al.*, 1981
H. mothi Khan and Husain, 1965
H. oryzae Luc *et al.*, 1961
H. oryzicola Rao and Jayaprakash, 1978
H. pakistanensis Maqbool and Shahina, 1986
H. pratensis Gäbler *et al.*, 2000
H. sacchari Luc and Merny, 1963
H. schachtii Schmidt, 1971
H. sorghi Jain *et al.*, 1982
H. sturhani Subbotin, 2015
H. swarupi Sharma *et al.*, 1999
H. trifolii Goffart, 1932
H. zeae Koshy *et al.*, 1971

Among these, *H. avenae* (European cereal cyst nematode), *H. australis* (Australian cereal cyst nematode), *H. cajani* (pigeon pea cyst nematode), *H. ciceri* (chickpea cyst nematode), *H. carotae* (carrot cyst nematode), *H. cruciferae* (cabbage cyst nematode), *H. elachista* (rice cyst nematode), *H. filipjevi* (Filipjev's cereal cyst nematode), *H. hordecalis* (barley cyst nematode), *H. glycines* (soybean cyst nematode), *H. goettingiana* (pea cyst nematode), *H. latipons* (Mediterranean cereal cyst nematode), *H. medicaginis* (alfalfa cyst nematode), *H. mediterranea* (Mediterranean cyst nematode), *H. mothi* (moth cyst nematode), *H. oryzae* (rice cyst nematode), *H. oryzicola* (rice cyst nematode), *H. sacchari* (sugarcane cyst nematode), *H. schachtii* (sugar beet cyst nematode), *H. sorghi* (sorghum cyst nematode), *H. sturhani* (Chinese cereal cyst nematode), *H. trifolii* (clover cyst nematode) and *H. zeae* (corn cyst nematode) are of economic importance and cause significant damage to their respective host crops (Siddiqi, 2000; Subbotin *et al.*, 2010; Turner and Subbotin, 2013; Perry *et al.*, 2018).

Systematic position of *Heterodera* species

	Classical (after Siddiqi, 2000)	Molecular (after Decraemer and Hunt, 2013)
Phylum	Nematoda	Nematoda
Class	Secernentea	Chromadorea
Order	Tylenchida	Rhabditida
Suborder	Hoplolaimina	Tylenchina
Infra-order	–	Tylenchomorpha
Superfamily	Holpolaimoidea	Tylenchoidea
Family	Heteroderidae	Hoplolaimidae
Subfamily	Heteroderinae	Heteroderinae
Genus	*Heterodera*	*Heterodera*

22 *Heterodera avenae*

22.1 Common names

Cereal cyst nematode; European cereal cyst nematode

22.2 Scientific name

Heterodera avenae Wollenweber, 1924

22.3 Synonyms

Bidera avenae (Wollenweber, 1924) Krall and Krall, 1978
Bidera ustinovi (Kirjanova, 1969) Krall and Krall, 1978
Heterodera major Schmidt, 1930
Heterodera schachtii major Schmidt, 1930
Heterodera schachtii var. *avenae* Wollenweber, 1924
Heterodera ustinovi Kirjanova, 1969

22.4 EPPO code

HETDMA

22.5 Diagnosis

At present 12 species belong to the Avenae group, including ten species of the *Heterodera avenae* complex (*H. arenaria*, *H. auklandica*, *H. australis*, *H. avenae*, *H. filipjevi*, *H. mani*, *H. pratensis H. riparia*, *H. sturhani* and *H. ustinovi*) as well as two other species, *H. latipons* and *H. hordecalis* (Subbotin *et al.*, 2010; Smiley *et al.*, 2017). Morphological differentiation of species of the *Heterodera avenae* complex is quite confusing and may fail to distinguish sibling species. Hence, biochemical and molecular diagnosis, aided by the morphological approach, is strongly recommended for precise identification of *H. avenae*.

22.5.1 Morphological diagnosis

The morphological descriptions of *H. avenae* after Williams and Siddiqi (1972) and Handoo (2002) are summarized below. For morphometric measurements, see Table 24.

Eggs: Cylindrical with rounded edges.
Gravid female: Body swollen, lemon-shaped, pearly-white and with protruding neck and vulva. Cuticle bears zigzag pattern and tends to run concentrically in neck and vulval regions. Head offset; single prominent labial disc with single distinct annule. Stylet sloping knobs. Ovaries paired and convoluted. Vulva is slit-like, protruding posteriorly. Anus is distinct.
Cysts: Cyst walls dark brown to black, partially covered with white subcrystalline layer. Cyst wall has zigzag patterns and irregularly arranged punctations. Cyst is bifenestrate; underbridge and bullae present (Fig. 6.10B).
Juveniles: Body cylindrical, tapering to rounded tail terminus. Head offset, with three annules; disc heavily sclerotized (Fig. 6.10H). Four lateral lines; stylet robust, with knobs slightly anteriorly directed. Tail has hyaline portion of more than 50% (Fig. 6.10J,K).
Male: Body cylindrical, twisted adjacent to spicules. Head offset, with 4–6 annules. Four lateral lines; outer lines crenate. Stylet robust, knobs directed slightly anteriorly. Spicules arcuate and bidentate (Fig. 6.10F). Gubernaculum simple.

22.5.2 Biochemical diagnosis

Several biochemical methods based on proteins are used to differentiate species, pathotypes or races of *H. avenae* (Mokabli *et al.*, 2001). *H. avenae* is morphologically very closely related to *H. filipjevi* and the two species differ in the protein patterns of white females as resolved by isoelectric focusing and malate dehydrogenase (MDH) isozyme patterns (Bishnoi *et al.*, 2004). *H. avenae* is also characterized on the basis of enzymes (Andrés *et al.*, 2001) and proteins (Subbotin *et al.*, 2002) using isoelectric focusing. Species-specific serological quantification of *H. avenae* and *H. glycines* using monoclonal antibodies (mAbs) have been achieved (Curtis *et al.*, 1997).

22.5.3 Molecular diagnosis

Various molecular techniques have been used to characterize *Heterodera* species. The *H. avenae* complex is identified on the basis of PCR-RFLP of ITS regions, RAPD-PCR, PCR-SSCP, ITS-rDNA and PCR with species-specific primers (Zheng *et al.*, 2000; Subbotin *et al.*, 2003; Tanha Maafi *et al.*, 2003; Waeyenberge *et al.*, 2009b; Dawabah *et al.* 2012; İmren *et al.*, 2015). Subbotin *et al.* (2003) characterized the *H. avenae* complex (*H. arenaria*, *H. auklandica*, *H. australis*, *H. avenae*, *H. filipjevi*, *H. mani*, *H. pratensis* and *H. ustinovi*) collected from different parts of the world using RAPD, PCR-RFLP and sequencing of the ITS-rDNA. Subbotin *et al.* (2018) made available a comprehensive phylogenetic analysis of 147 nematode populations (Avenae group) obtained from 26 countries using around 220 *COI* and ITS-rRNA gene sequences. The study revealed that the *COI* gene is a potential DNA barcoding marker for identification of species from the Avenae group or populations. Some other useful molecular diagnosis methods are discussed below.

ITS-RDNA SEQUENCES İmren *et al.* (2015) sequenced 41 populations of cereal cyst nematode (Avenae group) from the Mediterranean region using their ITS-rDNA regions. The DNA was extracted from one cyst of each population using worm lysis buffer (see Chapter 15, section 110.1.1). PCR amplification of the ITS-rDNA region was done using the following primers.

Forward primer: 5′-CGTAACAAGGTAGCTGTAG-3′
Reverse primer: 5′-TCCTCCGCTAAATGATATG-3′ (Ferris *et al.*, 1994),

The PCR amplification was performed using the following reagents and cycling parameters.

Table 24. Morphometric measurements of different *Heterodera* species.

Morphometric characters		Range from numerous populations													
		H. avenae (Handoo, 2002)	*H. cajani* (Koshy, 1967; Koshy *et al.* 1971)	*H. carotae* (Mathews, 1975)	*H. ciceri* (Vovlas *et al.*, 1985)	*H. elachista* (Nobbs *et al.*, 1992)[a]	*H. glycines* (Burrows and Stone, 1985)	*H. goettingiana* (Baldwin and Mundo-Ocampo, 1991)	*H. humuli* (Stone and Rowe 1977)	*H. latipons* (Franklin, 1969)	*H. oryzae* (Luc and Taylor, 1977)	*H. oryzicola* (Rao and Jayaprakash, 1978)	*H. sacchari* (Shahina and Maqbool, 1995)	*H. schachtii* (Baldwin and Mundo-Ocampo, 1991)	*H. zeae* (Asghari *et al.*, 2013)[b]
Females	Length (μm)	418–740	680–940	375–513	550–950	–	–	500	–	348–645	–	–	–	461	423–716
	Stylet (μm)	26–32	20–30	25–31	29–31	–	–	25.7±1	25.4±1.5	21–28	–	23–25	27	–	24.25
	Body width (μm)	285–534	363–488	168–238	300–520	–	–	400	–	277–510	–	–	–	494	325–684
	L/W ratio	–	1.4–2.1		1.5–2	–	–	1.25	–	–	–	–	–	–	1.0–1.5
Males	Length (μm)	1070–1610	844–1666	1090–1220	1235–1448	740	911–1400	1270±112	941±67	960–1406	790–970	896–980	1119	–	–
	Body width (μm)	–	–	27–33	29–30		–	24.7±0.8	27.9±2.1	25–32.5	–	28	–	28–42	–
	a	32–5	31.6–51.4	–	42–51	34.3	–	–	34±3.7		33.0–37.6	32–35	45.2–56.2	32–42	–
	b	7–12	5.8–11.8	–	5.8–6.4	6.7	–	–	–		7.1–7.5	7–9	4.5–5.6	–	–
	Stylet length (μm)	27–31	23–29	31–38	28–30	19.5	24-27	26.8±1	28.4±1.5	22–29	23–25	20–30	25.5–28	29	–
	DGO	3.6	–	5–7	5–6	3.5	–		–	–			–	–	–
	Lateral lines	4	4		4		–	4	4	4			3	4	–
	Spicule length (μm)	33–38	32–43	31-36	34–38	23	28–45	26.5±4.3	31.1±0.9	32–36	32–36	19–25	30.5–35.2	34–38	–
	Gubernaculum length (μm)	10–12	8.0–12.0	–	9–10	9.5	8–13	12.2±2	7.5 (7–8)	–	8–10	8	7–8[c]	10–11	–
Cysts	Length without neck (μm)	580–975	390–690	218–625	570–930	328–557	340–920	521±53	452±57	348–645	031-081	344–639	485–860	461	400–502
	Neck length (μm)	–			90–160	–	–	–	66±14	58–103			–	–	–
	Body width (μm)	390–725)	175–550	165–500	350–550	229–449	490±54	372±44	330±53	277–510	0.22–0.69	250–589	337–605	494	311–401
	L/W ratio	1.3	1–2.5		1.56–2.37	1.1–1.9	1.19–2.05	1.3–2.2	–	–	0.85–1.72	–	1.0–2.0	–	1.2–1.3
	Fenestral length (μm)	40–52	31–51	27–36	32–52	23–36	37–65	35.3±5.9	22.3±2.6	58–76	22–42	16–44	48.5–60	38	38–56
	Fenestral width (μm)	18–23	25–40	35–42	20–37	26–39	33–48	37.4±4.3	24.7±2.4	15–27	26–43	16–32	33.4–40	32	22–42
	Vulval bridge width (μm)	5–8	–	–	–	2–16	–	3.0±0.9		18–39		3–6	–	50	
	Vulval slit length μm	8–11	39–51	43–51	43–60	10–23	43–56	39.9±7.2	34.6±3.5	6–9	43–51	32–55	45–55.4	47	33–43
	Bullae	Present, heavy	Present, few	Absent	Present, heavy	Present	–	Present	Abbullate		Present	Present	Few peripheral	Bullate	Present
	Underbridge length	Absent	60–70	–	115–160	65–78	117.3±13.1	33.0±5.4	85.6±8.8	80–125	80–120	61–104	120–145.5	114	28–45

Morphometric characters		Range from numerous populations													
		H. avenae (Handoo, 2002)	*H. cajani* (Koshy, 1967; Koshy *et al.* 1971)	*H. carotae* (Mathews, 1975)	*H. ciceri* (Vovlas *et al.*, 1985)	*H. elachista* (Nobbs *et al.*, 1992)[a]	*H. glycines* (Burrows and Stone, 1985)	*H. goettingiana* (Baldwin and Mundo-Ocampo, 1991)	*H. humuli* (Stone and Rowe 1977)	*H. latipons* (Franklin, 1969)	*H. oryzae* (Luc and Taylor, 1977)	*H. oryzicola* (Rao and Jayaprakash, 1978)	*H. sacchari* (Shahina and Maqbool, 1995)	*H. schachtii* (Baldwin and Mundo-Ocampo, 1991)	*H. zeae* (Asghari *et al.*, 2013)[b]
Juveniles	Length (µm)	520–620	0.392	375–452	440–585	377–450	375–490	486±22	374±22	401–478	351–470	370–428	394–520	420–550	381–452
	Body width (µm)	20–24		19–21	19–22	15.5–18[d]	–	19.4±0.7	18.9±1.3	19–22	16–18	18–22	18–20[c]	22	16–19
	Stylet length (µm)	24–28	20–25	22–25	27–30	16–21	22.0–24.0	24.6±0.8	22.2±0.5	23–25	19–25	17–19	20–24	25.6	21–24
	DGO	5–6	–	5–6	5–6	–	–	–	5.2	–	–	–	5–6[c]	–	4–5
	Lateral line	4	4		4	–		3	4	4		–	3	3	
	a	21–27	22–23	18.2–22.7	–	21.3–28.1	–	–	–	–	22.6–28.0	19–20	23–26	–	21.8–27.0
	Tail length (µm)	58–70	39–54	43.5–58.6	53–72	47–70	42.0–59.4	60.1±5.3	50.3±3.5	42–54	51–64	41–75	45.5–60.2	49–55	35–49
	Length of hyaline terminus (µm)	35–45	14–34	20.3–31.8	31–42	26–36[d]	20.0–33.0	37.0±3.2	26.7±3.6	42–54	31–45	29–41	24.2–32.5	30	22–27
Eggs	Length (µm)	120–142	94–97	–	123–143		–	–	94±4.2	100–124	90–120	–	112–139	111–134	105–110
	Width (µm)	48–61	41–47	–	48–53		–	–	38.5±1.6	44–56	40–57	–	43–45	51–59	35–38
	L/W ratio	–	–	–	2.4–2.8		–	–	–	–	1.8–2.6	–	2.5–3.1	–	–

[a]Male from De Luca *et al.* (2013); [b]male not found, female and eggs from Koshy *et al.* (1970). [c]from Tanha Maafi *et al.*, 2007; [d]from original Ohshima, 1974.

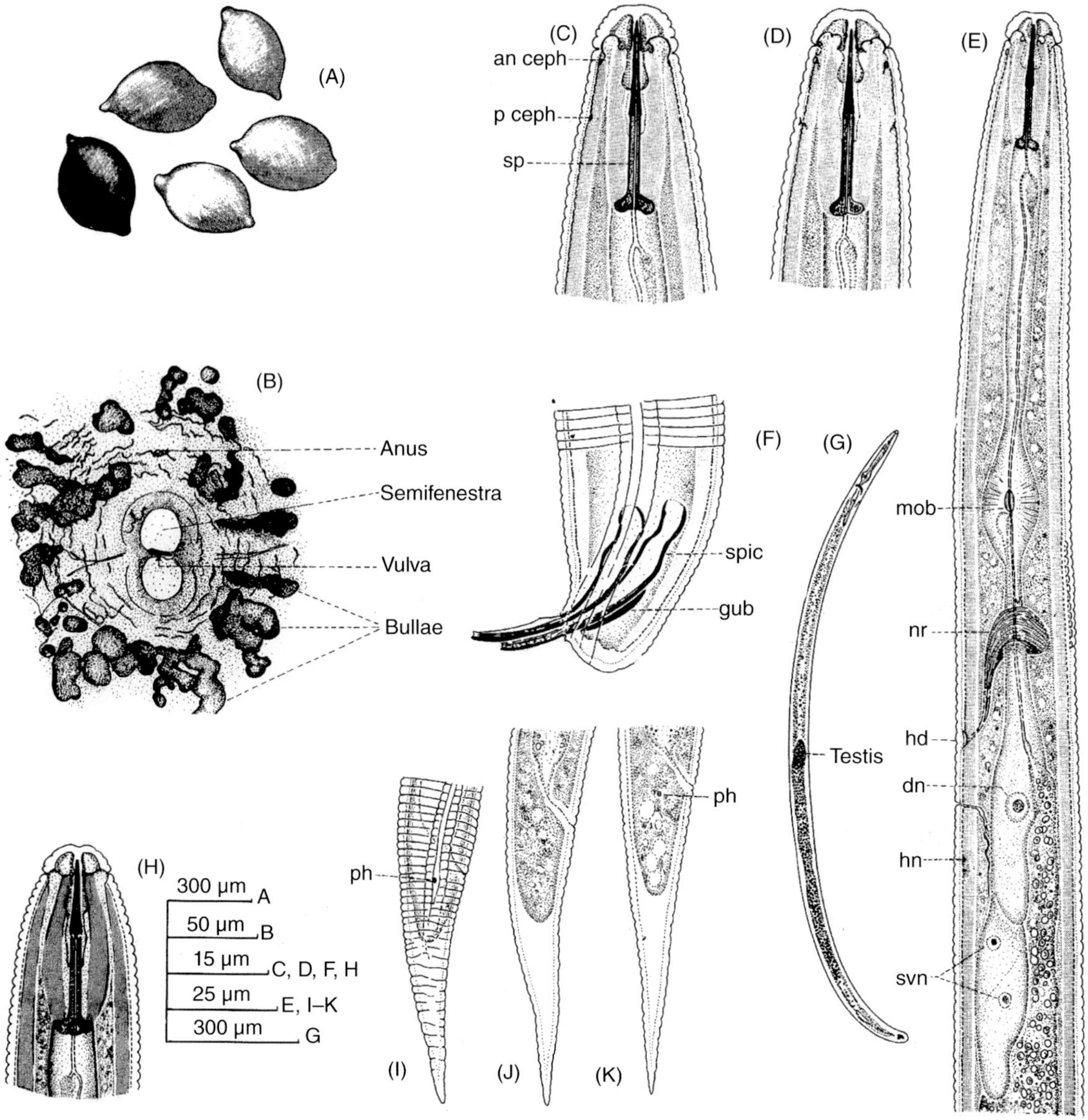

Fig. 6.10. Line drawing of *Heterodera avenae*. Cyst (A); *en face* view of vulval cone of cyst (B); head ends of male (C-D); oesophageal region of male (E); tail end of male (F); male (G); head end and tail ends of second stage juvenile, respectively (H, I–K); an ceph, anterior cephalid; dn, dorsal gland nucleus; gub, gubernaculum; hd, hemazonid; hn, hemizonion; mob, median oesophageal bulb; nr, nerve ring; p ceph, posterior cephalid; ph, phasmid; sp, spear; spic, spicule; svn, subventral gland nuclei. [Reproduced from Williams and Siddiqi, 1972; courtesy CAB International]

Reagent		Final concentration
PCR reaction	–	50 µl
Taq PCR Master Mix	–	25 µl
Forward primer	–	1 µM
Reverse primer	–	1 µM
DNA extract	–	1 µl
Cycling parameters		
Initial denaturation	–	95°C for 5 min
Number of cycles	–	40
Denaturation	–	94°C for 30 s
Annealing	–	48°C for 45 s
Extension	–	72°C for 45 s
Final extension	–	72°C for 8 min.

After PCR amplification the PCR product was purified using the manufacturer's instructions and sequenced. The sequences of all *Heterodera* populations were aligned with the *H. avenae* group and some other *Heterodera* groups available in GenBank using ClustalX2.0. A phylogenetic tree was constructed, and the study revealed that the *H. avenae* group divided into four groups based on the ITS-I, ITS-II and 5.8S rDNA sequences. Similarly,

Dawabah *et al.* (2012) characterized the ITS-rDNA of four populations of *H. avenae* from the arid zone of Saudi Arabia.

PCR-RFLP Yan and Smiley (2010) used polymerase chain reaction-restriction fragment length polymorphism (PCR-RFLP) of internal transcribed spacer (ITS)-ribosomal (r)DNA with seven restriction endonucleases to differentiate *H. avenae*, *H. filipjevi*, *H. latipons* and *H. schachtii*. The DNA was obtained from three gravid cysts. PCR amplification of ITS-rDNA was done using the primers described by Vrain *et al.* (1992):

18S (5′-TTGATTACGTCCCTGCCCTTT-3′)
26S (5′-TTTCACTCGCCGTTACTAAGG-3′)

The PCR amplification was performed using the following reagents and cycling parameters.

Reagent		Final concentration
PCR reaction	–	20 µl
$MgCl_2$	–	1.5 mM
dNTPs	–	0.2 mM
Taq polymerase	–	1.25 units
Primer	–	0.5 µM (each)
DNA template	–	3 µl
Cycling parameters		
Initial denaturation	–	95°C for 5 min
Number of cycles	–	40
Denaturation	–	94°C for 30 s
Annealing	–	48°C for 45 s
Extension	–	72°C for 45 s
Final extension		72°C for 8 min.

After PCR amplification, the PCR product was examined in a 1% standard agarose gel stained with ethidium bromide (0.1 µg/ml). The rest of the PCR product was purified according to the manufacturer's instructions and processed for RFLP analysis. Seven restriction endonucleases (*Taq*I, *Hin*fI, *Pst*I, *Hae*III, *Rsa*I, *Alu*I, and *Cfo*I) were analysed, using:

Forward primer: 5′-CGTAACAAGGTAGCTGTAG-3′
Reverse primer: 5′-TCCTCCGCTAAATGATATG-3′ (Ferris *et al.*, 1994),

Digestions were processed using a standard protocol (see Chapter 15, section 110.2.1). The species of *Heterodera* were determined by comparing the restriction pattern with those of the control species.

22.6 Geographical distribution

Cereal cyst nematode (CCN) was first recorded by Kuhn (1874) in Germany as a pest of cereal crops and later recorded from different countries (Smiley *et al.*, 2017). At present the *H. avenae* complex is widely distributed in temperate, tropical, subtropical and Mediterranean regions of the world (Fig. 6.11). It has been detected in the following countries (CABI-ISC, 2020).

Africa: Algeria, Egypt, Libya, Morocco, South Africa, Tunisia.
Asia: China, India, Iran, Israel, Japan, Kazakhstan, Pakistan, Saudi Arabia, Syria, Turkey.
Europe: Belgium, Bosnia and Herzegovina, Bulgaria, Czech Republic, Denmark, Estonia, France, Germany, Greece, Iran, Ireland, Italy, Latvia, Malta, Netherlands, Norway, Poland, Portugal, Russia, Serbia, Slovakia, Slovenia, Spain, Sweden, Switzerland, Ukraine, United Kingdom, former Yugoslavia.
North America: Canada, USA.
Oceania: Australia, New Zealand.
South America: Peru.

22.7 Hosts

The CCN has a limited host range. Barley (*Hordeum vulgare*), wheat (*Triticum aestivum* and *T. durum*), oats (*Avena sativa*), rye (*Secale cereale*) and triticale (wheat × rye) are the major crop hosts of *H. avenae* (Kanwar and Bajaj, 2010). However, Indian and Australian populations do not reproduce on cultivated varieties of oats (Cook and York, 1988). Maize (*Zea mays*) is an inefficient host of *H. avenae* as it supports a very limited reproduction of the nematode (Kanwar and Bajaj, 2010). Other hosts of *H. avenae* are listed in Table 25.

22.8 Symptoms

Disease symptoms are discernible within 10 days of first irrigation, which is generally applied 3–4 weeks after sowing. Above-ground symptoms caused by *H. avenae* are characterized by patches of poor plant growth in the field (Fig. 6.12A). These patches increase in size and tend to coalesce with monoculture of susceptible crops. With continuous cultivation of wheat for 3–4 years, the whole field may become uniformly infested (Fig. 6.12B). The affected plants exhibit stunted growth, yellowing, reduced tillering, fewer leaves with reduced leaf area, delayed emergence of ear heads and a smaller number of spikelets. Such effects can be attributed to nutrient and moisture stress resulting from improper functioning of the root system. Under severe infection plants are severely stunted coupled with overgrowth of weeds, leaving a crop with no economic value (Fig. 6.12C). In such a situation, farmers plough the field in mid-season (Fig. 6.12D).

On below-ground parts, at the initial stage of infection, root tips become swollen due to nematode penetration (Fig. 6.12E). The main root of an infected plant becomes elongated and has tufts of rootlets at the distal end (Fig. 6.12E). The root system becomes bushy and shallow due to excessive branching (Fig. 6.12F). Infected plants lack anchorage and can be easily pulled out of the soil. Since the above-ground symptoms are non-specific, presence of white females on the roots is the only confirmatory indication of nematode infection (Fig. 6.12F-G). White females can be seen on roots of wheat and barley in the months of January–February in India. A single root system may harbour numerous females (Kanwar and Bajaj, 2010).

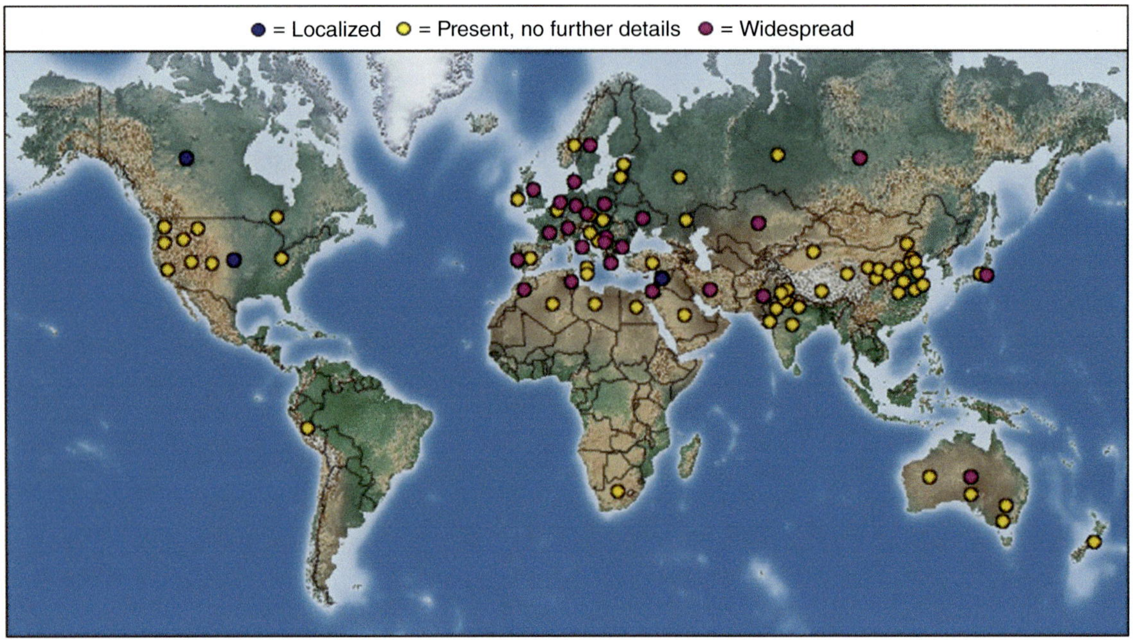

Fig. 6.11. Distribution map of *Heterodera avenae*. [Adapted from CABI-ISC, 2020]

22.9 Biology and life cycle

The life cycle of *H. avenae* is typical of a cyst nematode (Fig. 6.13). The nematode survives in the form of second-stage juveniles (J_2) enclosed in cysts (dead females) during the off-season. The J_2 starts coming out of diapause and cysts with favourable conditions (fall of temperature). It penetrates the growing roots and establishes at specific feeding sites, thereafter undergoing three moults to become an adult. The females become lemon shaped, whereas the mature males become vermiform and start moving out of the roots after 50–55 days. They live for around a further 15 days. During this period, they inseminate young white females whose vulval regions protrude after rupturing the cortex. Fertilized white females can be seen attached to the roots. The female starts laying eggs, which are deposited in its body (Turner and Subbotin, 2013) (Fig. 6.13A). As the crop matures, females die and their bodies turn from white to brown, called cysts. Eggs and juveniles inside the cysts undergo compulsive diapause induced by rising temperatures for about 6 months under Indian conditions during the summer period (Kanwar and Bajaj, 2010). The cysts get detached from roots into the soil, where they act as inoculum for the next season. A cyst, on average, contains 200–250 eggs. The nematode completes only one generation in a year (crop season). Life-cycle duration depends on temperature and development of white females, which may take 8–9 weeks in December-sown or 5 weeks in February-sown wheat in India (Kanwar and Bajaj, 2010).

The activity period of CCN coincides with the wheat crop that is grown during the winter season in India (Kanwar and Bajaj, 2010). In Australia, eggs remain in the quiescent stage during the hot days and early autumn months. Soon after the spring rains (mid-June–July), wheat is sown; concurrently hatching of eggs occurs and infection takes place; cysts attached to roots may be seen during August–October (Meagher, 1968). In Europe and Canada, the juveniles penetrate the roots of cereal hosts sown in autumn just after germination. During winter, nematode development remains suspended but restarts in spring. Females may be seen on the roots as early as April. In the case of spring-sown crops, the nematode overwinters in the egg stage and invades newly planted crops at temperatures as low as 4°C in mid-March (Decker, 1972). Moist and cool periods at the time of hatching and invasion and a dry period during the most active growing phase of the host crop cause the worst damage to crop (Kanwar and Baja, 2010).

22.10 Host–parasite relationship

The infective stage of *H. avenae* is the second-stage juvenile, which penetrate roots usually in the meristematic region behind the root cap and also other parts of the roots. Immediately after penetration, the nematode moves through the cortex both intercellularly and intracellularly and establishes itself for feeding in the stellar region by lying parallel to the stele. The cells surrounding the head begin to enlarge and their protoplasm becomes dense. Cell walls of adjoining cells are dissolved, due to mechanical and enzymatic action. As a result, elongate and multinucleate syncytia with thickened outer boundaries are formed (Williams, 1969). The cells involved in the formation of syncytium include those of endodermis, phloem and protoxylem. As the larva grows in size, disorganization of cortex, endodermis,

Table 25. Other hosts of *Heterodera avenae* (adapted from Kanwar and Bajaj, 2010).

Host name	Family
Agropyron repens	Poaceae
Agrostis stolonifera	Poaceae
A. castellana	Poaceae
A. capillaris	Poaceae
Apera spica-venti	Poaceae
Arrhenatherum elatius	Poaceae
Avena fetua	Poaceae
A. ludoviciana	Poaceae
A. brevis	Poaceae
A. byzantina	Poaceae
A. sterilis	Poaceae
Alopecurus geniculatus	Poaceae
A. pratensis	Poaceae
Brachypodium sp.	Poaceae
Bromus arvensis	Poaceae
B. mollis	Poaceae
B. secalinus	Poaceae
B. sterilis	Poaceae
B. tectorium	Poaceae
Cenchrus cilaris	Poaceae
Cynosurus cristatus	Poaceae
Dactilis glomerata	Poaceae
Elymus caninus	Poaceae
Echinochola colona	Poaceae
E. frumantacea	Poaceae
Festuca pratensis	Poaceae
F. longifolia	Poaceae
F. arundinacea	Poaceae
F. rubra	Poaceae
F. gashmiriana	Poaceae
Holocus lanatus	Poaceae
Keeleria spp.	Poaceae
Lolium multiflorum	Poaceae
L. perenne	Poaceae
L. temulentum	Poaceae
Melica ciliata	Poaceae
Phalaris canariensis	Poaceae
P. paradoxa	Poaceae
Phelum pratense	Poaceae
Poa pratensis	Poaceae
P. trivialis	Poaceae
P. bulbosa	Poaceae
Polypogon monospeliensis	Poaceae
Senebiera pinnatifida	Cruciferae
Setaria spp.	Poaceae
Sorghum vulgare	Poaceae
Stipa lagascae	Poaceae
Vulpia muralis	Poaceae
V. myuros	Poaceae
V. ciliata	Poaceae

pericycle, parenchymatous cells and the central vascular strands takes place. Syncytia may be formed within 6 days after inoculation (Johnson and Fushtey, 1966).

22.11 Similarities to other species and pathotype

Morphologically *H. filipjevi* is very closely related to *H. avenae* and was previously considered as the Gotland strain: British pathotype 3 and Ha 23 of *H. avenae*. *H. filipjevi* differs from *H. avenae* in having deeply cupped stylet knobs, shorter hyaline portion of tail in J_2, underbridge and fewer molar-shaped bullae in cysts (Bishnoi and Bajaj, 2004). Bajaj *et al.* (2005) conducted cross-hybridization tests between *H. avenae* and *H. filipjevi* and found that the populations of two species did not interbreed and that they therefore belong to distinct biological species. *H. australis*, previously considered as pathotype Ha 31, is morphologically indistinguishable from *H. avenae* and differs only in ITS-r DNA and RAPD data (Subbotin *et al.*, 2003).

Andersen (1959) recorded the occurrence of two pathotypes based on resistant and susceptible reactions of Danish populations on barley cv. 'Drost'. Thereafter, several pathotypes of CCN were reported from different countries (Cook and Rivoal, 1998). Andersen and Andersen (1982) indicated the presence of three pathotypes in India, i.e. Ha21 (similar to D of the Netherlands), Ha41 (similar to Fr1 of France) and Ha31, unique in its reaction. Cook and Rivoal (1998) identified 11 pathotypes of *H. avenae* world over (seven under group Ha1, one under group Ha2 and three under group Ha3) and discussed a decanary (decimal) coding system for distinguishing and designating pathotypes of *H. avenae*. They used four differentials each of barley, wheat and oats, and a particular differential was given a specific decanary notion. The pathotype scenario of *H. avenae* is far from complete. Some populations which were earlier considered to be its pathotypes are now recognized as different species (Bishnoi and Bajaj, 2004). Since the species of the *H. avenae* complex reproduce amphimictically, cross-breeding experiments should invariably be done to confirm the species status.

22.12 Interaction with other microorganisms and antagonism

A few studies have been conducted on the interaction of cereal cyst nematodes with plant-parasitic fungi. Meagher *et al.* (1978) found that *H. avenae* and *Rhizoctonia solani* interacted synergistically and together caused more growth reduction in wheat than either pathogen alone. Gill and Swarup (1977) reported that *Helminthosporium gramineum* and *Fusarium moniliforme* individually suppressed the multiplication of *H. avenae* but nematode multiplication was significantly greater when both fungi were present than their individual effects.

22.13 Economic importance

H. avenae is the principal nematode species on temperate cereals. In Europe, at least 50% of the fields in major cereal-cultivating areas are infected by this nematode (Rivoal and Cook, 1993). Crop damage depends upon the initial population

Fig. 6.12. *Heterodera avenae* infestation in wheat field. In patches (A); uniform (B); weeds overgrowing crop in severely infested field (C); severely infested field ploughed in mid-crop season, right (D); infected root showing bushy appearance (E); white female adheres to root (F-G). [Courtesy NASI, India]

of the nematode, soil type, crop variety and environmental factors. The annual yield loss in Europe may reach £3 million (Nicol and Rivol, 2008). In the USA losses of approximately $US3.4 million have been estimated (Turner and Subbotin, 2013). In Pakistan the nematode causes 15–20% yield losses (Maqbool, 1988), in Saudi Arabia 40–92% (Ibrahim *et al.*, 1999), in Australia 23–50% (Meagher,1972) and in China up to 70% (Turner and Subbotin, 2013). In India, *H. avenae* causes heavy losses to wheat and barley crops, sometimes resulting in complete crop failure as farmers have to plough their standing crops (Fig. 6.12D). Oats are resistant in India, as in Australia. Annual losses of US$4.5 million and US$5.3 million in wheat and barley, respectively, had been reported from Rajasthan state of India (Seshadri and Dasgupta, 1980). Kanwar and Paruthi (2000) reported 16.7% yield loss in wheat variety WH542 at initial inoculum of 13.3 eggs and J_2 per cm^3 of soil. Based on the prices in 2006, a conservative loss estimate of US$83 million (equal to 661.5 million Indian rupees) per annum in wheat was calculated due to *H. avenae* in Haryana state of India (Kanwar *et al.*, 2007).

ETL LEVEL Pathogenic levels vary with climatic and edaphic factors and crop variety. Gill and Swarup (1971) reported that 20 eggs or J2/g of soil is the critical threshold level for

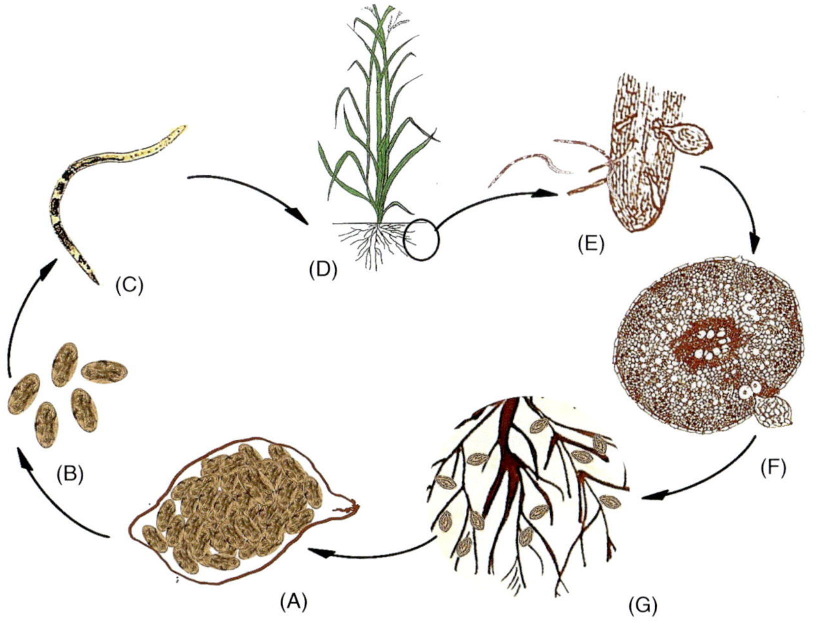

Fig. 6.13. An illustrated drawing of the life cycle of cereal cyst nematode. Cyst filled with eggs overwintering in soil (A); egg containing second-stage juvenile, J_2 (B); hatched J_2 free in soil (C); emerged plant roots host J_2 (D); young roots penetrated by J_2 and developing various stages (E); mature female form syncytium within root (F); mature female and cysts protruding on the root surface but still attached (G). [Author's own photograph]

causing economic damage. Meagher (1982) suggested that, in Australia, a population of 0.5 larva/g of soil can cause damage, whereas in West Siberia 2 J_2/g of soil was found to be the economic threshold level (ETL) for wheat crops (Shiabova, 1982). Based on field studies conducted for several years, 4–5 eggs or larvae/g of soil caused minimum economic damage in India (Kanwar and Bajaj, 2010).

22.14 Movement and means of dispersal

Maximum movement to new localities is facilitated by passive transport. The primary routes of spread are infested planting material and movement of soil (e.g. on farm machinery) from infested fields to other areas. Cysts can also be spread by wind and floodwater and cover long distances. The cysts could be carried 5–8 km in irrigation channels and canals (Odihirin, 1977).

22.14.1 Plant parts liable to carry the nematode in trade/transport

Roots, tuber, rhizome, seedlings, micro-propagated plants, growing medium accompanying plants, etc. can carry the eggs, juveniles, adults and cysts of the nematode.

22.14.2 Plant parts not known to carry the nematode in trade/transport

Stems (above ground), leaves, shoots, trunks, branches, bark, flowers, inflorescences, cones, calyx, fruits (including pods) and wood are not known to carry the nematode.

22.15 Pest risk analysis and looming threat from introduction

Because of the wide host range, damaging, dispersal and reproductive potential, CCN presents a major threat to cereal production in the countries where it is still absent. Establishment in all ecological zones of the world also concerns the environmental impact of this nematode on lowering biodiversity, disrupting natural communities or altering ecosystem processes. Introduction of this nematode into a new area will prompt additional treatment programmes, which lead to a hazardous impact on the environment. Introduction of the nematode shall also significantly impact agronomical and cultural practices. Hence, precautionary steps should be taken to prevent the introduction of the CCN into new areas and to check its further spread from known areas. The nematode has a very high fecundity rate and the cysts remain viable for a longer period. They can easily disseminate through infested soil and contaminated farm implements. Any mode of spread of infested soil can be a means of dissemination, including water, wind, migratory birds, and peds in seed lots.

22.16 Invasiveness rating

Based on the consequences of the introduction and pest risk analysis of the *H. avenae* complex, the following invasiveness rating has been assigned.

Climate/host interaction	–	High (3)
Hosts range	–	High (3)
Dispersal potential	–	Medium (2)
Economic impact	–	High (3)
Environmental impact	–	High (3)
Post-entry distribution	–	Low (–1)
Final rating (score)	–	High (13)

Based on the above invasiveness rating, *H. avenae* receives a score of 13 out of 15 and is rated as HIGHLY INVASIVE NEMATODE. The consequences of introduction into new areas are very high.

22.17 Region-wise status of invasiveness

Based on its worldwide geographical distribution and likely introduction, the *H. avenae* species complex attains highly invasive status in the following wheat-producing countries (FAOSTAT, 2019) where the nematode is absent.

APPPC: Afghanistan, Iraq, Turkmenistan, Uzbekistan
COSAVE: Argentina, Brazil
EPPO: Hungary, Lithuania, Romania
IAPSC: Ethiopia
NAPPO: Mexico

22.18 Management measures

22.18.1 Cultural practices

CROP ROTATION The cereal cyst nematode *H. avenae* has a narrow host range, hence it offers good chances of its management by crop rotation. With a rotation of two or more years with non-hosts in heavily infested areas, the nematode can be managed successfully. Meagher and Rooney (1966) reported that wheat yields improved when a legume rotation was used. Cultivation of resistant barley or mustard crops reduced the nematode population by 57–58% and 50%, respectively (Singh, 1985). Singh *et al.* (1987) observed population declines of 87–100% after growing *Brassica campestris* var. *dichotoma*, *B. campestris* var. *sarson*, *B. campestris* var. *toria*, *B. juncea* and *Eruca sativa* for one season.

SUMMER PLOUGHING Populations of cereal cyst nematode decline when soil is kept exposed to solar heat by ploughing during hot summers of May–June in India (Kanwar and Bajaj, 2010). Bhatti and Dabur (1982b) also reported that one, two and three ploughings at 10-day intervals increased wheat yield by 2.7%, 5.6% and 11.2%, respectively, over controls in CCN-infested fields. Summer ploughings may have the additional benefit of destroying other pests.

EARLY SOWING In Australia, early-sown crops in CCN-infested areas have yielded greater grains than later-sown crops (Meagher, 1972). In India, late (December) sown crops suffer more damage than early (November) sown crops (Brigbhan and Kanwar, 2004). Hence, early sowing is a recommended practice to minimize nematode damage.

TRAP CROP AND RE-SOWING Bhatti and Dabur (1982a) suggested an early wheat cv. ‘C 306’ as a trap crop before taking the main crop of wheat. The economics of this practice is questionable, as growing a susceptible crop and then ploughing it in the field involves extra cost. Bajaj and Kanwar (2005) indicated the possibility of winter maize as a trap crop for the control of *H. avenae* and *H. filipjevi* in India.

HOST RESISTANCE Several sources of resistance were identified in *Triticum aestivum* and *T. durum*, and the cultivars ‘Katyil’ and ‘Cocamba’ were specifically bred with resistance to CCN in Australia (Archer *et al.*, 1984). However, the cv. ‘Katyil’ was found susceptible to *H. avenae* in India (Bishnoi and Bajaj, 2004). Genes imparting resistance to different populations of *H. avenae* in wheat have been identified (Martin *et al.*, 2004). Montes *et al.* (2008) reported a higher level of resistance against Spanish pathotype Ha71 in wheat lines carrying Cre1, Cre4 or Cre7 genes than carrying Cre8 or Cre3 genes. In India a resistant cv. ‘Raj Molya Rodhak1’ (‘Raj MR1’) was developed for cultivation in Rajasthan state (Sharma *et al.*, 2004). Resistance in this variety is governed by a single dominant gene and pre- as well as post-infection resistance operates (Kanwar and Bajaj, 2009).

22.18.2 Chemical control

Many fumigant and non-fumigant nematicides have been used against cereal cyst nematode. For economic and environmental reasons, use of nematicides is generally not preferred. However, when other options of control have been exhausted, carbofuran at 1.0–1.5 kg a.i./ha are recommended against cereal cyst nematode in India (Kanwar and Bajaj, 2010).

22.18.3 Biological control

Several fungi, such as *Nematophthora gynophila*, *Pochonia chlamydosporia*, *Allomyces anomalus* and *Catenaria auxiliaris*, are known to suppress populations of *H. avenae* in fields and play an important role in the natural control of CCN (Holgado and Crump, 2003). Bhattacharya and Swarup (1988) reported parasitization of *H. avenae* by *Pasteuria penetrans*. Davies *et al.* (1990) reported a population of *Pasteuria* infecting J_2 of *H. avenae* which prevented infected juveniles from invading plant roots. A high-temperature resistant mutant strain HT-54 of *Azotobacter chroococcum* reduced CCN infection and promoted plant growth of wheat (Bansal *et al.*, 1999).

22.18.4 *Integrated nematode management*

Depending upon availability and feasibility, two or more control practices can be combined for achieving effective control of *H. avenae*. Increase in yield of wheat and barley was recorded when nitrogenous fertilizer was combined with nematicides (Kanwar and Bajaj, 2010). Integration of early November sowing with aldicarb at 2 kg a.i./ha gave a better yield and reduction in cyst population (Bhatti *et al.*, 1980). An increased yield of wheat was recorded from integration of seed treatment with nematicide, resistant cultivars, and crop rotation with legume and growing wheat with minimum tillage (Rovira and Simon, 1982). Kanwar and Bajaj (2010) suggested neem seed powder, farmyard manure and *Trichoderma viride* with nematicides for effective control of *H. avenae* in wheat.

22.18.5 *Phytosanitary measures*

See Chapter 16, section 115.

23 *Heterodera cajani*

23.1 Common name

Pigeon pea cyst nematode

23.2 Scientific name

Heterodera cajani Koshy, 1967

23.3 Synonym

Heterodera vigni Edward and Misra, 1968

23.4 EPPO code

HETDCJ

23.5 Diagnosis

Heterodera cajani was first reported by Swarup *et al.* (1964) as *Heterodera trifolii* which was later considered and described as a new species by Koshy (1967) on pigeon pea (*Cajanus cajan*). Morphological and molecular diagnosis approaches are described below.

23.5.1 *Morphological diagnosis*

The morphological and morphometric details (after Koshy, 1967; Koshy *et al.*, 1971) are as follows. For morphometric data, see Table 24.

ADULT FEMALE: Body swollen, pearl-white in colour. Lemon-shaped with protruding neck and vulva. Cuticle thick, with zigzag pattern externally. Head offset from neck, with two prominent annules. Second annule larger than first and usually disc-shaped. Cephalic sclerotization weak. Slight dorsal curvature to rather delicate stylet, knobs slope posteriorly. Median bulb large, inhabiting half of body width across valve. Oesophageal glands deceptively contained in single lobe, inconstant in size and shape. Excretory pore distinct, situated at level posterior to oesophageal glands end, generally about 127 µm from anterior end. Two ovaries, convoluted and almost indistinguishable due to existence of eggs. Vulva exposed, protrudes posteriorly. Egg sac present. Anus distinct, located approximately 81–84 µm anterior to vulva. Female turns into brown-coloured cyst upon death.

CYSTS: Light to dark brown colour, lemon-shaped with protruding neck and vulva (Fig. 6.14A). Occasionally, vulval cone tapers slowly into cyst; in others offset, tapers more abruptly, prominent. On young cysts, subcrystalline layer present. Cyst wall patterns on main body short, broken, angularly arranged lines, showing typical zigzag pattern. Vulval cone numerous concentric lines surrounding vulval slit and fenestrae. Punctations uneven and scanty, evident only under high magnification at lower depth than zigzag marking. Fenestrae ambifenestrate type on vulval cone-top with two semifenestrae detached by vulval bridge. Anus indistinct. Dark brown bullae present, prominent in vulval cone region below underbridge (Fig. 6.14B).

EGG SAC: Size varies between 0.5 and 2 times size of cyst (note: other species of Schachtii group have egg sacs not more than one cyst size). Yellow, seldom purple. On average 54 (few to 200) eggs in egg sacs (Fig. 6.14A).

EGGS: Oval. Eggshell hyaline, without surface markings.

SECOND-STAGE JUVENILE: Body elongate, cylindrical, tapering anteriorly and posteriorly but more posteriorly (Fig. 6.14C). Head offset, bearing four annules; first two more prominent than others situated anterior to cephalic constriction. Cuticular annulation distinct, annules around 1.6 µm apart at mid-body. Lateral field with four incisures (Fig. 6.14F), covering almost one-quarter body width. Cephalic framework sclerotized heavily (Fig. 6.14D). Anterior cephalids one annule behind cephalic constriction; posterior cephalids at 8th annule. Stylet well developed, strong, with anteriorly directed knobs, wider than the length (Fig. 6.14D). Dorsal oesophageal gland opening 4.0–5.0 µm behind stylet base. Procorpus a slim tube ending in median bulb with prominent valve plates. Basal part of oesophagus extends, glandular lobe overlapping intestine in anterior region, mostly laterally. Hemizonid distinct, one annule anterior to distinct anal excretory pore. Tail tip and length of hyaline tail-terminal vary, bluntly rounded narrow tip (Fig. 6.14E). Stylet/hyaline tail terminal ratio 1.0 (0.8–1.2), inconspicuous phasmids.

MALE: Adult resembles juvenile in anterior ends. Head offset from body with 4–5 distinct annules on cephalic region. Cephalic sclerotization heavy. Anterior cephalids generally on 2nd annule from head constriction; posterior cephalid on 8th–10th annules. Cuticular annulation distinct on body; nearly 2 µm apart. Lateral field marked by four incisures without prominent areolation, inhabiting about one-quarter to one-fifth of body

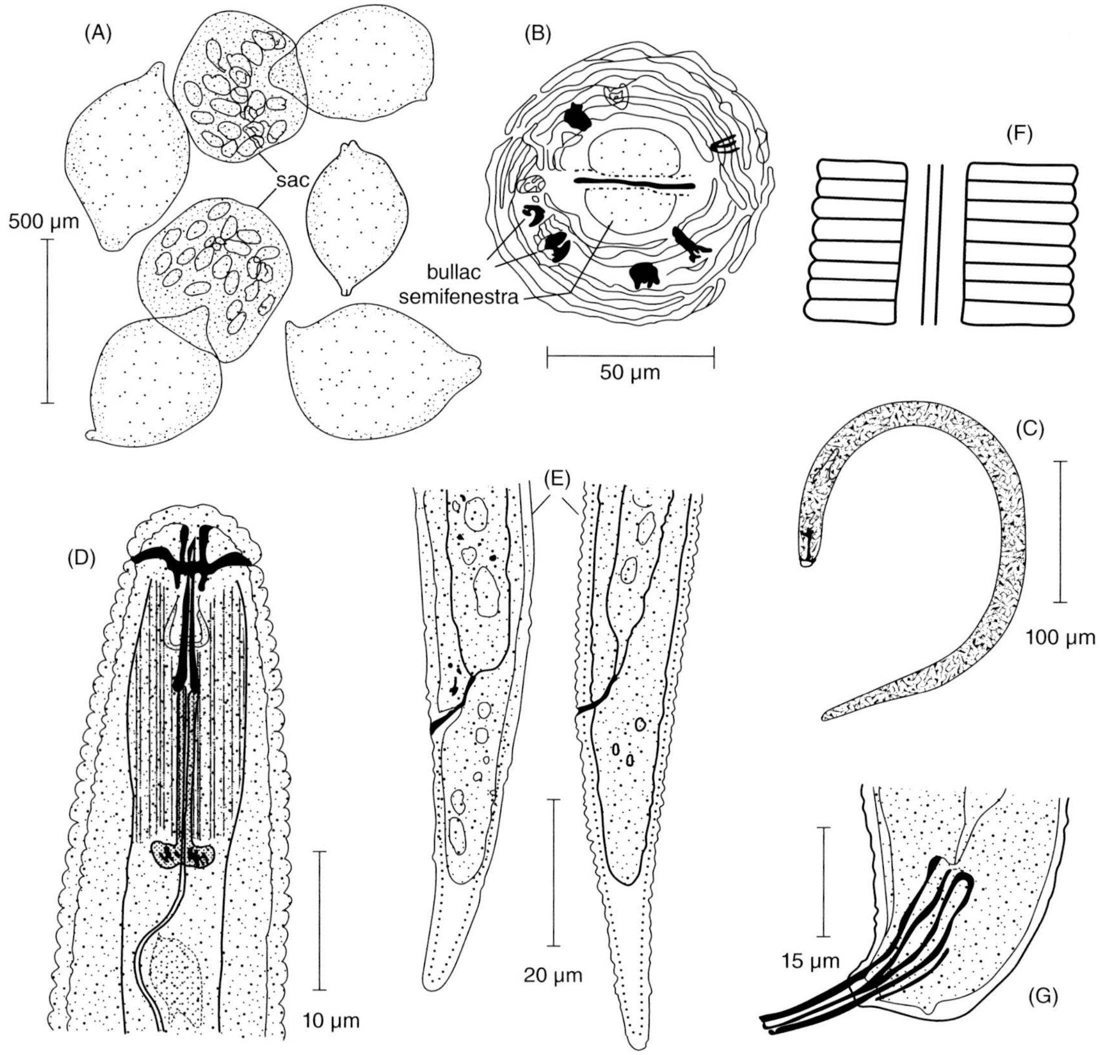

Fig. 6.14. Line drawing of *Heterodera cajani*. Cysts with egg sacs (A); vulval cone of a cyst (B); second-stage juvenile, J_2 (C); J_2 head end (D); tail of J_2 (E); lateral fields of J_2 (F); male spicules with bidentate tip (G). [Source: EcoPort (http://www.ecoport.org), C. Githure]

width. Stylet strong with slightly anteriorly directed knobs. Dorsal oesophageal gland opening 4–6 μm behind stylet base. Procorpus and intestine overlapping, similar to that of juvenile. Nerve ring encircles isthmus. Excretory pore usually opposite centre of posterior oesophageal gland. Hemizonid one annule anterior to excretory pore. Single testis, outstretches anteriorly. Paired bidentate spicules 31–41 μm in length (Fig. 6.14G). Simple rod-shaped gubernaculum, 8–12 μm in length. Tail very short, bluntly rounded with slight dorsal curvature. Phasmids not clear. May also be abnormally short males (480–650 μm) present in population that still can reproduce.

23.5.2 Biochemical and molecular diagnosis

Shashi *et al.* (1998) used the isozyme patterns of esterases and malate dehydrogenase to separate two races of *H. cajani* and three races of *H. zeae*. Shashi *et al.* (1999) also characterized *H. cajani*, *H. graminis* and *H. zeae* based on proteins, using one-dimensional native gel electrophoresis. Molecular diagnoses of *H. cajani* are available by sequencing the ITS-rRNA genes and by PCR-RFLP profiles (Waeyenberge *et al.*, 2009b; Subbotin *et al.*, 2010). For detailed molecular diagnosis, see section 22.5.3 of *H. avenae* (above).

23.6 Geographical distribution

H. cajani has limited distribution and has been reported only from two Asian countries, India and Pakistan (Koshy, 1967; Subbotin *et al.*, 2010), and one African country, Egypt (Aboul-Eid and Ghorab, 1974).

23.7 Hosts

H. cajani has a wide host range, including 18 plant species from Fabaceae and *Sesamum indicum* from Pedaliaceae (Koshy and

Swarup, 1973). Some discrepancies may occur in host preference due to the presence of physiological races of *H. cajani*. Major and minor hosts are listed below (CABI-ISC, 2020).

Major hosts: *Cajanus cajan* (pigeon pea), *Cajanus crassus*, *Cajanus platycarpus* (wild pigeon pea), *Cicer arietinum* (chickpea), *Cyamopsis tetragonoloba* (cluster bean, guar), *Desmodium triflorum* (three-flower tick trefoil), *Flemingia strobilifera* (wild hops), *Glycine max* (soybean), *Lablab purpureus* (hyacinth bean, lablab bean), *Macrotyloma uniflorum* (horsegram), *Macroptilium atropurpureum* (purple bush bean), *Macroptilium lathyroides* (wild bush bean), *Phaseolus acutifolus* (tepary bean), *Phaseolus lunatus* (sieve bean), *Phaseolus vulgaris* (kidney bean), *Phyllanthus maderaspatensis* (Madera leaf flower), *Pisum sativum* (common pea), *Sesamum indicum* (Indian sesame), *Sesamum orientale* (sesame), *Sesbania aculeata* (dhaincha), *Vicia narbonensis* (purple broad vetch), *Vicia sativa* (garden vetch), *Vigna aconitifolia* (moth bean), *Vigna mungo* (black gram), *Vigna radiata* (green gram, mungbean), *Vigna trilobata* (African gram), *Vigna umbellata* (rice bean), *Vigna unguiculata* (cowpea).

Minor hosts: *Crotalaria juncea* (sunn hemp), *Rhyncosia* spp., *Sesbania cannabina* (corkwood tree).

23.8 Symptoms

The main symptoms of *H. cajani* infection are stunted growth and reduced yield due to a damaged root system. A reduction in overall plant growth of the infected plants can be distinguished by careful comparison with non-infected and healthy plants. Yellowing (chlorosis) is common after an extended infection. Stunting, abridged leaf lamina size and chlorosis on cotyledonary leaves may or may not be readily visible. The size and number of flowers and pods are also reduced, with poorly developed root systems. The first symptoms of nematode infection are evident after 30–45 days; subsequently, the pearly white lemon-shaped females start protruding on the roots of infected plants (Sharma *et al.*, 1992b). Brown cysts can also be observed in the soil. Damage due to *H. cajani* infection is caused by the J_2s as they feed intracellularly on the elongation region of the growing root. The formation of syncytial cells around the head region interrupts xylem and phloem elements, thus hampering root efficiency (Aboul-Eid and Ghorab, 1974; Sharma *et al.*, 1992b).

In *Sesamum* (sesame), suppression in plant growth and leaf chlorophyll content are specifically observed in soils infested with *H. cajani* in India. Poor seed setting in the locule is also seen,with plants devoid of productive branching. In most cases, plants are found with a small, single mummified capsule at the plant tip with few seed (Balasubramanian and Vadivelu, 2003). In cowpea, *H. cajani* can cause delay in leaf emergence, delayed differentiation of flowers and pods, and a reduced number of flowers and pods per plant (Aboul-Eid and Ghorab, 1974).

23.9 Biology and life cycle

The life cycle of the pigeon pea cyst nematode, *H. cajani*, is similar to that in other cyst nematodes. The second-stage juvenile (J_2) hatches from the egg. The J_2, the infective stage, hunts for a host root and invades the cortex (usually within 48 h). Then the vermiform larva starts to become sedentary and feeds through the specialized trophic syncytial cells formed by the host in response to secretions from the nematode in the stele of the roots. Developing nematodes are continuously obese and moult to the J_3 usually around the 6th day. Finally, the J_4 stage is reached which moults to an adult. If the J_4 moults to female, it remains located within the root cortex. If J_4 moults to the vermiform male, it escapes from the J_4 cuticle and the root, and hunts for females to mate. Though males are encountered in large numbers, the nematode can reproduce even in the absence of males (Sharma and Swarup, 1984). Under stress conditions, plenty of adult males are developed in place of saccate females. These males can be differentiated on the basis of diarchic gonads in place of the monoarchic gonad present in natural males (Khan, 2008).

The life cycle of *H. cajani* is completed within 16 days at 29°C, and there can be many generations (up to nine) in a single growing season (Koshy and Swarup, 1979). At 25.5°C, 28–29 days are required by this nematode to complete a life cycle (Senthamizh *et al.*, 2005). Under cool conditions (10–25°C), *H. cajani* takes 45–80 days to complete one generation (Koshy and Swarup, 1971a). *H. cajani* prefers vertisol soils (Elyas and Sharma, 1997). Rao *et al.* (2011) indicate that the nematode is particularly widespread in sandy loams in northern India and vertisols in southern India.

23.10 Physiological races

Intraspecific variation and the occurrence of three races based on the host's differential reaction have been reported with this nematode. The pigeon pea (*Cajanus cajan*) race (race A) develops on all plants, whereas the other races do not reproduce either on cluster bean (*Cyamopsis tetragonolobus*) (race B) or on cluster bean and sun hemp (*Crotalaria juncea*) (race C) (Walia and Bajaj, 2000). Another study distinguished three races among 14 populations from seven districts of Uttar Pradesh, India (Siddiqi and Mahmood, 1993). Rao *et al.* (2011) used an amplified fragment length polymorphism (AFLP) and morphometric procedures to compare interspecific variation in *H. cajani* populations in India.

23.11 Similarities to other species

Morphologically *H. cajani* is similar to many other cyst nematodes; even *H. cajani* itself was originally described incorrectly as *H. trifoli* (Koshy *et al.*, 1971). *H. cajani* resembles species of the Schachtii group, including *H. trifolii*, *H. glycines*, *H. sacchari*, *H. schachtii*, *H. mothi*, *H. galeopsidis* and *H. lespedezae* (Subbotin *et al.*, 2010). The presence of males separates *H. cajani* from *H. galeopsidis*, and a different host range separates *H. cajani* from *H. lespedezae*.

23.12 Economic importance

In India, *H. cajani* is an economic pest of cowpea and pigeon pea and considered to be one of the four most important cyst nematodes. Although accurate estimates of economic crop loss are not available, high populations may cause up to 30% yield decline in pigeon pea (Ali *et al.*, 2010), with the highest reported yield loss up to 80% (Kannan and Lingaraju, 1999). *H. cajani* infection may suppress plant growth of pigeon pea by 28–34% and grain yield by 24%; and in mung bean, plant growth by 42% and grain yield by 68% (Sharma *et al.*, 1992b). Concomitant infection of *H. cajani* and *Fusarium udum* results in much greater yield loss and is difficult to manage (Hasan, 1984). Meena *et al.* (2009) worked out the economic threshold level as 100 second-stage juveniles per plant of pigeon pea.

23.13 Movement and means of dispersal

For details, see section 22.14, above.

23.14 Pest risk analysis and looming threat from introduction

Legume crops are important staple crops worldwide. *H. cajani* could have detrimental effects on production and yield if introduced into legume fields. The wide host range of *H. cajani* among leguminous crops coupled with the ability of this pest to become established in tropical and subtropical areas makes this nematode a highly invasive species. The presence of *H. cajani* in Egypt led to concerns that it could become established in tropical, subtropical and even Mediterranean countries (Luc, 1986).

23.15 Invasiveness rating

Considering the consequences of the introduction and pest risk analysis of *H. cajani*, the following invasiveness rating has been assigned.

Climate/host interaction	–	High (3)
Hosts range	–	High (3)
Dispersal potential	–	Medium (2)
Economic impact	–	High (3)
Environmental impact	–	Medium (2)
Post-entry distribution	–	Very low (–1)
Final rating (score)	–	High (12)

Based on the above invasiveness rating, *H. cajani* receives a score of 12 out of 15 and is rated as HIGHLY INVASIVE NEMATODE.

23.16 Region-wise status of invasiveness

H. cajani is listed as a harmful organism in Brazil and Ecuador (USDA-PCIT, 2014). There may be trade implications for leguminous crops from countries where this nematode enters and becomes established. The nematode attains highly invasive status in the following legume-producing countries.

APPPC: Bangladesh, Cambodia, China, Indonesia, Iraq, Israel, Japan, Jordan, Kazakhstan, Kyrgyzstan, Lebanon, Malaysia, Myanmar, Philippines, South Korea, Sri Lanka, Syria, Tajikistan, Thailand, Vietnam, Uzbekistan
CAHFSA: Barbados, Belize, Cuba, Dominican Republic, Haiti, Jamaica, Saint Kitts & Nevis
CAN: Bolivia, Colombia, Ecuador, Peru, Venezuela
COSAVE: Argentina, Brazil, Chile, Guyana, Paraguay, Suriname, Uruguay
EPPO: Austria, Belarus, Belgium, Bulgaria, Croatia, Cyprus, Czech Republic, Denmark, Estonia, Finland, France, Georgia, Germany, Greece, Guernsey, Hungary, Ireland, Israel, Italy, Jersey, Jordan, Kazakhstan, Kyrgyzstan, Latvia, Lithuania, Luxembourg, Macedonia, Malta, Moldova, Morocco, Netherlands, Norway, Poland, Portugal, Romania, Russia, Serbia, Slovakia, Slovenia, Spain, Sweden, Switzerland, Tunisia, Turkey, Ukraine, United Kingdom, Uzbekistan
IAPSC: Algeria, Angola, Benin, Botswana, Burkina Faso, Burundi, Cameroon, Cape Verde, Central African Republic, Chad, Comoros, Congo, Côte d'Ivoire, Eritrea, Ethiopia, Gabon, Gambia, Ghana, Guinea, Guinea-Bissau, Kenya, Liberia, Madagascar, Malawi, Mali, Mauritania, Mauritius, Mozambique, Namibia, Niger, Réunion, Rwanda, Senegal, Sierra Leone, Somalia, South Africa, Sudan, Swaziland, Togo, Uganda, Zambia, Zimbabwe
NAPPO: Canada, Mexico, USA
NEPPO: Jordan, Libya, Malta, Sudan, Syria, Tunisia
OIRSA: El Salvador, Guatemala, Honduras, Nicaragua
PPPO: Australia, Fiji, New Zealand, Papua New Guinea, Tonga, Vanuatu.

23.17 Management measures

23.17.1 Cultural control

The most effective management measure for cyst nematode is rotation with non-host crops (Sikora *et al.*, 2005). In India, it was revealed that little resistance is currently available within pigeon pea germplasm. India has begun to employ extensive intercropping for *H. cajani* management with sorghum, maize (corn), pearl millet and guar (Dwivedi and Upadhyay, 2001). Seed coating with neem formulations has been shown to be effective against *H. cajani* in pigeon pea (Mojumder and Mishra, 2001).

23.17.2 Chemical control

Seed treatment with carbofuran, phorate and fensulfothion significantly reduced the cyst population, with phorate exhibiting the best response (Zaki and Bhatti, 1986). The application of fensulfothion, phorate, vydate and carbofuran either in the form of seed treatment or as granules or a drench application resulted in better plant growth than with non-inoculated control (Panda and Seshadri, 1979). Application of phenamiphos (fenamiphos), oxamyl, phosalone and Danusban (chlorpyriphos) at 1000 ppm/plant (soil, spray, and soil + spray) resulted in a marked reduction in *H. cajani* populations infesting pigeon pea (Mishra and Prasad, 1990).

23.17.3 Phytosanitary measures

See Chapter 16, section 115.

24 *Heterodera carotae*

24.1 Common name

Carrot cyst nematode

24.2 Scientific name

Heterodera carotae Jones, 1950

24.3 EPPO code

HEMCAR

24.4 Diagnosis

Heterodera carotae is a member of the Goettingiana group and close relatives are *H. circae*, *H. cruciferae*, *H. goettingiana*, *H. mediterranea*, *H. scutelarae* and *H. persica*.

24.4.1 Morphological diagnosis

Morphological and morphometric descriptions (after Mathews, 1975) are summarized below. For measurements, see Table 24.

FEMALE: Lemon-shaped, white with large egg sac nearly size of the female. Head small, comprising labial plate and one annule. Median oesophageal bulb oval with discrete valve. Excretory pore behind base of median bulb and 81–119 µm from anterior end. Ovaries paired, almost cover entire body cavity. Vulval slit in cleft on cone-terminus. Cuticular pattern reticulate at mid-body.

CYST: Mature cysts light brown, lemon-shaped with distinct neck and small vulval cone (Fig. 6.15A); neck frequently twisted. No transitional yellow stage. Wall-pattern forming close network comprising irregular zigzag lines. Subcrystalline

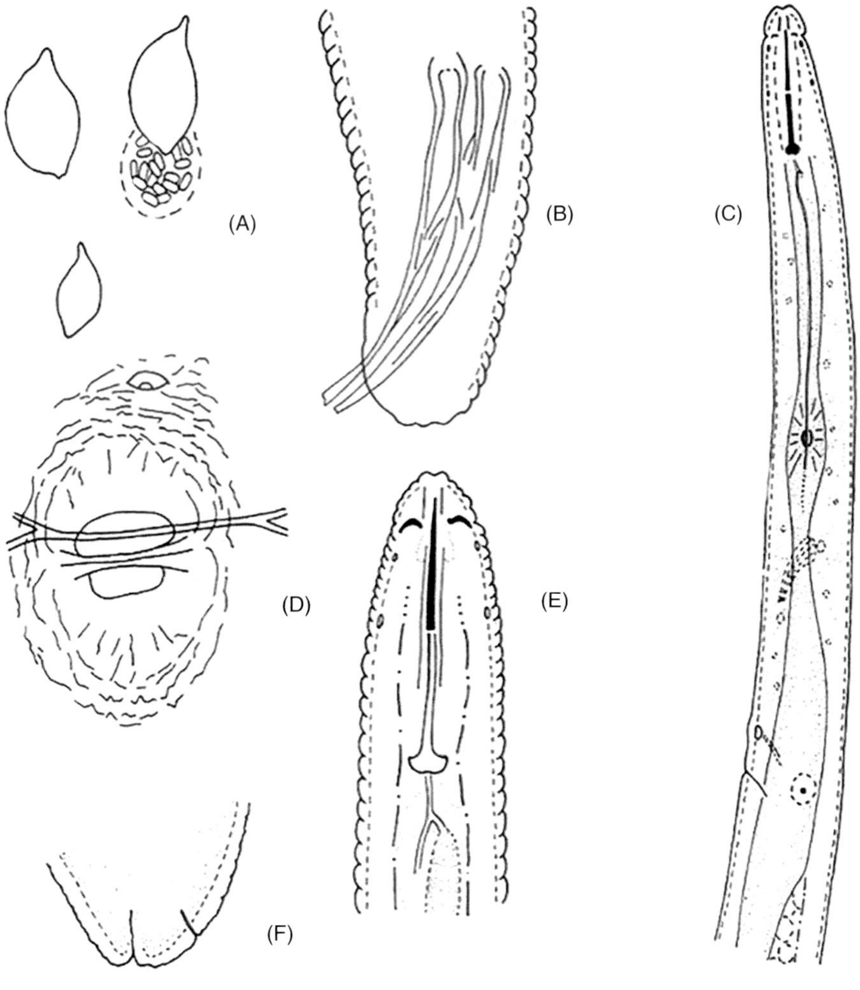

Fig. 6.15. Line drawing of *Heterodera carotae*. Cysts with egg sac (A); male posterior end (B); male anterior region (C); vulval cone of cyst (D); second-stage juvenile with stylet (E); tail of second-stage juveniles (F). [Source: Nemaplex; H. Ferris]

layer found, but delicate. Vulval cone protruding with slightly sloping sides merging smoothly with body contour. Bullae not present. Underbridge long slender, about 90 μm, bifurcate, unsclerotized. Vulval slit, 43–51 μm long, arises in dipped cleft on cone top and usually appears partly open. Vulval lips not sclerotized. Fenestration ambifenestrate but blurry. Large egg sac present usually filled with eggs.

MALE: Vermiform, short with bluntly rounded tail tip (Fig. 6.15B). Head offset with indistinct post-labial annules (Fig. 6.15C). Cephalic framework strong. Anterior cephalids at level of second body annule and posterior at sixth body annule. Stylet robust with rounded basal knobs. Dorsal oesophageal gland orifice 5–7 μm behind stylet knobs. Median oesophageal bulb rounded with weak valve plates. Excretory pore 148–161 μm from anterior end. Hemizonid noticeable, 2–3 annules long, situated 6–9 annules at level of excretory pore. Hemizonion discrete. Lateral field with four incisures producing three bands; outer lines crenate and outer bands areolated. Testis single, around 60% of total body length, consistently filled with sperm. Spicules arcuate with bulbous forward part and tubular mid-part, tapering into twisted posterior part, bidentate tip (Fig. 6.15B). Gubernaculum curved slightly. Phasmids present and adanal in position.

SECOND-STAGE JUVENILE: Head slightly offset with weak cephalic framework. Stylet well developed and stylet knobs with concave anterior appearances. Cephalids discrete; anterior at level of 3rd body annule and posterior at 8th–9th body annule. Dorsal oesophageal gland orifice 5–6 μm behind stylet knobs. Median bulb rounded, but weakly developed. Median valve plates 61–72 μm from anterior end. Lateral field similar to male. Hemizonion obscure. Phasmids unclear, 3-4 annules anterior to excretory pore. Genital primordium located 60% of body length from head, apparently consisting of two cells. Tail conical, typical heteroderid type in shape with hyaline region with rounded tip (Fig. 6.15F).

24.4.2 Molecular diagnosis

PCR-RFLP of internal transcribed space regions of ribosomal DNA (ITS-rDNA) is a very reliable diagnostic approach for many cyst nematodes, including *H. carotae* (Subbotin *et al.*, 2013). Madani *et al.* (2004) characterized 15 populations of cyst nematodes from a region near the Mediterranean Sea using PCR–RFLP and ITS-rDNA sequences. Eight restriction enzymes, *Alu*I, *Ava*I, *Bsh1236*I, *Hae*III, *Hin6*I, *Mva*I, *Pst*I and *Rsa*I, were used to generate RFLP profiles of *H. carotae*, *H. ciceri*, *H. fici*, *H. filipjevi*, *H. goettingiana*, *H. hordecalis*, *H. humuli*, *H. mediterranea*, *H. ripae* and *H. schachtii*. The PCR-RFLP and ITS-rDNA sequences were carried out using the protocol described by Tanha Maafi *et al.* (2003) (see Chapter 4, section 15.5.2, and Chapter 15, section 110.2.1). The obtained sequences are deposited in GenBank (accession numbers AY347917–AY347927). The RFLP patterns were similar to those previously described by other workers (Subbotin *et al.*, 2000; Tanha Maafi *et al.*, 2003). Recently Madani *et al.* (2018) used ITS, D2–D3 region of the 28S gene fragment of rRNA, cytochrome c oxidase I (COI) of mitochondrial DNA, and a heat shock protein gene (*hsp90*) to characterize *H. carotae* from carrot fields in Canada. The PCR-RFLP of ITS regions revealed dissimilar restriction patterns for *Rsa*I than *H. carotae* but the other six restriction profiles were similar to those reported in other studies. Escobar-Avila *et al.* (2018) recorded *H. carotae* for the first time in Mexico and the identification was confirmed by sequencing of ITS-rRNA and partial COI genes.

24.5 Geographical distribution

H. carotae is widespread but with localized distribution and mainly recorded from Cyprus, Czech Republic, Germany, France, Hungary, Ireland, Italy, Netherlands, Poland, Portugal, Russia, Sweden, Switzerland, United Kingdom and the USA (Berney and Bird, 1992; CABI-ISC, 2020). Recently the nematode has been reported from Canada (Madani *et al.*, 2018) and Mexico (Escobar-Avila *et al.*, 2018).

24.6 Hosts

H. carotae is a polyphagous nematode but has a narrow host range (Escobar-Avila *et al.*, 2018) that comprises beetroot (*Beta vulgaris*), broccoli (*Brassica oleracea* var. *italica*), Brussels sprouts (*B. oleracea* var. *gemmifera*), cabbage (*B. oleracea* var. *capitata*), cauliflower (*B. oleracea* var. *botrytis*), carrot (*Daucus carota* and *D. pulcherrima*), hedge parsley (*Torilis leptophylla*), lettuce (*Lactuca sativa*), radish (*Raphanus sativus*) and wheat (*Triticum aestivum*). Parasitization of olive trees has been reported, but not confirmed (Mathews, 1975).

24.7 Symptoms

Symptoms are similar to those of other cyst nematodes. They include patches of the crop with reduced growth (Fig. 6.16A), undersized individual plants with bronzing of leaves, small distorted roots system (Fig. 6.16B), a twisted overgrowth of rootlets (Fig. 6.16C) with the presence of cysts. The root system is reduced and smaller than healthier ones. Occasionally, forking of taproots is also observed. The conspicuous presence of a white female (cyst) on the root surface is an indication of nematode infection (Fig. 6.16D).

24.8 Biology and life cycle

Similar to other cyst-forming nematodes, *H. carotae* has sedentary endoparasitic habits. Infective juveniles retained in cysts do not emerge until 2–3 months after reaching the brown stage. Juveniles from the egg mass hatch within a few days, even in the absence of a host plant, and may begin a second generation by invading young rootlets and establishing a specialized feeding site (syncytium) in the stele. Root invasion takes place in 36 h at 18–20°C. Reproduction is amphimictic and development

Fig. 6.16. Symptoms of *Heterodera carotae* (A-D) and *H. ciceri* (E-F). Stunted carrot plants encircled (A); damaged and healthy carrot (B); infested roots (C); white cysts on carrot root surface (D); lentil root showing numerous females, f (E); close-up of chickpea root heavily infested with several females, 'f' and cysts, 'c' of the nematode, scale bar = 450 μm (F). [Photo courtesy A–C: Jones, BASF, source HYPP Zoology; D: Christopher Hogger, Swiss Federal Research Station for Agroecology and Agriculture, Bugwood.org; E: Nicola Greco; F: N. Vovlas]

of the egg sac begins about 4 weeks later. Some females remain embedded within the root even at maturity. The egg sac rapidly fills with unembryonated extruded eggs, but later holds fully developed juveniles. At the end of the reproductive phase, females die and become small lemon-shaped brown cysts. Cysts persist in soil for many years. Dehydrated egg masses may also persist in the soil, either free or adhering to the cyst or to pieces of root. In England, carrots sown in March and harvested in July have only one generation of the nematode, but two generations can develop on the main crop sown in May and harvested in November (Mathews, 1975). The dehydrated cysts remain infective for up to 10 years.

24.9 Economic importance

H. carotae is considered as a potential pest and might cause serious damage to the carrot crop. In Italy infestation of this nematode in carrot caused 20–90% yield loss (Greco *et al.*, 1994). In light soils and unmanaged, crop losses may reach up to 80%. If consecutive carrot crops are grown on the same infested site, the nematode populations may increase by tenfold per year (Berney and Bird, 1992).

24.10 Movement and means of dispersal

See section 22.14, above.

24.11 Pest risk analysis and looming threat from introduction

The damage caused by *H. carotae* in Europe and its establishment in the USA necessitate the enactment of regulatory measures for import from commercial carrot-producing countries to prevent the spread of *H. carotae* into non-infested carrot-growing areas. A complete pest risk analysis is recommended and appropriate actions should be evaluated by the quarantine department to minimize the risks of the introduction of this nematode from host countries.

24.12 Invasiveness rating

Considering the consequences of the introduction and pest risk analysis of *H. carotae*, the following invasiveness rating has been assigned.

Climate/host interaction	–	Medium (2)
Hosts range	–	Medium (2)
Dispersal potential	–	Medium (2)
Economic impact	–	High (3)
Environmental impact	–	Low (1)
Post-entry distribution	–	Not known (0)
Final rating (score)	–	Medium (10)

Based on the above invasiveness rating, *H. carotae* receives a score of 10 out of 15 and is rated as MEDIUM INVASIVE NEMATODE.

24.13 Region-wise status of invasiveness

The nematode attains medium invasive status in all the host countries where *H. carotae* has not yet been detected.

24.14 Management measures

To prevent further spread of the cyst nematode into non-infested areas, movement of nursery stock, planting material and soil should be restricted by a legislative method. For further details, see section 21.18 above.

25 *Heterodera ciceri*

25.1 Common name

Chickpea cyst nematode

25.2 Scientific name

Heterodera ciceri Vovlas et al., 1985

25.3 EPPO code

HETDCI

25.4 Diagnosis

Heterodera ciceri belongs to the Schachtii group along with *H. betae*, *H. cajani*, *H. deverti*, *H. galeopsidis*, *H. glycines*, *H. medicaginis*, *H. mediterranea* *H. sonchophyla*, *H. schachtii* and *H. trifolii* (Subbotin *et al.*, 2010).

25.4.1 Morphological diagnosis

The morphological and morphometric description (after Vovlas *et al.*, 1985) with minor modification are discussed below. For morphometric data see Table 24.

EGGS: Embryonated, eggshell hyaline and unsculptured. Second-stage juvenile crumpled four times inside the egg shell.
SECOND-STAGE JUVENILE: Vermiform. Head hemispherical, slightly offset; Stylet strong, stylet knobs with concave anterior surfaces (Fig. 6.17A). Oesophageal glands well developed. Hemizonid distinct, two annules long and 1–2 annules anterior to excretory pore. Genital primordium oval, 210–232 μm anterior to tail terminus. Cuticular annulations distinct. Lateral field with four lines, with outer bands aerolated (Fig. 6.17D). Tail irregularly annulated, uniformly tapering and ending in finely rounded tip. Terminal hyaline portion 48–58% of tail length (Fig. 6.17F).
ADULT FEMALE: Body lemon-shaped, opaque-white in colour with well-developed neck (Fig. 6.17S) and protruding terminal cone. At maturity, turning yellow and then yellowish-brown at start of tanning. Sometimes thin subcrystalline layer covering entire body. Gelatinous matrix present in gravid females, but only very seldom containing 1–2 eggs. Vulva a transverse slit terminally positioned. Anus subterminal, located in depression 7–10 μm in diameter. Cuticle ornamented by zigzag external patterns at mid-body, but on neck and cone region ornamentation consists of series of unbroken lines, appears as several concentric circles surrounding the vulva area in scanning electron microscope.

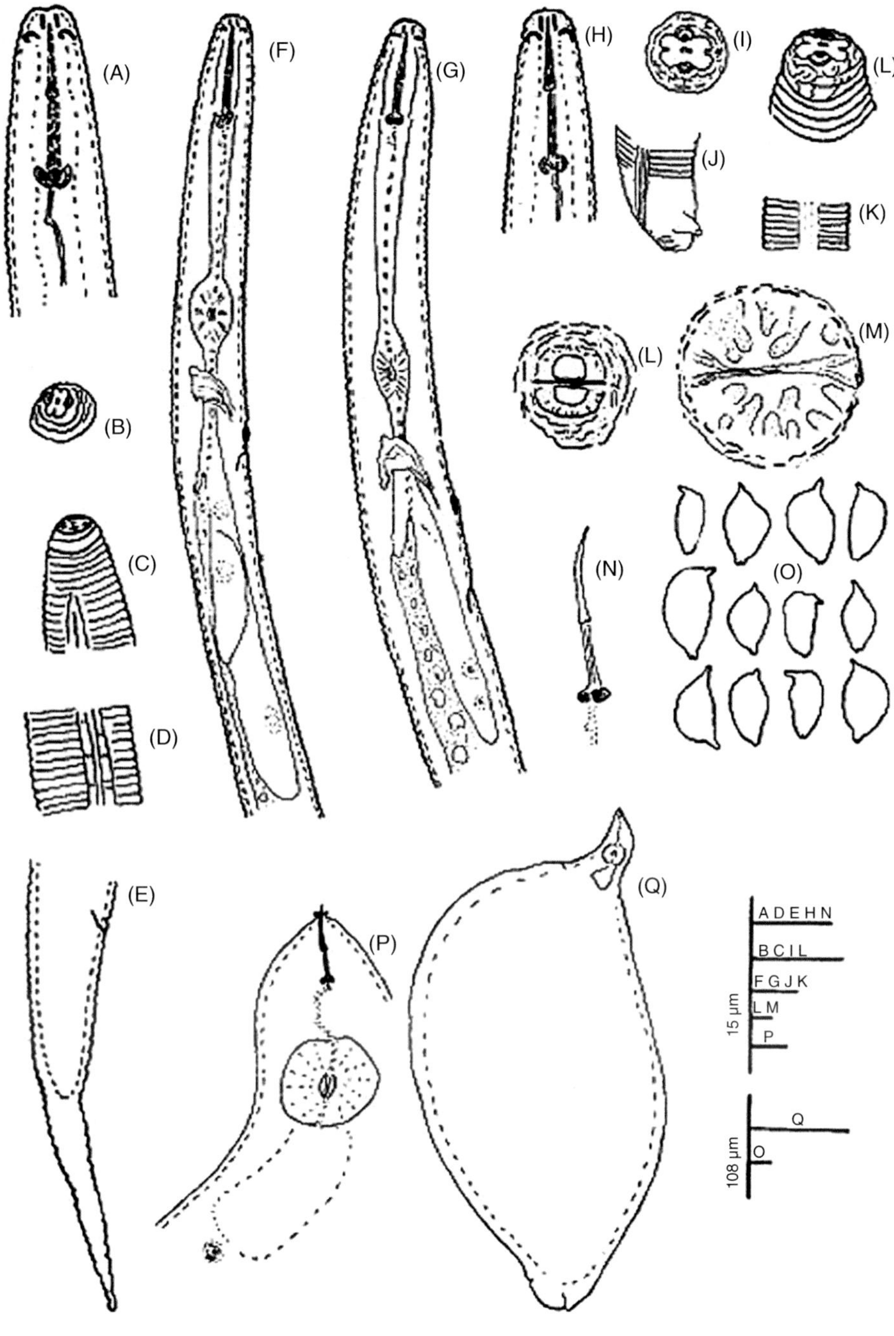

Fig. 6.17. Line drawing of *Heterodera ciceri*. Anterior end of second-stage juvenile, J_2 (A); J_2 lip pattern and profile (B,C); J_2 mid-body lateral field (D); J_2 tail (E); oesophageal region of J_2 and male, respectively (F,G); anterior end of male (H); lip pattern and profile of male tail (J); male mid-body lateral field (K); fenestrae and vulval slit (L); bullae and underbridge (M); female stylet (N); outline drawing of cysts (O); anterior portion of female (P); Outline drawing of a gravid female body (Q). [Courtesy Vovlas *et al.*, 1985; Nematologia Mediterranea]

CYSTS: Oval to lemon-shaped with distinct neck and prominent cone. Neck may be offset and posteriorly curved (Fig. 6.17R). Cysts with well-ornamented irregular zigzag patterns, creating fine network over body. Circular pattern of 15–20 straight ridges on top of vulval cone, including semifenestral areas before fenestration. Excretory pore 10–14 µm transversely within depression and surrounded by 3–4 continuous cuticular ridges. Cysts may be covered by thin subcrystalline layer entirely or partially. Numerous protruding dark brown bullae unevenly distributed at edge of vulval cone at and below underbridge present in most cysts. Vulval cones ambifenestrate; semifenestrae semicircular and sub-equal (Fig. 6.17P). Underbridge well developed lying 78–86 µm below top of vulval cone, with furcate ends and central thickening.

Male: Abundant in this species. Vermiform, typically open C-shaped after heat relaxation, with short, bluntly rounded tail, about one-third of body-width length, with long cloacal tube ending in circular opening. Head region hemispherical, offset from body (Fig. 6.17I). Prestoma rectangular-shaped, located in centre of slightly raised oral disc plate. Large amphidial apertures between ellipsoidal lateral sectors and labial disc. Cephalic framework sclerotized; strong stylet with rounded basal knobs 5–6 μm wide. Dorsal gland orifice opens 5–6 μm behind stylet knobs. Median oesophageal bulb oval with moderately developed valve 98–117 μm from head end. Hemizonid conspicuous, two annules long, located 9–12 annules anterior to excretory pore. Excretory pore 153–180 μm from anterior extremities. Lateral field with four incisures, about one-fourth of body width at mid-body, anteriorly begins as three lines at the 6th–8th body annules forming two regularly areolated bands. Testis single, not reflexed, occupies 61–68% of body length. Spicules arcuate, distally tapering. Gubernaculum ventrally curve slightly, not ornamented.

25.4.2 Molecular diagnosis

Molecular characterization of *H. ciceri* has been done by different workers using PCR–RFLP and ITS-rDNA sequences (Subbotin *et al.*, 2000; Tanha Maafi *et al.*, 2003; Madani *et al.*, 2004). Imren *et al.* (2012) characterized the ITS-rDNA of *H. ciceri* populations from Turkey. For detailed molecular diagnosis, see *H. carotae* section 23.4.2, above.

25.5 Geographical distribution

The chickpea cyst nematode, *H. ciceri*, has limited distribution and is mainly reported from Mediterranean countries such as Jordan, Lebanon, Syria (Greco *et al.*, 1992a), Turkey, Italy and Spain (EPPO, 2009b).

25.6 Hosts

Chickpea (*Cicer arietinum*) and lentil (*Lens culinaris* ssp. *culinaris*) are the main hosts of *H. ciceri* (Greco *et al.*, 1986). Other leguminous hosts are haricot bean (*Phaseolus vulgaris*), crimson clover (*Trifolium incarnatum*), red clover (*Trifolium pratense*), broad bean (*Vicia faba*), lupin (*Lupinus albus*), alfalfa (*Medicago sativa*) and vetch (*Vicia sativa*). Large numbers of *H. ciceri* specimens may also develop on wild annual *Cicer* spp. *C. bijugum*, *C. chorassanicum*, *C. echinospermum*, *C. judaicum*, *C. pinnatifidum*, *C. reticulatum* and *C. yamashitae* (Singh *et al.*, 1989; Di Vito *et al.*, 1996). Only one non-leguminous plant, carnation (*Dianthus caryophyllus*), supports multiplication, and few females develop on roots (Greco *et al.*, 1986).

25.7 Symptoms

The major symptoms on the roots of chickpea (Fig. 6.16F) and lentil (Fig. 6.16E) are the presence of white and yellow females at the flowering to early podding stage and of brown cysts at later stages (late podding to senescence). At earlier plant stages and in the case of heavy infestations, small necrotic spots (2–3 mm long and 0.5 mm wide), caused by juvenile penetration, appear longitudinally on the roots, from the centre of which the fourth and adult stage of the nematode protrudes. Rhizobium nodulation of heavily infested roots is reduced. Symptoms on above-ground parts are not specific; they are those that are characteristic of a plant whose root system is damaged so that water and nutrient uptake is impaired. Plants with severely infested roots have stunted growth, yellowish foliage, premature senescence and limited numbers of flowers and pods, which may be empty. The severity of these symptoms is correlated with intensity of the population density of the nematode in the soil; and generally infestations become obvious from early flowering onwards (Greco *et al.*, 1992a)

25.8 Biology and life cycle

The nematode survives in the soil as a cyst in the absence of host plants or during unsuitable environmental conditions. When a host plant is sown, the nematode is stimulated to hatch by root exudates. Second-stage juveniles, emerging from eggs, migrate in the soil and when a root is encountered they penetrate near the tip. After getting a suitable site, they become sedentary and stimulate the formation of syncytial cells. The nematode then undergoes three moults, becoming third- and fourth-stage juveniles and then adults. The lifespan of the different stages is greatly influenced by environmental conditions. Fourth-stage juveniles or adult females rupture the root cortex and become externally visible. Males are sedentary and endoparasitic until the fourth juvenile stage and then become free-living in the soil as adults. After fertilization, the female produces a small egg sac, made of a gelatinous matrix, and starts to produce eggs. All eggs are retained within the female body and very few (one to two) are laid in the egg sac. The female turns into a cyst by thickening its cuticle, which also becomes brown, and thus protects the eggs contained inside. Eggs undergo embryogenic development and second-stage juveniles remain coiled inside the eggshell. Cysts detach easily from the roots and persist in the soil. Eggs can remain viable in cysts for 5–10 years (Greco *et al.*, 1992b).

The highest rate of hatching occurs when the temperature is 15–25°C, whilst hatching is nil at 5°C or 30°C and negligible at 10°C. A few juveniles can emerge from eggs contained inside females. Penetration of second-stage juveniles within chickpea roots occurs equally well at 10°C and 25°C. At 8°C juveniles also invade the roots but they do not develop further. The minimum basal temperature for development is 10°C, while at 30°C root invasion by juveniles is suppressed (Kaloshian *et al.*, 1986a). At 20°C, the day-degrees required to reach different developmental stages within roots of chickpea are 80, 120, 180, 190 and 370 for third- and fourth-stage juveniles, adults, eggs and cyst formation, respectively. First- and second-stage juveniles are visible within the eggs after 15 and 17 days from their formation, respectively (Kaloshian *et al.*, 1986b). Generally, the nematode develops only one generation on spring chickpea, while there is evidence of a second generation, comprising a small proportion of the offspring of the first generation, which occurs on winter chickpea and lentil sown in fields infested with < 2 eggs/g of soil (Greco *et al.*, 1992b).

25.9 Similarities to other species

Morphologically, *H. ciceri* resembles *H. trifolii*, *H. rosii* and *H. daverti*. However, it differs from *H. trifolii* by host specification.

25.10 Economic importance

H. ciceri is a major problem in chickpea and lentil in north Syria, where severe stunting and occasionally complete crop failure have been known to occur (Sikora *et al.*, 2005). Severe symptoms have also been observed, but sporadically, on chickpea in Jordan, Lebanon and Turkey. Lentil, another pulse extensively grown in the same areas, is less severely affected by the nematode. Microplot experiments demonstrated that yield losses of chickpea may occur whenever the nematode soil population density exceeds 1 egg/cm^2 of soil. Yield losses of 20%, 50% and 100% are expected with nematode densities of 8, 16 and > 32 eggs/cm^2 of soil, respectively. The tolerance limit of lentil to the nematode is 2.5 eggs/cm^2 of soil, and yield losses of 20% and 50% may occur at about 25 and 128 eggs/cm^2 of soil, respectively (Greco *et al.*, 1993). Seed quality is also affected by the nematode.

25.11. Movement and means of dispersal

See section 22.14, above.

25.12 Pest risk analysis and looming threat from introduction

The damage caused by this nematode in Mediterranean countries presents a threat to chickpea and other legume production in other host countries. The nematode is not seed-borne and only attacks roots, therefore the chance of the introduction of cysts with seeds from infested areas to new areas is low, but due to the potential economic damage, a complete pest risk analysis should be enacted against this nematode. Preventive steps should be taken to check its introduction in the new area and to control further spread once the nematode is intercepted. There may be trade implications if this nematode is introduced and becomes established in the host countries.

25.13 Invasiveness rating

Considering the consequences of the introduction and pest risk analysis of *H. ciceri*, the following invasiveness rating has been assigned.

Climate/host interaction	–	Medium (2)
Hosts range	–	High (3)
Dispersal potential	–	Low (1)
Economic impact	–	High (3)
Environmental impact	–	Low (1)
Post-entry distribution	–	Not known (0)
Final rating (score)	–	Medium (10)

Based on the above invasiveness rating, *H. ciceri* receives a score of 10 out of 15 and is rated as MEDIUM INVASIVE NEMATODE.

25.14 Region-wise status of invasiveness

H. ciceri is not considered to be a pest warranting specific quarantine regulation. However, it can spread easily via infested soil and contaminated machinery and has attained low invasiveness status in the following major legume producing countries (FAOSTAT, 2019), where it is absent.

APPO: China, India, Indonesia, Iran, Myanmar, Nepal, Pakistan, South Korea, Vietnam
COSAVE: Argentina, Brazil, Peru
EPPO: Belarus, France, Germany, Morocco, Poland, Russian Federation, Ukraine, United Kingdom
IAPSC: Angola, Burkina Faso, Cameroon, Egypt, Ethiopia, Guatemala, Kenya, Malawi, Niger, Nigeria, Rwanda, Sudan, Tanzania, Uganda
NAPPO: Canada, Mexico, USA
PPPO: Australia

25.15 Management measures

H. ciceri can be controlled effectively by crop rotation, because of its restricted host ranges. Cultivation of a non-host plant once in a 3–4-year cycle reduces nematode soil population density to non-damaging levels and greatly increases the yield of chickpea (Saxena *et al.*, 1992). Deep ploughing of the soil in summer may speed up the decline of the nematode population. Solarization of the soil for at least 6 weeks gives satisfactory nematode control (Di Vito and Greco, 1986). The use of nematicides would be very effective in controlling *H. ciceri*, but these chemicals are expensive and may cause pollution. At present no cultivar with resistance to *H. ciceri* is available. However, resistance has been found in lines of the wild *Cicer* species *C. bijugum*, *C. pinnatifidum* and *C. reticulatum* (Di Vito *et al.*, 1996). For phytosanitary measures, see Chapter 16, section 115.

26 *Heterodera glycines*

26.1 Common name

Soybean cyst nematode

26.2 Scientific name

Heterodera glycines Ichinohe, 1952

26.3 EPPO code

HEMCAR

26.4 Diagnosis

Heterodera glycines is a member of the Schachtii group along with *H. betae*, *H. cajani*, *H. ciceri*, *H. deverti*, *H. galeopsidis*, *H. medicaginis*, *H. mediterranea*, *H. sonchophyla*, *H. schachtii* and *H. trifolii* (Subbotin *et al.*, 2010). The morphological and molecular diagnosis approach is recommended for correct species identification of *H. glycine*.

26.4.1 *Morphological diagnosis*

The morphological and morphometric descriptions (after Burrows and Stone, 1985) are presented below. For measurements, see Table 24.

FEMALE: Body lemon-shaped with prominent neck comprising oesophageal parts. Body annulation and lateral incisures not present, but covered with framework of ridges. Head with hexaradiate skeleton, stylet slender with posteriorly protrusive knobs. Median bulb subspherical and large. Females initially white, later turning pale yellow when eggs develop. Egg sac or gelatinous matrix present, holding up to 200 eggs (Fig. 6.18J). Subcrystalline layer present. Vulva and anus carried on obtuse cone-shaped apex with projection opposite neck. Vulva a transverse slit on vulval cone top, surrounded ventrally and dorsally by thin-walled crescent-shaped parts, the semifenestrae.

CYST: On death, female body wall tans to form brown tough-walled cyst. Morphologically similar to white obese female. Outer wall of cyst containing rugose zigzag pattern. Vulva cone ambifenestrate. Vulval region usually intact in younger cysts, but lost in older specimens, leaving open fenestra crossed by vulval bridge bearing vulval slit and dividing fenestra into two semifenestrae. Bullae present, well developed, elongate, at or just under level of prominent underbridge (Fig. 6.18I).

MALE: Vermiform with regularly annulated cuticle. Four incisures in lateral field. Head with strong head skeleton, offset with 4–5 annules. Stylet strong with knobs projecting laterally to anteriorly. Dorsal oesophageal gland orifices 4 µm posterior to base of stylet. Excretory pore funnel-shaped and 14.5 µm from head end. Dorsal oesophageal gland lobe ventrally overlapping intestine. Tail short, bluntly rounded. Spicules well developed, gubernaculum present (Fig. 6.18E).

SECOND-STAGE JUVENILE: Vermiform. Lateral field with four incisures. Incisures falling to three anteriorly and posteriorly. Head offset with 3 or 4 annules. Labial disc dumbbell-shaped. Stylet strong with knobs directed anteriorly (Fig. 6.18G). Anterior cephalids located 2nd–3rd annules and posterior 7th–9th annules. Tail tapering uniformly to finely rounded tip; hyaline portion about half of tail length (Fig. 6.18H).

26.4.2 Biochemical and molecular diagnosis

Several PCR-based approaches are described for diagnosis of *H. glycines*. For example, a PCR-ITS-RFLP test was designed by Subbotin *et al.* (2000) that differentiated *H. glycines* from other cyst species (*H. ciceri*, *H. medicaginis*, *H. schachtii* and *H. trifolii*) of the Schachtii group by using the restriction enzyme *Ava*I (Zheng *et al.*, 2000; Tanha Maafi *et al.*, 2003). Reliable identification can also be achieved by PCR-based species-specific primer to identify second-stage juveniles of *H. glycines* in mixed samples with other cyst and soil-inhabiting nematodes (Subbotin *et al.*, 2001; Ou *et al.*, 2008). A rapid Duplex real-time PCR method was developed by Ye (2012) using species-specific primer for accurate identification of *H. glycines* alone or in mixed samples with other cyst nematodes.

Subbotin *et al.* (2001) developed a method for the rapid identification of juveniles and cysts of *H. glycine* based on PCR with species-specific primers. A single cyst or second-stage juveniles of *H. glycines* alone or in a mixture with other soil-inhabiting nematodes will be detectable by this method. The DNA is extracted from single or several cysts, or single juveniles alone or in mixture with other nematode species using nematode lysis buffer (see Chapter 15, section 110.1.1). DNA suspension (2 µl) was used as a template for PCR. Four primers were used in the PCR reaction. The first set contained the universal primer to amplify the D3 expansion region of the large subunit ribosomal gene (28S rDNA gene):

D3A: 5′-GACCCGTCTTGAAACACGGA-3′
D3B: 5′-TCGGAAGGAACCAGCTACTA-3′

The second set primers included:

Species-specific primer GlyF1: 5′-TTACGGACCGTAACT-CAA-3′
Universal primer rDNA2: 5′-TTTCACTCGCCGTTAC-TAAGG-3′

The PCR amplification was performed using the following reagents and cycling parameters.

Reagent		Final concentration
PCR reaction	–	20 µl
10X Qiagen PCR buffer	–	2.0 µl
$MgCl_2$	–	1.6 µl (25 mM)
Deoxynucleotides	–	0.8 µµ (10 mM)
Taq DNA polymerase	–	5 U
Primer	–	0.2 µl (each 1.0 µg/µl)
DNA template	–	3 µl
Cycling parameters		
Initial denaturation	–	94°C for 4 min
Number of cycles	–	35
Denaturation	–	94°C for 30 s
Annealing	–	55°C for 30 s
Extension	–	72°C for 1.5 min
Final extension	–	72°C for 10 min.

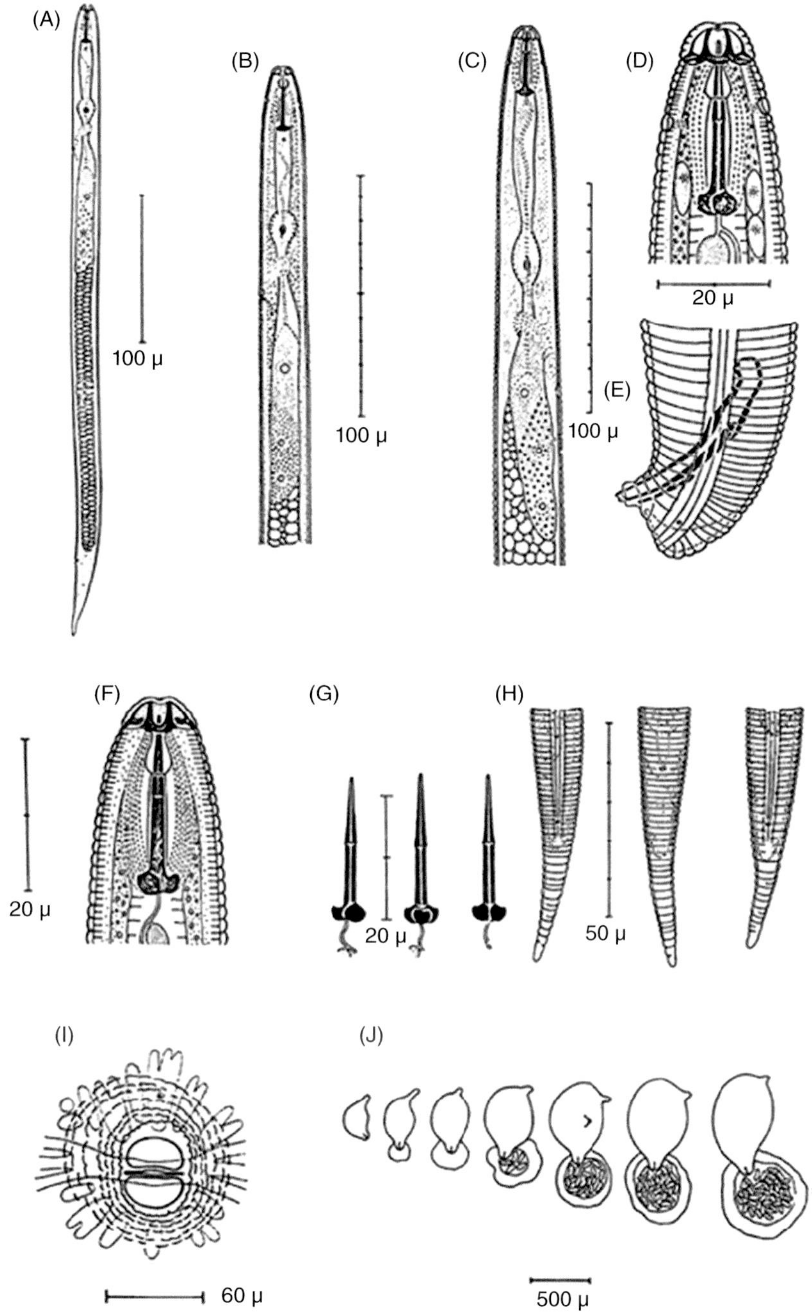

Fig. 6.18. Line drawing of *Heterodera glycines*. Body of second-stage juvenile, J_2 (A); J_2 anterior part (B); male anterior part (C); male head (D); J_2 tail (E); J_2 head region (F); J_2 stylets (G); J_2 tails (H); cyst cone front view (I); female development (J). [A–H,J: after Hirschmann, 1956, Helminthological Society of Washington; I: after Ohshima *et al.*, 1981, Japanese Journal of Nematology; courtesy EPPO Bulletin]

The amplified PCR product was visualized and photographed under UV light. All the *H. glycines* populations yielded two distinct fragments of 181 bp and 345 bp, whereas other species produced only one fragment of 345 bp.

Ye (2012) designed sequence-characterized amplified region (SCAR)-marker genomic sequences for detection of *H. glycines* cysts, second-stage juveniles and eggs using the species-specific primers/probe:

Forward primer SCNrtF: 5′-AAATTC CAGGCCGCTATC TC-3′

Reverse primer SCNrtR: 5′-CGTGGACTGAACTGGAC AAAG-3′

The double-quenched probe:

SCNrtP: 6.5′/56–FAM/TGGGCTGGG/ZEN/TGCTTCT AGAACTTTT/3IABkFQ/3′

Universal oligonucleotides for a nematode endogenous control to detect the presence of 18S rDNA gene:

Forward primer Ne18Sf: 5′-ATT GAC GGA AGG GCAC CA C-3′
Reverse primer Ne18Sr: 5′-GAA CGG CCATGC ACC AC-3′
Probe Ne18Sp: 5′/5-TET/TGCGGCTTAATTTGACTCA ACACGGG/3IABkFQ/3′

The real-time PCR cycling parameters are: denaturation at 95°C for 2 min, followed by 50 cycles of initial template denaturation at 95°C for 10 s, annealing and extension phase at 62°C for 45 s. The expected fragment length for species-specific primer is 83 bp.

Kovaleva *et al.* (2005) characterized full-length coding sequence for aldolase (a key enzyme in glycolysis) gene from *H. glycine* and *G. rostochiensis*. Gibson *et al.* (2011) sequenced the complete coding region of the mitochondrial genome of *H. glycines* (approximately 21–22 kb). The organization of the mitochondrial genome of *H. glycines* shows many similarities with *Radopholus similis*, but fewer similarities when compared with *Meloidogyne javanica*. Very few gene boundaries are shared with *Globodera pallida* or *G. rostochiensis*.

26.5 Geographical distribution

The soybean cyst nematode (SCN) occurs in most of the soybean-producing region of the world (Fig. 6.19) and is reported from the countries below (CABI-ISC, 2020). In India, SCN has been reported (EPPO, 2011) but has localized distribution.

Africa: Egypt.
Asia: China, India, Indonesia, Iran, Japan, Mongolia, North Korea, South Korea, Taiwan.
Europe: Italy, Russia.
North America: Canada, Puerto Rico, USA.
South America: Argentina, Brazil, Colombia, Ecuador, Paraguay.

26.6 Hosts

H. glycines is polyphagous and infects a wide range of plants, mainly from the Fabaceae family. Some plants of Caryophyllaceae and Scrophulariaceae are also recorded as hosts. Around 63 plant species from 50 genera of 22 families have been reported as non-fabaceous hosts (Riggs and Wrather, 1992). The extended host list of *H. glycines* is given in Table 26.

26.7 Symptoms

The symptoms of SCN are not specific. Typically, it causes stunting of the host and this may be combined with chlorosis in patches (Fig. 6.20A). Sometimes the plants in these patches show wilting or loss of leaves with reduced seed production. There is usually a sharp dividing line between affected and non-affected areas in a field of the soybean crop. The nematode causes yellow dwarf disease of soybean, the symptoms appearing in the field about 2 months after sowing. Diseased plants may be stunted with yellowed foliage and have fewer lateral roots than normal (Davis and Tylka, 2000). Reduction in seed yield is usually the first sign of infestation. The presence of

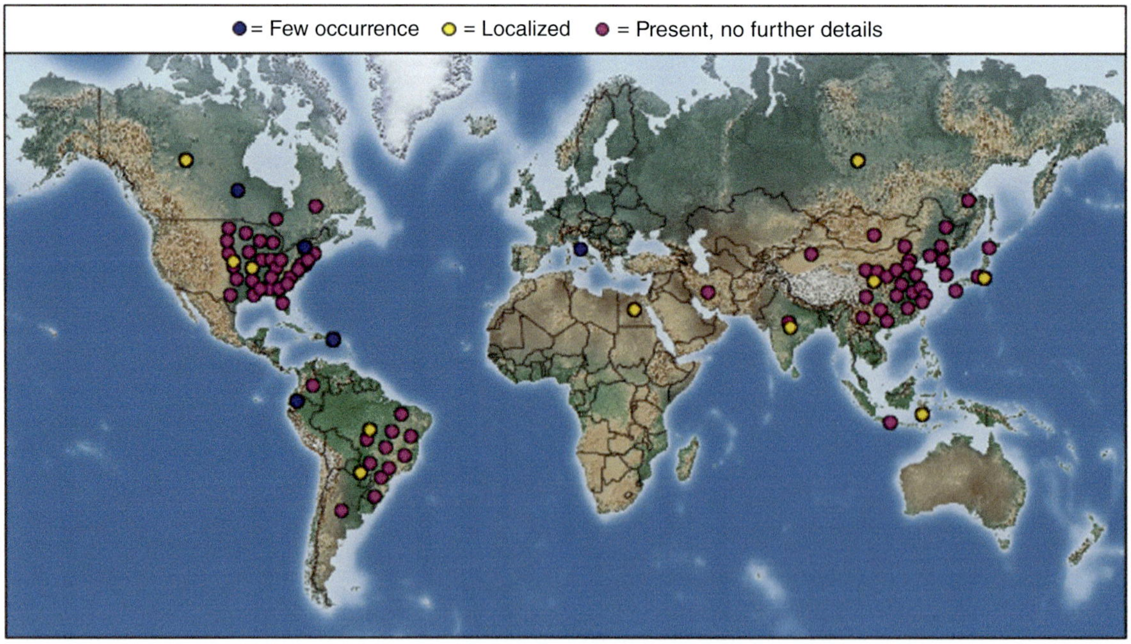

Fig. 6.19. Distribution map of *Heterodera glycine*. [Adapted from CABI-ISC, 2020]

Table 26. List of hosts of *Heterodera glycines* (adapted from CABI-ISC, 2020).

Major hosts
Glycine max (soybean)
Minor hosts
Beta vulgaris (beetroot)
Cajanus cajan (pigeon pea)
Glycine sp.
Kummerowia striata (Japanese lespedeza)
Lespedeza cuneata (Sericea lespedeza)
Lupinus (lupins)
Lupinus albus (white lupin)
Lycopersicon esculentum (tomato)
Phaseolus vulgaris (common bean)
Pisum sativum (pea)
Vicia villosa
Vigna aconitifolia (moth bean)
Vigna angularis (adzuki bean)
Vigna mungo (black gram)
Vigna radiata (mung bean)
Wild hosts
Aeschynomene indica (Indian jointvetch)
Fabaceae (leguminous plants)
Geranium (cranesbill)
Lamium amplexicaule (henbit deadnettle)
Penstemon sp.
Sesbania exaltata (coffeebean)
Stellaria media (common chickweed)
Verbascum thapsus (Aaron's-rod)

white females (cysts) on the root surface is confirmation of the disease (Fig. 6.20B,C). Yield is substantially reduced and, in severe cases, the plant may wilt and die. The most severe damage is often in the centre of the affected area. Damaged areas are frequently located near the field entrance where machinery moves into the field, or in areas where soil from another field is deposited by wind or water. There can also be reduced *Rhizobium* nodulation (Sikora *et al.*, 2005).

26.8 Biology and life cycle

The life cycle is typical of other cyst nematodes. Adult females protrude from plant roots. Reproduction is amphimictic and a single female may produce up to 600 eggs. Eggs are retained inside the cyst and are also laid in an external egg sac. The eggs remain viable in the cyst for up to 11 years (Schmitt *et al.*, 2004). Embryonic development occurs and the first moult is inside the egg. The second-stage juvenile emerges from the egg after hatching. It is presumed that the root exudates of soybean give hatching signals for dormant cysts. The J_2 migrates in soil, seeks out a suitable host root and penetrates the cortex. The optimum temperature for emergence of the J_2 from the egg and for root penetration is recorded to be 24°C. After finding a suitable host, the nematode goes through successive moults to reach pre-adult stage (J_4). The J_4 moults either to the obese female or to the vermiform male, which discharges from the J_4 cuticle and escapes the root to search for females to mate. The nematode takes 3–4 weeks to complete the life cycle and may complete six to seven generations per year, depending upon environmental conditions (Riggs and Wrather, 1992). The optimum temperature for development is 28–31°C; development ceases at below 15°C or above 33°C (Schmitt *et al.*, 2004).

26.9 Interaction with other organisms

Infected plants are predisposed to *Fusarium* wilt and sudden death syndrome of soybean is caused by interaction of *Fusarium udum* and *H. glycine* (Turner and Subbotin, 2013),

26.10 Economic importance

The soybean cyst nematode *H. glycines* is the most important constraint on the yield of soybean worldwide (Subbotin *et al.* 2001; Wrather *et al.*, 2001). Reported yield losses on soybean vary from 10% to 70% in Japan (Inagaki, 1977). All soybean-growing areas in the USA are at risk and the nematode is still spreading into previously non-infested areas. Losses in south-eastern USA were estimated at US$88.4 million in 1990 (Sciumbato, 1991). In both the USA and Canada, the reduction in yield in 2002 was 4.2 million tons, amounting to US$1960 million (Wrather *et al.*, 2003). Wrather *et al.* (2001) provided estimates that the total yield loss due to *H. glycines* in the top ten soybean-producing countries (USA, Brazil, China, Argentina, India, Canada, Paraguay, Indonesia, Italy and Bolivia) in 1998 was 28.5×10^6 t, valued at US$$6.29 \times 10^9$.

26.11. Movement and means of dispersal

See section 22.14, above.

26.12 Pest risk analysis and looming threat from introduction

Introduction of this nematode in non-native regions such as China, Brazil, Japan, India, Italy and Puerto Rico poses a threat to all other regions of the world where soybeans are grown and the nematode has not been detected. Establishment in different ecological zones of the world also concerns the environmental impact of this nematode on lowering biodiversity, disrupting natural communities or changing ecosystem processes. Introduction of the nematode into a new area will precipitate additional treatment programmes, which also leads to a hazardous impact on the environment. Hence steps should be taken to prevent the introduction in the first instance and to control spread once the nematode is known to occur. The nematode is spread most easily via infested root stock, soil and contaminated machinery.

Fig. 6.20. Symptoms of *Heterodera glycines* (A–D) and *H. goettingiana* (E–F). Soybean field with a small area of stunted plants (A); cysts on host plant roots (B,C); single cyst containing eggs (D); patches of stunted growth and chlorotic plants of pea infested with *H. goettingiana* (E); cysts on lateral root of pea (F). [Photos courtesy A: Paul Bachi, University of Kentucky Research and Education Center, Bugwood.org; B: Elizabeth Bush, Virginia Polytechnic Institute and State University, Bugwood.org; C: Gregory Tylka, Iowa State University, Bugwood.org; D: Agroscope FAL Reckenholz Archive, Swiss Federal Research Station for Agroecology and Agriculture, Bugwood.org; E: Di Vito (1991), source Nemaplex; F: Nigel Cattlin/Holt Studios International, source CABI-CPC-2010]

26.13 Invasiveness rating

Based on the consequences of the introduction and pest risk analysis of *H. glycines*, the following invasiveness rating has been assigned.

Climate/host interaction	–	High (3)
Hosts range	–	High (3)
Dispersal potential	–	Medium (2)
Economic impact	–	High (3)
Environmental impact	–	High (3)
Post-entry distribution	–	Medium (–2)
Final rating (score)	–	High (12)

Based on the above invasiveness rating, *H. glycines* receives a score of 12 out of 15 and is rated as HIGHLY INVASIVE NEMATODE. The consequences of introduction into new areas are high.

26.14 Region-wise status of invasiveness

SCN is a highly invasive pest in the following soybean-producing countries, where it is absent. There may be trade implications if this nematode is introduced and established in these countries.

APPO: Azerbaijan, Bhutan, Cambodia, Iraq, Jordan, Kazakhstan, Kyrgyzstan, Malaysia, Myanmar, Nepal, Pakistan, Philippines, Sri Lanka, Syria, Tajikistan, Thailand, Turkey, Vietnam

COSAVE: Bolivia, Chile, French Guiana, Guyana, Peru, Suriname, Uruguay

EPPO: Albania, Austria, Bulgaria, Croatia, Czech Republic, France, Germany, Greece, Hungary, Latvia, Moldova, Poland, Romania, Serbia, Slovakia, Slovenia, Spain, Switzerland, Ukraine

IAPSC: Benin, Burkina Faso, Burundi, Cameroon, Côte d'Ivoire, Ethiopia, Gabon, Kenya, Liberia, Madagascar,

Malawi, Mali, Morocco, Nigeria, Rwanda, Senegal, South Africa, Uganda, Zambia, Zimbabwe
NAPPO: Mexico
OIRSA: Belize, Costa Rica, El Salvador, Guatemala, Honduras, Nicaragua, Panama
PPPO: Australia, New Zealand

26.15 Management measures

Prevention is the best management strategy as the SCN, once established in a field, cannot be eradicated. *H. glycines* has a relatively limited host range and can thus be managed by appropriate crop rotation with non-hosts (Sikora *et al.*, 2005). Growing resistant soybean cultivars is another effective management measure. The cyst is a protective structure and no nematicide will kill all cysts in the soil. Chemical control may also be applied, as soybean is an economically high-return crop in some countries, including India. Only a few nematicides are used against SCN, such as 1,3-dichloropropene, aldicarb and oxamyl. However, yields and economic benefits are not certain, as the nematicides are expensive (Davis and Tylka, 2000). For phytosanitary measures, see Chapter 16, section 115.

27 *Heterodera goettingiana*

27.1 Common name

Pea cyst nematode

27.2 Scientific name

Heterodera goettingiana Liebscher, 1892

27.3 EPPO code

HETDGO

27.4 Diagnosis

Heterodera goettingiana belongs to the Goettingiana group that includes *H. carotae*, *H. circae*, *H. cruciferae*, *H. mediterranea*, *H. scutelarae* and *H. persica*. Morphologically they are similar, hence identification needs to be aided by molecular diagnosis methods.

27.4.1 Morphological diagnosis

The morphological and morphometric descriptions of *H. goettingiana* given below are after Baldwin and Mundo-Ocampo (1991) and Di Vito (1991). For measurements, see Table 24.

Second-stage juvenile: Vermiform with annulated cuticle. Lateral field with three bands at mid-body and not areolated (Fig. 6.21H). Head offset, bearing two or three head annules, similar to male, head skeleton hexaradiate with heavy sclerotization. Cephalids located at level of 2nd and 8th annules from anterior end. Stylet well-developed with rounded to moderately hook-shaped knobs. Oesophageal glands extended outside excretory pore for around one-third of body length, from the head. Dorsal oesophageal glands with prominent nucleus, subventral gland with obscure nuclei. Nerve ring encloses oesophagus between gland lobes and median bulb. Hemizonid located one annule anterior to excretory pore and two annules long. Hemizonion found in 6–8 annules posterior to the excretory pore and less than one annule long. Genital primordium can be seen very distinctly, two-celled, about halfway along body. Tail tapers to rounded tip, hyaline portion about two-thirds of tail length (Fig. 6.21E).
Female: Swollen, lemon-shaped. Head bearing 1–3 annules, weak skeleton, hexaradiate. Conus represents half of stylet length, basal knobs rounded. Excretory pore located at base of neck. Median oesophageal bulb spherical to sub-spherical with prominent valve. Oesophageal glands often displaced by large well-developed ovaries. Vulva situated at terminal end of vulval cone. Vulval bridge bisects fenestra to form two almost equally sized semifenestrae. Anus dorsal to bridge. Young females white, no yellow phase; after death of female, body tanned to form a cyst (Fig. 6.21B).
Cyst: Ambifenestrate with two closely depressed semifenestrae (Fig. 6.21A). Lateral muscles on inner cuticle wall can be seen under vulval bridge. Bullae not frequently present. Cuticle covered by ridges and folds to form characteristic zigzag pattern around cone and body wall. An egg sac may be attached to the cyst (Fig. 6.21B).
Male: Vermiform. On heat fixation, acquires 90° twist from mid-body to tail along length of nematode (Fig. 6.21G). Similar to second-stage juveniles in morphology. Tail short, blunt and without bursa. Testis single, contains plenty of round spermatogonia and nearly half of total body length. Cloacal opening small, surrounded by circular ring of muscle tissue. Spicule tips broad with tridentate tips (Fig. 6.21C). Gubernaculum simple, rod-like. Phasmids and caudalids not present.

27.4.2 Molecular diagnosis

Different molecular approaches are applied to characterize *H. goettingiana*, such as polymerase chain reaction–restriction fragment length polymorphism (PCR–RFLP) and internal transcribed spacer (ITS)-rDNA sequences (Tanha Maafi *et al.*, 2003; Madani *et al.*, 2004). For details, see section 15.5.2 in Chapter 4 and section 23.5.2, this chapter.

27.5 Geographical distribution

The pea cyst nematode (PCN) is widely distributed but in limited regions (Fig. 6.22) and is reported from the following countries (CABI-ISC, 2020).

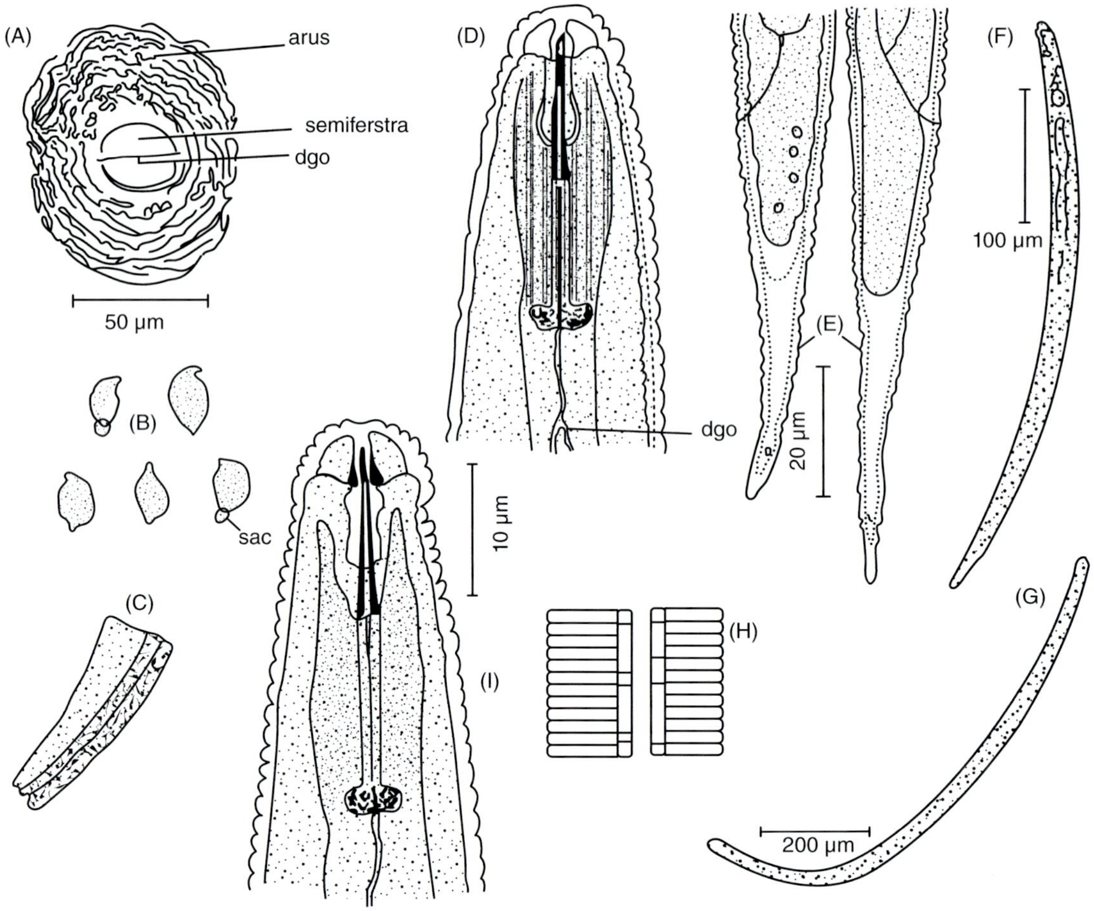

Fig. 6.21. Line drawing of *Heterodera goettingiana*. Cyst showing vulval slit, anus and semifenestra (A); cysts with egg sac (B); spicule with tridentate tip (C); second-stage juvenile with stylet (D); tail of second-stage juveniles (E); second-stage juvenile (F); male (G); lateral field (H); stylet (I). [Source: EcoPort (http://www.ecoport.org): C. Githure]

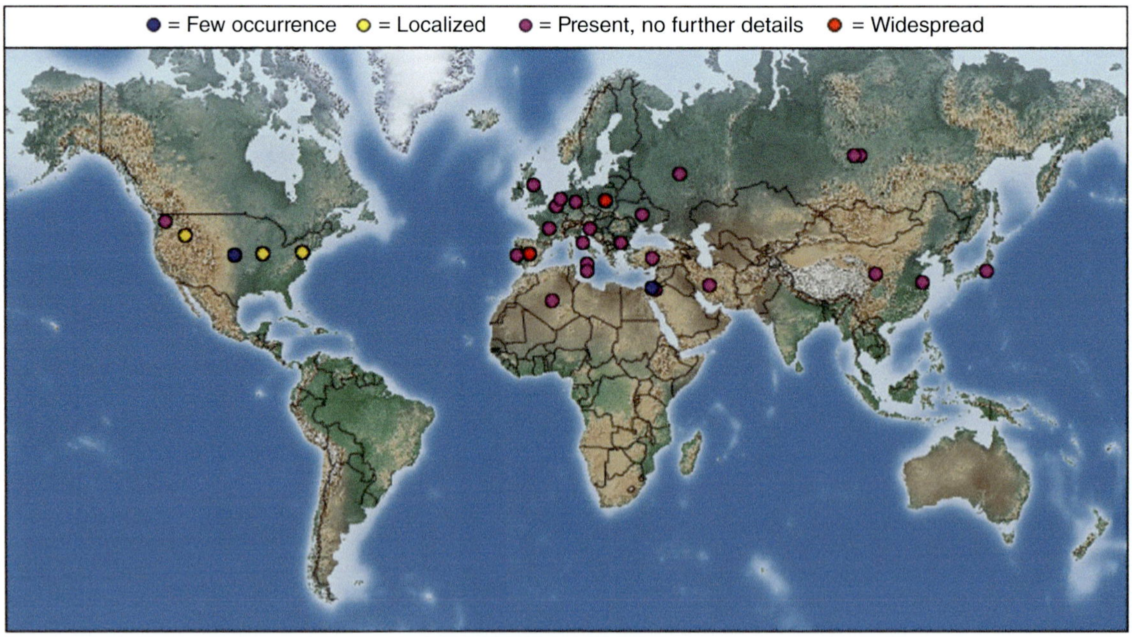

Fig. 6.22. Distribution map of *Heterodera goettingiana*. [Adapted from CABI-ISC, 2020]

Africa: Algeria.
Asia: China, Iran, Israel, Japan, Jordan, Turkey.
Europe: Belgium, Bulgaria, France, Germany, Italy, Malta, Netherlands, Poland, Portugal, Russia, Slovenia, Spain, Ukraine, United Kingdom.
North America: USA.

27.6 Hosts

Pisum sativum (pea) and *Vicia faba* (broad bean) are major hosts of *H. goettingiana*. Other minor and wild hosts (CABI-ISC, 2020) are as follows.

Minor hosts: *Cicer arietinum* (chickpea), *Glycine max* (soybean), *Lathyrus odoratus* (sweet pea), *Lathyrus sativus* (grass pea), *Lens culinaris* ssp. *culinaris* (lentil), *Lupinus albus* (white lupin), *Lupinus luteus* (yellow lupin), *Medicago sativa* (lucerne), *Pisum sativum* var. *arvense* (Austrian winter pea), *Vicia* (vetch), *Vicia benghalensis* (purple vetch), *Vicia cracca* (tufted vetch), *Vicia ervilia* (bitter vetch).
Wild hosts: *Lathyrus cicera* (flat-podded vetchling), *Vicia calcarata*.

27.7 Symptoms

Pea cyst nematode infestation becomes obvious in the field when plants in patches become stunted and yellowish. Heavily infected pea plants are severely stunted, chlorotic and die off at partially filled pod stage. Infested pea fields show abruptly outlined patches (Fig. 6.20E) (Turner and Subbotin, 2013). The chlorotic patches may enlarge during the growing season. This is due to the slower progress of symptoms in less heavily infected plants. In general, patches in fields enlarge slowly unless the nematode is moved during farming operations. Remarkably, plant roots infected with *H. goettingiana* do not show excessive numbers of lateral rootlets and the attached cyst can be noticeable to the naked eyes (Fig. 6.20F). The nematode is capable of eradicating *Rhizobium* nodules from roots. In such cases, plants display severe nitrogen deficiency, chlorosis, dwarfing, restriction of the number of flowering nodes and early senescence (Perry and Beane, 1983).

27.8 Biology and life cycle

The life cycle and biology of *H. goettingiana* are typical of the genus. The second-stage juvenile (J_2) possibly hatches in response to root exudates from host roots and environmental stimuli such as temperature. Once the J_2 has entered the host roots it will pass through three more successive moults to become an adult. Males are common and their development is possibly environmentally influenced (Guevara-Benitez *et al.*, 1970). Females enlarge in the host root tissue and finally break through the cortical root cells to become exposed on the root surface and are subsequently mated by males. Once mated, the white female begins to tan and become a cyst. There is no yellow transitional female phase. The optimum temperature for development is 13–15°C. A single egg sac usually contains 100 eggs. The number of generations produced in a year is influenced by environmental factors. One or two generations occur during the growing season in the UK and up to three in Italy. Viable cysts can survive for 12 years without a host, although each year the viability of the cyst contents falls (Turner and Subbotin, 2013).

27.9 Similarities to other species

H. goettingiana has many similarities with other cyst nematodes, particularly those that also belong to Mulvey's Group Five (before the Goettingiana group) (Subbotin *et al.*, 2010), such as *H. cruciferae*, *H. carotae and H. elachista*. It differs from several other members of the Goettingiana group by having a longer average juvenile body, longer tail and hyaline region (Turner and Subbotin, 2013)

27.10 Economic importance

In the field, yield suppression as high as 75% has been recorded (Di Vito and Greco, 1986). When peas are grown in a monocrop in an infested field, the successive crop will ultimately fail to give an economic yield. Nematode population densities and effects on plant yield are both inconstant. Nematode damage to field pea is serious and compounded by concomitant *Fusarium oxysporum* infections.

27.11 Movement and means of dispersal

See section 22.14, above.

27.12 Pest risk analysis and looming threat from introduction

The nematode presents a threat to pea production wherever it occurs worldwide and steps should be taken to prevent introduction into new host areas. The nematode is spread most easily via infested soil and contaminated machinery.

27.13 Invasiveness rating

Considering the consequences of the introduction and pest risk analysis of *H. goettingiana*, the following invasiveness rating has been assigned.

Climate/host interaction	–	Medium (2)
Hosts range	–	Medium (2)
Dispersal potential	–	Medium (2)
Economic impact	–	High (3)
Environmental impact	–	Low (1)
Post-entry distribution	–	Not known (0)
Final rating (score)	–	Medium (10)

Based on the above invasiveness rating, *H. goettingiana* receives a score of 10 out of 15 and is rated as MEDIUM INVASIVE NEMATODE.

27.14 Region-wise status of invasiveness

The invasiveness status is based on the absence of PCN in the following major pea-producing countries.

APPO: India, Indonesia, Myanmar, Nepal, North Korea, Pakistan, Vietnam
COSAVE: Argentina, Brazil, Peru
EPPO: Belarus, Morocco
IAPSC: Angola, Burkina Faso, Cameroon, Egypt, Ethiopia, Guatemala, Kenya, Malawi, Niger, Nigeria, Rwanda, Sudan (former), Tanzania, Uganda
NAPPO: Canada, Mexico
PPPO: Australia

27.15 Management measures

Control of *H. goettingiana* is difficult, as the hatching is slow and probably takes place over several weeks, and the production of a new cyst is influenced by the host. The application of granular nematicides to infested soil at the time of drilling in spring is effective in preventing crop damage but, as summer approaches in temperate regions, nematode densities are almost equal in treated and untreated plots (Green *et al.*, 1981). Crop rotation is another option in which growing non-host crops for 3–6 years can help in lowering the population densities (Di Vito and Greco, 1986). Leguminous weeds should be removed from fields as they can act as alternative hosts. In tropical and subtropical countries, summer solarization is an effective method of control. In Germany, a *Pasteuria* isolate (HGP) was found to attack 93% of all free juveniles in the field soil (Winkelheide and Sturhan, 1993). For regulatory measures, see section 25.15, above.

28 *Heterodera humuli*

28.1 Common name

Hop cyst nematode

28.2 Scientific name

Heterodera humuli Filipjev 1934

28.3 EPPO code

HETDHU

28.4 Diagnosis

Heterodera humuli is the representative member of the Humuli group, which also includes *H. fici*, *H. litoralis*, *H. ripae*, *H. turcomanica* and *H. vallicola* (Subbotin *et al.*, 2010).

28.4.1 Morphological diagnosis

The morphological and morphometric descriptions after Stone and Rowe (1977) are summarized below. For measurements, see Table 24.

FEMALE: Swollen, plump, lemon-shaped with neck frequently projecting at an angle, comprising oesophagus and part of oesophageal glands (Fig. 6.23H). Head with distinct 2–3 annules. Body without annulation or lateral field, covered by zigzag pattern of ridges. Head hexaradiate with weak skeleton. Stylet slender, basal knob with backward rake. Anterior part of oesophagus flask-shaped with massive spherical median bulb having distinct valve plate. Lumen convoluted within anterior part of oesophagus. Excretory pore prominent, situated at base of neck. Cuticle with thick hypodermis. Vulval cone obtuse, opposite neck. Vulva with long transvers slit and longer than vulval bridge, set in cleft below level of cone top. Fenestral region occupied by two thin-walled, circular areas of cuticle (the semifenestrate) with broad band of thicker cuticle (vulval bridge) between. Vagina thick-walled, supported by lateral underbridge. Anus posterior to vulval region in vermiform pre-adults, coming to lie dorsally as female shape changes. Female remains white until death. Small egg sac (gelatinous matrix) protruding from vulva but does not contain eggs.
CYSTS: Abullate, thin-walled, light coloured, darkening with age and tanning to brown. Small size (smallest among *Heterodera* cysts). Bifenestrate, broad vulval bridge, semifenestrate circular or subcircular and occasionally obscured by thin membrane with fine fingerprint-like pattern (Fig. 6.23G). Cyst wall with zigzag ridges, sometimes fine punctation also present. Cuticle smooth around fenestral area. Vulva set in transverse groove, similar to female.
MALE: Vermiform and morphologically similar to other genera of *Heterodera*. Body ventrally curved after heat fixation (Fig. 6.23J), frequently twisted around longitudinal axis. Cuticle annulated and lateral field with four incisures. Head offset, heavily sclerotized, hexaradiated. Anterior and posterior cephalid obscure, at about 3 and 8 body annules, respectively. Stylet strong, well developed, basal knob backwardly raked (Fig. 6.23E). Median bulb elliptical with distinct oesophageal gland lobe, overlapping one-fourth of body length from head. Dorsal and ventral cells prominent, subventral section longer. Oesophago-intestine junction obscure. Nerve ring encircling oesophagus posterior to median bulb. Hemizonid present, 2 annules long, 4 annules anterior to excretory pore. Hemizonion not present. Single testis, unreflexed, occupying 45% of body length. Cloaca raised, collar-like lip. Spicule massive, ventrally curved (Fig. 6.23F). Gubernaculum present. Phasmid not observed.
SECOND-STAGE JUVENILE: Vermiform, similar to other *Heterodera* species (Fig. 6.23D). Folded three times within egg. Body short and comparatively stout (shortest among *Heterodera*

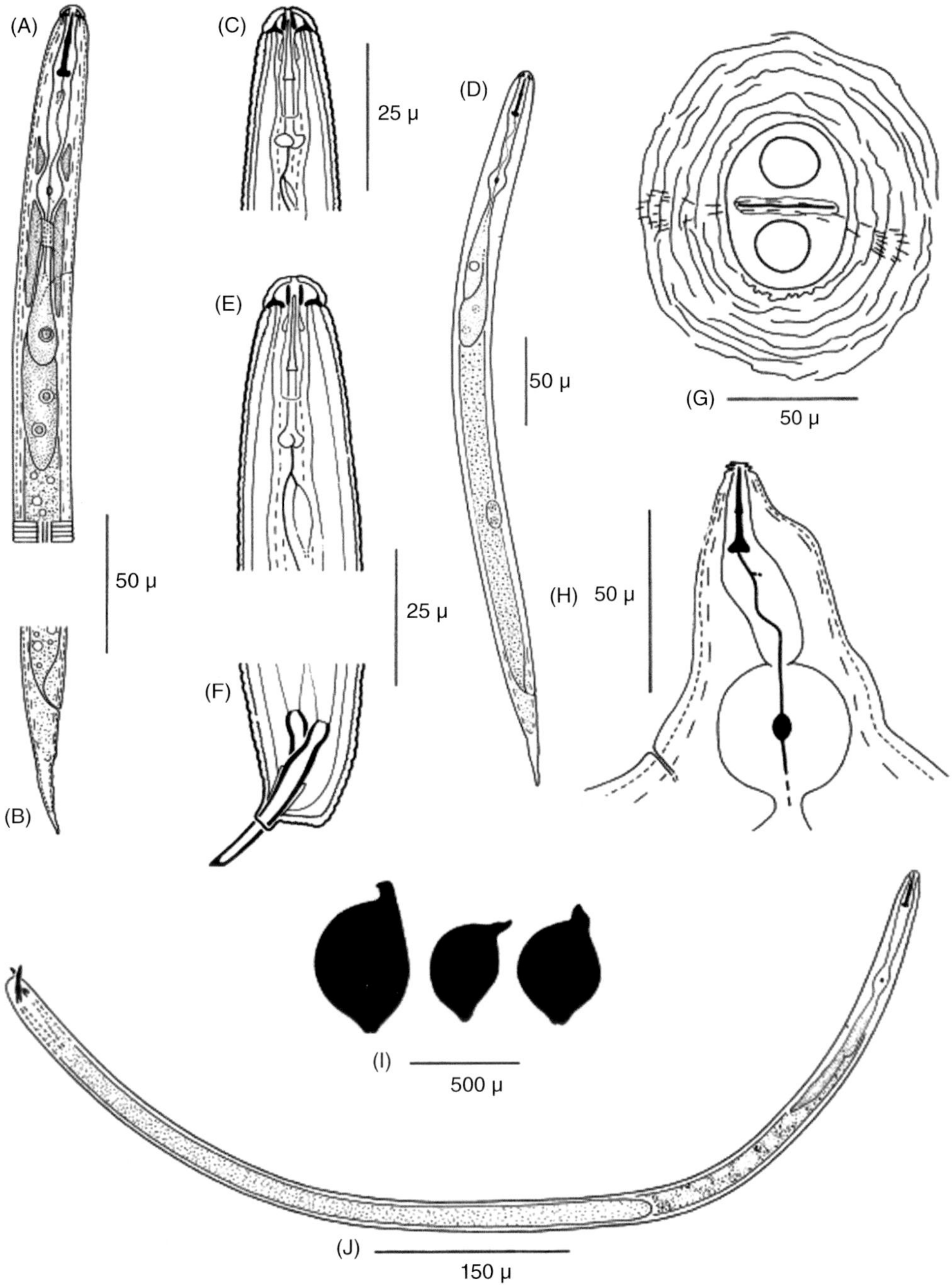

Fig. 6.23. Line drawing of *Heterodera humuli*. Anterior body of second-stage juvenile (A); tail of second-stage juvenile (B); head of second-stage juvenile (C); entire second-stage juvenile (D); head of male (E); tail of male (F); mature female terminal region (G); mature female head and neck (H); cysts (I); entire male (J). [Reproduced form CIH description, Set 7 No. 105, after Stone and Rowe (1977); courtesy CAB International]

juveniles). Cuticle annulated, thicker in first 7–8 annules. Lateral field not aerolated, with four incisures, reducing to three in anterior and posterior ends. Head with massive skeleton, offset, hexaradiated, comprising 2–4 annules. Cephalids obscure. Stylet robust with large basal knobs, anteriorly concave. Oesophagus similar to that of male. Hemizonid two annules wide, one annule anterior to excretory pore. Hemizonion not observed. Tail with tapering terminus often constricted and irregular shape. Hyaline portion about half of tail length, ratio with stylet length around 1:2. Phasmid obscure, around halfway between anus and limit of pseudocoelom.

28.4.2 Molecular diagnosis

Akyazi *et al.* (2019) characterized *H. humuli* from common nettle in Turkey based on ITS1, 5.8S gene and ITS2 regions using standard PCR. The genomic DNA is extracted from a single juvenile using a nematode lysis buffer and the PCR amplification of ITS-rDNA was done using the primers as described by Tanha Maafi *et al.* (2003) (see Chapter 4, section 15.5.2). The PCR products gave a fragment of 1100 bp and were sequenced. The obtained sequences were submitted to the GenBank database to compare with the available sequences of *H. humuli* and revealed a 100% similarity. The sequences were submitted under the GenBank accession numbers AY347926 and AF498384.

28.5 Geographical distribution

The hop cyst nematode (HCN) is the most common cyst nematode found on hops (*Humulus lupulus*). HCN is reported from Russia, Australia, the USA, Canada, New Zealand, South Africa and several European countries (England, Belgium, Germany, Italy, Spain, Greece and the Czech Republic) and recently from Turkey (Akyazi *et al.*, 2019).

28.6 Hosts

The host range of HCN is noted to be predominantly in the Urticaceae family (Akyazi *et al.*, 2019) and includes hop (*Humulus lupulus*), common nettle (*Urtica dioica*) and hemp (*Cannabis sativa*).

28.7 Symptoms

Hop plants are usually quite vigorous and can tolerate some nematode feeding. However, high levels of HCN can cause symptoms such as chlorosis and reduced length of vines, accompanied by wilting (Fig. 6.24A). A characteristic symptom of HCN infection is the presence of protruding white females on the root surface which later tan and turn brownish.

28.8 Biology and life cycle

The biology and life cycle of HCN are similar to those of other cyst nematodes. A detailed bio-ecological study was conducted by Danilova (1996). Reproduction is amphimictic. After insemination, the female produces eggs which remain inside. The female turns into cysts after the death. A single generation lasts 34–56 days, depending on environmental conditions. The optimum temperature for development is around 25°C. The life cycle is completed in 45 days at 20°C (Von Mende and McNamara, 1995). The nematode prefers heavy soils and the soil profile influences its occurrence. In England, the largest numbers of *H. humuli* cysts were found in the top 15 cm of the soil, but some were found at much greater depths (Von Mende and McNamara, 1995).

28.9 Similarities to other species

H. humuli is most closely matched with *H. fici*, but the latter fails to develop in hops (Stone and Rowe, 1977).

28.10 Economic importance

The nematode causes damage to the hop, *Humulus lupulus*, and can also attack the stinging nettle, *Urtica dioica*. Infection with this nematode can lead to a poor-quality crop and a decrease of 50% in hop yield. The yield losses depend on nematode population density and plant growth stage. Infestation also causes a significant decrease in the durability of hop cuttings (Danilova,1996).

28.11 Movement and means of dispersal

See section 22.14, above.

28.12 Pest risk analysis and looming threat from introduction

Hop (*Humulus lupulus*) has significant use in brewing. It is widely cultivated throughout the northern hemisphere. The USA and Germany are the two largest producers and the rest is produced mainly in other European countries. HCN is already present in the USA, Germany and most European countries. It still presents a threat to hop production in other countries and steps should be taken to prevent its introduction into new areas and also to check its spread once the nematode has been introduced. The nematode can disseminate through nursery stock, infested soil and contaminated machinery.

28.13 Invasiveness rating

Considering the consequences of the introduction and pest risk analysis of *H. humuli*, the following invasiveness rating has been assigned.

Climate/host interaction	–	Medium (2)
Hosts range	–	Low (1)
Dispersal potential	–	Medium (2)
Economic impact	–	Medium (2)
Environmental impact	–	Medium (2)
Post-entry distribution	–	Not known (0)
Final rating (score)	–	Medium (9)

Based on the above invasiveness rating, *H. humuli* receives a score of 9 out of 15 and is rated as MEDIUM INVASIVE NEMATODE.

Fig. 6.24. Symptoms of *Heterodera humuli* on hop plant (A); field of Syrian durum infested with *H. latipons* (B); Cyprus barley infested with *H. latipons* (C); cysts of *H. sacchari* (D); sugar beet field showing damage caused by *Heterodera schachtii* (E); *H. schachtii* infected beet (F); all stages of *H. schachtii* cyst nematode attached on roots (G); healthy and infected cabbage plant by *H. schachtii* (H); white female of *H. schachtii* attached on root (I); cysts of *H. schachtii* (J); *Heterodera zeae* cysts (K); white females and cyst of *Punctodera chalcoensis* on maize root (L). [Photo courtesy A: Gus'kova L.A. http://www.agroatlas.ru/content/pests/Heterodera_humuli; B,C: N. Greco; D: J. Bridge/CABI BioScience; E,I,K: J.D. Eisenback/Nemapix Vol.1; F: Mactode Publications, Bugwood.org; G: U. Zunke/Nemapix Vol. 1; H: DAFF Archive, Bugwood.org; J: F.A. Gray, University of Wyoming, source CABI; L: Janet A. Rowe, source CABI]

28.14 Region-wise status of invasiveness

Hop cyst nematode is a quarantine pest in several countries and requires phytosanitary certification. Considering the absence of *H. humuli* in the following hop-cultivating countries, a medium invasiveness status has been given.

APPPC: China, Japan, South Korea
COSAVE: Argentina
EPPO: Albania, Austria, Bulgaria, France, Hungary, Ireland, Poland, Portugal, Romania, Serbia, Slovakia, Slovenia, Switzerland, Ukraine
IAPSC: Ethiopia

28.15 Management measures

See section 22.18, above.

29 *Heterodera latipons*

29.1 Common name

Mediterranean cereal cyst nematode

29.2 Scientific name

Heterodera latipons Franklin, 1969

29.3 EPPO code

HETDLA

29.4 Diagnosis

Heterodera latipons belongs to the cereal cyst nematode complex, Avenae group, which includes 11 other species (*H. arenaria*, *H. auklandica*, *H. australis*, *H. avenae*, *H. filipjevi*, *H. hordecalis*, *H. mani*, *H. pratensis*, *H. riparia*, *H. sturhani* and *H. ustinovi*) (Subbotin *et al.*, 2010; Smiley *et al.*, 2017). They are morphologically very close to each other, hence morphometric data aided by molecular techniques is suggested for correct species identification of *H. latipons*.

29.4.1 Morphological diagnosis

The morphological and morphometric description (after Franklin, 1969) are given below. For measurements, see Table 24.

FEMALE: Lemon-shaped, dark to mid-brown. White subcrystalline layer present. Fenestration semifenestrae and separated by distance larger than fenestral width. Vulval slit short, underbridge strong with marked thickening in middle and splayed ends (Fig. 6.25O,P). Bullae often absent; if present, at level of underbridge.

MALE: Head offset with four postlabial annules. Amphidial apertures small. Basal annule with 18–19 longitudinal furrows. Lateral field with four longitudinal incisures, outer bands unevenly areolated all over body. Anterior part of body twisted in heat-fixed specimens (Fig. 6.25F). Stylet well-developed, knobs concave anteriorly: conus part nearly equal in length to shaft and knobs together. Cephalids present, anterior at level of second or third annule and posterior at mid-stylet level. Dorsal oesophageal gland duct orifices 3–5 μm behind knob of stylet. Hemizonid around three annules wide and at two body widths behind median bulb. Excretory pore 3–6 annules behind median bulb. Hemizonian not observed. Spicules somewhat bow-shaped, with anterior end broad, but narrow posterior part and apparently twisted (Fig. 6.25I,J). Gubernaculum 8 μm long, trough-shaped. Phasmids adanal in position and tail short, less than one body-width long.

JUVENILE: Body slightly curved dorso-ventrally after heat fixation (Fig. 6.25B). Head offset with three postlabial annules. Stylet similar to male. Oesophageal gland lobe overlying intestine latero-ventrally Cuticle in neck region appears slightly inflated in some specimens. Four longitudinal incisures in lateral field, about one-fourth to one-fifth body width, beginning at near mid-stylet level and ending midway along tail; forming three bands, outer bands areolated. Hemizonid discrete but hemizonion not seen. Excretory pore opening at same level as hemizonid. Cephalids obscure, posterior probably at 8th annule and anterior not observed. Phasmids located at 2–3 annules behind anus. Rectum about as long as anal body width. Genital primordium initially consisting of two cells, located at around 60% of body length from anterior end.

29.4.2 Molecular diagnosis

Ferris *et al.* (1999) used ribosomal DNA sequence analysis of *H. latipons* and *H. filipjevi* isolated from Russia and compared them with other cyst nematode isolates. The sequence data for two geographically separated isolates of *H. latipons* from Russia were 98.1% similar to the ITS rDNA sequence of an isolate from Israel. Several other molecular approaches have been implemented to characterize cereal cyst nematode including *H. latipons*, such as PCR-ITS-RFLP ITS-rRNA and COI gene sequence (Tanha Maafi *et al.*, 2003; Yan and Smiley, 2010; Subbotin *et al.*, 2010; see Chapter 4, section 15.5.2, and section 22.5.3, above).

29.5 Geographical distribution

Mediterranean cereal cyst nematode (MCCN) is known to occur in most of Mediterranean Europe, West Asia and North African regions (Subbotin *et al.*, 2010). This nematode has been recorded from Algeria (Mokabli *et al.*, 2002), the Azores archipelago (Mor and Sturhan 2000), Bulgaria, Canada, Cyprus, Greece, Japan, Jordan, Poland, Spain, Syria, United Kingdom (Greco *et al.*, 2002), former Czechoslovakia (Sabova *et al.*, 1988), Iran (Tanha Maafi *et al.*, 2003), Israel (Franklin, 1969), Italy (Nicol, 2002), Lebanon, Libya, Tunisia, Turkey (Inserra *et al.* 2003), Russia (Subbotin *et al.*, 2010) and South Africa (Fourie *et al.*, 2001).

29.6 Hosts

The MCCN *H. latipons* is polyphagous and parasitizes mainly cereals (Greco *et al.*, 2002). Wheat (*Triticum aestivum*) and barley (*Hordeum vulgare*) are the most seriously affected cereals, but the nematode also attacks oats (*Avena sativa*), rye (*Secale cereale*), carrot (*Daucus carota*), groundnut (*Arachis hypogaea*) and sugar beet (*Beta vulgaris* subsp. *vulgaris*).

29.7 Symptoms

Disease symptoms are discernible within 10 days of first irrigation, which is generally given 3–4 weeks after sowing. Above-ground

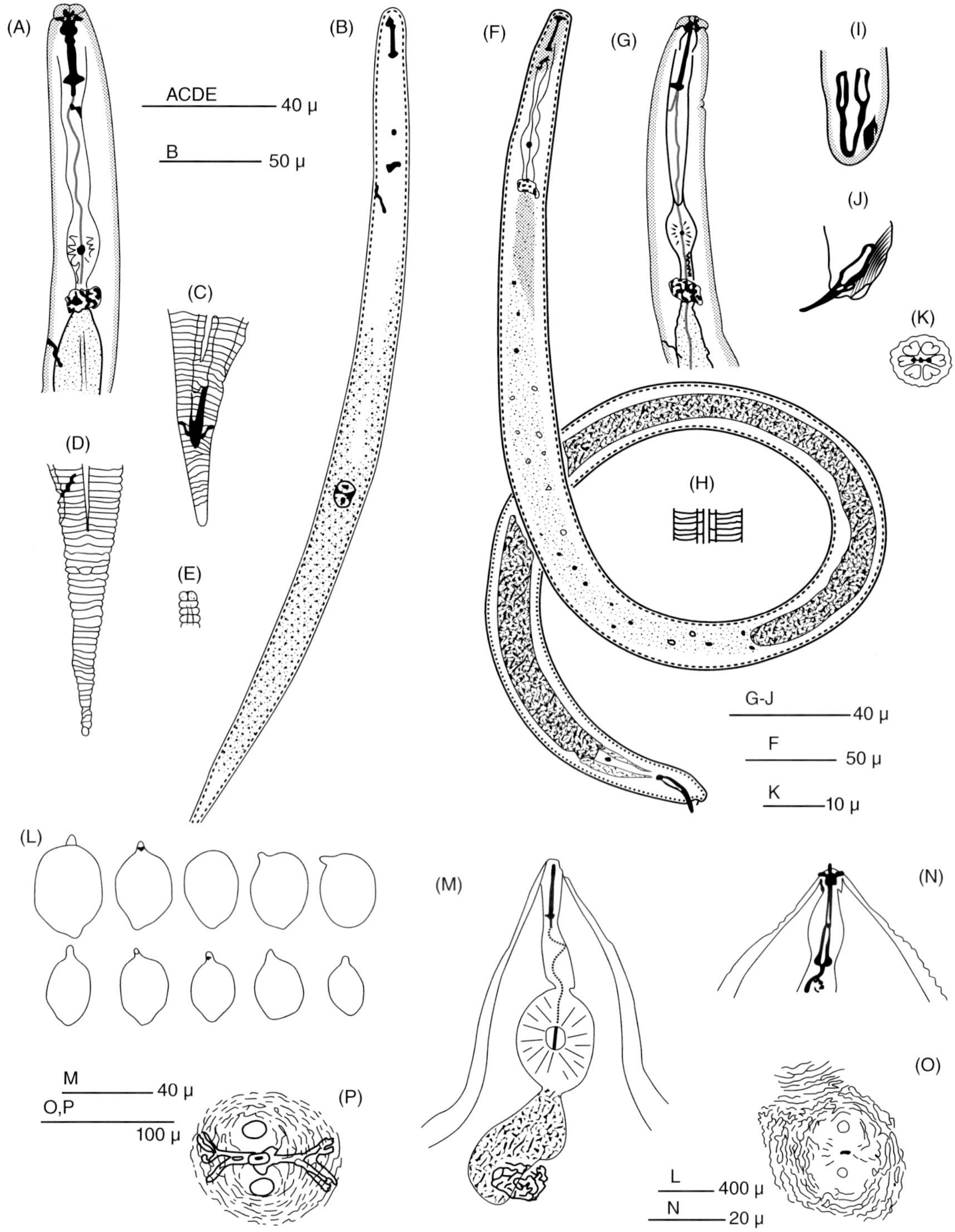

Fig. 6.25. Line drawing of *Heterodera latipons*. Second-stage juvenile J_2 anterior region (A); whole J_2 larva (B); J_2 tail (C); J_2 tail of *H. avenae* larva for comparison (D); anterior end of lateral field (E); whole male (F); male oesophageal region (G); male lateral field at mid body (H); spicules, ventral (I); spicules, lateral (J); *en face* view of male head at level of fourth annule (K); mature females (L); female anterior end with excretory pore (M); female head and stylet, lateral (N); female cuticle of vulval region and anus in white cyst (O); fenestrae and underbridge in brown cyst (P). [Franklin, 1969; courtesy Nematologica]

symptoms are recognized by patches of plants with poor growth irregularly scattered in the field (Fig. 6.24B,C). These patches increase in size and tend to coalesce in monoculture of susceptible crops. The affected plants exhibit stunted growth, yellowing, reduced tillering, fewer leaves with reduced leaf area, delayed emergence of ear heads and a smaller number of spikelets. At the initial stage of infection, root tips become swollen due to nematode penetration. The main root of the infected plant becomes elongated and has a tuft of rootlets at the distal end. The root system becomes bushy and shallow, due to excessive branching. Since the above-ground symptoms are non-specific, the presence of cyst females on the roots is the only confirmatory indication of nematode infection (Greco *et al.*, 2002).

29.8 Biology and life cycle

The biology and life cycle of *H. latipons* are similar to those of other cyst nematodes. The eggs are retained inside the cysts and a few in the egg sac. Cysts persist for many years in the soil. Hatching of second-stage juveniles (J_2) and host root penetration do not require root diffusate. The juveniles develop into swollen females, which retain eggs. Reproduction is amphimictic; at the end of the reproductive phase, females die and convert into lemon-shaped dark to mid-brown cysts (Inserra *et al.*, 2003). *H. latipons* prefers cool climates with wet winters and warm dry summers. The nematode is active from November to February. It exhibits high water-hatching ability. The optimum temperature for hatching in Syria is 10°C and the peak density of hatched juveniles is from January to February (Smiley *et al.*, 2017). Typically, one generation is completed in the Mediterranean region (Greco *et al.*, 2002). The presence of diapause has been reported in *H. latipons* populations in the dry zone of Syria and appeared to exhibit a facultative quiescence (Smiley *et al.*, 2017).

29.9 Similarities to other species

H. latipons closely resembles *H. hordecalis*, but the latter has a much bigger vulval slit (Smiley *et al.*, 2017). It can also be confused with other cyst nematode species that attack cereals (*H. avenae*, *H. bifenestra*, *H. filipjevi*, *H. mani*, *H. pakistanensis*, *H. zeae*) and *Punctodera punctata* (Nicol, 2002; Cook, 2004). Two pathotypes appear to have been distinguished in Israel (Smiley *et al.*, 2017)

29.10 Economic importance

On barley, a yield suppression as high as 50% was reported in Cyprus (Nicol, 2002). In the Mediterranean region, severe stunting of wheat and barley is common in infested fields (Franklin, 1969; Sikora, 1988). In Syria, an assessed 24% yield loss was reported in barley in a field with 28 eggs and second-stage juveniles per gram of soil (Greco *et al.*, 2002).

29.11 Movement and means of dispersal

See section 22.14, above.

29.12 Pest risk analysis and looming threat from introduction

H. latipons presents a threat to cereal production throughout the world. This nematode occurs chiefly in the Mediterranean region but has also been documented in eastern and northern Europe, the Middle and Near East, North and South Africa and Japan. Establishment of the MCCN outside its native region invokes a complete pest risk analysis and steps should be taken to prevent its introduction into new areas.

29.13 Invasiveness rating

Considering the consequences of the introduction and pest risk analysis of *H. latipons*, the following invasiveness rating has been assigned.

Climate/host interaction	–	Medium (2)
Hosts range	–	Medium (2)
Dispersal potential	–	Medium (2)
Economic impact	–	High (3)
Environmental impact	–	Medium (2)
Post-entry distribution	–	Low (–1)
Final rating (score)	–	Medium (10)

Based on the above invasiveness rating, *H. latipons* receives a score of 10 out of 15 and is rated as MEDIUM INVASIVE NEMATODE.

29.14 Region-wise status of invasiveness

Taking into consideration the economic impact of this nematode on cereals in MCCN-infested countries, medium invasiveness status is assigned in the following cereal-producing countries (FAOSTAT, 2019) where *H. latipons* is not known to occur.

APPPC: Afghanistan, India, Iraq, Kazakhstan, Pakistan, Uzbekistan
COSAVE: Argentina, Brazil
EPPO: Belgium, Belarus, Czech Republic, Denmark, France, Hungary, Germany, Greece, Lithuania, Morocco, Romania, Serbia, Sweden
IAPSC: Egypt, Ethiopia
NAPPO: Mexico, USA
PPPO: Australia

29.15 Management measures

For management including phytosanitary measures, see Chapter 16, section 115.

30 *Heterodera sacchari*

30.1 Common name

Sugarcane cyst nematode

30.2 Scientific name

Heterodera sacchari Luc and Merny, 1963

30.3 EPPO code

HETDSA

30.4 Diagnosis

Heterodera sacchari is a member of the Sacchari group and its closest known member is *H. goldeni* (Subbotin *et al.*, 2010).

30.4.1 Morphological diagnosis

The morphological description (after Luc, 1974) and morphometric measurements (after Shahina and Maqbool, 1995) are given below. For morphometric data see Table 24.

MATURE FEMALE: Lemon-shaped, white. Cephalic region truncated, offset with two retrorse annules, sclerotization weak (Fig. 6.26M). Stylet weak, with small spherical knobs. Dorsal oesophageal gland orifice 4–5 μm behind stylet. Procorpus cylindrical, median bulb heavy with prominent valvular apparatus. Isthmus short, basal glandular part almost globular. Nerve ring surrounding isthmus. Eggs retained in body.

CYSTS: Lemon-shaped, brown to dark brown with projecting vulval cone and medium-size neck. Cuticle with zigzag pattern, on the inner layer several 'perforations' evenly but irregularly distributed. Thick subcrystalline layer in young cysts. Vulval cone ambifenestrate, fenestrae relatively obscure demarcated, vulval slit as long as vulval bridge. Underbridge strongly developed with finger-like projections. Few peripheral bullae present.

MALE: Not common. Body cylindrical, almost straight to C-shaped after heat fixation, twisted at posterior end, slightly tapering anterior with rounded terminus (Fig. 6.26I). Cuticle annulated. Lateral field with three incisures, faintly marked by irregularly crossing by annule. Cephalic region spherical, heavily sclerotized, with 4–5 (rarely 6) annules (Fig. 6.26G). Cephalids obscure. Stylet strong, knobs posteriorly rounded. Dorsal oesophageal gland orifice 4 μm behind stylet. Median oesophageal bulb oval, dorsal oesophageal gland with large nucleus, anterior to subventral glands each with small nucleus (Fig. 6.26H). Nerve ring, encircling isthmus, strongly marked. Hemizonid smooth, extending over one annule, 7–15 μm anterior to excretory pore. Hemizonion not observed. Testis single. Spicules bowed, notched at the terminus (Fig. 6.26I). Spicular sheath found. Gubernaculum lamellate. Phasmids not seen.

SECOND-STAGE JUVENILE: Body straight or ventrally curved slightly after heat fixation, slightly tapering at anterior end, more attenuated at posterior one (Fig. 6.26A). Body cuticle and head-end morphologically similar to those of male. Dorsal oesophageal gland orifice 5–8 μm behind stylet. Median bulb with strong valvular plates. Dorsal oesophageal gland wide with large nucleus, anterior to subventrals in position, subventral glands each with small nucleus, not fully covering body cavity. Nerve ring well marked, located immediately behind median bulb. Excretory secretory pore 104–130 μm from anterior extremity. Hemizonid lenticular, closely in front of excretory pore, extending over two annules. Genital primordium located at the mid-body with two nuclei. Tail elongated, conical, with pointed tip. Phasmids pore-like, 45–47 μm from posterior end.

30.4.2 Molecular diagnosis

Tanha Maafi *et al.* (2007) characterized closely related members of the *H. sacchari* species complex (*H. sacchari*, *H. leuceilyma* and *H. goldeni*) using ITS-rRNA gene sequences. One to four cysts were processed for nucleic acid extraction by nematode lysis buffer (Tanha Maafi *et al.*, 2003). PCR amplification was done using the following primers.

Forward primer TW81 (5′-GTTTCCGTAGGTGAACC TGC-3′)
Reverse primer AB28 (5′-ATATGCTTAAGTTCAGCGGGT-3′)

The PCR amplification and cycling parameters are performed using the protocol described by Tanha Maafi *et al.* (2003) (see Chapter 4, section 15.5.2). The newly obtained sequences were aligned with ClustalX and submitted to the GenBank database (EF143604– EF143610) to match the DNA sequences of cyst nematodes from the Sacchari group. The study revealed that *H. hordecalis* and *H. latipons* are outgroup taxa (Subbotin *et al.*, 2001).

30.5 Geographical distribution

H. sacchari has limited distribution and is reported mainly from Asian countries (India, Pakistan and Thailand) and African (Benin, Burkina Faso, Chad, Congo, Côte d'Ivoire, Gambia, Ghana, Liberia, Nigeria and Senegal) (CABI-CPC, 2010). It is also recorded in Jamaica (Ferris, 2005). Earlier reports from Trinidad and Tobago have been not validated and it is considered as absent (CABI-ISC,2020).

30.6 Hosts

Rice (*Oryza sativa*) and sugarcane (*Saccharum officinarum*) are the primary hosts of *H. sacchari*. Some cereal crops are also attacked by *H. sacchari*, such as maize, millet and sorghum (Coyne and Plowright, 1999b). This nematode also parasitizes several Cyperaceae and Gramineae plants (Lamberti *et al.*, 2007). Some wild hosts are Bermuda grass (*Cynodon dactylon*), carpet grass (*Axonopus compressus*), *Brachiaria brizantha*, goose grass (*Eleusine indica*), jungle rice (*Echinochloa colona*), palisade-grass (*Urochloa brizantha*), sour grass (*Paspalum conjugatum*) and wild sugarcane (*Saccharum spontaneum*) (CABI-CPC, 2010).

30.7 Symptoms

Symptoms are similar to those caused by other cyst nematodes and include severely stunted plants, chlorotic foliage with a smaller number of tillers, reduction in root vigour, and black or brown discoloration on roots. A declined root system with proliferation of lateral roots at feeding points is characteristic of *H. sacchari* infections (Babatola, 1983). In heavily infested soils, seedlings may be killed (Bridge and Starr, 2007). In upland rice fields in Côte d'Ivoire, the nematode causes physiological drought symptoms (Audebert *et al.*, 2000).

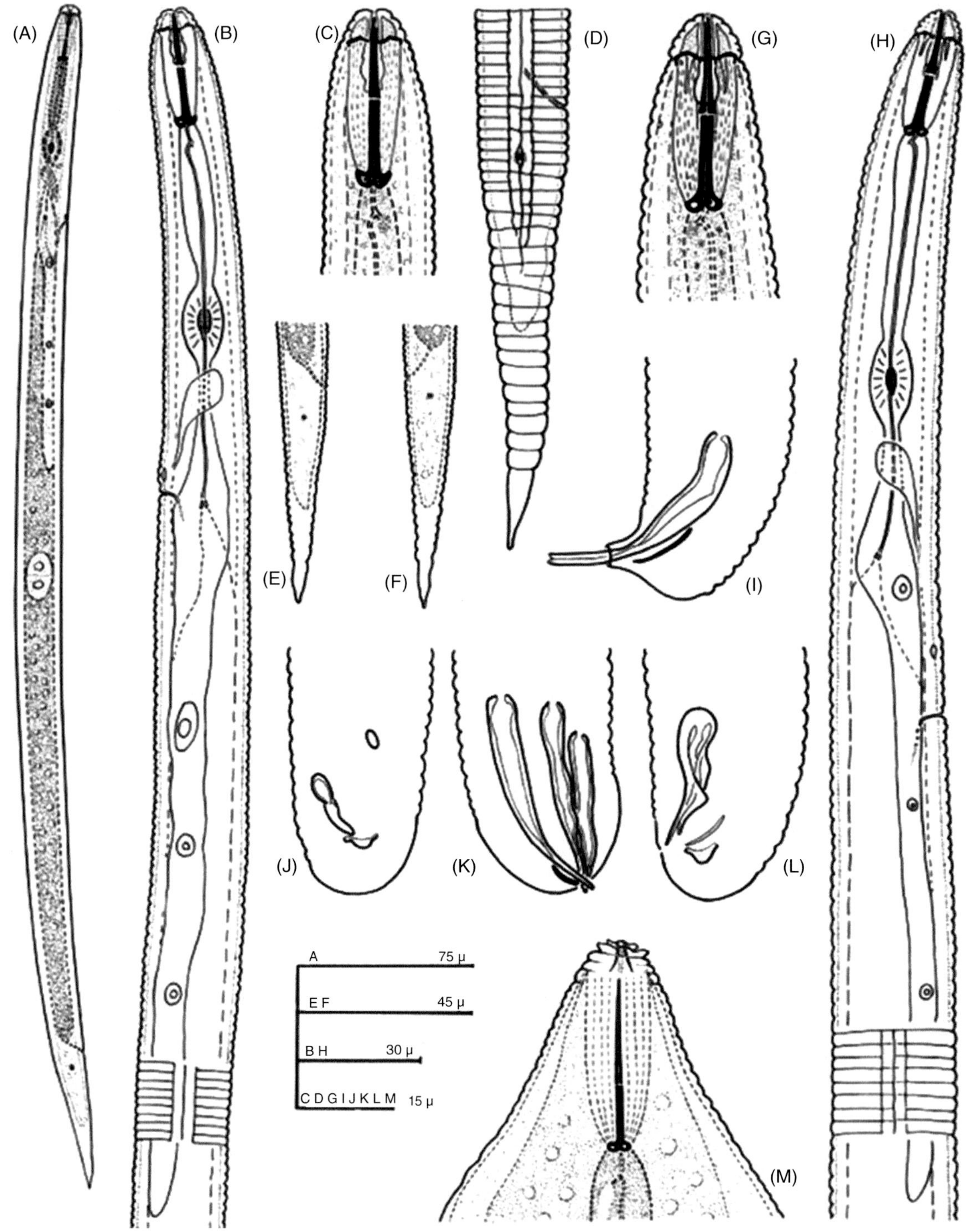

Fig. 6.26. Line drawing of *Heterodera sacchari*. Entire second-stage juveniles J_2 (A); oesophageal region of J_2 (B); head end of J_2 (C); tail ends of J_2 (D–F); male head end (G); male oesophageal region (H); male normal tail end, lateral view (I); male tail ends showing abnormal copulatory apparatus (J– L); anterior end of young female (M). [Source Nemaplex, H. Ferris]

Nematode-infested sugarcane shows fewer secondary roots. Sometimes the roots are congregated, creating patches of rotten root material in which cysts may be present (Jerath, 1968). Salawu (1992) recorded a significant reduction in shoot weight and severe root necrosis in sugarcane. Jerath (1968) reported that infected sugarcane plants were 30–50% shorter in height than uninfected plants. One of the most apparent signs of the nematode is the presence of swollen lemon-shaped white females protruding from the root surface. The mature cysts are dark brown and easily get detached from roots into soil (Fig. 6.24D).

30.8 Biology and life cycle

The J_2s of *H. sacchari* hatch over a prolonged period, allowing the nematode to frequently infect the available host (Coyne and Plowright, 2000). Hatching is water-dependent and not on root exudates as for other *Heterodera* species (Ibrahim *et al.*, 1993). After hatching, J_2s penetrate host roots and establish a particular feeding site (known as syncytium) in the stele (Khan, 2008). The J_2s start feeding and moult through third and fourth juvenile stages, taking about 30–35 days at 24–35°C to become adults (Salawu, 1994). Males are rare and reproduction is mainly through mitotic parthenogenesis, so the majority of individuals develop into females (Sullivan *et al.*, 2014). Females retain the eggs in the body (Siddiqi, 2000). After reproduction, females die and develop into brown to dark brown cysts. Cysts can persist in the soil for many years and can be dispersed through soil, wind, or irrigation water (Salawu, 1994). *H. sacchari* reproduces better on upland rice cultivars but it can multiply on both upland rice and lowland rice (Babatola, 1983). The nematode is more frequent in sandy soils than in clay soils (Jerath, 1968). *H. sacchari* usually cannot withstand prolonged flooding (Bridge and Starr, 2007).

30.9 Similarities to other species

H. sacchari is morphologically similar to other cyst nematodes that infect rice, such as *H. elachista*, *H. oryzae* and *H. oryzicola* (Sullivan *et al.*, 2014). Morphologically *H. sacchari* resembles *H. goldeni* and *H. leuceilyma* (Tanha Maafi *et al.*, 2007).

30.10 Economic importance

H. sacchari is one of the most economically significant cyst nematodes in the hot tropics (Evans and Rowe, 1998) and is a particular pest of rice in Western Africa (Lamberti *et al.*, 2007) and Trinidad (Bridge and Starr, 2007). Audebert *et al.* (2000) reported that *H. sacchari* is emerging as a potentially serious pest of upland rice in West Africa. In South Africa, this species causes severe plant growth suppression in susceptible upland rice and may result in up to 50% yield loss (Coyne and Plowright, 1999a). In the USA *H. sacchari* may substantially reduce yields of rice and sugarcane and would prompt probable regulatory measures by other host countries (USDA, 2012).

30.11 Movement and means of dispersal

See section 22.14, above.

30.12 Pest risk analysis and looming threat from introduction

Rice is the world's most important cereal crop and globally it ranks second in production (FAOSTAT, 2019). Sugarcane is also widely cultivated throughout the world. Considering *H. sacchari* as an important pest of these two important crops, regular interception of this nematode indicates its invasiveness and probability of introduction into new areas (AQAS, 2014). There would be trade restrictions if this nematode is introduced and becomes established in new host countries.

30.13 Invasiveness rating

Considering the consequences of the introduction and pest risk analysis of *H. sacchari*, the following invasiveness rating has been assigned.

Climate/host interaction	–	High (3)
Hosts range	–	Medium (2)
Dispersal potential	–	High (3)
Economic impact	–	High (3)
Environmental impact	–	Low (1)
Post-entry distribution	–	Not known (0)
Final rating (score)	–	High (12)

Based on the above invasiveness rating, *H. sacchari* receives a score of 12 out of 15 and is rated as HIGHLY INVASIVE NEMATODE.

30.14 Region-wise status of invasiveness

H. sacchari is listed as a harmful organism in Brazil, Indonesia, Japan, Taiwan and Timor-Leste (USDA-PCIT, 2014). Below is the list of major sugarcane- and rice-producing countries where *H. sacchari* is absent and attains high invasive status.

APPPC: Afghanistan, Bangladesh, Cambodia, China, Indonesia, Iran, Japan, Lao People's Democratic Republic, Malaysia, Myanmar, Nepal, North Korea, Philippines, South Korea, Sri Lanka, Vietnam
CAHFSA: Barbados, Belize, Cuba, Dominican Republic, Haiti, Saint Kitts & Nevis
COSAVE: Argentina, Bolivia, Brazil, Colombia, Ecuador, Guyana, Paraguay, Peru, Surinam, Uruguay, Venezuela
EPPO: Italy, Portugal, Russian Federation, Spain, Turkey
IAPSC: Algeria, Angola, Botswana, Burundi, Cameroon, Cape Verde, Central African Republic, Comoros, Egypt, Eritrea, Ethiopia, Gabon, Guinea, Guinea-Bissau, Kenya, Madagascar, Malawi, Mali, Mauritania, Mauritius, Mozambique, Namibia, Niger, Réunion, Rwanda, Sierra Leone, Somalia, South Africa, Sudan, Swaziland, Togo, Uganda, Zambia, Zimbabwe
NAPPO: Mexico, USA
OIRSA: El Salvador, Guatemala, Honduras, Nicaragua
PPPO: Australia, Fiji, Melanesia, Micronesia, New Zealand, Papua New Guinea, Samoa

30.15 Management measures

Crop rotation with non-host plants and chemicals may prove quite effective against the nematode, resulting in an increase of over 60% in yield (Bridge and Starr, 2007). Nematicidal effects on *H. sacchari* have been investigated in pot experiments but not validated in tropical field conditions. Oxamyl, aldicarb and carbofuran prevented the hatch of *H. sacchari* (Garabedian and Hague, 1984). Carbofuran and oxamyl delayed invasion (Mohammad and Hague, 1987) and subsequent development of *H. sacchari* (Mohammad, 1988). The treatment also reduced fecundity and increased the proportion of males (Mohammad, 1988). Dipping of sugarcane setts and foliar and soil application of nematicides suppressed the nematode invasion in sugarcane roots (Garabedian and Hague, 1982). For phytosanitary measures, see Chapter 16, section 115.

31 *Heterodera schachtii*

31.1 Common name

Sugar beet cyst nematode

31.2 Scientific name

Heterodera schachtii Schmidt, 1871

31.3 Synonyms

Heterobolbus schachtii (Schmidt, 1871) Railliet, 1896
Heterodera schachtii minor O. Schmidt, 1930
Tylenchus schachtii (Schmidt, 1871) Oerley, 1880

31.4 EPPO code

HETDSC

31.5 Diagnosis

Heterodera schachtii is a representative of the Schachtii group, other members of are *H. betae*, *H. cajani*, *H. ciceri*, *H. deverti*, *H. galeopsidis*, *H. glycines*, *H. medicaginis*, *H. mediterranea*, *H. sonchophyla* and *H. trifolii* (Subbotin *et al.*, 2010). Morphologically they are very similar, hence both morphological and molecular diagnosis approaches are recommended for authentic identification of *H. schachtii*.

31.5.1 Morphological diagnosis

The morphological and morphometric descriptions from Baldwin and Mundo-Ocampo (1991) are summarized below. For measurements, see Table 24.

Female: Body spherical with small head rapidly expanding into cylindrical neck, terminating in vulval cone, with dorsally subterminal anus (Fig. 6.27A). Head skeleton weak. Excretory pore at 'shoulder' where body swells. Stylet slender with small knobs. Median oesophageal bulb spherical, prominent. Oesophageal glands overlapping intestine latero-ventrally. Ovaries paired, extended and much coiled. Eggs retained in body with few deposited in a gelatinous matrix (Fig. 6.27A). Body cuticle three-layered, covered superficially with a reticulum of ridges.

Male: Body generally straight with posterior part spirally twisted through 90–180 degrees after heat fixation (Fig. 6.27H). Body annules distinct. Anteriorly body tapering slowly until at neck. Posterior terminus bluntly rounded with tail less than half body-width long. Phasmids adanal in position. Lateral field with four longitudinal lines but not areolated. Head dome-shaped (Fig. 6.27C), offset with three or four annules, skeleton hexaradiate (Fig. 6.27D) with amphid apertures small slit-like in lateral sectors near oral orifice (Fig. 6.27E). Anterior cephalids at 2nd annule and posterior at 6th–8th annules. Stylet strong with well-developed anteriorly concave knobs. Oesophagus cylindrical, median bulb fusiform with valve, nerve ring encircled isthmus (Fig. 6.27J). Oesophageal glands overlapping intestine ventro-laterally. Excretory pore behind median bulb, 2–3 body widths. Hemizonid pore-like, 6–10 annules. Single testis, blunt-ended. Spicules bowed, anteriorly slightly knobbed with notched tip (Fig. 6.27G). Gubernaculum simple.

Cysts: Egg sac present (Fig. 6.27B). Vulval bridge about as long as terminal vulval slit which is edged on either side by a kidney-shape. Semifenestrae separated by vulval bridge. Vagina joined to side walls by underbridge and several irregularly arranged. Bullae dark brown located short distance below vulval bridge.

Second-stage juvenile: Head hemispherical, offset with four annules. Skeleton hexaradiated, robust (Fig. 6.27K). Amphidial apertures small in lateral sectors close to oral opening. Lateral field with four incisures. Stylet moderate with prominent knobs forwardly directed (Fig. 6.27K). Oesophagus similar to that of male, but median bulb more prominent. Dorsal oesophageal gland duct opening 3–4 μm behind stylet base. Genital primordium slightly behind mid-body with two nuclei. Anus obscure, around four anal bodywidths from tip. Tail acutely conical with rounded terminus. Hyaline terminal section distinct, 1–1.25 times stylet length (Fig. 6.27I). Phasmids just post-anal, obscure.

31.5.2 Molecular diagnosis

Numerous molecular techniques have been used to characterized *H. schachtii* from different parts of the world. These approaches include RAPD (Amiri *et al.*, 2003), AFLP of ITS region (Madani *et al.*, 2007), PCR-RFLP of ITS-rDNA (Amiri *et al.* 2003; Madani *et al.*, 2004, see section 23.4.2, above; Yan and Smiley, 2010, see section 22.5.3, above) and quantitative detection with real-time PCR (Madani *et al.*, 2005).

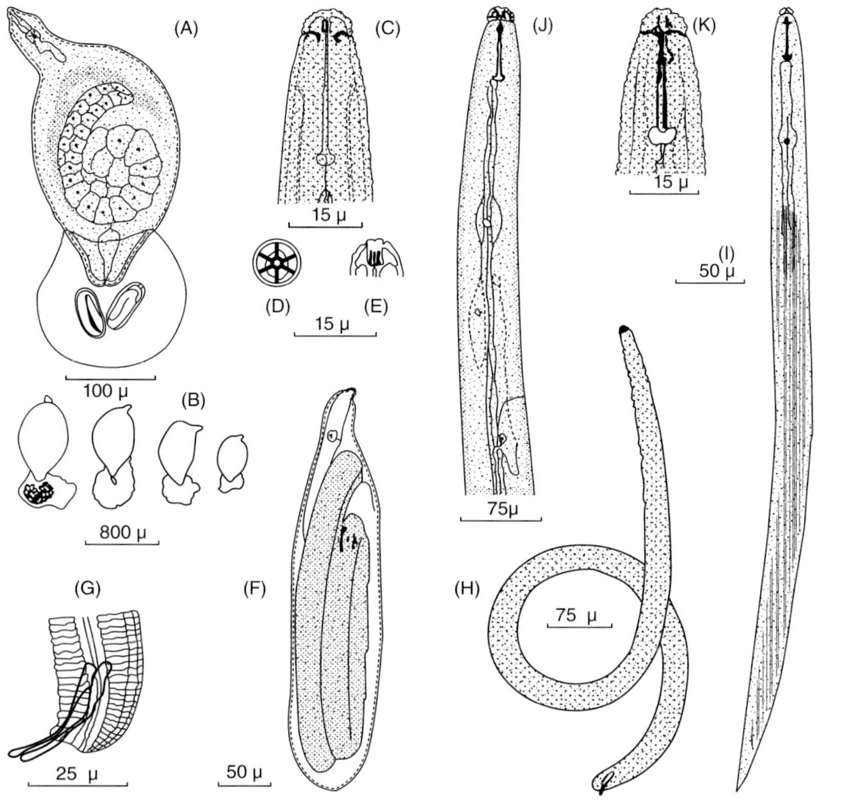

Fig. 6.27. Line drawing of *Heterodera schachtii*. Adult female with egg sac (A); cysts with egg sacs (B); head of male showing anterior and posterior cephalids (C); end-on view of male head (D); dorso-ventral view of male head showing amphid openings (E); fourth-stage male moulting (F); male tail (G); male (H); second-stage juvenile (I); male oesophageal region (J); head of second-stage larva (K). [Reproduced from CABI-CPC, 2010; Courtesy: CAB International]

31.6 Geographical distribution

H. schachtii is widely distributed throughout sugar beet-cultivating areas (Fig. 6.28). The nematode is known to occur in the following countries (CABI-ISC, 2020).

Africa: Algeria, Cape Verde, Egypt, Gambia, Libya, Morocco, Senegal, South Africa, Tunisia.
Asia: Azerbaijan, China, Iran, Iraq, Israel, Japan, Jordan, Kazakhstan, Kyrgyzstan, Pakistan, South Korea, Syria, Turkey.
Europe: Albania, Austria, Belgium, Bulgaria, Croatia, Czech Republic, Denmark, Estonia, Finland, France, Germany, Greece, Hungary, Ireland, Italy, Latvia, Moldova, Netherlands, Poland, Portugal, Romania, Russia, Serbia, Slovakia, Spain, Sweden, Switzerland, Ukraine, United Kingdom, former Yugoslavia.
North America: Canada, Mexico, USA.
Oceania: Australia, New Zealand.
South America: Chile, Peru, Uruguay.

31.7 Hosts

The sugar beet cyst nematode (SCN) has a very large and varied host range comprising several plant families such as Chenopodiaceae, Cruciferae, Polygonaceae, Caryophyllaceae, Amaranthaceae, Labiatae, Portulaceae, Scrophulariaceae, Fabaceae and Solanaceae. Escobar-Avila *et al.* (2019) reported broccoli as a host of *H. schachtii* from Mexico. The extended list of hosts of SCN is presented in Table 27.

31.8 Symptoms

The nematode infestation may be uniformly spread in the fields or scattered in one or more localized areas (Fig. 6.24E). Infestations may result in patches of plants showing poor stands and growth (Fig. 6.24H). Plants of all ages can be parasitized by *H. schachtii* but seedlings are severely damaged or killed. The nematode invasion in older plants causes less injury. Young plants infected by *H. schachtii* have extended petioles and are stunted, which persists until harvest. The outer leaves of attacked plants frequently show day wilting and have pronounced chlorosis. Early infection of roots often results in severe branching of storage roots. Severely attacked plants have small storage roots with excessive fibrous roots (Fig. 6.24F) usually referred as 'bearded' or 'whiskered' root (Jones and Jones, 1964). The above-ground symptoms described for SCN can be verified by pulling out the plants for a visual presence of young females and cysts on the roots (Fig. 6.24G,I). Soil

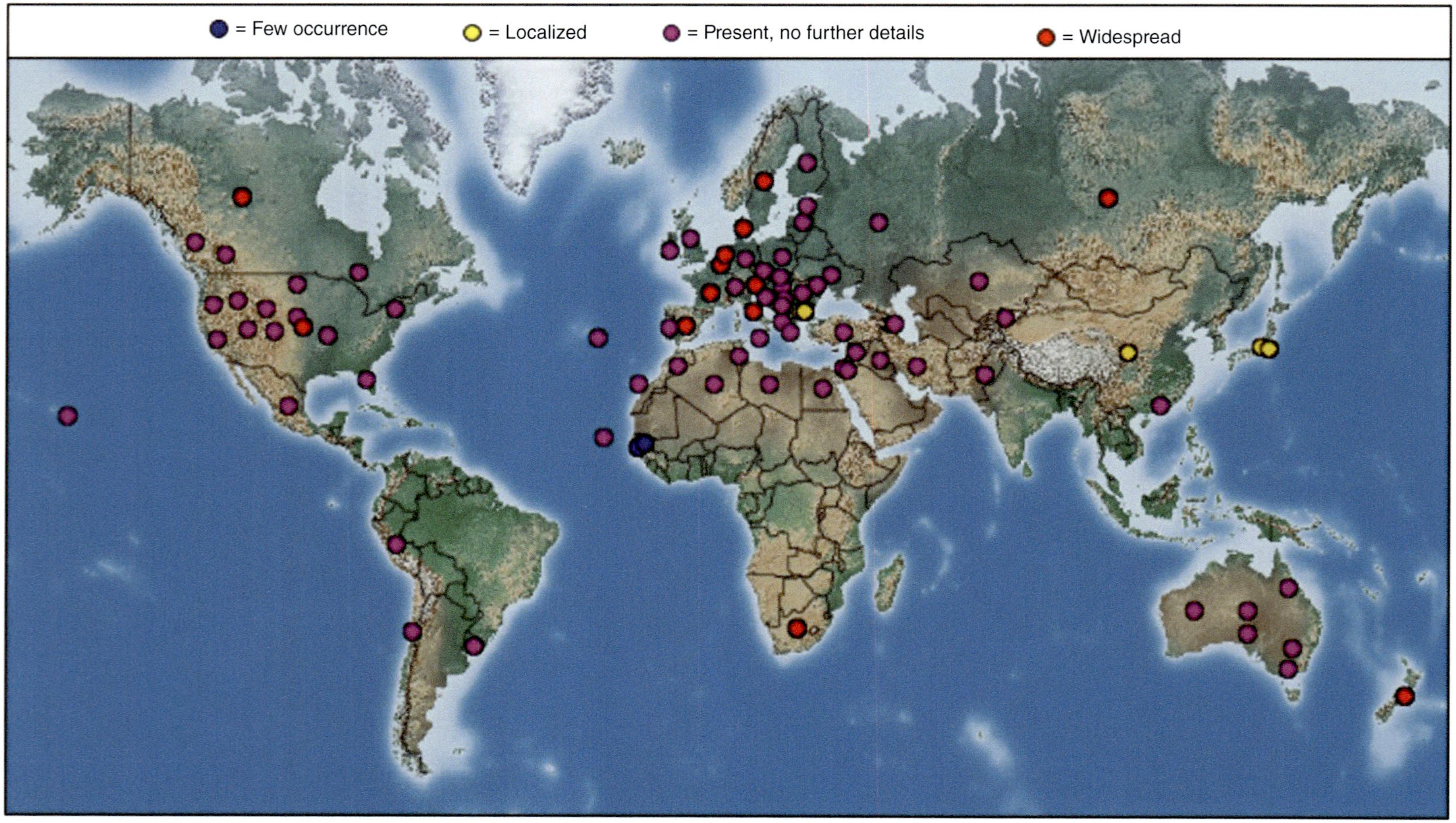

Fig. 6.28. Distribution map of *Heterodera schachtii*. [Adapted from CABI-ISC, 2020]

Table 27. List of hosts of *Heterodera schachtii* (adapted from CABI-ISC, 2020).

Main hosts
Alliaria petiolata (garlic mustard)
Alyssum argentium
Amaranthus carneus
Amaranthus caudatus (love-lies-bleeding)
Amaranthus graecizans (prostrate pigweed)
Amaranthus hybridus (smooth pigweed)
Amaranthus palmeri (Palmer amaranth)
Amaranthus quitensis
Apium dulce
Arabis alpina
Arabis arenosa
Arabis caucasica (snow-in-summer)
Arabis hirsuta (hairy rockcress)
Arabis muralis
Arabis perfoliata
Arabis turrita (tower rockcress)
Arabis verna
Armoracia rusticana (horseradish)
Atriplex confertifolia (shadscale saltbush)
Atriplex hastata
Atriplex hortensis (garden orache)
Atriplex littoralis (grass-leaved orache)
Atriplex patula (common orache)
Atriplex rosea (red orache)
Aubrieta (rainbow rock-cress)
Aubrieta columnae
Beta atriplicifolia
Beta corolliflora
Beta foliosa
Beta intermedia
Beta lomatogona
Beta macrocarpa
Beta macrorhiza
Beta nana
Beta patellaris
Beta trigyna
Beta vulgaris (beetroot)
Beta vulgaris subsp. *maritima* (sea beet)
Beta vulgaris var. *saccharifera* (sugar beet)
Brassica spp.
Brassica cernua
Brassica juncea var. *juncea* (Indian mustard)
Brassica napus var. *napobrassica* (swede)
Brassica napus var. *napus* (rape)
Brassica nigra (black mustard)
Brassica oleracea (cabbages, cauliflowers)
Brassica oleracea var. *viridis* (collards)
Brassica rapa ssp. *oleifera* (turnip rape)
Dianthus (carnation)
Dianthus barbatus (sweet william)
Dianthus caryophyllus (carnation)
Dianthus deltoides (maiden pink)
Dianthus plumarius (common pink)
Gypsophila acutifolia
Gypsophila elegans (baby's breath)
Phaseolus vulgaris (common bean)
Phytolacca acinosa
Pisum sativum (pea)
Polygonum (knotweed)
Polygonum convolvulus (black bindweed)
Portulaca oleracea (purslane)
Raphanus maritimus
Raphanus raphanistrum (wild radish)
Raphanus sativus (radish)
Rheum hybridum (rhubarb)
Rorippa amphibia (great yellowcress)
Rorippa islandica (yellow marshcress)
Rumex acetosa (sour dock)
Rumex acetosella (field sorrel)
Rumex alpinus (alpine dock)
Rumex hydrolapathum (water dock)
Rumex maritimus (golden dock)
Rumex palustris
Rumex patienta
Rumex sanguineus (red-veined dock)
Saponaria ocymoides (rock soapwort)
Saponaria officinalis (soapwort)
Sesbania exaltata (coffeebean)
Sesbania macrocarpa
Silene armeria (garden catchfly)
Silene maritima
Silene nutans (Nottingham catchfly)
Silene quadrifida
Silene saxifraga
Sinapis arvensis (wild mustard)
Sinapis eruca
Sisymbrium austriacum
Sisymbrium irio (London rocket)
Sisymbrium loeseli
Sisymbrium officinale (hedge mustard)
Sisymbrium orientale (eastern rocket)
Solanum sarrachoides (green nightshade)
Spinacia glabra
Spinacia oleracea (spinach)
Tetragonia tetragonioides (New Zealand spinach)
Thlaspi arvense (field pennycress)
Tropaeolum perigrinum
Vaccaria hispanica (bladder soapwort)
Vicia benghalensis (purple vetch)
Vigna catjang
Wild host
Chenopodium (goosefoot)
Stellaria media (common chickweed)

samples can also be taken for isolation of cysts that are visible to the naked eye as small, round, light to dark brown pin-head structures (Fig. 6.24J).

31.9 Biology and life cycle

H. schachtii overwinters as eggs and juveniles which remain dormant inside the cyst. Root exudates stimulate hatching and juveniles emerge from the cyst. Second-stage juveniles (J_2) migrate to the fibrous roots and penetrate near the root terminals. After penetration, J_2s move a short distance and reach the root cortex, where they start developing syncytium and moult into obese third-stage juveniles. After a fourth moult the female becomes swollen and lemon-shaped, protruding outside the root, and can be observed as a small white dot attached to fibrous roots (Fig. 6.24G). Adult males migrate out from the root and mate with the females. After fertilization around 200 eggs are produced by a single female inside the body and a few are deposited in the gelatine matrix of the egg sac. Subsequently the female dies, the body wall begins to harden and is transformed into a lemon-shaped, light brown to reddish-brown cyst. The life cycle takes 4–6 weeks, depending on environmental conditions (temperature 21–27°C). Three generations have been recorded in a growing season. Cysts may remain viable for several years in irrigated fields (Cooke, 1993).

31.10 Similarities to other species

Morphologically *H. glycines*, *H. trifolii* and *H. galeopsidis* are close relatives of *H. schachtii* (Mulvey, 1972).

31.11 Economic importance

The population density of the sugar beet cyst nematode increases dramatically under monocropping, causing significant losses in Germany. The sugar content of beet (*Beta vulgaris*) may be reduced by up to 30%, or more in many cases (CABI-CPC, 2010). Microplots were infested with *H. schachtii* juveniles at rates of 18,000/kg of soil, which reduced the marketable yield of table beet (*Beta vulgaris*), rutabagas (*Brassica napus*), cabbage and spinach by 30%, 35%, 24% and 49%, respectively. The juvenile nematode population was highest at harvest, except in spinach, and cyst numbers had also increased at harvest in relation to the pre-plant population density per root system and per gram of fresh root (Olthof *et al.*, 1974). An initial soil population density of 2–4 eggs or juveniles/cm^3 of soil may result in economic yield loss.

31.12 Movement and means of dispersal

See section 21.14, above.

31.13 Pest risk analysis and looming threat from introduction

H. schachtii has been detected from almost all sugar beet-growing areas and is likely to establish in the remaining areas, if not protected. Due to its polyphagous nature, high dispersal and reproductive potential and the subsequent economic impact, *H. schachtii* poses a very high risk to sugar beet production. Establishment in all ecological zones of the world also concerns the environmental impact of this nematode on lowering biodiversity, disrupting natural communities or changing ecosystem processes. The introduction of this nematode into a new area will trigger additional treatment programmes, which leads to a hazardous impact on the environment. Introduction of this nematode also significantly impacts agronomical and cultural practices. Hence, regulatory steps should be taken to prevent its introduction. The nematode is spread most easily via root stock, infested soil and contaminated machinery.

31.14 Invasiveness rating

Considering the consequences of the introduction and pest risk analysis of *H. schachtii*, the following invasiveness rating has been assigned.

Climate/host interaction	–	High (3)
Hosts range	–	High (3)
Dispersal potential	–	High (3)
Economic impact	–	High (3)
Environmental impact	–	Medium (2)
Post-entry distribution	–	Low (–1)
Final rating (score)	–	High (13)

Based on the above invasiveness rating, *H. schachtii* receives a score of 13 out of 15 and is rated as HIGHLY INVASIVE NEMATODE. The consequences of introduction into new areas are very high.

31.15 Region-wise status of invasiveness

Considering the wide host range, *H. schachtii* is a highly invasive nematode species in the following countries where this nematode does not occur.

APPPC: Bangladesh, Cambodia, India, Indonesia, Lebanon, Malaysia, Myanmar, Philippines, Sri Lanka, Tajikistan, Thailand, Vietnam, Uzbekistan
CAHFSA: Barbados, Belize, Cuba, Dominican Republic, Haiti, Jamaica, Saint Kitts & Nevis
CAN: Bolivia, Colombia, Ecuador, Venezuela
COSAVE: Argentina, Brazil, Guyana, Paraguay, Suriname
EPPO: Belarus, Cyprus, Georgia, Guernsey, Jersey, Lithuania, Luxembourg, Macedonia, Malta, Norway, Slovenia, Uzbekistan
IAPSC: Angola, Benin, Botswana, Burkina Faso, Burundi, Cameroon, Central African Republic, Chad, Comoros, Congo, Côte d'Ivoire, Eritrea, Ethiopia, Gabon, Ghana, Guinea, Guinea-Bissau, Kenya, Liberia, Madagascar, Malawi, Mali, Mauritania, Mauritius, Mozambique, Namibia, Niger, Réunion, Rwanda, Sierra Leone, Somalia, Swaziland, Togo, Uganda, Zambia, Zimbabwe
NEPPO: Malta, Sudan
OIRSA: El Salvador, Guatemala, Honduras, Nicaragua
PPPO: Fiji, Papua New Guinea, Tonga, Vanuatu.

31.16 Management measures

The control of *H. schachtii* is attainable if appropriate methods are integrated. Crop rotation with non-hosts for 5–6 years can affect the rate of natural attrition and therefore reduction of nematode inoculum between susceptible crops. In sugar beet, crop rotation with barley reduces *H. schachtii* more significantly than with wheat (Sikora *et al.*, 2018). Trap crops (*Sinapsis alba*, *Raphanus sativus* ssp. *oleifera*) and cover crops (*Fagopyrum esculentum* and *Phaceilia tanacetifolia*) are used effectively for the management of SCN (Sikora *et al.*, 2005). Crops grown at higher altitudes and in winter are not damaged severely by this nematode (Sikora and Fernandez, 2005). Biological control using *Nectria radicicola*, *Pochonia chlamydosporia*, *Rhizobacteria* and *Hirsutella rhossiliensis* may also prove effective against SCN (Khan, 2016). Application of non-fumigant nematicides such as phorate, carbofuran and oxamyl also control the nematode (Roberts and Thomason, 1981).

31.16.1 Phytosanitary measures

See Chapter 16, section 115.

32 *Heterodera zeae*

32.1 Common name

Corn cyst nematode

32.2 Scientific name

Heterodera zeae Koshy *et al.*, 1971

32.3 EPPO code

HETDZE

32.4 Diagnosis

Heterodera zeae belongs to the Humuli group with *H. turcomanica* and *H. salixophila* (Subbotin *et al.*, 2010). The nematode closely resembles other corn-infesting cyst nematodes such as *H. oryzae*, *H. sacchari* and *H. avenae*. Hence, to identify *H. zeae*, morphological diagnosis assisted by a molecular approach are recommended.

32.4.1 Morphological diagnosis

Morphological descriptions (after Koshy *et al.*, 1970; Golden and Mulvey, 1983; Shahina and Maqbool, 1995) are presented below. Morphometric data from Asghari *et al.* (2013) are summarized in Table 24.

ADULT FEMALE: Lemon-shaped, pearly white with a short neck and terminal vulval cone (Fig.6.29B). Body cuticle thin with lace-like or zigzag pattern anteriorly and wavy near vulval cone. Head consists of two annules. Stylet moderately developed, slender with flat knobs. Median bulb rounded with prominent valve plates. Excretory pore at level of oesophageal gland lobe. Ovaries paired. Vulva a long slit, vulval cone well developed. Anus small, obscure, near vulva. Egg sac present, small (Fig. 6.29A). Eggs retained in body and only few in egg sac.

CYSTS: Hard-walled, lemon-shaped, light brown with protruding neck (Fig. 6.29C). Cuticle thin with zigzag pattern, subcrystalline layer absent. Vulval fenestration present, ambifenestrate with two semifenestrae divided by vulval bridge. Vulva terminal. Anus indistinct, anal fenestration absent. Underbridge generally present, simple. Bullae present with finger-like projection.

SECOND-STAGE JUVENILE: Body slender, tapering at both ends. Lip region generally continuous, moderately developed, labial disc indistinct. Stylet slender, slightly backwards anteriorly. Dorsal oesophageal gland orifice 4–5 µm from stylet base. Lateral field with four incisures. Median bulb well developed with prominent valve plates. Glandular lobe overlapping intestine in anterior region, mostly laterally and dorsally. Hemizonid distinct. Phasmids conspicuous but small. Tail short with hyaline part and pointed or rounded terminus.

MALE: Very rare.

EGGS: Hyaline shell, without cuticular markings.

32.4.2 Biochemical and molecular diagnosis

Singh *et al.* (1998) used the isozyme patterns of esterases and malate dehydrogenase to separate three races of *H. zeae* and two races of *H. cajani*. Shashi *et al.* (1999) also characterized *H. zeae*, *H. graminis* and *H. cajani* based on proteins, using one-dimensional native gel electrophoresis. *H. zeae* has been characterized on a molecular basis using restriction fragment length polymorphism (RFLP) patterns generated by several enzymes, *Alu*I, *Ava*I, *Msp*I, *Hha*I and *Hinf*I (Umarao *et al.*, 2008) and using ITS (1 and 2) rDNA 28S large ribosomal subunit (LSU) D2–D3 expansion segment, and partial 18S small ribosomal subunit (SSU) followed by RFLP of ITS-rDNA using *Alu*I, *Ava*I, *Bsr*I, *Hha*I, *Bst*NI, *Rsa*I and *Scr*FI enzymes and heat shock protein 90 (*Hsp*90) sequence (Skantar *et al.*, 2012). In the method described by Skantar *et al.* (2012), the individual second-stage juveniles were mechanically disrupted and processed for DNA extraction by nematode lysis buffer (Tanha Maafi *et al.*, 2003) (see Chapter 15, section 110.1.1). The primers used in this study are as follows.

For ITS1, 5.8S rRNA coding sequence, and ITS2

Forward primer TW81 (5′-GTT TCC GTA GGT GAA CCT GC-3′)
Reverse AB28 (5′-ATA TGC TTA AGT TCA GCG GGT-3′)

Fig. 6.29. Line drawing of *Heterodera zeae*. Females with egg sac (A); females (B); cysts (C). [Koshy *et al.*, 1970, courtesy Nematologica]

For 28S LSU D2-D3 expansion segment

D2A forward primer (5′-ACAAGTACCGTGAGGGAAAGTT-3′)
D3B reverse primer (5′-TCGGAAGGAACCAGCTACTA-3′)

For partial 18S SSU sequence

18s1.2 (5′-GGCGATCAGATACCGCCCTAGTT-3′)
18sr2b (5′-TACAAAGGGCAGGGACGTAAT-3′)

For *Hsp*90 fragments degenerate primers

U288 (5′-GAYACVGGVATYGGNATGACYAA-3′)
L1110 (5′-TCRCARTTVTCCATGATRAAVAC-3′)

The PCR amplification of ITS1, 5.8S rRNA coding sequence and ITS2 were performed using the procedure described by Tanha Maafi *et al.* (2003) (see Chapter 4, section 15.5.2). The

28S LSU D2–D3 expansion segment was amplified as described by De Ley *et al.* (2005) and Ye and Giblin-Davis (2013) (see Chapter 2, section 12.5.3). The partial 18S SSU sequence was amplified using the cycling conditions: 1 cycle of 94°C for 2 min, followed by 40 cycles of 94°C for 20 s, 59°C for 30 s, and 72°C for 30 s, and finally with 1 cycle of 72°C for 5 min. The *Hsp*90 is as described by Skantar and Carta (2004). All the PCR products were visualized using the gel documentation system. PCR–RFLP of ITS amplifications were performed as described by Umarao *et al.* (2008) (see Chapter 15, section 110.2.1). The PCR products were sequenced and submitted to the GenBank database under the accession numbers: ITS sequences (GU145611–GU145616), 28S sequence (GU145612), 18S rDNA sequence (HQ724313) and *Hsp*90 sequences (JQ316172–JQ316193, MG019814–MG019907). The newly obtained sequences were aligned with the selected gene sequences of *Heterodera* species. The phylogenetic relationships placed *H. zeae* in a clade with *H. turcomanica* and *H. salixophila* species of the Humuli group, congruent with Subbotin *et al.* (2010).

32.5 Geographical distribution

H. zeae was first reported from roots of maize (*Zea mays*) collected from Rajasthan, India (Koshy *et al.*, 1971). Subsequently, it was detected from Pakistan (Maqbool, 1980), Egypt (Aboul-Eid and Ghorab, 1981), the USA (Sardanelli *et al.*, 1981), Thailand (Chinnasri *et al.*, 1995) and Nepal (Sharma *et al.*, 2001). More recently, this nematode has been reported from Afghanistan (Asghari *et al.*, 2013), China (CABI-ISC, 2020), Indonesia, Iran, Iraq (EPPO, 2004), Greece (Skantar *et al.*, 2012) and Portugal (Correia and Abrantes, 2005).

32.6 Hosts

H. zeae has been reported to be mostly confined to plants of family Poaceae. Maize (*Zea mays*) is the primary host, but the nematode also invades oats (*Avena sativa*), barley (*Hordeum vulgare*), rice (*Oryza sativa*), foxtail millet (*Setaria italica*), sorghum (*Sorghum bicolor*), Sudan grass (*Sorghum sudanense*), wheat (*Triticum aestivum*), teosinte (*Zea mexicana*) and some wild grasses of Poaceae (CABI-CPC, 2010). Tomato (Shazad and Ghaffar 1986) and *Prunus amygdalus* (Quasim and Ghaffar, 1986) are also recorded as hosts of *H. zeae*.

32.7 Symptoms

The disease is characterized by patches of plants showing poor unthrifty growth, stunting and yellowing (Koshy and Swarup, 1971b). The root system is also poorly developed. As well as being stunted, the infested maize plants have been reported to have retarded leaf emergence with reduced fresh and dry weights (Aboul-Eid and Ghorab, 1981). Stunted growth of tomato plants infested with *H. zeae* was described by Shazad and Ghaffar (1986) and of *Prunus amygdalus* by Quasim and Ghaffar (1986). As with other cyst nematodes, the presence of cysts of *H. zeae* (Fig. 6.24K) on the surface of the roots is the most important and characteristic symptom of nematode infection.

32.8 Biology and life cycle

The adult females of *H. zeae* are sedentary endoparasites and remain within the root or attached to the roots as semi-endoparasitic. The relevant hosts may be attacked either alone or in combination with other nematodes. *H. zeae* is bisexual, but only sporadic occurrence of males has been reported. It mainly reproduces through parthenogenesis (Shahina and Maqbool, 1989). Each cyst may contain approximately 200–250 eggs. The life cycle of *H. zeae* is very short; it takes 19–20 days at 25–30°C on maize (Shahina and Maqbool, 1989) and is able to be completed in about 6–7 generations during one cropping season (Srivasta and Sethi, 1986). Temperature plays an important role in the biology of *H. zeae*. At lower or higher temperatures, the life cycle was prolonged and took 40–50 days to complete (Bajaj *et al.*, 1986). *H. zeae* was able to survive for a period of over 20 months in the fields up to depths of 30 cm. Some cysts were recovered as long as 51 months after moist storage of naturally infested field soil at 2°C and 24°C (Krusberg and Sardanelli, 1989).

32.9 Similarities to other species

H. zeae can be easily confused with several other cyst nematodes that parasitize cereals, particularly *H. avenae*, *H. bifenestra*, *H. filipjevi*, *H. mani* and *H. pakistanensis*.

32.10 Economic importance

H. zeae is widespread in Egypt, India and Pakistan and is currently gaining more importance since its presence in the USA, where maize is cultivated over large areas. Amongst the many poaceous plants infested by this nematode, maize is the primary host. *H. zeae* is considered to be one of the four most important cyst nematodes and an economic pest of maize in India. Adequate data are not available on the losses caused by *H. zeae* but indirect evidence, as judged by the increase in growth parameters after nematicidal treatments, is available (Parihar and Siddiqi, 2010). Hirling (1974) reported poor plant growth of maize with yield loss of 40% due to infection by *H. zeae*. McGregor (1978) estimated a yield loss in maize due to *H. zeae* as approximately US$320 in the USA.

32.11. Movement and means of dispersal

See section 22.14, above.

32.12 Pest risk analysis and looming threat from introduction

The species is currently distributed in more than 15 countries in Asia, Africa, Europe and North America and is one of the most economically important pests on maize and other cereal crops in India and Pakistan. Considering the world's second-most important crop, the nematode possesses a direct risk to maize production. Karadjova *et al.* (2015) assessed the probability of entry and establishment of *H. zeae* in a non-native country (Bulgaria). Three main pathways of entry of *H. zeae* into the country were considered. The highest risk is associated with the import of bulbs, rhizomes, tubers and other underground plant parts with attached soil, intended for direct consumption and processing, from countries where the pest is distributed. For this pathway, the probability of entry has been assessed as moderate.

32.13 Invasiveness rating

Considering the consequences of the introduction and pest risk analysis of *H. zeae*, the following invasiveness rating has been assigned.

Climate/host interaction	–	High (3)
Hosts range	–	Medium (2)
Dispersal potential	–	High (3)
Economic impact	–	High (3)
Environmental impact	–	Medium (2)
Post-entry distribution	–	Low (–1)
Final rating (score)	–	High (12)

Based on the above invasiveness rating, *H. zeae* receives a score of 12 out of 15 and is rated as HIGHLY INVASIVE NEMATODE.

32.14 Region-wise status of invasiveness

Based on the worldwide geographical distribution of *H. zeae* and likely introduction in the major cereal (including corn) producing countries (FAOSTAT, 2019) where this nematode is still absent, the region-wise status of invasiveness is evaluated. Region-wise top cereal-producing countries where *H. zeae* may attain invasive status are:

APPPC: Bangladesh, Cambodia, Japan, Kazakhstan, Myanmar, Philippines, Uzbekistan, Vietnam
NAPPO: Canada, Mexico
COSAVE: Argentina, Brazil
EPPO: Denmark, France, Hungary, Germany, Italy, Morocco, Poland, Romania, Russian Federation, Serbia, Spain, Turkey, Ukraine, United Kingdom
IAPSC: Algeria, Ethiopia, Nigeria, South Africa, Tanzania
PPPO: Australia

32.15 Management measures

Once an infestation with cyst nematodes has taken place in the field, it is practically impossible to eliminate the nematode. Hence prevention is the best management strategy. However, the following management measures are useful.

32.15.1 Cultural practices

Since the nematode has a very limited host range, there is good scope for using crop rotation as a control strategy for *H. zeae*. Monoculture of maize and other host crops in the same field should be avoided. Two years of rotation with non-host crops like vegetables, pulses and oilseeds can be fruitful, as this would bring down the nematode populations to below economic threshold levels (Srivastava and Chawla, 2005). Two or three deep ploughings at intervals of 10–15 days during April–May in a hot summer may also reduce the nematode population densities to a considerable extent (Srivastava and Chawla, 2005). There are a few cultivars available that have expressed some resistance against *H. zeae*. The maize cultivars 'Ageti-76' and 'Karnal-1' are moderately resistant to *H. zeae* in India (Kali Ram, 1989). The use of oil cakes and other soil amendments have also been proved effective in managing the population of *H. zeae* (Parihar and Siddiqi, 2010).

32.15.2 Biological control

Certain biocontrol fungi such as *Cataneria*, *Pochonia* and *Trichoderma* spp. have been isolated from *H. zeae* cysts (Srivastava and Chawla, 2005). El-Sherif *et al.* (1995) reported the antagonistic potential of some *Bacillus* spp., *Corynebacterium* spp., *Serratia* spp., *Arthrobacterium* spp. and *Streptomyces* spp. against *H. zeae*.

32.15.3 Chemical control

Soil application of carbofuran or phorate at 2 kg a.i./ha is very effective in reducing nematode populations and subsequently increasing the crop yield (Parihar and Siddiqi, 2010). Seed treatment with carbosulfan at 2–3% (w/w) also provided a significant decline in the population of *H. zeae* (Kaushal *et al.*, 2007). Soil application of systemic nematicides against *H. zeae* has shown that aldicarb and fenamiphos treatments most effectively suppress larval emergence (Kaul and Sethi, 1988).

32.15.4 Phytosanitary measures

See Chapter 16, section 115.

Punctodera Spp.

The genus *Punctodera* was proposed by Mulvey and Stone (1976) in the subfamily Heteroderinae (Skarbilovich, 1947). Luc *et al.* (1978) rejected subfamilies in the Heteroderidae. Subsequently, *Punctodera* has been raised to subfamily Punctoderinae (Krall and Krall, 1978). Coomans (1979) recognized Heteroderini as a tribe in Heteroderinae and *Punctodera* has been placed in Heteroderinae (Siddiqi, 1986). The tribes within Heteroderinae could also be treated as subfamilies within the Heteroderidae. This view has been adopted by accepting the genus *Punctodera* in the subfamily Heteroderinae. The following four species of *Punctodera* are described and validated with *Punctodera punctata* as type species (Siddiqi, 2000; Subbotin *et al.*, 2010).

Punctodera chalcoensis Stone, Sosa-Moss and Mulvey, 1976
P. matadorensis Mulvey and Stone, 1976
P. punctata (Thorne, 1928) Mulvey and Stone, 1976
P. stonei Brzeski, 1998

Two species, *Punctodera chalcoensis* and *P. punctata*, are of agricultural importance and are reviewed for their invasiveness in this chapter. *P. chalcoensis* mainly infects maize crops and is only recorded from Mexico. *P. punctata* is a widespread species but has a narrow host range, primarily attacking Poaceae plants.

Systematic position of *Punctodera* species

	Classical (after Siddiqi, 2000)	Molecular (after Decraemer and Hunt, 2013)
Phylum	Nematoda	Nematoda
Class	Secernentea	Chromadorea
Order	Tylenchida	Rhabditida
Suborder	Hoplolaimina	Tylenchina
Infra-order	–	Tylenchomorpha
Superfamily	Holpolaimoidea	Tylenchoidea
Family	Heteroderidae	Hoplolaimidae
Subfamily	Heteroderinae	Heteroderinae
Genus	*Punctodera*	*Punctodera*

33 *Punctodera chalcoensis*

33.1 Common name

Mexican corn cyst nematode

33.2 Scientific name

Punctodera chalcoensis Stone, Sosa-Moss and Mulvey, 1976

33.3 Synonym

Heterodera punctata Thorne 1928

33.4 EPPO code

PUNCCH

33.5 Diagnosis

33.5.1 Morphological diagnosis

These morphological descriptions and measurements of *Punctodera chalcoensis* are adapted from Stone *et al.* (1976), Sosa-Moss (1977) and Baldwin and Mundo-Ocampo (1991).

MATURE FEMALE: Globose body white, with distinct head and neck (Fig. 6.30A). Head with one to two prominent annules. Stylet slender, tends to be curved with rounded basal knobs. Oesophagus well developed with large ovoid median bulb and very prominent valve (Fig. 6.30A). Oesophageal glands with distinct dorsal and subventral gland nuclei. Excretory pore situated at base of neck. Globular body cavity contains two large ovaries and can produce up to 400 eggs.
Measurements: Vulva fenestra length = 30.0 ± 5.0 μm, width = 31.5 ± 7.0 μm; length of vulval slit = 4.0 ± 0.5 μm; anal fenestra length = 30.0 ± 5.0 μm, width = 31 ± 5.0 μm; length of anal slit = 3.0 ± 0.7 μm; distance between fenestrae = 64.5 ± 16 μm.
CYSTS: Spherical to sub-spherical, light brown to deep brown. Vulval cone absent. Both vulval fenestra and anal fenestra break down in old cysts. Scattered bullae or occasionally grouped below vulval fenestra.
Measurements: Length excluding neck = 510 μm; maximum width = 470 μm; neck length = 88 μm; length excluding neck/width ratio = 1.1; vulval fenestra length = 18.5 ± 3 μm, width = 20 ± 2.5 μm; length of vulval slit = 3.9 ± 0.5 μm; anal fenestra length = 22.0 ± 3.1 μm, width = 23.3 ± 2.7 μm; length of anal slit = 2.9 ± 0.6 μm; distance between the fenestrae = 52.5 ± 7.3 μm.
SECOND-STAGE JUVENILES: Folded four times within egg. Body vermiform, tail conical with distinctive long hyaline region. Head offset with three head annules (Fig. 6.30F). Labial disc elongated with surrounding amphids. Anterior cephalid located at level of 2nd and posterior at 7th body annules. Stylet robust with large posteriorly rounded basal knobs, flat to concave anteriorly (Fig. 6.30J). Conus half of shaft in length. Lateral field with four incisures, areolated in mid-body region reducing to three near anterior and posterior ends. Median oesophageal bulb well developed (Fig. 6.30F). Oesophageal gland lobes overlap ventrally. Hemizonid 1–2 annules anterior to excretory pore. Genital primordium lying at around 40% of body length from tail terminus. Phasmids located two-thirds along tail length.
MEASUREMENTS: Length = 530 ± 35 μm; body width = 20 ± 0.7 μm; stylet length = 25 ± 0.4 μm; stylet base to dorsal oesophageal gland duct junction = 5.4 ± 0.8 μm; head tip to median bulb valve = 74 ± 3.0 μm; head tip to excretory pore = 110 ± 5.5 μm; tail length = 64 ± 3.0 μm; hyaline tail length = 40 ± 4.2 μm; hyaline tail length/stylet length = 1.6.
MALE: Body vermiform, twisted after heat fixation. Head offset with 5–7 annules. Head skeleton heavy sclerotized. Cephalids at level of 2nd and 8th body annules. Stylet strong, basal knobs flat to concave anteriorly. Lateral field with four lateral lines, ending on tail, areolated. Median bulb ellipsoidal, oesophageal gland lobe ventrally overlaps intestine. Hemizonid two annules wide, situated one annule opposite excretory pore. Gonad single, extending to half body length. Spicules paired, tapering distally with ventral curvature and rounded blunt

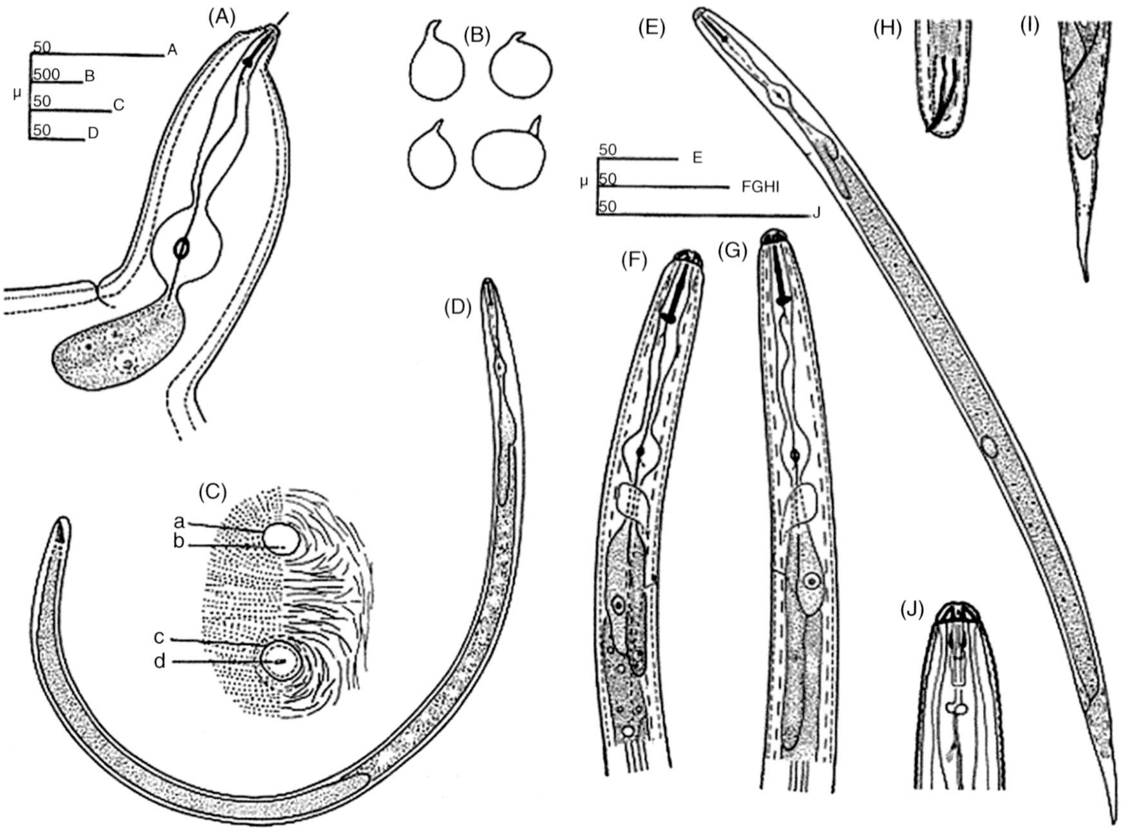

Fig. 6.30. Line drawing of *Punctodera chalcoensis*. Head and neck of female (A); entire female (B); perineum of female, (a) anal fenestra, (b) anus, (c) vulval fenestra, (d) vulval slit (C); entire male (D); entire second-stage juvenile, J_2 (E); anterior region of J_2 (F); anterior of male (G); male tail (H); tail of J_2 (I); head of J_2 (J). [Courtesy Stone *et al.*, 1976; Nematologica]

terminus. Gubernaculum single. Tail bluntly rounded. Phasmids not observed.
Measurements: Length = 973 ± 72 µm; body width = 26 ± 1.8 µm; stylet length = 27.5 ± 0.5 µm; stylet base to dorsal oesophageal gland duct junction = 3.8 ± 0.8 µm; head tip to oesophageal median bulb valve = 83 ± 3.5 µm; head tip to excretory pore = 129.7 ± 9.5 µm; tail length = 3.0 ± 1.6 µm; spicule length (across chord) = 32 ± 2.0 µm; gubernaculum length = 7.4 ± 0.7 µm; gonad length = 473 ± 105 µm.

33.5.2 Biochemical molecular diagnosis

See section 34.5.2, below.

33.6 Geographical distribution

P. chalcoensis has only been reported in Mexico (Baldwin and Mundo-Ocampo, 1991; CABI-CPC, 2010).

33.7 Hosts

This cyst-forming nematode is host specific and the primary agronomic host is maize (*Zea mays*). It also infects teosinte (*Zea mexicana*) (Muñoz and Sosa-Moss, 1983).

33.8 Symptoms

Plants severely attacked by *P. chalcoensis* display stunted growth and death of seedlings. Plants are smaller with pale stripes on the leaves and unhealthy-looking in the field. Symptoms on roots are more conspicuous, such as reduced root systems that produce smaller lateral roots and resemble bottle brushes (Stone *et al.*, 1976) with cysts visible on the roots (Fig. 6.24L).

33.9 Biology and life cycle

The life cycle of *P. chalcoensis* is similar to that of other cyst nematodes. The second-stage juveniles (J_2) hatch in large numbers after the stimulus of maize root diffusate (Muñoz and Sosa-Moss, 1983). After emergence from the cysts, J_2s penetrate host roots and establish syncytial cells in the stele. They develop into swollen females, rupture root cortex and protrude from the root surface. At the end of the reproductive phase, the fertilized female retains eggs in a sac-like body until they are embryonated. The females die and tan to become dark brown cysts. The nematode completes only one generation per year (Sosa-Moss, 1987), which synchronizes with the maize growing season. Cysts (containing embryonated juveniles) left in the soil need a period of hibernation to overwinter. Under laboratory conditions, Jeronimo (1988) recorded a typical life cycle of about 30 days, although it may take up to 50 days (Sosa-Moss, 1987). Nematodes in the soil survive as cysts.

33.10 Similarities with other organisms

P. chalcoensis cysts are spherical and dark brown and have the same dimensions as *Globodera* spp. In Mexico, maize and potatoes are sometimes grown in the same areas and these nematodes do not have common hosts, so the only confusion between *Punctodera* and *Globodera* would arise from a soil sample. Another cyst nematode, *Heterodera zeae*, is lemon-shaped and not a problem in Mexico, the host country (Sosa-Moss, 1987).

33.11 Economic importance

P. chalcoensis is a serious pest of maize in Mexico, where it can reduce the yield by 90% (Sosa-Moss, 1987). It causes severe damage to maize grown in temperate upland valleys of Central Mexico (Mundo-Ocampo *et al.*, 1987). *P. chalcoensis* on maize is as big problem to farmers in Mexico as are potato cyst nematodes (*Globodera* spp.) on potatoes in the same regions. In the most severe cases of nematode damage, as many as 500 cysts per 100 g of soil have been recorded (Vazquez, 1971). The maize straw is also so stunted in these areas that it is unusable for cattle (Becerra and Sosa-Moss, 1978).

33.12 Movement and means of dispersal

See section 21.14, above.

33.13 Pest risk analysis and looming threat from introduction

The specific host range of *P. chalcoensis* would not facilitate this nematode in establishing on other hosts if the nematode is introduced outside its native area. However, due to the economic importance of maize and direct economic impact of this nematode on the corn industry, it cannot be ignored. The production of maize and related industries would be seriously affected if this nematode were to be introduced into new areas and quarantine restrictions implemented. Planting material can only be imported for scientific purposes from Mexico, particularly the central part of the country.

33.14 Invasiveness rating

Based on the consequences of the introduction and probable economic damage caused by *P. chalcoensis*, the following invasiveness rating has been assigned.

Climate/host interaction	–	Medium (2)
Hosts range	–	Low (1)
Dispersal potential	–	Low (1)
Economic impact	–	High (3)
Environmental impact	–	Medium (2)
Post-entry distribution	–	Not known (0)
Final rating (score)	–	Medium (9)

Based on the above invasiveness rating, *P. chalcoensis* receives a score of 9 out of 15 and is rated as MEDIUM INVASIVE NEMATODE.

33.15 Region-wise status of invasiveness

P. chalcoensis is a nematode species of quarantine importance as it occurs only in Mexico and infects maize. Due to the tremendous economic importance of the maize crop and reported economic losses in Mexico, this cyst nematode is given a medium invasiveness rating in all the maize-growing countries of the world, except Mexico.

33.16 Management measures

The management options of *P. chalcoensis* are described below.

33.16.1 Prevention

It is assumed that *P. chalcoensis* co-evolved with maize in Mexico (Stone *et al.*, 1976). Cyst nematodes are difficult to control once established, because of their longevity and their survival strategies, such as diapause. Hence prevention of introduction into other areas is the best way to keep the nematode under check.

33.16.2 Cultural control

Crop rotation is a practical way to keep cyst levels down in the field, because *P. chalcoensis* is apparently host-specific to maize and teosinte (not cultivated on a commercial basis). One fallow in 3 years is an effective measure, but subsistence farmers (who rely on maize) cannot afford to leave the ground fallow and have few alternative crops. The time of sowing is critical. It is preferable to sow the crop early, before the seasonal rains begin, which is usually in late May in Mexico. After the rains, the dormant juveniles hatch much more freely and so seedlings have less chance of survival. The application of adequate fertilizer helped to overcome yield loss (Sosa-Moss and Gonzalez, 1973).

33.16.3 Chemical control

Carbofuran and fenamiphos have been reported to give good control in some instances (Jeronimo, 1988). Some increase in yield has been obtained by applying very small amounts of systemic nematicide to the seeds (de Santiago *et al.*, 1984).

33.16.4 Phytosanitary measures

See Chapter 16, section 115.

34 *Punctodera punctata*

34.1 Common name

Grass cyst nematode

34.2 Scientific name

Punctodera punctata (Thorne, 1928) Mulvey and Stone, 1976

34.3 Synonym

Heterodera punctata Thorne 1928

34.4 EPPO code

HETDPU

34.5 Diagnosis

34.5.1 Morphological diagnosis

The following morphological description and measurements of *Punctodera punctata* are adapted from Webley and Lewis (1977), Wouts (1985) and Turner and Subbotin (2013).

FEMALE: White, spherical, flask- or pear-shaped, posteriorly ovoid. Vulval cone not present. Labial disc elevated, oblong to rectangular, surrounded by circular plate, set off from and formed by fusion of lips. Stylet strong, well developed. Vulval slit surrounded by sub-circular area without normal cuticle markings. Subcrystalline layer present. Anus a slit, less than 4 μm long, enclosed by concentric cuticular markings. Ovaries paired.
Measurements: Length excluding neck = 330–420 (370) μm; neck = 110–150 (123) μm; width = 170–320 (236) μm; L/W = 1.22–2.05 (1.59); stylet = 20–25 (23.2) μm.
Cysts: Cuticle wall with lace-like pattern, annulations and ridges or striations, at times spaced by perforations, all arrangements occasionally present on same cyst, generally with sub-surface pattern of uneven lines of punctations. Underbridge and bullae absent. Cysts produce up to 250 eggs, retained inside the body.
SECOND-STAGE JUVENILE: Body vermiform, tail conoid with a distinctive long hyaline region (Fig. 6.31H). Head offset with three head annules (Fig. 6.31G). Labial disc ovoid, lip region with four annules. Body annules distinct. Stylet robust with forwardly projecting basal knobs (Fig. 6.31G). Lateral field with four incisures, areolated in the mid-body region reducing to three near anterior and posterior ends. Median oesophageal bulb ellipsoid, well developed. Nerve ring just behind bulb, obscure. Hemizonid one annule anterior to excretory pore. Rectum and anus not prominent. Phasmids located about 5–7 annules anterior to end of lateral field.
Measurements: Length = 510–570 (540) μm; width = 20–24 (22) μm; stylet = 24–32 (27) μm; tail = 68–93 (80) μm; anal body width = 13–16 (14.2) μm; hyaline tail terminus = 51–60 (54) μm; dorsal oesophageal gland duct opening from base of stylet = 4–7 (5.4) μm.
MALE: Vermiform, slender, developing by metamorphosis from swollen fourth-stage larva. Head offset with 5–7 annules. Lip cap around spherical, submedial lips fused, in majority of individuals also fused with lateral lips to form continuous annule around lip cap. Stylet robust, basal knobs well developed, projecting forward (Fig. 6.31C). Body tapers conically anterior to stylet base. Lateral field with four incisures delineating three longitudinal bands, fades near tail terminus. Phasmids not seen. Tail bluntly rounded, triangular in end view (Fig. 6.31E). Spicules paired, distally tapering, ventrally curved with blunt terminus. Gubernaculum single.
MEASUREMENTS: Length = 970–1270 (1088) μm; width = 25–32 (27) μm; a = 32–47 (40.3); stylet = 27–33 (30) μm; spicules = 28–36 (34) μm; dorsal oesophageal gland duct opening from base of stylet = 5–6 (5.5) μm.

34.5.2 Biochemical diagnosis

P. punctata was biochemically characterized on the basis of proteins and enzymes, using isoelectric focusing (Rumpenhorst, 1985).

34.5.3 Molecular diagnosis

Fleming *et al.* (2000) characterized *P. punctata* by performing PCR–RFLP of ITS regions (see section 20.5.3, above).

34.6 Geographical distribution

P. punctata is mainly distributed in the temperate parts of the world (Fig. 6.32) and widely reported in Canada and the USA. This cyst nematode is known to occur in the following countries (CABI-CPC, 2010).

Asia: Israel, Japan.
Africa: Morocco.
Europe: Austria, Belgium, Czech Republic, Denmark, Finland, France, Germany, Greece, Iceland, Ireland, Italy, Latvia, Netherlands, Norway, Poland, Portugal, Russia, Slovenia, Spain, Sweden, Switzerland, Ukraine. United Kingdom, former Yugoslavia.
North America: Canada, Mexico, USA.
South America: Argentina, Chile.

34.7 Hosts

P. punctata has a narrow host range and mainly attacks Poaceae plants such as *Agrostis capillaris* (brown bentgrass), *Agrostis stolonifera* (creeping bentgrass), *Agrostis stolonifera* var. *palustris*

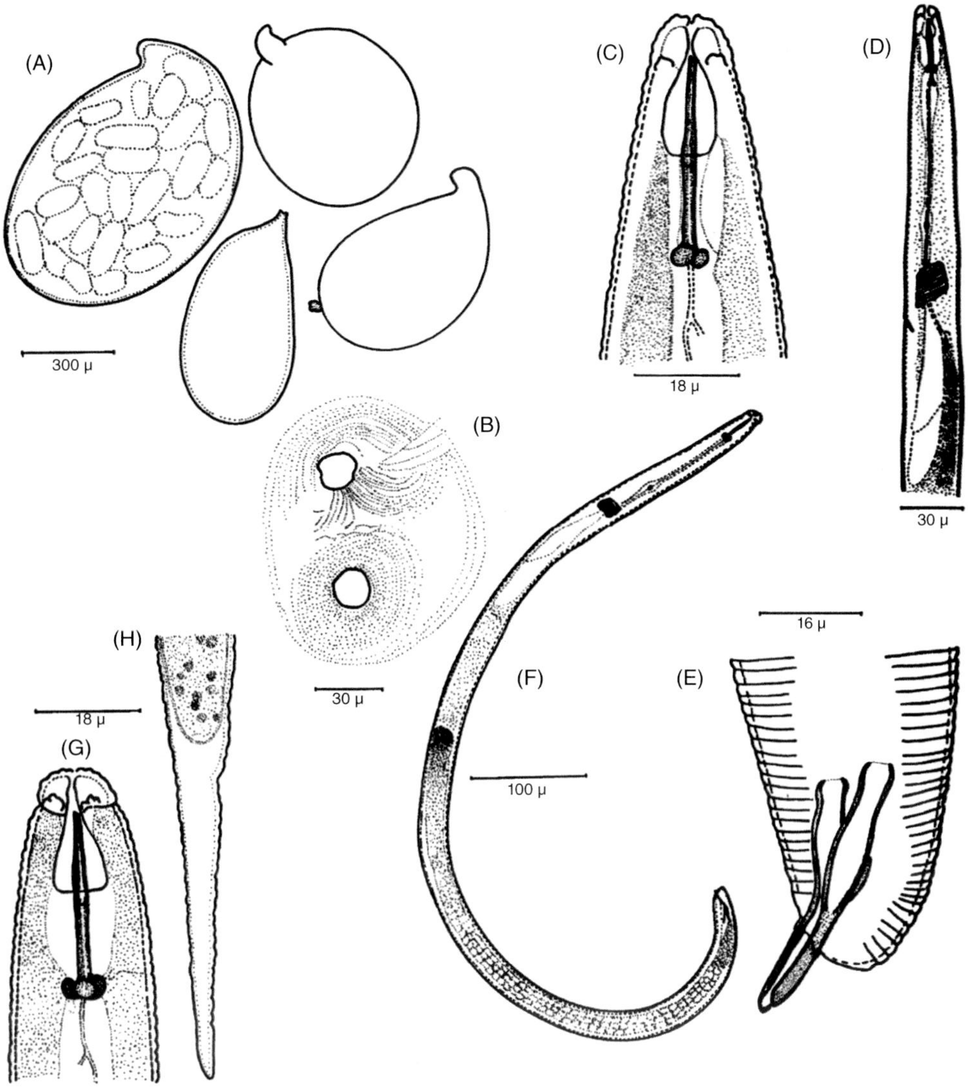

Fig. 6.31. Line drawing of *Punctodera punctata*. Cysts (A); surface of the cysts with vulval and anal fenestrae (B); head of male (C); oesophageal region of male (D); tail region of male (E); entire male (F); head region of second-stage juvenile (G); tail of second-stage juvenile (H). [Reproduced from Webley and Lewis, 1977, CIH Descriptions of Plant-parasitic Nematodes, Set 7, No. 102; courtesy CAB International]

(bentgrass), *Festuca rubra* (red fescue), *Lolium perenne* (perennial ryegrass), *Poa annua* (annual meadowgrass) and *Poa pratensis* (smooth-stalked meadowgrass). Other hosts include *Avena sativa* (oats), *Hordeum vulgare* (barley) and *Triticum* (wheat). However, European populations of *P. punctata* do not infect wheat (CABI-CPC, 2010).

34.8 Symptoms

Symptoms caused by *P. punctata* are not specific on its hosts. However, Brzeski *et al.* (1971) detected a distinct yellowing and lack of vigour of infected *Poa annua* lawns in Michigan (USA). The presence of cysts on the root system or through soil extraction is the only confirmatory symptom. The nematode usually occurs where moisture is abundant, so if the host is not subjected to water stress, as on golf courses or sports fields, the plant usually remains symptomless (Nischwitz *et al.* 2013).

34.9 Biology and life cycle

Hatching of *P. punctata* is stimulated by host root exudates. The life cycle is similar to that of *P. chalcoensis* (see section 33.9, above). Reproduction occurs sexually in *P. punctata* through amphimixis (Radice *et al.*, 1985). A single female lays up to 250 eggs and one generation occurs a year.

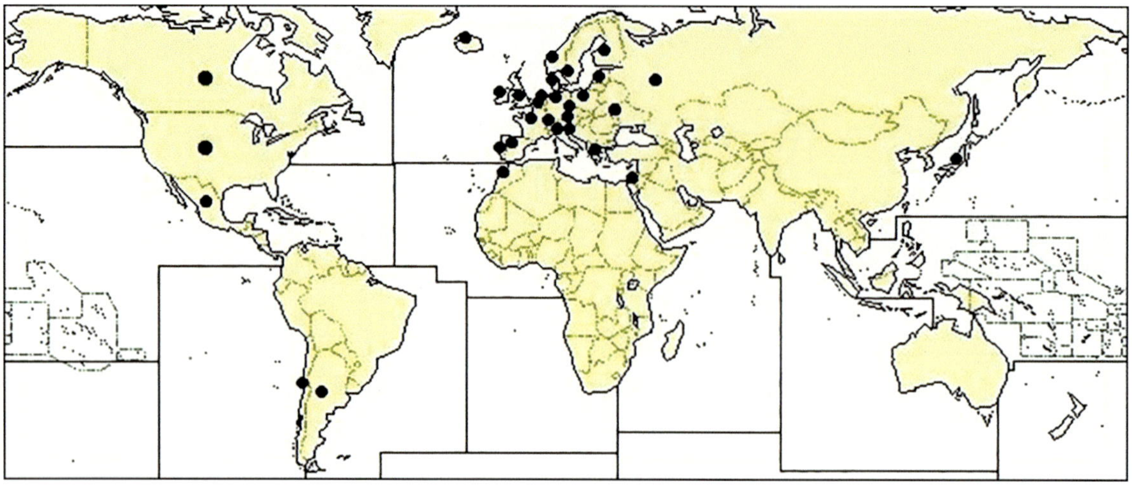

Fig. 6.32. Distribution map of *Punctodera punctata*. [Adapted from CABI-ISC, 2020]

34.10 Similarities with other organisms

Punctodera has a distinct fourth layer in the cuticle of the female, unlike *Heterodera* species, which seems to indicate a closer relationship of *Punctodera* to *Globodera* than to *Heterodera*. *P. punctata* differs from other *Punctodera* species by the pear-shaped cysts and absence of bullae (Turner and Subbotin, 2013).

34.11 Economic importance

There are no records of *P. punctata* causing economic damage in the agronomic field (Webley and Lewis, 1977).

34.12 Movement and means of dispersal

See section 22.14, above.

34.13 Pest risk analysis and looming threat from introduction

The specific and narrow host range, particularly the grasses, along with lack of information on economic damage makes *P. punctata* an insignificant pest in its present habitat. However, its hosts (Poaceae) are often moved between countries as seeds or slips. Other main means of transport for long-distance distribution include contaminated soil attached to non-host plants. Complete pest risk information for this nematode is missing and it may even be incapable of spreading to areas and countries it has not yet invaded. Hence a complete pest risk analysis is suggested for *P. punctata* and precluding procedures should be implemented to check the spread of this nematode.

34.14 Invasiveness rating

Considering the consequences of introduction and probable economic damage caused by *P. punctata*, the following invasiveness rating has been assigned.

Climate/host interaction	–	Medium (2)
Hosts range	–	Medium (2)
Dispersal potential	–	Medium (2)
Economic impact	–	Medium (2)
Environmental impact	–	Medium (2)
Post-entry distribution	–	Not known (0)
Final rating (score)	–	Medium (10)

Based on the above invasiveness rating, *P. punctata* receives a score of 10 out of 15 and is rated as MEDIUM INVASIVE NEMATODE.

34.15 Region-wise status of invasiveness

As a pest of Poaceae plants, *P. punctata* has been given a medium invasiveness rating in all the countries of the world where the nematode is not known to occur.

34.16 Management measures

P. punctata is not well researched as a pest of economic importance. Control measures to protect pastures from this nematode are lacking. However, systemic nematicides such as phorate or carbofuran through soil application may suppress this nematode. For phytosanitary measures, see Chapter 16, section 115.

7

Hoplolaimidae: *Aphasmatylenchus*, *Helicotylenchus*, *Hoplolaimus* and *Scutellonema* Species

The family Hoplolaimidae contains agricultural and economically important genera, viz. *Aphasmatylenchus*, *Helicotylenchus*, *Hoplolaimus* and *Scutellonema*.

Aphasmatylenchus Spp.

The genus *Aphasmatylenchus* belongs to the family Hoplolaimidae, subfamily Hoplolaiminae. As the name suggests, the genus is notably different because of the absence of phasmids. Four species are described and validated with *Aphasmatylenchus nigeriensis* as type species (Siddiqi, 2000). Some important species are:

Aphasmatylenchus liberiensis Baujard, Vovlas, Mounport and Martiny, 1998
A. nigeriensis Sher, 1965
A. straturatus Germani 1970
A. variabilis Germani and Luc, 1984

Among these, *A. straturatus* is an obligate migratory endo/ectoparasitic nematode and attacks some economically important crops, such as groundnut (Baujard and Martiny, 1995), and is considered for invasiveness.

Systematic position of *Aphasmatylenchus* species

	Classical (after Siddiqi, 2000)	Molecular (after Decraemer and Hunt, 2013)
Phylum	Nematoda	Nematoda
Class	Secernentea	Chromadorea
Order	Tylenchida	Rhabditida
Suborder	Hoplolaimina	Tylenchina
Infra-order	–	Tylenchomorpha
Superfamily	Holpolaimoidea	Tylenchoidea
Family	Hoplolaimidae	Hoplolaimidae
Subfamily	Aphasmatylenchinae	Hoplolaiminae
Genus	*Aphasmatylenchus*	*Aphasmatylenchus*

35 *Aphasmatylenchus straturatus*

35.1 Common name

Agent de la chlorose voltaïque des légumineuses (French)

35.2 Scientific name

Aphasmatylenchus straturatus Germani 1970

35.3 EPPO code

APHSYT

35.4 Diagnosis

35.4.1 Morphological diagnosis

Morphological and morphometric descriptions of *Aphasmatylenchus straturatus* after Germani (1970, 1977) and Baujard *et al.* (1998) are summarized below. For measurements, see Table 28.

Female: Body ventrally curved to form open C-shape, occasionally forming closed curve when heat relaxed. Cuticular markings prominent, forming maize cob-like arrangement and each transverse body annule separated by fine striae. Lateral fields with four incisures, three longitudinal bands, occupying about one-fourth of body width, outer bands with cross-striae; central band irregularly areolated. Cephalic region truncated, conical, subhexagonal with 8–13 annules and slightly offset from rest of body by distinct constriction (Fig. 7.1B). Cephalic framework well developed. Amphidial aperture small, elongate, located between oral disc and lateral lip sectors. Oesophagus with procorpus occasionally narrowing slightly at posterior end. Dorsal oesophageal gland duct orifice 5–9 μm behind spear base. Oesophageal glands with three distinct nuclei, overlapping intestine latero-ventrally or ventrally (Fig. 7.1A). Excretory secretory pore 138–208 μm from head end, 1–3 annules behind hemizonid. Vulva a transverse slit, epiptygma present, slightly protruding

DOI: 10.1079/9781789247367.0008

Table 28. Morphometric measurements of *Aphasmatylenchus* and *Helicotylenchus* species.

Morphometric characters		Range, numerous populations				
		A. straturatus (after Germani, 1970)	*H. dihystera* (Sher, 1966)	*H. multicinctus* (Sher,1966)	*H. pseudorobustus* (Sher, 1966)[a]	*H. variocaudatus* (Vovlas *et al.*, 1995)
Females	Length (μm)	1180–1750	590–790	470–530	600–820	431–500
	Stylet (μm)	30–36	25–28	22–24	26–30	24–27
	a	26.0–35.6	27–35	24–30	27–34	23–29
	b	–	5.8–6.9	4.7–5.4	6.0–7.2	3.9–4.3
	b'	6.0–9.1	4.4–5.9	3.4–4.1	–	–
	c	18.0–27.5	35–49	35–46	32–52	25–30
	c'	–	0.8–1.2	0.8–1.0	0.9–1.4	1.3–2.0
	V	50–56	60–65	65–69	59–64	87–89
	Tail length (μm)	–	–	–	–	15–22
	Tail annuli number	–	–	–	7–12	3–4
Males	Length (μm)	990–1260	590–650	470	–	392–437
	Stylet (μm)	28–30	22–27	22	–	19–22
	a	29.1–38.0	25–32	29–30	–	29–36
	b	6.1–7.2	5.1–6.1	5.2	–	3.7–5.2
	b'	–	4.2–4.9	3.9	–	–
	c	14.6–17.4	31–33	28–31	–	27–33
	c'	–	1.0–1.3	–	–	1.3–1.8
	Tail length (μm)	28–29.4	–	–	–	12–18
	Spicules (μm)	37–57	20–23	17–18	–	14–18
	Gubernaculum (μm)	14–21	7–8	5–6	–	4.3–6.0

[a]Male not found.

out of vagina or not protruding, looking like flattened tube. Spermathecae nearly rectangular, comprising numerous small refringent globular sperm-like bodies. Post-anal sac extending for most of tail. Tail cylindro-conical, tapering to round or bluntly rounded tip (Fig. 7.1E). Phasmids absent.

MALE: Cephalic region nearly hemispherical, with 9–12 cephalic annules, distinctly offset or continuous with body contour (Fig. 7.1D). Labial disc well developed, sub-hexagonal, conspicuous in apical view. Lip annule divided into six sectors; two lateral sectors much smaller than others. Amphidial openings narrow slits between labial disc and lateral lip sectors. Stylet and oesophagus less developed than female's, only one oesophageal gland nucleus seen (Fig. 7.1D). Cuticular marking similar to female's. Excretory secretory pore situated 130–206 μm from head end. Single testis, outstretched. Spicules long, curved (Fig. 7.1C). Gubernaculum with titillate. Anterior cloacal lip with several filiform appendices. Bursa enveloping tail with annulated edges (Fig. 7.1C). Tail conical, pointed. Phasmids absent.

35.4.2 Molecular diagnosis

No data are available on molecular approaches.

35.5 Geographical distribution

A. straturatus is known to have very restricted distribution recorded only from an African country, Burkina Faso, particularly in south-west region (Germani, 1977).

35.6 Hosts

A. straturatus has limited host range, particularly annual leguminous plants (Baujard and Martiny, 1995). The principal host is groundnut (*Arachis hypogaea*). Other hosts are cowpea (*Vigna unguiculata*), pigeon pea (*Cajanus cajan*), soybean (*Glycine max*) and bambara groundnut (*Voandzeia subterranea*). Hoary-pea (*Tephrosia* spp.) is recorded as wild host.

35.7 Symptoms

Symptoms produced by *A. straturatus* on leguminous crops appear as chlorotic foliage around 35–45 days after planting (Germani *et al.*, 1982). Chlorosis leads to a decrease in vegetative growth, root development and nodule production. The impairment of nodule formation results in nitrogen deficiency. This is probably associated with the damage caused by *A. straturatus* and by poor nitrogen fixation related to lesions on the nodules and/or to inhibition of *Rhizobium* activity (Germani and Luc, 1982). Symptoms may appear at a population density of 600 nematodes/cm^3. In laboratory conditions, nearly 2000 nematodes/cm^3 are required to reproduce these symptoms (Baujard and Martiny, 1995).

35.8 Biology and life cycle

A. straturatus is a migratory endo/ectoparasitic nematode and infects groundnut and other leguminous crops. In the localized

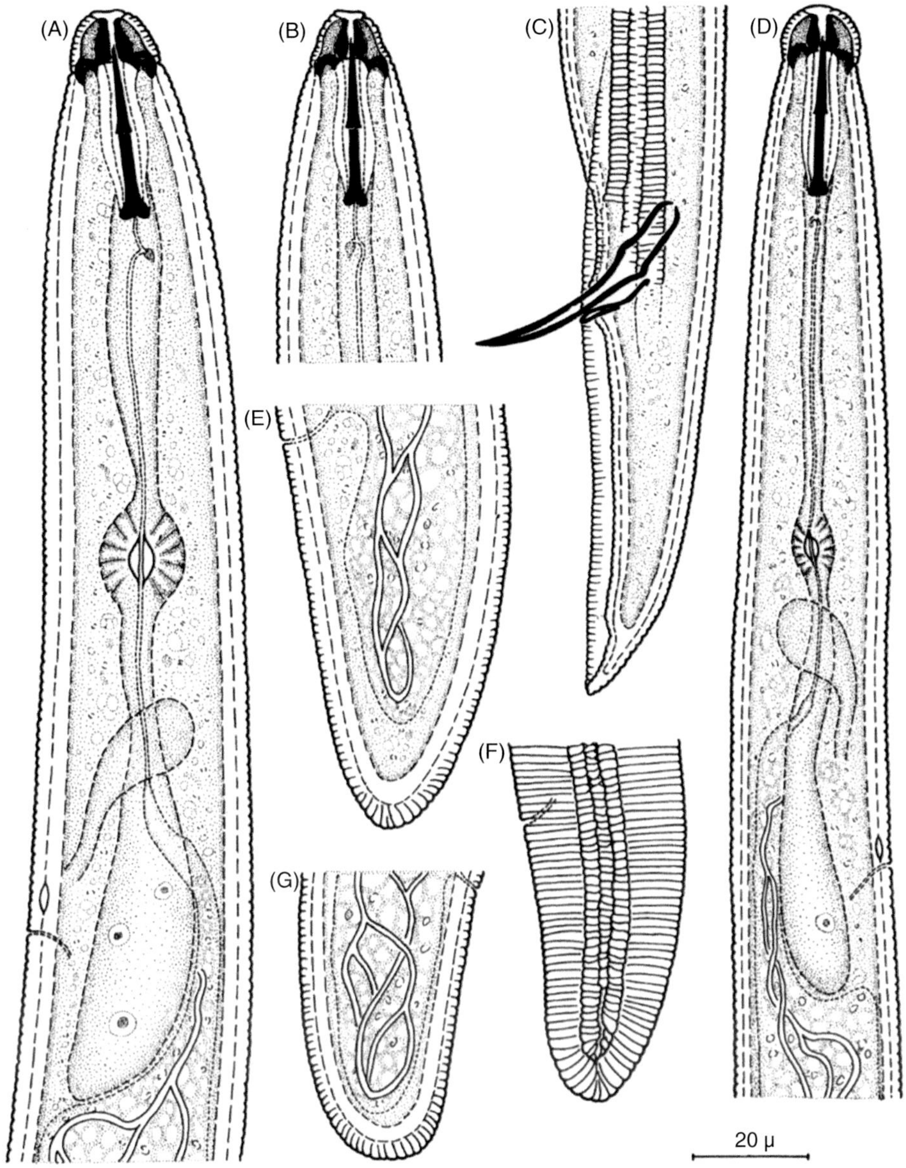

Fig. 7.1. Line drawing of *Aphasmatylenchus straturatus*. Female, anterior end (A,B); male, posterior end (C); male, anterior end (D); female, posterior end (E–G). [Reproduced from Germani, 1977, CIH Description of Plant-parasitic Nematodes, Set 7. No. 104; courtesy CAB International]

area, the host (groundnut) is interplanted with karite trees (*Butyrospermum parkii*) and the nematodes spend the dry season at a depth of 40–60 cm in the rhizosphere or in the roots of the karite trees. Thereafter, they migrate from the tree roots and enter groundnut roots at the beginning of the rainy season. The nematodes are most abundant in groundnut roots around 40–70 days after seeding. At 100–110 days after seeding, they again move from groundnut roots to the karite trees. Under laboratory conditions, nematode multiplication was higher at temperatures of 30–32°C and 9–11% soil moisture (Baujard and Martiny, 1995). The nematode seems to be able to reproduce on cereals and to survive it also appears to be able to enter anhydrobiosis at egg stage during the dry period (Baujard and Martiny, 1995).

35.9 Similarities to other species and conditions

A. straturatus shares some features which are characteristic of the family Hoplolaimidae (Fortuner, 1987). However, in both sexes, there is a tube-like structure or serpentine canal within the intestinal walls winding over the entire length of the intestine, including the post-anal sac. Within the genus, *A. straturatus* can easily be distinguished from the three species *A. nigeriensis*, *A. variabilis* and *A. liberiensis* by the ornamentation of the cuticle, i.e. without longitudinal divisions outside the lateral fields. Since *A. straturatus* is known to be restricted to a specific region of Burkina Faso, confusion concerning its identification is improbable.

35.10 Economic importance

A. straturatus causes significant yield reductions in groundnut that range from 30% to 70%. On pigeon pea, this nematode may cause complete crop failure (Germani, 1979). In Burkina Faso, the nematode infested nearly 4% of groundnut production and infestation had risen to 25% in 4 years (Dhéry *et al.*, 1975).

35.11 Movement and means of dispersal

The dispersal of *A. straturatus* and other members of family Hoplolaimidae is facilitated primarily via infected planting material. The nematodes can also be spread by the movement of contaminated farm machinery, harvesting combines and nematode-infected plant residues. Its endoparasitic habits also allow the nematode to be transported long distances inside infected roots.

35.11.1 Plant parts known to carry the nematode in trade/transport

Roots, bulbs, tubers, corms, rhizomes (sucker), growing medium accompanying plants, seedlings and micro-propagated plants.

35.11.2 Plant parts not known to carry the nematode in trade/transport

Bark, flowers, inflorescences, cones, calyx, fruits (including pods), leaves, stems (above ground), shoots, trunks, branches, wood and true seeds (including grain).

35.12 Pest risk analysis and looming threat from introduction

A. straturatus caused substantial yield loss to the groundnut crop in its localized area. The nematode also parasitizes other economically important leguminous plants and its rapid spread poses a threat to the production of groundnut and other leguminous crops. It can be easily spread through infested vegetative planting material. A complete pest risk analysis is suggested for all the host countries where this nematode is absent.

35.13 Invasiveness rating

Considering the consequences of introduction and probable economic damage caused by *A. straturatus*, the following invasiveness rating has been assigned.

Climate/host interaction	–	Low (1)
Hosts range	–	High (3)
Dispersal potential	–	Low (1)
Economic impact	–	High (3)
Environmental impact	–	Low (1)
Post-entry distribution	–	Not known (0)
Final rating (score)	–	Medium (9)

Based on the above invasiveness rating, *A. straturatus* receives a score of 9 out of 15 and is rated as MEDIUM INVASIVE NEMATODE.

35.14 Region-wise status of invasiveness

Considering the economic damage potential of *A. straturatus* on important leguminous crops, a medium invasiveness rating has been given for all the legume-growing countries except Burkina Faso, where *A. straturatus* is widespread.

35.15 Management measures

Limited information is available on the management of *A. straturatus* for groundnut and other crops. Application of fumigant nematicides, dibromochloropropane, at the time of planting gave satisfactory control (Dhéry *et al.*, 1975; Germani, 1979). Pre-plant application with fensulfothion may also reduce the soil population of the nematode.

35.15.5 Phytosanitary measures

See Chapter 16, section 115.

Helicotylenchus Spp.

The genus *Helicotylenchus* contains more than 200 species, excluding several synonyms, with *H. nannus* as its type species (Siddiqi, 2000; Uzma *et al.*, 2015). Keys to *Helicotylenchus* spp. are given by Fortuner and Wong (1984), Fotedar and Kaul (1985), Firoza and Maqbool (1994) and Subbotin *et al.* (2011c). The important species from *Helicotylenchus* are:

Helicotylenchus abunaamai Siddiqi, 1972
H. affinis (Luc, 1960) Fortuner, 1984
H. astriatus Khan and Nanjappa, 1972
H. brevis (Whithead, 1958) Fortuner, 1984
H. cavenessi Sher, 1966
H. crenacauda Sher, 1966
H. digitiformis Ivanova, 1967
H. digonicus Perry *et al.*, 1959
H. dihystera (Cobb, 1893) Sher, 1961
H. egyptiensis Tarjan, 1964
H. erythrinae (Zimmermann, 1904) Golden, 1956
H. indicus Siddiqi, 1963
H. intermedius (Luc, 1960) Siddiqi and Husain, 1964
H. microcephalus Sher, 1966
H. mucronatus Siddiqi, 1963
H. multicinctus (Cobb, 1893) Golden, 1956
H. neopaxilli Inserra Vovlas and Golden, 1975
H. oleae Inserra Vovlas and Golden, 1979
H. paracanalis Sauer and Winoto, 1975
H. pseudorobustus (Steiner, 1914) Golden, 1956
H. serenus Siddiqi, 1963
H. sharafati Mulk and Jairajpuri, 1975

H. varicaudatus Yuen, 1964
H. variocaudatus (Luc, 1960) Fortuner, 1984

Among these, *H. dihystera*, *H. multicinctus*, *H. pseudorobustus* and *H. variocaudatus* are associated with plant growth suppression as ecto- or semi-endoparasites and cause severe crop damage throughout the world (Siddiqi, 2000; Subbotin *et al.*, 2015; Hunt, 2018).

Systematic position of *Helicotylenchus* species

	Classical (after Siddiqi, 2000)	Molecular (after Decraemer and Hunt, 2013)
Phylum	Nematoda	Nematoda
Class	Secernentea	Chromadorea
Order	Tylenchida	Rhabditida
Suborder	Hoplolaimina	Tylenchina
Infra-order	–	Tylenchomorpha
Superfamily	Holpolaimoidea	Tylenchoidea
Family	Holpolaimidae	Holpolaimidae
Subfamily	Rotylenchoidinae	Holpolaiminae
Genus	*Helicotylenchus*	*Helicotylenchus*

36 *Helicotylenchus dihystera*

36.1 Common name

Spiral nematode

36.2 Scientific name

Helicotylenchus dihystera (Cobb, 1893) Sher, 1961

36.3 Synonyms

Aphelenchus dubius var. *peruensis* Steiner, 1920
Helicotylenchus crenatus Das, 1960
Helicotylenchus flatus Roman, 1965
Helicotylenchus membranatus Xie and Feng 1993
Helicotylenchus nannus Steiner, 1945
Helicotylenchus olaae (Cobb) Siddiqi, 1986
Helicotylenchus paraconcavus Rashid and Khan, 1974
Helicotylenchus punicae Swarup and Sethi, 1968
Helicotylenchus reversus Sultan, 1985
Helicotylenchus spiralis (Cassidy) Siddiqi 1986
Tylenchus dihystera Cobb, 1893
Tylenchus olaae Cobb, 1906
Tylenchus spiralis Cassidy, 1930
Tylenchorhynchus olaae (Cobb) Micoletzky, 1922

36.4 EPPO code

HELYDH

36.5 Diagnosis

Morphological, biochemical and molecular diagnoses of *Helicotylenchus dihystera* are described below.

36.5.1 Morphological diagnosis

This morphological diagnosis is based on descriptions by different authors (Sher, 1966; Rashid and Khan, 1974; Van Den Berg and Heyns, 1975; Fortuner *et al.*, 1981; Siddiqi, 2000). For measurements see Table 28.

Female: Body spirally curved (Fig. 7.2B), more so in posterior region after relaxation or death (Fig. 7.2C). Lateral fields with four incisures, not areolated. Cephalic region or head hemispherical, with 4–5 annules; framework heavily sclerotized (Fig. 7.2A), outer margins extending posteriorly 2–3 body annules. Posterior cephalids distinct, 8–11 annules behind cephalic region. Stylet fairly strong, usually 24–28 μm long, conus measuring 11.0–12.5 μm long, basal knobs well developed and anteriorly indented. Median oesophageal bulb oval, very muscular, 6–8 body annules long. Orifice of dorsal oesophageal gland at less than half stylet length behind stylet base. Oesophageal glands moderately covering anterior end of intestine; subventrals a little longer than dorsal gland. Excretory pore almost opposite oesophago-intestinal junction. Hemizonid 0–2 annules behind excretory pore. Vulva a transverse slit, at 60–66% of body length. Reproductive organs outstretched in two branches in opposite directions. Ovary with oocytes in row. Spermatheca offset, empty. Tail dorsally convex-conoid, with or without slight ventral projection, with 8–12 annules ventrally (Fig. 7.2D-H). Phasmids pore-like, preanal, 5–12 annules anterior to anal level, normally centrally placed in lateral field; inner incisures of lateral field on tail not fusing distally for around distance.

Male: Extremely rare. Body ventrally arcuate. Head, stylet and oesophagus as in female. Bursa enveloping tail, which is conoid.

Juvenile: Four juvenile stages, first occurring within egg; similar to female in structure of head, stylet, oesophagus and tail. Different juvenile stages recognized by degree of development of reproductive system: genital primordia of II-, III- and IV-stage juveniles have 4, 14 and 108 nuclei, respectively (Hirschmann and Triantaphyllou, 1968).

36.5.2 Molecular diagnosis

Species diagnosis of *Helicotylenchus* is quite challenging because of existence of significant interspecific variability. At present around 20% of *Helicotylenchus* species have been characterized molecularly frequently using ribosomal DNA fragments (18S, ITS, 28S rDNA) and mitochondrial cytochrome c

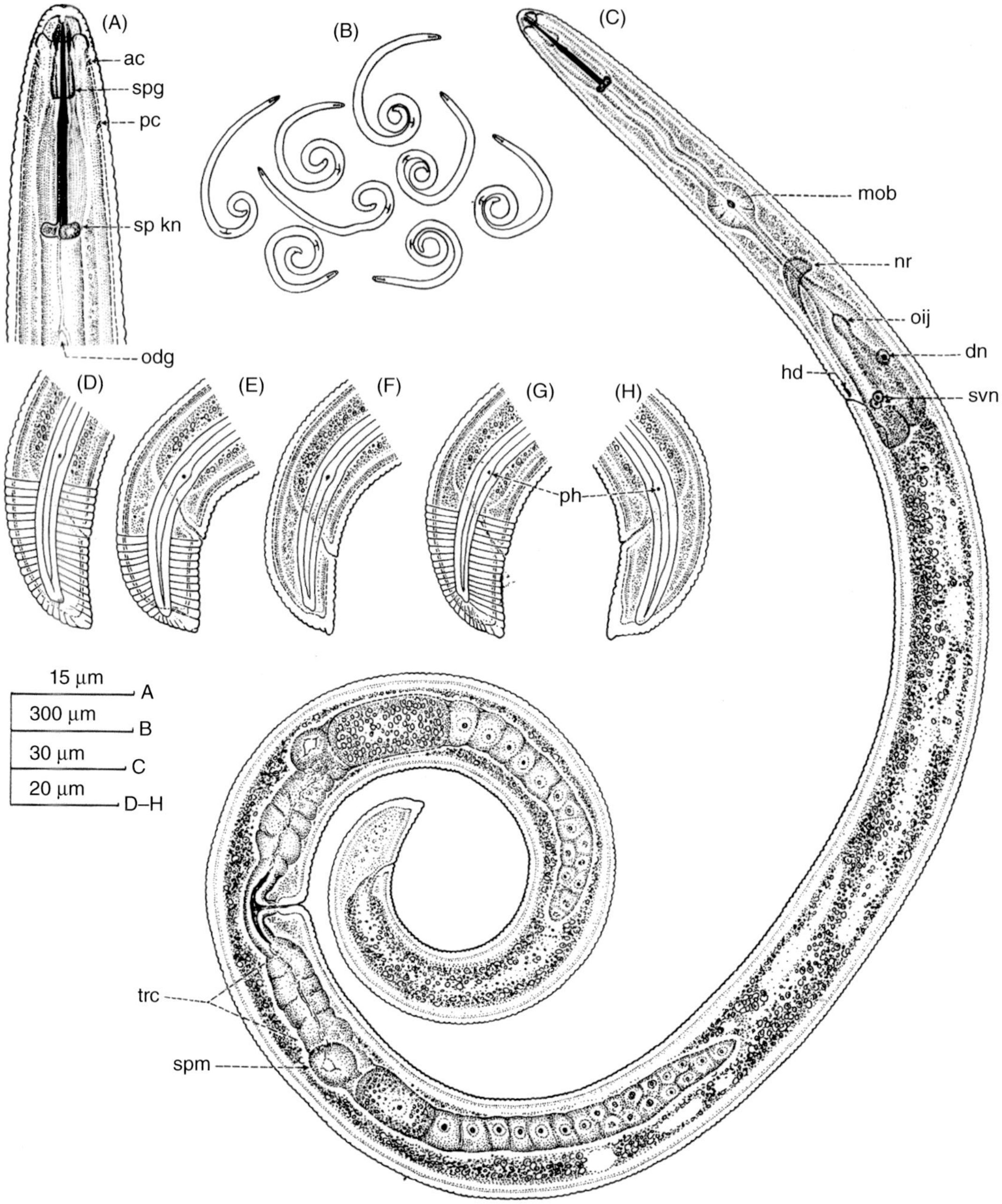

Fig. 7.2. Line drawing of *Helicotylenchus dihystera*. Head end (A); females (B,C); tail end (D–H); Ac, anterior cephalid; dn, dorsal oesophageal gland nucleus; hd, hemizonid; mob, median oesophageal bulb; nr, nerve ring; odg, orifice of dorsal gland nucleus; oij, oesophageal-intestine junction; pc, posterior cephalid; spg, stylet guide; sp kn, stylet knob; spm, spermatheca; svn, subventral gland nuclei; trc, tricolumella. [Reproduced from Siddiqi, 2000, courtesy CAB International]

oxidase subunit I (mtCOI) sequences (Palomares-Rius *et al.*, 2018; Rybarczyk-Mydłowska *et al.*, 2019). Brücher *et al.* (2019) characterized *H. dihystera* parasitizing maize from Argentina based on molecular analysis of D2–D3 of the 28S rDNA and the 5.8S ITS2 regions using standard PCR. The genomic DNA is extracted from 20 *H. dihystera* females using worm lysis buffer. Amplification of the D2–D3 expansion segment of the 28S rDNA region and 5.8S ITS2 region produces amplicon sizes of approximately 800 pb and 1100 pb, respectively. For detailed protocol, see Chapter 4, section 15.5.2.

Subbotin *et al.* (2015) provided morphological and molecular characterizations of numerous *Helicotylenchus* species such as *H. broadbalkiensis*, *H. digonicus*, *H. dihystera*, *H. microlobus*, *H. paxilli* and *H. pseudorobustus*, collected from various

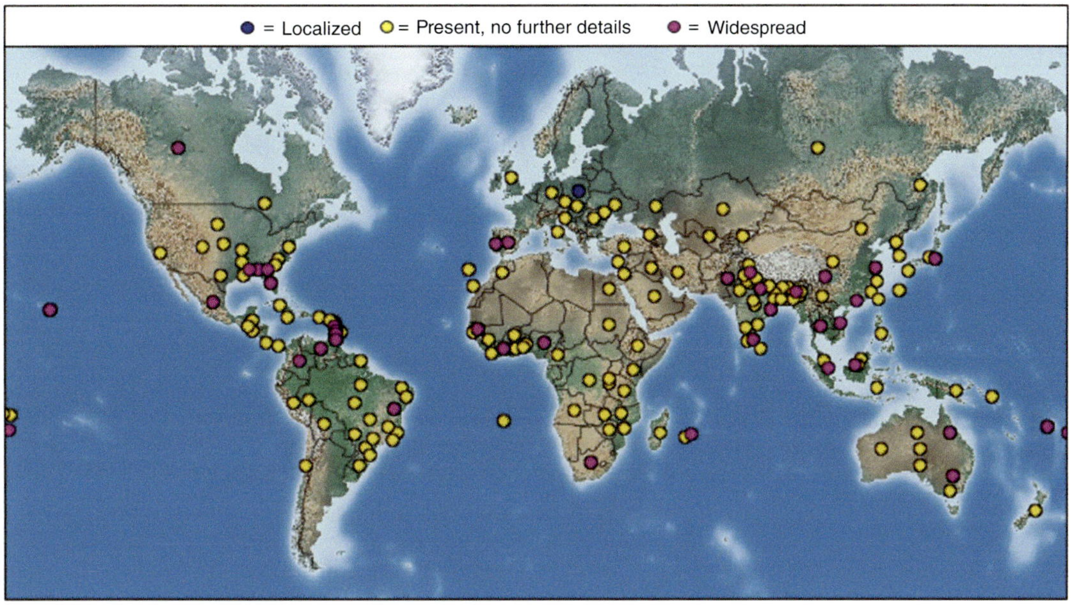

Fig. 7.3. Distribution map of *Helicotylenchus dihystera*. [Adapted from CABI-ISC, 2020]

geographical areas of Italy, New Zealand, South Korea, Spain, Switzerland, Russia, the USA and the UK. The phylogenetic analyses of 154 sequences of the D2–D3 of 28S rRNA gene and 37 sequences of ITS-rRNA gene in the study suggested that *H. microlobus* and *H. pseudorobustus* were morphologically and molecularly different (Subbotin *et al.*, 2015).

36.6 Geographical distribution

H. dihystera is a near-cosmopolitan species and is distributed throughout the world (Fig. 7.3). However, the nematode is more concentrated in tropical and subtropical countries. It is mainly distributed in the following countries (CABI-ISC, 2020).

Africa: Angola, Benin, Burkina Faso, Burundi, Cameroon, Congo Democratic Republic, Côte d'Ivoire, Egypt, Ethiopia, Gambia, Ghana, Guinea, Kenya, Liberia, Madagascar, Malawi, Mauritius, Morocco, Nigeria, Réunion, Rwanda, Saint Helena, Senegal, South Africa, Sudan, Tanzania, Togo, Zambia, Zimbabwe.

Asia: Bangladesh, Bhutan, Brunei, China, Georgia, India, Indonesia, Iran, Iraq, Israel, Japan, Kazakhstan, Kyrgyzstan, Malaysia, Nepal, North Korea, Pakistan, Saudi Arabia, Singapore, South Korea, Sri Lanka, Taiwan, Thailand, Turkey, Uzbekistan, Vietnam.

Europe: Cyprus, Czech Republic, Germany, Italy, Moldova, Netherlands, Poland, Portugal, Romania, Russia, Slovakia, Slovenia, Spain, Ukraine, United Kingdom.

North America: Barbados, Belize, Canada, Costa Rica, Cuba, Dominica, El Salvador, Grenada, Guadeloupe, Guatemala, Jamaica, Martinique, Mexico, Montserrat, Panama, Puerto Rico, Saint Kitts & Nevis, Saint Lucia, Saint Vincent and the Grenadines, Trinidad and Tobago, USA.

Oceania: American Samoa, Australia, Fiji, New Zealand, Niue, Papua New Guinea, Samoa, Solomon Islands, Tonga.

South America: Bolivia, Brazil, Chile, Colombia, French Guiana, Peru, Venezuela.

36.7 Hosts

H. dihystera is a polyphagous species and attacks a wide variety of hosts (Azmi, 1990). The host list comprising main, other and wild plants is presented in Table 29.

36.8 Symptoms

H. dihystera feeds ecto- or endoparasitically on the outer cortical cells, causing small wounds. The wounds obstruct uptake of water and nutrients by roots, leading to weakening and stunting of plants with undersize fruits and reduced yields. The vegetative cycle of infected plant is prolonged and the productive life of the farm is reduced. Root necrosis occurs, resulting in dieback and eventual debilitation of the entire plant. The nematode incites discrete, relatively shallow necrotic lesions on roots. Generally, they feed on the superficial layers to a few cells depth into the root cortex. In the infected cells, the nematode withdraws cytoplasm from surrounding cells and the cell walls are distorted or ruptured. The nucleus, if intact, is enlarged. Evacuated cells together with those near the 'nest' of eggs, soon become discoloured and necrotic. The necrosis and dieback of the root system facilitate entry for invasion by soil bacteria and fungi, which exaggerate the damage (Khan, 1993). On soybean, *H. dihystera* causes discernible stunting of plants. The cross-section of the infected plants showed adults, larvae and eggs within the root cortex accompanied by small brown lesions near the nematodes (Orbin, 1973).

Table 29. List of hosts of *Helicotylenchus dihystera* (adapted from CABI-CPC, 2010).

Main hosts
Abelmoschus esculentus (okra)
Acalypha hispida (copperleaf)
Aechmea fasciata
Agrostis capillaris (brown bentgrass)
Allium cepa (onion)
Allium chinense (spring onion)
Allium schoenoprasum (chives)
Ananas comosus (pineapple)
Andropogon pseudapricus
Annona muricata (soursop)
Apium graveolens (celery)
Arachis hypogaea (groundnut)
Araucaria
Araucaria cunninghamii (colonial pine)
Artocarpus altilis (breadfruit)
Avena sativa (oats)
Begonia
Beta vulgaris (beetroot)
Brassica
Broussonetia papyrifera (paper mulberry)
Buxus sempervirens (common boxwood)
Cajanus cajan (pigeon pea)
Calopogonium
Camellia sinensis (tea)
Canna indica (canna lily)
Capsicum annuum (bell pepper)
Carica
Carica papaya (papaw)
Cassia alata (Ringworm senna)
Casuarina (beefwood)
Cenchrus ciliaris (buffelgrass)
Cenchrus echinatus (southern sandbur)
Centrosema pubescens (centro)
Chamaecyparis (false cypress)
Chloris gayana (Rhodes grass)
Cinnamomum verum (cinnamon)
Citrullus lanatus (watermelon)
Citrus aurantiifolia (lime)
Citrus jambhiri (rough lemon)
Citrus limon (lemon)
Citrus sinensis (navel orange)
Citrus × paradisi (grapefruit)
Cocos nucifera (coconut)
Coffea arabica (arabica coffee)
Colocasia esculenta (taro)
Crossandra undulifolia
Crotalaria juncea (sunn hemp)
Crotalaria spectabilis (showy rattlepod)
Cryptanthus bromelioides
Ctenanthe oppenheimiana
Cucumis sativus (cucumber)
Cucurbita moschata (pumpkin)
Cucurbita pepo (ornamental gourd)
Cymbidium
Cynodon dactylon (Bermuda grass)
Datura (thorn-apple)
Daucus carota (carrot)
Delonix regia (gold mohar)
Desmodium tortuosum (Florida beggarweed)
Digitaria (crabgrass)
Dioscorea (yam)
Dioscorea alata (white yam)
Elaeis guineensis (African oil palm)
Eleusine coracana (finger millet)
Eucalyptus deanei
Falcataria moluccana (batai wood)
Ficus benjamina (weeping fig)
Ficus elastica (rubber plant)
Fortunella × crassifolia (meiwa kumquat)
Fragaria ananassa (strawberry)
Fragaria vesca (wild strawberry)
Gardenia
Glycine max (soybean)
Gossypium hirsutum (Bourbon cotton)
Helianthus annuus (sunflower)
Hibiscus (rose mallows)
Hordeum vulgare (barley)
Humulus lupulus (hop)
Hyptis pectinata (comb hyptis)
Ilex cornuta (Chinese holly)
Ilex crenata (Japanese holly)
Impatiens (balsam)
Indigofera hirsuta (hairy indigo)
Ipomoea batatas (sweet potato)
Ipomoea brasiliensis
Jasminum sambac (Arabian jasmine)
Juglans (walnuts)
Lablab purpureus (hyacinth bean)
Lactuca sativa (lettuce)
Lolium (ryegrasses)
Macadamia tetraphylla (macadamia nut)
Malus domestica (apple)
Mangifera indica (mango)
Manihot esculenta (cassava)
Medicago sativa (lucerne)
Mimosa pudica (sensitive plant)
Miscanthus floridulus (Japanese silvergrass)
Momordica charantia (bitter gourd)
Mucuna pruriens (buffalobean)
Musa (banana)
Musa acuminata (wild banana)
Musa × paradisiaca (plantain)
Nicotiana tabacum (tobacco)
Olea europaea subsp. *europaea* (olive)
Oryza sativa (rice)
Pandanus (screw-pine)
Panicum maximum (Guinea grass)
Panicum sumatrense (little millet)
Paspalum conjugatum (sour paspalum)
Paspalum notatum (bahiagrass)
Passiflora edulis (passionfruit)
Pelargonium graveolens (rose geranium)

Table 29. Continued.

Pennisetum clandestinum (kikuyu grass)
Pennisetum glaucum (pearl millet)
Pennisetum pedicellatum (deenanath grass)
Peony
Persea americana (avocado)
Petunia
Phaseolus vulgaris (common bean)
Phoenix dactylifera (date-palm)
Phytolaccaceae
Pinus caribaea (Caribbean pine)
Pinus elliottii (slash pine)
Piper nigrum (black pepper)
Pisum sativum (pea)
Poncirus trifoliata (trifoliate orange)
Prunus domestica (plum)
Prunus dulcis (almond)
Prunus persica (peach)
Psidium guajava (guava)
Pueraria phaseoloides (tropical kudzu)
Pyrus communis (European pear)
Raphanus sativus (radish)
Rosa (roses)
Rosa multiflora (multiflora rose)
Rubus idaeus (raspberry)
Saccharum edule
Saccharum officinarum (sugarcane)
Schefflera (umbrella tree)
Setaria magna (giant bristlegrass)
Setaria splendida
Solanum lycopersicum (tomato)
Solanum melongena (aubergine)
Solanum torvum (turkey berry)
Solanum tuberosum (potato)
Sorghum bicolor (sorghum)
Sorghum halepense (Johnson grass)
Spinacia oleracea (spinach)
Stylosanthes gracile
Swietenia macrophylla (mahogany)
Syzygium aromaticum (clove)
Syzygium malaccense (malay-apple)
Tectona grandis (teak)
Theobroma cacao (cocoa)
Tillandsia cyanea
Trifolium alexandrinum (berseem clover)
Trifolium repens (white clover)
Triticum aestivum (wheat)
Ulmus parvifolia (lacebark elm)
Urena lobata (hibiscus burr)
Urochloa decumbens (signal grass)
Urochloa mutica (tall panicum)
Vaccinium (blueberries)
Vanda
Vanilla planifolia (vanilla)
Vigna mungo (black gram)
Vigna unguiculata (cowpea)
Vitex trifolia
Vitis vinifera (grapevine)
Washingtonia (Washington-palm)
Wisteria
Xanthosoma sagittifolium (yautia)
Zea mays (maize)
Zingiber officinale (ginger)
Other hosts
Ceiba pentandra (kapok)
Prunus salicina (Japanese plum)
Wild hosts
Coffea (coffee)
Eucalyptus (Eucalyptus tree)
Hevea brasiliensis (rubber)
Leucaena leucocephala (leucaena)
Liriodendron tulipifera (tuliptree)
Pinus echinata (shortleaf pine)
Pinus patula (Mexican weeping pine)
Pinus taeda (loblolly pine)
Platanus occidentalis (sycamore)
Populus heterophylla (Swamp cottonwood)

36.9 Biology and life cycle

The spiral nematode is an ecto- or semi-endoparasite of roots and large populations are found in the rhizosphere. It usually reproduces by mitotic parthenogenesis (Hirschmann and Triantaphyllou, 1968). However, on pineapple, the species is bisexual and reproduces by cross-fertilization or amphimixis (Baujard and Martiny, 1995). Hence, the biology and life cycle slightly vary with the host. On soybean, the adults and juveniles are both endoparasitic and semi-endoparasitic, and the life cycle is completed within the roots (Orbin and Cairns, 1971). After penetration, feeding on a single cell continues for several days, with alternating salivation and ingestion (Jones, 1978). On sugarcane, both adults and juveniles invade the roots, the anterior end of the body entering the cortical tissues and the rest remaining outside, but often they enter the roots completely (Rao and Swarup, 1974). On pineapple, all stages of the nematode can be found within roots and, possibly, the entire life cycle is completed endoparasitically there (Mohilal and Khan, 2012).

The population levels of *H. dihystera* on sugarcane are directly correlated with rainfall and soil moisture content (Gomes Carneiro *et al.*, 1980). However, in *Cenchrus ciliaris* pastures in India, *H. dihystera* was least affected by environmental stress (Azmi, 1990). In the USA, the population of *H. dihystera* on maize, cotton, groundnut and soybean was stable during January to June but increased significantly from August to September (Johnson *et al.*, 1974). In wheat, the nematode population was relatively stable from January to May, but when soybeans were planted, a drastic increase in the nematode population occurred within one month (Kinloch *et al.*, 1971). The maximum activity of *H. dihystera* occurred at temperatures of 20–30°C and was much reduced between 5–10°C and 35–40°C (Azmi, 1979). The nematode survived for 6 months in soil kept in plastic bags both at room temperatures (18–24°C) and in the refrigerator at 1.1–4.4°C (Ferris, 1960). Studies have also shown that anhydrobiosis might induce a strong effect on the physiology of *H. dihystera* and numbers of the nematode were higher after soil desiccation during the dry season (Baujard and Martiny, 1995).

36.10 Economic importance

The spiral nematodes *H. dihystera*, *H. multicinctus* and *H. pseudorobustus* are widely distributed in different parts of world. Among these, *H. dihystera* is the most widely occurring species and causes economic damage to agricultural crops. Firoza and Maqbool (1995) reported populations of around 1000 *H. dihystera* per 250 g of soil in aubergine, tomato and wheat fields. The spiral nematode may cause up to 49% suppression in tomato growth (Slabaugh, 1974). On potato, *H. dihystera* reduced the dry weights of shoot and root of potato cv. 'Kufri Jyoti' by 24–45% while the tuber yields per plant were reduced by 9–17% at 90 days (Pradesh and Singh, 1987). In South Africa, *Helicotylenchus*, *Scutellonema* and *Rotylenchus*, have been reported to cause notable damage to pineapple (Keetch and Purdon, 1979). In India, Khan (1983) also found that, among different nematodes associated with pineapple, *Helicotylenchus* spp. were the most abundant nematode. Nath *et al.* (1997) reported that *Helicotylenchus* spp. had the highest frequency, prominence and importance values among the nematode species infesting pineapple in Tripura (India). In guava orchards in South Africa, 96% of plantings were infected by *H. dihystera* and the height of plants was suppressed by 53% (Willers and Grech, 1986).

The nematodes severely damaged the root systems of sugarcane plants (Rao and Swarup, 1974). On sorghum, an inoculum level of 10,000 nematodes per plant caused a significant reduction in plant growth parameters (Jain and Hasan, 1987). On cocoa, 1000 *H. dihystera* per plant caused stunting and there was a significant decrease in dry root weight (Campelo and Galli, 1980). This nematode is also pathogenic to groundnut and millet (Baujard and Martiny, 1995), seedlings of cottonwood (*Populus heterophylla*), American sycamore (*Platanus occidentalis*) and yellow poplar (*Liriodendron tulipifera*) (Ruehle, 1971) and olive seedlings (Diab and El-Eraki, 1968). *H. dihystera* was also associated with soybean-growing areas in Brazil, and 47% soil samples were found infested (Machado *et al.*, 2019).

36.11. Movement and means of dispersal

See section 35.11, above.

36.12 Pest risk analysis and looming threat from introduction

Taking into consideration that *H. dihystera* has a very wide host range, regulatory agencies must adopt adequate measures to prevent the spread of this nematode from infested countries to new areas. Pest risk analysis is recommended and appropriate actions should be adopted by quarantine departments to minimize the risks.

36.13 Invasiveness rating

Considering the consequences of introduction and pest risk analysis of *H. dihystera*, the following invasiveness rating has been assigned.

Climate/host interaction	–	High (3)
Hosts range	–	High (3)
Dispersal potential	–	High (3)
Economic impact	–	High (3)
Environmental impact	–	Medium (2)
Post-entry distribution	–	High (–3)
Final rating (score)	–	Medium (11)

Based on the above invasiveness rating, *H. dihystera* receives a score of 11 out of 15 and is rated as MEDIUM INVASIVE NEMATODE.

36.14 Region-wise status of invasiveness

Considering the polyphagous and cosmopolitan nature of *H. dihystera*, the nematode attained medium invasiveness status in all countries where it is absent. There may be trade implications if this nematode is introduced and established in these countries.

36.15 Management measures

Different management measures have been tested and recommended against spiral nematodes. Chemical controls have proved relatively more effective on *H. dihystera*, when the nematode is semi-endoparasite. However, the task becomes tough when it parasitizes as an endoparasite. The following management options may help to minimize the damage caused by this nematode.

36.15.1 Cultural practices and soil amendments

In beans (*Phaseolus vulgaris*) and maize, control of *H. dihystera* was attained by green manuring with *Crotalaria paulina*, followed by crop rotation with *Tagetes patula* under field conditions (Sharma and Scolari, 1984). On sorghum, two vesicular–arbuscular mycorrhizal (VAM) fungi, *Glomus fasciculatum* and *G. mosseae*, were associated with *H. dihystera*. Where roots had > 50% VAM infection the nematode numbers were lower (Jain and Hasan, 1986). In tomato, soil amendment with leaves of neem (*Azadirachta indica*), datura (*Datura fastuosa*) and calotropis (*Calotropis procera*) reduced the population of *H. dihystera* and improved plant growth (Firoza and Maqbool, 1996a,b). Aqueous extracts of *Citrullus colocynthis* and *Ricinus communis* killed 30% of *H. dihystera* (Korayem and Koura, 1993).

36.15.2 Biological control

Several biological control agents (BCA) have been successfully tested against *H. dihystera*. The fungal BCA *Catenaria anguillulae* restricted the growth of *H. dihystera* endobiotically (Singh *et al.*, 1996). It is well known that *Pasteuria penetrans* parasitizes *H. dihystera* (Mankau and Imbriani, 1975).

36.15.3 Host-plant resistance

Various cultivars have been tested for resistance against *H. dihystera*. Maize cultivars 'Seneca 110' and 'Seneca Explorer' were reported to be resistant (Johnson, 1975). In India, 21 cultivars of sorghum were screened against *H. dihystera* and two cultivars, 'Etawah-2' and 'JHS-822', were found to be the least susceptible (Jain and Hasan, 1987). Similarly, banana cultivars 'Pat Kapuria Mendhi', 'Kathia' and 'Athiakol' were found to be resistant to *H. dihystera* (Sheela and Sundraraju, 2012).

36.15.4 Chemical control

Immersing seeds of watermelon and beans (*Phaseolus vulgaris*) in oxamyl gave a reduction in the *H. dihystera* population of 80–90% (Rodriguez-Kabana *et al.*, 1977). In Brazil, cocoa seedlings infested with *H. dihystera* that were treated with fenamiphos, fensulfothion, aldicarb or fosthietan grew better than untreated seedlings (Sharma and Ferraz, 1977). Soil application with oxamyl, cadusafos and carbofuran in a tomato field reduced *H. dihystera* infestation by 44.8–70% (Cadet, 1990). On cotton in the USA, fenamiphos was effective in controlling *H. dihystera* (Rodriguez-Kabana and King, 1978). Preplant dipping of pineapple planting material in oxamyl or fenamiphos distinctly suppressed *H. dihystera* populations (Milne *et al.*, 1977). Application of dazomet, dichlorvos, phosphamidon, dimethoate and parathion killed *H. dihystera* larvae in the soil (Saxena *et al.*, 1990).

36.15.5 Phytosanitary measures

See Chapter 16, section 115.

37 *Helicotylenchus multicinctus*

37.1 Common name

Spiral nematode

37.2 Scientific name

Helicotylenchus multicinctus (Cobb, 1893) Golden, 1956

37.3 Synonyms

Anguillulina multicincta (Cobb) Goodey, 1932
Helicotylenchus iperoiguensis (Carvalho) Andrássy, 1958
Rotylenchus multicinctus (Cobb) Filipjev, 1936
Rotylenchus iperoiguensis Carvalho, 1956
Tylenchus multicinctus Cobb, 1893
Tylenchorhyncus multicinctus (Cobb) Micoletzky, 1922

37.4 EPPO code

HELYMU

37.5 Diagnosis

A morphological and molecular diagnosis approach is strongly recommended for correct species identification of *Helicotylenchus multicinctus*.

37.5.1 Morphological diagnosis

This morphological description is based on Sher (1966). For measurements see Table 28.

Female: Body arcuate or C-shaped when heat-killed (Fig. 7.4A,B); distinct annules, lateral field with four incisures but not areolated. Lip region hemispherical, framework heavily sclerotized with 3–5 annules; cephalids usually unclear. Stylet strong, with basal knobs 5–6 µm wide, looking anteriorly concave or flattened (Fig. 7.4D). Oesophagus with ovoid metacorpus, and glandular bulb overlapping intestine ventrally, with subventral glands posterior to dorsal (Fig. 7.4E). Dorsal oesophageal gland outlet around 26–35% of total length of spear. Excretory pore near the oesophago-intestinal junction; hemizonid present, 0–3 body annules anterior to excretory pore. Reproductive structure comprises two symmetrical ovaries; occasionally posterior ovary may appear reduced, but is still functional. Spermathecae rounded, generally filled with sperms. Vulva a transverse slit, depressed on ventral side of body. Tail a little tapering (Fig. 7.4I-J), with anus marked by minor depression; tail terminus annulated, hemispherical, with dorsal margin more curved than ventral; 6–13 cuticular annules on ventral side. Phasmids present, punctate and 1–6 annules anterior to anus.
Male: Abundant in number, morphologically similar to females (Fig. 7.4C), except for reproductive characters. Testis single, monoarchic, outstretched. Bursa crenate, enveloping tail but not protruding beyond it in lateral view (Fig. 7.4G). Spicules cephalated weakly (Fig. 7.4H). Gubernaculum simple.

37.5.2 Molecular diagnosis

Detailed molecular characterization of *Helicotylenchus* spp. is discussed in section 36.5.2, above.

37.6 Geographical distribution

H. multicinctus is known to occur in almost all banana-growing countries and is concentrated in tropical and subtropical regions (Fig. 7.5). The nematode has been reported from the following countries (CABI-ISC, 2020).

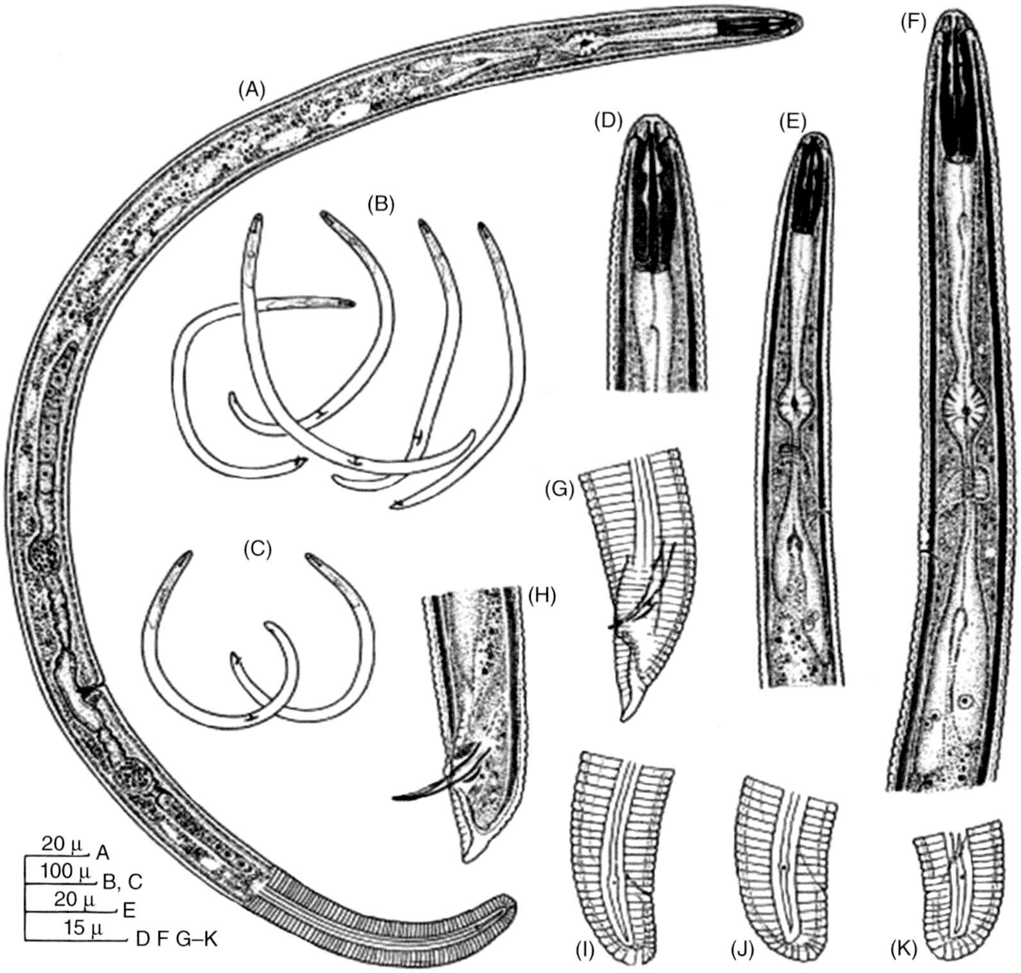

Fig. 7.4. Line drawing of *Helicotylenchus multicinctus*. Entire female (A); adult females (B); adult males (C); female head end (D); female oesophageal region (E); male oesophageal region (F); male tail ends (G,H); female tail ends (I–K). [Reproduced from Sher, 1966, CIH Descriptions, Set 2, No. 23; courtesy CAB International]

Africa: Angola, Benin, Burkina Faso, Burundi, Cameroon, Congo Democratic Republic, Côte d'Ivoire, Egypt, Ethiopia, Ghana, Kenya, Madagascar, Malawi, Mauritius, Mozambique, Nigeria, Réunion, Rwanda, São Tomé and Principe, Seychelles, Somalia, South Africa, Sudan, Tanzania, Tunisia, Uganda, West Africa, Zambia, Zimbabwe.

Asia: Bangladesh, Brunei, China, Cocos Islands, India, Iran, Israel, Jordan, Lebanon, Malaysia, Oman, Pakistan, Philippines, Sri Lanka, Thailand, Uzbekistan, Vietnam.

Europe: Belgium, Bulgaria, Cyprus, Czech Republic, Greece, Hungary, Italy, Lithuania, Moldova, Portugal, Romania, Russia, Slovakia, Slovenia, Spain, United Kingdom.

North America: Antigua and Barbuda, Barbados, Belize, Costa Rica, Cuba, Dominica, Dominican Republic, El Salvador, Grenada, Guadeloupe, Guatemala, Honduras, Jamaica, Martinique, Mexico, Montserrat, Nicaragua, Panama, Saint Lucia, Saint Vincent and the Grenadines, Trinidad and Tobago, USA.

Oceania: American Samoa, Australia, Cook Islands, Fiji, Kiribati, Niue, Papua New Guinea, Samoa, Solomon Islands, Tonga, Tuvalu, Vanuatu.

South America: Argentina, Bolivia, Brazil, Colombia, French Guiana, Peru, Suriname, Venezuela.

37.7 Hosts

Banana (*Musa* sp.) is the primary host of *H. multicinctus*, which is considered as an important nematode parasite of bananas in nearly every banana-cultivating region of the world (Moens *et al.*, 2006). Besides banana, it has also been detected on other plants (Gowen and Quénéhervé, 1990; Ploetz *et al.*, 2003). The extended host list of *H. multicinctus* is given in Table 30.

37.8 Symptoms

The damage symptoms produced by *H. multicinctus* are similar to those caused by other spiral and endoparasitic nematodes of

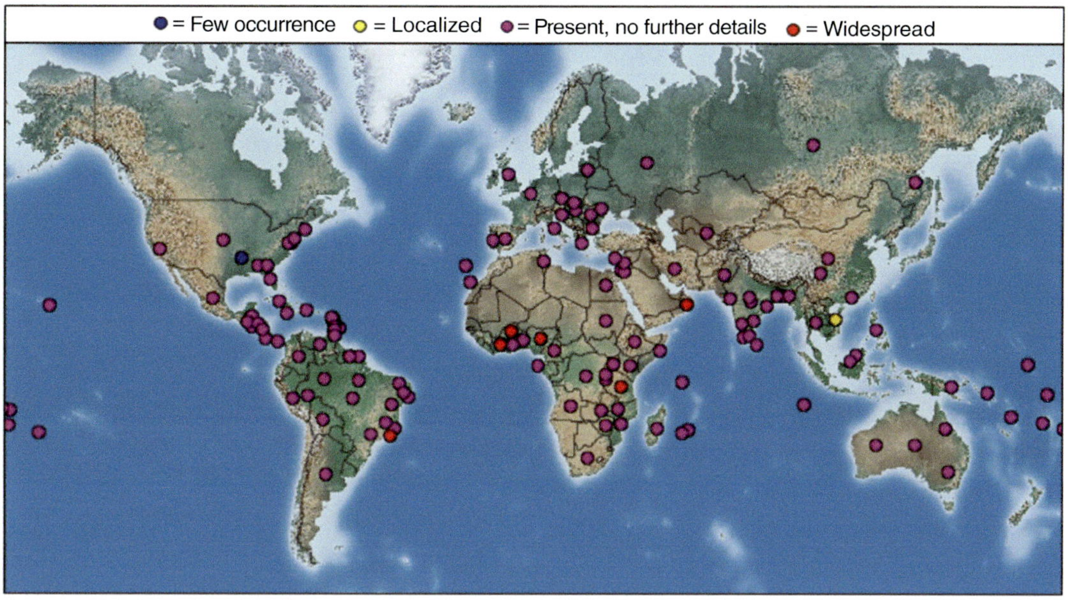

Fig. 7.5. Distribution map of *Helicotylenchus multicinctus*. [Adapted from CABI-ISC, 2020]

the family Pratylenchidae. The nematode infects the outer cortical part of the root, producing characteristic necrotic lesions (Fig. 7.6B), which are at first pale in colour and then turn reddish-brown to black (Fig. 7.6C). Infected plants have stunted growth with chlorotic foliage. In severe infestation the plant may topple (Fig. 7.6A). On banana, nematode infestation causes extensive root necrosis, dieback and disfunctional physiological activity leading to the death of the plant (Sheela and Sundraraju, 2012).

This nematode also occurs in the sugarcane root system along with other spiral nematodes, and feeds ectoparasitically, semi-endoparasitically or endoparasitically in the root cortex, causing distortion and collapse of the cells (Brathwaite, 1980). The primary root becomes blunt and malformed with fewer lateral roots (Rao and Swarup, 1975). Continuous feeding may also result in the formation of brownish-red lesions or a general discoloration as a consequence of secondary infection by bacteria and fungi (Khan, 1993).

37.9 Biology and life cycle

H. multicinctus feeds ecto- or endoparasitically on the outer cortical cells of plant roots and may complete its life cycle within the cortical layer of the roots, where all stages of the nematode except the eggs can be observed (Sheela and Sundraraju, 2012). Groups of 8–26 eggs, positioned parallel to the long cell axes, are often found in discoloured cortical tissues. The gravid females, held in tap water at 30°C, laid eggs within 24 h, which hatched in 48–51 h (Zuckerman and Strich-Harari, 1963). Gowen and Quénéhervé (1990) reported that the nematode takes 35–37 days to complete the life cycle from egg to adult at a soil temperature ranging from 22.6°C to 33°C. Feeding on host roots by second-stage larva was essential for completion of the life cycle. Stoyanov (1967) reported that the nematode can also survive for 4 months in the absence of a host plant.

37.10 Economic importance

H. multicinctus severely attacks banana and causes a serious decline in all varieties of banana throughout the tropics and subtropics. The nematode is a major problem in Jordan Valley, Israel (Minz *et al.*, 1960), Australia (Blake, 1961), Ivory Coast (Luc and Vilardebo, 1961), Uganda (Barekye *et al.*, 2000), São Tomé (Vovlas *et al.*, 1994), and Argentina, Cuba, South Africa, Cyprus, India and Pakistan (McSorley and Parrado, 1986; Mohilal and Pramodini, 2012). In subtropical regions, *H. multicinctus* may be the major nematode problem, particularly in regions where *Radopholus similis* is absent.

H. multicinctus was responsible for 34% yield loss, 56% loss in the number of fruits per bunch and delayed fruiting by 134 days (Vadivelu *et al.*, 1987). In India, the nematode is responsible for 20–50% loss to banana or even greater. Losses of up to 100% may occur in heavy infestations (Mohilal and Pramodini, 2012). In American Samoa, *H. multicinctus* was the most prevalent species and more than 23,000 nematodes per 100 g of roots were recorded (Brooks, 2004). *H. multicinctus* suppressed root growth by 13% and the bunch weight reduction was 4% (Moens *et al.*, 2006). In plants infected with *H. multicinctus*, Speijer and Fogain (1999) found reductions in bunch weight of 19–34%. A severe reduction (about 75%) in root system size at flower emergence was associated with nearly 20% reduction in corm and shoot growth (Blomme *et al.*, 2004). A density of 10,000 nematodes per 100 g of roots was considered damaging in Central America (Volkers and Gamboa, 1988).

Table 30. List of hosts of *Helicotylenchus multicinctus* (adapted from CABI-ISC, 2020).

Main hosts
Musa (banana)
Musa × paradisiaca (plantain)
Other hosts
Allium sativum (garlic)
Beta vulgaris var. *saccharifera* (sugar beet)
Brassica napus var. *napus* (rape)
Brassica oleracea var. *capitata* (cabbage)
Camellia sinensis (tea)
Carica papaya (papaw)
Citrullus lanatus (watermelon)
Citrus spp.
Citrus reticulata (mandarin)
Citrus sinensis (navel orange)
Coffea (coffee)
Cola acuminata (cola)
Cucumis melo (melon)
Daucus carota (carrot)
Dioscorea alata (white yam)
Elaeis guineensis (African oil palm)
Ficus carica (fig)
Hevea brasiliensis (rubber)
Ipomoea batata (sweet potato)*Lactuca sativa* (lettuce)
Mangifera indica (mango)
Manihot esculenta (cassava)
Oryza sativa (rice)
Persea americana (avocado)
Phaseolus vulgaris (common bean)
Pisum sativum (pea)
Saccharum officinarum (sugarcane)
Theobroma cacao (cocoa)
Vigna unguiculata (cowpea)
Vitis vinifera (grapevine)
Zea mays (maize)
Wild hosts
Cynodon dactylon (Bermuda grass)
Picea (spruces)
Pinus (pines)
Taxus (yew)

37.11 Movement and means of dispersal

See section 35.11, above.

37.12 Pest risk analysis and looming threat from introduction

Infected planting material is the main source of passive dissemination. The introduction of *H. multicinctus* into new areas usually takes place by suckers or rhizome of banana. For further details see section 36.12, above.

37.13 Invasiveness rating

Based on the consequences of introduction and pest risk analysis of *H. multicinctus*, the following invasiveness rating has been assigned.

Climate/host interaction	–	High (3)
Hosts range	–	High (3)
Dispersal potential	–	High (3)
Economic impact	–	High (3)
Environmental impact	–	Medium (2)
Post-entry distribution	–	High (–3)
Final rating (score)	–	Medium (11)

Based on the above invasiveness rating, *H. multicinctus* receives a score of 11 out of 15 and is rated as MEDIUM INVASIVE NEMATODE.

37.14 Region-wise status of invasiveness

Considering banana and other host crops and the worldwide distribution of *H. multicinctus*, the nematode is assigned medium invasiveness status in the following banana-producing countries, where it is still absent. There may be trade implications if this nematode is introduced and established in these countries.

APPPC: Bahrain, Bhutan, Cambodia, Indonesia, Japan, Kuwait, Maldives, Nepal, Singapore, South Korea, Syria, United Arab Emirates, Yemen
CAHFSA: Bahamas, Bermuda, British Virgin Islands, Cayman Islands, Haiti, Puerto Rico, US Virgin Islands
CAN: Ecuador
COSAVE: Guyana, Paraguay
EPPO: Turkey
IAPSC: Algeria, Cape Verde, Central African Republic, Comoros, Equatorial Guinea, Gabon, Guinea, Guinea-Bissau, Liberia, Mali, Morocco, Senegal, Swaziland, Togo.
PPPO: French Polynesia, Guam, Micronesia (Federated states of), New Caledonia, Tokelau, Wallis and Futuna Islands.

37.15 Management measures

The control measures suggested for *H. dihystera* may also prove effective against *H. multicinctus*, except for the use of resistant varieties. The banana hybrids H.74 (Matti × Pisang Lilin), H.94 (Matti × Tongat), H.100, H.106, H.109 (all Matti × Tongat) and cultivars 'Ney Vannan', 'Sirumalai', 'Peyan', 'Gross Michel', 'Karpuravalli', 'Robusta', 'Rasthali',' Kunnan', 'Kullan' and 'Vennettu Kunnan' are tolerant/resistant to *H. multicinctus* (Sheela and Sundraraju, 2012). For more management options, see section 36.15 of *H. dihystera*, above.

Fig. 7.6. Infestation symptoms caused by *Helicotylenchus multicinctus* (A-C), *Hoplolaimus columbus* (D) and *Hoplolaimus indicus* (E). Toppling and uprooting of banana (A); lesions on banana roots (B); damage to epidermis and outer root cortex (C); *Hoplolaimus columbus* inside the cotton root (D); *H. indicus* infestation on rice crop (E). [Photo courtesy A,B: Posadas *et al.*, 2016, Revista mexicana de fitopatología; C: SJ Collins and HF Hunter, DAFWA 2009; D: Source Nemaplex, H. Ferris; E: Author's own photograph]

38 *Helicotylenchus pseudorobustus*

38.1 Common name

Spiral nematode

38.2 Scientific name

Helicotylenchus pseudorobustus (Steiner, 1914) Golden, 1956

38.3 Synonyms

Helicotylenchus bradys Thorne and Malek, 1968
Helicotylenchus microlobus Perry in Perry *et al.*, 1956
Helicotylenchus phalerus Anderson, 1974
Tylenchus pseudorobustus Steiner, 1914
Tylenchus (*Tylenchorhynchus*) *pseudorobustus* Steiner, 1914
Tylenchorynchus robustus var. *pseudorobustus* (Steiner, 1914) Micoletzky, 1922

38.4 EPPO code

HELYPS

38.5 Diagnosis

38.5.1 Morphological diagnosis

This morphological description is based on Fortuner *et al.* (1984) and Subbotin *et al.* (2015). For measurements see Table 28.

Female: Body spiral in shape when heat killed (Fig. 7.7A); body annules distinct, lateral field with four incisures, not areolated, except in the caudal region of specimens of a few populations. Lip region hemispherical (Fig. 7.7E), with 4–5 annuli. Cephalids usually indistinct. Stylet developed, 26–28 μm long, with conspicuous, flattened basal knobs (Fig. 7.7F). Oesophagus glandular part

overlapping intestine ventro-laterally, with subventral glands slightly posterior or equal to dorsal. Dorsal oesophageal gland outlet 9–12 µm behind stylet base. Excretory pore at level of oesophago-intestinal junction; hemizonid 0–2 body annules anterior to it. Gonad double, symmetrical. Vulva a transverse slit, at 59–62% of body length. Spermathecae rounded, without sperms (Fig. 7.7B-D). Tail tapering at end, with dorsal profile more curved than ventral; terminus characterized by rounded ventral projection, mucro in some specimens (Fig. 7.7H-O), variable in length. Tail bears 7–12 annuli on ventral side. Phasmids 2–7 annules anterior to anus.
Male: Males are not known.

38.5.2 Molecular diagnosis

Shokoohi *et al.* (2018) characterized three *Helicotylenchus* species, *H. digonicus*, *H. pseudorobustus* and *H. vulgaris* from Iran, using 28S rDNA. Rybarczyk-Mydłowska *et al.* (2019) characterized three *Helicotylenchus* species, *H. canadensis*, *H. pseudorobustus* and *H. varicaudatus*, based on 28S rDNA and mtCOI sequence data. The following primers were used.

H. pseudorobustus (GenBank accession no. MG663105), approximate amplicon size 700 bp.

Forward primer (5′-3′) M3.5F: GGAGTGGiACARGiTGAAC
Reverse primer (5′-3′) M8aRa: GCAACiACATAATAAGWATCATG

H. pseudorobustus (MG663104) and *H. varicaudatus* (MG663116), amplicon size 450 bp. Reverse primer (5′-3′) M6.9R: ACCiACARTAAAiATATGATG

H. canadensis (MG663099, MG663100), *H. pseudorobustus* (MG663106-MG663112) and *H. varicaudatus* (MG663115), approximate amplicon size 600 bp.

Forward primer (5′-3′) M2Fb: ATTGGiGSTTTTGGTAATT
Reverse primer (5′-3′) RH1R: CCAACAATGAATATATGATG

H. canadensis (MG663098, MG663101) and *H. varicaudatus* (MG663114, MG663118), 350 bp.

Forward primer (5′-3′) RH2F: GGTGGAAGAATTAATTTYTG
Approximate amplicon size: 350

For 28S rDNA and mtCOI fragments, the PCR amplifications are performed using the following reagents and cycling parameters.

Reagent		**Final concentration**
PCR reaction	–	25 µl
Taq ready mix	–	12.5 µl
Primer	–	1 µl (5 µM each)
DNA template	–	1–3 µl
Annealing temperatures 28S rDNA	–	58–60°C
Annealing temperatures mtCOI	–	40–43°C
Cycling parameters		
Initial denaturation	–	96°C for 1 min
Number of cycles	–	25
Denaturation	–	96°C for 10 s
Annealing	–	50°C for 5 s
Extension	–	60°C for 105 s

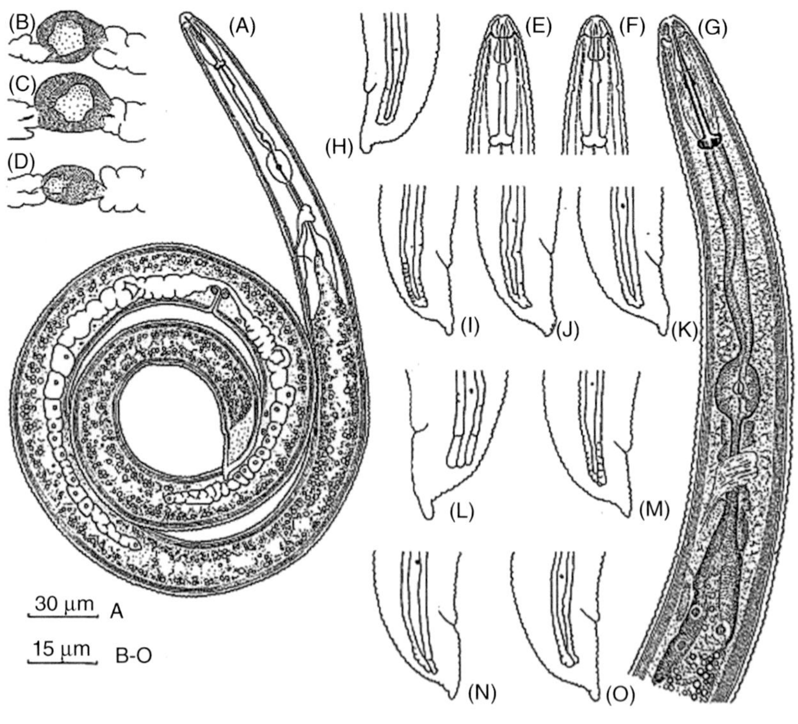

Fig. 7.7. Line drawing of *Helicotylenchus pseudorobustus*. Whole female (A); spermatheca (B–D); heads end (E–F); anterior end (G); tails end (H–O). [Courtesy Fortuner *et al.*, 1984; Revue Nématol]

Phylogenetics revealed that *H. canadensis* and *H. varicaudatus* are members of a clade that is a sister group to all other *Helicotylenchus* species; however, the closest known sister group to *H. canadensis* is *H. vulgaris* type A, which excludes *H. pseudorobustus*.

38.6 Geographical distribution

H. pseudorobustus is distributed worldwide (Fig. 7.8). The nematode is known to occur in the following countries (CABI-ISC, 2020).

Africa: Algeria, Benin, Cameroon, Congo Democratic Republic, Côte d'Ivoire, Egypt, Gambia, Ghana, Guinea, Kenya, Liberia, Madagascar, Malawi, Nigeria, South Africa, Sudan, Tanzania, Uganda, Zambia.
Asia: Afghanistan, China, India, Iran, Iraq, Israel, Jordan, Malaysia, Nepal, North Korea, Pakistan, Philippines, Saudi Arabia, Singapore, South Korea, Taiwan, Thailand, Turkey, Vietnam.
Europe: Austria, Belgium, Czech Republic, Finland, France, Germany, Greece, Italy, Netherlands, Poland, Portugal, Slovakia, Slovenia, Spain, Switzerland, United Kingdom.
North America: Belize, Canada, Cuba, Dominica, Dominican Republic, Martinique, Saint Lucia, Saint Vincent and the Grenadines, Trinidad and Tobago, USA.
Oceania: American Samoa, Australia, Fiji, New Zealand, Papua New Guinea.
South America: Argentina, Bolivia, Brazil, Chile, Colombia, French Guiana, Peru, Venezuela

38.7 Hosts

H. pseudorobustus is regarded as a root parasite of a wide range of plants, both herbaceous and arboreal (Table 31). Cereal crops such as oats (*Avena sativa*), barley (*Hordeum vulgare*), rye (*Secale cereale*), wheat (*Triticum*) and maize (*Zea mays*) are the main hosts of this nematode (Subbotin *et al.*, 2015).

38.8 Symptoms

H. pseudorobustus is a semi-endoparasite, living essentially on the root cortex of wild graminaceous plants. Symptoms similar to those of other spiral nematodes include dwarfing and stunting of leaves and root rot. The roots may also swell slightly. Damage is not easily detected unless the population density is high (more than 10,000 nematodes per plant).

38.9 Biology and life cycle

The nematode is usually found partially embedded in the cortical tissue, but there are records of females being completely inside the roots (Bridge, 1972). Vovlas and Inserra (1985) observed the anterior half of the nematode body inside maize roots, feeding on a single modified 'nurse cell'. Population of *H. pseudorobustus* was found to be higher in fine-textured soils with plenty of silt and clay (McSorley, 1978) and is significantly correlated with other edaphic factors, such as pH and organic matter. In native Iowa prairies, *H. pseudorobustus* mostly occurred in well-drained sites (Schmitt and Norton, 1972). Rãssner (1972) reported that the nematode preferred moist soil to dry soils, particularly the top 15 cm, but it survived throughout the year.

38.10 Economic importance

H. pseudorobustus is considered a weak pathogen although it is widely distributed throughout the world. It causes economic

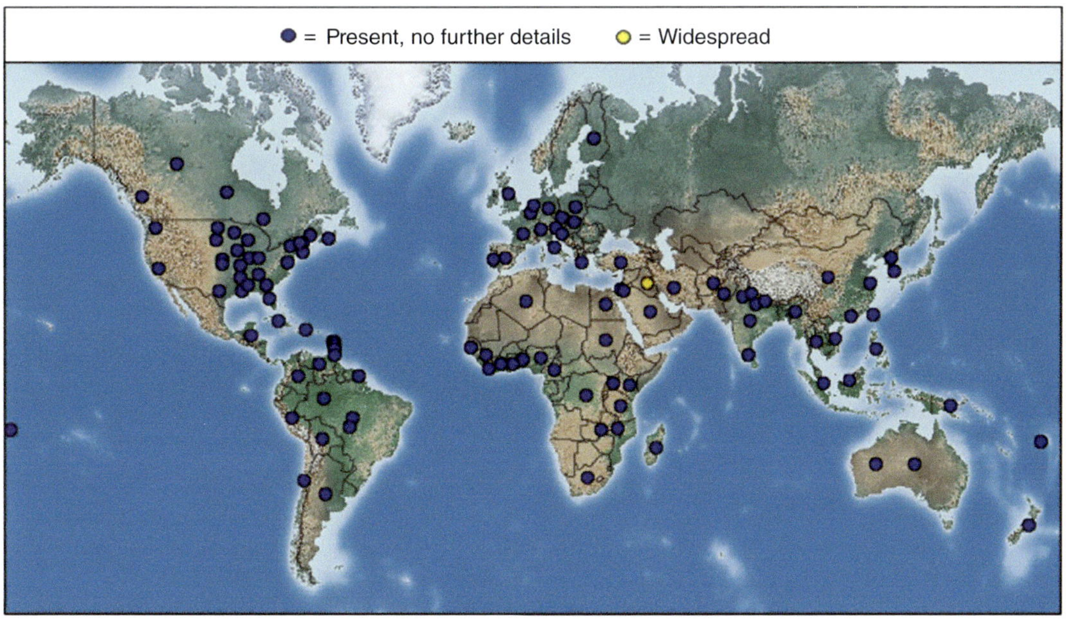

Fig. 7.8. Distribution map of *Helicotylenchus pseudorobustus*. [Adapted from CABI-ISC, 2020]

Table 31. List of hosts of *Helicotylenchus pseudorobustus* (adapted from CABI-CPC, 2010).

Main hosts
Avena sativa (oats)
Hordeum vulgare (barley)
Secale cereale (rye)
Triticum (wheat)
Zea mays (maize)
Other hosts
Acacia (wattles)
Actinidia chinensis (Chinese gooseberry)
Allium cepa (onion)
Allium sativum (garlic)
Bambusa vulgaris (common bamboo)
Carica papaya (papaw)
Cucumis melo (melon)
Daucus carota (carrot)
Dianthus caryophyllus (carnation)
Diospyros kaki (persimmon)
Elaeis guineensis (African oil palm)
Fragaria ananassa (strawberry)
Ginkgo
Gladiolus hybrids (sword lily)
Glycine max (soybean)
Gossypium (cotton)
Helianthus annuus (sunflower)
Juncus effusus (Soft rush)
Lactuca sativa (lettuce)
Lathyrus sativus (grasspea)
Linum usitatissimum (flax)
Maranta arundinacea (West Indian arrowroot)
Musa (banana)
Nicotiana tabacum (tobacco)
Oryza sativa (rice)
Pastinaca sativa (parsnip)
Persea americana (avocado)
Phaseolus vulgaris (common bean)
Pinus (pines)
Pisum sativum (pea)
Populus (poplars)
Prunus persica (peach)
Punica granatum (pomegranate)
Rosa (roses)
Saccharum officinarum (sugarcane)
Solanum lycopersicum (tomato)
Solanum melongena (eggplant)
Solanum tuberosum (potato)
Sorghum bicolor (sorghum)
Trifolium pratense (purple clover)
Vigna unguiculata (cowpea)
Vitis vinifera (grapevine)
Wild hosts
Epilobium angustifolium (rosebay willowherb)
Poaceae (grasses)
Pteridophyta (ferns)

damage only at a very high population level. For details see the other *Helicotylenchus* species, section 36.10 above.

38.11 Movement and means of dispersal

See section 35.11, above.

38.12 Pest risk analysis and looming threat from introduction

See section 36.12, above.

38.13 Invasiveness rating

Considering the consequences of the introduction and pest risk analysis of *H. pseudorobustus*, the following invasiveness rating has been assigned.

Climate/host interaction	–	High (3)
Hosts range	–	High (3)
Dispersal potential	–	High (3)
Economic impact	–	Medium (2)
Environmental impact	–	Low (1)
Post-entry distribution	–	High (–3)
Final rating (score)	–	Medium (9)

Based on the above invasiveness rating, *H. pseudorobustus* receives a score of 9 out of 15 and is rated as MEDIUM INVASIVE NEMATODE.

38.14 Region-wise status of invasiveness

Due to its wide host range and distribution, *H. pseudorobustus* is assigned medium invasiveness status in all the countries where it is still absent.

38.15 Management measures

For detail management options, see section 36.15 of *H. dihystera*, above.

Hoplolaimus Spp.

The genus *Hoplolaimus* was created by Daday in 1905. An identification key of 29 valid species of the genus is given by Handoo and Golden (1992). At present, the following 31 species have been described in the genus, with *Hoplolaimus tylenchiformis* as type species (Siddiqi, 2000; Handoo and Golden, 1992; Decraemer and Geraert, 2013).

Hoplolaimus abelmoschi Tandon and Singh, 1973
H. aegypti Shafiee and Koura, 1969
H. aorolaimoides Siddiqi, 1972

H. bachlongviensis Nguyen, Bui and Trinh, 2015
H. californicus Sher, 1963
H. capensis Van den Berg and Heyns, 1970
H. casparus Van den Berg and Heyns, 1970
H. cephalus Mulk and Jairajpuri, 1976
H. chambus Jairajpuri and Baqri, 1973
H. citri Khan and Khan, 1985
H. clarissimus Fortuner, 1973
H. columbus Sher, 1963
H. concaudajuvencus Golden and Minton, 1970
H. dimorphicus Mulk and Jairajpuri, 1976
H. dubius Chaturvedi, Singh and Khera,1979
H. galeatus (Cobb, 1913) Thorne, 1935
H. imphalensis Khan and Khan, 1985
H. indicus Sher, 1963
H. jalalabadiensis Khan and Khan, 1985
H. magnistylus Robbins, 1982
H. pararobustus (Schuurmans Stekhoven and Teunissen, 1938) Sher, 1963
H. puertoricensis Ramirez, 1964
H. sacchari (Shamsi, 1979) Luc, 1981
H. seinhorsti Luc, 1958
H. seshadrii Mulk and Jairajpuri, 1976
H. sheri Suryawanshi, 1971
H. smokyensis Ma *et al.*, 2019
H. singhi Das and Shivaswamy, 1977
H. stephanus Sher, 1963
H. tabacum Firoza and Maqbool, 1990
H. tylenchiformis (von Daday, 1905) Lewis and Fassuliotis, 1982

Hoplolaimus spp. (lance nematodes) are vermiform, comparatively long and robust with a prominent cephalic region and well-developed massive stylets (Fortuner, 1991) that parasitize on the roots of a wide range of monocotyl and dicotyl plants. Although all species are economically important, in this chapter *Hoplolaimus columbus*, *H. galeatus*, *H. indicus*, *H. pararobustus* and *H. seinhorsti* are considered for their invasiveness, as they can cause severe damage to agronomic and horticultural crops, including banana (*Musa*), cotton (*Gossypium hirsutum*), maize (*Zea mays*), rice (*Oryza sativa*), soybean (*Glycine max*) and turf grasses (Noe, 1993; Ma *et al.*, 2011; Decraemer and Geraert, 2013).

Systematic position of *Hoplolaimus* species

	Classical (after Siddiqi, 2000)	Molecular (after Decraemer and Hunt, 2013)
Phylum	Nematoda	Nematoda
Class	Secernentea	Chromadorea
Order	Tylenchida	Rhabditida
Suborder	Hoplolaimina	Tylenchina
Infra-order	–	Tylenchomorpha
Superfamily	Holpolaimoidea	Tylenchoidea
Family	Hoplolaimidae	Hoplolaimidae
Subfamily	Hoplolaiminae	Hoplolaiminae
Genus	*Hoplolaimus*	*Hoplolaimus*

39 *Hoplolaimus columbus*

39.1 Common name

Columbia lance nematode

39.2 Scientific name

Hoplolaimus columbus Sher, 1963

39.3. Synonym

Basirolaimus columbus Siddiqi, 1986

39.4 EPPO code

HOLLCO

39.5 Diagnosis

Morphologically *Hoplolaimus columbus* is similar to other species of the genus. Hence for precise species identification, both morphological and molecular diagnosis approaches are recommended.

39.5.1 Morphological diagnosis

The morphological and morphometric descriptions from Fassuliotis (1976) are given below. For measurements, see Table 32.

FEMALE: Lip region offset, frequently with three annules, often fourth annule observed on one side of lip region. Basal plate has six arms. Basal annule of cephalic region with 10–15 longitudinal lines. Stylet robust, knobs projecting two processes anteriorly (Fig. 7.9A). Oesophogeal glands dorsally overlap intestine (Fig. 7.9C). Lateral field with only one indistinct longitudinal line. Excretory pore located behind level of oesophago-intestinal valve. Hemizonid present, comprising 2–5 annules, posterior to excretory pore. Hemizonian not observed. Intestine extended into tail, overlaps rectum partly. Tail bluntly rounded to hemispherical, with 16–22 annules (Fig. 7.9B). Phasmid present, anterior 38% from the head end on right side; posterior 81% from the head end on left side. Ovaries outstretched, spermatheca absent. Diovarial reproductive system fully developed in adult and occupies approximately 61% of body length. Eggs laid individually with stalk up to 15 µm long.
MALE: Very rare. Morphologically similar to female except for sexual dimorphism. Longitudinal incisure not present on body. Spicules robust, 37–53 µm long, slightly arcuate with very thin velum, observed in extended spicules (Fig. 7.9F). Telamon discrete, lying between spicules. Gubernaculum trough-like, with distinct titillate. Bursa extends around tail terminus (Fig. 7.9F).

Table 32. Morphometric data for *Hoplolaimus* species.

Morphometric characters		Range, numerous populations				
		H. columbus (after Sher, 1963)[a]	*H. galeatus* (after Sher, 1963)	*H. indicus* (after Khan and Chawla, 1975)	*H. pararobustus* (after Sher, 1963)	*H. seinhorsti* (after Luc, 1958)[b]
Females	Length (µm)	1260–1800	1240–1940	1020–1280	940–1370	1060–1560
	Stylet (µm)	40–48	43–52	33–43	38–44	40–49
	a	30–38	25–34	22–29	26.7–34.6	25–34
	b	9.1–12.4	7.6–10.8	8.4–9.0	7.9–10.0	8.8–10.1
	b'	6.3–9.7	6.0–8.8	7.8–8.1	–	–
	c	39–57	42–82	51–67	53–91	6.0–10.1
	c'		–	0.66–0.68	0.40–0.75	38–74
	V	51–60	52–60	50–53	54–61	52–60
	m		–	–	–	–
	o	9–13	10–17	–	–	–
	Lateral incisures	1	4	2 (1–3)	1 (2–3)	1
	Oesophageal gland nuclei	6	3	6	3	6
	Anterior phasmid	34–47% of body	30–46% of body	–	–	31–44% of body
	Posterior phasmid	80–90% of body	75–88% of body	–	–	74–83% of body
	Tail annules	16–22	10–16	14 (18–22)	7–15	10–15
	Lips annuli number	3	5	3–4	4–5	4
Males	Length (µm)	1150–1400	1050–1560	940–1090	940–1150	–
	Stylet (µm)	40.2–43.7	40–48	30–38	38.0–41.5	–
	a	25.9–39.2	25–32	31–36	25–36	–
	b	9.58–12.18	8.3–10.3	8.9–9.9	7.0–8.5	–
	b'	–	6.0–8.4	7.2–8.4	–	–
	c	26.8–33.1	28–40	31–36	29.7–31.5	–
	c'	–	–	–	1.5–1.7	–
	m	–		–	–	–
	o	4.8–5.2	10–17	–	–	–
	Anterior phasmid	38% of body	29–45% of body	–	–	–
	Posterior phasmid	82% of body	75–89% of body	–	–	–
	Tail annules	–	–	–	–	–
	Spicules (µm)	36.6–52.5	40–52	38–48	44–51	–
	Gubernaculum (µm)	19.5–23.2	20–28	12–20	21.0–22.5	–

[a]Male rare, description from Fassuliotis, 1974; [b]male not known.

39.5.2 Biochemical and molecular diagnosis

Lawson *et al.* (1984) used isoelectric focusing of proteins for taxonomic identification of *H. columbus*. Numerous molecular approaches have been applied to characterize *Hoplolaimus* spp., including *H. columbus* (Bae *et al.*, 2008, 2009a,b; Holguin *et al.*, 2016). Bae *et al.* (2008) used first internal transcribed spacer (ITS1) and the D2–D3 expansion segments of 28S ribosomal DNA for molecular identification of the lance nematode, *Hoplolaimus* spp. (for protocol, see Chapter 4, section 15.5.2). Further, they applied duplex PCR, multiplex PCR and PCR–RFLP approaches using DNA sequences of the ITS1-5.8S-ITS2 ribosomal DNA (rDNA) and identified *H. columbus*, *H. galeatus*, *H. concaudajuvencus*, *H. magnistylus*, *H. seinhorsti* and three undescribed *Hoplolaimus* species (Bae *et al.*, 2009a). They also studied the genetic variability of *H. columbus* using PCR–RFLP analysis of nuclear rDNA ITS regions (Bae *et al.*, 2009b).

39.6 Geographical distribution

H. columbus has limited distribution and has only been reported from Egypt, India, Pakistan, Trinidad and Tobago and the USA (CABI-ISC, 2020). It is widespread in different states of the USA and one of the most common lance nematodes (Holguin *et al.*, 2016).

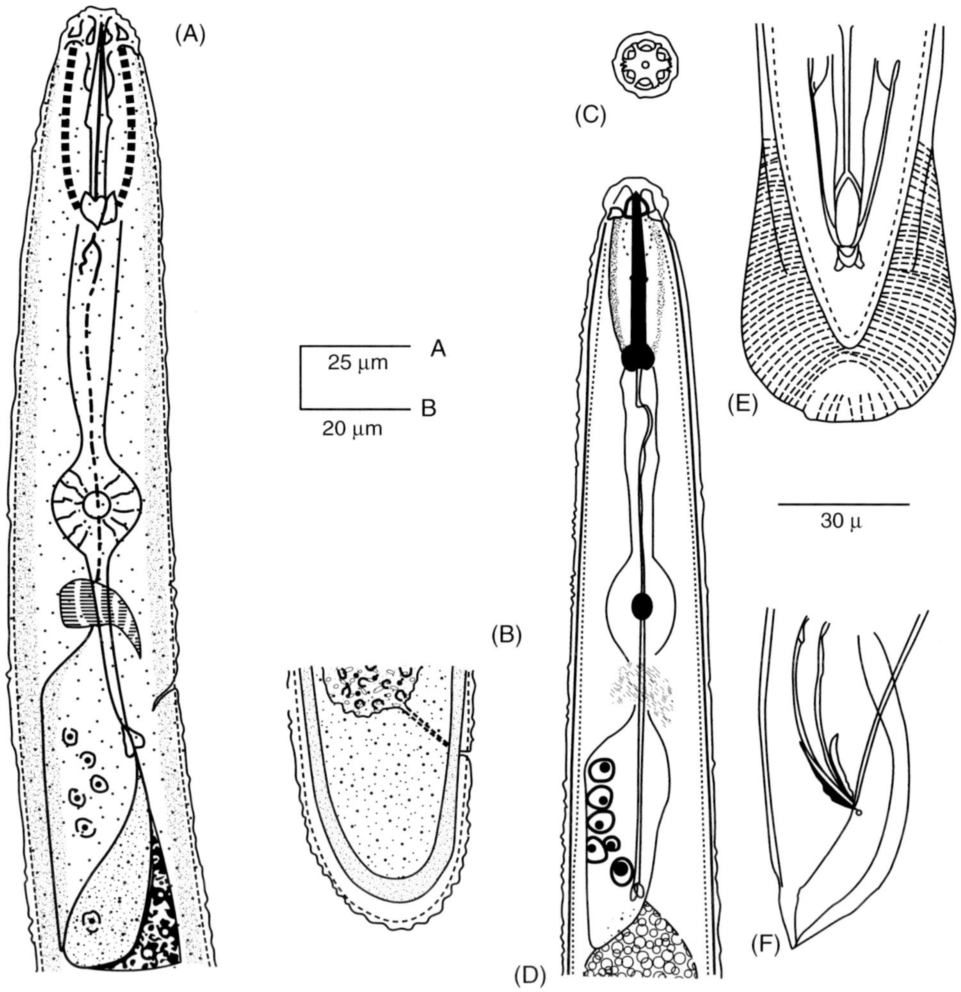

Fig. 7.9. Line drawing of *Hoplolaimus columbus*. Anterior region of female (A); tail end of female (B); cross-section through basal annule of lip region of male (C); anterior region of male (D); ventral view of posterior end of male (E); lateral view of posterior end of male (F). [Adapted from Fassuliotis, 1974; courtesy Journal of Nematology]

39.7 Hosts

H. columbus was first described in the USA from a soybean field (Sher, 1963). This nematode is primarily associated with the infestation of cotton and soybean (Mueller and Sanders, 1987). Maize, sugarcane and wheat are also reported to be hosts of *H. columbus*. Other hosts include banana, cucumber, cowpea, cantaloupe, lima bean, okra, millet, snap bean, watermelon, bermudagrass, St Augustine grass, zoysia, etc. Some weeds are also recorded as wild hosts, such as henbit, nutsedge, crimson clover, pigweed, sicklepod, and showy crotalaria (Koenning *et al.*, 2004).

39.8 Symptoms

H. columbus is ecto- and endoparasitic on cotton and soybean and causes severe damage to the cortical parenchyma and to the endodermal–vascular region of the roots of both plants, resulting in the hosts with stunted and chlorotic growth (Fassuliotis, 1976; Lewis *et al.*, 1976). The nematode may also cause lesions on the surface and discolour cortical cells in the maturation zone of the root. In some soybean varieties, *H. columbus* penetrates endodermis and pericycle and reaches up to phloem, resulting in stunted growth, leaf yellowing, wilting and fewer pods on soybeans. On cotton, it feeds as a migratory endoparasite, penetrating the cortex but not the endodermis (Fig. 7.6E); it produces symptoms similar to those on soybean and the affected plants exhibit nutritional deficiency with purple leaves. *H. columbus* also attacks sugarcane, penetrates the roots, damages parenchyma and instigates cell necrosis that results in coarse and exhausted root systems (Astudillo and Birchfield, 1980).

39.9 Biology and life cycle

All stages are infective. Reproduction in *H. columbus* is mainly by parthenogenesis, as males are very rare. The life cycle on cotton and soybean is completed in 45–49 days. The females lay eggs individually, not in masses (Noe, 1993). However, 20–50 eggs may be placed in a localized area. *H. columbus* can tolerate osmotic stress and extremely high desiccation. It can survive under adverse conditions and has been reported to survive under dry soil for up to 5 years (Ferris, 2019).

39.10 Similarities to other species

H. columbus is found associated with other lance nematode species under field conditions. *H. seinhosti* is the most closely related species and it can be distinguished by the lower position of the excretory pore and the noticeable overlap of the intestine over the rectum. *H. stephanus* is also a dominant species in soybean, but the two species are never recorded together (Holguin *et al.*, 2016).

39.11 Economic importance

H. columbus causes economic damage to agriculturally important crops such as cotton, soybean, sugarcane, etc. Losses on cotton are typically 10–25% and losses on soybean can be as high as 70% (Noe, 1993). In sugarcane, *H. columbus* infestation results in depleted root systems (Astudillo and Birchfield, 1980). Maximum probable yield reductions of 18% in cotton and 48% in soybean were forecast because of function damage developed by *H. columbus* (Noe, 1993). The soil populations of the nematode in a soybean field increased up to 4000 per 100 ml of soil. The economic threshold for cotton has been calculated as Pi 60/100 cm^3 of soil (Noe, 1993).

39.12 Movement and means of dispersal

See section 36.11, above.

39.13 Pest risk analysis and looming threat from introduction

Considering the multi-host range of *H. columbus*, its damage potential and occurrence in different ecological zones (Africa, Asia, America and Caribbean region), a complete pest risk analysis for this nematode should be instigated. Due to its limited distribution, preventive measures should be taken to avoid introduction into new areas. The nematode is endo- and ectoparasitic and can easily spread through nursery stock, infested soil and contaminated machinery.

39.14 Invasiveness rating

Considering the consequences of introduction and pest risk analysis of *H. columbus*, the following invasiveness rating has been assigned.

Climate/host interaction	–	Medium (2)
Hosts range	–	Medium (2)
Dispersal potential	–	Medium (2)
Economic impact	–	High (3)
Environmental impact	–	Low (1)
Post-entry distribution	–	Low (–1)
Final rating (score)	–	Medium (9)

Based on the above invasiveness rating, *H. columbus* receives a score of 9 out of 15 and is rated as MEDIUM INVASIVE NEMATODE.

39.15 Region-wise status of invasiveness

H. columbus attained medium invasiveness status in major cotton- and soybean-producing countries. However, due to its wide host range, it may be considered as a invasive nematode in all the areas or countries where it is still absent.

39.16 Management measures

Crop rotation with non-host crops or the use of cover crops is an effective management strategy. The use of tolerant or resistant cultivars is a credible management option. Chemical controls have proved the most effective means of nematode management. However, in the case of lance nematode, chemicals are ineffective due to its endoparasitic nature. Fumigant nematicides are quite effective, as they can penetrate the plant system and may result in a fivefold increase in yield of cotton (Ferris, 2019). For phytosanitary measures, see Chapter 16, section 115.

40 *Hoplolaimus galeatus*

40.1 Common name

Lance nematode

40.2 Scientific name

Hoplolaimus galeatus (Cobb, 1913) Thorne, 1935

40.3 Synonyms

Hoplolaimus coronatus Cobb, 1923
Nemonchus galeatus Cobb, 1913

40.4 EPPO code

HOLLGA

40.5 Diagnosis

Morphologically *Hoplolaimus galeatus* is similar to other species of the genus. Hence, for accurate species identification of *H. galeatus*, the molecular diagnosis approach aided by morphometric data is recommended.

40.5.1 Morphological diagnosis

This morphological and morphometric descriptions are from Orton Williams (1973b). For measurements, see Table 32.

FEMALE: Lip region offset, generally with five annules, forming low cone usually with flattened sides (Fig. 7.10D). Basal annule with around 32–36 longitudinal striations (Fig. 7.10C). Cephalic framework massive. Stylet strong, knobs with processes anteriorly directed. Lateral field with four incisures, usually areolated but sometimes partially areolated. Oesophageal gland with three nuclei (Fig. 7.10D). Hemizonid around two annules long just anterior to excretory pore which is close to level of oesophago-intestinal valve. Hemizonion around five annules posterior to excretory pore. Caudalid present, about eight annules anterior to anus. Ovaries outstretched, spermathecae round to ovoid. Epiptygma single or double, generally conspicuous. Intestine usually extending into tail and overlapping rectum. Tail rounded, comprising 10–16 annules (Fig. 7.10F).

MALE: Similar to female in morphology. Lip region raised but less conoid than female, with hemispherical and convex sides (Fig. 7.10G). Spicule slightly arcuate (Fig. 7.10H). Gubernaculum with titillate sometimes bent distally in a few specimens.

40.5.2 Molecular diagnosis

Holguin *et al.* (2015) used molecular diagnosis through sequences of the cytochrome c oxidase subunit 1 (COX1) and the internal transcribed spacer (ITS1) on 23 populations of *Hoplolaimus* to identify *H. galeatus* and morphologically similar species. The DNA was extracted from individual nematodes using the worm

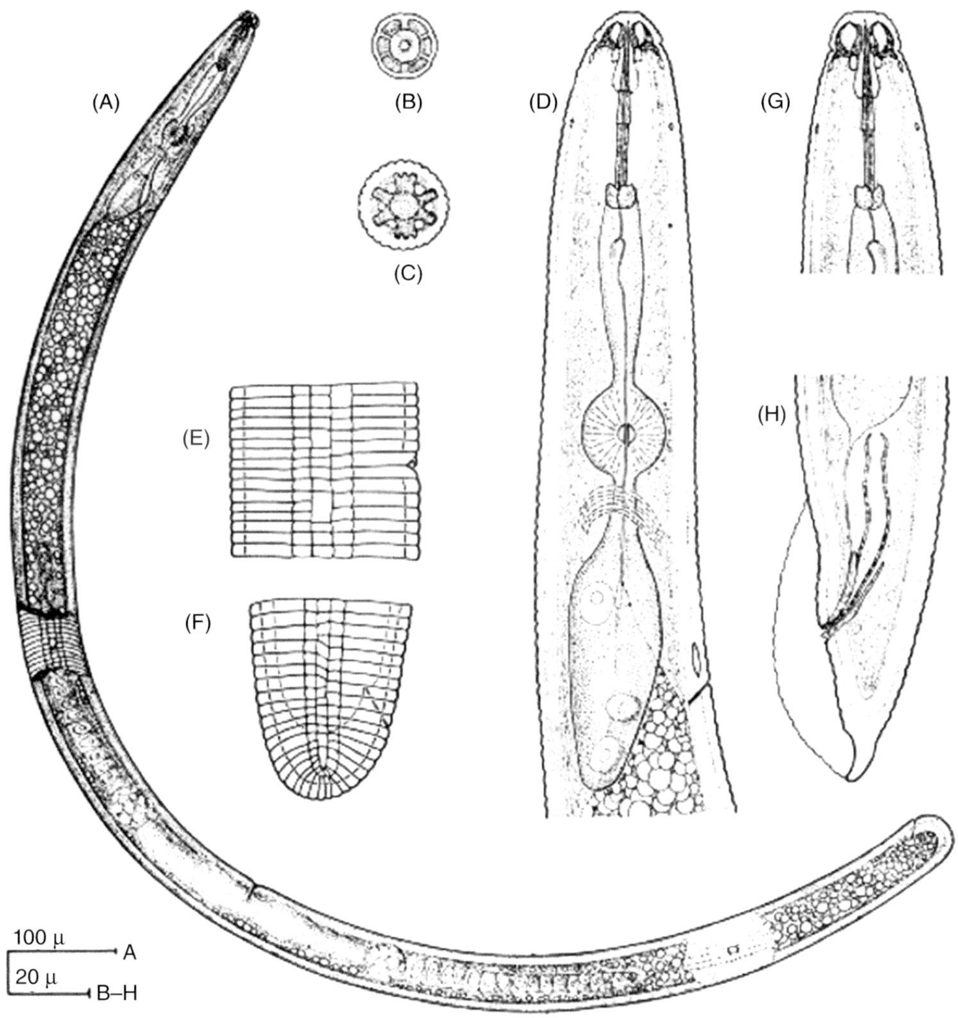

Fig. 7.10. Line drawing of *Hoplolaimus galeatus*. Entire female (A); female, face view (B); female cross-section through basal annule of cephalic region (C); female anterior region (D); female surface view at vulva (E); female posterior region (F); male anterior region (G); male tail (H). [Adapted from Orton Williams, 1973b, CIH Descriptions set 2, CABI; After Sher, 1961, courtesy Nematologica]

lysis buffer (see Chapter 15, section 110.1.1). The PCR amplification and sequencing were performed using the following primers.

For ITS region newly designed primers

F1-F (5′-CTGACGACCAGTTAGGCGTT-3′)
F1-R (5′-CGTGCCAAAGGATGTCACTC-3′)
F5-F (5′-CTTGATTGGAAAGCGCCCAC-3′)
F5-R (5′-ATGTCACTCCAATGGCGCA-3′)

For the COI gene and DNA barcoding

Forward primer JB3 (5′ TTTTTTGGGCATCCTGAGGTT TAT 3′) (Hu *et al.* 2002)

Reverse primer JB5 (5′ AGCACCTAAACTTAAAACATAA TGAAA 3′) (Derycke *et al.*, 2005)

For all primers, PCRs were performed using the following reagents and cycling parameters.

Reagent		Final concentration
PCR reaction	–	20 µl
Ready Mix Taq PCR	–	10 µl
$MgCl_2$	–	20 mmol/l
Tris–HCl (pH 8.3)	–	100 mmol/l
KCl	–	3 mmol/l
Gelatin	–	0.002%,
dNTP mixture	–	0.4 mmol/l
Taq DNA polymerase	–	0.06 units/ml
Each primer	–	0.5 µl
DNA template	–	1 µl
Cycling parameters (for the ITS marker)		
Initial denaturation	–	95°C for 3 min
Number of cycles	–	33
Denaturation	–	95°C for 45 s
Annealing	–	59°C for 75 s
Extension	–	72°C for 2 min
Final extension	–	72°C for 10 min

Cycling parameters for COI portion were the same except annealing at 50°C for 1 min. The PCR products were purified and were sequenced in both directions. The newly obtained sequences were aligned using ClustalW and submitted to the GenBank database under accession numbers KP303595–KP303681 for ITS region gene and KP230554–KP230657 for COI mtDNA gene. The study identified five morphospecies: *H. galeatus*, *H. magnistylus*, *H. concaudajuvencus*, *H. stephanus* and an undescribed species.

40.6 Geographical distribution

H. galeatus is reported from Canada, Egypt, India, Tanzania, Sumatra, the USA and Central and South America. In the USA, the nematode is known to occur along the east coast from New England to Florida and the Mississippi River basin from Minnesota and Wisconsin to Louisiana, as well as in Colorado, California and Texas (Crow and Brammer, 2015).

40.7 Hosts

H. galeatus is known as a primary pest of St Augustine grass, turfgrasses and lawn grasses (Henn and Dunn, 1989; Crow and Brammer, 2015). In addition, many crops, grasses and trees are parasitized, including alfalfa, bananas, beans, cabbage, cotton, clover, maize, peas, groundnuts, sweet potatoes, chrysanthemums, wheat, apple, pine, oak, etc. (Crow and Brammer, 2015; Holguin *et al.*, 2015). Recently Abd-Elgawad *et al.* (2018) also reported *H. galeatus* from forage legume fields in Egypt.

40.8 Symptoms

Damage from *H. galeatus* may show as yellow patches of grass with dying or unthrifty growth (Fig. 7.11A,B). Similar symptoms are produced by nutrient deficiency or drought. *H. galeatus* may feed on the roots ecto- or endo-parasitically, destroying the root system. Close observation of the roots of a nematode-infested lawn reveals exhaustive damage to the root system and often outer layers of the dead roots slough off. The damaged root appears dark and missing small feeder roots, with dead root tips (Fig. 7.11C). Destruction of the root system results in yellowing or drying of patches of grass (Crow and Brammer, 2015).

40.9 Biology and life cycle

The life cycle and biology of *H. galeatus* are similar to those of other lance nematodes. It acts as an ectoparasite as well as an endoparasite, but more frequently it is endoparasitic on cotton, alfalfa, St Augustine grass and Bermuda grass (Henn and Dunn, 1989). Reproduction of *H. galeatus* on seven St Augustine grass cultivars revealed Pf/Pi 5.7–8.6, indicating their high reproduction potential on suitable host grasses (Henn and Dunn, 1989).

40.10 Economic importance

H. galeatus is of concern for causing damage in turf and lawn grasses. The nematode is highly pathogenic to St Augustine grass and Bermuda grass putting greens. However, exact damage due to lance nematodes is difficult to quantify. In the USA, *H. galeatus* is next to *Belonolaimus longicaudatus* as the worst nematode pest of turfgrass. Both nematodes are major pests of home lawns and golf courses. Lance nematodes have a much larger distribution and are much more difficult to control (Crow and Brammer, 2015).

40.11 Movement and means of dispersal

See section 35.11, above.

40.12 Invasiveness rating

Considering the consequences of the introduction and pest risk analysis of *H. galeatus*, the following invasiveness rating has been assigned.

Fig. 7.11. Symptoms of *Hoplolaimus galeatus* (A–C), *Scutellonema brachyurus* (D–E) and *Scutellonema bradys* (F–G). *H. galeatus* infesting bermudagrass (A); patches of yellowing, dying or unthrifty growth on St. Augustine grass (B); healthy St. Augustine grass roots on the left and nematode-damaged roots (C); healthy and root lesioned soybean plants with darkening of roots inoculated with *S. brachyurus* (D); *S. brachyurus* infected soybean roots stained with fuchsin acid (E); damage yam tuber showing dry rot and cracking caused by *S. bradys* (F); dry rot caused by *S. bradys* on yam tubers (G). [Courtesy A-C: R.A. Dunn, University of Florida, Source: entnemdept.ufl.edu/creatures/nematode/lance_nematode.htm; D-F: Machado *et al.*, 2019, PlosOne; F-G: John Bridge]

Climate/host interaction	–	Medium (2)
Hosts range	–	High (3)
Dispersal potential	–	Medium (2)
Economic impact	–	Medium (2)
Environmental impact	–	Low (1)
Post-entry distribution	–	Low (–1)
Final rating (score)	–	Medium (9)

Based on the above invasiveness rating, *H. galeatus* receives a score of 9 out of 15 and is rated as MEDIUM INVASIVE NEMATODE.

40.13 Region-wise status of invasiveness

H. galeatus attains medium invasiveness status in its host countries. Due to its wide host range, it may be considered as a medium invasive nematode in all the areas or countries where it is still absent.

40.14 Management measures

Careful turf management may help minimize the turf damage caused by this nematode. Healthy roots are more resistant to nematode damage and deep watering is recommended to encourage the growth of deep roots. Use of balanced fertilizer is also helpful, but too much nitrogen can increase the growth of roots and make them more susceptible to *H. galeatus*. Mowing at the recommended height rather than too short can prevent the problem of there being insufficient foliage to provide food for the root system (Crow and Brammer, 2015). For phytosanitary measures, see Chapter 16, section 115.

41 *Hoplolaimus indicus*

41.1 Common name

Lance nematode

41.2 Scientific name

Hoplolaimus indicus Sher, 1963

41.3 Synonyms

Basirolaimus arachidis (Maharaju and Das, 1982) Siddiqi, 1986
Basirolaimus indicus (Sher, 1963) Shamsi, 1979
Hoplolaimus arachidis Maharaju and Das, 1982

41.4 EPPO code

HOLLIN

41.5 Diagnosis

41.5.1 Morphological diagnosis

The morphological and morphometric descriptions below are adapted from Khan and Chawla (1975). For measurements, see Table 32.

Female: Body elongate-cylindroid, slightly C-shaped when heat relaxed (Fig. 7.12C). Lip region hemispherical with 3–4 distinct annules. Basal annule with 6–12 longitudinal striations. Cephalic framework strongly sclerotized. Stylet massive, basal knobs with one to three forwardly directed tooth-like processes. Dorsal oesophageal gland opening 3–4 μm behind basal knobs. Median oesophageal bulb spheroid, oesophageal glands overlapping intestine dorsally, dorsal gland having four nuclei, thus giving total complement of six. Excretory secretory pore situated above level of oesophago-intestinal valve and six (4–7) annules anterior to hemizonid. Scutella not opposite one another, one being anterior and one posterior to vulva. Lateral field with single, occasionally two or three, weakly developed and indistinct lines (Fig. 7.12E). Epiptygma indistinct, single or double. Vulva median with amphididelphic genital tracts. Spermathecae present and usually with sperms. Intestine partially overlapping rectum. Tail bluntly rounded with 8–13 annules (Fig. 7.12G).

Male: Similar to female in general respects. Spicules cephalated and arcuate. Gubernaculum with titillae. Tail conoid, bursa enveloped tail tip (Fig. 7.12A).

41.5.2 Molecular diagnosis

Mansourabad *et al.* (2016) characterized *H. indicus* and *H. seinhorsti* from Iran using D2–D3 expansion segments of 28S rRNA gene. The PCR amplifications were performed using the primers and protocol described by Tanha Maafi *et al.* (2003) and Subbotin *et al.* (2007; see Chapter 4, section 15.5.2). The phylogenetic analyses revealed that the Iranian populations of *H. indicus* are located in the same clade with *H. seinhorsti* and *H. columbus*.

41.6 Geographical distribution

H. indicus has a restricted distribution and is only reported from Bangladesh (Mian and Tsuno, 1988), Canada (Anderson, 1983), China (Chang, 1993), India (Khan and Chawla, 1975), Iran (Tanha Maafi and Kheiri, 1993) and Pakistan (Maqbool and Ghazala, 1988). In India, the nematode is widespread and known to occur in almost every state (Rahman, 2010).

41.7 Hosts

The major host of *H. indicus* is rice. Other hosts include field crops, vegetable crops, fruit crops, forest trees, ornamental plants, plantation crops, grasses and few other plants (Rahman, 2010), which are listed in Table 33.

41.8 Symptoms

Lance nematode is a parasite of rice in India, and particularly causes damage to upland rice. *H. indicus* may restrict normal growth and development of rice roots (Singh and Misra, 1976; Astudillo and Birchfield, 1980). Purple-red necrotic lesions may appear on rice roots invaded by *H. indicus*, and severe infestations may lead to girdling of the root. *H. indicus* infestation decreased the shoot and root biomass (Singh and Misra, 1976) but seedling growth was hardly affected (Nath *et al.*, 1975).

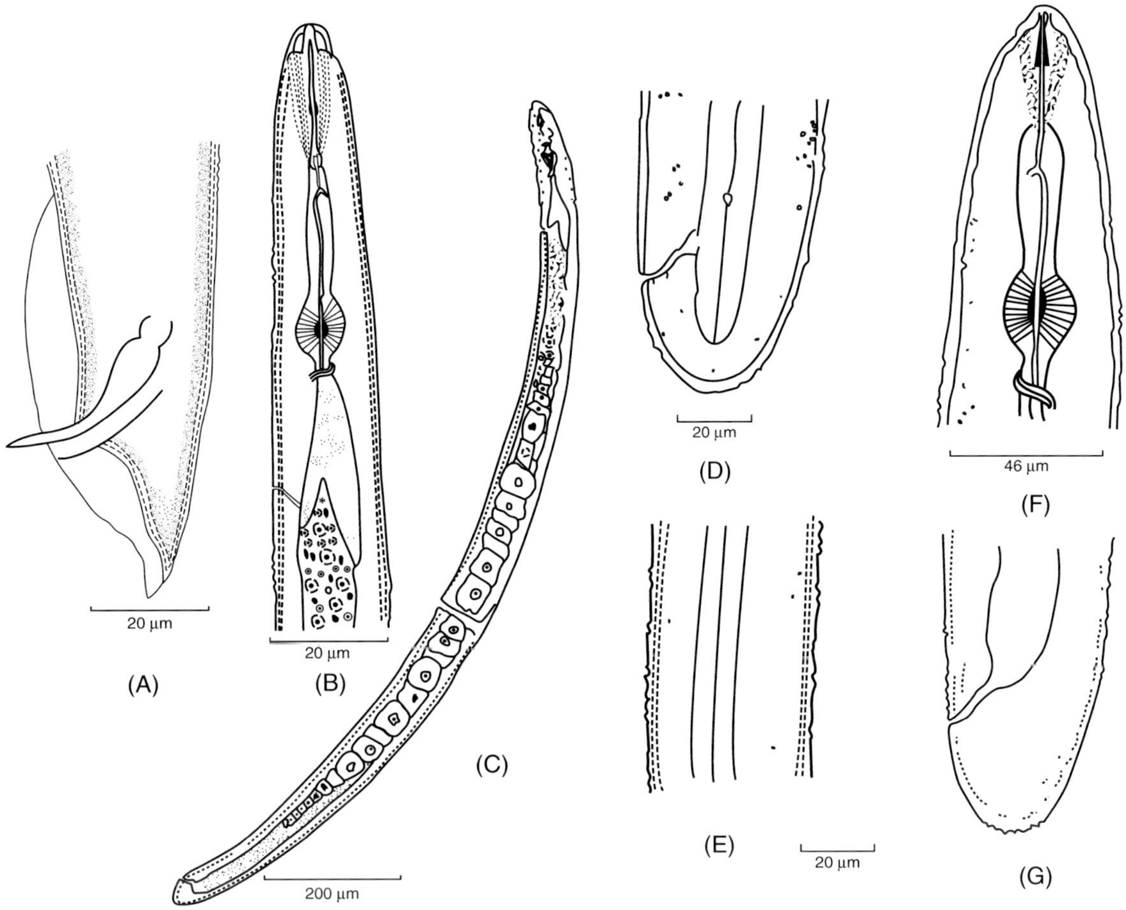

Fig. 7.12. Line drawing of *Hoplolaimus indicus*. Male posterior region (A); male anterior region (B); entire female (C); female posterior region (D); female anterior (E); female mid body with lateral incisures (F); tail end (G). [Courtesy Singh, 1995; A.P.S. University, India]

The nematode causes discernible symptoms at the different developmental stages of rice plants. Seedlings in nursery beds show suppressed growth and yellowing that resemble nitrogen deficiency symptoms. Infected seedlings in the field show yellowish leaves, which later turn brown and brittle with ash-coloured tips. Young plants become stunted with shortened upper internodes and the new leaves may curl. Chlorosis, curling of new leaves and wilting of the tip of older leaves occur in the early stages of crop development. In old plants, drooping of leaves takes place and the formation of ear heads is suppressed. The above-ground symptoms can be less apparent in the later stages of rice plants (Fig. 7.6E). On infected roots, brown lesions appear at the point of invasion. Due to the nematode's migration and feeding, cavities are formed in the cortex, cells lose their turgidity, vascular elements become distorted and consequently the roots become flaccid (soft and weak).

41.9 Biology and life cycle

The lance nematode *H. indicus* is a migratory endoparasite of roots (Decraemer and Geraert, 2013). Occasionally it is seen as semi-endoparasitic when the nematode inserts the anterior part of its body deeply, while the posterior part remains protruding outside in the soil (Nandakumar and Khera, 1974). Males and females of *H. indicus* remain vermiform throughout life. Reproduction is amphimictic type. Eggs laid in soil take about 8 days to hatch, giving rise to second-stage juveniles. First moulting occurs within the egg. Durations of second-, third- and fourth-stage larvae are 6, 8 and 10 days, respectively (Das and Rao, 1970). Root exudates seem to help the nematode in locating the roots for feeding (Khan, 2008). The fourth-stage larvae and adults puncture the epidermal cells of roots and rootlets at regions above the root caps and at places where branches of roots arise. They gradually move into the cortex. The nematode moves within the cortex intercellularly and intracellularly and, as a result, galleries and tunnels are formed in the cortical tissue pointing towards the stele. In the cortex, the nematode larvae move indiscriminately and cause damage to the cells of the vascular parenchyma. In heavy infestations, the cortical cells lose turgidity, due to nematode migration and feeding. The parenchyma of the stele also gets damaged and the vascular bundles become distorted (Rahman, 2010). Fully developed females feed for at least 3 days before they start laying eggs inside the roots. A single female on average may lay

Table 33. List of hosts of *Hoplolaimus indicus* (adapted from Rahman, 2010).

Food crops
Rice (*Oryza sativa*) – main host
Jowar (*Sorghum vulgare*)
Maize (*Zea mays*)
Pearl millet (*Pennisetum typhoides*)
Ragi (*Eleucine coracana*)
Wheat (*Triticum aestivum*)
Black gram (*Vigna mungo*)
Chickpea (*Cicer arietinum*)
Cluster bean (*Cyamopsis tetragonoloba*)
Cowpea (*Vigna unguiculata*)
Mung bean (*Vigna radiata*)
Pea (*Pisum sativum*)
Pigeon pea (*cajanus cajan*)
Vegetable crops
Bottle gourd (*Lagenaria siceraria*)
Brinjal (*Solanum melongena*)
Bitter gourd (*Momordica charantia*)
Chillies (*Capsicum frutescens*, *C. annum*)
Cauliflower (*Brassica oleracea* var. *botrytis*)
Cabbage (*Brassica oleracea* var. *capitata*)
Carrot (*Daucas carota*)
Cucumber (*Cucumis sativus*)
French bean (*Phaseolus vulgaris*)
Okra (*Abelmoschus esculentum*)
Potato (*Solanum tuberosum*)
Radish (*Raphanus sativus*)
Sugar beet (*Beta vulgaris*)
Sweet potato (*Ipomoea batatas*)
Tomato (*Lycopersicon esculentum*)
Ornamental crops
Bougainvillia spectabilis
China rose (*Hibiscus rosasinensis*)
Chrysanthemum spp.
Gerbera sp.
Nerium (*Nerium indicum*)
Rose (*Rosa indica*, *R. damascena*)
Thevetia peruvian
Forest trees
Acacia mangium
A. nilotica,
Azadirachta indica
Bombyx ceiba
Casuarina equisetifolia
Cedrus deodera
Dalbergia sisso
Eucalyptus citriodera
Gmelina arberia
Pinus roxberghii
Pongamia pinnata
Prosopis juliflora
Santalum album
Tamarinda indica
Fruit crops
Almond (*Prunus amygdalus*),
Apple (*Pyrus malus*)
Apricot (*Prunus armeniaca*)
Banana (*Musa sapientum*)
Grapevine (*Vitis vinifera*)
Guava (*Psidium guajava*)
Hill lemon (*Citrus pseudolimon*)
Indian gooseberry (*Embilica officinalis*)
Jujuba (*Zizyphus jujuba*)
Jambolan (*Syzygium cumini*)
Kiwi fruit (*Actinidia deliciosa*)
Lemon (*Citrus limon*)
Lychee (*Litchi chinensis*)
Mango (*Mangifera indica*)
Olive (*Olea europaea*)
Orange (*Citrus reticulata*)
Papaya (*Carica papaya*)
Peach (*Prunus persica*)
Pineapple (*Ananas comosus*)
Plum (*Prunus domestica*)
Rough lemon (*Citrus jambhiri*)
Sweet orange (*Citrus sinensis*)
Trifoliate orange (*Poncirus trifoliata*)
Industrial crops
Cotton (*Gossypium* spp.)
Jute (*Corchorus* spp.)
Poi (*Basella rubra*)
Roselle (*H. sabdariffa*)
Sugarcane (*Saccharum officinarum*)
Tobacco (*Nicotiana* sp.)
Oilseed crops
Groundnut (*Arachis hypogea*)
Linseed (*Linum usitatissimum*)
Mustard (*Brassica campestris*)
Soybean (*Glycine max*)
Plantation crops
Betelvine (*Piper betle*)
Black pepper (*Piper nigrum*)
Mulberry (*Morus indica*)
Tea (*Camellia sinensis*)
Aromatic and spices
Ginger (*Zingiber officinales*)
Mesta (*Hibiscus cannabinus*)
Mint (*Mentha* sp.)
Tulsi (*Ocimum sanctum*)
Turmeric (*Cucurma longa)*
Fodder and grasses
Cynodon dactylon
Chrysopogon aciculatus
Eriochloa chinensis
Eragrostis lenella
Medicago hispida
Panicum antidotale

14 eggs in 4 days. The eggs inside the roots hatch in 3–4 days. One generation from egg to egg is completed in 26–27 days. The nematode prefers well-drained soils rather than poorly drained soils. Sandy loam soil and a temperature of 21–27°C favour lance nematode infection (Rahman, 2010).

41.10 Economic importance

Limited information is available on quantitative and qualitative losses to paddy due to infestation by *H. indicus*. Rahman (2010) reported 5–19% yield reductions in rice. The nutritive value of the rice is also affected, as the grains from infected plants contain significantly less protein. Haider *et al.* (1978) reported that 100 individuals/500 g of soil caused significant growth reduction and stunting of rice plants in sandy loam soil. The entire growth parameters showed a declining trend with increasing population levels. *H. indicus* is common in uplands and areas where other cereals are grown in rice-based cropping system and causes up to 18% losses in rice (Prasad *et al.*, 1987).

41.11 Movement and means of dispersal

See section 35.11, above.

41.12 Pest risk analysis and looming threat from introduction

Considering rice as a host of *H. indicus*, its damage potential and probable introduction into other rice-producing countries, a complete pest risk analysis for this nematode should be instigated. Due to its restricted distribution, preventive steps should be taken to avoid introduction into new host areas. The nematode can easily spread through rice transplanting (seedlings), infested soil and contaminated machinery.

41.13 Invasiveness rating

Based on the consequences of the introduction and pest risk analysis of *H. indicus*, the following invasiveness rating has been assigned.

Climate/host interaction	–	Medium (2)
Hosts range	–	Medium (2)
Dispersal potential	–	Medium (2)
Economic impact	–	High (3)
Environmental impact	–	Low (1)
Post-entry distribution	–	Not known (0)
Final rating (score)	–	Medium (10)

Based on the above rating, *H. indicus* receives a score of 10 out of 15 and is rated as MEDIUM INVASIVE NEMATODE.

41.14 Region-wise status of invasiveness

H. indicus attains moderate invasiveness status in major rice-producing countries (FAOSTAT, 2019) where it is absent. There may be trade implications if this nematode is introduced and becomes established in these countries.

APPPC: Afghanistan, Cambodia, Indonesia, Japan, Lao People's Democratic Republic, Malaysia, Myanmar, Nepal, North Korea, Philippines, South Korea, Sri Lanka, Thailand, Vietnam
COSAVE: Argentina, Brazil, Colombia, Ecuador, Peru, Uruguay, Venezuela
EPPO: Italy, Russian Federation, Spain, Turkey
IAPSC: Côte d'Ivoire, Egypt, Guinea, Madagascar, Mali, Nigeria, Senegal, Sierra Leone, Tanzania
NAPPO: USA
PPPO: Australia, New Zealand

41.15 Management measures

In view of the economic importance of *H. indicus* in rice-based cropping systems and its potential to cause significant damage, appropriate management strategies need to be incorporated in the paddy fields.

41.15.1 Cultural methods

Crop rotation is one of the most effective methods of controlling *H. indicus*. In paddy-based cropping sequences, maximum reduction in the population of *H. indicus* was observed in paddy–gram rotation in two consecutive years followed by paddy–mustard or paddy–garlic (Haider *et al.*, 2001). Deep ploughing before a cropping sequence of wheat–chilli followed by fallow and thereafter lentil–cowpea–mung-bean cropping significantly reduced the total population of *H. indicus* (Wani, 2005). Use of green manuring and organic amendment has proved effective against *H. indicus*. Soil amendments with neem products such as leaf powder, sawdust and oilseed cake and urea resulted in a significant decrease in the population of *H. indicus* and other nematodes compared with control (Akhtar, 1998).

41.15.2 Chemical control

Biswas and Rao (1971) recorded effective control of *H. indicus* in rice with soil application of diazinon at 250 ppm, dimethoate at 500 ppm or DBCP at 1000 ppm. Diazinon and DBCP were also effective as root dip treatment in rice nurseries at 500 ppm for 10 min and reduced invasion and development of the nematode inside the rice roots. About 82% reduction in the population of *H. indicus* was recorded due to the application of fenamiphos at 10.0 kg/ha of soil (Sethi and Meher, 1991).

41.15.3 Integrated management

H. indicus populations were markedly reduced by the application of castor oil cake, mustard oil cake, carbofuran, sugarcane bagasse, sawdust, horse manure, poultry manure, neem leaves and farmyard manure (Khan and Shaukat, 2000). A reduction of 100% in the population of *H. indicus* was observed after combined treatment with carbofuran at 500 g a.i./ha + neem cake at 250 kg/ha followed by carbofuran at 500 g a.i/ha alone and carbofuran at 250 g a.i/ha + neem cake at 125 kg/ha (Goswami and Mishra, 1994). Deep ploughing combined with neem cake at 110 kg/ha was found to be superior in reducing *H. indicus* populations compared with normal ploughing combined with neem cake (Tiyagi *et al.*, 2002).

41.15.4 Phytosanitary measures

See Chapter 16, section 115.

42 *Hoplolaimus pararobustus*

42.1 Common name

Lance nematode

42.2 Scientific name

Hoplolaimus pararobustus (Sch. Stek. and Teun., 1938) Sher, 1963

42.3 Synonyms

Gottholdsteineria pararobusta (Sch. Stek. and Teun., 1938) Andrássy, 1958
Hoplolaimus angustulatus Whitehead, 1959
Hoplolaimus capensis Van den Berg and Heyns, 1970
Hoplolaimus casparus Van den Berg and Heyns, 1970
Hoplolaimus kittenbergeri Andrássy, 1961
Hoplolaimus pararobustus (Sch. Stek. and Teun., 1938) Sher, 1963
Hoplolaimus proporicus Goodey, 1957
Rotylenchus pararobustus (Sch. Stek. and Teun., 1938) Filipjev and Sch. Stek., 1941
Tylenchorhynchus pararobustus Schuurmans Stekhoven and Teunissen, 1938
Tylenchorhynchus robustus Schuurmans Stekhoven, 1936

42.4 EPPO code

HOLLPA

42.5 Diagnosis

42.5.1 Morphological diagnosis

Morphological descriptions after Siddiqi (1974d) after Sher (1963) in Coomans (1963) are presented below. For measurements, see Table 32.

FEMALE: Body straight to slightly arcuate ventrally when heat relaxed. Body annules distinct, about 2 μm wide. Cephalic region hemispherical, offset from rest of body by a constriction (Fig. 7.13A) and with terminal disc and generally four distinct annules (three may be present on one side and four on the other and as many as six have been recorded). Lip annules with 7–25 longitudinal striae (Fig. 7.13E). Cephalic framework hexaradiate, strongly cuticularized. Lateral fields reduced; characterized by interruption of body annules or a single incisure towards both ends; two or three incomplete incisures sometimes seen. Phasmids adapted as enlarged scutella, around 5 μm diameter; located at 24–52% anterior on right side and 58–89% posterior on left side of body length from head end. Excretory pore distinct, generally opposite to median bulb (Fig. 7.13B), but may be anterior or posterior, yet always anterior to hemizonid. Spear robust, about 38–49 μm long with tulip-shaped basal knobs (i.e. longitudinally oval with deeply dentate anterior margin). Oesophagus typical of genus with rounded median bulb and three nuclei in dorsally overlapping gland lobe (Fig. 7.13A). Intestine overlapping rectum. Vulva a transverse slit; epiptygma single, positioned anterior or posterior; vagina at right angles to body axis (Fig. 7.13K). Genital tracts amphididelphic; spermathecae spherical with sperm. Tail short and rounded; usually hemispherical and with 7–15 annules (Fig. 7.13D).

MALE: Commonly occurs. Body arcuate ventrally. Cephalic region raised and higher than female with 3–5 (frequently four) annules (Fig. 7.13I). Excretory pore near median oesophageal bulb, 8–23 annules anterior to hemizonid. Spear, oesophagus and lateral field same as in female. Spicules slightly cephalated, ventrally arcuate with large flanges in distal half (Fig. 7.13J). Gubernaculum trough-like, protrusible with distal titillae. Caudal alae large, crenate; enclosing conoid tail (Fig. 7.13J).

42.5.2 Molecular diagnosis

See section 39.5.2, above.

42.6 Geographical distribution

H. pararobustus is reported from several tropical and subtropical countries of Asia, Africa and the Caribbean region (Fig. 7.14). It is more dominant in African regions (CABI-ISC, 2020) and recorded from the following countries.

Africa: Angola, Burkina Faso, Cameroon, Côte d'Ivoire, Congo (Democratic Republic of), Egypt, Gambia, Guinea, Kenya, Madagascar, Malawi, Mozambique, Namibia, Niger, Nigeria, Réunion, São Tomé and Principe, Senegal, South Africa, Tanzania, Togo, Uganda, Zimbabwe.
Asia: China, Pakistan, Sri Lanka.
North America: Dominica, Grenada, Saint Lucia, Saint Vincent and the Grenadines.

42.7 Hosts

H. pararobustus is a polyphagous nematode infesting several higher plants. Banana (*Musa*) is considered as the main host, as the nematode is most frequently associated with this crop (CABI-CPC,

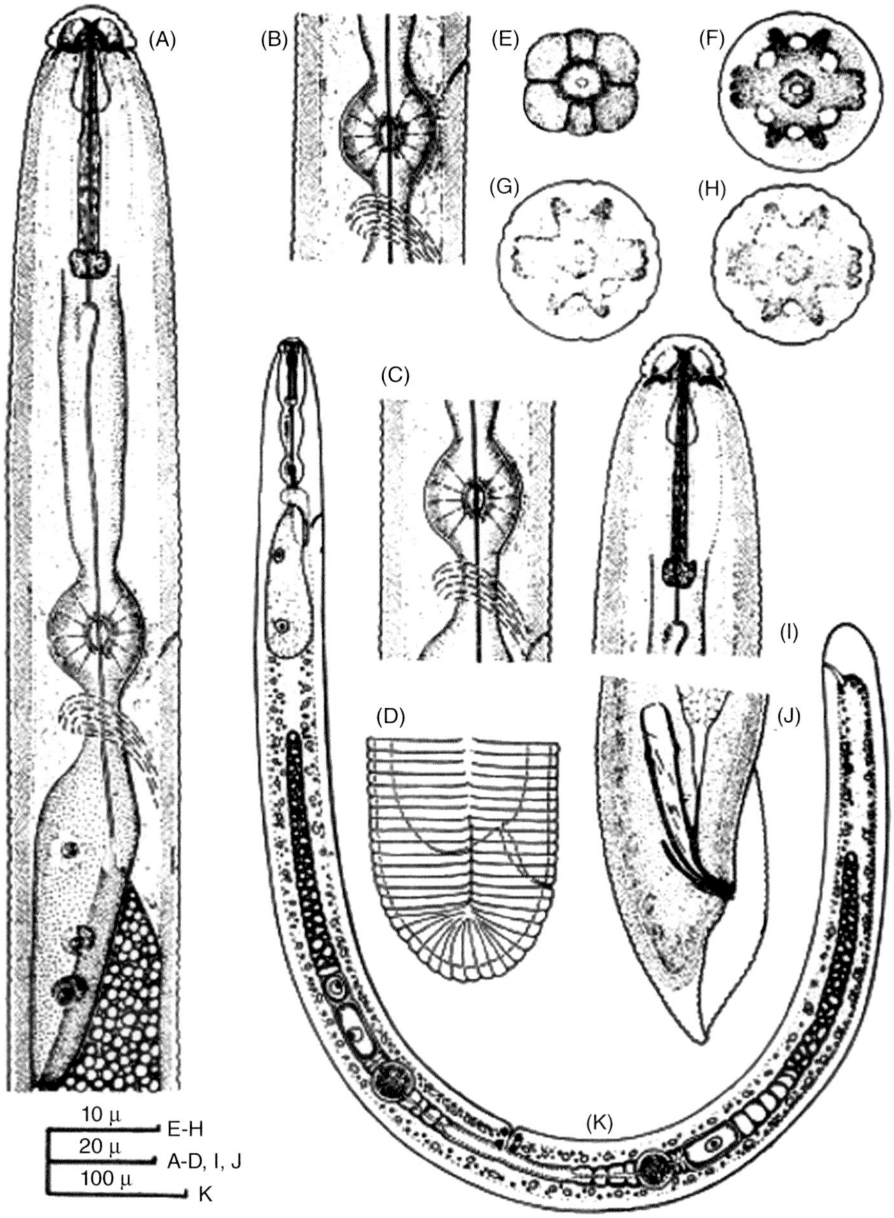

Fig. 7.13. Line drawing of *Hoplolaimus pararobustus*. Oesophageal region of female (A); position of excretory pore with respect to median oesophageal bulb of female (B,C); tail end of female (D); *en face* view of female (E); cross-section through basal bulb annula of lip region (F–H); head end of male (I); tail end of male (J); entire female (K). [Reproduced from Siddiqi, 1974d, CIH Descriptions set 3, No. 33; courtesy CAB International]

2010). The extended list of hosts of *H. pararobustus* is presented in Table 34.

42.8 Symptoms

General symptoms of *H. pararobustus* are similar to those of other lance nematode species attacking agricultural crops. Damage on banana roots may show dark brown narrow pustules about 1–2 mm long and parallel to the root axis. As the nematodes move deeper into the cortex, the necrosis becomes more extensive, giving rise to elongate, ulcerated lesions (Whitehead, 1959).

42.9 Biology and life cycle

The biology and life cycle of *H. pararobustus* are similar to those of other lance nematodes. The second-stage juvenile hatches from the egg. The first moult takes place inside the egg.

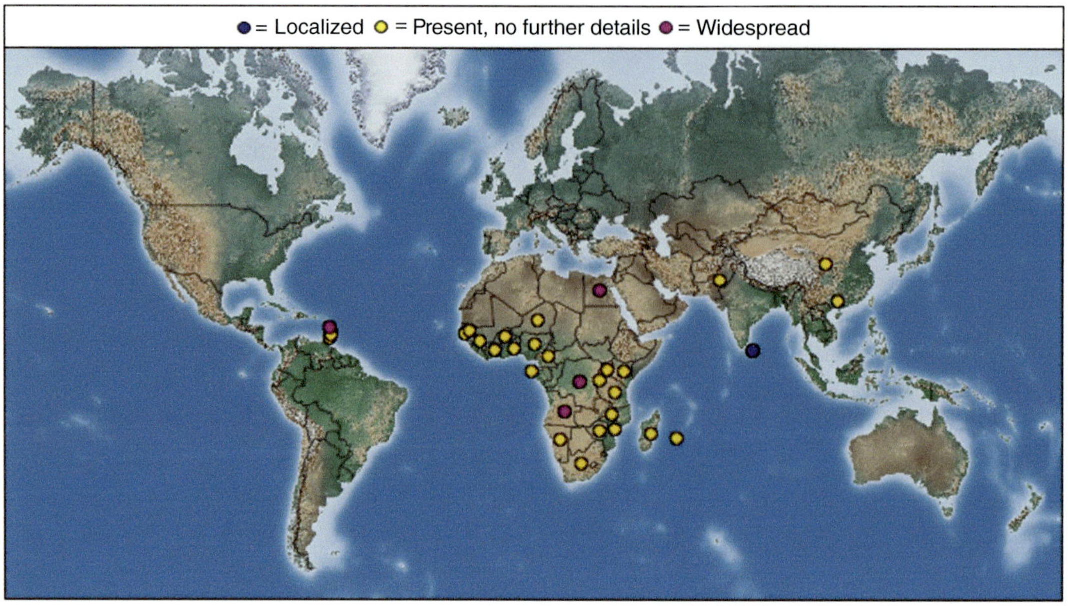

Fig. 7.14. Distribution map of *Hoplolaimus pararobustus*. [Adapted from CABI-ISC, 2020]

Table 34. List of hosts of *Hoplolaimus pararobustus* (adapted from CABI-CPC, 2010).

Main host	
Banana (*Musa* sp.).	Papaya (*Carica papaya*)
Other hosts	Pineapple (*Ananas comosus*)
African oil palm (*Elaeis guineensis*)	Rice (*Oryza sativa*)
Cotton (*Gossypium hirsutum*)	Roses (*Rosa* sp.)
Citrus (*Citrus* sp.)	Sorghum (*Sorghum bicolor*)
Cocoa (*Theobroma cacao*)	Sugarcane (*Saccharum officinarum*)
Coffee (*Coffea arabica*)	Tea (*Camellia sinensis*)
Cowpea (*Vigna unguiculata*)	Wheat (*Triticum aestivum*)
Finger millet (*Eleusine coracana*)	**Wild hosts**
Guava (*Psidium guajava*)	Crabgrass (*Digitaria abyssinica*) Tropical kudzu
Mango (*Mangifera indica*)	(*Pueraria phaseoloides*)

The hatched J_2 moults three times to become an adult. Males are common and reproduction is amphimictic. All stages of *H. pararobustus* can be found in the soil and plant roots. The nematode may be completely endoparasitic within the root tissue of the host, or it may often be ectoparasitic with only the anterior region penetrating the root tissue (Baujard and Martiny, 1995).

42.10 Economic importance

H. pararobustus is common in tropical African countries, but its economic importance appears to be low compared with other more damaging nematodes, which are usually found in much larger numbers, particularly on banana and plantain. In Dominica, populations of *H. pararobustus* on banana were recorded as up to 1000 nematodes per 100 g of roots (Hunt, 1977). Price (1994) was unable to show that *H. pararobustus* was pathogenic to banana in Cameroon. Baujard and Martiny (1995) demonstrated a reduction in root weight of infested cowpea.

42.11 Movement and means of dispersal

See section 35.11, above.

42.12 Pest risk analysis and looming threat from introduction

H. pararobustus has low economic significance but due to its polyphagous nature and with banana as a major host, its introduction into new areas, through any means of movement and dispersal such as nursery stock, banana corm, roots and soil, should be regulated. No contaminated consignment should be permitted from the host countries and a phytosanitary certificate should be attached before processing for export.

42.13 Invasiveness rating

Considering the consequences of the introduction and pest risk analysis of *H. pararobustus*, the following invasiveness rating has been assigned.

Climate/host interaction	–	Medium (2)
Hosts range	–	High (3)
Dispersal potential	–	High (3)

Continued

Economic impact	–	Medium (2)
Environmental impact	–	Low (1)
Post-entry distribution	–	Low (–1)
Final rating (score)	–	Medium (10)

Based on the above invasiveness rating, *H. pararobustus* receives a score of 10 out of 15 and is rated as MEDIUM INVASIVE NEMATODE.

42.14 Region-wise status of invasiveness

H. pararobustus is a quarantine nematode in many countries, including India, where it is listed as a schedule VI pest (PQ Order-2003, 2015). This nematode may be considered as a medium invasive pest in the following main banana-producing countries, where it is still absent.

APPPC: Bahrain, Bangladesh, Bhutan, Brunei, Cambodia, India, Indonesia, Israel, Japan, Jordan, Kuwait, Lebanon, Malaysia, Maldives, Nepal, Oman, Philippines, Singapore, South Korea, Syria, Thailand, United Arab Emirates, Vietnam
CAHFSA: Antigua and Barbuda, Bahamas, Barbados, Belize, Bermuda, British Virgin Islands, Cayman Islands, Costa Rica, Cuba, Dominican Republic, El Salvador, Guadeloupe, Guatemala, Haiti, Honduras, Jamaica, Martinique, Montserrat, Nicaragua, Panama, Puerto Rico, Trinidad and Tobago, US Virgin Islands
COSAVE: Argentina, Bolivia, Brazil, Colombia, Ecuador, French Guiana, Guyana, Paraguay, Suriname
EPPO: Cyprus, Greece, Italy, Portugal, Spain
IAPSC: Algeria, Benin, Burundi, Cape Verde, Central African Republic, Comoros, Ethiopia, Gabon, Ghana, Guinea-Bissau, Liberia, Mali, Mauritius, Morocco, Rwanda, Seychelles, Somalia, Sudan, Swaziland
NAPPO: Mexico, USA
PPPO: American Samoa, Australia, Cook Islands, Fiji, French Polynesia, Guam, Kiribati, Micronesia (Federated states of), New Caledonia, Niue, Papua New Guinea, Samoa, Solomon Islands, Tokelau, Tonga, Tuvalu, Vanuatu, Wallis and Futuna Islands.

42.15 Management measures

See section 39.16, above.

43 *Hoplolaimus seinhorsti*

43.1 Common name

Lance nematode

43.2 Scientific name

Hoplolaimus seinhorsti Luc, 1958

43.3 Synonyms

Basirolaimus seinhorsti Siddiqi, 1986
Hoplolaimus sheri Suryawanshi, 1971

43.4 EPPO code

HOLLSE

43.5 Diagnosis

Morphologically *Hoplolaimus seinhorsti* is differentiated from other species of the genus by the presence of extra gland nuclei and Siddiqi (1986) validated it as *Basirolaimus*. However, Fortuner (1987) rejected *Basirolaimus* and considered it as *Hoplolaimus*.

43.5.1 Morphological diagnosis

The morphological descriptions below are adapted from Luc (1958), Sher (1963) and Timm (1965). Morphometric measurements are after Sher, 1963 (see Table 32).

FEMALE: Body cylindrical, ventrally curved, tapering slightly towards anterior end (Fig. 7.15A). Cuticle annule distinct, subcuticular striations prominent. Cephalic region hemispherical, offset by deep constriction with 4–6 annules, separated into six sectors, central and dorsal sectors usually smaller than subdorsal and subcentrals. Lip disc squarish with oval corners and round oral opening (Fig. 7.15B). Amphidial apertures just behind lip disc. Basal lip annule with 8–12 longitudinal striations. Cephalic framework well developed, heavily sclerotized. Lateral fields characterized by single irregular incisure. Scutella 4–5 μm in diameter, one anterior to vulva, another between vulva and anus (Fig. 7.15E). Hemizonid located posterior to excretory pore. Hemizonion 5–10 annules posterior to hemizonid. Stylet massive, well developed; knobs with irregular anterior margins (Fig. 7.15D). Oesophagus with six gland nuclei; dorsal oesophageal gland with four nuclei; oesophagus overlaps dorsally. Intestine not overlapping rectum (Fig. 7.15D). Gonads amphidelphic, outstretched; vulva at mid-body or slightly posterior. Epiptygma single or double (Fig. 7.15G). Spermatheca not present. Tail hemispheroid, shorter than body diameter, with 10–15 annules (Fig. 7.15F).
MALE: Not known

43.5.2 Molecular diagnosis

See sections 39.5.2 and 41.5.2, above.

43.6 Geographical distribution

H. seinhorsti was first described in Madagascar from the roots of cotton by Luc (1958). The nematode has a wide distribution and

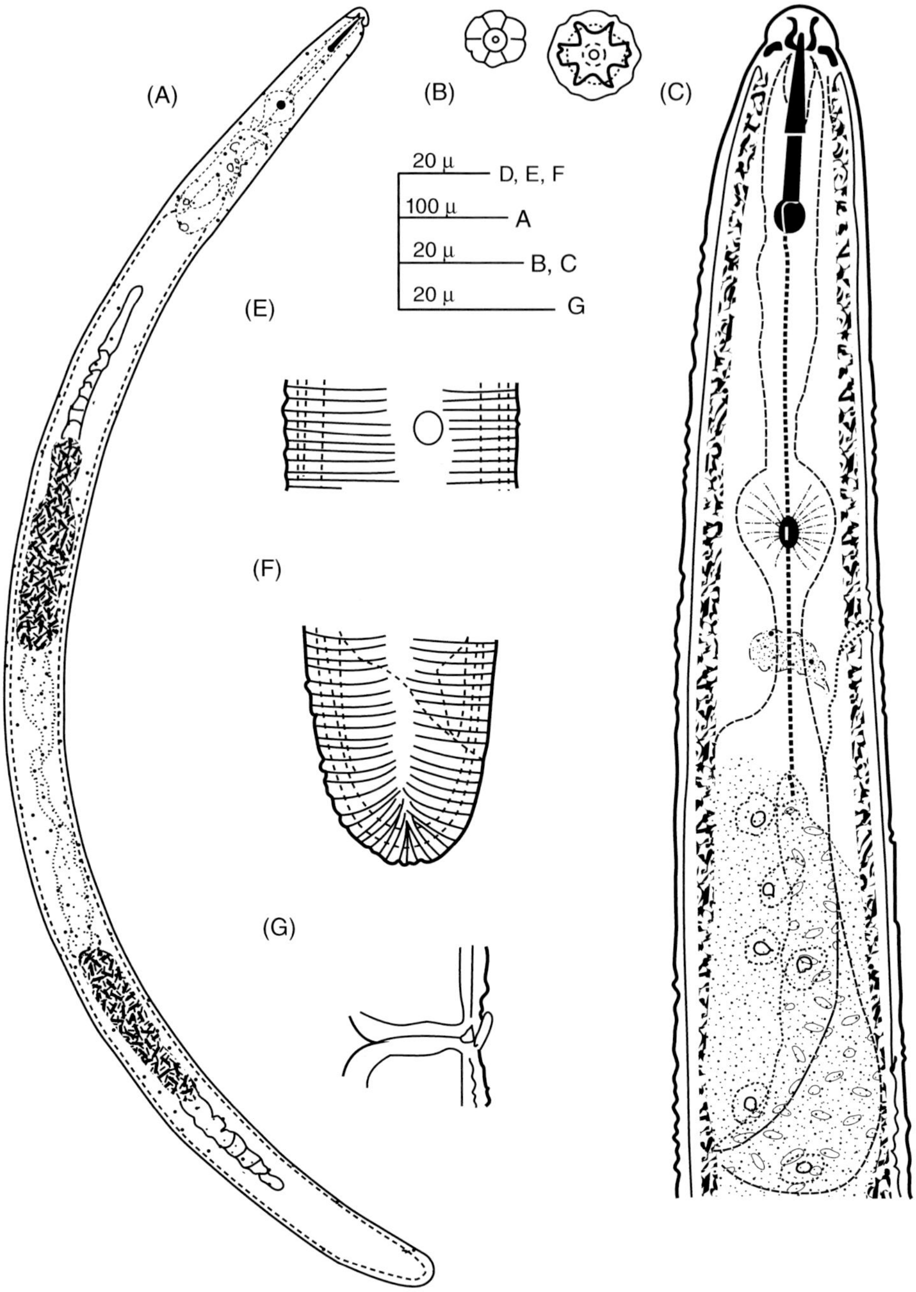

Fig. 7.15. Line drawing of *Hoplolaimus seinhorsti*. Entire female (A); female, face view (B); female cross-section through basal annule of cephalic region (C); female anterior region (D); scutella (E); female posterior region (F); female surface view at vulva (G). [Reproduced from Van den Berg, 1976, CIH Descriptions set 6, No. 76; courtesy CAB International]

is reported from Australia, Brunei, China, Egypt, Fiji, French Guiana, Guatemala, Indonesia, India, Iran, Kenya, Nigeria, Madagascar, Malaysia, Martinique, Pakistan, Papua New Guinea, Philippines, Réunion, Samoa, Sri Lanka, Sudan, Thailand, Trinidad and Tobago and Venezuela (Van den Berg and Kirby, 1979; Siddiqi, 2000; CABI-CPC, 2010). In India, *H. seinhorsti* is widespread and found in almost all states (Azmi, 1993).

43.7 Hosts

H. seinhorsti is a polyphagous nematode and attacks several economically important hosts comprising food crops, vegetables, ornamental plants and fruit trees (CABI-CPC, 2010). The extended list of hosts is given in Table 35.

Table 35. List of hosts of *Hoplolaimus seinhorsti* (adapted from CABI-CPC, 2010).

Ananas comosus (pineapple)
Arachis hypogaea (groundnut)
Brassica oleracea var. *botrytis* (cauliflower)
Cajanus cajan (pigeon pea)
Capsicum (peppers)
Cicer arietinum (chickpea)
Cucumis melo (melon)
Fabaceae (leguminous plants)
Glycine max (soybean)
Gossypium (cotton)
Hordeum vulgare (barley)
Jasminum (jasmine)
Leucaena leucocephala (leucaena)
Mangifera indica (mango)
Morus alba (mora)
Musa (banana)
Oryza sativa (rice)
Pennisetum glaucum (pearl millet)
Phaseolus (beans)
Piper nigrum (black pepper)
Poaceae (grasses)
Rosa (roses)
Saccharum officinarum (sugarcane)
Solanaceae plants
Solanum lycopersicum (tomato)
Solanum melongena (aubergine)
Sorghum bicolor (sorghum)
Theobroma cacao (cocoa)
Vigna unguiculata (cowpea)
Vitis vinifera (grapevine)
Zea mays (maize)

43.8 Symptoms

H. seinhorsti is an ecto- and endo-parasitic nematode and feeds on cortical cells. Continuous feeding causes yellowish-brown discoloration of root tissue. The nematode moves from one portion to another after destroying the feeding sites, which results in the formation of large cavities in the root cortex and the appearance of tissue peeling off. Only the central core of vascular elements remains enacted. On cowpeas, trails of necrosed cells are visible on either side of the migrating nematodes. The nematode also feeds on *Rhizobium* nodules occasionally. All the developing stages, adults and eggs can be seen inside the roots after 1 week of the germination of the host seedling. In 6–8 weeks, conspicuous necrosis of both lateral and secondary lateral roots is observed. The feeder root completely vanishes 9 weeks after germination and is badly necrosed. The infested plants are stunted with shorter shoot length, reduced fresh and dry root and decreased shoot weights (Bridge, 1973).

43.9 Biology and life cycle

H. seinhorsti can feed as an ectoparasite and endoparasite on cortical cells and moves from one portion to another portion within the root tissue. Very little information is available on the biology and life cycle of this nematode. In Nigeria, an initial population of 100 nematodes per litre of soil increased to 3800 nematodes per litre in 9 weeks on cowpeas (Bridge, 1973), while in 49 weeks the nematode population increased from 741 to 12,883 per pot in India (Muthukrishnan *et al.*, 1979). The highest density of *H. seinhorsti* usually occurs in the area of the lateral roots (Rajaram, 1982).

43.10 Economic importance

H. seinhorsti is a polyphagous nematode and has been reported to be associated with substantial yield decline on different hosts. However, exact data are not available on yield losses due to this nematode. In pigeon pea, a population of 204 nematodes per 500 g of soil caused 25% reduction in dry root weight and 25% reduction in shoot length when densities were higher than 1500 nematodes per 500 g soil (Sharma and Nene, 1988).

43.11 Movement and means of dispersal

See section 35.11, above.

43.12 Pest risk analysis and looming threat from introduction

Considering the wide host range of *H. seinhorsti*, its damage potential and presence in different ecological zones, a complete pest risk analysis for this nematode should be instigated. Due to its wide distribution, preventive steps should be taken to avoid its introduction into new areas. As an endoparsite, the nematode can easily spread through nursery stocks, infested soil and contaminated machinery.

43.13 Invasiveness rating

Considering the consequences of the introduction and pest risk analysis of *H. seinhorsti*, the following invasiveness rating has been evaluated.

Climate/host interaction	–	Medium (2)
Hosts range	–	High (3)
Dispersal potential	–	Medium (2)
Economic impact	–	Medium (2)
Environmental impact	–	Low (1)
Post-entry distribution	–	Low (–1)
Final rating (score)	–	Medium (9)

Based on the above invasiveness rating score, *H. seinhorsti* receives a score of 9 out of 15 and is rated as MEDIUM INVASIVE NEMATODE.

43.14 Region-wise status of invasiveness

Although *H. seinhorsti* attained medium invasiveness status, due to the polyphagous nature, however, the nematode may be considered as an invasive nematode in all the regions where it is still absent or not recorded.

43.15 Management measures

In *Capsicum anuum* fields, the crop rotation sequence of wheat–fallow–maize–capsicum–wheat–fallow–onions was found effective in reducing the nematode population (Khan *et al.*, 1987). On wheat, the application of carbofuran and neem leaves provided satisfactory control of *H. seinhorsti* (Khan and Shaukat, 1993). Neem and tobacco leaf extracts were also effective in hampering the nematode in wheat fields (Khan *et al.*, 1994). For further detail, see section 39.16, above.

Scutellonema Spp.

The genus *Scutellonema* was created by Andrássy in 1958 and included all *Rotylenchus* species with large phasmids (scutella) situated either opposite each other or closely so and at the level of the cloacal aperture or anus. *Scutellonema* spp. are migratory ectoparasites and/or endoparasites of roots of a wide range of agricultural and horticultural crops (Van Den Berg *et al.*, 2017). The genus was revised by different workers (Sher, 1964; Siddiqi, 2000; Van Den Berg *et al.*, 2013; Kolombia *et al.*, 2017). Currently, approximately 50 species of *Scutellonema* are described and validated with *Scutellonema blaberum* as type species (Siddiqi, 2000; Van Den Berg *et al.*, 2013; Kolombia *et al.*, 2017). Some of the *Scutellonema* species are:

Scutellonema africanum Smit, 1971
S. bizanae Van den Berg and Heyns, 1973
S. blaberum (Steiner, 1937) Andrássy, 1958
S. brachyurus (Steiner, 1938) Andrássy, 1958
S. bradys (Steiner and LeHew, 1933) Andrássy, 1958
S. cavenessi Sher, 1964
S. clathricaudatum Whitehead, 1959
S. clavicaudatum Van Den Berg *et al.*, 2017
S. commune Van den Berg and Heyns, 1973
S. conicephalum Sivakumar and Selvasekaran, 1982
S. dreyeri Van den Berg and De Waele, 1990
S. grande Sher, 1964
S. magniphasma Sher, 1964
S. nigermontanum Van den Berg, 1990
S. siamense Timm, 1965
S. sofiae Van den Berg and Heyns, 1973
S. sorghi Van den Berg and De Waele, 1990
S. transvaalense Van den Berg, 1981
S. truncatum Sher, 1964
S. tsitsikamense Van den Berg, 1976
S. unum Sher, 1964

Among these species, *S. brachyurus* and *S. bradys* are damaging pathogens of economically important agricultural and horticultural crops in tropical and subtropical regions and considered for their invasiveness in this chapter. Other species cause little or no known economic damage to agriculture crops (Van Den Berg *et al.*, 2017).

Systematic position of *Scutellonema* species

	Classical (after Siddiqi, 2000)	Molecular (after Decraemer and Hunt, 2013)
Phylum	Nematoda	Nematoda
Class	Secernentea	Chromadorea
Order	Tylenchida	Rhabditida
Suborder	Hoplolaimina	Tylenchina
Infra-order	–	Tylenchomorpha
Superfamily	Holpolaimoidea	Tylenchoidea
Family	Hoplolaimidae	Hoplolaimidae
Subfamily	Hoplolaiminae	Hoplolaiminae
Genus	*Scutellonema*	*Scutellonema*

44 *Scutellonema brachyurus*

44.1 Common name

Carolina spiral nematode

44.2 Scientific name

Scutellonema brachyurus (Steiner, 1938) Andrássy,1958

44.3 Synonyms

Rotylenchus brachyurus Steiner, 1938
Rotylenchus coheni Goodey, 1952
Scutellonema bangalorense Khan and Nanijappa, 1972
Scutellonema boocki Lordello, 1957
Scutellonema brachyurum (Steiner, 1938) Andrássy, 1958
Scutellonema coheni (Goodey, 1952) Andrássy, 1958
Scutellonema conicaudatum Sivakumar and Selvasekaran, 1982
Scutellonema orientale Rashid and Khan, 1974

Scutellonema ramai Verma, 1972
Scutellonema sheri Edward and Rai, 1970

44.4 EPPO code

SCUNBC

44.5 Diagnosis

44.5.1 *Morphological diagnosis*

The morphological and morphometric descriptions of *Scutellonema brachyurus* below are based on Van den Berg *et al.* (2013). For measurements, see Table 36.

Female: Body form an open to closed spiral after heat killed. Cephalic region round, offset slightly with discrete lip disc with usually three or occasionally four lip annuli, lip disc oval with small amphid which opens laterally. Cephalic framework moderate. Cephalids absent. Stylet strong with rounded stylet knobs posteriorly and a little hollow anteriorly (Fig. 7.16A). Telenchium longer than metenchium, seldom of identical length. Median bulb slightly longer than width, not covering body cavity. Pharyngeal gland dorsally overlapping with three gland nuclei. Excretory pore located mostly opposite basal pharyngeal lobe. Hemizonid two annules long, located behind excretory–secretory pore to five annules anterior to excretory pore. Hemizonion obscure if present, one annule long and positioned 13 annules posterior to hemizonid. Body annuli distinct. Spermatheca conspicuous but small, oval and empty. Vaginal glands oblong, small, not very discrete. Epiptygma double and folded over vulval opening or into vagina. Intestine mostly not overlapping rectum. Scutellum moderate in size, located three annules from anterior and three annules posterior to anus (Fig. 7.16D–J). Lateral fields with four longitudinal lines, areolated anteriorly, opposite oesophageal region and generally opposite scutellum, seldom without any areolation near scutellum. Tail rounded, more quickly dorsal side with 9–17 annules, those on posterior terminus slightly larger than rest of body.
Male: Absent or extremely rare, not required for reproduction. Van den Berg and Heyns (1973) reported males in very small numbers. Similar to females in most morphology. Spicules slightly cephalated, arcuate, 25–27 µm long. Gubernaculum large, protrusible. Caudal alae crenate, terminal. Tail conoid, completely enveloped by bursa.

44.5.2 *Molecular diagnosis*

Machado *et al.* (2019) identified *S. brachyurus* from soybean fields in Brazil by characterizing ITS-1 and D2–D3 rDNA gene fragments using the universal primers for 18S (5′-TTGATTACGTCCCTGCCCTTT-3′) and RN58SR (5′-ACGAGCCGAGTGATCCACCG-3′) for ITS-1 region of rDNA, and D2A (5′-TCGGAAGGAACCAGCTACTA-3′) and D3B (5′-ACAAGTACCGTGAGGGAAAGTTG-3′) for D2-D3 region.

44.6 Geographical distribution

S. brachyurus is known to occur in tropical and subtropical regions of the world (Fig. 7.17). The nematode is widespread in China and, unlike *S. bradys*, it is also reported from temperate countries such as Japan, Germany, Italy, Netherlands, Poland, Sweden and United Kingdom (CABI/EPPO, 2006). Recently, Tzortzakakis *et al.* (2016) reported *S. brachyurus* from Crete, Greece. The nematode has been detected from the following countries (CABI-ISC, 2020).

Africa: Congo (Democratic Republic), Egypt, Eswatini (formerly Swaziland), Ethiopia, Kenya, Madagascar, Malawi, Mauritius, Mozambique, Nigeria, Réunion, South Africa, Tanzania, Uganda, Zambia, Zimbabwe.
Asia: Brunei, China, India, Japan, Malaysia, Nepal, Pakistan, Singapore, South Korea, Sri Lanka, Thailand.
Europe: Germany, Greece, Italy, Netherlands, Poland, Sweden, United Kingdom.

Table 36. Morphometric measurements of *Scutellonema* species.

		Range, numerous populations	
Morphometric characters		*S. brachyurus* (Van den Berg *et al.*, 2013)[a]	*S. bradys* (after Siddiqi, 1972c)
Females	Length (µm)	656–805	880–1110
	Stylet length (µm)	28.5–30.5	26–30
	a	19.5–22.9	27–32
	b	4.6–5.6	6.7–8.7
	b'	5.5–6.8	5.2–6.0
	c	59.4–95.3	32–56
	c'	0.4–0.6	–
	DGO	5–7	
	V	58–61	54–59
	Excretory pore from anterior end (µm)	111–136	–
	Mid body diameter (µm)	31.5–38	–
	Tail length (µm)	7.5–12.5	–
	Scutellum length (µm)	3.5–5.0	–
	o	17.9–26	–
Males	Length (µm)	–	850–1000
	Stylet length (µm)	–	25–28
	a	–	26–36
	b	–	6.6–9.0
	b'	–	5.2–6.6
	c	–	27–32
	Tail length (µm)	–	11–16
	Spicule (µm)	–	29–33
	Gubernaculum (µm)	–	14–17
	Capitulum (µm)	–	10
	Excretory pore from anterior end (µm)	–	–

[a]Male rare or absent.

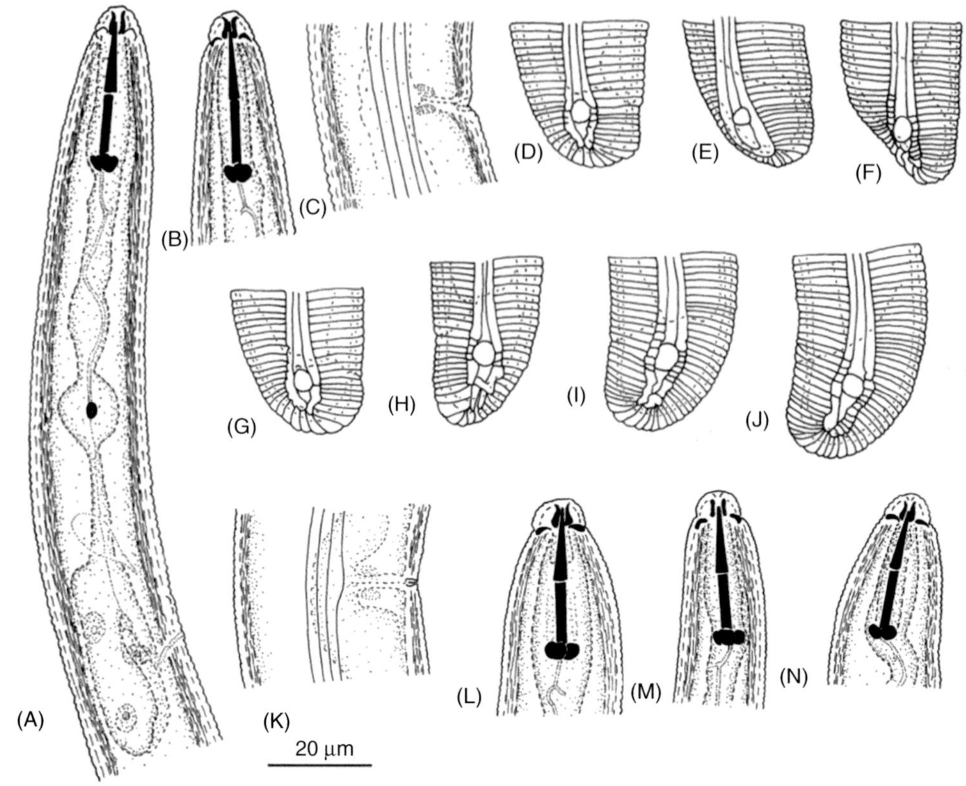

Fig. 7.16. Line drawing of *Scutellonema brachyurus*. Female, anterior end (A); lip regions of various females (B, L–N); body at vulva with lateral field (C,K); tails of various females showing areolation at scutellum (D–J). [Adapted from Van den Berg *et al.*, 2013; courtesy Nematology]

North America: Barbados, Canada, Costa Rica, Cuba, Jamaica, USA.
Oceania: Australia, Fiji, New Zealand, Samoa, Tonga.
South America: Brazil, Colombia, Falkland Islands, Peru, Venezuela.

44.7 Hosts

S. brachyurus is a polyphagous nematode and parasitizes many agronomical and horticultural crops (Siddiqi, 1974c). The major hosts of this nematode are banana, cotton, peach, soybean and tobacco. An extended list of hosts is presented in Table 37.

44.8 Symptoms

The plant roots parasitized by *S. brachyurus* usually produce lesions due to chemical stimulation as well as mechanical injury to the tissue (Schuerger and McClure, 1983). On *Hippeastrum* sp., the lesions are initially reddish and later turn brown (Fig. 7.11D). The nematode also penetrates the root and lives in the cavities formed in the root cortex. In soybean fields, plants showed stunting. The root injuries are apparently associated with drought in early-sown cultivars. Populations can reach up to 5000 nematodes/100 cm^3 of soil in soybean fields in Brazil (Maceda *et al.*, 2009). *S. brachyurus* is found inside roots, where it may multiply and causes root lesions (Fig. 7.11E) (Machado *et al.*, 2019). Inoculated amaryllis seedlings showed reduced leaf growth and slightly discoloured roots (Nong and Weber, 1964). Symptoms exhibited on *Aloe vera* include leaf stunting and yellowing, rough root surface and a reduction in leaf harvesting (Esser *et al.*, 1986). On potato, dark brown root lesions were formed after 12 h feeding and these lesions ultimately prolonged axially in the root (Schuerger and McClure, 1983). *S. brachyurus* is also associated with peach tree short-life syndrome in South Carolina (USA) along with *Macroposthonia xenoplax* (Nesmith *et al.*, 1981).

44.9 Biology and life cycle

S. brachyurus is mainly an ectoparasite, but it may invade the deeper cortical layers of roots and may become an endoparasite. *S. brachyurus* reproduces parthenogenetically, as males are extremely rare or absent. First-stage juveniles moult within the egg. The eggs are laid within the ectodermal tissue of the root and the juveniles develop within the root tissue. The nematode builds up large populations on suitable hosts. A higher population increase occurs at 28°C, while at 18°C reproduction reduces significantly (Malek and Jenkins, 1964). The nematode

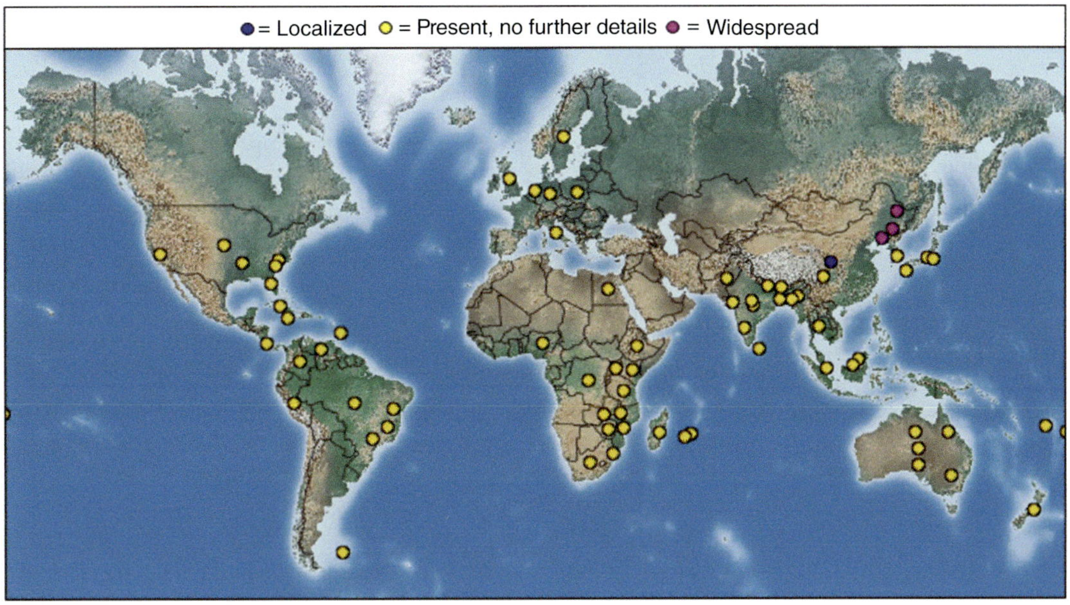

Fig. 7.17. Distribution map of *Scutellonema brachyurus*. [Adapted from CABI-ISC, 2020]

survives in dry soil by coiling and undergoing anhydrobiosis (Demeure *et al*., 1979).

44.10 Economic importance

S. brachyurus parasitizes many agriculturally important crops. In South Carolina (USA), *S. brachyurus* was recovered from 48% of cotton soil samples and found to be associated with poor growth of cotton (Kraus-Schmidt and Lewis, 1981). On tobacco, the nematode caused severe stunting of 'Golden Cure' tobacco and plant growth was reduced by 40–57% with soil populations of 4000–5000 per gallon of soil (Graham, 1955). Machado *et al*. (2019) regarded *S. brachyurus* as a potential pathogen for soybean in Brazil. The nematode spreads rapidly and its population densities increase in soybean-growing regions. In Poland, *S. brachyurus* is a serious pest of asparagus and a distinct reduction in plant growth was recorded, with a damage threshold at 400–500 nematodes per 100 m^2 initial population. At 1600 *S. brachyurus* per 100 m^2 of substrate, plant growth was reduced by 20% (Wopjtowicz and Szczygiel, 1990). On aloe, the nematode caused a severe reduction in leaf harvest in Florida, USA (Esser *et al*., 1986). In association with *Macroposthonia xenoplax*, it causes peach tree short-life syndrome and declines in tree vigour, and ultimately trees die in South Carolina, USA (Nesmith *et al*., 1981). A green-manuring crop of hairy vetch (*Vicia villosa*) harbouring a high population of *S. brachyurus* did not show any significant effects on shoot growth, but the root growth was insignificantly reduced (Malek and Jenkins, 1964).

44.11 Movement and means of dispersal

See section 35.11, above.

44.12 Pest risk analysis and looming threat from introduction

S. brachyurus is a potential pest of various economically important agricultural and horticultural crops. It poses a considerable phytosanitary risk because of its survival in different climatic zones, including temperate areas, and ease of dissemination within planting material and tubers. The occurrence of *S. brachyurus* in six European countries and recent detection in Greece suggested that the nematode was introduced through imported plant material.

44.13 Invasiveness rating

Based on the consequences of the introduction and probable economic damage caused by *S. brachyurus* the following invasiveness rating has been assigned.

Climate/host interaction	–	High (3)
Hosts range	–	High (3)
Dispersal potential	–	High (3)
Economic impact	–	High (3)
Environmental impact	–	High (3)
Post-entry distribution	–	Medium (–2)
Final rating (score)	–	High (13)

Based on the above invasiveness rating, *S. brachyurus* receives a score of 13 out of 15 and is rated as HIGHLY INVASIVE NEMATODE.

44.14 Region-wise status of invasiveness

On the basis of economic damage potential, establishment outside its native region and wide host range, *S. brachyurus* is

Table 37. List of host plants of *Scutellonema brachyurus* (adapted from CABI-CPC, 2010).

Main hosts
Aloe (grey alder)
Asparagus officinalis (asparagus)
Glycine max (soybean)
Gossypium (cotton)
Musa (banana)
Nicotiana tabacum (tobacco)
Prunus persica (peach)
Other hosts
Ananas comosus (pineapple)
Anthurium
Asparagus densiflorus (asparagus fern)
Beta vulgaris (beetroot)
Camellia sinensis (tea)
Carica papaya (papaw)
Citrus
Citrus maxima (pummelo)
Crotalaria pallida (smooth crotalaria)
Cucumis sativus (cucumber)
Cyamopsis tetragonoloba (guar)
Dianthus caryophyllus (carnation)
Eucalyptus (eucalyptus tree)
Eucalyptus spp.
Fragaria ananassa (strawberry)
Helianthus annuus (sunflower)
Hemerocallis (daylilies)
Hemerocallis spp.
Hippeastrum hybrids (amaryllis)
Hippeastrum vittatum (Barbados lily)
Hordeum vulgare (barley)
Hoya carnosa (wax plant)
Humulus lupulus (hops)
Ipomoea batatas (sweet potato)
Lactuca sativa (lettuce)
Liriodendron tulipifera (tuliptree)
Notocactus schumannianus
Oryza sativa (rice)
Passiflora edulis (passionfruit)
Prunus armeniaca (apricot)
Saccharum officinarum (sugarcane)
Saintpaulia ionantha (African violet)
Solanum lycopersicum (tomato)
Sorghum bicolor (sorghum)
Trifolium pratense (purple clover)
Vigna unguiculata (cowpea)
Vitis vinifera (grapevine)
Zea mays (maize)
Wild host
Acer pseudoplatanus (sycamore)
Arachniodes adiantiformis (leatherleaf fern)
Eleusine indica (goose grass)
Hairy vetch
Pinus patula (Mexican weeping pine)
Pippeastrum sp.
Platanus occidentalis (sycamore)
Ruscus hypophyllum
Thevetia peruviana (exile tree)
Thuja sp. (thuja)
Vicia villosa

given a high invasiveness rating for the following countries where it does not occur.

APPPC: Bangladesh, Cambodia, Indonesia, Iran, Iraq, Israel, Kazakhstan, Kyrgyzstan, Lebanon, Myanmar, Philippines, Saudi Arabia, Taiwan, Tajikistan, Vietnam, Uzbekistan
CAHFSA: Belize, Dominican Republic, Haiti, Martinique, Puerto Rico, Saint Kitts & Nevis, Trinidad and Tobago
CAN: Bolivia, Ecuador
COSAVE: Argentina, Chile, Guyana, Paraguay, Suriname, Uruguay
EPPO: Austria, Belarus, Belgium, Bulgaria, Croatia, Cyprus, Czech Republic, Denmark, Estonia, Finland, France, Georgia, Guernsey, Hungary, Ireland, Jersey, Latvia, Lithuania, Luxembourg, Macedonia, Malta, Moldova, Morocco, Norway, Portugal, Romania, Russia, Serbia, Slovakia, Slovenia, Spain, Switzerland, Turkey, Ukraine
IAPSC: Algeria, Angola, Benin, Botswana, Burkina Faso, Burundi, Cameroon, Cape Verde, Central African Republic, Chad, Comoros, Côte d'Ivoire, Eritrea, Gabon, Gambia, Ghana, Guinea, Guinea-Bissau, Liberia, Mali, Mauritania, Namibia, Niger, Senegal, Sierra Leone, Somalia, Togo
NAPPO: Mexico
NEPPO: Jordan, Libya, Malta, Sudan, Syria, Tunisia
OIRSA: El Salvador, Guatemala, Honduras, Nicaragua, Panama
PPPO: Papua New Guinea, Vanuatu

44.15 Management measures

44.15.1 Cultural practices

Various cultural practices are employed to control *S. brachyurus* in field crops. In Zimbabwe, crop rotation with grass decreased populations of *S. brachyurus* in tobacco (Shepherd, 1968). In South Africa, crop rotations of pineapple with Rhodes grass (*Chloris gayana*), *Desmodium uncinatum* and bare fallow lowered the population of various nematodes, including *S. brachyurus*, *Rotylenchus unisexus* and *Helicotylenchus dihystera* (Keetch and Dalldorf, 1980).

44.15.2 Chemical control

Use of fumigant nematicides was quite common in nematode management, but most fumigant nematicides have been superseded. Systemic nematicides such as carbofuran, ethoprophos, fenamiphos and oxamyl are alternatives for fumigants. These nematicides are not recommended for control of *S. brachyurus* in asparagus cultivation due to their high toxic impact (Wopjtowicz and Szczygiel, 1990). Chemical modules effective against *S. bradys* may also prove effective against this nematode (see section 45.15, below).

44.15.3 Phytosanitary measures

See Chapter 16, section 115.

45 *Scutellonema bradys*

45.1 Common name

Yam nematode

45.2 Scientific name

Scutellonema bradys (Steiner and LeHew, 1933) Andrássy, 1958

45.3 Synonyms

Anguillulina bradys (Steiner and Le Hew, 1933) Goodey, 1975
Hoplolaimus bradys Steiner and Le Hew, 1933
Rotylenchus blaberus Steiner, 1937
Rotylenchus bradys (Steiner and Le Hew, 1933) Filipjev, 1936
Scutellonema blaberum (Steiner, 1937) Andrássy, 1958
Scutellonema dioscorea Lordello, 1959

45.4 EPPO code

SCUNBR

45.5 Diagnosis

Baujard and Martiny (1995) considered *S. bradys* as a complex species including *S. bradys*, *S. cavenessi* and *S. clathricaudatum*. However, phylogenetic analysis by Kolombia *et al.* (2017) and Van den Berg *et al.* (2017) confirmed them as separate species.

45.5.1 Morphological diagnosis

Morphological and morphometric descriptions of *S. bradys* after Siddiqi (1972b) are summarized below. For measurements, see Table 36.

Female: Body straight to arcuate slightly when heat relaxed. Labial region knob-like, offset by a constriction. Cephalic framework heavily sclerotized (Fig. 7.18A). Stylet well developed with rounded to large oval basal knobs. Lateral fields with four incisures, about one-fifth body width, areolated anteriorly and at phasmids. Hemizonid distinct, 2–3 annules long close to oesophago-intestinal junction, 0–3 annules anterior to excretory pore. Hemizonion distinct, one annule long, around 8 annules behind the excretory pore. Oesophageal glands elongate, overlapping intestine dorso-laterally and dorsally (Fig. 7.18E). Paired ovaries with one or two rows of oocytes. Spermathecae ovoid, sometimes round, generally filled with sperms (Fig. 7.18D). Vulva a transverse slit with exposed cuticular thickenings towards ends. Epiptygma ordinary. Intestine overlapping rectum partially. Tail with obtusely rounded striated terminus (Fig. 7.18F-H). Phasmids present, with pore-like aperture, around 4 μm diameter.

Male: Similar to female in morphology except for sexual dimorphism. Testis single, outstretched with three to four rows of spermagonia, sperm around 4 μm in diameter. Caudal alae large, crenate, enclosing tail (Fig. 7.18I). Spicules ventrally arcuate and slightly cephalated with prominent distal flanges (Fig. 7.18J,K). Capitulum (= telamon) large, around 10 μm long. Cuticular, non-protoplasmic portion of tail 11–16 μm long at terminus (Fig. 7.18J). Phasmids present, generally postanal.

45.5.2 Molecular diagnosis

S. bradys-specific primers have been developed by Humphreys-Pereira *et al.* (2014) for ITS region. The *S. bradys*-specific primers are SBVF1 (5′-CCTCTCCA TGTGTCCCACTT-3′) and SBLR (5′-TGCACAAG GCACACATCT-3′). Etchiha Afoha *et al.* (2018) sequenced ITS1 and COI regions of 138 *Scutellonema* spp. samples from Benin using the protocol and primers developed by Humphreys-Pereira *et al.* (2014). The phylogenetic analysis based on 51 population sequences did not reveal any genetic variation between populations by cultivar, locality, cropping system and agroecological zone. Kolombia *et al.* (2017) also characterized *S. bradys* D2–D3 expansion segments of LSU rDNA (28S rDNA gene) and COI gene fragments using the protocol described in Van den Berg *et al.* (2013) (see section 44.5.2).

45.6 Geographical distribution

S. bradys was first reported from Jamaica infecting yam tubers (Steiner and Lehew, 1933). After that it was recorded from different parts of the world but mainly confined to tropical and subtropical regions (Fig. 7.19). The nematode has been detected from the following countries (CABI-ISC, 2020).

Africa: Benin, Burkina Faso, Cameroon, Côte d'Ivoire, Gambia, Ghana, Guinea, Kenya, Mali, Nigeria, Senegal, Sudan, Tanzania, Togo.
Asia: India, Pakistan.
North America: Barbados, Costa Rica, Cuba, Dominica, Dominican Republic, Guadeloupe, Guatemala, Haiti, Jamaica, Martinique, Puerto Rico, Trinidad and Tobago, USA.
South America: Brazil, Venezuela.

45.7 Hosts

S. bradys parasitizes many agricultural and horticultural crops but the primary hosts are yam (*Dioscorea* spp.), including white

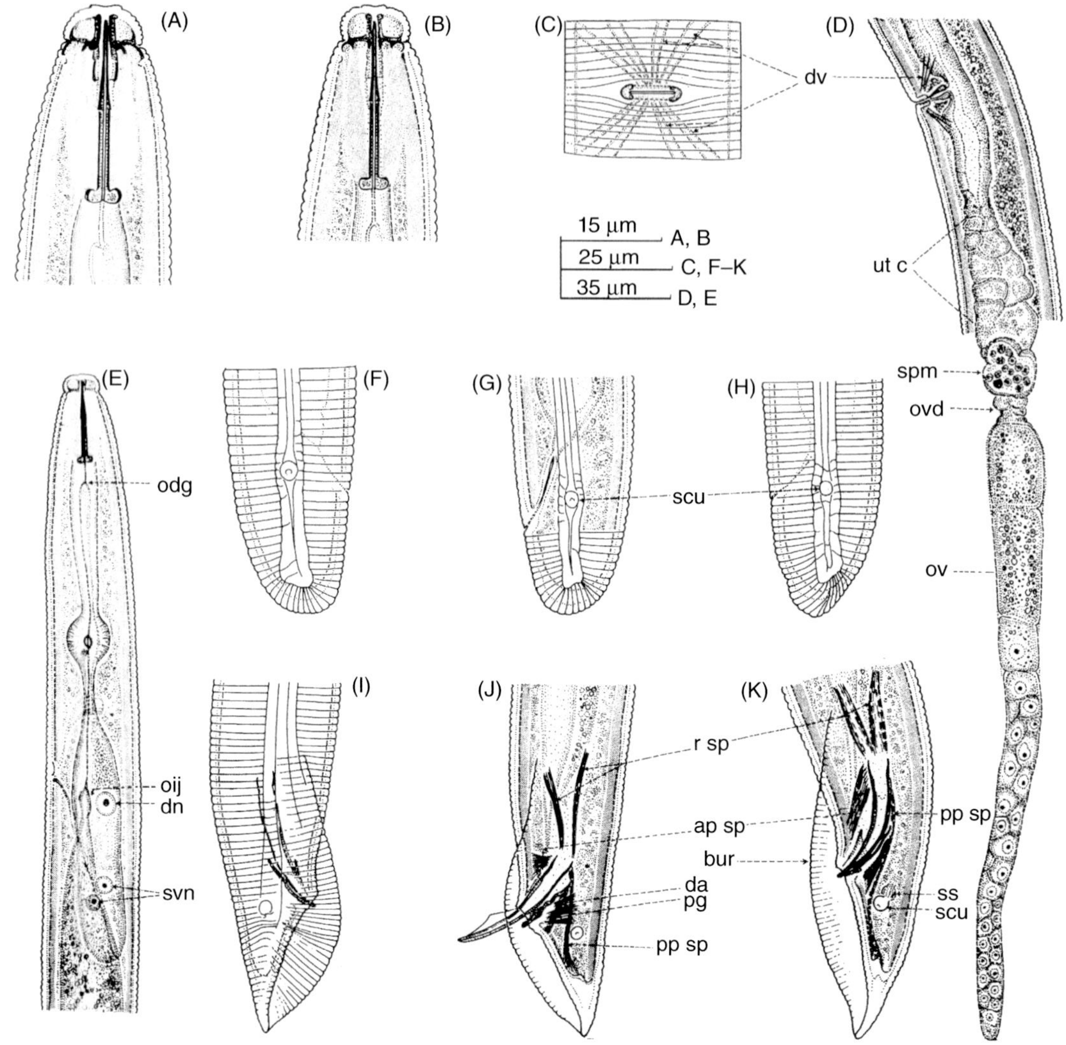

Fig. 7.18. Line drawing of *Scutellonema bradys*. Female, anterior end (A); male, anterior end (B); vulval region (C); posterior gonad (D); female oesophageal region (E); female tails (F–H); male tails (I–K). Ap sp, anterior protractor of spicule; bur, bursa; da, dilator of anus; dn, dorsal gland nucleus; dv, dilator of vulva; odg, orifice of dorsal oesophageal gland; oij, oesophago-intestinal junction; ov, ovary; ovd, oviduct; pg, protractor of gubernaculum; pp sp, posterior protractor of spicule; r sp, retractor of spicule; scu, scutellum; spm, spermatheca; ss, sensilla sac; svn, subventral gland nucleus; ut c, uterine columella. [Reproduced from Siddiqi, 2000 (after Siddiqi, 1972c); courtesy CAB International]

yam, air-potato and Asiatic yam. It also attacks and reproduces on cowpea (*Vigna unguiculata*). An expanded list of hosts is presented in Table 38.

45.8 Symptoms

S. bradys is directly associated with the dry rot disease of yam tubers (*Dioscorea* spp.). The characteristic symptom of dry rot appears in the outer skin of tubers up to 2 cm deep. At the initial stage, cream and light-yellow lesions below the outer skin of the tuber occur. When disease advances and reaches the tuber, lesions of usually 2 cm deep but occasionally deeper develop (Fig. 7.11G). At this stage, the lesions become light brown first and then turn dark brown to black. The skin of the tubers also shows external cracks (Fig. 7.11F), exposing patches of dark brown dry rot tissues, due to flaking off of infected parts. The dry rot symptoms are most severe in mature tubers, particularly in storage, and lead to decay of tubers. On above-ground parts, no specific symptoms are produced on yams growing in soil infested with *S. bradys* (Jatala and Bridge, 1990).

45.9 Biology and life cycle

S. bradys is a migratory endoparasite and attacks roots and tubers. In yams, *S. bradys* penetrates young and developing tubers through the growing point of tissues, alongside emerging roots

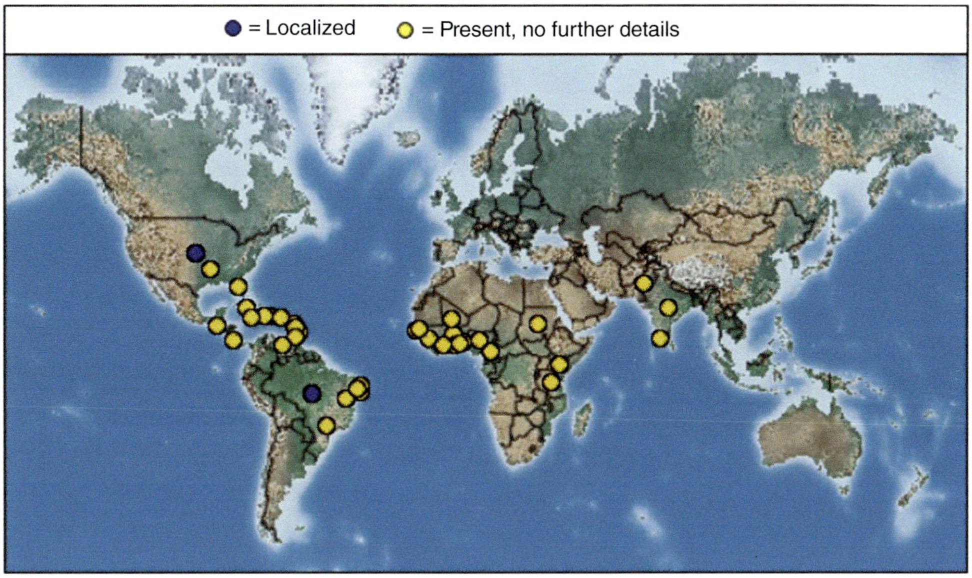

Fig. 7.19. Distribution map of *Scutellonema bradys*. [Adapted from CABI-ISC, 2020]

and shoots. The nematode also invades through roots, cracks or injured areas in the tuber skin (Bridge, 1972). *S. bradys* feed intracellularly in tissues of yam tuber, resulting in breakage of cell walls, loss of cell contents and the formation of cavities (Adesiyan, 1977). Nematodes are mainly confined to the subdermal, peridermal and underlying parenchymatous tissues in the outer 1–2 cm of the skin of tuber. The nematode continues to feed and multiply in yams during storage after harvesting. The populations can upsurge 9–14-fold in *D. rotundata* tubers over a storage period of 5–6 months, and 5 and 8 fold in *D. alata* and *D. cayenensis*, respectively, over the same period (Adesiyan, 1977). More nematodes are found in the oldest apical portions, adjacent to the stems of the partially infected tuber (Adesiyan, 1977).

Table 38. List of hosts affected by *Scutellonema bradys* (adapted from CABI-ISC, 2020).

Main hosts
Dioscorea alata (white yam)
Dioscorea bulbifera (air-potato)
Dioscorea esculenta (Asiatic yam)
Vigna unguiculata (cowpea)
Other hosts
Abelmoschus esculentus (okra)
Cajanus cajan (pigeon pea)
Celosia argentea (celosia)
Cocos nucifera (coconut)
Corchorus olitorius (jute)
Dioscorea trifida (cushcush yam)
Hibiscus cannabinus (kenaf)
Hibiscus sabdariffa (Jamaica sorrel)
Luffa aegyptiaca (loofah)
Lycopersicon esculentum (tomato)
Manihot esculenta (cassava)
Pachyrhizus erosus (yam bean)
Pueraria phaseoloides (tropical kudzu)
Sesamum indicum (sesame)
Sorghum bicolor (sorghum)
Synedrella nodiflora (synedrella)
Urena lobata (hibiscus burr)
Vigna radiata (mung bean)
Wild hosts
Catharanthus roseus (pink periwinkle)
Dioscorea (wild yam)

45.10 Economic importance

The primary economic importance of *S. bradys* is on yams, as it causes direct damage to the tubers resulting in dry rot disease (Jatala and Bridge, 1990). The nematode significantly reduces the quality, edible portions of tubers and marketable value. These reductions are more devastating in stored yam. The weight of yam is also decreased by 20–30% (Bridge, 1982) and 0–29% (Wood *et al.*, 1980). In Nigeria, approximately 47% of yam tubers in markets were infested with *S. bradys* (Bridge, 1973) and both wet rot and dry rot diseases have been observed (Adesiyan, 1977). In West Africa, *S. bradys* is also reported as a potential risk on potato, on which it can cause substantial damage (Etchiha Afoha *et al.*, 2018).

45.11 Movement and means of dispersal

See section 35.11, above.

45.12 Pest risk analysis and looming threat from introduction

S. bradys poses a substantial phytosanitary risk to yam production because of its survival and ease of dissemination within yam tubers. The principal means of dissemination of *S. bradys* is usually infected yams tuber, as whole tubers or pieces of the tuber are used for propagation. Relatively low populations of *S. bradys* in tubers do not produce any external symptom of the disease and therefore the risk of dissemination by this means is greater from the yam-growing areas to other parts of the world.

45.13 Invasiveness rating

Based on the consequences of the introduction and probable economic damage caused by *S. bradys*, the following invasiveness rating has been assigned.

Climate/host interaction	–	Medium (2)
Hosts range	–	High (3)
Dispersal potential	–	Medium (2)
Economic impact	–	High (3)
Environmental impact	–	Medium (2)
Post-entry distribution	–	Not known (0)
Final rating (score)	–	High (12)

Based on the above invasiveness rating, *S. bradys* receives a score of 12 out of 15 and is rated as HIGHLY INVASIVE NEMATODE.

45.14 Region-wise status of invasiveness

S. bradys has been listed as a quarantine nematode in many countries, including India, where it is restricted for cassava crop (PQ Order-2003, 2015). Based on economic damage potential, this nematode has given a high invasiveness rating for all the countries where *S. bradys* does not occur.

45.15 Management measures

45.15.1 Prevention and control

Using nematode-free planting material of yam is the most economic and practical means of preventing damage by *S. bradys* and also its dissemination. Hot-water treatment of tubers at 50–55°C for up to 40 min gives effective control of *S. bradys* without harming the tubers. Jatala and Bridge (1990) suggested the following management measures to control *S. bradys*: (i) controlling nematodes in field soil by cultural or chemical means; (ii) use of planting material that is naturally free of nematodes, or treatment of seed material (tubers and setts with yams) before planting to reduce or eliminate nematodes from propagative material; (iii) in the case of yams, treatment of tubers after harvesting to prevent storage losses; and (iv) hot-water treatment can reduce or eliminate *S. bradys* from yam tubers.

45.15.2 Phytosanitary measures

See Chapter 16, section 115.

8

Longidoridae: *Longidorus*, *Paralongidorus* and *Xiphinema* Species

Longidorus and *Paralongidorus* Spp.

The genera *Longidorus* and *Paralongidorus* belong to the family Longidoridae and subfamily Longidorinae (Thorne, 1935) along with genus *Longidoroides* (Khan *et al.*, 1978). Since the description of the first species as *Longidorus elongatus* by de Man (1876), nearly 170 valid species of *Longidorus* and 75 *Paralongidorus* (including 14 *Longidoroides*) have been described (Decraemer and Geraert, 2013; Barsalote *et al.*, 2018). The polytomous key was revised by Chen *et al.* (1997). Some important species of *Longidorus* and *Paralongidorus* species are:

Longidorus africanus Merny, 1966
L. americanum Handoo *et al.*, 2005
L. attenuatus Hooper, 1961
L. diadecturus Eveleigh and Allen, 1982
L. elongatus Hooper, 1961
L. fangi Xu *et al.*, 2017
L. glycines Ye and Robbins, 2004
L. juglans Xu *et al.*, 2017
L. hangzhouensis Zheng *et al.*, 2001
L. helveticus Lamberti *et al.*, 2001
L. leptocephalus Hooper, 1961
L. kheirii Pedram *et al.*, 2008
L. macrosoma Hooper, 1961
L. orientalis Loof, 1982
L. pinus Xu *et al.*, 2018
L. pisi Edward, Misra and Singh, 1964
L. polyae Lazarova *et al.*, 2019
L. sturhani Rubtsova, Subbotin, Brown and Moens, 2001
L. tabrizicus Niknam *et al.*, 2010
Paralongidorus australis Stirling and McCulloch, 1984
P. maximus (Bütschli, 1874) Siddiqi, 1964

Eleven *Longidorus* and one *Paralongidorus* species are of potentially greater importance because they act as vectors of nepoviruses (nematode-transmitted viruses with polyhedral particles) and can transmit 13 nepoviruses (Decraemer and Geraert, 2013; Xu and Zhao, 2019). Among these, *Longidorus attenuatus*, *L. diadecturus*, *L. elongatus*, *L. leptocephalus*, *L. macrosoma*, *Paralongidorus australis* (not a vector) and *P. maximus* are considered for their invasiveness as they are of economic importance and associated with plant growth suppression in different agricultural crops (Hunt, 1993; Decraemer and Geraert, 2013).

Systematic position of *Longidorus and Paralongidorus* species

	Classical (after Hunt, 1993)	Molecular (after Decraemer and Hunt, 2013)
Phylum	Nematoda	Nematoda
Class	Adenophorea	Enoplea
Subclass	Dorylaimia	Dorylaimia
Order	Dorylaimida	Dorylaimida
Suborder	Dorylaimina	Dorylaimina
Superfamily	Dorylaimoidea	Dorylaimoidea
Family	Longidoridae	Longidoridae
Subfamily	Longidorinae	Longidorinae
Genus	*Longidorus* *Paralongidorus*	*Longidorus* *Paralongidorus*

46 *Longidorus attenuatus*

46.1 Common name

Needle nematode

46.2 Scientific name

Longidorus attenuatus Hooper, 1961

46.3 EPPO code

LONGAT

46.4 Diagnosis

46.4.1 Morphological diagnosis

Morphological and morphometric descriptions after Brown and Boag (1977) are presented below. For measurements, see Table 39.

DOI: 10.1079/9781789247367.0009

Female: Body long and narrow, C-shaped with slightly greater curvature in posterior half when heat relaxed (Fig. 8.1B). Lip region flattened, prolonged and offset from neck contour (Fig. 8.1A), about three-quarters of width of neck at guiding ring. Amphids large with characteristic symmetrical bilobed base, extending halfway to guiding ring and nearly encircling head. Pores amphid very minute. Odontostyle large but narrow, twice length of odontophore with simple junction (Fig. 8.1C). Anterior oesophagus slender tube, joins wider muscular posterior oesophageal–intestinal valve, rounded. Lateral cords around half mid-body width, narrowing to thin line in pharyngeal region where lateral pores arranged serially. Oesophageal region 5–8 ventral body pores. Vulva a transverse slit one-quarter of mid-body width. Tail conoid, dorsally convex. Pair of caudal pores on each side of tail (Fig. 8.1J-K).

Male: Rare. Spicule long, lateral guiding piece with bifid distal ends. Adanal pair supplements plus ventral series of ten present (Fig. 8.1E). Tail similar to female, dorsally convex and narrowly conoid (Fig. 8.1E).

Juveniles: Lip region and amphids similar to adult. Tail of first stage conical, about four anal body widths long (Fig. 8.1F–I).

46.4.2 Molecular diagnosis

Over the past two decades, more than 45% of species from family Longidoridae including *Longidorus*, *Paralongidorus* and *Xiphinema* species have been molecularly characterized by various approaches, such as the sequencing of ribosomal RNA genes (rDNA) of partial 18S, ITS regions, or the D2 and D3 expansion segments of the 28S rRNA gene or large subunit (LSU) rRNA nuclear gene and protein-coding mitochondrial gene cytochrome c oxidase subunit 1 (COX1) (Handoo *et al.*, 2005; Archidona-Yuste *et al.*, 2016a; Susulovska *et al.*, 2016; Xu *et al.*, 2018; Amrei *et al.*, 2019).

Palomares-Rius *et al.* (2017) used COX1 as a barcode as well as a phylogenetic marker for species identification of Longidoridae nematodes, including *L. aetnaeus*, *L. leptocephalus* and *Xiphinema americanum*. The DNA was extracted from a single

Table 39. Morphometric data for *Longidorus* and *Paralongidorus* species.

Morphometric characters		Range, numerous populations			
		L. attenuatus (after Hooper, 1961)	*L. elongatus* (after Hooper, 1961)[a]	*P. australis* (after Stirling and McCulloch, 1984)	*P. maximus* (after Sturhan, 1963)
Females	Length (mm)	6.0	4.5–6.4	7.60–10.60	7.60–12.35
	Odontostyle (μm)	75	81–102	146–170	152–187
	Odontophore (μm)	43	34–71	101–124	42–70
	Oral aperture to guiding ring	29	30–36	58–70	37–47
	a	201	76–123	85.5–116.0	77–133
	b	18.9	9.7–17.5	11.2–18.0	12.8–23.7
	c	107	73–141	255.0–426.0	178–320
	c'	–	–	0.48–0.59	–
	V	497	45–53	46–56	36–40
	Tail length (μm)	55	–	21–35	26–38
Males	Length (mm)	5.5–6.3	–	7.65–10.57	8.55–10.70
	Odontostyle (μm)	80–84	–	146–171	171–173
	Odontophore (μm)	30–40	–	112–123	52–56
	Oral aperture to guiding ring	30–32	–	56–73	39.5–43.0
	a	119–171	–	86.9–120.1	97–110
	b	12.9–17.1	–	11.7–16.4	12.9–14.1
	c	99–114	–	215.5–336.0	190–237
	c'	–	–	0.48–0.68	–
	Spicule (μm)	–		112–134	100–106
Second stage juvenile	Length (μm)	1500–1800	1.5–2.1	–	–
	Odontostyle (μm)	43–52	59–67	–	–
	Odontophore (μm)	27–32	34–49	–	–
	Oral aperture to guiding ring	57–67	70–77	–	–
	a	54–72	47–63	–	–
	b	6.2–6.5	5.3–8.4	–	–
	c	31–35	27–43	–	–

[a]Male very rare, usually absent.

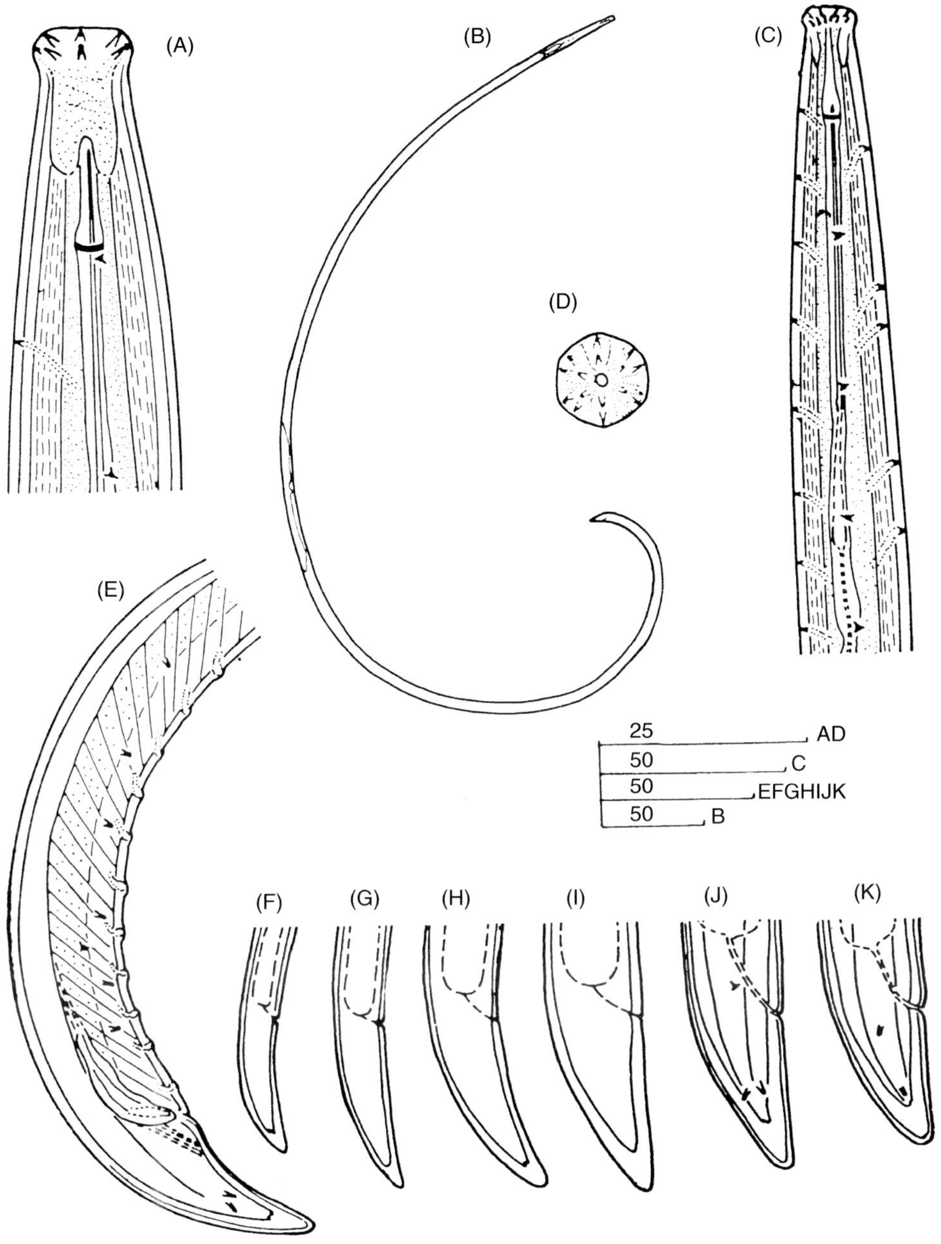

Fig. 8.1. Line drawing of *Longidorus attenuatus*. Head lateral (A); adult female (B); anterior oesophageal region ventral (C); lip region (D); male tail (E); larva tails (F-I); female tail (J, K). [Reproduced from Brown and Boag, 1977, CIH Descriptions of Plant-parasitic Nematodes, Set 7, No. 101; all except B after Hooper, 1961; courtesy CAB International]

nematode as described by Tanha Maafi *et al.* (2003) (see Chapter 15, section 110.1.1) and PCR assays were performed using the following primers.
For partial COX1 gene (Lazarova *et al.*, 2006)

COIF (5′-GATTTTTTGGKCATCCWGARG-3′)
COIR (5′-CWACATAATAAGTATCATG-3′)
XIPHR1 (5′-ACAATTCCAGTTAATCCTCCTACC-3′) or
XIPHR2 (5′-GTACATAATGAAAATGTGCCAC-3′)
JB3 (5′-TTTTTTGGGCATCCTGAGGTTTAT-3′) (Bowles *et al.*, 1992)
JB4 (5′-TAAAGAAAGAACATAATGAAAATG-3′)

The partial COX1 gene for mtDNA was amplified using PCR cycle conditions: 1 cycle for 1 min at 94°C, a further 1 min at 50°C and 2 min at 72°C; followed by 40 cycles for 1 min at 94°C, 1 min at 45°C and 2 min at 72°C. Final extension for 1 min at 94°C, 1 min at 45°C and 5 min at 72°C. The D2–D3 region was obtained using a protocol and primers described by Subbotin *et al.* (2007) (see Chapter 4, section 15.5.2) and Archidona-Yuste *et al.* (2016a). PCR products were purified and directly sequenced in both directions and the sequences were submitted to the GenBank database under the accession numbers KY816588–KY816723.

He *et al.* (2005) used the D2–D3 expansion region to sequence the large subunit (LSU) rRNA nuclear gene and construct the phylogeny of longidorids including *L. attenuatus*, *L. diadecturus*, *L. elongatus*, *L. leptocephalus*, *L. macrosoma* and *P. maximus*. Susulovska *et al.* (2016) also characterized *L. attenuatus* based on D2–D3 28S rDNA region sequence. Hübschen *et al.* (2004) developed species-specific primers for diagnosis of virus-vector longidorid nematodes, particularly the species associated with German viticulture.

46.5 Geographical distribution

L. attenuatus is primarily distributed in European countries and one African country (Nigeria). It has been recorded from Bulgaria, France, Germany, Italy, Nigeria, Poland, Russia, Spain, Ukraine and the United Kingdom (Cooke, 1993; Susulovska *et al.*, 2016).

46.6 Hosts

L. attenuatus mainly parasitizes field crops (Brown and Boag, 1977), namely barley (*Hordeum vulgare*), clover (*Trifolium* spp.), potato (*Solanum tuberosum*), ryegrass (*Lolium* spp.), sugar beet (*Beta vulgaris*) and wheat (*Triticum* spp.). Other crops infected directly by the nematode or by its virus vector include alfalfa (*Medicago sativa*), cabbage (*Brassica* spp.), carrot (*Daucus carota*), cucurbits (*Cucurbita* spp.), strawberry (*Fragaria* × *Ananassa*) and grape (*Vitis* spp.).

46.7 Vector

L. attenuatus is a vector of English and German strains of tomato black ring virus, TomBRV (Brown and Boag, 1977), which makes it a significant problem in the viticulture industry (Hübschen *et al.*, 2004).

46.8 Symptoms

L. attenuatus is ectoparasitic, feeds on root tips and causes small apical galls. The symptoms are similar to those of other *Longidorus* species. The nematodes feed exclusively at, or just behind, root tips, usually causing galls. Feeding by needle nematodes generally leads to an extensive reduction in the root systems, with lateral and sometimes tap roots being severely stunted. Seedlings and young cuttings are highly sensitive to damage caused by *Longidorus* species, appearing as cessation of growth of the main root system and short, swollen lateral roots usually with galls on root tips. Crop damage caused by needle nematodes is typically visible as 'disease patches', usually extending slowly each year and eventually these may extend to cover a whole field. The affected plants become susceptible to drought and may exhibit leaf malformations or discoloration as chlorosis. On sugar beet, *L. attenuatus* causes docking disorder. It induces galls on root tips, but sometimes galls are also formed on the later roots and exhibit a distal necrotic spot (Cooke, 1993). The nematode also causes stunting of strawberry in Germany (Wyss, 1970).

46.9 Biology and life cycle

L. attenuatus is reproduced through parthenogenesis. Larvae emerge from eggs and there are four life stages. The basic life cycle pattern of *L. attenuatus* is similar to that of other *Longidorus* nematodes and consists of six stages: the egg (or embryo), four juvenile stages and the adult. Hatching usually occurs as soon as the development of the first-stage juvenile is completed, depending upon environmental conditions such as moisture, temperature and host. Being a dorylaim, first moulting in *L. attenuatus* larvae takes place after hatching. The life cycles of *Longidorus* species are estimated to be completed in a few months to several years, depending on the species (Flegg, 1968). *L. attenuatus* takes more than 1 year to complete its life cycle. The nematode survives in varied types of soils, but prefers those containing more than 90% sand (Brown and Boag, 1977).

46.10 Economic importance

L. attenuatus is a destructive pest of agronomic and horticultural crops. The nematode is associated with docking disorder of sugar beet, which causes a significant decline in crop yield (Cooke, 1993). This nematode can transmit TomBRV, which is a devastating pathogen of grape (Brown *et al.*, 1993).

46.11 Movement and means of dispersal

L. attenuatus is an ectoparasite and usually spread with irrigation water, or soil attached to tools, shoes and farm machinery. Dissemination may also occur when infested seedlings are transplanted.

46.11.1 Plant parts known to carry the pest in trade/transport

Bulbs, corms, rhizomes (sucker), roots, tubers, growing medium accompanying plants, seedlings and micro-propagated plants.

46.11.2 Plant parts not known to carry the pest in trade/transport

Bark, branches, cones, calyx, flowers, inflorescences, fruits (including pods), leaves, stems (above ground), shoots, trunks, true seeds (including grain) and wood.

46.12 Pest risk analysis and looming threat from introduction

Considering the economic damage caused by this nematode in field crops, a complete pest risk analysis for *L. attenuatus* is suggested. The nematode can be easily detected and excluded because of its ectoparasite nature. The primary pathway of entry would be soil, which at present cannot be imported along with the plants. However, due to its vector role and restricted distribution, preventive steps should be taken to avoid introduction into new host areas.

46.13 Invasiveness rating

Based on the consequences of the introduction and probable pest risk analysis of *L. attenuatus,* the following invasiveness rating has been assigned.

Climate/host interaction	–	Medium (2)
Hosts range	–	Medium (2)
Dispersal potential	–	Medium (2)
Economic impact	–	High (3)
Environmental impact	–	Low (1)
Post-entry distribution	–	Not known (0)
Final rating (score)	–	Medium (10)

Based on the above invasiveness rating, *L. attenuatus* receives a score of 10 out of 15 and is rated as MEDIUM INVASIVE NEMATODE.

46.14 Region-wise status of invasiveness

L. attenuatus is an EPPO A2 List pathogen, mainly because of the virus transmission capability of TomBRV (EPPO, 2012), and it should be considered as a quarantine pest if detected. The nematode is given a medium invasiveness rating for the following countries where it is still not known to occur.

APPPC: Bangladesh, Cambodia, China, India, Indonesia, Iran, Iraq, Israel, Japan, Kazakhstan, Kyrgyzstan, Lebanon, Malaysia, Myanmar, Nepal, Pakistan, Philippines, Saudi Arabia, Singapore, South Korea, Sri Lanka, Taiwan, Tajikistan, Thailand, Vietnam, Uzbekistan

CAHFSA: Barbados, Belize, Cuba, Dominican Republic, Haiti, Jamaica, Puerto Rico, Saint Kitts & Nevis, Trinidad and Tobago

CAN: Bolivia, Colombia, Ecuador, Peru, Venezuela

COSAVE: Argentina, Brazil, Chile, Guyana, Paraguay, Suriname, Uruguay

EPPO: Austria, Belarus, Belgium, Croatia, Cyprus, Czech Republic, Denmark, Estonia, Finland, Georgia, Greece, Guernsey, Hungary, Ireland, Jersey, Kazakhstan, Kyrgyzstan, Latvia, Lithuania, Luxembourg, Macedonia, Malta, Moldova, Morocco, Netherlands, Norway, Portugal, Romania, Serbia, Slovakia, Slovenia, Sweden, Switzerland, Turkey

IAPSC: Algeria, Angola, Benin, Botswana, Burkina Faso, Burundi, Cameroon, Cape Verde, Central African Republic, Chad, Comoros, Congo, Côte d'Ivoire, Eritrea, Ethiopia, Gabon, Gambia, Ghana, Guinea, Guinea-Bissau, Kenya, Liberia, Madagascar, Malawi, Mali, Mauritania, Mauritius, Mozambique, Namibia, Niger, Réunion, Rwanda, Senegal, Sierra Leone, Somalia, South Africa, Swaziland, Tanzania, Togo, Uganda, Zambia, Zimbabwe

NAPPO: Canada, Mexico, USA

NEPPO: Egypt, Jordan, Libya, Malta, Sudan, Syria, Tunisia

OIRSA: Costa Rica, El Salvador, Guatemala, Honduras, Nicaragua, Panama

PPPO: Australia, Fiji, New Zealand, Papua New Guinea, Tonga, Vanuatu.

46.15 Management measures

46.15.1 Prevention and control

Earlier, application of fumigant nematicides such as DD (1,3-dichloropropene, 1,2-dichloropropane) mixture or chloropicrin or DBCP (1,2-dibromo-3-chloropropane) were found quite effective against the needle nematode and increased the yield of sugar beet (Brown and Boag, 1977). However, currently most of the soil fumigants are superseded and good control can be achieved by the application of organophosphate or carbamate nematicides, although their cost-effectiveness may be poor.

46.15.2 Phytosanitary measures

See Chapter 16, section 115.

47 *Longidorus elongatus*

47.1 Common name

Needle nematode

47.2 Scientific name

Longidorus elongatus (de Man, 1876) Micoletzky, 1922

47.3 Synonyms

Dorylaimus elongatus de Man, 1876
Dorylaimus tenuis von Linstow, 1879
Longidorus menthasolanus Konicek and Jensen, 1961
Longidorus monohystera Altherr, 1953
Trichodorus elongatus (de Man, 1876) Filipjev, 1921

47.4 EPPO code

LONGEL

47.5 Diagnosis

47.5.1 Morphological diagnosis

This morphological description of *Longidorus elongatus* is adapted from Hunt, 1993 (after Hooper, 1973b). For morphometric measurements (after Hooper, 1961), see Table 39.

FEMALE: Female body long (4.5–6.4 mm) and narrow, open C-shaped to slightly twisted when heat relaxed. Cuticle generally smooth, inner layer thickens in neck region and on tail and has very faint transverse striations that have radial appearance on tail. Cephalic region flattened, anteriorly narrows, slightly offset from neck contour (Fig. 8.2C). Eight pairs of papillae directed away from oral opening and present at conspicuous guiding ring, around half to two-thirds the body width. Amphids prominent, pouch-like, covering around half length from oral opening to guiding ring. Base of amphidial pouch a little bilobed in lateral view. Amphid apertures pore-like, obscure, located at base of labial region. Odontostyle long and narrow with modest junction with odontophore (Fig. 8.2B). Odontophore not very conspicuous, about half the odontostyle length and widening and thickening posteriorly. Oesophagus anteriorly a narrow tube often bent or looped as it joins wider muscular posterior oesophageal bulb that is approximately five times as long as wide (Fig. 8.2A). Oesophago-intestinal valve bluntly convex-conoid. Dorsal oesophageal and subventral gland openings and associated glands at about 15% and 59% from anterior end of oesophageal bulb. Second subventral opening at 85% of its length also suggested. Glandular hypodermal lateral chords present around one-quarter of mid-body width wide but narrowing to thin line in oesophageal region. Lateral pores arise from chords throughout length of body and mainly in staggered double row but merging to single row in oesophageal region. Four to six ventral pores in oesophageal region and occasionally single dorsal pore somewhat posterior to guiding ring. One nerve ring encircling anterior oesophagus just behind base of odontophore; second nerve ring about a body width behind first occasionally noticeable. Pre-rectum about ten and rectum less than one anal body width. Tail dorsally and ventrally flattened, convex or concave very slightly, around 1–1.3 anal body widths long with roundly conoid end and two or three lateral pores on each side of tail (Fig. 8.2O, P). Vulva a transverse slit around one-third of mid-body width (Fig. 8.2H). Genital tracts paired, opposed, reflexed; reflexed part contains mainly ovary in which larger oocytes are in single row, the rest multiple.

MALE: Usually very rare or absent in most of populations, but may be almost as common as females in some populations. Anterior region similar to female, posterior third ventrally curved when heat killed. Spicules paired, prominent, long (58 μm), separate, arcuate with bluntly rounded distal ends (Fig. 8.2G). Lateral guiding pieces present with furcate distal ends. Tail conoid to bluntly conoid, dorsally convex, ventrally concave with two or three lateral pores on both sides (Fig. 8.2N). One adanal pair plus a ventral series of 7–10 supplementary papillae. Paired testes, opposed, anterior testis nearly reaching to middle of body.

JUVENILE: Anterior region similar to adults but replacement odontostyle always present in anterior oesophagus. In first stage, anterior end lies inside odontophore of functional odontostyle. Juvenile tails become blunter and shorter relative to anal body width as they progress from first to fourth stage, but actual length of tail remains same (Fig. 8.2J-M).

47.5.2 Molecular diagnosis

See section 46.4.2, above.

47.6 Geographical distribution

L. elongatus is a northern species and widespread in cooler regions of northern Europe and Russia, with outlying populations further south (Decraemer and Geraert, 2013). Besides Europe, it is also reported to occur in some Asian countries, North America, South Africa and New Zealand (Fig. 8.3). *L. elongatus* has been detected in the following countries (CABI-ISC, 2020).

Africa: South Africa
Asia: India, Kazakhstan, Pakistan, Tajikistan, Uzbekistan, Vietnam.
Europe: Austria, Belgium, Bulgaria, Czech Republic, Denmark, Estonia, Finland, France, Germany, Greece, Hungary, Ireland, Italy, Latvia, Moldova, Netherlands, Norway, Poland, Portugal, Romania, Russia, Slovakia, Slovenia, Spain, Sweden, Switzerland, Ukraine, United Kingdom.
North America: Canada, USA.
Oceania: New Zealand.

47.7 Hosts

L. elongatus is a polyphagous species with a wide host range, including many agronomic crops and some weeds (Hooper, 1973b; Decraemer and Geraert, 2013). This needle nematode is frequently associated with sugar beet, carrot, peppermint, raspberry, redcurrant, strawberry, grasses and fruit trees. The expanded list of hosts is presented in Table 40.

47.8 Vector

L. elongatus transmits raspberry ringspot virus (RRV, or RRSV), tomato black ring virus (TomBRV) and spoon leaf virus. RRV is easily transmitted to raspberry although the plant is a poor host of *L. elongatus* (Hooper, 1973b).

47.9 Symptoms

L. elongatus is a polyphagous ectoparasite attacking the roots of numerous plants (see Table 40). This needle nematode tends to invade near and just behind the root tips where it feeds ectoparasitically, which results in the formation of galls on the root tip (Khan, 2008). Overall, the plant shows stunting of the root system. Above-ground symptoms are

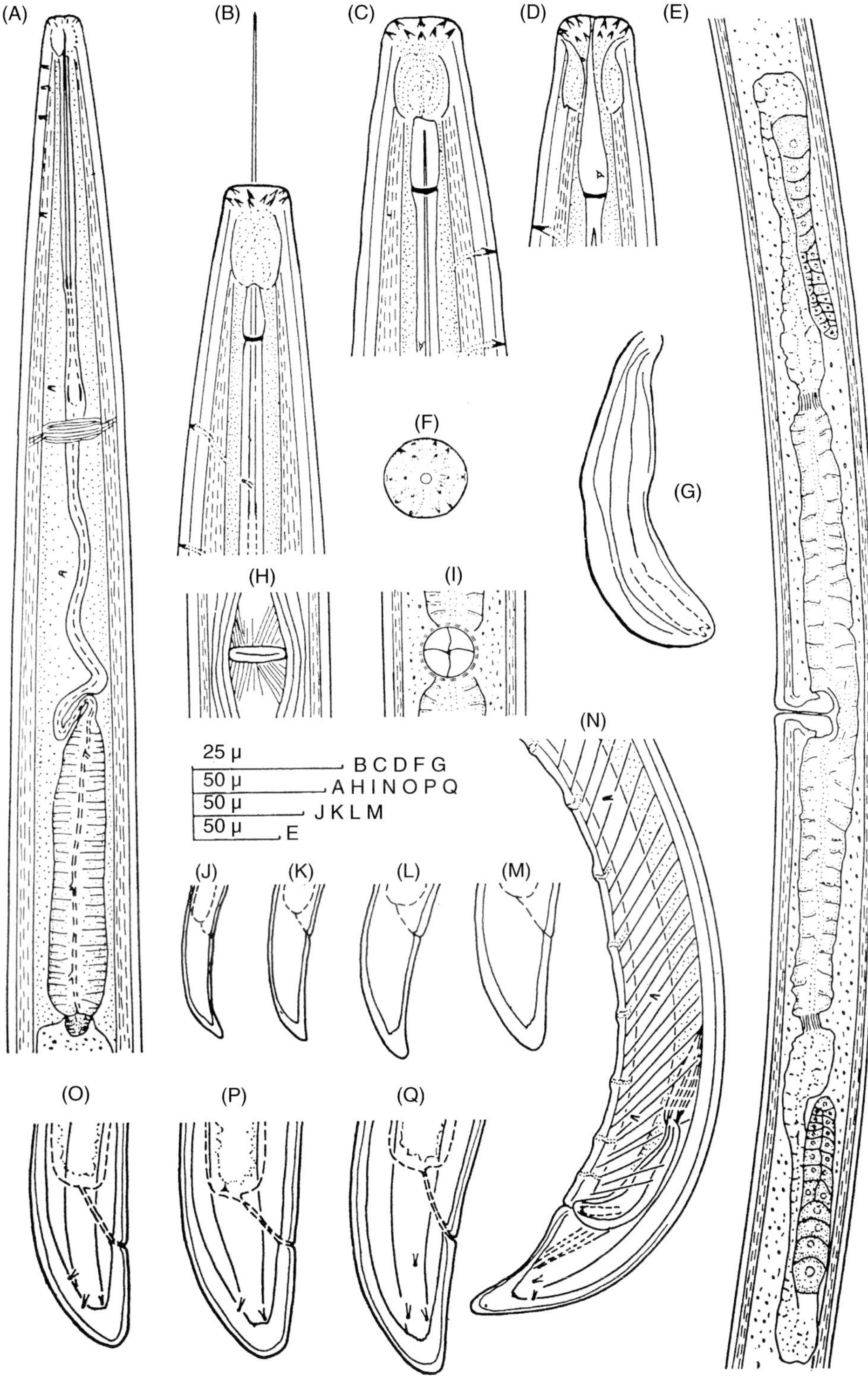

Fig. 8.2. Line drawing of *Longidorus elongatus*. Oesophageal region, dorso-lateral (A); head, lateral (B); female head, lateral (C); female head, ventral (D); female gonad, lateral (E); lip region *en face* view (F); spicules and guiding piece, lateral (G); vulva, ventral (H); vagina, ventral (I); larval tail, lateral (J–M); male tail, lateral (N); female tail, lateral (O–Q). [Reproduced from Hooper, 1973; CIH Descriptions of Plant-parasitic Nematodes, Set 2, No. 30; courtesy CAB International]

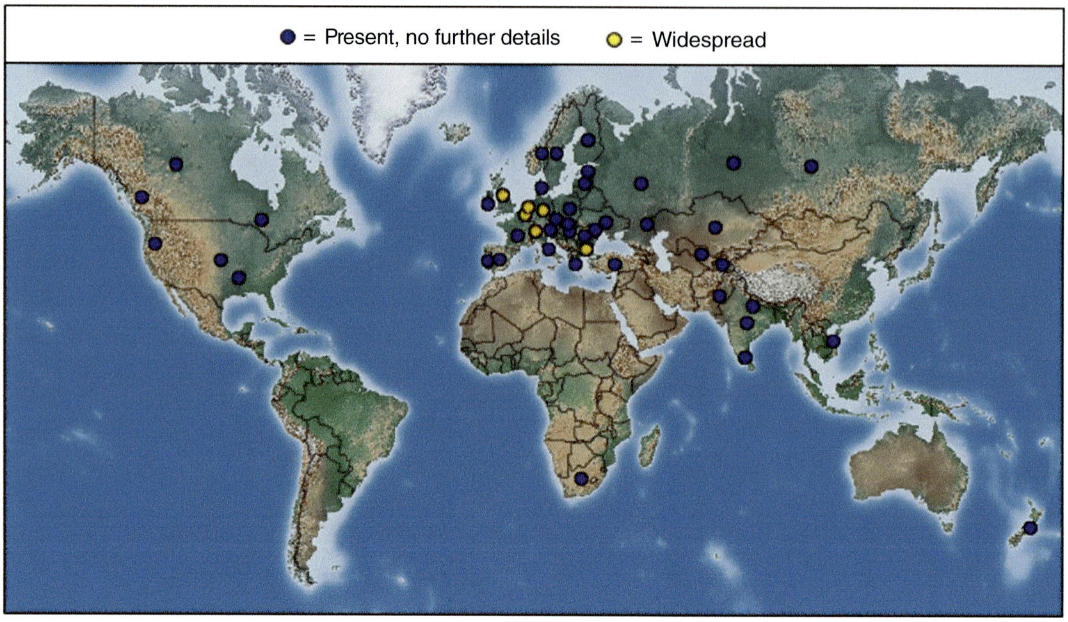

Fig. 8.3. Distribution map of *Longidorus elongatus*. [Adapted from CABI-ISC, 2020]

Table 40. List of hosts of *Longidorus elongatus* (adapted from CABI-CPC, 2010).

Main hosts
Beta vulgaris (beetroot)
Daucus carota (carrot)
Fragaria ananassa (strawberry)
Lolium perenne (perennial ryegrass)
Mentha piperita (peppermint)
Other hosts
Allium spp. (onions, garlic, leek, etc.)
Apium graveolens (celery)
Arachis hypogaea (groundnut)
Avena sativa (oats)
Brassica oleracea var. *capitata* (cabbage)
Brassica oleracea var. *gemmifera* (Brussels sprouts)
Brassica rapa subsp. *rapa* (turnip)
Ficus carica (fig)
Hordeum vulgare (barley)
Lactuca sativa (lettuce)
Malus domestica (apple)
Medicago sativa (lucerne)
Polygonum convolvulus (black bindweed)
Pyrus communis (European pear)
Ranunculus acris (meadow buttercup)
Raphanus sativus (radish)
Ribes nigrum (blackcurrant)
Ribes rubrum (red currant)
Rosa (roses)
Rubus idaeus (raspberry)
Solanum lycopersicum (tomato)
Trifolium pratense (purple clover)
Urtica dioica (stinging nettle)
Vitis vinifera (grapevine)
Unknown hosts
Anthemis cotula (dog fennel)
Chenopodium album (fat hen)
Chenopodium quinoa (quinoa)
Capsella bursa-pastoris (shepherd's purse)
Dactylis glomerata (orchardgrass)
Fragaria vesca (wild strawberry)
Lolium multiflorum (Italian ryegrass)
Poa annua (annual meadowgrass)
Stellaria media (common chickweed)
Senecio vulgaris
Symphytum officinale (blackwort)
Tussilago farfara (colt's-foot)
Urtica urens (annual nettle)

correlated to the obliteration of the root system (Hooper, 1973b). Specific plant viruses may also be transmitted during the feeding process, leading to typical above-ground virus symptoms (Fig. 8.4B). The long odontostyle is used to pierce deep into the root, with feeding taking place at all levels of host tissue within reach of the odontostyle. Direct feeding activities can severely damage carrot, peppermint, strawberry and sugar beet roots. The affected root systems become stunted, with galled root tips (Fig. 8.4A). *L. elongatus* has been implicated in docking disorder of sugar beet (Whitehead and Hooper, 1970).

47.10 Biology and life cycle

The biology of *L. elongatus* is similar to that of *L. attenuatus* (see section 46.9, above). Males are rare and reproduction is

Fig. 8.4. Symptoms of *Longidorus* spp. Root-tip galling on sugar beet caused by the feeding of *L. elongatus* (A); strawberries in eastern Scotland affected by tomato black ring and raspberry ringspot nepoviruses, transmitted by *L. elongatus* (B); *L. macrosoma* on *Petunia hybrida* showing root-tip galls (C); *L. macrosoma* in soil (D). [Courtesy Derek J.F. Brown]

usually through parthenogenesis. The nematode occurs in various soil types but mainly prefers coarse and well-drained soil (Decraemer and Geraert, 2013). The soil depth also influences the occurrence of longidorids and is associated with the host root system and environmental condition (soil moisture, temperature, etc.). The life cycle may be completed in around 65 days at 30°C, and usually one generation is completed in a year under field conditions. Large populations may build up in uninterrupted soils in perennial crops (Boag *et al.*, 1991).

47.11 Similarities with other organisms

One of the problems in identifying *L. elongatus* is the wide morphometric variation that occurs over its reported range; often these variations may indicate the existence of a species complex. *L. africanus* cohabits with nearly all the populations of *L. elongatus* in North Africa (Brown and Taylor, 1987).

47.12 Economic importance

The major economic importance of *L. elongatus* is because of its ability to transmit several important plant viruses, including raspberry ringspot virus (RRV), spoon leaf virus (SLV) and tomato black ring virus (TomBRV). Direct feeding on the root tips may cause severe damage, even death, to seedlings of crops such as sugar beet and grass (Hooper, 1973b).

47.13 Movement and means of dispersal

See section 46.11, above.

47.14 Pest risk analysis and looming threat from introduction

Most countries now have rigorous legislative regulations that prohibit the entry of plants, plant products and commodities that include nematodes. Considering the economic damage caused by *L. elongatus* and its virus transmission ability, EPPO has assigned it to the A2 pest list concerning fruit-plant importations into Europe (EPPO, 2012). Due to its vector role and confined distribution, mainly in Europe, preventive steps should be taken to avoid introduction into new host areas.

47.15 Invasiveness rating

Based on the consequences of the introduction and probable pest risk analysis of *L. elongatus*, the following invasiveness rating has been assigned.

Climate/host interaction	–	High (3)
Hosts range	–	Medium (2)
Dispersal potential	–	Medium (2)
Economic impact	–	High (3)
Environmental impact	–	Medium (2)
Post-entry distribution	–	Not known (0)
Final rating (score)	–	High (12)

Based on the above invasiveness rating, *L. elongatus* receives a score 12 out of 15 and is rated as HIGHLY INVASIVE NEMATODE.

47.16 Region-wise status of invasiveness

The European and Mediterranean Plant Protection Organization (EPPO) includes raspberry ringspot nepovirus in its A2 pest list. The virus is mainly transmitted by *L. elongatus*, which practically has quarantine status with regard to fruit-plant importations into Europe (EPPO, 2012) (see section 47.14 above). This nematode has been given a high invasiveness rating for the following countries where it is still not known to occur.

APPPC: Bangladesh, Cambodia, China, Indonesia, Iran, Iraq, Israel, Japan, Kyrgyzstan, Lebanon, Malaysia, Myanmar, Nepal, Philippines, Saudi Arabia, Singapore, South Korea, Sri Lanka, Taiwan, Thailand

CAHFSA: Barbados, Belize, Cuba, Dominican Republic, Haiti, Jamaica, Puerto Rico, Saint Kitts & Nevis, Trinidad and Tobago

CAN: Bolivia, Colombia, Ecuador, Peru, Venezuela

COSAVE: Argentina, Brazil, Chile, Guyana, Paraguay, Suriname, Uruguay

EPPO: Belarus, Croatia, Cyprus, Denmark, Georgia, Guernsey, Jersey, Lithuania, Luxembourg, Macedonia, Malta, Morocco, Serbia, Turkey

IAPSC: Algeria, Angola, Benin, Botswana, Burkina Faso, Burundi, Cameroon, Cape Verde, Central African Republic, Chad, Comoros, Congo, Côte d'Ivoire, Eritrea, Ethiopia, Gabon, Gambia, Ghana, Guinea, Guinea-Bissau, Kenya, Liberia, Madagascar, Malawi, Mali, Mauritania, Mauritius, Mozambique, Namibia, Niger, Nigeria, Réunion, Rwanda, Senegal, Sierra Leone, Somalia, Swaziland, Tanzania, Togo, Uganda, Zambia, Zimbabwe

NAPPO: Mexico

NEPPO: Egypt, Jordan, Libya, Malta, Sudan, Syria, Tunisia

OIRSA: Costa Rica, El Salvador, Guatemala, Honduras, Nicaragua, Panama

PPPO: Australia, Fiji, Papua New Guinea, Tonga, Vanuatu.

47.17 Management measures

47.17.1 *Cultural control*

In Scotland's raspberry plantations, fruiting canes were cut out at the end of the season and then they were thrashed regularly and rotary cultivated into the soil between the rows. This procedure reduced the numbers of *L. elongatus* in plantations (Taylor and Murant, 1966), but similar treatment failed to give good results in strawberry plants (Taylor and Murant, 1966). Boag (1983) also found that rotary cultivations of the soil killed more *L. elongatus* than tylenchid nematodes. Weeds should be removed, as they serve as alternative hosts and increase the virus burden on the crop, particularly if the crop itself is a poor host for the vector, as found with *L. elongatus* and raspberry ringspot nepovirus in raspberry (Taylor, 1967).

47.17.2 *Chemical control*

Systemic nematicides are generally regarded as unsuitable for control of vector-nematodes in perennial crops, but applications of oxamyl gave considerable control of raspberry ringspot transmitted by *L. elongatus* in raspberry plantations (Trudgill and Alphey, 1976). Taylor and Gordon (1970) compared the effectiveness of fungicide quintozene with DD (1,2-dichloropropane, 1, 3-dichloropropene) and methomyl to control *L. elongatus*. The nematicidal effect of methomyl lasts for only a short time, while quintozene acted slowly and the population density of *L. elongatus* was reduced to the level achieved by DD treatment. The treatment was equally efficient in preventing the transmission of raspberry ringspot nepovirus and tomato black ring nepovirus to strawberry crops (Taylor and Murant, 1968).

47.17.3 *Phytosanitary measures*

See Chapter 16, section 115.

48 *Longidorus macrosoma*

48.1 Common name

Needle nematode

48.2 Scientific name

Longidorus macrosoma Hooper, 1961

48.3 EPPO code

LONGMA

48.4 Diagnosis

48.4.1 Morphological diagnosis

FEMALE: Body very long, 6.8–12.0 mm in length, sometimes visible to naked eye (Fig. 8.4D). Open C-shaped to slightly coiled when heat relaxed. Cuticle generally smooth, inner layer thickens in neck region and on tail and has very faint transverse striations that have radial appearance on tail. Cephalic region flattened, anteriorly narrows, slightly offset from neck contour. Tail dorsally convex and conoid.
MALE: similar to female in shape. Spicule long, lateral guiding piece with bifid distal ends (Brown and Boag, 1977).

48.4.2 Molecular diagnosis

See section 46.4.2, above.

48.5 Geographical distribution

L. macrosoma is known to occur in Austria, Belgium, Bulgaria, Czech Republic, France, Germany, Ireland, Italy, Netherlands, Russia, Slovakia, Spain, Switzerland, United Kingdom and former Yugoslavia (Buser, 1990; Xu and Zhao, 2019).

48.6 Hosts

The needle nematode *L. macrosoma* is a polyphagous species and parasitizes apple (*Malus sylvestris*), *Fagus sylvatica*, *Prunus avium*, *Pyrus communis*, oak (*Quercus robur*) and red raspberry (*Rubus idaeus*) (Xu and Zhao, 2019).

48.7 Vector

L. macrosoma is a vector of the English strain of raspberry ringspot nepovirus (Buser, 1990).

48.8 Symptoms

Symptoms are typical of the genus, including galls or swelling on root tip (Fig. 8.4C).

48.9 Biology and life cycle

L. macrosoma is reproduced parthenogenetically, in the manner of other *Longidorus* spp. Juveniles hatch from eggs and there are four juvenile stages and adults. The nematode prefers light soils. The depth of soil also influences its occurrence; as the depth of soil increased, the number of *L. macrosoma* increased up to 70 cm depth (Flegg, 1968), and more than 800 *L. macrosoma* were isolated at a depth of 60 cm in raspberry plantations (Cotten, 1975). In Germany, Italy and the United Kingdom, *L. macrosoma* occurs in soils with a large clay fraction overlying chalk (calcareous) outcrops (Taylor *et al.*, 1978). Individuals of *L. macrosoma* can survive for up to 5 years in plant-free soil under natural conditions (Buser, 1990).

48.10 Economic importance

This needle nematode species is a damaging pest of fruit trees. Besides, it transmits raspbery ringspot nepovirus, which is devastating plant virus, and may cause additional damage to host crops (Brown *et al.*, 1993).

48.11 Movement and means of dispersal

See section 46.11, above.

48.12 Pest risk analysis and looming threat from introduction

Considering the economic damage caused by this needle nematode to fruit trees, a complete pest risk analysis is suggested. Due to its vector role and restricted distribution, preventive steps should be taken to avoid introduction into new host areas.

48.13 Invasiveness rating

Based on the consequences of probable introduction and pest risk analysis of *L. macrosoma*, the following invasiveness rating has been assigned.

Climate/host interaction	–	Medium (2)
Hosts range	–	Medium (2)
Dispersal potential	–	Medium (2)
Economic impact	–	High (3)
Environmental impact	–	Low (1)
Post-entry distribution	–	Not known (0)
Final rating (score)	–	Medium (10)

Based on the above invasiveness rating, *L. macrosoma* receives a score of 10 out of 15 and is rated as MEDIUM INVASIVE NEMATODE.

48.14 Region-wise status of invasiveness

Practically, *L. macrosoma* has A2 pest list status with regard to virus transmission of TomBRV to fruit-plant importations to

Europe. It is given a medium invasiveness rating for the countries where it is still not detected.

48.15 Management measures

It is necessary to propagate stock cherry in soil that is free from the virus-vector *L. macrosoma*. For further management measures, see section 47.17 of *L. elongatus*, above. For phytosanitary measures, see Chapter 16, section 115.

49 *Paralongidorus maximus*

49.1 Common name

Needle nematode

49.2 Scientific name

Paralongidorus maximus (Bütschli, 1874) Siddiqi, 1964

49.3 EPPO code

LONGMX

49.4 Diagnosis

49.4.1 Morphological diagnosis

Morphological and morphometric descriptions (after Siddiqi, 1964c) are summarized below. For measurements (after Sturhan, 1963), see Table 39.

Female: Very long, 7–12 mm in length. Ventrally curved in open spiral when heat killed (Fig. 8.5I). Cephalic region rounded and offset from body contour with deep constriction (Fig. 8.5C). Amphids stirrup-shaped. Amphidial fovea wide slit-like opening two-thirds of lip region width. Odontostyle and odontophores 152–187 and 42–71 μm long, respectively (Fig. 8.5B). Guiding ring far anteriorly (37–47 μm from anterior end). Dorsal oesophageal gland nucleus at anterior end of bulb. Cardia convex-conoid. Vulva a transverse slit with muscular vagina (Fig. 8.5F), around one-quarter the corresponding body width in length. Genital tracts amphidelphic, reflexed (Fig. 8.5E). Pre-rectum variable, 4–17 anal body widths in length. Tail short and bluntly rounded (Fig. 8.5G,H), about half of anal body width long, with two or three pairs of subterminal caudal papillae.
Male: Rare, no role in reproduction. Spicule strong, 100–106 μm long (Fig. 8.5J). Lateral guiding pieces 32–35 μm. Copulatory supplements comprising adanal pair and ventral series of 14–15 supplements (Fig. 8.5K).

49.4.2 Molecular diagnosis

See section 46.4.2, above.

49.5 Geographical distribution

P. maximus is a European species reported in Austria, France, Germany, Hungary, Poland, Portugal and United Kingdom (Heyns, 1975; Bravo and Lemos, 1997) and also in South Africa (Swart *et al.*, 1996).

49.6 Hosts

P. maximus is a polyphagous species and parasitizes numerous agronomic crops, ornamentals and forest trees (Heyns, 1975; Decraener and Geraert, 2013). There is evidence that *P. maximus* transmits raspberry ringspot virus (Brown *et al.*, 2004). An expanded list of hosts of *P. maximus* is given in Table 41.

49.7 Symptoms

Symptoms are typical of the genus. The nematode severely attacks coniferous and deciduous trees, currant, grape, potato and tobacco, resulting in the suppression of plant growth. Severe stunting was observed on potato, while infested tobacco plants show retarded growth in patches (Decker, 1989).

49.8 Biology and life cycle

Male are extremely rare and reproduction is parthenogenetic. The duration of the life cycle is long and is favoured by warm summers (Sturhan, 1963). The nematode mostly occurs in open-textured sandy or sandy loam soils. The highest population densities occur at 20–30 cm depth (Hunt, 1993). The nematode is known to transmit raspberry ringspot virus (Brown *et al.*, 2004).

49.9 Economic importance

The damage induced by this nematode has been established by its association with diseased plants and also by inoculation tests in the greenhouse. However, no information is available on quantitative yield losses.

49.10 Movement and means of dispersal

See section 46.11, above.

49.11 Pest risk analysis and looming threat from introduction

Limited information is available on damage to agricultural crops and forest trees. However, the risk posed by *P. maximus* is high

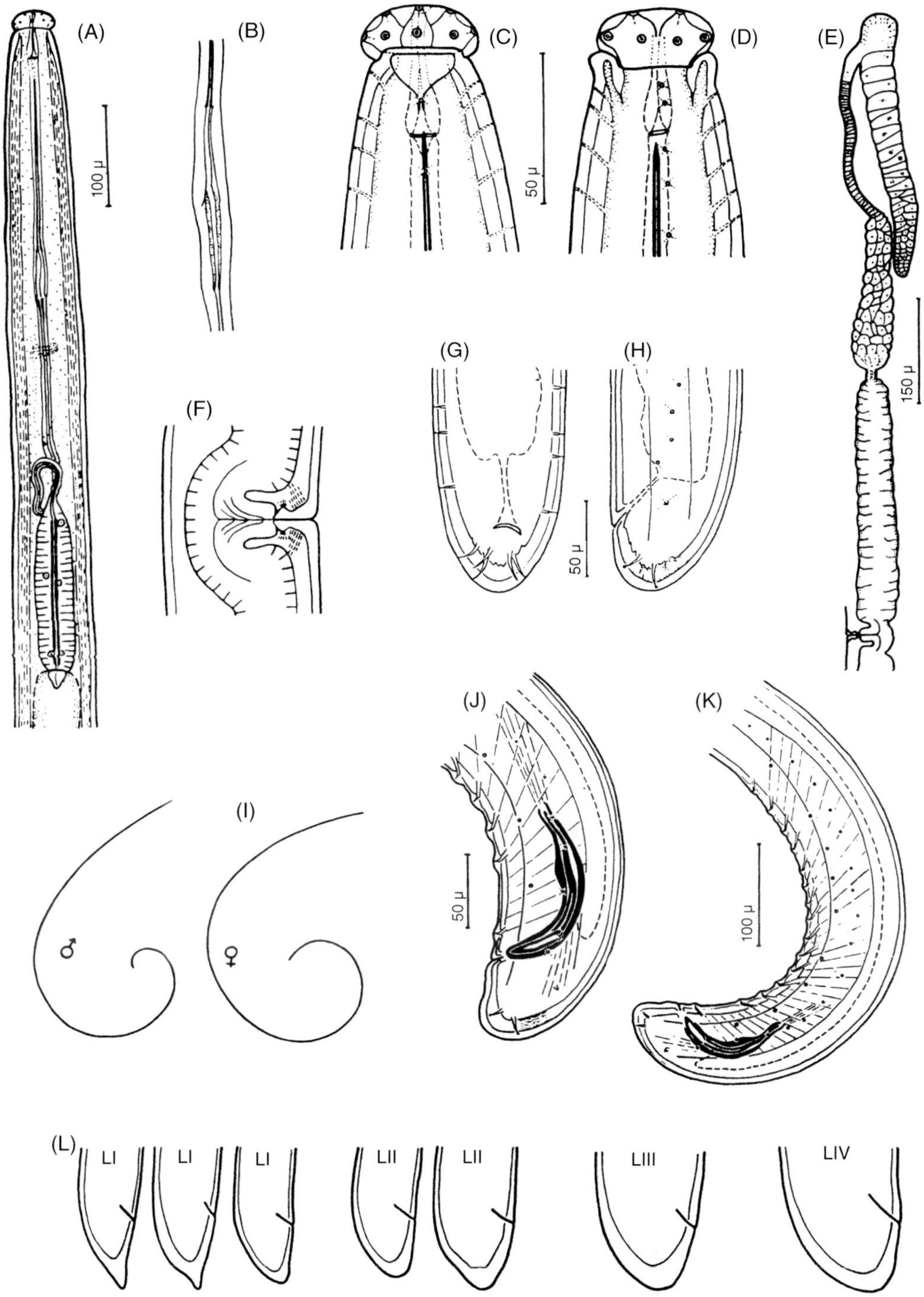

Fig. 8.5. Line drawing of *Paralongidorus maximus*. Pharyngeal region (A); odontophore (B); head region in lateral view and dorso-ventral view, respectively (C,D); anterior genital branch of female reproductive system (E); vagina region (F); female tail ventral view and view of left side, respectively (G, H); habitus of male and female upon thermal death (I); posterior region of male (J, K); tails in lateral view of the four juvenile stage, L_I–L_{IV} (L). [Reproduced from Hunt, 1993 (after Sturhan, 1963); courtesy CAB International]

Table 41. List of hosts of *Paralongidorus maximus*.

Alder (*Alnus* sp.)
Arborvitae (*Thuja* sp.)
Asparagus (*Asparagus officinalis*)
Beans (*Phaseolus* sp.)
Carrot (*Daucus carota*)
Cauliflower (*Brassica oleracea*)
Cowpea (*Vigna unguiculata*)
Celery (*Apium graveolens* and *A. rapaceum*)
Chrysanthemums (*Chrysanthemum* sp.)
Cucumber (*Cucumis sativus*)
Dahlias (*Dahlia* sp.)
Fir (*Abies* sp.)
Garden beet (*Beta* sp.)
Gladioli (*Gladiolus* sp.)
Grapevine (*Vitis vinifera*)
Hazel (*Corylus* sp.)
Hornbeam (*Carpinus* sp.)
Larch (*Larix* sp.)
Leek (*Allium ampeloprasum*)
Locust (*Robinia* sp.)
Onion (*Allium cepa*)
Oak (*Quercus* sp.)
Maple (*Acer* sp.)
Parsley (*Petroselinum crispum*)
Periwinkle (*Vinca rosea*)
Phlox (*Phlox* sp.)
Pine (*Pinus sylvestris*)
Rose (*Rosa* sp.)
Scorzonera (*Scorzonera* sp.)
Spruce (*Picea* sp.)
Sunflower (*Helianthus annuus*)
Strawberry (*Fragaria* × *Ananassa*)
Tobacco (*Nicotiana tabacum*)

due to its wide host range and nematode transmission ability. A complete pest risk analysis is suggested and preventive steps should be taken to avoid the introduction of this nematode into new host areas.

49.12 Invasiveness rating

Considering the consequence of probable introduction and pest risk analysis of *P. maximus*, the following invasiveness rating has been assigned.

Climate/host interaction	–	Medium (2)
Hosts range	–	Medium (2)
Dispersal potential	–	Medium (2)
Economic impact	–	High (3)
Environmental impact	–	Low (1)
Post-entry distribution	–	Not known (0)
Final rating (score)	–	Medium (10)

Based on the above invasiveness rating, *P. maximus* receives a score of 10 out of 15 and is rated as MEDIUM INVASIVE NEMATODE.

49.13 Region-wise status of invasiveness

P. maximus is given a medium invasiveness rating for the countries where it does not occur.

49.14 Management measures

See section 46.15, above.

Xiphinema spp.

The genus *Xiphinema* Cobb, 1913 belongs to the family Longidoridae and subfamily Xiphinematinae. *Xiphinema* has great morphological diversity and currently more than 260 valid species with *Xiphinema americanum* as types species, are described (Coomans *et al.*, 2001; Palomares-Rius *et al.*, 2017). Further, the genus *Xiphinema* has been divided into two species groups: *Xiphinema americanum* group and *Xiphinema* non-*americanum* group (Loof and Luc, 1990; Lamberti *et al.*, 2000; He *et al.*, 2005). The *Xiphinema americanum* group comprises a complex of around 61 species and the *Xiphinema* non-*americanum* group comprises a complex of more than 215 species (Gutiérrez-Gutiérrez *et al.*, 2012; Archidona-Yuste *et al.*, 2016b; Palomares-Rius *et al.*, 2017). Polytomous keys of *Xiphinema* identification have been given by different workers (Coomans *et al.*, 2001; Lamberti *et al.*, 2004; Barsi and De Luca, 2008; Gutiérrez-Gutiérrez *et al.*, 2012; Maria *et al.*, 2018). In addition to direct damage to the plant, some putative *Xiphinema* species are known to transmit nepoviruses (family Comoviridae). Some virus-transmitting *Xiphinema* species are as follows.

From *Xiphinema* non-*americanum* group:

X. diversicaudatum (Micoletzky, 1927) Thorne, 1939
X. index Thorne and Allen, 1950
X. italiae Meyl, 1953
X. vuittenezi Luc, Lima, Weischer and Flegg, 1964

From *X. americanum* group:

X. americanum Cobb, 1913
X. bricolense Ebsary Vrain and Graham, 1989
X. brevicolle Lordello and Da Costa 1961 (= *X. brevicollum* Luc *et al.*, 1998)
X. californicum Lamberti and Bleve-Zacheo, 1979
X. inaequale Khan and Ahmad, 1977
X. intermedium Lamberti and Bleve-Zacheo, 1979
X. rivesi Dalmasso, 1969
X. tarjanense Lamberti and Bleve-Zacheo, 1979

The dagger nematodes, *Xiphinema* species are root ectoparasites and polyphagous, feeding on a wide range of cultivated and wild plants. Nematode feeding causes extensive mechanical damage to plants roots and causes root galls, especially in young roots (Özturk *et al.*, 2018). The major economic impact of this nematode is due to vectoring nepoviruses (Hunt, 1993; Decraemer and Robbins, 2007). In this chapter, six virus-transmitting species,

X. americanum sensu stricto, *X. diversicaudatum*, *X. index*, *X. italiae* and *X. rivesi*, and one non-virus-transmitting species, *X. ifacolum*, are considered for invasiveness because of their economic importance and quarantine significance.

Systematic position of *Xiphinema* species

	Classical (after Hunt, 1993)	Molecular (after Decraemer and Hunt, 2013)
Phylum	Nematoda	Nematoda
Class	Adenophorea	Enoplea
Subclass	Dorylaimia	Dorylaimia
Order	Dorylaimida	Dorylaimida
Suborder	Dorylaimina	Dorylaimina
Superfamily	Dorylaimoidea	Dorylaimoidea
Family	Longidoridae	Longidoridae
Subfamily	Xiphinematinae	Xiphinematinae
Genus	*Xiphinema*	*Xiphinema*

50 *Xiphinema americanum*

50.1 Common name

American dagger nematode

50.2 Scientific name

Xiphinema americanum Cobb, 1913

50.3 Synonyms

Tylencholaimus americanus (Cobb, 1913) Micoletzky, 1922
Xiphinema taylori Lamberti *et al.*, 1991

50.3 EPPO code

XIPHAA (*sensu stricto*), XIPHAM (*sensu lato*)

50.4 Diagnosis

X. americanum is a complex group and largely referred to as *X. americanum sensu lato* (*s.l.*), and consisting of more than 61 nominal species (Palomares-Rius *et al.*, 2017). The majority of the members of this group are difficult to distinguish based on morphological and biochemical characterization. Hence, integrative taxonomy should be applied for an accurate diagnosis.

50.4.1 Morphological diagnosis

The following morphological and morphometric descriptions of *X. americanum sensu stricto* (*s.s.*) are after Siddiqi (1973b). For measurements, see Table 42.

FEMALE: Body posture open spiral shape on heat death. Cephalic region slightly expanded and appears hemispherical. Amphidial slits around half as long as width of cephalic region (Fig. 8.6E). Odontostyle long, needle-like. Odontophore large with three basal flanges (Fig. 8.6A). Basal stylet guiding ring close to base of retracted odontostyle. Oesophagus typical dorylamid, two parts; anterior part slender, posterior part bulbar. Vulva approximately equatorial (Fig. 8.6H). Ovaries opposed, paired reflexed, harbouring bacteria. Z-organ not present (Fig. 8.6B). Tail 1.6–1.8 times anal body width, conoid with a more dorsal curvature, ending in hemispherical tip, two pairs of lateral pores (Fig. 8.6I-K).
MALE: Extremely rare, no role in reproduction. Body with larger curvature in posterior end (Fig. 8.6F). Basal stylet guiding ring long, 51 μm from anterior end of body (Fig. 8.6C). Spicules relatively straight, long, cephalated (Fig. 8.6G). Preanal pair of supplements present at 7 μm anterior to anus, and five ventromedian papillae present and located anterior to anus. Tail length nearly twice anal body width, with hemispherical tail terminus (Fig. 8.6G), two pairs of lateral pores present just anterior to its middle.
JUVENILE: Four juvenile stages documented by comparative body length and odontostyle length. Stage can be differentiated by length of replacement odontostyle in one stage and corresponding functional odontostyle length in next stage.

50.4.2 Molecular diagnosis

Recently, more than 96 *Xiphinema* species (approximately 35% of total described species), including *Xiphinema americanum*, have been characterized (Archidona-Yuste *et al.*, 2016b), using ribosomal genes such as D2–D3 expansion segments of 28S rRNA, ITS1 rRNA and partial 18S (He *et al.*, 2005; Lazarova, *et al.*, 2006; Orlando *et al.*, 2016; Palomares-Rius *et al.*, 2017; Maria *et al.*, 2018). For details, see section 46.4.2 (above). Some species-specific primers are also developed to characterize *X. americanum s.s.* Özturk *et al.* (2018) designed *X. americanum*-specific primers for ITS-1 region of rDNA. The PCR amplification with the primer Forward (GTCGCCACTGTGAATGAATG) and Reverse (GATCCCGAACGTCTGAGATAA) yielded an amplicon of 183 bp.

50.5 Geographical distribution

X. americanum s.l. has been reported from several countries of the world (Fig. 8.7) and is widespread in Canada and the USA but rare in Australasia and Oceania (Hockland and Prior, 2009). It is known to be detected in the following countries (CABI-ISC, 2020).

Africa: Nigeria, Senegal.
Asia: China, India, Japan, Kazakhstan, North Korea, Oman, Pakistan, Philippines, Sri Lanka, Tajikistan, Turkey, Uzbekistan.

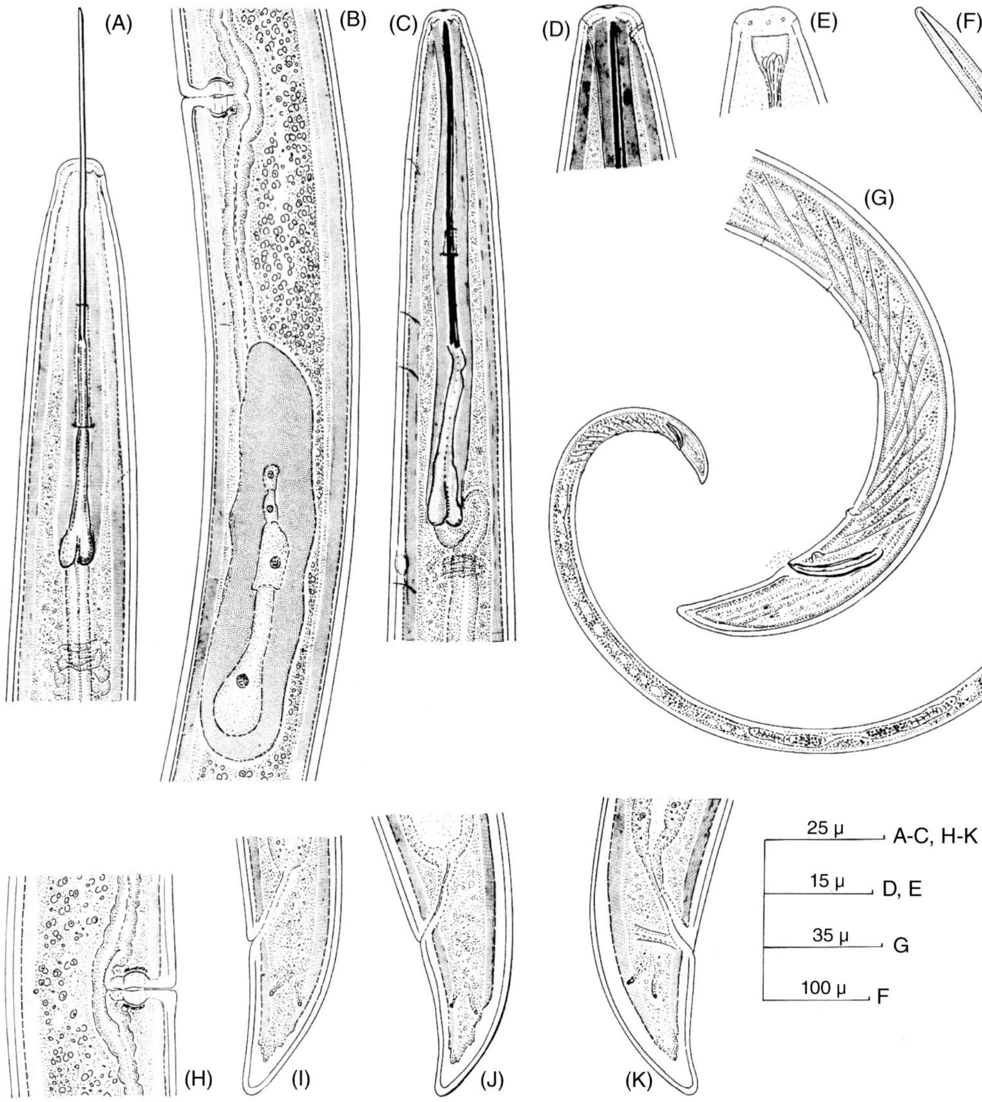

Fig. 8.6. Line drawing of *Xiphinema americanum sensu stricto.* Female anterior end (A); posterior female gonad (B); male anterior ends (C); female head end (D,E); entire male (F); male tail end (G); female vulval region (H); female tail ends (I–K). [Reproduced from Siddiqi, 1973. CIH Descriptions of Plant-parasitic Nematodes. Set 2, No. 29, CAB International; specimen courtesy A.C. Tarjan, University of Florida, USA]

Europe: France (intercepted only), Hungary, Netherlands (intercepted only), Poland, Russia, Spain, United Kingdom.
North America: Canada, Guatemala, Mexico, Panama, USA.
Oceania: Australia.
South America: Argentina, Brazil, Chile, Ecuador, Peru, Uruguay, Venezuela.

50.6 Hosts

X. americanum is extremely polyphagous and parasitizes numerous plants, including field crops, ornamentals and forest trees (Siddiqi, 1973b; MacGowan, 1980; Hunt, 1993; Decraemer and Geraert, 2013). The extended list of hosts of *X. americanum* is given in Table 43.

50.7 Vector

X. americanum s.l. is a well-organized virus-vector and transmits the four major American nepoviruses, namely, cherry rasp leaf nepovirus (CRLV), peach rosette mosaic nepovirus (PRMV), tomato ringspot nepovirus (TomRSV) and tobacco ringspot nepovirus (TRSV or TobRSV). Other members of *X. americanum s.l.* such as *X. americanum s.s.*, *X. californicum* and *X. rivesi* transmit TomRSV, TobRSV and CRLV, while *X. intermedium* and *X. tarjanense* transmit TomRSV and TobRSV. *X. bricolense* is

Table 42. Morphometric measurements of *Xiphinema* species.

Morphometric characters		Range, numerous populations					
		X. americanum s.s. (after Siddiqi, 1973b)[a]	*X. diversicaudatum* (after Goodey *et al.*, 1960)	*X. ifacolum* (after Luc, 1961)	*X. index* (after Heyns, 1971)[b]	*X. italiae* (after Heyns, 1974)	*X. rivesi* (after Dalmasso, 1969)
Females	Length (mm)	1.66–1.71	4.0–5.5	3.12–3.7	2.87–3.58	2.3–2.8	1.68–2.11
	Odontostyle (µm)	70–74	130–157	185–197	123–134	80–91	90–101
	Odontophore (µm)	45–47	70–97	52–79	74–81	58–66	48–57
	a	47–52	57–92	50.3–62.0	54–61	60–73	37–49
	b	6.3–7.5	6.6–11.4	7.3–11.3	6.2–8.0	6.1–8.4	5.7–6.9
	c	48–53	61–134	45.4–59.1	0.7–1.1	32–45	51–59
	c'	1.6–1.8			40–42	2.3–2.8	1.3–1.5
	V	51–54	39–46	48.3–53.3		44–49	51–54
Males	Length (mm)	1.5–1.7	41–6.2	3.20	3.6	2.3–2.4	1.92
	Odontostyle (µm)	–	131–153	–	–	88–90	94
	Odontophore (µm)	–	72–90	–	–	58–66	51
	a	39.7–51.6	57–96	62.9	63	62–63	43
	b	6.1–6.3	7.4–11.3	6.7	7.3	6.7–6.8	6.3
	c	37.8–50.1	55–100	43.9	88	46–49	60
	c'	–	–	–	–	1.8–1.9	1.1
	Spicule (µm)	27–30	69–81	53		44–48	50
	T%	40–54	47–67	–	49	–	–

[a]Male rare, after Tarjan (1956); [b]male after Thorne and Allen (1950).

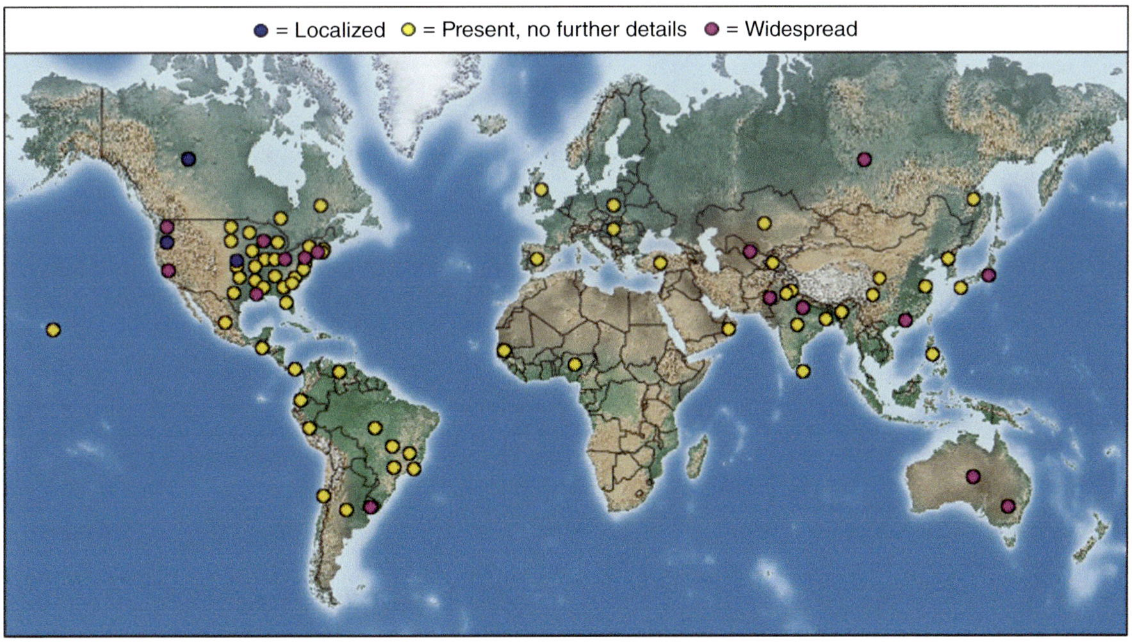

Fig. 8.7. Distribution map of *Xiphinema americanum*. [Adapted from CABI-ISC, 2020]

only known to transmit TomRSV (Brown *et al.*, 1994; Decraemer and Geraert, 2013). Brown *et al.* (1994) also reported that these North American populations have broad-spectrum virus transmission capabilities compared with the European population, which has relatively narrow vector species.

50.8 Symptoms

In the absence of virus infection, the plants parasitized by *X. americanum s.l.* generally do not show any noticeable symptoms on aerial parts of the plant unless population

Table 43. List of hosts of *Xiphinema americanum* (adapted from CABI-ISC, 2020).

Main hosts
Avena sativa (oats)
Citrus sp.
Citrus sinensis (navel orange)
Cocos nucifera (coconut)
Coffea (coffee)
Coffea canephora (robusta coffee)
Cucurbita pepo (ornamental gourd)
Ficus carica (fig)
Fragaria ananassa (strawberry)
Glycine max (soybean)
Gossypium hirsutum (Bourbon cotton)
Juglans regia (walnut)
Linum usitatissimum (flax)
Litchi chinensis (lychee)
Malus domestica (apple)
Malus sylvestris (crab-apple tree)
Mangifera indica (mango)
Medicago sativa (lucerne)
Nicotiana tabacum (tobacco)
Oryza sativa (rice)
Pistacia vera (pistachio)
Prunus (stone fruit)
Prunus armeniaca (apricot)
Prunus avium (sweet cherry)
Prunus domestica (plum)
Prunus dulcis (almond)
Prunus persica (peach)
Prunus salicina (Japanese plum)
Psidium guajava (guava)
Rubus spp. (blackberry, raspberry)
Saccharum officinarum (sugarcane)
Solanum lycopersicum (tomato)
Solanum tuberosum (potato)
Sorghum bicolor (sorghum)
Vaccinium spp. (blueberries)
Vitis labrusca (fox grape)
Vitis rupestris (sand-grape)
Vitis vinifera (grapevine)
Zea mays (maize)
Other hosts
Acer platanoides (Norway maple)
Agropyron cristatum (crested wheatgrass)
Agropyron desertorum (crested wheatgrass)
Anethum graveolens (dill)
Bambusa (bamboo)
Bambusa tuldoides (punting-pole bamboo)
Callicarpa americana (American beautyberry)
Cichorium intybus (chicory)
Codiaeum variegatum (croton)
Cydonia oblonga (quince)
Manilkara zapota (sapodilla)
Mentha (mints)
Mentha piperita (peppermint)
Mentha spicata (spearmint)
Nerium oleander (oleander)
Onobrychis viciifolia (sainfoin)
Petunia hybrida
Picea spp. (spruce)
Pinus (pines)
Pinus echinata (shortleaf pine)
Pinus elliottii (slash pine)
Pinus resinosa (red pine)
Pinus sylvestris (Scots pine)
Pinus taeda (loblolly pine)
Plantago (plantain)
Poncirus trifoliata (trifoliate orange)
Potentilla (Cinquefoil)
Rosa (roses)
Shorea robusta (sal)
Tectona grandis (teak)
Trifolium (clovers)
Trifolium pratense (purple clover)
Trifolium repens (white clover)
Zoysia japonica (zoysiagrass)
Wild hosts
Acer rubrum (red maple)
Agrostis spp.
Ambrosia artemisiifolia (ragweed)
Borrichia frutescens (sea-oxeye)
Chenopodium giganteum (lambsquarters)
Chenopodium quinoa (quinoa)
Cornus (dogwood)
Cynodon dactylon (Bermuda grass)
Erigeron annuus (annual fleabane)
Fraxinus pennsylvanica (downy ash)
Grewia asiatica (phalsa)
Pennisetum clandestinum (kikuyu grass)
Pittosporum tobira (Japanese pittosporum)
Populus deltoides (poplar)
Quercus laurifolia (laurel oak)
Salix alba (white willow)
Stellaria media (common chickweed)
Taraxacum officinale complex (dandelion)
Thevetia peruviana (exile tree)
Ulmus (elms)
Vinca minor (common periwinkle)

levels are high. In high population density, a general reduction in plant vigour appears in patches in the field, similar to those that occur when a plant is under stress conditions. In orchard trees, overall growth may drastically reduce and trees look stunted (Fig. 8.8A). Under heavy infestation, nematode feeding causes extensive mechanical damage to plant roots and incites root galls in young roots (Fig. 8.8B) (Özturk *et al.*, 2018). However, when nematode feeding involves the transmission of the virus, distinctive virus symptoms of the specific virus develop in the host crop. After transmission to the roots, viral symptoms usually appear first on aerial plant parts in the growing season, for example peach rosette mosaic virus-infected grapevine (Fig. 8.8C).

Fig. 8.8. Damaging effect *Xiphinema americanum* (A-C), *X. diversicaudatum* (D-G) and *X. index* (H-J). Whole infected plant of *Ulmus parviflora* (A); swollen and stubby root tips (B); peach rosette mosaic virus infected grapevine (C); raspberry leaves infected with ArMV (D); symptoms on raspberry plantation severely stunted plants (E); symptoms 'yellow disease' of barley crop (F); single infected barley plant (G); typical root symptoms showing swellings of root tips in comparison with an uninfected root (H); multinucleate cells induced by *X. index* in *Vitis* rootstock (I); grape fan leaf virus infected grapevine leaf (J). [Courtesy A-B: J.D. Stanley, DPI; C: W.R. Allen, Agriculture and Agri-Food Canada, Bugwood.org D: Scottish Crop Research Institute; E-G: Derek J.F. Brown; H-I: P. Castillo and N. Vovlas; J: Canadian food inspection agency]

50.9 Biology and life cycle

X. americanum s.s. is an ectoparasite and lives completely in the soil. All stages invade the roots. The larvae and adults appear to be attracted towards newly growing roots and puncture several successive layers of cells through a long stylet, feeding by extracting cytoplasm. Males are rare or absent and females produce eggs parthenogenetically. The best temperatures for reproduction are 20–24°C (Hunt, 1993). In the majority of populations, three juvenile stages are formed; however, particularly in the North American populations which transmit viruses, four juvenile stages are found (Brown *et al.*, 1994). The life cycle takes around 1 year to complete. The nematode does not survive for long periods in sub-zero soil, or in areas with low winter temperatures. The nematode overwinters mainly in the egg stage. All stages have been found to survive where the soil is not frozen and there is no specialized survival stage. They can mature in the soil even without a host, but multiplication will not occur.

X. americanum prefers well-drained sandy soil, but also occur in clay soil with an optimum pH 5.6–7.4 (Decraemer and Geraert, 2013). Vertical distribution of the nematode was recorded for two grapevine sites infested with peach rosette mosaic nepovirus in Michigan, USA. The population of nematodes was 0.3–2.5 per 100 cc soil at one site (up to a depth of 213.4 cm) and 0.3–10.5 per 100 cc soil at the other site (152.4 cm) (Ramsdell and Myers, 1978). In Uttar Pradesh, India, *X. americanum* population density was highest in March, then gradually declined with soil moisture and reached a minimum in September. These population fluctuations were greatest in the upper soil level (0–25 cm depth) and lowest in the lower level (50–75 cm). Soil moisture seemed to be a more important factor than soil temperature in determining population fluctuation of the nematode (Malik and Jairajpuri, 1983).

50.10 Vector transmission

X. americanum s.l. is a potential vector of several nepoviruses (CRLV, PRMV, TobRSV, TomRSV). However, the ability to transmit viruses may vary among different populations of the same nematode species. In North America, both adults and juvenile stages of *X. americanum s.l.* are able to transmit virus (Teliz *et al.*, 1966). *X. americanum s.s.* transmits CRLV, TobRSV and TomRSV (Decraemer and Geraert, 2013). The nematode acquires the virus particles when feeding on a virus-infected plant. The virus particles adhere to the lining of the stylet and oesophagus. When the nematode attacks the next healthy plant, it injects virus particles into the host. The nematode is able to transmit the acquired virus particles up to 2 years (Bitterlin and Gonsalves, 1987).

50.11 Economic importance

In the USA, direct damage by *X. americanum s.s.* appears to be economically important in many states on strawberries, fruit trees, forage legumes and forest trees (CABI, 2013). *X. americanum* is believed to be a major factor in the unthriftiness of shelterbelt trees of South Dakota (Malek, 1968) and red clover (*Trifolium pratense*) in Iowa (Norton, 1967). Yields of some hybrid grapevine cultivars were significantly reduced by four species of nematode, including *X. americanum*, in Michigan (Ramsdell *et al.*, 1996).

The overall importance of this nematode group is due to the ability of some species to transmit economically important nepoviruses (Taylor and Brown, 1981). PRMV causes severe crop loss and death to *Vitis labrusca* cultivars (Ramsdell *et al.*, 1995). Soybean severe stunt virus has been recorded in 30 of 50 soybean fields in Delaware, USA. The disease was transmitted via soil and *X. americanum* is strongly connected with the incidence of the disease in the field (Weldekidan *et al.*, 1992).

50.12 Movement and means of dispersal

See section 46.11, above.

50.13 Pest risk analysis and looming threat from introduction

The main phytosanitary risk is that these nematodes are known to be natural vectors of four economically important nepoviruses, namely cherry rasp leaf virus (CRLV, *Cheravirus*), peach rosette mosaic virus (PRMV, *Nepovirus*), tobacco ringspot virus (TobRSV, *Nepovirus*) and tomato ringspot virus (TomRSV, *Nepovirus*). The European populations of *X. americanum s.l.* have not been shown to transmit the quarantine-listed viruses, but Sirca *et al.* (2007) and Auger *et al.* (2009) reported transmission of TobRSV and TomRSV to host plants in Slovenia and Chile, respectively, by *Xiphinema rivesi* (a member of *X. americanum s.l.*). These nematodes could thus act as a direct agent for the introduction of these nepoviruses. If the viruses are introduced or spread into new areas of host countries, it would create an additional phytosanitary risk and the prerequisite more complex procedures for the certification of virus-free planting material for fruit crops.

50.14 Invasiveness rating

Considering the economic impact and pest risk analysis of *X. americanum*, the following invasiveness rating has been assigned.

Climate/host interaction	–	High (3)
Hosts range	–	High (3)
Dispersal potential	–	High (3)
Economic impact	–	High (3)
Environmental impact	–	High (3)
Post-entry distribution	–	Medium (–2)
Final rating (score)	–	High (13)

Based on the above invasiveness rating, *X. americanum* receives a score of 13 out of 15 and is rated as HIGHLY INVASIVE NEMATODE.

50.15 Region-wise status of invasiveness

X. americanum s.l. is a quarantine pest in European Union (EPPO, 2012) and other countries, including India (PQ Order-2003, 2015) and is recommended for regulation. It is given a high invasiveness rating for the following countries where it is still not present.

APPPC: Bangladesh, Cambodia, Indonesia, Iran, Iraq, Israel, Kyrgyzstan, Lebanon, Malaysia, Myanmar, Nepal, Saudi Arabia, Singapore, South Korea, Taiwan, Thailand, Vietnam

CAHFSA: Barbados, Belize, Cuba, Dominican Republic, Haiti, Jamaica, Martinique, Puerto Rico, Saint Kitts & Nevis, Trinidad and Tobago

CAN: Bolivia, Colombia

COSAVE: Guyana, Paraguay, Suriname,

EPPO: Austria, Belarus, Belgium, Bulgaria, Croatia, Cyprus, Czech Republic, Denmark, Estonia, Finland, France, Georgia, Germany, Greece, Guernsey, Ireland, Italy, Jersey, Latvia, Lithuania, Luxembourg, Macedonia, Malta, Moldova, Morocco, Netherlands, Norway, Portugal, Romania, Serbia, Slovakia, Slovenia, Sweden, Switzerland, Ukraine

IAPSC: Algeria, Angola, Benin, Botswana, Burkina Faso, Burundi, Cameroon, Cape Verde, Central African Republic, Chad, Comoros, Congo, Côte d'Ivoire, Eritrea, Ethiopia, Gabon, Gambia, Ghana, Guinea, Guinea-Bissau, Kenya, Liberia, Madagascar, Malawi, Mali, Mauritania, Mauritius, Mozambique, Namibia, Niger, Réunion, Rwanda, Sierra Leone, Somalia, South Africa, Swaziland, Tanzania, Togo, Uganda, Zambia, Zimbabwe

NEPPO: Egypt, Jordan, Libya, Malta, Sudan, Syria, Tunisia

OIRSA: Costa Rica, El Salvador, Honduras, Nicaragua

PPPO: Fiji, New Zealand, Papua New Guinea, Tonga, Vanuatu.

50.16 Management measures

50.16.1 Cultural control

It has been shown that rotation crops and green manure treatments can have a significant impact on *X. americanum* population development, and that nematode populations may be suppressed by incorporation of crops as a green manure treatment or by the application of a nematicide (Halbrendt, 1995). Almost 60 plant species were tested for their effect on *X. americanum* populations in Russia. Among these, only rye, maize and barley inhibited the development of *X. americanum* by 89–95% (compared with numbers in fallow plots). Wheat, millet, oats, chicory, ryegrass, timothy, mustard, calendula and some others reduced the numbers of nematodes, whereas clover, mint, *Stellaria* sp., *Chenopodium* sp. and dill favoured nematode development (Boldyrev and Borzykh, 1983). *Brassica* spp. (rape and mustard) and *Tagetes* spp. (marigold) were the most effective in reducing *X. americanum* populations in replanted orchards in Pennsylvania, USA (Halbrendt, 1993).

50.16.2 Host resistance

American grapevine (*Vitis labrusca*) cultivars ('Concord', 'Delaware' and 'Niagara'), French-American hybrid rootstock cultivars ('Couderc' 1202 and 1616, 'Teleki' 5A and 5C) and French-American hybrid scion cultivars ('Chancellor', 'Foch', 'Seyval' and 'Vignoles') were resistant to *X. americanum* (Ramsdell and Gillett, 1985).

50.16.3 Biological control

Treatment of *C. anguillulae, Lagenidium caudatum, Aphanomyces* and *Leptolegnia* parasitized *X. americanum* and *X. rivesi* (Jaffee, 1986). Parasitism of *X. americanum* by *Pseudomonas denitrificans* was reported by Sayre (1980). *Predatory nematodes Mononchoides longicaudatus* and *Mononchoides fortidens* attacked and fed on *X. americanum* (Bilgrami and Jairajpuri, 1989).

50.16.4 Chemical control

Soil application of nematicides may give over 95% control. Preplant soil fumigation eliminated *X. americanum* and increased *Prunus* tree growth, providing effective control of stem pitting induced by tomato ringspot nepovirus (Stouffer and Mowery, 1980). Application of fenamiphos, carbofuran or carbosulfan to peach trees infested with *X. americanum* and *X. rivesi* resulted in a significant increase in yield over controls (Jaffee and Shaffer, 1985). Application of prophos (ethoprophos) directly to lawn grass (*Zoysia japonica*) provided a marked enhancement in the condition of lawns within 2–4 weeks, and the population density of nematodes was greatly reduced (Feldmesser and Golden, 1972). Application of oxamyl, fenamiphos or ethylene dibromide to apple orchards in South Carolina, USA, reduced populations of *X. americanum* and *X. rivesi* (CABI-CPC, 2010).

50.16.5 Phytosanitary measures

See Chapter 16, section 115.

51 *Xiphinema diversicaudatum*

51.1 Common name

European dagger nematode

51.2 Scientific name

Xiphinema diversicaudatum (Micoletzky, 1927) Thorne, 1939

51.3 Synonyms

Dorylaimus (*Longidorus*) *elongatus apud* Micoletzky, 1923
Dorylaimus (*Longidorus*) *diversicaudatus* Micoletzky, 1927

Longidorus diversicaudatus (Micoletzky, 1927) Thorne and Swanger, 1936
Xiphinema paraelongatum Altherr, 1958
Xiphinema seredouense apud Luc, 1958
Xiphinema israeliae apud Cohn, 1969, Cohn and Mordechai, 1969
Xiphinema amarantum Macara, 1970
Xiphinema sahelense apud Riffle, 1968, 1970
Diversiphinema diversicaudatum (Micoletzky, 1927) Cohn and Sher, 1972
Xiphinema basiri apud Javed, 1983

51.4 EPPO code

XIPHDI

51.5 Diagnosis

Xiphinema diversicaudatum belongs to the *Xiphinema non-americanum* group.

51.5.1 Morphological diagnosis

The morphological description of *X. diversicaudatum* is after Hunt (1993). For morphometric measurements (after Goodey *et al.*, 1960), see Table 42.

Female: Body shape, cephalic region, odontostyle, odontophore and oesophagus similar to those of male. Vulva a transverse slit at around 40% of body length from anterior (Fig. 8.9D). Genital tracts amphidelphic, ovary reflexed, symmetrical. Uteri close to vagina with well-developed ovijector. Oviduct and uterus united through a sphincter-Z. Prominent pseudo-Z organ comprising 10–20 irregular globular bodies found in each uterus (Fig. 8.9J).
Male: Body curved after heat fixation (Fig. 8.9C). Body cuticle smooth, outer layer thinner than inner. Lip region low, finely rounded, continuous with body contour. Amphid stirrup-shaped, as wide as lip region, just post-labial with crescentic slit. Odontostyle long, distinctly furcate at base (Fig. 8.9A). Odontophore long with prominent tripartite basal flanges attached to retracted muscles. Basal guiding ring 125 µm from anterior end. Oesophagus typical of longidorids, with narrow tubular enlarging to form muscular cylindrical bulb. Male genital tract diarchic, two testes, one reflexed and other outstretched anteriorly. Vas deferens generally filled with sperm which are spindle-shaped. Copulatory supplementary papillae paired, slightly anterior to the anal opening followed by 2–5 well-developed supplements. Copulatory muscle strong, located in region of supplements. Tail with strong curvature, digitate peg-like tip (Fig. 8.9H). Spicules robust, well developed and ventrally curved near middle. Short, lateral guiding pieces (Fig. 8.9H).
Juvenile: Three or four juvenile stages documented by comparative body length and odontostyle length. Stage can be differentiated by length of replacement odontostyle in one stage and corresponding functional odontostyle length in next stage. Pre-adult juveniles have digitate tail like adults but have more tapering tails and missing the digitate peg. Tail of first-stage juvenile elongate, ventrally arcuate conoid with hyaline fifth terminal.

51.5.2 Biochemical diagnosis

A few *Xiphinema* species, including *X. diversicaudatum*, *X. index* and *X. italiae*, are characterized based on enzymes, using isoelectric focusing (Lamberti *et al.*, 1999).

51.5.3 Molecular diagnosis

Chizhov *et al.* (2014) characterized D2–D3 expansion segments of 28S rRNA, ITS rRNA and COX1 mtDNA gene sequences of *X. diversicaudatum* isolated from Russia. They also developed species-specific primers of ITS rRNA region:

TW81 (5′-GTTTCCGTAGGTGAACCTGC-3′) (amplicon length approx. 990–1010 bp)
Xip5.8S (5′-CACCGCTTAGAATGGAATCGC-3′)
TW81 (5′-GTTTCCGTAGGTGAACCTGC-3′) (amplicon length approx. 864 bp)
Xip_diver_ITS (5′-GAATAAACACCTTTCAACGCTC-3′)

The PCR amplification was performed using the protocol described by Tanha Maafi *et al.* (2003). For further details, see Chapter 4, section 15.5.2. Wang *et al.* (2003) also developed *X. diversicaudatum*-specific primers of ITS1 region of rDNA with multiplex PCR (for primers, see section 53.5.2, below).

51.6 Geographical distribution

X. diversicaudatum is widely distributed in several European countries (Chizhov *et al.* 2014) (Fig. 8.10). However, the occurrence of *X. diversicaudatum* is relatively rare in Asia, Africa, America and Oceania, although it has been reported in South Africa (Heyns and Coomans, 1984), India (Singh and Kumar, 2003) and several states of the USA (Pitcher *et al.*, 1974). The nematode is known to occur in the following countries (CABI-ISC, 2020).

Africa: South Africa
Asia: India
Europe: Austria, Belgium, Bulgaria, Croatia, Czech Republic, Denmark, France, Germany, Ireland, Italy, Moldova, Netherlands, Norway, Poland, Portugal, Russia, Slovakia, Spain, Sweden, Switzerland, Ukraine, United Kingdom, former Yugoslavia
North America: USA
Oceania: New Zealand

51.7 Hosts

X. diversicaudatum is a polyphagous species and attacks numerous woody and herbaceous plants (Chizhov *et al.* 2014). The extended list of hosts of *X. diversicaudatum* is given in Table 44.

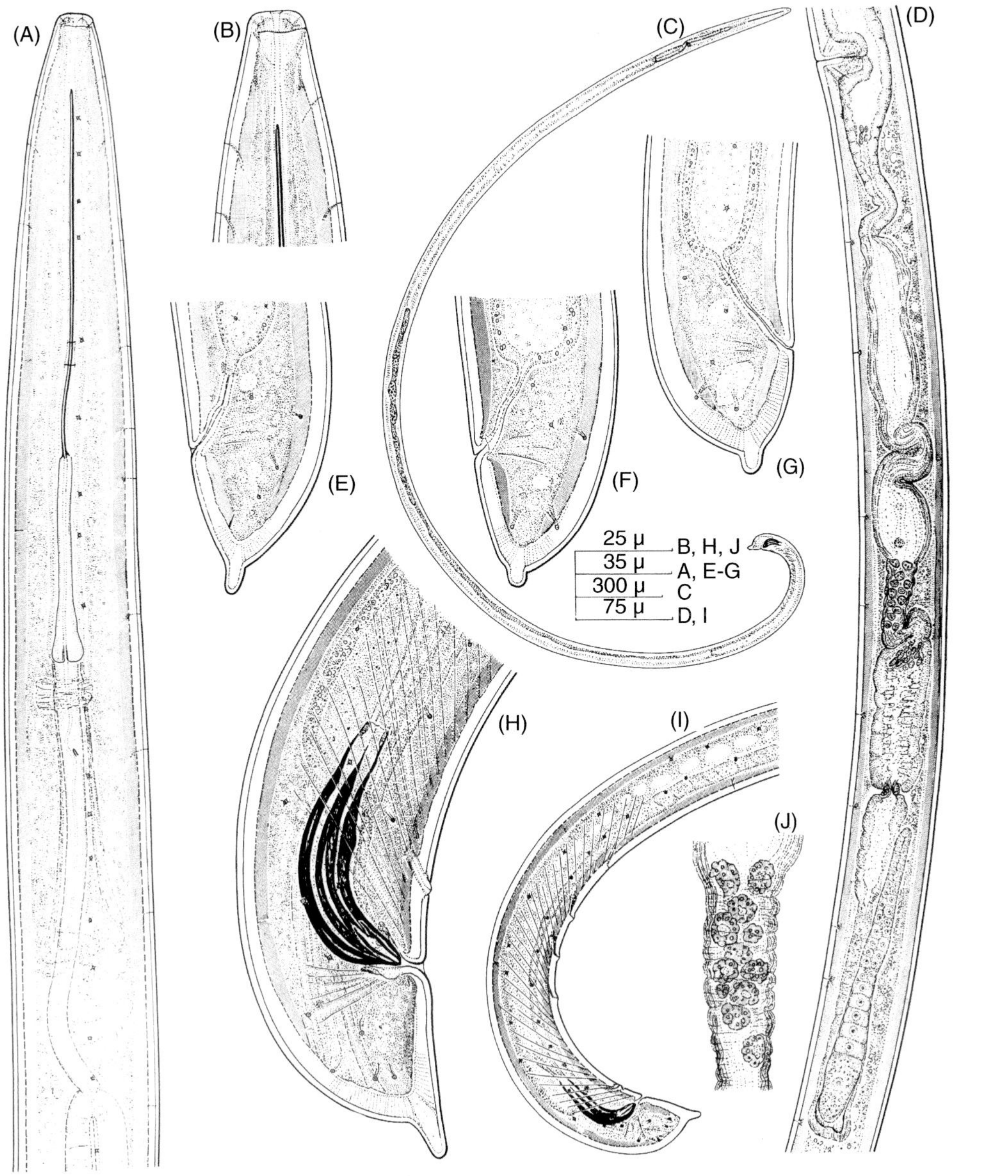

Fig. 8.9. Line drawing of *Xiphinema diversicaudatum.* Male anterior end (A,B); entire male (C); posterior female reproductive system (D); female tails (E-G); male tail end (H,I); pseudo Z-organ with globular bodies (J). [Reproduced from Hunt, 1993 (after Siddiqi); courtesy CAB International]

51.8 Vector

X. diversicaudatum is known to transmit two nepoviruses: Arabis mosaic (ArMV) and strawberry latent ringspot (SLRSV) (Pitcher *et al.*, 1974).

51.9 Symptoms

X. diversicaudatum causes direct damage in several plants by inducing severe root galling, which can result in significant suppression of plant yield. The nematode has a very long stylet and feeding causes characteristic root-tip galling, resulting in severe dwarfing of the entire plant (Fig. 8.8E). The nematode also vectors Arabis mosaic nepovirus and strawberry latent ringspot virus, causing indirect damage (Pitcher *et al.*, 1974). The distinctive symptoms of the specific virus, such as yellowing, chlorotic mosaic and vein clearing in raspberry leaves infected with ArMV (Fig. 8.8D) and bright yellowing of the barley foliage known as 'yellow disease' as a result of infection by Arabis mosaic nepovirus, only appear after transmission of the virus by *X. diversicaudatum* (Fig. 8.8F,G).

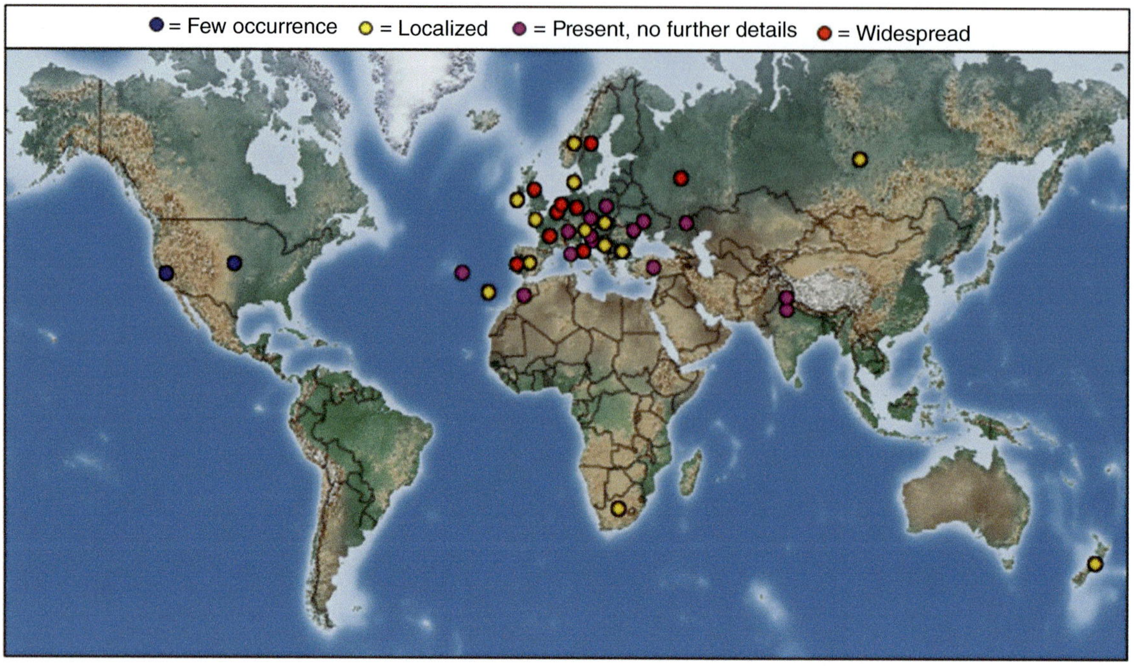

Fig. 8.10. Distribution map of *Xiphinema diversicaudatum* (Adapted from CABI-ISC, 2020).

Table 44. List of hosts of *Xiphinema diversicaudatum* (adapted from CABI-ISC, 2020).

Main hosts
Beta vulgaris (beetroot)
Fragaria ananassa (strawberry)
Prunus domestica (plum)
Prunus persica (peach)
Prunus spinosa (blackthorn)
Rosa (roses)
Rosa canina (dog rose)
Rubus fruticosus (blackberry)
Rubus idaeus (raspberry)
Vitis vinifera (grapevine)
Habitat/association
Humulus lupulus (hop)
Other hosts
Prunus salicina (Japanese plum)
Wild hosts
Acer pseudoplatanus (sycamore)
Allium porrum (leek)
Brassica oleracea (cabbages, cauliflowers)
Chamaecyparis lawsoniana (Orford cedar)
Chamomilla suaveolens (rounded chamomile)
Chrysanthemum coronarium (chrysanthemum)
Crataegus laevigata
Cucumis sativus (cucumber)
Daucus carota (carrot)
Fagus sylvatica (common beech)
Fraxinus excelsior (ash)
Hordeum vulgare (barley)
Lactuca sativa (lettuce)
Malus sylvestris (crab-apple tree)
Mentha arvensis (corn mint)
Pisum sativum (pea)
Pyrus communis (European pear)
Sambucus nigra (elder)
Senecio vulgaris
Solanum lycopersicum (tomato)
Solanum tuberosum (potato)
Trifolium pratense (purple clover)
Tussilago farfara (colt's-foot)
Veronica (speedwell)

51.10 Biology and life cycle

Typical of the genus, *X. diversicaudatum* has four juvenile stages. The female lays eggs in the soil and hatching takes place after the complete development of the first-stage juvenile. *X. diversicaudatum* may take almost 2 years to develop from egg to adult and the adults have a lifespan of 3–5 years (Flegg, 1968). The nematode has a reproductive span of around 12–13 months and produces approximately 180–200 eggs, laying one egg every 21 days with a threshold temperature above 5°C (Brown and Coiro, 1983). Distribution of *X. diversicaudatum* is also influenced by the soil profile and associated host plant species (Flegg, 1968). In England, *X. diversicaudatum* was detected to a depth of 60–100 cm, but numbers decreased below a depth of 20 cm (Taylor and Thomas, 1968).

51.11 Similarities to other species/conditions

Loof and Luc (1990) placed *X. diversicaudatum* in the species group that has a pseudo-Z organ and they are morphologically similar, such as *X. coxi europaeum* and *X. artemisiae*. Some other species, without a pseudo-Z organ, are also similar to *X. diversicaudatum*, such as *X. index*.

51.12 Economic importance

The European dagger nematode *X. diversicaudatum* can directly damage a wide range of fruit, ornamental and vegetable crops by feeding with its long stylet. However, the nematode is frequently associated with two nepoviruses, ArMV and SLRSV, throughout Europe and localized in New Zealand, South Africa and some states of the USA. The economic importance of *X. diversicaudatum* is mainly because of its ability to transmit ArMV and SLRSV. These viruses are able to reduce the quality as well as quantity of infected crops and make them unmarketable. Sometimes the infected plants are killed directly by the virus, or through secondary invasion by other pathogens.

51.13 Movement and means of dispersal

See section 46.11, above.

51.14 Pest risk analysis and looming threat from introduction

The main phytosanitary risk is because the nematode is known to be a natural vector of the two economically important nepoviruses, ArMV and SLRSV. Robbins and Brown (1991) suggested that *X. diversicaudatum* was probably introduced relatively recently in North America, having been imported with plant material from Europe, particularly roses from infested nurseries. As a vector of nepoviruses, these nematodes could act as a direct agent for the introduction of the viruses. If the viruses are introduced or spread into new areas or into the host countries, this would create an additional phytosanitary measure for the certification of virus-free material of numerous crops.

51.15 Invasiveness rating

Considering the consequences of probable introduction and pest risk analysis of *X. diversicaudatum*, the following invasiveness rating has been assigned.

Climate/host interaction	–	Medium (2)
Hosts range	–	High (3)
Dispersal potential	–	High (3)
Economic impact	–	High (3)
Environmental impact	–	Medium (2)
Post-entry distribution	–	Low (–1)
Final rating (score)	–	High (12)

Based on the above invasiveness rating, *X. diversicaudatum* receives a score of 12 out of 15 and is rated as HIGHLY INVASIVE NEMATODE.

51.16 Region-wise status of invasiveness

In several countries *X. diversicaudatum* has a quarantine status, including in India (PQ Order-2003, 2015), mainly because it acts as a vector of arabis mosaic virus and strawberry latent ringspot virus (Decraemer and Geraert, 2013), and is recommended for quarantine regulations. It has been given a high invasiveness rating for all the countries where it is absent.

51.17 Management measures

51.17.1 Chemical control

Treatments with chemical nematicides have proved effective in controlling virus-vector nematodes, including *X. diversicaudatum*. Earlier, the use of fumigant nematicides was highly effective to control *X. diversicaudatum* (Pitcher and McNamara, 1973), but now most of these fumigants have been banned for pest control. Application of non-fumigant nematicides with a commercially recommended dose may provide up to 80–90% suppression of nematodes in the top 40–60 cm of soil (Thomason and McKenry, 1975). Application of carbamate chemicals such as oxamyl prevented virus transmission by *X. diversicaudatum* in pot experiments (Taylor and Gordon, 1970). However, chemical control is not economical when the plants are infected by either of the nepoviruses transmitted by *X. diversicaudatum*.

51.17.2 Phytosanitary measures

See Chapter 16, section 115.

52 *Xiphinema ifacolum*

52.1 Common name

Dagger nematode

52.2 Scientific name

Xiphinema ifacolum Luc, 1961

52.3 EPPO code

XIPHIF

52.4 Diagnosis

52.4.1 *Morphological diagnosis*

The morphological and morphometric descriptions of *Xiphinema ifacolum* below are after Luc (1961). For measurements, see Table 42.

FEMALE: Body slender, cylindrical, open C-shaped after heat fixation. Body with very fine transversely striated cuticle and tapering at either end. Lateral chord covering one-quarter of corresponding body width near mid-body. Body pores present in latero-subventral, latero-subdorsal, cervical region and ventral pores in posterior region of body. Cephalic region ovoid, separated from body contour by a weak constriction (Fig. 8.11A). Amphidial aperture opening at close to depression and covering about two-thirds of corresponding body width (Fig. 8.11B). Odontostyle and odontophore very long. Oesophagus typical of the genera. Vulva median (Fig. 8.11D). Genital tracts paired, amphidelphic (Fig. 8.11I) with very prominent Z-organs (Fig. 8.11J). Tail conoid irregularly narrowed towards the terminus with a sub-digitate peg (Fig. 8.11H). Three pairs of caudal pores. A narrow, typically shaped protoplasmic process spreads down sub-digitate part of tail tip.

MALE: Morphologically similar to female in general except sexual structure. Spicules robust, broad and arcuate. Supplements arranged just before cloaca as one ventral pair and three others ventral plus five subventral pairs. Tail similar to female's with three pairs of caudal pores.

52.4.2 *Molecular diagnosis*

For details, see section 46.4.2, above.

52.5 Geographical distribution

X. ifacolum has been reported from West Africa, where it is widely distributed in Benin, Cameroon, Côte d'Ivoire, Guinea, Liberia, Sierra Leone and Togo (CABI-CPC, 2010). It is also recorded from Brazil (Rashid *et al.*, 1986) and Sri Lanka (Lamberti *et al.*, 1983).

52.6 Hosts

X. ifacolum is polyphagous, feeding on numerous plants, but it mainly parasitizes rice, *Oryza sativa* (Lamberti *et al.*, 1991). Other hosts are *Abelmoschus esculentus* (okra), *Ananas comosus* (pineapple), *Capsicum annuum* (bell pepper), *Citrus limon* (lemon), *Coffea* (coffee), *Glycine max* (soybean), *Hevea brasiliensis* (rubber), *Lycopersicon esculentum* (tomato), *Musa* (banana), *Musa* × *paradisiaca* (plantain), *Piper nigrum* (black pepper), *Solanum melongena* (aubergine), *Theobroma cacao* (cocoa) and *Vigna unguiculata* (cowpea) (CABI-CPC, 2010).

52.7 Vector

X. ifacolum is not known to transmit viruses.

52.8 Symptoms

As this species is not a vector of nepoviruses, the symptoms appear only due to direct damage by ectoparasitic feeding on roots. Vigorous feeding of the nematode by its long stylet causes mechanical damage with confined discoloration/necrosis and with the root tip showing swelling and galling. Root growth and vigour are suppressed. Conspicuous symptoms on the above-ground parts do not appear, but rice plants show deteriorating foliar growth and patchy appearance of plants in the field, with lower grain yields (Lamberti *et al.*, 1991).

52.9 Biology and life cycle

The biology and life cycle are typical of the genus.

52.10 Economic importance

Lamberti *et al.* (1991) regarded *X. ifacolum* as a major pest of rice in Liberia. The nematode is also capable of damaging black pepper in Sri Lanka and it reduced vine height in pot experiments (Lamberti *et al.*, 1983). Another pot trial showed 30% growth reduction in rice (Lamberti *et al.*, 1987). Lamberti *et al.* (1993) also reported *X. ifacolum* as pathogenic to soybean, cowpea, okra and eggplant in Liberia. However, assessments on economic loss are not available.

52.11 Movement and means of dispersal

See section 46.11, above.

52.12 Pest risk analysis and looming threat from introduction

Considering the economic damage caused by this dagger nematode to rice and other probable losses in other host crops, a complete pest risk analysis is suggested. However, this nematode can be excluded if it is detected because it is an ectoparasite and the primary entry pathway would be soil, which is not allowed to be carried with plants. Due to its restricted distribution, preventive steps should be taken to avoid introduction into new host areas.

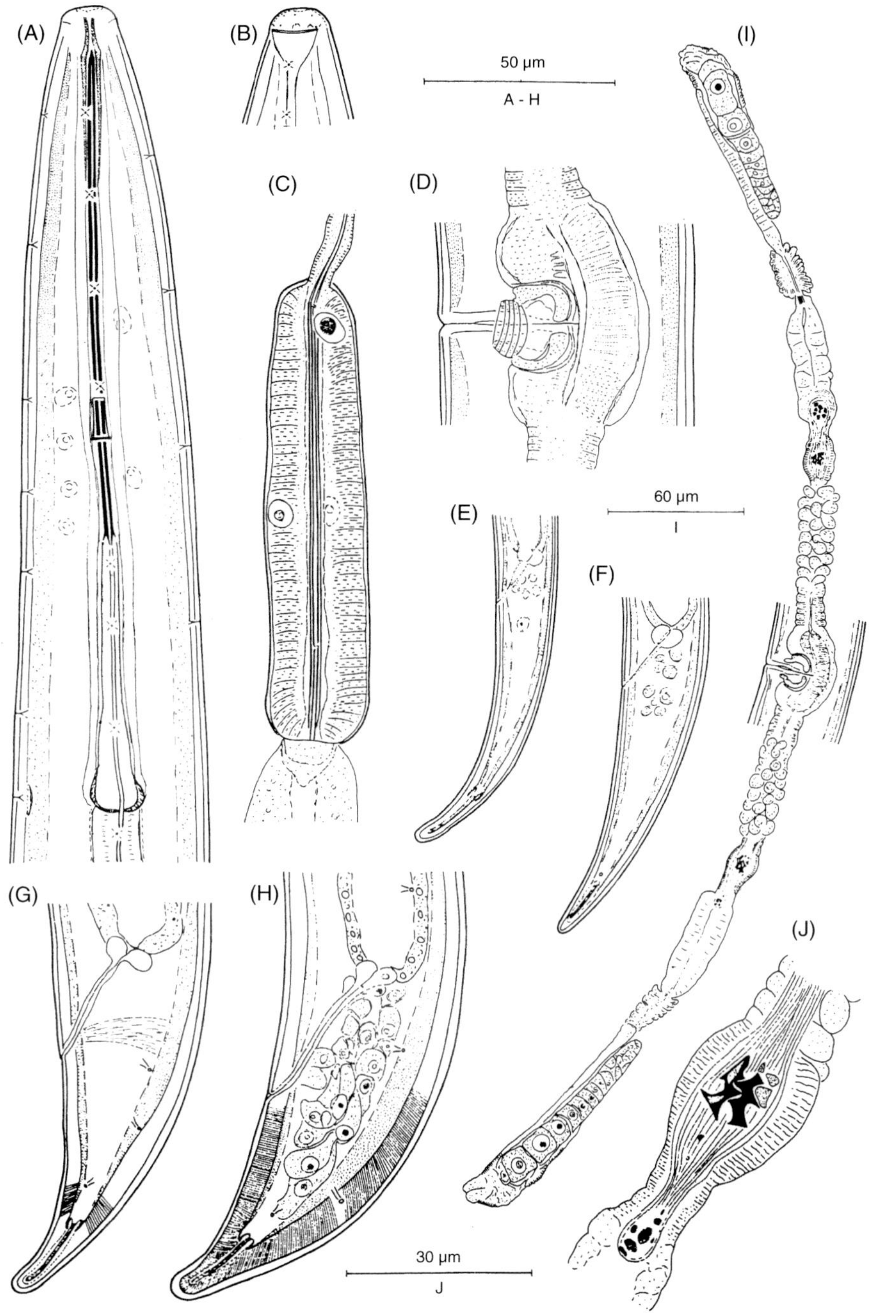

Fig. 8.11. Line drawing of *Xiphinema ifacolum*. Female anterior region (A); head end, surface view (B); oesophageal bulb (C); vagina and vulva region (D); tail of first-stage juvenile (E); tail of second-stage juvenile (F); tail of fourth-stage juvenile (G); female tail (H); female reproductive system (I); Z organ (J). [Reproduced from Hunt, 1993 (after Rashid *et al.*, 1986; courtesy Nematologia Mediterranea]

52.13 Invasiveness rating

Based on the probable consequences of introduction and the economic impact of *X. ifacolum* following invasiveness, the following rating has been assigned.

Climate/host interaction	–	Medium (2)
Hosts range	–	Medium (2)
Dispersal potential	–	Medium (2)
Economic impact	–	Medium (2)
Environmental impact	–	Medium (2)
Post-entry distribution	–	Not known (0)
Final rating (score)	–	Medium (10)

Based on the above invasiveness rating, *X. ifacolum* receives a score of 10 out of 15 and is rated as MEDIUM INVASIVE NEMATODE.

52.14 Region-wise status of invasiveness

X. ifacolum is a quarantine pest in India for pineapple (*Ananas comosus*) (PQ Order-2003, 2015). However, taking into consideration the damaging effect and distribution of this dagger nematode, a medium invasive rating has been given for all the countries where it is still not detected.

52.15 Management measures

The application of carbofuran significantly reduced *X. ifacolum* populations under rice in Liberia (Lamberti *et al.*, 1987). Carbofuran also checked the nematode in nurseries. In the field, carbofuran and ethoprophos can be used to control the nematode (Lamberti *et al.*, 1992). For further management options and phytosanitary measures, see Chapter 16, section 115.

53 *Xiphinema index*

53.1 Common name

Fan-leaf virus nematode

53.2 Scientific name

Xiphinema index Thorne and Allen, 1950

53.3. Synonym

Diversiphinema index Thorne and Allen (Cohn and Sher, 1972)

53.4 EPPO code

XIPHIN

53.5 Diagnosis

Xiphinema index belongs to the *Xiphinema* non-*americanum* group.

53.5.1 Morphological diagnosis

The morphological and morphometric descriptions of *Xiphinema index* after Heyns (1971) are presented below. For measurements, see Table 42.

FEMALE: Body open C-shape when heat relaxed. Cephalic region not offset, continuous with body contour. Amphidial aperture nearly as long as width of head end (Fig. 8.12D). Odontostyle long needle-like. Odontophore with three large basal flanges (Fig. 8.12A). Basal spear guiding ring 100–114 μm from head end. A short mucro similar to stylet tip existing about 30 μm behind odontophore. Oesophagus typical of genus, slender anterior and bulbar posterior part. Dorsal oesophageal gland (DOG) cell spreads along length of basal oesophageal bulb on dorsal side and anterior half of ventral side. DOG cell with six ducts moulded by deep infolds cell membrane limiting. Ovaries paired, amphidelphic, reflexed. Z-organ not present (Fig. 8.12A). Prerectum 320–390 μm long. Tail about as long as anal body width, mammiform, generally with a digitate terminal or ventral peg 9–13 μm long (Fig. 8.12F). In some females, tail is pegless.
MALE: Extremely rare, not important for reproduction. Spicules robust and strong with lateral guiding pieces 12 μm long (Fig. 8.12C). Ventromedian arrangements of seven supplementary papillae are found (Fig. 8.12C).
JUVENILES: Three or four juvenile stages documented by comparative body length and odontostyle length. Stage can be differentiated by length of replacement odontostyle (Fig. 8.12E) in one stage and corresponding functional odontostyle length in next stage.

53.5.2 Molecular diagnosis

Species-specific primers have been developed by Wang *et al.* (2003) to differentiate *X. index* from other closely related species (*X. diversicaudatum*, *X. italiae* and *X. vuittenezi*). They used multiplex PCR to identify single individuals by amplifying one region (ITS1) spanning the 18S and 5.8S ribosomal genes with the primers specific for:

X. index (GAGTCGTAACGTTTCTCGTCTATCAGG) fragment size 340 bp
X. diversicaudatum (GAGATATAAAGCGAAAACCGCGAG) fragment size 813 bp
X. italiae (GAAATAAGAACCCTGAAAAAGATAGG) fragment size 414
X. vuittenezi (GTGGAACGAAAAGACCTC) fragment size 591 bp.

Molecular studies of *X. index* have also been conducted by a few workers (Demangeat *et al.*, 2010; Gutiérrez-Gutiérrez *et al.*, 2011). For other molecular details, see section 46.4.2 above.

53.6 Geographical distribution

X. index occurs worldwide (Fig. 8.13) and has been reported from almost all viticulture regions of the world. The nematode is known to have been detected in the following countries CABI-ISC, 2020.

Africa: Algeria, South Africa, Tunisia
Asia: Armenia, Azerbaijan, Georgia, India, Iran, Iraq, Israel, Lebanon, Pakistan, Tajikistan, Turkey, Turkmenistan, Uzbekistan

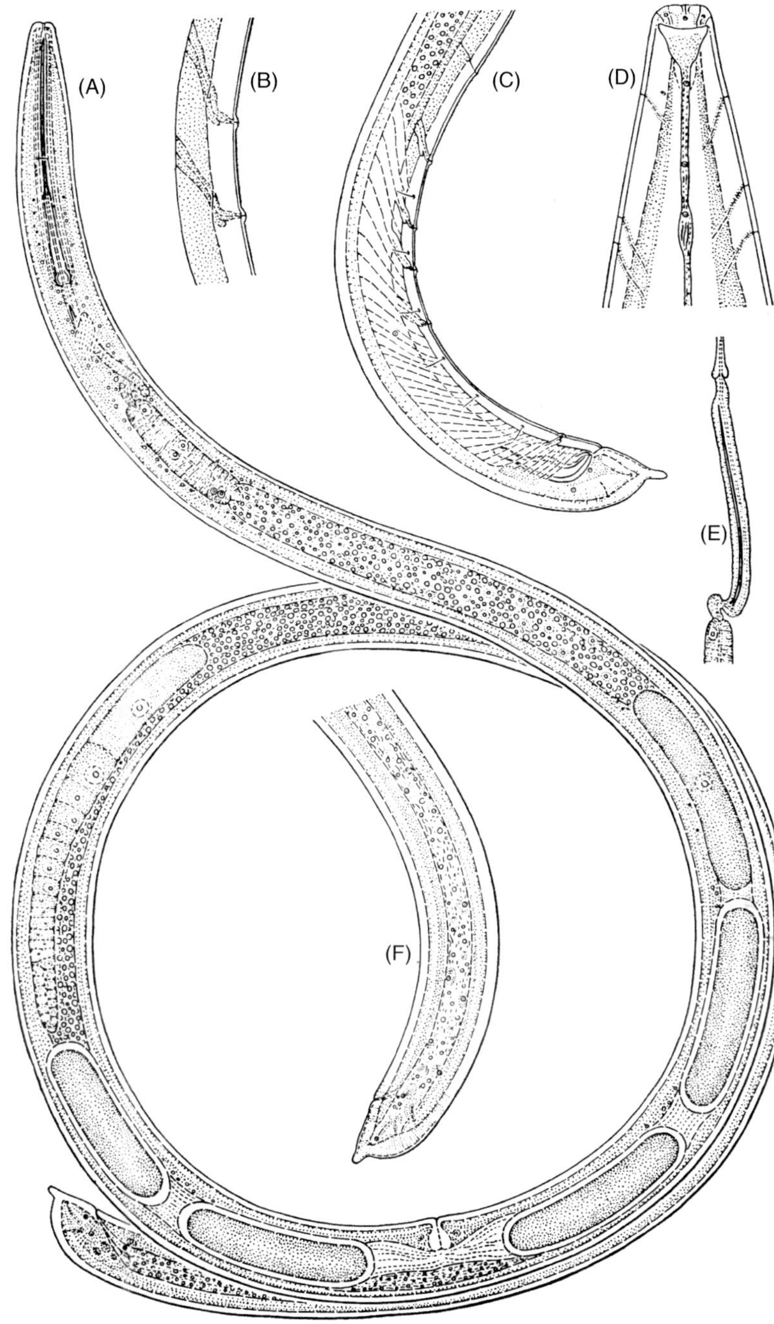

Fig. 8.12. Line drawing of *Xiphinema index*. Female (A); details of supplements (B); male posterior ends (C); head end showing amphid (D); replacement of odontostyle in the anterior portion of oesophagus of juvenile (E); female posterior region (F). [Reproduced from Hunt, 1993 (after Thorne and Allen, 1950; courtesy Helminthological Society of Washington]

Europe: Albania, Austria, Bulgaria, Croatia, Cyprus, Czech Republic, France, Germany, Greece, Hungary, Italy, Malta, Moldova, Poland, Portugal, Romania, Russia, Serbia and Montenegro, Slovakia, Slovenia, Spain, Switzerland, Ukraine, former Yugoslavia.
North America: USA
Oceania: Australia, New Zealand
South America: Argentina, Brazil, Chile, Peru

53.7 Hosts

X. index is polyphagous, with grapevine (*Vitis vinifera*) as the main host. It is also known to parasitize numerous woody plants, including citrus, fig, mulberry, *Prunus* and *Pyrus* spp. (Decraemer and Geraert, 2013). The extended list of hosts of *X. index* is given in Table 45.

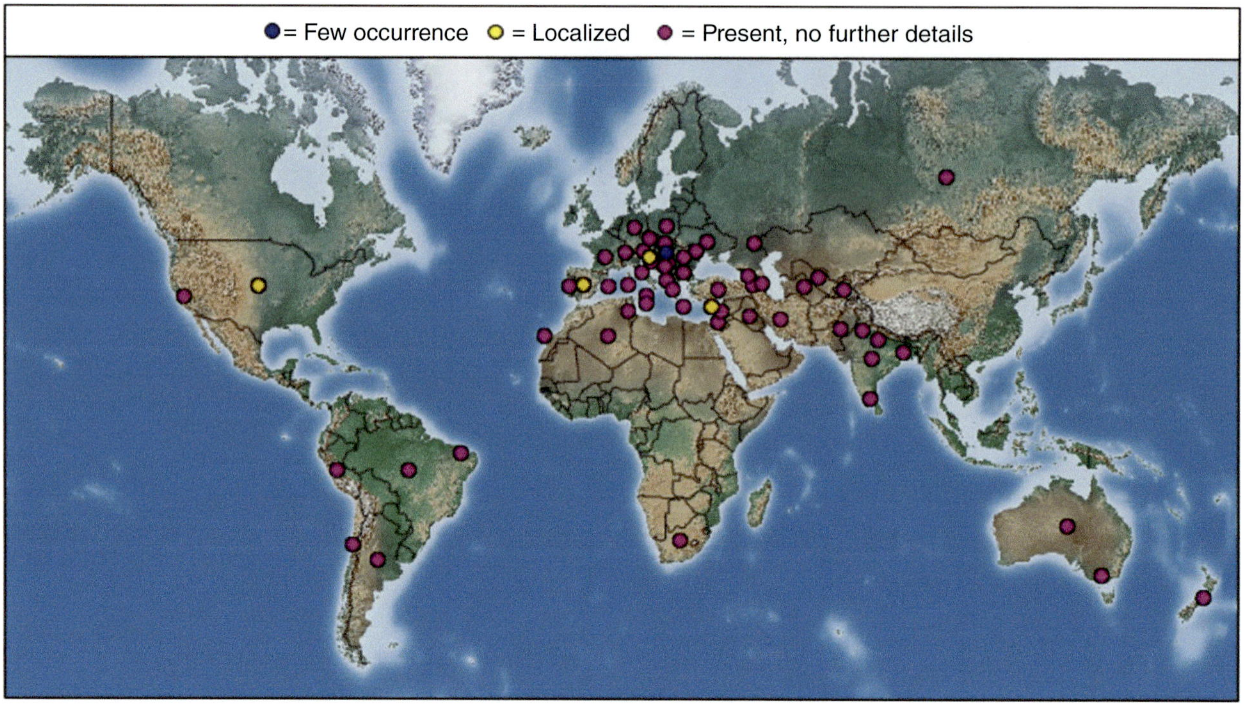

Fig. 8.13. Geographical distribution of *Xiphinema index* (adapted from CABI-ISC, 2020).

Table 45. List of hosts of *Xiphinema index* (adapted from CABI-ISC, 2020).

Main hosts
Ficus carica (fig)
Vitis arizonica (canyon grape, USA)
Vitis candicans (mustang grape)
Vitis vinifera (grapevine)
Vitis vinifera × *Muscadinia rotundifolia*
Other hosts
Ampelopsis aconitifolia (monkhood-vine)
Anacardium occidentale (cashew nut)
Citrus
Citrus aurantium (sour orange)
Cydonia oblonga (quince)
Jasminum officinale
Juglans regia (walnut)
Malus domestica (apple)
Malus sylvestris (crab-apple tree)
Morus alba (mora)
Olea europaea subsp. *europaea* (olive)
Opuntia sp. (pricklypear)
Parthenicissus tricuspidata
Petunia hybrida
Phaseolus vulgaris (common bean)
Pistacia mutica
Pistacia vera (pistachio)
Prunus amygdalus
Prunus armeniaca (apricot)
Prunus avium (sweet cherry)
Prunus domestica (plum)
Prunus salicina (Japanese plum)
Pyrus communis (European pear)
Pyrus persica
Rosa (roses)
Solanum lycopersicum (tomato)
Solanum tuberosum (potato)
Wild hosts
Betula (birches)
Chenopodium giganteum (large lambsquarters)
Chenopodium quinoa (quinoa)
Cupressus pyramidalis
Fagus silvaticus
Gomphrena globosa (globe amaranth)
Nicotiana rustica (wild tobacco)
Pinus halepensis (Aleppo pine)
Pinus sp. (pine)
Populus sp. (poplar)
Quercus lucombeana
Solanum nigrum (black nightshade)
Thuja
Urtica urens (annual nettle)
Vitis rupestris (sand-grape)

53.8 Vector role

X. index is a vector of grapevine fanleaf virus (GFLV) and it was first demonstrated by Hewitt *et al.* (1958). GFLV is one of the most devastating viruses of grapevine (Jelly *et al.*, 2012). The virus particles adhere to the lumen and cuticular lining of the odontophore and also the pharynx. The virus particles can be retained for up to 4 years. However, the virus is not transferred by nematodes in eggs or when the nematode moults (Demangeat *et al.*, 2005).

53.9 Symptoms

X. index causes both direct and indirect damage symptoms. Direct symptoms are due to extensive feeding activity resulting in retarded root extension and swelling or galls on root tips. Heavily parasitized roots of grapevine and figs turn brownish, swell at the tips and form terminal galls (Fig. 8.8H). The epidermal and outer cortical cells of the root tip break down at the feeding point and show necrotic lesions. Beneath the layer of necrotic lesions, cells are formed within the gall tissue that are multinucleate (Fig. 8.8I); they are substantially enlarged and comprise dense cytoplasm (Vovlas *et al.*, 1978). Damage to the root system eventually leads to a reduction in plant growth. However, the most important impact of *X. index* is indirect damage by vectoring grapevine fanleaf virus (GFLV). Attacked grapevines and roses show vein clearing (Fig. 8.8J) and the root breaks in the cortex appear darker. In grapevines, a layer of phellogen tissue develops which consists of three to five cells from the feeding site, and slowly extends to encircle the whole cross-section of the root (Cohn and Orion, 1970).

53.10 Biology and life cycle

X. index is a migratory ectoparasite; all stages are biotrophic and occur in soil. The nematode reproduces by meiotic parthenogenesis and an individual juvenile is capable of raising a population (Dalmasso, 1970). Sexual reproduction seldom occurs even though males are abundant (Villate *et al.*, 2010). Single females produced 55–118 eggs (Wyss, 1978). Embryogenic development occurs in 10–12 days at 24 ± 2°C (Vovlas and Larizza, 1994). Hatching occurs in 6–8 days and the first moult takes place outside the egg 24–48 h after hatching. Thereafter second, third and fourth moults occur at 6-day intervals (Fisher and Raski, 1967). The life cycle has been recorded as being as short as 22–27 days at 24°C in the USA (Radewald and Raski, 1962) and up to 7–9 months at 20–23°C in Israel (Cohn and Mordechai, 1969).

The nematode occurs in a wide range of soils, including heavy soils, but it has a preference for soils of light to medium texture with pH 6.5–7.5 (Decraemer and Geraert, 2013). *X. index* can survive up to 9–10 months in moist sterile soil without a food source (Taylor and Raski, 1964) and 4.5 years in the soil after the grapevine has been harvested but not the roots (Raski *et al.*, 1965). It can survive in freezing temperatures in England (Cotten *et al.*, 1971). The population dynamic of *X. index* is also influenced by the soil profile. In grapevine fields, the lowest number of nematodes was observed at up to 25 cm depth in soil having a high clay content, and up to 40 cm in sandier soil textures (Esmenjaud *et al.*, 1992). In south-eastern France the nematode is found at depths of up to 1.5 m in heavy soils but is more abundant between 20 cm and 60 cm (Scotto La Massese *et al.*, 1988).

53.11 Economic importance

X. index causes significant damage to viticulture throughout the world, alone or as a vector of GFLV (Decraemer and Geraert, 2013). High population levels negatively affected grape yields in the USA (Anwar and Van Gundy, 1989) and Spain (Decraemer and Geraert, 2013). *X. index* also significantly reduced plant growth and delayed bud break in California (Anwar and Van Gundy, 1989). Inoculation of 500 *X. index* in grapevine plants increased abscission of oldest leaves in the first year by 23% and suppressed top and root weights by 65% and 38%, respectively; in the second year the inflorescences and fruit size were reduced by 60% and 89%, respectively (Boubals *et al.*, 1971). In Cyprus, 22.2% of soil samples revealed the occurrence of *X. index* (Philis, 1993). *X. index* was also reported from 105 of 170 vineyards from eight provinces in Iran (Mojtahedi *et al.*, 1980).

53.12 Movement and means of dispersal

See section 46.11, above.

53.13 Pest risk analysis and looming threat from introduction

The main phytosanitary risk is due to *X. index* being an established vector of the nepovirus, grapevine fanleaf virus (GFLV). The nematode could thus act as a direct agent for the introduction of this virus. If the virus is introduced or spread into new host countries, it would require an additional phytosanitary measure for the certification of virus-free material of particularly grapevine and other fruit crops.

53.14 Invasiveness rating

Considering the probable consequences of introduction and pest risk analysis of *X. index*, the following invasiveness rating has been assigned.

Climate/host interaction	–	High (3)
Hosts range	–	Medium (2)
Dispersal potential	–	High (3)
Economic impact	–	High (3)
Environmental impact	–	High (3)
Post-entry distribution	–	Low (–1)
Final rating (score)	–	High (13)

Based on the above invasiveness rating, *X. index* receives a score of 13 out of 15 and is rated as HIGHLY INVASIVE NEMATODE.

53.15 Region-wise status of invasiveness

X. index is a quarantine pest in the European Union (EPPO, 2012) and is currently targeted in regulatory programmes worldwide (CABI-ISC, 2020). It has been given a high invasiveness rating for all the countries where it still does not occur.

53.16 Management measures

53.16.1 Cultural practices

To prevent *X. index* in grapevine nurseries, the site should be located on non-infested soil. Dalmasso (1969) advised a rotation with non-hosts for 6–7 years, followed by a 3-year rotation with non-hosts and the use of nematicides. If *X. index* has been detected, grapevine should not be cultivated for 5–7 years. The soil should be disinfested and replanting should be done in spring, 3–6 months after treatment (Esmenjaud, 1983). In Cyprus, fallow alone or fallow rotated with barley can suppress the populations in 3–4 years after uprooting grapevine plantations (Philis, 1994). Application of botanical pesticides has also been found effective in decreasing the *X. index* population. Treatment of macerated *Capsicum annuum* pods in water may cause significant mortality to the nematode (Sasanelli and Catalano, 1991). Aqueous extracts from leaves of *Ruta graveolens* are also reported to have a nematicidal effect against *X. index*, and the mortality increased with the increase of the concentration and exposure time (Sasanelli, 1992). In the vineyard nursery, hot-water treatment for 10 min at 52°C controlled *X. index* (Vega, 1978). The hot-water treatment has also proved effective in grapevines in South Africa (Orffer, 1977).

53.16.2 Host-plant resistance

Some *Vitis* species, such as *V. arizonica*, *V. candicans*, *V. rufotomentosa*, *V. smalliana* and *V. solonis*, were found to be highly resistant to *X. index* (Kunde *et al.*, 1968). Among 31 grapevine rootstocks of *V. longii* tested by Malan and Meyer (1993), the rootstocks 'Harmony', 'Freedom' and '1613C' recorded no root damage and a low reproduction rate of *X. index*. Similarly, 19 Californian hybrid rootstocks were tested as resistant, including 'Harmony', 'Freedom', 'Schwarzmann' and '3309' (Harris, 1983). *V. candicans* was reported as a source of resistance to *X. index* and the resistance of the cultivar 'Dog Ridge' was confirmed (Coiro *et al.*, 1990).

53.16.3 Chemical control

Effective control of *X. index* by application of fumigant nematicides is well documented in earlier works. However, these fumigant nematicides have been banned in crop protection due to their hazardous impact. Application of non-fumigant nematicides such as carbofuran and fenamiphos gave good control of *X. index* on grapevines in California, USA (Raski *et al.*, 1981). Application of oxamyl also effectively controlled the nematode (Hafez *et al.*, 1981a). In Switzerland, a solution of glyphosate, applied to grapevines 3–6 months before grubbing, suppressed *X. index* populations to below economic threshold (Boubals, 1994).

53.16.4 Phytosanitary measures

See Chapter 16, section 115.

54 *Xiphinema italiae*

54.1 Common name

Mediterranean dagger nematode

54.2 Scientific name

Xiphinema italiae Meyl, 1953

54.3 EPPO code

XIPHIT

54.4 Diagnosis

X. italiae belongs to the *Xiphinema* non-*americanum* group.

54.4.1 Morphological diagnosis

The morphological descriptions of *X. italiae* from Hunt (1993) are presented below. For morphometric measurements (after Heyns, 1974), see Table 42.

FEMALE: Body slender, about 3 mm long. Almost straight in first two-thirds of body, then ventrally arcuate (Fig. 8.14L). Cephalic region gently rounded, offset by a slight constriction (Fig. 8.14A). Amphidial stirrup-shaped, post labial, slit-like about one-half and two-thirds body width, about 70–74 μm long. Odontostyle needle-like, 87–112 μm long; odontophore with prominent tripartite basal flanges, about 54–68 μm long. Oesophagus typical of genus with anterior slender and posterior bulbar part. Nucleus of dorsal oesophageal gland near anterior extremity of bulb. Vulva a transverse slit, slightly premedian leading to vagina which extends one-third into body. Genital tract amphidelphic, reflexed at oviduct (Fig. 8.14M). Z-organ absent. Tail shape variable (Fig. 8.14B-F), usually elongate-conoid with slight dorsal and ventral constrictions towards tip. Occasionally bluntly conoid and even subdigitate.
MALE: Extremely rare, no role in reproduction. Body shape similar to that of female. Spicules arcuate, 44–48 μm long (Fig. 8.14H). A pre-anal pair of supplements present and one or two additional ventromedian ones. Copulatory muscles strongly developed.

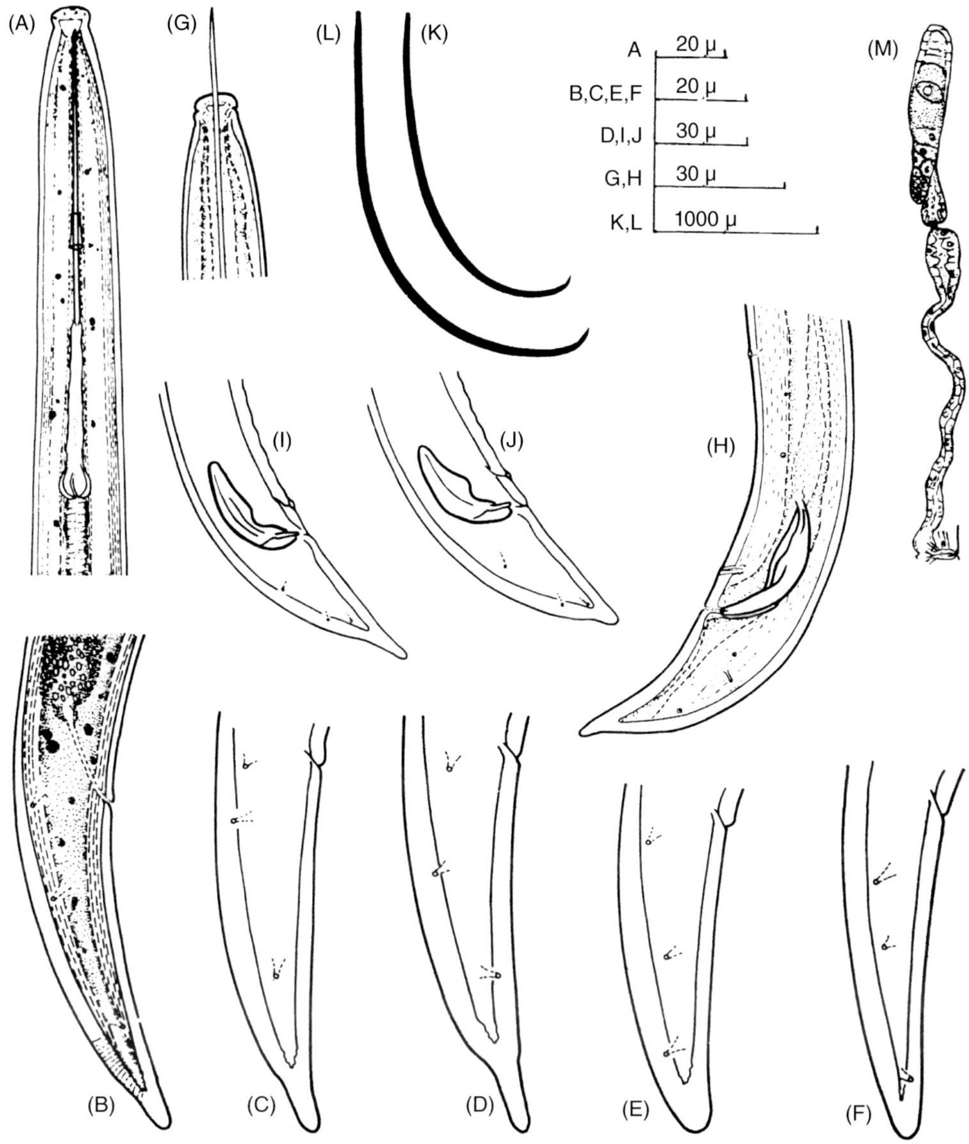

Fig. 8.14. Line drawing of *Xiphinema italiae*. Female anterior end (A); posterior end of female (B); tail shape of different populations (C-F); head end (G); male posterior end (H); male tails (I,J); female body postures (K, L); female reproductive system (M). [Reproduced from Hunt, 1993; A-H, K-L from Martelli *et al*., 1966, courtesy Nematologica; I,J from Heyns, 1974; M from Dalmasso, 1969]

JUVENILE: Three or four juvenile stages documented by comparative body length and odontostyle length. Stage can be differentiated by length of replacement odontostyle in one stage and corresponding functional odontostyle length in next stage.

54.4.2 Molecular diagnosis

Species-specific primers are available to characterize *X. italiae* (Wang *et al*., 2003; Kumari and Lišková, 2009). For details, see section 46.4.2, above.

54.5 Geographical distribution

X. italiae has limited distribution and is concentrated near the Mediterranean region. The nematode is known to occur in some African countries (Cameroon, Egypt, Libya, Nigeria, South Africa and Tunisia) and is also reported from some European countries (Greece, Hungary, Italy, Moldavia, Romania, Russia, Slovakia and Spain) (Bonta *et al*., 2013; CABI-ISC, 2020). Outside Europe and Africa, *X. italiae* has been recorded in Cuba (Fernandez Díaz-Silveira and Ortega Herrera, 1995).

54.6 Hosts

The main host of *X. italiae* is grapevine (*Vitis* sp.). Other hosts include almond (*Prunus amygdalus*), apricot (*Prunus armeniaca*), eucalyptus (*Eucalyptus* sp.), mulberry (*Morus* sp.), olive (*Olea europaea*), peach (*Prunus persica*), plum (*Prunus domestica*), sour orange (*Citrus aurantium*) and a few conifers (Cohn, 1977).

54.7 Vector

X. italiae has been reported to be a vector of grapevine fanleaf virus in Israel (Cohn *et al.*, 1970).

54.8 Symptoms

X. italiae attacks feeder roots and its feeding causes root discoloration and stunting with some terminal galling (Hunt, 1993). When the nematode is involved in the transmission of GFLV, the typical symptoms of the virus develop in the host plant (Brown *et al.*, 1993).

54.9 Biology and life cycle

X. italiae has ectoparasitic habits and takes 6 months to complete its life cycle at 28°C (Cohn, 1977). *X. italiae* prefers sandy soil rather than loamy soil, and the highest population occurs in top 30 cm of soil (Hunt, 1993). In Romania, the nematode is mainly associated with sandy soil (Bonta *et al.*, 2013)

54.10 Economic importance

The main economic impact is associated with the ability of *X. italiae* to transmit GFLV, which has been shown to have a significant damaging impact on grapevine (Brown and Trudgill, 1997).

54.11 Movement and means of dispersal

See section 46.11, above.

54.12 Pest risk analysis and looming threat from introduction

See section 46.12, above.

54.13 Invasiveness rating

Considering the probable consequences of introduction and the economic impact of *X. italiae*, the following invasiveness rating has been assigned.

Climate/host interaction	–	Medium (2)
Hosts range	–	Medium (2)
Dispersal potential	–	Medium (2)
Economic impact	–	Medium (2)
Environmental impact	–	Medium (2)
Post-entry distribution	–	Low (–1)
Final rating (score)	–	Medium (9)

Based on the above invasiveness rating, *X. italiae* receives a score of 9 out of 15 and is rated as MEDIUM INVASIVE NEMATODE.

54.14 Region-wise status of invasiveness

X. italiae is widespread in the Mediterranean region, but it has been reported to transmit grapevine fanleaf virus only in Israel and the virus vector status also needs to be confirmed. Due to its uncertain vector status in different countries, the nematode is given a medium invasiveness rating in the countries where it is absent.

54.15 Management measures

No specific management options are available for this nematode, but the management measures practised to control other dagger nematodes, particularly the vector of GFLV (*X. index*), may be employed. For further management and phytosanitary measures, see Chapter 16, section 115.

55 *Xiphinema rivesi*

55.1 Common name

Dagger nematode

55.2 Scientific name

Xiphinema rivesi Dalmasso, 1969

55.3 EPPO code

XIPHRI

55.4 Diagnosis

Xiphinema rivesi is a member of the *Xiphinema americanum* group.

55.4.1 Morphological diagnosis

The morphological and morphometric descriptions of *Xiphinema rivesi* after Dalmasso (1969) are summarized below. For measurements, see Table 42.

Female: Body tapering gradually, closed C-shape when heat relaxed. Cephalic region rounded, continuous not offset, 10 µm wide. Amphid aperture slit-like, stirrup-shaped, about half as wide as cephalic region. Odontostyle robust, 90 µm long, with flanged odontophore 50 µm long. Basal guiding sheath 3–4 µm long with double guiding ring, situated 69 µm from anterior end. Basal bulb of oesophagus occupies about one-quarter of total oesophageal length, 75–85 µm long and 16–25 µm wide. Cardia bluntly conical, almost as long as wide. Vulva around equatorial. Vagina occupying one-quarter corresponding body width. Gonads paired, opposed and reflexed. Spermatheca and Z organ not seen. Prerectum 50–90 µm long. Rectum around half of anus diameter. Tail conoid, 33 µm long, generally with small terminal bulge.
Male: Very rare. Similar to female but with more body curvature in posterior region. Spicules dorylaimid type, 49 µm long. Tail slightly slenderer than in female. An adanal pair of copulatory supplements present with six supplements.
Juvenile: Three or four juvenile stages documented by comparative body length and odontostyle length. Stage can be differentiated by length of replacement odontostyle in one stage and corresponding functional odontostyle length in next stage.

55.4.2 Molecular diagnosis

As *X. rivesi* belongs to the *Xiphinema americanum* group, the molecular characterization employed for this group may be equally effective for this nematode. For details, see section 46.4.2, above.

55.5 Geographical distribution

X. rivesi has limited distribution but is widespread in some regions of the world (Fig. 8.15). The nematode is known to occur in the following countries (CABI-ISC, 2020).

Africa: Egypt.
Asia: Iran, Pakistan.
Europe: France, Germany, Italy, Portugal, Slovakia, Slovenia, Spain.
Oceania: Australia, Samoa, Togo.
North America: Canada, USA.
South America: Argentina, Chile, Peru.

55.6 Hosts

X. rivesi is polyphagous and parasitizes several fruit plants. The main hosts are apple, blueberry, grapevine, *Chenopodium quinoa*, peach, raspberry, sweet cherry and stone fruit (CABI-CPC, 2010). The extended hosts list is given in Table 46.

55.7 Vector

Xiphinema rivesi is the principal vector of three American nepoviruses, namely tomato ringspot virus (TomRV), tobacco ringspot virus (TobRV) and peach rosette mosaic virus (PRMV) (Brown *et al.*, 1993).

55.8 Symptoms

X. rivesi is a member of *X. americanum* group and symptoms are typical of the group. In the absence of virus, generally no distinctive symptoms show on the above-ground parts of plants. With high populations, the roots show swellings near the root tips due to nematode feeding (Perry, 1958). Sometimes extensive necrosis and destruction of feeder roots occur. On strawberry, *X. rivesi* feeding causes sunken, reddish-brown lesions that became dark brown to black, with reduction of the root

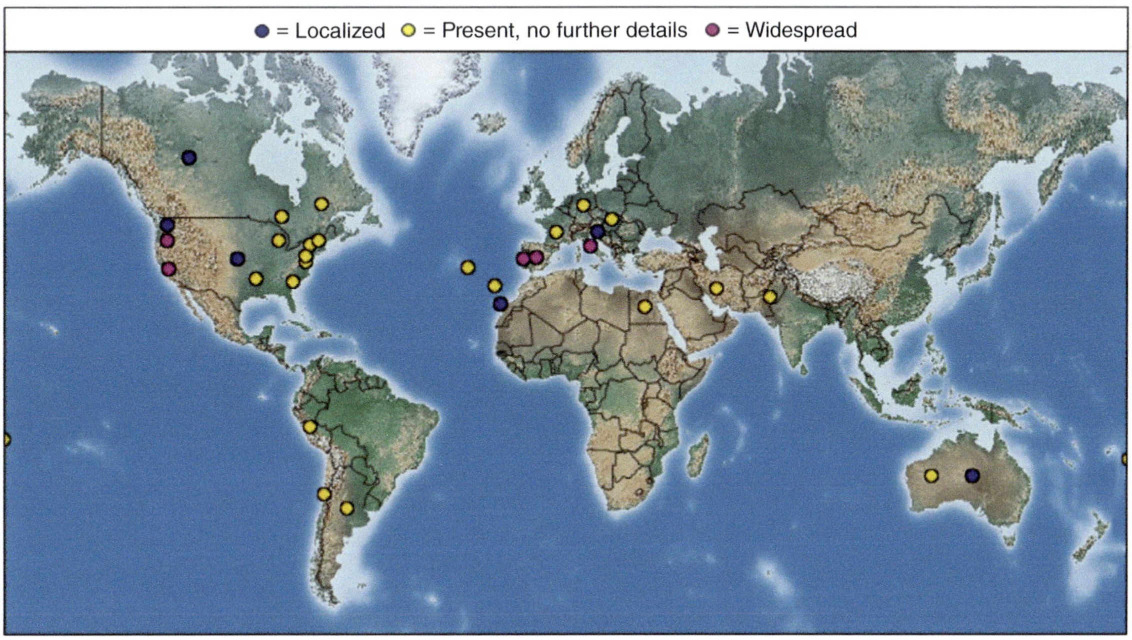

Fig. 8.15. Distribution map of *Xiphinema rivesi*. [Adapted from CABI-ISC, 2020]

Table 46. List of hosts of *Xiphinema rivesi* (adapted from CABI-CPC, 2010).

Main hosts
Acer (maples)
Avena sativa (oats)
Betula pubescens (downy birch)
Chenopodium quinoa (quinoa)
Cucumis sativus (cucumber)
Fagus (beeches)
Fragaria ananassa (strawberry)
Juniperus (junipers)
Lonicera (honeysuckles)
Malus domestica (apple)
Malus sylvestris (crab-apple tree)
Medicago sativa (lucerne)
Nicotiana tabacum (tobacco)
Picea (spruces)
Picea glauca (white spruce)
Picea pungens (blue spruce)
Pinus koraiensis (fruit pine)
Prunus avium (sweet cherry)
Prunus domestica (plum)
Prunus persica (peach)
Rosa (roses)
Rubus (blackberry, raspberry)
Rubus idaeus (raspberry)
Sorghum bicolor (sorghum)
Taraxacum (dandelion)
Trifolium pratense (purple clover)
Trifolium repens (white clover)
Trifolium spp.
Tsuga (hemlocks)
Turfgrasses
Ulmus (elms)
Vaccinium (blueberries)
Vitis vinifera (grapevine)
Other hosts
Prunus salicina (Japanese plum)

system (Perry, 1958). Infected seedlings of spruce (*Picea pungens* and *Picea glauca* var. *densata)* had abridged root growth and were more susceptible to winter kill than non-infected plants (Griffin and Epstein, 1964).

55.9 Biology and life cycle

X. rivesi is a migratory ectoparasite and its life cycle is typical of *X. americanum s.l.* The populations of *X. americanum s.l.* (including *X. rivesi* and *X. californicum*) have three juvenile stages (Halbrendt and Brown, 1992). All stages of *X. rivesi* appear to overwinter (Jaffee *et al.*, 1987). On apple, the population density of *X. rivesi* may increase from January to June and decline from October to January (Khan *et al.*, 1997). The population density increases dramatically in the soil with sand and silt amendments (Jaffee *et al.*, 1988).

55.10 Economic importance

X. rivesi causes economic loss to peach and apple due to direct damage. Its economic importance is greatly increased when the nematode is associated with virus transmission, particularly TomRV and PRMV in Canada (Allen *et al.*, 1988). In Canada, PRMV causes substantial damage in *Prunus* species, grapes and blueberry (Brown *et al.*, 1993).

55.11 Similarities with other species

X. rivesi is included in the *X. americanum s.l.* complex. However, unlike *X. rivesi*, the head of *X. americanum* is slightly offset, the stylet is shorter and the tail shape is different (more pointed).

55.12 Movement and means of dispersal

See section 46.11, above.

55.13 Pest risk analysis and looming threat from introduction

X. rivesi is a parasite of peach and apple and causes economic loss due to direct damage. Its economic importance is greatly increased because of its vector role for important plant nepoviruses, TomRV, TobRV and PRMV. *X. rivesi* is a member of the *Xiphinema americanum* group; for further details see section 50.13 of *X. americanum*, above.

55.14 Invasiveness rating

Considering the probable consequences of introduction and pest risk analysis of *X. rivesi*, the following invasiveness rating has been assigned.

Climate/host interaction	–	High (3)
Hosts range	–	High (3)
Dispersal potential	–	High (3)
Economic impact	–	High (3)
Environmental impact	–	High (3)
Post-entry distribution	–	High (–3)
Final rating (score)	–	High (12)

Based on the above invasiveness rating, *X. rivesi* receives a score of 12 out of 15 and is rated as HIGHLY INVASIVE NEMATODE.

55.15 Region-wise status of invasiveness

Both *X. rivesi* and the vector viruses are of quarantine significance in EPPO countries and are included in the EPPO A2 list (EPPO, 2012). They are also of quarantine significance in many other countries, including India for peach, *Prunus persica* (PQ Order-2003, 2015). Due to its economic damage potential and vector status, *X. rivesi* is given a high invasiveness rating in the countries where it is not known to occur.

55.16 Management measures

See section 53.16, above.

9

Meloidogynidae: *Meloidogyne* Species

Meloidogyne Spp.

The genus *Meloidogyne* was proposed by Göldi (1887). At present more than 100 species of *Meloidogyne* have been described and validated with *Meloidogyne exigua* as type species (Subbotin, 2018). The recent key for the genus has been reviewed by numerous authors (Hunt and Handoo, 2009; Karssen *et al.*, 2013). The following *Meloidogyne* species are considered in this chapter for their invasiveness, as they are of economic importance and related to plant growth destruction on different agricultural crops (Siddiqi, 2000; Perry *et al.*, 2009; Karssen *et al.*, 2013).

Meloidogyne acronea Coetzee, 1956
M. arabicida Lopez and Salazar, 1989
M. ardenensis Santos, 1967
M. arenaria (Neal, 1889) Chitwood, 1949
M. artiellia Franklin, 1961
M. brevicauda Loos, 1953
M. chitwoodi Golden, O'Bannon, Santos and Finely, 1980
M. citri Zhao, Gao and Weng, 1990
M. coffeicola Lordello and Zamith, 1960
M. decalineata Whitehead, 1968
M. donghaiensis Zheng, Lin and Zheng, 1990
M. enterolobii Yang and Eisenback, 1983
M. ethiopica Whitehaed, 1968
M. exigua Goldi, 1887
M. fallax Karssen, 1996
M. fujianensis Pan, 1985
M. graminicola Golden and Birchfield, 1965
M. hapla Chitwood, 1949
M. incognita (Kofois and White, 1919) Chitwood, 1949
M. javanica (Treub, 1885) Lavergne, 1901
M. jianyangensis Yang, Hu, Chen and Zhu, 1990
M. kikuyensis De Grisse, 1960
M. kongi Yang, Wang and Feng, 1988
M. mali Itoh, Ohshima, and Ichinohe, 1969
M. marylandi Jepson and Golden, 1987
M. mingnanica Zhang, 1993
M. naasi Franklin, 1965
M. paranaensis Carneiro *et al.*, 1996
M. salasi Lopez, 1984

Systematic position of *Meloidogyne* species

	Classical (after Siddiqi, 2000)	Molecular (after Decraemer and Hunt, 2013)
Phylum	Nematoda	Nematoda
Class	Secernentea	Chromadorea
Order	Tylenchida	Rhabditida
Suborder	Hoplolaimina	Tylenchina
Infra-order	–	Tylenchomorpha
Superfamily	Holpolaimoidea	Tylenchoidea
Family	Meloidogynidae	Hoplolaimidae
Subfamily	Meloidogyninae	Meloidogyninae
Genus	*Meloidogyne*	*Meloidogyne*

56 General Characteristics of *Meloidogyne* Species

56.1 Diagnosis

Morphological identification based on perennial patterns is still the most accurate diagnostic method for *Meloidogyne* species. Due to the substantial diversity among the *Meloidogyne* species, various diagnostic approaches are employed to identify the genus at species level. These approaches include morphological, cytogenetics, isozymes and molecular methods of characterization of root-knot nematodes.

56.1.1 Morphological diagnosis

The morphology of the genus *Meloidogyne* has been reviewed by numerous authors (Whitehead,1968; Taylor and Sasser, 1978; de Guiran and Ritter, 1979; Hirschmann, 1985; Eisenback and Triantaphyllou, 1991; Hunt and Handoo, 2009; Karssen *et al.*, 2013). Morphological characters are important diagnostic features and are used in primary identification of *Meloidogyne* species. With the increasing number of described species in the genus, a general standard for morphometric characters should be followed to facilitate precise comparisons of differential diagnoses (Eisenback and Hunt, 2009). The list of characters presented in Table 47 should be considered and at least 20

DOI: 10.1079/9781789247367.0010

Table 47. Morphometrics of *Meloidogyne* spp. All measurements in micrometres (µm) and in the form: mean ± standard deviation (range) (Eisenback and Hunt, 2009).

	Female		Male	Second-stage juvenile (J_2)
Characters/stages	Holotype	Paratype	Paratype	Paratype
n	–	20	20	20
L	√	√	√	√
a	√	√	√	√
c	–	–	√	√
T	–	–	√	–
Maximum body diameter	√	√	√	√
Neck length	√	√	–	–
Stylet length	√	√	√	√
Stylet knob height	√	√	√	√
Stylet knob width	√	√	√	√
DGO	√	√	√	√
Excretory pore to anterior end	√	√	√	√
Inter-phasmidial distance	–	√	–	–
Vulva length	–	√	–	–
Vulva–anus distance	–	√	–	–
Tail length	–	–	√	√
Spicule length	–	–	√	–
Gubernaculum length	–	–	√	–
Testis length	–	–	√	–
Hyaline tail terminus (h)	–	–	–	√
H% (h/tail length × 100)	–	–	–	√

specimens each of the second-stage juvenile (J_2), male and mature female stages should be measured.

Based on the descriptions after Siddiqi (2000) and Hunt and Handoo (2009), modified morphological diagnoses of the genus *Meloidogyne* are summarized below.

MATURE FEMALE: Pearly white, body round to pear-shaped with short projecting neck, sedentary endoparasite. No cyst stages. Vulva and anus located close together at terminal. Perineal pattern with fingerprint-like cuticular marking, usually flattened, rarely elevated. Phasmids present, dot-like, slightly anterior, on either side of anus. Cuticle with fine to coarse striae. Stylet slender, well developed, 12–15 µm long with small basal knobs. Excretory secretory pore anterior to median bulb, often just posterior to base of stylet. Genital tracts paired, convoluted, prodelphic. Six large rectal glands present, secreting gelatinous material in which eggs are laid, usually not retained in body except moribund females.

MALE: Body vermiform, developed by metamorphosis within a swollen juvenile, up to 2 mm long, posterior end twisted. Cuticle strongly annulated. Lateral field with four longitudinal lines or incisures. Labial region smoothly offset, with distinct lip disc consisting few (1–3) annules; lateral sectors wider than submedian sectors, appearing as 'cheeks'. Stylet robust (18–25 µm), with large basal knobs. Pharyngeal glands lying mostly ventral to intestine. Spicules slender, usually 25–33 µm long; gubernaculum simple, 7–11 mm long. Testis single, but paired when sex reversal occurs. Tail rounded. Phasmids dot-like, located near cloacal opening, subterminal. Bursa absent.

JUVENILE: First stage with blunt tail tip, moults within egg; second and third moults occur within cuticle of second stage. Second stage infective, vermiform, migratory, straight to arcuate habitus when heat relaxed. Labial region with 1–4 coarse annules with distinct lip disc. Cephalic framework lightly sclerotized, lateral sectors wider than submedian sectors. Stylet robust, slender, less than 20 µm. Excretory secretory pore posterior to hemizonid. Median bulb large, rounded, with refractive thickenings. Tail with hyaline region, tip narrow with irregular edge. Third stage sedentary, stylet absent, swollen, sausage-shaped with short blunt tail. Fourth stage sedentary, swollen, with terminal anus. Stylet absent.

56.1.2 Cytogenetic diagnosis

The cytogenetic status of genus *Meloidogyne* is complex but important in understanding the overall biology and evolution. Within the genus *Meloidogyne* there are three modes of reproduction: amphimixis, facultative meiotic parthenogenesis (automixis) and obligatory mitotic parthenogenesis (apomixis). The chromosome number and mode of reproduction of some invasive nematodes are given in Table 48.

56.1.3 Biochemical diagnosis

Numerous new approaches, such as the use of isozyme with polyacrylamide gel electrophoresis (PAGE) of crude protein extracts and histochemical markers/stains for non-specific

Table 48. The mode of reproduction and chromosome number of some invasive *Meloidogyne* species (adapted from Karssen *et al.*, 2013).

Mode of reproduction	Meloidogyne species	Chromosome number	
		n	*2n*
Amphimixis	*M. kikuyensis*	7	
Facultative meiotic parthenogenesis (automixis)	*M. chitwoodi*	14–18	–
	M. exigua	18	–
	M. fallax	18	–
	M. graminicola	18	–
	M. hapla (race A)	13–17	–
	M. naasi	18	–
Obligatory mitotic parthenogenesis (apomixis)	*M. ardenensis*	–	51–54
	M. arenaria	–	30–38
	M. decalineata (= *M. africana*)	–	21
	M enterolobii (= *M. mayaguensis*)	–	44–46
	M. ethiopia	–	36–44
	M. hapla	–	30–32
	M. hapla (race B)	–	43–48
	M. incognita	–	32–38
	M. indica	–	41–46
	M. javanica	–	42–48
	M. oryzae	–	51–55
	M. paranaensis	–	50–52
	M. salasi	–	36

esterases, superoxide dismutase, malate dehydrogenase, and glutamate oxaloacetate transaminase, have provided useful tools in the identification of *Meloidogyne* species (Esbenshade and Triantaphyllou, 1985a,b; Carneiro *et al.*, 2000; Xu *et al.*, 2004). Among these, esterases and malate dehydrogenase are the most useful for differentiating the four common *Meloidogyne* species, such as phenotypes A2 and A3 for *M. arenaria*, HI for *M. hapla*, II for *M. incognita* and J3 for *M. javanica* (Esbenshade and Triantaphyllou, 1985b). For protocol details, see Chapter 15, section 109.1.

Polyclonal and monoclonal antibodies have also been produced for root-knot nematode identification by ELISA, dot blot and western blots (Tastet *et al.*, 2001; Kapur-Ghai *et al.*, 2014). Davies *et al.* (1996) selected three monoclonal antibodies (mAbs) to differentiate females of *M. incognita*, *M. javanica* and *M. arenaria*. Antisera produced to purified species-specific esterase bands permitted variation of *M. incognita* from *M. javanica* but the mAbs cross-reacted with other root-knot species (Davies *et al.*, 1996; Ibrahim *et al.*, 1996). Ibrahim *et al.* (1996) also raised a mAb to purified esterase from *M. incognita* and were able to differentiate *M. incognita* from *M. javanica*. Tastet *et al.* (2001) used two-dimensional electrophoresis and were able to differentiate *M. chitwoodi* and *M. fallax* from eight other *Meloidogyne* species in a dot-blot hybridization with soluble proteins extracted from a single female. Kapur-Ghai *et al.* (2014) developed an ELISA-based technique for the detection of *M. incognita* antigens. The use of antibodies as diagnostic tools for *Meloidogyne* spp. is limited to a few examples, having mainly been superseded by DNA-based diagnostics, which generally have greater sensitivity and specificity.

56.1.4 Molecular diagnosis

Molecular approaches based on polymerase chain reaction (PCR), PCR–RFLP of internal transcribed spacer (ITS) region of rDNA, ribosomal DNA of intergenic spacer (IGS) region, species-specific PCR-RAPD, satellite DNA probes and AFLP, sequencing of SSU rRNA, D2–D3 expansion segments of LSU rRNA, mitochondrial DNA (mtDNA), sequence characterized amplified regions (SCARs), etc. are successfully used to distinguish different species of *Meloidogyne*. Álvarez-Ortega *et al.* (2019) studied the phylogenetic analyses of sequences of the partial 18S rRNA, D2–D3 of 28S rRNA, ITS of rRNA, mitochondrial COI and COII genes of 16S rRNA intergeneric fragment from 56 valid *Meloidogyne* species using Bayesian inference and maximum likelihood methods (Fig. 9.1). From these results, they inferred 11 distinct clades among studied species of *Meloidogyne*. Seventy-five per cent of these species belonged to seven clades within the *Meloidogyne* superclade.

Details of DNA-based methods are described in section 110 of Chapter 15. Here we are only providing the species-specific primers used to differentiate the invasive *Meloidogyne* species (Table 49).

56.2 Biology, life cycle and ecology

The life cycle starts as the females lay eggs into gelatinous sac, which are produced by rectal glands in the female and composed of a glycoprotein matrix. This protects the eggs from external environmental extremes and predation. Usually,

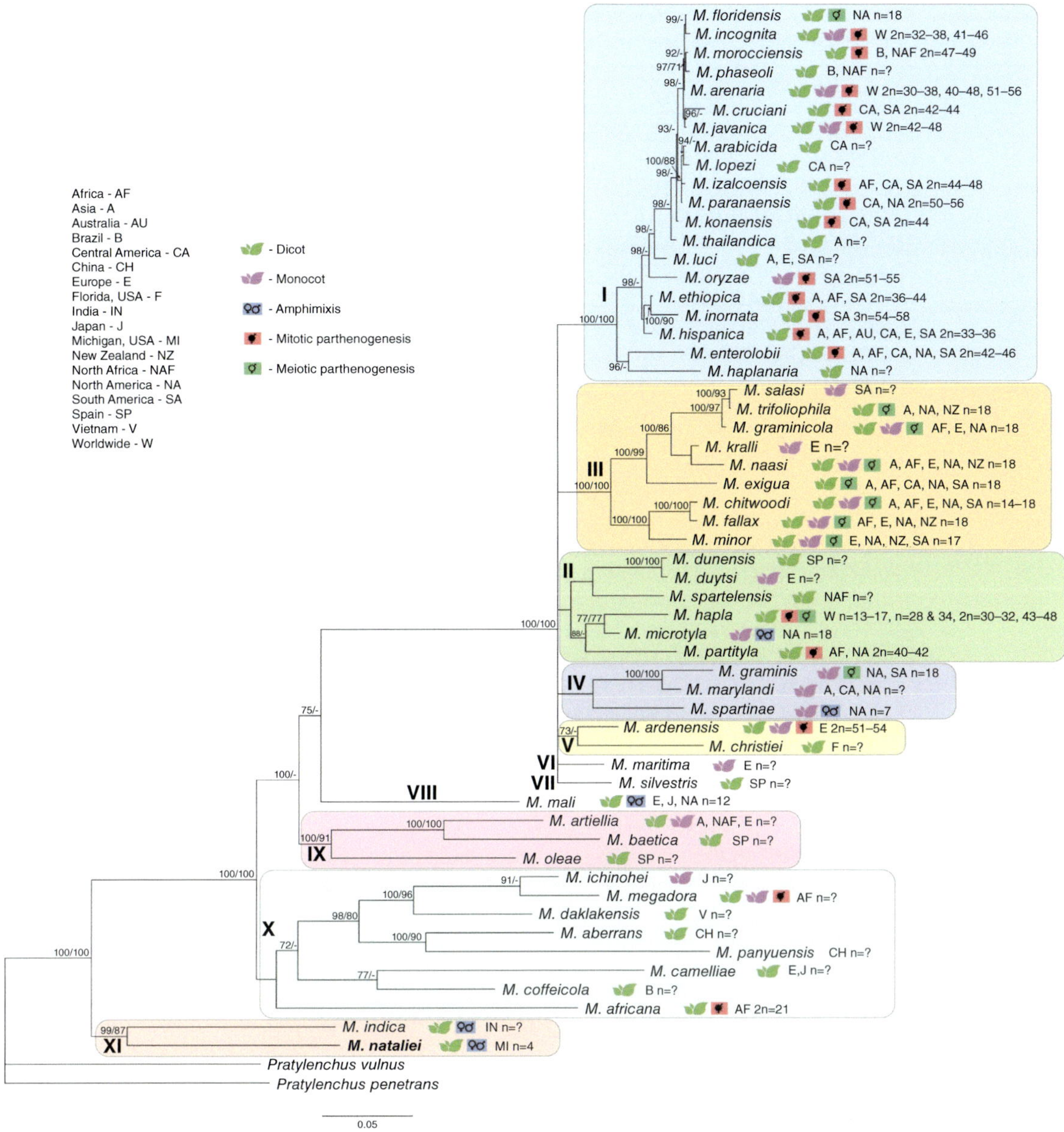

Fig. 9.1. Bayesian 50% majority rule consensus tree as inferred from 18S rRNA, ITS1 rRNA, D2–D3 expansion segments of 28S rRNA, COI gene and COII-16S rRNA sequence alignment under the GTR + I + G model. Branch support of over 70% is given for appropriate clades and it is indicated as: posterior probabilities value in Bayesian inference analysis/bootstrap value from maximum-likelihood analysis. (*n* = ?, chromosome number information unknown). [Courtesy Álvarez-Ortega *et al.*, 2019; Scientific Reports]

the egg masses are found on the surface of galled roots. However, in some species such as *M. graminicola*, egg masses may also be embedded within the gall. Inside the egg, embryogenesis proceeds to the first-stage juvenile, which moults to become an infective second-stage juvenile (J_2). The J_2 hatches from the egg; hatching is mainly dependent on temperature and moisture, though it is also influenced by other factors, like root diffusate (Curtis *et al.*, 2009). The hatched J_2 is vulnerable and needs to locate a host as rapidly as possible (Moens *et al.*, 2009). The J_2 is attracted to roots, and carbon dioxide is considered as a most important factor for the nematode attraction to roots (Karssen *et al.*, 2013). The J_2 penetrates the root, usually behind the root tip, and moves through the root to initiate and develop a permanent feeding site on protoxylem and protophloem cells. The feeding of the J_2 induces the formation of specialized multinucleate nurse cells, which are called giant cells. With the formation of giant cells, the nematode becomes sedentary and obese assuming a flask shape.

Under suitable conditions, the J_2 stage moults to the third-stage juvenile (J_3) after 14 days, then to the fourth-stage juvenile (J_4),

Table 49. Species-specific primers for *Meloidogyne* identification (adapted from Blok and Powers, 2009). See references for species and isolates used for validation of these primers.

Species	Primer set (5'-3')	Amplicon length	Reference
M. arabicida	GGTCAATGTTCAGAAATTTGTGG TACCTTTGACCAATCACGCT	1.5–1.6 kb COII/16S	Okimoto *et al.*, 1991
M. ardenensis	CTCAAAGATTAAGCCATGC CTGCGTGAGAGGTGAAAT	≈1.700 bp	Holterman *et al.*, 2009
	GCTTGTCTCAAAGATTAAGCC TGATCCWKCYGCAGGTTCAC	1819 bp	Janssen *et al.*, 2017b
M. arenaria	TCGGCGATAGAGGTAAATGAC TCGGCGATAGACACTACAACT	420 bp	Zijlstra *et al.*, 2000
	TCGAGGGCATCTAATAAAGG GGGCTGAATAATCAAAGGAA	950 bp	Dong *et al.*, 2001b
M. artiellia	TAGCCTTACCGGTGGATCAC CTGGTTCAGGGTCTTATC	531 bp	Damme *et al.*, 2013
M. chitwoodi	CCAATGATAGAGATAGGAAC CTGGCTTCCTCTTGTCCAAA	400 bp	Williamson *et al.*, 1997
	GATCTATGGCAGATGGTATGGA AGCCAAAACAGCGACCGTCT AC	900 bp	Petersen and Vrain, 1996
	TGGAGAGCAGCAGGAGAAAGA GGTCTGAGTGAGGACAAGAGTA	800 bp	Zijlstra, 2000
M. coffeicola	GCTTGTCTCAAAGATTAAGCC TGATCCWKCYGCAGGTTCAC	1819 bp	Janssen *et al.*, 2017b
M. decalineata (= *M. africana*)	CTTTGTACACACCGCCCGTCGCT TTTCACTCGCCGTTACTAAGG GAATC	18S, 28S & Cox1-2847 bp	Janssen *et al.*, 2017b
M enterolobii (= *M. mayaguensis*)	GAAATTGCTTTATTGTTACTAAG TAGCCACAGCAAAATAGTTTTC	322 bp	Blok *et al.*, 2002
	GATAACTTTTGTGAAAGTGCCGCTG ACATCAGTTCAGGCAGGATCAACC	240 bp	Subbotin, 2018
M. ethiopia	GCTTGTCTCAAAGATTAAGCC TGATCCWKCYGCAGGTTCAC	1819 bp	Janssen *et al.*, 2017b
	GGTCAATGTTCAGAAATTTGTGG TACCTTTGACCAATCACGCT	1.5–1.6 kb COII/16S	Okimoto *et al.*, 1991
M. exigua	CATCCGTGCTGTAGCTGCGAG CTCCGTGGGAAGAAAGACTG	562 bp	Randig *et al.*, 2002a
M. fallax	TGGGTAGTGGTCCCACTCTG AGCCAAAACAGCGACCGTC TAC	1100 bp	Petersen and Vrain, 1996
	CCAAACTATCGTAATGCATTATT GGACACAGTAATTCATGAGCTAG	515 bp	Zijlstra, 2000
M. graminicola	TTGATTACGTCCCTGCCCTTT ACGAGCCCGAGTGATCCACCG	450 bp	Pokharel *et al.*, 2010
	GCTTGTCTCAAAGATTAAGCC TGATCCWKCYGCAGGTTCAC	1819 bp	Janssen *et al.*, 2017b
	TTGATTACGTCCCTGCCCTTT TTTCACTCGCCGTTACTAAGG	450 bp	Khan and Ahamad, 2020
M. hapla	CAGGCCCTTCCAGCTAAAGA CTTCGTTGGGGAACTGAAGA	960 bp	Williamson *et al.*, 1997
	TGACGGCGGTGAGTGCGA TGACGGCGGTACCTCATAG	610 bp	Zijlstra, 2000
	GGCTGAGCATAGTAGATGATGTT ACCCATTAAAGAGGAGTTTTGC	1500 bp	Dong *et al.*, 2001b
	GGATGGCGTGCTTTCAAC AAAAATCCCCTCGAAAAATCCACC	440 bp	Wishart *et al.*, 2002

Continued

Table 49 Continued.

Species	Primer set (5'-3')	Amplicon length	Reference
M. incognita	CTCTGCCCAATGAGCTGTCC CTCTGCCCTCACATTAGG	1200 bp	Zijlstra *et al.*, 2000
	TAGGCAGTAGGTTGTCGGG CAGATATCTCTGCATTGGTGC	1350 bp	Dong *et al.*, 2001b
	GTGAGGATTCAGCTCCCCAG ACGAGGAACATACTTCTCCGTCC	955 bp	Meng *et al.*, 2004
M. indica	GCTTGTCTCAAAGATTAAGCC TGATCCWKCYGCAGGTTCAC	1819 bp	Janssen *et al.*, 2017b
M. javanica	CCTTAATGTCAACACTAGAGCC GGCCTTAACCGACAATTAGA	1650 bp	Dong *et al.*, 2001b
	GGTGCGCGATTGAACTGAGC CAGGCCCTTCAGTGGAACTATAC	670 bp	Zijlstra *et al.*, 2000
	ACGCTAGAATTCGACCCTGG GGTACCAGAAGCAGCCATGC	517 bp	Meng *et al.*, 2004
	ATCGGGGTTTAATAATGGG AAATTCAATTGAAATTAATAGC	740 bp	Brito *et al.*, 2018
M. mali	GGTCAATGTTCAGAAATTTGTGG TACCTTTGACCAATCACGCT	520–540 bp	Okimoto *et al.*, 1991
M. marylandi	GGCGATCAGATACCGCCCTAGTT TACAAAGGGCAGGGACGTAAT	721 bp	McClure *et al.*, 2012
M. naasi	CTCTTTATGGAGAATAATCGT CCTCCGCTTACTGATATG	433 bp	Zijlstra *et al.*, 2004
M. paranaensis	GCCCGACTCCATTTGACGGA CCGTCCAGATCCATCGAAGTC	208 bp	Randig *et al.*, 2002b

and finally to the adult stage. The J_3 and J_4 stages are much shorter (4–6 days); they lack a functional stylet and do not feed. The last moult is true metamorphosis for the male, which appears as long vermiform, and there is no evidence that males feed (Moens *et al.*, 2009). The life cycle is summarized in Fig. 9.2. The adult females enlarge as they mature and become pyriform. There are three types of reproduction in *Meloidogyne* species: amphimixis, facultative meiotic parthenogenesis (automixis) and obligatory mitotic parthenogenesis (apomixis). In amphimictic species (such as *M. carolinensis*. *M. microlyla*, *M. pim*), the males mate and females secrete a gelatinous matrix into which they lay large numbers of eggs, usually 30–80 eggs per day. In parthenogenetic species (*M. incognita*, *M. enlerolobii* and *M. oryzae*), the proportion of males in a population varies under unfavourable conditions, such as when population densities are very high and presumably there is a limitation of the food supply. Sometimes female juveniles become adult males under such conditions, with true intersexes (Karssen *et al.*, 2013). A normal egg mass contains 200–500 eggs, exceptionally as high as 2900 eggs. Duration of the life cycle varies with host and environmental conditions, especially temperature. At 25–30°C a J_2 may take 20 days to become a mature female and lay eggs, otherwise 3–4 weeks is a normal period for the life cycle (Khan, 2008). The number of generations varies with the species and availability of food. Generally there are many generations, but in some species, for example *M. naasi*, there is only one generation per year. The infective juveniles (J_2) can survive in the soil in a quiescence stage for a longer period. However, during that period, they consume food reserves present in the intestine (Karssen *et al.*, 2013).

Root-knot nematode species demonstrate a large diversity in various aspects of their life cycles with respect to the temperature requirements, *Meloidogyne* species can be divided into two distinct groups of species, thermophiles and cryophiles, which can be separated by their ability to survive lipid-phase transitions that occur at 10°C (Evans and Perry, 2009). *M. chitwoodi*, *M. hapla* and probably *M. naasi* are cryophilic and able to survive soil temperatures below 10°C; *M. arenaria*, *M. javanica* and *M. exigua* are thermophilic and do not have extended survival at temperatures below 10°C. Like survival, hatching is primarily controlled by temperature (Curtis *et al.*, 2009). Thermotypes exist within species (Moens *et al.*, 2009).

56.3 Host–parasite relationship

The nematode develops a very specialized and complex relationship with the host. As a perfect example of highly adapted root parasitism, these nematodes enjoy a guaranteed continuous supply of food and water from giant cells and protection within the gall for the females and their progeny. The giant cells are highly specialized cellular adaptations induced and maintained by the developing *Meloidogyne* female in susceptible hosts. Formation of giant cells is essential for successful parasitism. If a juvenile fails to induce the formation of giant

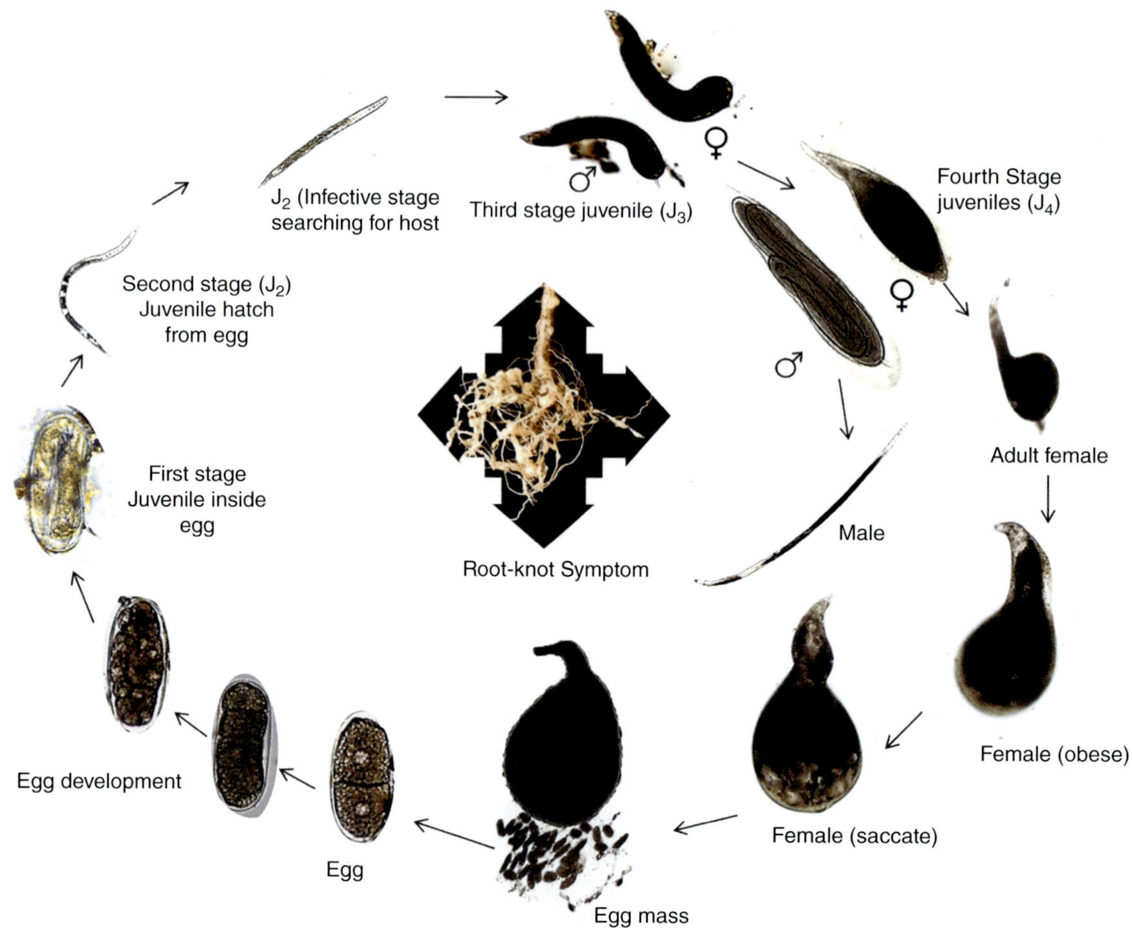

Fig. 9.2. Diagram of the life cycle of the root-knot nematode, *Meloidogyne* species. [Author's own photograph]

cells in roots, it will either starve or migrate out and locate another root, if it has sufficient energy reserves. The roots containing giant cells shift their metabolism in the direction of increased synthesis of proteins, their diversion towards roots and reduced transport of nutrients to shoots. The giant cells are formed through repeated endomitosis without cytokinesis (Fig. 9.3). The giant cells are formed around the head region of the female. They are 3–6 (2–12) in number and many times bigger than a normal cell (Fig. 9.3). They are multinucleate and size and shape of nuclei vary greatly. Giant cells are rich in the cytoplasm, and Golgi apparatus, mitochondria, ribosomes, polysomes and endoplasmic reticulum are abundant. The central vacuole disappears but small vacuoles remain in the cytoplasm. Wall ingrowths traverse the giant cell cytoplasm that greatly increase the surface area of the plasmalemma. These structures facilitate the transport of photosynthates, which are largely transported from the surrounding tissues to giant cells to meet the nutritional requirements of the female nematode (Khan, 2008).

Concurrent with the establishment of giant cells, root tissue around the nematode and its feeding site undergo hyperplasia (repeated mitotic divisions) and hypertrophy (enlargement), giving rise to swellings, and eventually galls are formed on the roots (Fig. 9.3). Stellar parenchymatous and pericycle cells undergo hyperplasia while the surrounding cortical cells become hypertrophied. Galls usually develop 1–2 days after juvenile penetration. The galling is induced by growth regulators released from the sub-ventral glands of the second-stage juveniles or when tryptophan is released upon hydrolysis of plant proteins by the juveniles. The tryptophan reacts with endogenous phenolic acids to yield auxins, which incite hyperplasia and hypertrophy in the root tissue (Huang, 1985).

56.4 Symptoms

Root-knot nematodes are among the few phytonematodes that cause specific symptoms on plants (Khan, 2008). The symptoms produced on crop plants due to infection by *Meloidogyne* appear on above-ground and below-ground plant parts.

ABOVE-GROUND SYMPTOMS Specific symptoms do not appear on above-ground parts. However, due to impairment of absorption and conduction of water and minerals through the infected root system, reduction in the top growth occurs and plants show stunted growth, yellowing and chlorosis of leaves,

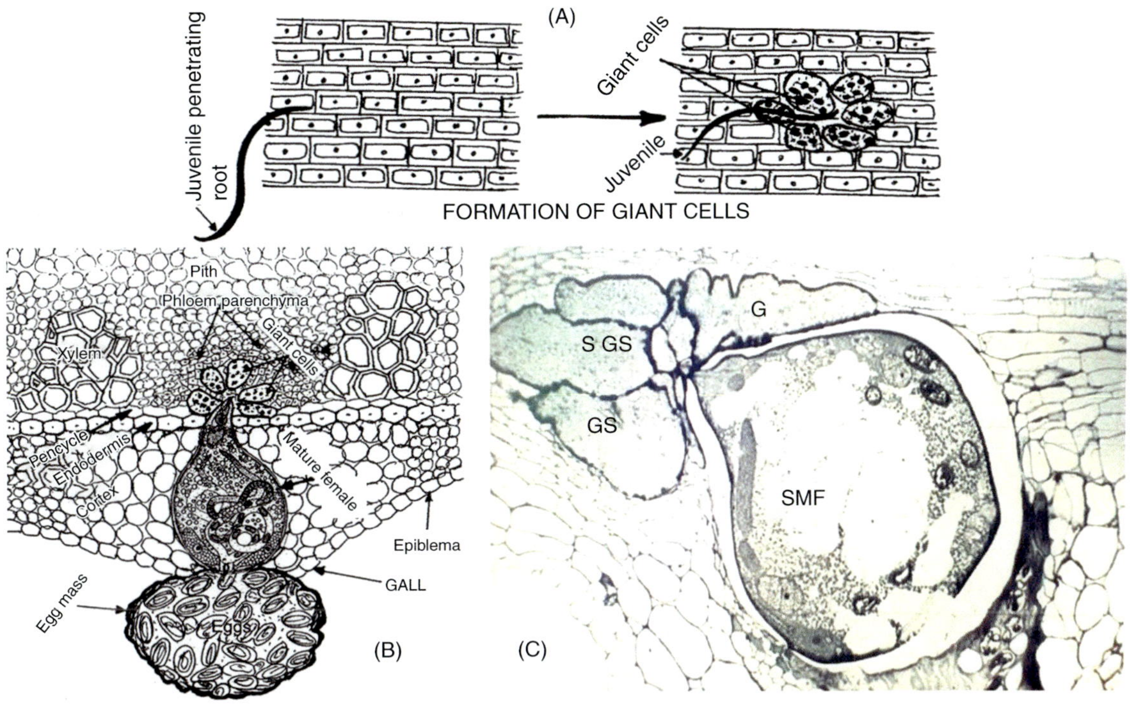

Fig. 9.3. Schematic presentation and microphotograph of host–parasite relationship of root-knot nematodes. Formation of giant cells (A); establishment of host–parasite relationship (B); transverse section of galled tissue showing saccate *Meloidogyne* female, SMF and giant cells, GS (C). [courtesy A,B: Khan (2008), reprinted with permission; C: IMP, USDA]

loss of vigour and premature leaf drop (Fig. 9.4). These symptoms usually appear in patches of plants in a field; plants in rest of the field may appear healthy. The patches are irregularly visible in the field, but in a case of heavy infection, yellowing and other deficiency symptoms and wilting are evenly visible throughout the entire field. Due to nematode infection, an imbalance between absorption of water and transpiration takes place; as a result, plants show a marked tendency to wilt during periods of rapid transpiration (hotter part of the day). Plants recover from the incipient wilting during the night and it recurs the next day. Overall, above-ground symptoms resemble those of mineral deficiencies and/or drought injury, even in the presence of adequate fertilizer and moisture. The rate of photosynthesis is also retarded due to nematode infection. These adverse effects cumulatively manifest as stunted plant growth, with small and chlorotic leaves, and an overall bushy appearance of the foliage. Fruits ripen prematurely and remain smaller or undersized with diminished natural attraction, colour and flavour. Hence, their market value is greatly reduced.

BELOW-GROUND SYMPTOMS The presence of galls or knots, which are cancerous outgrowths on the infected roots, are primary symptoms of the root-knot disease (Fig. 9.5). Size and shape of the galls vary with the plant and the nematode species. On cucurbitaceous crops, large and fleshy galls are extensively formed and their appearance somewhat resembles to beads in a string, especially on bottle gourd and pumpkin (Fig. 9.5). Stem galls can also be seen on some cucurbits, bottle-gourd, ridge gourd, etc. In most of the other vegetables and legumes, such as eggplant, tomato, chickpea and beans, galls are large and firm (Fig. 9.5). Occasionally very small galls develop and, in some cases, galls are not visible. Symptoms of *Meloidogyne* on monocots like onion and leek are very discrete and the gall appears because of protruding egg mass. Galls on sweet and chilli pepper are frequently small. Among the species, the symptoms caused by *M. hapla* differ from those of other species and only small and spherical galls are formed, with profuse root-branching originating from gall tissue causing a 'bearded' root system.

56.5 Host races concept of root-knot nematodes

Pathotype or physiological races are the populations of a number of species which give a differential response with regard to susceptibility or resistance of a plant cultivar (Khan, 1997). It is well known that races are present in certain species of *Meloidogyne*. The differential host test is based on the susceptibility/resistance response of certain host cultivars to species and races of *Meloidogyne* (Taylor and Sasser, 1978). Quite a few hosts are available that can be used to identify the species and races. The North Carolina Host Differential Test (NCHDT) is one of the most reliable and widely used host differential tests to identify the four major species of root-knot nematodes and differentiation of the races (Hartman and Sasser, 1985). This test can also separate particular species and races from an unknown or mixed population. The test uses six standard host

Fig. 9.4. Above-ground symptoms caused by *Meloidogyne* species on different crops. Rice nursery (A); chickpea (B); pigeon pea (C); cotton (D); wheat (E); Bermudagrass (F); cucumber (G); okra (H); tomato (I). [Courtesy D: J. Bridge and S.L.J. Page/CABI BioScience; E: S.S. Vaish; F: William T. Crow, University of Florida; rest author's own photographs]

plants: cotton cv. 'Deltapine-16', tobacco cv. 'NC-95', pepper cv. 'California Wonder', watermelon cv. 'Charleston Grey', groundnut cv. 'Florunner' and tomato cv. 'Rutgers' (Hartman and Sasser, 1985). One more cultivar of pepper ('Surya Mukhi Green') has been included in the NCHDT to recognize two races in *M. javanica* (Khan *et al.*, 2003).

56.6 Interaction with other organisms

In addition to causing direct damage to plants, root-knot nematodes possess great potential to synergize bacterial and fungal plant pathogens, leading to the development of disease complexes (Khan, 1993). The first report of synergistic interaction between nematodes and other plant pathogens was recorded with *Meloidogyne acrita* (= *Meloidogyne incognita*) and *Fusarium oxysporum* on cotton that resulted in breaking of resistance of the cotton cultivar against wilt (Atkinson, 1892). The presence of root-knot nematodes may also break the resistance of cultivars against wilts caused by species of *Fusarium*, *Xanthomonas*, *Pseudomonas*, etc. Root-rots are also aggravated by the presence of *Meloidogyne* spp. Root-knot disease complexes involving *Meloidogyne* species and *Fusarium*, *Rhizoctonia*, *Macrophomina*, *Verticillium*, *Pythium* spp., etc. are quite common in a large number of agricultural crops (Khan and Sharma, 2020).

56.7 Management measures

56.7.1 Chemical control

Fumigant nematicides (halogenated hydrocarbons) have been found highly effective for controlling soil populations of *Meloidogyne* spp., but due to their adverse effects on ecology, all these chemicals have been banned in crop protection. Granular nematicides such as phorate, carbamate, carbofuran and thionazin are relatively safer and can provide satisfactory decline in the root-knot nematode population at a dose of 4–5 kg a.i./ha. Granular nematicides can also be applied at the post-planting stage. In transplanting a crop, it is always advisable to disinfest the planting material by root-dip treatment with 50–100 ppm carbofuran or phorate (Khan *et al.*, 2014). When plants are small and disease has been recognized, a foliar spray with oxamyl or fenamiphos at 5 l/ha can effectively decrease the disease severity. There are several new nematicides such as fluensulfone and fluopyram which can offer effective control of root-knot nematodes. Application of nematicides, however, should be restricted to serious or epidemic situations.

56.7.2 Non-chemical control

Management of root-knot nematodes can also be achieved through non-chemical methods. Solarization and steaming can

Fig. 9.5. Root-knot symptoms caused by *Meloidogyne* species on different crops. Tomato (A); citrus (B); cotton (C); pigeon pea (D); rice (E); chickpea (F); peanut (G); lettuce (H); sponge gourd (I); eggplant (J); okra (K); broad bean (L); hollyhock (M); tea (N); potato (O); carrot (P); tobacco (Q); guava (R); grapevine (S); coffee (T); banana (U). [Photo courtesy C,T: J. Bridge and S.L.J. Page/CABI BioScience; G–H: H.J. Jensen and K.R. Barker/Nemapix; N: N.C. Gnanapragasam, source CABI-CPC, 2010; O: DAFF Archive, Bugwood.org; P: Walter Peraza Padilla, National University of Costa Rica, Bugwood.org; Q: J.D. Eisenback; R: Philippe Castagnone-Sereno; S: RMDG Carneiro, Embrapa, BR; U: NASI, India; rest author's own photographs]

reduce the infestation level in the soil. Solarization in tropical and subtropical countries may become very effective when fields are deep ploughed immediately after harvest to expose roots and deeper soil layers to the sun continuously for a week or two. This practice will result in the reduction of nematode populations to below the economic injury level. The efficiency of solarization can be further enhanced if fields are irrigated before the ploughing and later covered by polythene sheets. In temperate countries, steaming through permanently or temporarily buried pipes and covering the field with polythene sheets proves to be an efficient technique to prevent nematode build-up.

Removal of stubbles immediately after harvest (e.g. tomato, tobacco, eggplant, okra, cucurbits) deprives many females of the opportunity to lay eggs, leading to a check in the expected population increase. Crop rotation of cereals or oil-seed crucifers in place of vegetables, legumes and cucurbits or the use of resistant/tolerant cultivars is the most effective and easy method of the nematode management. With recent public awareness of the health risks and environmental contamination involved in the use of synthetic chemicals, there is an increasing demand for food materials produced through organic farming and biopesticide application (Khan, 2005). A few bionematicides based on *Pochonia chlamydosporia*, *Pupureocillium lilacinum*, *Trichoderma harzianum* and *Pseudomonas fluorescens* have been developed recently (Khan *et al.*, 2011; Khan and Haque, 2011; Haque *et al.*, 2018). Their application at 2g/kg of seed or 2 ml/l of microbial suspension can decrease the galling, resulting in a 20–50% increase in yield of legumes, vegetables and cereals. These bionematicides can also be applied to other crops. The bacterium *Pasteuria penetrans* has also proved to be an effective antagonist to root-knot nematodes.

56.7.3 Phytosanitary measures

See Chapter 16, section 115.

57 *Meloidogyne acronea*

57.1 Common name

African cotton root-knot nematode

57.2 Scientific name

Meloidogyne acronea Coetzee, 1956

57.3 Synonyms

Hypsoperine acronea Sledge and Golden 1964
Hypsoperine (Hypsoperine) acronea Siddiqi 1986

57.4 EPPO code

MELGAC

57.5 Diagnosis

57.5.1 Morphological diagnosis

Morphological and morphometric descriptions (after Coetzee, 1956; Page, 1985) are presented below. For measurements, see Table 50.

Female: Body spherical to oval with projecting neck, tapering towards head end (Fig. 9.6I); cuticle thick with protruding perineal region. Head truncate with one annule behind head cap. Cuticle thickening abruptly behind stylet (Fig. 9.6G); annulation diminishes with increased cuticle thickening, particularly in gravid females. Stylet knobs rounded. Excretory secretory pore posterior to stylet knobs. Perineal pattern circular with faint and smooth intermittent striae, arch low often broken on either side of vulval slit. Vulva and anus located on flattened protuberance. Eggs laid in gelatinous matrix. Moribund females retain embryonated eggs and hatched juveniles in the body (Fig. 9.6H).

Male: Body vermiform. Cephalic region not offset, flattened with one annule behind head cap (Fig. 9.6A). Cephalic framework sclerotized (Fig. 9.6B). Stylet short and strong with prominent pear-shaped knobs (Fig. 9.6B). Lateral field with four incisures but not areolated. Testis single, frequently reflexed. Spicules curved. Tail bluntly rounded (Fig. 9.6D).

Second-stage juvenile: Infective stage. Head continuous, not offset; cone shaped with one annule behind flattened head (Fig. 9.6E). Stylet slender with rounded knobs (Fig. 9.6E). Lateral field with four longitudinal lines. Tail tapering with short hyaline portion and rounded tip (Fig. 9.6F).

57.6 Geographical distribution

M. acronea is only recorded from South Africa (Coetzee, 1956) and Malawi (Bridge *et al.*, 1976).

57.7 Hosts

M. acronea parasitizes mainly *Gossypium* (cotton) and *Gossypium hirsutum* (Bourbon cotton). Other hosts includes *Arachis hypogaea* (groundnut), *Cajanus cajan* (pigeon pea), *Cassia* (sennas), *Chloris gayana* (Rhodes grass), *Croton tiglium* (purging croton), *Cyamopsis tetragonoloba* (guar), *Eleusine coracana* (finger millet), *Eragrostis* (lovegrasses), *Helianthus annuus* (sunflower), *Hibiscus sabdariffa* (Jamaica sorrel), *Lycopersicon esculentum* (tomato), *Pennisetum glaucum* (pearl millet), *Phaseolus* (beans), *Phaseolus vulgaris* (common bean), *Solanum tuberosum* (potato) and *Sorghum bicolor* (sorghum) (CABI-ISC, 2020).

57.8 Symptoms

Stunting, chlorosis and wilting are common above-ground symptoms of infection. On the roots of *Sorghum vulgare* var. Radar, *M. acronea* caused inconspicuous galls and growth of the plants was not visibly affected due to infestation. In Malawi, cotton grown on land after two seasons of planting with sorghum showed severe stunting and loss of vigour in patches due to heavy infestation with *M. acronea* (Bridge *et al.*, 1982); the nematodes attacked the fine, secondary roots and caused their distortion (Fig. 9.5-C) and stunting of plant growth (Fig. 9.4-D). The nematode generally causes slight swellings, which apparently occur due to an aggregation of large numbers of females, males, egg sacs and eggs on the roots, and the stimulation of cell division and development at feeding sites (Bridge *et al.*, 1982).

Table 50. Morphometric data of *Meloidogyne* species.

Characters		Range, numerous populations										
		M. acronea (after Page, 1985)	*M. ardenensis* (after Santos, 1967)	*M. arenaria* (Hunt and Handoo, 2009)	*M. artiellia* (after Franklin, 1961)	*M. brevicauda* (after Loos, 1953)	*M. chitwoodi* (Hunt and Handoo, 2009)	*M. coffeicola* (after Lordello and Zamith, 1960)	*M. decalineata* (after Whitehead, 1968)	*M. enterolobii* (Rashidifard, 2019)	*M. ethiopia* (Hunt and Handoo, 2009)	*M. exigua* (Flores and Lopez, 1989a, b)
Female	L (μm)	597–1243	472–795	500–1000	650–760	680–1860	430–740	992–1348	649–1041	–	–	–
	a	–	–	–	–	–	–	–	1.2–2.1	–	–	–
	Max. body diameter (μm)	–	309–520	400–600	340–460	310–1060	344–518	310–387	–	–	–	–
	Neck length (μm)	9.9–12.6	153–300	–	–	–	–	–	–	130–210	–	–
	Stylet length (μm)	–	–	13–17	12–16	22.1	11–12.5	15.3–19.0	12–17	16–24	12–15	12–14
	Stylet knob height (μm)	1.6–3.0	–	–	–	–	–	–	–	1–3	–	–
	Stylet knob width (μm)	2.2–4.4	–	–	–	4–5	–	–	3–4	3–4	–	–
	DGO (μm)	2.7–5.5	–	3–7	–	–	–	3.8–4.6	4–6	4–6	–	–
	Excretory pore to anterior end (μm)	14.9–41.4	–	–	–	–	–	–	–	–	–	14–33 annules
	Vulva length (μm)	–	–	–	15–22	–	–	–	–	26–33	–	–
Male	L (μm)	946–1419	–	900–2300	820–1370	970–1440	887–1268	1279–1596	1630–1700	1010–1837	–	830–1090
	a	–	–	–	31–40	26–44	–	32.2–40.1	29.6–42.5	41–60	–	24–32
	c	–	–	–	60–100	–	–	58.3–80.2	–	82–148	–	–
	Max. body diameter (μm)	–	–	27–48	23–36	28–52	22–37	38–44	–	23–34	–	–
	Stylet length (μm)	15.9–18.6	–	20–28	17–27	19–21	17–19	23–26	19–20	20–26	23–27	18–20
	Stylet knob height (μm)	2.7–4.3	–	–	–	–	–	–	–	2–3	–	–
	Stylet knob width (μm)	3.2–5.4	–	–	–	5.2	–	–	–	4–6	–	–
	DGO (μm)	2.1–6.3	–	4–8	–	–	2.2–3.4	–	4	4–6	3–5	–
	Excretory pore to anterior end (μm)	100.0–132.2	–	–	–	–	–	–	–	124–185	–	–
	Tail length (μm)	–	–	–	–	–	4.7–9.0	–	–	12–17	–	–
	Spicule length (μm)	32–34[a]	–	–	–	34–42.5	–	20–29	33–37	27–33	–	20–26
	Gubernaculum length (μm)	–	–	–	–	10–10.5	–	6–9.2	7	6–10	–	7–8
	Testis length (μm)	–	–	–	–	333–511	–	–	–	515–1072	–	–

Second stage Juvenile (J2)	L (µm)	341–370	392–605	300–370	460–590	336–417	336–424	471–573	363–460	326–510	290–370
	a		–	–	23–33	–	22–25	32.8–40.0	24–35	–	–
	c		–	10–16	21–29	–	9.5–13.9	10.3–12.2	7–10	–	–
	Max. body diameter (µm)		13–18	–	147–206	12.5–15.5	–	–	13–18	–	–
	Stylet length (µm)	10.0–14.6	10–12	14–16	14–15	–	9.2–10.7	10.7–13.7	11–15	–	8.6–11.4
	Stylet knob height (µm)	0.6–1.7	–	–	–	–	–	–	–	–	–
	Stylet knob width (µm)	1.7–3.1	–	–	–	–	–	–	–	–	–
	DGO (µm)	2.2–4.5	3–5	–	–	–	2–4	–	2–4	–	2.5–4.1
	Excretory pore to anterior end (µm)	61.8–85.5	–	–	–	–	–	–	62–95	–	–
	Tail length (µm)	33.2–48.6	44–69	18–26	17.5–28	39–47	–	44–52	42–60	52–72	39–50
	Hyaline tail terminus	4.4–6.7	6–13	4–6	–	9–14	–	15.5±3.1[b]	5–15	12–15	–

From [a]Coetzee (1956), [b]Jepson (1987).

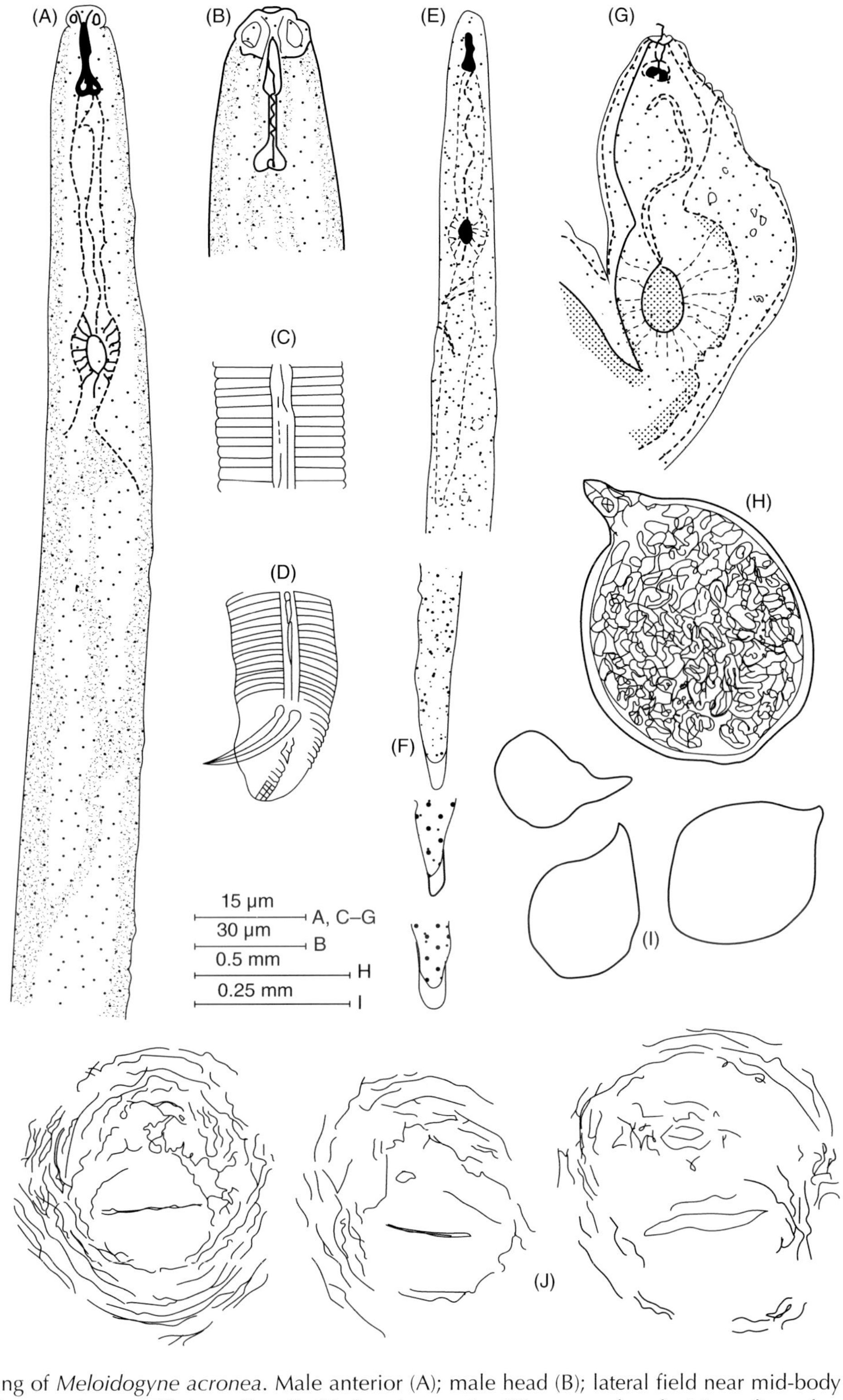

Fig. 9.6. Line drawing of *Meloidogyne acronea*. Male anterior (A); male head (B); lateral field near mid-body (C); male tail (D); second-stage juvenile, anterior end (E); juvenile tails (F); female anterior (G); mature female (H); outline of mature females (I); perineal patterns (J). [Reproduced from Hunt and Handoo, 2009; after Page, 1985; courtesy of CAB International]

57.9 Biology and life cycle

Mature females are semi-endoparasitic and secrete gelatinous matrices directly into the soil. Males are abundant, up to nine being found coiled within each egg sac. Reproduction is predominantly by amphimixis (Page, 1983). Eggs are laid about 2 days after copulation and up to 1023 eggs may be produced. A proportion of eggs are usually retained within the body of the moribund female, where they hatch and await the rotting of the protective maternal cuticle before being liberated into the soil. The rest of the life cycle is typical of the genus *Meloidogyne* and is completed in 19 days at 33°C. The juveniles rapidly die at temperatures below 10°C or at relative humidity (RH) less than 100% (Page, 1983). In southern Malawi, *M. acronea* was found to be restricted by soil type, preferring alluvial soils, and does not survive in vertisols. It survives 6–7 months throughout the dry season as eggs within moribund females and/or gelatinous egg sacs in soils with the wilting point above 97.7% RH (Page, 1983).

57.10 Economic importance

Pathogenicity studies showed that *M. acronea* infection may lead to a delay in flowering by up to 2 weeks and may reduce the yield of *Sorghum bicolor* by up to 56%. The nematode may also severely retard growth and delay flowering in cotton, *Gossypium hirsutum*, and reduce the yield by up to 90% (Page, 1983).

57.11 Movement and means of dispersal

57.11.1 Plant parts known to carry the pest in trade/transport

Roots, bulbs, tubers, corms, rhizomes (sucker), growing medium accompanying plants, seedlings, transplanting/nursery stock and micro-propagated plants.

57.11.2 Plant parts not known to carry the pest in trade/transport

Bark, flowers, inflorescences, cones, calyx, fruits (including pods), leaves, stems (above ground), shoots, trunks, branches, wood and true seeds (including grain).

57.12 Pest risk analysis and looming threat from introduction

The distribution of *M. acronea* is limited, but the isolated discoveries of this nematode in South Africa and Malawi indicate that other climatically similar areas of southern Africa may also harbour this nematode. The crops, sorghum and cotton, which are mainly affected by this nematode, are sown directly so there would be less chance of dispersal of this nematode from infested soil and roots. If a tentative infection with the nematode on coffee in Tanzania and tobacco in Kenya is confirmed, there is the possibility of the spread of infection on transplanted material from nurseries.

57.13 Invasiveness rating

Considering the consequences of probable introduction and pest risk analysis of *M. acronea*, the following invasiveness rating has been assigned.

Climate/host interaction	–	Low (1)
Hosts range	–	Medium (2)
Dispersal potential	–	Medium (2)
Economic impact	–	High (3)
Environmental impact	–	Low (1)
Post-entry distribution	–	Not known (0)
Final rating (score)	–	Medium (9)

Based on the above invasiveness rating, *M. acronea* receives a score of 9 out of 15 and is rated as MEDIUM INVASIVE NEMATODE.

57.14 Region-wise status of invasiveness

M. acronea is on the quarantine pest list in India (PQ Order-2003, 2015). Due to the wide host range, the nematode is considered as invasive; however, due to limited distribution and lack of information, the nematode is given a medium invasiveness rating for the countries where it still does not occur.

57.15 Management measures

See section 56.7, above.

58 *Meloidogyne arenaria*

58.1 Common name

Peanut root-knot nematode

58.2 Scientific name

Meloidogyne arenaria (Neal, 1889) Chitwood, 1949

58.3 Synonyms

Anguillula arenaria Neal, 1889
Heterodera arenaria (Neal, 1889) Marcinowski, 1909
Meloidogyne arenaria (Neal, 1889) Chitwood, 1949
Meloidogyne arenaria thamesi Chitwood in Chitwood *et al.*, 1952
Meloidogyne thamesi (Chitwood *et al.*, 1952) Goodey, 1963
Tylenchus arenarius (Neal, 1889) Cobb, 1890

58.4 EPPO code

MELGAR

58.5 Diagnosis

58.5.1 Morphological diagnosis

The morphological description and morphometric data based on Hunt and Handoo (2009) are presented below. For measurements, see Table 50.

FEMALE: Pear-shaped, without posterior terminal protuberance. Stylet strong, cone dorsally curved, gradually tapering anteriorly to blunt tip; shaft broad, cylindrical, posteriorly widening gradually; basal knobs teardrop-shaped to rounded, offset (Fig. 9.7A–D). Perineal pattern variable, rounded to ovoid with fine to coarse striae. Dorsal arch low, flattened with smooth striae or slightly wavy, continuous or broken, somewhat bent towards tail terminus at lateral line; usually creating shoulders on lateral portion of arch. Dorsal and ventral striae often jointing at an angle at lateral lines; lateral field distinct, slightly irregular (Fig. 9.7I).

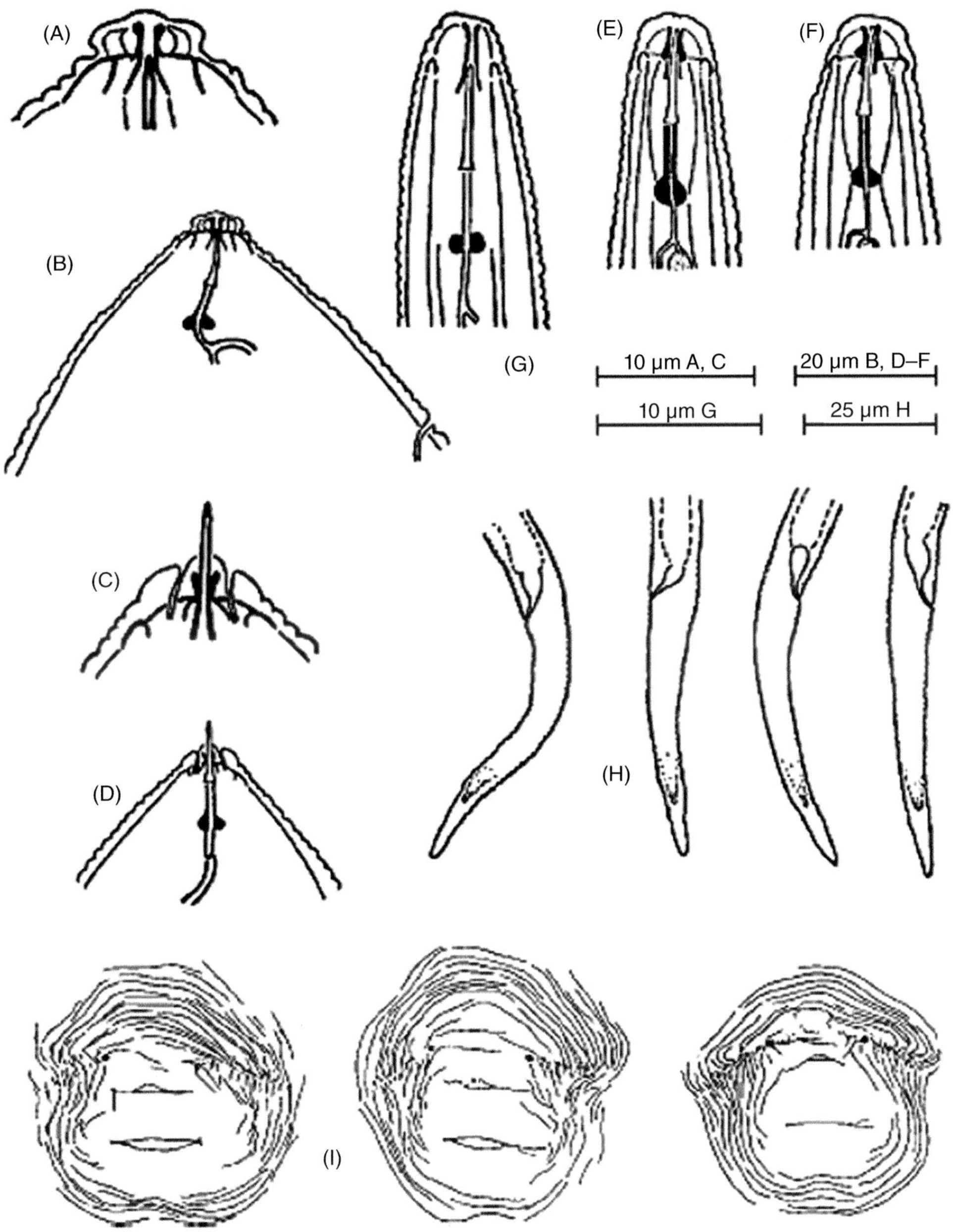

Fig. 9.7. Line drawing of *Meloidogyne arenaria*. Female anterior region (A–D); male anterior region (E, F); J_2 anterior region (G); J_2 tail regions (H); perineal patterns (I). [A–H, after Whitehead (1968), courtesy Journal of Zoology, John Wiley and Sons; I, after Orton Williams (1975), courtesy of CAB International]

Male: Cephalic region continuous, smooth, not offset, rarely with one or two incomplete annulations (Fig. 9.7E–F). Lip disc more or less oval, slightly raised above level of medial lips, lateral lips usually absent (remnants occasionally present). Stylet long, basal knobs offset, angular or more amalgamated (Fig. 9.7E).

Second-stage juvenile: Hemizonid 1–3 annules anterior to excretory secretory pore. Tail with rounded to pointed tail terminus and indistinct hyaline region (Fig. 9.7H).

58.5.2 Cytogenetic diagnosis

Two cytological races of *M. arenaria* have been designated (Triantaphyllou, 1979). Race A is triploid ($3n$ = 51–56) and is the most commonly occurring population. Less common populations comprise Race B, which is diploid ($2n$ = 34–37). See also Table 48.

58.5.3 Biochemical diagnosis

Esterase phenotype A1, A2 and A3, a typical esterase phenotype S1-M, S2-M1, M3-F1 and malate dehydrogenase isozyme phenotypes N1 and N3 are associated with *M. arenaria* (Esbenshade and Triantaphyllou, 1985a).

58.5.4 Molecular diagnosis

See the primers in Table 49.

58.6 Geographical distribution

M. arenaria is cosmopolitan and distributed worldwide (Fig. 9.8). It occurs in tropical, subtropical and temperate climates with an average temperature of 36°C in the hottest month or lower and at 3°C in the coldest month. In temperate regions, it is mainly a problem in greenhouses (Karssen *et al.*, 2013).

58.7 Hosts

M. arenaria has a very wide host range. This root-knot species parasitizes several food crops from dicotyledons, monocotyledons, herbaceous and woody plants grown in tropical, subtropical and temperate climates (Taylor *et al.*, 1982). An expanded list of hosts is presented in Table 51.

58.8 Symptoms

Above-ground symptoms are non-specific and typical of the genus. On below-ground parts, *M. arenaria* populations frequently produce several small bead-like galls that do not form short lateral roots (Eisenback *et al.*, 1981) (Fig. 9.5-Q). Small wart-like protuberances may occur on infected corms, tubers and peanut pods. These galls result in abnormal formation and function of the root system and blockage of the vascular cylinder.

58.9 Biology and life cycle

The biology and life cycle are typical of the genus, with obligatory mitotic parthenogenesis (apomixis). The length of one generation of the nematode is greatly affected by temperature. The life cycle takes approximately 3 weeks at very high temperatures (> 29°C), but at very cool temperatures it can be

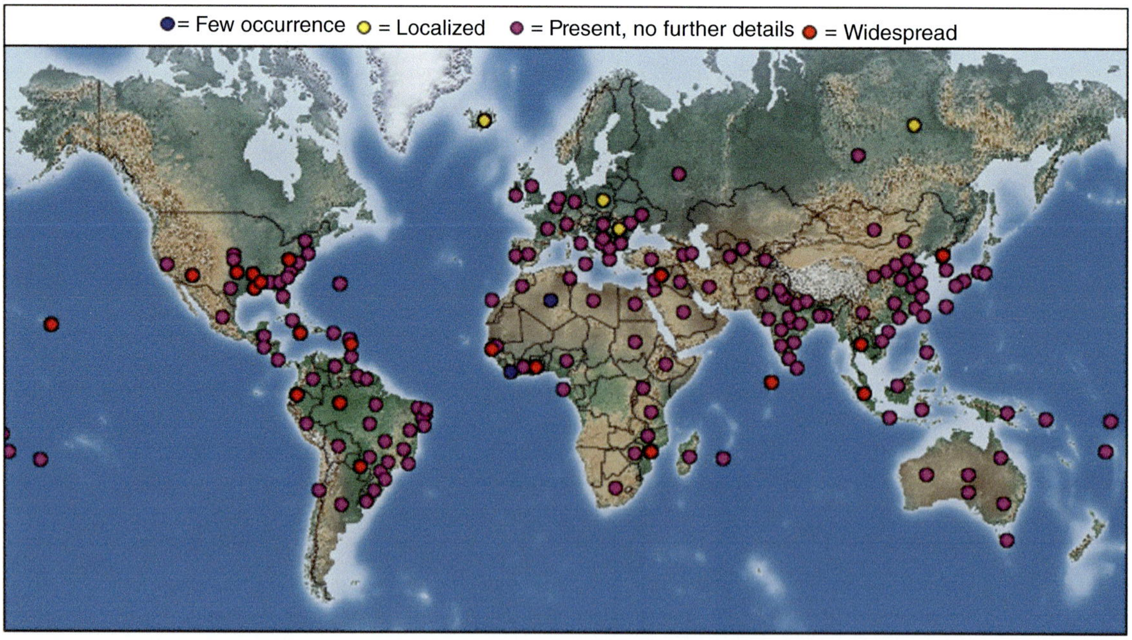

Fig. 9.8. Distribution map of *Meloidogyne arenaria*. [Adapted from CABI-ISC, 2020]

Table 51. List of hosts of *Meloidogyne arenaria* (adapted from CABI-ISC, 2020).

Main hosts
Abelmoschus esculentus (okra)
Arachis hypogaea (groundnut)
Avena sativa (oats)
Capsicum frutescens (chilli)
Citrullus lanatus (watermelon)
Coffea arabica (arabica coffee)
Cucumis sativus (cucumber)
Cucurbita pepo (ornamental gourd)
Daucus carota (carrot)
Glycine max (soybean)
Gossypium hirsutum (Bourbon cotton)
Hordeum vulgare (barley)
Ipomoea batatas (sweet potato)
Medicago sativa (lucerne)
Musa (banana)
Nicotiana tabacum (tobacco)
Oryza sativa (rice)
Passiflora edulis (passionfruit)
Phaseolus vulgaris (common bean)
Pisum sativum (pea)
Prunus persica (peach)
Psidium guajava (guava)
Saccharum officinarum (sugarcane)
Solanum lycopersicum (tomato)
Solanum tuberosum (potato)
Trifolium repens (white clover)
Vigna unguiculata (cowpea)
Vitis vinifera (grapevine)
Zea mays (maize)
Wild hosts
Chenopodium album (fat hen)
Cyperus esculentus (yellow nutsedge)
Datura stramonium (jimsonweed)
Desmodium tortuosum (Florida beggarweed)
Deutzia gracilis (slender deutzia)
Digitalis purpurea (foxglove)
Eleusine indica (goosegrass)
Festuca arundinacea (reed fescue)
Lespedeza stipulacea (Korean lespedeza)
Lolium multiflorum (Italian ryegrass)
Raphanus sativus (radish)
Ricinus communis (castor bean)
Rosa multiflora (multiflora rose)
Solanum nigrum (black nightshade)
Trifolium pratense (purple clover)

extended to 2–3 months. *M. arenaria*, like *M. javanica* and *M. exigua*, is thermophilic and does not have prolonged survival at temperatures below 10°C.

58.10 Economic importance

The peanut root-knot nematode is one of the most economically important nematodes that attacks thousands of plant species worldwide. *M. arenaria* significantly reduces the quantity and quality of food and fibre production, and on average 5% loss is associated with this root-knot nematode. In high infestations, some fields exhibit complete crop failure (Sasser and Carter, 1984; Khan, 1997).

58.11 Movement and means of dispersal

See section 57.11, above.

58.12 Pest risk analysis and looming threat from introduction

M. arenaria is a phytosanitary risk, but because it is so widespread it is not specifically quarantined.

58.13 Invasiveness rating

Considering the consequences of economic impact and pest risk analysis of *M. arenaria*, the following invasiveness rating has been assigned.

Climate/host interaction	–	High (3)
Hosts range	–	High (3)
Dispersal potential	–	High (3)
Economic impact	–	High (3)
Environmental impact	–	High (3)
Post-entry distribution	–	High (–3)
Final rating (score)	–	High (12)

Based on the above invasiveness rating score, *M. acronea* receives a score of 12 out of 15 and is rated as HIGHLY INVASIVE NEMATODE.

58.14 Region-wise status of invasiveness

M. arenaria is a highly invasive nematode, but the phytosanitary risk is low because it is widely distributed and detected in almost every country. There are no quarantine measures for this nematode.

58.15 Management measures

See section 56.7, above.

59 *Meloidogyne artiellia*

59.1 Common name

British root-knot nematode

59.2 Scientific name

Meloidogyne artiellia Franklin, 1961

59.3 EPPO code

MELGAT

59.4 Diagnosis

59.4.1 Morphological diagnosis

Juveniles relatively small with short tail. Cuticular perineal pattern of female very distinct inner area that contains the vulva and anus. Pattern marked by a few coarse figure-of-eight-shaped striae with small top and large base. Top arch usually angular (Greco *et al.*, 1992c). For measurements, see Table 50.

59.4.2 Biochemical diagnosis

Esterase isozymes phenotype M2-VF12 and malate dehydrogenase (mdh) phenotype N1b can be used to identify *M. artiellia* on a biochemical basis (Karssen and van Hoenselaar, 1998).

59.4.3 Molecular diagnosis

De Giorgi *et al.* (2002) and Davis and Venette (2004b) used restriction fragment length polymorphism (PCR–RFLP) of ribosomal DNA, and De Luca *et al.* (2009) used heat shock protein (hsp) 90 gene to distinguish *M. artiellia* from other root-knot nematodes. For primers, see Table 49.

59.5 Geographical distribution

M. artiellia is a Mediterranean species. It has been known to occur in Algeria, Belgium, China, France, Greece, Italy, Israel, Morocco, Russia, Siberia, Spain, Syria, Tunisia, Turkey and United Kingdom (Shiabova, 1977; Greco *et al.*, 1992c; Damme *et al.*, 2013).

59.6 Hosts

M. artiellia infects and damages cereal, crucifer and legume crops (Greco *et al.*, 1992c; Davis and Venette, 2004b) such as barley (*Hordeum vulgare*), sorghum (*Sorghum vulgare*), wheat (*Triticum durum* and *T. vulgare*), cabbage (*Brassica oleravea*), cauliflower (*B. oleracea* var. *Botrytis*), radish (*Raphanus sativus*), rashad (*Nasturtium fontanum*), turnip (*Brassica rapa*), alfalfa (*Medicago sativa*), broad bean (*Vicia faba*), chickpea (*Cicer arietinum*), clovers (*Trifolium incarnatum* and *T. repens*), dogtooth pea (*Lathyrus sativus*) and vetch (*Vicia sativa*). Experimental studies show that nearly all members of Brassicaceae, Fabaceae and Poaceae are hosts of *M. artiellia* (Karssen *et al.*, 2013).

59.7 Symptoms

Symptoms are similar to those for other root-knot nematodes. The infested chickpea plant produces very few pods and seeds are small. On wheat, spikes are sparse and reduced in size. The root galls are very small (Greco *et al.*, 1992c).

59.8 Biology and life cycle

The biology and life cycle are typical of the genus; it is a meiotic parthenogenetic species (Janssen *et al.*, 2017b).

59.9 Economic importance

In Mediterranean countries where wheat is continuously grown, regular damage to wheat has been reported (Di Vito and Greco, 1988a). When wheat is rotated with chickpea, damage on the latter is also observed in fields (Sikora and Greco, 1990). Microplot studies revealed up to 80% yields loss in chickpea and wheat (DiVito and Greco, 1988b). Complete crop failure has also been observed with levels of 8 eggs/cc of soil (Greco *et al.*, 1992c).

59.10 Movement and means of dispersal

See section 57.11, above.

59.11 Pest risk analysis and looming threat from introduction

M. artiellia is an important nematode pest of chickpea and wheat in the Mediterranean region. Since both are one of the major crops, pest risk analysis is recommended for assessment of the risk posed by this nematode to the cereal- and pulse-growing areas. Appropriate actions should be evaluated by quarantine departments in order to minimize the risks.

59.12 Invasiveness rating

Considering the economic impact and pest risk analysis of *M. artiellia*, the following invasiveness rating has been assigned.

Climate/host interaction	–	Medium (2)
Hosts range	–	Medium (2)
Dispersal potential	–	Low (1)
Economic impact	–	High (3)
Environmental impact	–	Low (1)
Post-entry distribution	–	Not known (0)
Final rating (score)	–	Medium (9)

Based on the above invasiveness rating, *M. artiellia* receives a score of 9 out of 15 and is rated as MEDIUM INVASIVE NEMATODE.

59.13 Region-wise status of invasiveness

Due to the economic importance of the hosts, wheat and chickpea, the nematode is considered as invasive. However, due to limited distribution and less chance of dispersal, the nematode is given a low invasiveness rating for the countries where it is still not reported.

59.14 Management measures

See section 56.7, above.

60 *Meloidogyne brevicauda*

60.1 Common name

Indian root-knot nematode

60.2 Scientific name

Meloidogyne brevicauda Loos, 1953

60.3 EPPO code

MELGBR

60.4 Diagnosis

60.4.1 Morphological diagnosis

The females of *Meloidogyne brevicauda* are very large and can be easily differentiated from the four most common species of root-knot nematodes: *M. arenaria*, *M. hapla*, *M. incognita* and *M. javanica*. Although the internal morphology is similar to that of most *Meloidogyne* species, the morphology of the anterior end is different (Eisenback, 1985). The perineal pattern is oblong, made up of transverse striae, which are widely spaced and interrupted close to the perineal region by the lateral field. The dorsal arch is high and flat; the ventral arch is also flat with some striae, which are continuous between the anus and vulva. Phasmids are fairly widely spaced, almost in line with the tail end; the lateral field is indistinct, indicated by breaks in some striae. For measurements, see Table 50.

60.5 Geographical distribution

M. brevicauda is confined to Sri Lanka (Loos, 1953) and India (Mukherjee and Dasgupta, 1982). Outside Sri Lanka and India, the nematode has only been reported in Azerbaijan (Kasimova and Atakishieva, 1980). In India, it is reported from localized pockets in the Nilgiris (Muraleedharan, 1992) and Darjeeling area (Mukherjee and Dasgupta, 1982).

60.6 Hosts

M. brevicauda primarily parasitizes tea plants (*Camellia sinensis*) in India and Sri Lanka. The only other host reported is saffron (*Crocus sativus*) in Azerbaijan (Kasimova and Atakishieva, 1980).

60.7 Symptoms

Above-ground symptoms on tea can be easily confused with those caused by other tea nematodes such as *Pratylenchus loosi*. However, the characteristic swelling and pitting of the roots caused by *M. brevicauda* are easily identifiable. Tea bushes attacked by this root-knot nematode are stunted with pale, dull leaves. Severe infestation may result in defoliation. Damage is more apparent in bushes after pruning and often severely infected bushes may fail to recover (Sivapalan, 1972). Galls are large on the roots, with characteristic pinhole pits (Fig. 9.5N).

60.8 Biology and life cycle

Males are extremely rare. The life history of *M. brevicauda* is otherwise very similar to that of other *Meloidogyne* species.

60.9 Economic importance

M. brevicauda causes severe damage to tea, but no data are available on the quantified damage. It is a localized nematode pest of two major tea-producing countries, Sri Lanka and India.

60.10 Movement and means of dispersal

See section 57.11, above.

60.11 Pest risk analysis and looming threat from introduction

Although *M. brevicauda* poses a significant threat to tea production, the looming threat from its introduction is limited due to its restricted distribution and host range (tea and saffron).

60.12 Invasiveness rating

Considering the probable consequences of introduction and pest risk analysis of *M. brevicauda*, the following invasiveness rating has been assigned.

Climate/host interaction	–	Medium (2)
Hosts range	–	Low (1)
Dispersal potential	–	Low (1)
Economic impact	–	Medium (3)
Environmental impact	–	Medium (2)
Post-entry distribution	–	Not known (0)
Final rating (score)	–	Medium (9)

Based on the above invasiveness rating, *M. brevicauda* receives a score of 9 out of 15 and is rated as MEDIUM INVASIVE NEMATODE.

60.13 Region-wise status of invasiveness

M. brevicauda has a limited host range and distribution; however, because of the economic damage potential and lack of information, the nematode is considered as medium invasive and should be prevented in the following tea- and saffron-producing countries where it is absent.

APPPC: Bangladesh, China, Indonesia, Iran, Japan, Malaysia, Myanmar, Nepal, South Korea, Thailand, Turkey, Vietnam
CAN: Bolivia, Peru
COSAVE: Argentina, Brazil, Colombia, Ecuador
EPPO: Portugal, Russian Federation
IAPSC: Burundi, Cameroon, Ethiopia, Kenya, Madagascar, Malawi, Mali, Mauritius, Mozambique, Réunion, Rwanda, Seychelles, South Africa, Uganda, Zambia, Zimbabwe
OIRSA: Guatemala, Panama
PPPO: Papua New Guinea

60.14 Management measures

See section 56.7, above.

61 *Meloidogyne chitwoodi*

61.1 Common name

Columbia root-knot nematode

61.2 Scientific name

Meloidogyne chitwoodi Golden, O'Bannon, Santo and Finley, 1980

61.3 EPPO code

MELGCH

61.4 Diagnosis

61.4.1 Morphological diagnosis

Male vermiform, elongated with slight taper at both ends (Fig. 9.9A). Tail short with rounded terminus. Body annules distinct and more prominent near anterior and posterior end (Fig. 9.9A). Second-stage juveniles with short tail, scarcely tapered and hyaline (Fig. 9.9G) (Hunt and Handoo, 2009). Female pearly-white and pear-shaped (Fig. 9.9J). Perineal pattern oval to round, dorsal arch low and round to high and angular. Striae near perineal area broken, twisted, curved with weakly visible lateral lines (Fig. 9.9K) (Karssen *et al.*, 2013). For measurements, see Table 50.

61.4.2 Biochemical diagnosis

Esterase isozymes phenotype S1 and malate dehydrogenase phenotype N1a are used to identify *M. chitwoodi* on a biochemical basis (Esbenshade and Triantaphyllou, 1985b).

61.4.3 Molecular diagnosis

For primers, see Table 49.

61.5 Geographical distribution

M. chitwoodi is localized in different parts of the world and is recorded from Argentina, Belgium, France, Germany, Italy, Mexico, Mozambique, Netherlands, South Africa, Sweden and the USA (Karssen *et al.*, 2013; CABI-ISC, 2020). The nematode is widespread in the USA. Earlier reports of its occurrence in Canada, Bulgaria, Denmark, Portugal, Tunisia and United Kingdom have been rejected after survey (CABI-ISC, 2020).

61.6 Hosts

The nematode parasitizes economically important crops such as carrot (*Daucus carota*), lucerne (*Medicago sativa*), potato (*Solanum tuberosum*) and tomato (*Lycopersicon esculentum*). Other hosts include common bean (*Phaseolus vulgaris*), pea (*Pisum sativum*), maize (*Zea mays*), oyster plant (*Scorzonera hispanica*) and quinoa (*Chenopodium quinoa*) (CABI-ISC, 2020).

61.7 Symptoms

Symptoms are similar to those of several other root-knot species, including inciting galls, but the galls are usually relatively smaller and appear as numerous small, pimple-like, raised areas on the surface of infected potato tubers (Fig. 9.5O).

61.8 Biology and life cycle

The biology and life cycle are typical of the genus. Reproduction is automictic, with facultative meiotic parthenogenesis (n = 14–18).

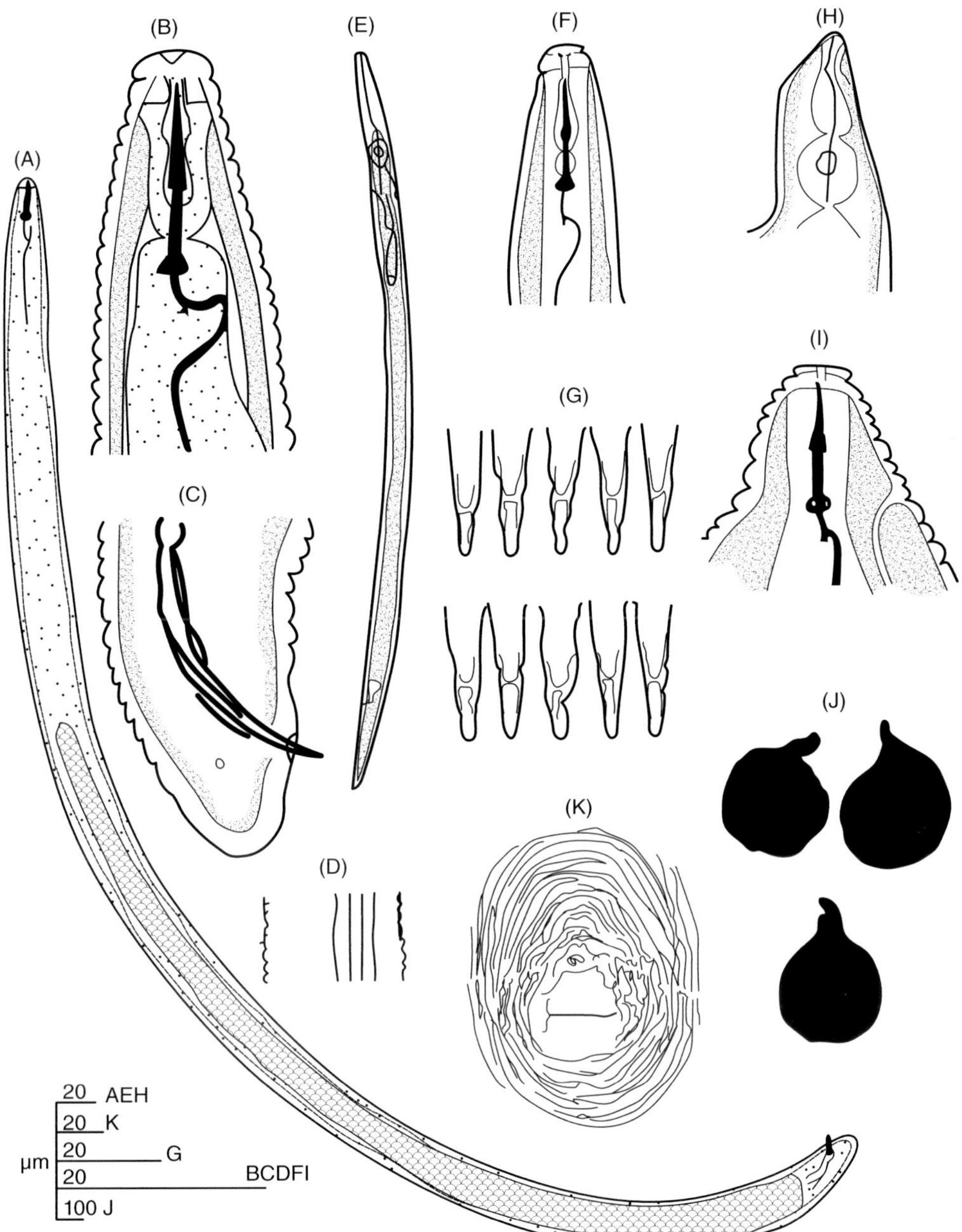

Fig. 9.9. Line drawing of *Meloidogyne chitwoodi*. Entire male (A); male anterior region (B); male tail region (C); male longitudinal lines (D); entire J_2 (E); J_2 anterior region (F); J_2 tail regions (G); female anterior region, lateral view (H,I); entire females (J); perineal pattern (K). [Reproduced from Eisenback and Hunt, 2009; after Jepson (1985), courtesy of CAB International]

61.9 Economic importance

M. chitwoodi causes significant damage to potato and reduces tuber yields as well as market value, due to internal necrosis and external galling. Necrotic spots in the flesh of the tuber of as little as 5% of a crop make it commercially undesirable. The nematode is considered as one of the major pests of potatoes in the Pacific Northwest states of the USA. There would be an annual predicted loss of US$40 million if management measures were not practised (Santo, 1994). The damaging impact of the nematode in other countries on potato (Muller *et al*., 1996) and on other host crops (Santo and O'Bannon, 1981; Chaves and Torres, 2000) has also been documented.

61.10 Movement and means of dispersal

See section 57.11, above.

61.11 Pest risk analysis and looming threat from introduction

M. chitwoodi represents a greater threat to the EPPO region than any other *Meloidogyne* species, except *M. hapla*, which is already widespread in the region. The main pest risk of *M. chitwoodi* is for potato crops, particularly in Europe, which includes main potato-growing countries. Braasch *et al.* (1996) carried out a detailed pest risk analysis in Germany and concluded that the nematode was most likely to be damaging in potato-growing regions. The nematode also has very high fecundity and under suitable climatic conditions as many as four generations per year can be completed. It has a wide host range. The soil temperature requirements of *M. chitwoodi* for population development are similar to those of *Globodera rostochiensis* (Tiilikkala *et al.*, 1995), indicating that the former species may occupy the same geographical distribution as the latter.

61.12 Invasiveness rating

Considering the consequences of introduction and probable economic impact of *M. chitwoodi*, the following invasiveness rating has been assigned.

Climate/host interaction	–	High (3)
Hosts range	–	High (3)
Dispersal potential	–	High (3)
Economic impact	–	High (3)
Environmental impact	–	Medium (2)
Post-entry distribution	–	Medium (–2)
Final rating (score)	–	High (12)

Based on the above invasiveness rating, *M. chitwoodi* receives a score of 12 out of 15 and is rated as HIGHLY INVASIVE NEMATODE.

61.13 Region-wise status of invasiveness

M. chitwoodi was included in the European list of quarantine organisms in 1998 to stop further distribution within the EPPO region. It is also listed as a quarantine organism in many countries (Karssen *et al.*, 2013) where it is absent. Considering its potential as a pest of potato, *M. chitwoodi* has a high invasiveness rating in the following potato-producing countries.

APPPC: Afghanistan, Armenia, Azerbaijan, Bahrain, Bangladesh, Bhutan, China, India, Indonesia, Iran, Iraq, Israel, Japan, Jordan, Kuwait, Lebanon, Malaysia, Myanmar, Nepal, Mongolia, Oman, Pakistan, Philippines, Qatar, South Korea, Sri Lanka, Syria, Saudi Arabia, Tajikistan, Thailand, Turkmenistan, United Arab Emirates, Vietnam, Yemen

CAHFSA: Bahamas, Bermuda, Cuba, Dominica, Dominican Republic, Guadeloupe, Haiti, Honduras, Jamaica, Martinique, St Kitts & Nevis

CAN: Bolivia, Chile, Colombia, Ecuador, Peru, Venezuela

COSAVE: Paraguay, Uruguay

EPPO: Albania, Austria, Belarus, Bosnia and Herzegovina, Bulgaria, Croatia, Cyprus, Czech Republic, Denmark, Estonia, Faroe Islands, Finland, Greece, Hungary, Iceland, Ireland, Kazakhstan, Kyrgyzstan, Latvia, Liechtenstein, Lithuania, Luxembourg, Malta, Moldova, Morocco, Norway, Poland, Romania, Russia, Serbia, Slovakia, Slovenia, Spain, Switzerland, Ukraine, United Kingdom, Uzbekistan

IAPC: Algeria, Benin, Burkina Faso, Burundi, Cameroon, Cape Verde, Central African Republic, Chad, Comoros, Congo, Côte d'Ivoire, Eritrea, Ethiopia, Guinea, Kenya, Lesotho, Libya, Madagascar, Malawi, Mali, Mauritania, Mauritius, Niger, Nigeria, Réunion, Rwanda, Senegal, Sierra Leone, Sudan, Swaziland, Uganda, Zambia

NAPPO: Canada

OIRSA: Guatemala, Nicaragua

PPPO: Belize, Costa Rica, El Salvador, New Caledonia, Papua New Guinea

PPPO: Australia, New Zealand.

61.14 Management measures

For details of management, including phytosanitary measures, see Chapter 16, section 115.

62 *Meloidogyne citri*, *M. donghaiensis*, *M. fujianensis*, *M. indica*, *M. jianyangensis*, *M. kongi* and *M. mingnanica*

62.1 Common name

Asian citrus root-knot nematodes (ACRKN)

62.2 Scientific names

Meloidogyne citri Zhao, Gao and Weng, 1990
Meloidogyne donghaiensis Zheng, Lin and Zheng, 1990
Meloidogyne fujianensis Pan, 1985
Meloidogyne indica Whitehead, 1968
Meloidogyne kongi Yang, Wang and Feng, 1988
Meloidogyne jianyangensis Yang Wang and Feng, 1988
Meloidogyne mingnanica Zhang, 1993

62.3 EPPO codes:

MELGFU (*M. fujianensis*), MELGJI (*M. jianyangensis*), MELGKO (*M. kongi*) and MELGMG (*M. mingnanica*).

62.4 Diagnosis

62.4.1 Morphological diagnosis

The morphological diagnoses of seven species comprising the complex of Asian citrus root-knot nematodes (ACRKN) are given below.

M. CITRI: Female with oval cuticular perineal pattern, marked by fine and coarse striae. Lateral lines indistinct and dorsal arch moderately high or low (Zhang *et al.*, 1990).
M. DONGHAIENSIS: Female with rounded cuticular perineal pattern, marked by fine and coarse striae. Lateral lines indistinct and dorsal arch moderately high or low (Zheng *et al.*, 1990)
M. FUJIANENSIS: Female with rounded cuticular perineal pattern, marked by fine and coarse striae. Lateral lines indistinct and dorsal arch moderately high. A small swelling with a central pit may be found outside vulva edge (Pan, 1985).
M. INDICA: Female with rounded cuticular perineal pattern, dorsal arch low, indistinct lateral lines or any forking. Striae indistinct, creating whorl in tail region spreading between anus and vulva (Whitehead, 1968).
M. JIANYANGENSIS: Female with rounded cuticular perineal pattern, marked by fine and coarse striae. Lateral lines indistinct and dorsal arch moderately high or low (Yang *et al.*, 1990).
M. KONGI: Female with ovoidal cuticular perineal pattern, marked by fine striae. Lateral lines indistinct and dorsal arch moderately high or low (Yang *et al.*, 1988).
M. MINGNANICA: Female with ovoidal cuticular perineal pattern, marked by a few coarse striae in figure-of-eight shape, large base and small top marked by fine striae. Lateral lines indistinct and dorsal arch moderately high or low (Zhang, 1993).

62.4.2 Molecular diagnosis

Information is lacking, except for *M. indica*. See Table 49 for primers for *M. indica*.

62.5 Geographical distribution

All ACRKN species are recorded from China (*M. citri*, *M. donghaiensis*, *M. fujianensis*, *M. indica*, *M. jianyangensis*, *M. kongi* and *M. mingnanica*) except *M. indica*, which is reported from India.

62.6 Hosts

Citrus is the main host for all ACRKN species. Other experimental hosts include pepper and tomato. Some species prefer certain *Citrus* species over others (Davis and Venette, 2004a).

M. CITRI: Citrus, mandarin/tangerine orange, satsuma orange, sour orange and tomato.
M. DONGHAIENSIS: Mandarin/tangerine orange.
M. FUJIANENSIS: Cogon grass (*Imperata cylindrical*) and mandarin/tangerine orange.
M. INDICA: Citrus, lime, orange and morinda (*Morinda officianalis*).
M. JIANYANGENSIS: Citrus, mandarin/tangerine orange.
M. KONGI: Citrus and pepper.
M. MINGNANICA: Citrus, satsuma citrus, trifoliate or hardy orange (*Poncirus trifoliate*), sour orange (experimental host).

62.7 Symptoms

The ACRKN produces symptoms similar to those of other *Meloidogyne* species. The galls are produced on the infected roots. Galls may occur singly or coalesce to form compounded root swellings. Above-ground symptoms are general and typical of an impaired root system caused by biotic or abiotic factors. Visible symptoms of infection usually include unthriftiness, yellowing of leaves, wilting, defoliation, reduced growth and even death of the host (Chitambar, 2016).

62.8 Biology and life cycle

The ACRKN life cycle is typical of other root-knot nematode species. The eggs are produced within roots or in the gelatinous matrix on the root surface. Little information is available on the biology of ACRKN species. *M. fujianensis* takes 33–35 days from root penetration to egg production on citrus and the life cycle is completed in 55–60 days at 25°C (Pan *et al.*, 1988). *M. indica* requires 28–35 days from root penetration to egg production on sour orange, satsuma citrus and tomato (Vovlas and Inserra, 2000).

62.9 Economic importance

Knowledge of the exact economic damage caused by individual ACRKN species is lacking. However, the impact of mixed populations in China and India was estimated to be 20–50% yield losses in citrus production (Pan, 1985; Vovlas and Inserra, 2000).

62.10 Movement and means of dispersal

See section 57.11, above.

62.11 Pest risk analysis and looming threat from introduction

Considering citrus as host, the introduction of ACRKN in citrus-growing countries may cause devastation in citrus orchards. Most citrus is cultivated in warm, tropical or subtropical climates. Based on the current geographical distribution, the ACRKN species are likely to establish and become widespread in areas with subtropical and tropical climates (Davis and Venette, 2004a). *M. jianyangensis* and *M. kongi* prefer temperate climates and can establish in temperate areas. Davis and Venette (2004a) suggested that *M. jianyangensis* may find a suitable climate in the northern cooler region of the USA. The nematode can be introduced through nursery stock and there is also possibility that introduction into new areas would support populations of the nematode, especially areas under citrus, tomato and pepper production.

62.12 Invasiveness rating

Considering the probable consequences of introduction and pest risk analysis of the ACRKN complex, the following invasiveness rating has been assigned.

Climate/host interaction	–	Medium (2)
Hosts range	–	Medium (2)
Dispersal potential	–	Medium (2)
Economic impact	–	High (3)
Environmental impact	–	Medium (2)
Post-entry distribution	–	Not known (0)
Final rating (score)	–	Medium (11)

Based on the above invasiveness rating, the ACRKN complex receives a score of 11 out of 15 and is rated as MEDIUM INVASIVE NEMATODE.

62.13 Region-wise status of invasiveness

Among the seven species of ACRKN, *M. fujianensis* and *M. kongi* are on Taiwan's Harmful Organisms list, and *M. indica* is listed as a quarantine nematode in Indonesia and Timor-Leste. Introductions of *M. citri* and *M. fujianensis* are also restricted into Florida, USA (USDA-PCIT, 2014). However, due to its limited and localized distribution in India and China, the nematode is given a medium invasiveness rating for the following countries where these nematode species do not occur.

APPPC: Bangladesh, Cambodia, Indonesia, Iran, Iraq, Israel, Japan, Kazakhstan, Kyrgyzstan, Lebanon, Malaysia, Myanmar, Nepal, Pakistan, Philippines, Saudi Arabia, Singapore, South Korea, Sri Lanka, Taiwan, Tajikistan, Thailand, Vietnam, Uzbekistan
CAHFSA: Barbados, Belize, Cuba, Dominican Republic, Haiti, Jamaica, Puerto Rico, Saint Kitts & Nevis, Trinidad and Tobago
CAN: Bolivia, Colombia, Ecuador, Peru, Venezuela
COSAVE: Argentina, Brazil, Chile, Guyana, Paraguay, Suriname, Uruguay
EPPO: Austria, Belarus, Belgium, Croatia, Cyprus, Czech Republic, Denmark, Estonia, Finland, Georgia, Greece, Guernsey, Hungary, Ireland, Jersey, Kazakhstan, Kyrgyzstan, Latvia, Lithuania, Luxembourg, Macedonia, Malta, Moldova, Morocco, Netherlands, Norway, Portugal, Romania, Serbia, Slovakia, Slovenia, Spain, Sweden, Switzerland, Turkey.
IAPSC: Algeria, Angola, Benin, Botswana, Burkina Faso, Burundi, Cameroon, Cape Verde, Central African Republic, Chad, Comoros, Congo, Côte d'Ivoire, Eritrea, Ethiopia, Gabon, Gambia, Ghana, Guinea, Guinea-Bissau, Kenya, Liberia, Madagascar, Malawi, Mali, Mauritania, Mauritius, Mozambique, Namibia, Niger, Réunion, Rwanda, Senegal, Sierra Leone, Somalia, South Africa, Swaziland, Tanzania, Togo, Uganda, Zambia, Zimbabwe
NAPPO: Canada, Mexico, USA
NEPPO: Egypt, Jordan, Libya, Malta, Sudan, Syria, Tunisia
OIRSA: Costa Rica, El Salvador, Guatemala, Honduras, Nicaragua, Panama
PPPO: Australia, Fiji, New Zealand, Papua New Guinea, Tonga, Vanuatu.

62.14 Management measures

See section 56.7, above.

63 *Meloidogyne coffeicola*

63.1 Common name

Coffee root-knot nematode

63.2 Scientific name

Meloidogyne coffeicola Lordello and Zamith, 1960

63.3 EPPO code

MELGCO

63.4 Diagnosis

63.4.1 Morphological diagnosis

The morphological description and morphometric data below are after Lordello and Zamith, 1960. For measurements, see Table 50.

FEMALE: White or brownish, obese, oval, with long neck (in contrast to other species) and slight posterior protuberance of body. Perineal pattern roughly rounded, rectangular or oval with flattened dorsal arch, striae in dorsal sector smooth to slightly wavy. Tail tip region surrounded by concentric striae which form distinct tail tip pattern. Striae smooth, closely spaced and usually faint; striae present between vulva and anus. Lateral field area not marked, no breaks or space in lateral area. Distance between phasmids one vulva length or shorter.
MALE: Body vermiform, slightly ventrally arcuate when heat relaxed; posterior part twisted by 180°. Cuticle strongly annulated. Lateral fields with four incisures, outer bands areolated. Stylet strongly developed, round or longitudinally oval, posteriorly sloping knob. Oesophageal glands elongate, mostly ventral to

intestine. Testis single, usually reflexed. Spicules slightly arcuate. Gubernaculum fixed. Phasmids pore-like, located in front of anal level.

SECOND-STAGE JUVENILE: Head slightly offset, with one large annule and flat labial disc. Cuticle finely striated. Lateral field 4.6 µm wide, with four incisures, outer bands areolated. Stylet small, with small rounded knobs. Median oesophageal bulb oval, muscular, with refractive inner thickenings. Oesophageal glands long, mostly ventral to intestine. Phasmids pore-like, at one-third to middle of tail. Tail conoid, with rounded terminus; terminal hyaline region distinct.

63.4.2 Biochemical diagnosis

The esterase isozyme phenotype C2 and malate dehydrogenase phenotype C1 are associated with *M. coffeicola* (Carneiro *et al.*, 2000).

63.4.3 Molecular diagnosis

For primers of *M. coffeicola*, see Table 49.

63.5 Geographical distribution

M. coffeicola is only recorded from Brazil (Campos and Melles, 1987).

63.6 Hosts

Coffee is the primary crop parasitized by this root-knot nematode. Major hosts are from the genus *Coffea* in the family Rubiaceae (Lordello and Zamith, 1960), including arabica coffee (*Coffea arabica*), robusta coffee (*Coffea canephora*) and Liberian coffee (*Coffea liberica*). The nematode has also been detected on native plants such as *Eupatorium pauciflorum* and *Psychotria nitidula* (Jaehn *et al.*, 1980).

63.7 Symptoms

M. coffeicola produces slight swellings on the lateral roots, similar to *M. exigua* (Fig. 9.5-T); distinct galls are absent. Roots are not typically galled but thickened and cracked. Cortical tissues become detached, resulting in 'rough root' symptoms. The nematode inhabits the root cortex damaging it badly, and, in advanced stages, the infected cortex, sloughs off as a powdery mass of cells. Severely infested coffee plants showed dieback symptoms, killing both the main and feeder roots, and eventually results in plant death (Curi, 1968).

63.8 Biology and life cycle

The biology and life cycle are typical of the genus. Eggs and juveniles are found in egg masses in root cavities, on the outside of roots and in the surrounding soil (Lordello and Zamith, 1960).

63.9 Economic importance

M. coffeicola has a damaging impact on coffee plants and is a serious problem in coffee plantations in Brazil (Schmidt, 1969). The nematode is directly associated with severe decline and dieback of coffee trees; its infestations resulted in around 5% yield loss and the death of 15,000 coffee trees in Parana State, Brazil (Campos *et al.*, 1990).

63.10 Movement and means of dispersal

See section 57.11, above.

63.11 Pest risk analysis and looming threat from introduction

M. coffeicola is an economic pest of coffee plants in Brazil and would pose a threat to other coffee-producing countries like India if introduced. Pest risk analysis is recommended to assess the risk posed by this nematode to new areas, especially coffee-growing regions, and appropriate actions should be evaluated by quarantine departments in order to minimize the risks.

63.12 Invasiveness rating

Based on the consequences of introduction and probable pest risk analysis of *M. coffeicola*, the following invasiveness rating has been assigned.

Climate/host interaction	–	Medium (2)
Hosts range	–	Low (1)
Dispersal potential	–	Medium (2)
Economic impact	–	High (3)
Environmental impact	–	Low (1)
Post-entry distribution	–	Not known (0)
Final rating (score)	–	Medium (9)

Based on the above invasiveness rating, *M. acronea* receives a score of 9 out of 15 and is rated as MEDIUM INVASIVE NEMATODE.

63.13 Region-wise status of invasiveness

Considering coffee as a major host of this nematode and its potential to cause economic damage, *M. coffeicola* is considered as an invasive nematode. Due to limited distribution and lack of information, the nematode is given a medium invasiveness rating for the following coffee-producing countries where it is absent.

APPPC: Cambodia, China, India, Indonesia, Malaysia, Myanmar, Nepal, Philippines, Saudi Arabia, Sri Lanka, Thailand, Vietnam, Yemen

CAHFSA: Belize, Cuba, Dominican Republic, Haiti, Jamaica, Puerto Rico, Saint Lucia, Trinidad and Tobago
CAN: Bolivia, Colombia, Ecuador, Peru
COSAVE: Guyana, Paraguay, Suriname
EPPO: Spain
IAPSC: Angola, Benin, Burundi, Cameroon, Cape Verde, Central African Republic, Comoros, Congo, Côte d'Ivoire, Equatorial Guinea, Ethiopia, Gabon, Ghana, Guinea, Kenya, Liberia, Madagascar, Malawi, Mauritius, Mozambique, Nigeria, Rwanda, São Tomé and Principe, Sierra Leone, Togo, Uganda, Zambia, Zimbabwe
NAPPO: Mexico, USA
OIRSA: Costa Rica, El Salvador, Guadeloupe, Guatemala, Haiti, Honduras, Martinique, Nicaragua, Panama
PPPO: Cook Islands, Fiji, French Polynesia, New Caledonia, Papua New Guinea, Samoa, Tonga, Vanuatu.

63.14 Management measures

See section 56.7, above.

64 *Meloidogyne decalineata*

64.1 Common name

African coffee root-knot nematode

64.2 Scientific name

Meloidogyne decalineata Whitehead, 1968

64.3 Synonyms

Meloidogyne africana (Whitehead, 1959) Janssen *et al.*, 2017
Meloidogyne oteifae Elmiligy 1968

64.4 EPPO code

MELGDE

64.5 Diagnosis

64.5.1 Morphological diagnosis

Janssen *et al.* (2017b) considered *M. decalineata* as a junior synonym of *M. africana* along with *M. oteifae*. However, in this section, the morphological description and morphometric data of *M. decalineata* (after Whitehead, 1968) are presented, until further validation of the species. For measurements, see Table 50.

Female: Body pyroid, posterior end without or with very slight protuberance (Fig. 9.10F). Spear knobs drawn out laterally with rounded or backwardly sloping anterior margins. Posterior cuticular pattern with fairly close striae and evenly spaced, often broken especially at sides of pattern (Fig. 9.10G). Distinct tail whorl and vulva, rudimentary lateral field in some patterns. Phasmids close to tail terminus, cuticle often folded in ventral region of pattern (Fig. 9.10G).
Male: Head not offset, truncated cone shape with small head-cap (Fig. 9.10A). Head framework thin. Stylet knobs laterally rounded with back-sloped anterior margins. Oesophago-intestinal junction about half length of median bulb behind median bulb. Posterior oesophageal region a long, narrow glandular lobe running mostly ventral to intestine. Lateral field arising about level of base of relaxed stylet. Dierids and phasmids not observed. Testis single. Spicules with large offset heads, blades thick walled, tapering to subacute termini (Fig. 9.10C). Gubernaculum crescentic. Tail somewhat rostrate.
Second-stage juvenile: Head high with slightly inflated, truncated cone shape. Stylet with rounded knobs occasionally with flattened anterior margins. Hemizonid 2–4 annules long opposite posterior end of isthmus. Posterior oesophageal region ventrally overlapping intestine. Lateral field with four incisures, often three secondary incisures between them. Rectum not inflated. Tail tapering smoothly or irregularly to subacute tip with fine annulation to just before last section of tail (Fig. 9.10D).

64.5.2 Biochemical diagnosis

M. decalineata, a junior synonym of *M. africana*, is characterized by a unique esterase isozyme phenotype AF2 and malate dehydrogenase phenotype H1 (Janssen *et al.*, 2017b)

64.5.3 Molecular diagnosis

For primers of *M. decalineata* (= *M. africana*), see Table 49.

64.6 Geographical distribution

M. decalineata is a localized nematode parasite in Tanzania (Whitehead, 1969) and São Tomé Island, Africa (Lordello and Fazuoli, 1980). However, considering *M. decalineata* as a junior synonym of *M. africana*, the geographical distribution includes Kenya, Democratic Republic of Congo, Sudan and India (Janssen *et al.*, 2017b).

64.7 Hosts

M. decalineata primarily parasitizes arabica coffee (*Coffea arabica*) and robusta coffee (*Coffea canephora*) (Whitehead, 1969). If *M. decalineata* is considered as a junior synonym of *M. africana*, the new host range includes pepper (*Capsicum annuum*) and maize (*Zea mays*) (Janssen *et al.*, 2017b).

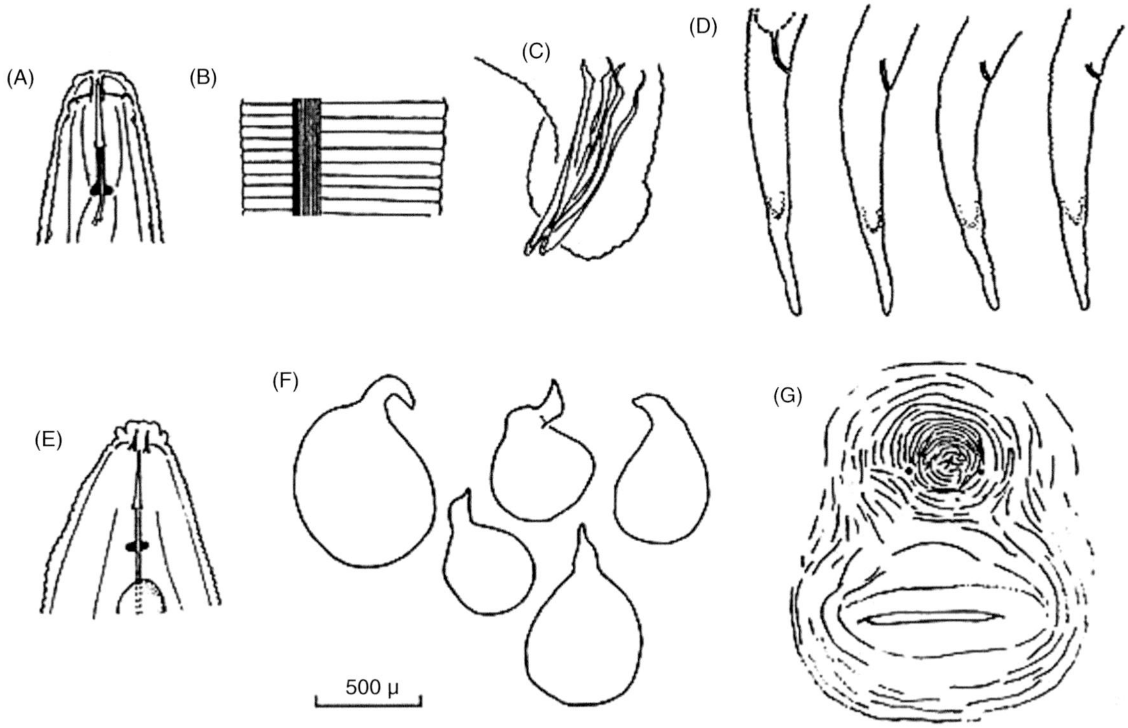

Fig. 9.10. Line drawing of *Meloidogyne decalineata*. Anterior end of male (A); lateral mid-body cuticle markings of male (B); posterior end of male (C); juvenile tails (D); anterior end of ovigerous female (E); shapes of ovigerous females (F); typical posterior cuticular pattern of female (G). [Courtesy: J.E. Machon/CABI BioScience]

64.8 Symptoms

M. decalineata occurs sympatrically with other *Meloidogyne* spp. on coffee in Africa, usually known as African coffee root-knot nematodes. The nematode causes galling on the fine tertiary lateral or feeder roots. The galls are characteristically rounded, occurring on the root terminals, but sometimes they are elongate behind the tip along the roots. On severely infested roots, the galls appear as a mass of fine, bead-like structures 1–3 mm in size, along the main lateral root (Campos *et al.*, 1990). The nematodes may hamper plant growth and cause malformation and blinding of the roots. Infected seedlings are stunted and chlorotic and are often killed as a direct result of nematode infestation (Lordello and Fazuoli, 1980).

64.9 Biology and life cycle

The biology and life cycle are typical of the genus, with obligatory mitotic parthenogenesis (Janssen *et al.*, 2017b).

64.10 Economic importance

In Tanzania, dieback and yield decline of coffee trees are associated with *M. decalineata* infection, but damage caused by this nematode on coffee has not been quantified. In Tanzania, *M. decalineata* is found along the equatorial coffee belt, and around 25% of plants sampled were found to be infested with the nematode. In other parts of Tanzania, 12% of plants sampled were infested (Whitehead, 1969).

64.11 Movement and means of dispersal

See section 57.11, below.

64.12 Pest risk analysis and looming threat from introduction

Taking into consideration *M. decalineata* as a junior synonym of *M. africana*, there is a high chance of its introduction into new areas and it may be considered as a medium invasive nematode. The risk posed by this root-knot nematode to the coffee industry is very high, as coffee is a transplanted crop there is always the possibility of spreading the nematode through infected planting material.

64.13 Invasiveness rating

Considering the consequences of probable introduction and pest risk analysis of *M. decalineata*, the following invasiveness rating has been assigned.

Climate/host interaction	–	Medium (2)
Hosts range	–	Medium (2)
Dispersal potential	–	Medium (2)
Economic impact	–	High (3)
Environmental impact	–	Low (1)
Post-entry distribution	–	Not known (0)
Final rating (score)	–	Medium (10)

Based on the above invasiveness rating, *M. decalineata* receives a score of 10 out of 15 and is rated as MEDIUM INVASIVE NEMATODE.

64.14 Region-wise status of invasiveness

M. decalineata is on the quarantine pest list in India under schedule VI (PQ Order-2003, 2015) and movement of nursery stock requires phytosanitary certification from the exporting country. However, considering *M. decalineata* as junior synonym of *M. africana*, it is already present in India and infesting maize (Chitwood, 1960), which requires an urgent revision of pest risk analysis of *M. decalineata* in India and other countries. Due to the localized distribution of this nematode in different countries, medium invasiveness rating has been given for the coffee-producing countries (see section 63.13, above) except where this nematode already occurs.

64.15 Management measures

See section 56.7, above.

65 *Meloidogyne enterolobii*

65.1 Common name

Guava root-knot nematode / Pacara earpod tree root-knot nematode

65.2 Scientific name

Meloidogyne enterolobii Yang and Eisenback, 1983

65.3 Synonym

Meloidogyne mayaguensis Rammah and Hirschmann, 1988

65.4 EPPO code

MELGMY (*Meloidogyne mayaguensis*)

65.5 Diagnosis

The nematode *M. enterolobii*, formerly known as *M. mayaguensis*, is now regarded as a junior synonym (Xu *et al*., 2004; Hunt and Handoo, 2009; Karssen *et al*., 2013). Hence, all the diagnostic characters related to *M. mayaguensis* are considered for *M. enterolobii*.

65.5.1 Morphological diagnosis

FEMALE: Pear-shaped, no posterior protuberance. Stylet with basal knobs reniform, indented, offset (Fig. 9.11A,B). Perineal pattern round to ovoid, dorsal arch rounded; striae fine, widely spaced; lateral field absent or with single line occurring at junction of dorsal and ventral arches (Fig. 9.11G).
MALE: Cephalic region not offset, labial disc not elevated, lateral lips absent (Fig. 9.11C). Stylet with basal knobs rounded, offset, sloping posteriorly (Fig. 9.11C).
SECOND-STAGE JUVENILE: Hemizonid two annules anterior to excretory pore, tail with bluntly pointed tail tip and medium-long hyaline region (Fig. 9.11F) (Hunt and Handoo, 2009). For measurements, see Table 50.

65.5.2 Biochemical diagnosis

M. enterolobii is characterized by esterase isozyme phenotype VS1-S1, M2 and malate dehydrogenase phenotype N1a and N3c (Esbenshade and Triantaphyllou, 1985b; Rammah and Hirschmann, 1988; Carneiro *et al*., 2000)

65.5.3 Molecular diagnosis

Subbotin (2018) used recombinase polymerase amplification (RPA) assays targeting the IGS rRNA gene for rapid diagnosis of *M. enterolobii*. For primers of *M. enterolobii*, see Table 49.

65.6 Geographical distribution

M. enterolobii is widespread and recorded from different parts of the world (Fig. 9.12). The nematode is known to occur in the following countries (Hunt and Handoo, 2009; CABI-ISC, 2020).

Africa: Burkina Faso, Congo, Côte d'Ivoire, Kenya, Malawi, Mozambique, Niger, Nigeria, Senegal, South Africa, Togo.
Asia: China, India, Thailand, Vietnam.
Europe: Portugal, Switzerland.
North America: Costa Rica, Cuba, Guadeloupe, Guatemala, Martinique, Mexico, Puerto Rico, Trinidad and Tobago, USA.
South America: Brazil, Venezuela.

65.7 Hosts

M. enterolobii parasitizes eggplant (*Solanum melongena*), basil (*Basillum* sp.), bell pepper (*Capsicum annuum*), coffee (*Coffea* spp.), guava (*Psidium guajava*), potato (*Solanum tuberosum*), pumpkin (*Cucurbita* spp.), soybean (*Glycine max*), sweet potato (*Ipomoea batata*), tobacco (*Nicotiana tabacum*), tomato

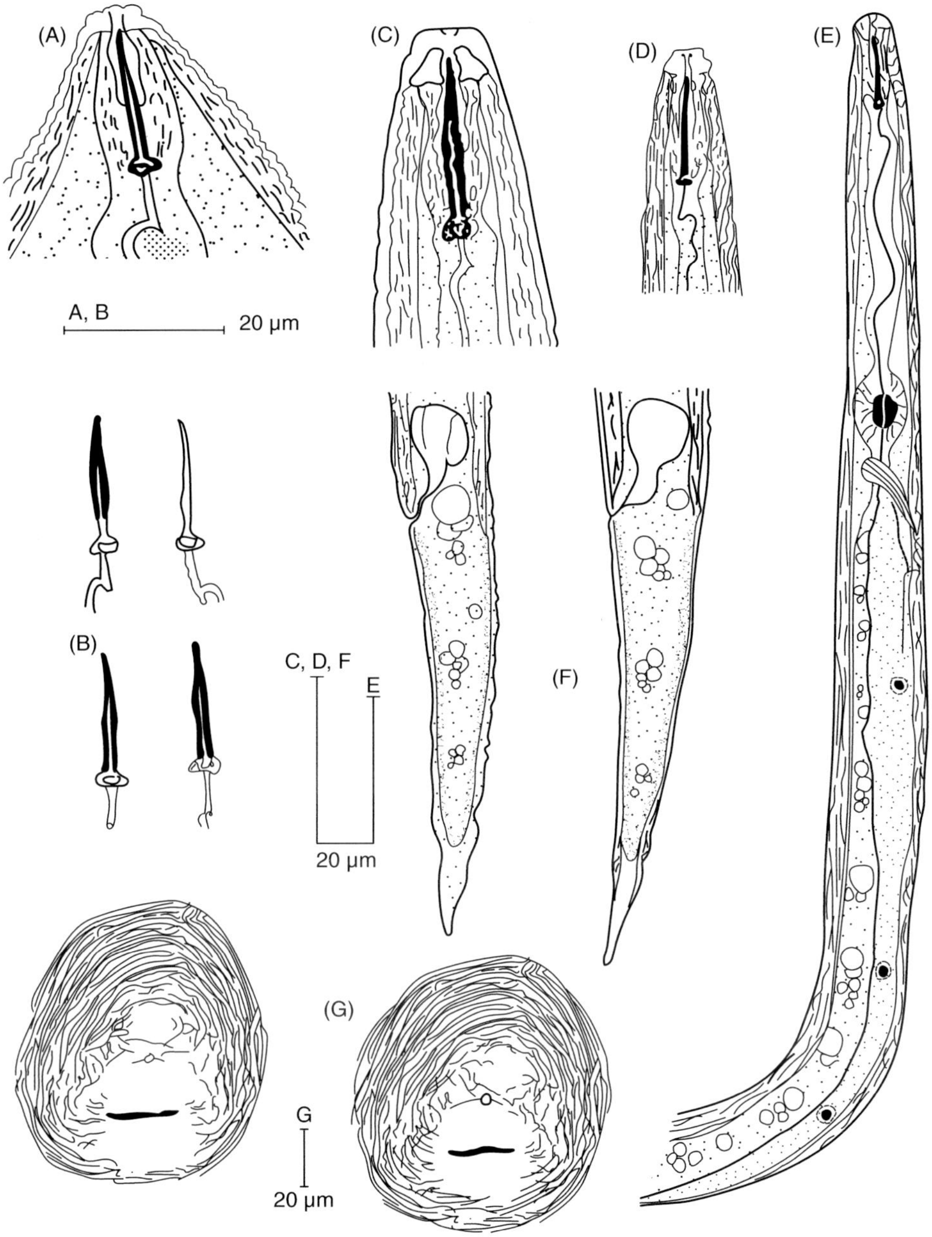

Fig. 9.11. Line drawing of *Meloidogyne enterolobii*. Female anterior region (A); female stylets (B); male anterior region (C); J_2 anterior region (D); J_2 pharyngeal region (E); J_2 tail regions (F); perineal patterns (G). [Rammah and Hirschmann, 1988; courtesy Journal of Nematology]

(*Lycopersicon esculentum*) and watermelon (*Citrullus lanatus*). Other hosts include Spanish needle, bean, beet, broccoli, celery, horsebean, parsley, American black nightshade, wild poinsettia, angel trumpet, glory bush, ajuga, glory flower and blackjack (Hunt and Handoo, 2009; Moens *et al.*, 2009).

65.8 Symptoms

M. enterolobii heavily attacks its host plants or trees and induces severe root galling (Fig. 9.5-R). Infestation of the trees causes a drastic reduction in overall growth, leaf size and yield decline, both quantitative and qualitative. Severely attacked guava trees die within 3 years due to wilting. Wilting of trees occurs mainly in patches surrounded by relatively healthy trees (Willers, 1997).

65.9 Biology and life cycle

The biology and life cycle are typical of the genus. The nematode is apomictic and reproduces through mitotic parthenogenesis ($2n$ = 44–46).

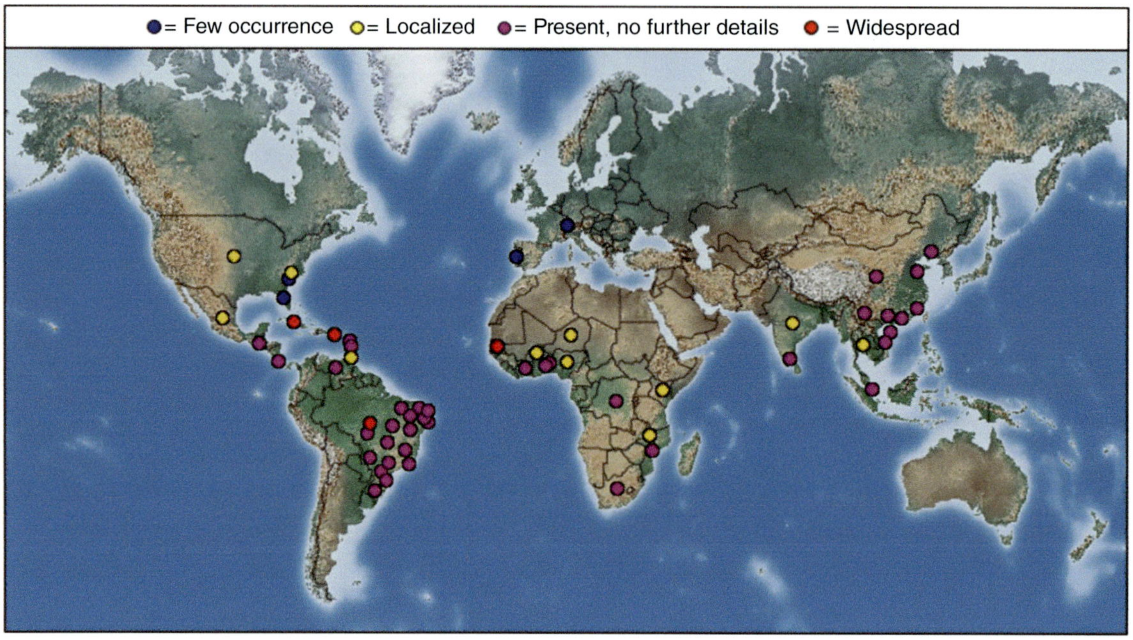

Fig. 9.12. Distribution map of *Meloidogyne enterolobii*. [Adapted from CABI-ISC, 2020]

65.10 Economic importance

M. enterolobii is an economically important root-knot nematode species because of its global distribution and the commercial value of most of its hosts (Moens *et al.*, 2009). This nematode parasitizes coffee in Cuba and occurs sympatrically with other root-knot species (Rodriguez *et al.*, 1995). On guava, severe root galling was observed in commercial orchards in Nelspruit, South Africa (Willers, 1997) and in Brazil (Carneiro *et al.*, 2001). The nematode is more proficient at breaking resistance than other common tropical root-knot species such as *M. arenaria*, *M. javanica* and *M. incognita* that carry the *Mi* gene, which confers resistance to the genotypes of tomato, pepper and some agronomic crops (Fargette *et al.*, 1996; Moens *et al.*, 2009).

65.11 Movement and means of dispersal

See section 57.11, above.

65.12 Pest risk analysis and looming threat from introduction

M. enterolobii is one of the most economically important root-knot nematodes because of its global distribution, wide host range, high reproduction rate and ability to overcome the resistance of important crop plants. A complete pest risk analysis is recommended to assess the risk posed by this nematode to the areas where it still does not occur. Appropriate preventive measures should also be employed by quarantine departments in order to minimize the risks.

65.13 Invasiveness rating

Considering the consequences of introduction and probable economic impact of *M. enterolobii*, the following invasiveness rating has been assigned.

Climate/host interaction	–	High (3)
Hosts range	–	Medium (2)
Dispersal potential	–	High (3)
Economic impact	–	High (3)
Environmental impact	–	Medium (2)
Post-entry distribution	–	Low (–1)
Final rating (score)	–	High (12)

Based on the above invasiveness rating, *M. enterolobii* receives a score of 12 out of 15 and is rated as HIGHLY INVASIVE NEMATODE.

65.14 Region-wise status of invasiveness

M. enterolobii is considered as a quarantine pest in different countries and listed as an EPPO A2 list (No. 361) pest for the EPPO region. In California (USA) the nematode is under the highest A rating as a nematode pest (Chitambar, 2016). Due to the wide host range, the nematode is considered as highly invasive and given a high invasiveness rating for all the countries where it is not present.

65.15 Management measures

See section 56.7, above.

66 *Meloidogyne ethiopica*

66.1 Common name

Tropical root-knot nematode

66.2 Scientific name

Meloidogyne ethiopica Whitehead, 1968

66.3 EPPO code

MELGET

66.4 Diagnosis

66.4.1 *Morphological diagnosis*

FEMALE: Elongate to pyriform, terminal protuberance absent. Stylet basal knobs rounded, tapering gradually into shaft (Fig. 9.13A). Perineal pattern oval to squarish; dorsal arch moderately high to high; striae coarse, widely separated, smooth to wavy; lateral field indistinct (Fig. 9.13C).
MALE: Labial region not offset, labial disc distinct, lateral lips present. Stylet basal knobs rounded to pear-shaped, offset (Fig. 9.13D).
SECOND-STAGE JUVENILE: Hemizonid anterior to excretory pore, tail long, hyaline region distinct, tail tip finely rounded to pointed (Fig. 9.13E) (Hunt and Handoo, 2009). For measurements, see Table 50.

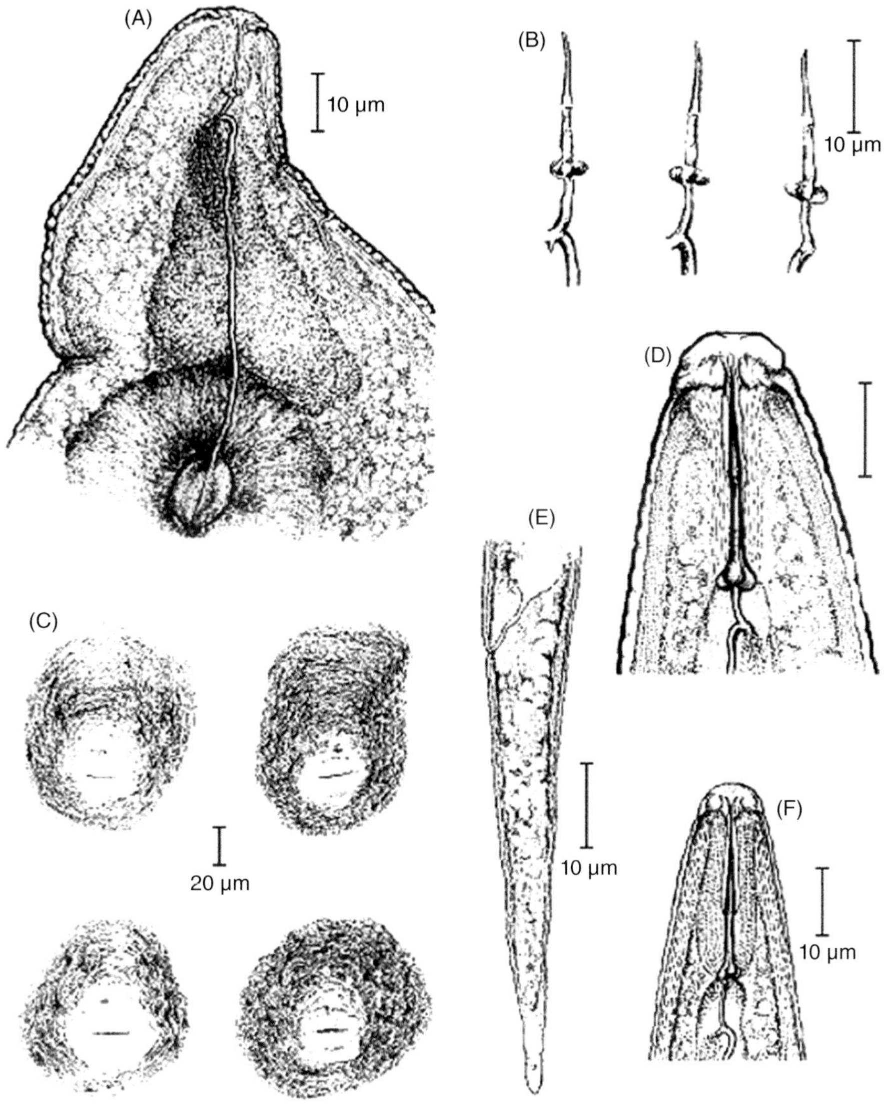

Fig. 9.13. Line drawing of *Meloidogyne ethiopica*. Female pharyngeal region, lateral view (A); female stylets (B); perineal patterns (C); male labial region, lateral view (D); J_2 tail, lateral view (E); J_2 anterior region, lateral view (F). [Adapted from Carneiro *et al.* (2004), courtesy Nematology]

66.4.2 *Biochemical diagnosis*

The esterase phenotype E3 (Ki3) and malate dehydrogenase N1 phenotype are species-specific and this is the most useful character for differentiating *M. ethiopica* from other species (Carneiro *et al.*, 2007).

66.4.3 *Molecular diagnosis*

RAPD profiles were used by Carneiro *et al.* (2004). For primers, see Table 49.

66.5 Geographical distribution

M. ethiopia is recorded mainly in East and Southern Africa (Ethiopia, Kenya, Mozambique, South Africa, Tanzania, Zimbabwe) and South America. Outside Africa, the nematode is known to occur in Brazil, Chile, Greece, Peru, Slovenia and Turkey (Hunt and Handoo, 2009; CABI-ISC, 2020).

66.6 Hosts

This root-knot nematode is polyphagous (Hunt and Handoo, 2009) and parasitizes cabbage (*Brassica oleracea* var. *capitata*), bell pepper (*Capsicum annuum*), common bean (*Phaseolus vulgaris*), cowpea (*Vigna unguiculata*), gourd (*Cucurbita pepo*), grapevine (*Vitis vinifera*), kiwi (*Actinidia deliciosa*), lettuce (*Lactuca sativa*), soybean (*Glycine max*), sugarcane (*Saccharum officinarum*), tomato (*Lycopersicon esculentum*) and yacon tubers (*Polymia sonchifolia*) (Lima *et al.*, 2009).

66.7 Symptoms

Symptoms are similar to those of other root-knot nematodes. *M. ethiopica* induces small and large multiple galls such as on grapevine (Fig. 9.5R). The affected plant has a distorted root system and is devoid of fine roots. The nematode is considered to be particularly destructive to several crops (e.g. beans, cucumber, tomatoes) and causes very large galls (Širca *et al.*, 2004).

66.8 Biology and life cycle

The biology and life cycle are typical of the genus. The nematode is apomictic, reproducing through mitotic parthenogenesis (Karssen *et al.*, 2013). Despite its tropical origins, *M. ethiopica* has shown the potential to survive in cold winters, even in regions where soil temperatures fall below 0°C, as well as under a sub-Mediterranean climate (hot summers and mild winters). In addition, the nematode could survive under glasshouse conditions across the region (Širca *et al.*, 2004).

66.9 Economic importance

M. ethiopica is an aggressive nematode pest of beans, cucumber and tomatoes, causing significant yield loss to these crops (Širca *et al.*, 2004)., The nematode is also considered as a destructive species on kiwi and grapevine in Brazil and Chile, with heavy infestations resulting in substantial suppression of plant growth, fruit size and quality (Hunt and Handoo, 2009). However, estimated data on the extent of damage on its different host plants is lacking.

66.10 Movement and means of dispersal

See section 57.11, above.

66.11 Pest risk analysis and looming threat from introduction

M. ethiopica is a polyphagous nematode and damages many economically important crops, including vegetable, ornamental and fruit crops. The recent incursion of this nematode into European countries demonstrated its potential to survive in temperate regions and to establish in new areas outside its native region.

66.12 Invasiveness rating

Considering the consequences of probable introduction and economic damage of *M. ethiopica*, the following invasiveness rating has been assigned.

Climate/host interaction	–	High (3)
Hosts range	–	High (3)
Dispersal potential	–	High (3)
Economic impact	–	High (3)
Environmental impact	–	Medium (2)
Post-entry distribution	–	Low (–1)
Final rating (score)	–	High (13)

Based on the above invasiveness rating, *M. ethiopica* receives a score of 13 out of 15 and is rated as HIGHLY INVASIVE NEMATODE.

66.13 Region-wise status of invasiveness

The Panel on Quarantine Nematodes endorsed that *M. ethiopica* should be added to the EPPO Alert List. This nematode is already on quarantine pest lists in many countries, including India (PQ Order-2003, 2015). Due to the wide host range, the nematode is considered as highly invasive for the countries where it is absent.

66.14 Management measures

See section 56.7, above.

67 *Meloidogyne exigua*

67.1 Common name

Coffee root-knot nematode

67.2 Scientific name

Meloidogyne exigua Goldi, 1887

67.3 Synonym

Heterodera exigua (Göldi, 1887) Marcinowski, 1909

67.4 EPPO code:

MELGEX

67.5 Diagnosis

67.5.1 Morphological diagnosis

M. exigua is the type species of the genus *Meloidogyne* (see detailed morphology in section 57.1.1). Important diagnostic morphological characters are as below. For measurements, see Table 50.

FEMALE: Pear-shaped, no posterior terminal protuberance (Fig. 9.14A). Stylet with basal knobs rounded, offset. Perineal pattern rounded to oval, low dorsal arch. Striae smooth, widely spaced, coarse, broken and folded in lateral regions; lateral field absent (Fig. 9.14B).

MALE: Labial region slightly offset, with distinct elevated labial disc and one annule, lateral lips present, trapezoidal or almost triangular. Stylet long, basal knobs rounded, posteriorly sloping, offset.

SECOND-STAGE JUVENILE: Excretory pore opposite to posterior end of isthmus, tail with narrowly rounded tip (Fig. 9.14H) (Hunt and Handoo, 2009).

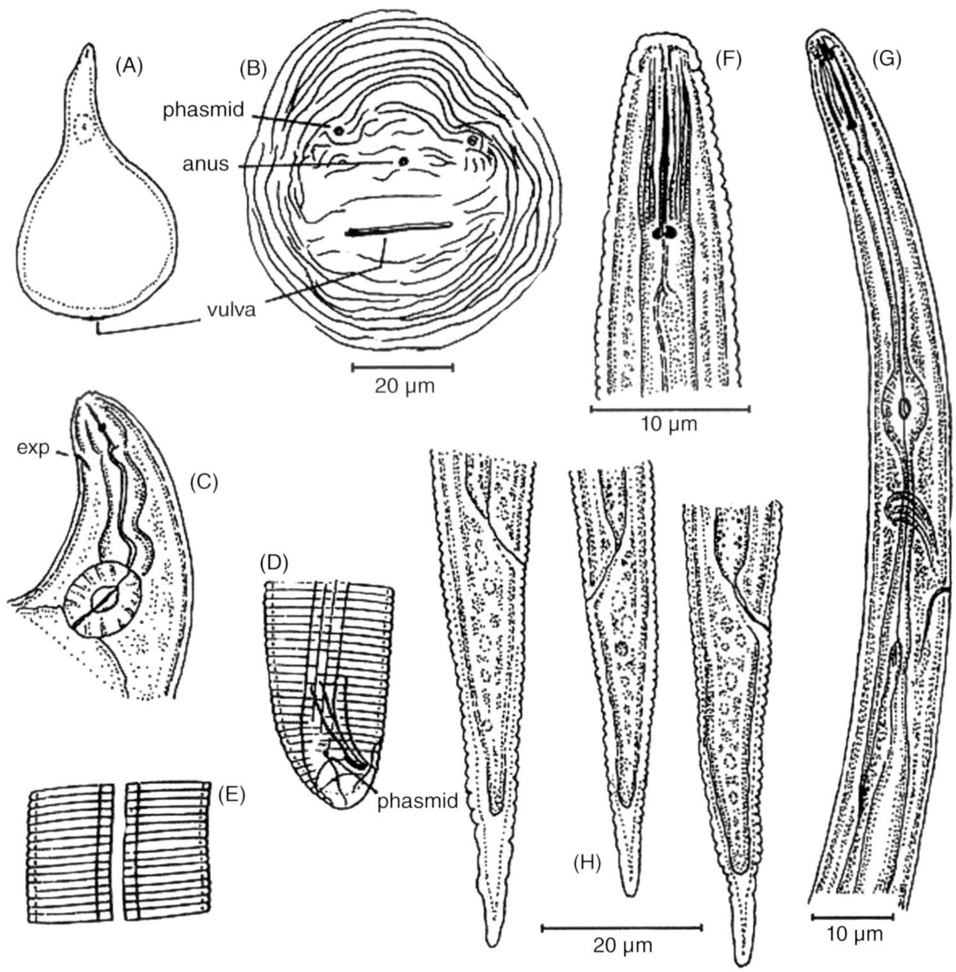

Fig. 9.14. Line drawing of *Meloidogyne exigua*. Entire female (A); perineal pattern of female (B); anterior end of female (C); posterior end of male (D); lateral field of male (E); anterior end of second-stage juvenile (F); oesophageal region of second-stage juvenile (G); tails of second-stage juveniles (H). [Courtesy CAB International]

67.5.2 *Biochemical diagnosis*

Several esterase phenotypes, such as V81 (Esbenshade and Triantaphyllou, 1990), E1 (VF1) and E1b (VF1) (Carneiro *et al.*, 2000) and E2, E2a, E3 (Blok and Powers, 2009) have been used for *M. exigua*. The VF1 esterase phenotype of *M. exigua* is shared with *M. naasi*, although it can be differentiated from the latter by its N1 malate dehydrogenase phenotype (Hunt and Handoo, 2009).

67.5.3 *Molecular diagnosis*

Randig *et al.* (2002a) characterized *M. exigua* using a cloned BglII satellite and used it as a radioactively labelled probe to detect single individuals. For primers, see Table 49.

67.6 Geographical distribution

M. exigua is widely distributed in humid coffee-producing areas and recorded from the following countries (Hunt and Handoo, 2009; CABI-ISC, 2020).

Asia: Turkey
North America: Costa Rica, Dominican Republic, El Salvador, Guadeloupe, Guatemala, Honduras, Martinique, Nicaragua, Puerto Rico, Trinidad and Tobago.
South America: Bolivia, Brazil, Colombia, French Guiana, Peru, Suriname, Venezuela.

67.7 Hosts

M. exigua is a polyphagous nematode and parasitizes mostly coffee (*Coffea* spp.), but also common weeds in coffee plantations (Lopez and Vilchez, 1991). The host range includes *Allium cepa* (onion), *Bidens pilosa* (blackjack), *Capsicum annuum* (bell pepper), *Citrullus lanatus* (watermelon), *Citrus aurantium* (sour orange), *Coffea arabica* (arabica coffee), *Coffea eugenioides* (coffee), *Cucumis sativus* (cucumber), *Cyphomandra betacea* (tree tomato), *Ipomoea acuminate*, *Musa* (banana), *Oryza sativa* (rice), *Pile* sp., *Poncirus trifoliata* (trifoliate orange), *Saccharum officinarum* (sugarcane), *Solanum lycopersicum* (tomato), *Solanum nigrum* (black nightshade) and *Stachys arvensis* (staggerweed). Three races have been described for *M. exigua*: race 1 attacks pepper and coffee; race 2 attacks tomato, pepper and coffee; and race 3 attacks rubber but not coffee (Muniz *et al.*, 2008).

67.8 Symptoms

M. exigua typically incites severe root galling in heavily infested coffee and other plants. In humid valleys and near rivers and streams, coffee trees are most susceptible. The symptoms on coffee include rapid yellowing and leaf fall spread along lines of trees. In a case of severe infestation, trees may die and are simply uprooted. Roots show small galls about the size of a hemp seed (Fig. 9.5S), and root hairs and rootlets are destroyed (Dias *et al.*, 1996). Mature, pyroid females are found in fibrovascular bundles and in parenchymatous tissue, where they form lesions that are later invaded by microorganisms (De Souza, 1978).

67.9 Biology and life cycle

The biology and life cycle are typical of the genus and reproduction is by facultative parthenogenesis ($2n = 36$). *M. exigua* is thermophilic and may not survive at temperatures below 10°C.

67.10 Economic importance

M. exigua attacks and damages mainly coffee, severely reducing coffee yield in several countries in Central and South America (Hunt and Handoo, 2009). In Brazil, it infests coffee in almost all states and yield losses have been estimated at 45% (Barbosa *et al.*, 2004) and 10–15% in Costa Rica (Muniz *et al.*, 2008). In Colombia, 48% of coffee root samples were found infested with this nematode (Gomez, 1980). Maize, cotton or soybean plants grown for 1 or 2 years after coffee in plantations infested with *M. exigua* exhibited significant reductions in yield (Moraes, 1977), likewise rubber (Muniz *et al.*, 2008).

67.11 Movement and means of dispersal

See section 57.11, above.

67.12 Pest risk analysis and looming threat from introduction

M. exigua is a polyphagous species and primarily parasitizes coffee in Latin America and a few locations outside this area. Due to potential economic damage, it is a threat to all coffee- and rubber-growing regions and should be targeted in regulatory programmes. There is a very high chance of dispersal of this nematode through nurseries from infested soil and root stock.

67.13 Invasiveness rating

Considering the probable introduction and economic damage of *M. exigua*, the following invasiveness rating has been assigned.

Climate/host interaction	–	High (3)
Hosts range	–	High (3)
Dispersal potential	–	Medium (2)
Economic impact	–	High (3)
Environmental impact	–	Medium (2)
Post-entry distribution	–	Low (–1)
Final rating (score)	–	High (12)

Based on the above invasiveness rating, *M. exigua* receives a score of 12 out of 15 and is rated as HIGHLY INVASIVE NEMATODE.

67.14 Region-wise status of invasiveness

Because of the wide host range, the nematode is considered as highly invasive for all the countries where it still does not occur.

67.15 Management measures

See section 56.7, above.

68 *Meloidogyne fallax*

68.1 Common name

False Columbia root-knot nematode

68.2 Scientific name

Meloidogyne fallax Karssen, 1996

68.3 Synonym

Meloidogyne chitwoodi (Baexem) B-type van Meggelen *et al.*, 1994

68.4 EPPO code

MELGFA

68.5 Diagnosis

Meloidogyne fallax and *M. chitwoodi* are sympatric and morphologically closely related to each other (Hunt and Handoo, 2009).

68.5.1 Morphological diagnosis

FEMALE: Globular to pear-shaped, with slight posterior protuberance (Fig. 9.15C). Stylet long with basal knobs large, rounded to transversally ovoid, offset (Fig. 9.15B). One or two large and several smaller vesicle-like structures located along lumen lining. Perineal pattern ovoid to oval-shaped, or rectangular; dorsal arch low to moderately high, striae coarse; tail terminus and lateral field indistinct, resulting in a relatively large area without striae (Fig. 9.15G,H).
MALE: Labial region slightly offset, labial disc rounded, elevated, fused with medial lips, lateral lips present. Stylet long, basal knobs offset, large rounded (Fig. 9.15E).
SECOND-STAGE JUVENILE: Hemizonid at same level as excretory pore, tail with broadly rounded tail tip and long hyaline region (Fig. 9.15D) (Karssen *et al.*, 2013). For measurements, see Table 52.

68.5.2 Biochemical diagnosis

M. fallax is characterized by a unique malate dehydrogenase N1b phenotype (Karssen and van Hoenselaar, 1998) and lacks any major esterase band. All populations share this rare malate dehydrogenase phenotype and '0' esterase phenotype (Hunt and Handoo, 2009). Prolonged esterase staining (60 min) showed a very weak, three-banded pattern named F3 (van der Beek and Karssen, 1997).

68.5.3 Molecular diagnosis

For primers, see Table 49.

68.6 Geographical distribution

M. fallax was recorded for the first time in the Netherlands (Karssen, 1994). After that, it was detected in France (Daher *et al.*, 1996), Belgium (Waeyenberge and Moens, 2001), Germany (Schmitz *et al.*, 1998), Sweden, Switzerland and the United Kingdom (CABI-ISC, 2020). Outside Europe, it has been recorded in New Zealand (Marshall *et al.*, 2001) and Australia (Nobbs *et al.*, 2001).

68.7 Hosts

M. fallax parasitizes mainly monocotyledons and dicotyledons. Important hosts include potato (*Solanum tuberosum*) beetroot (*Beta vulgaris*), carrots (*Daucus carota*), globe artichoke (*Cynara cardunculus* var. *scolymus*), lettuce (*Lactuca sativa*), lucerne (*Medicago sativa*), oyster plant (*Scorzonera hispanica*), strawberry (*Fragaria ananassa*) and tomatoes (*Solanum lycopersicum*) (CABI-ISC, 2020) and cereals (Moens *et al.*, 2009).

68.8 Symptoms

The nematode infection causes stunting, chlorosis and wilting symptoms on foliage. On the roots, galling is a typical symptom. Relatively small galls are formed and secondary roots do not emerge from the gall. On potato tubers, numerous small, blister-like raised areas on the surface are formed (similar to *M. chitwoodi*) (Fig. 9.5O).

68.9 Biology and life cycle

The biology and life cycle are typical of the genus. *M. fallax* is automictic and reproduction is by facultative meiotic parthenogenesis ($n = 18$). Unlike *M. chitwoodi*, *M. fallax* did not need hatch stimulation from root diffusate (Moens *et al.*, 2009).

68.10 Economic importance

M. fallax is a major pest of potatoes, oyster plant and carrots in the Netherlands (Korthals *et al.*, 2000). It is of economic importance on tomatoes on a single site in France and also has a more significant effect on yields of artichoke and lettuce than *M. chitwoodi*.

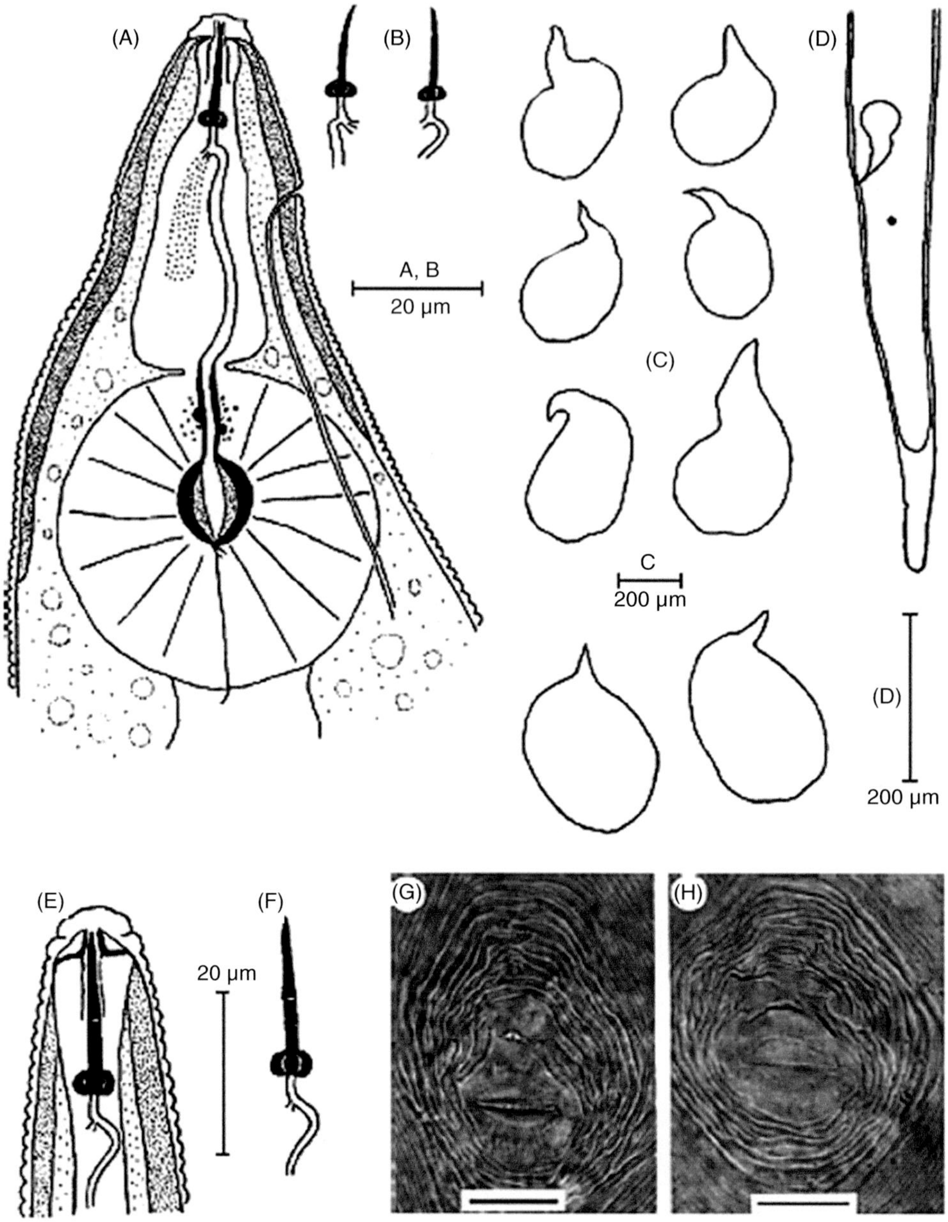

Fig. 9.15. Line drawing of *Meloidogyne fallax*. Female pharyngeal region, lateral view (A); female stylets (B); entire females (C); J_2 tail region, lateral view (D); male anterior region (E); male stylets (F); perineal patterns, scale bar = 25 μm (G,H). [Courtesy Karssen, 1996; Fundamental and Applied Nematology]

68.11 Movement and means of dispersal

See section 57.11, above.

68.12 Pest risk analysis and looming threat from introduction

The distribution of *M. fallax* is limited within Europe; outside Europe, it is recorded in New Zealand and Australia. This nematode is polyphagous and affects economically important crops, including potato, and there would be a high chance of dispersal of this nematode from infested areas to new areas through seed tubers and contaminated soil. *M. fallax* should be targeted in regulatory programmes to check its further spread in Europe and other countries.

68.13 Invasiveness rating

Considering the consequences of probable introduction and pest risk analysis of *M. fallax*, the following invasiveness rating has been assigned.

Table 52. Morphometries data of *Meloidogyne* species.

Characters		Range, numerous populations							
		M. fallax (Karssen, 1996; Karssen *et al.*, 2013)	*M. graminicola* (after Golden and Birchfield, 1965)	*M. hapla* (after Whitehead, 1968)	*M. incognita* (after Whitehead, 1968)	*M. javanica* (after Orton Williams, 1972)	*M. mali* (after Itoh *et al.*, 1969)	*M. naasi* (Krassen *et al.*, 2013)	*M. paranaensis* (Hunt and Handoo, 2009)
Female	L (µm)	404–720	445–765	419–845	500–723	541–804	–	–	–
	a	–	1.2–1.8	–	–	–	–	–	–
	Max. body diameter (µm)	256–464	275–520	311–561	331–520	311–581	–	–	–
	Stylet length (µm)	13.5–15.5	10.64–11.20	10–13	13–16	14–18	13–17	13–15	15–17.5
	Stylet knob height (µm)	–	–	–	–	–	–	–	–
	Stylet knob width (µm)	–	–	2–3	3–5	2–5	–	–	–
	DGO (µm)	–	2.8–3.9	4–6	2–4	2–5	–	–	–
Male	L (µm)	736–1520	1020–1428	791–1432	1108–1953	757–1297	–	–	–
	a	–	72.8–215	33.3–47.0	31.4–55.4	17.5–42.9	–	–	–
	c	–	–	73–283	97–255	50–144	–	–	–
	Max. body diameter (µm)	27–44	–	–	–	–	–	–	–
	Stylet length (µm)	19–21	16.24–17.36	17.3–22.7	23.0–32.7	20–23	18–22	16–18.5	20–27
	Stylet knob height (µm)	–	–	–	–	–	–	–	–
	Stylet knob width (µm)	–	–	2.5–5.0	4.7–6.8	3.6–5.4	–	–	–
	DGO (µm)	3–6	3.5–4.8	2.5–3.2	1.4–2.5	2.2–4.7	6–13	2.5–3.5	3.5–5.0
	Tail length (µm)	7.6–12.1	6.2–15.1	–	–	–	–	–	–
	Spicule length (µm)	–	27.4–29.1	21.6–28.1	28.8–40.3	20.9–31.7	–	–	–
	Gubernaculum length (µm)	–	5.6–6.7	7.2–9.4	9.4–13.7	7.2–9.4	–	–	–
Second stage Juvenile (J_2)	L (µm)	381–435.2	415–484	312–355	337–403	387–459	390–450	410–460	389–513
	a	–	22.3–27.3	20.1–26.6	24.9–31.5	27.1–35.9	–	–	–
	c	–	5.5–6.7	7.3–10.2	6.9–10.6	7.3–11.1	–	–	–
	Max. body diameter (µm)	13–16.4	–	–	–	–	–	–	–
	Stylet length (µm)	–	11.20–12.32	7.9–10.9	9.6–11.7	9.4–11.4	–	–	–
	DGO (µm)	–	2.8–3.4	–	–	–	–	–	–
	Tail length (µm)	46–55.6	67–76	33–48	43–65[a]	36–56	30–34	60–75	48–51
	Hyaline tail terminus (h)	12–16	14.0–21.2	–	6–14[a]	9–18[a]	4–12	16–25	9–10

[a]From Hunt and Handoo (2009).

Climate/host interaction	–	High (3)
Hosts range	–	Medium (2)
Dispersal potential	–	High (3)
Economic impact	–	High (3)
Environmental impact	–	Medium (2)
Post-entry distribution	–	Low (–1)
Final rating (score)	–	High (12)

Based on the above invasiveness rating, *M. fallax* receives a score of 12 out of 15 and is rated as HIGHLY INVASIVE NEMATODE.

68.14 Region-wise status of invasiveness

M. fallax is of quarantine importance in a number of countries (Blok and Powers, 2009). The nematode was added to the

European list of quarantine organisms to prevent further spread within Europe (Hunt and Handoo, 2009). Due to the wide host range, this nematode is considered highly invasive for all the countries where it is absent.

68.15 Management measures

See section 56.7, above.

69 *Meloidogyne graminicola*

69.1 Common name

Rice root-knot nematode

69.2 Scientific name

Meloidogyne graminicola Golden and Birchfield, 1965

69.3 EPPO code

MELGGC

69.4 Diagnosis

69.4.1 Morphological diagnosis

Morphological descriptions by Golden and Birchfield (1965), Mulk (1976) and Hunt and Handoo (2009) are summarized below. For morphometric measurements, see Table 52.

FEMALE: Adult pear-shaped to spheroid with elongated neck and usually embedded within root tissue. Stylet long with three small, prominent, dorsally curved basal knobs (Fig. 9.16G). Oesophageal glands overlap anterior part of intestine. Ovaries paired, cover most of body cavity. Vulva with typically terminal anus, slightly raised or flush with body contour and encircled by cuticular striae. Perineal pattern egg-shaped or oval (Fig. 9.16H). Striae smooth, thick in outer region without lateral line. In dorsal arch, striae anastomose into pyramidal web or tetragonal that joins at tail terminus. Semicircular whorl consisting of two to three striae present at each corner of vulvar slit. Vulvar lips smooth without invaginations.

MALE: Labial region usually not offset, or slightly offset by a constriction, labial disc not raised, lateral lips usually present. Stylet with rounded posteriorly sloping knobs (Fig. 9.16B). Hemizonid 1–3 annules anterior to excretory secretory pore, 1–2 annules wide. Hemizonion inconspicuous, a few annules behind excretory pore. Spicules arcuate or slightly curved ventrally near middle (Fig. 9.16D). Gubernaculum rod-shaped. Tail with smooth tip. Phasmids postanal, small, situated near middle of tail.

SECOND-STAGE JUVENILE: J_2 short with weak cephalic framework. Stylet with posteriorly sloping rounded knobs. Oesophageal gland lobe overlaps intestine ventrally (Fig. 9.16E). Tail terminus rounded, often slightly clavate. Tail terminus tapers to long hyaline region with fine point (Fig. 9.16F).

69.4.2 Biochemical diagnosis

The esterase VS1 phenotype with one slow band with a large drawn-out area of enzymatic activity is associated with *M. graminicola* (Esbenshade and Triantaphyllou, 1985a,b). The malate dehydrogenase N1a phenotype is similar to that of *M. chitwoodi* and *M. salasi* (Hunt and Handoo, 2009).

69.4.3 Molecular diagnosis

Khan and Ahamad (2020) characterized several isolates of *M. graminicola* from Northern India. For primers, see Table 49.

69.5 Geographical distribution

The rice root-knot nematode *M. graminicola* has been recorded from major rice-growing countries (Fig. 9.17), in particular in South-East Asia (Haque *et al.*, 2019). The nematode has been detected from the following countries (CABI-ISC, 2020).

Africa: Madagascar, South Africa.
Asia: Bangladesh, China, India, Indonesia, Laos, Malaysia, Myanmar, Nepal, Pakistan, Philippines, Singapore, Sri Lanka, Thailand, Vietnam.
Europe: Italy.
North America: USA.
South America: Brazil, Colombia, Ecuador.

69.6 Hosts

M. graminicola is primarily a pest of rice. Besides rice, it also parasitizes onion, cabbage and tomato, and many common weeds of rice (Haque *et al.*, 2019). MacGowan and Langdon (1989) recorded more than 100 host plants of *M. graminicola*. The nematode has also been found to be damaging wheat and barley (Vaish *et al.*, 2016). The extended list of host plants is presented in Table 53.

69.7 Symptoms

The symptoms incited by *M. graminicola* are produced in the form of distinctive hook, spindle or horseshoe-like terminal galls on the roots (Fig. 9.5E). Due to formations of galls, the absorption capacity of roots is impaired, leading to stunting

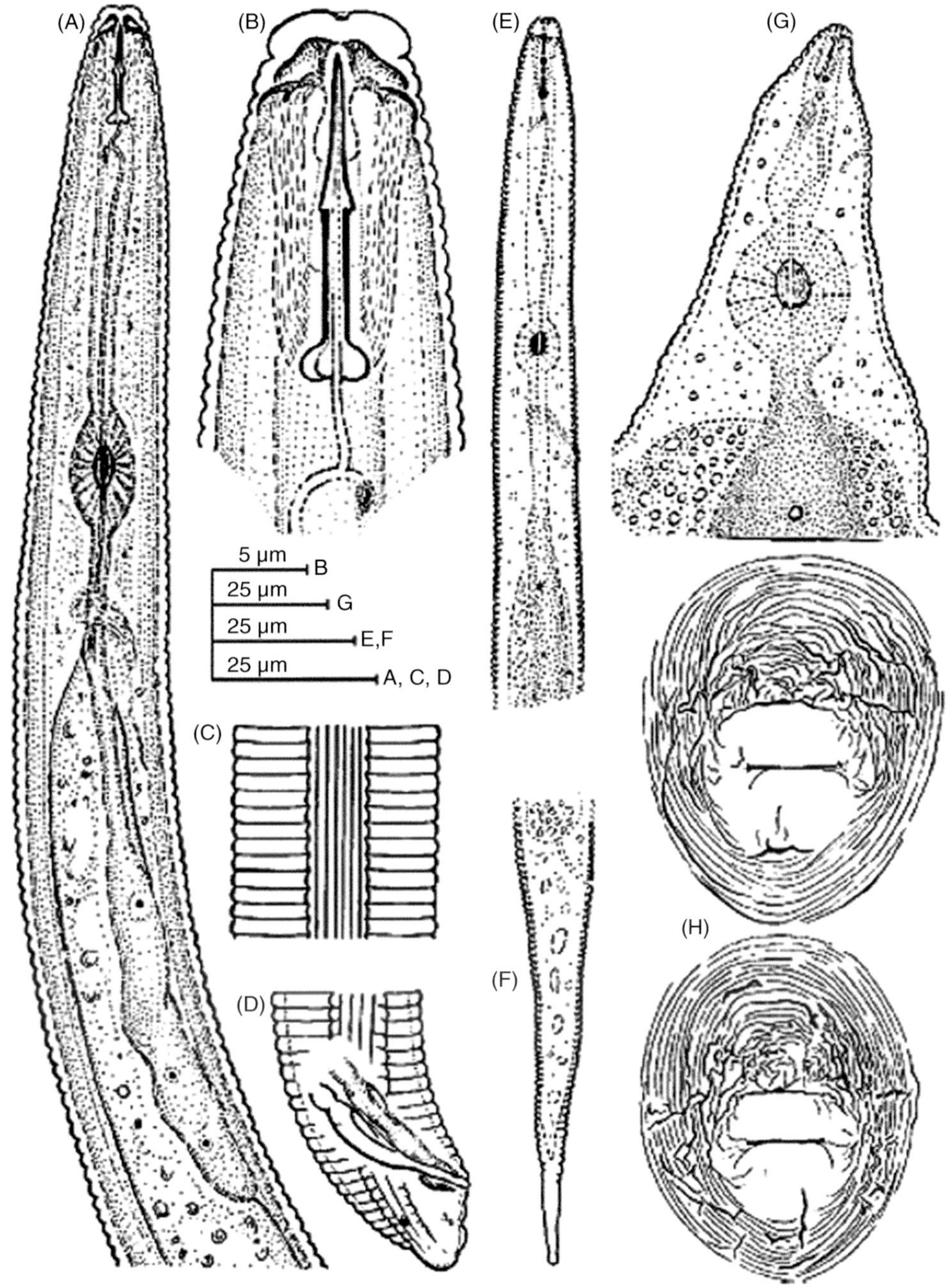

Fig. 9.16. Line drawing of *Meloidogyne graminicola*. Male pharyngeal region (A); male anterior region (B); male lateral field (C); male tail region (D); J_2 pharyngeal region (E); J_2 tail region (F); female anterior region (G); perineal patterns (H). [Reproduced from Hunt and Handoo, 2009; A–D,H after Mulk, 1976; courtesy CAB International; E–G after Golden and Birchfield, 1965; courtesy of Proceedings of the Helminthological Society of Washington]

of plant growth and yellowing of leaves (Khan and Ahamad, 2020). These symptoms appear in patches on the plants within a nursery (Fig. 9.4A) or in the field. The newly emerged leaves become distorted and crinkled along the margins. Severely infested plants mature early (Haque *et al*. 2018).

69.8 Biology and life cycle

The biology and life style are typical of the genus. *M. graminicola* is automictic and reproduction is by facultative meiotic parthenogenesis (n =18). In *M. graminicola*, generally egg masses are not formed on the root surface, rather the masses are

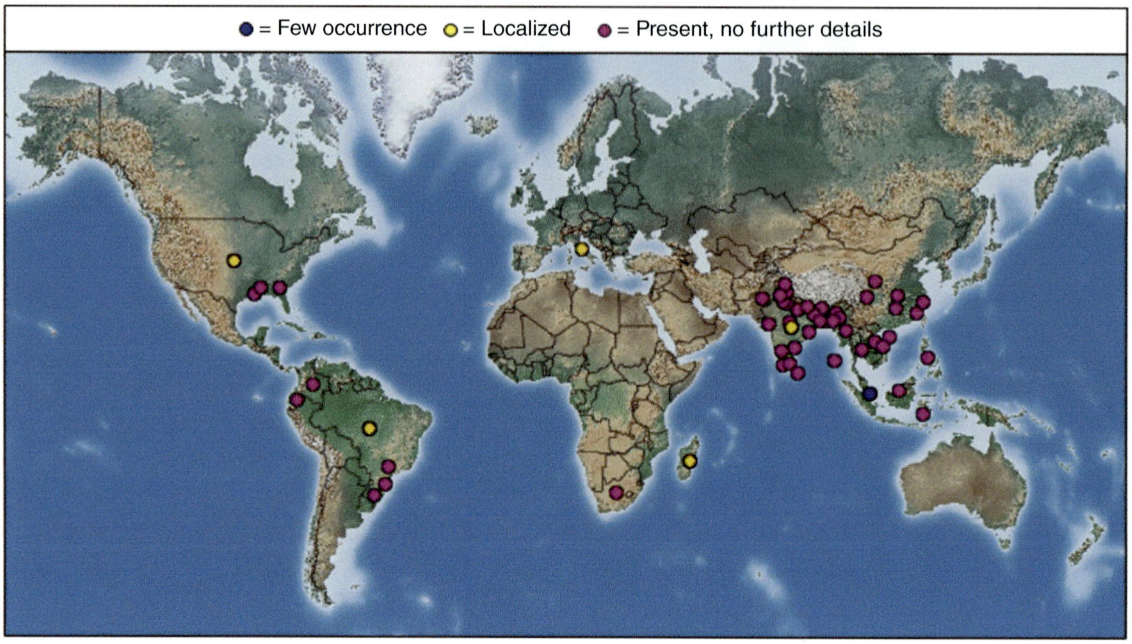

Fig. 9.17. Distribution map of *Meloidogyne graminicola*. [Adapted from CABI-ISC, 2020]

Table 53. List of host plants of *Meloidogyne graminicola* (Haque *et al*., 2019).

Main hosts
Oryza sativa (rice)
Poaceae (grasses)
Other hosts
Allium cepa (onion)
Brassica
Eleusine coracana (finger millet)
Glycine max (soybean)
Panicum miliaceum (millet)
Sorghum bicolor (sorghum)
Triticum aestivum (wheat)
Zea mays (maize)
Wild hosts
Alopecurus (foxtails)
Ammania pentandra
Andropogon (beardgrass)
Blumea
Cynodon dactylon (Bermuda grass)
Cyperus imbricatus
Cyperus procerus
Cyperus pulcherrimus
Cyperus rotundus (purple nutsedge)
Digitaria filiformis (slender crabgrass)
Echinochloa colona (junglerice)
Echinochloa crus-galli (barnyard grass)
Eleusine indica (goosegrass)
Fimbristylis complanata
Fimbristylis dichotoma (tall fringe rush)
Fimbristylis littoralis (lesser fimbristylis)
Fuirena glomerata
Grangea madraspatensis
Ischaemum rugosum (saramollagrass)
Ludwigia (water primrose)
Monochoria vaginalis (pickerel weed)
Panicum repens (torpedo grass)
Paspalum scrobiculatum (ricegrass paspalum)
Phyllanthus urinaria (leafflower)
Poa annua (annual meadowgrass)
Ranunculus (buttercup)
Scirpus articulatus
Sphaeranthus
Sphenoclea zeylanica (wedgewort)

well embedded in the galled tissue (Khan *et al*., 2014). The progeny moves internally to invade the fresh root tissue (Haque *et al*., 2018) but a considerable population also migrates to the soil. *M. graminicola* is well adapted to waterlogged soil and can survive in flooded conditions for longer periods as eggs in egg masses or as juveniles. The nematode prefers sandy loam soil (Khan and Haque, 2012). Variations in the duration of the life cycle of *M. graminicola* have been reported; in Bangladesh the life cycle was completed within 19 days at 22–29°C (Bridge and Page, 1982), whereas in the USA it took 25–27 days at 26.8°C (Yik and Birchfield 1979).

69.9 Economic importance

Infestation by this root-knot nematode has been found to be prevalent in upland (Prot *et al*., 1994), rain-fed lowland (Padgham *et al*., 2004) and irrigated rice crops (Khan *et al*., 2020). The nematode attacks rice plants in both the nursery bed and the field, resulting in 21–34% yield loss to the crop (Haque *et al.*, 2018). Sometimes transplanting fails to establish in highly infested field and farmers have no option but to plough the field one month after planting. Globally, *M. graminicola* has been found to cause serious damage to rice in various countries (Prot *et al.* 1994; Padgham *et al.* 2004; Bridge *et al.* 2005a; Peng *et al.*, 2018). Khan and Ahamad (2020) reported 5–47% relative yield loss to irrigated rice in northern India. The nematode has also emerged as a notorious pest of wheat and barley grown after rice (Vaish *et al.*, 2016).

69.10 Movement and means of dispersal

See section 57.11, above.

69.11 Pest risk analysis and looming threat from introduction

M. graminicola poses a serious threat to rice cultivation world over. In recent decades, the nematode has emerged as an important nematode pest in nursery and field, due to continuous and extensive cultivation of rice in South-East Asia and other major rice-growing countries. The nematode prefers sandy loam soil but it has adapted well to clay soil and can survive in water-logged soil for long periods. *M. graminicola* is easily dispersed from infested fields to non-infested fields through infested nurseries and irrigation water. Hence, the nematode should be targeted in regulatory programmes to check its further spread from infested areas to new areas or regions.

69.12 Invasiveness rating

Considering the consequences of probable introduction and economic damage of *M. graminicola*, the following invasiveness rating has been assigned.

Climate/host interaction	–	High (3)
Hosts range	–	Medium (2)
Dispersal potential	–	High (3)
Economic impact	–	High (3)
Environmental impact	–	High (3)
Post-entry distribution	–	Low (–1)
Final rating (score)	–	High (13)

Based on the above invasiveness rating, *M. graminicola* receives a score of 13 out of 15 and is rated as HIGHLY INVASIVE NEMATODE.

69.13 Region-wise status of invasiveness

Because of the economic damage potential to the world's second most important crop, rice, *M. graminicola* is considered as highly invasive nematode, particularly for the following rice-producing countries in which the nematode does not occur.

APPPC: Afghanistan, Cambodia, Iran, Japan, North Korea, South Korea
COSAVE: Argentina, Peru, Uruguay, Venezuela
EPPO: Russia, Spain, Turkey
IAPSC: Côte d'Ivoire, Egypt, Guinea, Mali, Nigeria, Senegal, Sierra Leone, Tanzania
PPPO: Australia, New Zealand.

69.14 Management measures

See section 56.7, above.

70 *Meloidogyne hapla*

70.1 Common name

Northern root-knot nematode

70.2 Scientific name

Meloidogyne hapla Chitwood, 1949

70.3 EPPO code

MELGHA

70.4 Diagnosis

70.4.1 Morphological diagnosis

The morphological description and morphometric data from Karssen *et al.* (2013) are presented below. For measurements, see Table 52.

FEMALE: Pear-shaped, without terminal protuberance (Fig. 9.18C). Stylet long, ranging in length, basal knobs small, rounded, offset (Fig. 9.18A,B). Perineal pattern with fine striae, rounded, low dorsal arch, characteristic punctations usually present above the anus, lateral field present (Fig. 9.18J).
MALE: Cephalic region offset from body. Labial disc not usually elevated. Lateral lips present. Head cap rounded. Stylet long, basal knobs relatively small, rounded, offset (Fig. 9.18D).
SECOND STAGE JUVENILE: Hemizonid anterior to excretory pore, not adjacent. Tail slender, tail hyaline region often irregular in form, anterior region not clearly enclosed, tail terminus finely rounded (Fig. 9.18I).

70.4.2 Biochemical diagnosis

Most populations show the H1 esterase phenotype (Esbenshade and Triantaphyllou, 1985a,b). One population from France did not have any major esterase activity, and another population from Minnesota, USA, had an A1 esterase phenotype

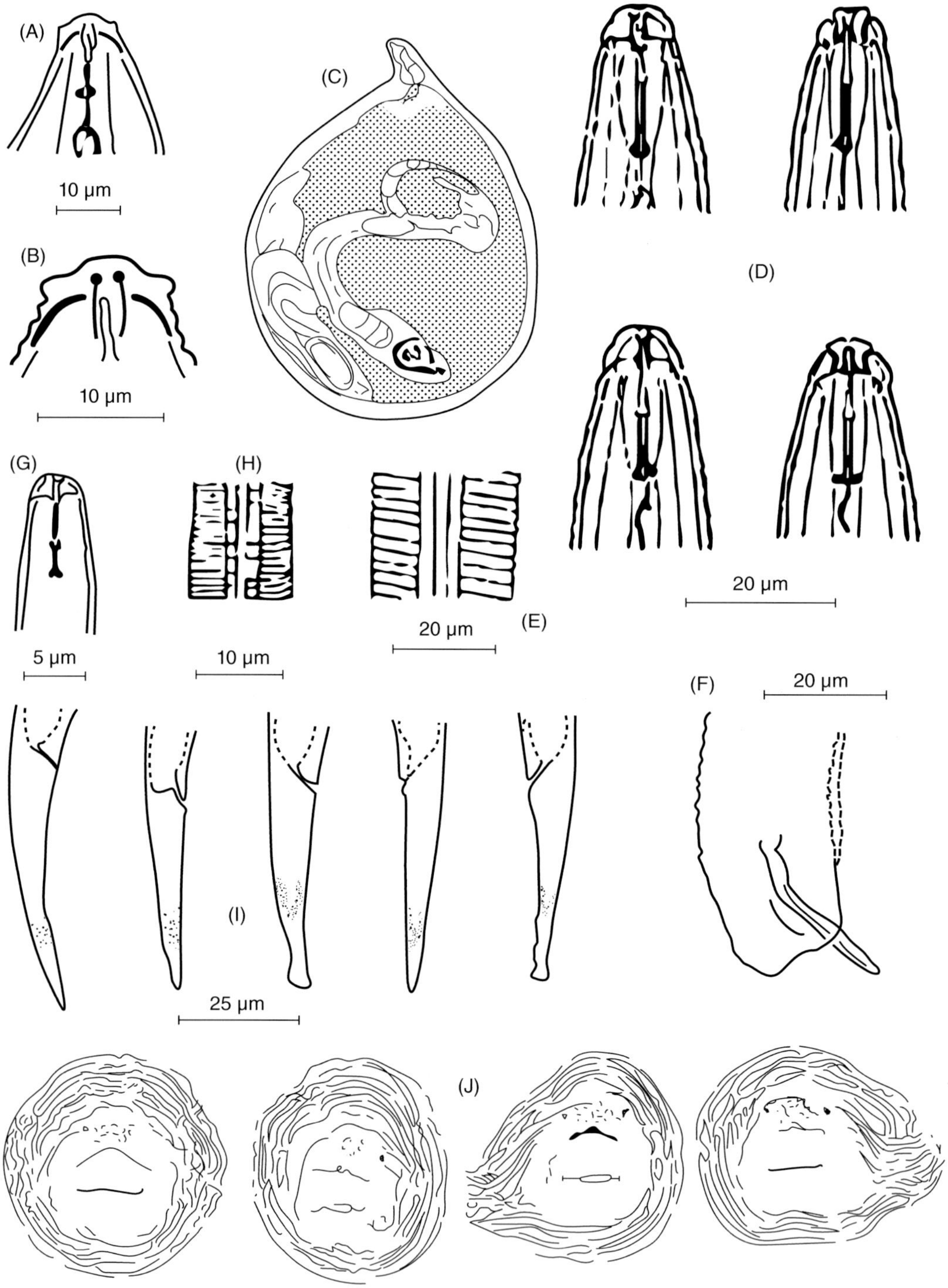

Fig. 9.18. Line drawing of *Meloidogyne hapla*. Female anterior region (A,B); entire female (C); male anterior region (D); male lateral field (E); male tail (F); J_2 anterior region (G); J_2 lateral field at mid-body (H); J_2 tail regions (I); perineal patterns (J). [A–I, after Whitehead, 1968; courtesy Journal of Zoology, John Wiley and Sons. J after Orton Williams (1974); courtesy of CAB International]

identical to *M. arenaria*. The malate dehydrogenase H1 phenotype is unique for this species (Hunt and Handoo, 2009).

70.4.3 Molecular diagnosis

Piotte *et al.* (1995) used satDNA as a diagnostic probe that included radioactively labelled repeat sequences from *M. hapla*. The probe had sufficient sensitivity to detect DNA from a single female of *M. hapla* even within root tissue (Dong *et al.*, 2001a). For primers, see Table 49.

70.5 Geographical distribution

M. hapla is one of the four commonest species and has been recorded worldwide, mainly from temperate regions (Fig. 9.19) but also reported from tropical and subtropical areas at high altitudes (Karssen *et al.*, 2013). The nematode has been found to occur in the following countries (CABI-ISC, 2020).

Africa: Algeria, Côte d'Ivoire, Egypt, Ethiopia, Kenya, Libya, Malawi, Morocco, Nigeria, South Africa, Tanzania, Uganda, Zimbabwe.
Asia: Armenia, China, India, Indonesia, Iran, Israel, Japan, Kazakhstan, Kyrgyzstan, Malaysia, Mongolia, Pakistan, Philippines, South Korea, Sri Lanka, Taiwan, Tajikistan, Thailand, Turkey, Turkmenistan, Uzbekistan.
Europe: Belarus, Belgium, Bulgaria, Czech Republic, Estonia, Finland, France, Germany, Greece, Hungary, Italy, Latvia, Lithuania, Moldova, Netherlands, North Macedonia, Norway, Poland, Portugal, Romania, Russia, Serbia and Montenegro, Slovakia, Slovenia, Spain, Switzerland, Ukraine, United Kingdom, former Yugoslavia.
North America: Canada, Costa Rica, Guatemala, Mexico, Panama, USA.
Oceania: Australia, New Zealand, Norfolk Island, Papua New Guinea.
South America: Argentina, Brazil, Chile, Colombia, Ecuador, Paraguay, Peru, Uruguay, Venezuela.

70.6 Hosts

M. hapla is a polyphagous nematode that parasitizes mainly dicotyledons and rarely monocotyledons. An expanded list of hosts is given in Table 54.

70.7 Symptoms

Typical symptoms of *M. hapla* infection include small galls on the root systems (Fig. 9.5G,H). The galls are subspherical, often marked with a small proliferation of roots at the site of infection (Moens *et al.*, 2009). On potato tubers, brown spots may appear and help to identify the infection sites. The spots are due to gelatinous matrix secretion by the females to deposit eggs in the tubers. Severely attacked roots become impaired, resulting in disruption to physiological function. Eventually, the above-ground parts exhibit stunting and result in a significant reduction in the yield of crops.

70.8 Biology and life cycle

The biology and life cycle are typical of the genus. Race A is automictic and reproduces by facultative meiotic parthenogenesis, whereas Race B and others are apomictic, reproducing by obligatory mitotic parthenogenesis. *M. hapla* is cryophilic and

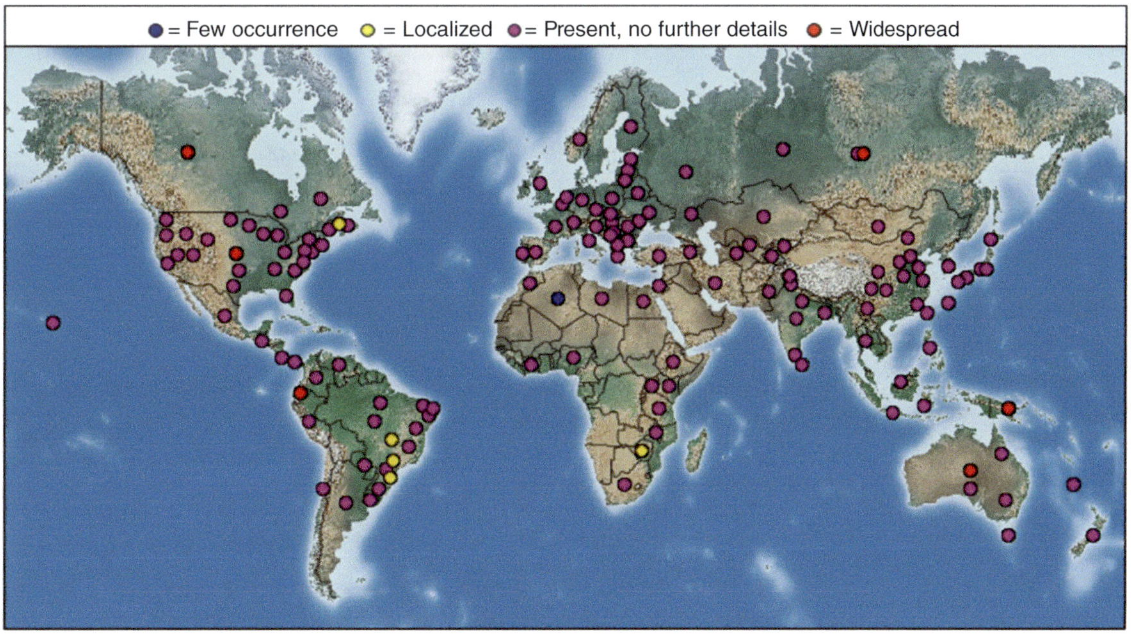

Fig. 9.19. Distribution map of *Meloidogyne hapla*. [Adapted from CABI-ISC, 2020]

Table 54. List of host plants of *Meloidogyne hapla* (adapted from CABI-ISC, 2020).

Main hosts
Actinidia chinensis (Chinese gooseberry)
Arachis hypogaea (groundnut)
Beta vulgaris var. *saccharifera* (sugar beet)
Cichorium intybus (chicory)
Daucus carota (carrot)
Fragaria ananassa (strawberry)
Glycine max (soybean)
Medicago sativa (lucerne)
Pelargonium (pelargoniums)
Rosa (roses)
Solanum lycopersicum (tomato)
Solanum tuberosum (potato)
Tanacetum cinerariifolium (pyrethrum)
Trifolium (clovers)
Other hosts
Allium cepa (onion)
Anemone (windflower)
Beta
Brassica napus var. *napus* (rape)
Brassica oleracea var. *capitata* (cabbage)
Cajanus cajan (pigeon pea)
Camellia sinensis (tea)
Capsicum annuum (bell pepper)
Chenopodium quinoa (quinoa)
Chrysanthemum (daisy)
Coffea (coffee)
Cucumis (melons, cucumbers, gherkins)
Cyclamen
Dianthus caryophyllus (carnation)
Dioscorea batatas (Chinese yam)
Fabaceae (leguminous plants)
Lactuca sativa (lettuce)
Linum usitatissimum (flax)
Mentha (mints)
Nicotiana tabacum (tobacco)
Olea europaea subsp. *europaea* (olive)
Phaseolus (beans)
Raphanus sativus (radish)
Rubus (blackberry, raspberry)
Sinapis alba (white mustard)
Solanum (nightshade)
Vicia (vetch)
Vitis vinifera (grapevine)

survives in the soil where temperatures are below 10°C (Moens *et al.*, 2009).

70.9 Economic importance

M. hapla parasitizes nearly all economically important temperate crops. For example, carrot, groundnut, lucerne, onion, potato, pyrethrum, sugar beet and strawberry are severely affected (Luc *et al.*, 1990). The nematode is well capable of causing substantial reductions in yield, or even total crop loss in the field as well as in the greenhouse (Evans *et al.*, 1993).

70.10 Movement and means of dispersal

See section 57.11, above.

70.11 Pest risk analysis and looming threat from introduction

M. hapla is polyphagous and has worldwide distribution, but there is no legislative measure to check the introduction of this nematode. It attacks economically important crops, so there would be a high chance of dispersal of this nematode from infested areas to new areas through nursery stock and contaminated soil. Hence, regulatory programmes must be applied to check its further spread from infested areas to new areas or regions.

70.12 Invasiveness rating

Considering the economic damage and pest risk analysis of *M. hapla*, the following invasiveness rating has been assigned.

Climate/host interaction	–	High (3)
Hosts range	–	High (3)
Dispersal potential	–	High (3)
Economic impact	–	High (3)
Environmental impact	–	High (3)
Post-entry distribution	–	High (–3)
Final rating (score)	–	High (12)

Based on the above invasiveness rating, *M. hapla* receives a score of 12 out of 15 scores and is rated as HIGHLY INVASIVE NEMATODE.

70.13 Region-wise status of invasiveness

M. hapla is a highly invasive nematode, but the phytosanitary risk is low because it is widely distributed and detected in almost every temperate country. There are no quarantine measures for this nematode.

70.14 Management measures

See section 56.7, above.

71 *Meloidogyne incognita*

71.1 Common name

Southern root-knot nematode

71.2 Scientific name

Meloidogyne incognita (Kofoid and White, 1919) Chitwood, 1949

71.3 Synonyms

Heterodera incognita (Kofoid and White, 1919) Sandground, 1923
Meloidogyne acrita Chitwood, 1949
Meloidogyne elegans da Ponte, 1977
Meloidogyne grahami Golden and Slana, 1978
Meloidogyne incognita acrita Chitwood, 1949
Meloidogyne incognita grahami Golden and Slana, 1978
Meloidogyne incognita incognita (Kofoid and White, 1919) Chitwood, 1949
Meloidogyne incognita wartellei Golden and Birchfield, 1978
Meloidogyne kirjanovae Terenteva, 1965
Meloidogyne wartellei Golden and Birchfield, 1978
Oxyuris incognita Kofoid and White, 1919

71.4 EPPO code

MELGIN

71.5 Diagnosis

71.5.1 Morphological diagnosis

Morphological descriptions of *Meloidogyne incognita* from Hunt and Handoo (2009) are summarized below. For measurements, see Table 52.

FEMALE: Pear-shaped, no posterior protuberance at terminal end. Stylet medium long with rounded basal knobs, offset (Fig. 9.20A). Cuticular perineal pattern rounded to oval, typically with raised, squared dorsal arch. Striae generally wavy, lateral field weakly demarcated by forked striae or absent (Fig. 9.20K).
MALE: Cephalic region not offset, lip disc elevated and lateral lips usually absent. Stylet long with rounded to transversely elongate basal knobs, offset (Fig. 9.20F,G).
SECOND-STAGE JUVENILE: Hemizonid anterior or near to excretory pore, tail long with hyaline region and rounded tail tip (Fig. 9.20J).

71.5.2 Biochemical diagnosis

The unique esterase I1 type and malate dehydrogenase N1 type have been described for *M. incognita*. Variability of esterase activity is low, only one population not having the typical phenotype. The phenotype N1 of malate dehydrogenase is similar to that of *M. exigua*, *M. javanica* and a few populations of *M. arenaria* (Esbenshade and Triantaphyllou 1985a,b).

71.5.3 Molecular diagnosis

Various molecular approaches have been employed to characterize *M. incognita*. For primers, see Table 49.

71.6 Geographical distribution

M. incognita is cosmopolitan and distributed worldwide (Fig. 9.21). It is more common in tropical and subtropical countries. The nematode has been recorded from the following countries (CABI-ISC, 2020).

Africa: Algeria, Angola, Burkina Faso, Cameroon, Congo, Côte d'Ivoire, Egypt, Ethiopia, Gambia, Ghana, Guinea, Kenya, Liberia, Libya, Madagascar, Malawi, Mauritania, Mauritius, Morocco, Mozambique, Niger, Nigeria, Réunion, Senegal, Seychelles, Somalia, South Africa, Sudan, Tanzania, Tunisia, Uganda, Zambia, Zimbabwe.
Asia: Armenia, Bangladesh, Brunei, China, India, Indonesia, Iran, Iraq, Israel, Japan, Jordan, Kazakhstan, Lebanon, Malaysia, Mongolia, Myanmar, Nepal, Oman, Pakistan, Philippines, Saudi Arabia, Singapore, South Korea, Sri Lanka, Syria, Taiwan, Tajikistan, Thailand, Turkey, Turkmenistan, Uzbekistan, Vietnam, Yemen.
Europe: Albania, Belarus, Belgium, Bulgaria, Cyprus, Czech Republic, Estonia, France, Germany, Greece, Hungary, Iceland, Italy, Latvia, Lithuania, Macedonia, Malta, Moldova, Montenegro, Netherlands, Poland, Portugal, Romania, Russia, Serbia and Montenegro, Slovenia, Spain, Switzerland, Ukraine, United Kingdom, former Yugoslavia.
North America: Antigua and Barbuda, Barbados, Belize, Bermuda, Canada, Costa Rica, Cuba, Dominica, Dominican Republic, El Salvador, Guadeloupe, Guatemala, Haiti, Honduras, Jamaica, Martinique, Mexico, Montserrat, Nicaragua, Panama, Puerto Rico, Saint Lucia, Saint Vincent and the Grenadines, Trinidad and Tobago, USA.
Oceania: American Samoa, Australia, Fiji, Kiribati, New Caledonia, New Zealand, Niue, Norfolk Island, Papua New Guinea, Samoa, Solomon Islands, Tonga, Tuvalu, Vanuatu.
South America: Argentina, Bolivia, Brazil, Chile, Colombia, Ecuador, French Guiana, Guyana, Paraguay, Peru, Suriname, Uruguay, Venezuela.

71.7 Hosts

M. incognita is highly polyphagous, attacking and parasitizing both monocotyledons and dicotyledons (Hunt and Handoo, 2009). Four races (Race 1, Race 2, Race 3 and Race 4) are known to occur in the *M. incognita* population (Taylor and Sasser, 1978). Various hosts of this nematode are given in Table 55.

71.8 Symptoms

Above-ground symptoms are similar to other root-knot nematodes and include stunted growth, chlorotic leaves and wilting in patches of plants (Fig. 9.4). Damage to the root system may also occur without showing any above-ground symptoms. The nematode causes characteristic galls or knots on roots. Galls caused by *M. incognita* are warty in appearance or can become

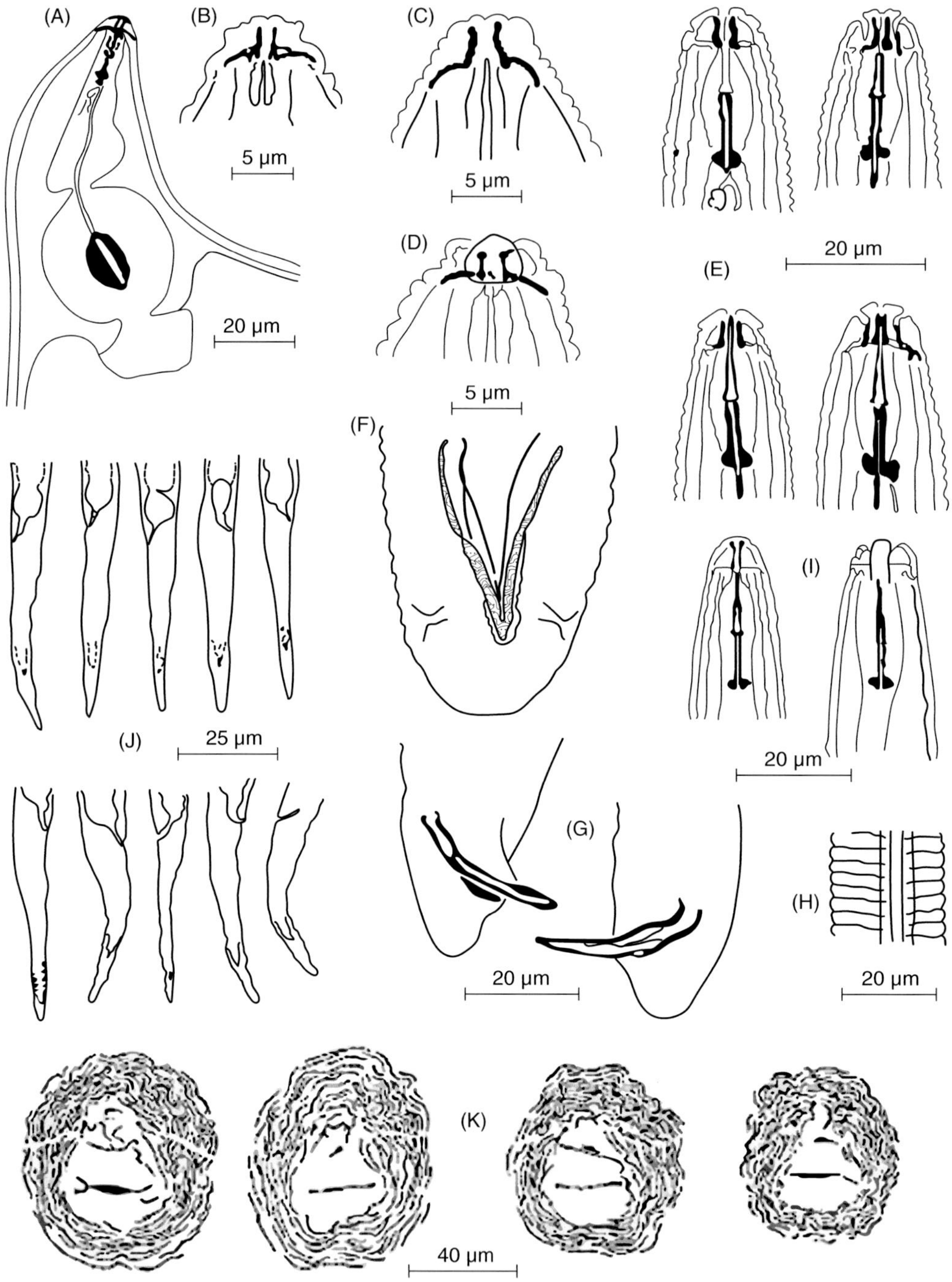

Fig. 9.20. Line drawing of *Meloidogyne incognita*. Female pharyngeal region (A); female anterior region (B–D); male anterior region (E); male tail region (F–G); male lateral field (H); J_2 anterior region (I); J_2 tail regions (J); perineal patterns (K). [A–J after Whitehead, 1968; courtesy Journal of Zoology, John Wiley and Sons. K after Orton Williams, 1973; courtesy of CAB International]

completely deformed on the root surface such as on tomato (Fig. 9.5A), pigeon pea (Fig. 9.5D), chickpea (Fig. 9.5F), cucurbits (Fig. 9.5I), eggplant (Fig. 9.5J), okra (Fig. 9.5K), broad bean (Fig. 9.5L), ornamental plants (Fig. 9.5M) and banana (Fig. 9.5U).

71.9 Biology and life cycle

The biology and life cycle are typical of the genus. *M. incognita* is apomictic and reproduces by obligatory mitotic parthenogenesis.

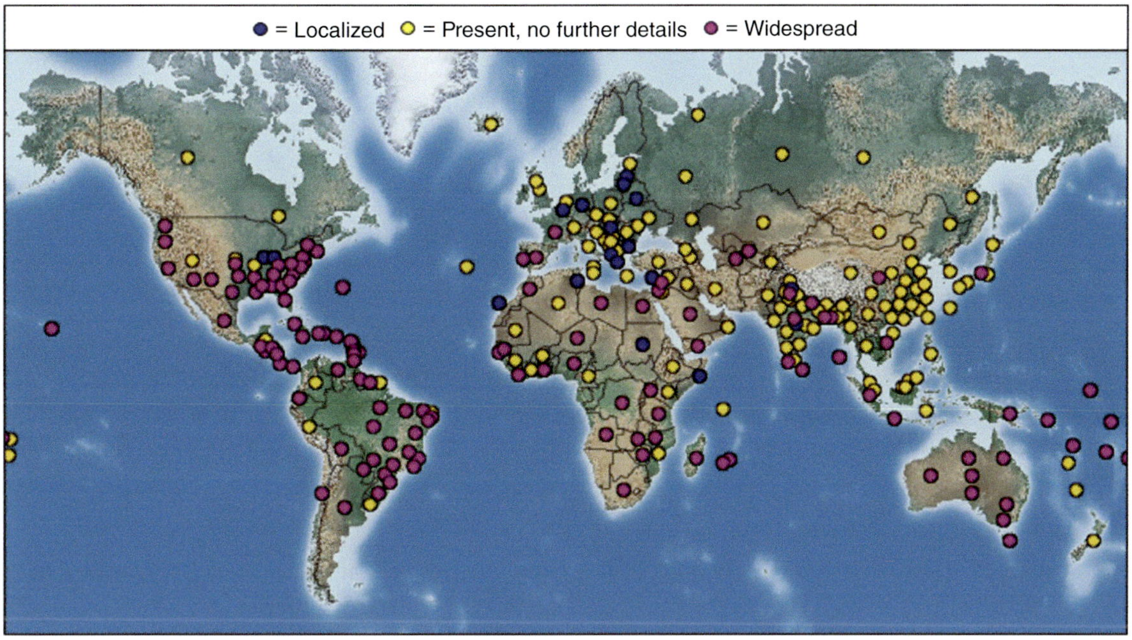

Fig. 9.21. Distribution map of *Meloidogyne incognita*. [Adapted from CABI-ISC, 2020]

Table 55. List of host plants of *Meloidogyne incognita* (adapted from CABI-ISC, 2020).

Main hosts
Abelmoschus esculentus (okra)
Acacia confusa
Brassica oleracea var. *gongylodes* (kohlrabi)
Capsicum annuum (bell pepper)
Carica papaya (papaw)
Coffea spp. (coffee)
Coffea arabica (arabica coffee)
Cucurbitaceae (cucurbits)
Fabaceae (leguminous plants)
Hibiscus cannabinus (kenaf)
Lactuca sativa (lettuce)
Luffa aegyptiaca (loofah)
Mangifera indica (mango)
Medicago sativa (lucerne)
Oryza sativa (rice)
Phaseolus (beans)
Phaseolus vulgaris (common bean)
Prunus salicina (Japanese plum)
Rosmarinus officinalis (rosemary)
Samanea saman (rain tree)
Solanaceae crops
Solanum lycopersicum (tomato)
Solanum melongena (aubergine)
Other hosts
Actinidia deliciosa (kiwifruit)
Amaranthus (grain amaranth)
Ananas comosus (pineapple)
Brassicaceae (cruciferous crops)
Calendula officinalis (pot marigold)
Canna
Cannabis sativa (hemp)
Chrysanthemum (daisy)
Cichorium (chicory)
Clitoria ternatea (butterfly-pea)
Coffea canephora (robusta coffee)
Colocasia
Cucumis auguria (gherkin)
Cucumis melo (melon)
Cucumis sativus (cucumber)
Curcuma alismatifolia
Cyperus (flatsedge)
Daucus carota (carrot)
Dioscorea (yam)
Dioscorea batatas (Chinese yam)
Dioscorea rotundata
Gossypium (cotton)
Helianthus annuus (sunflower)
Hemerocallis (daylilies)
Ipomoea batatas (sweet potato)
Jacquemontia pentantha
Jasminum sambac (Arabian jasmine)
Manihot esculenta (cassava)
Musa (banana)
Musa × *paradisiaca* (plantain)
Portulaca quadrifida
Prunus (stone fruit)
Saccharum officinarum (sugarcane)
Solanum tuberosum (potato)
Spinacia oleracea (spinach)
Vigna mungo (black gram)
Vigna radiata (mung bean)
Xanthosoma (cocoyam)
Zea mays (maize)

71.10 Economic importance

M. incognita is perhaps the most notorious plant-parasitic nematode pest in tropical and subtropical regions of the world (Sasser, 1989; Khan, 1998). It can severely damage a very wide range of agricultural crops and cause significant yield losses (Khan and Jairajpuri, 2010). Impacts on some important crops are described below;

VEGETABLES Vegetables like tomato, eggplant, chilli, okra, cucumber, melon, carrot, gourds, lettuce, etc. suffer severe damage due to *M. incognita* infection, especially in the tropics and subtropics. Estimated crop losses due to root-knot nematodes in major geographical regions of the tropics range from 5% to 43% or even greater (Sasser, 1989; Khan *et al.*, 2019a). In the areas where management practices are not in effect, average yield losses are assessed to be about 25%, while damage in the individual field may be as high as 60%. In tomato, brinjal and chilli, yield losses of 30–50%, 14–27% and 7–40% due to root-knot nematode infestation have been reported (Khan *et al.*, 2012a). On cabbage and cauliflower, the nematode may cause yield losses of 22–60% (Khan and Jairajpuri, 2012). In India, farmers may lose at least 10–25% of cucurbitaceous produce due to attack by *M. incognita* (Khan *et al.*, 2012b). In potato crops, losses mainly due to *M. incognita* are estimated at 25% or more (Devrajan *et al.*, 2012). *M. incognita* is also a serious pest in vegetable legumes such as cowpea, pea, French bean, broad bean, lima bean and other beans and causes yield losses of 5–30%, depending on nematode population, legume species and soil conditions (Khan *et al.*, 2019b).

FOOD LEGUMES *M. incognita* is one of the main economic pests of chickpea, green gram, black gram, lentil, pea, etc. in the tropics and subtropics (Sikora and Greco, 1990). Yield losses in food legumes depend upon nematode population levels, host and environmental conditions. Root-knot nematodes may inflict a 20–35% yield decline in pulses (Ali *et al.*, 2010).

SPICES *M. incognita* is frequently found a serious pest of black pepper (*Piper nigrum*) in Brazil and India (Koshy and Bridge, 1990). The nematode caused 32–47% yield losses to cardamom (*Elettaria cardamomum*) in India (Ali, 1986). *M. incognita* and *M. javanica* are important pests of ginger and cause yield losses of up to 57–74% (Bora and Choudhury, 2010).

COFFEE *M. incognita* is frequently found associated with the coffee plant and has its most catastrophic impact in Brazil, where farmers have been forced to leave the coffee crop and grow alternative crops (Campos *et al.*, 1990). The nematode also causes significant yield loss to coffee plantations in India (Subramanian *et al.*, 2010).

YAMS Damage caused by *M. incognita* to yams results in a significant reduction in marketable quality of the tubers (Jatala and Bridge, 1990). The nematode may reduce the price of tubers by 39–52% (Nwauzor and Fawole, 1982).

COTTON *M. incognita* Race 3 has become a major threat to cotton cultivation in the USA, with an estimated annual yield loss of about 2.2% (Blasingame and Patel, 2005). In Arkansas, yield losses were estimated at 1.5% annually with *M. incognita* (Kirkpatrick, 1989). In Brazil, a yield loss of 17% due to root-knot nematode alone was reported by Sasser (1979). In India, the nematode has been a major cause of damage to cotton (Banu, 2010b).

TOBACCO Root-knot nematodes are one of the economically important pests of tobacco, particularly where temperature and soil type favour the nematodes (Shepherd and Barker, 1990). *M. incognita* was estimated to cause yield losses of 50–60% in tobacco in some parts of Turkey (Shepherd and Barker, 1990). In Iraq, more than 40% of the tobacco crop was reported to be infested, with up to 100% infestation in the field and a 50% loss if the infestation started in the seedbed (Johnson and Chinheya, 2018). In India, yield losses of 25–30% were reported from field infestations (Khan *et al.*, 2010). In Cuba, losses were estimated at 27.5% in high-quality cigar tobacco (Fernandez Díaz-Silveira and Ortega Herrera, 1998).

71.11 Movement and means of dispersal

See section 57.11, above.

71.12 Pest risk analysis and looming threat from introduction

Because of the endemic nature of *M. incognita* in subtropical and tropical regions, quarantine regulations would be redundant and none are known to be in place. However, this nematode is polyphagous and damages almost all economically important crops, so there would be a high chance of dispersal of this nematode from infested areas to new areas through nursery stock and contaminated soil. The existence of four races further increases the risk of introduction in exotic areas. Hence, regulatory programmes must be applied to check its further spread from infested areas to new areas or regions.

71.13 Invasiveness rating

Considering the economic damage and pest risk analysis of *M. incognita*, the following invasiveness rating has been assigned.

Climate/host interaction	–	High (3)
Hosts range	–	High (3)
Dispersal potential	–	High (3)
Economic impact	–	High (3)
Environmental impact	–	High (3)
Post-entry distribution	–	High (–3)
Final rating (score)	–	High (12)

Based on the above invasiveness rating, *M. incognita* receives a score of 12 out of 15 and is rated as HIGHLY INVASIVE NEMATODE.

71.14 Region-wise status of invasiveness

M. incognita is a highly invasive nematode, but the phytosanitary risk is low because it is widely distributed and detected in almost every country. There are no quarantine measure for this nematode.

71.15 Management measures

See section 56.7, above.

72 *Meloidogyne javanica*

72.1 Common name

Javanese root-knot nematode; sugarcane root-knot nematode

72.2 Scientific name

Meloidogyne javanica (Treub, 1885) Chitwood, 1949

72.3 Synonyms

Anguillula javanica (Treub, 1885) Lavergne, 1901
Heterodera javanica Treub, 1885
Meloidogyne bauruensis Lordello, 1956b
Meloidogyne javanica bauruensis Lordello, 1956a
Meloidogyne javanica javanica (Treub, 1885) Chitwood, 1949
Meloidogyne lordelloi da Ponte, 1969
Meloidogyne lucknowica Singh, 1969
Tylenchus (*Heterodera*) *javanicus* (Treub, 1885) Cobb, 1890

72.4 EPPO code

MELGJA

72.5 Diagnosis

72.5.1 Morphological diagnosis

For measurements, see Table 52.

FEMALE: Pear-shaped, no posterior protuberance at terminal (Fig. 9.22E). Stylet long with ovoid basal knobs, offset (Fig. 9.22A). Perineal pattern rounded with low dorsal arch. Striae smooth, often distinct tail whorl. Lateral field distinct, clearly drawn from striae by more or less parallel lines (Fig. 9.22F).

MALE: Labial region continuous, not offset, lip disc not elevated. Lateral lips not present. Stylet long with ovoid basal knobs, offset (Fig. 9.22B).

SECOND-STAGE JUVENILE: Hemizonid anterior or near to excretory pore. Tail long with moderate hyaline region, tail tip finely rounded (Fig. 9.22D) (Hunt and Handoo, 2009).

72.5.2 Biochemical diagnosis

The esterase phenotype J3, J2 or J2a (Tomaszewski *et al.*, 1994; Castro *et al.*, 2003) and malate dehydrogenase phenotype N1 type (Esbenshade and Triantaphyllou, 1985a,b) are associated with *M. javanica*. The majority of *M. javanica* populations have a malate dehydrogenase phenotype N1, similar to that of *M. exigua*, *M. incognita* and a few populations of *M. arenaria*. One population of *M. javanica* from Bangladesh and one from Korea had the N3 phenotype, similar to some populations of *M. arenaria* (Hunt and Handoo, 2009).

72.5.3 Molecular diagnosis

Several molecular approaches are applied to characterize *M. javanica*. For primers, see Table 49.

72.6 Geographical distribution

M. javanica is distributed worldwide (Fig. 9.23), mainly in warm and tropical climates where the nematode is a predominant root-knot species. The nematode is known to occur in the following countries (CABI-ISC, 2020).

Africa: Algeria, Angola, Botswana, Burundi, Comoros, Congo Democratic Republic, Côte d'Ivoire, Egypt, Eritrea, Ethiopia, Gabon, Gambia, Ghana, Kenya, Liberia, Libya, Madagascar, Malawi, Mauritania, Mauritius, Morocco, Mozambique, Nigeria, Réunion, Rwanda, Senegal, South Africa, Sudan, Tanzania, Tunisia, Uganda, Zambia, Zimbabwe.
Asia: Armenia, Bangladesh, Brunei, China, India, Indonesia, Iran, Iraq, Israel, Japan, Jordan, Kazakhstan, Lebanon, Malaysia, Myanmar, Nepal, Oman, Pakistan, Philippines, Saudi Arabia, Singapore, South Korea, Sri Lanka, Syria, Taiwan, Tajikistan, Thailand, Turkey, Turkmenistan, Uzbekistan, Vietnam, Yemen.
Europe: Bosnia and Herzegovina, Bulgaria, Cyprus, France, Germany, Greece, Hungary, Iceland, Italy, Macedonia, Malta, Moldova, Poland, Portugal, Russia, Serbia and Montenegro, Spain, Ukraine, former Yugoslavia.
North America: Costa Rica, Cuba, El Salvador, Guadeloupe, Guatemala, Honduras, Jamaica, Martinique, Mexico, Nicaragua, Panama, Puerto Rico, Trinidad and Tobago, USA.
Oceania: Australia, Cook Islands, Fiji, Kiribati, Niue, Papua New Guinea, Samoa, Solomon Islands, Tonga.

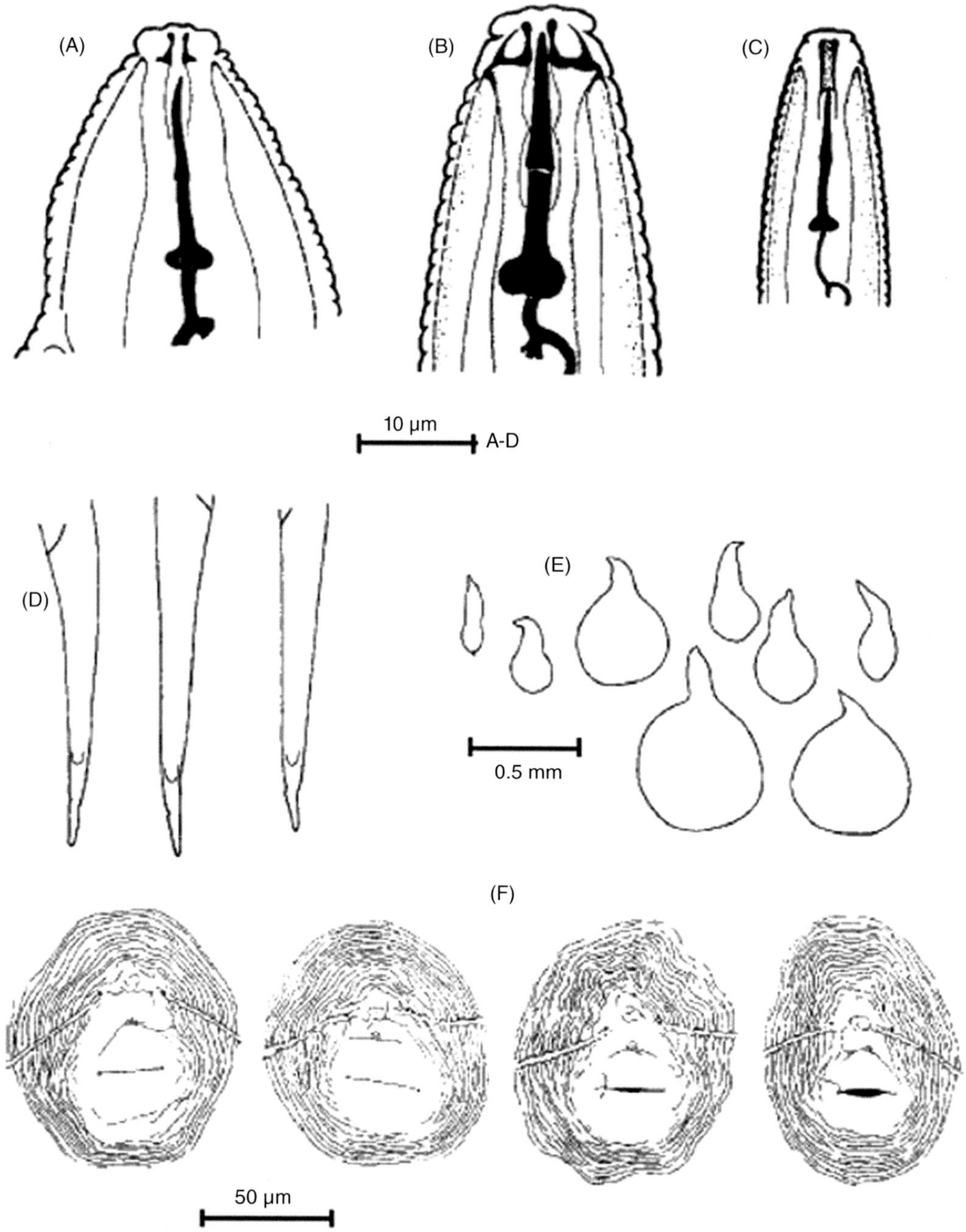

Fig. 9.22. Line drawing of *Meloidogyne javanica*. Female pharyngeal region, lateral view (A); male anterior region (B); J_2 anterior region (C); J_2 tail regions (D); entire females (E); perineal pattern variation (F). [A–D after Jepson, 1987 and E, F after Orton Williams, 1972; courtesy of CAB International]

South America: Argentina, Bolivia, Brazil, Chile, Colombia, Ecuador, Paraguay, Peru, Suriname, Uruguay, Venezuela.

72.7 Hosts

M. javanica is highly polyphagous and parasitizes both monocotyledons and dicotyledons. *M. javanica* is sympatric with other root-knot species (*M. arenaria*, *M. hapla* and *M. incognita*) and they all infect several common host crops (Moens *et al.*, 2009). An expanded list of hosts is presented in Table 56.

72.8 Symptoms

M. javanica is sympatric with *M. incognita* and other root-knot species. Symptoms resemble those of *M. incognita* infection (see section 71.8, above).

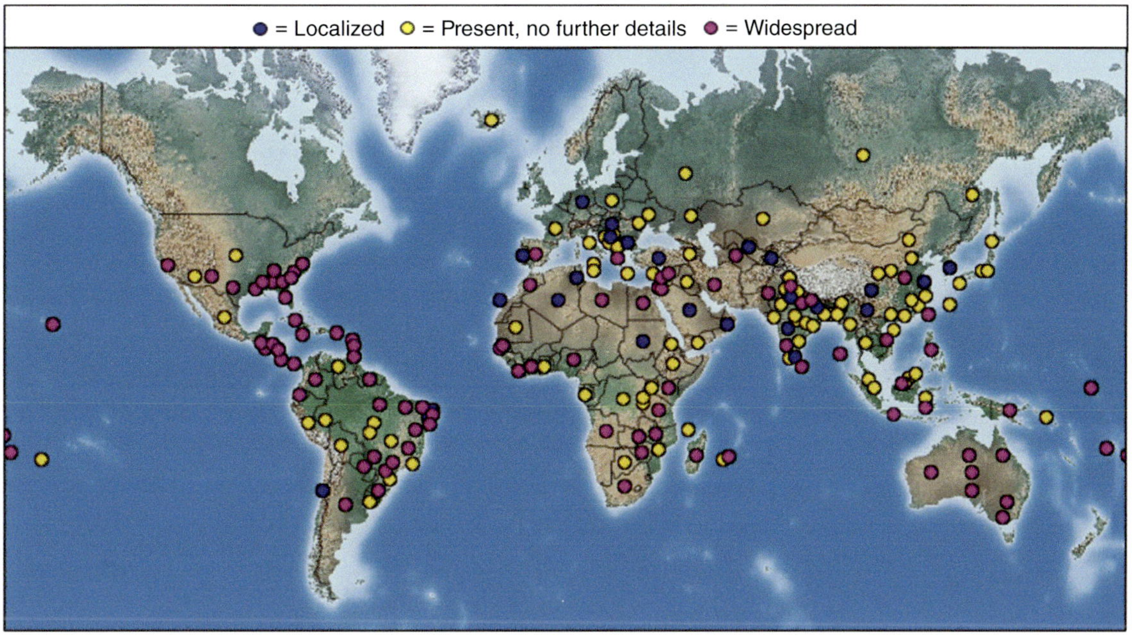

Fig. 9.23. Distribution map of *Meloidogyne javanica*. [Adapted from CABI-ISC, 2020]

Table 56. List of hosts affected by *Meloidogyne javanica* (adapted from CABI-ISC, 2020).

Main hosts
Capsicum annuum (bell pepper)
Cucurbitaceae (cucurbits)
Fabaceae (leguminous plants)
Nicotiana tabacum (tobacco)
Prunus salicina (Japanese plum)
Solanaceae crops
Solanum lycopersicum (tomato)
Solanum melongena (aubergine)
Solanum tuberosum (potato)
Other hosts
Abelmoschus esculentus (okra)
Amaranthus (grain amaranth)
Ananas comosus (pineapple)
Brassicaceae (cruciferous crops)
Canna
Cannabis sativa (hemp)
Chrysanthemum (daisy)
Cichorium (chicory)
Clitoria ternatea (butterfly-pea)
Colocasia
Cyperus (flatsedge)
Daucus carota (carrot)
Dioscorea (yam)
Gossypium (cotton)
Humulus lupulus (hops)
Ipomoea batatas (sweet potato)
Jasminum sambac (Arabian jasmine)
Manihot esculenta (cassava)
Musa × paradisiaca (plantain)
Oryza sativa (rice)
Prunus (stone fruit)
Raphanus (radish)
Saccharum officinarum (sugarcane)
Xanthosoma (cocoyam)

72.9 Biology and life cycle

The biology and life cycle are typical of the genus. *M. javanica* is apomictic and reproduces by obligatory mitotic parthenogenesis ($2n = 42–48$). *M. javanica* is thermophilic and does not survive for long periods in temperatures below 10°C (Moens *et al.*, 2009).

72.10 Economic importance

Most of the host crops that are susceptible to root-knot nematodes are probably susceptible to both *M. incognita* and *M. javanica* (Khan, 1998) and it is difficult to quantify the exact yield loss. However, usually over 30% yield losses in various vegetable crops occur due to infection with *Meloidogyne* spp. (Sikora and Fernandez, 2005; Khan *et al.*, 2019a). Besides vegetables, *M. javanica* is a common pest of pulse crops in various countries (Sikora *et al.*, 2005; Ali *et al.*, 2010; Khan *et al.*, 2019b) and is a notorious problem on tobacco in some parts of Africa and Asia (Khan *et al.*, 2010; Johnson and Chinheya, 2018). The nematode is also one of the main pests of pineapple in many countries (Mohilal and Khan, 2012).

72.11 Movement and means of dispersal

See section 57.11, above.

72.12 Pest risk analysis and looming threat from introduction

Because of wide distribution in tropical and subtropical regions, no quarantine measures against *M. javanica* are known to be in force in any specific region or country. This nematode is polyphagous and, like *M. incognita*, it severely attacks almost all economically important crops, so there would be a high chance of dispersal of this nematode from infested areas to new areas through nursery stock and contaminated soil. Hence, regulatory programmes must be applied to check its further spread from infested areas to new areas or regions.

72.13 Invasiveness rating

Considering the economic damage potential and pest risk analysis of *M. javanica*, the following invasiveness rating has been assigned.

Climate/host interaction	–	High (3)
Hosts range	–	High (3)
Dispersal potential	–	High (3)
Economic impact	–	High (3)
Environmental impact	–	High (3)
Post-entry distribution	–	High (–3)
Final rating (score)	–	High (12)

Based on the above invasiveness rating, *M. javanica* receives a score of 12 out of 15 and is rated as HIGHLY INVASIVE NEMATODE.

72.14 Region-wise status of invasiveness

M. javanica is a highly invasive nematode, but the phytosanitary risk is low because it is endemic in the tropics and subtropics region. There are no quarantine measures implemented for this nematode.

72.15 Management measures

See section 56.7, above.

73 *Meloidogyne mali*

73.1 Common name

Dutch elm root-knot nematode

73.2 Scientific name

Meloidogyne mali Itoh, Ohshima, and Ichinohe, 1969

73.3 Synonym

Meloidogyne ulmi (Palmisano and Ambrogioni, 2000) Ahmed *et al.*, 2013

73.4 EPPO code

MELGMA

73.5 Diagnosis

73.5.1 Morphological diagnosis

Morphological descriptions (after Itoh *et al.*, 1969; Palmisano and Ambrogioni, 2000; Ahmed *et al.*, 2013) are summarized below. For measurements, see Table 52.

FEMALE: Spherical with projecting neck (Fig. 9.24M). Stylet with dorsally curved cone, shaft straight and rounded to pear-shaped stylet knobs slightly backwardly sloping (Fig. 9.24I–L). Perineal pattern oval shape, dorsal arch rounded to square-shaped with distinct phasmids. Lateral field indistinct or marked by broken lines or with striae folds (Fig. 9.24U).
MALE: Stylet straight with backwardly sloping rounded knobs (Fig. 9.24N,O). Cephalic region weakly offset with low head cap. Post-labial incisures absent. Distance from stylet base to dorsal oesophageal gland orifice comparatively long.
SECOND-STAGE JUVENILES: Stylet knobs small with rounded and slightly backwards sloping (Fig. 9.24B,C). Hemizonid located behind excretory pore. Tail short, conical with small hyaline region with finely pointed tip (Fig. 9.24F–H).

73.5.2 Biochemical diagnosis

The esterase isozymes and malate dehydrogenase phenotype are used to identify young females of *M. mali* (Esbenshade and Triantaphyllou, 1985a).

73.5.3 Molecular diagnosis

M. mali has been characterized by sequencing of 18S rDNA (SSU) and 28S rDNA (LSU) (Okimoto *et al.*, 1991;

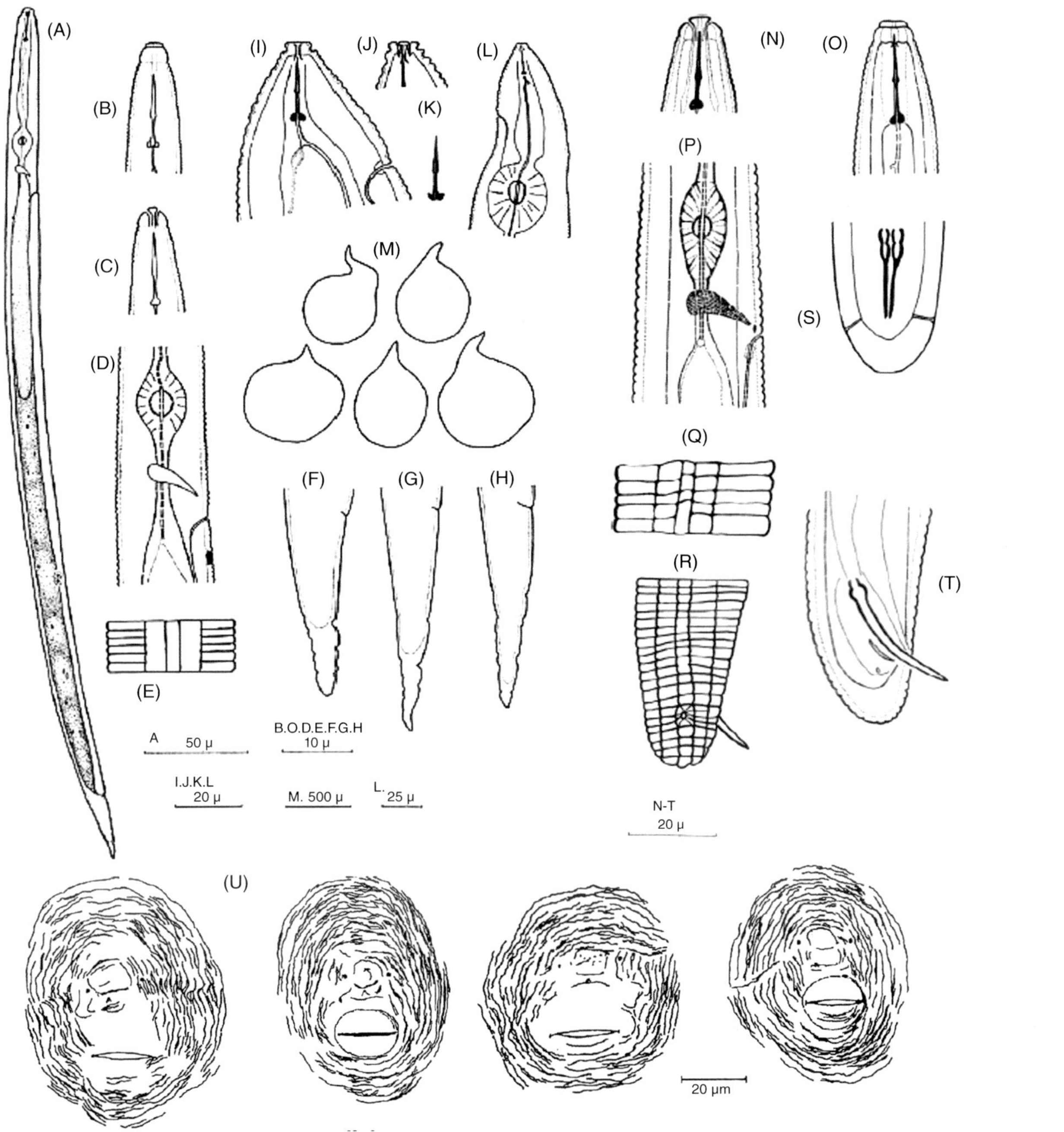

Fig. 9.24. Line drawing of *Meloidogyne mali*. Second-stage juvenile J_2 body (A); J_2 anterior region, lateral (B); J_2 anterior region dorsal (C); J_2 metacorpus region (D); J_2 lateral field (E); J_2 tails (F–H); females anterior region (I,J,L); females stylet (K); females body shape (M); male anterior region, lateral (N); male anterior region, dorsoventral (O); male metacorpus region (P); male lateral field (Q); male tail regions lateral, ventral, lateral, respectively (R–T); perineal pattern (U). [A–T after Itoh *et al.*, 1969, Applied Entomology and Zoology, Tokyo; U- after Palmisano and Ambrogioni, 2000, Nematologia Mediterranea; courtesy EPPO Bulletin]

Holterman *et al.*, 2009; Ahmed *et al.*, 2013). Zhou *et al.* (2017) developed a loop-mediated isothermal amplification with lateral flow device (LAMP-LFD) for rapid, highly sensitive and accurate diagnosis of *M. mali*. For primers, see Table 49.

73.6 Geographical distribution

M. mali was recorded for the first time from Japan in apple orchards (Itoh *et al.*, 1969) and is widespread in Japan. After that, it was detected in the Netherlands and Italy (Heybroek, 1993).

At present *M. mali* has been recorded in France, Italy, Japan, the Netherlands and the USA (EPPO, 2018).

73.7 Hosts

M. mali has been found to parasitize a large number of hosts, including trees, shrubs and herbaceous plants. Apple and Dutch elm (*Ulmus chenmoui*) are the main hosts of *M. mali* (EPPO, 2017a).

73.8 Symptoms

On young roots, the galls induced by *M. mali* are similar to those of other root-knot species (up to 0.5 cm in diameter). However, on older roots relatively large galls (1–2 cm in diameter) develop. Above-ground symptoms on the trees are non-specific and only noticeable in heavy infestations. The symptoms include early leaf fall and an overall reduction in growth. In heavily infested Dutch elm, the trees topple down during storms (EPPO, 2017a).

73.9 Biology and life cycle

The biology and life cycle are typical of the genus, but the mode of reproduction is different. Males are common and reproduce sexually through amphimixis (n = 12) (Janssen *et al.*, 2017b).

73.10 Economic importance

M. mali is a major pest of apple and Dutch elm. However, information on exact yield loss caused by this nematode is not known.

73.11 Movement and means of dispersal

See section 57.11, above.

73.12 Pest risk analysis and looming threat from introduction

Since the first report from Japan, *M. mali* has been detected in the Netherlands and is considered to have been introduced from Asia (Japan) as part of a breeding programme when large amounts of elm material (seeds, cuttings and sometimes rooted material) were brought into the country to develop disease resistance in elm. At the end of the programme, in 1992, the newly developed resistant rooted elm seedlings were sent to different European countries and probably spread the nematode (Heybroek, 1993).

73.13 Invasiveness rating

Considering the consequences of introduction and pest risk analysis of *M. mali*, the following invasiveness rating has been assigned.

Climate/host interaction	–	Medium (2)
Hosts range	–	Medium (2)
Dispersal potential	–	Medium (2)
Economic impact	–	Medium (2)
Environmental impact	–	Medium (2)
Post-entry distribution	–	Low (–1)
Final rating (score)	–	Medium (9)

Based on the above invasiveness rating, *M. mali* receives a score of 9 out of 15 and is rated as MEDIUM INVASIVE NEMATODE.

73.14 Region-wise status of invasiveness

M. mali has been added to the A2 pest list of quarantine organisms in the EPPO region (EPPO, 2018) to prevent further spread within Europe. Being a pest of apple and elm, which are important European crops, this nematode is considered as medium invasive for all the European and other countries where it is absent.

73.15 Management measures

See section 56.7, above.

74 *Meloidogyne naasi*

74.1 Common name

Root-knot nematode

74.2 Scientific name

Meloidogyne naasi Franklin, 1965

74.3 EPPO code

MELGNA

74.4 Diagnosis

74.4.1 Morphological diagnosis

For measurements, see Table 52.

FEMALE: Globular with slight posterior protuberance (Fig. 9.25F). Stylet long with ovoid and backwardly sloping basal knobs (Fig. 9.25E). Excretory pore near style knob level. Perineal pattern round to oval-shaped with prominent phasmid, relatively coarse striae without lateral lines (Fig. 9.25L).

MALE: Head offset from body, labial disc not elevated and lateral lips absent (Fig. 9.25B). Stylet long, basal knobs similar to female. Small vesicles usually present in anterior lumen lining of metacarpus.

SECOND-STAGE JUVENILE: Anterior metacarporial lumen lining with several small vesicles (Fig. 9.25H). Hemizonid anterior

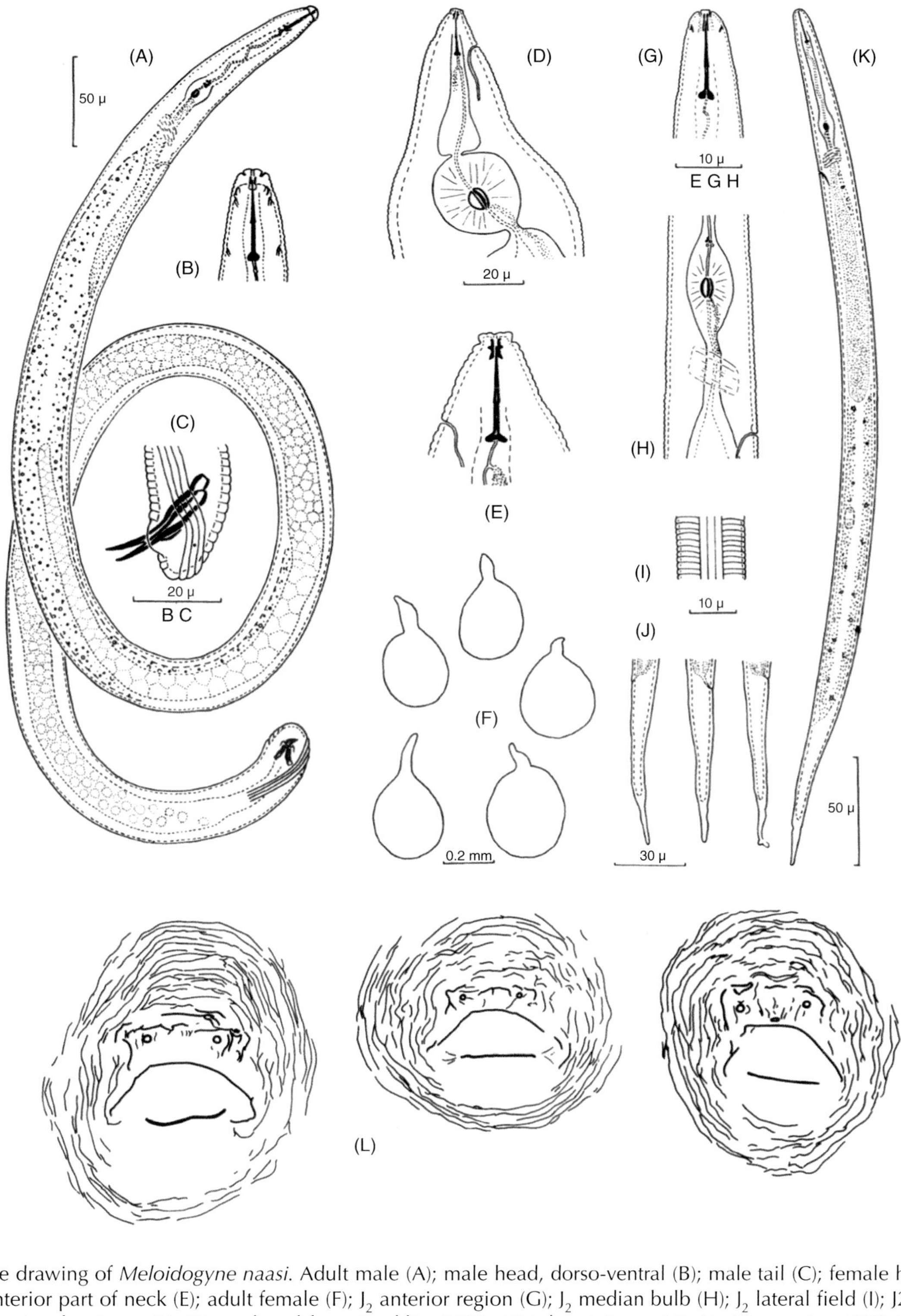

Fig. 9.25. Line drawing of *Meloidogyne naasi*. Adult male (A); male head, dorso-ventral (B); male tail (C); female head and neck (D); female anterior part of neck (E); adult female (F); J_2 anterior region (G); J_2 median bulb (H); J_2 lateral field (I); J2 tails (J); whole J_2 (K); perineal patterns (L). [Reproduced from Franklin, 1965, CIH description set 2, No.19; courtesy CAB International]

adjacent to the excretory pore. Tail long and slender with fine rounded tail tip and long hyaline region (Fig. 9.25J) (Karssen *et al.*, 2013).

74.4.2 Biochemical diagnosis

The esterase isozyme phenotype VF1 and malate dehydrogenase phenotype N1a are associated with *M. naasi* (Esbenshade and Triantaphyllou, 1985b).

74.4.3 Molecular diagnosis

For primers, see Table 50.

74.5 Geographical distribution

M. naasi is widely distributed in Europe. The nematode is also recorded in North and South America and occasionally in Asia (Karssen *et al.*, 2013).

74.6 Hosts

M. naasi reproduces on several grasses, including turfgrasses, golf courses, lawn grasses and some cereals, but rarely on dicotyledons.

74.7 Symptoms

M. naasi also produces small galls, but usually on the terminal or hooked or with lateral roots of infected plants (Karssen *et al.*, 2013).

74.8 Biology and life cycle

The biology and life cycle are typical of the genus. *M. naasi* is an automictic species reproducing through facultative meiotic parthenogenesis (Karssen *et al.*, 2013). The female is generally completely embedded within the root system (McClure *et al.*, 2012). *M. naasi* is cryophilic and survives at soil temperatures below 10°C (Moens *et al.*, 2009).

74.9 Economic importance

Turfgrass, lawn grasses, golf courses and sports fields are reported to be damaged by *M. naasi*. The nematode is one of the most common species from southern California to Washington state (USA) and was found in 58 golf courses (McClure *et al.*, 2012).

74.10 Movement and means of dispersal

See section 57.11, above.

74.11 Pest risk analysis and looming threat from introduction

M. naasi is widely distributed but confined to cooler regions. This nematode is a major problem in turfgrass, golf courses and lawn grasses, so there would be a high chance of dispersal of this nematode from infested areas to new areas through grass slips and contaminated soil. *M. naasi* should be targeted in regulatory programmes to check its further spread in other countries.

74.12 Invasiveness rating

Considering the consequences of probable introduction and pest risk analysis of *M. naasi*, the following invasiveness rating has been assigned.

Climate/host interaction	–	High (3)
Hosts range	–	Low (1)
Dispersal potential	–	High (3)
Economic impact	–	Medium (2)
Environmental impact	–	Low (1)
Post-entry distribution	–	Low (–1)
Final rating (score)	–	Medium (9)

Based on the above invasiveness rating, *M. naasi* receives a score of 9 out of 15 and is rated as MEDIUM INVASIVE NEMATODE.

74.13 Region-wise status of invasiveness

Due to the limited host range and distribution mainly in temperate regions, *M. naasi* is considered as medium invasive nematode for all the countries where it is absent.

74.14 Management measures

See section 56.7, above.

75 *Meloidogyne paranaensis*

75.1 Common name

Coffee root-knot nematode

75.2 Scientific name

Meloidogyne paranaensis Carneiro, Carneiro, Abrantes, Santos and Almeida, 1996

75.3 EPPO code

MELGPA

75.4 Diagnosis

75.4.1 Morphological diagnosis

For measurements, see Table 52.

FEMALE: Ovoid to pear-shaped, without posterior terminal protuberance. Stylet long with broad basal knobs, offset (Fig. 9.26A). Perineal pattern rectangular to oval, high dorsal arch; striae fine/coarse, smooth/wavy. Lateral field absent (Fig. 9.26F).

MALE: Labial region continuous, not offset, labial disc elevated, lateral lips not present (Fig. 9.26B). Stylet long with round to transversally elongate basal knobs, offset.

SECOND-STAGE JUVENILE: Hemizonid anterior to excretory pore. Tail long with rounded tail tip and medium hyaline region (Fig. 9.26D,E) (Hunt and Handoo, 2009).

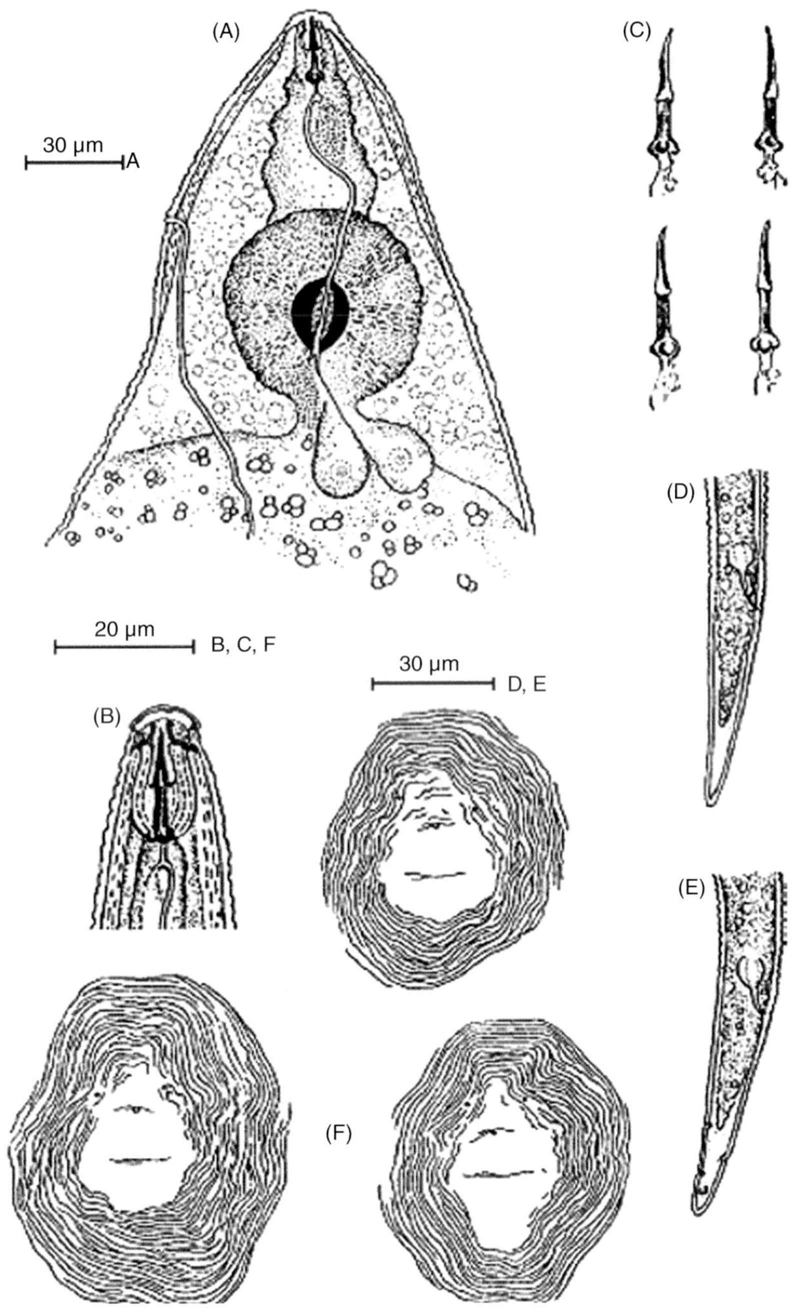

Fig. 9.26. Line drawing of *Meloidogyne paranaensis*. Female pharyngeal region, lateral view (A); male anterior region, lateral view (B); female stylets (C); J_2 tail region (D,E); perineal patterns (F). [Reproduced from Hunt an Handoo, 2009; modified after Carneiro *et al.*, 1996; courtesy of Journal of Nematology]

75.4.2 Biochemical diagnosis

The esterase phenotype F1 has one fast-migrating band of *M. paranaensis* and the malate dehydrogenase phenotype is N1 (Carneiro *et al.*, 2000).

75.4.3 Molecular diagnosis

For primers, see Table 49.

75.5 Geographical distribution

M. paranaensis is recorded from Brazil (Carneiro *et al.*, 1996), the USA and Guatemala (Moens *et al.*, 2009)

75.6 Hosts

Coffee (*Coffea* spp.) is the main host of *M. paranaensis*. Besides coffee, the nematode reproduces well on soybean, tobacco (*Nicotiana tabacum*), tomato (*Lycopersicon esculentum*), watermelon (*Citrullu lanatus*) and mate (*Ilex paraguariensis*). It also parasitizes other species of Solanaceae, Cucurbitaceae and Aquifoliaceae plants (Hunt and Handoo, 2009).

75.7 Symptoms

Plants infected by *M. paranaensis* display cracking and splitting of roots without apparent galling (Carneiro *et al.*, 1996). Root necrosis also appears where females have induced giant cells. Above-ground symptoms include yellowing of leaves and plant growth decline, leading to death (Moens *et al.*, 2009).

75.8 Biology and life cycle

The biology and lifecycle are typical of the genus. *M. paranaensis* is apomictic and reproduces by mitotic parthenogenesis (Karssen *et al.*, 2013).

75.9 Economic importance

Infestation by *M. paranaensis* leads to up to 50% yield suppression, which is usually associated with dieback. In the Paraná state of Brazil, damage on coffee was attributed erroneously to *M. incognita* (Carneiro *et al.*, 1996).

75.10 Movement and means of dispersal

See section 57.11, above.

75.11 Pest risk analysis and looming threat from introduction

M. paranaensis has a wide host range and is a potential pest of coffee. The economic damage caused by the nematode in Brazil instigated a complete pest risk analysis for all the coffee-growing regions and to evaluate the consequences of its introduction and probable impact on other host crops.

75.12 Invasiveness rating

Considering the economic damage potential and pest risk analysis of *M. paranaensis*, the following invasiveness rating has been assigned.

Climate/host interaction	–	Medium (2)
Hosts range	–	Medium (2)
Dispersal potential	–	Medium (2)
Economic impact	–	High (3)
Environmental impact	–	Medium (2)
Post-entry distribution	–	Low (–1)
Final rating (score)	–	Medium (10)

Based on the above invasiveness rating, *M. paranaensis* receives a score of 10 out of 15 and is rated as MEDIUM INVASIVE NEMATODE.

75.13 Region-wise status of invasiveness

Because of the wide distribution, extensive host range and economic damage caused by this nematode in Brazil, *M. paranaensis* has been given a medium invasiveness rating for all the coffee-producing countries (see the country list in section 63.13, above) where this nematode is absent.

75.14 Management measures

See section 56.7, above.

10

Pratylenchidae: *Achlysiella*, *Hirschmanniella*, *Nacobbus*, *Pratylenchus*, *Radopholus* and *Zygotylenchus* Species

The family Pratylenchidae contains six agriculturally important genera: *Achlysiella*, *Hirschmanniella*, *Nacobbus*, *Pratylenchus*, *Radopholus* and *Zygotylenchus*.

Achlysiella Spp.

Achlysiella williamsi was originally described as *Radopholus similis*, parasitizing sugarcane in Mauritius (Siddiqi, 1964b). Hunt *et al.* (1989) created a new genus, *Achlysiella*, and accommodated *Radopholus williamsi*, which has some distinguishing morphological and biological features that separate it from *Radopholus* species. *A. williamsi* is a potential nematode pest of sugarcane in tropical countries, particularly in Pacific regions.

Systematic position of *Achlysiella* species

	Classical (after Siddiqi, 2000)	Molecular (after Decraemer and Hunt, 2013)
Phylum	Nematoda	Nematoda
Class	Secernentea	Chromadorea
Order	Tylenchida	Rhabditida
Suborder	Hoplolaimina	Tylenchina
Infra-order	–	Tylenchomorpha
Superfamily	Holpolaimoidea	Tylenchoidea
Family	Pratylenchidae	Pratylenchidae
Subfamily	Pratylenchinae	Radopholinae
Genus	*Achlysiella*	*Achlysiella*
Species	*A. williamsi*	*A. williamsi*

76 *Achlysiella williamsi*

76.1 Common name

Lesion nematode

76.2 Scientific name

Achlysiella williamsi (Siddiqi, 1961) Hunt, Bridge and Machon, 1989

76.3 Synonyms

Radopholus similis apud. Williams, 1959
Radopholus williamsi Siddiqi, 1961

76.4 EPPO code:

ACLYWI

76.5 Diagnosis

The vermiform stages of *A. williamsi* morphologically resemble with *Radopholus* species. However, the mature female is sausage-shaped and resembles with *Rotylenchulus* females.

76.5.1 Morphological diagnosis

The morphological and morphometric descriptions of *A. williamsi* are based on Siddiqi (1964b).

Immature vermiform female: Body slightly arcuate ventrally after heat relaxed. Cephalic region hemispherical, a little flattened apically with 3–4 annules. Head slightly offset. Cuticle annulated. Lateral field with four incisures, outer two crenate, usually without areolation. Stylet moderately strong with round basal knobs, often concave or flattened anteriorly. Dorsal oesophageal gland orifice 3–4 μm posterior to stylet base. Oesophagus with ellipsoid median bulb and long basal lobe. Oesophageal gland cells separate and well developed. Nerve ring behind bulb (Fig. 10.1-A). Hemizonid posterior to nerve ring, about 4–5 annules and directly anterior to excretory pore. Hemizonion 8–9 annules further posterior. Vulva slightly protuberant lip, post-median in position. Genital tract amphididelphic and outstretched (Fig. 10.1-B). Spermathecae prominent with rod-shaped sperm. Tail conoid, convex dorsally with smoothly rounded tip. Phasmids 2–9 (usually 7) annules posterior to anus.

DOI: 10.1079/9781789247367.0011

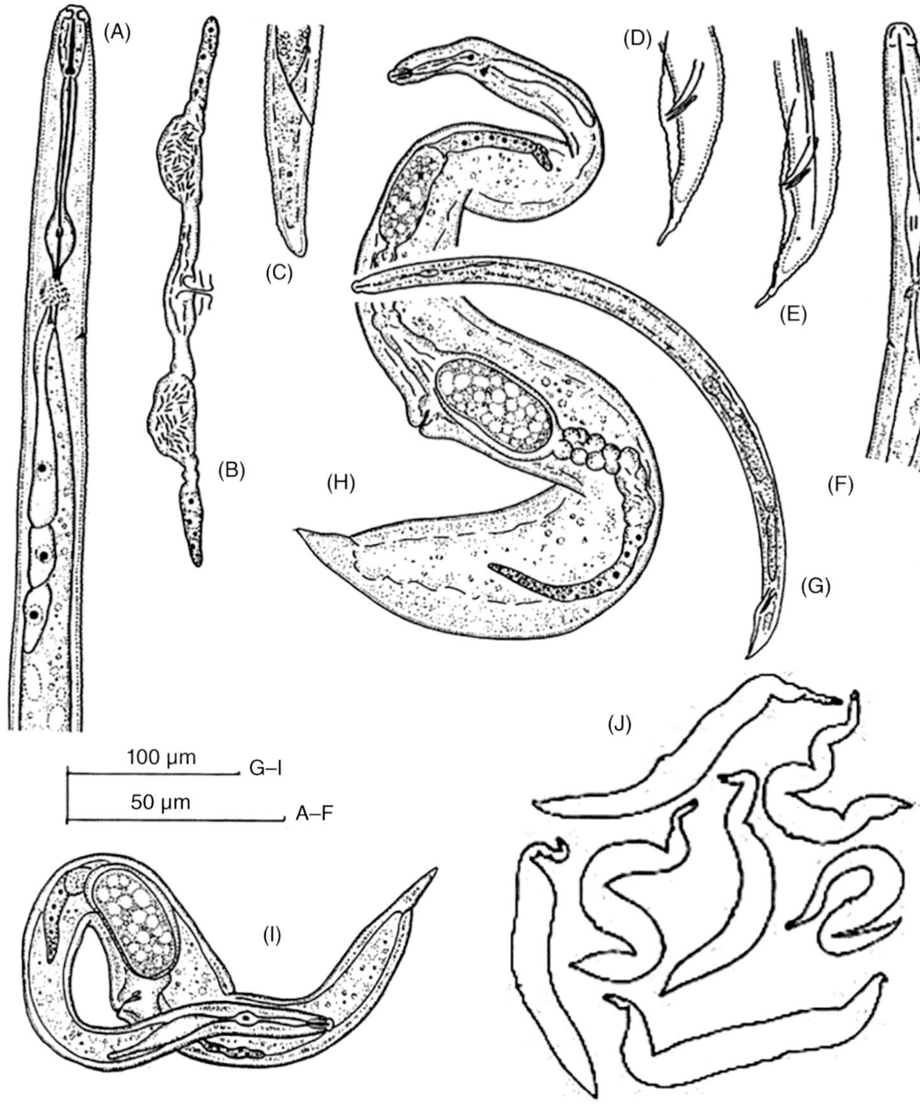

Fig. 10.1. Line drawing of *Achlysiella williamsi.* Oesophagus of vermiform immature female (A); genital tracts of vermiform immature female (B); tail of vermiform immature female (C); male tail (D,E); male oesophagus (F); entire male (G); mature swollen female (H–J). [Courtesy J.E. Machon/CABI BioScience]

Measurements: L = 0.35–0.45 mm; a = 22.4–30.2; MB = 6.5–8.0; b' = 2.6–3.4; c = 11.6–15.5; c' = 2.3–3.4; V = 60.3–63.8; body diameter = 13–20 μm; tail = 21–35 μm; anal body diameter = 9–11 μm; stylet = 13.1–15.4 μm; h = 5–7 μm.

MATURE SWOLLEN FEMALE: Body swollen, sausage-shaped (Fig. 10.1-H). Endo- or semi-endoparasitic in the root. Neck region slenderer than rest of body and mobile (Fig. 10.1-I). Vulval lips bulging strongly, reproductive system well developed (Fig. 10.1-H). Anterior genital tract extending up to oesophageal gland lobe. Posterior genital tract may extend to anus, reflexed one or more times.

Measurements: L = 0.56–0.89 mm; a = 8.1–11.2; V = 51–61; maximum diameter = 64–107 μm.

MALE: Body open C-shaped after heat relaxed (Fig. 10.1-G). Head distinctively offset with 5–6 annules and as high as wide. Stylet and oesophagus degenerate. Caudal alae crenate, reaching nearly to tail terminus (Fig. 10.1-D). Spicules arcuate, paired (Fig. 10.1-F). Gubernaculum simple and protrusible. Tail almost digitiform tip.

Measurements: L = 0.35–0.44 mm; a = 25.2–30.2; c = 11.2–13.6; c' = 2.7–3.3; body diameter = 12–17 μm; tail = 26–34 μm; anal body diameter = 10–12 μm; stylet = 9–11.4 μm; spicules = 13.7–15.4 μm; gubernaculum = 8–10 μm; h = 8–10.7 μm.

JUVENILES: Stylet fully developed but non-feeding. Frequently enclosed in one or two cast cuticles.

76.6 Geographical distribution

A. williamsi was originally reported from Mauritius (Williams, 1961). Subsequently, it has been recorded in many localities in Papua New Guinea and a wide area of the Pacific region such as Fiji, Queensland (Australia), Samoa, Tonga, Tuvalu and Vanuatu (De Waele and Elsen, 2007).

76.7 Hosts

The nematode is only known to infect sugarcane (*Saccharum officinarum*). This nematode also occurs in the rhizosphere of several uncultivated plants (Orton Williams, 1980).

76.8 Symptoms

A. williamsi does not produce swelling or galling on the root like sedentary nematodes and also the nematode does not form any permanent feeding sites or specialized trophic cells like other sausage-shaped nematode females, such as *Rotylenchulus*. The nematode induces necrotic lesions similar to those induced by pratylenchid nematodes; the lesions are purplish in colour and apparent on the root surface at the site of the swollen females (De Waele and Elsen, 2007). Partly or fully swollen mature females are found, surrounded by necrotic cells, within a cavity in the root cortex (Fig. 10.2B). Sometimes the posterior section of the nematode body can be exposed in smaller roots (Hunt *et al.*, 1989).

76.9 Biology and life cycle

Juveniles, males and immature females of *A. williamsi* are vermiform. The mature females become sausage-shaped on entering host roots and produce a gelatinous egg sac. Females lay 15–91 eggs within the sac. In distilled water, the eggs develop into vermiform adults in 14–16 days at 25°C. Moulting is usually superimposed and J_4 are enclosed in J_3 and J_2 cuticle. Vermiform juveniles and males, and sometimes inseminated females, are recorded in the soil. However, the females do not become sausage-shaped until they penetrate the root and start feeding (Hunt *et al.*, 1989). Eggs are laid during the final sedentary phase of the life cycle of the female.

76.10 Economic importance

The nematode is localized in the Pacific region, but no details of the direct economic impact of *A. williamsi* are available. De Waele and Elsen (2007) considered it as a potentially damaging species on sugarcane in tropical countries, particularly in the Pacific region.

76.11 Movement and means of dispersal

In general, the endoparasitic habits of Pratylenchids nematodes (including *A. williamsi*) allow them to be transported long distances inside infected roots, propagative plant materials and tubers, rhizomes and corms as eggs, juveniles and adults. Infected soil debris, run-off and irrigation water and human activity also disperse the nematode from infested to non-infested sites.

76.11.1 Plant parts known to carry the pest in trade/transport

Roots, bulbs, tubers, corms, rhizomes (sucker), growing medium accompanying plants, seedlings and micro-propagated plants.

76.11.2 Plant parts not known to carry the pest in trade/transport

Bark, flowers, inflorescences, cones, calyx, fruits (including pods), leaves, stems (above ground), shoots, trunks, branches, wood and true seeds (including grain).

76.12 Pest risk analysis and looming threat from introduction

A. williamsi is considered as a potential pest of sugarcane in the Pacific region and it is currently localized there. However, there is a high probability that it could infest sugarcane in other subtropical and tropical areas. *A. williamsi* can be easily spread through infested vegetative planting material such as sugarcane cuttings. A complete pest risk analysis is suggested for all the sugarcane-growing countries where this nematode is absent.

76.13 Invasiveness rating

Considering the probable consequences of introduction and pest risk analysis of *A. williamsi*, the following invasiveness rating has been assigned.

Climate/host interaction	–	Medium (2)
Hosts range	–	Low (1)
Dispersal potential	–	Medium (2)
Economic impact	–	Medium (2)
Environmental impact	–	Medium (2)
Post-entry distribution	–	Not known (0)
Final rating (score)	–	Medium (9)

Based on the above invasiveness rating, *A. williamsi* receives a score of 9 out of 15 and is rated as MEDIUM INVASIVE NEMATODE.

76.14 Region-wise status of invasiveness

Information on the host range and economic importance of this nematode is lacking. However, taking into consideration that it is a pest of sugarcane and the potential pathway of movement through infected sugarcane cuttings to various part of the world,

Fig. 10.2. Symptoms of *Achlysiella williamsi* (A,B), *Hirschmanniella miticausa* (C–E), *Hirschmanniella oryzae* (F,G) and *Nacobbus aberrans* (H,J). Ruptured egg sac attached to sugarcane root (A); swollen female in cavity of sugarcane root (B); patch of wilting taro in a swamp (C); early symptom of miti-miti disease showing red streaks and rotting on outer part of corm (D); late symptom showing complete decay of corm (E); infested rice plant showing patches of stunted growth and reduced ear head formation (F); stunted rice root system (G); beet seedling infected by the nematode (H); galls on tomato (I); beads like galls on root (J). [Photo courtesy AB: J.E. Machon, D.J. Hunt and J. Bridge/CABI BioScience; C-E: Helen Tsatsia and Grahame Jackson, Source Pacific Pests and Pathogens - Fact Sheets; G: Nemaplex, H. Ferris; H: R.N. Inserra; I: Carmen Triviño, INIAP, Ecuador; J: J. Bridge/CABI BioScience]

a medium invasive rating has been given for *A. williamsi* for all sugarcane-growing countries where this nematode is absent.

76.15 Management measures

No information on control of *A. williamsi* has been found, but the nematode would be susceptible to soil nematicides.

76.16 Phytosanitary measures

See Chapter 16, section 115.

Hirschmanniella Spp.

The genus *Hirschmanniella* belongs to the family Pratylenchidae and currently includes at least 33 species, with *Hirschmanniella spinicaudata* as holotype (Siddiqi, 2000; Khun *et al.*, 2015). Identification keys of *Hirschmanniella* spp. within the family Pratylenchidae were given by Razjivin *et al.* (1981), Ebsary and Anderson (1982), Sivakumur and Khan (1982) and Loof (1991). A recent polytomous key to the genus was given by Khun *et al.* (2015). Below is a list of all valid and some new species of *Hirschmanniella*.

Hirschmanniella anchoryzae Ebsary and Anderson, 1982
H. areolata Ebsary and Anderson, 1982
H. behningi (Micoletzky, 1923) Luc and Goodey, 1964
H. belli Sher, 1968
H. brassicae Duan, Liu, Liu and Zhao, 1996
H. caribbeana Van den Berg and Quénéhervé, 2000
H. caudacrena Sher, 1968
H. dicksoni Habteweld *et al.*, 2019
H. diversa Sher, 1968
H. furcata Razjivin, Fernandez, Ortega and Quincosa, 1981
H. gracilis (de Mann, 1880) Luc and Goodey,1964
H. halophila Sturhan and Hallmann, 2010
H. imamuri Sher, 1968
H. kwazuna Van den Berg, Subbotin, Handoo and Tiedt, 2009
H. loofi Sher, 1968
H. magna Siddiqi, 1966
H. marina Sher, 1968
H. mexicana (Chitwood. 1961) Sher, 1968
H. microtyla Sher, 1968
H. miticausa Bridge, Mortimer and Jackson, 1984
H. mucronata (Das, 1960) Luc and Goodey,1964
H. nghetinhiensis (Eroshenko and Chau) Eroshenko *et al.*, 1985
H. obesa Razjivin, Fernandez, Ortega and Quincosa, 1981
H. ornata (Eroshenko and Chau) Eroshenko *et al.*, 1985
H. oryzae (Van Breda de Haan, 1902) Luc and Goodey,1964
H. pisquidensis Ebsary and Pharoah, 1982
H. pomponiensis Abdel-Rahman and Maggenti, 1987
H. santarosae Tandingan De Ley *et al.*, 2007
H. shamimi Ahmad, 1972
H. spinicaudata (Schuurmans Stekhoven, 1944) Luc and Goodey, 1964
H. thornei Sher, 1968
H. truncata Razjivin, Fernandez, Ortega and Quincosa, 1981
H. zostericola (Allgén, 1934) Luc and Goodey, 1964

Hirschmanniella spp. are uniquely adapted to aquatic environments (Jeger *et al.*, 2018). Most are tropical and their main hosts are monocotyledonous aquatic plants. Around 60% of rice fields are infested with this genus, hence they are collectively called rice root nematodes (Duncan and Moens, 2013). Besides rice, the nematode also parasitizes diverse groups of plants such as taro (Bridge *et al.*, 1983), cabbage (Duan *et al.*, 1996), maize, sugarcane (Bridge *et al.*, 2005a), tomato (Tandingan *et al.*, 2007) and many aquatic plants, like lotus (Chong Bin *et al.*, 2001) and waterlily (Chitamber *et al.*, 2018). Several *Hirschmanniella* species pose a serious threat to lowland rice fields in Asia (Bridge *et al.*, 2005a; Jeger *et al.*, 2018). Among the above-listed species, *H. belli*, *H. diversa*, *H. miticausa*, *H. oryzae* and *H. spinicaudata* are considered for their invasiveness as they are of economic importance and associated with plant growth suppression in different agricultural crops (Khun *et al.*, 2015; Jeger *et al.*, 2018).

Systematic position of *Hirschmanniella* species

	Classical (after Siddiqi, 2000)	Molecular (after Decraemer and Hunt, 2013)
Phylum	Nematoda	Nematoda
Class	Secernentea	Chromadorea
Order	Tylenchida	Rhabditida
Suborder	Hoplolaimina	Tylenchina
Infra-order	–	Tylenchomorpha
Superfamily	Holpolaimoidea	Tylenchoidea
Family	Pratylenchidae	Pratylenchidae
Subfamily	Hirschmanniellinae	Pratylenchinae
Genus	*Hirschmanniella*	*Hirschmanniella*

77 *Hirschmanniella diversa*

77.1 Common name

Lotus root nematode

77.2 Scientific name

Hirschmanniella diversa Sher, 1968

77.3 EPPO code

HIRSDI

77.4 Diagnosis

77.4.1 Morphological diagnosis

Morphologically *Hirschmanniella diversa* is similar to other species of the genus. For measurements, see Table 57.

FEMALE: Body long, slender, 1.7–2.8 mm in length. No sexual dimorphism in lip region and oesophageal region. Cuticle with distinct annules. Lateral field with four incisures, areolated towards anterior and posterior end. Deirids absent. Lip region, not offset, composed of 3–7 annules. Stylet strong, 22–25 µm long, with prominent stylet knobs. Oesophageal glands ventrally overlapping intestine. Reproductive system paired, equally branched. Tail elongate-conoid, usually with terminal mucro.

MALE: Tail with crenate, subterminal bursa. Phasmids pore-like situated near tail tip (Fortuner and Maggenti, 1991; Sun *et al.*, 2013).

77.4.2 Molecular diagnosis

Habteweld *et al.* (2019) characterized *Hirschmanniella* spp. by sequencing of D2–D3 expansion segments of 28S and the ITS rRNA region. The DNA isolation and PCR amplifications were performed using the procedure described by Tanha Maafi *et al.* (2003) (see Chapter 4, section 15.5.2). The PCR products are purified and sequenced (GenBank accession numbers KY773577 for D2–D3 28S gene and KY773578 for ITS rRNA). Koyama *et al.* (2013) developed specific primers to identify *Hirschmanniella* spp. infesting lotus in Tokushima (Japan). PCR primers were designed based on the ITS regions specific to *H. diversa* (designated as divF and divR), *H. imamuri* (imaF and imaR) and both species (HdiF and HdiR).

77.5 Geographical distribution

H. diversa was first described from Hawaii (USA) by Sher (1968). The nematode was later reported from China (Hu *et al.*, 2004), Japan (Koyama *et al.*, 2013) and Vietnam (Jeger *et al.*, 2018).

77.6 Hosts

Lotus (*Nelumbium nelumbo*) is considered as the primary host of this nematode, as it was first isolated from lotus rhizomes (Sher, 1968). There are numerous reports of detection of *H. diversa* from roots of rice (*Oryza sativa*). Recently it has been recorded in waterlily in the USA (Chitambar *et al.*, 2018).

Table 57. Morphometric data for *Hirschmanniella* species.

Morphometric characters		Range, numerous populations			
		H. diversa (after Sher, 1968; Fortuner and Maggenti, 1991)[a]	*H. miticausa* (after Bridge *et al.*, 1983)	*H. oryzae* (after Sher,1968)	*H. spinicaudata* (after Luc and Fortuner, 1975)
Females	Length (µm)	1740–2850	1720 (1600–1860)	1440 (1140–1630)	2840 (2440–3016)
	Stylet (µm)	22–25	20 (19–21)	17 (16–19)	46 (44–48)
	a	51–87	53.5 (49.2–58.8)	60 (50–67)	56 (49–63)
	b	12–20	11.6 (10.5–12.6)	10.7 (8.8–12.1)	14.9 (12.7–17.0)
	b'	5.4–9.9	4.4 (3.8–4.8)	5.7 (4.5–7.2)	5.8 (5.0–7.6)
	c	15–20	17 (15.2–18.9)	17 (15–19)	23.9 (20.4–28.3)
	c'	4.8–6.8	4.1 (3.4–4.8)		3.3 (3.0–3.8)
	V	45–51	53.7 (49.4–57.2)	52 (50–55)	54.5 (51.5–58.5)
	m	–	48.5 (47–50)	–	–
	o	–	18.4–20.5	17 (15–19)	–
	Tail length (µm)	–	–	–	–
	Lips annuli number	5	5–6	4 (3–5)	5 (4–6)
Males	Length (µm)	–	1580 (1520–1650)	1170 (1010–1400)	2370 (2120–2640)
	Stylet (µm)	–	19 (18.5–20)	17 (16–18)	41.5 (40–44)
	a	–	52.1 (49.1–54.7)	59 (52–61)	56.2 (48.7–60.7)
	b	–	11.2 (10.7–12.3)	10 (9.1–11.3)	13.3 (11.3–15.8)
	b'	–	4.3 (3.7–4.6)	5 (4.6–5.7)	5.2 (4.6–6.3)
	c	–	18 (17–19.2)	17 (16–18)	20.3 (18.0–23.3)
	c'	–	4.5 (4.1–4.9)	–	4.7 (3.2–5.7)
	m	–	47 (44–50)	17 (13–18)	–
	o	–	19 (17–22)	–	–
	Spicules (µm)	30–39	33 (31–37)	23 (18–26)	48.5 (41–51)
	Gubernaculum (µm)	–	9.5 (9–11)	8 (7–9)	15 (13–16)

[a]From Khun *et al.*, 2015

77.7 Symptoms

There are no easily identifiable above-ground symptoms of damage from this nematode in the field. Reduction of plant growth occurs, especially when plants are attacked in the early stages. Yellowing of plants is occasionally observed and flowering can be delayed. Roots invaded by *H. diversa* turn yellowish-brown and rot (Peng *et al.*, 2018). On lotus roots, various life stages of this nematode were present in the root cortex cells, which were reduced in size (Koyama *et al.*, 2013; Uematsu *et al.*, 2019).

77.8 Biology and life cycle

H. diversa is an obligate migratory endoparasite of plant roots and it inhabits moist soils. In Japan, the fourth-stage juveniles or adults overwinter in lotus roots, weed roots, or soil. The overwintered adults mate during April to May and the females lay eggs in June. Second-stage juveniles emerge at the beginning of August and continue developing thereafter. The peak occurrence of *H. diversa* has been recorded in late August or early September and thereafter goes down. The seasonal breeding period, determined by quantifying the number of *H. diversa* females with spermatheca filled with sperm, peaks from April until mid-May and decreases subsequently (Takagi *et al.*, 2019). However, the length of the life cycle could be as short as one month and a tenfold increase in the number of individuals per generation is possible (Jeger *et al.*, 2018).

77.9 Economic importance

H. diversa is recognized as a major concern in lotus production, due to infestations, blackening and deforming lotus tubers; this reduces their economic value by more than US$1 million per year in Japan. Some farmers in Japan have even given up lotus cultivation because of serious and severe damage by this nematode (Takagi *et al.*, 2020).

77.10 Movement and means of dispersal

See section 76.11, above.

77.11 Pest risk analysis and looming threat from introduction

Considering the multi-host range of *H. diversa*, its damage potential and distribution in North America and Asia, a complete pest risk analysis for this nematode should be instigated. Due to its restricted distribution, preventive steps should be taken to avoid introduction into new host areas. The nematode can easily spread through nursery stock, infested tuber, soil and contaminated machinery.

77.12 Invasiveness rating

Based on the consequences of introduction and pest risk analysis of *H. diversa*, the following invasiveness rating has been given.

Climate/host interaction	–	Medium (2)
Hosts range	–	Low (1)
Dispersal potential	–	Medium (2)
Economic impact	–	Medium (2)
Environmental impact	–	Medium (2)
Post-entry distribution	–	Not known (0)
Final rating (score)	–	Medium (9)

Based on the above invasiveness rating, *H. diversa* receives a score of 9 out of 15 and is rated as MEDIUM INVASIVE NEMATODE.

77.13 Region-wise status of invasiveness

H. diversa is on the list of harmful organisms for South Korea and the A1 pest list for Turkey (PExD, 2019). The nematode is also recommended for the quarantine pest list in the European Union along with *H. imamuri*, *H. miticausa*, *H. mucronata*, *H. oryzae* and *H. spinicaudata* (Jeger *et al.*, 2018).

77.14 Management measures

Application of carbamate (Benfuracarb), organophosphate (Fenthion), nereistoxin (Cartap hydrochloride) and cyanamide had significant nematicidal effects against *H. diversa*. Among these, Benfuracarb at 1 μg/ml has been found most effective and is recommended to be used for *H. diversa* control in lotus cultivation (Takagi *et al.*, 2020). For phytosanitary measures, see Chapter 16, section 115.

78 *Hirschmanniella miticausa*

78.1 Common name

Miti-miti or taro nematode

78.2 Scientific name

Hirschmanniella miticausa Bridge, Mortimer and Jackson, 1983

78.3 EPPO code

HIRSMI

78.4 Diagnosis

78.4.1 Morphological diagnosis

The morphological and morphometric measurement descriptions from Bridge *et al.* (1983) are summarized below. For measurements, see Table 57.

Female: Body straight or slightly curved after heat fixation (Fig. 10.3A). Lateral field with four incisures with indistinct areolation at extremities. Cephalic region not offset, hemispherical with elevated lip disc. Head with prominent skeleton. Stylet with rounded knobs, anterior part slightly shorter or same as posterior. Hemizonid comprising 3–4 annules, anterior to oesophago-intestinal junction about 3–4 annules. Oesophageal glands elongate, ventrally overlapping intestine with posterior tapering. Intestine not overlapping rectum. Genital tract paired, outstretched. Tail tip non-annulated, with pointed ventral projection or rounded, occasionally with mucro (Fig. 10.3G). Phasmids indistinct, small, around one-third tail length.

Male: No sexual dimorphism except reproductive structures. Spicule robust, gubernaculum non-protruding. Tail with same variable terminus. Bursa not enveloping tail tip (Fig. 10.3C).

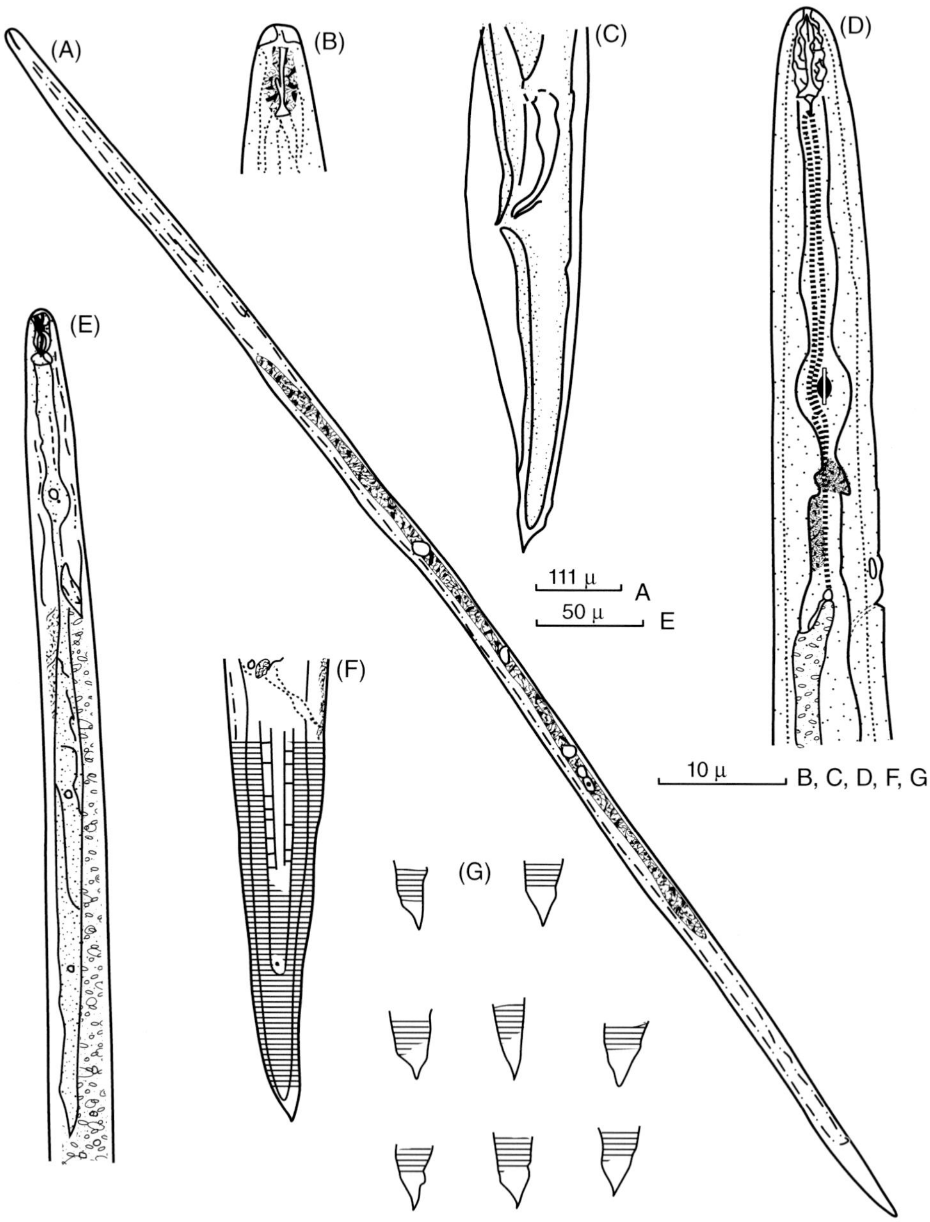

Fig. 10.3. Line drawing of *Hirschmanniella miticausa*. Whole female (A); male head (B); posterior part of male showing spicules and tail (C); anterior part of the body (D,E); posterior part of body (F); tail tips (G). [Courtesy Bridge *et al.*, 1983; Revue de Nematologie]

78.4.2 Molecular diagnosis

Molecular characterization of *H. miticausa* is lacking. However, the molecular approaches employed for other *Hirschmanniella* species may also be used for this nematode. For details, see sections 77.4.2, 79.5.2 and 80.5.2.

78.5 Geographical distribution

H. miticausa has only been found in Papua New Guinea and the Solomon Islands (Bridge, 1988). Reports by Jeger *et al.* (2018) on the occurrence of *H. miticausa* in Hawaii (USA) need to be verified.

78.6 Hosts

Taro (*Colocasia esculenta*) is the primary host for *H. miticausa*; other hosts are not known (Bridge *et al.*, 2005b).

78.7 Symptoms

H. miticausa causes the corm rot disease of taro known as 'miti-miti' disease. Only the roots and corms of taro are attacked by *H. miticausa* but secondary or referred symptoms are produced on the leaves. The initial foliar symptoms of miti-miti disease are wilting of the older leaves; these leaves become chlorotic, while the new central leaf remains straight instead of bending (Fig. 10.2C). Diseased plants die prematurely as a result of corm damage. When cutting longitudinally, diseased corms at first show red streaks which radiate from the base of the corm (Fig. 10.2D). These streaks later become irregular zones of dry brown rot, 1–10 mm wide, with the advancing diseased tissues remaining red. The basal portions of severely diseased corms are often completely decayed due to brown soft rot (Fig. 10.2E). The number of cormels is reduced in plants affected by the disease (Mortimer *et al.*, 1981; Bridge *et al.*, 2005b).

78.8 Biology and life cycle

H. miticausa is a migratory endoparasite and causes miti-miti disease of *C. esculenta* in dry-land soils, rainfed mountain slopes and in flooded swamp pits. The population of *H. miticausa* is highest in corms of infected plants and lower in the roots. Only a few nematode larvae occur in the surrounding soil (Jatala and Bridge, 1990). Nematodes are found in or in the area immediately around the red necrotic tissues of the basal portion of the corm. Relatively few larvae are found in the white tissues at the centre and are rarely in the crown (the top 1 cm). The nematode population commonly exceeds 1000/10 g of corm tissue and sometimes over 3000/10 g of corm tissue (Bridge *et al.*, 2005b).

78.9 Economic importance

H. miticausa is the causal organism of the corm rot disease of *Colocasia esculenta* known as 'miti-miti' in the Solomon Islands. Miti-miti disease has been reported from four islands in the Solomon Islands group (Mortimer *et al.*, 1981) and the highlands of Papua New Guinea (Bridge and Page, 1984). Miti-miti disease hampers the taro plant so much that the corms become inedible. In severe infestations, almost all the consumable corm tissue of the crop is destroyed. In some parts of the Solomon Islands, the disease is so devastating that the cultivation of *C. esculenta* has been almost entirely abandoned, particularly in areas where the crop was continuously cultivated in swamp pits (Bridge *et al.*, 2005b).

78.10 Movement and means of dispersal

See section 76.11, above.

78.11 Pest risk analysis and looming threat from introduction

The nematode is known to occur in four islands of the Solomon Islands and Papua New Guinea. A complete pest risk analysis is recommended for assessment of the looming threat posed by this nematode to other parts of the world, particularly where taro is cultivated.

78.12 Invasiveness rating

Considering the probable consequences of introduction and pest risk analysis of *H. miticausa*, the following invasiveness rating has been given.

Climate/host interaction	–	Medium (2)
Hosts range	–	Low (1)
Dispersal potential	–	Medium (2)
Economic impact	–	High (3)
Environmental impact	–	Medium (2)
Post-entry distribution	–	Not known (0)
Final rating (score)	–	Medium (10)

Based on the above invasiveness rating, *Hirschmanniella miticausa* receives a score of 10 out of 15 and is rated as MEDIUM INVASIVE NEMATODE.

78.13 Region-wise status of invasiveness

H. miticausa is listed as a quarantine pest in the Pacific island countries, South-East Asia and the European Union (Jeger *et al.*, 2018). *H. miticausa* attains medium invasiveness status in major taro-producing countries where it is absent. There may be trade restrictions if the nematode is introduced and established in these countries.

78.14 Management measures

The disease is, at present, restricted to those areas of the Pacific where *C. esculenta* is a subsistence crop. Hence effective management measures are required to control this menace. The best management practice is the use of nematode-free planting material, as the main source of inoculum is infested planting material. Nematodes can be eliminated from normal planting material (the corm top and 40 cm of leaf base), without damaging the tissues, by immersing in hot water at 50°C for 15 min (Mortimer *et al*., 1981). For phytosanitary measures, see Chapter 16, section 115.

79 *Hirschmanniella oryzae*

79.1 Common name

Rice root nematode

79.2 Scientific name

Hirschmanniella oryzae (van Breda de Haan,1902) Luc and Goodey,1964

79.3 Synonyms

Anguillulina apapillata (Imamura, 1931) Goodey, 1932
Anguillulina oryzae (van Breda de Haan, 1902) Goodey, 1932
Hirschmannia oryzae (vanBreda de Haan 1902) Luc and Goodey, 1962
Hirschmanniella abnormalis Renubala, Dhanachand and Gambhir, 1992
Hirschmanniella apapillata (Imamura, 1931) Siddiqi, 1986
Hirschmanniella exacta Kakar, Siddiqi and Khan, 1994
Hirschmanniella exigua Khan, 1972
Hirschmanniella nana Siddiqi, 1966
Radopholus oryzae (van Breda de Haan, 1902) Thorne, 1949
Rotylenchus apapillatus (Imamura, 1931) Filipjev,1936
Rotylenchus oryzae (van Breda de Haan, 1902) Filipjev and Schuurmans, 1941
Tylenchus apapillatus Imamura, 1931
Tylenchus oryzae van Breda de Haan, 1902

79.4 EPPO code

HIRSOR

79.5 Diagnosis

The presence of several synonymous species indicates that they are morphologically very similar. Hence morphological characters aided with molecular approaches are recommended for correct species identification of *H. oryzae*.

79.5.1 Morphological diagnosis

The present descriptions are based on Dey and Baqri (1990) and measurements after Sher (1968). For measurements, see Table 57.

FEMALE: Body slightly curved upon fixation (Fig. 10.4A). Body annules 1.2–2.0 μm apart in middle region. Lateral fields 5–7 μm or one-quarter to one-third of body width near middle, marked with four incisures of which outer ones crenate (Fig. 10.4I). In some specimens, incomplete areolation found in posterior region of body. Head continuous with body, flat at apex, marked with 3–4 annules, with well-developed cephalic framework. Stylet robust, about 1.8–2.2 times head width. Metanchium (anterior part of stylet) 7–8 μm long, 41–42% of stylet length. Stylet knobs rounded. Oesophagus typical of genus. Median oesophageal bulb at 50–60% of oesophageal length from anterior end. Oesophageal glands lobed, overlapping intestine ventrally. Orifice of dorsal oesophageal gland 2.5–3.5 μm below base of stylet. Nerve ring situated at 72–80% of oesophageal length from anterior extremity. Excretory pore at 112–131 μm of oesophageal length from anterior end, located slightly anterior to oesophago-intestinal junction. Thorneian cells joining intestinal cells with hypodermis also observed (Siddiqi, 1973a) (Fig. 10.4G). Hemizonid 2–4 annules long, situated 0–5 annules anterior to excretory pore. Vulva post-equatorial in position. Vagina about half of corresponding body width. Reproductive system amphidelphic. Spermatheca filled with sperms (Fig. 10.4J). Oocytes arranged in single row except in growth region. Tail cylindrical, with mucronate terminus (Fig. 10.4K–O), 5.5–6.5 anal body width long. Phasmid located in posterior half of tail, 26.4–42.8% of tail length from tail terminus.

MALE: Similar to female in general shape and morphology except reproductive system and tail shape. Bursa subterminal, gubernaculum slightly curved (Fig. 10.4F). Spicule 18–26 μm long when measured along median line. Tail slightly curved, 60–90 μm or 4.5–6 anal body width long, with pointed ventral mucro (Fig. 10.4F).

79.5.2 Molecular diagnosis

Khun *et al*. (2015) characterized all valid *Hirschmanniella* species, including *H. oryzae*, using three molecular markers: SSU, D2–D3 LSU and ITS1-5.8S-ITS2 rDNA. The study revealed that the D2–D3 region was the easiest, fastest and most successful region to be amplified. Based on phylogenetic analysis, different *Hirschmanniella* species were re-evaluated and *H. abnormalis* and *H. exacta* are considered to be junior synonyms of *H. oryzae*.

79.6 Geographical distribution

The rice root nematode *H. oryzae* is cosmopolitan in rice crops and has been reported from nearly all rice-growing countries (Fig. 10.5), such as Thailand, India, Pakistan, Philippines, Bangladesh, Cuba, Madagascar, Sri Lanka, the USA, East Asia (including China and Taiwan), New Zealand, Nigeria and Vietnam (Bridge *et al*., 2005a; Maung *et al*., 2010). *H. oryzae* is usually sympatric (co-occurring) on rice with a few more species throughout South and South-East Asia (Ghosh and Manna, 2010). The nematode is known to occur in the following countries (CABI-ISC, 2020).

Africa: Côte d'Ivoire, Egypt, Gambia, Ghana, Guinea, Madagascar, Mauritania, Niger, Nigeria, Senegal, Sierra Leone, South Africa.
Asia: Bangladesh, China, India, Indonesia, Iran, Japan, Malaysia, Myanmar, Nepal, Pakistan, Philippines, Singapore, South Korea, Sri Lanka, Taiwan, Thailand, Vietnam.

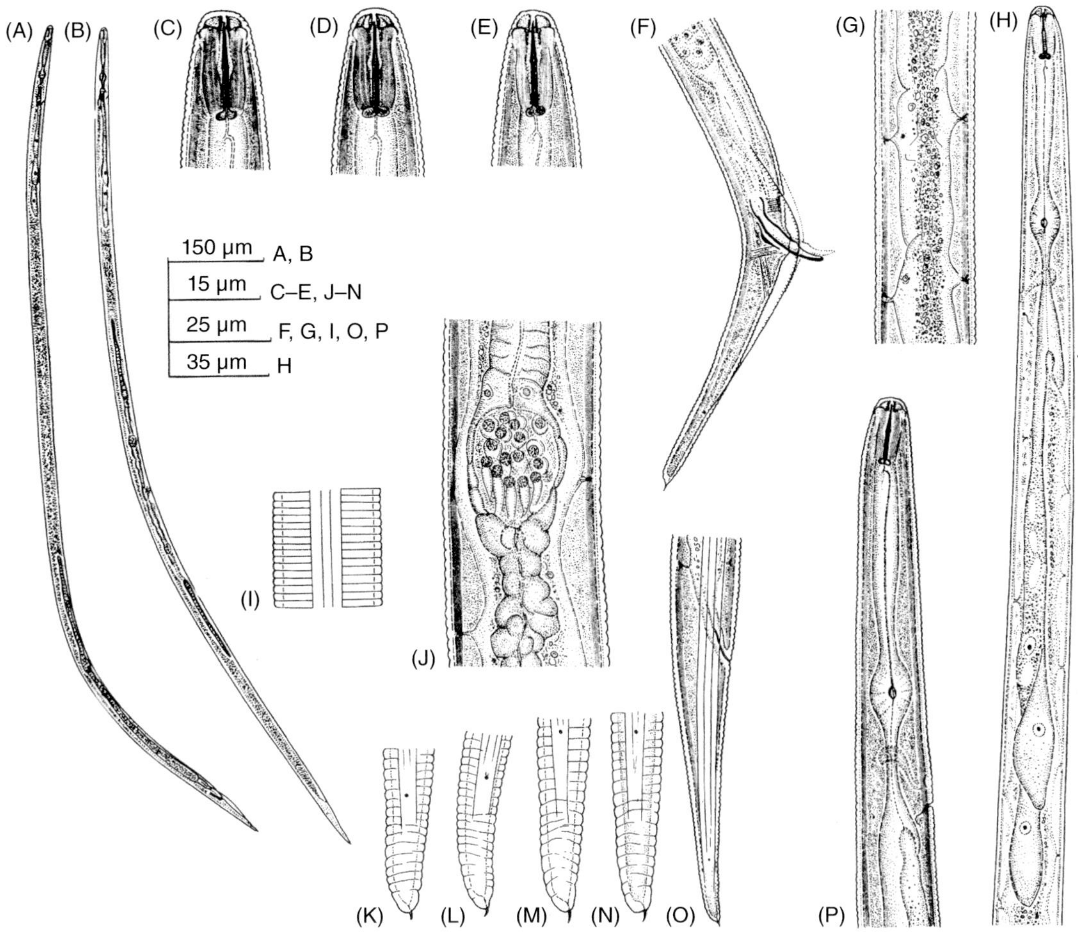

Fig. 10.4. Line drawing of *Hirschmanniella oryzae*. Adults (A,B); head region (C–E); tail region (F,K–O); mid-body showing 'Thorneian cells' joining intestinal cells with hypodermis (G); oesophageal region (H,P); lateral field at mid-body (I); spermatheca with sperms (J). [Reproduced from M. R. Siddiqi (2000), after Siddiqi, 1973a; courtesy CAB International]

Europe: Portugal
North America: Costa Rica, El Salvador, USA.
South America: Argentina, Brazil, Guyana, Venezuela.

79.7 Hosts

Rice (*Oryza sativa*) is a primary host of *H. oryzae*. However, *H. oryzae* has also been found to infect nearly 30 other plant species, mostly members of Cyperaceae and Gramineae (Anwar *et al.*, 1992; Ghosh and Manna, 2010). An extended list of hosts of *H. oryzae* is presented in Table 58.

79.8 Symptoms

Rice root nematodes (RRNs) do not cause specific symptoms on rice plants. In Java, Indonesia, they are associated with an important disease of rice known as mentek (Ghosh and Manna, 2010). The nematode attack is not easily identifiable; generally, foliar growth of plants is retarded. The infected plants show stunted growth with fewer and shorter tillers (Fig. 10.2F). Flowering is delayed and ear head numbers are decreased, with reduced dry weight. Roots show stunted growth (Fig. 10.2G) and sometimes a yellow to brown discoloration of the cortex with extensive necrosis and large cavities, which are formed due to internal migration and feeding of the nematode. Root necrosis does not occur in the absence of secondary pathogenic microorganisms, but it is greatly increased in the presence of *H. oryzae*. Mentek disease is a root rot, although the name is applied to the above-ground manifestations, i.e. reddening of leaves. Their abnormal colour indicates that the roots are not healthy and this can be confirmed on closer examinations.

79.9 Biology and life cycle

H. oryzae is a migratory endoparasitic nematode moving up and down within the rice roots. The adults and larvae enter roots at some distance from the tip and move in both directions, so that they may be distributed in the entire root, from the base

Fig. 10.5. Distribution map of *Hirschmanniella oryzae*. [Adapted from CABI-ISC, 2020]

Table 58. List of hosts of *Hirschmanniella oryzae* (adapted from CABI-CPC, 2010).

Main hosts
Oryza sativa (rice)
Other hosts
Abelmoschus esculentus (okra)
Gossypium hirsutum (Bourbon cotton)
Pennisetum glaucum (pearl millet)
Saccharum officinarum (sugarcane)
Solanum lycopersicum (tomato)
Triticum aestivum (wheat)
Zea mays (maize)
Wild hosts
Alternanthera sessilis (sessile joyweed)
Boerhavia diffusa (red spiderling)
Cyperus difformis (small-flower nutsedge)
Cyperus iria (rice flatsedge)
Cyperus rotundus (purple nutsedge)
Digitaria ciliaris (southern crabgrass)
Echinochloa colona (junglerice)
Echinochloa crus-galli (barnyard grass)
Eclipta prostrata (eclipta)
Eichhornia crassipes (water hyacinth)
Eleocharis spiralis
Eleusine indica (goosegrass)
Fimbristylis ferruginea
Fimbristylis globulosa (globular fimbristylis)
Fimbristylis littoralis (lesser fimbristylis)
Ischaemum rugosum (saramollagrass)
Leptochloa chinensis (Chinese sprangletop)
Leptochloa fascicularis (bearded sprangletop)
Ludwigia perennis
Monochoria vaginalis (pickerel weed)
Nelumbo nucifera (sacred lotus)
Paspalum distichum (knotgrass)
Polygonum plebeium (small knotweed)
Scirpus maritimus (saltmarsh bulrush)
Urochloa ramosa (browntop millet)
Vallisneria spiralis (eelweed)

to tip. The opening caused by the entrance of one individual is likely to be used by others, resulting in a concentration of nematodes in that particular part of the root. After entering, the nematode moves through the air channels between the radial lamellae of the parenchyma. A few days after the invasion, the females lay eggs inside the root and these hatch within 4–6 days (Bridge *et al.*, 2005a; Peng *et al.*, 2018). Development to the adult stage takes at least a month, passing through third-stage juveniles. The nematodes remain alive in the soil for more than 10 weeks in the absence of host plants but the mature eggs producing females cannot survive without a host, as they have not been reported from dry fields without the rice crop (Ghosh and Manna, 2010). *H. oryzae* requires high soil moisture levels. The nematode can survive in both high (25–45°C) and low (8–12°C) temperatures (Bridge *et al.*, 2005a). A rise in temperature during the paddy growing period has a significant correlation with the nematode population. The nematode can also survive in anaerobic condition in a wide range of pH (Babatola, 1981).

79.10 Economic importance

In general, *Hirschmanniella* species infest 58% of the world's rice fields, resulting in yield losses of nearly 25% (Bridge *et al.*,

2005a; Peng *et al.*, 2018). In Asia and some rice fields in Africa, *H. oryzae* alone is responsible for yield losses of 10–36% (Ghosh and Manna, 2010). In India, the yield loss due to different *Hirschmanniella* species on rice has been estimated as 25–35% (Ahmad *et al.*, 1984; Ghosh and Manna, 2010), in which RRN may induce up to 32% yield loss.

79.11 Movement and means of dispersal

See section 76.11, above.

79.12 Pest risk analysis and looming threat from introduction

Considering the importance of rice as host, the damaging potential of the nematode and its presence in all ecological zones, *H. oryzae* poses a high threat to rice cultivation in the countries where the nematode is still absent. Since the nematode is an endoparasite, it can easily spread through rice transplanting (seedlings). Preventive measures should be taken to check the introduction of this nematode into new host areas.

79.13 Invasiveness rating

Based on the consequences of introduction and economic impact of *H. oryzae*, the following invasiveness rating has been given.

Climate/host interaction	–	High (3)
Hosts range	–	Medium (2)
Dispersal potential	–	High (3)
Economic impact	–	High (3)
Environmental impact	–	Medium (2)
Post-entry distribution	–	Medium (2)
Final rating (score)	–	Medium (11)

Based on the above rating, *H. oryzae* receives a score of 11 out of 15 and is rated as MEDIUM INVASIVE NEMATODE.

79.14 Region-wise status of invasiveness

H. oryzae is listed as a quarantine pest in the European Union (Jeger *et al.*, 2018). The nematode attains medium invasiveness status in the following major rice-producing countries (FAOSTAT, 2019), where it is still absent.

APPPC: Afghanistan, Cambodia, Lao People's Democratic Republic
COSAVE: Colombia, Ecuador, Peru, Uruguay
EPPO: Italy, Russian Federation, Spain, Turkey
IAPSC: Mali, Tanzania
PPPO: Australia, New Zealand

79.15 Management measures

79.15.1 Cultural methods

Crop rotation of rice with Jalmasta jute (*Hibiscus subdariffa*) followed by 2-month fallow then rice reduced the soil population of *H. oryzae* by 50% (Ghosh and Manna, 2008). Soil amendments with fresh leaves of neem, *Sesbania aculeata* or water hyacinth or its compost at the rate of 60 kg N/ha were found effective to some extent in the management of *H. oryzae* populations and brought subsequent yield enhancement (Prasad *et al.*, 2004).

79.15.2 Host resistance

Use of resistant cultivars is the most effective and economic method of RRN control. Ghosh and Manna (2008) reported TKM–9 as a rice variety resistant to *H. oryzae*. The cultivars 'Annapurna', 'CR 52', 'Teenpakhia', 'N 136' and 'Basmati' 370, 385 have also been found moderately resistant to the nematode in India (Ghosh and Manna, 2010).

79.15.3 Chemical controls

Root dipping with carbofuran and soil application of non-fumigant nematicides like diazinon, fensulfothion, aldicarb, carbofuran, phenamiphos and phorate in a split dose of 1.5 kg a.i./ha at planting and on the 40th and 60th day after planting or 1 kg a.i./ha on the 30th day after transplanting significantly decreased the population of *H. oryzae* in a rice field (Muthukrishnan *et al.*, 1979; Ghosh and Manna, 2010).

79.15.4 Phytosanitary measures

See Chapter 16, section 115.

80 *Hirschmanniella spinicaudata*

80.1 Common name

Rice root nematode

80.2 Scientific name

Hirschmanniella spinicaudata (Schuurmans, 1944) Luc and Goodey

80.3 Synonyms

Hirschmannia spinicaudata (Schuurmans Stekhoven,1944) Luc and Goodey, 1962
Hirschmanniella lavabri (Luc, 1957) Sher, 1968
Radopholus lavabri Luc, 1957
Tylenchorhynchus spinicaudatus Schuurmans Stekhoven, 1944

80.4 EPPO code:

HIRSSC

80.5 Diagnosis

Hirschmanniella spinicaudata is usually sympatric with other *Hirschmanniella* species that infest rice crops. Hence precise identification using morphological and molecular techniques is suggested.

80.5.1 Morphological diagnosis

Morphological and morphometric descriptions after Luc and Fortuner (1975) are as below. For measurements, see Table 57.

FEMALE: Body slender, ventrally curved or straight after heat relaxation (Fig. 10.6A). Body annule transversely striated. Lateral fields

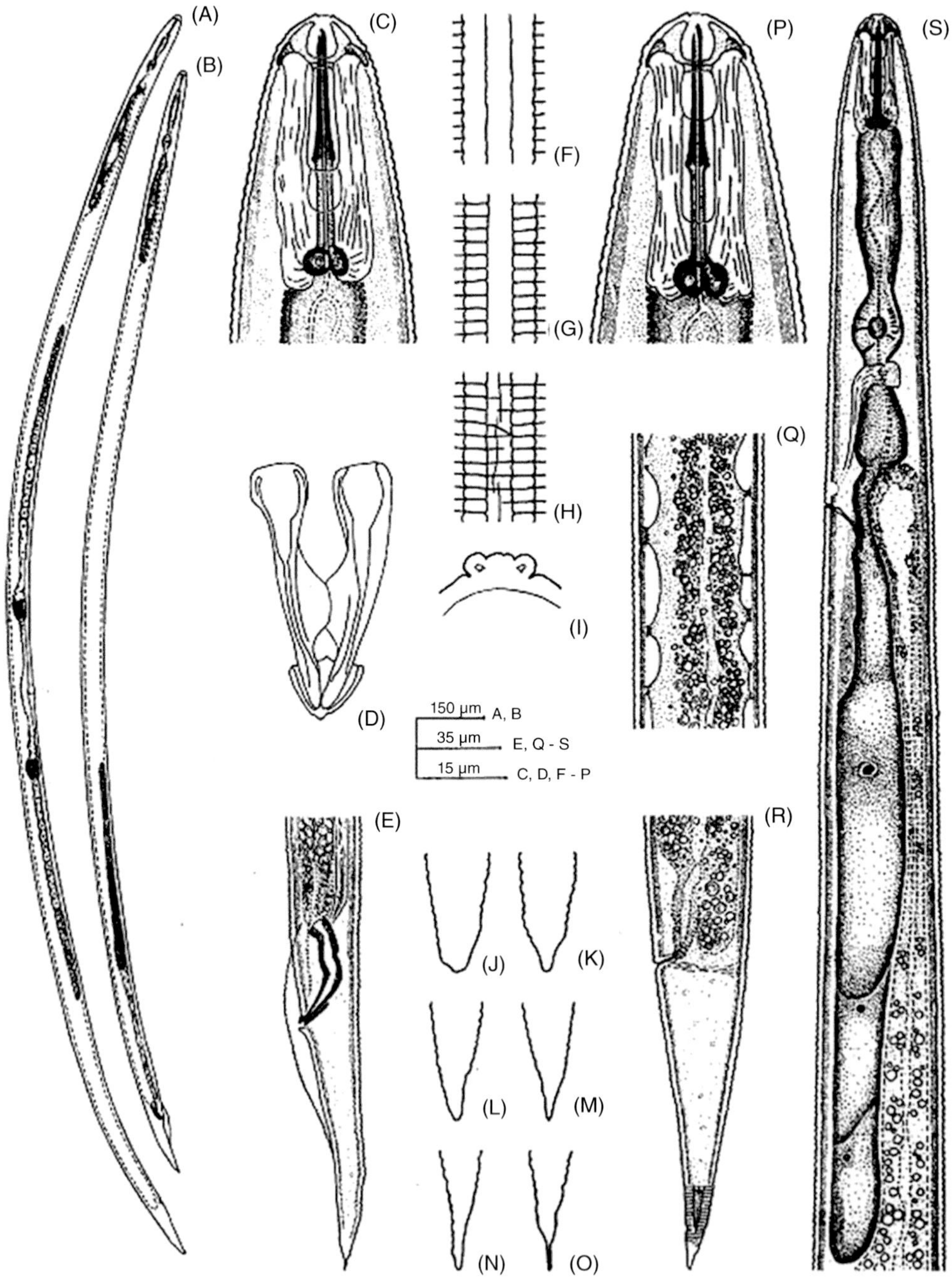

Fig. 10.6. Line drawing of *Hirschmanniella spinicaudata*. Female (A); male (B); male head end (C); ventral view of spicules and gubernaculum (D); male tail (E); female lateral field (F-H); cross-section of female lateral field (I); female tail tips (J–O); female head end (P); female mid-body portion showing 'Thornean cells' (Q); female tail (R); female oesophageal region (S). [Reproduced from CABI-CPC, 2010; Courtesy CAB International]

with four longitudinal incisures, transverse striae crossing the three bands or only the two outer ones (Fig. 10.6F–H). Lip region hemispherical, raised, not offset, with 4–6 (usually 5) annules. Cephalic framework with heavy sclerotization (Fig. 10.6P). Hemizonid consisting of two annules, located 2–4 annules at level of excretory pore. Hemizonion obscure, posterior to excretory pore around 15–18 annules. Caudalid present, located 4–5 annules in front of anus. Stylet strong, over 40 μm long with rounded knobs. Median bulb oval. Oesophageal glands elongate, ventrally or ventro-laterally overlapping intestine. Intestine slightly extending into rectum. Genital tracts branched, paired, ovaries with single row of oocytes. Spermatheca spherical, with sperm. Tail elongate-conoid, extremely variable from effilate to rounded or with distinct mucro (Fig. 10.6J–O). Phasmids pore-like, located 17–24 annules from the tip.

MALE: Similar to female in general respects except for reproductive organs (Fig. 10.6B). Spicules cephalated, arcuate, flanged on inner edges (Fig. 10.6F). Gubernaculum simple, curved. Bursa crenate, covering proximal end of spicules to posterior end of tail (Fig. 10.6E). Tail elongate-conoid, tip as in female.

80.5.2 *Molecular diagnosis*

Van Den Berg *et al.* (2009) used sequence analysis of D2–D3 expansion segments of 28S, partial 18S and ITS rRNA region to characterize *H. spinicaudata*. The DNA was extracted from several specimens of a sample using the nematode lysis buffer (Tanha Maafi *et al.*, 2003) (see Chapter 4, section 15.5.2). The following primer sets were used for amplification and sequencing of rRNA gene fragments.

For D2–D3 expansion segments of the 28S rRNA

D2A forward primer (5′-ACA AGT ACC GTG AGG GAA AGT TG-3′)

D3B reverse primer (5′-TCG GAA GGA ACC AGC TAC TA-3′),

For ITS1-5.8S-ITS2 of rRNA gene

Forward primer TW81 (5′-GTT TCC GTA GGT GAA CCT GC-3′)

Reverse AB28 (5′-ATA TGC TTA AGT TCA GCG GGT-3′)

For partial 18S rRNA gene

Forward G18SU (5′-GCTTGTCTCAAAGATTAAGCC-3′) (Dorris and Blaxter, 2000)

Reverse R18Tyl1 (5′-GGTCCAAGAATTTCACCTCTC-3′) (Chizhov *et al.*, 2006)

The PCR amplifications were performed using the procedure described by Tanha Maafi *et al.* (2003, see Chapter 4, section 15.5.2). Each gene fragment was sequenced with two clones. The sequences were submitted to the GenBank database (for 18S rRNA gene EU620460–EU620463 and D2–D3 of 28S gene EU620464–EU620469) and for the ITS rRNA (EU620470–EU620475).

80.6 Geographical distribution

H. spinicaudata mainly occurs in tropical West Africa, where it is widespread and reported from Benin, Burkina Faso, Cameroon, Congo Democratic Republic, Côte d'Ivoire, Gambia, Ghana, Nigeria, Senegal, South Africa, Togo and Zambia. Outside Africa, the nematode has been recorded from Brazil, Cuba, China, Costa Rica, Pakistan, the USA and Venezuela (CABI-CPC, 2010). Van Den Berg *et al.* (2009) also reported *H. spinicaudata* from South Africa.

80.7 Hosts

Rice (*Oryza sativa*) is the main host of *H. spinicaudata*. Other hosts are recorded from the Juncaceae and Cyperaceae (Babatola, 1980). Besides rice, the nematode has been detected in the rhizosphere of sugarcane (Sher, 1968), tomato and okra (Netscher, 1970), but not in the roots of these plants. Babatola (1980) recorded 45 weed species as being good hosts for *H. spinicaudata*, most of these being grasses and sedges. A wild rice, *Oryza longistaminata*, is highly susceptible to the nematode. Fernandez Díaz-Silveira and Ortega Herrera (1982) recorded eight weed hosts for this nematode in a rice-growing area in Cuba.

80.8 Symptoms

H. spinicaudata is predominantly a parasite of rice, where it is endoparasitic within the roots. The infected roots turn yellowish-brown in colour and rot. Lesions on the root surface may be visible where the nematodes have entered the tissue. Aboveground symptoms may involve chlorosis of the foliage, retarded growth, decrease in tillering, delayed flowering, reduced grain yield and an increase in grain chaffiness (Babatola and Bridge, 1979).

80.9 Biology and life cycle

H. spinicaudata is an obligate endoparasite of rice, where it feeds and reproduces within the cortical tissues. The life cycle is similar to that of *H. oryzae*. The second-stage juvenile (J_2) emerges from the egg and moults three times to become an adult. The species is amphimictic. The nematode survives between crops either in the soil or in the roots of other hosts or intact roots of rice stubble. The nematode can survive in prolonged flooding and anaerobic condition in a wide range of pH (Bridge *et al.*, 2005a). *H. spinicaudata* may also become quiescent in dry soil. The species is restricted to areas with wet or flooded soils.

80.10 Economic importance

H. spinicaudata, along with other species of the genus, is a common pest in rice fields throughout the world and may cause a yield loss of 25% (Hollis *et al.*, 1984). In Côte d'Ivoire, nematicide treatments against *H. spinicaudata* increased yield by 20–53%, yet there was no significant correlation between nematode numbers and yield (Cadet and Quénéhervé, 1982). This nematode may cause yield losses of

16–32% when the population level is 1000 nematodes per plant (Bridge *et al.*, 2005a).

80.11 Movement and means of dispersal

See section 76.11, above.

80.12 Pest risk analysis and looming threat from introduction

In view of the importance of rice as host, the damage potential of the nematode and its presence in different ecological zones, introduction of the nematode into new host areas should be prevented. A complete pest risk analysis is recommended for assessment of the looming threat posed by this nematode to other parts of the world where it is absent.

80.13 Invasiveness rating

Considering the consequences of introduction and probable economic impact of *H. spinicaudata*, the following invasiveness rating has been given.

Climate/host interaction	–	High (3)
Hosts range	–	Medium (2)
Dispersal potential	–	Medium (2)
Economic impact	–	High (3)
Environmental impact	–	Medium (2)
Post-entry distribution	–	Medium (–2)
Final rating (score)	–	Medium (10)

Based on the above invasiveness rating, *H. spinicaudata* receives a score of 10 out of 15 and is rated as MEDIUM INVASIVE NEMATODE.

80.14 Region-wise status of invasiveness

H. spinicaudata is listed as a quarantine pest in many countries and as I / A1 pest in the European Union (EPPO, 2009b; Jeger *et al.*, 2018). *H. spinicaudata* has attained medium invasiveness status in major rice-producing countries (FAOSTAT, 2019) where it is absent. However, there would be trade implications if this nematode is introduced and established in these countries.

APPPC: Afghanistan, Bangladesh, Cambodia, India, Indonesia, Iran, Japan, Lao People's Democratic Republic, Malaysia, Myanmar, Nepal, North Korea, Philippines, South Korea, Sri Lanka, Thailand, Vietnam
COSAVE: Argentina, Colombia, Ecuador, Peru, Uruguay
EPPO: Italy, Russian Federation, Spain, Turkey
IAPSC: Egypt, Guinea, Madagascar, Mali, Sierra Leone, Tanzania
PPPO: Australia, New Zealand.

80.15 Management measures

Management measures are similar to those for *H. oryzae*. For details, see section 79.15. For phytosanitary measures, see Chapter 16, section 115.

Nacobbus Spp.

The false root-knot nematodes, *Nacobbus* species, were first observed by Thorne (1935) as *Anguillulina aberrans*. Later they were transferred to the new genus *Nacobbus* by Thorne and Allen (1944). The genus was assigned to the family Nacobbidae by Golden (1971). Six species of *Nacobbus* have been described, namely: *N. aberrans* (Thorne, 1935) Thorne and Allen, 1944; *N. dorsalis* Thorne and Allen, 1944 (type species); *N. batatiformis* Thorne and Schuster, 1956; *N. serendipiticus* Franklin, 1959; *N. serendipiticus bolivianus* Lordello, Zamith and Boock, 1961; and *N. bolivianus* Lordello, Zamith and Boock, 1961. Siddiqi (2000) validated only two species, *N. aberrans* and *N. dorsalis*, and synonymized the rest with *N. aberrans* (Siddiqi, 2000; Manzanilla-López *et al.*, 2004). Molecular studies of several populations of *N. aberrans* confirmed that the species consists of different pathotypes (Manzanilla-López *et al.*, 2004). Therefore, for practical reasons, *N. aberrans sensu lato* is considered to include *N. batatiformis* from North America and *N. serendipiticus*, *N. serendipiticus bolivianus*, *N. bolivianus* and *N. aberrans sensu stricto* (Reid *et al.*, 2003).

Systematic position of *Nacobbus* species

	Classical (after Siddiqi, 2000)	Molecular (after Decraemer and Hunt, 2013)
Phylum	Nematoda	Nematoda
Class	Secernentea	Chromadorea
Order	Tylenchida	Rhabditida
Suborder	Hoplolaimina	Tylenchina
Infra-order	–	Tylenchomorpha
Superfamily	Holpolaimoidea	Tylenchoidea
Family	Pratylenchidae	Pratylenchidae
Subfamily	Nacobbinae	–
Genus	*Nacobbus*	*Nacobbus*

81 *Nacobbus aberrans*

81.1 Common name

False root-knot nematode

81.2 Scientific name

Nacobbus aberrans (Thorne,1935) Thorne and Allen, 1944

81.3 Synonyms

Anguillulina aberrans Thorne, 1935
Nacobbus batatiformis Thorne and Schuster, 1956
Nacobbus bolivianus Lordello, Zamith and Boock, 1961
Nacobbus serendipiticus Franklin, 1959
Nacobbus serendipiticus bolivianus Lordello, Zamith and Boock, 1961
Pratylenchus aberrans (Thorne, 1935) Filipjev, 1936

81.4 EPPO code

NACOBA

81.5 Diagnosis

Nacobbus aberrans sensu lato is a species complex and the presence of several pathotypes make it very difficult to identify at the species level. Hence, both morphological and molecular analysis aided by biochemical diagnosis (use of isoenzyme markers) are suggested for an accurate diagnosis.

81.5.1 Morphological diagnosis

Morphological and morphometric descriptions after Lordello *et al.* (1961) and Sher (1970) are presented below.

Mature female: Body obese, oval to spindle-shaped with tapering extremities, posterior end usually nipple-like (Fig. 10.7C). Cephalic framework sclerotized. Stylet knobs rounded. Oesophageal glands dorsally overlap intestine. Vulva and anus conspicuous and situated near posterior region (Fig. 10.7B). Ovary single and coiled within body, spreading to median oesophageal bulb. Eggs average 80 × 42 µm and do not fill body cavity (Fig. 10.7C).
Measurements: L = 0.8–1.2 mm; stylet = 22–24 µm.
Immature female: Body vermiform, J-shaped or ventrally arcuate when heat relaxed. Lateral fields with four incisures with incomplete areolation towards extremities. Head slightly

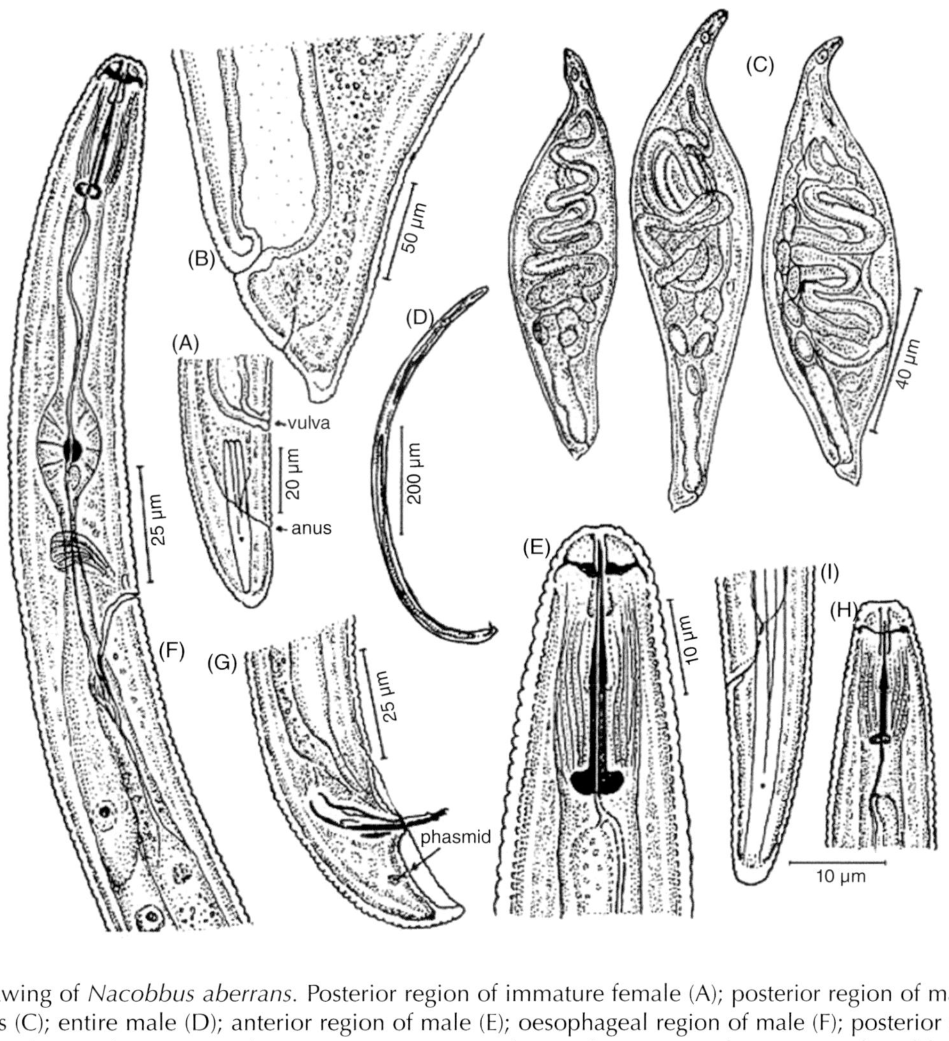

Fig. 10.7. Line drawing of *Nacobbus aberrans*. Posterior region of immature female (A); posterior region of mature female (B); mature females (C); entire male (D); anterior region of male (E); oesophageal region of male (F); posterior region of male (G); anterior region of second-stage juvenile (H); posterior region of second-stage juvenile (I). [Reproduced from CABI-CPC, 2010; Courtesy CAB International]

offset with 3–4 annules, with round labial disc. Cephalic framework hexaradiate and strongly sclerotized. Stylet strong with rounded knobs. Dorsal gland opening 2–3 µm behind stylet. Median bulb oval with distinct muscular valve plates. Oesophageal glands overlap intestine dorsally. Dorsal gland anterior to subventral. Vulva located posteriorly (Fig. 10.7A). Ovary single with numerous oocytes arranged in row. Tail subcylindroid with smooth rounded tip (Fig. 10.7A). Phasmids pore-like and situated just behind anus.

Measurements: L = 0.71–0.93 mm; stylet = 21–25 µm; a = 23–40; b' = 2.8–4.1; c = 24–40; c' = 0.9–1.5; V = 92–94%; distance from vulva to anus = 15–24 µm.

MALE: Body ventrally arcuate (Fig. 10.7D). Cephalic region slightly offset or continuous, with 3–4 annules. Lateral fields with four incisures. Stylet well developed with rounded knobs (Fig. 10.7E). Oesophageal glands overlap intestine dorsally. Spicules medial. Gubernaculum simple, trough-shaped in lateral view, fixed (Fig. 10.7G). Bursa ventrally arcuate, enveloping tail (Fig. 10.7G). Phasmids just anterior to middle of tail.

Measurements: L = 0.788–0.90 mm; a = 18.8–25.4; b = 7.2–8.5; c = 34.3–46.8; stylet = 21.4–22.9 µm; spicules = 26.0–30.6 µm; gubernaculum = 7.6–9.2 µm.

SECOND-STAGE JUVENILES: Body vermiform, slightly ventrally arcuate to straight when heat relaxed. Head hemispherical, slightly offset or continuous with 2–4 annules. Cephalic framework heavily sclerotized (Fig. 10.7H). Lateral fields with four incisures. Stylet with rounded or backwardly sloping knobs (Fig. 10.7H). Excretory secretory pore lies 63–80 µm from head end. Tail end with small rounded smooth terminus (Fig. 10.7H), 1.8–4.0 times anal body width. Phasmids indistinct, 9–15 annules from anus.

Measurements: L = 0.298–0.364 mm; a = 19.5–23.8; b = 3.3–4.1; c = 9.7–13.0; stylet = 10.7–12.3 µm.

81.5.2 Molecular diagnosis

Numerous molecular studies have been carried out by different workers (Reid *et al.*, 2003; Vovlas *et al.*, 2007; Castillo *et al.*, 2012) and provided molecular evidence to support the existence of at least two major clades within South American *Nacobbus* species. Kerry *et al.* (2005) developed molecular tools with specific primers targeting the ITS rDNA and could detect *N. aberrans* in soil and in samples of tuber skin.

81.6 Geographical distribution

N. aberrans is recorded only in temperate and tropical latitudes of North and South America, mainly in the USA, Mexico, Argentina, Bolivia, Chile, Ecuador and Peru (EPPO, 2009b). Outside America, it was intercepted in Finland, Italy, the Netherlands, Russia and United Kingdom (EPPO, 2009b). Saxena (1986) reported *N. aberrans sensu lato* on vegetable crops in India but this is considered an invalid record (EPPO, 2009b).

81.7 Pathotype

Based on different host preferences, *N. aberrans* populations can be separated into three main pathotype groups (Inserra *et al.*, 1985; Manzanilla *et al.*, 2002). ***Sugar beet pathotype*** infects sugar beet, but not potato, and occurs in the USA, Argentina and Ecuador. ***Potato pathotype*** infects potato and many other crops, including sugar beet, but not chilli pepper. It does not occur in the USA and is common in Argentina, Bolivia, Chile, Peru and Mexico. ***Bean pathotype*** infects bean, but not potato and sugar beet, and occurs in Mexico.

81.8 Hosts

N. aberrans sensu lato is a polyphagous nematode and has a very wide host range spreading to 17 plant families other than the Poaceae (PROINPA, 1991). The nematode parasitizes 69 plant species, such as potatoes, tomatoes, sugar beet, cabbages, broccoli, turnips, lettuces, cucumbers, peas (Tacconi and Ambrogioni, 1995), eggplant and sweet peppers (Manzanilla *et al.*, 2002). Among these, potato and sugar beet are the most preferred host. An expanded list of hosts of *N. aberrans sensu lato* is presented in Table 59.

81.9 Symptoms

N. aberrans sensu lato causes root galls mainly along the root axis and at the root tips. The galls resemble those caused by *Meloidogyne* spp. However, the galls caused by false root-knot nematode tend to be more distinct and rounded, giving a beaded appearance (Fig. 10.2J), while those caused by *Meloidogyne* often coalesce to form elongated swellings along the root. Root swellings occur within 5 days of invasion (Altman and Thomason, 1971). The root knots are in a bead-like arrangement; hence it is named as a rosary root nematode. Mature galls tend to be flat with a rough corky surface (Fig. 10.2I) and bear many lateral roots (Inserra *et al.*, 1983; Manzanilla *et al.*, 2002). In severely infested plants, the above-ground symptoms include retarded growth, stunting and chlorosis of leaves (Fig. 10.2H).

81.10 Biology and life cycle

All the pathotypes of *N. aberrans sensu lato* have peculiar parasitic habits and have both endoparasitic sedentary and migratory feeding habits. The vermiform stages, juveniles and immature adults, are migratory endoparasites. Like lesion nematodes, they invade the feeder root tissues or tubers where they feed, migrate and cause lesions, necrosis and cavities. In contrast, the vermiform immature females leave the necrotic tissues and re-invade appropriate healthy roots, where they become sedentary and obese and induce specialized feeding sites, resulting in gall formations and egg-mass production, as with root-knot nematodes. Vermiform stages (third- and fourth-stage juveniles) undergo

Table 59. List of host plants of *Nacobbus aberrans sensu lato* (adapted from CABI-CPC, 2010).

Amaranthus (grain amaranth)
Atriplex confertifolia (shadscale saltbush)
Beta vulgaris (beetroot)
Beta vulgaris var. *saccharifera* (sugar beet)
Brassica napus var. *napobrassica* (swede)
Brassica nigra (black mustard)
Brassica oleracea (cabbages, cauliflowers)
Brassica oleracea var. *gemmifera* (Brussels sprouts)
Brassica oleracea var. *gongylodes* (kohlrabi)
Brassica oleracea var. *italica* (broccoli)
Brassica oleracea var. *viridis* (collards)
Brassica rapa subsp. *pekinensis*
Capsicum annuum (bell pepper)
Chenopodium album (fat hen)
Chenopodium quinoa (quinoa)
Cucumis sativus (cucumber)
Cucurbita maxima (giant pumpkin)
Cucurbita pepo (ornamental gourd)
Daucus carota (carrot)
Gaillardia
Kochia
Lactuca sativa (lettuce)
Opuntia (pricklypear)
Pisum sativum var. *arvense* (Austrian pea)
Sisymbrium irio (London rocket)
Solanum lycopersicum (tomato)
Solanum melongena (aubergine)
Solanum nigrum (black nightshade)
Solanum tuberosum (potato)
Spergula arvensis (corn spurry)
Spinacia oleracea (spinach)
Tragopogon porrifolius (oysterplant)
Tribulus terrestris (puncture vine)
Ullucus tuberosus (ulluco)

dormancy to survive under adverse environmental conditions and overwinter under the skin of dry roots or stored or marketed tubers of potato (Costilla, 1985; Inserra *et al.*, 1985).

81.11 Economic importance

The false root-knot nematode *N. aberrans* is an important pest of potato and estimated losses on potato are as high as 65% in South America. Besides potato, the nematode also suppresses the yield of tomato by 55% and bean by 36% in Mexico (Otazú *et al.*, 1985; Manzanilla *et al.*, 2002). In the USA, *N. aberrans* is an economic pest of sugar beet, inflicting crop losses of 10–20% (Inserra *et al.*, 2005).

81.12 Movement and means of dispersal

See section 76.11, above.

81.13 Pest risk analysis and looming threat from introduction

The broad host range of *N. aberrans* would facilitate its establishment on alternative hosts other than main hosts like potato and sugar beet. The direct economic damage of this nematode on the potato industry can be easily assessed in Latin America. The indirect economic and regulatory impact on the native areas should not be undervalued. If the nematode is introduced outside its native area, associated industries would be severely affected by the quarantine restrictions imposed against *N. aberrans* on plant shipments originating from host countries.

81.14 Invasiveness rating

Considering the probable consequences of introduction and pest risk analysis of *N. aberrans*, the following invasiveness rating has been assigned.

Climate/host interaction	–	High (3)
Hosts range	–	High (3)
Dispersal potential	–	High (3)
Economic impact	–	High (3)
Environmental impact	–	High (3)
Post-entry distribution	–	Low (–1)
Final rating (score)	–	High (14)

Based on the above invasiveness rating, *N. aberrans* receives a score of 14 out of 15 and is rated as HIGHLY INVASIVE NEMATODE.

81.15 Region-wise status of invasiveness

N. aberrans is an important quarantine nematode in many countries and in 2011 at least 52 countries regulated this nematode (Hockland *et al.*, 2013). These countries included Argentina, Brazil, Bulgaria, Colombia, the European Union, Iceland, Indonesia, Japan, Morocco, Norway, Paraguay, South Korea, Thailand, Tunisia, the USA (California), Uruguay and former Yugoslavia (Manzanilla *et al.*, 2002; Lehman, 2004). The European and Mediterranean Plant Protection Organization (EPPO) also includes *N. aberrans* on its A1 list of harmful organisms whose introduction and spread within all member states are banned (EPPO, 2012). *N. aberrans* is also regarded as a high-risk nematode species in Taiwan (Tsay, 1997).

81.16 Management measures

The management options of *N. aberrans sensu lato* are described below.

81.16.1 Cultural control

Cultural control includes prolonged crop rotations (for 4–6 years) and control of host weed species (Ambrogioni and Tacconi, 1992). Mixed cropping with leguminous crops (*Concanavalia ensiformis*) and tomatoes reduced the galling

(Marban-Mendoza *et al.*, 1989). *Tagetes erecta* and tomato intercropping reduced the infection of *N. aberrans sensu lato* on tomato roots compared with tomatoes grown alone (Zavaleta Mejia and Gomez, 1995).

81.16.2 Biological control

Pupureocillium lilacinum (=*Paecilomyces lilancinus*) is a good biological control agent for *N. aberrans sensu lato* (Eguiguren-Carrion, 1995). The trapping capability shown by *Arthrobotrys conoides* was higher than 90% for *N. aberrans sensu lato* second-stage juveniles (Mendoza-de Gives *et al.*, 1994). Experiments in greenhouses showed that *Beauveria brogniartii* can parasitize and reduce the numbers of *N. aberrans sensu lato* on potatoes (Brodie, 1993).

81.16.3 Chemical control

Soil fumigation and granular nematicides gave good control of false root-knot nematode (Inserra *et al.*, 1985; Brodie, 1993). In sugar beet fields, aldicarb reduced galls by nearly 50% compared with untreated controls, although the reduction was not significant. Substantial reductions of nematode galling were observed after combined drench treatment with aldicarb and iprodionem (Beaupre *et al.*, 1989). Application of carbofuran or Telone in sugar beet fields significantly reduced the infection level of the nematode (Kerr, 1991). Jatala (1979) suggested that chemical dips or hot-water treatment of tubers may facilitate control of *N. aberrans sensu lato*.

81.16.4 Phytosanitary measures

See Chapter 16, section 115.

Pratylenchus Spp.

Species of the genus *Pratylenchus* are migratory endoparasitic nematodes, also known as root-lesion nematodes as they produce lesions by feeding on feeder roots and occasionally on other underground plant parts (Castillo *et al.*, 2012). The first described *Pratylenchus* species was *Tylenchus pratensis* (de Man, 1880). Filipjev (1936) named the genus *Pratylenchus* with *P. pratensis* as the type species. Sher and Allen (1953) first described the taxonomy of the genus. Keys to the species are given by different workers (Handoo and Golden, 1989; Loof, 1991; Castillo and Vovlas, 2007). The monograph of Castillo and Vovlas (2007) provides comprehensive information on *Pratylenchus*. At present the genus *Pratylenchus* comprises more than 100 valid species (Siddiqi, 2000; Castillo and Vovlas, 2007; Janssen *et al.*, 2017a). Some important *Pratylenchus* species are:

Pratylenchus alleni Ferris, 1961
P. brachyurus (Godfrey, 1929) Filipjev and Schuurmans Stekhoven, 1941
P. coffeae (Zimmermann, 1898) Filipjev and Schuurmans Stekhoven, 1941
P. crenatus Loof, 1960
P. fallax Seinhorst, 1968
P. floridensis De Luca *et al.*, 2010
P. goodeyi Sher and Allen, 1953
P. haiduongensis Nguyen *et al.*, 2017
P. hispaniensis Palomares-Rius *et al.*, 2010
P. lentis Troccoli, De Luca, Handoo and Di Vito, 2008
P. loosi Loof, 1960
P. neglectus (Rensch, 1924) Filipjev and Schuurmans Stekhoven, 1941
P. oleae Palomares-Rius *et al.*, 2014
P. parafloridensis De Luca *et al.*, 2010
P. penetrans (Cobb, 1917) Filipjev and Schuurmans Stekhoven, 1941
P. pratensis (de Man, 1880) Filipjev, 1936
P. rwandae Singh *et al.*, 2018
P. scribneri Steiner in Sherbakoff and Stanley, 1943
P. speijeri De Luca *et al.*, 2012
P. thornei Sher and Allen, 1953
P. vulnus Allen and Jensen, 1951
P. zeae Graham, 1951

Systematic position of *Pratylenchus* species

	Classical (after Siddiqi, 2000)	Molecular (after Decraemer and Hunt, 2013)
Phylum	Nematoda	Nematoda
Class	Secernentea	Chromadorea
Order	Tylenchida	Rhabditida
Suborder	Hoplolaimina	Tylenchina
Infra-order	–	Tylenchomorpha
Superfamily	Holpolaimoidea	Tylenchoidea
Family	Pratylenchidae	Pratylenchidae
Subfamily	Pratylenchinae	Pratylenchinae
Genus	*Pratylenchus*	*Pratylenchus*

82 General Characteristics of *Pratylenchus* species

Pratylenchus species are highly damaging plant-parasitic nematodes worldwide and rank second only to root-knot and cyst nematodes (Castillo and Vovlas, 2007). They are predominantly recognized as polyphagous nematodes, parasitizing a broad variety of plant species, including cereals, vegetables, forage crops, industrial crops, banana, cotton, coffee, maize, potatoes and ornamental plants, as well as weed species (Bélair *et al.*, 2007; Castillo *et al.*, 2012). Host preferences can differ significantly among species, such as: *P. brachyurus* on maize, cotton, groundnut, pineapple, potato and tobacco; *P. coffeae* on coffee, citrus, sugarcane and tea; *P. goodeyi* on banana; *P. neglectus* on cereals and legumes; *P. penetrans* on grasses, forages, fruit trees and strawberries; *P. scribneri* on potato, soybean and strawberries; *P. thornei* on cereals and legumes; *P. vulnus* on pome, stone fruit, ornamentals and roses; and *P. zeae* on maize, rice, sugarcane and wheat (Castillo and Vovlas, 2007). In this volume we review *Pratylenchus alleni*, *P. brachyurus*, *P. coffeae*, *P. crenatus*, *P. fallax*, *P. goodeyi*, *P. loosi*, *P. neglectus*, *P. penetrans*, *P. scribneri*, *P. thornei*, *P. vulnus* and *P. zeae* for their invasiveness, as they are a limiting factor for the production of several crops of economic importance.

82.1 Diagnosis

Morphologically *Pratylenchus* species show substantial variability intra- and interspecifically, depending on the host, feeding and geographically distant populations (Castillo *et al.*, 2012). Identification of *Pratylenchus* species based on only morphology and morphometry can be difficult to confirm, thus biochemical, molecular and phylogenetic analyses are powerful tools for precisely identifying a *Pratylenchus* species. Hence, to obtain a correct identification, an integrative approach combining classical taxonomy aided with molecular methods is strongly suggested.

82.1.1 Morphological diagnosis

The general morphological characters of *Pratylenchus* species are summarized below.

Female: Small to medium size (less than 900 μm). Lip region anteriorly flattened, not offset, usually continuous with body contour and framework strongly sclerotized. Stylet typical, strong and around 15–20 μm long with varied basal shape. Number of lateral fields incisures four to six. Deirids not present. Phasmids near middle of tail. Metacorpus oval to round and muscular, pharyngeal glands ventrally overlap intestine. No sexual diamorphism in anterior region of body. Genital tract with single ovary and spermatheca only in anterior region (monoprodelphic). Postvulval uterine sac usually present. Vulva in posterior position, generally at 70–80% of body length. Tail subcylindrical to conoid with smooth or annulated terminus.

Male: Similar to female in overall morphology, except reproductive structure. Testis usually single outstretched. Spicules arcuate and with well-marked manubrial in some species. Gubernaculum simple and does not protrude. Tail pointed with caudal alae extending to tip.

82.1.2 Biochemical diagnosis

P. brachyurus (Payan and Dickson, 1990), *P. coffeae* (Andrés *et al.*, 2000), *P. crenatus* (Ibrahim *et al.*, 1995), *P. fallax* (Ibrahim *et al.*, 1995), *P. goodeyi* (Andrés *et al.*, 2000), *P. neglectus* (Ibrahim *et al.*, 1995), *P. penetrans* (Andrés *et al.*, 2000), *P. scribneri* (Andrés *et al.*, 2000), *P. thornei* (Andrés *et al.*, 2000) and *P. vulnus* (Jaumot *et al.*, 1997) have been characterized based on proteins and isozymes using one-dimensional sodium dodecyl sulfate polyacrylamide gel electrophoresis (SDS PAGE) (Abrantes *et al.*, 2004). Protein polymorphism and genetic diversity have been studied using isoelectric focusing (IEF), such as for *P. scribneri* and *P. brachyurus* (Payan and Dickson 1990). Most protein- or isoenzyme-based diagnostic techniques have now been superseded by DNA- and PCR-based techniques.

82.1.3 Molecular diagnosis

Various regions such as ITS rDNA (Bélair *et al.*, 2013; Oliveira *et al.*, 2017), D2–D3 expansion segment of 28S rDNA (Al-Banna *et al.*, 2004; Baidoo *et al.*, 2017; Dauphinais *et al.*, 2018) and β-1,4-endoglucanase gene (Mokrini *et al.*, 2014; Peetz and Zasada, 2016) are a potential target for detection and identification of several *Pratylenchus* species. Species-specific primers for different *Pratylenchus* species have also been developed for conventional polymerase chain reaction (PCR) and quantitative PCR (qPCR). Details of species-specific primers for all the *Pratylenchus* species reviewed in this chapter are given in Table 60.

82.2 Biology, life cycle and ecology

Generally, the life cycle of root-lesion nematodes starts with the first moult within the egg and the first-stage juvenile (J_1) becomes a second-stage juvenile (J_2). The J_2 remains inside the egg until hatch, typically after 1 week from egg deposition. After hatching, the J_2 moults into J_3 and subsequently J_4, within 35–40 days. Thereafter there is the final moult to become adults, either female or male (Castillo and Vovlas, 2007) (Fig. 10.8).

Males are rare or absent in some species, such as *P. crenatus*, *P. neglectus* and *P. thornei*, and common in others, such as *P. coffeae* and *P. penetrans*. When males are present, reproduction is amphimictic; and in their absence, reproduction is by parthenogenesis (Castillo and Vovlas, 2007; Duncan and Moens, 2013). Females lay eggs (15–35) in clusters that can be found inside roots or also in the surrounding soil (Castillo and Vovlas, 2007). The duration of the life cycle among the species varies with temperature and host plant, in the ranges of 25–35 days (*P. thornei*) and 54–65 days (*P. penetrans*). Temperature is an important factor to determine the duration of the life cycle (Turner and Chapman, 1972; Castillo *et al.*, 1995). Temperatures above 30°C or below 17°C affect the life cycle and 25°C is optimal to complete the life cycle in 26 days (Mizukubo and Adachi, 1997).

82.3 Host–parasite relationship

Pratylenchus species are typical migratory endoparasitic nematodes (Fig. 10.8). The invasion processes involve recognition of the root, probing, penetration, ectoparasitic feeding, root entry and feeding on cells in the cortex, followed by moulting and reproduction (Linsell *et al.*, 2014). All mobile stages can penetrate roots and usually move through epidermal and cortical cells by piercing cell walls along the migration pathway (Fig. 10.8). After entering into the root behind the zone of elongation, the nematode feeds on parenchyma cells (Castillo *et al.*, 1995). In the host roots, cortical parenchyma cells are usually fed upon and evacuated, and cells walls are broken, leaving elongated cavities. Such cells may collapse and become necrotic, producing lesions and an elongated tunnel through a section of the cortex. The nematode moves both inter- and intra-cellularly within the cortex (Khan, 2008). Formation of lesions begins with small, faint yellow spots associated with feeding sites, later developing into elliptical, brownish, necrotic cortical lesions. The changes in the colour of cells and tissues are due to chemical alterations within the affected parts and can be seen at various stages during the parasitism. The degradation of pectic and cellulose materials in cell walls is a probable reason for these colour changes, as evidence exists for the presence of pectic and cellulolytic enzymes in the nematode saliva. *Pratylenchus* species are known to secrete

Table 60. Species-specific primers for *Pratylenchus* diagnosis. See references for species and molecular assay technique for validation of these primers (adapted from Orlando *et al*., 2020).

Species	DNA region	Primer set (5'–3')	Molecular assay	Amplicon length (bp)	Reference
P. alleni	ITS rDNA	F -TTGATTACGTCCCTGCCCTTT R - TTTCACTCGCCGTTACTAAGG	Conventional PCR	1080	Bélair *et al*. (2013)
P. brachyurus	18S rDNA ITS rDNA	F - TTGATTACGTCCCTGCCCTTT* R - GCWCCATCCAAACAAYGAG	Conventional PCR	267	Machado *et al*. (2007)
P. coffeae	ITS rDNA	F - ATGCGCACATTGCATTCA R - GAGCGAGAAACACCTCTCAC	Conventional PCR	632	Uehara *et al*. (1998a)
		F - ATGCCCACATTGCATTCAGC R - GAGAGAGAAACACCTCTCAC	Multiplex PCR	638	Saeki *et al*. (2003)
P. crenatus	28S rDNA	F - AAAGCCTGAATGCCCTGAG R - AATTGAAAGAGGTCGGTCGT	Conventional PCR	610	Mekete *et al*. (2011)
	ITS rDNA	F - TTCTTGACAAGTTCATTGCTTC R - CACTCACGATGTGCTTCTG	TaqMan qPCR	116	Oliveira *et al*. (2017)
	β-1,4-endoglucanase	F - TCTCCTGGACGGACGTGCTC R - AGGCCGTCCAGGAAGGTGTAC	Conventional PCR	381	Peetz and Zasada (2016)
P. fallax	28S rDNA	F - ACAAGTACCGTGAGGGAAAGTTG R - TCGGAAGGAGCTACTA	Conventional PCR	838	Subbotin *et al*. (2008)
P. goodeyi	ITS rDNA	F - GTGGATCACTCGGCTCGTAG R - CAGCGGGTATTCACGTCTGA	Conventional PCR	493	Mgonja *et al*. (2019)
	28S rDNA	F - ACAAGTACCGTGAGGGAAAGTTG R - TCGGARGGAACCAGCTACTA	Conventional PCR	781	Mgonja *et al*. (2019)
P. loosi	28S rDNA	F - ACAAGTACCGTGAGGGAAAGTTG R - TCGGAAGGAGCTACTA	Conventional PCR	838	Subbotin *et al*. (2008)
	ITS rDNA	F - ATGCGCACATTGCATTCA R - GAGCGAGAAACACCTCTCAC	Conventional PCR	632	Uehara *et al*. (1998a)
P. neglectus	28S rDNA	F - ATGAAAGTGAACATGTCCTC R - TCGGAAGGAACCAGCTACTA*	Conventional PCR	290	Al-Banna *et al*. (2004)
	28S rDNA	F - CGCAATGAAAGTGAACAATGTC R - AGTTCACCATCTTTCGGGTC	SYBR green qPCR	144	Yan *et al*. (2008)
	ITS rDNA	F - GGCACTGTGCGAAGTGTCCG R - TTAACACCTCAGGCGTCATGTAC	SYBR green qPCR	234	Yan *et al*. (2013)
	ITS rDNA	F - ACTGTGCGAAGTGTCCG R - GATCCACCGATAAGGCTAGA	TaqMan qPCR	121	Oliveira *et al*. (2017)
	β-1,4 endoglucanase	F - TGACCACAACGCGCAGAACCAC R - GCCACGTCCACGTCCTGGGA	Conventional PCR	293	Peetz and Zasada (2016)
P. penetrans	28S rDNA	F - TAAAGAATCCGCAAGGATAC R - TCGGAAGGAACCAGCTACTA*	Conventional PCR	278	Al-Banna *et al*. (2004)
	28S rDNA	F - ACATGGTCGACACGGTGATA R - TGTTGCGCAAATCCTGTTTA	Conventional PCR	520	Mekete *et al*. (2011)
	28S rDNA	F - GGTTTTCGGGCTCATATGGGTTC R - TTTACGCCGAGAGTGGGATTGTG	SYBR green qPCR	111	Baidoo *et al*. (2017)
	28S rDNA	F - GAGACTTTCGAGAAGGCGATATG R - AGGACCGAATTGGCAGAAG	TaqMan qPCR	176	Dauphinais *et al*. (2018)
	ITS rDNA	F - ATGATGGAAGTGTCCGCCT R - CCCAAACGACGGTCAAAAGG	Conventional PCR	462	Uehara *et al*. (1998b)
	ITS rDNA	F - ATTCCGTCCGTGGTTGCTATG R - GCCGAGTGATCCACCGATAAG	SYBR green qPCR	134	Sato *et al*. (2007)
	ITS rDNA	F -TGACTATATGACACATTTRAACTTG R - ATATGCTTAAGTTCAGCGGGT	Duplex PCR	660	Waeyenberge *et al*. (2009a)
	ITS rDNA	F - AATGTGTCTCGCCCTGAGG R - GCAACCACGGACGGAATAC	TaqMan qPCR	80	Oliveira *et al*. (2017)
	β-1,4 endoglucanase	F - CCAACCTCTGCTACACTA R - CAGTGCCGTATTCAGTGA	TaqMan qPCR	–	Mokrini *et al*. (2013)
	β-1,4 endoglucanase	F - GCATTTATGTG(A/C)TCGTGGATTGGC R - GTTGCCATCAGCGCTGACAGTG	Conventional PCR	528	Peetz and Zasada (2016)

Continued

Table 60. Continued.

Species	DNA region	Primer set (5'–3')	Molecular assay	Amplicon length (bp)	Reference
P. scribneri	28S rDNA	F - AAAGTGAACGTTTCCATTTC R - TCGGAAGGAACCAGCTACTA*	Conventional PCR	286	Al-Banna *et al.* (2004)
	28S rDNA	F - ATGTGTTGCCATGCATCTG R - GTCCAGAACCCATTTGGACT	Conventional PCR	750	Mekete *et al.* (2011)
	ITS rDNA	F - GTGTTGCTATAATTCATGTAAAGTTGC R - TGGCCAGATGCGATTCGAGAGGTGT	SYBR green qPCR	136	Huang and Yan (2017)
P. thornei	28S rDNA	F - GAAAGTGAAGGTATCCCTCG R - TCGGAAGGAACCAGCTACTA*	Conventional PCR	288	Al-Banna *et al.* (2004)
	ITS rDNA	F - GTGTGTCGCTGAGCAGTTGTTGCC R - GTTGCTGGCGTCCCCAGTCAATG	SYBR green qPCR	131	Yan *et al.* (2012)
	β-1,4 endoglucanase	F - GGATGCGGTCATCAAGGC R - TTGGCTCTGGTGGTTCTG	TaqMan qPCR	88	Mokrini *et al.* (2014)
	β-1,4 endoglucanase	F - GGCTGGTCAGGAGTGAAGTCC R - GCCAGTTCCACCACTCGTTGG	Conventional PCR	364	Peetz and Zasada (2016)
P. vulnus	β-1,4-endoglucanases	F - AGTCGAGCGGCGATGGCAC R - GGCACTGTCGACACTTCGG	qRT-PCR	900	Fanelli *et al.* (2014)
	cDNA	F - GAACAATCTTCGAAGGACT R - ACCCCCTCCCATTGTCTAGC	qRT-PCR	1228	Fanelli *et al.* (2019)
	18S rDNA ITS	F - TTGGATAACTGTGGTTTAACTA R - ATTTCACCTCTCACGCAACA	Conventional PCR	–	Chihani-Hammas *et al.* (2018)
	28S rDNA	F - ACAAGTACCGTGAGG GAAAGTTG R - TCGGAAGGAACCAGCTACTA	Conventional PCR	–	
P. zeae	28S rDNA	F - GTTTCCGTAGGTGAACCTGC* R - TACGCATACRGTTCTGCTCAT	Conventional PCR	560	Troccoli *et al.* (2016)
	LSU rDNA	F - GACCCGTCTTG AAACACGGA R - TCGGAAGGAACCAGCTACTA	Conventional PCR	345	Carta *et al.* (2001)
Zygotylenchus guevarai	28S rDNA	F - GTT TCCGTAGGTGAACCTGC R - ATATGCTTAAGTTCAGCGGGT	Conventional PCR	774	Majd-Taheri *et al.* (2013)

F - forward; R - reverse; *universal primer.

a beta-glucosidase in the saliva, which can free bound phenols from phenolic glucosides (Acedo and Rohde, 1971).

82.4 Symptoms

Pratylenchus species are migratory endoparasites and feed and reproduce within the root system (Khan, 2008). The nematode may also feed on the root surface as an ectoparasite without entering the root tissue (Duncan and Moens, 2013). The nematode directly feeds on roots, tubers and cortical cells. All stages, including adults and juveniles, infect roots and degrade cell walls with mechanical movements of the stylet but also by secreting enzymes, which degrade the cytoplasm within cells, resulting in brown lesions at the points of entry and root cavities (Zunke, 1990a,b). Such lesions lead to cell death and an apparent reduction in root growth (Castillo and Vovlas, 2007). Damaged roots obstruct the uptake of nutrients and water, thus showing above-ground symptoms with stunted plants and leaf yellowing (Duncan and Moens, 2013). The above-ground symptoms are usually non-specific, but in some cases infestation of *Pratylenchus* occurs in patches in fields, as with other plant-parasitic nematodes (Khan, 2008). Generally, leaves show chlorosis and appear pale green to yellowish. Infected plants become stunted and weak, with a tendency to wilt in drought situations. Fruit size may also reduce. Depending on the species, various symptoms can be observed on the above-ground parts of the plant (Fig. 10.9).

On below-ground parts, the primary symptom of *Pratylenchus* attack is the presence of characteristic lesions on host plant feeder roots. The colour of the lesion may be varied, depending upon species and hosts (Fig. 10.10). The lesions, sometimes similar to 'cat scratches', appear as streaked yellowish to brownish stretch marks parallel to the long axis of the root. On severely damaged roots, the lesions are dark-coloured, normally yellow-brown or 'rusty'. At high populations, the lesions may girdle the root and the cortical tissues of the girdle may slough off in a sleeve-like fashion (Khan, 2008). The root systems may also show 'witch's-broom' symptoms in which the feeder roots proliferate and form clusters or may be sparse as a result of root-terminal meristem killing (Fig. 10.10). Secondary root rots may also result, due to invasion by fungi or bacteria entering through the wounds created by the nematode.

82.5 Management measures

Crop protection strategies such as cultural practices including the use of resistant varieties, crop rotation, organic

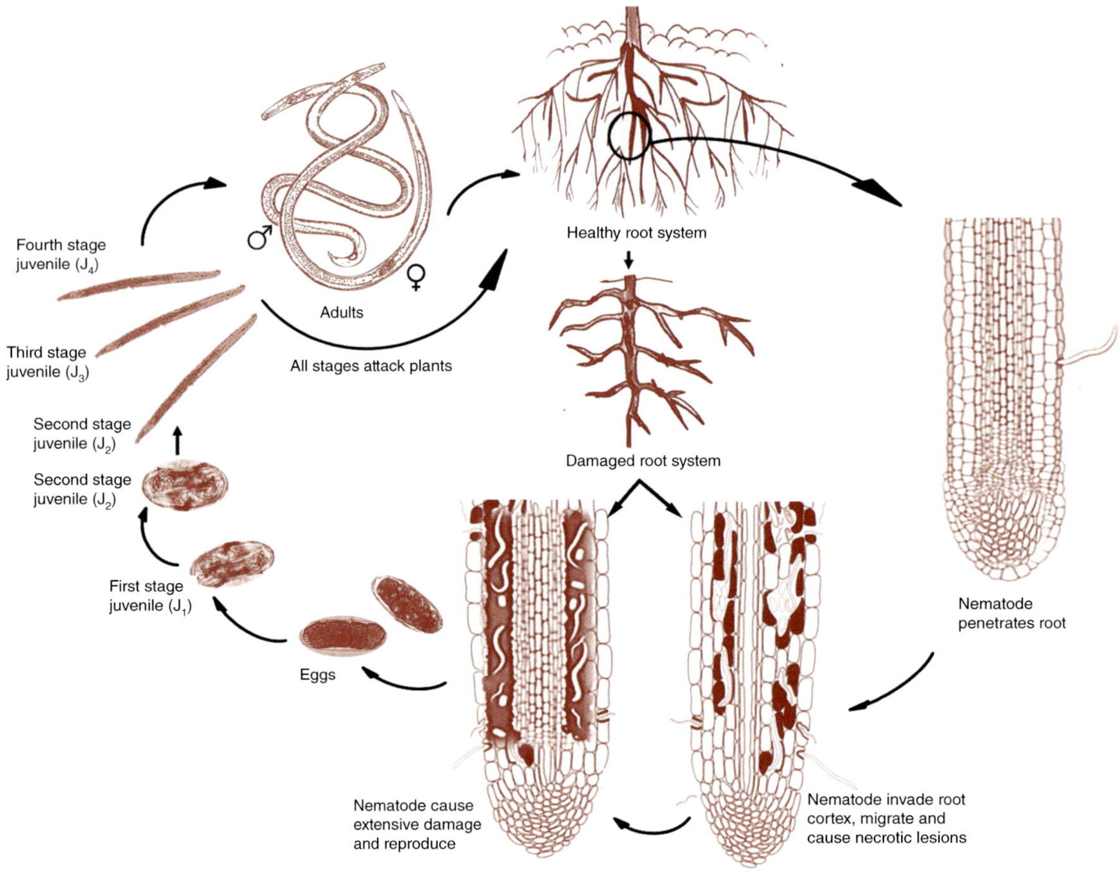

Fig. 10.8. A diagrammatic representation of the life and disease cycle of *Pratylenchus*. [Author's own photographs]

amendments, cover crops, removal of weeds, physical, chemical and biological control are well practised for root-lesion nematodes (Castillo *et al.*, 2012; Duncan and Moens, 2013).

82.5.1 Cultural practices

Crop rotation with marigold is reported to effectively suppress *P. penetrans* (Hooks *et al.*, 2010). Cover crops such as alfalfa (*Medicago sativa*), marigold (*Tagetes* spp.) and redtop (*Agrostis alba*), grown as inter crops, also help to reduce the nematode population (Castillo and Vovlas, 2007). Organic amendment with oil cakes of neem (*Azadirachta indica*), cotton (*Gossypium* sp.), karanj (*Pongamia pinnata*), Indian mustard (*Brassica juncea*), mahua (*Madhuca longifolia*) and polanga (*Calophyllum inophyllum*) have been reported to suppress *Pratylenchus* spp. (Alam, 1991). Green manuring with *Crotalaria paulina* followed by crop rotation with *Tagetes patula* and the application of carbofuran effectively controlled *P. brachyurus* in bean (*Phaseolus vulgaris*) and maize (Sharma and Scolari, 1984).

82.5.2 Host resistance

The potato cultivar 'Butte' in the USA is resistant to *P. penetrans* and *P. neglectus* (Davis *et al.*, 1992). Similarly, the pineapple clone 'Perolera 103-104' showed tolerance against *P. brachyurus* in Côte d'Ivoire (Gnonhouri *et al.*, 2000). Groundnut cultivars '295233', 'PI 290606' and 'PI 365553' (Starr, 1984), clones of potato (Dunn, 1973b), maize (Lordello *et al.*, 1985), coffee (Oliveira *et al.*, 1999) and soybean (Ferraz, 2006) are reported to be resistant to *P. brachyurus*. The banana varieties 'Kunnan', 'Paka' and 'Yamgambi KM5' (Collingborne *et al.* 1998) are resistant or tolerant to *P. coffeae* in India. Robusta coffee genotypes BP961, BP959, BP973, BP991 and BP993 show resistance to *P. coffeae* (Wiryadiputra, 1996). Cowpeas (*Vigna unguiculata*) and mungbean (*Vigna mungo*) are resistant to *P. zeae* (Aung and Prot, 1990).

82.5.3 Biological control

Root-lesion nematodes are difficult targets for biological control because of their endoparasitism (Stirling, 2014). The nematode-trapping fungus *Hirsutella rhossiliensis* showed effective parasitism on *P. penetrans* (Timper and Brodie, 1994). The endophytic bacterial isolates, *Microbacterium esteraromaticum* and *Kocuria varians* showed nematicidal activity against *P. penetrans* (Sturz and Kimpinski 2004). An isolate of *Bacillus thuringiensis*, tested in the field and greenhouse against other nematodes, was shown to reduce *P. penetrans* populations in roots of strawberry in a greenhouse (Zuckerman *et al.*, 1993).

Fig. 10.9. Above-ground symptoms of *Pratylenchus* spp. on different corps. Potato (A); wheat (B); field pea (C); coffee (D); banana (E); tea (F); tobacco (G); saffron (H). [Photo courtesy A: Nicole Viaene; B: Sarah Collins, Department of Agriculture and Food, Government of Western Australia (DAFWA); C: V. Vanstone; D: Moura *et al.*, 2002; E: John Bridge; F: N.C. Gnanapragasam; G: Agriculture & Agri-Food Canada]

82.5.4 Chemical control

Various studies show that the root-lesion nematodes can be managed with nematicides. Non-fumigant nematicides like cadusafos, carbofuran, disulfoton, ethoprophos, fenamiphos, fensulfothian, isazofos, oxamyl and thionazin have been reported to control *Pratylenchus* species in various crops (Olthof, 1986; Rabie *et al.*, 1997; Orlando *et al.*, 2020).

82.5.5 Phytosanitary measures

See Chapter 16, section 115.

83 *Pratylenchus brachyurus*

83.1 Common name

Root-lesion nematode

83.2 Scientific name

Pratylenchus brachyurus (Godfrey, 1929) Filipjev and Schuurmans Stekhoven, 1941

83.3 Synonyms

Anguillulina brachyura (Godfrey, 1929), Goodey, 1932
Anguillulina (*Pratylenchus*) *brachyura* (Godfrey, 1929) Goodey, 1932
Pratylenchus leiocephalus Steiner, 1949
Pratylenchus pratensis Thorne, 1949
Pratylenchus steineri Lordello, Zamith and Boock, 1954
Tylenchus brachyurus Godfrey, 1929
Tylenchus (*Chitinotylenchus*) *brachyurus* Godfrey, 1929 (Filipjev, 1934)

83.4 EPPO code

PRATBR

83.5 Diagnosis

83.5.1 Morphological diagnosis

The morphological descriptions of *P. brachyurus* given by Sher and Allen (1953), Corbett (1976) and Loof (1978) are summarized below. For morphometric measurements, see Table 61.

Fig. 10.10. Below-ground symptoms of *Pratylenchus* spp. on different crops. Yam corm (A); maize (B); coffee (C); strawberry (D); legumes (E); plantain (F); sugarcane (G); tea (H); saffron (I); Missing in figures. [Photo courtesy A,D,F: John Bridge/CABI BioScience; B: Agriculture & Agri-Food Canada; C: Moura *et al.*, 2002; E,G,I: NASI, India; H: N.C. Gnanapragasam, source CABI; J: Sarah Collins; K: Machado *et al.*, 2019, PlosOne; L: V. Vanstone]

Female: Body slender. Head region low and usually flat, frame sclerotized. Lateral field with four incisures. Cephalic region angular, offset with two distinct annules. Stylet with large, round basal knobs. Hemizonid about two body annules long just above excretory pore. Ovary single uterus prodelphic, not extending to oesophageal gland, consists of single row of oocytes, post-uterine sac short. Tail rounded, no striae near terminus. Phasmids situated near middle of tail.

Male: Extremely rare. Testis single, outstretched. Phasmids slightly posterior to middle of tail. Spicule arcuate, hafted. Gubernaculum simple, trough shaped. Tail conical and rounded to flattened at terminus. Bursa delicate.

83.5.2 Molecular diagnosis

For primers, see Table 60.

83.6 Geographical distribution

P. brachyurus is widely distributed throughout the tropical and subtropical regions of the world (Fig. 10.11). The

Table 61. Morphometric measurements of *Pratylenchus* species.

Characters		Range, numerous populations												
		P. alleni (after Ferris, 1961)	*P. brachyurus* (after Loof, 1960)[a]	*P. coffeae* (after Sher and Allen, 1953)	*P. crenatus* (after Kumari, 2015)[a]	*P. fallax* (after Seinhorst 1977)	*P. goodeyi* (after Sher and Allen, 1953)	*P. loosi* (Loof, 1960)	*P. neglectus* (after Kumari, 2015)[a]	*P. penetrans* (after Sher and Allen, 1953; Loof, 1991)	*P. scribneri* (Yan *et al.*, 2016)	*P. thornei* (after Loof, 1960)[a]	*P. vulnus* (after Allen and Jensen, 1951)	*P. zeae* (after Fortuner, 1976)[a]
Female	L (µm)	370–482	390–750	450–700	415–483	420–560	640–680	480–640	413–573	430–650	407–532	408–708	460–910	373–506
	a	20.6–23.0	15–29	25–35	23.0–30.1	24–33	27–37	28–36	19.2–27.9	17–34	17.7–24.6	25.3–36.4	26.6–39.5	20–30
	b	4.7–5.5	5–10	5–7	4.2–5.1	5.2–6.7	5.5–6.1	5.7–7.1	3.7–5.0	4.1–8.1	4.0–5.2	5.4–8.3	5.3–7.7	4.9–6.1
	c	19.3	13–28	17–22	17.2–23.1	18–24	16–18	18–25	16.0–26.7	14–26	16.2–22.4	16.8–25.1	14.2–27.7	15–19
	Stylet length (µm)	14–15	17–22	15–18	13–15	16–17		14–18	14–19	15–17	15.0–15.5	15–19	16–18	15.5–16.5
	V	79.0–79.7	82–89	76–83	81–84	77–81	73–75	20–40	78–84	75–77	75.5– 78.7	74.4–79	78.0–84.1	68.6–73.9
	Tail length (µm)	25	51–53%[b]	–	18–28	–	27.7–41[c]		19–33	–	25–28	–	–	–
	DGO (µm)	–	2	–	2–3	–	2.3–4.3[c]		2–4	–	–	–	–	3
	Excretory pore to anterior end (µm)	62.5–82.5	57–108	–	64–74	–			75–94	74–101	–	–	–	–
	Body diameter at mid body (µm)	18–21	–	–	13–20	–	17.8–28.7[c]		17–22	–	21–28	–	–	–
	Vulva-anus length (µm)	76	–	–	46–63	–			59–85	–	–		–	–
Male	L (µm)	–	–	450–700	–	400–500	550–570	38–58	–	305–574	–	–	460–730	–
	a	–	–	26–40	–	26–33	26	28–41	–	23–34	–	–	28.3–39.2	–
	b	–	–	6–7	–	5.0–6.2	5.4–5.8	5.4–6.7	–	5.4–7.3	–	–	5.3–7.4	–
	c	–	–	17–24	–	16–25	17–18	18.5–23.2	–	16–22	–	–	17.5–29.4	–
	Stylet length (µm)	–	–	15–17	–	15	12.7–16.0[c]	12–16	–	13–16	–	–	15–18	–
	Tail length (µm)	–	–	45–52	–	–	28.4–35[$]	36–52	–	36–58	–	–	35.8–66	–
	Spicule length (µm)	–	–	16–20	–	14–16	–	16–20	–	14–17	–	–	14–20	14–15
	Gubernaculum length (µm)	–	–	4–7	–	4	–	4–7	–	3.9–4.2	–	–	4–6	4–5

[a]Males are extremely rare or absent; [b]from Sher and Allen (1953); [c]from Mgonja *et al.* (2019)

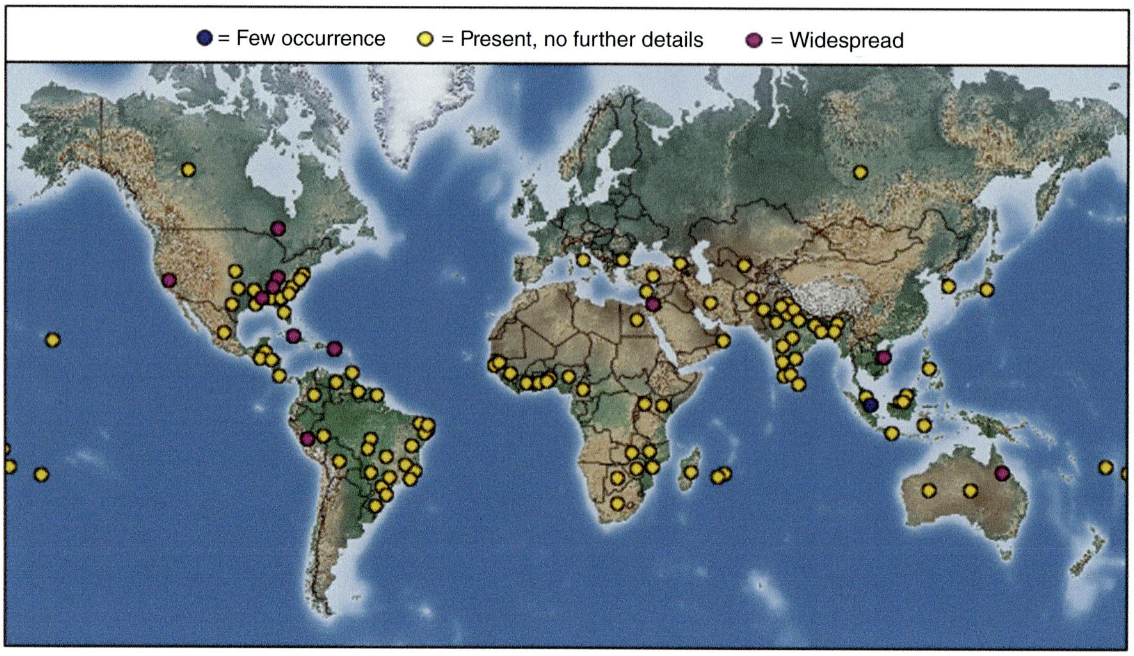

Fig. 10.11. Distribution map of *Pratylenchus brachyurus*. [Adapted from CABI-ISC, 2020]

nematode has been recorded in the following countries (CABI-ISC, 2020).

Africa: Benin, Botswana, Cameroon, Côte d'Ivoire, Egypt, Gambia, Ghana, Guinea, Kenya, Madagascar, Malawi, Mauritius, Mozambique, Nigeria, Réunion Senegal, South Africa, Togo, Uganda, Zambia, Zimbabwe.
Asia: Afghanistan, Brunei, Georgia, India, Indonesia, Iran, Israel, Japan, Malaysia, Oman, Pakistan, Philippines, Singapore, South Korea, Sri Lanka, Turkey, Uzbekistan, Vietnam.
Europe: Bulgaria, Cyprus, Italy, Russia.
North America: Belize, Canada, Costa Rica, Cuba, Guatemala, Honduras, Mexico, Puerto Rico, Trinidad and Tobago, USA.
Oceania: Australia, Cook Islands, Fiji, Niue, Samoa, Tonga.
South America: Bolivia, Brazil, Colombia, French Guiana, Guyana, Peru, Venezuela.

83.7 Hosts

P. brachyurus is extremely polyphagous, with a very wide host range, including fruit trees, forest trees, cereals, vegetables, tuber crops, oilseed crops, fibre crops, woody ornamentals, plantation crops, grasses and medicinal plants (CABI-CPC, 2010). An expanded list of hosts of *P. brachyurus* is presented in Table 62.

83.8 Symptoms

P. brachyurus produces characteristic typical necrotic lesions on the surface and throughout the cortex of infected roots. The type, colour and size of the lesions vary according to the level of infestation, age of the lesion and the host species.

GROUNDNUTS *P. brachyurus* mainly infects groundnut pods and pegs. The nematode causes brown tunnels in groundnut shells which enlarge into purplish-brown lesions with discrete boundaries. Roots are also attacked, resulting in stunting of the plant and chlorosis.

POTATOES In potatoes, the damage is only to the cortical tissues. Tubers are often attacked and small lesions are found on the surface of roots and the outer peel of tubers. The lesions are initially light brown in colour, becoming purplish-brown with raised centres. They are shallow and rarely penetrate more than 0.5 mm deep. Lesions, scabs, pustules or pimples are also formed on the tubers, reducing their market value.

PINEAPPLE The nematode burrows through the root cortex and induce necrosis after 4 days. Growth of pineapple roots ceases, resulting in vessel blocking.

83.9 Biology and life cycle

The biology and life cycle are typical of the genus. *P. brachyurus* is monosexual: males are very rare and reproduction is by mitotic parthenogenesis. Reproduction is more rapid at higher temperatures (26.7–32.2°C) and the optimum temperature for reproduction is 26°C (Boswell, 1968).

83.10 Economic importance

Severe crop damage and significant yield and quality reduction have been reported in many groundnut fields in the USA and

Table 62. List of host plants of *Pratylenchus brachyurus* (adapted from CABI-CPC, 2010).

Main hosts
Acer rubrum (red maple)
Ananas comosus (pineapple)
Arachis hypogaea (groundnut)
Cajanus cajan (pigeon pea)
Camellia sinensis (tea)
Casuarina equisetifolia (casuarina)
Chrysanthemum coronarium (chrysanthemum)
Citrus
Coffea (coffee)
Cymbopogon citratus (citronella grass)
Dianthus barbatus (sweet william)
Digitaria decumbens (pangolagrass)
Dioscorea (yam)
Eucalyptus (eucalyptus tree)
Fragaria (strawberry)
Glycine max (soybean)
Gossypium (cotton)
Hevea brasiliensis (rubber)
Liriodendron tulipifera (tuliptree)
Nicotiana tabacum (tobacco)
Oryza sativa (rice)
Persea americana (avocado)
Phaseolus lunatus (lima bean)
Prunus persica (peach)
Rosa multiflora (multiflora rose)
Saccharum officinarum (sugarcane)
Solanum tuberosum (potato)
Theobroma cacao (cocoa)
Tipuana tipu (tipu tree)
Vigna radiata (mung bean)
Zea mays (maize)
Other hosts
Abelmoschus esculentus (okra)
Acalypha hispida (copperleaf)
Allium cepa (onion)
Allium fistulosum (Welsh onion)
Avena sativa (oats)
Bambusa vulgaris (common bamboo)
Bidens pilosa (blackjack)
Citrullus
Cucumis sativus (cucumber)
Cucurbita (pumpkin)
Feijoa sellowiana (horn of plenty)
Ficus carica (fig)
Gossypium barbadense (Gallini cotton)
Helianthus annuus (sunflower)
Hibiscus cannabinus (kenaf)
Hordeum vulgare (barley)
Hyparrhenia rufa (jaraguagrass)
Impatiens (balsam)
Ipomoea batatas (sweet potato)
Leucaena leucocephala (leucaena)
Mangifera indica (mango)
Manihot esculenta (cassava)
Mentha piperita (peppermint)
Mucuna pruriens (buffalobean)
Pennisetum purpureum (elephant grass)
Phaseolus vulgaris (common bean)
Pinus palustris (longleaf pine)
Poaceae (grasses)
Secale cereale (rye)
Sesamum indicum (sesame)
Setaria italica (foxtail millet)
Solanum lycopersicum (tomato)
Solanum melongena (aubergine)
Solanum torvum (turkey berry)
Sorghum bicolor (sorghum)
Sorghum sudanense (Sudan grass)
Triticum aestivum (wheat)
Urochloa decumbens (signal grass)
Vanilla
Vigna unguiculata (cowpea)
Zingiber officinale (ginger)

other countries due to *P. brachyurus* infestation (Wheeler and Starr, 1987; Minton and Baujard, 1990). Pineapple is an important host for this species in the equatorial tropics and yield losses as high as 30% for the plant crop and 80% for the ratoon crop are reported in Côte d'Ivoire (Sarah, 1986). In Nigeria, a 28.5% reduction in yield of maize was correlated with a 50% increase in *P. brachyurus* density (Egunjobi, 1974). *P. brachyurus* is also of economic importance to cotton and decreases fibre strength, fibre maturity and fibre length (Lordello and Sabino, 1985). *P. brachyurus* is a pathogen of economic importance to potatoes (Koen, 1967). Infestation in potato not only causes losses in yield but also reduces the marketable quality of the tuber. Very low initial population levels of *P. brachyurus* significantly reduced the growth of coffee plants in the greenhouse (Winslow and Willis, 1972). On soybean, *P. brachyurus* caused yield losses of 30–50%, with higher damage in sandy soils (Machado *et al.*, 2019).

83.11 Movement and means of dispersal

See section 76.11, above.

83.12 Pest risk analysis and looming threat from introduction

The wide host range of *P. brachyurus* would facilitate the establishment of this nematode on any alternative hosts in a new host area. The direct economic impact of this nematode on cotton, groundnut, pineapple, potato and soybean production is well established. The indirect economic and regulatory impacts due to its introduction should not be underestimated. The production of host crops would be seriously affected by quarantine restrictions imposed against *P. brachyurus*.

83.13 Invasiveness rating

Considering the consequences of introduction and probable economic damage caused by *P. brachyurus*, the following invasiveness rating has been assigned.

Climate/host interaction	–	High (3)
Hosts range	–	High (3)
Dispersal potential	–	High (3)
Economic impact	–	High (3)
Environmental impact	–	High (3)
Post-entry distribution	–	Medium (–2)
Final rating (score)	–	High (13)

Based on the above invasiveness rating, *P. brachyurus* receives a score of 13 out of 15 and is rated as HIGHLY INVASIVE NEMATODE.

83.14 Region-wise status of invasiveness

P. brachyurus is listed as one of the quarantine species in many countries, including India where it is restricted for the ginger crop (PQ Order-2003, 2015). Because of the polyphagous nature and damaging potential of the nematode, *P. brachyurus* has been given a highly invasive rating for all the countries where this nematode does not occur.

83.15 Management measures

See section 82.5, above.

84 *Pratylenchus coffeae*

84.1 Common name

Coffee root-lesion nematode

84.2 Scientific name

Pratylenchus coffeae (Zimmermann 1898) Filipjev and Schuurmans Steckhoven, 1941

84.3 Synonyms

Anguillulina mahogani (Cobb, 1920) Goodey, 1932
Pratylenchus mahogani (Cobb, 1920) Filipjev, 1936
Pratylenchus musicola (Cobb, 1920) Filipjev, 1936
Tylenchus coffeae Zimmermann, 1898
Tylenchus mahogani Cobb, 1920
Tylenchus musicola Cobb, 1919

84.4 EPPO code

PRATCO

84.5 Diagnosis

84.5.1 Morphological diagnosis

Morphological descriptions of *P. coffeae* after Siddiqi (1972a) are summarized below. For morphometric measurements (after Sher and Allen, 1953), see Table 61.

FEMALE: Body slender with distinct annule (Fig. 10.12G). Lateral fields usually with four, occasionally five or six. Cephalic region offset slightly, consisting of two distinct annules. Stylet with round to oblong basal knobs (Fig. 10.12A). Post-uterine sac present with terminal rudimentary ovary. Spermathecae large, rounded to oval, frequently with sperms. Tail with indented tip, occasionally smoothly rounded, irregularly crenate or truncate (Fig. 10.12M,N).
MALE: Abundant in number. Spicules with well-marked manubria and ventrally arcuate slender shaft (Fig. 10.12E). Gubernaculum 4–7 µm long. Hypoptygma prominent. Bursa delicate with faintly crenate margins (Fig. 10.12D).

84.5.2 Molecular diagnosis

For primers, see Table 60.

84.6 Geographical distribution

P. coffeae is widely distributed throughout the tropical and subtropical regions of the world (Fig. 10.13). The nematode has been recorded from the following countries (CABI-ISC, 2020).

Africa: Cameroon, Congo Democratic Republic, Côte d'Ivoire, Eswatini (Swaziland), Ghana, Kenya, Madagascar, Malawi, Mauritius, Morocco, Mozambique, Nigeria, Seychelles, South Africa, Tanzania, Uganda, Zambia, Zimbabwe.
Asia: Afghanistan, Bangladesh, Bhutan, Brunei, China, Georgia, India, Indonesia, Iran, Japan, Malaysia, North Korea, Oman, Pakistan, Philippines, South Korea, Sri Lanka, Taiwan, Thailand, Turkey, Vietnam.
Europe: Austria, Bulgaria, Italy, Slovenia, Spain.
North America: Barbados, Belize, Costa Rica, Cuba, Dominica, Dominican Republic, El Salvador, Grenada, Guadeloupe, Guatemala, Honduras, Jamaica, Martinique, Mexico, Nicaragua, Panama, Puerto Rico, Trinidad and Tobago, USA.
Oceania: Australia, Cook Islands, Fiji, Kiribati, Niue, Palau, Papua New Guinea, Samoa, Solomon Islands, Tonga, Vanuatu.
South America: Brazil, Colombia, Ecuador, French Guiana, Suriname, Venezuela.

84.7 Hosts

P. coffeae is highly polyphagous; its host range is spread over 250 plant species covering almost all plant families

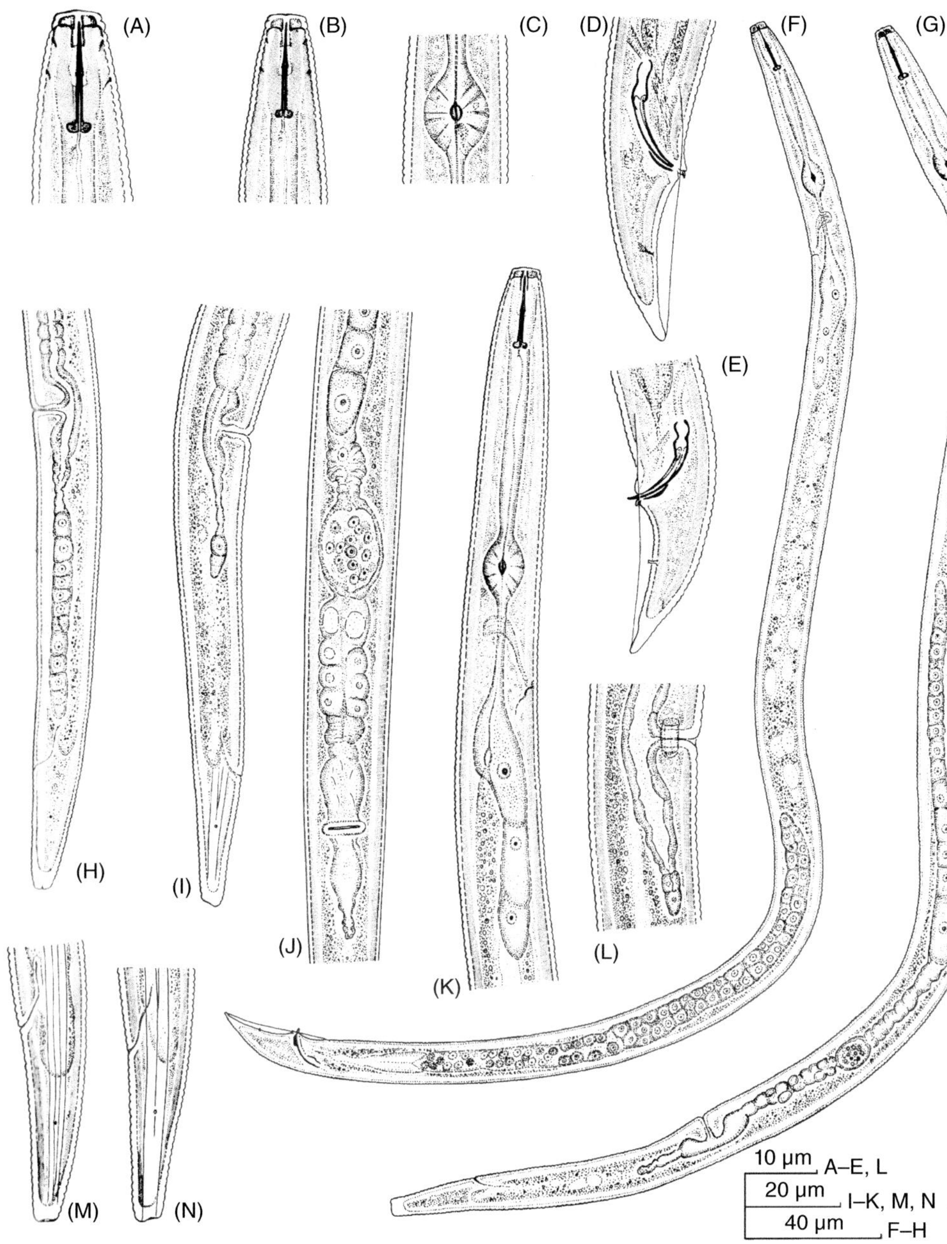

Fig. 10.12. Line drawing of *Pratylenchus coffeae*. Female head end (A); male head end (B); median oesophageal bulb, male (C); male tail end (D,E); entire male (F); entire female (G); female tail end (H,I); female vulval region (J,L); female oesophageal region (K); female tail end (M,N). [Reproduced from M.R. Siddiqi, 2000; courtesy CAB International]

(Jackson *et al.*, 2003). The main hosts of *P. coffeae* include Araceae, *Citrus*, *Coffea* (coffee), *Colocasia esculenta* (taro), cucurbits, *Curcuma longa* (turmeric), *Dioscorea* (yam), *Elettaria cardamomum* (cardamom), *Ficus carica* (fig), *Musa* (banana), *Musa textilis* (abacá, manila hemp), *Musa* × *paradisiaca* (plantain), *Piper nigrum* (black pepper), *Solanum tuberosum* (potato), *Theobroma cacao* (cocoa), *Zea mays* (maize) and *Zingiber officinale* (ginger) (CABI-ISC, 2020).

84.8 Symptoms

P. coffeae attacks root, corms, rhizome and tuber of host plants. Symptoms on different host plants and plant parts are described below.

BANANAS, PLANTAINS AND ABACÁ The nematode causes stunting, extended vegetative growth period, decreased size and number of leaves and bunch weight, and toppling similar to that of

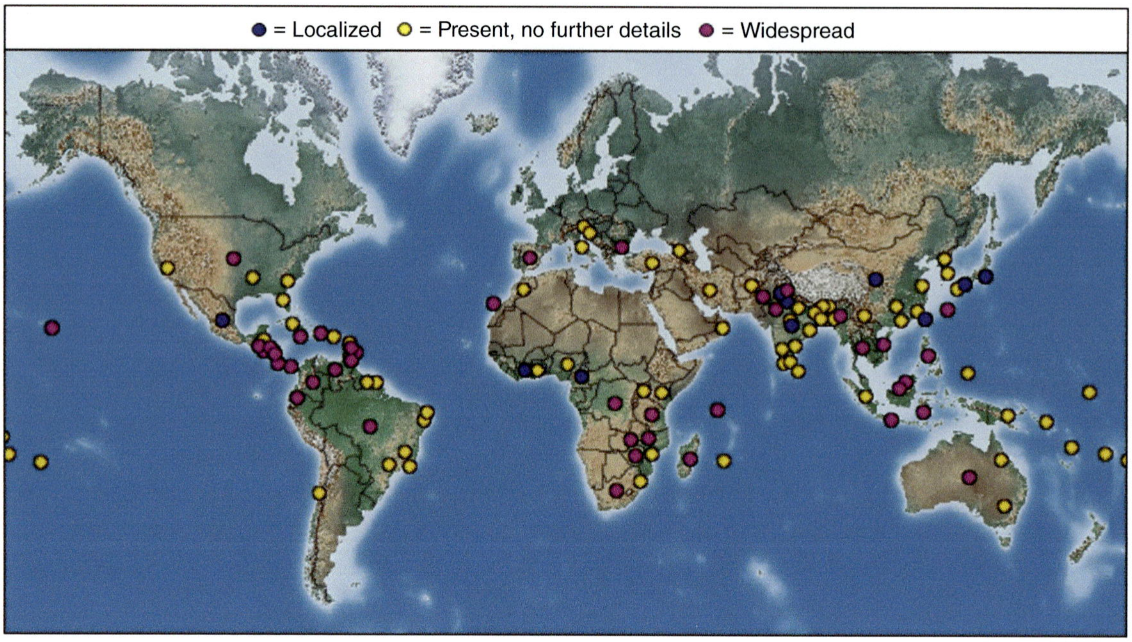

Fig. 10.13. Distribution map of *Pratylenchus coffeae* (adapted from CABI-ISC, 2020).

Radopholus similis. Infected roots and outer parts of the corm have extensive purple or black necrotic lesions, frequently resulting in secondary rotting and root breakage (Luambano *et al.*, 2019).

YAMS In yam tubers, *P. coffeae* causes dry rot symptoms with brown, irregular dry rot extending 1–2 cm into the outer tissues of tubers, sometimes as deep as 5 cm (Fig. 10.10A). The symptoms are more obvious in the oldest apical portions of the tubers. External cracks with a corky appearance exposing dark-brown rotted areas on diseased tubers have also been observed (Bridge and Page, 1984).

COFFEE Infected coffee roots initially turn yellow, then brown, with rotting of most of the lateral roots. Diseased plants are stunted, with a few small chlorotic leaves, and the wilt gradually sets, leading to the death of the whole plant. Heavily infected plants may die prematurely (Campos *et al.*, 1990).

GINGER AND TURMERIC The infected rhizome of ginger is rotted, with yellowing of the leaves. Turmeric rhizomes also show discoloration and rotting. In severe infestations, turmeric rhizomes become less turgid; they become deep-red to brown in colour and wrinkled with dark-brown necrotic lesions (Koshy and Bridge, 1990).

84.9 Biology and life cycle

The biology and life cycle are typical of the genus. *P. coffeae* is bisexual, males are abundant and reproduction is amphimictic. The eggs hatch in 6–8 days at 28–30°C. The average life cycle on potato is about 27 days at 25–30°C (Siddiqi, 1972a). On yam roots and tubers, the life cycle takes 21–28 days. The nematode can survive in moist soil for up to 8 months without host plants (Bridge and Page, 1984).

84.10 Economic importance

BANANAS, PLANTAINS AND ABACÁ *P. coffeae* is of economic importance on bananas in the Pacific island countries and several parts of Africa, and on bananas, plantains and abacá (*Musa textilis*) in Cuba and in Central and South America (Gowen and Quénéhervé, 1990). In South Africa, estimated crop losses can be as high as 80% (Sarah, 1989). In Cuba, high population densities cause toppling of the plants and limit the production of the crop (Fernández Díaz-Silveira and Ortega Herrera, 1998). *P. coffeae* has been reported to cause 24% losses in bunch weight and the toppling of banana plants (Moens *et al.*, 2006). Similarly, in Ghana *P. coffeae* caused severe death and necrosis of roots and suckers, and lesion on corm in bananas (Brentu *et al.*, 2004). Luambano *et al.* (2019) discovered the presence of *P. coffeae* in around 80% of the samples collected from banana plantations in Tanzania, but there are no research data on the effect of this nematode on banana production.

YAMS *P. coffeae* is an important parasite of yam tubers and reduces their edible portions, marketable value and mainly their storage qualities. Around 67–100% of *Dioscorea cayenensis* and *D. rotundata* tubers were found to be infested with *P. coffeae* in Jamaica and over 50% of *D. alata* tubers in Papua New Guinea (Thompson *et al.*, 1973). The infested tuber may contain over 60,000 nematodes per 50 g of tissues (Bridge and Page, 1984).

COFFEE *P. coffeae* is an important pest of coffee in Brazil, Costa Rica, Cuba, Dominican Republic, El Salvador, Guatemala and Puerto Rico. The nematode also infects coffee in India, Barbados, Martinique, Madagascar and Tanzania. In Java and India, it attained the status of a major pest in coffee (Campos *et al.*, 1990).

GINGER AND TURMERIC: In India, *P. coffeae* is widely distributed in ginger and turmeric growing parts of the country; it causes 'ginger yellows' and it damages both roots and rhizomes (Bora and Chaudhury, 2010).

84.11 Movement and means of dispersal

See section 76.11, above.

84.12 Pest risk analysis and looming threat from introduction

P. coffeae has a very wide host range and its direct economic damage to the production of bananas, citrus, coffee, ginger, plantains, yams, ginger, turmeric and other host crops is well documented. The indirect economic and regulatory impacts due to its introduction should not be underestimated. The production of host crops would be seriously affected by quarantine restrictions imposed against *P. coffeae*.

84.13 Invasiveness rating

Considering the consequences of introduction and pest risk analysis of *P. coffeae*, the following invasiveness rating has been assigned.

Climate/host interaction	–	High (3)
Hosts range	–	High (3)
Dispersal potential	–	High (3)
Economic impact	–	High (3)
Environmental impact	–	High (3)
Post-entry distribution	–	Medium (–2)
Final rating (score)	–	High (13)

Based on the above invasiveness rating score, *P. coffeae* receives a score of 13 out of 15 and is rated as HIGHLY INVASIVE NEMATODE.

84.14 Region-wise status of invasiveness

Because of the polyphagous nature of the nematode and its damaging potential, *P. coffeae* poses a phytosanitary risk in all tropical and subtropical countries. It has been given a highly invasive rating for all the countries where this nematode does not occur.

84.15 Management measures

See section 82.5, above.

85 *Pratylenchus crenatus*

85.1 Common name

Root-lesion nematode

85.2 Scientific name

Pratylenchus crenatus Loof, 1960

85.3 EPPO code

PRATCR

85.4 Diagnosis

85.4.1 Morphological diagnosis

The female bears a clavate tail with annulated tip shape; there is strong musculature at the secretory excretory pore (Karssen and Bolk, 2000). Males are extremely rare or absent. For morphometric measurements (Kumari, 2015), see Table 61.

85.4.2 Molecular diagnosis

For primers, see Table 60.

85.5 Geographical distribution

P. crenatus is distributed worldwide, occurring in 27 countries, especially in temperate and subtropical zones (Bonfim *et al.*, 2016). The nematode has also been reported in tropical Africa (Castillo and Vovlas, 2007) and recently detected in Brazil (Bonfim *et al.*, 2016).

85.6 Hosts

P. crenatus is polyphagous and mostly recorded on cereals and grasses (Loof, 1991). The nematode has also been reported to associate with potatoes, common bean and soybean (Olthof and Wolynetz, 1991; Esteves *et al.*, 2015).

85.7 Symptoms

P. crenatus produces characteristic typical necrotic lesions, similar to other *Pratylenchus* species (Fig. 10.10). The type, colour and size of the lesions vary according to the host species.

85.8 Biology and life cycle

The biology and life cycle are typical of the genus. *P. crenatus* is monosexual; males are absent and reproduction is by mitotic parthenogenesis. The nematode can be found in a wide range of soils from heavy clays to sandy soils but occurs more frequently in acidic than in neutral or alkaline soils (Kumari, 2015).

85.9 Economic importance

P. crenatus causes significant yield losses in barley, common bean, potato and soybean (Bonfim *et al.*, 2016; Orlando *et al.*, 2020).

85.10 Movement and means of dispersal

See section 76.11, above.

85.11 Pest risk analysis and looming threat from introduction

The nematode can be easily disseminated through infected potato seed tubers and nursery stock. The recent introduction of *P. crenatus* into Brazil is likely to have originated in Europe, possibly through either contaminated seed potatoes or in association with international vehicle movements (Bonfim *et al.*, 2016).

85.12 Invasiveness rating

Considering the consequences of introduction and probable economic damage caused by *P. crenatus*, the following invasiveness rating has been assigned.

Climate/host interaction	–	Medium (2)
Hosts range	–	Medium (2)
Dispersal potential	–	High (3)
Economic impact	–	High (3)
Environmental impact	–	Medium (2)
Post-entry distribution	–	Low (–1)
Final rating (score)	–	Medium (11)

Based on the above invasiveness rating, *P. crenatus* receives a score of 11 out of 15 and is rated as MEDIUM INVASIVE NEMATODE.

85.13 Region-wise status of invasiveness

Due to the polyphagous nature and damaging potential of *P. crenatus*, it has been given a medium invasive rating for all the countries where this nematode does not occur.

85.14 Management measures

See section 82.5, above.

86 *Pratylenchus goodeyi*

86.1 Common name

Banana root-lesion nematode

86.2 Scientific name

Pratylenchus goodeyi Sher and Allen, 1953

86.3 Synonyms

Anguillulina musicola Apud Goodey, 1932
Tylenchus musicola Apud Goodey, 1928

86.4 EPPO code

PRATGO

86.5 Diagnosis

86.5.1 Morphological diagnosis

Morphological descriptions of *P. goodeyi* after Sher and Allen (1953) and de Guiran and Vilardebó (1962) are summarized below. For morphometric measurements, see Table 61.

FEMALE: Body slender, nearly straight but ventrally curved slightly in posterior region after heat killing (Fig. 10.14a). Lip region with four annules. Stylet with anteriorly flattened knobs (Fig. 10.14b). Lateral field with four incisures, inconspicuous. Genital tract monoprodelphic with vulva anteriorly located at 71–78%. Postvulval sac present. Spermatheca large, subrectangular, filled with sperm (Fig. 10.14c). Tail conoid, ventrally concave with sinuate dorsal contour just prior to tail terminus (Fig. 10.14d). Phasmid generally conspicuous, 10–14 annules from tail terminus.
MALE: Abundant. Spicules well marked and ventrally arcuate (Fig. 10.14e). Gubernaculum simple. Bursa delicate, extending to tail terminus (Fig. 10.14e).

86.5.2 Molecular diagnosis

For primers, see Table 60.

86.6 Geographical distribution

P. goodeyi has limited distribution; it is mainly spread in Africa, particularly in upland areas known as the afro-montane

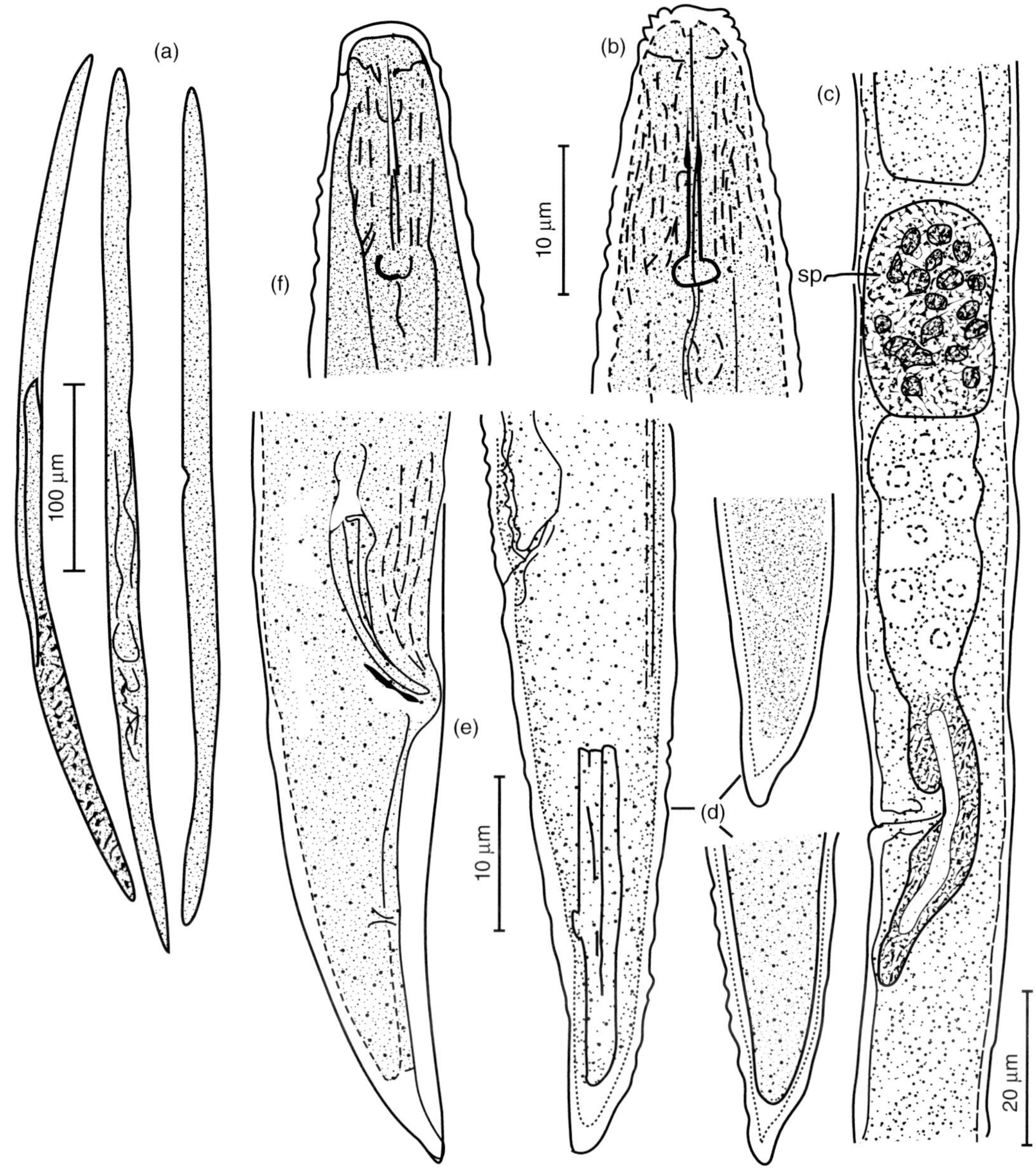

Fig. 10.14. Line drawing of *Pratylenchus goodeyi.* female body (a); cephalic region (b); female showing vulva, post-uterine branch and spermatheca (c); tail (d); spicule shaft (e); rounded male head (f). [Source: EcoPort (http://www.ecoport.org)]

highlands (Price and Bridge, 1995), and is reported from Burundi, Cameroon, Egypt, Ethiopia, Kenya, Rwanda, Tanzania and Uganda. Outside Africa, it has been detected in Australia, China, Greece, Pakistan, Portugal and Spain (CABI-ISC, 2020).

86.7 Hosts

P. goodeyi is polyphagous species and major hosts are *Musa* (banana) and *Musa* × *paradisiaca* (plantain). Other hosts include *Commelina benghalensis* (wandering jew), *Leucaena leucocephala* (leucaena), *Lycopersicon esculentum* (tomato), *Phaseolus vulgaris* (common bean), *Solanum tuberosum* (potato), *Sorghum bicolor* (sorghum) and *Zea mays* (maize). The nematode also reproduces on a few wild hosts like *Hyparrhenia rufa* (jaraguagrass) (CABI-ISC, 2020)

86.8 Symptoms

Plants infected by *P. goodeyi* show characteristic purplish necrotic lesions throughout the cortical tissue of the roots. The lesions are initially yellow, later turning to purple and then brown. The surface of the root turns black over necrosed tissues. On banana and plantain, the symptoms can be observed after longitudinal cutting of root (Fig. 10.10F), similar to those caused by *Pratylenchus coffeae* and *Radopholus similis*. Feeding by the nematode damages cortical tissues but the damage does not extend into the stele. As a result, the roots break off near the necrotic tissue, leading to toppling or uprooting of plants (Fig. 10.9E) (Mgonja *et al.*, 2019).

86.9 Biology and life cycle

The biology and life cycle are typical of the genus. *P. goodeyi* reproduces by amphimixis and, unlike *P. brachyurus*, reproduction rates are quite high at lower temperatures (16–21°C) and significantly low at higher temperatures (25°C) (Pinochet *et al.*, 1995).

86.10 Economic importance

P. goodeyi is among the major damaging pests of banana and plantain crops in afro-montane highlands of Africa, particularly in Tanzania, Uganda, Kenya and Cameroon (Sikora *et al.*, 2018). The nematode also damages ensete (false banana) in Ethiopia (Peregrine and Bridge, 1992). Yield losses on banana are estimated to be up to 50% annually in East Africa (Coyne, 2009; Mgonja *et al.*, 2019). In Tanzania, *P. goodeyi* was found in 94.5% of the banana-growing area and up to 84% damage was directly associated with toppling and other symptoms incited by the nematode (Bridge, 1988; Mgonja *et al.*, 2019).

86.11 Movement and means of dispersal

See section 76.11, above.

86.12 Pest risk analysis and looming threat from introduction

The type locality of *P. goodeyi* in the United Kingdom indicated that the nematode was disseminated through banana vegetative planting material from European greenhouses, introduced into the Canary Islands in the 19th century during the Cavendish banana breeding programme. The occurrence of *P. goodeyi* in the Canary Islands, Greece and Egypt is most probably due to introduction of the nematode through planting material.

86.13 Invasiveness rating

Considering the consequences of introduction and pest risk analysis of *P. goodeyi*, the following invasiveness rating has been assigned.

Climate/host interaction	–	High (3)
Hosts range	–	Medium (2)
Dispersal potential	–	High (3)
Economic impact	–	High (3)
Environmental impact	–	High (3)
Post-entry distribution	–	Low (–1)
Final rating (score)	–	High (13)

Based on the above invasiveness rating, *P. goodeyi* receives a score of 13 out of 15 and is rated as HIGHLY INVASIVE NEMATODE.

86.14 Region-wise status of invasiveness

P. goodeyi has a narrow host range but it includes economically important crops: banana and plantain. This nematode is listed as a quarantine species in many countries. Based on the damaging potential of *P. goodeyi* to banana production, it has been given a highly invasive rating for the following banana-producing countries where this nematode is not known to occur.

APPPC: Bahrain, Bangladesh, Bhutan, Brunei, Cambodia, Indonesia, Israel, Japan, Jordan, Kuwait, Lebanon, Malaysia, Maldives, Nepal, Oman, Philippines, Singapore, South Korea, Syria, Thailand, United Arab Emirates, Vietnam, Yemen

CAHFSA: Antigua and Barbuda, Bahamas, Barbados, Belize, Bermuda, British Virgin Islands, Cayman Islands, Costa Rica, Cuba, Dominica, Dominican Republic, El Salvador, Grenada, Guadeloupe, Guatemala, Haiti, Honduras, Jamaica, Martinique, Montserrat, Nicaragua, Panama, Puerto Rico, Saint Lucia, Trinidad and Tobago, US Virgin Islands

CAN: Colombia, Ecuador

COSAVE: Argentina, Bolivia, Brazil, French Guiana, Guyana, Paraguay, Suriname

EPPO: Cyprus, Italy

IAPSC: Algeria, Angola, Benin, Cape Verde, Central African Republic, Comoros, Congo, Côte d'Ivoire, Equatorial Guinea, Gabon, Ghana, Guinea, Guinea-Bissau, Liberia, Madagascar, Malawi, Mali, Mauritius, Morocco, Mozambique, Réunion, São Tomé and Principe, Senegal, Seychelles, Somalia, South Africa, Sudan, Swaziland, Togo, Zambia, Zimbabwe

NAPPO: Mexico, USA

PPPO: American Samoa, Cook Islands, Fiji, French Polynesia, Guam, Kiribati, Micronesia (Federated States of), New Caledonia, Niue, Papua New Guinea, Samoa, Solomon Islands, Tokelau, Tonga, Tuvalu, Vanuatu, Wallis and Futuna Islands

86.15 Management measures

See section 82.5, above.

87 *Pratylenchus loosi*

87.1 Common name

Root-lesion nematode

87.2 Scientific name

Pratylenchus loosi Loof, 1960

87.3 EPPO code

PRATLO

87.4 Diagnosis

87.4.1 Morphological diagnosis

Morphological descriptions of *P. loosi* after Seinhorst (1977) are summarized below. For morphometric measurements (after Loof, 1960), see Table 61.

FEMALE: Body very slender, almost straight after heat killing. Lips with two annules, rounded. Posterior edges of cephalic framework extend one to two body annules into body. Stylet knobs rounded and broadly fused to shaft (Fig. 10.15A). Broad lateral field with four, seldom five or even six incisures in anterior part and three posterior to anus (Fig. 10.15I). Just posterior to vulva, body narrowing fairly at about 0.8 times width anterior to vulva, then cylindrical to anus. Spermatheca oval, up to 20 µm long, located 50–90% of distance from vulva to anus in front of vulva (Fig. 10.15E). Postvulvar uterine sac present and branching 30–50% of distance between vulva and anus. Tail about 2.5 anal body diameters long, tapering gradually to smooth conical tip, usually with narrowly rounded tip (Fig. 10.15F–I).
MALE: Abundant in number. Spicules slender, with well-marked manubria and ventrally arcuate shaft, 16–20 µm long; gubernaculum 4–7 µm long, bursal margins faintly crenate (Fig. 10.15D).

87.4.2 Molecular diagnosis

For primers, see Table 60.

87.5 Geographical distribution

P. loosi is widely distributed throughout the tea-cultivating parts of the world and has been recorded in the following countries (CABI-ISC, 2020).

Africa: Senegal.
Asia: Bangladesh, China, India, Iran, Japan, South Korea, Sri Lanka, Taiwan, Turkey.
Europe: Bulgaria.
North America: Guadeloupe, USA.
Oceania: American Samoa, Australia, Cook Islands.
South America: Brazil, Chile.

87.6 Hosts

The primary host of *P. loosi* is tea (*Camellia sinensis*). However, the nematode is polyphagous and can reproduce on natural grasslands, orchards and forests (CABI-CPC, 2010). An expanded list of hosts of *P. loosi* is presented in Table 63.

87.7 Symptoms

Typical above-ground symptoms of *P. loosi* appear on both young and mature tea, and include patches of unthrifty plants showing spindly growth with sparse foliage (Fig. 10.9F). The leaves are initially dull and yellowish in colour. Heavily infested plants tend to enter into the reproductive phase at an earlier stage by flowering and setting fruit prematurely (Dasgupta *et al.*, 2010). Below ground, infested plants show a marked decrease in the growth of feeder roots. The remaining roots appear brown and dried up when compared with normal healthy roots, which are succulent and whitish in colour (Fig. 10.10H). When the bark is peeled, the larger storage roots display dark brown necrotic patches or lesions of varying size. Although the invasion of *P. loosi* into the epidermis and cortex caused thickening of cell walls, necrosis and cavities, there was no evidence of damage to the vascular cylinder (Sivapalan, 1972). Severely infested plants either recover very poorly from pruning or die.

87.8 Biology and life cycle

P. loosi is amphimictic. The nematode takes 15–17 days for the eggs to hatch, 15–16 days as juveniles, and 15 days as adults before egg-laying. The life cycle is completed in 45–48 days and the optimum temperature range for reproduction and development is 18–24°C (Gnanapragasam and Manuelpillai, 1984).

87.9 Economic importance

P. loosi is the most serious pest of tea in Sri Lanka (Sivapalan, 1972) and Japan (Takaji, 1969). In Sri Lanka, the annual decline in yield has been estimated at around 225–350 kg of made tea per hectare (Visser, 1959), with a variation from 4% to 40% depending on the type of host and environmental conditions (Dasgupta *et al.*, 2010). The extent of damage is greater in young infested tea clearings and nurseries, where 60-100% damage may occur if proper sanitary measures are not followed. In Bangladesh, *P. loosi* causes damage to tea nurseries (Campos *et al.*, 1990). The damage threshold of *P. loosi* was

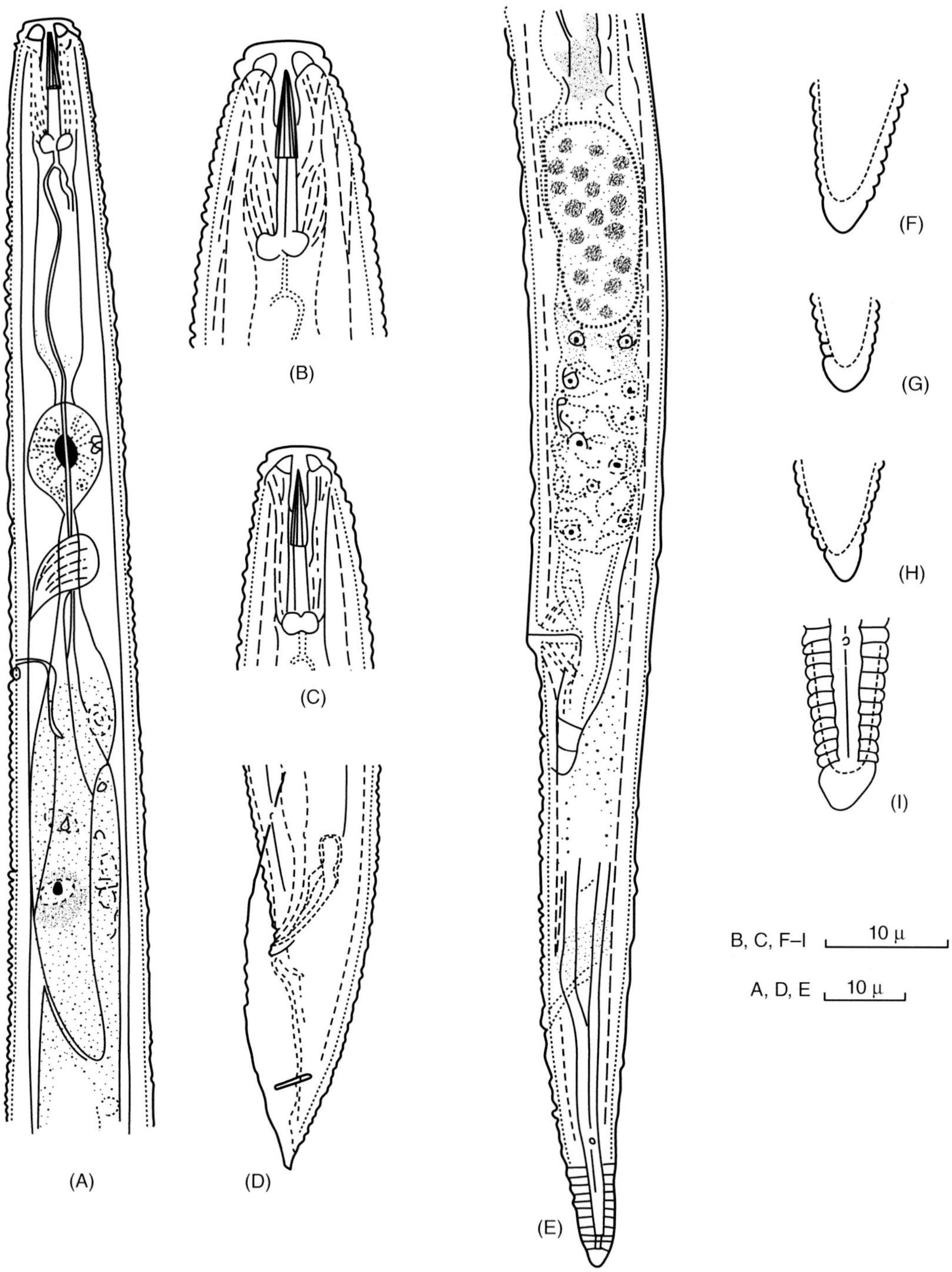

Fig. 10.15. Line drawing of *Pratylenchus loosi.* Anterior part of female (A); head of female (B); head of male (C); tail of male (D); posterior part of female (E); tail of female (F-I). [Reproduced from Seinhorst, 1977, CIH description of Plant Parasitic Nematodes Set 7, No. 98; Courtesy CAB International]

estimated to be 40 nematodes per 100 g of soil at 24°C (Gnanapragasam and Manuelpillai, 1984).

87.10 Movement and means of dispersal

See section 76.11, above.

87.11 Pest risk analysis and looming threat from introduction

P. loosi is a phytosanitary risk in all tea-growing regions of the world. The spread of the nematode could be prevented by restricting the transport of infested nursery stock and using proper soil conservation measures in tea nurseries and infested

Table 63. List of host plants of *Pratylenchus loosi* (adapted from CABI-CPC, 2010).

Main host
Camellia sinensis (tea)
Other hosts
Abelmoschus esculentus (okra)
Cinnamomum camphora (camphor laurel)
Citrus
Coffea (coffee)
Convallaria
Dioscorea (yam)
Dioscorea rotundata
Diospyros kaki (persimmon)
Fragaria ananassa (strawberry)
Malus domestica (apple)
Mangifera indica (mango)
Musa × paradisiaca (plantain)
Panicum hemitomon
Paspalum notatum (bahiagrass)
Pisum sativum (pea)
Poncirus trifoliata (trifoliate orange)
Prunus avium (sweet cherry)
Pyrus communis (European pear)
Saccharum officinarum (sugarcane)
Solanum tuberosum (potato)
Sorghum bicolor (sorghum)
Vigna unguiculata (cowpea)
Zea mays (maize)
Wild hosts
Acacia decurrens (green wattle)
Alternanthera sessilis (sessile joyweed)
Artemisia vulgaris (mugwort)
Cassia alata (ringworm senna)
Catharanthus roseus (pink periwinkle)
Cestrum (jessamine)
Cymbopogon citratus (citronella grass)
Cyperus (flatsedge)
Cyperus rotundus (purple nutsedge)
Grevillea robusta (silky oak)
Hibiscus rosa-sinensis (China-rose)
Imperata cylindrica (satintail)
Panicum repens (torpedo grass)
Sesbania cannabina (corkwood tree)
Solanum nigrum (black nightshade)
Tagetes (marigold)
Tecoma stans (yellow bells)
Tephrosia (hoary-pea)
Tithonia diversifolia (Mexican sunflower)

areas. Besides tea, its wide host range would also facilitate the establishment of *P. loosi* on other alternative hosts, if introduced into a new area. The direct economic impact of this nematode on tea production is well established. The indirect economic and regulatory impacts due to the introduction should not be underestimated.

87.12 Invasiveness rating

Considering the economic damage caused by *P. loosi* and pest risk analysis, the following invasiveness rating has been assigned.

Climate/host interaction	–	High (3)
Hosts range	–	Medium (2)
Dispersal potential	–	High (3)
Economic impact	–	High (3)
Environmental impact	–	Medium (2)
Post-entry distribution	–	Low (–1)
Final rating (score)	–	High (12)

Based on the above invasiveness rating, *P. loosi* receives a score of 12 out of 15 and is rated as HIGHLY INVASIVE NEMATODE.

87.13 Region-wise status of invasiveness

P. loosi poses a high threat to tea cultivation and is listed as a quarantine species in many countries. Due to its damaging potential and polyphagous nature, *P. loosi* has been given a highly invasive rating for all the countries where this nematode does not occur.

87.14 Management measures

See section 82.5, above.

88 *Pratylenchus neglectus*

88.1 Common name

Root-lesion nematode

88.2 Scientific name

Pratylenchus neglectus (Rensch, 1924) Filipjev and Schuurmans Stekhoven, 1941

88.3 Synonym

Pratylenchus minyus Sher and Allen

88.4 EPPO code

PRATNE

88.5 Diagnosis

88.5.1 Morphological diagnosis

Morphologically, *Pratylenchus neglectus* resembles *P. crenatus*. However, the former generally has a smooth tail tip and musculature is poorly developed at the excretory pore (Mizukubo and Minagawa, 1991). For morphometric measurements (after Kumari, 2015), see Table 61.

88.5.2 Molecular diagnosis

For primers, see Table 60.

88.6 Geographical distribution

P. neglectus is widely distributed and known to occur in the following countries (Taylor, 2000; Kumari, 2015; CABI-ISC, 2020).

Africa: Algeria, Tunisia, Morocco, South Africa, Sudan.
Asia: India, Iran, Japan, Oman, Pakistan, Turkey.
Europe: Bulgaria, Croatia, Czech Republic, Estonia, France, Germany, Hungary, Italy, Netherlands, Poland, Portugal, Russia, Slovakia, Slovenia, Spain, former Yugoslavia.
North America: Canada, Mexico, USA.
Oceania: Australia.
South America: Argentina.

88.7 Hosts

P. neglectus is polyphagous and infects both monocot and dicot plants (Castillo and Vovlas, 2007). Hosts include cereals: wheat (*Triticum aestivum*), alfalfa (*Medicago sativa*), barley (*Hordium vilgare*), chickpea (*Cicer arietinum*), oats (*Avena sativa*), maize (*Zea mays*), sorghum (*Sorghum vulgare*), vetch (*Vicia sative*); vegetables: potato (*Solanum tuberosum*); fruit trees: apple (*Malus silvestris*), peach (*Prunus persica*), pear (*Pyrus communis*); brassicas: canola (*Brassica napus*), camelina (*Camelina sativa*), mustard (*B. juncea*); tobacco (*Nicotiana tabacum*) and grasses: sudangrass (*Sorghum bicolor*), *Agropyron* spp. and *Elymus* sp. (Taylor, 2000; Smiley *et al.*, 2014).

88.8 Symptoms

Field symptoms associated with *P. neglectus* are nonspecific and can be confused with the symptoms of nutritional deficiencies or water shortage (Castillo and Vovlas, 2007). Infested roots with this nematode show dead cortical and epidermal cells and loss of lateral roots and root hairs (Vanstone *et al.*, 1998). Infested plants appear dwarf, with premature yellowing of older leaves, and many other growth and productivity parameters are negatively affected (Smiley *et al.*, 2014).

88.9 Biology and life cycle

P. neglectus reproduces by mitotic parthenogenesis; males are very rare and not required for reproduction (Kumari, 2015). The optimum soil temperature for embryogenesis of *P. neglectus* is around 25°C (Vanstone and Nicol, 1993). The life cycle is completed in 30–45 days under natural conditions, depending on environmental conditions (Taylor, 2000). On tobacco, greater reproduction occurs at 38°C and the life cycle is completed in 28 days (Townshend and Anderson, 1976). The nematode can survive at 2°C, but not at below-zero temperatures.

88.10 Economic importance

P. neglectus severely hampers wheat production worldwide (Castillo and Vovlas, 2007), with documented yield reductions in Asia, Africa, Australia, Europe and the USA (Castillo and Vovlas, 2007). In Turkey, *P. neglectus* has been detected in 11.6% of samples from wheat and causes significant yield loss (Dababat *et al.*, 2019). Damage threshold levels of 0.05–30 nematodes/cm^3 of soil were estimated on wheat, depending on the host crop and other existing *Pratylenchus* spp. (Castillo and Vovlas, 2007). *P. neglectus* is also one of the most common *Pratylenchus* species infecting potato roots (found in 35% of samples) and results in severe yield losses to potato production (Esteves *et al.*, 2015).

88.11 Movement and means of dispersal

See section 76.11, above.

88.12 Pest risk analysis and looming threat from introduction

P. neglectus infests two important hosts, wheat and potato, besides several other host plants. The direct economic impact of this nematode is well documented. As an endoparasitic pest of potato, the nematode easily disseminates from seed tubers and instigates regulatory impact due to introduction. The production of both crops would be seriously affected by quarantine restrictions if imposed against *P. neglectus*.

88.13 Invasiveness rating

Considering the consequences of introduction and pest risk analysis of *P. neglectus*, the following invasiveness rating has been assigned.

Climate/host interaction	–	High (3)
Hosts range	–	High (3)

Dispersal potential	–	High (3)
Economic impact	–	High (3)
Environmental impact	–	Medium (2)
Post-entry distribution	–	Medium (–2)
Final rating (score)	–	High (12)

Based on the above invasiveness rating, *P. neglectus* receives a score of 12 out of 15 and is rated as HIGHLY INVASIVE NEMATODE.

88.14 Region-wise status of invasiveness

As a pest of wheat and potato, *P. neglectus* has been given a highly invasive rating for all the countries where this nematode does not occur.

88.15 Management measures

See section 82.5, above.

89 *Pratylenchus penetrans*

89.1 Common name

Northern root-lesion nematode

89.2 Scientific name

Pratylenchus penetrans (Cobb, 1917) Filipjev and Schuurmans Stekhoven, 1941

89.3 Synonyms

Pratylenchus gulosus (Kühn, 1890) Filipjev and Schuurmans Stekhoven, 1941
Tylenchus penetrans Cobb, 1917

89.4 EPPO code

PRATPE

89.5 Diagnosis

89.5.1 Morphological diagnosis

Morphological descriptions of *P. penetrans* based on Corbett (1973) and Loof (1991) are summarized below. For morphometric measurements (after Sher and Allen, 1953; Loof, 1991), see Table 61.

FEMALE: Body medium slender, nearly straight when heat killed (Fig. 10.16J). Body annules fine. Lateral field usually with four incisures, outer bands occasionally areolate, not spreading to tail tip. Lip region low, slightly offset, flat in front, with oval outer margins, three annules. Head with strong, conspicuous skeleton. Stylet with broadly rounded knobs, occasionally concave anteriorly (Fig. 10.16A). Oesophagus lobe ventrally overlapping intestine about 1.5 body widths long. Excretory secretory pore opposite oesophago-intestinal junction. Hemizonid consisting about two body annules immediately in front of excretory pore. Post-uterine sac short, identical, about 1 or 1.5 vulval body widths long. Spermatheca round, or nearly so, generally filled with spermatozoa (Fig. 10.16C). Tail tapering to broad, evenly rounded smooth terminus, with 15–27 annules on ventral surface (Fig. 10.16I).
MALE: Abundant in number. Similar to female but slightly smaller (Fig. 10.16H). Lateral field with four incisures, extending up to bursa, sometimes with sloping striae in mid-body. Spicules slender, ventrally arcuate shafts, with well-marked manubria and 14–17 μm long; gubernaculum simple (Fig. 10.16F,G). Tail about twice as long as anal body diameter. Caudal alae with irregularly crenate margin, enveloping tail terminus (Fig. 10.16G).

89.5.2 Molecular diagnosis

For primers, see Table 60.

89.6 Geographical distribution

P. penetrans is cosmopolitan and widely distributed throughout the world (Fig. 10.17). The nematode has been recorded from the following countries (CABI-ISC, 2020).

Africa: Algeria, Egypt, Kenya, Libya, Morocco, Namibia, Nigeria, South Africa, Tanzania, Tunisia, Zimbabwe.
Asia: Azerbaijan, China, India, Iran, Japan, Kuwait, Kyrgyzstan, Pakistan, Philippines, Saudi Arabia, Singapore, South Korea, Taiwan, Turkey, Uzbekistan, Vietnam.
Europe: Albania, Austria, Belgium, Bulgaria, Cyprus, Czech Republic, Denmark, Estonia, Finland, France, Germany, Greece, Hungary, Italy, Moldova, Netherlands, Norway, Poland, Portugal, Romania, Russia, Serbia and Montenegro, Slovakia, Spain, Sweden, Switzerland, United Kingdom.
North America: Canada, Costa Rica, Mexico, Trinidad and Tobago, USA.
Oceania: Australia, Cook Islands, New Zealand.
South America: Argentina, Brazil, Venezuela.

89.7 Hosts

P. penetrans is extremely polyphagous, and probably the nematode with the highest number (< 400) of hosts after *Meloidogyne*

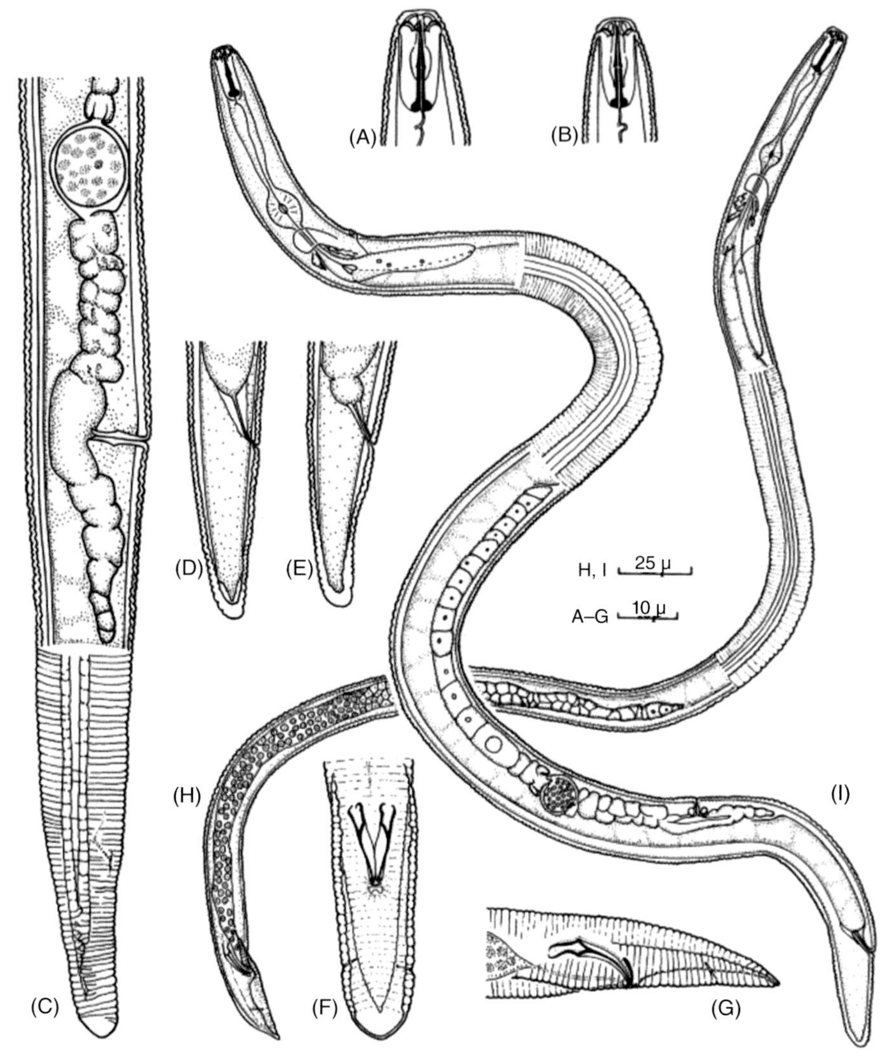

Fig. 10.16. Line drawing of *Pratylenchus penetrans*. Female head (A); male head (B); female vulva region and tail (C); female tail tips (D,E); male tail ventral view (F,G); male tail lateral view (F); male (H); female (I). [Reproduced from Corbettt, 1973, CIH Descriptions of Plant-parasitic Nematodes Set 2, No. 25; A-G after M.W. Allen; Courtesy CAB International]

incognita and *M. javanica* (Vieira *et al.*, 2015). The host range includes all kinds of plant species from fruit trees, forest trees, cereals, vegetables, tuber crops, oilseed crops, fibre crops, woody ornamentals, plantation crops, grasses and medicinal plants. An expanded list of hosts of *P. penetrans* is presented in Table 64.

89.8 Symptoms

Symptoms are typical of the genus. The presence of lesions on host-plant feeder roots is a characteristic symptom of *P. penetrans*. The lesions sometimes resemble 'cat scratches', and generally appear as yellowish to brownish elongate discoloured marks parallel to the long axis of the root. On severely damaged roots, lesions are 'rusty' in appearance. Under severe infestation, the lesion may girdle the lateral root and upon rotting the cortical tissue may slough off in a sleeve-like fashion (Khan, 2008). For detail see section 82.4, above.

89.9 Biology and life cycle

The biology and life cycle are typical of the genus. *P. penetrans* is an obligate plant-parasite, which reproduces sexually. The life cycle is temperature-dependent, most often requiring about 30 days. The life cycle in lucerne was completed in 30 days at 30°C, 37 days at 25°C and 92 days at 15°C (Dunn 1973a).

89.10 Economic importance

P. penetrans is considered one of the most economically important species within the genus (Vieira *et al.*, 2015). Significant yield

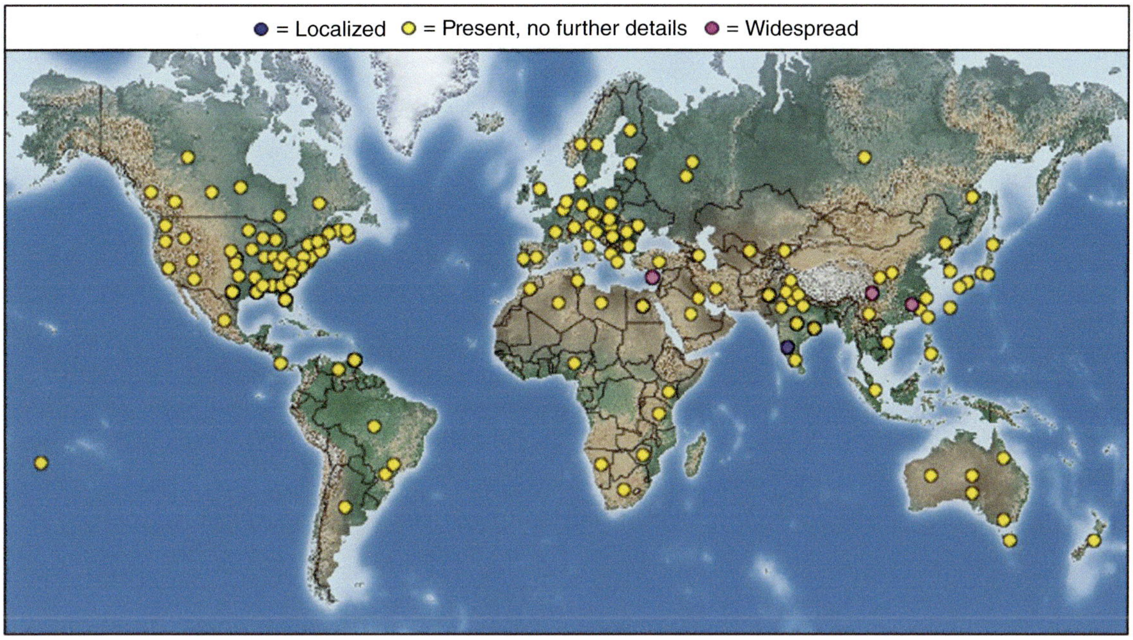

Fig. 10.17. Distribution map of *Pratylenchus penetrans*. [Adapted from CABI-ISC, 2020]

Table 64. List of hosts affected by *Pratylenchus penetrans* (adapted from CABI-CPC, 2010).

Main hosts

Abelmoschus esculentus (okra)
Allium cepa (onion)
Allium cepa var. *aggregatum* (shallot)
Allium fistulosum (Welsh onion)
Allium porrum (leek)
Allium sativum (garlic)
Apium graveolens (celery)
Arachis hypogaea (groundnut)
Arachniodes adiantiformis (leatherleaf fern)
Arctium (burdock)
Astragalus sinicus (Chinese clover)
Avena sativa (oats)
Beta vulgaris (beetroot)
Brassica juncea var. *juncea* (Indian mustard)
Brassica napus var. *napus* (rape)
Brassica nigra (black mustard)
Brassica oleracea (cabbages, cauliflowers)
Brassica oleracea var. *botrytis* (cauliflower)
Brassica oleracea var. *capitata* (cabbage)
Brassica oleracea var. *gemmifera* (Brussels sprouts)
Brassica oleracea var. *viridis* (collards)
Brassica rapa subsp. *pekinensis*
Brassica rapa subsp. *rapa* (turnip)
Brassica rapa var. *glabra* (Japanese turnip)
Cajanus cajan (pigeon pea)
Calendula officinalis (pot marigold)
Camellia sinensis (tea)
Capsicum annuum (bell pepper)
Carica papaya (papaw)
Chrysanthemum (daisy)
Chrysanthemum vestitum
Citrus aurantiifolia (lime)
Citrus limon (lemon)
Citrus sinensis (navel orange)
Colocasia esculenta (taro)
Cryptotaenia canadensis (honewort)
Cucumis sativus (cucumber)
Cucurbita moschata (pumpkin)
Cucurbita pepo (ornamental gourd)
Cydonia oblonga (quince)
Cynara cardunculus var. *scolymus* (globe artichoke)
Daucus carota (carrot)
Dioscorea japonica (Japanese yam)
Echinochloa frumentacea (Japanese millet)
Eleusine indica (goosegrass)
Fragaria ananassa (strawberry)
Glycine max (soybean)
Gossypium (cotton)
Hordeum vulgare (barley)
Humulus lupulus (hop)
Ipomoea batatas (sweet potato)
Lactuca indica (Indian lettuce)
Lactuca sativa (lettuce)
Lespedeza
Lespedeza cuneata (sericea lespedeza)
Leucanthemum vulgare (oxeye daisy)
Linum usitatissimum (flax)
Lolium multiflorum (Italian ryegrass)
Lotus corniculatus (bird's-foot trefoil)
Luffa aegyptiaca (loofah)
Lupinus angustifolius (lupin)

Continued

Table 64. Continued.

Malus domestica (apple)
Malus sylvestris (crab-apple tree)
Mentha (mints)
Mentha piperita (peppermint)
Mentha spicata (spear mint)
Musa (banana)
Musa × *paradisiaca* (plantain)
Nerium oleander (oleander)
Nicotiana tabacum (tobacco)
Nigella damascena (love-in-a-mist)
Oryza sativa (rice)
Pastinaca sativa (parsnip)
Petasites japonicus (creamy butterbur)
Petroselinum crispum (parsley)
Phaseolus coccineus (runner bean)
Phaseolus vulgaris (common bean)
Phleum pratense (timothy)
Phlox paniculata (summer perennial phlox)
Phoenix dactylifera (date-palm)
Phormium tenax (New Zealand flax)
Pistacia
Pistacia vera (pistachio)
Pisum sativum (pea)
Pisum sativum var. *arvense* (Austrian winter pea)
Poa pratensis (smooth-stalked meadowgrass)
Polygonum aviculare (prostrate knotweed)
Prunus (stone fruit)
Prunus amygdalus
Prunus armeniaca (apricot)
Prunus avium (sweet cherry)
Prunus cerasifera (myrobalan plum)
Prunus cerasus (sour cherry)
Prunus domestica (plum)
Prunus mahaleb (mahaleb cherry)
Prunus persica (peach)
Prunus salicina (Japanese plum)
Prunus serotina (black cherry)
Pyrus communis (European pear)
Rheum hybridum (rhubarb)
Rhododendron indicum (Satsuki azalea)
Ribes nigrum (blackcurrant)
Ribes rubrum (red currant)
Ribes uva-crispa (gooseberry)
Rorippa indica (Indian marshcress)
Rosa (roses)
Rosa corymbifera
Rosa multiflora (multiflora rose)
Rubus idaeus (raspberry)
Saccharum officinarum (sugarcane)
Secale cereale (rye)
Sesamum indicum (sesame)
Solanum lycopersicum (tomato)
Solanum tuberosum (potato)
Sorghum sudanense (Sudan grass)
Spinacia oleracea (spinach)
Tanacetum cinerariifolium (pyrethrum)
Tanacetum coccineum (common pyrethrum)
Theobroma cacao (cocoa)
Tradescantia fluminensis (wandering Jew)
Trifolium hybridum (alsike clover)
Trifolium repens (white clover)
Trifolium subterraneum (subterranean clover)
Tulipa (tulip)
Ulmus americana (American elm)
Vaccinium corymbosum (blueberry)
Vicia hirsuta (hairy tare (UK))
Vicia sativa (common vetch)
Vicia villosa
Zea mays (maize)
Other hosts
Abies magnifica (red fir)
Abies sachalinensis (Sakhalin fir)
Acer saccharum (sugar maple)
Achillea filipendulina (fern-leaf yarrow)
Actinidia chinensis (Chinese gooseberry)
Ageratum houstonianum (blue billygoatweed)
Agropyron cristatum (crested wheatgrass)
Agrostis gigantea (black bent)
Agrostis stolonifera (creeping bentgrass)
Agrostis stolonifera var. *palustris* (bent grass)
Akebia quinata (five-leaf akebia)
Alyssum (alison)
Amaranthus blitum (livid amaranth)
Ambrosia artemisiifolia (common ragweed)
Ambrosia trifida (giant ragweed)
Ammi majus (bishop's-weed)
Anagallis arvensis (scarlet pimpernel)
Andropogon gerardii (big bluestem)
Andropogon virginicus (broomsedge bluestem)
Anemone coronaria (poppy anemone)
Anethum graveolens (dill)
Antirrhinum majus (snapdragon)
Arctium lappa (burdock)
Arracacia xanthorrhiza (arracacha)
Artemisia absinthium (wormwood)
Artemisia dracunculus (tarragon)
Asparagus officinalis (asparagus)
Avena barbata (slender oat)
Avena fatua (wild oat)
Avena sterilis (winter wild oat)
Azadirachta indica (neem tree)
Basella alba (Malabar spinach)
Begonia
Begonia cucullata var. *hookeri* (perpetual begonia)
Begonia tuberhybrida
Bellis perennis (common daisy)
Berberis julianae (evergreen barberry)
Berberis thunbergii (Japanese barberry)
Borago officinalis (borage)
Bouteloua curtipendula (side-oats grama)
Brassica cernua
Brassica rapa subsp. *oleifera* (turnip rape)
Bromus inermis (awnless brome)
Buxus microphylla (little-leaf box)
Buxus sempervirens (common boxwood)

Continued

Table 64. Continued.

Callistemon (bottle brush)
Callistephus chinensis (China aster)
Camellia japonica (camellia)
Canavalia gladiata (horse bean)
Cannabis sativa (hemp)
Capsicum frutescens (chilli)
Carum carvi (caraway)
Celosia argentea (celosia)
Centaurea cyanus (cornflower)
Chaenomeles japonica (Japanese quince)
Chamaecyparis obtusa (hinoki cypress)
Chamaemelum nobile (common chamomile)
Chenopodium ambrosioides (Indian goosefoot (UK))
Chlorophytum comosum (ribbon plant)
Chrysanthemum indicum (chrysanthemum)
Cichorium endivia (endives)
Cichorium intybus (chicory)
Citrullus lanatus (watermelon)
Clematis
Clematis vitalba (old man's beard)
Coffea canephora (robusta coffee)
Coleus blumei (common coleus)
Commelina communis (common dayflower)
Consolida ambigua (rocket larkspur)
Convallaria majalis (lily-of-the-valley)
Cosmos atrosanguineus (black cosmos)
Cosmos bipinnatus (garden cosmos)
Cotoneaster simonsii (khasiaberry)
Crambe abyssinica
Crocosmia crocosmiiflora
Crocus
Crocus sativus (saffron)
Crotalaria spectabilis (showy rattlepod)
Cryptomeria japonica (Japanese cedar)
Cyperus rotundus (purple nutsedge)
Dactylis glomerata (orchardgrass)
Dahlia
Dahlia hybrids
Daphne odora (winter daphne)
Datura metel (Hindu datura)
Dianthus barbatus (sweet william)
Dianthus caryophyllus (carnation)
Dianthus chinensis (china pink)
Digitalis purpurea (foxglove)
Digitaria ciliaris (southern crabgrass)
Doronicum (leopard's bane)
Echinacea purpurea (purple coneflower)
Echinops ritro (small globe-thistle)
Elymus repens (quackgrass)
Eriobotrya japonica (loquat)
Erysimum cheiri (wallflower)
Eschscholzia californica (California poppy)
Euonymus japonicus (Japanese spindle)
Festuca arundinacea (reed fescue)
Festuca rubra (red fescue)
Foeniculum vulgare (fennel)
Forsythia intermedia (golden bells)
Fragaria chiloensis (Chilean strawberry)
Gaillardia
Galinsoga parviflora (gallant soldier)
Gladiolus hybrids (sword lily)
Glechoma hederacea (ground ivy)
Gomphrena globosa (globe amaranth)
Grevillea robusta (silky oak)
Guizotia abyssinica (niger)
Hedera helix (ivy)
Helianthus annuus (sunflower)
Helleborus niger (Christmas rose)
Hibiscus rosa-sinensis (China-rose)
Hibiscus syriacus (shrubby althaea)
Hippeastrum
Hyacinthus (hyacinth)
Iberis sempervirens (edging candytuft)
Ilex crenata (Japanese holly)
Ilex glabra (gallberry)
Impatiens balsamina (garden balsam)
Impatiens walleriana (busy-lizzie)
Ipomoea quamoclit (Cupid's-flower)
Iris germanica (German iris)
Juniperus virginiana (eastern red cedar)
Lablab purpureus (hyacinth bean)
Lagenaria siceraria (bottle gourd)
Lantana camara (lantana)
Larix kaempferi (Japanese larch)
Lathyrus odoratus (sweet pea)
Liatris spicata (spike gayfeather)
Ligustrum japonicum (Japanese privet)
Ligustrum ovalifolium (California privet)
Ligustrum vulgare (privet)
Lilium (lily)
Lilium longiflorum (Easter lily)
Lilium regale (regal lily)
Lilium speciosum (showy lily)
Linaria (toadflax)
Lolium (ryegrasses)
Lolium perenne (perennial ryegrass)
Lupinus (lupins)
Lupinus luteus (yellow lupin)
Lychnis (campion)
Mahonia (holly grape)
Mangifera indica (mango)
Matthiola incana (stock)
Medicago sativa (lucerne)
Melia azedarach (chinaberry)
Melilotus alba (honey clover)
Morus alba (mora)
Mucuna pruriens (buffalobean)
Myosotis sylvatica (wood forget-me-not)
Narcissus (daffodil)
Narcissus pseudonarcissus (wild Lent lily)
Nemesia strumosa (cape-jewels)
Nephrolepis exaltata (Boston fern)
Onobrychis viciifolia (sainfoin)
Ornithopus sativus (bird's foot)
Oxalis corniculata (creeping woodsorrel (USA))

Continued

Table 64. Continued.

Paeonia (peonies)
Paeonia lactiflora (Chinese peony)
Paeonia suffruticosa (tree peony)
Panax ginseng (Asiatic ginseng)
Panax quinquefolius (American ginseng)
Panicum maximum (Guinea grass)
Papaver rhoeas (common poppy)
Papaver somniferum (opium poppy)
Parthenium argentatum (guayule)
Parthenium hysterophorus (Parthenium weed)
Persea americana (avocado)
Phacelia tanacetifolia (California bluebell)
Phalaris arundinacea (reed canary grass)
Phalaris paradoxa (awned canary grass)
Pharbitis nil (Japanese morning glory)
Phaseolus lunatus (lima bean)
Phlox
Picea jezoensis (Yezo spruce)
Pieris japonica (lily-of-the-valley shrub)
Pinus caribaea (Caribbean pine)
Pinus densiflora (Japanese umbrella pine)
Pinus merkusii (Tenasserim pine)
Pinus ponderosa (ponderosa pine)
Pinus radiata (radiata pine)
Pittosporum tobira (Japanese pittosporum)
Platycodon grandiflorus (balloonflower)
Poa annua (annual meadowgrass)
Poa compressa (Canada bluegrass)
Poa trivialis (rough bluegrass)
Polygonum blumei (tufted knotweed (USA))
Populus sp. (poplar)
Primula veris (cowslip)
Pyracantha (firethorn)
Ranunculus (buttercup)
Raphanus raphanistrum (wild radish)
Raphanus sativus (radish)
Rosa canina (dog rose)
Rosa chinensis (China rose)
Rosa glauca (red-leaved rose)
Rosa rubiginosa (sweet briar)
Rosa rugosa (rugosa rose)
Rosmarinus officinalis (rosemary)
Rubus (blackberry, raspberry)
Rubus loganobaccus (loganberry)
Rubus strigosus (American red raspberry (USA))
Rudbeckia (coneflower)
Rudbeckia laciniata (thimbleweed)
Rumex acetosa (sour dock)
Saintpaulia (African violet)
Saintpaulia ionantha (African violet)
Salvia officinalis (common sage)
Scabiosa atropurpurea (pincushion)
Scabiosa caucasica
Scorzonera hispanica (oyster plant)
Senecio vulgaris
Sequoia sempervirens (coast redwood)
Sequoiadendron giganteum (giant sequoia)
Setaria italica (foxtail millet)
Setaria pumila (yellow foxtail)
Setaria viridis (green foxtail)
Sicyos angulatus (burcucumber)
Sinapis alba (white mustard)
Sinapis arvensis (wild mustard)
Solanum melongena (aubergine)
Solanum nigrum (black nightshade)
Sorghastrum nutans (Indian grass)
Sorghum bicolor (sorghum)
Spergula arvensis (corn spurry)
Spiraea
Stellaria media (common chickweed)
Syringa vulgaris (lilac)
Tagetes (marigold)
Tagetes erecta (African marigold)
Tagetes minuta (stinking Roger)
Tagetes patula (French marigold)
Tanacetum parthenium (feverfew)
Trifolium alexandrinum (berseem clover)
Trifolium incarnatum (crimson clover)
Triticum aestivum (wheat)
Trollius (globeflower)
Vaccinium angustifolium (lowbush blueberry)
Viburnum carlesii
Vicia faba (broad bean)
Vigna angularis (adzuki bean)
Vigna radiata (mung bean)
Vigna unguiculata (cowpea)
Viola (violet)
Viola cornuta (horned violet)
Viola wittrockiana (wild pansy)
Vitis labrusca (fox grape)
Vitis vinifera (grapevine)
Washingtonia filifera (desert fanpalm)
Xanthium strumarium (common cocklebur)
Zinnia elegans (zinnia)
Zoysia japonica (zoysiagrass)
Wild hosts
Abies alba (silver fir)
Acer (maples)
Acer campestre (field maple)
Acer negundo (box elder)
Acer platanoides (Norway maple)
Acer pseudoplatanus (sycamore)
Acer rufinerve (grey snake-bark maple)
Aesculus hippocastanum (buckeye)
Alnus glutinosa (European alder)
Alnus incana (grey alder)
Alopecurus pratensis (meadow foxtail)
Amaranthus retroflexus (redroot pigweed)
Amelanchier laevis (Allegheny serviceberry)
Angelica archangelica (angelica)
Anthemis cotula (dog fennel)
Anthoxanthum odoratum (sweet vernalgrass)
Arctium minus (common burdock)
Asclepias syriaca (common milkweed)
Betula pendula (common silver birch)
Calluna vulgaris (heather)

Continued

Table 64. Continued.

Capsella bursa-pastoris (shepherd's purse)
Cedrus atlantica (Atlas cedar)
Cercis siliquastrum (Judas tree)
Chamaecyparis lawsoniana (Port Orford cedar)
Chamomilla suaveolens (rounded chamomile)
Chenopodium (goosefoot)
Chenopodium album (fat hen)
Cirsium arvense (creeping thistle)
Convolvulus arvensis (bindweed)
Conyza canadensis (Canadian fleabane)
Cornus sanguinea (dogwood)
Coronilla varia (crown vetch)
Corylus avellana (hazel)
Cotinus coggygria (fustet)
Crataegus (hawthorns)
Crataegus laevigata
Cytisus nigricans
Cytisus scoparius (broom)
Dactylis (orchardgrass)
Daphne mezereum (mezereon)
Digitaria (crabgrass)
Epilobium (willowherbs)
Equisetum arvense (field horsetail)
Erigeron annuus (annual fleabane)
Erysimum cheiranthoides (treacle mustard)
Fagus sylvatica (common beech)
Fragaria (strawberry)
Fragaria vesca (wild strawberry)
Fraxinus excelsior (ash)
Galanthus nivalis (snowdrop)
Gaultheria procumbens (aromatic wintergreen)
Ginkgo biloba (kew tree)
Hippophae rhamnoides (sea buckthorn)
Hyssopus officinalis (hyssop)
Laburnum anagyroides (laburnum)
Lamium amplexicaule (henbit deadnettle)
Lepidium campestre (field cress)
Liatris (gayfeathers)
Liriodendron tulipifera (tuliptree)
Lupinus polyphyllus (perennial lupin)
Medicago lupulina (black medick)
Phalaris canariensis (canarygrass)
Physostegia virginiana (American dragonhead)
Picea abies (common spruce)
Pinus nigra (black pine)
Pinus sylvestris (Scots pine)
Polygonum convolvulus (black bindweed)
Populus canadensis (hybrid black poplar)
Portulaca oleracea (purslane)
Prunus spinosa (blackthorn)
Pseudotsuga menziesii (Douglas-fir)
Ranunculus asiaticus (garden crowfoot)
Rhamnus cathartica (buckthorn)
Rhamnus frangula (alder buckthorn)
Robinia pseudoacacia (black locust)
Rorippa sylvestris (creeping yellowcress)
Rumex crispus (curled dock)
Salix alba (white willow)
Sisymbrium altissimum (tall rocket)
Sonchus arvensis (perennial sowthistle)
Sonchus oleraceus (common sowthistle)
Sorbus aucuparia (mountain ash)
Taraxacum officinale complex (dandelion)
Taxodium distichum (bald cypress)
Thlaspi arvense (field pennycress)
Thuja occidentalis (eastern white cedar)
Trifolium fragiferum (strawberry clover)
Ulmus pumila (dwarf elm)
Valeriana officinalis (common valerian)
Veronica arvensis (corn speedwell)
Viburnum lantana (wayfaring tree)
Viburnum opulus (guelder rose)
Weigela florida (old-fashioned weigela)

losses due to *P. penetrans* have been reported on various crops such as potato, tobacco, tree-fruit and berries, legumes, cereals and garden vegetables (Olthof, 1986; MacGuidwin and Rouse, 1990). Olthof *et al.* (1973) reported 11% and 27.5% yield losses in tobacco at high nematode densities. Zepp and Szczygiel (1986) recorded 45% and 75% decreased yield of Antonovka apple in soils formerly planted with pear and cherry, respectively. Willis and Thompson (1972) reported substantial yield losses in legumes. Olthof and Potter (1973) determined crop losses at pre-defined lesion nematode population densities (PD); losses in yield of maize were 30% (PD = 666 nematodes/kg soil) and 49% (PD = 18,000 nematodes/kg soil), and in other crops the losses at the same two levels were, respectively: onion 14% and 71%; cabbage 17% and 25%; lettuce 18% and 33%; cauliflower 19% and 59%; potatoes 25% and 43%. At a pre-plant density of 18,000 *P. penetrans*/kg soil, losses in weight of marketable produce were 27% for beet, 43% for lettuce and 21% for spinach. At 14,360 nematodes/kg of soil there was a 44% loss in total fruit weight of tomato.

89.11 Movement and means of dispersal

See section 76.11, above.

89.12 Pest risk analysis and looming threat from introduction

P. penetrans is an extremely polyphagous and highly invasive nematode and has a huge economic impact on agricultural crops. Its distribution is cosmopolitan; it has been reported from all over the world and there is no direct threat from this nematode. However, it can impact native populations of nematodes and is a significant threat to native biodiversity.

89.13 Invasiveness rating

Considering the consequences of introduction on the native population and economic damage caused by *P. penetrans*, the following invasiveness rating has been assigned.

Climate/host interaction	–	High (3)
Hosts range	–	High (3)
Dispersal potential	–	High (3)
Economic impact	–	High (3)
Environmental impact	–	High (3)
Post-entry distribution	–	High (–3)
Final rating (score)	–	High (12)

Based on the above invasiveness rating, *P. penetrans* receives a score of 12 out of 15 and is rated as HIGHLY INVASIVE NEMATODE.

89.14 Region-wise status of invasiveness

P. penetrans is a highly invasive nematode. However, because of its cosmopolitan distribution and extensive host range, there are no known quarantine restrictions on this nematode.

89.15 Management measures

See section 82.5, above.

90 *Pratylenchus scribneri*

90.1 Common name

Root-lesion nematode

90.2 Scientific name

Pratylenchus scribneri Steiner in Sherbakoff and Stanley, 1943

90.3 Synonyms

Pratylenchus agilis (Thorne and Malik, 1968) Subbotin *et al.*, 2008

90.4 EPPO code

PRATSC

90.5 Diagnosis

90.5.1 Morphological diagnosis

For morphometric measurements (Yan *et al.*, 2016), see Table 61.

FEMALE: Cephalic region slightly higher or offset lip region, with two lip annules. Stylet short with mostly rounded knobs, with anterior indentions (cup-shaped anteriorly) observed in some specimens. Median bulb massive and generally rounded. Valve of median bulb most often situated centrally but anterior position observed in some specimens. Lateral field most often consists of four lines, but fifth and sixth lines observed in some specimens, particularly in vulval region. Posterior uterine sac length approximately equal to mid-body width. Tail shape varies from slightly tapering (sub-hemispherical) to broad (hemispherical) with smooth terminus (Ozbayrak, 2019).
MALE: Morphologically similar to female (Orlando *et al.*, 2020).

90.5.2 Molecular diagnosis

For primers, see Table 60.

90.6 Geographical distribution

P. scribneri is widely distributed throughout the world (Orlando *et al.*, 2020).

90.7 Hosts

P. scribneri is polyphagous, with potato, maize and soybean as its main hosts (Ozbayrak, 2019). Other hosts include wheat, barley, alfalfa, dry beans, grapevines and sugar beet (Huang and Yan, 2017; Ozbayrak, 2019).

90.8 Symptoms

P. scribneri produces characteristic necrotic lesions on the surface of potato tubers. The infected potato tuber has a scabby appearance (Orlando *et al.*, 2020). Seed potatoes are also infected by *P. scribneri* (Olthof and Wolynetz, 1991). Population densities of 1–2 nematodes/g of soil inflict economic damage to potatoes (Olthof, 1986).

90.9 Biology and life cycle

The biology and life cycle are typical of the genus. *P. scribneri* reproduces through amphimixis.

90.10 Economic importance

Limited information is available on quantified damage caused by *P. scribneri*. However, the nematode significantly reduces the yield and quality of potato tubers (Orlando *et al.*, 2020).

90.11 Movement and means of dispersal

See section 76.11.

90.12 Pest risk analysis and looming threat from introduction

Considering potato, maize and soybean as the main hosts of *P. scribneri*, the nematode has high economic impact. It can disseminate through seed tubers. The direct economic impact of this nematode on potato is well recognized. The indirect economic and regulatory impacts due to its introduction should not be underestimated. The production of potato, maize and soybean and related industries would be seriously affected if quarantine restrictions are imposed against *P. scribneri*.

90.13 Invasiveness rating

Considering the consequences of introduction and probable economic damage caused by *P. scribneri*, the following invasiveness rating has been assigned.

Climate/host interaction	–	Medium (2)
Hosts range	–	Medium (2)
Dispersal potential	–	High (3)
Economic impact	–	Medium (2)
Environmental impact	–	Medium (2)
Post-entry distribution	–	Low (–1)
Final rating (score)	–	Medium (10)

Based on the above invasiveness rating, *P. scribneri* receives a score of 10 out of 15 and is rated as MEDIUM INVASIVE NEMATODE.

90.14 Region-wise status of invasiveness

Information on the direct economic impact is lacking. However, being a pest of three important crops (maize, potato and soybean) and the probable economic consequences of its introduction, *P. scribneri* has been given a medium invasive rating for all the countries where this nematode does not occur.

90.15 Management measures

See section 82.5, above.

91 *Pratylenchus thornei*

91.1 Common name

Root-lesion nematode

91.2 Scientific name

Pratylenchus thornei Sher and Allen, 1953

91.3 EPPO code

PRATTH

91.4 Diagnosis

91.4.1 *Morphological diagnosis*

Morphological descriptions of *P. thornei* after Fortuner (1977) are summarized below. For morphometric measurements (after Loof, 1960), see Table 61.

FEMALE: Body slender and long, assuming an open C-shape on heat killing (Fig. 10.18F). Body conspicuously narrower immediately posterior to vulva. Cuticle with inconspicuous transverse striae, 1 µm apart. Lateral line with four incisures, outer ones straight, occasionally crenate faintly, extending past phasmids. Lip region slightly high with three annules but not offset. Stylet broadly rounded to almost anteriorly flattened basal knobs (Fig. 10.18D). Orifice of dorsal oesophageal gland around 3 µm behind basal knobs. Nerve ring right behind median bulb. Hemizonid occupying two annules and one annule posterior to excretory pore. Ovary not outstretching to oesophagus; oocytes developing in single row (Fig. 10.18B). Spermatheca obscure, without spermatozoa. Post-uterine sac present, around 1.5 vulval body widths long. Phasmids slightly posterior to middle of tail. Tail dorsally convex-conoid, terminus bluntly rounded to truncate and unstriated (Fig. 10.18H–L).
MALE: Very rare.

91.4.2 *Molecular diagnosis*

For primers, see Table 60.

91.5 Geographical distribution

P. thornei is widely distributed throughout the temperate and subtropical regions of the world (Fig. 10.19). It is known to occur in the following countries (CABI-ISC, 2020).

Africa: Algeria, Egypt, Kenya, Libya, Malawi, Morocco, South Africa, Sudan.
Asia: Afghanistan, India, Iran, Israel, Japan, Jordan, Pakistan, Saudi Arabia, South Korea, Syria, Tajikistan, Turkey.
Europe: Belgium, Bulgaria, Croatia, Cyprus, Denmark, Germany, Greece, Italy, Netherlands, Poland, Portugal, Romania, Russia, Slovakia, Slovenia, Spain, United Kingdom, former Yugoslavia.
North America: Canada, Mexico, USA.
Oceania: Australia.
South America: Argentina, Chile, Venezuela.

91.6 Hosts

P. thornei is highly polyphagous and attacks cereals, mainly wheat but also barley and maize, along with a number of legumes

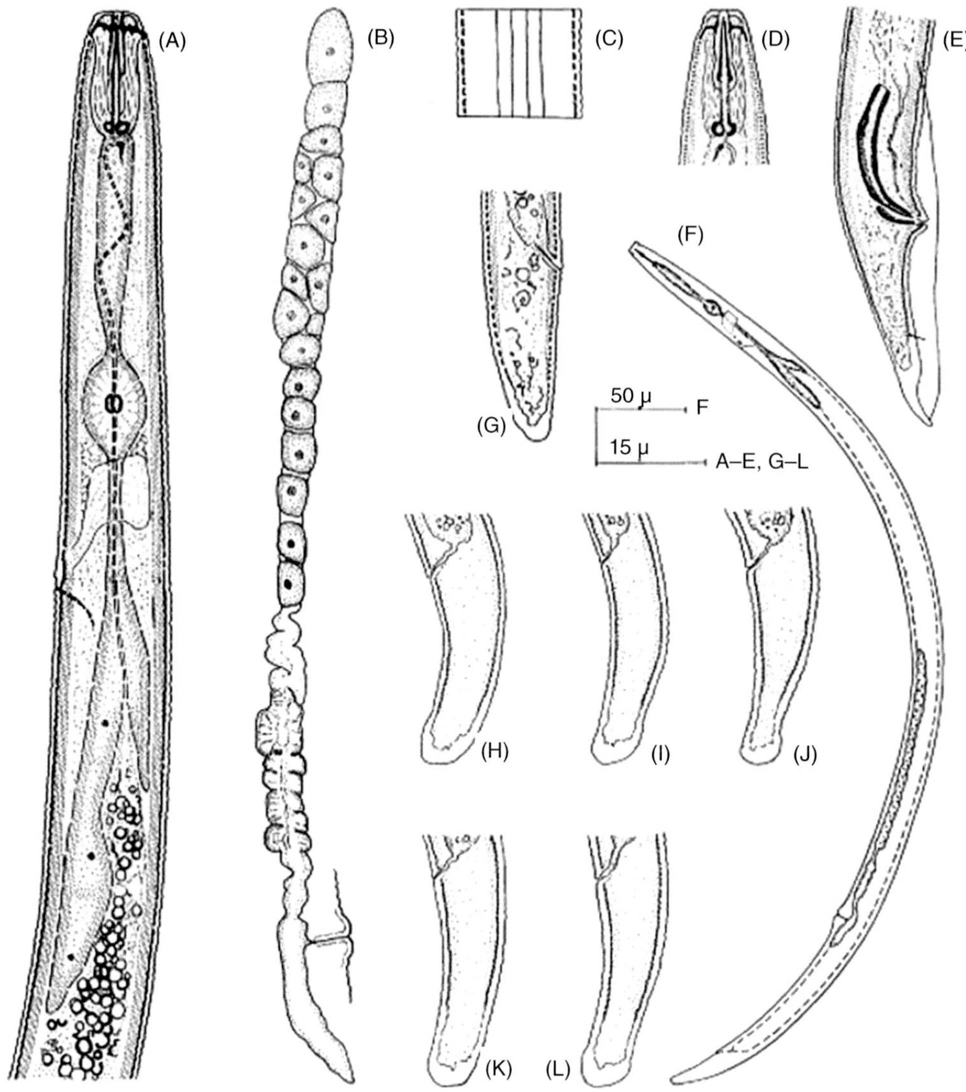

Fig. 10.18. Line drawing of *Pratylenchus thornei*. Oesophagous region (A); ovary (B); lateral field (C); anterior region of male (D); posterior region of male (E); entire female (F); tails (G-L). [Reproduced from Fortuner, 1977, CIH Descriptions of Plant-parasitic Nematodes Set 7, No. 93; courtesy CAB International]

(predominantly chickpea) and woody plants, including fruit trees and grapevine. An expanded list of hosts is presented in Table 65.

91.7 Symptoms

P. thornei produces characteristic lesions on infected roots and secondary necrosis usually follows. Root systems may be reduced in size and other damage includes stunting, growth retardation, chlorosis, reduced tillering (in cereals) and lower yield. In former Yugoslavia, *P. thornei* was reported to cause heavy crop losses in wheat by inducing ear sterility (Grujicic, 1969). Patchy growth in infested fields may also be correlated with the presence of the nematode.

91.8 Biology and life cycle

The biology and life cycle are typical of the genus. *P. thornei* reproduces by mitotic parthenogenesis. Males rarely occur. Heavy soils, such as clay loams, frequently reveal a higher population of the nematode. *P. thornei* completes the life cycle in 6–8 weeks and the reproductive rate is highest at temperatures of 20–25°C (Smiley *et al.*, 2014).

91.9 Economic importance

P. thornei is regarded as an important pest of wheat in many countries but is also known to attack potato in Australia. Estimated yield losses caused by *P. thornei* can be as high as

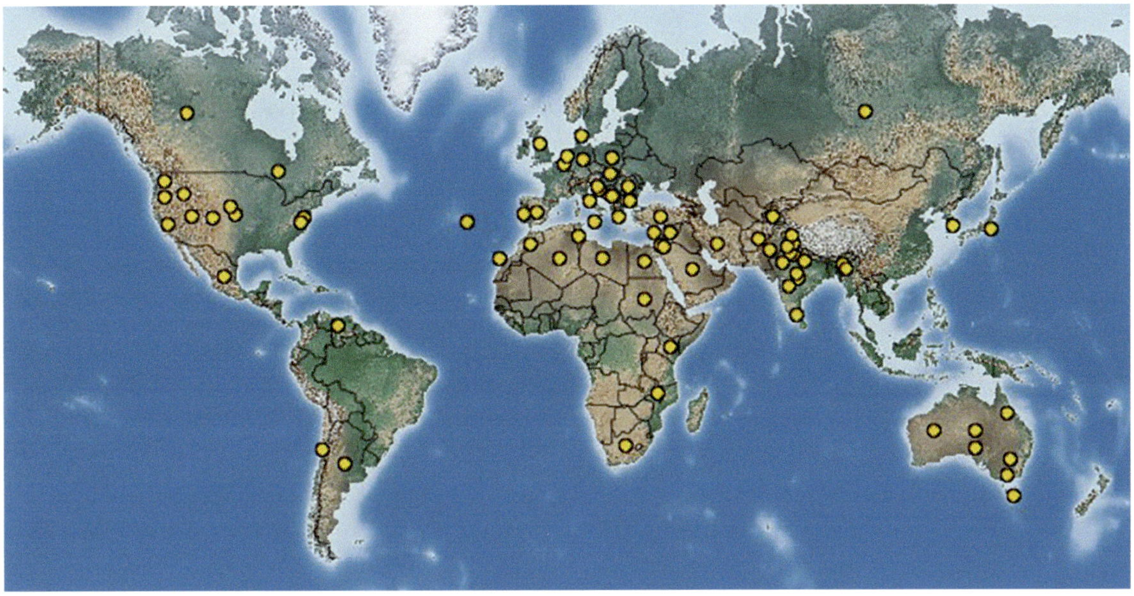

Fig. 10.19. Distribution map of *Pratylenchus thornei*. [Adapted from CABI-ISC, 2020]

Table 65. List of host plants of *Pratylenchus thornei* (adapted from CABI-CPC, 2010).

Main hosts
Cicer arietinum (chickpea)
Triticum aestivum (wheat)
Triticum turgidum (durum wheat)
Other hosts
Agrostis (bentgrasses)
Allium cepa (onion)
Apium graveolens (celery)
Arachis hypogaea (groundnut)
Avena sativa (oats)
Beta vulgaris (beetroot)
Citrullus lanatus (watermelon)
Cupressus (cypresses)
Cynara cardunculus var. *scolymus* (globe artichoke)
Daucus carota (carrot)
Fabaceae (leguminous plants)
Fragaria ananassa (strawberry)
Glycine max (soybean)
Hordeum vulgare (barley)
Iberis (candytuft)
Juglans regia (walnut)
Lathyrus odoratus (sweet pea)
Lens culinaris subsp. *culinaris* (lentil)
Medicago sativa (lucerne)
Mentha piperita (peppermint)
Mentha spicata (spear mint)
Phaseolus vulgaris (common bean)
Pinus (pines)
Pisum sativum (pea)
Prunus (stone fruit)
Prunus amygdalus
Prunus dulcis (almond)
Rosa (roses)
Saccharum officinarum (sugarcane)
Solanum tuberosum (potato)
Sorghum bicolor (sorghum)
Trifolium alexandrinum (berseem clover)
Trifolium fragiferum (strawberry clover)
Trifolium repens (white clover)
Vicia faba (broad bean)
Vigna unguiculata (cowpea)
Vitis vinifera (grapevine)
Zea mays (maize)
Wild host
Poaceae (grasses)

85% in wheat in Australia (Thompson *et al.*, 2008), 70% in Israel and 37% in Mexico (May *et al.*, 2016). The nematode is also commonly reported damaging legume crops, ornamental roses and other hosts.

91.10 Movement and means of dispersal

See section 76.11, above.

91.11 Pest risk analysis and looming threat from introduction

The wide and important host range of the *P. thornei* may result in significant economic damage if introduced into new areas. Alternative hosts also facilitate the establishment of the nematode on hosts other than the main host. The direct and indirect economic impacts of this nematode on the production of main host crops like wheat and chickpea should not be underestimated. The production of host crops would be seriously affected if quarantine restrictions are not imposed against *P. thornei*.

91.12 Invasiveness rating

Considering the economic damage and pest risk analysis of *P. thornei*, the following invasiveness rating has been assigned.

Climate/host interaction	–	Medium (2)
Hosts range	–	High (3)
Dispersal potential	–	High (3)
Economic impact	–	High (3)
Environmental impact	–	Medium (2)
Post-entry distribution	–	Low (–1)
Final rating (score)	–	High (12)

Based on the above invasiveness rating, *P. thornei* receives a score of 12 out of 15 and is rated as HIGHLY INVASIVE NEMATODE.

9.13 Region-wise status of invasiveness

Considering wheat as the main host, *P. thornei* has been given a highly invasive rating for the following wheat-producing countries where this nematode is absent.

APPPC: Iraq, Kazakhstan, Turkmenistan, Uzbekistan
COSAVE: Brazil
EPPO: Belarus, Czech Republic, France, Hungary, Lithuania, Romania, Serbia, Sweden, Ukraine
IAPSC: Algeria, Ethiopia

91.14 Management measures

See section 82.5, above.

92 *Pratylenchus vulnus*

92.1 Common name

Walnut root-lesion nematode

92.2 Scientific name

Pratylenchus vulnus Allen and Jensen, 1951

92.3 EPPO code

PRATVU

92.4 Diagnosis

92.4.1 *Morphological diagnosis*

Morphological descriptions of *P. vulnus* given by different workers (Allen and Jensen, 1951; Sher and Allen, 1953; Loof, 1960; Corbett, 1974) are summarized below. For morphometric measurements (after Allen and Jensen, 1951), see Table 61.

FEMALE: Body slender, almost straight after heat killed. Head continuous with 3–4 annules, not offset. Cephalic framework heavily sclerotized. Stylet with rounded basal knobs occasionally cupped anteriorly (Fig. 10.20A). Lateral field with four incisures. Excretory secretory pore near to opposite oesophago-intestinal junction. Median oesophageal bulb elliptical, comparatively narrow. Oesophageal lobe ventrally overlaps intestine. Spermatheca oblong with swollen sperm. Postvulval sac about twice vulval body width long with rudimentary ovary (Fig. 10.20E). Tail tapering, with narrowly rounded to subacute smooth or irregular terminus (Fig. 10.20F,G).
MALE: Abundant in number. Morphologically similar to female in appearance of head and oesophagus (Fig. 10.20H). Lateral field extends to bursa. Spicules curved, cephalated; gubernaculum simple. Bursa envelops tail (Fig. 10.20C).

92.4.2 *Molecular diagnosis*

The complete mitochondrial genome sequences (21,656 bp) of *P. vulnus* was carried out by Sultana *et al.* (2013). For species-specific primers, see Table 60.

92.5 Geographical distribution

P. vulnus is widely distributed throughout the temperate, tropical and subtropical regions of the world (Fig. 10.21). Recently the nematode has been detected in apple orchards in Tunisia (Chihani-Hammas *et al.*, 2018). The nematode has been recorded from the following countries (CABI-ISC, 2020).

Africa: Cameroon, Egypt, Kenya, Réunion, Tunisia.
Asia: China, India, Iran, Israel, Japan, Kyrgyzstan, North Korea, Pakistan, South Korea, Sri Lanka.
Europe: Belgium, Bulgaria, Denmark, Finland, France, Germany, Greece, Italy, Netherlands, Norway, Poland, Slovenia, Spain, former USSR, United Kingdom.
North America: Canada, Cuba, USA.
Oceania: Australia, New Zealand.
South America: Argentina, Brazil, Uruguay.

92.6 Hosts

P. vulnus is highly polyphagous and attacks several fruit plants and ornamental crops all over the world. It is a common parasite of almond, apple, grapevine, olive, plum, peach, sour orange, walnut, etc. (Castillo and Vovlas, 2007; Fanelli *et al.*, 2019) but is seldom associated with agronomic crops (Askary *et al.*, 2012). An expanded list of hosts of *P. vulnus* is presented in Table 66.

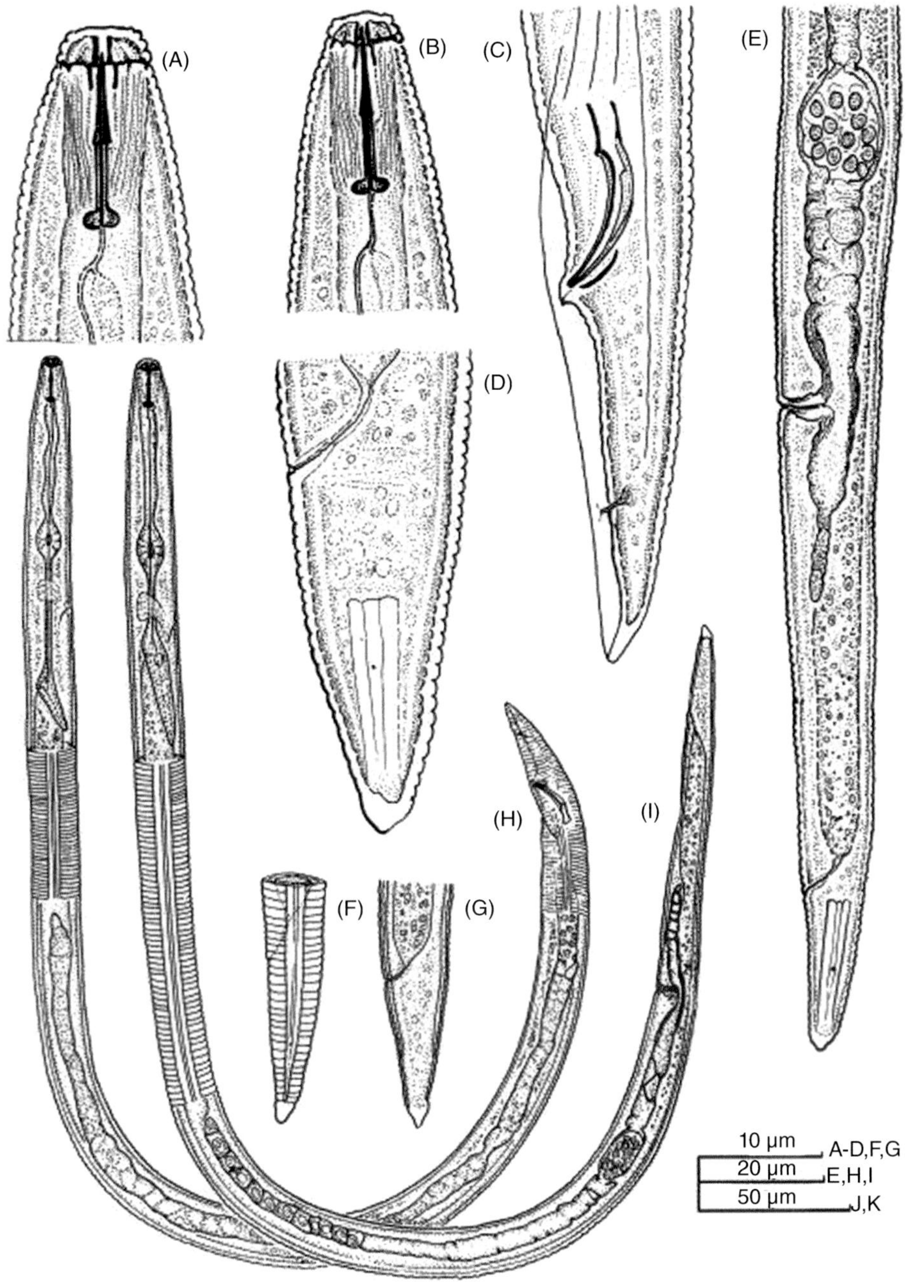

Fig. 10.20. Line drawing of *Pratylenchus vulnus*. Head end of female (A); head end of male (B); tail end of male (C): tails of female (D,F,G); posterior region of female showing vulva, uterus, spermatheca, posterior uterine branch with vestigial ovary and tail (E); male (H); female (I). [Adapted from CABI-CPC, 2010; F-H from CIH description, Set 3, No. 37 courtesy D.C.M. Corbett; rest from M.R. Siddiqi]

92.7 Symptoms

Pratylenchus vulnus causes symptoms similar to those of other lesion nematodes. Characteristic blackened necrotic lesions develop on the root surface (Fig. 10.10D). Above-ground symptoms include stunting of growth, nutrient deficiency-like chlorosis and dieback.

92.8 Biology and life cycle

The biology and life cycle are typical of the genus. Males are abundant and reproduction is amphimictic. Reproductive potential increases with increasing temperature from 23°C to 28°C, but ceases at 33°C (Fanelli *et al.*, 2019).

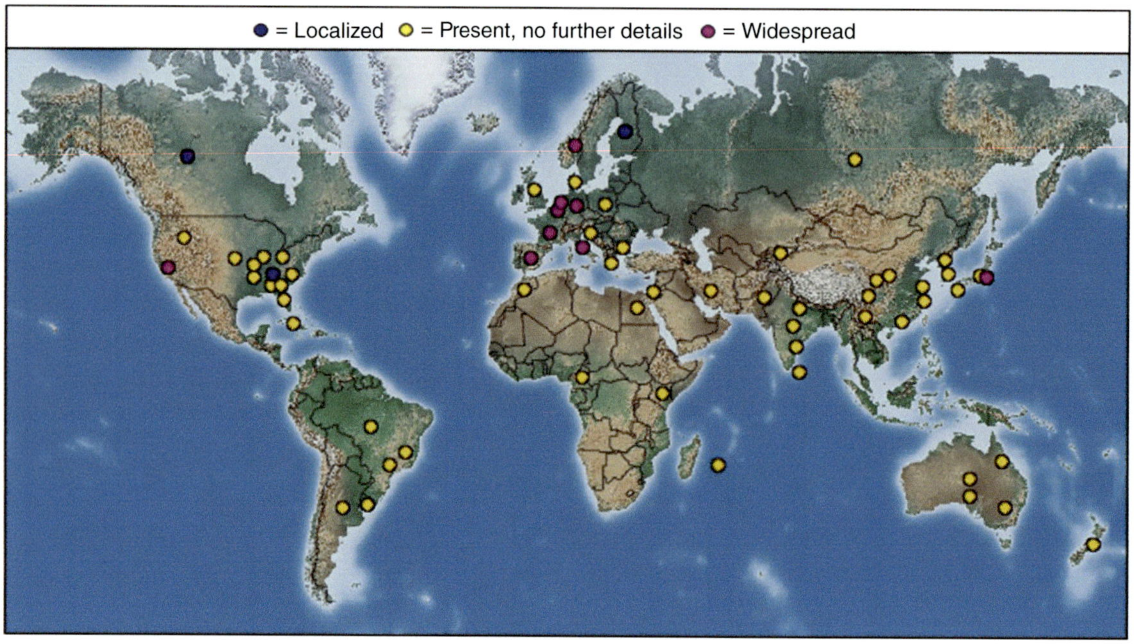

Fig. 10.21. Distribution map of *Pratylenchus vulnus*. [Adapted from CABI-ISC, 2020]

Table 66. List of host plants of *Pratylenchus vulnus* (adapted from CABI-CPC, 2010).

Main hosts

Actinidia arguta (tara vine)
Actinidia deliciosa (kiwifruit)
Araucaria araucana (monkey puzzle)
Berberis thunbergii (Japanese barberry)
Brassica oleracea var. *capitata* (cabbage)
Buxus sempervirens (common boxwood)
Carya illinoinensis (pecan)
Citrus aurantium (sour orange)
Citrus sinensis (navel orange)
Citrus × *paradisi* (grapefruit)
Corylus avellana (hazel)
Crotalaria juncea (sunn hemp)
Crotalaria spectabilis (showy rattlepod)
Cydonia oblonga (quince)
Cytisus scoparius (broom)
Daucus carota (carrot)
Eriobotrya japonica (loquat)
Ficus carica (fig)
Forsythia intermedia (golden bells)
Fragaria ananassa (strawberry)
Gossypium hirsutum (Bourbon cotton)
Hamamelis (witch hazel)
Impatiens balsamina (garden balsam)
Juglans hindsii (Californian black walnut)
Juglans major (Arizona walnut)
Juglans microcarpa (river walnut tree)
Juglans nigra (black walnut)
Juglans regia (walnut)
Juniperus chinensis (Chinese juniper)
Lathyrus odoratus (sweet pea)
Ligustrum ovalifolium (California privet)
Lilium longiflorum (Easter lily)
Lilium speciosum (showy lily)
Lotus corniculatus (bird's-foot trefoil)
Mahonia (holly grape)
Malus sylvestris (crab-apple tree)
Medicago sativa (lucerne)
Melilotus alba (honey clover)
Nicotiana tabacum (tobacco)
Olea europaea subsp. *europaea* (olive)
Papaver somniferum (opium poppy)
Persea americana (avocado)
Phaseolus lunatus (lima bean)
Phaseolus vulgaris (common bean)
Pistacia vera (pistachio)
Pisum sativum (pea)
Prunus amygdalus
Prunus armeniaca (apricot)
Prunus avium (sweet cherry)
Prunus besseyi (bessey cherry)
Prunus cerasifera (myrobalan plum)
Prunus domestica (plum)
Prunus dulcis (almond)
Prunus mahaleb (mahaleb cherry)
Prunus persica (peach)
Pterocarya stenoptera (Chinese wing nut)
Pyrus communis (European pear)
Rosa (roses)
Rosa canina (dog rose)
Rosa chinensis (China rose)
Rosa multiflora (multiflora rose)
Rubus (blackberry, raspberry)
Rubus idaeus (raspberry)
Taxus baccata (English yew)

Continued

Table 66. Continued.

Vicia benghalensis (purple vetch)
Vicia faba (broad bean)
Vigna catjang
Vitis rupestris (sand-grape)
Vitis vinifera (grapevine)
Zea mays (maize)
Other hosts
Hedera helix (ivy)
Prunus salicina (Japanese plum)
Sequoia sempervirens (coast redwood)
Habitat association
Pinopsida (conifers)
Pinus koraiensis (fruit pine)

92.9 Economic importance

P. vulnus is reported to attack at least 37 commercial fruit tree species, including almond, apple, avocado, cherry, grape, hazelnut, loquat, peach, pear, pistachio, plum, walnut, grape and citrus rootstocks tested in Spain (Pinochet *et al.*, 1996). In the USA, *P. vulnus* is pathogenic to boxwood; it suppressed the growth by 50% and plants died within 15–20 months at populations of 7100 nematodes/500 cc of soil (Benson *et al.*, 1976). *P. vulnus* is also associated with growth decline of peach trees (Amici and Lalatta, 1971; Simeone, 1993). In China, the nematode caused 81–100% infection in peach plants in some nurseries and orchards and also infected strawberry (Gao *et al.*, 1989). It also extensively colonized roots of papaya (Pinochet *et al.*, 1993) and poppy (Zyubin, 1968).

P. vulnus can cause significant damage to roses. It appears to be destructive even in small numbers (Winfield, 1974); root fresh weight was reduced by 24–31% and fresh weight of rose flowers by 24–25% (Amsing, 1991). The hairy root of roses caused by *Agrobacterium rhizogenes* is of minor importance, but in presence of *P. vulnus* the disease becomes very severe (Sitaramaiah and Pathak, 1993). Lily is another crop that is prone to *P. vulnus* infestation. In Japan, serious losses to *Lilium speciosum rubrum* cv. 'Uchida' were recorded (Ohkawa and Saigusa, 1972). Inoculation of *P. vulnus* in grapevine, alone or in combination with *Xiphinema index*, suppressed vine shoot and root growth under glasshouse conditions (Pinochet *et al.*, 1976). *P. vulnus* and *Meloidogyne incognita* also caused substantial reduction in shoot and root growth of grapes compared with controls (Anwar, 1995).

92.10 Movement and means of dispersal

See section 76.11, above.

92.11 Pest risk analysis and looming threat from introduction

Wide distribution and high adaptability of *P. vulnus* to various temperature regimes revealed that the nematode is able to establish in temperate as well as tropical regions. It has a wide and diverse host range that includes several economically important fruit crops. The direct and indirect economic impacts of this nematode on the productivity of its host crops should not be underestimated. This nematode poses a definite threat to the production of these crops in host countries. Introduction of *P. vulnus* into new areas or countries would seriously affect the productivity of host crops.

92.12 Invasiveness rating

Considering the economic damage and pest risk analysis of *P. vulnus*, the following invasiveness rating has been assigned.

Climate/host interaction	–	Medium (2)
Hosts range	–	High (3)
Dispersal potential	–	High (3)
Economic impact	–	High (3)
Environmental impact	–	Medium (2)
Post-entry distribution	–	Low (–1)
Final rating (score)	–	High (12)

Based on the above invasiveness rating, *P. vulnus* receives a score of 12 out of 15 and is rated as HIGHLY INVASIVE NEMATODE.

92.13 Region-wise status of invasiveness

P. vulnus is listed as a quarantine species in many countries, including Italy for pome and stone fruits and vegetables (Tacconi and Talame, 1995) and India, where it is restricted to protect the lily crop (PQ Order-2003, 2015). Due to the wide host range, *P. vulnus* has been given a highly invasive rating for all the countries where this nematode is absent.

92.14 Management measures

See section 82.5, above.

93 *Pratylenchus zeae*

93.1 Common name

Root-lesion nematode

93.2 Scientific name

Pratylenchus zeae Graham, 1951

93.3 Synonym

Pratylenchus indicus Das 1960

93.4 EPPO code

PRATZE

93.5 Diagnosis

93.5.1 Morphological diagnosis

Morphological descriptions of *P. zeae* after Fortuner (1976) are presented below. For measurements, see Table 61.

FEMALE: Body slender, almost straight after heat relaxation. Body annules indistinct. Lateral field with four incisures, spreading up to tail terminus. Lip region continuous, bearing three annules. Labial framework strongly sclerotized and extending into body about one annule. Stylet knobs broad, anteriorly flattened. Dorsal oesophageal gland orifice around 3 µm behind stylet base. Hemizonid, just in front of excretory pore, about two body annules long. Hemizonion 9–11 annules behind hemizonid. Ovary with oocytes arranged in double row, not extending to oesophagus. Uterus short, oviduct indistinct, spermatheca small and round without sperm. Vulva located at 68–76% of body length. Postvulval uterine sac short with rudiments of ovary. Phasmids slightly posterior to middle of tail. Tail tapering, unstriated, with variable terminus, usually pointed to narrowly rounded to subacute.
MALE: Extremely rare, not required for reproduction. Morphologically similar to female except for genital structure. Spicules slender, arcuate ventrally; gubernaculum simple. Bursa with weak crenate margins.

93.5.2 Molecular diagnosis

For species-specific primers, see Table 60.

93.6 Geographical distribution

P. zeae is widely distributed throughout the tropical and subtropical regions of the world (Fig. 10.22). The nematode has been recorded from the following countries (CABI-CPC, 2010).

Africa: Benin, Burkina Faso, Cameroon, Côte d'Ivoire, Egypt, Ethiopia, Gambia, Ghana, Guinea, Kenya, Madagascar, Malawi, Mozambique, Namibia, Nigeria, Senegal, Somalia, South Africa, Tanzania, Togo, Uganda, Zambia, Zimbabwe.
Asia: Afghanistan, India, Indonesia, Iraq, Japan, Malaysia, Oman, Pakistan, Philippines, Sri Lanka, Turkey, Vietnam.
Europe: Bulgaria.
North America: Barbados, Belize, Costa Rica, Cuba, El Salvador, Jamaica, Martinique, Panama, Puerto Rico, Saint Vincent and the Grenadines, Trinidad and Tobago, USA.
Oceania: Australia, Fiji, Papua New Guinea, Samoa, Tonga.
South America: Argentina, Brazil, Colombia, Ecuador, Venezuela.

93.7 Hosts

P. zeae has a wide host range and attacks several gramineous crops, mainly rice, maize, sorghum and sugarcane (Fortuner, 1976). An expanded list of hosts of *P. vulnus* is presented in Table 67.

93.8 Symptoms

P. zeae causes symptoms similar to those of other lesion nematodes. There are no specific above-ground symptoms. In fields, the infected plants exhibit stunted growth with chlorotic foliage in patches. On roots, depending upon host crop, the characteristic necrosis varies marginally. On rice, the nematode causes light-brown lesions in the root cortex. In severe

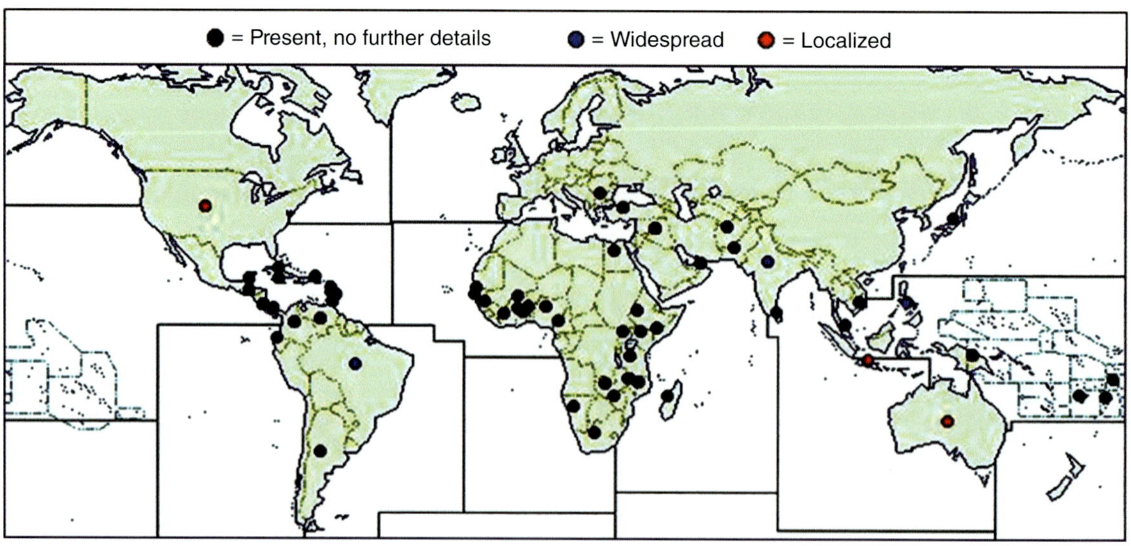

Fig. 10.22. Distribution map of *Pratylenchus zeae*. [Adapted from CABI-CPC, 2010]

Table 67. List of hosts affected by *Pratylenchus zeae* (adapted from CABI-CPC, 2010).

Main hosts
Arachis hypogaea (groundnut)
Capsicum annuum (bell pepper)
Feijoa sellowiana (horn of plenty)
Nicotiana tabacum (tobacco)
Oryza sativa (rice)
Psidium guajava (guava)
Saccharum officinarum (sugarcane)
Solanaceae
Sorghum bicolor (sorghum)
Zea mays (maize)
Other hosts
Allium cepa (onion)
Allium sativum (garlic)
Amaranthus hybridus (pigweed)
Borrichia frutescens (sea-oxeye)
Caryocar brasiliense
Cocos nucifera (coconut)
Crotalaria pallida (smooth crotalaria)
Crotalaria spectabilis (showy rattlepod)
Crotalaria sphaerocarpa
Lactuca sativa (lettuce)
Lespedeza
Lupinus (lupins)
Musa × paradisiaca (plantain)
Pennisetum glaucum (pearl millet)
Prunus persica (peach)
Zingiber officinale (ginger)
Wild hosts
Amaranthus spinosus (spiny amaranth)
Andropogon virginicus (broomsedge)
Chenopodiaceae
Chloris gayana (Rhodes grass)
Cynodon dactylon (Bermuda grass)
Cyperus iria (rice flatsedge)
Dactyloctenium aegyptium (crowfoot grass)
Datura stramonium (jimsonweed)
Digitaria sanguinalis (large crabgrass)
Diodia teres (poorjoe)
Echinochloa crus-galli (barnyard grass)
Eleusine indica (goosegrass)
Eragrostis curvula (weeping lovegrass)
Heterotheca subaxillaris (camphorweed)
Panicum maximum (Guinea grass)
Poaceae (grasses)
Solidago gigantea (giant goldenrod)
Tribulus terrestris (puncture vine)
Urochloa mutica (tall panicum)
Xanthium strumarium (common cocklebur)

conditions, the lesions become necrotic, darken and coalesce. Root size also decreases and plants remain dwarf. On maize and sorghum, necrosis of cortical tissues and breakage of root cells results in cavity formation (Olowe and Corbett, 1976). On sugarcane, the nematode causes severe necrosis with conspicuous red lesions within the root cortex (Fig. 10.10G). Heavy infestation leads to yellowing of the leaves and reduction in top and stalk length (Sankaranarayanan, 2010).

93.9 Biology and life cycle

The biology and life cycle are typical of the genus and *P. zeae* reproduces amphimictically (Troccoli *et al.*, 2016). Highest reproduction rates occur at 30°C. The life cycle is completed in around 21 days on maize and 30 days on rice (Olowe and Corbett, 1976). The nematode thrives well in lighter, sandy soils (Plowright and Hunt, 1994).

93.10 Economic importance

P. zeae mainly attacks rice crops along with other *Pratylenchus* spp. under upland conditions (Bridge *et al.*, 2005a). The nematode may infest 50% of the upland rice in South and South-East Asia and inflict yield losses of up to 30% (Prot, 1993). *P. zeae* also parasitizes other gramineous crops, such as maize, sorghum and sugarcane (Fortuner, 1976). On sugarcane, it is major nematode problem in Panama (Pinochet, 1987), north-east Argentina (Costilla, 1973), Burkina Faso., South Africa (Spaull and Cadet, 1990) and India (Sankaranarayanan, 2010).

93.11 Movement and means of dispersal

See section 76.11, above.

93.12 Pest risk analysis and looming threat from introduction

P. zeae is a major problem on four economically important crops: rice, maize, sorghum and sugarcane, confined to subtropical and tropical regions. The direct and indirect economic impacts of this nematode on the production of its host crops are well documented. The nematode is not recorded in temperate regions and poses a definite threat to the production of host crops in these countries. Introduction of *P. zeae* and subsequent damage would seriously affect the production of host crops.

93.13 Invasiveness rating

Based on the economic damage and pest risk analysis of *P. zeae*, the following invasiveness rating has been assigned.

Climate/host interaction	–	Medium (2)
Hosts range	–	High (3)
Dispersal potential	–	High (3)
Economic impact	–	High (3)
Environmental impact	–	Medium (2)
Post-entry distribution	–	Low (–1)
Final rating (score)	–	High (12)

Based on the above invasiveness rating, *P. zeae* receives a score of 12 out of 15 and is rated as HIGHLY INVASIVE NEMATODE.

93.14 Region-wise status of invasiveness

Considered as an important pest of rice, maize, sorghum and sugarcane, *P. zeae* has been given a highly invasive rating for all the host countries where this nematode is not known to occur.

93.15 Management measures

See section 82.5, above.

Radopholus Spp.

Cobb (1893) described males and females of the *Radopholus* as two different species, *Tylenchus similis* and *Tylenchus granulosus*, respectively. Later, he described a third species, *Tylenchus biformis*, on sugarcane (Cobb, 1906). Thorne (1961) proposed the genus *Radopholus* (1949) and placed it in the family Pratylenchidae. Sher (1968) recorded the similarities in the three species proposed by Cobb and merged them as *Radopholus similis*. Siddiqi (2000) place the genus in the superfamily Hoplolaimoidea and the family Pratylenchidae and this hoplolaimoid nexus is now supported by molecular analysis (Subbotin *et al.*, 2006; Haegeman *et al.*, 2010). Currently, more than 30 species are described and validated with *Radopholus similis* as type species (Siddiqi, 2000; De Waele and Elsen, 2007; Duncan and Moens, 2013). Numerous new species have been described in the past two decades. Some of the recognized and important *Radopholus* species are:

Radopholus arabocoffeae Trinh, Nguyen, Waeyen, Subbotin, Krassen and Moens, 2004
R. bridgei Siddiqi and Hahn, 1995
R. citri Machon and Bridge, 1996
R. daklakensis Trinh, Waeyenberge, Nguyen and Moens, 2012
R. duriophilus Nguyen, Subbotin, Madani, Trinh and Moens, 2003
R. musicola Stanton, Mundo-Ocampo, Baldwin and Kaplan, 2001
R. nativus Sher, 1968
R. neosimilis Sauer, 1958
R. vacuus Colbran, 1971
R. vangundyi Sher, 1968

The burrowing nematode, *Radopholus* is a migratory endoparasite. It feeds on underground parts of plants, similar to root-lesion nematodes, and produces lesions on roots. *R. similis* is an important parasite of bananas and citrus, but it also attacks other fruits, vegetables and plantation crops. *R. citri* is a pest of citrus and has been localized in Java, Indonesia.

Systematic position of *Radopholus* species

	Classical (after Siddiqi, 2000)	Molecular (after Decraemer and Hunt, 2013)
Phylum	Nematoda	Nematoda
Class	Secernentea	Chromadorea
Order	Tylenchida	Rhabditida
Suborder	Hoplolaimina	Tylenchina
Infra-order	–	Tylenchomorpha
Superfamily	Holpolaimoidea	Tylenchoidea
Family	Pratylenchidae	Pratylenchidae
Subfamily	Pratylenchinae	Radopholinae
Genus	*Radopholus*	*Radopholus*

94 *Radopholus citri*

94.1 Common name

Citrus burrowing nematode

94.2 Scientific name

Radopholus citri Machon and Bridge, 1996

94.3 EPPO code

RADOCT

94.4 Diagnosis

94.4.1 Morphological diagnosis

Morphological descriptions of *Radopholus citri* after Machon and Bridge (1996) are summarized below. For measurements, see Table 68.

FEMALE: Body almost straight to ventrally arcuate after heat relaxed (Fig. 10.23K). Annules distinct. Lateral field with four incisures, reducing to three in region of phasmids, outer bands often marked by transverse striae towards posterior extremity. Head low, slightly offset, rounded and slightly flattened apically (Fig. 10.23D). Stylet moderately strong with distal surface of dorsal knob extending anteriorly (Fig. 10.23B). Orifice of dorsal oesophageal gland 4–5 μm behind stylet knobs. Procorpus cylindroid, median bulb oval to round and slightly offset from rest of oesophagus. Oesophageal gland dorsally overlapping intestine (Fig. 10.23A). Nerve ring just posterior to median bulb. Excretory secretory pore immediately posterior to hemizonid and situated posterior to bulb. Vulva postmedian with distinct protuberant lips. Genital tract amphididelphic, branched,

Table 68. Morphometric measurements of *Radopholus* and *Zygotylenchus* species.

Morphometric characters		Range, numerous populations		
		R. citri (after Machon and Bridge, 1996)	*R. similis* (after Sher, 1968)	*Z. guevarai*(after Siddiqi, 1975a)
Females	Length (μm)	620–810	520–880	480–550
	Stylet length (μm)	17.4–19.4	17–20	16–17
	a	5.1–32.5	22–30	27–30
	b	6.9–8.3	4.7–7.4	5.1–5.7
	b'	3.8–4.9	3.5–5.2	–
	c	12.3–16.7	8–13	18–20
	c'	2.5–3.6	2.9–4.0	–
	Body diameter (μm)	–	–	–
	V	54.1–62.3	55–61	59–62
	Tail length (μm)	43.6–57.0	52–77[a]	–
	h (μm)	5.4–8.7	–	–
	0	–	12–20	–
Males	Length (μm)	470–580	590–670	500–520
	Stylet length (μm)	12.7–14.7	12–17	15.5–16.0
	a	28.4–41.7	31–44	31–36
	b	5.4–8.0	6.1–6.6	5.2–5.5
	b'	–	4.1–4.9	–
	c	12.4–15.9	8–10	16–18
	c'	2.6–3.9	5.1–6.7	–
	Body diameter (μm)	–	–	–
	Tail length (μm)	31.5–43	64–86[a]	40–45
	Spicule (μm)	14.1–17.4	19–22	16–21
	Gubernaculum (μm)	7.4–10.7	8–12	5–7
	h	5.4–8.0	–	–

[a] From Taylor (1969); h = length of terminal, non-protoplasmic portion of tail.

outstretched, with oval to rod-shaped sperm in axial spermathecae (Fig. 10.23M). Tail tapering, conoid to rounded tip annulated regularly. Phasmids situated around middle of tail, at 12–17 annules posterior to anus.

MALE: Body with marked sexual dimorphism in anterior parts. Head elevated, knob-like, offset with four or five annules (Fig. 10.23C). Stylet conus strongly developed, shaft and knobs distinct but weakly developed. Oesophagus degenerate and deceptively non-functional. Spicules paired, arcuate ventrally (Fig. 10.23J). Gubernaculum rod-like, protrusible. Bursa crenate, nearly reaching tail terminus. Phasmids situated near middle of tail, about 16 annules posterior to cloaca. Tail conoid, tapering to a finely pointed and partially offset tip (Fig. 10.23I).

94.4.2 Biochemical and molecular diagnosis

Phosphate glucose isomerase, lactate dehydrogenase and malate dehydrogenase isoenzyme patterns with isoelectric focusing may be used to compare the *Radopholus* isolates and species (Fallas *et al.*, 1996; Kaplan and Opperman, 1997). Early RAPD profiles confirmed that *R. citri* is a distinct species (Hahn *et al.*, 1996; Kaplan and Opperman, 1997). However, molecular approaches employed to characterize *R. similis* may be useful. For further details, see section 95.5.4, below.

94.5 Geographical distribution

This burrowing nematode has been reported only in Java, Indonesia, and is localized there (Machon and Bridge, 1996).

94.6 Hosts

R. citri is a monophagous nematode and infects only citrus. Currently, there is no record of other hosts (Machon and Bridge, 1996). Since *R. citri* can be cultured on excised maize roots and carrot discs, these may act as a host.

94.7 Symptoms

R. citri causes very severe stunting, cortical necrosis and root destruction in citrus. The nematode remains confined to the cortex and is often associated with complete destruction of the cortical tissues. In pot trials, 1000, 5000 and 10,000 larvae/pot caused a highly significant reduction in seedlings height, fresh and dry weights of tops, number of leaf nodes and length of stems and branches (Fig. 10.24A). Fresh root weights were also

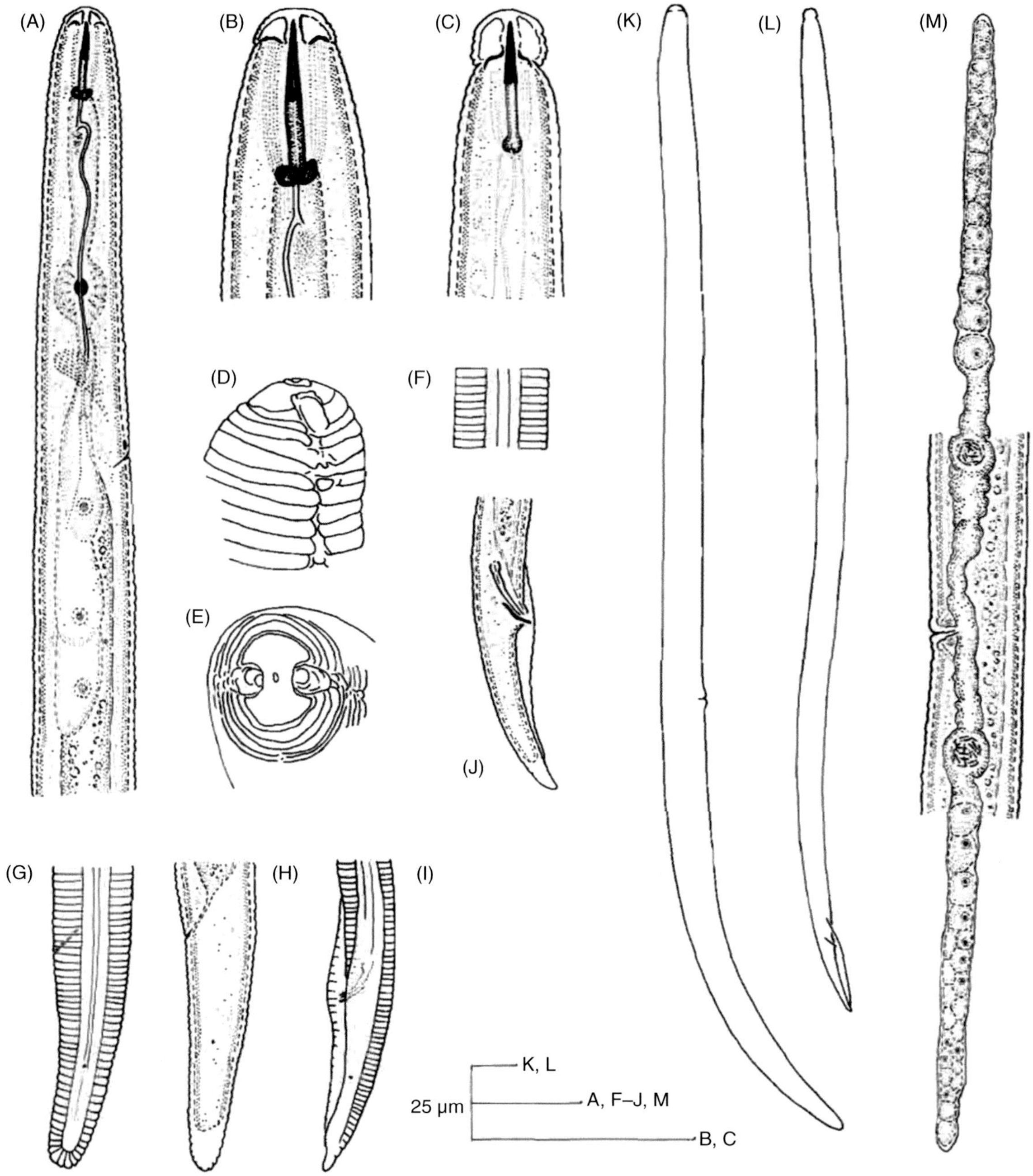

Fig. 10.23. Line drawing of *Radopholus citri.* Female oesophagus (A); female head (B); male head (C); female head (D); female head, enface (E); lateral field, mid-body (F); female tails (G,H); male tails (I,J); entire female (K); entire male (L); female genital tracts (M). [Courtesy Machon and Bridge, 1996; Fundamental and Applied Nematology]

significantly reduced in all nematode treatments (Machon and Bridge, 1996).

94.8 Biology and life cycle

R. citri is a migratory endoparasite which can be found in the host's roots as well as in soil. The nematode completes its life cycle within the root cortex. Females and juveniles are infective but males, although having a relatively strongly developed stylet, have an apparently non-functional, degenerate oesophagus. Eggs are laid singly or in groups in the root tissues and hatch in around 10 days at 25°C. The juveniles and females feed on the cortical cells, causing cell death and necrosis of the root (Machon and Bridge, 1996).

Fig. 10.24. Symptoms of burrowing nematode on different crops. (A) caused by *Radopholus citri*, rest caused by *R. similis*. Citrus seedlings of the same age with different populations of *R. citri*: 1=Zero, 2=1000, 3=5000, 4=10000 (A); infested banana field (B); necrotic lesions and rot (C); LS of infected root pieces showing root necrosis and decay (D); toppling of banana trees (E); lesion on cortex exposing stele, arrow (F); citrus plant showing die back symptom (G); ginger rhizome showing rotting (H); yellow disease of black peppers (I). [Courtesy A,F,I: J. Bridge & J.E. Machon/CABI BioScience; B-E: NASI, India; F: Fred E. Brooks, https://www.apsnet.org; G: S.A. Lewis, Clemson, Bugwood.org]

94.9 Similarities to other species/conditions

R. citri is morphologically similar to the genus *Pratylenchus* but can be differentiated by having a median vulva with two genital tracts. The nematode can be distinguished from other *Radopholus* species by the male's stronger stylet and from the female of *R. similis* by a shorter and more rounded tail terminus.

94.10 Economic importance

Machon and Bridge (1996) demonstrated that the nematode is potentially a destructive pest of citrus. Very severe root destruction and growth reduction were observed in controlled pathogenicity experiments.

94.11 Movement and means of dispersal

See section 76.11, above.

94.12 Pest risk analysis and looming threat from introduction

R. citri is an important pest of citrus and is currently localized in Java, Indonesia. However, there is a high probability that it could infest citrus in other subtropical and tropical areas, particularly those with sandy soils similar to the type locality. *R. citri* can be spread through infested vegetative planting material such as nursery rootstocks and seedlings.

94.13 Invasiveness rating

Based on the consequences of introduction and probable pest risk analysis of *R. citri*, the following invasiveness rating has been assigned.

Climate/host interaction	–	Low (1)
Hosts range	–	Low (1)
Dispersal potential	–	Medium (2)
Economic impact	–	High (3)
Environmental impact	–	Medium (2)
Post-entry distribution	–	Not known (0)
Final rating (score)	–	Medium (9)

Based on the above invasiveness rating, *R. citri* receives a score of 9 out of 15 and is rated as MEDIUM INVASIVE NEMATODE.

94.14 Region-wise status of invasiveness

Considering the economic importance of citrus in various parts of the world, the perennial nature of the host crop and the limited information available on the host range and pathways of entry of *R. citri*, a medium invasive rating has been given for all the citrus-growing countries except Java, Indonesia, where this nematode localized.

94.15 Management measures

Management measures include the production of nematode-free citrus planting material. Growing seedlings in fields free of nematodes or use of sterilized soil for pots or containers is advised. For further details, see the management of citrus in section 95.15 below, and phytosanitary measures in Chapter 16, section 115.

95 *Radopholus similis*

95.1 Common name

Burrowing nematode

95.2 Scientific name

Radopholus similis (Cobb, 1893) Thorne, 1949

95.3 Synonyms

Anguillulina acutocaudatus (Zimmermann, 1898) Goodey, 1932
Anguillulina biformis (Cobb, 1909) Goodey, 1932
Anguillulina granulosa (Cobb, 1893) Goodey, 1932
Anguillulina similis (Cobb, 1893) Goodey, 1932
Radopholus acutocaudatus (Zimmermann, 1898) Siddiqi, 1986
Radopholus biformis (Cobb, 1909) Siddiqi, 1986
Radopholus citrophilus Huettel, Dickson and Kaplan, 1984
Radopholus granulosus (Cobb, 1893) Siddiqi, 1986
Rotylenchus similis (Cobb, 1893) Filipjev, 1936
Radopholus similis citrophilus Huettel, Dickson and Kaplan, 1984
Tetylenchus granulosus (Cobb, 1893) Filipjev, 1936
Tylenchorhynchus acutocaudatus (Zimmermann, 1898) Filipjev, 1934
Tylenchus biformis Cobb, 1909
Tylenchus granulosus Cobb, 1893
Tylenchus similis Cobb, 1893

95.4 EPPO code

RADOCI

95.5 Diagnosis

Previously *Radopholus citrophilus* was regarded as a separate species from *Radopholus similis* but it is now accepted as being synonymous (Kornobis, 1999; Xu *et al.*, 2014).

95.5.1 Morphological diagnosis

Morphological descriptions of *R. similis* after Orton Williams and Siddiqi (1973) are summarized below. For morphometric measurements (after Sher, 1968), see Table 68.

FEMALE: Body straight to slightly arcuate ventrally. Annules distinct. Lateral field with four incisures, not areolated except towards extremities. Cephalic region strongly sclerotized, hemispherical, occasionally offset, bearing 3–4 annules (Fig. 10.25A). Anterior cephalids immediately posterior to lip sclerotization. Stylet well developed, basal knobs round, usually indented anteriorly (Fig. 10.25B). Median bulb well developed, oval to round with prominent valvular apparatus. Oesophageal glands three, in separate lobes, dorsally and dorso-laterally overlapping intestine; dorsal gland anterior (Fig. 10.25O). Excretory pore just behind level of oesophago-intestinal valve. Hemizonid just anterior to excretory pore, three annules long. Genital tract paired, opposed branches, outstretched. Vulva prominent, just postequatorial. Spermathecae globular, generally packed with rod-shaped small sperm. Ovaries usually with single row of oocytes. Intestine indistinctly overlapping rectum, filled with spherical granules. Tail slightly elongate-conoid with indented terminus or narrow rounded (Fig. 10.25K,L).
MALE: Distinct sexual dimorphism in anterior region. Stylet degenerate without distinct knobs (Fig. 10.25C,D). Oesophagus non-functional, with indistinct median bulb and valvular apparatus. Labial region high, four-lobed, reduced lateral lips, weakly sclerotized. Testis single, anteriorly outstretched (Fig. 10.25P). Spermatozoa rod-like. Spicules strongly cephalated with pointed distal ends. Gubernaculum protrusible, rod-like, with distinct sharp claw-like titillae at distal end. Caudal alae coarsely crenate, enveloping around two-thirds of tail (Fig. 10.25N).

95.5.2 Cytogenetic diagnosis

The presence of two races in *R. similis* can be distinguished by cytogenetic characters (Xu *et al.*, 2014). Kaplan and Opperman

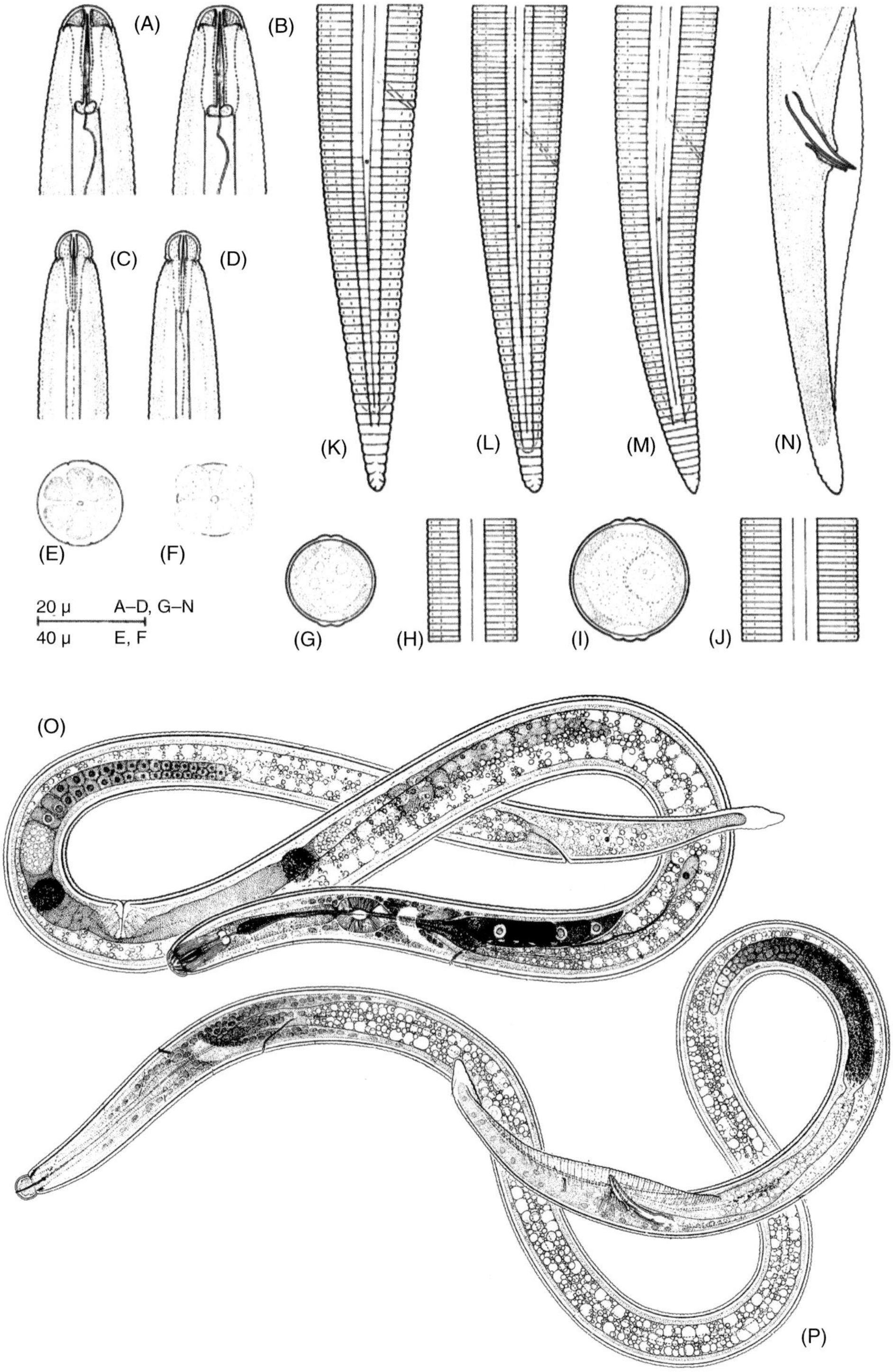

Fig. 10.25. Line drawing of *Radopholus similis*. Anterior region of female (A,B); anterior region of male (C,D); female face view (E); male face view (F); juvenile cross-section near mid-body (G); juvenile lateral view (H); female cross-section near mid-body (I); female lateral view (J); posterior region of female (K,L); posterior region of male (M,N); entire female (O); entire male (P). [Adapted from Ortan Williams and Siddiqi, 1973. CIH Descriptions of Plant-parasitic Nematodes. Set 2, No. 27; courtesy CAB International]

(2000) reported that the haploid chromosome numbers of the karyotype of two physiological races of *R. similis* were $n = 4$ (banana race) and $n = 5$ (citrus race) and demonstrated that the citrus race and the banana race could mate and produce offspring which had chromosome number $n = 5$. However, Goo and Sipes (1999) found that the chromosome number ranged from $n = 4$ to $n = 7$ from six isolates of *R. similis*. Xu *et al.* (2014) studied the karyotype of 19 populations of *R. similis* and confirmed that the karyotype of all the 20 populations was $n = 5$. The study supported the proposal that *R. citrophilus* is a synonym of *R. similis*.

95.5.3 Biochemical diagnosis

R. similis was characterized based on proteins and isozymes using one-dimensional native gel electrophoresis (Jaumot *et al.*, 1997; Sarah, 1997). Phosphate glucose isomerase, lactate dehydrogenase and malate dehydrogenase isoenzyme patterns using isoelectric focusing were also used to compare the various *R. similis* isolates (Fallas *et al.*, 1996; Kaplan and Opperman, 1997).

95.5.4 Molecular diagnosis

Several molecular approaches have been employed to characterize *R. similis* (Elbadri *et al.*, 2002; Jacob *et al.*, 2008; Mgonja *et al.*, 2020). Fallas *et al.* (1996) used 18S and 26S rDNA fragment amplified by the polymerase chain reaction (PCR) to characterize ten isolates of *Radopholus* spp. from different part of the world using the universal primer (Vrain *et al.*, 1992) as follows:

18S primer: 5'-TTGATTACGTCCCTGCCCTTT-3'
26S primer: 3'-GGAATCATTGCCGCTCACTTT-5'.

Restriction fragment length polymorphism (RFLP) analysis using each of six restriction enzymes (*Hha* I, *Hpa* II, *Rsa* I, *Hae* III, *Taq* I and *Alu* I) revealed little variation among the isolates tested. Mgonja *et al.* (2020) designed *R. similis*-specific primers to amplify the ITS1 and ITS2 rDNA and large subunit (LSU) rDNA regions.

For ITS1 and ITS2 rDNA region

RD1F - ACTGAGCCGATTCGAGAAATC
RD1R - ATGATTTGGAAAAGCTGCCAATTT
RS3ITSF - CTGTGAGTCGTGGAGCAGTT
RS3IT SR - ATGATTTGGAAAAGCTGCCAAT
RS4ITSF - TGTAGTCC ATGTCCGTGGC
RS4ITSR - TGATTTGGAAAAGCTGCCAATTT)

For LSU rDNA regions

RS8LSUF - AGGACGTGAAACCGGTGAGG
RS8LSUR - T ATACCCAAGTCAGACGATCG
RS6LSUF - CTGG CGTATCTAGCCTGCAT
RS6LSUR - TTTACACCGAGGATTGGCGT

The ITS1 and ITS2 and the LSU rDNA region produced amplicon sizes ranging 308 to 750 bp. The phylogenetic analysis revealed two clades: clade-I, which contained *R. similis* isolates; and clade-II, which contained *R. arabocoffeae*.

95.6 Geographical distribution

R. similis was first reported by Cobb in the necrotic tissue of the roots of *Musa* sp. in New South Wales from Fiji in July 1891. This burrowing nematode has since been reported throughout the world and is widely distributed in warmer regions (Fig. 10.26), including Asia, Africa, the Caribbean, South America and the Pacific. The nematode has been detected in the following countries (CABI-ISC, 2020).

Africa: Benin, Burkina Faso, Burundi, Cameroon, Central African Republic, Congo Democratic Republic, Congo

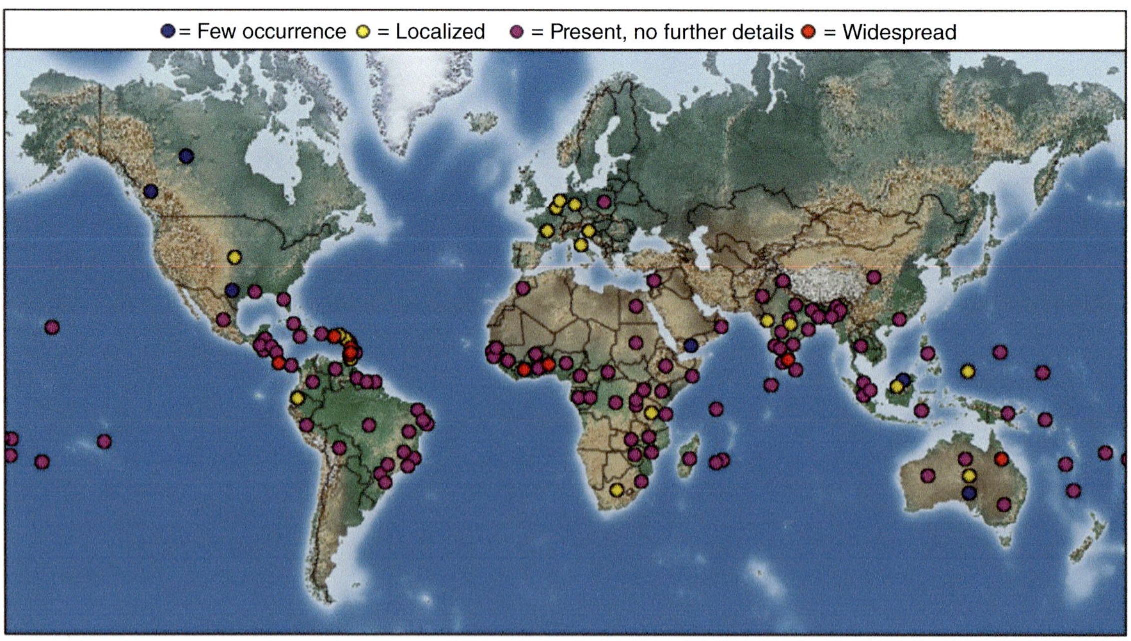

Fig. 10.26. Distribution map of *Radopholus similis*. [Adapted from CABI-ISC, 2020]

Republic, Côte d'Ivoire, Egypt, Eswatini, Ethiopia, Gabon, Gambia, Ghana, Guinea, Guinea-Bissau, Kenya, Madagascar, Malawi, Mauritius, Morocco, Mozambique, Nigeria, Réunion, Rwanda, Senegal, Seychelles, Somalia, South Africa, Sudan, Tanzania, Uganda, Zambia, Zimbabwe.

Asia: Brunei, China, India, Indonesia, Japan, Lebanon, Malaysia, Maldives, Oman, Pakistan, Philippines, Singapore, Sri Lanka, Taiwan, Thailand, Turkey, Yemen.

Europe: Belgium, France, Italy, Netherlands, Slovenia.

North America: Barbados, Belize, Canada, Costa Rica, Cuba, Dominica, Dominican Republic, El Salvador, Grenada, Guadeloupe, Guatemala, Honduras, Jamaica, Martinique, Mexico, Nicaragua, Panama, Puerto Rico, Saint Kitts & Nevis, Saint Lucia, Saint Vincent and the Grenadines, Trinidad and Tobago, USA, US Virgin Islands.

Oceania: American Samoa, Australia, Fiji, French Polynesia, Guam, Micronesia (Federal States of), New Caledonia, Niue, Norfolk Island, Palau, Papua New Guinea, Samoa, Solomon Islands, Tonga.

South America: Bolivia, Brazil, Colombia, Ecuador, French Guiana, Guyana, Peru, Suriname, Venezuela.

95.7 Hosts

R. similis is a polyphagous nematode and infects more than 250 host plants from 16 families, including monocot and dicot plant species (Duncan and Moens, 2013). DuCharme and Birchfield (1956) reported for the first time the existence of two races of *R. similis* in Florida, based on their ability to parasitize citrus and banana crops. The 'banana race' attacked banana but not citrus, whereas the 'citrus race' infested both banana and citrus, and other plant species belonging to diverse families. The banana race of *R. similis* is known to infect *Musa* spp., *Ipomea batatas*, *Pueraria phaseoloides javanica*, coconut, areca nut, black pepper, cardamom, ginger and turmeric, tea, vegetables, ornamentals, trees, sugarcane, coffee, maize, avocado, weeds, etc. Some wild *Musa* types are also infested by this nematode. Venkitesan (1976) reported that a population of *R. similis* isolated from pepper roots did not infest *Citrus sinensis*, *C. reticulata*, *Coffea arabica*, coconut and areca nut; it only infested banana. The banana population infected pepper, coconut and areca nut; the coconut population infected pepper and banana but not areca nut; while the areca nut population infected pepper and banana but not coconut. The citrus race is highly pathogenic to citrus grown in Florida, Mediterranean regions or other parts of the world where citrus is cultivated. It attacks banana, citrus, black pepper, coconut, tea, *Strelitzia*, *Philodendron*, *Anthurium* and *Calathea* spp. An expended list of hosts is given in Table 69.

95.8 Symptoms

In general, *R. similis* causes severe stunting, cortical necrosis and root destruction in the host plants. Symptoms on different hosts are discussed below.

Table 69. List of host plants of *Radopholus similis* (adapted from CABI-ISC, 2020).

Main hosts
Ananas comosus (pineapple)
Anthurium andreanum
Arachis hypogaea (groundnut)
Areca catechu (betelnut palm)
Camellia sinensis (tea)
Citrus
Cocos nucifera (coconut)
Coffea (coffee)
Coffea arabica (arabica coffee)
Coffea canephora (robusta coffee)
Curcuma longa (turmeric)
Daucus carota (carrot)
Dioscorea (yam)
Musa (banana)
Musa textilis (manila hemp)
Musa × *paradisiaca* (plantain)
Persea americana (avocado)
Piper betle (betel pepper)
Piper nigrum (black pepper)
Pyrus (pears)
Saccharum officinarum (sugarcane)
Solanum lycopersicum (tomato)
Zea mays (maize)
Zingiber officinale (ginger)
Other host
Illicium verum (star anise)
Wild host
Solanum nigrum (black nightshade)

BANANA Banana plants infested by the burrowing nematode display general decline, premature defoliation, unthriftiness, stunting and small bunches of fruit (Fig. 10.24B). Symptoms on roots and corms are more specific, exhibiting small cuticular sunken lesions on young cord roots (Fig. 10.24C), whereas surface cracks appear on older roots. Such roots show extensive reddish-brown lesions in the cortex when cut longitudinally (Fig. 10.24D). The nematode causes decay and death of distal cells. This deadly pathogen is well established in banana tracts where soil and temperature conditions are favourable. The infested trees have half the quantity of functional feeder roots compared with healthy trees. The resulting disruption in the uptake of nutrients eventually leads to suppression in plant growth and in production of fruits. The formation of lesions of the primary roots combined with the girdling and death of the anchor roots makes the plant amenable to toppling by wind action, especially during the reproductive phase (Fig. 10.24E) (Sheela and Sundraraju, 2012). This is followed by retarded growth and extensive root and rhizome necrosis, resulting in serious economic consequences like ineffective utilization of fertilizers, delay in harvest from the time of planting, failure to attain maximum bunch weight and impoverished fruit quality. Heavy nematode infestation affects plant vigour, resulting in undersized fruits and longer periods intervening between the ratoon crops (Gowen, 1979). Infested fields have to be replanted every 2–3 years because of the drastic reduction of plant numbers.

CITRUS *R. similis* (= *R. citrophilus*) causes spreading decline, a serious disease of citrus in Florida. Above-ground symptoms generally appear about a year after primary infection. A typical spreading decline area consists of a discrete group of unthrifty-appearing trees (Fig. 10.24G) which expands much more rapidly than the decline caused by *Tylenchulus semipenetrans*, in all directions regardless of rows or direction of cultivation. The diameter of the decline area increases by approximately 10–20 m/year. Infested trees have sparse foliage, stunted terminal growth resembling the symptom of twig dieback, general unthriftiness, reduced foliage size with less K and N and fewer fruits on the trees. Plants may wilt at midday and rejuvenate with rain and irrigation. Trees sometimes bloom abundantly but bear few fruits, which are of small size. Below ground, feeding on roots by nematodes causes cessation of apical meristem activity and destruction of cortical cells, leading to the development of necrotic lesions which may coalesce and eventually die. Reduction in the number of feeder roots in deeper soil profile occurs. At depths of 25–50 cm, 25% of the feeder roots are destroyed; and below this depth, almost all of them are destroyed (Agrios, 2005).

BLACK PEPPER *R. similis* is associated with pepper yellows (slow-wilt) disease of black pepper. The disease appears as whitish-yellow or pale yellow drooping leaves on the vines, followed by shedding of leaves, cessation of growth and dieback symptoms (Fig. 10.24I). Within 3–5 years of initiation of disease, all the leaves are shed and the vine is killed (Koshy and Bridge, 1990).

GINGER Infected plants exhibit stunting, chlorosis, reduced vigour and tillering. Affected plants mature early and rapidly dry out. Initial infections of the rhizomes are evidenced by small, shallow, sunken, water-soaked lesions (Fig. 10.24H). In advanced stages, large infection channels or galleries are formed within the rhizomes.

COCONUT In coconut, *R. similis* causes symptoms of general decline such as stunting, yellowing, button shedding and reduced yield. On roots, infestation induces small, elongate, orange-coloured lesions and causes tender creamy-white roots to become spongy (Griffith and Koshy, 1990).

95.9 Biology and life cycle

R. similis is a migratory endoparasitic nematode. The juveniles and adult females infest the roots and enter the cortex, where they feed, resulting in the development of lesions and burrows/cavities. Males are unable to infest roots. Several cell wall-degrading enzymes, such as cellulases (Haegeman *et al.*, 2008), pectate lyases and xylanases (Haegeman *et al.*, 2009), have been identified from *R. similis*. The nematode may also invade the stellar zone involving phloem and cambium. Although all stages remain vermiform throughout, sexual dimorphism is apparent in adult males having somewhat degenerate stylets and probably being non-parasitic. Females are larger than males and are very active. Sexual reproduction can occur but, in the absence of males, the females become hermaphrodites (Kaplan and Opperman, 2000). Females lay their eggs within infected tissues, with an average of 7–8 eggs per day for 2 weeks. More than one generation can occur within the roots. The duration of the life cycle varies among the hosts. On banana, the life cycle from egg to egg completes in 20–25 days at a temperature range of 24–32°C. On black pepper, it takes about 25–30 days at 21–31°C (Koshy and Bridge, 1990), areca nut 25–30 days at 21–31°C (Koshy and Sosamma, 1978) and coconut about 25 days at 25–28°C.

Although the nematode is found active at 12–32°C of soil temperature, its biological activity remains intense at an optimum temperature of 24°C. Soil texture also plays an important role in parasitism. The population builds up most rapidly in sandy soil, followed by gravelly or loamy soil. There is hardly any build-up in clay soils. The nematode is more pathogenic to citrus in light-textured soil owing to its greater mobility in such soils (Tarjan, 1971). In the absence of hosts, *R. similis* can survive in the soil for more than 6 months without becoming anhydrobiotic (Volcy, 2011).

95.10 Economic importance

R. similis infestation in banana causes gross reduction in the quantity and quality of yield (Sheela and Sundararaju, 2012). In Central America, an increase of 17,000 lb of fruit production per acre (19,000 kg (19 t)/ha) per year occurred in burrowing nematode-free banana plots (Wehunt and Edwards, 1968). Similarly, in Suriname, the burrowing nematode reduced the banana yield to 30 t/ha per year whereas plantain with 3% nematode infestation had yielded 73 t/ha per year. In India, *R. similis* infestation is reported to reduce the yield of banana by 30–41% (Sheela and Sundararaju, 2012). The nematode also causes a severe decline in citrus and resulted in 40–70% yield losses for oranges and 50–80% for grapefruit (Duncan and Cohn, 1990). In Fiji, more than 50% of ginger fields were found infested with *R. similis* resulting in about 40% yield loss (Vilsoni *et al.*, 1976).

95.11 Movement and means of dispersal

See section 76.11, above.

95.12 Pest risk analysis and looming threat from introduction

R. similis is a tropical nematode, native to Indo-Malaya or Australasia. However, it has become a worldwide pest, particularly in subtropical and tropical climates. The nematode has also been introduced into temperate parts of the world, where it may cause damage to glasshouse crops when populations adapt to a lower temperature (Haegeman *et al.*, 2010). RAPD analysis of isolates from different parts of the world by Fallas *et al.* (1996) supported the hypothesis that *R. similis* has been spread with infected plant materials such as rootstocks, corms and tubers, particularly banana, and tea sets transported across several countries for commercial purposes.

95.13 Invasiveness rating

Considering the consequences of introduction and pest risk analysis of *R. similis*, the following invasiveness rating has been assigned.

Climate/host interaction	–	High (3)
Hosts range	–	High (3)
Dispersal potential	–	High (3)
Economic impact	–	High (3)
Environmental impact	–	High (3)
Post-entry distribution	–	High (–3)
Final rating (score)	–	High (12)

Based on the above invasiveness rating, *R. similis* receives a score of 12 out of 15 and is rated as HIGHLY INVASIVE NEMATODE.

95.14 Region-wise status of invasiveness

The *R. similis* citrus race (attacking citrus and other plants) is an EPPO A1 list pest and the *R. similis* banana race (not attacking citrus) is an A2 list pest (EPPO, 2008). Due to the economic importance of *R. similis* in various part of the world, its wide host range and pathways of entry for this nematode, a high invasive rating has been given for all the countries where the nematode is absent.

95.15 Management measures

Prevention is the most effective means of controlling the spread of *R. similis* into new areas. Saplings raised in known infested areas should not be planted. Creating a buffer zone 15–20 ft (4.5–6 m) wide around infested orchards to check the spread of this nematode is highly effective. The use of trenches and plastic barriers may also prove effective in contending the nematode. Nursery certification may serve as a secondary screening to check the spread of this nematode. Site certification is most effective for the production of nematode-free nursery stock. The movement of soil and water from infested fields to healthy ones should be avoided and farm implements used in infested orchards should be disinfested before being used in nematode-free orchards. Several other effective prophylactic measures are of considerable importance for reducing the initial inoculum level and for curbing the progressive multiplication of *R. similis*. Appropriate cultural practices, host resistance, chemical methods and biological control on different host crops are discussed below.

BANANA The use of nematode-free planting materials is an important measure against nematodes. Fallowing for 3 months or flood fallowing for 5 months after banana harvest effectively suppressed *R. similis* populations (Rajendran *et al.*, 1979). Organic amendment with oil cakes of neem, mahua, castor, karanj, etc. have also shown considerable effectiveness in reducing nematode populations. Crop rotation with paddy, sugarcane, green gram, cotton or turmeric may suppress the nematode population (Rajendran *et al.*, 1979). Several cultivars, such as 'Pisang Batuau', 'Pisang Jari Buaya' and 'Pisang Edor Kuda', 'Prata', 'Mysore', 'Hybrid SH-3142', 'Kadali', 'Pedalimoongil', 'Kunnan', 'Ayrianka poovan', 'Pisang Seribu', 'Tongat', 'Venettu Kunnan', 'Anaikomban', 'Yelakkibale', 'Ayiranka poovan', 'Karpuravally' and 'Palayankodan' have been found to be moderately resistant to *R. similis* infestation (Pinochet and Rowe, 1979; Binks and Gowen, 1996; Harish and Nanjegowda, 2000).

Paring and treating with carbofuran at 1.2 g a.i./sucker or dipping the suckers in a mixture of DBCP and Bordeaux mixture lowers the incidence of *R. similis* and increases yields. Complete disinfestation is achieved by double paring followed by a dip in monocrotophos solution at 0.5% for 30 min and later dried under shade for 72 h before planting (Sheela and Sundraraju, 2012). Nair (1979) showed a reduction in the population of *R. similis* with the application of carbofuran, phorate and fenamiphos at 2 g a.i./plant. Soil application of fenamiphos at 3 kg a.i./ha was effective in reducing nematode populations, as well as increasing plant height, plant girth, the number of leaves and yield (Reddy *et al.*, 1991). Paring combined with hot-water treatment of suckers at 50°C for 15 min followed by application of neem cake or carbofuran at 0.5 g a.i./plant significantly reduced the nematode population and increased yield in banana (Sheela *et al.*, 1996; Shreeinivasa *et al.*, 2005).

CITRUS Crop rotation with pangola grass, use of *Crotalaria spectabilis* and *C. striata* as a cover crop, flood-fallowing for 5–6 months and hot-water root dip treatment of saplings at 55°C for 20 min are highly effective in suppressing nematode populations. The rootstocks of 'Milam lemon' and 'Ridge Pineapple sweet orange' are resistant against Biotype 1 of *R. citrophilus*. 'Milam', 'Carign', 'Rough Lemon' and 'Ridge Pineapple' are varieties of *Citrus sinensis* which possess resistance, but none are immune to this nematode (Duncan and Cohn, 1990). Application of systemic and granular nematicides such as phorate or fenamiphos are also effective (Koshy *et al.*, 1985). A higher dose of chemicals is recommended so as to hold the nematode population at 1 nematode/g of root, a level at which little damage occurs to the feeder roots.

BLACK PEPPER Integrated methods of nematode management that can be suggested include: (i) the planting of nematode-free rooted cuttings; (ii) uprooting of affected vines and replanting after 9–12 months; (iii) use of non-living supports or standards; (iv) exclusion of *R. similis*-susceptible trees as standards for trailing black pepper vines; (v) exclusion of susceptible intercrops such as banana, ginger and turmeric; (vi) application of organic amendments, such as neem oil cake (*Azadirachta indica*), green foliage, or farm-yard manure; and (vii) earthing-up after application of nematicides, NPK fertilizers and organic amendments.

95.15.1 Phytosanitary measures

See Chapter 16, section 115.

Zygotylechus Spp.

Zygotylenchus guevarai was described as a new species by three workers, as *Pratylenchoides guevarai* by Tobar Jiménez (1963), as *Zygotylenchus browni* by Siddiqi (1963a) and as *Mesotylus gallicus* by de Guiran (1964). Only one other species of the genus has been described by Siddiqi (2000): *Z. taomasinae* (de Guiran, 1964) Braun and Loof, 1966. *Z. guevarai* is a serious threat to cultivated violets, lucerne, oats, beans and grapevines in some Asian and European countries (Siddiqi, 2000).

Systematic position of *Zygotylenchus* species

	Classical (after Siddiqi, 2000)	Molecular (after Decraemer and Hunt, 2013)
Phylum	Nematoda	Nematoda
Class	Secernentea	Chromadorea
Order	Tylenchida	Rhabditida
Suborder	Hoplolaimina	Tylenchina
Infra-order	–	Tylenchomorpha
Superfamily	Holpolaimoidea	Tylenchoidea
Family	Pratylenchidae	Pratylenchidae
Subfamily	Pratylenchinae	Pratylenchinae
Genus	*Zygotylenchus*	*Zygotylenchus*
Species	*Z. guevarai*	*Z. guevarai*

96 *Zygotylenchus guevarai*

96.1 Common name

Root-lesion nematode

96.2 Scientific name

Zygotylenchus guevarai (Tobar Jiménez, 1963) Braun and Loof, 1966

96.3 Synonyms

Mesotylus gallicus de Guiran, 1964
Pratylenchoides guevarai Tobar Jiménez, 1963
Zygotylenchus browni Siddiqi, 1963

96.4 EPPO code

ZYGTGU

96.5 Diagnosis

96.5.1 Morphological diagnosis

This morphological description of *Z. guevarai* is after Tobar Jiménez (1963) and Siddiqi (1963a and 1975a). For morphometric measurements (after Siddiqi, 1975a), see Table 68.

FEMALE: Body ventrally arcuate upon relaxation or death (Fig. 10.27B); annules 1.1–1.7 μm wide near middle. Lateral fields about one-fourth of body width, with four incisures, not areolated. Cephalic region low, rounded, continuous with body contour, with 3–4 distinct annules; framework strongly sclerotized (Fig. 10.27C), with outer margins extending back 1–2 body annules. Stylet strong, conus 43–49% of its total length; basal knobs rounded (Fig. 10.27C). Orifice of dorsal oesophageal gland 2–3 μm behind stylet base. Median oesophageal bulb muscular, oval, extending over 7–9 body annules. Oesophageal glands extending over intestine ventrally and ventro-laterally. Deirids absent. Excretory pore near base of isthmus, 80–88 μm from anterior end of body. Vulva a submedian, transverse slit. Vagina straight, one-third of body width long. Genital tracts paired, outstretched in opposite direction. Ovaries two, with oocytes in single row. Spermatheca rounded, often with round sperm (Fig. 10.27F). Intestine partially extending over rectum. Serpentine canals present in intestinal region. Tail subcylindrical, with slightly concave dorsal surface and a broadly rounded (Fig. 10.27J,K), unstriated terminus, about 2–3 anal body widths long, with 16–23 annules. Phasmids pore-like, near middle of tail.
MALE: Abundant. No marked sexual dimorphism in anterior region, but cephalic region slightly higher and stylet slightly shorter with smaller basal knobs than in female (Fig. 10.27A). Testis single, outstretched. Bursa completely enveloping tail (Fig. 10.27I), with coarsely crenate margins; phasmids near middle of tail, slightly extending into bursa. Spicules ventrally arcuate, slightly cephalated, distally pointed, 16–21 (average 18) μm long. Gubernaculum fixed, trough-shaped, 5–7 μm long.

96.5.2 Molecular diagnosis

For species-specific primers, see Table 60.

96.6 Geographical distribution

Z. guevarai has limited distribution but occurs widely in Germany (Sopp, 1994) and Spain (Tobar Jiménez *et al.*, 1974). The nematode

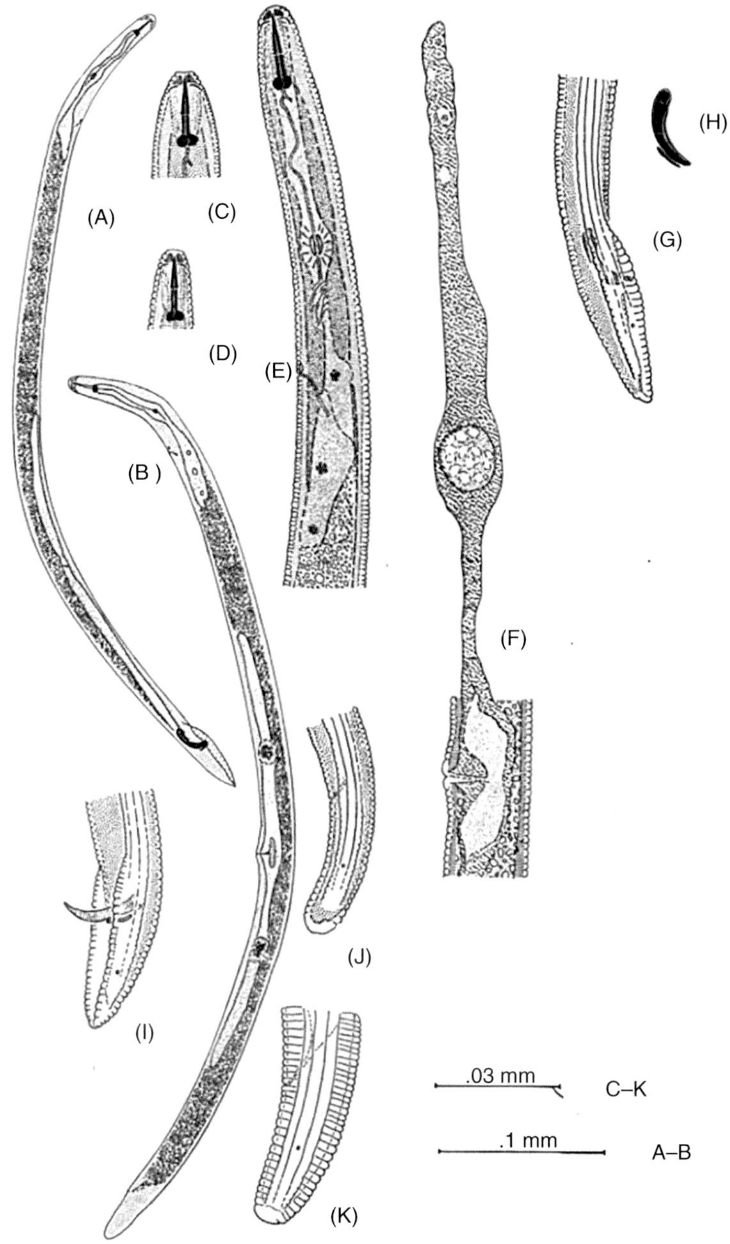

Fig. 10.27. Line drawing of *Zygotylenchus guevarai.* Whole male (A); whole female (B); head end of female (C); head end of male (D); anterior end of female (E); anterior gonad of female (F); male tails (G,I); spicule and gubernaculum (H); female tails (J,K). [Tarjan and Weischer, 1965; courtesy Nematologica]

is mainly confined to the temperate region and recorded in the following countries (Siddiqi, 1975a; CABI-ISC, 2020).

Africa: Algeria, Morocco, South Africa, Tunisia.
Asia: Iran, Jordan, Pakistan, Syria, Tajikistan, Turkey, Uzbekistan.
Europe: Austria, Bulgaria, Croatia, former Czechoslovakia, France, Germany, Greece, Hungary, Italy, Malta, Netherlands, Poland, Slovakia, Spain, Switzerland, United Kingdom, former Yugoslavia.
North America: USA.
Oceania: New Zealand.

96.7 Hosts

Z. guevarai is a polyphagous nematode species and attacks all kind of plants, including cereals, vegetables, ornamentals and fruit. An expanded list of hosts of *Z. guevarai* is presented in Table 70.

Table 70. List of host plants of *Zygotylenchus guevarai* (adapted from CABI-CPC, 2010).

Main hosts
Apium graveolens var. *rapaceum* (celeriac)
Avena sativa (oats)
Cicer arietinum (chickpea)
Cupressus sempervirens (Mediterranean cypress)
Lathyrus cicera (flat-podded vetchling)
Phaseolus vulgaris (common bean)
Pisum sativum (pea)
Vicia faba (broad bean)
Viola odorata (English violet)
Viola wittrockiana (wild pansy)
Vitis vinifera (grapevine)
Zea mays (maize)
Other hosts
Allium cepa (onion)
Alstroemeria (Inca lily)
Cannabis sativa (hemp)
Capsicum annuum (bell pepper)
Cucumis melo (melon)
Daucus carota (carrot)
Medicago sativa (lucerne)
Petroselinum crispum (parsley)
Solanum lycopersicum (tomato)
Solanum tuberosum (potato)
Wild hosts
Poaceae (grasses)
*Prunus dulcis (*almond)
Vicia (vetch)

96.8 Symptoms

Brownish or brownish-black root lesions indicating large cavities in the root cortex where *Z. guevarai* feeds and reproduces were observed in chickpea (*Cicer arietinum*), lesser pea (*Lathyrus cicera*), oat (*Avena sativa*), and vetch (*Vicia sativa*) (Varo Alcala and Tobar Jiménez, 1970). Diseased young cypress (*Cupressus sempervirens*) and violet (*Viola odorata*) showed general unthriftiness, significantly reduced growth and serious root damage (Tobar Jiménez and Martinez Sierra, 1966). In bean roots (*Phaseolus vulgaris*), *Z. guevarai* invaded and damaged the cortical parenchyma and the necrosis also extended to the endodermis and the pericycle (Vovlas and Inserra, 1977). Sections of parasitized roots of celeriac showed that *Z. guevarai* preferentially invaded the cortex, where it induces cavities and destroys the tissue. Damage to the epidermis and vascular tissue also occurs (Vovlas *et al.*, 1976).

96.9 Biology and life cycle

The biology and life cycle are typical of the genus. Males are abundant and reproduction is amphimictic. A temperature above 7°C is required for reproduction and low winter temperatures reduce the population (Tobar Jiménez, 1973). In culture, the *Z. guevarai* life cycle is completed in 45 days at 25°C (Verdejo and Pinochet, 1991).

96.10 Economic importance

Z. guevarai is a serious threat to common beans, cultivated violets, lucerne, oats and grapevines. In Spain, *Z. guevarai* caused serious damage to oats but not to wheat or barley (Tobar Jiménez, 1973). It is also a serious threat to violets and cypress (Tobar Jiménez and Martinez, 1966). In Italy, *Z. guevarai* and *Pratylenchus thornei* seriously damaged beans (Vovlas and Inserra, 1977). In France, *Z. guevarai* has been found to be the most important nematode attacking violet roots (Deportes *et al.*, 1979).

96.11 Movement and means of dispersal

See section 76.11, above.

96.12 Pest risk analysis and looming threat from introduction

Considering the economic damage caused by *Z. guevarai* on host crops and its establishment in various countries, regulatory agencies particularly in vegetable-producing countries should adopt preventive measures to check the introduction of this nematode.

96.13 Invasiveness rating

Based on the consequences of economic damage and pest risk analysis of *Z. guevarai*, the following invasiveness rating has been assigned.

Climate/host interaction	–	Medium (2)
Hosts range	–	High (3)
Dispersal potential	–	High (3)
Economic impact	–	High (3)
Environmental impact	–	Medium (2)
Post-entry distribution	–	Low (–1)
Final rating (score)	–	High (12)

Based on the above invasiveness rating, *Z. guevarai* receives a score of 12 out of 15 and is rated as HIGHLY INVASIVE NEMATODE.

96.14 Region-wise status of invasiveness

Because of the wide host range and limited distribution, particularly in temperate regions, *Z. guevarai* has been given highly invasive rating for all the countries where this nematode is still absent.

96.15 Management measures

See section 82.5, above.

11

Rotylenchulidae: *Rotylenchulus* Species

Rotylenchulus Spp.

The reniform nematodes of the genus *Rotylenchulus* are sedentary and semi-endoparasites of plants. The genus was created by Linford and Oliveira (1940) with the description of *Rotylenchulus reniformis*. Siddiqi (2000) placed *Rotylenchulus* in the superfamily Hoplolaimoidea, family Rotylenchulidae and subfamily Rotylenchulinae. However, a new classification based on molecular studies diminished the family Rotylenchulidae and placed the genus *Rotylenchulus* in the family Hoplolaimidae (Decraemer and Hunt, 2013). The key to the genus has been given by different workers (Dasgupta *et al.*, 1968; Siddiqi, 1986; Lehman and Inserra, 1990; Siddiqi, 2000). Currently, at least 11 species are described and validated with *Rotylenchulus reniformis* as its type species (Siddiqi, 2000; Van den Berg *et al.*, 2016).

Rotylenchulus anamictus Dasgupta, Raski and Sher, 1968
R. borealis Loof and Oostenbriiik, 1962
R. clavicaudatus Dasgupta, Raski and Sher, 1968
R. leptus Dasgupta, Raski and Sher, 1968
R. macrodoratus Dasgupta, Raski and Sher, 1968
R. macrosomoides Van den Berg *et al.* 2016
R. macrosomus Dasgupta, Raski and Sher, 1968
R. parvus (Williams, 1960) Sher, 1961
R. reniformis Linford and Oliveira, 1940
R. sacchari Van den Berg and Spaull, 1981
R. variabilis Dasgupta, Raski and Sher, 1968

Among these species, *R. macrodoratus*, *R. parvus* and *R. reniformis* are important pathogens of agricultural crops and are considered in this chapter for their invasiveness. The other species have limited distribution and are of minor economic importance (Robinson *et al.*, 1997; Jones *et al.*, 2013). *R. reniformis* is a parasite of more than 350 plant species, including a large number of economically important vegetables, ornamentals, fruit and fibre crops (Jones *et al.*, 2013). In the USA the annual losses attributed to reniform nematodes in cotton production are estimated to exceed US$100 million (Robinson *et al.*, 2005; Blasingame, 2006). *R. parvus* is a major pest of cereals and leguminous crops, while *R. macrodoratus* parasitizes many fruit trees and ornamental plants.

Systematic position of *Rotylenchulus* species

	Classical (after Siddiqi, 2000)	Molecular (after Decraemer and Hunt, 2013)
Phylum	Nematoda	Nematoda
Class	Secernentea	Chromadorea
Order	Tylenchida	Rhabditida
Suborder	Hoplolaimina	Tylenchina
Infra-order	–	Tylenchomorpha
Superfamily	Holpolaimoidea	Tylenchoidea
Family	Rotylenchulidae	Holpolaimidae
Subfamily	Rotylenchulinae	Rotylenchulinae
Genus	*Rotylenchulus*	*Rotylenchulus*

97 *Rotylenchulus macrodoratus*

97.1 Common name

Mediterranean reniform nematode

97.2 Scientific name

Rotylenchulus macrodoratus Dasgupta, Raski and Sher, 1968

97.3 EPPO code

ROTYMA

97.4 Diagnosis

97.4.1 Morphological diagnosis

Morphological descriptions of *Rotylenchulus macrodoratus* (after Dasgupta *et al.*, 1968) are summarized below. For measurements, see Table 71.

DOI: 10.1079/9781789247367.0012

Table 71. Morphometric measurements of *Rotylenchulus* species (after Dasgupta *et al.*, 1968).

Morphometric characters		Range, numerous populations		
		R. macrodoratus	*R. parvus*	*R. reniformis*
Immature females	Length (μm)	400–490	250–340	340–420
	Stylet length (μm)	22–26	12–14	16–18
	a	22–28	20–26	22–27
	b	3.2–4.7	3.1–3.7	3.6–4.3
	b′	2.8–3.6	2.1–3.0	2.4–3.5
	c	16–24	16–20	14–17
	c′	1.8–2.4	2.0–2.7	2.6–3.4
	V	64–72	60–66	68–73
	h (μm)	6–12	< 0 3	7–11[b]
	0	56–82	90–107	81–106
Males	Length (μm)	450–530	380–460	380–430
	Stylet length (μm)	17–20	10	12–15
	a	27–32	28–32	24–29
	b′	3.8–4.6	3.6–4.0	2.8–4.8
	c	21–28	14–23	12–17
	Tail length (μm)	–	32–41	35–45
	Spicule (μm)	20	16	19–23
	Gubernaculum (μm)	7	5–7	7–9
	h	6–10	3	4–8
Mature female	Length (μm)	410–530	250–360	380–520
	Stylet length (μm)	–	12–15	13–15[b]
	a	4–6	4–7	4–5
	V	64–71	61–66	68–73
	Width of vulva (μm)	60–90	40–80	–
	Length of the swollen portion plus tail (μm)	170–340	150–250	190–470
Juvenile	Length (μm)	400–500	250–340	350–410
	Stylet length (μm)	20–22	12–13	13–15
	a	21–28	21–25	20–24
	b	3.0–4.7	–	–
	b′	–	2.5–3.5	3.5–4.1
	c	15–20	14–18	12–16

From [a]Van den Berg *et al.* (2016); [b]Palomares-Rius *et al.* (2018)

Immature female: Cephalic region conoid with fine annules. Second and third incisures in lateral field weaker than two outer lines, slowly diminishing near flexure level of posterior ovary. Cephalic sclerotization strong, arches downwards heavily (Fig. 11.1A). Stylet knobs large and anchor-shaped (Fig. 11.1A). Excretory pore about 100 μm from head end. Metacorpus oblong, valve around 3 μm long. Oesophageal glands ventrally overlapping intestine. Tail about 21 μm in length with bluntly rounded terminus (Fig. 11.1E–G).

Mature female: Body curved ventrally, body contour sloping gradually anterior and abruptly posterior to vulva, tapering behind vulva to an intermediate degree. Metacorpus ovoid. Vulva elevated with pronounced lips. Tail tip with spike-like process.

Male: Cephalic region raised, hemispherical (Fig. 11.1B). Spear knobs and oesophagus much degenerated. Metacorpus and median valve indistinct.

Juvenile: Resembling immature females. Tail more rounded than immature females (Fig. 11.1I).

97.4.2 Molecular diagnosis

Palomares-Rius *et al.* (2018) designed species-specific PCR primers, Forward –GGTGAACCTGCTGCTGGATC ATTAC and Reverse – CRCGTCTGAGYTCAGGTCG for ITS region sequence of *R. macrodoratus*. The phylogenetic study revealed that *R. macrodoratus* has two distinct rRNA gene types in its genomes in the populations of different Mediterranean countries.

97.5 Geographical distribution

The Mediterranean reniform nematode occurs in Crete, France, Greece, Israel, Italy and Malta (Robinson *et al.*, 1997) and South Africa (Van den Berg, 1998).

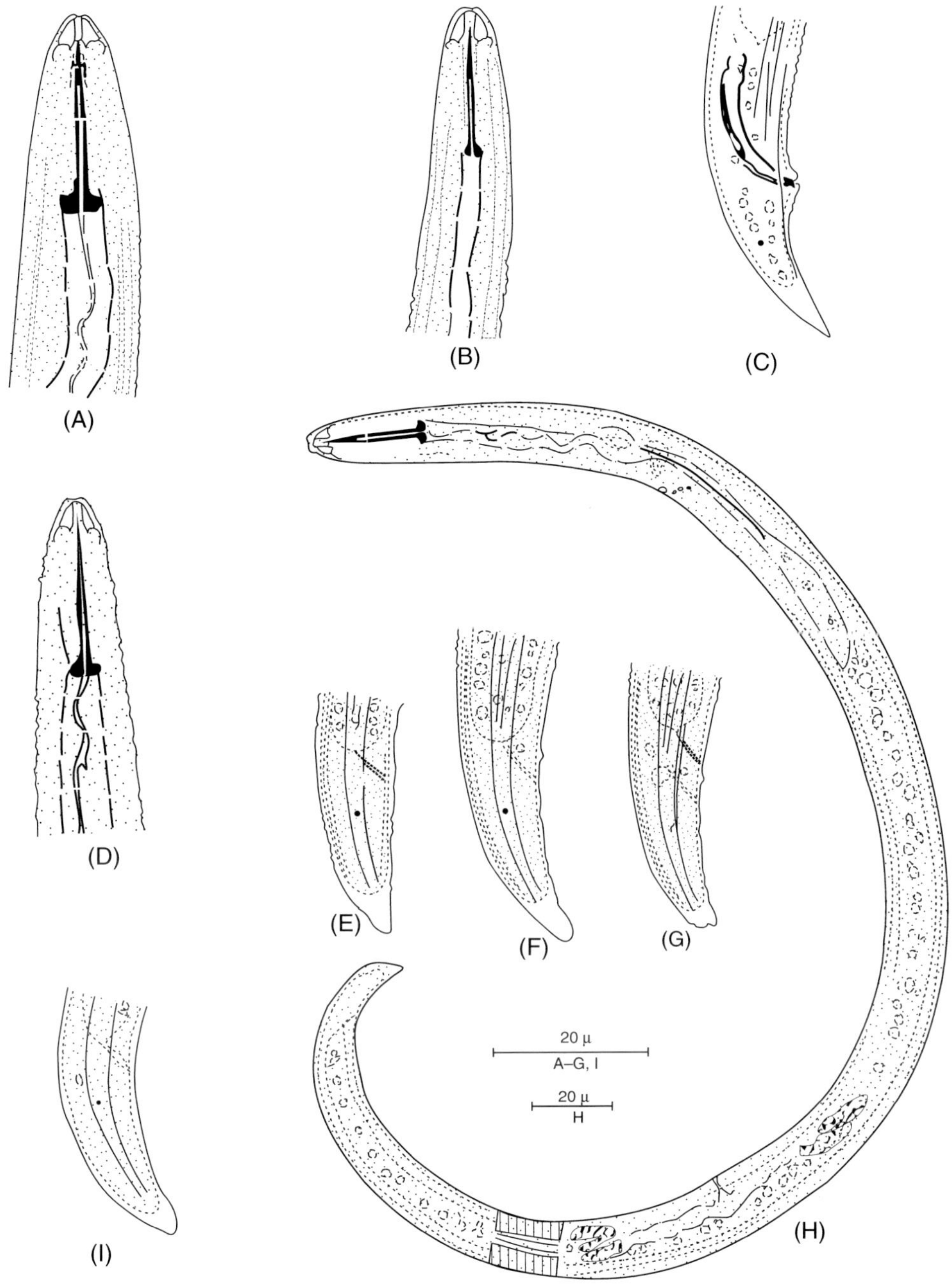

Fig. 11.1. Line drawing of *Rotylenchulus macrodoratus*. Immature female, anterior end (A); male, anterior end (B); male, tail (C); juvenile, anterior end (D); immature female, tail (E–G); immature female (H); juvenile, tail (I). [Courtesy Dasgupta *et al.*, 1968; The Helminthological Society of Washington]

97.6 Hosts

R. macrodoratus parasitizes many fruit trees and ornamental plants, including olive and grapevine (Robinson *et al.*, 1997). The reported host fruit trees are almond (*Prunus amygdalus*), apricot (*Prunus armeniaca*), carob (*Ceratonia siliqua*), fig (*Ficus carica*), grape (*Vitis vinifera*), laurel (*Laurus nobilis*), loquat (*Eurybotria japonica*), oleander (*Nerium oleander*), olive (*Olea europaea*), oak (*Quercus calliprinos* and *Q. farnetto*), pistachio (*Pistacia vera*) and plum (*Prunus domestica*). Other hosts include carnation (*Dianthus caryophyllus*), ivy (*Hedera ilex*), large-flowered sweet william (*Dianthus barbatus*), pellitory (*Parietaria officinalis*), phlomis (*Phlomis fruticosa*) and soybean (*Glycine max*).

97.7 Symptoms

The above-ground symptoms of *Rotylenchulus* infection resemble those of other plant-parasitic nematodes, such as moisture and nutrient deficiencies. Severely infected plants may be stunted with chlorotic foliage. On below-ground parts, root growth is abridged with a smaller number of secondary roots. The presence of many gelatinous egg sacs on the root surface, which appear as dirty root symptoms, can be confirmed after close observation (Jones *et al.*, 2013).

97.8 Biology and life cycle

The biology and life cycle of *R. macrodoratus* are similar to those of other reniform nematodes (Robinson *et al.*, 1997), but these nematodes have an exceptional life cycle. The second-stage juvenile (J_2) hatches from the egg in 8–10-days and undergoes three consecutive moults, developing into a vermiform adult female or male within 7–9 days without feeding. These adult stages are relatively smaller than the J_2. The cuticles of the previous moults are retained, which helps to reduce the rate of water loss and improves the chances of anhydrobiotic survival in drier soil (Jones *et al.*, 2013). When the nematodes migrate in moist soil, these cuticular sheaths are shed with the friction of soil particles. The infective stage is the immature young female, which penetrates the roots of plants, injecting around one-third of the anterior body, and establishes feeding sites on endodermal and pericycle cells. Usually, reniform nematodes share the same type of feeding cell induction in host plant roots as a syncytium (Castillo *et al.*, 2003). However, in contrast to this, *R. macrodoratus* induces a uninucleate giant cell (Cohn and Mordechai, 1977). After 2–3 days of feeding, the posterior portion of the female starts to become obese near the vulval region and in around 7 days the body assumes a bean or kidney shape. A gelatinous matrix is produced by the uterine glands and 40–100 eggs are laid within 7–9 days under optimal environmental conditions. Males are abundant in most populations but they do not feed and they remain in the soil.

97.9 Similarities to other species/conditions

R. macrodoratus is close in morphology to *R. macrosomoide* because of the long stylet for immature females, but it can be separated by several other morphological characters (Van den Berg *et al.*, 2016).

97.10 Economic importance

The damage caused by *R. macrodoratus* to fruit trees and other crops has not been quantified. In Greece, *R. macrodoratus* was detected in cultivated olive with a prevalence of 8.2% (Palomares-Rius *et al.*, 2018). One to two females per cm^3 of soil can cause economic yield reduction (Jones *et al.*, 2013).

97.11 Movement and means of dispersal

The dispersal of *R. macrodoratus* is facilitated primarily via infected planting material. The nematodes can also spread by the movement of contaminated farm machinery, harvesting combines and nematode-infected plant residues. Its semi-endoparasitic habits also allow the nematode to be dispersed long distances inside infected plant material as eggs, juveniles and adults.

97.11.1 Plant parts known to carry the nematode in trade/transport

Roots, bulbs, tubers, corms, rhizomes (sucker), growing medium accompanying plants, seedlings and micro-propagated plants.

97.11.2 Plant parts not known to carry the nematode in trade/transport

Bark, flowers, inflorescences, cones, calyx, fruits (including pods), leaves, stems (above ground), shoots, trunks, branches, wood and true seeds (including grain).

97.12 Pest risk analysis and looming threat from introduction

R. macrodoratus is considered as a potential pest of many fruit trees and ornamental plants, including olive and grapevine, and it is currently widespread in the Mediterranean basin. Due to its wide host range, there is a high probability that it could infest other host crops in other subtropical and tropical areas. *R. macrodoratus* can be easily spread through infested vegetative planting material. A complete pest risk analysis is suggested for all the countries where this nematode is absent.

97.13 Invasiveness rating

Considering the consequences of introduction and pest risk analysis of *R. macrodoratus*, the following invasiveness rating has been assigned.

Climate/host interaction	–	Medium (2)
Hosts range	–	High (3)
Dispersal potential	–	Medium (2)
Economic impact	–	Medium (2)
Environmental impact	–	Medium (2)
Post-entry distribution	–	Low (–1)
Final rating (score)	–	Medium (10)

Based on the above invasiveness rating, *R. macrodoratus* receives a score of 10 out of 15 and is rated as MEDIUM INVASIVE NEMATODE.

97.14 Region-wise status of invasiveness

Information on crop losses caused by *R. macrodoratus* is lacking, but the nematode has the potential to cause economic damage to many fruit trees and other hosts. Based on its wide host range and limited distribution, the nematode has been given a medium invasiveness rating for all the countries where it is absent.

97.15 Management measures

Information on the management of *R. macrodoratus* is deficient, but the nematode would be susceptible to soil nematicides. The management options employed for *R. reniformis* may be useful to control *R. macrodoratus* as well. For further details, see section 99.15, below.

98 *Rotylenchulus parvus*

98.1 Common name

Reniform nematode

98.2 Scientific name

Rotylenchulus parvus (Williams, 1960) Sher, 1961

98.3 Synonym

Helicotylenchus parvus Williams, 1960

98.4 EPPO code

ROTYPA

98.5 Diagnosis

98.5.1 Morphological diagnosis

Morphological descriptions of *Rotylenchulus parvus* (after Dasgupta *et al.*, 1968) are summarized below. For measurements, see Table 71.

IMMATURE FEMALE: Body in open C to loose spiral shape after heat relaxation (Fig. 11.2F). Cephalic region low, rounded with fine annules. Spear knobs sloping backwards (Fig. 11.2C). Excretory pore 60–75µm from head end. Metacorpus ovoid. Oesophageal glands laterally and ventrally overlapping intestine, more frequently laterally. Tail short with short ventral projection in form of terminus (Fig. 11.2H,I).
MATURE SWOLLEN FEMALE: Body curved ventrally, body contour tapers at both ends, more abruptly beyond vulva, and posterior part often crosses neck (Fig. 11.2G). Postanal part slender but wider than immature females, not hemispherical. Anal body width 10–16µm. Vulva elevated with prominent lips. Vagina funnel-shaped, extending across half the length (Fig. 11.2G). Tail without projection or spike-like process.
MALE: Rare. Cephalic region raised, hemispherical with weak sclerotization (Fig. 11.2A). Prorhabdion not well marked, obscure. Oesophagus degenerated. Lumen of oesophagus, metacorpus and valve difficult to resolve. Tail tip rounded (Fig. 11.2D).
JUVENILE: Similar to immature female. Tail terminus bluntly rounded, without projections (Fig. 11.2E).

98.5.2 Molecular diagnosis

Van den Berg *et al.* (2016) characterized D2–D3 of 28S rRNA region of *R. parvus* species using the primers, Forward – ACAAGTACCGTGAGGGAAAGTTG and Reverse – TCGGAAGGAACCAGCTACTA with conventional PCR. The study revealed that *R. parvus* had a sister relationship with *R. clavicaudatus*.

98.6 Geographical distribution

R. parvus was originally reported from Mauritius (Williams, 1960). Since then, it has been recorded in many African countries. The nematode has been detected in the following countries (CABI-ISC, 2020).

Africa: Côte d'Ivoire, Egypt, Kenya, Malawi, Mauritius, Mozambique, Somalia, South Africa, Tanzania, Uganda, Zambia, Zimbabwe.
Asia: India, Iran, Pakistan.
Europe: Cyprus.
North America: Dominican Republic, USA, US Virgin Islands.
Oceania: Australia

98.7 Hosts

R. parvus is polyphagous and known to infect both monocot and dicot plants. The main hosts are common bean, cowpea, groundnut, maize, papaw, potato, Rhodes grass, sorghum, sugarcane, thyme and African star grass (CABI-CPC, 2010). An extended list of *R. parvus* hosts from different plant families is given in Table 72.

98.8 Symptoms

R. parvus-infested roots can exhibit severe distortion but not extensive necrosis, though occasional lesions can be observed (Roman and Grullon, 1975). On thyme plants, *R. parvus* caused stunting, chlorotic foliage and senescence at an early stage of growth, leading to an overall decline in plant growth (Chinappen *et al.*, 1988). In association with wilt fungus (*Fusarium udum*) in pigeon pea, *R. parvus* accelerated wilt incidence (Hillocks and Songa, 1993).

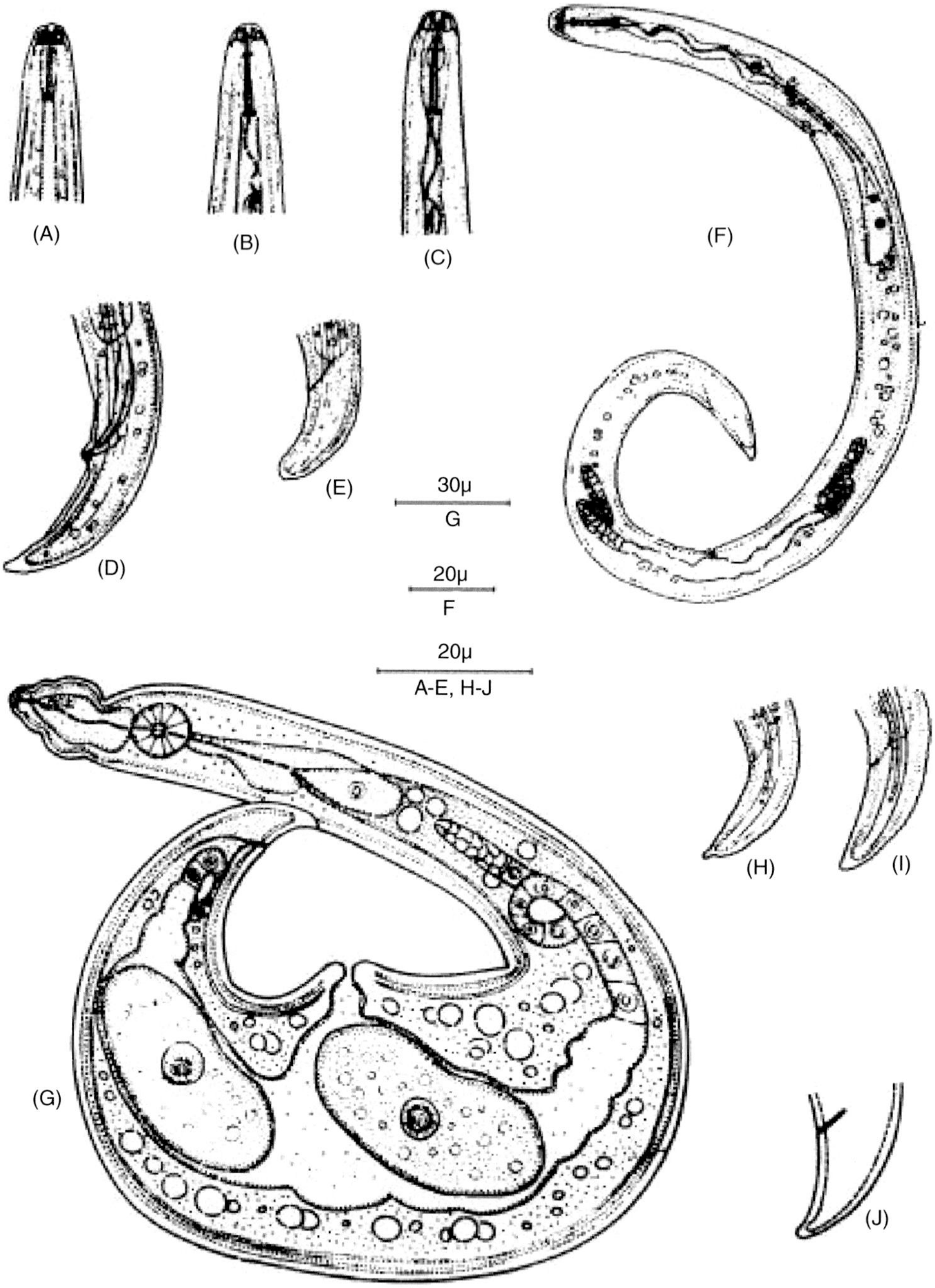

Fig. 11.2. Line drawing of *Rotylenchulus parvus*. Male, head end (A); juvenile, head end (B); immature female, head end (C); male tail (D); juvenile tail (E); immature female (F); mature female (G); tails of immature females (H,I); tail of mature female (J). [Courtesy CAB International]

98.9 Biology and life cycle

The biology and life cycle are typical of the genus. However, males are extremely rare and reproduction is mainly through parthenogenesis. Eggs are laid by mature females in a gelatinous matrix outside the host tissue. The number of eggs per egg mass varies from 14 to 37. Hatching occurs within 2 days. The entire life cycle is completed in around 27–36 days at 20–35°C (Dasgupta and Raski, 1968).

98.10 Similarities to other species/conditions

Mature females of *R. parvus* have a shorter stylet, like other *Rotylenchulus* species such as *R. anamictus*, *R. borealis* and *R. leptus*. In *R. anamictus* and *R. borealis*, males are rare (Van den Berg *et al.*, 2016). *R. parvus* can be differentiated from *R. reniformis* by the straight-line distance from head to tail terminus and from head to vulva.

Table 72. List of hosts affected by *Rotylenchulus parvus* (adapted from Rathore, 2018).

Monocots	
Cyperaceae:	*Cyperus* sp.
Poaceae:	*Cynodon dactylon*, *Hordeum vulgare*, *Pennisetum americanum*, *Saccharum officinarum*, *Sorghum bicolor*, *Zea mays*
Eudicots	
Proteaceae:	*Macadamia* sp.
Rosids – Brassicaceae:	*Brassica oleracea*
Caricaceae:	*Carica papaya*
Cucurbitaceae:	*Cucumis sativus*
Fabaceae:	*Crotalaria juncea*, *Glycine max*, *Medicago sativa*, *Phaseolus vulgaris*, *Vigna unguiculate*
Malvaceae:	*Gossypium hirsutum*
Rosaceae:	*Prunus persica*
Vitaceae:	*Vitis vinifera*
Superasterids – Amaranthaceae:	*Beta vulgaris*
Nectaginaceae:	*Bougainvillea* sp.
Asterids – Apiaceae:	*Apium graveolens*, *Daucus carota*
Apocynaceae:	*Carissa* sp.
Asteraceae:	*Lactuca sativa*
Lamiaceae:	*Thymus sp.*, *Thymus vulgaris*
Oleaceae:	*Oleaeuropaea*
Pittosporaceae:	*Pittosporum* sp.
Solanaceae:	*Nicotiana tabacum*, *Solanum lycopersicum*, *S. tuberosum*
Gymnosperm – Cupressaceae:	*Cupressus* sp.

98.11 Economic importance

Little information is available on losses attributed to *R. parvus* but it can be assumed that yield will be significantly affected in heavy infestations. Furstenberg (1974) found a 3000% increase in numbers of the nematode at the end of 3 years of maize monoculture in South Africa. The population peaks were recorded towards the end of the maize-growing season and in spring when the ground was fallow, possibly due to optimum temperatures at these times. During the 3 years under maize, the proportion of *R. parvus* rose from 16% to 96% of all plant-parasitic species.

98.12 Movement and means of dispersal

See section 97.11, above.

98.13 Pest risk analysis and looming threat from introduction

R. parvus is considered as a potential pest of cereals and legumes such as maize, sorghum, groundnut, cowpea, etc. Little information is available on its direct economic impact on these crops. *R. parvus* can be easily spread through infested vegetative planting material. Although these crops are directly grown *in situ*, nematodes can be transported on seedlings from infested nursery stock of transplanted hosts such as pawpaw and citrus. The nematode is indigenous to southern Africa but has been found in Cyprus, India and the USA. There is a high probability that it could infest crops in other subtropical and tropical areas. A complete pest risk analysis is suggested for all the subtropical and tropical countries where this nematode is absent.

98.14 Invasiveness rating

Considering the consequences of introduction and probable economic damage caused by *R. parvus*, the following invasiveness rating has been assigned.

Climate/host interaction	–	Medium (2)
Hosts range	–	High (3)
Dispersal potential	–	Medium (2)
Economic impact	–	Medium (2)
Environmental impact	–	Medium (2)
Post-entry distribution	–	Low (–1)
Final rating (score)	–	Medium (10)

Based on the above invasiveness rating, *R. parvus* receives a score of 10 out of 15 and is rated as MEDIUM INVASIVE NEMATODE.

98.15 Region-wise status of invasiveness

Taking into consideration that it is a pest of cereals, legumes, fruit, etc., and establishment of the nematode outside its native region, a medium invasive rating has been given for *R. parvus* for all the countries where this nematode is absent.

98.16 Management measures

Limited experiments on the control of *R. parvus* have been reported. Application of carbofuran significantly decreased soil populations and plants recovered completely in 6 weeks after treatment in Mauritius (Chinappen *et al.*, 1988). The chemical treatments that are effective against *R. reniformis* would also be effective against *R. parvus* (see section 99.15, below).

99 *Rotylenchulus reniformis*

99.1 Common name

Reniform nematode

99.2 Scientific name

Rotylenchulus reniformis Linford and Oliveira, 1940

99.3 Synonyms

Leiperotylenchus leiperi Das, 1960
Rotylenchulus leiperi (Das) Loof and Oostenbrunk, 1961
Rotylenchulus queirozi (Lordello and Cesnik) Sher, 1961
Rotylenchulus stakmani Husain and Khan, 1965
Spyrotylenchus queirozi Lordello and Cesnik, 1958

99.4 EPPO code

ROTYRE

99.5 Diagnosis

99.5.1 Morphological diagnosis

Morphological descriptions of *R. reniformis* after Dasgupta *et al.* (1968) and Siddiqi (1972b) are presented below. For measurements, see Table 71.

IMMATURE FEMALE: Body slender, vermiform and spiral to C-shaped after heat fixation (Fig. 11.3D). Stylet with rounded and posteriorly sloping knobs (Fig. 11.3A). Median oesophageal bulb with distinct valve, basal lobe overlaps intestine laterally and ventrally. Vulva not prominent, located at about 70% of body length (Fig. 11.3D). Ovaries paired, opposed with double flexure. Tail tapers to a narrow-rounded tip (Fig. 11.3F,G).
MATURE FEMALE: Body swollen, bean or kidney-shaped with irregular neck (Fig. 11.3E). Vulva postequatorial with raised lips. Body hemispherical beyond anus with tapering slender terminal portion (Fig. 11.3I,J). Stylet well developed with thick cuticle. Ovaries convoluted, very long. Eggs laid in a gelatinous matrix.
MALE: Vermiform, does not feed. Head end and stylet reduced. Oesophagus degenerate with reduced median bulb and valve. Spicules slender, elongated, curved ventrally. Bursa inconspicuous, not enveloping tail terminus (Fig. 11.3C).
JUVENILE: Similar to immature female. Oesophagus asymmetrical in posterior region, basal lobe overlaps intestine laterally and ventrally. Orifice of dorsal gland around one stylet length posterior to base of stylet knob. Genital primordium four-celled,

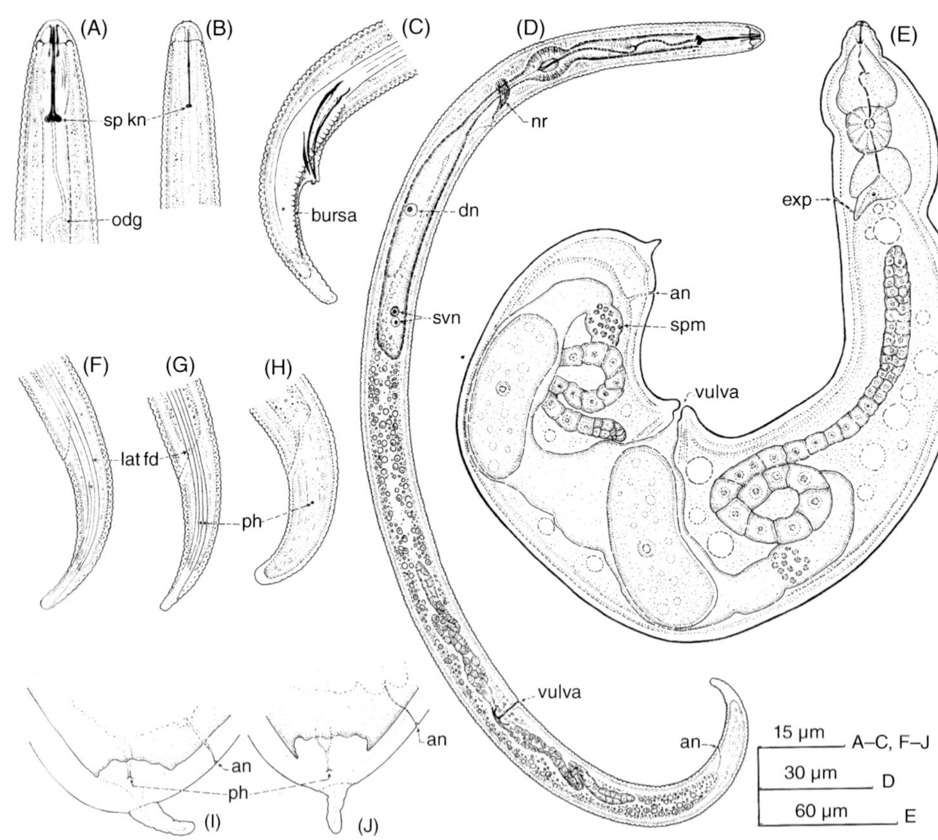

Fig. 11.3. Line drawing of *Rotylenchulus reniformis.* Immature female head (A); male head (B); male tail (C); immature female (D); mature female (E); immature female tail (F,G); juvenile tail (H); mature female tail (I,J). an, anus; dn, dorsal gland nucleus; exp, excretory pore; lat fd, lateral field; nr, nerve ring; odg, orifice of dorsal gland; ph, phasmid; sp kn, stylet knob; spm, spermatheca; svn, subventral gland nuclei. [Reproduced from Siddiqi, 2000; A-D and F-J after Siddiqi, 1972b; E after Dasgupta *et al.*, 1968; courtesy CAB International]

situated at same position as vulva. Tail similar to immature females but more rounded (Fig. 11.3H).

99.5.2 *Biochemical diagnosis*

R. reniformis was characterized on the basis of isozymes, using one-dimensional gel electrophoresis. Rao *et al.* (1998) observed variations in B-esterase profiles of two geographical populations of *R. reniformis*.

99.5.3 *Molecular diagnosis*

DNA-based molecular approaches have been successfully used for the diagnosis of *Rotylenchulus* species, including *R. reniformis* (Vovlas *et al.*, 2008; Van Den Berg *et al.*, 2016; Palomares-Rius *et al.*, 2018; Riascos-Ortiz *et al.*, 2019). Van Den Berg *et al.* (2016) developed *R. reniformis* species-specific primer for rapid diagnosis using D2–D3 expansion region of 28S rRNA. The primers are:

R_renif_R1A R - GAAAAGGCCTACCCAATGTG
D2A F - ACAAGTACCGTGAGGGAAAGTTG
R_renif_R2A R - CCCGATACCATTTCCATACAAG
D2A F - ACAAGTACCGTGAGGGAAAGTTG
R_renif_R1B R - CACAGACRCCCRAGCAGCCA

The study showed that *R. reniformis* and *R. macrosoma* have a sister relationship. The phylogenetic analysis also confirmed that this genus originated from the Afrotropical zoogeographical region.

Palomares-Rius *et al.* (2018) characterized *R. reniformis* using ITS region, D2–D3 expansion region of the large subunit (LSU) of ribosomal DNA (28S) and mitochondrial cytochrome c oxidase subunit 1 (COX1 mtDNA) genes using the following primers.

For ITS rRNA (Tanha Maafi *et al.*, 2003)

TW81 F - GTT TCC GTAGGTGAACCTGC
AB28 R - ATATGCTTAAGTTCAGCGGGT

For D2–D3 of 28S rRNA (De Ley *et al.*, 1999)

D2A F - ACAAGTACCGTGAGGGAAAGTTG
D3B R - TCGGAAGGAACCAGCTACTA
D2A_ROT F - ACAAGTACYGTGARGGAAAGTTG
D3B R - TCGGAAGGAACCAGCTACTA

For COX1 mtDNA (Bowles *et al.*, 1992)

JB3 F - TTTTTTGGGCATCCTGAGGTTTAT
JB4.5 R - TAAAGAAAGAACATAATGAAAATG

The phylogenetic study was similar to that obtained by Van Den Berg *et al.* (2016). Riascos-Ortiz *et al.* (2019) also used D2–D3 of LSU 28S rRNA and COX1 mtDNA to identify *R. reniformis* using the primers described in the previous paragraph.

99.6 Geographical distribution

R. reniformis is widespread in tropical and subtropical regions of the world (Fig. 11.4). It has also been known to occur in warm temperate regions of Europe, the USA, China and Japan. The nematode is reported from the following countries (CABI-ISC, 2020).

Africa: Angola, Burundi, Cameroon, Cape Verde, Côte d'Ivoire, Egypt, Ethiopia, Gambia, Ghana, Kenya, Liberia, Malawi, Mozambique, Nigeria, Réunion, Senegal, Somalia, South Africa, Sudan, Tanzania, Togo, Uganda, Zimbabwe.

Asia: Bangladesh, Brunei, China, India, Indonesia, Iraq, Israel, Japan, Lebanon, Malaysia, Oman, Pakistan, Philippines, Saudi Arabia, Singapore, Sri Lanka, Taiwan, Thailand, Vietnam.

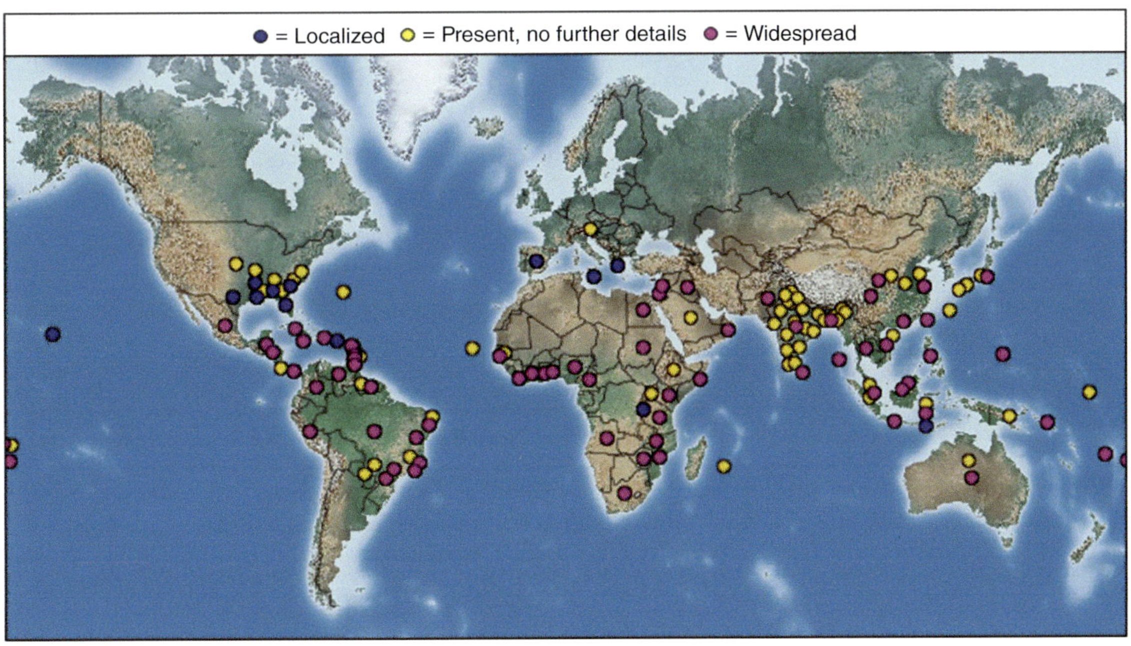

Fig. 11.4. Distribution map of *Rotylenchulus reniformis*. [Adapted from CABI-ISC, 2020]

Europe: Austria, Greece, Malta, Spain.
North America: Antigua and Barbuda, Barbados, Belize, Bermuda, Costa Rica, Cuba, Dominica, Dominican Republic, Grenada, Guadeloupe, Honduras, Jamaica, Martinique, Mexico, Montserrat, Panama, Puerto Rico, Saint Lucia, Saint Vincent and the Grenadines, Trinidad and Tobago, USA.
Oceania: American Samoa, Australia, Fiji, Guam, Kiribati, Niue, Papua New Guinea, Samoa, Solomon Islands, Tonga.
South America: Brazil, Colombia, Guyana, Peru, Suriname, Venezuela.

99.7 Hosts

R. reniformis is extremely polyphagous and feeds on many field and horticultural crops. Among 364 plant species screened against *R. reniformis*, 314 species belonging to 77 families have been reported as host (Robinson *et al.*, 1997). An expanded list of *R. reniformis* hosts is given in Table 73.

99.8 Symptoms

In general, above-ground symptoms of *R. reniformis* resemble moisture and nutrient deficiencies. The nematode can induce stunting, leaf curling and chlorosis on above-ground plant parts. On below-ground parts, females and their adhering egg masses can be easily observed under a dissecting microscope (Fig. 11.5B,D). Soil adhering to the gelatinous egg masses often gives them a dark appearance. The roots invaded by reniform nematode appear bulgy, due to the protruding posterior part of the nematode and attached egg mass, and look dirty, due to soil particles and organic matter adhering to the gelatinous matrix of the egg mass (Khan, 2008). Symptoms also vary with the hosts and may be explicit on different crops.

COTTON Symptoms of reniform nematode damage on cotton include dwarfing, chlorosis, premature decay and loss of secondary roots and plant mortality (Fig. 11.5A). Injury to cotton caused by *R. reniformis* becomes evident with a reduction in the emergence of seedlings. At the 3rd or 4th leaf stage, seedlings cease rapid growth and become light green to pale in colour. Severely infested plants die while others remain stunted, giving an uneven crop (Banu, 2010b). Roots of infected plants are fewer, smaller and sometimes show browning. Affected plants bear fewer and smaller bolls. Roots of affected plants show several semi-endoparasitic female nematodes, which are darkened (Fig. 11.5D). *R. reniformis* also causes a delay in maturity, a reduction in the size of boll and a reduction in lint percentage (Cook and Namken, 1992). Stunted patches with pale greenish leaves can be seen in *R. reniformis*-infested fields. A hybrid cotton DHB-115 exhibited a pinkish discoloration of leaves, leaf shedding and bursting of the affected boll when parasitized by this nematode (Lakshmanan *et al.*, 1989).

PINEAPPLE Reniform nematode infestation in pineapple causes foliar symptoms in the form of reddish or yellow coloration of leaves and retarded and stunted growth (Fig. 11.5C). Heavy infestation may result in the collapse and death of the plant. Plants appear dwarf, with curling leaves. Infected plants show less erectness, with reduced plant lifespan and poor growth (Ferreira and Boley, 1991). On below-ground parts, most of the feeder roots are pruned or decayed as a result of feeding by the nematode. The primary roots continue to elongate and give good anchorage, but secondary root formation is inhibited, thus the root system is poorly developed and leads to crop failure (Mohilal and Khan, 2012).

PULSE CROPS Infection with *R. reniformis* results in malformation of the root system in pigeon pea and is also called 'dirty root disease'. Symptoms of nematode infestation are very distinctly visible in early stages of the crop in chickpea, mungbean, urdbean and lentil, as patches of plants appear light green with uneven growth, but at maturity it is difficult to distinguish between infected and healthy plants as no visual symptoms are seen on above-ground parts. However, the presence of sedentary females protruding on rootlets may be observed on the whole root system after a mild wash (Ali *et al.*, 2010).

99.9 Biology and life cycle

The biology and life cycle are typical of the genus. Feeding by *R. reniformis* induces syncytia similar to that of cyst nematodes (Van Megen *et al.*, 2009). *R. reniformis* is bisexual and reproduces amphimictically, sometimes parthenogenetically (Siddiqi, 1972b). Duration of the life cycle depends upon the hosts. On cotton, it is completed in 17–23 days, on average 19 days at 29.5°C (Birchfield, 1962). The life cycle is completed in 27–32 days at 30–32°C on tobacco (Patel *et al.*, 1984), in 24–29 days on okra and within 20 days on cowpea at the optimum temperature of 25ºC. The reniform nematode can survive in moist soil in the absence of its host for 7 months and in dry soil for 6 months. After 4 months, 84% nematodes were still alive (Sivakumar and Sheshadri, 1979). Stoyanov (1971) reported that *R. reniformis* was able to survive 29 months in the absence of host plants.

99.10 Economic importance

R. reniformis is known to be a major pest of vegetables, cotton, pineapple, banana, tobacco, etc.

VEGETABLES Tomato, eggplant, okra and pepper are highly susceptible to *R. reniformis* and may exhibit 5–16% yield loss. Tomato yield was significantly reduced following inoculation with 100 juveniles per plant (Singh and Khera, 1979). The nematode has been found to damage a number of vegetable crops. A population of 6500 larvae/kg of soil caused 33% yield reduction in some crops in India (Khan *et al.*, 2019a).

COTTON *R. reniformis* is a serious threat to the production of cotton (*Gossypium hirsutum*) in the USA, with yield losses

Table 73. List of host plants of *Rotylenchulus reniformis* (adapted from CABI-CPC, 2010).

Main hosts
Abelmoschus esculentus (okra)
Ananas comosus (pineapple)
Brassica oleracea var. *capitata* (cabbage)
Cajanus cajan (pigeon pea)
Carica papaya (papaw)
Citrus
Cucumis melo (melon)
Cucurbitaceae (cucurbits)
Glycine max (soybean)
Gossypium (cotton)
Gossypium hirsutum (Bourbon cotton)
Ipomoea batatas (sweet potato)
Musa (banana)
Phaseolus (beans)
Solanum lycopersicum (tomato)
Solanum melongena (eggplant)
Syzygium aromaticum (clove)
Vigna unguiculata (cowpea)
Other hosts
Adansonia digitata (baobab)
Aeschynomene americana (American jointvetch)
Aeschynomene indica (Indian jointvetch)
Ageratum conyzoides (billygoat weed)
Amaranthus (grain amaranth)
Anagallis arvensis (scarlet pimpernel)
Annona squamosa (sugar apple)
Anoda cristata (anodaweed)
Arachis hypogaea (groundnut)
Argemone mexicana (Mexican poppy)
Artemisia vulgaris (mugwort)
Artocarpus altilis (breadfruit)
Averrhoa carambola (carambola)
Beaucarnea recurvata
Begonia cucullata var. *hookeri* (perpetual begonia)
Benincasa hispida (wax gourd)
Beta vulgaris (beetroot)
Bidens pilosa (blackjack)
Bixa orellana (annatto)
Brassica oleracea (cabbages, cauliflowers)
Caesalpinia pulcherrima (Paradise flower)
Cajanus
Calendula officinalis (pot marigold)
Callistephus chinensis (China aster)
Camellia sinensis (tea)
Carica
Centrosema pubescens (centro)
Chlorophytum comosum (ribbon plant)
Cicer arietinum (chickpea)
Citrullus lanatus (watermelon)
Citrus sinensis (navel orange)
Coccoloba uvifera (seaside grape)
Coffea arabica (arabica coffee)
Coffea liberica (Liberian coffee tree)
Coleus blumei (common coleus)
Colocasia esculenta (taro)
Commelina diffusa (spreading dayflower)
Convolvulus arvensis (bindweed)
Cordyline fruticosa (good-luck plant)
Crotalaria spectabilis (showy rattlepod)
Cucumis sativus (cucumber)
Cucurbita maxima (giant pumpkin)
Cucurbita moschata (pumpkin)
Cucurbita pepo (ornamental gourd)
Cynara cardunculus var. *scolymus* (globe artichoke)
Cyperus esculentus (yellow nutsedge)
Daucus carota (carrot)
Desmodium tortuosum (Florida beggarweed)
Digitaria sanguinalis (large crabgrass)
Dracaena draco (dragon tree)
Elettaria cardamomum (cardamom)
Eleusine indica (goosegrass)
Epipremnum pinnatum (hunters-robe)
Euphorbia milii (crown-of-thorns)
Euphorbia pulcherrima (poinsettia)
Ficus carica (fig)
Ficus elastica (rubber plant)
Ficus lyrata (fiddle-leaf fig)
Flemingia macrophylla (large-leaf flemingia)
Foeniculum vulgare (fennel)
Grewia asiatica (phalsa)
Helianthus annuus (sunflower)
Hemerocallis spp.
Hibiscus cannabinus (kenaf)
Hibiscus mutabilis (cottonrose)
Hibiscus rosa-sinensis (China-rose)
Hibiscus sabdariffa (Jamaica sorrel)
Impatiens balsamina (garden balsam)
Jacquemontia tamnifolia (small morning glory)
Jasmium officinale
Kalanchoe
Lactuca sativa (lettuce)
Lagenaria siceraria (bottle gourd)
Lepidium virginicum (Virginian peppercress)
Litchi chinensis (lychee)
Luffa acutangula (angled luffa)
Luffa aegyptiaca (loofah)
Malpighia glabra (acerola)
Malus sylvestris (crab-apple tree)
Malvaviscus arboreus (wax mallow)
Mangifera indica (mango)
Manihot esculenta (cassava)
Mentha (mints)
Momordica balsamina (common balsamapple)
Momordica charantia (bitter gourd)
Musa acuminata (wild banana)
Musa × *paradisiaca* (plantain)
Nerium oleander (oleander)
Nicotiana tabacum (tobacco)
Parthenium hysterophorus (Parthenium weed)
Paspalum conjugatum (sour paspalum)
Passiflora edulis (passionfruit)
Persea americana (avocado)

Continued

Table 73. Continued.

Phaseolus lathyroides (Phasey bean)
Phaseolus lunatus (lima bean)
Phaseolus vulgaris (common bean)
Philodendron bipinnatifidum (lacy tree)
Phlox drummondii (annual phlox)
Phyllanthus emblica (Indian gooseberry)
Physalis angulata (cutleaf groundcherry)
Piper betle (betel pepper)
Piper nigrum (black pepper)
Pisum sativum (pea)
Polyscias quilfoylei
Portulaca oleracea (purslane)
Prosopis glandulosa (honey mesquite)
Prunus armeniaca (apricot)
Prunus domestica (plum)
Prunus persica (peach)
Prunus salicina (Japanese plum)
Psophocarpus tetragonolobus (winged bean)
Pueraria phaseoloides (tropical kudzu)
Radermachera sinica
Raphanus sativus (radish)
Rhynchosia
Ricinus communis (castor bean)
Rottboellia cochinchinensis (itch grass)
Sansevieria trifasciata (snake plant)
Schefflera actinophylla (Queensland umbrella tree)
Schinus terebinthifolius (Brazilian pepper tree)
Sesamum indicum (sesame)
Sesbania exaltata (coffeebean)
Solanum nigrum (black nightshade)
Solanum tuberosum (potato)
Sonchus oleraceus (common sowthistle)
Sorghum halepense (Johnson grass)
Spathiphyllum
Stachys arvensis (staggerweed)
Syzygium malaccense (Malay-apple)
Tagetes erecta (African marigold)
Tagetes patula (French marigold)
Theobroma cacao (cocoa)
Trachycarpus fortunei (Chinese windmill palm)
Tribulus terrestris (puncture vine)
Trichosanthes dioica (pointed gourd)
Trifolium alexandrinum (berseem clover)
Trifolium incarnatum (crimson clover)
Trifolium pratense (purple clover)
Trifolium repens (white clover)
Trigonella foenum-graecum (fenugreek)
Triticum aestivum (wheat)
Urena lobata (hibiscus burr)
Vicia faba (broad bean)
Vicia villosa
Vigna aconitifolia (moth beans)
Vigna mungo (black gram)
Vigna radiata (mung bean)
Vitis vinifera (grapevine)
Washingtonia robusta (Mexican Washington-palm)
Xanthosoma caracu
Xanthosoma sagittifolium (yautia, yellow)
Zea mays (maize)
Unknown hosts
Allium sativum (garlic)
Amaranthus dubius (spleen amaranth)
Anacardium occidentale (cashew nut)
Areca catechu (betelnut palm)
Capsicum (peppers)
Capsicum annuum (bell pepper)
Catharanthus roseus (pink periwinkle)
Citrus limon (lemon)
Coffea (coffee)
Curcuma longa (turmeric)
Emilia sonchifolia (red tasselflower)
Euphorbia heterophylla (wild poinsettia)
Euphorbia hirta (garden spurge)
Gladiolus hybrids (sword lily)
Hemerocallis (daylilies)
Lablab purpureus (hyacinth bean)
Lagerstroemia indica (Indian crape myrtle)
Pennisetum glaucum (pearl millet)
Phoenix dactylifera (date-palm)
Polianthes tuberosa (tuberose)
Solanum torvum (turkey berry)
Sorghum bicolor (sorghum)
Zingiber officinale (ginger)

ranging from 10% to 70% in infested fields (Kirkpatrick and Lorenz, 1997). In Alabama (USA), the nematode is reported to be the most damaging nematode to cotton, accounting for 8–9% of the total yield loss in 2002 (Gazaway and McLean, 2003; Blasingame, 2004). The total annual loss to US cotton was estimated to exceed US$100 million (Blasingame, 2006). Reniform nematodes cause reduction in the lint yield, boll size, lint per cent, plant growth, seed index and fibre micronaire value (Cook and Namken, 1992). The damage threshold has been determined as 1000 young females per plant for *G. hirsutum* (Sud *et al.*, 1984).

PINEAPPLE The nematode has also been reported as a major problem in pineapple in the Philippines, Caribbean Islands (especially Puerto Rico), some areas in Thailand, North Queensland in Australia, Oaxaca in Mexico (Davide, 1988), South Africa (Keetch, 1982) and Hawaii, USA (Caswell *et al.*, 1990). Nematode infection reduced the crop yield by 60–74% and the marketable yield by 40–45%. Infestation by reniform nematodes may cause average yield losses of 21% (up to 38%) in first plant crops and 29% (up to 60%) in first ratoon crops (Mohilal and Khan, 2012).

BANANA In field plots where *R. reniformis* was the predominant parasitic nematode, yield reduction of 19.3 t/ha was reported in India (Sheela and Sundararaju, 2012).

PULSES In India, losses due to reniform nematode in pulses have been estimated as 23% in mungbean, 14–29% in pigeon pea and 11% in chickpea (Khan *et al.*, 2019b).

Fig. 11.5. Symptoms of *Rotylenchulus reniformis*. Reniform nematode infested cotton field showing stunted growth and leaf browning (A); posterior half of mature female protruding out of cotton root (B,D); left, infested pineapple plants whereas right plot treated with nematicide (C). [Courtesy A,B,C: NASI, India; D: J. Bridge/CABI BioScience]

TOBACCO *R. reniformis* is the second most important nematode problem in tobacco after root-knot in India. The nematode may reduce the tobacco yield by up to 22% (Khan *et al*., 2010).

99.11 Movement and means of dispersal

See section 97.11, above.

99.12 Pest risk analysis and looming threat from introduction

R. reniformis is widespread in many warm temperate regions. However, due to economic damaging potential to various agriculturally important crops, a quarantine risk prevails in all subtropical and tropical countries where the nematode is absent and susceptible host crops are grown. Roots of host plants, planting materials and contaminated soils facilitate dissemination of the nematode. It is a quarantine risk in the ornamental flower industries in the USA that have been established in areas where the nematode occurs and the movement of infected material from these areas into cotton-growing states is prohibited.

99.13 Invasiveness rating

Considering the economic damage and pest risk analysis of *R. reniformis*, the following invasiveness rating has been assigned.

Climate/host interaction	–	Medium (2)
Hosts range	–	High (3)
Dispersal potential	–	High (3)
Economic impact	–	High (3)
Environmental impact	–	High (3)
Post-entry distribution	–	Medium (–2)
Final rating (score)	–	High (12)

Based on the above invasiveness rating, *R. reniformis* receives a score of 12 out of 15 and is rated as HIGHLY INVASIVE NEMATODE.

99.14 Region-wise status of invasiveness

R. reniformis has a high invasive rating for all subtropical and tropical countries and also in warm temperate regions where susceptible host crops are grown and this nematode does not occur.

99.15 Management measures

99.15.1 Cultural management

Crop rotation of cotton with sorghum (Thames and Heald, 1974), maize, sugarcane and pangola grass (Heald and Thames, 1982) for 2 years was found quite effective in reducing nematode infestations. Soil amendment with animal manure and cotton seed cakes successfully controlled the nematode (Badra *et al.*, 1979). Soil solarization may reduce nematode populations by as much as 86% (Chellemi *et al.*, 1994).

99.15.2 Resistance

Numerous resistant varieties have been reported against *R. reniformis*, such as cowpea cultivars 'V-16', 'C-152' and 'Pusa Phaguni' (Thakur and Patel, 1984; Haque and Padmavathy, 1985), tomato accessions 'Radiant', 'SL-120', 'Patriot', 'Bresch', 'Ronita', 'Piersol',' F-38-E-2', 'Punjab NR-7' and 'Hisar Lalit' (Ramakrishnan *et al.*, 2013) and cotton cv. 'Deltapine-16' (Robinson *et al.*, 1997).

99.15.3 Biological control

Seed treatment with *Pseudomonas fluorescens* at 20 g/kg followed by soil application at 2.5 kg/ha significantly reduced the nematode population (Sivakumar *et al.*, 2005). Application of *Purpureocillium lilacinum* reduced nematode densities on tomato (Walters and Barker, 1994). Integration of biocontrol agent, *P. lilacinum* and carbofuran at 2 kg a.i./ha was found to be effective in managing *R. reniformis* in tomato (Reddy and Khan, 1988).

99.15.4 Chemical control

Application of oxamyl as a foliar spray has been found effective against *R. reniformis* in sweet potato (McSorley and Parrado, 1983). Foliar sprays with fenamiphos at 2–5 kg/ha at intervals of 1–3 months in pineapple reduced the nematode population and increased yield of pineapple (Tsai and Apt, 1979). Seed treatment with carbofuran at 700 g a.i./100 g of seeds (Lordello and Brancalion, 1986) or soil application of carbofuran at 0.5 kg a.i./ha (Brancalion and Lordello, 1982) provided good results against the reniform nematode. Soil application of either carbofuran, phorate, fenamiphos or carbosulfan, each at 1 kg/ha, gave maximum reduction in the nematode population at 30 days after germination and significantly increased cotton yield (Patel *et al.*, 1996).

99.15.5 Phytosanitary measures

See Chapter 16, section 115.

12

Telotylenchidae: *Tylenchorhynchus* Species

Tylenchorhynchus Spp.

The stunt nematodes of the genus *Tylenchorhynchus* are obligate migratory root ectoparasites and belong to the family Telotylenchidae (Siddiqi, 2000)/Dolichodoridae (Decraemer and Hunt, 2013). The subfamily Telotylenchinae constitutes a large group of plant-parasitic nematodes, including several related genera such as *Bitylenchus*, *Quinisulcius*, *Telotylenchus* and *Trophurus*. The genus has been revised by several workers (Handoo, 2000; Siddiqi, 2000; Geraert, 2011; Ganguly *et al.*, 2013). *Tylenchorhynchus* contains the highest number of species and more than 260 nominal species are described (Esser, 1991; Handoo, 2000; Decraemer and Geraert, 2013). Currently, at least 130 species are validated with *Tylenchorhynchus cylindricus* as type species (Handoo, 2000; Siddiqi, 2000; Geraert, 2011; Ganguly *et al.*, 2013). Among these, numerous species have been reported to damage agronomical and horticultural crops, such as *Quinisulcius acutus* (= *Tylenchorhynchus acutus*) on cassava, lucerne, soybean and tobacco, *T. annulatus* on rice and sugarcane and *T. claytoni* on maize, cotton, potato and soybean (Siddiqi, 2000). However, due to lack of information, only *T. claytoni* is considered in this volume for invasiveness.

Systematic position of *Tylenchorhynchus* species

	Classical (after Siddiqi, 2000)	Molecular (after Decraemer and Hunt, 2013)
Phylum	Nematoda	Nematoda
Class	Secernentea	Chromadorea
Order	Tylenchida	Rhabditida
Suborder	Hoplolaimina	Tylenchina
Infra-order	–	Tylenchomorpha
Superfamily	Dolichodoroidea	Tylenchoidea
Family	Telotylenchidae	Dolichodoridae
Subfamily	Telotylenchinae	Telotylenchinae
Genus	*Tylenchorhynchus*	*Tylenchorhynchus*

100 *Tylenchorhynchus claytoni*

100.1 Common name

Stunt nematode

100.2 Scientific name

Tylenchorhynchus claytoni Steiner, 1937

100.3 Synonyms

Billineelus claytoni (Steiner) Volkova, 1993
Dolichorhynchus claytoni (Steiner) Sultan, Singh and Sakhuja, 1991
Tessellus claytoni Jairajpuri and Hunt, 1984

100.4 EPPO code

TYLRCL

100.5 Diagnosis

100.5.1 Morphological diagnosis

Morphological descriptions of *Tylenchorhynchus claytoni* after Allen (1955), Loof (1974) and Golden *et al.* (1987) are summarized below. For morphometric measurements (after Golden *et al.*, 1987), see Table 74.

FEMALE: Body vermiform, elongate-cylindroid, with tapering at extremities, slightly ventrally curved when heat relaxed (Fig. 12.1C). Body cuticle tessellated in appearance, marked by transverse and longitudinal striae. Longitudinal lines 23–29, outside lateral fields. Annules coarse, 1.9–2.5 μm wide on mid-body, wider (2.9 μm) on distal part of tail. Lateral fields with four incisures forming three smooth bands. Lip region rounded, slightly offset bearing 3–4 annules. Cephalic sclerotization weak. Spear slender, moderate with sharply pointed tip with

DOI: 10.1079/9781789247367.0013

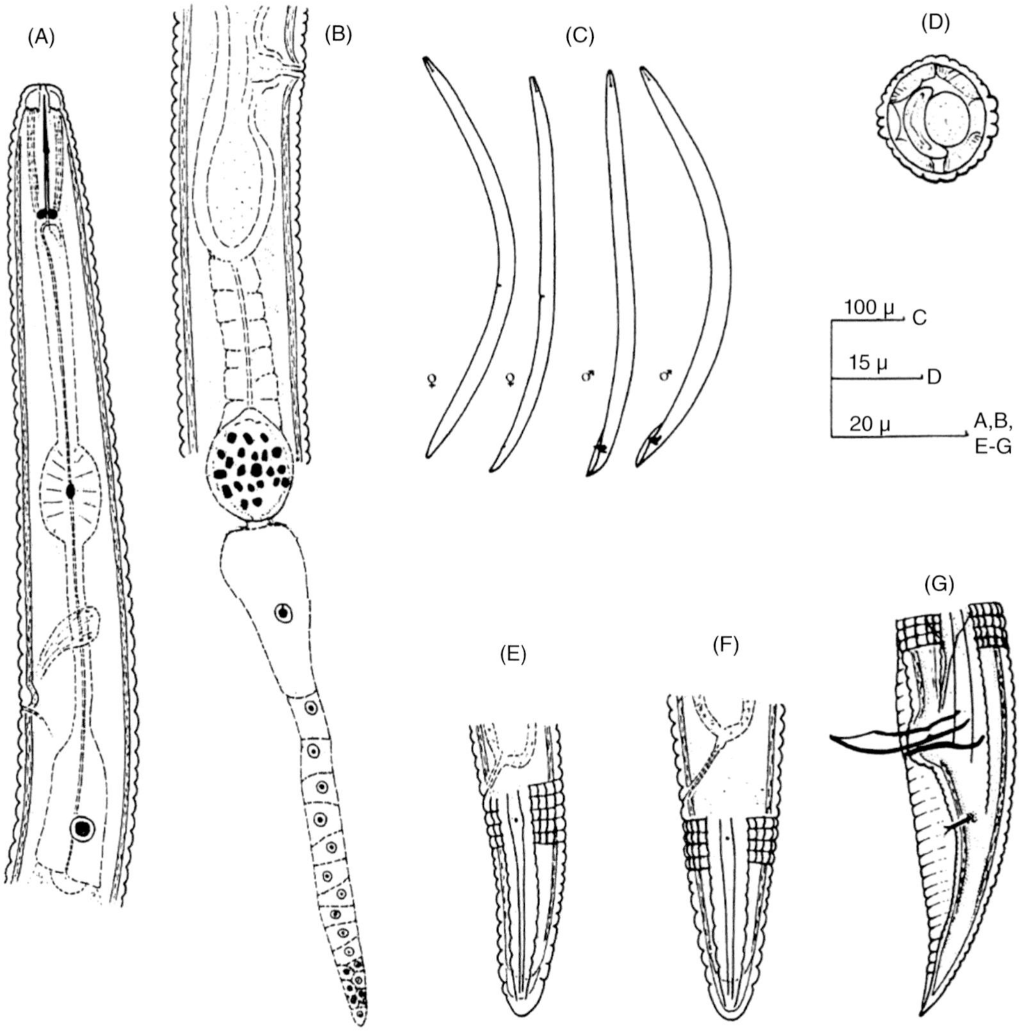

Fig. 12.1. Line drawing of *Tylenchorhynchus claytoni*. Female oesophageal region, lateral (A); female reproductive system (B); whole male and female (C); cross-section of body in mid-region (D); female tails, lateral (E,F); male tail, lateral (G). [Reproduced from Loof, 1974. CIH Descriptions of Plant-parasitic Nematodes. Set 3, No. 39; courtesy CAB International]

rounded basal knobs. Dorsal oesophageal gland orifice 1.7–2.5 μm from stylet base. Median oesophageal bulb muscular, oval. Isthmus slender, encircled by nerve ring. Basal lobe pyriform with flat base and distinct dorsal gland nucleus (Fig. 12.1A). Cardia small, rounded. Excretory pore 95–122 μm from head end. Hemizonid just anterior to excretory pore, two annules long. Deirids absent. Vulva a transverse slit, equatorial at around 55–57% of body length. Vagina straight. Genital tracts paired, amphidelphic, branches opposed, outstretched. Uterus with wide lumen. Spermathecae round, with sperm. Ovaries with single row of oocytes (Fig. 12.1B). Tail conoid, often tapering to rounded, smooth tip, bearing 9–16 annules (Fig. 12.1E,F). Phasmids pore-like, adanal to just postanal, 27–34 μm from tail terminus.

Male: Morphologically similar to female except reproductive system (Fig. 12.1C). Single testis, outstretched anteriorly. Spicules ventrally arcuate, cephaled, distally flanged. Gubernaculum rod-shaped with slightly curved and rounded proximal end. Tail conoid-pointed. Bursa completely enveloped tail, marked with transverse striae (Fig. 12.1G).

100.5.2 Molecular diagnosis

Handoo *et al.* (2014) characterized 18 known species of the family Telotylenchidae, including *T. annulatus*, *T. clarus*, *T. claytoni* and two new taxa (*T. mediterraneus* and *Bitylenchus hispaniensis*), using D2–D3 expansion segments of 28S rRNA, ITS rRNA and partial 18S rRNA sequences. The DNA was extracted from a single nematode and PCR assays were processed as described by Tanha Maafi *et al.* (2003) (see Chapter 4, section 15.5.2) using the following primers.

For D2–D3 expansion segments of 28S rRNA (Castillo *et al.*, 2003)

D2A (5′-ACAAGTACCGTGAGGGAAAGTTG-3′)

D3B (5′-TCGGAAGGAACCAGCTACTA-3′)

For ITS region (Subbotin *et al.*, 2001)

TW81 (5′-GTTTCCGTAGGTGAACCTGC-3′)

AB28 (5′-ATATGCTTAAGTTCAGCGGGT-3′)

Table 74. Morphometric measurements of *Quinisulcius* and *Tylenchorhynchus* species.

Morphometric characters		Range, numerous populations		
		Q. acutus (after Allen, 1955)[a]	*T. annulatus* (after Siddiqi, 1976)[a]	*T. claytoni* (after Golden *et al.*, 1987)
Females	Length (μm)	630–700	660–720	589–753
	Stylet length (μm)	16–17	17–18	18.4–21.0
	a	31–38	34–35	22.1–27.8
	b	4.8–5.3	5.0–5.6	4.3–5.7
	c	17–22	15–16	14.8–19.8
	c'	–	3.1–3.4	2.0–2.7
	DGO	–	1.5–2.0	1.7–2.5
	V	56–57	54–56	53.9–59.9
	Excretory pore from anterior end (μm)	–	–	95–122
	Maximum body width (μm)	–	–	21–30
	Tail length (μm)	–	–	27–34
	m	–	49–50	–
Males	Length (μm)	–	–	576–662
	Stylet length (μm)	–	–	18.0–19.3
	a	–	–	27.7–32.5
	b	–	–	4.6–5.5
	c	–	–	14.8–16.5
	c'	–	–	2.3–2.8
	Spicule (μm)	–	–	22–24
	Gubernaculum (μm)	–	–	12.9–14.1

[a]Male rare or absent.

For partial 18S rRNA (Chizhov *et al.*, 2006)

G18SU (5′-GCTTGTCTCAAAGATTAAGCC-3′)
F18Tyl1 (5′-GTGCCAGCAGCCGCGGTAATTCC-3′)

The phylogenetic reconstruction was performed with newly obtained sequences of 28S rRNA (705 bp), ITS rRNA (804 bp) and partial 18S rRNA (862 bp) and other published sequences of Telotylenchidae nematodes from GenBank. Mwamula *et al.* (2020) also characterized 28S rRNA and ITS rRNA gene sequences to identify *T. annulatus*, *T. claytoni* and *T. thermophilus*.

100.6 Geographical distribution

T. claytoni is distributed throughout the world (Fig. 12.2). The nematode is widespread in Canada, Japan and the USA and has been detected in the following countries (CABI-ISC, 2020).

Africa: South Africa.
Asia: Bangladesh, China, India, Japan, North Korea, South Korea, Turkey.
Europe: Belgium, Bulgaria, Cyprus, Denmark, Germany, Italy, Netherlands, Poland, Switzerland, United Kingdom.
North America: Canada, Honduras, USA.
Oceania: Australia.

100.7 Hosts

T. claytoni is highly polyphagous and parasitizes azaleas, maize, pea, pine, peach, sugarcane, tobacco and grasses (MacGowan, 1980; Benson and Barker, 1985). The rice weeds *Cynodon dactylon* and *Echinochloa colonum* are good alternative hosts of *T. claytoni* (Satyanarayana *et al.*, 1980). An expanded list of hosts of *T. claytoni* is given in Table 75.

100.8 Symptoms

T. claytoni feeds ectoparasitically on roots and may produce brownish lesions at the feeding point. The root damage can cause severe stunting and yellowing of foliage on azalea (Radewald *et al.*, 1972). Stunting and chlorosis are more severe when plants are under strain (Benson and Barker, 1985).

100.9 Biology and life cycle

T. claytoni is a migratory ectoparasite that feeds on epidermal cells of roots and inhabits rhizospheres. Its life cycle is similar to that of other stunt nematodes. However, males are abundant and more common than in other species of the genus and are necessary for reproduction, which is amphimictic type. Females lay eggs 8–10 days after the last moult. The first moult occurs inside the egg. At the third juvenile stage, sexes can be recognized by the structure and location of the gonad primordium. The life cycle takes 31–38 days to complete at 28°C (Wang, 1971). In India, *T. claytoni* populations showed a build-up in February to May in rice fields despite high temperatures and the nematode population levels were highest at the flowering stage of the rice crop, preceded by peaks of rainfall (Prasad and Rao, 1984). The nematode can survive in the soil for 10 months at 1.7–24°C (Loof, 1974).

100.10 Economic importance

T. claytoni is the main nematode problem for plants grown in peat soils in Germany. It can cause severe damage to azaleas (Radewald *et al.*, 1972). *T. claytoni* also causes economic losses to other ornamentals. *T. claytoni* and *T. annulatus* are considered as pests of sugarcane in Natal, South Africa (Anon., 1981). *T. claytoni* also parasitized maize and tobacco in New York State, USA (DiSanzo, 1981), tobacco in Bulgaria (Katalan-Gateva and Baicheva, 1976) and rice in India (Krishnaprasad and Krishnappa, 1982). Invasion of *T. claytoni* may predispose host plants to fungal infection, leading to disease complexes.

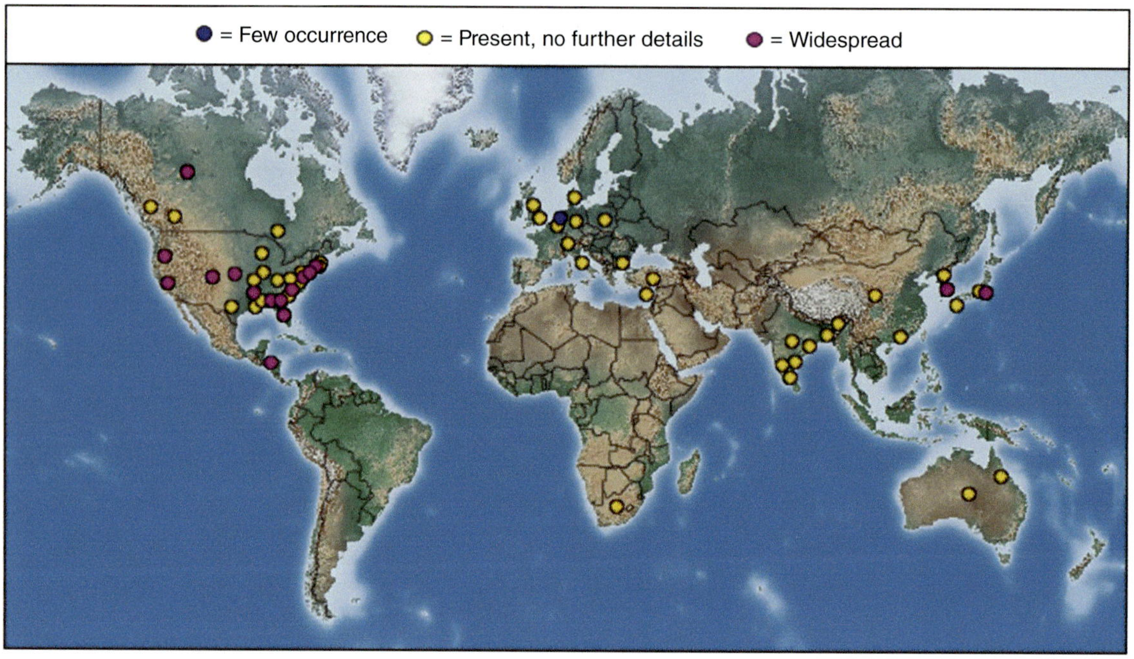

Fig. 12.2. Distribution map of *Tylenchorhynchus claytoni* (adapted from CABI-ISC, 2020).

Table 75. List of hosts affected by *Tylenchorhynchus claytoni* (adapted from CABI-CPC, 2010).

Main hosts
Arachniodes adiantiformis (leatherleaf fern)
Avena sativa (oats)
Azaleas
Glycine max (soybean)
Gossypium (cotton)
Lolium (ryegrasses)
Medicago sativa (lucerne)
Nandina
Nicotiana tabacum (tobacco)
Olea europaea subsp. *europaea* (olive)
Oryza sativa (rice)
Peony
Photinia glabra
Pinus resinosa (red pine)
Pisum sativum (pea)
Poaceae (grasses)
Prunus persica (peach)
Rhododendron (azalea)
Rhododendron indicum (Satsuki azalea)
Rhododendron japonicum (Japanese azalea)
Rhododendron obtusum (Kirishima azalea)
Saccharum officinarum (sugarcane)
Secale cereale (rye)
Solanum tuberosum (potato)
Trifolium spp.
Zea mays (maize)
Other hosts
Arachis hypogaea (groundnut)
Aucuba japonica (Japanese aucuba)
Cajanus cajan (pigeon pea)
Citrus aurantium (sour orange)
Citrus limon (lemon)
Citrus × *paradisi* (grapefruit)
Eleusine coracana (finger millet)
Ipomoea batatas (sweet potato)
Litchi chinensis (lychee)
Picea abies (common spruce)
Pinus echinata (shortleaf pine)
Pinus elliottii (slash pine)
Pinus palustris (longleaf pine)
Pinus taeda (loblolly pine)
Rubus (blackberry, raspberry)
Rubus fruticosus (blackberry)
Solanum lycopersicum (tomato)
Solanum melongena (aubergine)
Vaccinium myrtillus (blueberry)
Wild hosts
Agrostis stolonifera var. *palustris* (bentgrass)
Arctium lappa (burdock)
Bouteloua curtipendula (side-oats grama)
Bouteloua gracilis (blue grama)
Camellia japonica (camellia)
Chamaecyparis pisifera (Sawara false cypress)
Cynodon dactylon (Bermuda grass)
Echinochloa colona (junglerice)
Enkianthus perulatus (white enkianthus)
Erica (heaths)
Festuca (fescues)
Festuca arundinacea (reed fescue)
Osmanthus fragrans
Pieris japonica (lily-of-the-valley shrub)
Poa pratensis (smooth-stalked meadow grass)

100.11 Movement and means of dispersal

Tylenchorhynchus spp. are migratory ectoparasites and their dispersal is primarily facilitated via infected planting materials adhering contaminated soil. The nematodes can also be spread by the movement of contaminated farm machinery and the movement of soil.

100.11.1 Plant parts known to carry the nematode in trade/transport

Roots, bulbs, tubers, corms, rhizomes (sucker) and seedlings, growing medium accompanying plants and micro-propagated plants.

100.11.2 Plant parts not known to carry the nematode in trade/transport

Bark, flowers, inflorescences, cones, calyx, fruits (including pods), leaves, stems (above ground), shoots, trunks, branches, wood and true seeds (including grain).

100.12 Pest risk analysis and looming threat from introduction

T. claytoni is a polyphagous species and causes substantial damage to various agriculturally important crops. There is less chance of the nematode being introduced into new areas, because of its ectoparasite nature, but it has been detected in England on bonsai trees imported from Japan (Southey and Aitkenhead, 1972), indicating the possibility of introduction of this nematode into new areas.

100.13 Invasiveness rating

Considering the consequences of introduction and probable economic damage caused by *T. claytoni*, the following invasiveness rating has been assigned.

Climate/host interaction	–	High (3)
Hosts range	–	High (3)
Dispersal potential	–	Low (1)
Economic impact	–	High (3)
Environmental impact	–	Low (1)
Post-entry distribution	–	Low (–1)
Final rating (score)	–	Medium (10)

Based on the above invasiveness rating, *T. claytoni* receives a score of 10 out of 15 and is rated as MEDIUM INVASIVE NEMATODE.

100.14 Region-wise status of invasiveness

Because of the wide host range and economic damage potential, a medium invasiveness rating has been given for all the countries where *T. claytoni* is absent.

100.15 Management measures

In peat soils, control measures include prevention of transmission and rearing nematode-free seedlings (Braasch, 1976). Crop rotation using *Lolium multiflorum*, wheat, rye and a fallow to control nematodes on soybean has been found quite effective (Pedersen and Rodriguez-Kabana, 1991). Crop rotation of azalea with non-host plants also significantly controlled *T. claytoni* (Yamazaki *et al.*, 1981). Rice varieties 'Bala', 'CR-1014' and 'Supriya' (Krishnaprasad and Krishnappa, 1982) have shown significantly low population levels of *T. claytoni*. Applications of oil cakes or animal excrement as organic amendments are effective nematode suppressants (Rodriguez-Kabana, 1986). Several non-fumigant nematicides, such as prophos or fenamiphos in lawn grasses (Peacock and Dunn, 1986), cinnamic aldehyde in azalea (Saotome and Ishibashi, 1989), carbofuran and oxamyl in maize, tobacco and potato (DiSanzo, 1981, 1982), ethazole, phorate, tecnazene or disulfoton in cotton (Adams and Rodriguez-Kabana, 1977), fenamiphos in peach (Funt *et al.*, 1982) and fensulfothion and phorate in soybean (Schmitt and Corbin, 1981), were successfully used to control nematode complexes including *T. claytoni*. Immersion of potato seed, wheat and rye seeds in oxamyl, carbofuran or fenamiphos caused a sharp decline in numbers of *T. claytoni* in the soil (Rodriguez-Kabana *et al.*, 1977).

100.16 Phytosanitary measures

See Chapter 16, section 115.

13

Trichodoridae: *Nanidorus, Paratrichodorus* and *Trichodorus* Species

Nanidorus, Paratrichodorus and *Trichodorus* Spp.

The genera *Nanidorus*, *Paratrichodorus* and *Trichodorus* belong to the family Trichodoridae and subfamily Trichodorinae Thorne, 1935 along with three other genera, namely *Allotrichodorus*, *Ecuadorus* and *Monotrichodorus* (Decraemer and Geraert 2013; Asghari *et al.*, 2018; Subbotin *et al.*, 2019a). Cobb (1913) established the genus *Trichodorus*; Siddiqi (1974a) proposed the genus *Paratrichodorus* with *Nanidorus* as subgenus and described three new species. At present, the family Trichodoridae includes 116 valid species from six genera (Decraemer and Geraert, 2013; Asghari *et al.*, 2018; Subbotin *et al.*, 2019a). Members of the Trichodoridae are potential vectors of tobraviruses (= netuviruses, nematode-transmitted tubular particles). Two species of *Nanidorus* (six valid species), eight species of *Paratrichodorus* (29 valid species) and four species of *Trichodorus* (67 valid species) are known to transmit three tobraviruses, namely, tobacco rattle virus (TRV), pea early browning virus (PEBV) and pepper ringspot virus (PepRSV) (Brown *et al.*, 1989; Decraemer and Robbins, 2007; Subbotin *et al.*, 2019a). The virus-transmitting species are:

Nanidorus minor (Colbran, 1956) Siddiqi, 1974
N. nanus (Allen, 1957) Siddiqi, 1974
Paratrichodorus allius (Jensen, 1963) Siddiqi, 1974
P. anemones (Loof, 1965) Siddiqi, 1974
P. divergens Almeida *et al.*, 2005
P. hispanus Roca and Arias, 1986
P. pachydermus (Seinhorst, 1954) Siddiqi, 1974
P. porosus (Allen, 1957) Siddiqi, 1974
P. teres (Hooper, 1962) Siddiqi, 1974
P. tunisiensis (Siddiqi, 1963) Siddiqi, 1974
Trichodorus cylindricus Hooper, 1962
T. primitivus (deMann, 1880) Micoletzky, 1922
T. similis Seinhorst, 1963
T. viruliferus Hooper, 1963

Feeding by these nematodes can cause cessation of root elongation and abbreviation as a result the emerging lateral roots become short and wide, like stubs; hence are termed as stubby root disease and stubby nematode (Khan, 2008). Three species, *Nanidorus minor* (= *Paratrichodorus minor*), *Paratrichodorus porosus* and *Trichodorus viruliferus*, are considered in this chapter for their invasiveness, as they are of economic importance and associated with an important vector of plant viruses (Weischer, 1993; Xu and Zhao, 2019; Subbotin *et al.*, 2019a).

Systematic position of *Nanidorus, Paratrichodorus* and *Trichodorus* species

	Classical (after Hunt, 1993)	Molecular (after Decraemer and Hunt, 2013)
Phylum	Nematoda	Nematoda
Class	Adenophorea	Enoplea
Subclass	Dorylaimia	Enoplia
Order	Triplonchida	
Suborder	Diptherophorina	
Superfamily	Trichoroidea	Diptherophoroidea
Family	Trichodoridae	Trichodoridae
Subfamily	Trichodorinae	–
Genus	*Nanidorus*	*Nanidorus*
Species	*N. minor*	*N. minor*
Genus	*Trichodorus*	*Trichodorus*
Species	*T. viruliferus*	*T. viruliferus*
Genus	*Paratrichodorus*	*Paratrichodorus*
Species	*P. porosus*	*P. porosus*

101 *Nanidorus minor* (= *Paratrichodorus minor*)

101.1 Common name

Stubby nematode

101.2 Scientific name

Nanidorus minor (Colbran, 1956) Siddiqi, 1974

101.3 Synonyms

Nanidorus christiei (Allen, 1957) Siddiqi, 1974
Paratrichodorus christiei (Allen, 1957) Siddiqi 1974

DOI: 10.1079/9781789247367.0014

Paratrichodorus minor (Colbran, 1956) Siddiqi, 1974
Paratrichodorus obesus (Razj. and Pent., 1975) Rodriguez and Bell, 1978
Trichodorus christiei Allen, 1957
Trichodorus minor Colbran, 1956
Trichodorus obesus Razjivin and Penton, 1975.

101.4 EPPO code

TRIHMI (*Paratrichodorus minor*)

101.5 Diagnosis

In the literature, *Nanidorus minor* was generally referred to as *Paratrichodorus minor* until molecular diagnosis (Duarte *et al.*, 2010; Kumari and Subbotin, 2012), and now *Nanidorus* is widely accepted as a valid genus (Subbotin *et al.*, 2019a).

101.5.1 Morphological diagnosis

Morphological descriptions from Hunt (1993) are summarized below. For measurements, see Table 76.

FEMALE: Body almost straight when relaxed, tapering slightly at both ends (Fig. 13.1A). Lip region spherical with slightly protruding papillae. Amphidial aperture a large transverse slit appearing as a fovea or cup-shaped pouch, detached posteriorly from sensillum pouch or fusus by a constriction. Pharynx armed with rather short ventrally curved onchiostyle having inner onchium. Secretory-excretory pore generally near pharynx base. Oesophageal glands frequently overlapping intestine subventrally and ventrally, pharyngeal bulb seldom offset. Lateral body pores absent. Spermathecae not formed; very small sperm and spermatids dispersed throughout uterus, if present. Vagina with rod-shaped sclerotization, close, parallel to longitudinal body axis (Fig 13.1B). Vagina in side view quadrangular. Vulva a transverse slit, short (2–2.5 µm). Caudal pores usually absent.

MALE: Usually rare. Ventromedian cervical papillae and lateral cervical pores absent. Spicule mostly straight, slightly ventrally curved distally, capitulum slightly marked, shaft transversely striated, except at both ends (Fig. 13.1F). Gubernaculum with small dorsal keel. Single ventromedian pre-adanal supplement present, shortly anterior to cloacal opening. Bursa present (Fig. 13.1G) and extending immediately anterior to precloacal supplement posterior to postcloacal papillae.

101.5.2 Molecular diagnosis

Molecular characterization using various approaches such as the sequencing of ribosomal RNA genes (rDNA) of partial 18S, ITS regions, or the D2 and D3 expansion segments of the 28S rRNA gene or large subunit (LSU) rRNA nuclear gene and protein-coding mitochondrial gene cytochrome c oxidase subunit 1 (COX1) of the members of the family Trichodoridae including *Nanidorus*, *Paratrichodorus* and *Trichodorus* was carried out by several workers (Asghari *et al.*, 2018; Decraemer *et al.*, 2018; Xu and Zhao, 2019; Subbotin *et al.*, 2019a).

Subbotin *et al.* (2019a) characterized D2–D3 expansion fragments of 28S rRNA of 22 species of Trichodoridae, including *N. minor* and *P. porosus*, collected from Malaysia, Mexico, Russia and the USA. The DNA was extracted from several individual nematodes in a sample as described by Tanha Maafi *et al.* (2003, see Chapter 4, section 15.5.2) and PCR assays were performed using the following primers (Subbotin *et al.*, 2006).

Table 76. Morphometric data for *Nanidorus*, *Paratrichodorus* and *Trichodorus* species.

Morphometric characters		Range, numerous populations		
		N. minor (after Siddiqi, 1962 and 1963b)	*P. porosus* (after Chau, 1997)	*T. viruliferus* (after Hooper, 1976)
Females	Length (µm)	480–580	612–760	640–860
	Onchiostyle (µm)	26–30	47–49	40–48
	a	17–23	22.5–26.0	20–26
	b	4.3–5.8	4.4–5.4	4.7–7.0
	c	–	Subterminal	Subterminal
	V	52–57	56.0–58.8	51–59
Males	Length (mm)	600	585–723	620–890
	Onchiostyle (µm)	28	39–48	40–48
	a	25	20.6–26.0	23–33
	b	6	4.7–5.3	4.6–7.3
	c	40	–	47–71
	Excretory pore from anterior end (µm)	–	63–76	–
	Spicule (µm)	50	38–40	31–37
	Gubernaculum	17	11–13	13–18
	T	53	53–69	57–73

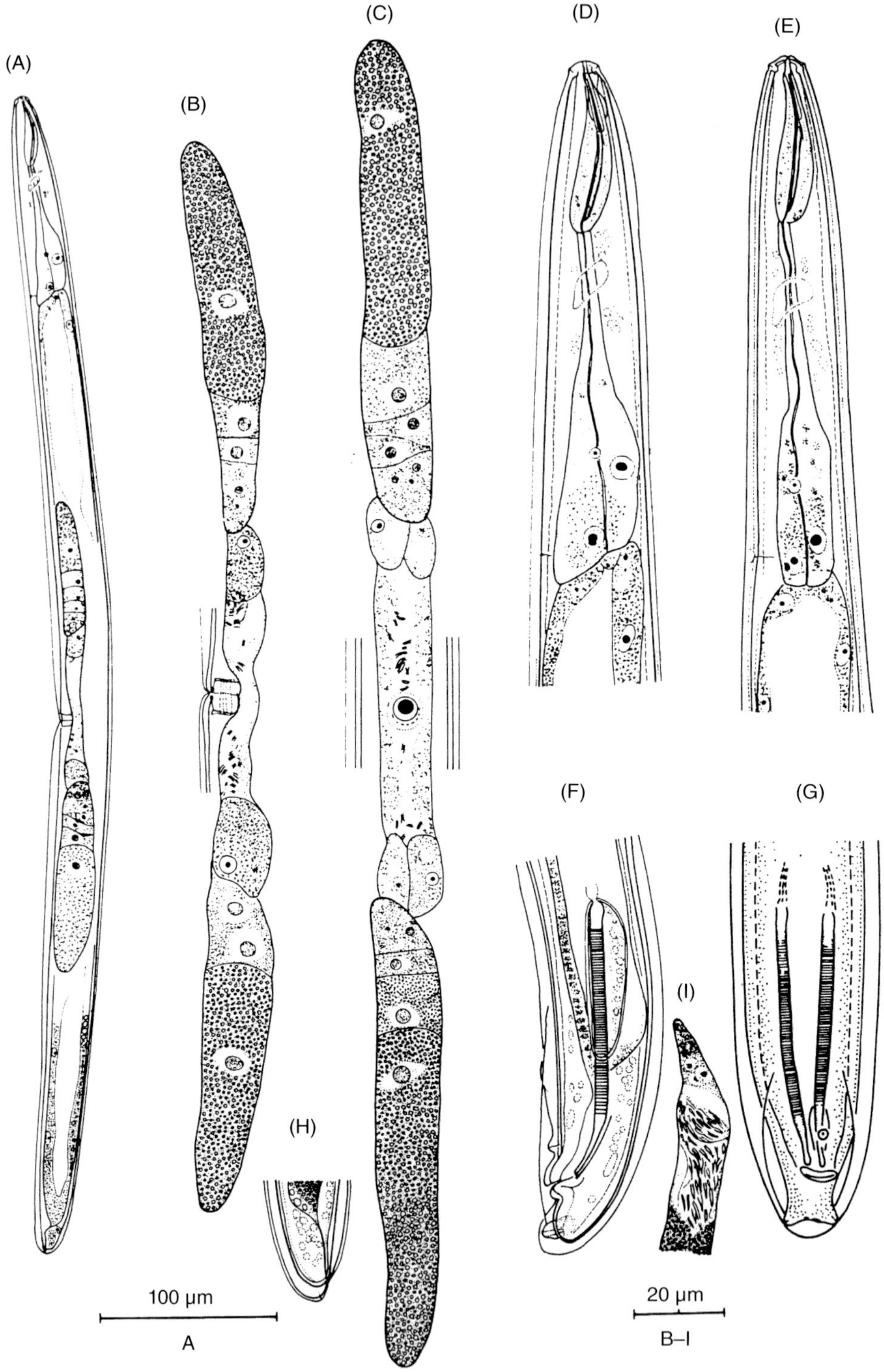

Fig. 13.1. Line drawing of *Nanidorus minor*. Female (A); reproductive system (B); reproductive system, ventral view (C); anterior end of female (D); anterior end of male (E); copulatory apparatus and tail (F); copulatory apparatus, ventral view (G); female tail (H); testis and part of vas deferens (I). [Reproduced from Hunt, 1993; after Decraemer, 1991; courtesy Taylor & Francis Group LLC]

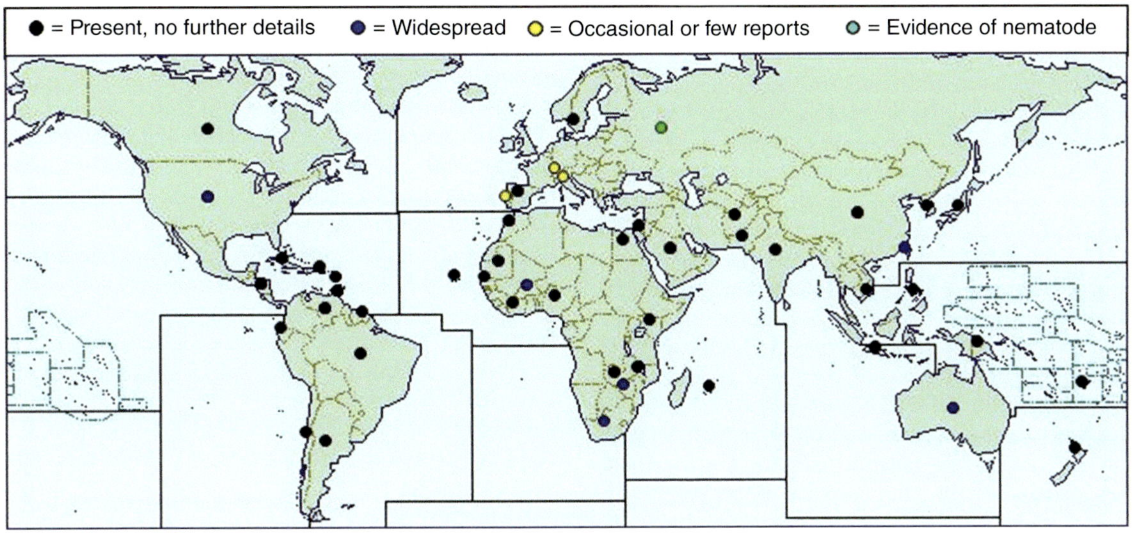

Fig. 13.2. Distribution map of *Nanidorus minor*. [Modified from CABI-CPC, 2010]

Forward D2A (5′-ACAAGTACCGTGAGGGAAA-GTTG-3′)
Reverse D3B (5′-TCG GAAGGAACCAGCTACTA-3′)

The D2–D3 of 28S rRNA gene sequence alignment was 754 bp long. PCR products were purified and directly sequenced and the sequences were submitted to the GenBank database under the accession numbers MK584753–MK584804 and MK615145–MK615150. The phylogenetic study of the stubby root nematodes revealed five major clades: (i) *Trichodorus* from Europe, Asia and North America; (ii) *Nanidorus* and *Trichodorus* from Asia; (iii) *Trichodorus* from California; (iv) *Paratrichodorus* from several distant geographical regions; and (v) *Monotrichodorus* from Central and South America. Xu and Zhao (2019) also characterized *N. minor* from New Zealand and the sequenced fragments of SSU, D2–D3 of LSU and ITS rDNA obtained approximately 800 bp, 800 bp and 600 bp, respectively. Huang *et al.* (2019) developed a one-step multiplex PCR assay for rapid detection of four stubby root nematode species, *N. minor*, *P. allius*, *P. porosus* and *T. obtusus* (for protocols, see Chapter 15, section 110.2.4).

101.6 Geographical distribution

N. minor is widely distributed throughout the world (Fig. 13.2). It has been reported from the following countries (CABI-ISC, 2020).

Africa: Burkina Faso, Cape Verde, Côte d'Ivoire, Egypt, Kenya, Malawi, Mauritania, Mauritius, Morocco, Nigeria, Senegal, South Africa, Zambia, Zimbabwe.
Asia: Afghanistan, China, Hong Kong, India, Indonesia, Israel, Java, Japan, Pakistan, Philippines, Saudi Arabia, South Korea, Taiwan, Vietnam.
Europe: Italy, Portugal, Spain, Sweden, Switzerland.
North America: Canada, Cuba, Martinique, Nicaragua, Puerto Rico, Trinidad and Tobago, USA.
Oceania: Australia, Fiji, New Zealand, Papua New Guinea.
South America: Argentina, Brazil, Chile, Ecuador, French Guiana, Venezuela.

101.7 Hosts

N. minor is polyphagous and parasitizes more than 100 plant species, including: *Acacia caven*, *Acer* sp., *Actinidia deliciosa*, *Agrostis stolonifera*, *Allium cepa*, *Beta vulgaris*, *Carica papaya*, *Chamaecyparis obtusa*, *Citrus* × *aurantium*, *Citrus limon*, *Citrus reticulata* , *Citrus* × *paradisi*, *Citrus sinensis*, *Citrus* sp., *Cotula* sp., *Cryptomeria japonica*, *Cydonia oblonga*, *Cyphomandra betacea*, *Dahlia* sp., *Elaeagnus angustifolia*, *Glycine max*, *Gossypium hirsutum*, grasses and pastures, *Larix kaempferi*, lawn grasses, *Leptinella* sp., *Lilium* sp., *Lolium multiflorum*, *Lolium perenne*, *Lycopersicon esculentum*, *Malus domestica*, *Medicago sativa*, *Musa acuminata*, *Musa* sp., *Nicotiana tabacum*, *Oryza sativa*, *Passiflora edulis*, *Passiflora mollissima*, *Phaseolus vulgaris*, *Phormium tenax*, *Pinus densiflora*, *Pinus thunbergii*, *Poa annua*, pome trees, *Prunus persica*, *Rheum rhaponticum*, *Rhododendron* sp., *Saccharum officinarum*, *Solanum betaceum*, *Solanum lycopersicum*, *Solanum melongena*, *Solanum tuberosum*, *Trifolium repens*, turfgrass, *Vitis vinifera* and *Zea mays* (CABI-CPC, 2010; Xu and Zhao, 2019).

101.8 Vector

N. minor is a primary vector of two tobraviruses: tobacco rattle virus (TRV, only known from Brazil) and pepper ringspot virus (PepRV) (Brown *et al.*, 1989).

101.9 Symptoms

N. minor is an ectoparasite and feeds on the epidermal cells, outermost cortical cells all along the root surface and root hairs, preferably near the root tip, and causes stubby root disease (Fig. 13.3A). The above-ground symptoms are non-specific. Severe infestation leads to retarded growth, wilted foliage and sensitivity to drought (Birchfield, 1984). In severe infestations, stunting and leaf-tip chlorosis of wheat plants take place under field conditions (Fig. 13.3B). On below-ground parts, the nematode starts feeding at the root tip, pierces the loose outer cells and apparently feeds on the apical meristem cells. Feeding causes damage to the meristem tissue, resulting in cessation of root growth; subsequently, the root becomes short and stunted, then new root tips can also be attacked and an overall suppression occurs in the root system. Seedlings are the most seriously attacked. Sometimes root-tip galls may also be incited by the nematode (Fig. 13.3A). Heavily attacked roots may have very short knob-like lateral branches, each with a stubby growth of branch rootlets (MacGowan, 1983). Parasitized roots generally show little or no necrosis, discoloration, or other nematode injury symptoms (Anwar, 1994). In severe infestations the stubby roots may become discoloured with necrotic lesions (Barker and Lucas, 1984). On cranberry roots, infection by *N. minor* showed as aborted lateral roots, normally with swollen tips and root discoloration (Bird and Jenkins, 1964). The associated virus, TRV, causes a corky ringspot disease on potato and yellow line patterns on leaves of herbaceous ornamentals such as daffodils, gladioli and hyacinths (Zwieg and Hudelson, 2010).

101.10 Biology and life cycle

N. minor is a migratory ectoparasite and four juvenile stages are recognized (Schneider and Ferris, 1987). Males are usually rare and reproduction occurs parthenogenetically (Winfield and Cooke, 1975). Eggs are laid in the single-cell stage. The first-stage juvenile is recognized after 96 h and hatches from the egg after 100–120 h. No moulting is observed in the egg (Bird *et al.*, 1968). On tomato, the life cycle takes 16–17 days at 30°C and 21–22 days at 22°C to complete (Rohde and Jenkins, 1957). For reproduction, the optimum temperatures were 16°C and 24°C on alfalfa (Ayala *et al.*, 1970) and 26°C on eggplant, cotton, wheat and strawberry (Schilt and Cohn, 1975).

Soil temperature and moisture may influence the population dynamics of the nematode. The lowest number of nematodes was recorded at 36°C (Baujard and Martigny, 1995) while the highest was at the maximum level of soil moisture. Soil profiles also affect the population densities. The highest densities occurred at 10–40 cm where soil moisture content was 20–29% in a potato field in Florida (Harrison and Smart, 1975). In a soybean field, the highest population was found at 30 cm depth in sandy soil (Brodie, 1976) and 15–30 cm depth on fine sandy soil with 9% moisture content (McSorley and Dickson, 1990). Collins (1972) found *N. minor* to be most abundant in non-fertilized and non-limed soils with pH less than 5.8.

101.11 Economic importance

N. minor is known to cause stubby root symptoms on a wide variety of plants and severe crop losses was reported in vegetable crops grown in lighter soils in subtropical regions (Perry and Rhoades, 1982b). Netscher (1970) reported 50% reduction in root weight of tomato in Senegal. *N. minor* is considered a major pest of maize (Norton, 1984) and sugarcane (Birchfield, 1984). Greenhouse tests showed substantial damage on 'Thompson Seedless' grape after inoculation with 100 individuals of *N. minor* (Hafez *et al.*, 1981b). Association of the nematode with the transmission of tobraviruses (TRV and PepRV) causes more devastating impacts and the nematode attains major pest status, especially for potatoes and other viral hosts (Davis *et al.*, 2014).

101.12 Movement and means of dispersal

The nematode is an ectoparasite and spreads with irrigation water and soil attached to tools, footwear and machinery. Dissemination can also occur when infested seedlings are transplanted.

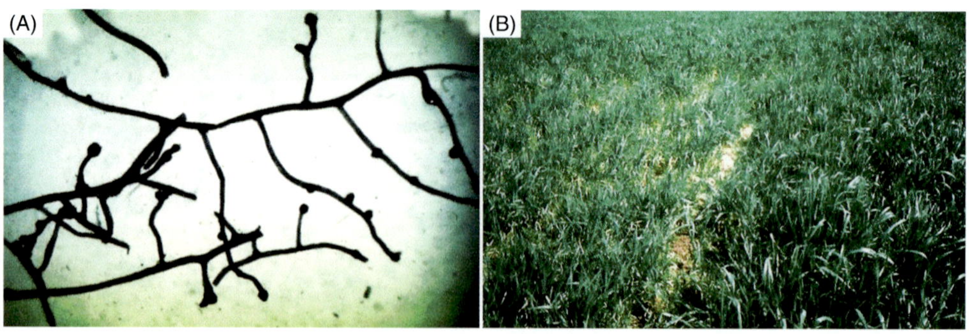

Fig. 13.3. Damage symptoms caused *Nanidorus* spp. on durum wheat. Stubby-root and root tip galling (A); field symptoms with stunting and leaf-tip chlorosis of affected plants (B). [Courtesy Derek J.F. Brown]

101.12.1 Plant parts known to carry the pest in trade/transport

Bulbs, tubers, corms, rhizomes (sucker), growing medium accompanying plants, roots, seedlings and micro-propagated plants.

101.12.2 Plant parts not known to carry the pest in trade/transport

Bark, flowers, inflorescences, cones, calyx, fruits (including pods), leaves, stems (above ground), shoots, trunks, branches, wood and true seeds (including grain).

101.13 Pest risk analysis and looming threat from introduction

Considering the economic damage caused by *N. minor* to vegetables and other crops, and its wide distribution in subtropical agricultural areas, *N. minor* poses a definite threat to other host countries. It is also vector of TRV, which causes additional economic loss. However, this nematode can be excluded if it is detected, because it is an ectoparasite and the primary entry pathway would be soil, which currently cannot be imported with plants. Due to its vector role and restricted distribution, preventive steps should be taken to avoid introduction into new host areas.

101.14 Invasiveness rating

Considering the economic damage and pest risk analysis of *N. minor*, the following invasiveness rating has been assigned.

Climate/host interaction	–	High (3)
Hosts range	–	High (3)
Dispersal potential	–	Medium (2)
Economic impact	–	High (3)
Environmental impact	–	High (3)
Post-entry distribution	–	Medium (–2)
Final rating (score)	–	High (12)

Based on the above invasiveness rating, *N. minor* receives a score of 12 out of 15 and is rated as HIGHLY INVASIVE NEMATODE.

101.15 Region-wise status of invasiveness

N. minor is considered a quarantine pest in many countries mainly because it is a vector of tobravirus. It is already on the Harmful Organisms list for Honduras, Jordan, Peru, South Korea and Taiwan (USDA-PCIT, 2019). The nematode is given a high invasiveness rating for the countries where it is absent.

101.16 Management measures

101.16.1 Cultural control

Due to the polyphagous nature of *N. minor*, crop rotation is generally not recommended. However, *Crotalaria spectabilis* is a non-host of *N. minor* and supports the lowest increase in nematode population when cultivated in the field followed by a fallow (Good *et al.*, 1973). Six weeks of fallow between tomato harvest and use of crotalaria or maize as cover crops were effective in reducing the population of *N. minor* (Brodie and Murphy, 1975). Overman (1964) reported that flooding for 2 weeks followed by 2 weeks of drying decreased nematode populations significantly. Sandy soil is preferred by *N. minor*; hence, growing sugarcane or wheat on soils other than sandy and loamy can significantly reduce the nematode population (Swarup and Sosa-Moss, 1990). Time of planting may also influence the effect of nematodes on the host plants: sugarcane is more susceptible to nematodes when planted in spring (Spaull and Cadet, 1990); and wheat sown in early autumn in sandy soils in Australia and the USA is highly susceptible to *N. minor* (Swarup and Sosa-Moss, 1990). Soil solarization, using clear plastic mulch or a photo-selective black polyethene sheet, reduced populations of *N. minor* (Chellemi *et al.*, 1994). *Asparagus officinalis* is known to be antagonistic to *N. minor*, due to toxic glycoside present in the root (Paxton, 1975).

101.16.2 Chemical control

Application of non-fumigant nematicides proved to be more effective in controlling *N. minor* than fumigant nematicides (Johnson and Chalfant, 1973). Treatment with fensulfothion, carbofuran, ethoprophos and phorate reduced population re-establishment on onion (Rhoades, 1969). Treatment with ethoprophos and fenamiphos on Bermudagrass (Tarjan and Frederick, 1984) and maize (Rhoades, 1985) significantly reduced the population of *N. minor*.

101.16.3 Phytosanitary measures

See Chapter 16, section 115.

102 *Paratrichodorus porosus*

102.1 Common name

Stubby nematode

102.2 Scientific name

Paratrichodorus porosus (Allen, 1957) Siddiqi, 1974

102.3 Synonyms

Atlantadorus porosus (Allen, 1957) Siddiqi, 1980
Paratrichodorus (*Atlantadorus*) *porosus* (Allen, 1957) Siddiqi, 1974

Trichodorus porosus Allen, 1957
Trichodorus bucrius Lordello and Zamith, 1958

102.4 EPPO code

TRIHPS

102.5 Diagnosis

102.5.1 Morphological diagnosis

Morphological descriptions from Hunt (1993) are summarized below. For measurements (after Chau, 1997), see Table 76.

FEMALE: Body cylindroid with blunt extremities, straight to slightly ventrally arcuate after heat relaxation (Fig. 13.4A). Excretory pore located near middle of oesophagus. Cuticle finely striated. Lip region slightly offset, papillae elevated distinctly. Amphids flask-shaped, with oval opening and just behind sensilla sac. Stoma tubular, sclerotized, forming 'collar' or stylet guiding ring in posterior part. Onchiostyle hair-like, solid, slightly arcuate (Fig. 13.4D). Oesophagus non-muscular, expanding posteriorly into large sac-like bulb offset from intestine bearing glands and outlets. Intestine slightly overlapping dorsal side of bulb (Fig. 13.4D). Paired ventromedian pores present close to vulva, one anterior and the other posterior. Vagina short with quadrangular vaginal mass, weakly sclerotized, inconspicuous, triangular-rounded in lateral view (Fig. 13.4F). Anus subterminal in position. Tail hemispherical to obtusely rounded with one pair of lateral pores.

MALE: Very rare. Anterior region similar to female. Single pair of lateral body pores often present at level of ventromedian cervical papilla. Single testis, outstretched. Spicules straight, striated transversely. Gubernaculum linear, slightly enlarged at distal end. Tail conical, short (Fig. 13.4J). Caudal alae well developed, enveloping tail terminus (Fig. 13.4K).

JUVENILE: Four juvenile stages present, morphologically similar to female in most characters except genital tracts.

102.5.2 Molecular diagnosis

See section 101.5.2, above.

102.6 Geographical distribution

P. porosus is distributed throughout the world and is widespread in Australia, Japan, South Africa and the USA (Fig. 13.5). It has been reported from the following countries (CABI-ISC, 2020).

Africa: South Africa.
Asia: China, India, Iran, Japan, South Korea, Sri Lanka, Uzbekistan, Vietnam.
Europe: Italy, Portugal, Russia, United Kingdom.
Oceania: Australia, New Zealand, Papua New Guinea.
North America: Costa Rica, USA.
South America: Brazil, Chile.

102.7 Hosts

P. porosus is highly polyphagous and parasitizes a diverse group of plants, including field crops, fruit trees, ornamentals, forest trees and nursery stock, shrubs, grasses and weeds (CABI-CPC, 2010). An extended list of hosts is given in Table 77.

102.8 Vector

P. porosus is a vector of tobacco rattle virus (TRV) (Brown *et al.*, 1989).

102.9 Symptoms

P. porosus is an ectoparasite and its feeding causes extensive damage to the host's root system, which becomes reduced, stunted and stubby, as with other trichodorids (see section 101.9 for further details). *P. porosus* also causes black-rot disease of Chinese yam (Zeng *et al.*, 2014). Extensive damage was observed on *Camellia* roots (Barriga, 1965), maize and sorghum roots (Chévres-Román *et al.*, 1971). The associated virus, TRV, produces necrotic spots and stripes on the leaves, leaf petioles and stems. Leaves may also become mottled and show yellow blotches. On potato plants, the virus may cause noticeable brown rings on the surface and brown flecking or discoloration inside the potato (Walkinshaw *et al.*, 1961).

102.10 Biology and life cycle

P. porosus is an ectoparasite of roots and is found in soil around the roots but never inside the roots. Males are usually rare and reproduction occurs parthenogenetically. Temperature affects reproduction: the optimum temperature for reproduction on maize was 24°C and the lowest reproduction rates were recorded at 12°C and 35°C (Ayala *et al.*, 1970). In Brazil, the population density of *P. porosus* on sugarcane was directly correlated with rainfall and soil moisture content, usually increasing with rainfall from November onwards and declining at the end of the rainy season in March or April (Gomes Carneiro *et al.*, 1980). *P. porosus* prefers sandy and sandy-loam soils and highest numbers usually occur at depths of 45–60 cm (Carneiro *et al.*, 1982).

102.11 Similarities with other species

P. porosus is morphologically similar to *P. paraporosus* (Khan *et al.*, 1989). However, *P. porosus* is the only species in the genus which has advulval ventromedian body pores. Absence of

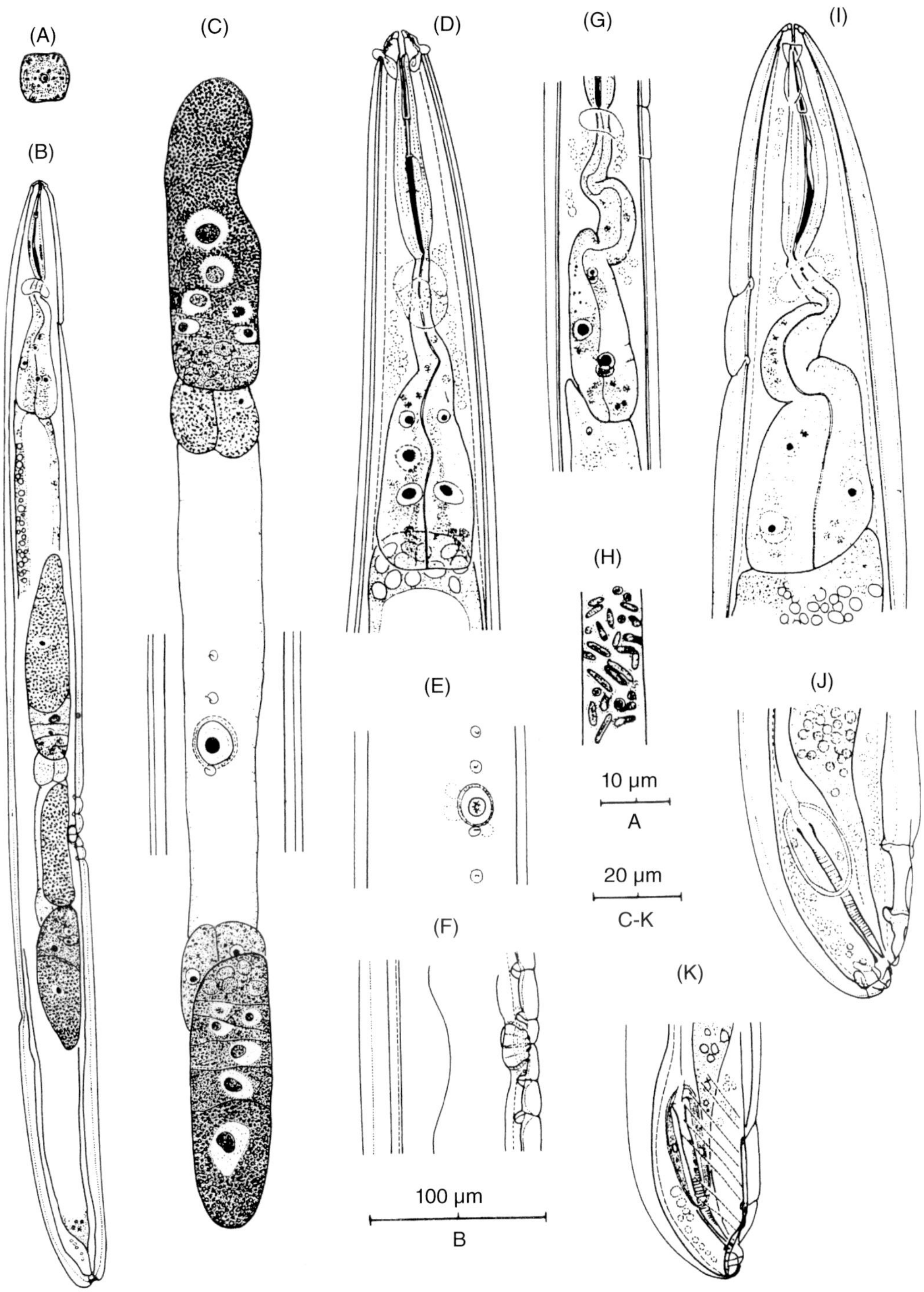

Fig. 13.4. Line drawing of *Paratrichodorus porosus*. *En face* view of female head (A); entire female (B); reproductive system of female, ventral view (C); anterior region of female (D); vulva, ventral view (E); vaginal region (F); oesophageal region male (G); spermatozoa (H); anterior region male (I); copulatory apparatus and tail (J,K). [Reproduced from Hunt, 1993; after Decraemer, 1991; courtesy Taylor & Francis Group LLC]

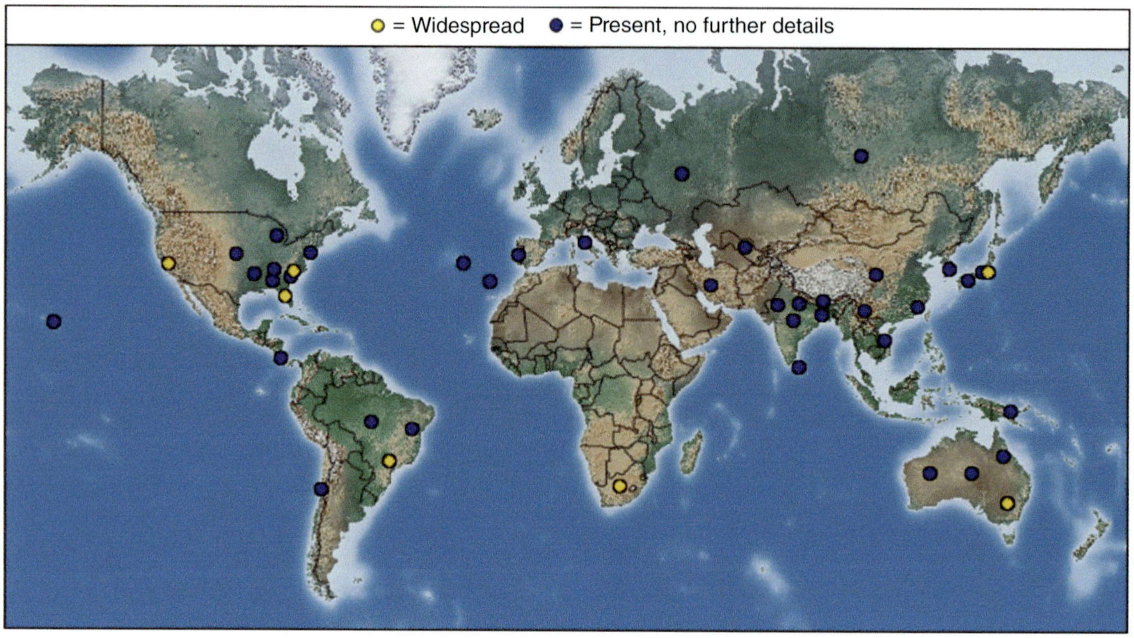

Fig. 13.5. Distribution map of *Paratrichodorus porosus*. [Adapted from CABI-ISC, 2020]

Table 77. List of hosts of *Paratrichodorus porosus* (adapted from CABI-CPC, 2010).

Main hosts
Allium cepa (onion)
Brassica oleracea var. *capitata* (cabbage)
Brassica rapa subsp. *chinensis* (Chinese cabbage)
Camellia
Camellia sinensis (tea)
Citrus jambhiri (rough lemon)
Citrus limon (lemon)
Citrus reticulata (mandarin)
Citrus sinensis (navel orange)
Citrus × *paradisi* (grapefruit)
Dioscorea (yam)
Juglans regia (walnut)
Lycopersicon peruvianum
Musa (banana)
Musa × *paradisiaca* (plantain)
Prunus armeniaca (apricot)
Prunus persica (peach)
Pyrus communis (European pear)
Rubus loganobaccus (loganberry)
Saccharum officinarum (sugarcane)
Sorghum bicolor (sorghum)
Tagetes erecta (African marigold)
Tagetes patula (French marigold)
Vitis vinifera (grapevine)
Zea mays (maize)
Other hosts
Abelmoschus esculentus (okra)
Allium sativum (garlic)
Bambusa vulgaris (common bamboo)
Cucurbita maxima (giant pumpkin)
Eucalyptus (eucalyptus tree)
Ficus carica (fig)
Gardenia
Glycine max (soybean)
Gossypium (cotton)
Lagerstroemia indica (crape myrtle)
Lupinus (lupins)
Malus domestica (apple)
Morus alba (mora)
Olea europaea subsp. *europaea* (olive)
Persea americana (avocado)
Poncirus trifoliata (trifoliate orange)
Prunus domestica (plum)
Prunus mahaleb (mahaleb cherry)
Prunus salicina (Japanese plum)
Rhododendron (azalea)
Rosa (roses)
Wild hosts
Acer palmatum (Japanese maple)
Aphananthe aspera (muku)
Aucuba japonica (Japanese aucuba)
Bambusa tuldoides (punting-pole bamboo)
Butia
Callicarpa japonica
Carpinus tschonoskii
Castanea crenata (Japanese chestnut)
Castanopsis cuspidata (chinkapin)
Casuarina (beefwood)
Casuarina equisetifolia (casuarina)

Table 77. Continued

Celtis sinensis (Chinese elm)
Cercidiphyllum japonicum (katsura)
Cinnamomum camphora (camphor laurel)
Cleyera japonica
Cryptomeria japonica (Japanese cedar)
Daphniphyllum macropodum
Fatsia japonica (Japanese aralia)
Ginkgo biloba (kew tree)
Idesia polycarpa (idesia)
Ilex crenata (Japanese holly)
Ilex integra (mochi)
Ilex rotunda (kurogane holly)
Juniperus (junipers)
Magnolia hypoleuca (Japanese big-leaved magnolia)
Musa acuminata (wild banana)
Neolitsea sericea
Pinopsida (conifers)
Pinus densiflora (Japanese umbrella pine)
Pinus pseudostrobus var. *pseudostrobus*
Pinus thunbergii (Japanese black pine)
Pittosporum tobira (Japanese pittosporum)
Pyrus amygdaliformis
Quercus acutissima (sawtooth oak)
Quercus glandulifera (glandbearing oak)
Quercus salicina
Rhododendron indicum (Satsuki azalea)
Taxus cuspidata (Japanese yew)
Ternstroemia gymnanthera
Viburnum odoratissimum

spermatheca and the rarity of males in *P. porosus* are other characters to differentiate it from *P. paraporosus*.

102.12 Economic importance

P. porosus is an important pest of banana, camellia, citrus, grapes, maize, pear, sorghum, sugarcane and yam.

102.13 Movement and means of dispersal

See section 101.12, above.

102.14 Pest risk analysis and looming threat from introduction

Risk analysis is similar to that for *N. minor*; see section 101.13, above.

102.15 Invasiveness rating

Considering the economic impact and pest risk analysis of *P. porosus*, the following invasiveness rating has been assigned.

Climate/host interaction	–	High (3)
Hosts range	–	High (3)
Dispersal potential	–	Medium (2)
Economic impact	–	Medium (2)
Environmental impact	–	Medium (2)
Post-entry distribution	–	Medium (–2)
Final rating (score)	–	Medium (10)

Based on the above invasiveness rating, *P. porosus* receives a score of 10 out of 15 and is rated as MEDIUM INVASIVE NEMATODE.

102.16 Region-wise status of invasiveness

P. porosus is considered as a quarantine pest, mainly because of its tobravirus transmission capability. It is already on the Harmful Organisms list for Colombia, Guatemala, Honduras, Mexico, Nicaragua, Panama, Peru, Taiwan and Thailand (USDA-PCIT, 2019). The nematode is given a medium invasiveness rating for the countries where it is absent.

102.17 Management measures

See section 101.16, above.

103 *Trichodorus viruliferus*

103.1 Common name

Stubby nematode

103.2 Scientific name

Trichodorus viruliferus Hooper, 1963

103.3 EPPO code

TRIHVI

103.4 Diagnosis

103.4.1 Morphological diagnosis

Morphological descriptions from Hunt (1993) are summarized below. For measurements (after Hooper, 1976) see Table 76.

FEMALE: Body cylindrical, tapering towards anterior end with slight ventral curvature when heat relaxed (Fig. 13.6G).

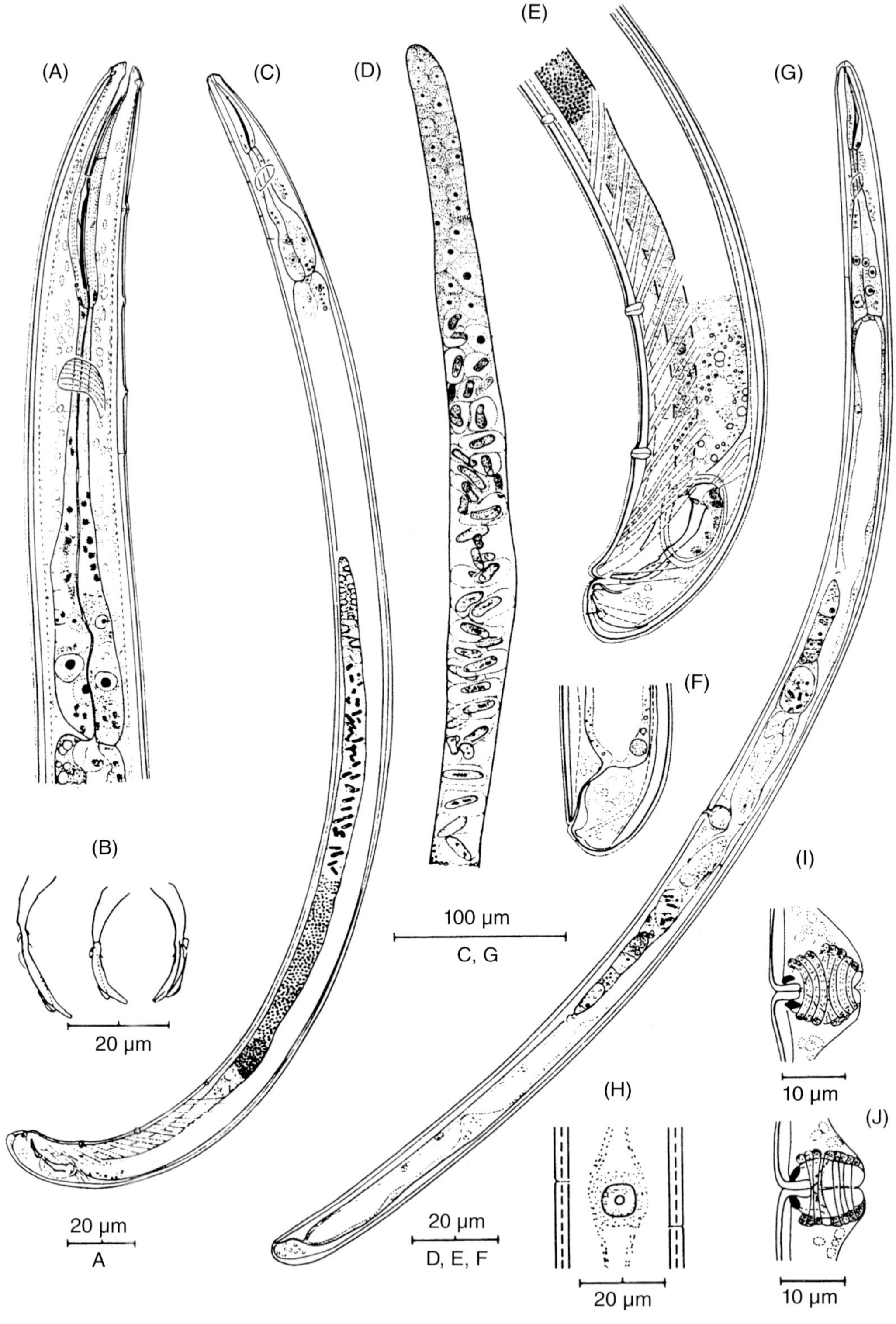

Fig. 13.6. Line drawing of *Trichodorus viruliferus*. Male anterior region (A); male copulatory apparatus (B); entire male (C); testis (D); male posterior region (E); female tail (F); entire female (G); vulva ventral view (H); vaginal region (I,J). [Reproduced from Hunt, 1993; after Decraemer, 1991; courtesy Taylor & Francis Group LLC]

Excretory pore just over one onchiostyle length posterior to onchiostyle base and about opposite middle of basal oesophageal bulb. One pore present on each side of body in vulval region. Genital tract amphidelphic, reflexed. Vulva a median, pore-shaped opening in ventral view. Spermatheca present at junction of oviduct and uterus. Refractive thickening at junction of vulva and vagina appears as ring in ventral view; and as two small rather indistinct oval-shaped pieces in lateral view directed towards vulva at angle of about 45°, their ends nearer vulva appearing slightly more pointed. Vagina as wide as long, rhomboid shape, surrounded by prominent ring muscle in innermost half (Fig. 13.6I,J).

MALE: Usually common. Body ventrally curved in posterior region after heat relaxation (Fig. 13.6E). Onchiostyle with guiding ring around one-third of its length. Oesophagus gradually expands into spathulate basal bulb that adjoins intestine, sometimes very slight ventromedian overlap (Fig. 13.6A). Three ventromedian cervical papillae situated anterior to excretory pore. Three ventromedian preanal copulatory supplements present at tail end. Spicule paired, without transverse striation, ventrally curved, proximal part wide and gradually tapering around halfway along, with distinct narrow bend (Fig. 13.6D).

103.4.2 Molecular diagnosis

Kumari and Subbotin (2012) characterized various members of Trichodoridae, including *T. viruliferus*, *N. minor* and *P. porosus* collected from different regions. The PCR assays of 18S rRNA, D2–D3 expansion fragments of 28S rRNA and ITS1 rRNA were performed using the following primers.

For 18S rRNA (Blaxter *et al.*, 1998)

SSU_F_04 Forward (5′-GCTTGTCTCAAAGATTAAGCC-3′)
SSU_R_09 Reverse (5′-AGCTGGAATTACCGCGGCTG-3′)
SSU_F_22 Forward (5′-TCCAAGGAAGGCAGCAGGC-3′)
SSU_R_13 Reverse (5′-GGGCATCACAGACCTGTTA-3′)
SSU_F_23 Forward (5′-ATTCCGATAACGAGCGAGA-3′)
SSU_R_81 Reverse (5′-TGATCCWKCYGCAGGTTCAC-3′)
1091-F Forward (5′-AGGAATTGACGGAAGGGCAC-3′) (Duarte *et al.*, 2010)
1671-R Reverse (5′-TCCTCTAAGTAAATCCCATTGG-3′)

For ITS1 rRNA (Boutsika *et al.*, 2004)

BL18 Forward (5′-CCCGTCGMTACTACCGATT-3′)
5818 Reverse (5′-ACGARCCGAGTGATCCAC-3′)
ITSA Forward (5′-ATCGATGAAGAACGCAGC-3′)
PXb481 Reverse (5′-TTTCACTCGCCGTTACTAAGG-3′) (Vrain *et al.*, 1992)

For 28S rRNA (Nunn, 1992)

D2A Forward (5′-ACAAGTACCGTGAGGGAAAGTTG-3′)
D3B Reverse (5′-TCG GAAGGAACCAGCTACTA-3′)

Based on the above study, ten trichodorid species were distinguished within collected samples: *Nanidorus minor*, *N. renifer*, *Paratrichodorus pachydermus*, *P. porosus*, *Trichodorus pakistanensis*, *T. primitivus*, *T. similis*, *T. sparsus*, *T. variopapillatus* and *T. viruliferus*.

103.5 Geographical distribution

T. viruliferus has limited distribution mainly in temperate Europe and the USA. The nematode has been reported from the following countries (CABI-ISC, 2020).

Europe: Belgium, Bulgaria, France, Germany, Hungary, Italy, Netherlands, Poland, Slovakia, Spain, Sweden, Switzerland, United Kingdom.
North America: USA.

103.6 Hosts

T. viruliferus is polyphagous and parasitizes a large number of plants, including apple (*Malus pumila*), barley (*Hordeum vulgare*), maize (*Zea mays*), pea (*Pisum sativum*), potato (*Solanum tuberosum*), rye (*Secale cereale*), sugar beet (*Beta vulgaris* var. *saccharifera*) and wheat (*Triticum aestivum*) (CABI-ISC, 2020).

103.7 Vector

T. viruliferus is known to transmit tobacco rattle virus (TRV, only known from Brazil) and pepper ringspot virus (PepRV) in Europe (Brown *et al.*, 1989).

103.8 Symptoms

Typical of the genus and produces stubby root symptom. The symptoms of the associated virus, TRV, causes a corky ringspot disease on potato and other host crops (Zwieg and Hudelson, 2010).

103.9 Biology and life cycle

The life cycle of *T. viruliferus* from egg to adult on apple roots was completed in about 45 days at 15–20°C in England (Pitcher and McNamara, 1970). *T. viruliferus* occurs most frequently in sandy or sandy-loam soils (Rau, 1976) with preferred distribution at depths of 10–40cm, with some specimens present down to 60 cm deep (Decraemer, 1995). Pitcher (1967) observed swarming behaviour of *T. viruliferus* on apple roots in an underground laboratory.

103.10 Economic importance

T. viruliferus is an important pest of apple, barley, maize, rye, pea, wheat, potato and sugar beet. However, information on estimated yield loss is lacking.

103.11 Movement and means of dispersal

See section 101.12, above.

103.12 Pest risk analysis and looming threat from introduction

Risk analysis is similar to that of *N. minor* (see section 101.13, above).

103.13 Invasiveness rating

Considering the probable consequences of introduction and pest risk analysis of *T. viruliferus*, the following invasiveness rating has been assigned.

Climate/host interaction	–	Low (1)
Hosts range	–	Medium (2)
Dispersal potential	–	Medium (2)
Economic impact	–	Medium (2)
Environmental impact	–	Medium (2)
Post-entry distribution	–	Not known (0)
Final rating (score)	–	Medium (9)

Based on the above invasiveness rating, *T. viruliferus* receives a score of 9 out of 15 and is rated as MEDIUM INVASIVE NEMATODE.

103.14 Region-wise status of invasiveness

T. viruliferus is a nematode of quarantine significance, mainly because of its limited distribution and virus transmission capability of two tobraviruses. The nematode is given a medium invasiveness rating for the countries where it is absent.

103.15 Management measures

See section 101.16, above.

14

Tylenchulidae: *Tylenchulus* Species

Tylenchulus Spp.

The family Tylenchulidae was proposed by Raski (1957) to include the subfamily Tylenchulinae that was created by Skarbilovich (1947) with the genera *Tylenchulus* (Cobb, 1913) and *Trophotylenchulus* (Raski, 1957). The family Tylenchulidae has been revised by several workers (Siddiqi, 2000; Decraemer and Hunt, 2006; Ghaderi *et al.*, 2016). For many years, *Tylenchulus semipenetrans* was noted as the only species of this genus. Currently, *Tylenchulus* contains five nominal species with *T. semipenetrans* as type species (Tanha Maafi *et al.*, 2012; Ghaderi *et al.*, 2016).

Tylenchulus furcus Van den Berg and Spaull, 1982
T. graminis Inserra, Vovlas, O'Bannon and Esser,1988
T. musicola Tanha Maafi *et al.*, 2012
T. palustris Inserra, Vovlas, O'Bannon and Esser, 1988
T. semipenetrans Cobb, 1913

T. semipenetrans, commonly known as citrus root nematode, is an important plant-parasitic nematode of citrus, widely distributed throughout the world in citrus orchards and causing slow decline disease (Siddiqi, 2000). It attacks most citrus relatives from the Rutaceae family, including *Citrus* species, *Poncirus trifoliata* and others (Tanha Maafi *et al.*, 2012). Some non-rutaceous plants are also described as good hosts, such as grape, olive and persimmon (Inserra *et al.*, 1988).

Systematic position of *Tylenchulus* species

	Classical (after Siddiqi, 2000)	Molecular (after Decraemer and Hunt, 2013)
Phylum	Nematoda	Nematoda
Class	Secernentea	Chromadorea
Order	Tylenchida	Rhabditida
Suborder	Criconematina	Tylenchina
Infra-order	–	Tylenchomorpha
Superfamily	Tylenchuloidea	Criconematoidea
Family	Tylenchulidae	Tylenchulidae
Subfamily	Tylenchulinae	Tylenchulinae
Genus	*Tylenchulus*	*Tylenchulus*

104 *Tylenchulus semipenetrans*

104.1 Common name

Citrus root nematode

104.2 Scientific name

Tylenchulus semipenetrans Cobb, 1913

104.3 EPPO code

TYLESE

104.4 Diagnosis

104.4.1 Morphological diagnosis

Morphological descriptions of *Tylenchulus semipenetrans* after Siddiqi (1974b, 2000) are summarized below. For morphometric measurements (Inserra *et al.*, 1988), see Table 78.

ADULT FEMALE: Sedentary, elongate-obese, ventrally arcuate, neck extended, enlarging mostly on dorsal side behind median bulb, tapering postvulval part elongated (Fig. 14.1G). Stylet with prominent knobs. Excretory pore at level of vulva about 68–85% body length from head end. Excretory system well marked, produces gelatinous matrix. Basal bulb offset from intestine. Vulva a transverse slit without postvulval uterine sac. Spermatheca present. Ovary with one to two flexures or coiled. Anus obscure. Tail tapering, terminus with a peg or rounded.
IMMATURE FEMALE: Body slender, straight to arcuate. Stylet and oesophagus similar to mature female (Fig. 14.1E). Ovary with few oocytes, immature.
MALE: Body slender, relatively short, slightly arcuate to straight when heat killed (Fig. 14.1I). Stylet and oesophagus degenerate (Fig. 14.1A). Excretory pore situated about 53–60% of body length. Spicules slender, arcuate. Gubernaculum simple, fixed. Bursa not present. Tail elongate-conoid with a peg or rounded terminus (Fig. 14.1B,C).

DOI: 10.1079/9781789247367.0015

Second-stage juvenile: Body slender, straight to slightly arcuate. Lateral field with two longitudinal lines. Stylet knobs rounded. Median oesophageal bulb round to fusiform with elongate valve-plates, isthmus slender, basal bulb offset from intestine (Fig. 14.1F). Tail conoid, elongate.

104.4.2 Molecular diagnosis

Molecular characterization of *T. semipenetrans* was carried out by PCR-ITS-RFLP profile (Wang *et al.*, 2004; Park *et al.*, 2009). Rashidifard *et al.* (2015) also characterized the D2–D3 segment of 28S rDNA to identify *T. semipenetrans* from citrus root samples collected from south-eastern Iran. Lin *et al.* (2016) and Song *et al.* (2017) developed a loop-mediated isothermal amplification (LAMP) assay for quick identification of *T. semipenetrans*. Song *et al.* (2017) directly extracted the DNA from soil and used the primers 18S/26S to obtain the rDNA-ITS sequence of *T. semipenetrans* (Vrain *et al.*, 1992).

18S TTGATTACGTCCCTGCCCTTT (amplicon size 21 bp)
26S TTTCACTCGCCGTTACTAAGG

The LAMP reaction of the rDNA-ITS1 sequence was performed using the following primers according to the protocol described by Notomi *et al.* (2000) and Nagamine *et al.* (2002), with some amendments.

For *T. semipenetrans*-specific Ts-LAMP (Song *et al.*, 2017)
Ts-F3 CCAGGTTGAGCAGAGTCCTT (amplicon size 20 bp)

Table 78. Morphometric measurements of *Trophotylenchulus* and *Tylenchulus* species.

Morphometric characters		Range, numerous populations	
		Tro. piperis (after Mohandas *et al.*, 1985)	*Tyl. semipenetrans* (after Inserra *et al.*, 1988)[a]
Adult females	Length (µm)	254–325	312–465
	Stylet length (µm)	13–14	11.0–12.5
	a	4.5–7.9	3.5–6.4
	b	2.5–3.1	2.4–4.3
	c	12–15	–
	c'	2.2–4.6	–
	DGO	9	–
	V	72–80	–
	Excretory pore from anterior end (µm)	110–134	77.7–89.8
	Maximum body width (µm)	35–58	68–100
	Tail length (µm)	18–32	
Immature females	Length (µm)	–	25–360
	Stylet length (µm)	–	12–15
	a	–	15–20
	b	–	2.15–3.00
	Excretory pore from anterior end (µm)	–	77–84
Males	Excretory pore from anterior end (µm)	349–392	346–380
	Stylet length (µm)	7–11	9.0–10.5
	a	26–44	28.4–34.7
	b	3.5–4.6	2.9–3.7
	c	4.0–8.4	7.7–10.1
	c'	4.0–6.4	–
	Excretory pore from anterior end (µm)	–	54.0–68.5
	Spicule (µm)	13–16	15–18
	Gubernaculum (µm)	3–4	3-–
	Tail length (µm)	47–99	34.5–44.5
Second-stage juvenile	Length (µm)	260–329	333.5–384.5
	Stylet length (µm)	12–13	12–13
	a	21–32	24.8–29
	b	2.3–3.1	3.0–3.7
	c	6.4–13.5	5.4–5.9
	c'	4.5–5.7	–
	DGO	4–5	4–6
	Excretory pore from anterior end (µm)	108–119	52.4–57.6
	Tail length (µm)	22–46	55–70

[a]Immature female after Van Gundy (1958); second-stage juvenile measurement from female.

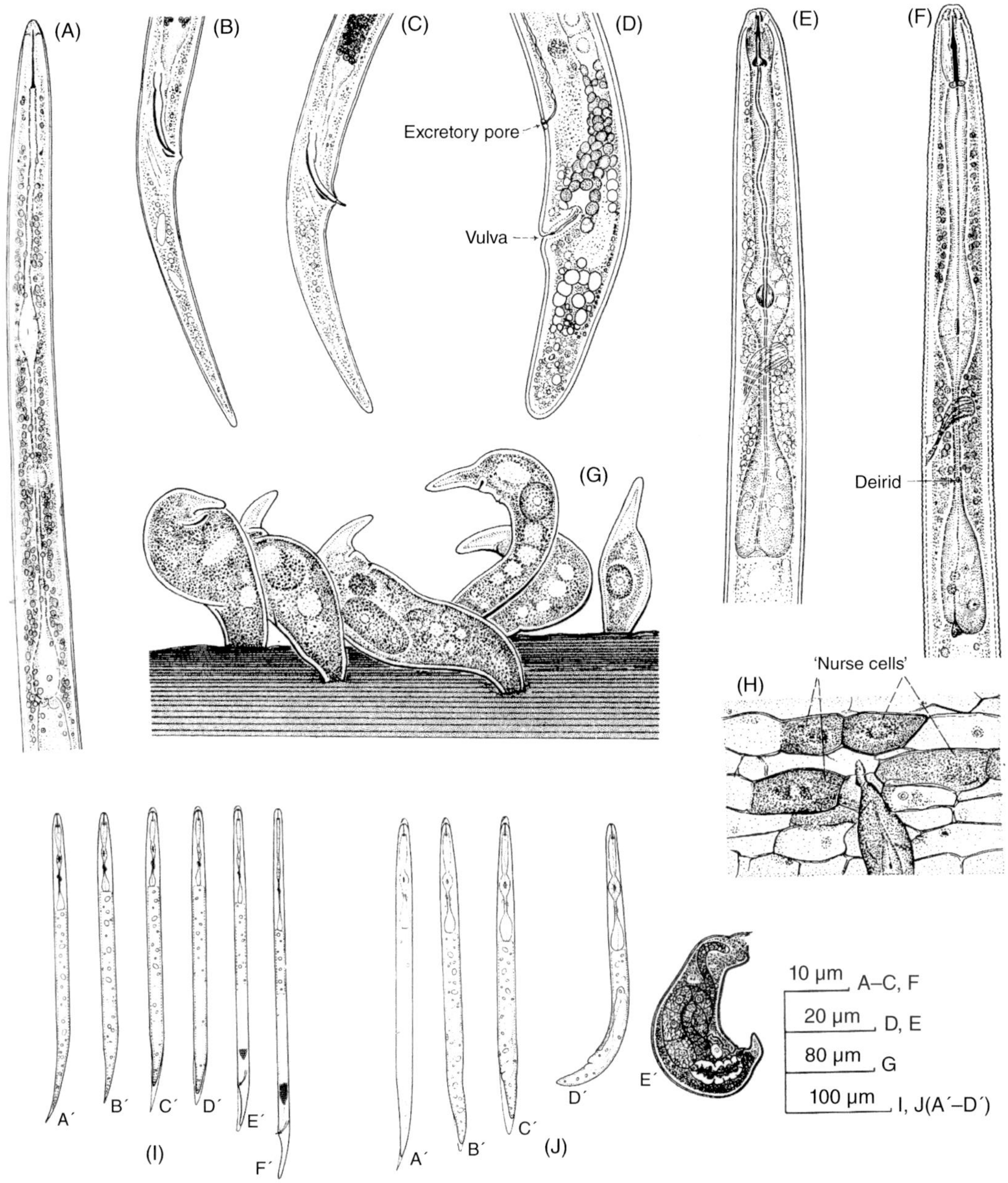

Fig. 14.1 Line drawing of *Tylenchulus semipenetrans.* Oesophageal region (A); tail region of male (B,C); posterior and anterior region of immature female (D,E); oesophageal region of female second-stage juvenile (F); female on citrus root (G); feeding site of female in cortex of a citrus root with nurse cells (H); developmental stages of male (I); developmental stages of female (J). [Reproduced from Siddiqi, 2000; D,E,G redrawn by M.R. Siddiqi from N.A. Cobb; I,J from Van Gundy; courtesy CAB International]

Ts-B3 GCCGTTTTCGCCTGTAGTC (amplicon size 19 bp)
Ts-FIP CAGATGCCAGAAGCAGCGACAGGCACATCGGG TGGAGAG (39 bp)
Ts-BIPTGGTCATACTTCCTCTGCCGCTAGGTAAGAGCC GAGAAGGA (41 bp)
Ts-LB GAGGAAGGGATACTGAGCTT (20 bp)

T. semipenetrans-specific primers Ts-F/Ts-R developed by Liu *et al.* (2011) were used to confirm the specificity of the LAMP assay:

Ts-F CTGCTGCTGGATCATTAC (amplicon size 18 bp)
Ts-R ATCTCGACTGAGTTCAGGT

Tanha Maafi *et al.* (2012) also designed *T. semipenetrans* species-specific primers and characterized ITS region and D2–D3 expansion region of 28S rRNA genes.

For ITS rRNA (Tanha Maafi *et al.*, 2003) (amplicon length 809–841 bp)

TW81 5′-GTT TCC GTAGGTGAACCTGC-3′
AB28 5′-ATATGCTTAAGTTCAGCGGGT-3′

For D2–D3 of 28S rRNA (Subbotin *et al.*, 2006) (amplicon length 774–777 bp)

D2A 5′-ACAAGTACCGTGAGGGAAAGTTG-3′
D3B 5′-TCGGAAGGAACCAGCTACTA-3′

Tylenchulus-specific primers for D2–D3 of 28S rRNA (amplicon length 361–364 bp)

D2A 5′-ACAAGTACCGTGAGGGAAAGTTG-3′
Tyl-specific 5′-TAGGYAAAGCKTGCTRTTRCCA -3′

Semipenetrans-specific primers ITS rRNA (amplicon length 113 bp)

TW81 5′-GTT TCC GTAGGTGAACCTGC-3′
Semipenetrans-specific 5′-GGACTCTGCTCAACCTGG-TAGA-3′

All these primer groupings were effectively tested in a multiplex PCR. The newly developed species-specific primer for *T. semipenetrans* is able to detect a single second-stage juvenile from a sample. The phylogenetic analysis revealed no grouping amongst *T. semipenetrans* with respect to location or host plant.

104.5 Geographical distribution

T. semipenetrans is distributed throughout the world where citrus is cultivated (Fig. 14.2). The nematode is widespread in the USA (Heald and O'Bannon, 1987), Iran (Tanha Maafi and Damadzadeh, 2008), China (Zhu *et al.*, 1992), Spain (Sorribas *et al.*, 2008), Pakistan (Ahmed and Khan, 1999) and Egypt (Abd-Elgawad *et al.*, 2016). In India, the occurrence of this nematode was recorded for the first time in 1961 (Siddiqi, 1964a). *T. semipenetrans* is known to occur in the following countries (CABI-ISC, 2020).

Africa: Algeria, Central African Republic, Congo (Democratic Republic of), Congo (Republic of), Côte d'Ivoire, Egypt, Ethiopia, Guinea, Kenya, Libya, Malawi, Morocco, Mozambique, Nigeria, Senegal, South Africa, Sudan, Tanzania, Tunisia, Uganda, Zimbabwe.
Asia: Afghanistan, Azerbaijan, Bahrain, Brunei, China, Georgia, India, Indonesia, Iran, Iraq, Israel, Japan, Jordan, Lebanon, Malaysia, Oman, Pakistan, Philippines, Saudi Arabia, South Korea, Sri Lanka, Syria, Taiwan, Thailand, Turkey, Uzbekistan, Vietnam, Yemen.
Europe: Albania, Cyprus, France, Greece, Italy, Malta, Portugal, Spain.
North America: Belize, Bermuda, Costa Rica, Cuba, Dominica, El Salvador, Grenada, Guadeloupe, Honduras, Jamaica, Martinique, Mexico, Panama, Puerto Rico, Saint Lucia, Saint Vincent and the Grenadines, Trinidad and Tobago, USA.
Oceania: Australia, Cook Islands, Fiji, French Polynesia, New Caledonia, Niue, Papua New Guinea, Samoa, Tonga.
South America: Argentina, Bolivia, Brazil, Chile, Colombia, Peru, Uruguay, Venezuela.

104.6 Hosts

T. semipenetrans has a narrow host range with citrus as primary hosts. Besides citrus, it may infect grapes (*Vitis vinifera*), lilac (*Syringia vulgaris*), olive (*Olea europeae*) and palmetto (*Sabal palmetto*). An extended list of hosts is given in Table 79.

Four biotypes or physiological races have been identified in *T. semipenetrans* populations: Citrus, Poncirus, Mediterranean and Grass (Inserra *et al.*, 1980; Gottlieb *et al.*, 1986).

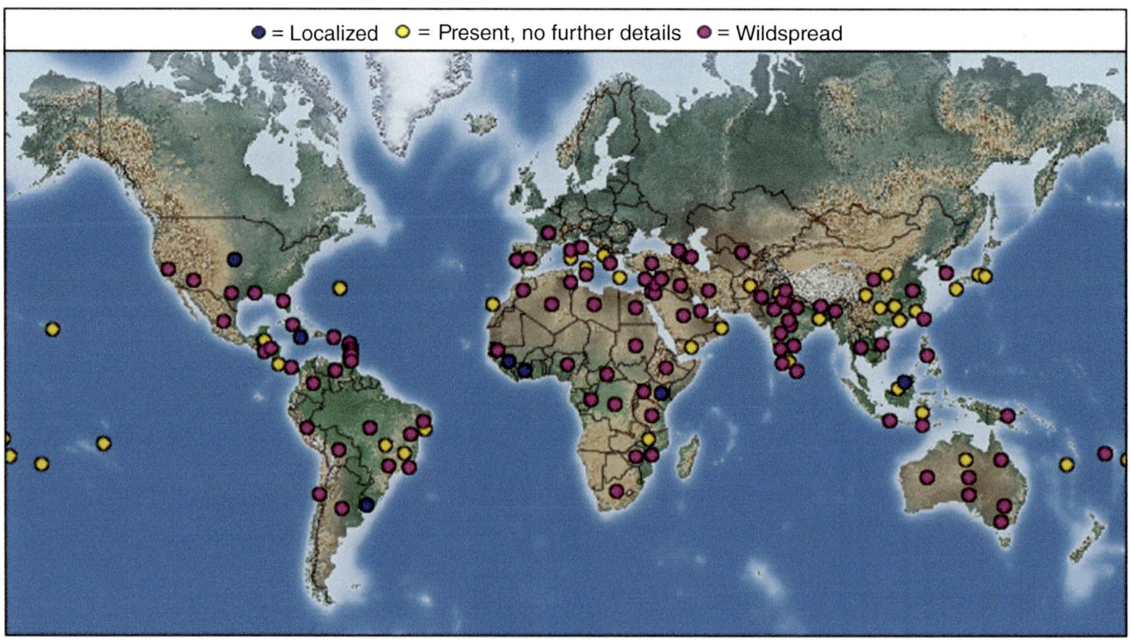

Fig. 14.2. Distribution map of *Tylenchulus semipenetrans*. [Adapted from CABI-ISC, 2020]

104.7 Symptoms

The symptoms induced by citrus nematode are non-specific and difficult to diagnose, and depend on overall orchard maintenance and age. Symptom development is very slow, following the build-up of the nematode population (Khan, 2008). There is a reduction in terminal growth, very slowly followed by reduced vigour of plants. The foliage become sparse, often dull grey-green or bronze-green, upright, coupled with reduced size and branches with dieback symptoms. As a result of poor emergence and slow growth of leaves, the premature terminal branches become exposed and this symptom can be recognized a few metres away from the infected tree (Duncan and Cohn, 1990). Fruit numbers and their size are reduced. The debilitating effects on the plant as a whole are referred as 'slow decline'. Plants may wilt during midday and respond poorly to irrigation or fertilization. In highly saline conditions, there is excessive accumulation of sodium in the leaves. The feeder roots of heavily infected plants are fewer, shortened and thicker (Fig. 14.3A) giving a dirty or encrusted appearance due to soil particles adhering to the gelatinous egg matrix on the root surface around the female's body (Heald and O'Bannon, 1987). The cortical cells of heavily infected roots may slough off and finally result in root death. Damage threshold levels of 20 juveniles/g of soil or 300 females/g of root have been recorded (Duncan and Cohn, 1990).

104.8 Biology and life cycle

The life cycle of *T. semipenetrans* consists of six development stages, starting with the egg, then four juvenile stages (J_1 to J_4) and finally adults (male and female). Around 100 eggs are laid by a female in the gelatinous matrix called the egg mass. This matrix is secreted by the excretory gland present in adult females. The development of eggs deposited in the egg mass proceeds and the J_1 stage develops within the egg; it undergoes first moult, resulting in the J_2 stage within the egg. The J_2 juveniles hatch in soil from the eggs. They are of two types: some are shorter and wider and others longer and slender. The shorter and wider J_2 develop into males. Such juveniles undergo three subsequent moults within 7–10 days without feeding and adult males develop, with weak stylet and degenerate oesophagus. They do not feed on the roots and thus cause no injury to the host. The longer and slender J_2 develop into females and infect the roots within 14 days. If the host and environment are unfavourable, J_2 juveniles survive in the soil for several months to years (Duncan and Cohn, 1990).

After infecting the roots, the longer and slender J_2 nematodes start feeding ectoparasitically on the epidermis and undergo three moults, resulting in the development of immature females. At this stage the last cuticle is moulted in the least time (fourth moulting) and the young female develops within 21 days. Within a week, the anterior part of the female's body penetrates deep into the cortex towards the pericycle and induces a highly specialized 'nurse cells' system in the cortex on which the female nematode feeds. The feeding of females on these nurse cells adversely affects the host's physiology, resulting in 'slow decline' disease (Shokoohi and Duncan, 2005). After the establishment of nurse cells, the female becomes sedentary (Fig. 14.3B). The posterior three-fourths of the body, remaining outside the root and becomes obese (enlarged and swollen). The feeding on nurse cells stimulates the reproductive phase and the female starts laying eggs without mating with a male. These eggs are unfertilized even though the population also consists of males. Reproduction is parthenogenetic. The life cycle from egg to egg is completed within 6–8 weeks at 24–26°C (Van Gundy, 1958).

Table 79. List of hosts affected by *Tylenchulus semipenetrans* (adapted from CABI-ISC, 2020).

Main hosts
Citrus
Cydonia oblonga (quince)
Diospyros (Malabar ebony)
Olea europaea subsp. *europaea* (European olive)
Philadelphus coronarius (mock orange)
Poncirus trifoliata (trifoliate orange)
Vitis vinifera (grapevine)
Other hosts
Andropogan rhizomatus (Grass)
Citrus aurantiifolia (lime)
Citrus reticulata (mandarin)
Citrus sinensis (navel orange)
Justicia adhatoda (Malabar nut)
Punica granatum (pomegranate)
Syringa vulgaris (lilac)
Toona ciliata (toon)

104.9 Similarities to other species/conditions

Morphologically *T. semipenetrans* resembles the other four species of the genus: *T. furcus*, *T. graminis*. *T. musicola* and *T. palustris* (Ghaderi *et al.*, 2016).

104.10 Economic importance

T. semipenetrans is a ubiquitous plant-parasitic nematode in citrus-growing countries and is associated with 90% infestation in the USA (Heald and O'Bannon, 1987), 89% soil samples in Iran (Tanha Maafi and Damadzadeh, 2008), 94.7% of groves in China (Zhu *et al.*, 1992), 77% of orchards in northeastern Spain (Sorribas *et al.*, 2008) and 98.4% of citrus orchards in Pakistan (Ahmed and Khan, 1999). Annual worldwide losses in citrus due to *T. semipenetrans* are estimated at 15–35% by different workers (Willers, 1979; Duncan *et al.*, 1995). In advanced stages of citrus decline, the losses may approach 50% or even more (Heald and O'Bannon, 1987). Besides citrus, *T. semipenetrans* is a mild to moderate pathogen of grape, mainly in California (USA) and Australia. No information is available on economic damage to other crops.

104.11 Movement and means of dispersal

The dispersal of *T. semipenetrans* is facilitated primarily via infected planting material. The nematodes can also spread by the movement

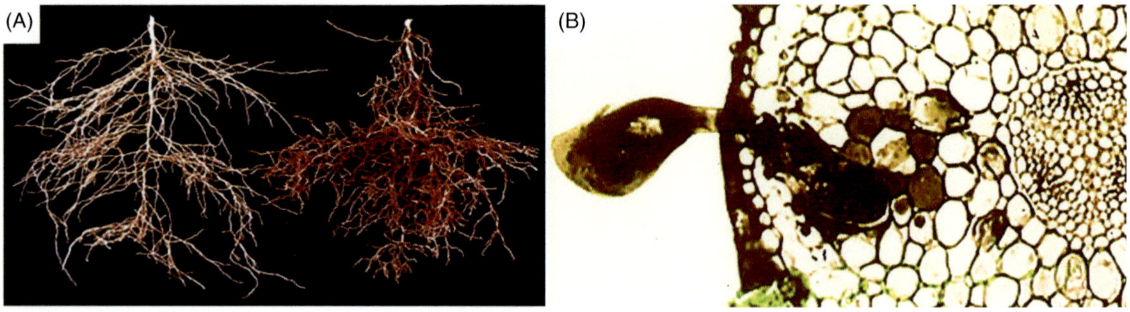

Fig. 14.3. Symptoms of *T. semipenetrans* on citrus fibrous roots infected, left and non-infected, right with the nematode (A); feeding site of female *T. semipenetrans* infecting citrus with visible nurse cells (B). [Courtesy Larry Duncan and Renato Inserra]

of contaminated agricultural machinery, farm implements and nematode-infected plant residues. Its semi-endoparasitic habits also allow the nematode to be transported long distances inside infected roots as eggs, juveniles and adults.

104.11.1 Plant parts known to carry the nematode in trade/transport

Roots, under ground plant parts, growing medium accompanying plants, seedlings and micro-propagated plants.

104.11.2 Plant parts not known to carry the nematode in trade/transport

Bark, flowers, inflorescences, cones, calyx, fruits (including pods), leaves, stems (above ground), shoots, trunks, branches, wood and true seeds (including grain).

104.12 Pest risk analysis and looming threat from introduction

T. semipenetrans is ubiquitous, distributed in citrus-growing regions of the world. The risk of introducing the nematode into new areas is high, particularly in regions where citrus plants are newly planted, as the nematode has only a few alternate native or cultivated hosts. *T. semipenetrans* can be transported on infected planting stock and infested soil and machinery. Strict phytosanitary measures should be implemented to prevent the spread of the nematode from an infested area to new areas.

104.13 Invasiveness rating

Considering the economic damage caused by *T. semipenetrans* and pest risk analysis, the following invasiveness rating has been assigned.

Climate/host interaction	–	High (3)
Hosts range	–	Medium (2)
Dispersal potential	–	High (3)
Economic impact	–	High (3)
Environmental impact	–	High (3)
Post-entry distribution	–	Medium (–2)
Final rating (score)	–	Medium (12)

Based on the above invasiveness rating, *T. semipenetrans* receives a score of 12 out of 15 and is rated as HIGHLY INVASIVE NEMATODE.

104.14 Region-wise status of invasiveness

T. semipenetrans is widely distributed in the world's tropical and subtropical regions but is not common in the temperate regions of Europe. Considering the economic damage caused by *T. semipenetrans* to citrus crops and possible damage to grape, a high invasiveness rating has been given to all the countries where the nematode is absent.

104.15 Management measures

Once *T. semipenetrans* has established in a citrus orchard it is virtually impossible to control; therefore, greater emphasis has to be given to exclusion and management of its populations below the threshold level. Management is crucial in the sense that the nematode could survive for more than 7 years after an old infested orchard has been removed.

104.15.1 Prevention and exclusion

Planting stock should be obtained from nurseries that are free from citrus nematode infestation. Nursery orchards should be monitored continuously, through: (i) regular soil sampling; (ii) avoiding run-off water from any adjacent infested orchard or nursery reaching a newly raised orchard; and (iii) avoiding the use of equipment, farm implements, footwear, etc. contaminated with nematode-infested soil from already infested nurseries or old orchards. These must be disinfested before being used in newly raised orchards free from this nematode.

104.15.2 Cultural control

Since damage leading to root dysfunction is more pronounced when plants are under stress, orchards should be properly irrigated and fertilized. Crop rotation with annual plants for 1–3 years before replanting with citrus may reduce the nematode population in the soil. Where clean nursery stock is not available, saplings should be subjected to hot-water treatment at 45°C for 25 min before planting. Pre-planting application of oil cakes as a source of organic amendment at 20–50 g per pit could be beneficial in suppressing nematode populations.

104.15.3 Resistant rootstocks

Rootstocks that are resistant ('Carrizo citrange', 'Citromela' and 'Grapefruit') or moderately resistant ('Swingle Citrumelo' and 'Trifoliate Orange') to *T. semipenetrans* could be used for grafting of commercial varieties. 'Troyer citrange', a resistant rootstock, should be avoided, due to the presence of resistance-breaking biotypes (Javed *et al.*, 2008).

104.15.4 Biological control

Pasteuria spp. affected seasonal fluctuations of the citrus nematode but only in one of four study orchards (Sorribas *et al.*, 2000). The egg-parasitic fungus *Pupureocillium lilacinum* reduced *T. semipenetrans* densities in pot experiments (Parvatha Reddy *et al.*, 1991) and the results were best when combined with organic amendments (oil cakes). Similar results were observed on rough lemon rootstock in an 8-year-old Valencia orchard (Le Roux *et al.*, 2000).

104.15.5 Chemical control

Treatment of orchards with granular and systemic nematicides of organophosphate (oxamyl, fenamiphos, phorate, etc.) or organo-carbamate group (carbofuran, etc.) is recommended and these nematicides are highly effective when applied at 5–11 kg a.i./ha. Tanha Maafi and Damadzadeh (2008) reported that the nematicides Vydate® and fenamiphos suppressed the nematode population when applied at 8 ppm 6 weeks after nematode inoculation. Aldicarb (Temik® 15G) when applied in-band along the dripline on both sides of tree rows effectively controlled citrus nematode (Duncan *et al.*, 2009). Application of fenamiphos 40 EC in irrigation water at 10 ml/m^2 was also effective (Mani and Dakshinamurthy, 1986). However, care has to be taken as the nematodes may occur up to a depth of 2 m and so nematicides should reach this depth.

104.15.6 Phytosanitary measures

See Chapter 16, section 115.

15

Techniques and Tools

To study and identify the nematode(s) present in soil or plant materials, proper sampling, extraction, killing and fixing of the nematode are important steps and these techniques are need to be followed carefully. A clean, good specimen showing most of the morphological details is vital for correct identification. Important nematological techniques have been described in details by several authors (Southey, 1986; Barker and Niblack, 1990; Khan, 2008; Manzanilla-López and Marbán-Mendoza. 2012; Hallman and Subbotin, 2018) and are discussed below.

105 Sampling

To obtain a reliable estimation of both the qualitative and quantitative occurrences of important plant-parasitic nematode species and their relative population levels, sampling should be done with an appropriate sampling technique(s).

105.1 Soil sampling technique

The technique for soil collection is simple and it does not require any special device. The classic cylindrical tube-type sampler, auger and polythene bags are standard equipment for the sampling (Khan, 2008). For soil collection, the top 5–6 cm of soil is discarded as it does not contain many nematodes. A standard depth of up to 25 cm is used for advisory purposes and good populations generally occur at depths of 5–15 cm. For virus-vector nematodes (longidorids and trichodorids), the samples are sometimes taken at a depth of 30–60 cm. However, the root system of the host is the most important factor influencing the vertical distribution of plant-parasitic nematodes, so the soil of the root zone needs to be collected for nematode assays.

105.1.1 Sampling pattern

There are different patterns of soil sampling, depending on the purpose and requirements. The relative size of the sample area and the number of cores collected per area for a nematode assay differ greatly with the purpose. For advisory purposes, a sample area of 1–2 ha or less is recommended, whereas for regulatory purposes a sample area of about one-third of a hectare and for research purposes the entire research plot is sampled (Khan, 2008). Different patterns are used to collect soil samples, depending on the objective, size of the sampling area and type of vegetation, which are as follows.

LARGE PLOT In large plots of 1–2 ha, 20–30 cores are made to collect soil samples. The cores can be made in a randomized (Fig. 15.1A) or in a stratified manner (Fig. 15.1B). In a stratified manner, cores are made in lines, while in a random manner the cores can be made anywhere in the plot, depending on the wishes of the researcher.

ROWED VEGETATION In a field where plants are grown in rows (e.g. cotton, potato, tobacco), cores are made in two central rows in the stratified manner if it is a four-row plot (Fig. 15.1C). In a multi-rowed plot, cores should be made after two rows (Fig. 15.1D).

PERENNIAL ORCHARD To collect soil samples from feeder root zones of established perennial plots, for example a plot having trees of citrus, peach, poplar, etc., pairs of cores are alternately made around each tree (Fig. 15.1E).

SINGLE-PLANT PLOT To collect soil from a single-plant plot, a total of ten cores are made in two circles (five each) around the plant in a pattern that gives an impression of a star; this is also known as a star sampling (Fig. 15.1F).

105.2 Storage of samples

Since plant-parasitic nematodes are thin-walled and short-lived microorganisms, improper handling such as exposure to the sun, extremes temperature, droughts or poor storage may result in high mortality rates that will lead to erroneous information on the nematode population.

105.2.1 Soil samples

After collection, each sample should be sealed in a plastic or polythene bag, and be kept at a place away from direct sun exposure, preferably at 20–25°C. The samples should not be stacked for too long and should not be given any forceful treatment (pressure, etc.) that might kill the nematodes. The samples should be transported to the laboratory as rapidly as

DOI: 10.1079/9781789247367.0016

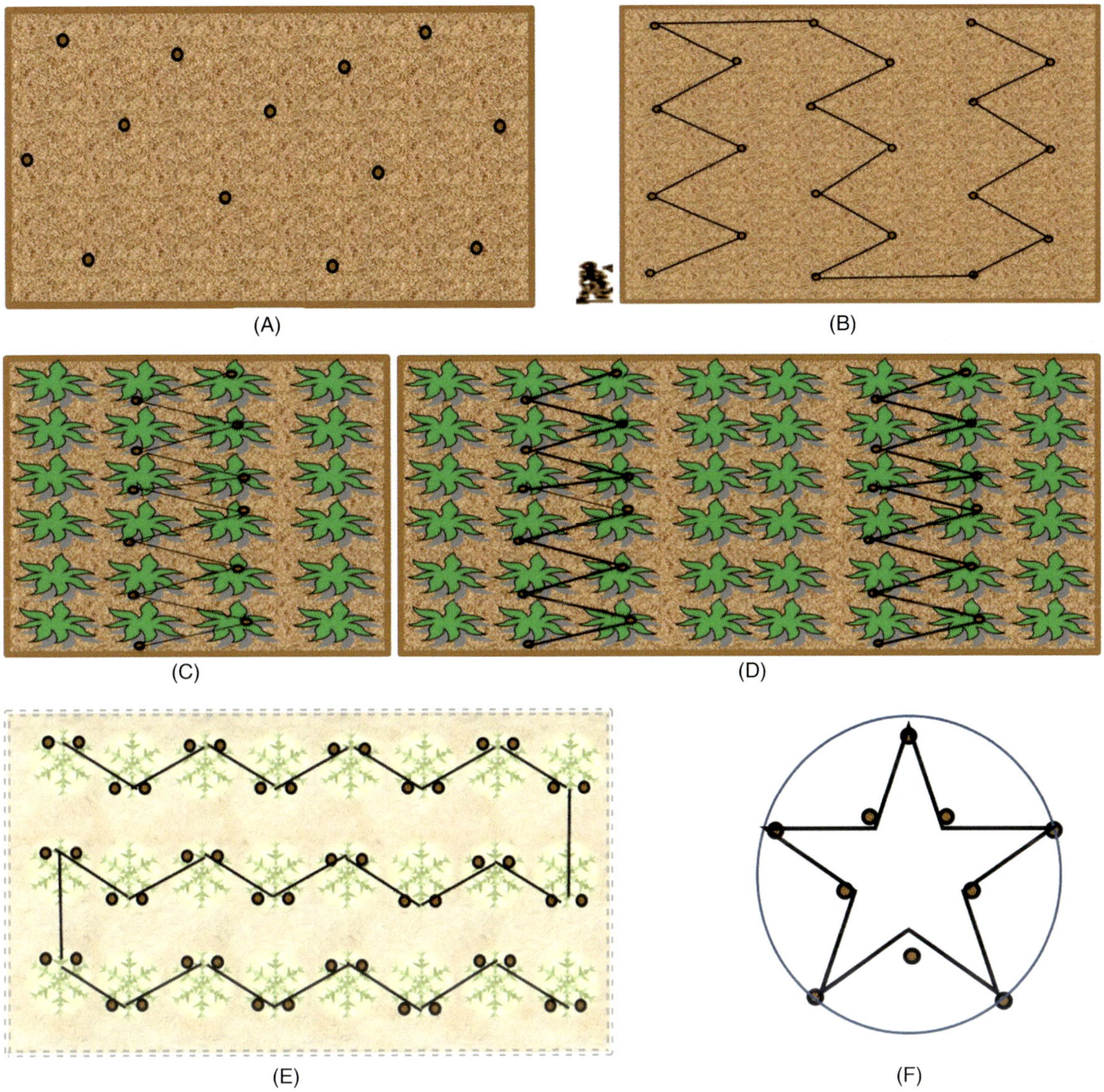

Fig. 15.1. Sampling patterns for large plots, fallow or broad cast vegetation (A,B); rowed vegetation (C,D); perennial orchards (E); single plant-plot or 'star' pattern (F). [Author's own photographs]

possible. Samples should be accompanied by proper labels, which might be placed between two bags.

105.2.2 Plant material

To isolate endo- or semi-endoparasitic nematodes from the plant material, the part of the plant invaded by the nematode is collected, such as roots for *Meloidogyne* spp. and *Rotylenchulus reniformis*, or bulbs, tubers and rhizomes for *Ditylenchus* spp. and *Radopholus similis*, leaves for *Aphelenchoides* spp., trunk for *Bursaphelenchus cocophilus* and *B. xylophilus*, or seeds for *Anguina tritici* and *Aphelenchoides besseyi*. The root samples should be rinsed gently in water before storage. Plant material containing nematodes should usually be kept cool and moist and examined as soon as possible (Khan, 2008). Whole plants (free of soil) are best stored. Shoots decompose more quickly than roots, so they should be kept in a separate polythene bag. The plant material should be made wet before being packed in polybags and should be stored at 10–15°C, especially in subtropical and tropical regions.

106 Detection and Isolation

106.1 Extraction of nematodes from soil

There are various methods to separate nematodes from soil samples. Although vermiforms and cysts can be isolated by a single method, recovery is good if a specific technique is used. Hence an appropriate isolation technique must be selected on the basis of the nematodes to be expected or isolated.

106.1.1 Extraction of vermiform nematodes

COBB'S DECANTING AND SIEVING (MODIFIED) METHOD Soil from a composite or simple sample (500 g or cc) is placed in 5–6 l of water in a bucket. Lumps are thoroughly broken and the suspension is left for 2 min to allow settling of large clumps, debris, etc. The suspension is decanted to a 35-mesh (500 μm) sieve placed over a 400-mesh (37 μm) sieve. If the soil is rich in humus, silt or clay, 100-mesh (149 μm) or 200-mesh (74 μm) sieves are used before the 35-mesh or 400-mesh sieves (Fig. 15.2C). For larger longidorid nematodes, 150 μm sieves are more suitable. The catch from the finest sieve is poured over a wire gauze sieve or coarse sieve (7–8 cm diameter) having two layers of tissue paper on the bottom. The sieve is kept on a modified Baermann funnel, holding a sufficient amount of water to remain in contact with the bottom of the wire gauze (Fig. 15.2A,B). The assembly is left overnight. Due to their wriggling/mobility and random movements, the nematodes gradually settle down in the lower end of the funnel. The nematode suspension is collected from the beak of the funnel through the stop valve.

SIEVING AND SUCROSE CENTRIFUGATION TECHNIQUE This technique is based on the density difference between sucrose solutions, and therefore nematodes that are lighter in density will float on the surface/upper layer. In this technique, 100 ml of the soil sample is taken and mixed in 4 l of water. The mixture is then passed through 60-mesh (250 μm) and 500-mesh (25 μm) sieves. The suspension is collected from the 500-mesh sieve and poured into 50 ml centrifuge tubes. The suspension is centrifuged at 3000 rpm for 4 min. The supernatant is discarded and the sucrose solution (454 g sucrose in 1 l of water) is added. The supernatant is centrifuged again at 3000 rpm for 1 min. Now, the supernatant is collected and the pellet is discarded. The suspension is washed on a 500-mesh (25 μm) sieve until the sugar content is completely removed. The catch is collected and the nematodes are observed under a stereomicroscope. This is a simple and less time-consuming technique and useful to isolate both sluggish and active nematodes. However, only a small quantity of samples can be processed at once.

106.1.2 Extraction of cysts of Heterodera *and* Globodera

SIMPLE METHOD Due to their size and flat shape, cysts float on water. The soil sample should be air dried before the isolation, so that the cysts can become lighter in weight. A small amount of air-dried soil (50 g) is put on a 60-mesh (250 μm) sieve and washed carefully with a slow stream of water (Khan, 2008). The catch at this sieve is poured into a white bowl with sloping edges. The cysts floating along the edge of the bowl are separated with a fine camel-hair brush.

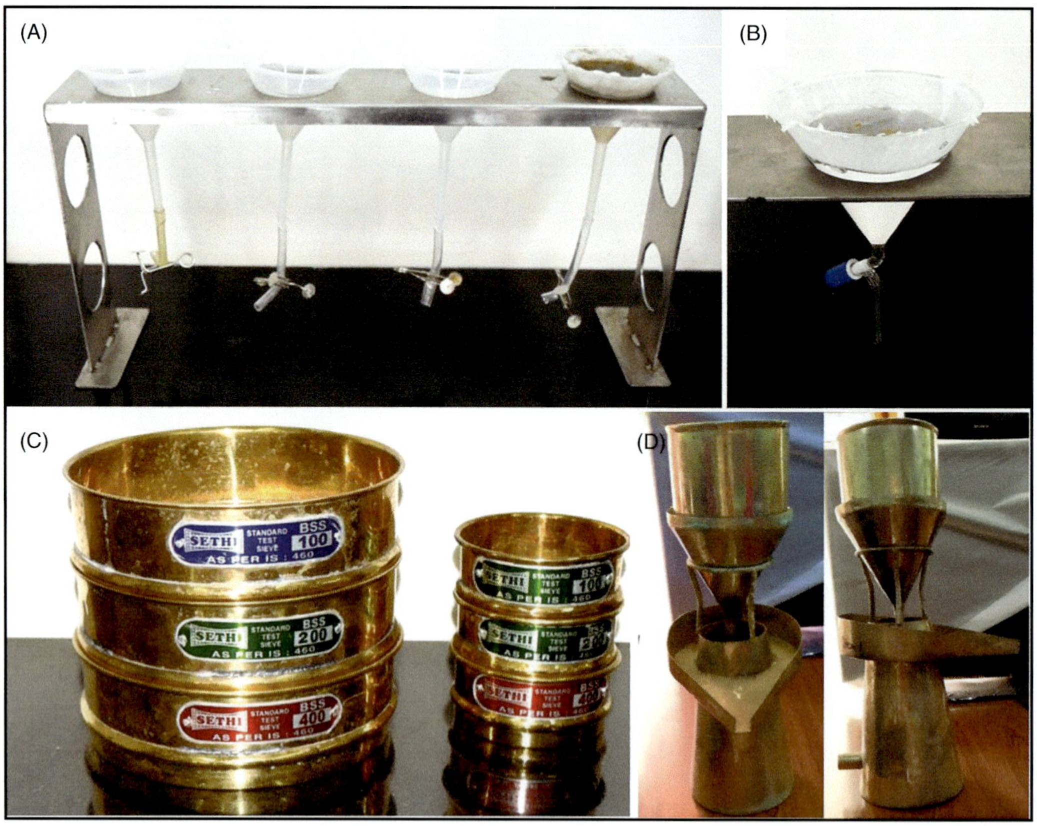

Fig. 15.2. Baermann's funnel in ring stand (A); an individual Baerman funnel (B); sets of sieves (C); Fenwick can (D). [Author's own photographs]

FENWICK CAN METHOD (MODIFIED) This is a simple but specially designed brass can for isolation of nematode cysts. Before the isolation, the can should be cleaned and filled with water. A sample of 100 g of air-dried soil is placed in the top sieve of 1 mm pore size. The sample is washed into the apparatus via the funnel (Fig. 15.2D). The coarse material is retained on the top, while heavy soil particles sink to the bottom of the apparatus. The floating cysts are carried off over the overflow collar. The cysts, root debris and other particles are collected on a 60-mesh sieve. The catch of this sieve is poured into a white bowl with sloping edges and the cysts can be collected by using a fine camel-hair brush and examined under a stereomicroscope (Khan, 2008).

106.2 Extraction of nematodes from plant material

106.2.1 Direct examination of plant material

If a small amount of material is available, it can be examined directly under a stereo microscope using transmitted and/or incident light for the nematode assay. The gently washed tissue is taken in Petri plates (without cover) and is teased with a stout needle under the microscope to release and collect the nematodes. The material can be re-examined after a few hours; often deeply embedded nematodes tend to migrate from the damaged tissue in 2–3 h. To isolate egg masses and females, the washed roots are placed under the stereo microscope with incident light. Egg masses (*Meloidogyne* spp.) and females (*Heterodera*, *Globodera*) are picked out with forceps. To collect females of *Meloidogyne* spp., the root tissue beneath the egg mass is gently torn to excise the creamy or transparent-looking shiny females (Khan, 2008).

106.2.2 Baermann funnel technique

The infested root, seeds, stem, leaves, bulb or plant tissues are soaked in water for 24 h and then macerated or cut into pieces (0.5–1 cm) and placed in muslin cloth or on a wire gauze with tissue paper in a coarse sieve. The sieve is immersed in water in a Baermann funnel. The funnel is incubated for 12 h at room temperature. During incubation, the larvae migrate from the tissue, pass through the muslin or tissue paper and aggregate in the beak of the funnel, from where they are collected. Recovery of *Aphelenchoides* and *Ditylenchus* species from shoot tissue would be better if 0.15% of H_2O_2 water is used in the funnel, which improves oxygenation.

106.2.3 Maceration-filtration technique

Roots or planting material (5 g) are chopped into pieces of 1 cm long and macerated with 100 ml of water in an electric blender for 10–15 s (Khan, 2008). The suspension is poured on a wire gauze sieve with tissue paper. The sieve is placed in a dish or Baermann funnel for 24–48 h. The larvae migrate down through the paper and are collected from the dish/funnel.

106.2.4 Root/shoot incubation technique

The root/shoot material is cut into pieces 5–10 cm long. The pieces are put into closed Petri plates or container. The pieces are either wetted well or immersed in shallow water and are incubated at 20–25°C for a few days to weeks (Khan, 2008). Water from the container is poured off daily and the root/shoot pieces are rinsed. Fresh water is added and containers are recapped. The collected water and wash out are poured on 35-mesh and 500-mesh sieves. The catch from the finer sieve is examined for nematodes. Most of the nematodes are recovered within 4–7 days.

107 Processing of Nematodes

107.1 Nematode counting

For counting the nematodes present in the suspension, special counting dishes are available with the basic principle of a transparent base with an imprinted grid on the bottom (Fig. 15.3). Usually, a counting dish with approximately 5 ml capacity is used for this purpose. A known volume (e.g. 10 ml) of extracted nematodes suspension from any of the procedures described above is taken. The suspension is stirred and aliquots (e.g. 1 or 2 ml) are carefully pipetted into the counting dish and left for a few seconds to settle the nematodes. The suspension is observed under a stereo microscope. If the nematode density is very high, the nematode suspension is diluted and another aliquot can be taken. Now all the nematodes present in aliquots in the counting dish are counted in a systematic way, following the grid pattern of the dish. It is better to use a tally counter. The counting is repeated using another two to three aliquots from the suspension and then the mean number of nematodes per aliquot is calculated. The mean is multiplied by the total volume of the suspension to estimate the total number of nematodes. This number represents the sample population that was extracted from the original sample (e.g. 100 ml of soil or 5 g of root).

107.2 Picking of vermiform nematodes

The collected nematode suspension is poured into a watch glass or cavity block and the desired nematode species is picked (fished out) with the help of a fine-pointed needle. A fine steel needle, bamboo splinter, an eyebrow or eyelash hair or horsetail hair are very effective tools to pick nematodes from the water suspension (Khan, 2008). The nematode suspension in a cavity block is placed under a stereo microscope. The picking device (needle) is held in the right hand while the left hand is placed on the focusing knob of microscope in the case of a right-hander. The needle is brought under the nematode larva perpendicular to its body length, then an upward push is given to bring the larva to the surface. With the left hand, the focusing knob is rotated to keep an eye on the larva. When the nematode has come on the surface, the needle is again brought perpendicular to the nematode body, which is lifted with a jerk (Khan, 2008). The nematode is held under surface tension on the needle. The picked larva is transferred to a drop of water on a glass slide and is further processed.

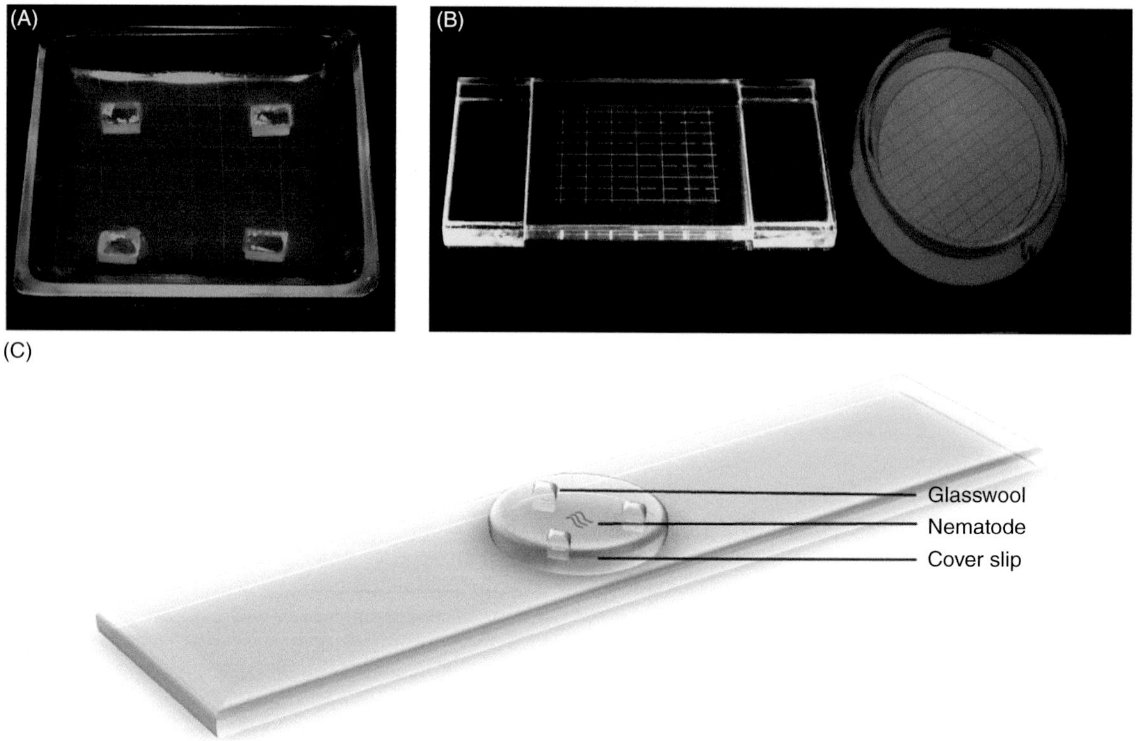

Fig. 15.3. Various nematode counting dish (A, B) and nematode mounted slide (C). [Author's own photographs]

107.3 Anaesthetizing or killing the nematode

Nematodes are anaesthetized or killed to immobilize them for microscopic examination. Nematodes are spoiled if transferred alive to a fixative. Some structures are more easily seen and clearly visible in an anaesthetized specimen on a temporary mount than in dead, fixed or processed nematodes. Sodium azide (0.01 M) is very effective for anaesthetizing nematodes within a few minutes; a 0.5–1% solution of propylene phenoxetol in water can also be used to immobilize nematodes. Some of the important methods are as follows.

107.3.1 Direct flame method

The best method to kill nematodes is gentle heating on a flame. A glass slide with a specimen in a drop of water is exposed to a spirit lamp flame for a few seconds. The temperature is assessed by touching the slide with the skin of the back of the palm. It is heated intermittently to an extent that it becomes unbearable to touch (Khan, 2008). During the heating, care must be taken not to let the water droplets evaporate completely. This method is very handy when a few specimens are to be killed.

107.3.2 Hot plate method

Instead of flame, a hot plate at 65°C can be used. The glass slide with the nematode specimen in one or two drops of water may be placed on the hot plate for 4–5 min. A small beaker (50 ml capacity) or cavity block may be used to kill the nematodes in a 5–10 ml suspension.

107.3.3 Hot water method

A safer method of killing a larger sample is hot-water treatment. First, the volume of the suspension is decreased by centrifuge at 1500 rpm. The concentrated specimen in a test tube is shaken and the tube is plunged into a hot-water bath at 65°C for 2–5 min or until the nematodes are killed. Thereafter, an equal volume of double-strength fixative is added.

107.3.4 Hot fixative method

One of the best methods to kill and fix nematodes by one process is the use of hot fixative. Formalin-acetic acid fixative (4:1) is heated at 100°C in a test tube or dropper in boiling water. The hot fixative is quickly added to the nematode suspension at double volume (Hallman and Subbotin, 2018). This method fixes glands and gonads and nuclei are clearly visible. Use of propionic acid in place of acetic acid followed by 4% formaldehyde may yield better killing and fixing.

107.4 Fixing the nematode

This is a process which prevents distortion or decay of the nematode body and preserves the organs/structures intact and in the original form (Khan, 2008). Nematodes can be stored in a fixative indefinitely. Vials containing nematodes should be labelled with identity (if known), locality, the fixative used and date of fixation. Fixatives containing alcohol should be avoided, as they cause some shrinkage to the body. Well-fixed specimens have a smooth outline. Various fixatives are used to fix nematodes and some of those that are commonly used are described below.

107.4.1 Formalin–glycerol

A solution of 5–10% formalin plus 2% glycerol (glycerine) is widely used. If a small amount of $CaCO_3$ is added to the stock solution of formalin, it will prevent darkening and granulation of tissue.

Formalin (40% formaldehyde)	8 ml
Glycerol (glycerine)	2 ml
Distilled water	up to 100 ml

107.4.2 Formalin–acetic (FA) or formalin–propionic (FP)

This is another widely used fixative for nematodes. The addition of 2% glycerol helps to prevent drying of specimens if the fixative evaporates slowly, hence the nematodes stored in vials will eventually end up in glycerol, which is the best-known mounting material for nematodes.

Formalin (40% formaldehyde)	10 ml
Glacial acetic acid or propionic acid	1 ml
Glycerol (glycerine)	2 ml
Distilled water	up to 100 ml

107.4.3 Formalin–aceto-alcohol (FAA)

This fixative contains 20% alcohol; as a result, a certain amount of shrinkage of the specimen always occurs. This quality is sometimes desired, to have the nematode structures plain as lateral lines and annulations.

Ethanol 95%	20 ml
Formalin (40% formaldehyde)	6 ml
Glacial acetic acid	1 ml
Distilled water	up to 100 ml

107.4.4 Triethanol amine formalin (TAF)

TAF has an advantage over FA or FP. With this fixative, the nematode retains its life-like appearance for several hours. However, during long storage in TAF, some degeneration in the cuticle may occur. To prevent further degeneration, the specimen should be transferred to FA or FP plus 2% glycerol. Addition of these fixatives to TAF with specimens in excess also works satisfactorily.

Formalin (40% formaldehyde)	7 ml
Triethanol amine	2 ml
Distilled water	up to 100 ml

107.5 Processing of nematodes for mounting

Internal body details, especially the gonad structure of nematodes in a fixative, may gradually diminish or obscure due to the granular appearance of the intestine. The specimen can be cleared by processing with a mountant such as lactophenol or glycerol

107.5.1 Rapid lactophenol method

Lactophenol mountant is prepared as follows. A cavity slide is filled with the mountant and is heated at 65–70°C. Already-fixed nematodes are transferred to the hot lactophenol. The slide is kept hot for 2–3 min and is checked under a stereo microscope for clearing the specimen.

Phenol (liquid)	500 ml
Lactic acid	500 ml
Glycerol	1000 ml
Distilled water with a pinch of methyl blue (cotton blue)	500 ml

107.5.2 Slow glycerol method

After fixation, nematodes are transferred to diluted glycerol (1.5% in distilled water, or in 75% ethanol) in a cavity block. A trace of copper sulfate or a little thymol can be added to prevent microbial growth. The cavity block is kept in a partially closed Petri plate inside a desiccator. The diluted glycerol (1.5%) should evaporate very slowly (3–5 weeks). After desiccation, the specimen remains in pure glycerol. An oven at 30–60°C can substitute the desiccator.

107.6 Mounting of nematodes

107.6.1 Temporary mounts

A few millilitres of nematode suspension are taken into a cavity block or watch glass. The desired number of nematodes are picked and transferred to a large drop of water in the centre of a clean glass slide. Water is used to mount unprocessed specimens, otherwise lactophenol or glycerol is used as a mountant (Khan, 2008). Three or four glass fibres or small pieces of coverslip are placed around the drop. After ensuring that the specimens are near the centre of the drop, the coverslip is applied gently. The coverslip is sealed properly from the edge using molten wax or nail polish (Fig. 15.3).

107.6.2 Permanent mounting

A drop of glycerol is kept on a clean glass slide. The drop should not be spread; rather, it should be kept as convex as possible to avoid air bubbles. The drop should be of such a size that it spreads just to the edge of the coverslip without any excess (Khan, 2008). Already fixed nematodes are picked and are mounted in glycerol in the way described above for temporary mounting. The coverslip can be sealed by Glyceel, DPX mountant or Canada balsam. The sealing should be done two or three times with an interval of 6–12 h.

107.7 Preparation of *en face* view, perineal pattern and vulval cone

107.7.1 Perineal pattern of Meloidogyne *species*

The galled roots are stained in acid fuchsin-lactophenol (see section 107.8.1, below). The females are dissected out from the roots and transferred in pure lactophenol. Working under a stereo microscope at 4×, the posterior end is cut off with a sharp blade and the body is pressed gently with the brush to avoid bursting during section cutting. Thereafter, the posterior end is cut having a perineal pattern (Fig. 15.4A). The posterior end is trimmed so that only the perineal area (the area around vulva and anus) remains in a piece of cuticle that is 5–10 times the perineal region. The inner tissues are removed carefully with a nylon bristle and transferred to a fresh drop of lactophenol on a glass slide. A coverslip is placed and is sealed with nail polish or DPX mountant.

107.7.2 Vulval cone of Heterodera *species*

The dry cysts are soaked in water for 24 h. A cyst is placed on a glass slide in a drop of water, and the posterior end is cut off so that the fenestral area is in the centre of the piece (Fig. 15.4B). The cut portion is trimmed so that it is not more than 5–10 times the fenestral area. The adhering body contents are removed and the cone is transferred in molten glycerine jelly on to a glass slide. A coverslip is placed and held in such a position that the fenestral area faces upwards when the jelly solidifies.

107.8 Staining of nematode in plant tissue

107.8.1 Cotton blue or acid fuchsin in lactophenol

The plant tissue is rinsed and made free of soil particles and other debris. Small pieces, 3–5 cm long, are cut and plunged into boiling cotton blue (0.1% cotton blue) or acid fuchsin stain for 3 min in a deep beaker. Multiple samples can be treated in one operation by wrapping them separately in muslin cloth. This also helps to immerse the leaves. After 3 min, the sample is taken out, washed in running water and placed in Petri plates with plain lactophenol. Liquid phenol is used to decolorize leaves and stem. Care should be taken to avoid fumes from phenol. The treated tissue is left for several hours or up to 2–3 days to allow the tissue to differentiate. Nematodes stain blue or red, whereas plant tissues except meristematic tissue remain largely stainless. The small parts of plant tissue may be pressed gently between two glass slides to make the surface plain and are examined under the microscope.

108 Morphological Examinations

Nematode taxonomy is based mainly on the morphological characters; hence a sound knowledge of the microscopic examination of nematodes is absolutely essential for their identification. The main morphological features that have application in the identification of plant-parasitic nematodes are: size and shape; head and tail; cuticular markings; stoma and stylet; position of oesophagus; male and female reproductive organs and so on. Morpho-anatomical characters are studied by using light microscope (LM), scanning electron microscope (SEM) and transmission electron microscope (TEM). The characteristic morphological features of each nematode species and their morphological measurements are described in the respective chapter for the species.

108.1 Morphometry

Besides qualitative morphological characters, certain quantitative characters (morphometric measurements) are the most important tools in the conventional taxonomy of nematodes (for specific morphometric characters, see respective chapter for the nematode species). Morphometry and allometric measurements often vary greatly under the influence of geographical and

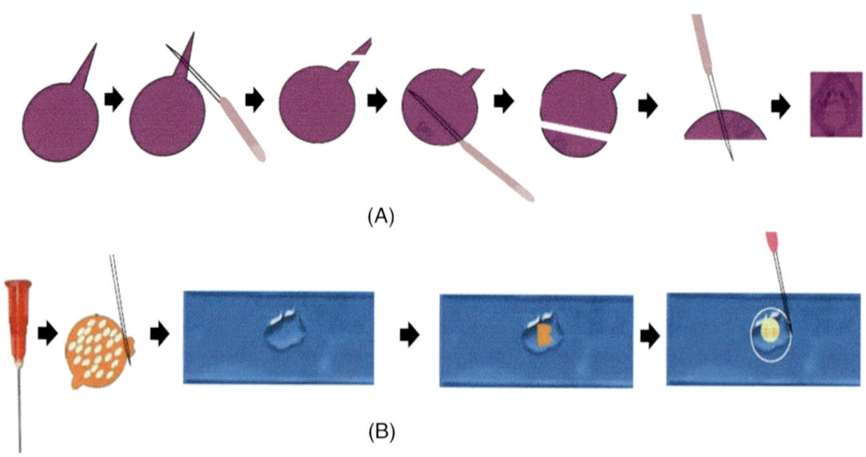

Fig. 15.4. Processing involved in the preparation of perineal pattern (A) and vulval cone top (B). [Adapted from H.K. Bajaj and R.K. Walia]

ecological conditions; thus care must be taken in using them. In addition, de Man (1884) proposed the following formula of body ratio (**de Man's ratios**) for nematode measurement.

L = Total body length

$$a = \frac{\text{body length}}{\text{maximum body width}}$$

$$b = \frac{\text{body length}}{\text{oesophageal length}}$$

$$b' = \frac{\text{body length}}{\text{distance from head end to posterior end of oesophageal gland}}$$

$$c = \frac{\text{body length}}{\text{tail length}}$$

$$c' = \frac{\text{tail length}}{\text{body width at anus}}$$

$$V = \frac{\text{distance from head end to vulva}}{\text{body length}} \times 100$$

$$T = \frac{\text{distance from the cloacal aperture to anterior end of testis}}{\text{body length}} \times 100$$

Measurement of specimens/objects involves the use of high-quality microscopes (LM, SEM and TEM) and the use of a camera lucida for line drawing.

108.1.1 Measurement of straight specimens

Straight structures can be measured directly and need not be drawn with a camera lucida. But prior calibration of the ocular micrometer is required for each set of eyepieces and objectives of a high-power research microscope. The specimen to be measured, is fixed in the microscope and aligned parallel to the ocular micrometer scale. The number of scale divisions along the specimen are counted and multiplied by the calibrated value from the table for that particular set of lenses.

108.1.2 Measurement of curled specimen

It is difficult to directly measure curled specimens accurately; therefore, a rough drawing or a median line along the curvatures is drawn with the help of a camera lucida. A variety of drawing apparatus is now available suited to different brands of microscopes, but two types are most common: the prism type and a mirror type. These are fitted on to a monocular tube. A **camera lucida** projects the picture of any object lying beside the microscope into it and can be viewed simultaneously with the specimen. Adjustment of illumination of the object outside the microscope and the specimen inside the microscope is required for sharp pictures. For measurements, first the stage microscope and the camera lucida are fixed on the microscope. A drawing sheet fixed on the bench beside the microscope is used. A straight line is marked on the sheet with a pencil. Viewing through the camera lucida, the illumination is adjusted to view this line clearly on the sheet and overlap it parallel with the micrometer scale. The divisions of the scale are marked on the line. Similarly, the scale divisions on other lines for different combinations of eyepieces and objectives (the smallest division represents 10 μm) are marked. The lines on the drawing sheet are neatly drawn and the scale is extended to prepare a permanent 'standard scale' for taking measurements. Then the specimen is placed on the stage of the microscope and a rough drawing of the object to be measured is made on another sheet. A simple median line along the curvatures may serve the purpose. A thread is run along the curvatures of the drawing from one end to another and straightened, then measurements are taken from the 'standard scale' for that particular set of eyepiece and objective. Flexible pieces of wire can be used in place of thread.

108.2 Camera lucida drawings

The description of a species is accompanied by illustrations of its morphological details which delineate it from other similar taxa. Line drawings using a camera lucida are often employed, if direct photographs obscure details. Rough pencil sketches of the specimens are prepared on paper sheets using a camera lucida as described above at desired magnifications. The entire specimen may not be visible in the microscope field and will have to be drawn in several sections, overlapping a little each time so that these can be joined accurately. Essential structures should be drawn in detail but unimportant ones can be indicated by a median line and width markings. These are then traced on to the drawing sheet. Several sketches are adjusted on the drawing sheet suitably within the dimensions of the plate. The finer details are worked out on the drawing sheet. To maintain scope for corrections during inking, good quality scratchable sheets or boards are preferred rather than ones with glazed surfaces. Magnification scales drawn from the 'standard scale' are also fitted into the plate.

Biochemical and Molecular Diagnosis

Advances in research have enhanced conventional morphological methods to precisely identify nematodes at generic and species levels and also to characterize races or pathotypes. A huge step towards nematode identification has been the use of molecular characterization based on polymerase chain reaction (PCR). Although numerous biochemical and molecular techniques are used to identify nematodes, this chapter describes the techniques that have potential application in nematode diagnosis.

109 Biochemical Techniques

A wide range of biochemical tools such as the protein-based enzyme-linked immunosorbent assay (ELISA), isoelectric focusing (IEF), one-dimensional gel electrophoresis (1-DGE), two-dimensional gel electrophoresis (2-DGE), sodium dodecyl sulphate–capillary gel electrophoresis (SDS–CGE) and matrix-assisted laser desorption/ionization–time-of-flight mass spectrometry (MALDI-TOF MS) have been developed to detect protein variation among nematodes. Serological techniques using monoclonal or polyclonal antibodies do not reach an application level. However, using 1-DGE for the separation of protein fractionation (general proteins and isozymes), IEF for general proteins and isozymes, 2-DGE for electrophoretic resolution of proteins and SDS–CGE for the separation of micro-amounts of proteins and peptides were more common in the past, with principal use of electrophoresis.

109.1 Protein (isozymes) electrophoresis

Electrophoresis is a technique that separates molecules in a matrix such as starch, agar, acrylamide or cellulose acetate. Total proteins can be analysed by staining the matrix containing the electrophoresis-treated protein extract with a protein-specific stain. The electrophoresic pattern of total proteins is complex, due to the presence of thousands of different proteins, but specific proteins such as enzymes or isozymes may give simple or less complex patterns which vary greatly interspecifically. Isozymes are enzymes that catalyse the reactions, which differ considerably in electrophoresic mobility and are more sensitive to stains than total proteins. This relative sensitivity allows identification of individual nematodes in a mixed population. Isozyme analysis has been extensively used to identify root-knot nematode species because of the larger size of a female that yields enough protein to undergo electrophoresis. However, isozyme electrophoresis has not been effective to characterize races of root-knot nematodes (Moens *et al.*, 2009). Species of *Meloidogyne* from numerous populations have been identified using esterase isozyme patterns (Carneiro *et al.*, 2000; Blok and Powers, 2009). Other isozymes such as malate dehydrogenase or glucose phosphate isomerase have also been found effective in identifying *Meloidogyne* spp. Isozyme patterns have also been used to identify *Heterodera glycines*, *Radopholus similis*, *Anguina tritici*, *A. agrostis* and other nematodes. The following steps are involved to prepare the isozyme patterns.

109.1.1 Sample preparation

Nematodes are isolated from soil or excised from roots, and are processed immediately or can be stored in 0.9% saline solution at room temperature for up to 3 weeks. An individual nematode, for example a female of a root-knot nematode, is crushed with a small glass rod in a small volume of extraction buffer (20% sucrose, 2% Triton X-100 and 0.01% bromophenol blue). The crushed nematode is immediately processed or can be stored at –20°C, as esterase activity remains stable for several months.

109.1.2 Gel casting

The gels may be prepared following a simple method. A thin (less than 0.7 mm) acrylamide or agarose gel is used in the electrophoresis system. Finer gels give a better resolution matrix for separation of isozymes. Starch or cellulose acetate paper, which is less expensive, can also be used. A detailed description of gel electrophoresis is given in section 110.2, below.

109.1.3 Running electrophoresis

The cassette and the gel are placed in the electrophoresis apparatus and the upper and lower tanks are filled with the electrophoresis buffer. A pipette is used to remove any air trapped in the gel wells so that the wells become filled with the buffer. The sample tubes containing nematode extracts are removed from the freezer and are placed into a centrifuge at 13,000× *g* for 15 min at 5°C. The centrifugation results in three fractions: the insoluble pellet at the bottom, a clear aqueous layer in the middle and an opaque layer at the top in the sample tube. The middle layer is transferred carefully into a sample well in the stacking gel. This sample contains sucrose or glycerol, and is more viscous and denser than the electrophoresis buffer, hence remaining in the well without mixing with the buffer. Thereafter a drop of bromophenol blue dye solution is added to the bridge buffer in the upper tank and an electric current at 80 volts is allowed to run for 30 min followed by 200 or 300 volts for another 15 min. By this time a definite band of bromophenol dye should be visible in the gel. When this band has migrated approximately 100 mm, the power supply is turned off and the cassette is removed from the electrophoresis apparatus.

109.1.4 Gel staining

Staining solutions are prepared using following reagents.

FOR ISOZYME ESTERASE	
0.1 Sodium phosphate buffer (pH 7.4)	25 ml
EDTA	7.5 mg
Fast blue R salt	15 mg
α Naphthyl acetate in 0.5 ml acetone	10 mg
FOR MALATE DEHYDROGENASE	
Water	38 ml
Solution A (sodium bicarbonate 10.6 g, L-malic acid 1.3 g and water up to 100 ml)	5.0 ml
Tris HCl (pH 7.1)	7.5 ml
NAD	25 mg
Tetrazolium	15 mg
Phenazine methosulfate	1.0 mg

A mixture of buffer substrates, chelators and coenzymes is poured on the stir plate. The dye and other photo-sensitive chemicals are added to the mixture just prior to staining the gels. The gel showing bands is cut from the unused/excess portion and is washed under a slow stream of water in a glass dish. The cassette is held sloping towards the dish to allow the gel to slide into the dish without tearing. The water from the dish is poured out and the staining solution is filled into the dish to cover the gel thoroughly. The dish is covered with a poly-sheet and placed inside an incubator on a gentle shaker (1 horizontal shaking/10 min) under darkness at 37°C for 30–60 min, depending on proper staining of enzyme bands. Thereafter, the gel is washed in a slow stream of water to remove staining solution and is stored in 7% acetic solution in an airtight container for 1–2 h. The enzyme pattern can now be studied and photographed. The sample's isozyme pattern of esterase and malate dehydrogenase can be compared with available isozyme data (Esbenshade and Triantaphyllou, 1985a; Carneiro *et al.*, 2000).

110 Molecular Techniques

Molecular characterization is potentially a more commanding method in nematode diagnosis and also to characterize inter- and intraspecific variability. Nowadays it is most frequently used and does not require expertise in nematode taxonomy. After the development of polymerase chain reaction (PCR) technology, the use of DNA-based techniques such as PCR–restriction fragment length polymorphism (PCR–RFLP), randomly amplified polymorphic DNA (RAPD) and amplified fragment length polymorphism (AFLP), multiplex PCR, real-time PCR, conventional PCR with specific primers, loop-mediated isothermal amplification (LAMP) and sequencing of DNA has improved the accuracy and speed of nematode identification and characterization. Other techniques such as microsatellites or single sequence repeats (SSRs), satellite DNA (satDNA), sequence characterized amplified regions (SCARs), DNA hybridization arrays and PCR–single-strand conformational polymorphism (PCR–SSCP) have also been used to characterize various plant-parasitic nematodes.

110.1 DNA-based techniques

Probably, for the first time Curran *et al.* (1985) reported the use of a DNA technique for taxonomic purposes by analysing fragments of genomic DNA with restriction enzymes to distinguish *Meloidogyne arenaria* from *M. javanica* and other non-plant-parasitic nematodes. However, this method was quite laborious until the development of PCR, which made it possible to diagnose even a small amount of material, such as single juveniles or eggs (Powers and Harris, 1993). Currently there is a wide range of DNA-based techniques that can be exploited or adapted for taxonomic and diagnostic purposes (Subbotin *et al.*, 2013). The amplification of specific genomic regions is an extremely effective way of distinguishing inter- and intraspecific variations between genera and species, and also within species, thus diagnosing the nematode. The two most commonly used repeated regions for taxonomic and diagnostic purposes are ribosomal DNA (rDNA) and mitochondrial DNA (mtDNA). Sequencing of rDNA fragments such as 18S, internal transcribed spacer (ITS) regions, 28S rDNA and mitochondrial cytochrome c oxidase subunit I (mtCOI) genes enables nematologists to identify a nematode more rapidly and accurately. For diagnosis of plant-parasitic nematodes, the most common techniques based on DNA involve four-step procedures: (i) DNA extraction; (ii) amplification of DNA by polymerase chain reaction; (iii) sequencing; and (iv) assembling of sequences and matching with available gene sequences in a database. In addition to these, molecular phylogenetic relationships can be constructed.

The first step in molecular diagnosis is extraction of DNA. Various DNA extraction methods have been developed using extracted nematodes or directly from the plant materials. However, nowadays most of DNA extraction methods are based on commercial kits. Some of the potential methods are described below.

110.1.1 DNA isolation from vermiform or saccate nematodes

WORM LYSIS BUFFER The protocol is described from Tanha Maafi *et al.* (2003). One to four eggs, juveniles, adults, saccate females or cysts are taken in 82 µl ddH_2O on a glass slide and cut under a dissecting microscope. The cut nematodes are transferred to an Eppendorf tube containing 12 µl worm/nematode lysis buffer (500 mM KCl, 100 mM Tris-HCl pH 8, 15 mM $MgCl_2$, 10 mM DTT, 4.5% Tween 20) and crushed with a micro-homogenizer (or micro-pestle). Two-microlitre proteinase K (600 µg/ml) is added and the tubes are frozen at –80°C for at least 10 min and then incubated at 65°C (1 h) and 95°C (10 min), consecutively. After incubation, the tubes are centrifuged for 2 min at 14,000 rpm and kept at –20°C until use.

PHENOL WITH CHLOROFORM EXTRACTION In this technique (Christoforou *et al.*, 2017), two steps are involved: (i) DNA extraction; and (ii) its purification. First, DNA is extracted from individual or mixed life stages (adults, juveniles). The nematode samples (1–30 individuals) are placed in 5 µl distilled water in Eppendorf tubes and frozen in liquid nitrogen until extraction. Using a micro-pestle, the frozen nematode is crushed to make a homogenized mixture. The mixture is transferred to a 50 ml tube and spun briefly to settle the tissues at the bottom of the tube. TNES buffer is added. The buffer is prepared by mixing 50 mM Tris-HCl pH 7.5, 100 mM EDTA pH 8.0, 400 mM NaCl, 0.5% sodium dodecyl sulfate-containing proteinase K at 20 mg/ml and bringing final volume to 10 ml. The tube is incubated for 12 h at 50°C. Then it is adjusted to 1.5 M NaCl by adding an equal volume of 2.6 M NaCl and vigorously shaking it for 30 s. Next it is centrifuged at 12,000× *g* for 5 min. The supernatant is decanted into another fresh tube and an equal volume of ethanol is added. Finally the DNA is spooled out and dissolved in 0.1 ml of distilled water.

For purification, an equal volume of phenol/chloroform/isoamyl alcohol is added to the DNA solution and vortexed vigorously for 10 s. Thereafter it is micro-centrifuged for 15 s at room temperature. Then the tube is carefully decanted and the aqueous (top) phase containing the DNA is taken. If a white precipitate is present at the aqueous/organic interface, the organic phase is re-extracted and the aqueous phases are pooled. One-tenth volume of 3 M sodium acetate (pH 5.2) is added to the DNA solution and mixed by vortexing briefly. An addition of 2–2.5 volumes of ice-chilled absolute ethanol is mixed in by vortexing. The tube is placed in dry ice (crushed) for 5 min or longer and then centrifuged at 13,000× *g* for 20 min in an ultra-centrifuge. The supernatant is removed and 1 ml of ethanol (70%) is added. The tube is shaken several times and again centrifuged for 5 min at 13,000× *g*. The supernatant is discarded and allowed to air dry for 15 min. Finally, the DNA pellet is re-suspended in 100 µl of TE buffer or ultra-pure water until further use.

*110.1.2 Isolation of DNA from cyst nematodes (*Cactodera, Globodera, Heterodera *and* Punctodera*)*

The protocol is described by Reid *et al.* (2015). The cyst material is suspended in 500 µl buffer comprising 5 M guanidine isothiocyanate, 10 mM EDTA, 50 mM Tris-HCl (pH 7.5) and 8% mercaptoethanol in a 2 ml Eppendorf tube and is crushed using a micro-pestle. After adding 200 µl of 0.5 mm silica/zirconia beads, the tube is placed in a bead beater for 30 s at 50,000 rpm and then incubated for up to 1 h at room temperature. The solution containing DNA is extracted by adding phenol and chloroform–isoamyl alcohol (24:1) in equal volumes. After adding the chloroform–isoamyl alcohol, the solution is precipitated with 0.3 M sodium acetate and isopropanol (two volumes). The DNA is re-suspended in 100 µl of distilled water until further use.

Another method was described by FERA (2009). In this method, DNA is extracted from a single cyst, or a cyst containing eggs, or single juveniles. Before extraction, the single cyst is dissected individually in water and homogenized in 180 µl of ATL buffer using a sterile micro-pestle until the complete cyst gets dissolved. Now the DNA can be extracted by using any of the protocols described above. The extracted DNA should be stored at –20°C until further use.

110.1.3 Isolation of DNA from live planting material (for Anguina, Aphelenchoides, Bursaphelenchus *and* Ditylenchus *species)*

A PCR diagnostic assay has been developed to identify nematodes from infected plant tissue (McCuiston *et al.*, 2007) and this PCR assay technique can be modified according to the plant material. In this technique the following steps are involved. First, plant tissue (leaf, stem, root, tuber, bulb, rhizome, etc.) exhibiting nematode symptoms is collected and symptomatic plant tissue is thinly sliced with scissors. A maximum 1.0 mg of sample tissue is placed in a 2 ml tube and two 5 mm glass beads are added. The tube is immersed in liquid nitrogen for 30 s and then the tissue is disrupted in a bead beater for 10 s at 2500 rpm. The process is repeated three times. Then the homogenized tissue is processed for DNA extraction using a DNeasy Plant Mini Kit (Qiagen, Inc.), following the manufacturer's instructions.

Another quick method is the NaOH extraction method. In this method, the plant tissue is crushed in 150 µl of 0.5N NaOH for 5–10 s using a micro-pestle, then 5 µl of ground suspension is taken into a 1.5 ml microcentrifuge tube with the addition of 495 µl of 100 mM Tris-HCl (pH 8.3). Total genomic DNA is eluted in a final volume of 100 µl elution buffer and stored at –20°C until further use. For PCR, the DNA suspension is diluted by adding 2 or 3 µl to 25 µl of PCR master mix with nematode-specific primers (for primers, see the respective nematode species). Since most of the nematodes are tissue-specific (foliar nematode, stem and bulb nematode, seed-borne nematode, etc.), a single nematode individual within a plant tissue can be detected by this PCR assay even with asymptomatic host plant tissue.

FROM WOOD CHIPS OR TIMBER LOGS Only two nematodes, *Bursaphelenchus cocophilus* and *B. xylophilus*, can be intercepted in wood chips or timber logs. The wood specimen suspected for these two nematode species should be processed as soon as possible. The DNA can be extracted from the specimen directly without isolating the nematode by the protocol described by Ye *et al.* (2013). First, the specimen is placed in DNA extraction buffer solution, which includes an enzyme (keratinolytic enzyme), and is incubated. The keratinolytic enzyme breaks down keratin (the main component of the *B. xylophilus* body surface), dissolves cells and elutes the DNA into the extraction buffer. Now the DNA can be extracted by using any of the protocols described above.

110.2 Polymerase chain reaction (PCR)

The PCR technique is one of the most widely used methods for molecular diagnosis. In a typical target PCR, prior DNA sequence information of the target site is required in order to amplify the site. It can detect and amplify any DNA fragments. At the start of the reaction, it requires a DNA template containing the region to be amplified, two oligonucleotide primers flanking this target region, DNA polymerase and four deoxynucleotide triphosphates (dATP, dCTP, dGTP, dTTP) mixed in a buffer containing magnesium ions (Mg^{2+}). The double-stranded DNA is denatured by heating. The single-stranded template DNA is then annealed to the oligonucleotide primers that are complementary to the sites on either side of the target locus. DNA polymerase is then used to synthesize the complementary strand. This entire cycle results in the synthesis of a replica of the target sequence.

PCR is performed in a tube in a thermal cycler with programmed heating and cooling. For nematodes, Subbotin *et al.* (2013) described a procedure consisting of a succession of three steps determined by temperature condition: (i) template denaturation at 95°C for 3–4 min; (ii) primer annealing at 55–60°C for 30 s to 2 min; and (iii) extension of DNA chain at 72°C for 30 s to 2 min (for specific PCR cycle parameters, see respective nematode sections). Repeating the cycles 25–40 times leads to more than a million-fold amplification in the copy number of the initial target sequence. Analysis of the PCR end product is done by **agarose gel electrophoresis**, which separates amplified products according to their size. The PCR product needs to be run on a 1% standard agarose gel or polyacrylamide gel stained with ethidium bromide (0.1 μg/ml). The gel is run at 50 volts for 1 h in electrophoresis. Ethidium bromide interacts with double-stranded DNA and causes it to fluoresce under UV radiation. The gel can be photographed using a gel documentation system. Approximate length of the bands, in base pairs, can be estimated by comparing with standard markers. In some cases, the amplicon size may serve as a diagnostic marker for a nematode group or species. However, the amplification can be characterized by various analyses, such as PCR–restriction fragment length polymorphism (PCR–RFLP), dot blotting or sequencing.

110.2.1 PCR–restriction fragment length polymorphism (PCR–RFLP)

Variations in sequences of PCR products can be revealed by restriction endonuclease digestion. PCR–RFLP of the internal transcribed spacer (ITS) region of rDNA is a very reliable method for diagnosis of various plant-parasitic nematodes. Different digestive enzymes (*Alu*I, *Bsh1236*I, *Cfo*I, *Hae*III, *Hinf*I, *Msp*I, *Pst*I, *Rsa*I and *Taq*I) can be used with universal or species-specific primers. Digestions are carried out in 5–15 μl reaction mixtures containing 2 units of the enzyme, 1× enzyme buffer, and 10 μl of the PCR product. The DNA fragments are separated using a standard agarose gel (1.8%) in 0.5× TBE buffer (45 mM Tris base, 1 mM EDTA and 45 mM boric acid). The length of fragments is estimated by a 100 bp DNA ladder and the band pattern is photographed under UV light using a gel documentation system. The expected species is determined by comparing the restriction pattern with those of the control species. The exact lengths of restriction fragments from the PCR products can also be obtained by virtual digestion of the sequences using WebCutter 2.0.

110.2.2 Randomly amplified polymorphic DNA (RAPD)

The RAPD technique uses a single random primer of around ten nucleotides long for creating genomic fingerprints. It is an extended DNA analysis to characterize a wide range of nematodes and is usually employed to estimate genetic diversity between individuals, populations or closely related species. RAPD PCR has been used for interspecific and more specifically intra-specific discrimination, particularly of *Globodera* and *Meloidogyne* species.

110.2.3 Amplified fragment length polymorphism (AFLP)

AFLP is one of the most popular fingerprinting techniques. It detects DNA sequence variation without the need for prior DNA sequence information. It produces a higher number of bands than RAPD. AFLP is a much more reliable technique; it is used to detect and evaluate genetic variation and can screen the whole genome. It consists of following steps (Subbotin *et al.*, 2013): (i) restriction of total DNA with two restriction enzymes; (ii) ligation of double-stranded adapter to the ends of the restriction fragment; (iii) amplification of a subset of the restriction fragment using 17–21 nucleotide primers complementary to the adapter and one that is 1–3 nucleotides adjacent to the restriction sites; and (iv) separation and visualization of AFLP PCR fragments with a variety of techniques, usually on denaturing polyacrylamide gels with further staining. Qualitative and quantitative studies have been conducted to assess genetic relationships among populations of cyst nematodes and root-knot nematodes.

110.2.4 Multiplex PCR

Multiplex PCR is a modification of a PCR process that enables the detection of one or several species of nematodes in a mixture sample by a single PCR test. Multiplex PCR includes pairs of multiple primers that are specific to different DNA sequences in a single amplification reaction; thus the process considerably decreases diagnostic time, labour, cost and the number of reactions for amplification. For a single nematode detection, multiplex PCR uses two primers. One set is to amplify the internal control confirming the presence of DNA in the sample; and the second set includes one species-specific primer to target the DNA of the desired nematode. This method uses the following steps (Devran *et al.*, 2018). DNA is extracted from the isolated nematode juveniles, adults and eggs using any of the methods described in previous sections. Multiplex PCR is carried out with two sets of primers: one with universal primer and the second with species-specific primers. Both PCR reactions are performed in a thermal gradient cycler with a total volume of 25 μl (containing 2.5 μl 10× PCR buffer, 2 mM $MgCl_2$, 200 μM dNTPs, 0.4 μM of each primer, 1 unit Platinum Taq DNA polymerase and 2.5 μl DNA). The cycling parameters are denaturing at 95°C for 3 min followed by 35 cycles at 95°C for 30 s, 56°C for 30 s and 72°C for 60 s, and a final extension step at 72°C for 7 min. Amplified products are electrophoresed on a 1.5% agarose gel in 1× TAE buffer (tris-acetate buffer, pH 8.3, and EDTA), stained with ethidium bromide and visualized. The PCR products are then purified and proceed for sequencing

110.2.5 Real-time PCR

Real-time PCR is a more rapid, sensitive and cost-efficient technique for identifying nematodes. Real-time PCR requires an instrument platform that consists of a thermal cycler with optics for fluorescents and provides real-time amplification of the DNA target sequence using direct analysis of the PCR products by integrating fluorescent probes or dyes into the reaction mix without the need for gel electrophoresis. In this method, the fluorescent molecule gives an estimate of the amount of DNA, as it is multiplied in each cycle as the fluorescent signal increases proportionally. Madani *et al.* (2005) described a real-time PCR method using SYBR green-I dye with melting curve analysis using the following steps. The nematode specimens are taken into an Eppendorf tube with 10 μl of double-distilled water and crushed with a micropestle or a Vibro Mixer. Then 50 μl of 1× PCR buffer and

5 μl of proteinase K (600 μg/ml) are added to the tube and it is incubated for 2 h at 65°C followed by 10 min at 95°C. The DNA suspension is serially diluted two-fold with distilled water. Afterwards, amplifications are carried out in real-time with 12 μl 1× SYBR Green I PCR Master Mix, 0.1 mM of each primer (using short amplicon species-specific primers for target nematode) and 4 μl DNA suspension in a total volume of 25 μl. Amplification reactions are performed using recommended universal thermal protocols: preheating for 2 min at 50°C, followed by initial template denaturation and activation of hot start polymerase for 10 min at 95°C, followed by 30 cycles of 95°C for 15 s each and combined annealing/extension phase of 65°C for 1 min. To generate the amplification curves for each reaction, sequence detection software (SDS) is used. The threshold cycle number (Ct) is also calculated with SDS and an automatic setting of the baseline. A disassociation curve is generated after each reaction to differentiate species amplicons from non-specific products. The DNA melting curve examination of the obtained amplicon is carried out at the end of amplification by cooling the samples at 1.6°C/s (ramp 100%) to 60°C and then increasing the temperature to 95°C at 0.03°C/s (ramp 2%). The emission of fluorescence is measured every 7 s (each data point collection in 0.2°C/s) over 20 min. The resulting PCR products can also be run on agarose gels to compare and estimate the PCR efficacy. Control samples without DNA template are included in each experiment with real-time PCR.

Quantitative real-time PCR (qPCR) is used for the detection and quantification of DNA present in a sample which is reproduced by the number of nematodes present in the sample. However, the stability of DNA from dead specimens in samples, particularly those extracted from cysts, appears to be an impediment. To avoid this, a specific dye, propidium monoazide, can be used, which enables the DNA to amplify only from viable/intact cells. The qPCR technique proves to be very useful for routine diagnosis of nematodes from field samples.

110.2.6 DNA microarrays

Microarrays provide a high-potential tool for diagnosing nematodes in multi-complex samples. It is next-generation diagnostics and can target multiple sequences simultaneously. Because of the specificity of the microarray method, it is able to detect unique sequences for each target species. Microarrays are composed of complementary DNAs (cDNAs) that can be sensed due to a fluorescence bind on the cDNA, on silicon chips or microscope slides, which comprise specific synthesized known DNA after hybridization of the cDNA. The microarrays have a high potential as a diagnostic tool for identifying nematodes, but little progress has been made in their use.

110.2.7 Other methods

Other techniques, such as microsatellites or single sequence repeats (SSRs), satellite DNA (satDNA), sequence characterized amplified regions (SCARs) and PCR–single strand conformational polymorphism (PCR–SSCP), have also been used to characterize *Bursaphelenchus*, *Ditylenchus*, *Globodera*, *Heterodera* and *Meloidogyne* species. Recently, a number of nucleic acid amplification techniques have been developed that do not require the use of thermal cycling equipment, such as loop-mediated isothermal amplification (LAMP) and recombinase polymerase amplification (RPA).

LOOP-MEDIATED ISOTHERMAL AMPLIFICATION (LAMP) LAMP is an innovative approach in nucleic acid amplification. The LAMP reaction requires a DNA polymerase with strand displacement activity (Bst polymerase) and a set of 4–6 specially designed primers based in a distinct region of the target DNA (Subbotin *et al.*, 2013). Because of the specific nature of the action of these primers, the amount of DNA produced in LAMP is significantly higher than PCR-based amplification. The reactions occur under isothermal conditions (60–65°C) and yield a large amount of product in a short time (30–60 min). The LAMP product can be visualized by the naked eye or gel electrophoresis by adding ethidium bromide or SYBR Green 1 dye in a tube. LAMP does not require a thermal cycler and optical detection and all the steps are conducted in a tube; thus it is less costly and holds potential for diagnosis in field conditions.

RECOMBINASE POLYMERASE AMPLIFICATION (RPA) The RPA technique represents a multipurpose alternative to PCR. RPA uses an extremely efficient displacement polymerase that rapidly amplifies a few copies of target nucleic acid (in 20 min) at a low constant operating temperature (37–42°C). RPA products can be carried out by using fluorescent probes in real-time or detected by lateral flow assay or agarose gel electrophoresis and can be used for the detection of plant-parasitic nematodes (Subbotin, 2018).

110.3 DNA sequencing

DNA sequencing is fundamental for modern gene manipulation. In any event, sequence information is a prerequisite for planning any substantial manipulation of the DNA. The amplified PCR product can be sequenced for determining the specific order of the bases in the DNA strands that are amplified. This particular sequence of A, T, G and C can then be compared with known sequences of an organism or strain to determine which organism's DNA matches the sequence of the amplified product obtained as a result of PCR. Sequencing of a purified PCR product can be carried out by a standard DNA sequencer or genetic analyser in any standard sequencing laboratory. However, if the facility is available, then the PCR products proceed with the following steps as described by Meldal *et al.* (2007). The PCR product is treated with alkaline phosphatase (1 U/μl) and exonuclease I (20 U/μl) for 15 min at 37°C, followed by inactivation of enzymes at 80°C for 15 min. Without any further purification, the material is then used for cycle sequencing using the ABI Prism BigDyeV 3.1 Terminator Cycle Sequencing kit. The sequencing conditions are 30 s at 96°C, 15 s at 50°C and 1 min at 60°C for 27 cycles. Primers used for sequencing are the same as for PCR amplification cases. Cycle sequence products are precipitated by adding 25 µl of 95% ethanol and 1 µl 3 M sodium acetate, pH 4.6, to each cycle sequencing reaction (10 µl). The samples are placed at room temperature for 15 min and centrifuged at 20,000× *g* for 15 min. After precipitation, an additional wash of the pellet is done with 125 µl of 70% ethanol and centrifuged at 20,000× *g* for 5 min. The pellet is dried in a Speedvac concentrator, redissolved in formamide and run on 50 cm capillaries with POP7 polymer. The exact lengths of restriction fragments from PCR products will be obtained by virtual digestion of sequence using WebCutter 2.0. Sequences are aligned with ClustalX and manually checked. Chromatograms are checked and corrected by using the Chromas Pro 1.7.6 software. Differences between sequences are counted using the BioEdit sequence alignment options. Additional sequences of interest for phylogenetic analyses can also be obtained from GenBank.

110.4 Phylogenetic analysis

All amplified DNA sequences are assembled with sequencing analysis software and aligned with ClustalX initially under the default alignment parameters. This is then manually upgraded in the PAUP program. All phylogenetic analyses can be performed using both distance- and character-based methods. The Kimura 2-parameter is generally used for distance methods. The maximum parsimony (MP) and maximum likelihood (ML) are used for character-based methods. Both MP and ML trees can be constructed through heuristic search in the PAUP 4.0 program (Swofford, 1999). Sequence variability within individuals and between samples can be determined by pairwise comparisons in the PAUP program with the available sequence in GenBank. Genetic diversity and isolate differentiation analysis can also be calculated by Software MEGA 6.0.

16

International Regulations

Today the modern transport systems and the unrestricted global trade of commodities across continents have opened the floodgate for the introduction of invasive species into exotic areas at a much faster rate and in a shorter time than ever before. While all species compete to survive, invasive species have traits that allow them to outcompete native species. The World Trade Organization (WTO) intends to supervise and liberalize international trade, but it is also the WTO's duty to review and propagate national trade policies, and to ensure the implementation of 'the agriculture agreement' and 'the Sanitary and Phytosanitary (SPS) measures agreement' (WTO-SPS). Member countries are encouraged to follow international standards, guidelines and recommendations where they exist. The WTO-SPS agreement allows countries to use different standards and different methods of inspection for products (IPPC, 2021). In order to achieve the targets, international standards need to be developed and the WTO-SPS agreement has assigned responsibilities to the International Plant Protection Convention (IPPC) to develop standards for plant health and International Standards for Phytosanitary Measures (ISPM).

111 International Plant Protection Convention (IPCC)

The WTO-SPS agreement recognizes the International Plant Protection Convention as the international standard-setting body for developing international standards, guidelines and recommendations with respect to phytosanitary measures related to trade in food and agriculture. The IPPC came into force in 1952 and currently 183 member countries adhere to it. The IPPC work programme involves three main activities (IPPC, 2021):

- the development of International Standards for Phytosanitary Measures (standard setting);
- the provision of information required by the IPPC and the facilitation of information exchange between contracting parties (information exchange); and
- the provision of technical assistance, especially for capacity building, to facilitate the implementation of the IPPC (technical assistance).

112 Regional Plant Protection Organizations (RPPOs)

Under article IX of the IPPC (revised in 1997), regional and inter-regional cooperation and harmonization in all the areas of plant protection are performed by recognized Regional Plant Protection Organizations (RPPOs are) and there are coordinating bodies in different continents to further implement the objectives of the Convention and to gather and disseminate information. Neither all contracting parties to the IPPC are members of RPPOs, nor all members of RPPOs are contracting parties to the IPPC. Moreover, certain contracting parties to the IPPC belong to more than one RPPO. There are the following ten organizations under the (IPPC, 2021).

1. Asia and Pacific Plant Protection Commission (APPPC)
2. Caribbean Plant Protection Commission (CPPC) – abolished in 2014 and reconstituted as Caribbean Agricultural Health and Food Safety Agency (CAHFSA)
3. European and Mediterranean Plant Protection Organization (EPPO)
4. North American Plant Protection Organization (NAPPO)
5. Near East Plant Protection Organization (NEPPO)
6. Pacific Plant Protection Organization (PPPO)
7. Inter-African Phytosanitary Council (IAPSC)
8. Comunidad Andina (CAN)
9. Comité de Sanidad Vegetal del Cono Sur (COSAVE)
10. Organismo Internacional Regional de Sanidad Agropecuaria (OIRSA)

112.1 Members countries (IPPC, 2021)

APPPC: Australia, Bangladesh, Cambodia, China, Fiji, French Polynesia, India, Indonesia, Lao, Malaysia, Myanmar, Nepal, New Zealand, North Korea, Pakistan, Papua New Guinea, Philippines, Samoa, Solomon Islands, South Korea, Sri Lanka, Thailand, Tonga, Vietnam.

CAHFSA: Anguilla, Bahamas, Barbados, Bermuda, Cayman Islands, Colombia, Costa Rica, Cuba, Dominica, Dominican Republic, French Guiana, Grenada,

DOI: 10.1079/9781789247367.0017

Guadeloupe, Guyana, Haiti, Honduras, Jamaica, Martinique, Mexico, Netherlands Antilles, Nicaragua, Panama, Puerto Rico, Saint Lucia, Saint Kitts & Nevis, Suriname, Trinidad and Tobago, Turks and Caicos Islands, Virgin Islands (British), Virgin Islands (US).

CAN: Bolivia, Colombia, Ecuador, Peru, Venezuela.

COSAVE: Argentina, Brazil, Chile, Paraguay, Peru, Uruguay.

EPPO: Albania, Algeria, Austria, Belarus, Belgium, Bulgaria, Croatia, Cyprus, Czech Republic, Denmark, Estonia, Finland, France, Georgia, Germany, Greece, Guernsey, Hungary, Ireland, Israel, Italy, Jersey, Jordan, Kazakhstan, Kyrgyzstan, Latvia, Lithuania, Luxembourg, Macedonia, Malta, Moldova, Morocco, Netherlands, Norway, Poland, Portugal, Romania, Russia, Serbia, Slovakia, Slovenia, Spain, Sweden, Switzerland, Tunisia, Turkey, Ukraine, United Kingdom, Uzbekistan.

IAPC: Algeria, Angola, Benin, Botswana, Burkina Faso, Burundi, Cameroon, Cape Verde, Central African Republic, Chad, Comoros, Congo (Democratic Republic of), Congo (Republic of), Côte d'Ivoire, Djibouti, Egypt, Equatorial Guinea, Eritrea, Ethiopia, Gabon, Gambia, Ghana, Guinea, Guinea-Bissau, Kenya, Lesotho, Liberia, Libya, Madagascar, Malawi, Mali, Mauritania, Mauritius, Morocco, Mozambique, Niger, Nigeria, Rwanda, São Tomé & Principe, Senegal, Seychelles, Sierra Leone, Somalia, Sudan, Swaziland, Tanzania, Togo, Tunisia, Uganda, Zaire, Zambia, Zimbabwe.

NAPPO: Canada, Mexico, USA.

NEPPO: Algeria, Egypt, Jordan, Libya, Malta, Morocco, Pakistan, Sudan, Syria, Tunisia.

OIRSA: Costa Rica, El Salvador, Guatemala, Honduras, Mexico, Nicaragua, Panama.

PPPO: American Samoa, Australia, Cook Islands, Fiji, French Polynesia, Guam, Kiribati, Marshall Islands, Micronesia, Nauru, New Caledonia, New Zealand, Niue, Northern Mariana Islands, Palau, Papua New Guinea, Pitcairn, Samoa, Solomon Islands, Tokelau, Tonga, Tuvalu, Vanuatu, Wallis and Futuna Islands.

113 National Plant Protection Organizations (NPPOs)

To meet the objective of preventing the introduction and spread of pests, exporting and importing contracting parties, and to implement phytosanitary measures, all member countries of IPPC need to establish a National Plant Protection Organization (NPPO). The NPPOs have responsibilities to implement Plant Quarantine (PQ) regulations that should be entrusted with public authority and legally establish barriers against the dissemination of injurious, invasive and exotic pests and diseases (IPPC, 2021). The need for PQ to be entrusted to a public authority is considered as one of the fundamental principles of PQ. Almost all countries now have PQ regulations (in some countries combined as Biosecurity) in force with certain adaptations according to the country's needs and barriers.

114 International Standards for Phytosanitary Measures (ISPMs)

International Standards for Phytosanitary Measures are the standards, guidelines and recommendations recognized as the basis for phytosanitary measures applied by members of the WTO under the Agreement on the Application of Sanitary and Phytosanitary Measures (the WTO-SPS Agreement). The Commission on Phytosanitary Measures (CPM), which is the governing body of the IPPC, has adopted the ISPMs set out below to provide guidance to countries and assist contracting parties to implement the objectives of the IPPC (IPPC, 2021). These standards in themselves are not regulatory instruments but come into force once countries have established requirements within their national legislation. Further information on import and export requirements should be requested from national plant protection organizations (NPPOs).

The ISPMs provide guidance to member countries in implementing national programmes and fulfilling the requirements of the IPPC. The ISPMs may be very general (e.g. Glossary, Principles, etc.) or highly specific (e.g. pest status, solid wood packaging material, treatment, pest risk analysis, etc.). As of now, there are 45 ISPMs adopted by CPM and a few more in draft status circulated for 'Member consultations' for comments from member countries before adopting the ISPMs. The ISPMs are as follows (each can be accessed through https://www.ippc.int).

ISPM 1 *Phytosanitary principles for the protection of plants and the application of phytosanitary measures in international trade* (adopted in 1993, revised in 2006)

ISPM 2 *Framework for pest risk analysis* (adopted in 1995, revised in 2007)

ISPM 3 *Guidelines for the export, shipment, import and release of biological control agents and other beneficial organisms* (adopted in 1995, revised in 2005)

ISPM 4 *Requirements for the establishment of pest-free areas* (adopted in 1995)

ISPM 5 *Glossary of phytosanitary terms* (updated as needed)

ISPM 6 *Surveillance* (adopted in 1997, revised in 2018)

ISPM 7 *Phytosanitary certification system* (adopted in 1997, revised in 2011)

ISPM 8 *Determination of pest status in an area* (adopted in 1998)

ISPM 9 *Guidelines for pest eradication programmes* (adopted in 1998)

ISPM 10 *Requirements for the establishment of pest-free places of production and pest-free production sites* (adopted in 1999)

ISPM 11 *Pest risk analysis for quarantine pests* (adopted in 2001, revised in 2004 and 2013)

ISPM 12 *Phytosanitary certificates* (adopted in 2001, revised in 2011)

ISPM 13 *Guidelines for the notification of non-compliance and emergency action* (adopted in 2001)

ISPM 14 *The use of integrated measures in a systems approach for pest risk management* (adopted in 2002)

ISPM 15 *Regulation of wood packaging material in international trade* (adopted in 2002, revised in 2009, 2013 and 2018)

ISPM 16 *Regulated non-quarantine pests: concept and application* (adopted in 2002)

ISPM 17 *Pest reporting* (adopted in 2002)
ISPM 18 *Guidelines for the use of irradiation as a phytosanitary measure* (adopted in 2003)
ISPM 19 *Guidelines on lists of regulated pests* (adopted in 2003)
ISPM 20 *Guidelines for a phytosanitary import regulatory system* (adopted in 2004, revised in 2017)
ISPM 21 *Pest risk analysis for regulated non-quarantine pests* (adopted in 2004)
ISPM 22 *Requirements for the establishment of areas of low pest prevalence* (adopted in 2005)
ISPM 23 *Guidelines for inspection* (adopted in 2005)
ISPM 24 *Guidelines for the determination and recognition of equivalence of phytosanitary measures* (adopted in 2005)
ISPM 25 *Consignments in transit* (adopted in 2006)
ISPM 26 *Establishment of pest-free areas for fruit flies (Tephritidae)* (adopted in 2006, revised in 2014 and 2015)
ISPM 27 *Diagnostic protocols for regulated pests* (adopted in 2006 and updated as needed)
ISPM 28 *Phytosanitary treatments for regulated pests* (adopted in 2007)
ISPM 29 *Recognition of pest-free areas and areas of low pest prevalence* (adopted in 2007)
ISPM 30 *(Revoked). Establishment of areas of low pest prevalence for fruit flies (Tephritidae)* (adopted in 2008; incorporated as an annexe to ISPM 35 in 2018)
ISPM 31 *Methodologies for sampling of consignments* (adopted in 2008)
ISPM 32 *Categorization of commodities according to their pest risk* (adopted in 2009)
ISPM 33 *Pest free potato* (Solanum *spp.*) *micropropagative material and minitubers for international trade* (adopted in 2010)
ISPM 34 *Design and operation of post-entry quarantine stations for plants* (adopted in 2010)
ISPM 35 *Systems approach for pest risk management of fruit flies (Tephritidae)* (adopted in 2012)
ISPM 36 *Integrated measures for plants for planting* (adopted in 2012)
ISPM 37 *Determination of host status of fruit to fruit flies (Tephritidae)* (adopted in 2016)
ISPM 38 *International movement of seeds* (adopted in 2017)
ISPM 39 *International movement of wood* (adopted in 2017)
ISPM 40 *International movement of growing media in association with plants for planting* (adopted in 2017)
ISPM 41 *International movement of used vehicles, machinery and equipment* (adopted in 2017)
ISPM 42 *Requirements for the use of temperature treatments as a phytosanitary measure* (adopted in 2018)
ISPM 43 *Requirements for the use of fumigation as a phytosanitary measure* (adopted in 2019).
ISPM 44 *Requirements for the use of modified atmosphere treatments as phytosanitary measures* (adopted in 2021)
ISPM 45 *Requirements for national plant protection organizations if authorizing entities to perform phytosanitary actions* (adopted in 2021)

115 ISPMs Applicable for Plant-parasitic Nematodes

The typical phytosanitary measures to prevent the spread of plant-parasitic nematodes, as described by Hockland *et al.* (2012), include: a specific quarantine treatment of the imported commodity; or prohibiting the entry of a host commodity likely to carry a particular nematode; or prohibiting the entry of non-treated media; or that the plants be imported with bare roots (free from media); or that plants should have been grown in approved growing media according to an officially agreed protocol; or that the commodities should originate from an area determined to be free of a particular nematode (nematode/pest-free area). In general, the phytosanitary measures impose restriction on the movement of soil, machinery, lorries, workers, etc. from fields where an infestation is already recognized. All soil contaminated articles should be thoroughly cleaned. Disposal of infested plant material and soil must be carried out in a proper manner. No contaminated consignment of nursery stock or plant parts should be permitted to be imported unless subjected to phytosanitary treatments by the exporting country and a phytosanitary certificate should be endorsed by the exporting country. If a contaminated consignment is detected at the port of entry, the same will be recommended for deportation or destruction at the port of entry. If a quarantine or regulated nematode is detected, immediate regulatory measures are taken to contain the potential spread and it is reported to the quarantine station concerned. The regulatory measures also include fumigation under vacuum (650 mm Hg) with hydrogen cyanide (initial dose, 4 g/m^3) for 1 h at over 10°C or treatments of seeds, planting material and cut flowers with methyl bromide fumigation. These recommendations are included in ISPM 28. Further, if the nematode is detected during post-entry quarantine inspection, the seeds or plants part should be destroyed by incineration.

Details of ISPMs related to plant-parasitic nematodes, how to determine their invasiveness and the subjects dealt with in this chapter are summarized below.

115.1 Pest risk analysis

In this book the common assessing criteria for the invasiveness of a nematode species are based on the pest risk analysis framework (ISPM 02), pest risk analysis for quarantine pest (ISPM 11) and pest risk analysis for regulated non-quarantine pests (ISPM 21).

ISPM 02 (1995, 2007, 2016) ***Framework for pest risk analysis:*** This standard provides a framework that describes the pest risk analysis (PRA) process within the scope of the IPPC. It introduces the three stages of pest risk analysis – initiation, pest risk assessment and pest risk management. Generic issues of information gathering, documentation, risk communication, uncertainty and consistency are addressed. The PRA process is a technical tool used for identifying appropriate phytosanitary measures. The PRA process is initiated in Stage 1 with the identification of an organism or pathway that may be considered for pest risk assessment, or as part of a review of existing phytosanitary measures, concerning a defined PRA area. The first step is to determine or confirm whether or not the organism is considered as a pest based on scientific evidence. If no pests are identified, the analysis does not need to continue. The analysis of pests identified in Stage 1 continues to Stages 2 and 3 using guidance provided in other standards. Information gathering, documentation and risk communication, as well as uncertainty and consistency, are issues common to all PRA stages.

ISPM 11 (2001, 2003, 2004, 2017) ***Pest risk analysis for quarantine pests, including analysis of environmental risks and living modified organisms:*** The standard provides details for the conduct of pest risk analysis (PRA) to determine if pests are quarantine pests. It describes the integrated processes to be used for risk assessment as well as the selection of risk management options. It also includes details regarding the analysis of risks of plant pests to the environment and biological diversity, including those risks affecting uncultivated or unmanaged plants, wild flora, habitats and ecosystems contained in the PRA area. It includes guidance on evaluating potential phytosanitary risks to plants and plant products posed by living modified organisms (LMOs).

ISPM 21 (2004) ***Pest risk analysis for regulated non-quarantine pests:*** This standard provides guidelines for conducting pest risk analysis (PRA) for regulated non-quarantine pests (RNQPs). It describes the integrated processes to be used for risk assessment and the selection of risk management options to achieve a pest tolerance level. The objectives of a PRA for RNQPs are, for a specified PRA area, to identify pests associated with plants for planting, to evaluate their risk and, if appropriate, to identify risk management options to achieve a tolerance level. PRA for RNQPs follows a process defined by three stages:

Stage 1. Initiating the process. This involves identifying the pest(s) associated with the plants for planting that are not quarantine pests but which may be of regulatory concern and that should be considered for risk analysis in relation to the identified PRA area.

Stage 2. Risk assessment. This implies or begins with the categorization of individual pests associated with the plants for planting and their intended use to determine whether the criteria for an RNQP are satisfied. Risk assessment continues with an analysis to determine if the plants for planting are the main source of the pest infestation and if the economic impacts of the pest on the intended use of those plants for planting are unacceptable.

Stage 3. Risk management. This involves identifying a pest tolerance level to avoid the unacceptable economic impact(s) identified at Stage 2 and management options to achieve that tolerance.

115.2 Pest surveillance (survey and monitoring)

Pest surveillance is an official process which collects and records data on pest occurrence or absence by survey, monitoring or other procedures, whereas survey is an official procedure conducted over a defined period of time to determine the characteristics of a pest population or to determine which pest species occurs in an area. Pest surveillance implies close and regular vigilance of the pest over the development of its population or incidence in an area or ecosystem. Following ISPMs cover the issues related to pest surveillance.

ISPM 06 (1997, 2018) ***Guidelines for surveillance:*** This standard describes the components of the survey and monitoring systems for pest detection and the supply of information for use in pest risk analyses, the establishment of pest-free areas and, where appropriate, the preparation of pest lists. The collecting and recording of pest information are fundamental to all these concepts. The implication is that National Plant Protection Organizations (NPPOs) should be in a position to validate declarations of the absence or limited distribution of quarantine pests. There are two major types of surveillance systems:

General surveillance: This involves a process whereby information on particular pests which are of concern for an area is gathered from many sources, wherever it is available and provided for use by the NPPO.

Specific surveys: The NPPOs obtain information on pests of concern on specific sites in an area over a defined period.

ISPM 08 (1998) ***Determination of pest status in an area:*** This standard describes the content of a pest record, and the use of pest records and other information in the determination of pest status in an area. Descriptions of pest status categories and recommendations for good reporting practices are provided.

115.3 Diagnostic protocols

Following standard provides guidance on the structure and content of the IPPC diagnostic protocols for regulated pests.

ISPM 27 (2006, updated as needed) and DPs 1–29 ***Diagnostic protocols for regulated pests:*** The protocols describe procedures and methods for the official diagnosis of regulated pests that are relevant for international trade. They provide at least the minimum requirements for a reliable diagnosis of regulated pests. Diagnostic protocols (DPs) for specific regulated pests are included as annexes (in original) to this standard. DPs contain the minimum requirements for a reliable diagnosis of the specified regulated pests and provide flexibility to ensure that methods are appropriate for use in the full range of circumstances. The methods included in DPs are selected on the basis of their sensitivity, specificity and reproducibility. The information related to these factors is provided for each of these methods. DPs are intended to be used by laboratories performing pest diagnosis as part of phytosanitary measures. To date, five DPs have been standardized for plant-parasitic nematodes, namely:

DP 8: Diagnostic protocol for *Ditylenchus dipsaci* and *Ditylenchus destructor* (2015)
DP 10: Diagnostic protocol for *Bursaphelenchus xylophilus* (2016)
DP 11: Diagnostic protocol for *Xiphinema americanum sensu lato* (2016)
DP 17: Diagnostic protocol for *Aphelenchoides besseyi*, *A. fragariae* and *A. ritzemabosi* (2016)
DP 18: Diagnostic protocol for *Anguina* spp. (2017)

115.4 Pest (nematode)-free areas, pest-free places of production and production sites and areas of low pest prevalence

A 'Nematode-free area' is: 'an area in which a specific nematode does not occur as demonstrated by scientific evidence and in which, where appropriate, this condition is being officially maintained'. To declare nematode-free areas, nematode-free places of production and production sites or areas of low nematode prevalence, there must be collection of surveillance data on the distribution of specific nematodes for which an additional declaration is imposed by the importing countries. To satisfy an importing country's requirement that a consignment is pest-free when the pest/nematode in question is known to be present in the exporting country, the exporting country may establish a pest-free area or pest-free place(s) of production within its country by following five ISPMs (ISPM 04, 10, 22, 29 and 33).

ISPM 04 (1995) ***Requirements for the establishment of pest-free areas:*** This standard describes the requirements for the establishment and use of pest-free areas (PFAs) as a risk management option for phytosanitary certification of plants and plant products and other regulated articles exported from the PFA or to support the scientific justification for phytosanitary measures taken by an importing country for protection of an endangered PFA.

ISPM 10 (1999) ***Requirements for the establishment of pest-free places of production and pest-free production sites:*** This standard describes the requirements for the establishment and use of pest-free places of production and pest-free production sites as risk management options for meeting phytosanitary requirements for the import of plants, plant products and other regulated articles. This standard uses the concept of 'pest freedom' to allow exporting countries to assure importing countries that plants, plant products and other regulated articles are free from a specific pest or pests and meet the phytosanitary requirements of the importing country when imported from a pest-free place of production. In circumstances where a defined portion of a place of production is managed as a separate unit and can be maintained pest free, it may be regarded as a pest-free production site.

ISPM 22 (2005) ***Requirements for the establishment of areas of low pest prevalence:*** This standard describes the requirements and procedures for the establishment of areas of low pest prevalence (ALPPs) for regulated pests in an area and, to facilitate export, for pests regulated by an importing country only. This includes the identification, verification, maintenance and use of those ALPPs. The establishment of an ALPP is a pest management option used to maintain or reduce a pest population below a specified level in an area. An ALPP may be used to facilitate exports or to limit pest impact in the area. Surveillance of the relevant pest should be conducted according to appropriate protocols. Additional phytosanitary procedures may be required to establish and maintain an ALPP.

ISPM 29 (2007) ***Recognition of pest-free areas and areas of low pest prevalence:*** This standard provides guidance and describes a procedure for the bilateral recognition of pest-free areas (PFAs) and areas of low pest prevalence (ALPPs). This standard does not include specified timelines for the recognition procedure. This standard also provides some considerations regarding pest-free places of production and pest-free production sites. Recognition of PFAs and ALPPs is a technical and administrative process to achieve acceptance of the phytosanitary status of a delimited area. Contracting parties to the IPPC should proceed with a recognition process without undue delay. The process should be applied without discrimination between contracting parties. Contracting parties should endeavour to maintain transparency in all aspects of the recognition process. Both exporting and importing contracting parties have specific responsibilities relating to the recognition of PFAs and ALPPs. The recognition process should be sufficiently documented by contracting parties. However, where the absence of the pest in an area and the PFA status can easily be determined, the procedure for recognition described in section 4 (see original standard at www.ippc.in) may not be required or very little supporting information may be necessary.

ISPM 33 (2010) ***Pest-free potato (*****Solanum** ***spp.) micropropagative material and minitubers for international trade:*** This standard guides the production, maintenance and phytosanitary certification of pest-free potato (*Solanum tuberosum* and related tuber-forming species) micropropagative material and minitubers intended for international trade. This standard does not apply to field-grown propagative material of potato, or potatoes intended for consumption or processing. Facilities used for the production of potato micropropagative material and minitubers for export should be authorized or operated directly by the National Plant Protection Organization (NPPO) of the exporting country. Pest risk analysis (PRA), carried out by the NPPO of the importing country, should justify establishing phytosanitary import requirements for regulated pests in the trade of potato micropropagative material and minitubers.

115.5 Regulatory system

The regulatory system provides guidelines to determine whether a pest is a regulated pest or regulated non-quarantine pests, and application aspects to prevent the introduction of quarantine pests or limit the entry of regulated non-quarantine pests with imported commodities and other regulated articles. Following ISPMS cover the issues on regulatory system (IPPC, 2021).

ISPM 16 (2002) ***Regulated non-quarantine pests – concept and application:*** This standard describes the concept of regulated non-quarantine pests and identifies their characteristics. It describes the application of the concept in practice and the relevant elements for regulatory systems. Pests that are not quarantine pests may be subject to phytosanitary measures because of their presence in plants for planting results in economically unacceptable impacts. They are defined in the IPPC (1997) as regulated non-quarantine pests (RNQPs). Where official control is applied to plants for planting produced within countries to protect them from such pests, then the same or equivalent phytosanitary measures may be applied to those pests on imported plants for planting of the same species for the same intended use (to avoid economic impact).

ISPM 19 (2003) ***Guidelines on lists of regulated pests:*** This standard describes the procedures to prepare, maintain and make available lists of regulated pests. Lists of regulated pests are established by an importing contracting party to specify all currently regulated pests for which phytosanitary measures may be taken. Quarantine pests, including those that are subject to provisional or emergency measures, and regulated non-quarantine pests should be listed. Required information associated with the listing includes the pest's scientific name, the pest category and commodities or other articles that are regulated for the pest. Lists should be made available to the IPPC Secretariat, to Regional Plant Protection Organizations (RPPOs) and, on request, to other contracting parties.

ISPM 20 (2004, 2017) ***Guidelines for a phytosanitary import regulatory system:*** This standard describes the structure and operation of a phytosanitary import regulatory system and the rights, obligations and responsibilities which should be considered in establishing, operating and revising the system. The objective of a phytosanitary import regulatory system is to prevent the introduction of quarantine pests or limit the entry of regulated non-quarantine pests with imported commodities and other regulated articles. An import regulatory system should consist of two components: a regulatory framework of phytosanitary legislation, regulations and procedures; and an official service, the NPPO, responsible for operation or oversight of the system. The legal framework should include: legal authority for the NPPO to carry out its duties; measures with which imported commodities should comply; other measures (including prohibitions) concerning imported commodities and other regulated articles; and actions that may be taken when incidents of non-compliance or incidents requiring emergency action are detected. In operating an import regulatory system, the NPPO has several responsibilities. These include the responsibilities identified in Article IV.2 of the IPPC (1997) relating to import including surveillance, inspection, disinfestation or disinfection, the conduct of pest risk analysis, and training and development of staff.

ISPM 25 (2006) ***Consignments in transit:*** This standard describes procedures to identify, assess and manage phytosanitary risks associated with consignments of regulated articles which pass through a country without being imported, in such a manner that any phytosanitary measures applied in the country of transit are technically justified and necessary to prevent the introduction into and/or spread of pests within that country.

ISPM 32 (2009) ***Categorization of commodities according to their pest risk:*** This standard provides criteria for NPPOs of importing countries on how to categorize commodities according to their pest risk when considering import requirements. This categorization should help in identifying whether further pest risk analysis (PRA) is required and if phytosanitary certification is needed. The first stage of categorization is based on whether the commodity has been processed and, if so, the method and degree of processing to which the commodity has been subjected before export. The second stage of categorization of commodities is based on their intended use after import. This allows pest risks associated with specific commodities to be assigned to categories. The objective of such categorization is to provide importing countries with criteria to better identify the need for a pathway-initiated PRA and to facilitate the decision-making process regarding the possible establishment of import requirements. Contaminating pests or storage pests that may become associated with the commodity after processing are not considered in this standard.

115.6 Inspection methodology and sampling

Following standards describe the procedures for the inspection of consignments of plants, plant products and other regulated articles at import and export, and guides NPPOs in selecting appropriate sampling methodologies for inspection or testing of consignments to verify compliance with phytosanitary requirements.

ISPM 23 (2005) ***Guidelines for inspection:*** This is focused on the determination of compliance with phytosanitary requirements, based on visual examination, documentary checks and identity and integrity checks. The results of inspection should allow an inspector to decide whether to accept, detain or reject the consignment, or whether further analysis is required. NPPOs may determine that consignments should be sampled during the inspection. The sampling methodology used should depend on the specific inspection objectives.

ISPM 31 (2008) ***Methodologies for sampling consignments:*** The sampling methodologies used by NPPOs in selecting samples for the inspection of consignments of commodities moving in international trade are based on several sampling concepts. These include parameters such as acceptance level, level of detection, confidence level, the efficacy of detection and sample size. In using sampling methodologies, NPPOs accept some degree of risk that non-conforming lots may not be detected. Inspection using statistically based methods can provide results with a certain level of confidence only and cannot prove the absence of a pest from a consignment. This standard does not give guidance on field sampling (for example, as required for surveys).

115.7 Post-entry quarantine

Following standard describes general guidelines for the design and operation of post-entry quarantine (PEQ) stations for holding imported consignments of plants, mainly plants for planting, in confinement to verify whether or not they are infested with quarantine pests.

ISPM 34 (2010) ***Design and operation of post-entry quarantine stations for plants:*** For certain commodities, the NPPO of the importing country may decide that PEQ is required to manage pest risks identified by PRA. Confinement of a consignment of plants in a PEQ station may be an appropriate phytosanitary measure in cases where a quarantine pest is difficult to detect, where it takes time for sign or symptom expression, or where testing or treatment is required. For a PEQ station to function successfully, its design and management should ensure that any quarantine pests that may be associated with consignments of plants are suitably confined, and do not move or escape from the station. The PEQ station should also ensure that consignments of plants are held in a manner that best facilitates observation, research, further inspection, testing or treatment of the plants. PEQ stations may consist of a field site, screen house, glasshouse and/or laboratory, amongst others. The type of facility to be used should be determined by the type of imported plants and the quarantine pests that may be associated with them.

115.8 Pest reporting and eradication

Following standard describes the responsibilities of and requirements for contracting parties in reporting the occurrence, outbreak and spread of pests in areas for which they are responsible. It also guides reporting successful eradication of pests and establishment of pest-free areas.

ISPM 17 (2002) ***Pest Reporting:*** IPPC (1997) requires countries to report on the occurrence, outbreak and spread of pests to communicate immediate or potential danger. NPPOs have the responsibility to collect pest information by surveillance and to verify the pest records thus collected. Occurrence, outbreak or spread of pests that are known (based on observation, previous experience, or pest risk analysis) to be of immediate or potential danger should be reported to other countries, in particular to neighbouring countries and trading partners. Pest reports should contain information on the identity of the pest, location, pest status, and nature of the immediate or potential danger. They should be provided without undue delay, preferably through electronic means, through direct communication, openly available publication and/or the International Phytosanitary Portal (IPP).

ISPM 09 (1998) ***Guidelines for pest eradication programmes:*** This standard describes the components of a pest eradication programme which can lead to the establishment or re-establishment of pest absence in an area. A programme for pest eradication may be developed by an NPPO as an emergency measure to prevent the establishment and/or spread of a pest following its recent entry (re-establish a pest-free area), or a measure to eliminate an established pest (establish a pest-free area). The eradication process involves three main activities: surveillance, containment and treatment and/or control measures.

115.9 Pest management

Following standards provide guidelines for the development and evaluation of integrated measures in a systems approach as an option for pest risk management under the relevant international standards for pest risk analysis designed to meet phytosanitary requirements for the import of plants, plant products and other regulated articles.

ISPM 14 (2002) ***The use of integrated measures in a systems approach for pest risk management:*** Systems approaches, which integrate measures for pest risk management in a defined manner, could provide an alternative to single measures to meet the appropriate level of phytosanitary protection of an importing country. They can also be developed to provide phytosanitary protection in situations where no single measure is available. A systems approach requires the integration of different measures, at least two of which act independently, with a cumulative effect. Exporting and importing countries may consult and cooperate in the development and implementation of a systems approach.

ISPM 36 (2012) ***Integrated measures for plants for planting:*** This standard outlines the main criteria for the identification and application of integrated measures at the place of production for the production of plants for planting (excluding seeds) for international trade. It provides guidance to help identify and manage pest risks associated with plants for planting as a pathway. Plants for planting are generally considered to pose a higher pest risk than other regulated articles. Integrated measures may be used to manage the pest risks that plants for planting pose as a pathway for regulated pests and to ensure they meet phytosanitary import requirements. Integrated measures may be developed and implemented by the NPPO of the exporting country.

115.10 Phytosanitary treatments

ISPM 15 (2002, 2009) ***Guidelines for regulating wood packaging material in international trade:*** This standard describes phytosanitary measures that reduce the risk of introduction and spread of quarantine pests associated with the movement in international trade of wood packaging material made from raw wood. Wood packaging material covered by this standard includes dunnage but excludes wood packaging made from wood processed in such a way that it is free from pests (e.g. plywood). Pests associated with wood packaging material are known to have negative impacts on forest health and biodiversity. Implementation of this standard is considered to reduce significantly the spread of pests and subsequently their negative impacts. Approved phytosanitary measures that significantly reduce the risk of pest introduction and spread via wood packaging material consist of the use of debarked wood (with a specified tolerance for remaining bark) and the application of approved treatments (methyl bromide fumigation and forced hot air treatment). NPPOs of exporting and importing countries have specific responsibilities. Treatment and application of the mark must always be under the authority of the NPPO.

ISPM 18 (2003) ***Guidelines for the use of irradiation as a phytosanitary measure:*** This standard provides technical guidance on the specific procedures for the application of ionizing radiation as a phytosanitary treatment for regulated pests or articles. Treatment with ionizing radiation (irradiation) may be used for pest risk management. NPPOs should be assured that the efficacy of the treatment is scientifically demonstrated for the regulated pest(s) of concern and the required response. Application of the treatment requires dosimetry and dose mapping to ensure that the treatment is effective in particular facilities and with specific commodity configurations. The NPPO is responsible for ensuring that facilities are appropriately designed for phytosanitary treatments. Procedures should be in place to ensure that the treatment can be conducted properly and commodity lots are handled, stored and identified to ensure that phytosanitary security is maintained.

ISPM 28 (2007) & PTs 1–32 *Phytosanitary treatments for regulated pests:* Harmonized phytosanitary treatments (PTs) support efficient phytosanitary measures in a wide range of circumstances and enhance the mutual recognition of treatment efficacy. This standard presents, in its Annexes, phytosanitary treatments evaluated and adopted by the CPM. The treatments are for the control of regulated pests on regulated articles, primarily those moving in international trade. The adopted treatments provide the minimum requirements necessary to control a regulated pest at a stated efficacy. NPPOs and Regional Plant Protection Organizations (RPPOs) may submit data and other information for the evaluation of efficacy, feasibility and applicability of treatments. Treatments that are eligible for evaluation include mechanical, chemical, irradiation, physical and controlled-atmosphere treatments. Submissions with complete information will be considered by the Technical Panel on Phytosanitary Treatments (TPPT), and if the treatment is deemed acceptable, it will be recommended to the CPM for adoption. At present 32 PTs have been approved by IPPC and available at www.ippc.int.

115.11 Phytosanitary certification system

Following standard describes the components of a national system for the issuance of phytosanitary certificates.

ISPM 07 (1997, 2011) *Export certification system:* IPPC requires its contracting parties to make arrangements to issue phytosanitary certificates certifying compliance with the phytosanitary regulations of other contracting parties. This standard describes an export certification system to produce valid and credible phytosanitary certificates. Exported consignments certified under these systems should meet the current phytosanitary requirements of the importing country. The basic elements of the phytosanitary certification process include ascertaining the relevant phytosanitary requirements of the importing country (including import permits if required), verifying that the consignment conforms to those requirements at the time of certification and issuing a phytosanitary certificate.

ISPM 12 (2001, 2011) *Guidelines for phytosanitary certificates:* This standard describes principles and guidelines for the preparation and issue of phytosanitary certificates and of phytosanitary certificates for re-export. The standard describes principles and guidelines to assist NPPOs with the preparation and issue of phytosanitary certificates and phytosanitary certificates for re-export. Model certificates are provided in the Annexe of the IPPC adopted in 1997 and are appended to this standard for reference.

116 Incursion Management

In spite of the impact of the multilateral trade agreement and phytosanitary measures, interception of regulated and non-regulated nematodes (including invasive ones) by various means of dispersal does occur (Hockland *et al.*, 2012). There is an imperative requirement of the legislative and institutional framework that clearly mandates basic aspects of incursion management, including prevention and management of invasive nematode species. The basic principles for dealing with incursion management of invasive nematode species are prevention, early detection, eradication and control.

116.1 Prevention

Prevention or exclusion is the first and most cost-effective option. Preventive measures include prohibiting introductions of a nematode's host plant and associated material that have a high risk of contamination. The lessons learnt from the incursion of numerous highly destructive and invasive nematode species, such as the pine wilt nematode, coconut red ring nematode, etc. in the past are to be taken as the guiding framework to devising emergency action plans in the event of pest incursions. There are many invasive nematodes which are likely to enter newer areas and may have a substantial economic impact on many crops. The most important measures for nematode prevention are restrictions of the import of soil, planting materials, growing media and packaging material.

116.2 Early detection

Early detection of a potential invasive nematode species is a vital step in preventing the establishment or spread of exotic or new nematode pests. It also helps in determining whether eradication of the species is practically feasible. For early detection, surveillance is an essential component, where information on an invasive nematode can be gathered from many sources, or it may be collected from specific surveys. Surveys and surveillance should have support from a nematode diagnostic laboratory for reliable and early identifications of the plant-parasitic nematodes.

116.3 Eradication

In cases of failure of prevention or exclusion, eradication of the invasive nematode species is the ideal course of management. However, eradication is a costly affair and it can only be successful when a cost-effective solution is implemented. The approach must be a vigilant analysis of the costs and likelihood of success, and adequate resources should be mobilized before eradication is attempted. Successful eradication programmes in the past have been done based on: (i) mechanical control, e.g. ear cockle diseases; (ii) chemical control, e.g. potato cyst nematode; (iii) habitat management, e.g. pine wilt nematode; and (iv) push and treat, e.g. citrus slow decline. However, maximum eradication programmes need to engage numerous different methods and the programme must evaluate the best methods under the specific circumstances.

116.4 Control

The last step in incursion management is the control of an invasive/exotic nematode species when other options have failed or are not feasible. However, the primary objective of control is to reduce the population density and abundance of an invasive nematode below the economic threshold level. Several control methods are practised to suppress the nematode population. Nematode control by the use of mechanical measures is labour-intensive but highly specific to the target. Chemicals are the most effective and rapid solution for nematode control, but their toxic and hazardous effects, high costs, non-target effects, etc. are major drawbacks. Other methods are biological control and integrated pest management. Biological control is cost-effective, highly target-specific, permanent, self-sustaining and ecologically safe and is suitable for all areas, including conserved ones.

116.5 Integrated Nematode Management

Integrated Nematode Management (INM) is 'an approach to reduce nematode damage to tolerable levels through a variety of techniques considering the agro-ecosystem, and the integrated strategy should have socio-economic and eco-friendly consequences. The basic strategies of INM includes:

- Natural control – maximum reliance on nature (biological control).
- Employing all principles of control in regulation, prevention, avoidance, eradication, protection, therapy and resistance.
- Selection of control tactics on the parameters of efficacy, economics and ecology, integrating them in a compatible manner.
- Application of the principles of systems and management in all decision-making processes, operations, optimization and analysis.

References

Abad, P., Tares, S., Brugier, N. and Deguiran, G. (1991) Characterisation of the relationships in the pinewood nematode species complex (PWNSC) (*Bursaphelenchus* spp.) using a heterologous unc-22 DNA probe from *Caenorhabditis elegans. Parasitology* 102, 303–308.

Abd-Elgawad, M.M.M., Koura, F.F.H., Montasser, S.A. and Hammam M.M.A. (2016) Distribution and losses of *Tylenchulus semipenetrans* in citrus orchards on reclaimed land in Egypt. *Nematology* 18, 1141–1150.

Abd-Elgawad, M.M.M., Eissa, M.F.M., El-Gindi, A.Y., Smart, G.C. and El-bahrawy, A. (2018) Description and identification of four species of plant-parasitic nematodes associated with forage legumes. *Egyptian Journal Agronematology* 17, 51–64.

Abelleira, A., Picoaga, A., Mansilla, J.P. and Aguin, O. (2011) Detection of *Bursaphelenchus xylophilus*, causal agent of pine wilt disease on *Pinus pinaster* in northwestern Spain. *Plant Disease* 95(6), 776–776.

Aboul-Eid, H.Z. and Ghorab, A.I. (1974) Pathological effects of *Heterodera cajani* on cowpea. *Plant Disease Reporter* 58(12), 1130–1133.

Aboul-Eid H.Z, and Ghorab, A.I. (1981) The occurrence of *Heterodera zeae* in maize fields in Egypt. *Egyptian Journal of Phytopathology* 13(1-2), 51–61.

Abrantes, I.M.D.O., Santos, M.C.V.D., Conceição, I. L.P.M. D., Cunha, M.J.M.D. and Santos, M.S.N.D.A. (2004) Biochemical and molecular characterization of plant-parasitic nematodes. *Phytopathologia Mediterrenea* 43, 232–258.

Acedo, J.R. and Rohde, R.A. (1971) Histochemical root pathology of Brassica oleracea capitata L. infected by Pratylenchus penetrans (Cobb) Filipjev and Schuurmans Stekhoven (Nematoda: Tylenchidae). *Journal of Nematology* 3, 62–68.

Acharya, A. and Das, S.N. (1983) Host range and pathogenicity of *Macroposthonia ornata* on ragi (Eleusine coracana Gin). *Indian Journal of Nematology* 13(1), 9–15.

Adam, W. (1932) Note sur *Heterodera schachtii* Schm. parasite des cactus. *Bulletin du Musée Royal d'Histoire Naturelle de Belgique* 8, 1–10.

Adams, I., Woodhall, J., Glover, R., Shah, F., Hockland, S., Boonham, N. and Marshall, J. (2009) Validation of quantitative DNA detection systems for PCN (Ref: R287). Agriculture and Horticulture Development Board (AHDB), Potato Council Ltd, UK.

Adams, J.R. and Rodriguez-Kabana, R. (1977) Effects of 2,3,4,5-tetrachloronitrobenzene on plant parasitic and free-living nematodes. *Proceedings of the American Phytopathological Society* 3(1976), 244.

Adesiyan, S.O. (1977) Penetration and multiplication of *Scutellonema bradys* in yams (*Dioscorea* spp.). *Nematologia Mediterranea* 5(2), 313–317.

Adylova, N.A. and Vasilevskii, V.N. (1983) The potato stem nematode at the collective farm 'Sokh' in the Fergana valley and its control. (Materialy simpoziuma, Voronezh, 27–29 Sentyabrya 1983 g.). *Voronezh, USSR: Vserossiiskii NII Zashchity Rastenii*, 59–61.

Agrios, G.N. (2005) *Plant Pathology*, 5th edn. Elsevier Academic Press, Burlington, Massachusetts.

Ahmad, N., Das, P.K. and Baqri, Q. H. (1984) Evaluation of crop losses in rice due to *Hirschmanniella gracilis* (de Mann, 1880) Luc and Goodey 1964 (Tylenchida: Nematoda) at Hooghly (West Bengal). *Bulletin of the Zoological Survey of India* 5, 85–91.

Ahmed, M., van de Vossenberg, B.T.L.H., Cornelissen, C. and Karssen, G. (2013) On the species status of the root-knot nematode *Meloidogyne ulmi* Palmisano and Ambrogioni, 2000 (Nematoda, Meloidogynidae). *ZooKeys* 362, 1–27.

Ahmed, R. and Khan, H. (1999) Citrus slow decline: problems and progress. In: *Punjab National Conference of Plant Pathology*. Sep. 27–29, University of Agriculture, Faisalabad, pp. 20–23.

Akhtar, M. (1998) Biological control of plant parasitic nematodes by neem products in agricultural soils. *Applied Soil Ecology* 7, 219–223.

Akyazi, F., Felek, A.F. and Cermak, V. (2019) First report of the hop cyst nematode *Heterodera humuli* from common nettle *Urtica diocia* in Turkey. *Plant Disease* 103(1), 161.

Alam, M.M. (1991) Control of plant-parasitic nematodes with oilseed cakes on some vegetables in field. *Pakistan Journal of Nematology* 9, 21–30.

Al-Banna. L., Ploeg, A.T., Williamson, V.M. and Kaloshian, I. (2004) Discrimination of six *Pratylenchus* species using PCR and species-specific primers. *Journal of Nematology* 36, 142–146.

Alderman, S.C., Bilsland, D.M., Griesbach, J.A., Milbrath, G.M., Schaad, N.W. and Postnikova, F. (2003) Use of a seed scarifier for detection and enumeration of galls of Anguina and Rathayibacter species in orchard grass seed. *Plant Disease*, 87, 320–323.

Ali, S.S. (1986) Root-knot nematode problem in cardamom and its management. *Proceedings of the Second Group Discussions on the Nematological Problems of Plantation Crops*, April 24–25, 1986. Central Coffee Research Station, Balehonnur, Karnataka, India, pp. 10–12.

Ali, S.S., Naimuddin, and Ali, M. (2010) Nematode infestation in pulses. In: Khan, M.R. and Jairajpuri, M.S. (eds) *Nematode Infestations – Part I: Food Crops*. National Academy of Sciences, Allahabad (Prayagraj), UP, India, pp. 227–257.

Allen, M.W. (1952) Taxonomic status of the bud and leaf nematodes related to *Aphelenchoides fragariae* (Ritzema Bos, 1891). *Proceedings of the Helminthological Society of Washington* 19, 108–120.

Allen, M.W. (1955) A review of the nematode genus *Tylenchorhynchus. University of California Publications Zoology* 61, 129–166.

Allen, M.W. and Jensen, H.J. (1951) *Pratylenchus vulnus* new species (Nematoda: Pratylenchinae), a parasite of trees and vines in California. *Proceedings of the Helminthological Society of Washington* 18, 47–50.

Allen, W.R., Stobbs, L.W., Schagen, J.G. van and Ebsary, B.A. (1988) Association of *Xiphinema* species with soil type and grapevines infected with tomato ringspot virus in Ontario, Canada. *Plant Disease* 72(10), 861–863.

Altman, I. and Thomason, I.J. (1971) Nematodes and their control. In: Johnson, R.T., Alexander, J.T., Rush, G.E. and Hawkes, G.R. (eds). *Advances in Sugarbeet Production: Principles and Practices*. Iowa State University Press, Ames, Iowa, pp. 335–370.

Álvarez-Ortega, S., Brito, J.A. and Subbotin, S.A. (2019) Multigene phylogeny of rootknot nematodes and molecular characterization of *Meloidogyne nataliei* Golden, Rose & Bird, 1981 (Nematoda: Tylenchida). *Scientific Reports* 9, 11788. doi: 10.1038/s41598-019-48195-0

Ambrogioni, L. and Tacconi, R. (1992) *Nacobbus aberrans* (Thorne and Allen) (*sensu lato*) Tylenchida, Pratylenchidae. *Informatore Fitopatologico* 42(9), 33–38.

Amici, A. and Lalatta, F. (1971) Studies on soil fumigation in replanting peaches. II. The relationship between *Pratylenchus vulnus* and soil exhaustion. *Annali dell'Istituto Sperimentale per la Frutticoltura* 2(1), 257–276.

Amiri, S., Subbotin, S.A. and Moens, M. (2003) Comparative morphometric and RAPD studies of *Heterodera schachtii* and *H. betae* population. *Russian Journal of Nematology* 11, 91–99.

Amrei, S.B., Peneva, V., Rakhshandehroo, F. and Pedram, M. (2019) Morphological and molecular description of *Longidorus behshahrensis* n. sp. (Nematoda: Longidoridae) from natural forests of Abbas Abad, north of Iran. *European Journal of Plant Pathology* 156(1). doi: 10.1007/s10658-019-01884-y

Amsing, J.J. (1991) Damage to soil-grown greenhouse roses from root lesion nematodes. Multic and Inermis equally sensitive. *Vakblad voor de Bloemisterij* 46(15), 62–63.

Andersen, S. and Andersen, K. (1982) Suggestions for determination and terminology of pathotypes and genes for resistance in cyst-forming nematodes, especially *Heterodera avenae*. *EPPO Bulletin* 12, 379–382.

Anderson, E.J. (1959) Summary of nematodes affecting crops in Hawaii. *Proceedings Shell Nematology Workshop*, 27–28, Portland Oregon, USA, pp. 73–76.

Anderson, R.V. (1983) Morphological characteristics of *Hoplolaimus indicus* Sher, 1963 in Canada, a parasite of wild rice. *Journal of Nematology* 15(3), 366–369.

Andrássy, I. (1958) *Hoplolaimus tylenchiformis* Daday, 1905 (syn. *H. coronatus* Cobb, 1923) und die Gattungen der Unterfamilie Hoplolaiminae Filip'ev, 1936. *Nematologica* 3, 44–56.

Andrássy, I. (2007) Free-living nematodes of Hungary (*Nematoda errantia*), Vol. II. In: Csuzdi, C. and Mahunka, S. (eds) *Pedozoologica Hungarica No. 4*. Hungarian Natural History Museum and Systematic Zoology Research Group of the Hungarian Academy of Sciences, Budapest, pp. 18–22.

Andrássy, I. and Farkas, K. (1988) Kertészeti növények fonálféreg kártevői. *Budapest, Mezőgazdasági Kiadó*. pp. 181–198.

Andrés, M.F., Pinochet, J., Hernández-Dorrego, A. and Delibes, A. (2000) Detection and analysis of inter- and intraspecific diversity of *Pratylenchus* spp. using isozyme markers. *Plant Pathology* 49, 640–649.

Andrés, M.F., Romero, M.D., Montes, M.J. and Delibes, A. (2001) Genetic relationships and isozyme variability in the *Heterodera avenae* complex determined by isoelectric focusing. *Plant Pathology* 50, 270–279.

Anonymous (1981) Annual Report of the Secretary for Agricultural Technical Services for the period 1 July 1979–31 March 1980. Republic of South Africa: Department of Agricultural Technical Services. *Plant Nematology* 25-26, 113–114.

Anwar, S.A. (1994) Effect of *Meloidogyne incognita* and *Paratrichodorus minor* on the growth of tomato. *Afro-Asian Journal of Nematology* 4(2), 155–160.

Anwar, S.A. (1995) Reproduction and influence of *Meloidogyne incognita* and *Pratylenchus vulnus* in the growth of grape. *Pakistan Journal of Phytopathology* 7(2), 118–123.

Anwar, S.A. and Van Gundy, S.D. (1989) Influence of four nematodes on root and shoot growth parameters in grape. *Journal of Nematology* 21(2), 276–283.

Anwar, S.A., Rauf, C.A. and Gorsi, S.D. (1992) Weeds as alternate hosts of phytonematodes. *Afro-Asian Journal of Nematology* 2(1-2), 41–47.

Anwar, S.A., McKenry, M.V., Riaz, A. and Khan, M.S.A. (2001) Evaluation of wheat cultivars for *Anguina tritici* resistance, development and influence of nematode on wheat growth. *International Journal of Nematology* 11, 150–156.

Apt, W.J., Austenson, H.M. and Courtney, W.D. (1960) The use of herbicides to break the life cycle of the bentgrass nematode *Anguina agrostis* (Steinbuch, 1799) Filipjev, 1936. *Plant Disease Reporter* 44(7), 524–526.

AQAS (2014) Agricultural Quarantine Activity Systems. Available at: https//aqas.aphis.usda.gov/aqas/ (accessed 10 July 2014)

Archer, M.J., Griffiths, J.B. and Pye, D.L. (1984) *Cocamba – a new wheat variety resistant to cereal cyst nematode*. Agnote, Department of Agriculture, Victoria.

Archidona-Yuste, A., Navas-Cortés, J.A., Cantalapiedra-Navarrete, C., Palomares-Rius, J.E. and Castillo, P. (2016a) Unravelling the biodiversity and molecular phylogeny of needle nematodes of the genus *Longidorus* (Nematoda: Longidoridae) in olive and a description of six new species. *PLoS ONE* 11(1), e0147689. doi: 10.1371/journal.pone.0147689

Archidona-Yuste, A., Navas Cortés, J., Cantalapiedra-Navarrete, C., Palomares-Rius, J. and Castillo, P. (2016) Remarkable diversity and prevalence of dagger nematodes of the genus *Xiphinema* Cobb, 1913 (Nematoda: Longidoridae) in olives revealed by integrative approaches. *PLoS ONE* 11(11), e0165412. doi:10.1371/journal.pone.0165412

Asghari, R., Pourjam, E., Heydari, R., Tanha Maffi, Z. and Latifi, A.M. (2013) First report of corn cyst nematode, *Heterodera zeae* in Afghanistan. *Australasian Plant Disease Notes*, 8, 93–96. doi: 10.1007/s13314-013-0104-z

Asghari, R., Eskandari, A., Tanha Maafi,.Z. and Decraemer, W. (2018) Characterisation of a new species of *Trichodorus* Cobb, 1913 (Triplonchida: Trichodoridae) from Iran based on morphological and molecular data. *Systematic Parasitology* 95, 257–269. doi: 10.1007/s11230-018-9780-6.

Askary, T.H., Banday, S.A. and Iqbal, U. (2012) Plant parasitic nematode diversity in pome, stone and nut fruits. In: Lichtfouse, E. (ed.) *Agroecology and Strategies for Climate Change. Sustainable Agriculture Reviews*, vol 8. Springer, Dordrecht, pp. 237–268.

Astudillo, G.E. and Birchfield, W. (1980) Pathology of *Hololaimus coulumbus* on sugarcane. *Phytopathology* 70, 565.

Atkinson, G.F. (1892) Some diseases of cotton. *Alabama Polytechnical Institute of Agriculture Experimental Station Bulletin* 41, 61–65.

Atkinson, K. (1964) *Aphelenchoides ritzemabosi* Schwartz on *Crassula coccinea*. *Plant Pathology* 13, 137.

Audebert, A., Coyne, D.L., Dingkuhn, M. and Plowright, R.A. (2000) The influence of cyst nematodes (*Heterodera sacchari*) and drought on water relations and growth of upland rice in Côte d'Ivoire. *Plant and Soil* 220, 235–242.

Auger, J., Leal, G., Magunacelaya, J.C. and Esterio, M. (2009) *Xiphinema rivesi* from Chile transmits Tomato ringspot virus to cucumber. *Plant Disease* 93, 971.

Aung, T. and Prot, J.C. (1990) Effects of crop rotations on *Pratylenchus zeae* and on yield of rice cultivar UPL Ri-5. *Revue de* 13(4), 445–447.

Ayala, A., Allen, M. and Noffsinger, E.M. (1970) Host range, biology, and factors affecting survival and reproduction of the stubby rootnematode. *Journal of Agriculture of University of Puerto Rico* 64, 341–369.

Azmi, M.I. (1979) Responses to temperature in nematodes. I. Mechanism of heat tolerance in *Helicotylenchus dihystera*. *Nematologia Mediterranea* 7(2), 163–175.

Azmi, M.I. (1993) Plant parasitic nematodes associated with su-babool. *Indian Journal of Forestry* 16(4), 384.

Azmi, M.J. (1990) Seasonal population fluctuation behaviour of *Helicotylenchus dihystera*, *Pratylenchus zeae* and *Tylenchorhynchus vulgaris* in buffalo grass pasture. *Indian Journal of Nematology* 20(1), 19–24.

Babatola, J.O. (1980) Studies on the weed hosts of the rice root nematode, *Hirschmanniella spinicaudata* Sch. Stek. 1944. *Weed Research* 20(1), 59–61.

Babatola, J.O. (1981) Effect of pH, oxygen and temperature on the activity and survival of *Hirschmanniella* spp. *Nematologica* 27(3), 285–291.

Babatola, J.O. (1983) Rice cultivars and *Heterodera sacchari*. *Nematologia Mediterranea* 11, 103–105.

Babatola, J.O. and Bridge, J. (1979) Pathogenicity of *Hirschmanniella oryzae*, *H. spinicaudata*, and *H. imamuri* on rice. *Journal of Nematology* 11(2), 128–132.

Babatola, J.O. and Bridge, J. (1980) Feeding behaviour and histopathology of *Hirschmanniella oryzae*, *H. imamuri*, and *H. spinicaudata* on rice. *Journal of Nematology* 12(1), 48–53.

Back, M., Haydock, P. and Jenkinson, P. (2006) Interaction between the potato cyst nematode *Globodera rostochiensis* and disease caused by *Rhizoctonia solani* AG3 in potatoes under field conditions. *European Journal of Plant Pathology* 114, 215–223.

Badra, T., Salem, M.A. and Oteifa, B.A. (1979) Nematicidal activity of some organic fertilizers and soil amendments. *Revue de Nematologie* 2, 29–36.

Bae, C.H., Szalanski, A.L. and Robbins, R.T. (2008) Molecular analysis of the lance nematode, *Hoplolaimus* spp., using the first internal transcribed spacer and the D2-D3 expansion segments of 28S ribosomal DNA. *Journal of Nematology*, 40, 201–209.

Bae, C.H., Robbins, R.T. and Szalanski, A.L. (2009a) Molecular identification of some *Hoplolaimus* species from the USA based on duplex PCR, multiplex PCR and PCRRFLP analysis. *Nematology* 11, 471–480.

Bae, C.H., Szalanski, A.L., and Robbins, R.T. (2009b) Genetic variation of *Hoplolaimus columbus* populations in the United States using PCR-RFLP analysis of nuclear rDNA ITS regions. *Journal of Nematology* 41, 201–209.

Baheti, B.L. and Verma, M.K. (2001) Varietal response against wheat seed gall nematode. *National Congress on Centenary Nematology in India. Appraisal and Future Plans*. Organised by IARI, New Delhi 5–7 Dec. 2001, pp. 130–131.

Baidoo, R., Yan, G., Nagachandrabose, S. and Skantar, A.M., (2017) Developing a Real-Time PCR assay for direct identification and quantification of *Pratylenchus penetrans* in soil. *Plant Disease* 101, 1432–1441.

Bajaj, H.K. and Kanwar, R.S. (2005) Parasitization of maize by *Heterodera avenae* and *H. filipjevi*. *Nematologia Mediterranea* 33, 203–207.

Bajaj, H.K., Gupta, D.C. and Dahiya, R.S. (1986) Development of *Heterodera zeae* Koshy et al, on wheat and maize. *Nematologica* 32, 209–215.

Bajaj, H.K., Kanwar, R.S. and Vats, R. (2005) Test for interbreeding between *Heterodera avenae* and *H. filipjevi*. In: *National Symposium on Recent Advances and Research Priorities in India. Nematology*, 9–10 Dec., 2005. IARI, New Delhi, p.16.

Balasubramanian, P. and Vadivelu, S. (2003) Assessment of avoidable yield los in Sesamum in Tamil Nadu due to pigeon-pea cyst nematode, *Heterodera cajani* Kosh, 1967. *Indian Journal Agricultural Research* 37(4), 310–312.

Baldwin, J.G. and Mundo-Ocampo, M. (1991) Heteroderinae, cyst and non-cyst forming nematodes. In: Nickle, W.R. (ed.) *Manual of Agricultural Nematology*. Marcel Dekker, New York, pp. 275–362.

Bansal, R.K., Dahiya, R.S., Lakshminarayana, K., Suneja, S., Anand, R.C. and Narula, N. (1999) Effect of rhizospheric bacteria on plant growth of wheat infected with *Heterodera avenae*. *Nematologia Mediterranea* 27, 311–314.

Banu, G. (2010a) Nematode infestation in coffee. In: Khan, M.R. and Jairajpuri, M.S. (eds) *Nematode Infestations – Part II: Industrial Crops*. National Academy of Sciences, Allahabad (Prayagraj), UP, India, pp. 149–170.

Banu, G. (2010b) Nematode infestation in cotton. In: Khan, M.R. and Jairajpuri, M.S. (eds) *Nematode Infestations – Part II: Industrial Crops*. National Academy of Sciences, Allahabad (Prayagraj), UP, India, pp. 207–232.

Barabashova, V.N. (1978) Karyological investigations of stem nematodes from the *Ditylenchus dipsaci* complex. *Nauchnye Doklady Vysshei Shkoly, Biologicheskie Nauki* 5, 109–114.

Barbercheck, M.E. and Broembsen, S.L. von (1986) Effects of soil solarization on plant-parasitic nematodes and *Phytophthora cinnamomi* in South Africa. *Plant Disease* 70(10), 945–950.

Barbosa, D.H.S.G., Vieira, H.D., Souza, R.M., Viana, A.P. and Silva, C.P. (2004) Field estimates of coffee yield losses and damage threshold by *Meloidogyne exigua*. *Nematologia Brasileira* 28, 49–54.

Barekye, A., Kashaija, I.N., Tushemereirwe, W.K. and Adipala, E. (2000) Comparison of damage levels caused by *Radopholus similis* and *Helicotylenchus multicinctus* on bananas in Uganda. *Annals of Applied Biology* 137, 273–278.

Barker, K.R. and Lucas, G.B. (1984) Nematode parasites of tobacco. In: Nickle, W.R. (ed.) *Plant and Insect nematodes*. Marcel Dekker, New York, pp. 218–242.

Barker, K.R., and Niblack, T.L. (1990) Soil sampling methods and procedures for field diagnosis, In: Zuckerman, B.M., Mai, W.F. and Krusberg, L.R. (eds) *Plant Nematology Laboratory Manual*. University of Massachusetts Agriculture Experiment Station Publ., pp. 10–19.

Barriga, R. (1965) Parasitism of Camellias by some nematode species. *Nematologica* 34, 34.

Barsalote, E.M., Pham, H.T., Lazarova, S., Peneva, V. and Zheng, J. (2018) Description of *Longidorus cheni* sp. n. (Nematoda, Longidoridae) from China. *ZooKeys* 744, 1–18.

Barsi, L. and De Luca, F. (2008) Morphological and molecular characterisation of two putative *Xiphinema americanum*-group species, *X. parasimile* and *X. simile* (Nematoda: Dorylaimida) from Serbia. *Nematology* 10, 15–25.

Basson, S., De Waele, D. and Meyer, A.J. (1993) Survival of *Ditylenchus destructor* in soil, hulls and seeds of groundnut. *Fundamental and Applied Nematology* 16, 79–85.

Baujard, P. (1989) Remarques sur les genres des sous-familles Bursaphelenchidae Paramanov, 1964 et Radinaphelenchinae Pramanov, 1964 (Nematoda: Aphelenchoididae). *Rev. Nématology* 12, 323–324.

Baujard, P. and Martiny, B. (1995) Ecology and pathogenicity of the Hoplolaimidae (Nemata) from the Sahelian zone of West Africa. 4. The genus *Aphasmatylenchus* Sher, 1965. *Fundamental and Applied Nematology* 18(4), 355–360.

Baujard, P., Vovlas, N., Mounport, D. and Martiny, B. (1998) Description of *Aphasmatylenchus liberiensis* n. sp.; and observations on the other species of the genus *Aphasmatylenchus* Sher, 1965 (Nematoda: Hoplolaimidae). *Fundamental and Applied Nematology* 21(2), 129–138.

Beaupre, C.M.S., Gray, F.A., Vincelli, P.C. and Miller, S.D. (1989) Effects of aldicarb, *Rhizoctonia* disease, and weeds on severity of the false root knot nematode of sugar beet. *Fungicide and Nematicide Tests* 45, 161.

Becerra, L.E.N. and Sosa-Moss, C. (1978) Relacion del nematodo enquistado del maiz *Punctodera chalcoensis* (Stone, Sosa-Moss y Mulvey) con otros microorganismo fitopatogenos del suelo. Tesis de Maestria en ciencias agricolas del primer autor. Colegio de Postgraduados, Chapingo, Mexico.

Becker, J.O. and Westerdahl, B.B. (2018) Onion and garlic nematodes. UCIPM Statewide Integrated Pest Management Program, University of California Agriculture and Natural Resources (updated 2/07). Available at: http://ipm.ucanr.edu/PMG/r584200111.html

Bélair, G., Dauphinais, N., Benoit, D.L. and Fournier, Y. (2007) Reproduction of *Pratylenchus penetrans* on 24 common weeds in potato fields in Quebec. *Journal of Nematology* 39, 321–326.

Bélair, G., Mimee, B., Duceppe, M.O. and Miller, S. (2013) First report of the root-lesion nematode *Pratylenchus alleni* associated with damage on soybean in Quebec, Canada. *Plant Disease* 97, 292. doi: 10.1094/PDIS-03-12-0309-PDN

Beniers, J.E., Been, T.H., Mendes, O., van Gent-Pelzer, M.P.E. and van der Lee, T.A.J. (2014) Quantification of viable eggs of the potato cyst nematodes (*Globodera* spp.) using either trehalose or RNA-specific Real-Time PCR. *Nematology* 16, 1219–1232.

Benson, D.M. and Barker, K.R. (1985) Nematodes – a threat to ornamental plants in the nursery and landscape. *Plant Disease* 69(2), 97–100.
Benson, D.M., Barker, K.R. and Aycock, R. (1976) Effects of density of *Helicotylenchus dihystera* and *Pratylenchus vulnus* on American boxwood growing in microplots. *Journal of Nematology* 8(4), 322–326
Benton, A.W. and Myers, R.F. (1966) Esterases, phosphatases, and protein patterns of *Ditylenchus triformis* and *Panagrellus redivivus*. *Nematologica* 12, 495–500.
Berney, M.F., and Bird, G.W. (1992) Distribution of *Heterodera carotae* and *Meloidogyne hapla* in Michigan carrot production. *Journal of Nematology* 24(4S), 776–778.
Bhattacharya, D. and Swarup, G. (1988) *Pasteuria penetrans*, a pathogen of the genus *Heterodera*, its effect on nematode biology and control. *Indian Journal of Nematology* 18(1), 61–70.
Bhatti, D.S. and Dabur, K.R. (1982a) Use of wheat as trap crop for the control of *Heterodera avenae*. *Journal of Nematology* 14, 429.
Bhatti, D.S. and Dabur, K.R. (1982b) Effect of summer plowings on *Heterodera avenae* population and wheat yield. *Journal of Nematology* 14, 429–430.
Bhatti, D.S., Dalal, M.R., Dahiya, R.S. and Dhawan, S.C. (1980) Integrated control of *Heterodera avenae* in wheat. *Journal of Nematology* 12, 214–215.
Bilgrami, A.L. and Jairajpuri, M.S. (1989) Resistance of prey to predation and strike rate of the predators, *Mononchoides longicaudatus* and *M. fortidens* (Nematoda: Diplogasterida). *Revue de Nématologie* 12(1), 45–49
Binks, R.H. and Gowen, S.R. (1996) Field evaluation of nematode infestations in *Musa* germplasm at FHIA (Lalime, Honduras). *Infomusa* 5, 15–17.
Birchfield, W. (1962) Host–parasite relationship of *Rotylenchulus reniformis* on *Gossypium hirsutum*. *Phytopathology* 52, 862–865.
Birchfield, W. (1984) Nematode parasite of sugarcane. In: Nickle, W.K. (ed.) *Plant and Insect Nematodes*. Marcel Dekker Inc., New York, pp. 571–588.
Bird, A.F. (1981) The *Anguina–Corynebacterium* association. *Plant Parasitic Nematodes* (III), 303–323.
Bird, A.F. (1985) The nature of the adhesion of *Corynebacterium rathayi* to the cuticle of the infective juveniles of *Anguina agrostis*. *International Journal for Parasitology* 15, 301–308.
Bird, A.F. and Riddle, D.L. (1984) Effect of attachment of *Corynebacterium rathayi* on movement of *Anguina agrostis* larvae. *International Journal for Parasitology* 14(5), 503–511.
Bird, A.F. and Stynes, B.A. (1981) The life cycle of *Anguina agrostis*: development in host plant. *International Journal for Parasitology* 11(6), 431–440.
Bird, A.F., Stynes, B.A. and Thomson, W.W. (1980) A comparison of nematode and bacteria-colonized galls induced by *Anguina agrostis* in *Lolium rigidum*. *Phytopathology* 70(11), 1104–1109.
Bird, G.W. and Jenkins, W.R. (1964) Occurrence, parasitism, and pathogenicity of nematodes associated with cranberry. *Phytopathology* 54, 677–680.
Bird, G.W., Goodman, R.M. and Mai, W.F. (1968) Observations on the embryogenesis of *Trichodorus christiei* (Nematoidea: Diphtherophoroidea). *Canadian Journal of Zoology* 46, 292–293.
Bishnoi, S.P. and Bajaj, H.K. (2004) On the species and pathotypes of *Heterodera avenae* complex of India. *Indian Journal of Nematology* 34, 147–152.
Bishnoi, S.P., Singh, S., Mehta, S. and Bajaj, H.K. (2004) Isozyme patterns of *Heterodera avenae* and *H. filipjevi* populations of India. *Indian Journal of Nematology* 34, 33–36.
Biswas, H. and Rao, Y.S. (1971) Toxicological assays of nematicides with rice nematodes. *Indian Phytopathology* 24, 159–165.
Bitterlin, M.W. and Gonsalves, D. (1987) Spatial distribution of *Xiphinema rivesi* and persistence of tomato ringspot virus and its vector in soil. *Plant Disease* 71(5), 408–411.
Blair, G.P. (1970) Studies on red ring disease of the coconut palm. *Oleagineux* 25, 19–22, 78–83.
Blake, C.D. (1961) Root rot of bananas caused by *Radopholus similis* (Cobb) and its control in New South Wales. *Nematologica* 6, 295–310.
Blank, W. (1985) Leaf and stem nematodes in strawberries – a serious problem. *Mitteilungen des Obstbauversuchsringes des Alten Landes* 40(6), 229–234.
Blasingame, D.J. (2004) Cotton pest loss database. National Cotton Council of America, 15 Jan. 2005. Available at: http://www.cotton.org/tech/pest/upload/cotton-disease-loss-Database.xls.
Blasingame, D.J. (2006) Cotton disease loss estimate. In: *Proceedings of the Beltwide Cotton Production Research Conference*, Memphis, TN. January 3–5, 2006. National Cotton Council of America, Memphis, Tennessee, pp.155–157.
Blasingame, D.J. and Patel, M.V. (2005) Cotton disease loss estimates committee report. In: *Proceedings of the Beltwide Cotton Conference*, New Orleans, LA. 4–7 Jan. 2005. National Cotton Council of America, Memphis, Tennessee, pp. 259–262.
Blaxter, M.L., De Ley, P., James, R. *et al.* (1998) A molecular evolutionary framework for the phylum Nematoda. *Nature* 392, 71–75.
Blok, V.C. and Powers, T.O. (2009) Biochemical and molecular identification. In: Perry, R.N., Moens, M. and Starr, J.L. (eds) *Root-knot Nematodes*, CABI, Wallingford, UK, pp. 98–118.
Blok, V.C., Wishart, J., Fargette, M., Berthier, K. and Phillips, M.S. (2002) Mitochondrial DNA differences distinguishing *Meloidogyne mayaguensis* from the major species of tropical root-knot nematodes. *Nematology* 4(7), 773–781.
Blomme, G., Beule, H. de, Swennen, R.L., Tenkouano, A. and Waele, D. de. (2004) Effect of nematodes on root and shoot growth of in vitro-propagated and sword sucker-derived plants of six *Musa* spp. genotypes. *Nematology* 6(4), 593–604.
Boag, B. (1983) Effect of rotary cultivation on plant parasitic nematodes. *Crop Research* 23(1), 33–37.
Boag, B., Crawford, J.W. and Neilson, R. (1991) The effect of potential climatic changes on the geographical distribution of the plant-parasitic nematodes *Xiphinema* and *Longidorus* in Europe. *Nematologica* 37(3), 312–323.
Bohmer, B. (1981) The harmful effect of *Aphelenchoides fragarip* and *A. ritzemabosi* on *Fragaria ananassa*. *Gesunde Pflanzen* 33(5), 113–117.
Boldyrev, M.I. and Borzykh, G.T. (1983) Plants in the control of nematodes. *Zashchita Rastenii*, Moscow, 9, 30.
Bolla, R.I., Weaver, C. and Winter, R. (1988) Genomic differences among pathotypes of *Bursaphelenchus xylophilus*. *Journal of Nematology* 20, 309–316.
Bonfim, J.M.F., Consoli, E.A., Inomoto, M.M. and de Oliveira, C.M.G. (2016) First report of *Pratylenchus crenatus* in Brazil. *New Disease Reports* 34, 7.
Bonta, M., Roşca, I. and Costache, C. (2013) New host plant for virus vector nematode *Xiphinema italiae* Meyl, 1953 (Nematoda: Longidoridae) in Romania. *Scientific Papers – Series B, Horticulture*, 159–161.
Bora, B.C. and Choudhury, B.N. (2010) Nematode infestation in ginger. In: Khan, M.R. and Jairajpuri, M.S. (eds) *Nematode Infestations – Part II: Industrial Crops*. National Academy of Sciences, Allahabad (Prayagraj), UP, India, pp. 96–115.
Bora, B.C. and Rahman, M.F. (2010) Stem nematode infestation in rice. In: Khan, M.R. and Jairajpuri, M.S. (eds) *Nematode Infestations – Part I: Food Crops*. National Academy of Sciences, Allahabad (Prayagraj), UP, India. pp. 97–110.
Bos, W.S. (1977a) *Aphelenchoides arachidis* n.sp. (Nematoda: Aphelenchoidea), an endoparasite of the testa of groundnuts in Nigeria. *Zeitschrift für Pflanzenkrankheiten und Pflanzenschutz* 84(2), 95–99.
Bos, W.S. (1977b) A preliminary report on the distribution and host-range of the nematode *Aphelenchoides arachidis* Bos. in the north of Nigeria. *Samaru Agricultural Newsletter* 19(1), 21–23.

Bossis, M., Andaloussi, F.A., Caubel, G., Esquibet, M. and Renault, L. (1998) Caractérisation par électrophorèse bidimensionelle de populations (marocaines et françaises) de *Ditylenchus dipsaci* collectés sur féverole. *Colloques de l'INRA* 88, 187–197.

Boswell, T.E. (1968) Pathogenicity of *Pratylenchus brachyurus* to Spanish peanuts. PhD thesis. Texas A&M University, College Station, Texas.

Boubals, D. (1994) Use of glyphosate (Round-up) to devitalize the root system of vines to be grubbed. *Progrès Agricole et Viticole*, 111(9), 205.

Boubals, D., Pistre, R., Dalmasso, A. and Bongiovani, M. (1971) Aspects of the attack of vine roots by the nematode *Xiphinema index*, vector of court-noue disease of the vine. *Progrès Agricole et Viticole*, 171(9), 4 pp.

Boutsika, K., Phillips, M.S., MacFarlane, S.A., Brown, D.J.F., Holeva, R.C. and Blok, V.C. (2004) Molecular diagnostics of some trichodorid nematodes and associated Tobacco rattle virus. *Plant Pathology* 53, 110–6.

Bowles, J., Blair, D. and McManus, D.P. (1992) Genetic variants within the genus *Echinococcus* identified by mitochondrial DNA sequencing. *Molecular and Biochemical Parasitology* 54, 165–174.

Boyd, F.T. and Perry, V.G. (1971) Effects of seasonal temperatures and certain cultural treatments on sting nematodes in forage grass. *Proceedings Soil and Crop Science Society Florida* 30, 360–365.

Braasch, H. (1976) Nematodes of plants grown in peat soils. *Nachrichtenblatt fur den Pflanzenschutz in der DDR*. 30(11), 228-231.

Braasch, H. (2001) *Bursaphelenchus* species in conifers in Europe: distribution and morphological relationships. *EPPO Bulletin* 31, 127–142.

Braasch, H. and Schönfeld, U. (2015) Improved morphological key to the species of the *xylophilus* group of the genus *Bursaphelenchus* Fuchs, 1937. *EPPO Bulletin* 45(1), 73–80.

Braasch, H., Wittchen, U. and Unger, J.G. (1996) Establishment potential and damage probability of *Meloidogyne chitwoodi* in Germany. *Bulletin OEPP* 26(3/4), 495–509.

Braasch, H., Tomiczek, Th., Metge, K., Hoyer, U., Burgermeister, W., Wulfert, I. and Schönfeld, U. (2001) Records of *Bursaphelenchus* spp. (Nematoda, Parasitaphelenchidae) in coniferous timber imported from the Asian part of Russia. *Forest Pathology* 31, 129–140.

Braasch, H., Burgermeister, W. and Gu, J. (2009) Revised intra-generic grouping of *Bursaphelenchus* Fuchs, 1937 (Nematoda: Aphelenchoididae). *Journal of Nematode Morphology and System* 12, 65–88.

Brancalion, A.M. and Lordello, L.G.E. (1982) Chemical control of *Rotylenchulus reniformis* on cotton. *Trabalhos apresentados a V Reuniao Brasileira de Nematol.*, 9–13 fevereiro de 1981, Londrina, PR, Brasil. Publicacao No. 5, 191–195.

Brathwaite, C.W.D. (1980) Plant parasitic nematodes associated with sugarcane in Trinidad. *FAO Plant Protection Bulletin* 28, 133–136.

Brathwaite, C.W.D. and Siddiqi, M.R. (1975) *Rhadinaphelenchus cocophilus*. *C.I.H. Descriptions of Plant-parasitic Nematodes, Set 5* (No. 72). Commonwealth Institute of Helminthology, St Albans, UK, 4 pp.

Bravo, M.A. and Lemos, R.M. (1997) *Longidorus*, *Paralongidorus* and *Xiphinema* in Portugal. In: Santos, M.S. de A., Abrantes, I.O., Brown, D.J.F. and Lemos, R.M. (eds) *An Introduction to Virus Vector Nematodes and their Associated Viruses*. Instituto do Ambiente e Vida, Universidade de Coimbra, Coimbra, pp. 421–441.

Brentu, C.F., Speijer P.R., Green K.R., Hemeng B.M.S, DeWaele, D. and Coyne, D.L. (2004) Micro-plot evaluation of the yield reduction potential of *Pratylenchus coffeae, Helicotylenchus multicinctus* and *Meloidogyne javanica* on plantain cv. Apantu-pa (Musa spp., AAB-group) in Ghana. *Nematology* 6, 455–462.

Bridge, J. (1972) Nematode problems with yams (*Dioscorea* spp). *PANS* 1, 89–91.

Bridge, J. (1973) Nematodes as pests of yams in Nigeria. *Mededelingen Fakulteit Landbouwwetenschappen* 38, 841–852.

Bridge, J. (1982) Nematodes of yams. In: Miege, J. and Lyonga, S.N. (eds) *Yams. Ignames*. Oxford University Press, New York, 253–264.

Bridge, J. (1988) Plant-parasitic nematode problems in the Pacific Islands. *Journal of Nematology* 20(2), 173–183.

Bridge, J. and Hunt, D.J. (1985) *Aphelenchoides arachidis*. *CIH Descriptions of Plant Parasitic Nematodes*, Set 8, No. 116. Commonwealth Institute of Helminthology, St Albans, UK.

Bridge, J. and Page, S.L.J. (1982) The rice root-knot nematode, *Meloidogyne graminicola*, on deep water rice (*Oryza sativa* subsp. *indica*). *Revue de Nematologie*, 5(2), 225–232.

Bridge, J. and Page, S.L.J. (1984) Plant nematode pests of crops in Papua New Guinea. *Journal of Plant Protection in the Tropics* 1, 99–109.

Bridge, J. and Starr, J.L. (2007) *Plant Nematodes of Agricultural Importance: A Color Handbook*. Academic Press, Boston, Massachusetts.

Bridge, J., Jones, E. and Page, L.J. (1976) *Meloidogyne acronea* associated with reduced growth of cotton in Malawi. *Plant Disease Reporter* 60(1), 5–7.

Bridge, J., Bos, W.S., Page, L.J. and McDonald, D. (1977) The biology and possible importance of *Aphelenchoides arachidis*, a seed-borne endoparasitic nematode of groundnuts from northern Nigeria. *Nematologica* 23(2), 253–259.

Bridge, J., Page, S.L.J. and Jordan, S.M. (1982) An improved method for staining nematodes in roots. *Rothamsted Report for 1981, Part 1*. Rothamsted Experimental Station, Harpenden, UK.

Bridge, J., Mortimer, J.J. and Jackson, G.V.H. (1983) *Hirschmanniella miticausa* n.sp. (Nematoda: Pratylenchidae) and its pathogenicity on taro (*Colocasia esculenta*). *Revue de Nematologie* 6(2), 285–290.

Bridge, J., Plowright, R.A. and Peng, D. (2005a) Nematode parasites of rice. In: Luc, M., Sikora, R.A. and Bridge, J. (eds) *Plant Parasitic Nematodes in Subtropical and Tropical Agriculture*, 2nd edn. CABI Publishing, Wallingford, UK, pp. 87–130.

Bridge, J., Coyne, D.L. and Kwoseh, C.K (2005b) Nematode parasites of tropical root and tuber crops (excluding potato). In: Luc, M., Sikora, R. A., and Bridge, J. (eds). *Plant Parasitic Nematodes in Subtropical and Tropical Agriculture*, 2nd edn. CABI Publishing, Wallingford, UK, pp. 221–258.

Brigbhan. and Kanwar, R.S. (2004) Effect of sowing time on multiplication of *Heterodera avenae* and performance of wheat crop. *Indian Journal of Nematology* 33, 172–174.

Brito, J.A., Subbotin, S.A., Desaeger J., Achinelly F. and Qiu, S. (2018) First report of the root-knot nematode (*Meloidogyne javanica*) infecting hops (*Humulus lupulus*) in Florida, USA. *Nematology* 50(4). doi: 10.21307/jofnem-2018-040

Brodie, B.B. (1976) Vertical distribution of three nematode species in relation to certain soil properties. *Journal of Nematology* 8(3), 243–247.

Brodie, B.B. (1993) Nematode parasites of potatoes. In: Evans, E., Trudgill, D.L. and Webster, J.M. (eds) *Plant Parasitic Nematodes in Temperate Agriculture*, CAB International, Wallingford, UK, pp. 87–132.

Brodie, B.B. (1998) Potato. In Barker, K.R., Pederson, G.A. and Windham, G.L. (eds) *Plant and Nematode Interactions*American Society of Agronomy, Inc., Crop Science Society of America, Inc. and Soil Science Society of America, Inc., Madison, Wisconsin, pp. 567–594.

Brodie, B.B. and Murphy, W.S. (1975) Population dynamics of plant nematodes as affected by combinations of fallow and cropping sequence. *Journal of Nematology* 7(1), 91–92.

Brooks, F.E. (2004) Plant-parasitic nematodes of banana in American Samoa. *Nematropica* 34, 65–72.

Brown, D., Macfarlane, S., Furlanetto, C., Oliveira, C. and Ferraz, L. (2004) Transmissão de vírus por nematóides parasitos de plantas. *Revisão Anual de Patologia de Plantas* 12, 201–242.

Brown, D.J.F. and Boag, B. (1975) *Longidorus macrosoma. CIH Descriptions of Plant-parasitic Nematodes*, Set 5, No. 67. Commonwealth Institute of Helminthology, St Albans, UK.

Brown, D.J.F. and Boag, B. (1977) *Longidorus attenuatus. CIH Descriptions of Plant-parasitic Nematodes*, Set 7, No. 101. Commonwealth Institute of Helminthology, St Albans, UK.

Brown, D.J.F. and Coiro, M.I. (1983) The total reproductive capacity and longevity of individual female *Xiphinema diversicaudatum* (Nematoda: Dorylaimida). *Nematologia Mediterranea* 11(1), 87–92.

Brown, D.J.F. and Taylor, C.E. (1987) Comments on the occurrence and geographical distribution of longidorid nematodes in Europe and the Mediterranean region. *Nematologia Mediterranea* 15, 333–373.

Brown, D.J.F. and Trudgill, D.L. (1997) Longidorid nematodes and their associated viruses. In: Santos, M.S.N. de A., Abrantes, I.M.O., Brown, D.J.F. and Lemos, R.M. (eds). *An Introduction to Virus Vector Nematodes and their Associated Viruses*. IAV, Departamento de Zoologia, Universidade de Coimbra, Portugal, pp.1–40.

Brown, D.J.F., Ploeg, A.T. and Robinson, D.J. (1989) A review of reported associations between *Trichodorus* and *Paratrichodorus* species (Nematoda: Trichodoridae) and tobraviruses with a description of laboratory methods for examining virus transmission by trichodorids. *Revue de Nematologie* 12(3), 235–241.

Brown, D.J.F., Halbrendt, J.M., Robbins, R.T. and Vrain, T.C. (1993) Transmission of nepoviruses by *Xiphinema americanum*-group nematodes. *Journal of Nematology* 25(3), 349–354.

Brown, D.J.F., Halbrendt, J.M., Jones, A.T., Vrain, T.C. and Robbins, R.T. (1994) Transmission of three North American nepoviruses by populations of four distinct species of the *Xiphinema americanum* group. *Phytopathology* 84, 646.

Brown, E.B. (1956) A seed-borne attack of chrysanthemum eelworm (*Aphelenchoides ritzema-bosi*) on the annual aster (*Callistephus chinensis*). *Journal of Helminthology* 30(2/3), 145–148.

Brown, R.H. (1987) Control strategies in low value crops. In: Brown, R.H. and Kerry, B.R. (eds) *Principles and Practice of Nematode Control in Crops*. Academic Press, NSW, Australia

Brücher, E., Vuletic, E.E., Guerra, F.A., De Rossi, R.L., Plazas, M.C.G.D., Guerra, M.G. *et al.* (2019) Characterization of a population of *Helicotylenchus dihystera* (Cobb, 1893) Sher, 1961, parasitizing maize roots, in southern Córdoba, Argentina. *Nematropica* 49, 49–58.

Brugni, N. and Chaves, E. (1994) Criconematids from a cypress forest of South Argentina. *Nematologica* 40(4), 467–473.

Brzeski, M.W. (1998) *Nematodes of Tylenchina in Poland and Temperate Europe*. Muzeum i Instytut Zoologii PAN, Warsaw, Poland.

Brzeski, M.W., Laughlin, C.W., Vargas, J.M., McCallum, D.K. and Hanna, M.K. (1971) Occurrence of cyst nematodes (*Heterodera*) in Michigan. *Plant Disease Reporter* 55, 399.

Bulman, S.R. and Marshall, J.W. (1997) Differentiation of Australasian potato cyst nematode (PCN) populations using the polymerase chain reaction (PCR). *New Zealand Journal of Crop and Horticultural Science* 25, 123–129.

Burgermeister, W., Braasch, H., Metge, K., Gu, J., Schröder, T. and Woldt, E. (2009) ITS-RFLP analysis, an efficient tool for identification of *Bursaphelenchus* species. *Nematology* 11, 649–668.

Burrows, P.R. and Stone, A.R. (1985) *Heterodera glycines. CIH Descriptions of Plant-Parasitic Nematodes*, Set 8, No. 118. Commonwealth Institute of Helminthology, St. Albans (UK).

Buser, A. (1990) Untersuchungen uber die Pfeffingerkrankheit der Susskirsche und deren vektor *Longidorus macrosoma* (Nematoda: Longidoridae). Dissertation. Eidgenossischen Technischen Hochschule Nr. 9194, Zurich.

Butler, E.J. (1913a) Diseases of rice. I. An eelworm disease of rice. *Agricultural Research Insitute Pusa Bull.* 34B, 1–27.

Butler, E.J. (1913b) Diseases of rice. I. *Agricultural Journal of India* 8 (3), 205–220.

Butler, E.J. (1919) The rice worm (*Tylenchus angustus*) and its control. *Botanical Series* X , 1–37.

CABI (2013) Datasheets for plant-parasitic nematodes: *Xiphinema americanum*. Crop Protection Compendium (CPC), CABI, Wallingford, UK. Available at http:// www.cabi.org/cpc/

CABI-CPC (2010) *Crop Protection Compendium*. CABI, Wallingford, UK.

CABI/EPPO (2006) *Scutellonema brachyurus. Distribution Maps of Plant Diseases*, No. 990. CAB International, Wallingford, UK.

CABI-ISC (2020) *Invasive Species Compendium*. CAB International, Wallingford, UK. Available at: http:// www.cabi.org/isc (accessed 1 March 2021)

Cadet, P. (1990) Effect of some nematicides on the yields and on the multiplication of nematode parasites of tomatoes in Martinique. *Bulletin Agronomique des Antilles et de la Guyane* 10, 42–48.

Cadet, P. and Albrecht, A. (1992) Hillock-levelling field in Martinique. 3. Influence on the sugarcane plant parasitic nematode community in relation with plant growth. *Cahiers ORSTOM, Série Pédologie* 27, 49–58.

Cadet, P. and Quénéhervé, P. (1982) Action of nematicides against *Hirschmanniella spinicaudata* in flooded rice fields. *Revue de Nematologie* 5(1), 93–102.

Campelo, A.M.F.L. and Galli, F. (1980) Pathogenicity of *Helicotylenchus dihystera* in *Theobroma cacao*. *Revista Theobroma* 10(1), 5–14.

Campos, P., Sivapalan, P. and Gnanapragasam, N.C. (1990) Nematode parasites of coffee, cocoa and tea. In: Luc, M., Sikora, R.A. and Bridge, J. (eds) *Plant Parasitic Nematodes in Sub Tropical and Tropical Agriculture*. CAB International, Wallingford, UK, pp. 387–430.

Campos, V.P. and Melles, C.C.A. (1987) Occurrence and distribution of *Meloidogyne* species in coffee plantations of Campos das Vertentes and in the south of Minas. *Nematologia Brasileira* 11, 233–241.

Carbonell, E.E. (1978) The nematodes, pests of sugarcane in Peru. *Proceedings International Society for Sugarcane Technologists* 16, 545–551.

Carneiro, R.M.D.G., Carneiro, R.G. and Monteiro, A.R. (1982) Vertical distribution of four sugarcane nematodes in relation to soil factors. In: Lordello, L.G.E. (ed.) *Trabalhos apresentados a VI reuniao Brasileira de nematologia*, 8–12 fevereiro de 1982, Fortaleza. Publicacao No. 6. Sociedade Brasileira de Nematologia, Piracicaba, Sao Paulo, Brazil, pp. 117–132.

Carneiro, R.M.D.G., Carneiro, R.G., Abrantes, I.M.O., Santos, M.S.NA and Almeida, M.R.A. (1996) *Meloidogyne paranaensis* n. sp. (Nemata: Meloidogynidae), a root-knot nematode parasitizing coffee in Brazil. *Journal of Nematology* 28, 177–189.

Carneiro, R.M.D.G., Almeida, M.R.A. and Quénéhervé, T.P. (2000) Enzyme phenotypes of *Meloidogyne* spp. populations. *Nematology* 2(6), 645–654.

Carneiro, R.M.D.G., Moreira, W.A., Almeida, M.R.A. and Gomes, A.C.M.M. (2001) First record of *Meloidogyne mayaguensis* on guava in Brazil. *Nematologia Brasileira* 25(2), 223–228.

Carneiro, R.M.D.G., Tigano, M.S., Randig, O., Almeida, M.R.A. and Sarah, J.L. (2004) Identification and genetic diversity of *Meloidogyne* spp. (Tylenchida: Meloidogynidae) on coffee from Brazil, Central America and Hawaii. *Nematology* 6(2), 287–298.

Carneiro, R.M.D.G., Almeida, M.RA, Cofcewicz, E.T., Magunacelaya, J.C. and Aballay, E. (2007) *Meloidogyne ethiopica*, a major root-knot nematode parasitising *Vitis vinifera* and other crops in Chile. *Nematology* 9, 635–641.

Carta, L.K., Skantar, A.M. and Handoo, Z.A. (2001) Molecular, morphological and thermal characters of 19 *Pratylenchu* spp. and relatives using the D3 segment of the nuclear LSU rRNA gene. *Nematropica* 31, 195–209.

Castagnone, C., Abad, P. and Castagnone-Sereno, P. (2005) Satellite DNA-based species-specific identification of single individuals of the pinewood nematode *Bursaphelenchus xylophilus* (Nematoda: Aphelenchoididae). *European Journal of Plant Pathology* 112, 191–193.

Castillo, P. and Vovlas, N. (2007) *Pratylenchus* (Nematoda: Pratylenchidae): diagnosis, biology, pathogenicity and management. Vol. 6 in: Hunt, D.J. and Perry, R.N. (series eds) *Nematology Monographs and Perspectives*. Brill, Leiden, Netherlands.

Castillo, P., Trapero-Casas, J.L. and Jiménez-Díaz, R.M. (1995) Effect of time, temperature, and inoculum density on reproduction of *Pratylenchus thornei* in carrot disk cultures. *Journal of Nematology* 27, 120–124.

Castillo, P., Vovlas, N. and Troccoli, A. (2003) The reniform nematode, *Rotylenchulus macrosoma*, infecting olive in southern Spain. *Nematology* 5, 23–29.

Castillo, P., Stanley, J., Inserra, R.N. and Manzanilla-López, R.H. (2012) Pratylenchidae – the lesion nematodes. In: Manzanilla-López, R.H. and Marbán-Mendoza, N. (eds) *Practical Plant Nematology*. Colegio de Postgraduados and Mundi-Prensa, Biblioteca Básica de Agricultura, Guadalajara, Jalisco, pp. 411–478.

Castro, J.M.C., Lima, R.D. and Carneiro, R.M.D.G. (2003) [Isoenzymatic variability in Brazilian populations of *Meloidogyne* spp. from soybean]. *Nematologia Brasileira* 27(1), 1–12 [in Portuguese].

Caswell, E.P., Sarah, J.L.and Apt, W.J. (1990) Nematode parasites of pineapple. In: Luc, M., Sikora, R.A. and Bridge, J. (eds) *Plant Parasitic Nematodes in Sub tropical and Tropical Agriculture*. CAB International, Wallingford, UK, pp. 519–537.

Catford, J.A., Jansson, R. and Nilsson, C. (2009) Reducing redundancy in invasion ecology by integrating hypotheses into a single theoretical framework. *Diversity and Distributions* 15, 22–40.

Catling, H.D., Cox, P.G., Islam, Z. and Rahman, L. (1979) Two destructive pests of deep water rice, yellow stem borer and ufra. *The ADAB News VI* 8, 16–21.

Caubel, G., Abbad Andaloussi, F., Bekal, S., Bossis, M., Esquibet, M. and Sellami, S. (1998) Variabilité de populations du nématode des tiges, *Ditylenchus dipsaci* sur quelques légumineuses a grosses graines. [The variability of populations of stem nematode, *Ditylenchus dipsaci* on some large grain legumes.] *Les légumineuses alimentaires méditerranéennes*. Contraintes biotiques et potentialités de développement, Rennes, France, 20–22 février 1997. Colloques de l'INRA, No. 88, 173–186.

Cayrol, J.C. and Dalmasso, A. (1975) Intraspecific relationships between three leaf nematodes (*A. fragarip*, *A. ritzemabosi* and *A. besseyi*). *Cahiers ORSTOM, Serie Biologie, nematologie* 10(3), 215–225.

CBD (2002) Sixth Conference of the Parties, The Hague, the Netherlands, 7–19 April 2002: Decision VI/23: *Alien species that threaten ecosystems, habitats or species* to which is annexed *Guiding principles for the prevention, introduction and mitigation of impacts of alien species that threaten ecosystems, habitats or species*. Available at: www.biodiv.org

Chakraborti, S. (2000) Evaluation of neem products on management of *ufra* disease of rice. *Indian Journal of Nematology* 30, 234–236.

Chakraborti, S. (2001) A biorational approach in the management of onion stem nematode. *Journal of Applied Zoological Researches* 12(2/3), 119–121.

Chakraborty, H.S., Nayak, D.K. and Pal, A. (1985) *Ufra* incidence in summer rice in West Bengal. *International Rice Research Newsletter* 10(1), 15–16.

Chang, S.S. (1993) The occurrence of nematodes on citrus in Fujian. *Fujian Fruit Tree* 3, 55–57.

Chau, N.N. (1997) On some trichodorids (Nematoda: Trichodoridae) from Vietnam. *International Journal of Nematology* 7(1), 80–84.

Chaves, E. and Torres, M.S. (2000) Nematode fauna associated with golf courses in the south-eastern region of the Buenos Aires province. *Revista de la Facultad de Agronomía* (Universidad de Buenos Aires) 20(3), 379–386.

Chaves, N., Cervantes, E., Zabalgogeazcoa, I. and Araya, C. (2013) *Aphelenchoides besseyi* Christie (Nematoda: Aphelenchoididae), agente causal del amachamiento del frijol común. *Tropical Plant Pathology* 38, 243-252.

Chellemi, D.O., Olson, S.M., Scott, J.W., Mitchell, D.J. and McSorley, R. (1994) Reduction of phytoparasitic nematodes on tomato by soil solarization and genotype. *Journal of Nematology* 25(Supp. 4), 800–805.

Chen, D.Y., Chan, H.Y., Yen, J.H., Chen, P. and Tsay, T.T. (2012) The first report of dragon fruit that parasitized by cactus cyst nematode, *Cactodera cacti* in Taiwan. *Plant Pathology Bulletin Taiwan* 21, 170. [Abstract in Chinese.]

Chen, Q.W., Hooper, D.J., Loof, P.A.A. and Xu, J.H. (1997) A revised polytomous key for the identification of species of the genus *Longidorus* Micoletzky, 1922 (Nematoda: Dorylaimoidea). *Fundamental and Applied Nematology* 20, 15–28.

Cherry, T., Szalanski, A.L., Todd, T.C. and Powers, T.O. (1997) The internal transcribed spacer region of *Belonolaimus* (Nemata: Belonolaimidae). *Journal of Nematology* 29(1), 23–29.

Chévres-Román, R., Gross, H.D. and Sasser, J.N. (1971) Influence of selected nematode species and number of consecutive plantings of corn and sorghum on forage production, chemical composition and plant and soil, and water use efficiency. *Nematropica* 1(2), 40–41.

Chihani-Hammas, N., Hajji-Hedfi, L., Regaieg, H., Larayedh, A., Badiss, A., Qing, Y. and Najet, H. (2018) First report of *Pratylenchus vulnus* associated with apple in Tunisia. *Journal of Nematology*, 4. doi: 10.21307/jofnem-2018-056

Chinappen, M., Jokhun, P. and Lamberti, F. (1988) A decline of thyme associated with *Rotylenchulus parvus* in Mauritius. *Nematologia Mediterranea* 16(2), 233–234.

Chinnasri, B., Tanchitsomkid, N. and Toida, Y. (1995) *Heterodera zeae* on maize in Thailand. *Japanese Journal of Nematology* 24, 35–38.

Chitambar, J.J. (2016) *California Pest Rating for Asian Citrus Root-knot Nematodes*. California Department of Food and Agriculture, Sacramento, California. Available at: https://blogs.cdfa.ca.gov/Section3162/?p=686 (accessed 25 February 2021).

Chitambar, J.J. (2018) *California Pest Rating for* Ditylenchus dipsaci *(Kühn, 1857) Filipjev, 1936*. California Department of Food and Agriculture, Sacramento, California. Available at: https://blogs.cdfa.ca.gov/Section3162/?p=5680 (accessed 25 February 2021).

Chitambar, J.J. and Subbotin, S.A. (2014) Systematics of the sheath nematodes of the superfamily Hemicycliophoroidea. Vol. 10 in: Hunt, D.J. and Perry, R.N. (series eds) *Nematology Monographs and Perspectives*. Brill, Leiden, Netherlands.

Chitambar, J.J., Westerdahl, B.B. and Subbotin, S.A. (2018) Plant Parasitic Nematodes in California Agriculture. In: Subbotin, S. and Chitambar, J. (eds) *Plant Parasitic Nematodes in Sustainable Agriculture of North America*. *Sustainability in Plant and Crop Protection*. Springer, Cham.

Chitwood, B.G. (1951) North American marine nematodes. *Texas Journal of Science* 4, 617–72.

Chitwood, B.G. and Birchfield, W. (1957) A new genus, *Hemicriconemoides* (Criconematidae: Tylenchina). *Proceedings Helminthological Society Washington* 24, 80–86.

Chitwood, B.G. and Toung, M. (1960) Meloidogyne from Taiwan and New Delhi. *Phytopathology* 50, 631–2.

Chizhov, V.N. and Subbotin, S.A. (1990) [Phytoparasitic nematodes of the subfamily Anguininae (Nematoda: Tylenchida). Morphology, trophic specialization, taxonomic.] *Zoologichesky Zhurnal*, 69(4), 15–26. [In Russian.]

Chizov, V.N., Chumakova, O.A., Subbotin, S.A. and Baldwin, J.G. (2006) Morphological and molecular characterization of foliar nematodes of the genus *Aphelenchoides*: *A. fragariae* and A. ritzemabosi (Nematoda: Aphelenchoididae) from the Main Botanical Garden of the Russian Academy of Science, Moscow. *Russian Journal of Nematology* 14, 179–184.

Chizhov, V.N., Borisov, B.A. and Subbotin, S.A. (2010) A new stem nematode, *Ditylenchus weischeri* sp.n. (Nematoda: Tylenchida), a parasite of *Cirsium arvense* (L) Scop. in the Central Region of the Non-Chernozem Zone of Russia. *Russian Journal of Nematology* 18, 95–102.

Chizhov, V.N., Pridannikov, M.V., Peneva, V. and Subbotin, S.A. (2014) Morphological and molecular characterisation of the Saratov population of the European dagger nematode, *Xiphinema diversicaudatum* (Nematoda: Dorylaimida), with notes on phylogeography of the species. *Nematology* 16, 847–862.

Chong Bin, L., Qi, Y.-H., Ke, Q.X. and Shun, Q.M. (2001) Incidence of *Hirshmanniella diversa* in *Nelumbo nucifera* Gaertn and its infection pattern. *Plant Protection*, 27, 26-27.

Christie, J.R. (1959) *Plant Nematodes: Their Bionomics and Control*. University of Florida, Gainesville, Florida.

Christoforou, M., Orford, M. and Tsaltas D. (2017) Molecular diagnostic tool for nematodes. *Intech Open*, DOI:http://dx.doi.org/10.5772/intechopen.69075

Cid Del Prado Vera, I. and Subbotin, S.A. (2012) *Belonolaimus maluceroi* sp. n. (Tylenchida: Belonolaimidae) from a tropical forest in Mexico and key to the species of *Belonolaimus*. *Nematropica* 42, 201–210.

Cid Del Prado Vera, I. and Subbotin, S.A. (2014) A new cyst nematode, *Cactodera torreyanae* sp. n. (Tylenchida: Heteroderidae), parasitising romerito, *Suaeda torreyana*, in Texcoco, Mexico. *Nematology* 16, 163–174.

CISEH (2019) Center for Invasive Species and Ecosystem Health. Available at: https://www.invasive.org/species/nematodes.cfm (accessed 28 October 2019)

Clapp, J.P., van der Stoel, C.D. and van der Putten, W.H. (2000) Rapid identification of cyst (*Heterodera* spp., *Globodera* spp.) and root-knot (*Meloidogyne* spp.) nematodes on the basis of ITS2 sequence variation detected by PCR-single-strand conformational polymorphism (PCR-SSCP) in cultures and field samples. *Molecular Ecology* 9, 1223–1232.

Clark, R.A., Esser, R.P. and O'Bannon, J.H. (1991) *Procedures to detect wheat seed gall nematode in Florida*. Nematology Circular No. 186. Florida Department of Agriculture and Consumer Services, Division of Plant Industry, Tallahassee, Florida.

Clerjeau, M., Rancillac, M. and Veschambre, D. (1983) The position regarding strawberry decline in France. *Pepinieristes Horticuteurs Maraichers – Revue Horticole* No. 237, 39–42.

Cobb, N.A. (1893) Nematodes, mostly Australian and Fijian. In: *Macleay Memorial Volume*. Linnean Society of New South Wales, pp. 252–308.

Cobb, N.A. (1906) *Fungus Maladies of the Sugar Cane with Notes on Associated Insects and Nematodes*. Bulletin, Division of Pathology and Physiology, Hawaiian Sugar Planters Association, Honolulu, Hawaii.

Cobb, N.A. (1913) Notes on *Mononchus* and *Tylenchulus*. *Journal of Washington Academic Science* 3, 287–288.

Cobb, N.A. (1919) A newly discovered nematode (*Aphelenchus cocophilus* n. sp.) connected with a serious disease of the coconut palm. *West Indian Bulletin* 17, 203–210.

Cobon, J. and O'Neill, W. (2011) *Aphelenchoides fragariae* (Ritzema Bos, 1890) Christie, 1932. *Australasian Plant Pathology Society Pathogen of the Month*, August. Available at: http://www.appsnet.org/publications/potm/pdf/Aug11.pdf (accessed 7 March 2016)

Coetzee, V. (1956) *Meloidogyne acronea*, a new species of root-knot nematode. *Nature* 177, 899–900.

Cohn, E. (1977) *Xiphinema italiae*. *CIH Descriptions of Plant-parasitic Nematodes*, Set 7, No. 95. Commonwealth Institute of Helminthology, St Albans, UK.

Cohn, E. and Mordechai, M. (1969) Investigations on the life cycles and host preference of some species of *Xiphinema* and *Longidorus* under controlled conditions. *Nematologica* 15(3), 295–302.

Cohn, E. and Mordechai, M. (1974) Experiments in suppressing citrus nematode populations by use of a marigold and a predacious nematode. *Nematologia Mediterranea* 1, 43–53.

Cohn, E. and Mordechai, M. (1977) Uninucleate giant cell induced in soybean by the nematode *Rotylenchulus macrodoratus*. *Phytoparasitica* 5, 83–93.

Cohn, E. and Orion, D. (1970) The pathological effect of representative *Xiphinema* and *Longidorus* species on selected host plants. *Nematologica* 16, 423–428.

Cohn, E., Tanne, E. and Nitzany, F.E. (1970) *Xiphinema italiae*, a new vector of grapevine fanleaf virus. *Phytopathology* 60, 181–182.

Coiro, M.I., Taylor, C.E., Borgo, M. and Lamberti, F. (1990) Resistance of grapevine rootstocks to *Xiphinema index*. *Nematologia Mediterranea* 18(2), 119–121.

Colbran, R.C. (1963) Studies of plant and soil nematodes. 6. Two new species from citrus orchards. *Queensland Journal of Agriculture and Animal Science* 20, 469–474.

Collingborn, F.M.B., Gowen, S.R. and Galan-Sauco, V. (1998) Screening Indian cultivars of Musa for resistance or tolerance to *Radopholus similis* and *Pratylenchus coffeae*. *Proceedings of the first international symposium on banana in the subtropics, Puerto de la Cruz*, Tenerife, Spain, 10–14 November, 1997. *Acta Horticulturae* 490, 369–372.

Collins, R.J. (1972) Relationship of fertilizer treatments and cropping sequence to populations of plant-parasitic and free-living nematodes. *Dissertation Abstracts International* 32B(12), 6770–6771.

Collis-George, N. and Blake, C.D. (1959) The influence of the soil moisture regime on the expulsion of the larval mass of the nematode *Anguina agrostis* from galls. *Australian Journal of Biological Sciences* 12(3), 247–256.

Cook, C.G. and Namken, L.N. (1992) Effect of reniform nematodes on growth, lint yield and fibre quality of 15 cotton cultivars. P201. In: Duggar, C.P. and Richtar, D.A. (eds) *1992 Proceedings Beltwide Cotton Conferences*, Jan. 6–10, Nashville, TN. National Cotton Council, Memphis, Tennessee.

Cook, R. (2004) Personal communication. Principal Research Scientist, Institute of Grassland and Environmental Research (IGER), Aberystwyth Research Station (UK).

Cook, R. and Rivoal, R. (1998) Genetics of resistance and parasitism. In: Sharma, S.B. (ed.) *The Cyst Nematodes*. Chapman and Hall, London, pp. 322–352.

Cook, R. and York, P.A. (1988) Reaction of some European and Australian oat cultivars to cereal cyst nematode. *Annals of Applied Biology* 112, 84–85.

Cook, R. and Yeates, G.W. (1993) Nematode pests of grassland and forage crops. In: Evans, K., Trudgill, D.L. and Webster, J.M. (eds) *Plant parasitic nematodes in temperate agriculture*. CAB International, Wallingford, UK, pp. 305–350.

Cooke, D. (1993) Nematode parasites of sugarbeet. In: Evans, K., Trudgill, D.L. and Webster, J.M. (eds) *Plant Parasitic Nematodes in Temperate Agriculture*. CAB International, Wallingford, UK, pp. 133–169.

Coomans, A. (1963) Observations on the variability of morphological structures in *Hoplolaimus pararobustus*. *Nematologica* 9, 241–254.

Coomans, A. (1979) General principles of systematics with particular reference to speciation. In: Lamberti, F. and Taylor, C.E. (eds) *Root-Knot Nematodes (*Meloidogyne *species): Systematics, Biology and Control*. Academic Press, London, pp. 1–19.

Coomans, A., Huys, R., Heyns, J. and Luc, M. (2001) Character analysis, phylogeny and biogeography of the genus *Xiphinema* Cobb, 1913 (Nematoda: Longidoridae). Tervuren, Belgium, Musée Royal de L'Afrique Centrale. *Annales Sciences Zoologiques* 287, 1–239.

Corbett, D.C.M. (1973) *Pratylenchus penetrans*. *CIH Descriptions of Plant-parasitic Nematodes*, Set 2, No. 25. Commonwealth Institute of Helminthology, St Albans, UK.

Corbett, D.C.M. (1974) *Pratylenchus vulnus*. *CIH Descriptions of Plant-parasitic Nematodes,* Set 3, No. 37. Commonwealth Institute of Helminthology, St Albans, UK.

Corbett, D.C.M. (1976) *Pratylenchus brachyurus*. *CIH Descriptions of Plant-parasitic Nematodes,* Set 6, No. 89. Commonwealth Institute of Helminthology, St Albans, UK.

Cordero, D.A. (1989) Comparative morphology and development of the cone of *Heterodera schachtii* and *Cactodera cacti*. PhD Thesis. University of California, Riverside, California.

Cordero, M.A., Robbins, R.T. and Szalanski, A.L. (2012) Taxonomic and molecular identification of *Mesocriconema* and *Criconemoides* species (Nematoda: Criconematidae). *Journal of Nematology* 44(4), 399–426.
Correia, F.J.S. and Abrantes, I.M. de O. (2005) Characterization of *Heterodera zeae* populations from Portugal. *Journal of Nematology* 37, 328–335.
Costilla, M.A. (1973) The nematode *Pratylenchus zeae* Graham in sugar cane. *Revista Industrial y Agricola de Tucuman* 50(2), 39–43.
Costilla, M. (1985) El falso nematodo del nudo *Nacobbus aberrans* (Thorne, 1935) Thorne & Allen, 1944 y su relación con el cultivo de la papa en el Noroeste Argentino. *Revista Internacional y Agrícola de Tucumán* 62, 79–97.
Cotten, J. (1975) Virus vector species of *Xiphinema* and *Longidorus* in relation to certification schemes for fruit and hops in England. *In:* Lamberti F, Taylor CE, Seinhorst JW, ed. *Nematode vectors of plant viruses*. Plenum Press. London & New York, USA UK, 283-285.
Cotten, J., Flegg, J.J.M. and Popham, A.M. (1971) Population studies with *Xiphinema diversicaudatum* and *X. index* maintained under two temperature regimes. *Nematologica* 16, 584–590.
Courtney, W.D. (1962) Stem nematode of red clover in the Pacific Northwest. *Bulletin of the Washington State Agricultural Experiment Station* 640, 1–17.
Cox, P.G. (1980) Symptoms of *ufra* disease in deep water rice in Bangladesh. *International Rice Research Newsletter* 5, 18.
Cox, P.G. and Rahman, L. (1979) The over winter decay of *Ditylenchus angustus*. *International Rice Research Newsletter* 4, 4.
Cox, P.G. and Rahman, L. (1980) Effect of *ufra* disease on yield loss of deep water rice in Bangladesh. *Tropical Pest Management* 26, 410–415.
Coyne, D.L. (2009) Pre-empting plant-parasitic nematode losses on banana in Africa: Which species do we target? *Acta Horticulture* 828, 227–235.
Coyne, D.L. and Plowright, R.A. (1999a) Effects of *Heterodera sacchari* population density on establishment and development of upland rice cv. IDSA6 under field and pot conditions. *IIRN* 24.2, 21–22.
Coyne, D.L. and Plowright, R.A. (1999b) Susceptibility of some cereal crops to cyst nematode *Heterodera sacchari* in West Africa. *International Rice Research Notes* 24(3), 17.
Coyne, D.L. and Plowright, R.A. (2000) *Heterodera sacchari*: field population dynamics and damage to upland rice in Côte d'Ivoire. *Nematology* 2(5), 541–550.
Cronin, D., Loccoz, Y.M., Febton, A., Dunne, C., Dowling, D.N and Gara, F.O. (1997) Role of 2,4-diacetyl phloroglucinol in the interactions of the biocontrol *Pseudomonas* strain F113 with the potato cyst nematode. *Applied Environmental Microbiology* 63, 1351–1361.
Crosse, J.E. and Pitcher, R.S. (1952) Studies in the relationship of eelworms and bacteria to certain plant diseases. I. The etiology of strawberry cauliflower disease. *Annals of Applied Biology* 39, 475–484.
Crow, W.T. (2013) *Nematode management for non-residential lawns, athletic fields, racetracks and cemeteries in Florida*. Extension publication ENY-038, Entomology and Nematology Department, IFAS, University of Florida, Gainesville, Florida.
Crow, W.T. and Brammer, A.S. (2015) *Lance Nematode,* Hoplolaimus galeatus *(Cobb, 1913) Thorne, 1935 (Nematoda: Secernentea: Tylenchida: Tylenchoidea: Hoplolaimidae)*. Extension publication ENY-234, Entomology and Nematology Department, UF/IFAS Extension, University of Florida, Gainesville, Florida. Available at: http://edis.ifas.ufl.edu/pdffiles/in/in395000 (accessed 16 February 2021)
Crozzoli, R., Lamberti, F., Greco, N. and Rivas, D. (1998) Plant parasitic nematodes associated with citrus in Venezuela. *Nematologia Mediterranea* 26, 31–58.
Cuany, A. and Ritter, M. (1977) New trends for control of nematodes in carnation crops. *Acta Horticulturae* 71, 201–207.
Cuc, N.T.T. (1982) New weed host of rice stem nematode identified in Vietnam. *International Rice Research Newsletter* 7, 15.
Cuc, N.T.T. (1989) Study on the life cycle of *Ditylenchus angustus* [unpublished].
Cuc, N.T.T. and Giang, L.T. (1982) Relative humidity and nematode number of survivals in rice seeds. *International Rice Research Newsletter* 7, 14.
Cunha, M.J.M., Bossis, M., Abrantes, I.M. de O., Santos, M.S.N. de A. and Mugniéry, D. (2000) Protein variability among Portuguese and other populations of *Globodera rostochiensis* revealed by two-dimensional gel electrophoresis with computed image analysis. *Nematology* 2, 461–471.
Curi, S.M. (1968) *Meloidogyne coffeicola* destruidor e insidiosa nematoide de cafeeiro. *O Biologico* 34(6), 144.
Curran J., Baillie D.L. and Webster, J.M. (1985) Use of genomic DNA restriction fragment length differences to identify nematode species. *Parasitology* 90, 137–144.
Curran, J., Driver, F., Ballard, J.W.O. and Milner, R.J. (1994) Phylogeny of *Metarhizium*: analysis of ribosomal DNA sequence data. *Mycological Research* 98, 547–555. doi: 10.1016/S0953-7562(09)80478-4
Curtis, R.H.C., Al-Hinai, M.S., Diggines, A.E.R. and Evans, K. (1997) Serological identification and quantification of *Heterodera avenae* from processed soil samples. *Nematologica* 43, 199–213.
Curtis, R.H.C., Robinson, A.F. and Perry, R.N. (2009) Hatch and host location. In: Perry, R.N., Moens, M. and Starr, J.L. (eds) *Root-knot Nematodes*. CABI, Wallingford, UK, pp. 139–162.
Dababat, A.A., Yildiz, Ş., Çiftçi, V., Duman, N. and İmren, M. (2019) Occurrence and seasonal variation of the root lesion nematode *Pratylenchus neglectus* on cereals in Bolu, Turkey. *Turkey Journal of Agriculture* 43, 21–27.
Daday, E. (1905) Untersuchungen über die Süßwasser-Mikrofauna Paraguays. *Zoologica* 18, 1–374.
Daher, S., Gillet, S., Mugniéry, D. and Marzin, H. (1996) Discovery in France and characteristics of the Dutch variant of *Meloidogyne chitwoodi*. *Proceedings of the Third International Nematology Congress*, Gosier, Guadeloupe, p. 188.
Dallimore, C.E. and Thorne, G. (1951) Infection of sugar beets by *Ditylenchus destructor* Thorne, the potato rot nematode. *Phytopathology* 41, 872–874.
Dalmasso, A. (1969) Méthodes de lutte contre un nématode vecteur du court noué de la vigne. *Phytoma* 21, 20–23.
Dalmasso, A. (1970) La gamétogenèse des genres *Xiphinema* et *Longidorus* (Nematoda: Dorylaimida). *Compte Rendu Hebdomadaire des Seances de l' Academie des Sciences Paris* 270, 824–827.
Damme, N., Waeyenberge, L., Viaene, N., van Hoenselaar, T. and Karssen, G. (2013) First Report of the Root-Knot Nematode *Meloidogyne artiellia* in Belgium. *Plant Disease* 97(1), 152. doi: 10.1094/PDIS-08-12-0720-PDN
Dan, S.D. (2008) Determination of parasitic nematode *Aphelenchoides* sp. causing leaf blotch disease of sambiloto *(Andrographis paniculata)*. *Journal Littri* 14(2), 61–66. [In Indonesian.]
Danilova, Yu. S. (1996) Bio-ecological justification of control measures of hop against hop cyst nematode *Heterodera humuli* Filipjev, 1934. PhD Thesis. TSKhA, Moscow. [In Russian]
Das, D. (2004) Chemical control of rice stem nematode, *Ditylenchus angustus* in deep ware rice in Assam. *Annales of Biology* 20, 43–45.
Das, P.K. and Rao, Y.S. (1970) Life history and pathogenesis of *Hoplolaimus indicus* incidence in rice. *Indian Pathology* 23, 459–464.
Das, S.N., Mohanty, K.C. and Ray, S. (1988) Population dynamics of *Criconemella ornata* under multiple cropping systems. *International Nematology Network Newsletter* 5(1), 27–30.
Dasgupta, D.R. and Raski, D.J. (1968) The biology of *Rotylenchulus parvus*. *Nematologica* 14, 429–440.

Dasgupta, D.R., Raski, D.J. and Sher, S.A. (1968) A revision of the genus *Rotylenchulus* Linford and Oliveira, 1940 (Nematoda: Tylenchidae). *Proceedings of the Helminthological Society of Washington* 35, 169–192.

Dasgupta, M.K. (1998) *Phytonematology*. Naya Prakash, Calcutta.

Dasgupta, M.K., Mukherjee, B., Bhattacharya, C., Khan, M.R. and Ghosh, S. (2010) Nematode infestation in tea. In: Khan, M.R. and Jairajpuri, M.S. (eds) *Nematode Infestations – Part II: Industrial Crops*. National Academy of Sciences, Allahabad (Prayagraj), UP, India, pp. 171–206.

Dauphinais, N., Vandal, M., Gagnon, A.E., Bélair, G., Véronneau, Y.P. and Mimee, B. (2018) Development and application of a multiplex qPCR method for the simultaneous detection and quantification of *Pratylenchus alleni* and *P. penetrans* in Quebec, Canada. *Plant Disease* 102, 970–976.

Davide, R.G. (1988) Nematode problems affecting agriculture in the Philippines. *Journal of Nematology* 20(2), 214–218.

Davies, K.G., Flynn, C.A., Laird, V. and Kerry, B.R. (1990) The life cycle, population dynamics and host specificity of a parasite of *Heterodera avenae*, similar to *Pasteuria penetrans*. *Review of Nematology* 13, 303–309.

Davies, K.G., Curtis, R.H. and Evans, K. (1996) Serologically based diagnostic and quantification tests for nematodes. *Pesticide Science* 47, 81–87.

Davis, E.E. and Venette, R.C. (2004a) *Mini risk assessment. Asian Citrus root-knot nematodes: Meloidogyne citri* Zhang, Gao & Weng; *M. donghaiensis* Zheng, Lin & Zheng; *M. fujianensis* Pan; *M. indica* Whitehead; *M. jianyangensis* Yang, Hu, Chen & Zhu; *M. kongi* Yang, Wang and Feng; and *M. mingnanica* Zhang [Nematoda: Meloidogynidae]. Cooperative Agricultural Pest Survey (CAPS), Animal and Plant Health Inspection Service, US Department of Agriculture, Riverdale, Maryland. Available at: http://www.aphis.usda.gov/plant_health/plant_pest_info/pest_detection/downloads/pra/asi ancitrusmmeloidogynepra.pdf

Davis, E.E. and Venette, R.C. (2004b) *Mini Risk Assessment. British root-knot nematode: Meloidogyne artiellia* Franklin [Nematoda: Meloidogynidae] Available at: . http://www.aphis.usda.gov/plant_health/plant_pest_info/pest_detection/downloads/pra/martielliapra.pdf

Davis, E.L. and Mitchum, M.G. (2005) Nematodes: sophisticated parasites of legumes. *Plant Physiology* 137, 1182–1188.

Davis, E.L. and Tylka, G.L. (2000) Soybean cyst nematode disease. *The Plant Health Instructor*. doi: 10.1094/PHI-I-2000-0725-01

Davis, J.R., Hafez, S.L. and Soressen, L.H. (1992) Lesion nematode suppression with the Butte potato and relationships to *Verticillium* wilt. *American Potato Journal* 69, 371–83.

Davis, R.F. and Webster, T.M. (2005) Relative host status of selected weeds and crops for *Meloidogyne incognita* and *Rotylenchulus reniformis*. *Journal of Cotton Science* 9, 41–46.

Davis, R.M., Nuñez, J., and Aegerter, B.J. (2014) *UC IPM: UC Management Guidelines for Corky Ringspot on Potato*. University of California, Davis, California.

Dawabah, A.A.M., Al-Hazmi, A.S., Al-Rehiayani, S.M., Abdel-Mawgood, A.L., Motawei, M.I., Al-Otayk, S., Sadder, M.T., Elgorban, A.M. *et al.* (2012) Morphological and molecular characterization of cereal cyst nematode (*Heterodera avenae*) populations from arid environments. *Australian Journal of Crop Science* 6(6), 970–979.

De Boer, J.M., Overmars, H.A., Pomp, H., Davis, E.L., Zilverentant, J.F., Goverse, A., Smant, G., Stokkermans, J.P.W.G., Hussey, R.S., Gommers, F.J., Bakker, J. and Schots, A. (1996) Production and characterization of monoclonal antibodies to antigens from second stage juveniles of the potato cyst nematodes, *Globodera rostochiensis*. *Fundamental and Applied Nematology* 19, 545–554.

De Giorgi, C., Veronico, P., De Luca, F., Natilla, A., Lanave, C. and Pesole, G. (2002) Structural and evolutionary analysis of the ribosomal genes of the parasitic nematode *Meloidogyne artiellia* suggests its ancient origin. *Molecular and Biochemical Parasitology* 124(1-2), 91–94.

De Guiran, G. (1964) *Mesotylus*, nouveau genre de Pratylenchinae (Nematoda: Tylenchoidea). *Nematologica* 9(4), 567–575.

De Guiran, G. and Ritter, M. (1979) Life cycle of *Meloidogyne* species and factors influencing their development. In: Lamberti, F. and Taylor, C.E. (eds) *Root-Knot Nematodes (*Meloidogyne *species). Systematics, Biology and Control*. Academic Press, London, pp. 173–191.

De Guiran, G. and Vilardebo, A. (1962) Le bananier aux Iles Canaries. IV. Les nématodes parasites du bananier. *Fruits* 17, 263–277.

De Ley, P., Félix, M.A., Frisse, L.M., Nadler, S.A., Sternberg, P.W. and Thomas, K.W. (1999) Molecular and morphological characterisation of two reproductively isolated species with mirror-image anatomy (Nematoda: Cephalobidae). *Nematology* 1, 591–612.

De Ley, P., Tandingan De Ley, I., Morris, K., Abebe, E., Mundo-Ocampo, M. *et al.* (2005) An integrated approach to fast and informative morphological vouchering of nematodes for applications in molecular barcoding. *Philosophical Transactions of the Royal Society* 360, 1945–1958.

De Luca, F., Di Vito, M., Fanelli, E., Reyes, A., Greco, N. and De Giorgi, C. (2009) Characterization of the heat shock protein 90 gene in the plant parasitic nematode *Meloidogyne artiellia* and its expression as related to different developmental stages and temperature. *Gene* 440 (1-2), 16–22.

De Luca, F., Vovlas, N., Lucarelli, G., Troccoli, A., Radicci, V. *et al.* (2013) *Heterodero elachista* the Japanese cyst nematode parasitizing corn in Italy: integrative diagnosis and bionomics. *European Journal of Plant Pathology* 136, 857–872.

de Man, J.G. (1876) Onderzoekingen over vrij in de aarde levende Nematoden. *Tijdschrift der Nederlandsche Dierkundige Vereeniging* 2, 78–196.

de Man, J.G. (1880) Die einheimischen, frei in der reinen erde und im süssen wasser lebenden nematoden. *Tijdschrift der Nederlandsche Dierkundige Vereeniging* 5, 1–104.

de Man, J.G. (1884) Die frei in der reinen Erde und im süssen Wasser lebenden Nematoden der Niederländischen Fauna. Eine systematisch-faunistische Monographie. *Tijdschrift der Nederlandsche Dierkundige Vereeniging* 5, 138–143.

de Santiago, S.A., Sosa-Moss, C. and Garcia-Gomez, F.A. (1984) Control de nemátodos del maiz *Zea mays* L., Crillo chalqueo, mediante immersfon de semillas en nematicidas sistemicos. *Revista Chapingo* 43-44, 93–96.

De Souza, P. (1978) A disease complex of coffee involving *Meloidogyne exigua* and *Rhizoctonia solani*. *Dissertation Abstracts International* 39B(2), 467–468.

De Waele, D. and Elsen, A. (2007) Challenges in Tropical Plant Nematology. *Annual Review of Phytopathology*, 45, 457–85.

De Waele, D., Jones, B.L., Bolton, C., and Van Den Berg, E. (1989) *Ditylenchus destructor* in hulls and seeds of peanut. *Journal of Nematology* 21, 10–15.

De Waele, D. and Wilken, R. (1990) Effect of temperature on the in vitro reproduction of *Ditylenchus destructor* isolated from peanut. *Revue de Nématologie* 13, 171–174.

Dean, C.G. (1979) *Red ring disease of* Cocos nucifera *L. caused by* Rhadinaphelenchus cocophilus *(Cobb, 1919) Goodey, 1960. An annotated bibliography and review*. Technical Communication No. 47 of Commonwealth Institute of Helminthology, Commonwealth Agricultural Bureaux, St Albans, UK.

Decker, H. (1972) Leaf parasitic nematodes. In: Sveshnikova, N.M. (ed.) *Plant Nematodes and their Control (Phytonematology)*. Amerind Publishing Co., New Delhi.

Decker, H. (1989) *Plant Nematodes and their Control (Phytonematology)*. E.J. Brill., New York.

Decraemer, W. (1991) Stubby root and virus vector nematodes: *Trichodorus*, *Paratrichodorus*, *Allotrichodorus* and *Monotrichodorus*. In Nickle, W.R. (ed.), Manual of Agricultural Helminthology, Marcel Dekker, New York, pp. 587–625.

Decraemer, W. (1995) *The Family Trichodoridae: Stubby Root and Virus Vector Nematodes*. Kluwer Academic Publishers, Dordrecht, The Netherlands.

Decraemer, W. and Geraert, E. (1996) On the taxonomic status of *Hemicriconemoides mangiferae* Siddiqi, 1961 and *H. strictathecatus* Esser, 1960 (Nematoda: Criconematidae). *Fundamental and Applied Nematology* 19, 608.

Decraemer, W. and Geraert, E. (2013) Ectoparasitic nematodes. In: Perry, R.N. and Moens, M. (eds) *Plant Nematology*, 2nd edn. CAB International, Wallingford, UK, pp. 199–202.

Decraemer, W. and Hunt, D. (2006) Structure and classification. In: Perry, R.N. and Moens, M. (eds) *Plant Nematology*. CAB International, Wallingford, UK, pp. 3–32.
Decraemer, W. and Hunt, D.J. (2013) Structure and classification. In: Perry, R.N. and Moens, M. (eds) *Plant Nematology*, 2nd edn. , CAB International, Wallingford, UK, pp. 3–39.
Decraemer, W. and Robbins, R. (2007) The who, what and where of *Longidoridae* and *Trichodoridae*. *Journal of Nematology* 39, 295–297.
Decraemer, W., Cantalapiedra-Navarrete, C., Archidona-Yuste, A., Varela-Benavides, I., Gutiérrez-Gutiérrez, C., Castillo, P. and Palomares-Rius, J.E. (2018) Integrative taxonomy unravels cryptic diversity in the *Paratrichodorus hispanus* group complex and resolves two new species of the genus and the molecular phylogeny of the family (Nematoda: Trichodoridae). *Zoological Journal of the Linnean Society* 185, 656–692. doi: 10.1093/zoolinnean/zly059
Deimi, A.M. and Mitkowski, N. (2010). Nematodes associated with vineyards throughout Markazi Province (Arak), Iran. *Australasian Plant Pathology* 39, 571–577.
Demangeat, G., Minot, J.C., Voisin, R., Bosselut, N., Fuchs, M. and Esmenjaud, D. (2005) Survival of *Xiphinema index* and retention of Grapevine fanleaf virus over extended periods of time in the absence of host plants. *Phytopathology* 95, 1151–1156.
Demangeat, G., Komar, V., Van-Ghelder, C., Voisin, R., Lemaire, O., Esmenjaud, D. and Fuchs, M. (2010) Transmission competency of single-female *Xiphinema index* lines for Grapevine fanleaf virus. *Phytopathology* 100(4), 384–389. doi:10.1094/PHYTO-100-4-0384
Demeure, Y., Freckman, D.W. and Van Gundy, S.D. (1979) Anhydrobiotic coiling of nematodes in soil. *Journal of Nematology* 11(2), 189–195.
Den, N.L.J.M.F. and Lock, C.A.M. (1990) Quantification of *Globodera rostochiensis* and *G. pallida* in mixed populations using species-specific thermostable proteins. *Netherlands Journal of Plant Pathology* 96, 179–186.
Deportes, L., Gilly, G., Cuany, A., Mercier, S., Poupet, A. and Marais, A. (1979) A contribution to the study of certain technical problems posed by violet culture. *Pepinieristes Horticulteurs Maraichers* 200, 13–21.
Derycke, S., Remerie, T.A., Vierstraete, T., Backeljau, J., Vanfleteren, M., Vincx, M. and Moens, T. (2005) Mitochondrial DNA variation and cryptic speciation within the free-living marine nematode *Pellioditis marina*. *Marine Ecology Progress Series* 300, 91–103.
Devrajan, K. and Balasubramanian, P. (2008) Management of *Globodera rostochiensis* and *G. pallida* in potato by antagonistic crop, mustard in Nilgiris. *Indian Journal of Nematology* 38, 15–17.
Devrajan, K., Seenivasan, N., Jonathan, E.I. and Prabhu, S. (2012) Nematode infestation in potato. In: Khan, M.R. and Jairajpuri, M.S. (eds) *Nematode Infestations – Part III: Horticultural Crops*. National Academy of Sciences, Allahabad (Prayagraj), UP, India, pp. 182–209.
Devrajan, K., Seenivasan, N., Selvaraj, N. and Rajendran, G. (2004) An integrated approach for the management of potato cyst nematodes, *Globodera rostochiensis* and *G. pallida* in India. *Nematologia Mediterranea* 32, 67–70.
Devran, Z., Polat, İ., Mıstanoğlu, İ. and Baysal, Ö. (2018) A novel multiplex PCR tool for simultaneous detection of three root-knot nematodes. *Australasian Plant Pathology* 47, 389–392.
Dey, S. and Baqri, Q.H. (1990) Nematodes from West Bengal (India). XXII. Morphometric and Allometric varieties in *Hirschmanniella oryzae* (Van Breda De Haan, 1902) Luc and Goodey 1963 (Pratylenchidae : Tylenchida). *Records of the Zoological Survey of India* 82(2), 319–327.
Dhéry, M., Germani, G. and Giard, A. (1975) Results of nematicidal treatments against chlorosis and clump disease of peanut in Upper Volta. *Cahiers ORSTOM, Serie Biologie, nematologie* 10(3), 161–167.
Di Vito, M. (1991) *The pea cyst nematode* Heterodera goettingiana. Nematology Circular Gainesville, No. 188. Florida Department of Agriculture and Consumer Services, Tallahassee, Florida.
Di Vito, M. and Greco, N. (1986) Pea cyst nematode. In: Lamberti, F. and Taylor, C.E. (eds) *Cyst Nematodes*. NATO ASI Series A: Life Sciences, Vol. 121. Plenum Press, New York and London.
Di Vito, M. and Greco, N. (1988a) Investigation on the biology of *Meloidogyne artiellia*. *Revue de Nématologie* 11, 221–225.
Di Vito, M. and Greco, N. (1988b) Effect of population densities of *Meloidogyne artiellia* on yield of wheat. *Nematologia Mediterranea* 16, 167–169.
Di Vito, M., Singh, K.B., Greco, N. and Saxena, M.C. (1996) Sources of resistance to cyst nematode in cultivated and wild *Cicer* species. *Genetic Resources and Crop Evolution* 43, 103–107.
Diab, K.A. and El-Eraki, S. (1968) Plant-parasitic nematodes associated with olive decline in the United Arab Republic. *Plant Disease Reporter* 52(2), 150–154.
Dias, W.P., Liberato, J.R. and da Fonseca, A.F.A. (1996) Survey of plant parasitic nematodes associated with coffee crop in the Espirito Santo State, Brazil. *Revista Ceres* 43(250), 808–812.
Dickson, D.W. and De Waele, D. (2005) Nematode parasites of peanut. In: Luc, M., Sikora, R.A. and Bridge, J. (eds) *Plant Parasitic Nematodes in Subtropical and Tropical Agriculture*, 2nd edn. CABI Publishing, Wallingford, UK, pp. 393–436.
Dickson, D.W., Sasser, J.N. and Huisingh, D. (1970) Comparative disc electrophoretic protein analysis of selected *Meloidogyne*, *Ditylenchus*, *Heterodera*, and *Aphelenchus* spp. *Journal of Nematology* 2, 286–293.
Dillon, A.B., Rolston, A.N., Meade, C.V., Downes, M.J. and Griffin, C.T. (2008) Establishment, persistence, and introgression of entomopathogenic nematodes in a forest ecosystem. *Ecological Applications* 18, 735–747.
DiSanzo, C.P. (1981) Effect of foliar application of carbofuran and a related compound on plant-parasitic nematodes under greenhouse and growth chamber conditions. *Journal of Nematology* 13(1), 20–24.
DiSanzo, C.P. (1982) Effect of foliar application of carbofuran and a related compound on plant-parasitic nematodes under microplot and field conditions. *Journal of Nematology* 14(2), 208–212.
Dominik, J. (1981) Summer control of xylophagous insect pests in Scots pine stands. *Sylwan* 125(7/8/9), 111–117
Dong, K., Dean, R.A., Fortnum, B.A. and Lewis, S.A. (2001a) A species-specific DNA probe for the identification of *Meloidogyne hapla*. *Nematropica* 31, 17–23.
Dong, K., Dean, A.A., Fortnum, B.A. and Lewis, S.A. (2001b) Development of PCR primers to identify species of root-knot nematodes: *Meloidogyne arenaria*, *M. hapla*, *M. incognita* and *M. javanica*. *Nematropica* 31, 271–280.
Dorris, M. and Blaxter, M. (2000) The small subunit ribosomal RNA sequence of *Strongyloides stercoralis*. *International Journal for Parasitology* 30, 939–941.
Drozdovskii, E.M., Blinov, B.A. and Yakovleva, V.A. (1971) *Study of Aphelenchoides–Corynebacterium complex*. Sbornik Nauchnykh Rabot Nauchno-Issledovatel'skogo Zonal'nogo Instituta Sadovodstva Nechernozemnoi Polosy (Plodovodstvo i yagodovodstvo nechernozemnoi polosy) Moscow, USSR, 3, 393–402.
Duan, Y., Liu, W., Liu, Y. and Zhao, H. (1996) A new species of *Hirschmanniella* (Pratylenchidae) associated with the Brassicae (*Brassica oleracea* var. *capitata*). *Journal of Shenyang Agricultural University* 3, 35–8.
Duan, Y.X., Wang, D. and Chen, L.J. (2012) First report of the cactus cyst nematode, *Cactodera cacti*, on cactus in Northern China. *Plant Disease* 96, 1385.
Duarte, I.M., De Almeida, M.T.M., Brown, D.J.F., Marques, I., Neilson, R. and Decraemer, W. (2010) Phylogenetic relationships, based on SSU rDNA sequences, among the didelphic genera of the family Tridhodoridae from Portugal. *Nematology*, 12, 171–80.

DuCharme, E.P. and Birchfield, W. (1956) Physiologic races of the burrowing nematode. *Phytopathology* 46, 615–616.

Duggan, J.J. (1969) Leaf and bud eelworms of strawberries. *Farm Research News* 10(6), 134–136.

Duggan, J.J. and Moore, J.F. (1962) Host-range of tuber rot eelworm (*Ditylenchus destructor*). *Irish Journal of Agricultural Research* 1, 109–114.

Duncan, L.W. and Cohn, E. (1990) Nematode parasites of citrus. In: Luc, M., Sikora, R.A. and Bridge, J. (eds) *Plant Parasitic Nematodes in Subtropical and Tropical Agriculture*. CAB International, Wallingford, UK, pp. 321–346.

Duncan, L.W. and Moens, M. (2013) Migratory endoparasitic nematodes. In: Perry, R.N. and Moens, M. (eds) *Plant Nematology*. CABI, Wallingford, UK, pp. 145–162.

Duncan, L.W. Mashela, P., Ferguson, J., Graham, J., Abou-Setta, M.M. and El-Morshedy, M.M. (1995) Estimating crop loss in orchards with patches of mature citrus trees infected by *Tylenchulus semipenetrans*. *Nematropica* 25(1), 43–51.

Duncan, L.W., Noling, J.W. and Inserra, R.N. (2009) *2009 Florida Citrus Pest Management Guide: Nematodes*. ENY-606 series of Entomology and Nematology Department, Florida Cooperative Extension Service, Institute of Food and Agricultural Sciences, University of Florida, Gainesville, Florida.

Dunn, R.A. (1973a) Effect of temperature on survival and reproduction of *Pratylenchus penetrans* (Cobb, 1917) Filipjev and Schurmans Stekhoven, 1941. *Dissertation Abstracts International*, 34B(1), 14.

Dunn, R.A. (1973b) Resistance in potato (*Solanum tuberosum*) to *Pratylenchus penetrans*. Abstracts of the papers. International Congress of Plant Pathology (2nd) Minneapolis, Minnesota, September 5–12, St Paul, Minnesota, USA. American Phytopathological Society Inc., No. 0860.

Dwinell, L.D. (1997) The pinewood nematode: regulation and mitigation. *Annual Review of Phytopathology* 35, 153–166.

Dwivedi, K. and Upadhyay, K.D. (2001) Population dynamics of *Heterodera cajani* associated with pigeon pea in various cropping systems of central Uttar Pradesh. *Indian Journal of Nematology* 31(2), 120–121.

Ebsary, B.A. and Anderson, R.V. (1982) Two new species of *Hirschmanniella* Luc & Goodey, 1963 (Nematoda: Pratylenchidae) with a key to the nominal species. *Canadian Journal of Zoology* 60, 530–535.

Eguiguren-Carrion, R. (1995) Control of *Meloidogyne incognita* and *Nacobbus* sp. with *Paecilomyces lilacinus* in the greenhouse and effect of nematicides on the fungus. *Biocontrol* 1(4), 41–49.

Egunjobi, O.A. (1974) Nematodes and maize growth in Nigeria. I. Population dynamics of *Pratylenchus brachyurus* in and about the roots of maize and its effects on maize production at Ibadan. *Nematologica* 20(2), 181–186.

Eisenback, J.D. (1985) Diagnostic characters useful in the identification of the four most common species of root-knot nematodes (*Meloidogyne* spp.). In: Sasser, J.N. and Carter, C.C. (eds) *An Advanced Treatise on Meloidogyne. Vol I. Biology and Control.* North Carolina State University Graphics, Raleigh, North Carolina, pp. 95–112.

Eisenback, J.D. and Hunt, D.J. (2009) General morphology. In: Perry, R.N., Moens, M. and Starr, J.L. (eds) *Root-knot Nematodes*. CABI, Wallingford, UK, pp. 19–54.

Eisenback, J.D. and Triantaphyllou, H. (1991) Root-knot nematodes: *Meloidogyne* species and races. In: Nickle, W.R. (ed.) *Manual of Agricultural Nematology*. Marcel Dekker, New York, pp. 191–274.

Eisenback, J.D., Hirschmann, H., Sasser, J.N. and Triantaphyllou, AC. (1981) *A Guide to the Four Most Common Species of Root-knot Nematodes (*Meloidogyne *species), with a Pictorial Key*. A Cooperative Publication of the Departments of Plant Pathology and Genetics, North Carolina State University and US Agency for International Development, Raleigh, North Carolina.

Elbadri, G.A.A., De Ley, P., Waeyenberge, L., Vierstraete, A., Moens, M. and Vanfleteren, J. (2002) Intraspecific variation in *Radopholus similis* isolates assessed with restriction fragment length polymorphism and DNA sequencing of the internal transcribed spacer region of the ribosomal RNA cistron. *International Journal of Parasitology* 32(2), 199–205.

El-Sherif, M.A., Ali, A.H. and Barakat, M.I. (1995) Suppressive bacteria associated with plant parasitic nematodes in Egyptian agriculture. *Japanese Journal of Nematology* 24(2), 55–59.

Elyas, Z. and Sharma, S.B. (1997) Mechanism of resistance to *Heterodera cajani* in *Cajanus platycarpus* accessions. *International Journal of Nematology* 7(2), 119–121.

EPPO (2004) European and Mediterranean Plant Protection Organization. PM 7/39(1) *Aphelenchoides besseyi*. *EPPO Bulletin* 34, 303–308.

EPPO (2008) European and Mediterranean Plant Protection Organization. PM 7/88(1) *Radopholus similis*. EPPO Bulletin 38, 374–378.

EPPO (2009a) European and Mediterranean Plant Protection Organization. PM 4/35 (1) Soil test for virus–vector nematodes in the framework of EPPO Standard PM 4 Schemes for the production of healthy plants for planting of fruit crops, grapevine, Populus and Salix. *EPPO Bulletin* 39, 284–288.

EPPO (2009b) PQR database. Paris, France: European and Mediterranean Plant Protection Organization. Available at: www.eppo.org (accessed February 2021)

EPPO (2011) European and Mediterranean Plant Protection Organization. EPPO Reporting Service. Paris, France. Available at: http://archives.eppo.org/EPPOReporting/Reporting_Archives.htm.

EPPO (2012) European and Mediterranean Plant Protection Organization. EPPO A1 and A2 lists of pests recommended for regulation as quarantine pests PM 1/2(21). Available at: http://www.eppo.org/QUARANTINE/quarantine.htm.

EPPO (2013) PQR: European and Mediterranean Plant Protection Organization (EPPO) Plant Quarantine Data Retrieval System. Available at http//www.eppo.org-DATABASES/pqr/pqr.htm

EPPO (2017a) *Pest Risk Analysis for Meloidogyne mali* (Tylenchida: Meloidogynidae). Available at: https://gd.eppo.int/taxon/MELGMA/documents (accessed 8 October 2019)

EPPO (2017b) *European and Mediterranean Plant Protection Organization (EPPO) Global Database*. Available at: https://gd.eppo.int

EPPO (2017c) European and Mediterranean Plant Protection Organization. Diagnostics PM 7/40 (4) *Globodera rostochiensis* and *Globodera pallida*. *EPPO Bulletin* 47(2), 174–197.

EPPO (2018) European and Mediterranean Plant Protection Organization. PM 7/136 (1) *Meloidogyne mali*. *EPPO Bulletin* 48(3), 438–445.

EPPO (2019) European and Mediterranean Plant Protection Organization. Available at: http://www.eppo.org/ (accessed 28 October 2019)

Eriksson, K.B. and Granberg, J. (1969) Studies of *Ditylenchus dipsaci* races using electrophoresis in acrylamide gel. *Nematologica* 15, 530–534.

Esbenshade, P. and Triantaphyllou, A. (1985a) Use of enzyme phenotypes for identification of *Meloidogyne* species. *Journal of Nematology* 17, 6–20.

Esbenshade, P.R. and Triantaphyllou, A.C. (1985b) Identification of major *Meloidogyne* species employing phenotypes as differentiating characters. In: Sasser, J.N. and Carter C.C. (eds) *An Advanced Treatise on Meloidogyne. Vol.I. Biology and Control.* North Carolina State Graphics, Raleigh, N Carolina, pp. 135–140.

Esbenshade, P.R. and Triantaphyllou, A.C. (1990) Isozyme phenotypes for the identification of *Meloidogyne* species. *Journal of Nematology* 22, 10–15.

Escobar-Avila, M., Lopex-Villegas, E.O., Subbotin, S.A. and Tovar-Soto, A. (2018) First report of carrot cyst nematode *Heterodera carotae* in Mexico: morphological, molecular characterization, and host range study. *Journal of Nematology* 50, 229–242. doi: 10.21307/jofnem-2018-021

Escobar-Avila, I.M., Cruz-Alvarado, Y., Tovar-Soto, A. and Subbotin, S.A. (2019) First report of sugar beet cyst nematode, *Heterodera schachtii* on beetroot and broccoli in Mexico. *Disease Notes*. doi: 10.1094/PDIS-11-18-2000-PDN

Eskandari, A., Karegar, A., Pourjam, E., vander Berg, E. and Tiedt, L. (2010) Additional data on some poorly known species of *Criconemoides* Taylor, 1936 (Nematoda: Criconematidae). *Nematology* 12, 505–518.

Esmenjaud, D. (1983) Grapevine infectious degeneration (court-noue): transmission by the nematode *Xiphinema index* and control measures. *Progres Agricole et Viticole* 100(20), 514–515.

Esmenjaud, D., Walter, B., Valentin, G., Guo, Z.T. and Cluzeau, D. (1992) Vertical distribution and infectious potential of *Xiphinema index* (Thorne et Allen, 1950) (Nematoda: Longidoridae) in fields affected by grapevine fanleaf virus in vineyards in the Champagne region of France. *Agronomie* 12(5), 395–399.

Esquibet, M., Grenier, E., Plantard, O., Andaloussi, F.A. and Caubel, G. (2003) DNA polymorphism in the stem nematode *Ditylenchus dipsaci*: development of diagnostic markers for normal and giant races. *Genome* 46, 1077–1083.

Esser, R.P. (1969) Rhadinaphelenchus cocophilus *a potential threat to Florida palms*. Nematology Circular No. 9. Division of Plant Industry, Florida Department of Agriculture, Gainesville, Florida.

Esser, R.P. (1991) *A computer-ready checklist of the genera and species of phytoparasitic nematodes, including a list of mnemonically coded subject categories*. Bulletin 13. Contribution No. 387. Florida Department of Agriculture and Consumer Services, Division of Plant Industry, Gainesville, Florida.

Esser, R.P. and Meredith, J.A. (1987) *Red ring nematode*. Nematology Circular No. 141. Division of Plant Industry, Florida Department of Agriculture and Consumer Services, Gainesville, Florida.

Esser, R.P., Riherd, C.C. and Harkcom, K.J. (1986) Pathogenicity of *Scutellonema brachyurum* to Aloe vera. *Nematropica* 16, 65–71.

Esteves, I., Maleita, C. and Abrantes, I. (2015) Root-lesion and root-knot nematodes parasitizing potato. *European Journal of Plant Pathology* 141, 397–406.

Etchiha, A.S.A.P., Affokpon A., Waeyenberge L., de Sutter, N., Agbangla C., Dansi, A., Coyne, D.L. and Viaene, N. (2018) Molecular diversity of *Scutellonema bradys* populations from Benin, based on ITS1 rDNA and COI mtDNA. *Tropical Plant Pathology* 43, 323–332 doi: 10.1007/s40858-018-0221-5

Evans, A.A.F. (1971) Taxonomic value of gel electrophoresis of proteins from mycophagous and plant-parasitic nematodes. *International Journal of Biochemistry* 2, 72–79.

Evans, H.F. and Perry, R.N. (2009) Survival mechanisms. In: RN Perry, R.N., Moens, M. and Starr, J.L (eds) *Root-knot Nematodes*, CAB International, Wallingford, UK pp. 201–219.

Evans, H.F., McNamara, D.G., Braasch, H., Chadoeuf, J. and Magnusson, C. (1996) Pest risk analysis (PRA) for the territories of the European Union (as PRA area) on *Bursaphelenchus xylophilus* and its vectors in the genus *Monochamus*. *EPPO Bulletin* 26, 199–249.

Evans, K. (1970) Longevity of males and fertilisation of females of *Heterodera rostochiensis*. *Nematologica* 16, 369–374.

Evans, K. and Brodie, B.B. (1980) The origin and distribution of the golden nematode and its potential in the USA. *American Potato Journal* 57, 79–90.

Evans, K. and Rowe, J.A. (1998) Distribution and economic importance. In: Sharma, S.B. (ed.) *The Cyst Nematodes*. Kluwer Academic Publishers, Dordrecht, pp. 1–30.

Evans, K. and Stone, A.R. (1977) A review of the distribution and biology of the potato cyst nematodes *Globodera rostochiensis* and *G. pallida*. *PANS* 23, 178–189.

Evans, K., Trudgill, D.L. and Webster, J.M. (1993) *Plant Parasitic Nematodes in Temperate Agriculture*. CABI, Wallingford, UK.

Fallas, G.A., Hahn, M.L., Fargette, M., Burrows, P.R. and Sarah, J.L. (1996) Molecular and biochemical diversity among isolates of *Radopholus* spp. from different areas of the world. *Journal of Nematology* 28(4), 422–430.

Fanelli, E., Troccoli, A., Picardi, E., Pousis, C. and De Luca, F. (2014) Molecular characterization and functional analysis of four β-1,4-endoglucanases from the root-lesion nematode *Pratylenchus vulnus*. *Plant Pathology* 63, 1436–1444.

Fanelli, E., Troccoli, A., Picardi, E. and De Luca, F. (2019) Functional variation of two novel cellulases, pv-eng-5 and pv-eng-8, and the heat shock 90 gene, pv-hsp-90, in *Pratylenchus vulnus* and their expression in response to different temperature stress. *International Journal of Molecular Sciences* 20, 107. doi: 10.3390/ijms20010107

FAO (1997) *Revised International Plant Protection Convention*. Report of the 29th Session of the FAO Conference, Rome, Italy, 7–18 November 1997. Available at: www.ippc.int

FAOSTAT (2019) *World Food and Agriculture – Statistical Pocketbook 2019*. Food and Agriculture Organization of the United Nations, Rome. Available at: http://www.fao.org/3/ca6463en/ca6463en.pdf

Fargette, M., Phillips, M.S., Blok, V.C., Waugh, R. and Trudgill, D.L. (1996) An RFLP study of relationships between species, populations and resistance-breaking lines of tropical species of *Meloidogyne*. *Fundamental and Applied Nematology* 19(2), 193–200.

Farkas, K., Hangya, L. and Nemeth, L. (1985) Nematological studies on chrysanthemums. *Novenyvedelem* 21(12), 529–537.

Fassuliotis, G. (1974) A Description of Males of *Hoplolaimus Columbus*. *Journal of Nematology* 6(3),116–8.

Fassuliotis, G. (1976) *Hoplolaimus columbus*. *CIH Descriptions of Plant-parasitic Nematodes*, Set 6, No. 81. Commonwealth Institute of Helminthology, St Albans, UK.

Feldmesser, J. and Golden, A.M. (1972) Control of nematodes damaging home lawngrasses in two counties in Maryland. *Plant Disease Reporter* 56(6), 476–480.

FERA (2009) *Validation of quantitative DNA detection systems for PCN*. Potato Council Project R287. Food and Environment Research Agency, York, UK.

Fernandez Díaz-Silveira, M. and Ortega Herrera, J. (1982) Weeds as hosts of rice nematodes. *Ciencias de la Agricultura* 12, 114–116

Fernandez Dias-Silveira, M. and Ortega Herrera, J. (1995) Principales problemas nematologicos en Cuba. In: *Proceedings Congresso International de Nematologia Tropical, Rio Quente*, Brazil, 161–171.

Fernandez Díaz-Silveira, M. and Ortega Herrera, J. (1998) An overview of nematological problems in Cuba. *Nematropica* 28(2), 151–164.

Ferraz, B.C.C.L. (2006) The nematode *Pratylenchus brachyurus* and soybean under tillage. *Tillage Magazine* 95, 23–27.

Ferreira, S.A. and Boley, R.A. (1991) *Rotylenchulus reniformis*. University of Hawaii Crop Knowledge Master, available at www.extento.hawaii.edu/kbase/crop/Type/r_renif.htm.

Ferris, H. (2005) *Heterodera sacchari*. Available at: http://plpnemweb.ucdavis.edu/nemaplex/Taxadata/G060s6.htm. (accesses 5 April 2005)

Ferris, H. (2013) *Nemaplex: The Nematode Plant Expert Information System: A Virtual Encyclopedia on Soil and Plant Nematodes: Anguina tritici*. University of California, Davis, California. Available at: http://plpnemweb.ucdavis.edu/nemaplex/Taxadata/G006S4.htm (accessed 7 December 2013)

Ferris, H. (2019) Nemaplex. Available at: http://plpnemweb.ucdavis.edu/nemaplex.

Ferris, J.M. (1960) Effect of storage temperatures on survival of plant-parasitic nematodes in soil. *Phytopathology* 50(9), 635.

Ferris, J.M. and Ferris, V.R. (1998) Biology of plant parasitic nematodes. In: Barker, K.R., Pederson, G.A. and Windham, G.L. (eds) *Plant and Nematode Interactions*. American Society of Agronomy, Inc., Crop Science Society of America, Inc. and Soil Science Society of America, Inc., Madison, Wisconsin, pp. 21–36.

Ferris, V.R. (1961) A new species of *Pratylenchus* (Nemata-Tylenchida) from roots of soybeans. *Proceedings of the Helminthological Society of Washington* 28, 109–111.

Ferris, V.R., Ferris, J.M. and Faghihi, J. (1993) Variation in spacer ribosomal DNA in some cyst-forming species of plant parasitic nematodes. *Fundamental and Applied Nematology* 16, 177–184.

Ferris, V.R., Ferris, J.M., Faghihi, J. and Ireholm, A. (1994) Comparisons of isolates of *Heterodera avenae* using 2-D PAGE protein patterns and ribosomal DNA. *Journal of Nematology* 26, 144–151.

Ferris, V.R., Subbotin, S.A., Ireholm A. Spiegel Y., Faghihi, J. and Ferris, J.M. (1999) Ribosomal DNA sequence analysis of *Heterodera filipjevi* and *H. latipons* isolates from Russia and comparisons with other nematode isolates Virginia. *Russian Journal of Nematology* 7(2), 121–125.
Filipjev, I.N. (1936) On the classification of Tylenchidae. *Proceedings of the Helminthological Society of Washington* 3, 80–82.
Filipjev, I.N. and Schuurmans, S. Jr (1941) *A Manual of Agricultural Helminthology*. E.J. Brill, Leiden, Netherlands.
Firoza, K. and Maqbool, M.A. (1994) A diagnostic compendium of the genus *Helicotylenchus* Steiner, 1945 (Nematoda: Hoplolaimidae). *Pakistan Journal of Nematology* 12, 11–50.
Firoza, K. and Maqbool, M.A. (1995) Numerical threshold for infection of the spiral nematode, *Helicotylenchus dihystera* (Cobb, 1893) Sher, 1961 on brinjal, tomato and wheat. *Pakistan Journal of Nematology* 13(2), 93–97.
Firoza, K. and Maqbool, M.A. (1996a) Effect of plant extracts in the control of *Helicotylenchus dihystera* (Cobb, 1893) Sher, 1961. *Pakistan Journal of Nematology* 14(1), 61–66.
Firoza, K. and Maqbool, M.A. (1996b) Nematicidal properties of leaves of some plant species against *Helicotylenchus dihystera* (Cobb, 1893) Sher, 1963 on tomato. *Pakistan Journal of Nematology* 14(2), 107–110.
Fisher, J.M. and Raski, D.J. (1967) Feeding of *Xiphinema index* and *X. diversicaudatum*. *Proceedings of the Helminthological Society of Washington* 34, 68–72.
Flegg, J.J.M. (1968) The occurrence and depth distribution of *Xiphinema* and *Longidorus* species in southeastern England. *Nematologica* 14, 189–196.
Fleming, C.C. and Powers, T.O. (1998) Potato cyst nematode diagnostics: morphology, differential hosts and biochemical techniques. In: Marks, R.J. and Brodie, B.B. (eds), *Potato Cyst Nematodes, Biology, Distribution and Control*. CAB International, Wallingford, UK, pp. 91–114.
Fleming, C.C., Rao, J., Moreland, B., Craig, D. and Turner, S.J. (2000) Diagnostics of cyst nematodes and tephritid fruit flies using mitochondrial and ribosomal DNA. *Bulletin OEPP* 30, 385–590.
Flores, L. and Lopez, R. (1989a) Morphological characterization of the root-knot nematode of coffee *Meloidogyne exigua* (Nemata: Heteroderidae) I. Females and eggs. *Turrialba* 39(3), 352–360.
Flores, L. and Lopez, R. (1989b) Morphological characterization of the root-knot nematode of coffee, *Meloidogyne exigua* (Nemata: Heteroderidae). III. Second stage juveniles. *Turrialba* 39(3), 369–376.
Fonseca, L., Cardoso, J.M.S., Lopes, A., Pestana, M., Abreu, F., Nunes, N., Mota, M. and Abrantes, I. (2012) The pinewood nematode, *Bursaphelenchus xylophilus*, in Madeira Island. *Helminthologia* 49(2), 96–103.
Fortuner, R. (1970) On the morphology of *Aphelenchoides besseyi* Christie, 1942 and *A. siddiqii* n. sp. *Journal of Helminthology* 44, 141–152.
Fortuner, R. (1976) *Pratylenchus zeae*. *CIH Descriptions of Plant-parasitic Nematodes*, Set 6, No. 77. Commonwealth Institute of Helminthology, St Albans, UK.
Fortuner, R. (1977) *Pratylenchus thornei*. *CIH Descriptions of Plant-parasitic Nematodes*, Set 7, No. 93. Commonwealth Institute of Helminthology, St Albans, UK.
Fortuner, R. (1987) A reappraisal of Tylenchina (Nemata) 8. The family of Hoplolaimidae Filipjev, 1934. *Revue de Nematologie* 10, 219–232.
Fortuner, R. (1991) The Hoplolaiminae. In: Nickle, W.R. (ed.) *Manual of Agricultural Nematology*. Marcel Dekker, Inc., New York, pp. 669–720.
Fortuner, R. and Maggenti, A.R. (1991) A statistical approach to the objective differentiation of *Hirschmanniella oryzae* from *H. belli* (Nemata: Pratylenchidae). *Revue de Nématologie* 14, 165–180.
Fortuner, R. and Wong, Y. (1984) Review of the genus *Helicotylenchus* Steiner, 1945. 1. A computer program for identification of the species. *Revue de Nématologie* 7, 385–392.
Fortuner, R., Merny, G. and Roux, C. (1981) Morphometrical variability in *Helicotylenchus Steiner*, 1945. 3: Observations on African populations of *Helicotylenchus dihystera* and considerations on related species. *Revue de Nematologie* 4(2), 235–260.
Fortuner, R., Maggenti, A.R. and Whittaker, L.M. (1984) Morphometrical variability in *Helicotylenchus* Steiner, 1945. 4: Study of field populations of *H. pseudorobustus* and related species. *Revue de Nématologie* 7, 121–135.
Fotedar, D.N. and Kaul, V. (1985) A revised key to the species of the genus *Helicotylenchus* Steiner, 1945 (Nematoda: Rotylenchoidinae). *Indian Journal of Nematology* 15, 138–147.
Fourie, H., Zijlstra, C. and McDonald, A.H. (2001) Identification of root-knot nematode species occurring in South Africa using the SCAR-PCR technique. *Nematology* 3(7), 675–680.
Francl, L.J. and Wheeler, T.A. (1993) Interaction of plant-parasitic nematodes with wilt inducing fungi. In: Khan, M.W. (ed.) *Nematode Interactions*. Chapman and Hall, New York, pp. 79–103.
Franco, J., Main, G. and Oros, R. (1999) Trap crops as a component for the integrated management of *Globodera* spp. in Bolivia. *Nematropica* 29, 51–60.
François, C., Castagnone, C., Boonham, N., Tomlinson, J., Lawson, R., Hockland, S., Quill, J., Vieira, P., Mota, M. and Castagnone-Sereno, P. (2007) Satellite DNA as a target for TaqMan real-time PCR detection of the pinewood nematode, *Bursaphelenchus xylophilus*. *Molecular Plant Pathology* 8, 803–809.
Franklin, M.T. (1961) A British root-knot nematode, *Meloidogyne artiellia* n. sp. *Journal of Helminthology R.T. Leiper supplement*, 85–92.
Franklin, M.T. (1965) A root-knot nematode, *Meloidogyne naasi* n. sp., on field crops in England and Wales. *Nematologica* 11, 79–86.
Franklin, M.T. (1969) *Heterodera latipons* n. sp., a cereal cyst nematode from the Mediterranean region. *Nematologica* 15, 535–542.
Franklin, M.T. and Siddiqi, M.R. (1972) *Aphelenchoides besseyi*. *CIH Descriptions of Plant-parasitic Nematodes*, Set 1, No. 4. Commonwealth Institute of Helminthology (CIH), St Albans, UK.
Franklin, M.T. and Stone, A.R. (1974) *Hemicycliophora arenaria*. *CIH Descriptions of Plant-parasitic Nematodes*, Set 3, No. 43. Commonwealth Institute of Helminthology, St Albans, UK.
Fujimura, T., Ichita, T. and Kimura, T. (1989) Occurrence of potato-rot nematode, *Ditylenchus destructor* Thorne, in garlic and control. 1. Evaluation of treatments applied before planting and after harvest for control. *Japanese Journal of Nematology* 18, 22–29.
Fullaondo, A., Vicario, A., Aguirre, A., Barrena, I. and Salazar, A. (2001) Quantitative analysis of two-dimensional gel electrophoresis protein patterns: a method for studying genetic relationships among *Globodera pallida* populations. *Heredity* 87, 266–272.
Funt, R.C., Krusberg, L.R., Ross, D.S. and Goulart, B.L. (1982) Effect of post-plant nematicides and trickle irrigation on newly planted peach trees. *Journal of the American Society for Horticultural Science* 107(5), 891–895
Furstenberg, P.J. (1974) The effect of cultivation on soil nematodes, especially *Rotylenchulus parvus*. In: *Simposio Internacional (XII) de Nematologia, Sociedad Europea de Nematologos*, 1–7 Septiembre, 1974, Granada, Spain. European Society of Nematologists, Rennes, France.
Futai, K. (2013) Pine wood nematode, *Bursaphelenchus xylophilus*. *Annual Review of Phytopathology* 51, 61–83.
Ganguly, S., Lal, M. and Rathour, K.S. (2013) A check-list of globally known species of *Tylenchorhynchus* Cobb, 1913 along with compendium of the Indian species. *Indian Journal of Nematology* 43, 47–60.
Gao, X.B., Cheng, H.R. and Fang, Z.D. (1989) Identification and determination of pathogenicity of lesion nematodes in peach roots. *Acta Phytopathologica Sinica* 19(2), 65–68.

Garabedian, S. and Hague, N. (1982) Comparison of methods of application of oxamyl for control of *Heterodera sacchari* on sugarcane plants. *Nematropica* 12(1), 39–44.
Garabedian, S. and Hague, N.G.M. (1984) The effect of weekly exposures to non-volatile nematicides and sugarcane root diffusate on the hatching of *Heterodera sacchari*. *Revue de Nematologie* 7(1), 95–96.
García, D., García, C., Montero, Z., Salazar, L., Brenes, A. and Gómez-Alpízar, L. (2009) Morphological and molecular identification of potato cyst-forming nematode *Globodera pallida* in soil samples from Costa Rica. *Revista Latinoamericana de la Papa* 15(1), 38–45.
Gazaway, W.S. and McLean,.S. (2003) A survey of plant parasitic nematodes associated with cotton in Alabama. *Journal of Cotton Science* 7, 1–7.
Geraert, E. (2010) *Criconematidae of the world – identification of the family Criconematidae* (Nematoda). Academia Press, Gent, Belgium, and New York.
Geraert, E. (2011) *The Dolichodoridae of the World. Identification of the family Dolichodoridae (Nematoda)*. Academia Press. Gent, Belgium.
Gerber, K. and Giblin-Davis, R.M. (1990) Association of the red ring nematode and other nematode species with the palm weevil, *Rhynchophorus palmarum*. *Journal of Nematology* 22, 143–149.
Gergon, E.B. and Prot, J.C. (1993) Effect of benomyl and carbofuran on *Aphelenchoides besseyi* on rice. *Fund Applied Nematology* 6, 563–566.
German, E.V. (1972) Control measures against the potato stem nematode. *Vestnik Sel'skokhozyaistvennoi Nauki, Alma-Ata* 11, 109–111.
Germani, G. (1970) *Aphasmatylenchus straturatus* n. sp. (Nematoda: Hoplolaimidae) from West Africa. *Proceedings of the Helminthological Society of Washington* 37, 48–51.
Germani, G. (1972) Une chlorose des legumineuses de Haute-Volta lie la presence d'un nematode. *Comptes rendus hebdomadaires des Seances de l'Academie-d'Agriculture de France* 58, 202–204.
Germani, G. (1977) *Aphasmatylenchus straturatus*. *CIH Descriptions of Plant-parasitic Nematodes*, Set 7, No. 104. Commonwealth Institute of Helminthology, St Albans, UK.
Germani, G. (1979) Nematicide application as a tool to study the impact of nematodes on plant productivity. In: Mongi, H.O. and Huxley, P.A, (eds) *Soils Research in Agroforestry*. (Procedings of an expert consultation, International Council for Research in Agroforestry, Nairobi, 26–30 Mar. 1979). ICRA. Nairobi, Kenya, pp. 297–313.
Germani, G. and Luc, M. (1982) Studies on 'voltaic chlorosis' of legumes caused by the nematode *Aphasmatylenchus straturatus*. *Revue de Nematologie* 5(1), 139–146.
Germani, G., Cuany, A. and Merny, G. (1982) Factorial analysis of the effect of 2 nematodes on the growth of Arachis and on the symbiotic fixation of nitrogen. *Revue de Nematologie* 5(1), 161–168.
Ghaderi, R., Geraert, E. and Karegar, A. (2016) *The Tylenchulidae of the World: Identification of the Family Tylenchulidae (Nematoda: Tylenchida)*. Academia Press, Ghent, Belgium.
Ghosh, S.C. and Manna, B. (2008) Studies on nematode parasites associated with paddy crop of West Bengal, India. *Records of the Zoological Survey of India*, Occasional paper No. 287, pp 1–144.
Ghosh, S.C. and Manna, B. (2010) Rice nematode infestation in paddy. In: Khan, M.R. and Jairajpuri, M.S. (eds) *Nematode Infestation – Part I: Food Crops*. National Academy of Sciences, Allahabad (Prayagraj), UP, India, pp. 65–73.
Giblin-Davis, R.M. (1990) Potential for biological control of phytoparasitic nematodes in Bermudagrass turf with isolates of the *Pasteuria penetrans* group. *Proceedings of the Florida State Horticultural Society* 103, 349–351.
Giblin-Davis, R.M. (2001) Borers of palms. In: Howard, F.W., Moore, D., Giblin-Davis, R.M. and Abad, R.G. (eds) *Insects on Palms*. CABI Publishing, Wallingford, UK, pp. 267–304.
Giblin-Davis, R.M., Hazir, S., Center, B.J., Ye, W., Keskin, N., Thorp, R.W. and Thomas, W.K. (2005) *Bursaphelenchus anatolius* n. sp. (Nematoda: Parasitaphelenchidae), an associate of bees in the genus *Halictus*. *Journal of Nematology* 37(3), 336–342.
Giblin-Davis, R.M., Mundo-Ocampo, M., Baldwin, J.G., Gerber, K. and Griffith, R. (1989) Observation of the morphology of the red ring nematode, *Rhadinaphelenchus cocophilus* (Nemata: Aphelenchoididae). *Revue de Nématologie* 12, 285–292.
Gibson, T., Farrugia, D., Barrett, J., Chitwood, D.J., Rowe, J., Subbotin, S. and Dowton, M. (2011) The mitochondrial genome of the soybean cyst nematode, *Heterodera glycines*. *Genome* 54, 565–574. doi: 10.1139/G11-024
Gill, H.S. (1971) Occurrence and reproduction of *Meloidogyne javanica* on three species of citrus in California. *Plant Disease Reporter* 55, 607–608.
Gill, J.S. and Swarup, G. (1971) On the host range of the cereal cyst nematode, *Heterodera avenae* Woll., 1924, the causal organism of 'molya' of wheat and barley in Rajasthan, India. *Indian Journal of Nematology* 1, 63–67.
Gill, J.S. and Swarup, G. (1977) Effect of interaction between *Heterodera avenae* Woll.1924, *Fusarium moniliforme* and *Helminthosporium gramineum* on barley plants and nematode reproduction. *Indian Journal of Nematology* 7(1), 4–45.
GISD (2019) *Global Invasive Species Database*. Available at: http://www.iucngisd.org/gisd/search.php (accessed 28 October 2019)
GISP (2019) *Global Invasive Species Programme*. Available at: https://www.gisp.org/ (accessed 30 October 2019)
Giudici, M.L., Villa, B., Callegarin, A.M. and Tamborini, L. (2004) Rice nematode; diffusion and experimentation in Italy. *Informatore Agrario* 60 (46), 69–73.
Gnanapragasam, N.C. and Manuelpillai, M.E.K. (1984) Influence of inoculum level and temperature on the population build-up and pathogenicity of the root-lesion nematode (*Pratylenchus loosi* Loof) of tea. *Tea Quarterly* 53(1-2), 19–22.
Gnonhouri, P.G., Ouya, A., Assi Tnan, B. and Ats, T.Y. (2000) Resistance and tolerance of *Ananas comosus* to *Pratylenchus brachyurus* in Ivory Coast. *Cahiers Agricultures* 9(2), 145–147.
Gokte, N. and Mathur, V.K. (1996) New fungal host records of *Aphelenchoides besseyi*. *Indian Journal of Nematology* (publ. 1997) 26(2), 255–256.
Gokte, N., Mathur, V.K., Ramasundaram, P. and Sabesh, M. (2001) Reproductive varaiations in *Aphelenchoides bessei* populations. *Indian Journal of Nematology* 31(2), 115–119.
Golden, A.M. (1971) Classification of the genera and higher categories of the Order Tylenchida (Nematoda). In: Zuckerman, B.M., Mai, W.F. and Rohde, R.A. (eds) *Plant Parasitic Nematodes. Volume 1*. Academic Press, New York, 191–234.
Golden, A.M. and Birchfield, W. (1965) *Meloidogyne graminicola* (Heteroderidae) a new species of root-knot nematode from grass. *Proceedings of the Helminthology Society of Washington* 32, 228–231.
Golden, A.M. and Mulvey, R.H. (1983) Redescription of *Heterodera zeae*, the corn cyst nematode, with SEM observations. *Journal of Nematology* 15(1), 60–70.
Golden, A.M. and Raski, D.J. (1977) *Heterodera thornei* n. sp. (Nematoda: Heteroderidae) and a review of related species. *Journal of Nematology* 9(2), 93–112.
Golden, A.M., Maqbool, M.A. and Handoo, Z.A. (1987) Descriptions of two new species of *Tylenchorhynchus* Cobb, 1913 (Nematoda: Tylenchida). with details on morphology and variation of T. claytoni. *Journal of Nematology* 19(1), 58–68.
Göldi, E.A. (1887) Relatório sôbre a moléstia do cafeeiro na provincia do Rio de Janeiro. Extrahido do VIII Vol. dos. *Archivos do Museu Nacional, Rio de Janeiro, Imprensa Nacional*, pp. 1–121 + 4 plates and a map.
Gomes Careiro, R.M.D., Lordello, L.G.E., Gomes Careiro, R. and Monteiro, A.R. (1980) Fluctuacao populacional de quatro especies de nematoides parasitos de cana-de-acucar (*Saccharum officinarum* L.) em relacao a pluviosidade e umidade do solo. *Trabalhos apresentados a IV Reuniao de Nematologia, Sociedade Brasileira de Nematologia*, Piricicaba, SP, Brazil, pp. 127–141.

Gomez, J. (1980) Preliminary results on the distribution of phytonematodes in the coffee areas of the departments of Caldas, Risaralda and Quindio in Colombia. *Nematropica* 10(2), 67.

Goo, M.Y.C. and Sipes, B.S. (1999) Chromosome number and reproductive isolates of *Radopholus similis* from Havaii. *International Journal of Nematology* 9, 43–46.

Good, J.M., Murphy, W.S. and Brodie, B.B. (1973) Population dynamics of plant nematodes in cultivated soil: length of rotation in newly cleared and old agricultural land. *Journal of Nematology* 5(2), 117–122.

Goodey, J.B. (1951) A secondary piliferous layer on the roots of Hippeastrum. *Nature* London, 167, 822–823.

Goodey, J.B. (1952) The influence of the host on the dimensions of the plant parasitic nematode Ditylenchus destructor. *Annals of Applied Biology* 39, 1468–474.

Goodey, J.B. (1960) Gall-forming nematodes of grasses in Britain. *Journal of the Sports Turf Research Institute* 10, 5–60.

Goodey, J.B., Peacock, F.C. and Pitcher, R.S. (1960) A redescription of *Xiphinema diversicaudatum* (Micoletzky, 1923and 1927) Thorne, 1939 and observations on its larval stages. *Nematologica* 5, 127–135.

Goodey, T. (1932) The genus *Anguillulina* Gerv. and v. Ben., 1859, vel *Tylenchus* Bastian, 1865. *Journal of Helminthology* 10, 75–180.

Gokte, N., Mathur, V.K., Ramasundaram, P. and Sabesh, M. (2001) Reproductive variations in *Aphelenchoides bessei* populations. *Indian Journal of Nematology* 31(2), 115–119.

Goswami, B.K. and Mishra, S.D. (1994) Comparative efficacy of neem cake and carbofuran on plant parasitic nematodes infesting pea. *Current Nematology* 5, 7–12.

Gottlieb, Y., Cohn, E. and Spiegel-Roy, P. (1986) Biotypes of the citrus nematode (*Tylenchulus semipenetrans* Cobb.) in Israel. *Phytoparasitica* 14, 193–198.

Gowen, S.R. (1979) Some considerations of problems associated with nematode pests of banana. *Nematropica* 9, 79–90.

Gowen, S. and Quénéhervé, P. (1990) Nematode parasites of bananas, plantains and abaca. In: Luc, M., Sikora, R.A. and Bridge, J. (eds) *Plant Parasitic Nematodes in Subtropical and Tropical Agriculture*. CAB International, Wallingford, UK, pp. 431–460.

Gozel, U., Adams, B.J., Nguyen, K.B., Inserra, R. N., Giblin-Davis, R.M. and Duncan, L.W. (2006) A phylogeny of *Belonolaimus* populations in Florida inferred from DNA sequences. *Nematropica* 36, 55–171.

Graham, T.W. (1955) Pathogenicity of *Rotylenchus brachyurus* on tobacco and of *Criconemoides* sp. on tobacco and peanuts. *Phytopathology* 45, 347.

Graney, L.S.O. and Bird, G.W. (1990) Descriptions and comparative morphology of *Cactodera milleri* n. sp. (Nematoda: Heteroderidae) and *Cactodera cacti* with a review and key to the genus *Cactodera*. *Journal of Nematology* 22(4), 457–480.

Gratwick, M. and Southey, J.F. (1972) *Hot-water treatment of plant material*. Bulletin 201, Ministry of Agriculture, Fisheries and Food. HMSO, London.

Greco, N. (1986) The carrot cyst nematode. In: Lamberti, F. and Taylor, C.E. (eds) *Cyst Nematodes*. Plenum Press, New York, pp. 333–346.

Greco, N., Di, V.M., Reddy, M.V. and Saxena, M.C. (1986) Effect of Mediterranean cultivated plants on the reproduction of *Heterodera ciceri*. *Nematologia Mediterranea* 14, 193–200.

Greco, N., Vito, M., Saxena, M.C. and Reddy, M.V. (1988) Effect of *Heterodera ciceri* on yield of chickpea and lentil and development of this nematode on chickpea in Syria. *Nematologica* 34(1), 98–114.

Greco, N., Di, V.M. and Saxena, M.C. (1992a) Plant parasitic nematodes of cool season food legumes in Syria. *Nematologia Mediterranea* 20, 37–46.

Greco, N., Vito, M. and Nombela, G. (1992b) The emergence of juveniles of *Heterodera ciceri*. *Nematologica* 38(4), 514–519.

Greco, N., Vovlas, N., Vito, M.Di. and Inserra, R.N. (1992c) Meloidogyne artiellia*: a root-knot nematode parasite of cereals and other field crops*. Nematology Circular No. 201, Florida Department of Agriculture and Consumer Services, Division of Plant Industry, Gainesville, Florida.

Greco, N., Vito, M., Singh, K.B. and Saxena, M.C. (1993) Effect of *Heterodera ciceri* on the growth of selected lines of *Cicer* species. *Nematologia Mediterranea* 21(1), 111–116.

Greco, N., D'Addabbo, T., Brandonisio, A. and Elia, F. (1994) Damage to Italian crops caused by cyst-forming nematodes. *Journal of Nematology* 25, 836–842.

Greco, N., Vovlas, N., Troccoli, A. and Inserra, R.N. (2002) *The Mediterranean cereal cyst nematode,* Heterodera latipons*: a menace to cool season cereals of the United States*. Nematology Circular 221. Florida Department of Agriculture & Conservation Services Division of Plant Industry, Gainesville, Florida.

Green, C.D., Williamson, K., Dennis, E.B. and McBurney, T. (1981) The effect of oxamyl on the growth of peas attacked by pea cyst nematode. *Annals of Applied Biology* 97(3), 303–309.

Griffin, G.D. (1985) Nematode parasites of alfalfa, cereals and grasses. In: Nickle, W.R. (ed.) *Plant and Insect Nematodes*. Marcel Dekker Inc., New York, pp. 243–322.

Griffin, G.D. and Epstein, A.H. (1964) Association of dagger nematode, *Xiphinema americanum*, with stunting and winterkill of ornamental spruce. *Phytopathology* 54(2), 177–180.

Griffith, R. and Koshy, P.K. (1990) Nematode parasites of coconut and other palms. In: Luc, M., Sikora, R.A. and Bridge, J. (eds) *Plant Parasitic Nematodes in Subtropical and Tropical Agriculture*. CAB International, Wallingford, UK, pp. 363–386.

Griffith, R., Giblin-Davis, R.M., Koshy, P.K. and Sosamma, V.K. (2005) Nematode parasites of coconut and other palms. In: Luc, M., Sikora, R.A. and Bridge, J. (eds) *Plant Parasitic Nematodes in Subtropical and Tropical Agriculture*. CAB International, Wallingford, UK, pp. 493–527.

Griffith, R., Giblin-Davis, R.M., Koshy, P.K., Sosamma, V.K. and Kanzaki, N. (2018) Nematode parasites of coconut and other palms. In: Sikora, R.A., Coyne, D., Hallmann, J., and Timper, P. (eds) *Plant Parasitic Nematodes in Subtropical and Tropical Agriculture*, 3rd edn. CABI Publishing, Wallingford, UK, pp. 504–535.

Grujicic, G. (1969) Contributions to the study of parasitic nematodes on wheat in Yugoslavia. *Savremena Poljoprivreda* 17(11/12), 531–539.

Gu, J. (ed.) (2014) *Identification of* Bursaphelenchus xylophilus *and its Closely Related Species*. Xiamen University Press, Xiamen, China. [In Chinese.]

Gu, J., Braasch, H., Burgermeister, W. and Zhang, J. (2006) Records of *Bursaphelenchus* spp. intercepted in imported packaging wood at Ningbo, China. *Forest Pathology* 36, 323–333.

Gu, J., Zheng, W., Braasch, H. and Burgermeister, W. (2008) Description of *Bursaphelenchus macromucronatus* sp. n. (Nematoda: Parasitaphelenchidae) in packaging wood from Taiwan and India – a new species of the *xylophilus* group. *Journal of Nematode Morphological Systems* 11(1), 31–40.

Gu, J., Wang, J., Braasch, H., Burgermeister, W. and Schröder, T. (2011) Morphological and molecular characterisation of mucronate isolates (M form) of *Bursaphelenchus xylophilus* (Nematoda: Aphelenchoididae). *Russian Journal of Nematology* 19(2), 103–120.

Guevera Benitez, D.C., Tobar Jiménez, A. and Guevera Pozo, D. (1970) Estudio cuantitativa del ciclo vital de *Heterodera goettingiana* Liebscher, 1892 y de la possibilitad de su control mediante cultivos cept. *Revista Iberica de Parasitologia* 30, 229.

Guntzel, O., Klingler, J. and Delucchi, V. (1987) Tylenchids (Nematoda) extracted from soil of Swiss vineyards north of the Alps. *Revue de Nematologie* 10(3), 361–368.

Gutiérrez-Gutiérrez, C., Palomares-Rius, J.E., Cantalapiedra-Navarrete, C., Landa, B.B. and Castillo, P. (2011) Prevalence, polyphasic identification, and molecular phylogeny of dagger and needle nematodes infesting vineyards in southern Spain. *European Journal of Plant Pathology*. doi: 10.1007/s10658-010-9705-y

Gutiérrez-Gutiérrez, C., Cantalapiedra-Navarrete, C., Decraemer, W., Vovlas, N., Prior, T., Palomares Rius, J.E. and Castillo, P. (2012) Phylogeny, diversity, and species delimitation in some species of the *Xiphinema americanum*-group complex (Nematoda: Longidoridae), as inferred from nuclear and mitochondrial DNA sequences and morphology. *European Journal of Plant Pathology*, 134, 561–597.

Habteweld, A.W., Akyazi, F., Joseph, S., Crow, W.T., Abebe, E. and Mekete, T. (2019) Description of *Hirschmanniella dicksoni* n. sp. (Nematoda: Pratylenchidae) from rhizosphere soil of limpograss from Florida, USA. *Journal of Nematology* 51. doi: 10.21307/jofnem-2019-083 e2019-83.

Haegeman, A., Jacob, J., Vanholme, B., Kyndt, T. and Gheysen, G.A. (2008) Family of GHF5 endo-1, 4-beta-glucanases in the migratory plantparasitic nematode *Radopholus similis*. *Plant Pathology* 57, 581–590.

Haegeman, A., Vanholme, B. and Gheysen, G. (2009) Characterization of a putative endoxylanase in the migratory plant-parasitic nematode *Radopholus similis*. *Molecular Plant Pathology* 10, 389–401.

Haegeman, A., Elsen, A., De Waele, D. and Gheysen, G. (2010) Emerging molecular knowledge on *Radopholus similis*, an important nematode pest of banana. *Molecular Plant Pathology* 11(3), 315–323.

Hafez, S.L., Raski, D.J. and Lear, B. (1981a) Action of systemic nematicides in control of *Xiphinema index* on grape. *Journal of Nematology* 13(1), 24–29.

Hafez, S.L., Raski, D.J. and Lownsbery, B.F. (1981b) Effects of *Paratrichodorus minor* (Colbran) Siddiqi on Thompson seedless grapevine. *Revue de Nematologie* 4(1), 115–120.

Haglund, W.A. (1983) Efficacy of selected nonvolatile nematicides on control of *Ditylenchus destructor* in iris. *Journal of Nematology* 15(1), 92–96.

Hague, N.G.M. (1963) Fumigation of agricultural products. XVIII. Effect of methyl bromide on the bent grass nematode, *Anguina agrostis* (Steinbuch, 1799) Filipjev, 1936, and on the germination of bent grass, *Agrostis tenuis*. *Journal of Science of Food and Agriculture* 14(8), 577–579.

Hahn, M.L., Burrows, P.R. and Wright, D.G. (1996) Genomic diversity between *Radopholus similis* populations from around the world detected by RAPD-PCR analysis. *Nematologica* 42(5), 537–545.

Haider, M.G., Askary, T.H. and Nath, R.P. (2001) Nematode population as influenced by paddy based cropping sequences. *Indian Journal of Nematology* 31, 68–71.

Haider, M.G., Nath, R.P. and Prasad, S.S. (1978) Studies on the lance nematode, *Hoplolaimus indicus* I. Pathogenicity and histopathogenisis on maize. *Indian Journal of Nematology* 8, 9–12.

Halbrendt, J.M. (1993) Evaluation of rotational crops for nematode management in orchards. *Pennsylvania Fruit News* 73(4), 62–66.

Halbrendt, J.M. (1995) Progress report: efficacy of crop rotation, green manure, and oxamyl as preplant nematode treatments. *Pennsylvania Fruit News* 75(2), 44–45.

Halbrendt, J.M. and Brown, D.J.F. (1992) Morphometric evidence for three juvenile stages in some species of *Xiphinema americanum* sensu lato. *Journal of Nematology* 24(2), 305–309.

Hallman, J. and Subbotin, S.A. (2018) Method of extraction, processing and detection of plant and soil nematodes. In: Sikora, R. A., Coyne, D., Hallmann, J., and Timper, P. (eds) *Plant Parasitic Nematodes in Subtropical and Tropical Agriculture*, 3rd edn. CABI Publishing, Wallingford, UK, pp. 62–86.

Han, H., Chung, Y.J. and Shin, S.C. (2008) First report of pine wilt disease on *Pinus koraiensis* in Korea. *Plant Disease* 92(8), 1251. doi: 10.1094/PDIS-92-8-1251A

Han, H.R., Jeyaprakash, A., Weingartner, D.P. and Dickson, D.W. (2006) Morphological and molecular biological characterization of *Belonolaimus longicaudatus*. *Nematropica* 36, 37–52.

Handoo, Z.A. (2000) A key and diagnostic compendium to the species of the genus *Tylenchorhynchus* Cobb, 1913 (Nematoda: Belonolaimidae). *Journal of Nematology*, 32(1), 20–34.

Handoo, Z.A. (2002) A key and compendium to species of the *Heterodera avenae* group (Nematoda: Heteroderidae). *Journal of Nematology* 34, 250–262.

Handoo, Z.A. and Golden, A.M. (1989) A key and diagnostic compendium to the species of the genus *Pratylenchus* Filipjev, 1936 (Lesion Nematodes). *Journal of Nematology* 21, 202–218.

Handoo, Z.A. and Golden, A.M. (1992) A key and diagnostic compendium to the species of the genus *Hoplolaimus* Daday, 1905. *Journal of Nematology* 24 (1), 45–53.

Handoo, Z.A., Carta, L.K., Skantar, A.M., Ye, W., Robbins, R.T., Subbotin, S.A., Fraedrich, S.W. and Cram, M.M. (2005) Morphological and molecular characterization of *Longidorus americanum* n. sp. (Nematoda: Longidoridae), a needle nematode parasitizing pine in Georgia. *Journal of Nematology* 37(1), 94–104.

Handoo, Z.A., Palomares-Rius, J.E., Cantalapiedra-Navarrete, C., Liébanas, G., Subbotin, S.A. and Castillo, P. (2014) Integrative taxonomy of the stunt nematodes of the genera *Bitylenchus* and *Tylenchorhynchus* (Nematoda, Telotylenchidae) with description of two new species and a molecular phylogeny. *Zoological Journal of the Linnean Society* 172(2), 231–264.

Hansen, T., Rasmussen, A.N. and Schadegg, E. (1972) Trials with plant protection materials in fruit growing and nurseries 1970. *Tidsskrift for Planteavl*, 76(1), 77-104.

Haque, M.M. and Padmavathy, P.V. (1985) Comparative studies on the penetration of roots by *Rotylenchulus reniformis* in relation to age of the seedling and time intervals on cowpea. *Indian Journal of Nematology* 15, 259–260.

Haque, Z., Khan, M.R. and Ahamad, F. (2018) Relative antagonistic potential of some rhizosphere biocontrol agent for the management of rice root-knot nematode, *Meloidogyne graminicola*. *Biological Control* 126, 109–116.

Haque, Z., Khan, M.R. and Ahamad, F. (2019) Rice root-knot nematode and its management through non-chemical approaches. In: Khan, M.R., Mukhopadhyay, A.N., Pandey, R.N., Thakur, M.P., Singh, D. *et al.* (eds) *Bio-intensive Approaches: Application and Effectiveness in Management of Plant Nematodes, Insects and Weeds*. Today & Tomorrow's Printers and Publishers, New Delhi, India, pp. 225–250.

Harish, M. and Nanjegowda, D. (2000) Screening of banana cultivars against the burrowing nematode, *Radopholus similis* (Cobb, 1893) Thorne, 1949. *Indian Journal of Nematology* 30, 239–240.

Harmey, J.H. and Harmey, M.A. (1994) DNA profiling of *Bursaphelenchus* species. *Gene* 145, 227–230.

Haroon, S. and Smart, G.C. Jr (1983) Effects of Pangola digitgrass on *Meloidogyne arenaria*, *M. javanica* and *M. hapla*. *Journal of Nematology* 15(4), 649–650.

Harper, R.S. (1975) Pests, diseases and weeds of crops in Indonesia. *World Crops* 27(3), 134–137.

Harris, A.R. (1983) Resistance of some Vitis rootstocks to *Xiphinema index*. *Journal of Nematology* 15(3), 405–409.

Harrison, R.E. and Smart, G.C. Jr (1975) Vertical distribution of *Trichodorus christiei* and *Trichodorus proximus* relative to soil moisture. *Journal of Nematology* 7(4), 324.

Hart, J.H. (1905) Coconut disease. *Bulletin of Miscellaneous Information, Royal Botanic Gardens, Trinidad* 6, 241–243.

Hartman, K. and Sasser, J. (1985) Identification of *Meloidogyne* species on the basis of differential host test and perineal-pattern morphology. In: Barker, K.R., Carter, C.C. and Sasser, J.N. (eds) *An Advanced Treatise on Meloidogyne. Volume 1: Methodology*. North Carolina State University Graphics, Raleigh, North Carolina, pp. 69–77.

Hasan, A. (1984) Synergism between *Heterodera cajani* and *Fusarium udum* attacking *Cajanus cajan*. *Nematology Mediterranean* 12, 159–162.

He, Y., Subbotin, S.A., Rubtsova, T.V., Lamberti, F., Brown, D.J.F. and Moens, M. (2005) A molecular phylogenetic approach to Longidoridae (Nematoda: Dorylaimida) *Nematology* 7(1), 111–124.

Heald, C.M. and O'Bannon, J.H. (1987) *Citrus decline caused by nematodes. V. Slow Decline*. Nematology Circular 143. Florida Department of Agriculture and Consumer Services, Division of Plant Industry, Gainesville, Florida.

Heald, C.M., and Thames, W.H. (1982) The reniform nematode, *Rotylenchulus reniformis*. In: Riggs, R.D. (ed.) *Nematology in the Southern Region of the United States*. Southern Cooperative Series Bulletin 276, Arkansas Agricultural Experiment Station, Fayetteville, pp. 139–143.

Henn, R.A. and Dunn, R.A. (1989) Reproduction of *Hoplolaimus galeatus* and growth of seven St Augustine grass (*Stenotaphrum secundatum*) cultivars. *Nematropica* 19, 81–87.

Hewitt, W.B., Raski, D.J. and Goheen, A.C. (1958) Nematode vector of soil-borne fanleaf virus of grapevines. *Phtopathology* 48, 586–595.

Heybroek, H.M. (1993) The Dutch elm breeding program. In: Sticklen, M.B. and Sherald, J.L. (eds) *Dutch Elm Disease Research: Cellular and Molecular Approaches*. Springer–Verlag, New York, pp. 16–25.

Heyns, J. (1971) *A Guide to the Plant and Soil Nematodes of South Africa*. A.A. Balkema. Cape Town, South Africa.

Heyns, J. (1974) The genus *Xiphinema* in South Africa. II. *X. elongatum* group (Nematoda: Dorylaimida). *Phytophylactica* 6, 249–260.

Heyns, J. (1975) *Paralongidorus maximus*. *CIH Descriptions of Plant-parasitic Nematodes*, Set 5, No. 75. Commonwealth Institute of Helminthology, St. Albans, UK.

Heyns, J. and Coomans, A. (1984) The genus *Xiphinema* in South Africa. VIII. *Xiphinema diversicaudatum* (Micoletzky, 1927) Thorne, 1939. *Phytophylactica* 16, 117–120.

Hillocks, R.J. and Songa, W. (1993) Root-knot and other nematodes associated with pigeonpea plants infected with *Fusarium udum* in Kenya. *Afro-Asian Journal of Nematology* 3, 143–147.

Hirling, V.W. (1974) Damage to maize by cereal cyst nematode, *Heterodera avenae* and the technique for examination. *Gesunde Pflanzen* 26 (3), 58–62.

Hirschmann, H. (1956) Comparative morphological studies on the soybean cyst nematode, *Heterodera glycines* and the clover cyst nematode, *H. trifolii* (Nematoda: Heteroderidae). *Proceedings of the Helminthological Society of Washington* 23, 140–151.

Hirschmann, H. (1985) The genus *Meloidogyne* and morphological characters differentiating its species. In: Sasser, J.N. and Carter, C.C. (eds) *An Advanced Treatise on Meloidogyne. Volume I: Biology and Control*. A cooperative publication of the Department of Plant Pathology and the United States Agency for International Development, North Carolina State University Graphics, Raleigh, North Carolina, pp. 79–93.

Hirschmann, H. and Triantaphyllou, A.C. (1968) Mode of reproduction and development of the reproductive system of *Helicotylnchus dihystera*. *Nematologica* 13(4), 558–574.

Hoberg, E.P. (2010) Invasive processes, mosaics and the structure of helminth parasite faunas. *Revue Scientifique et Technique–Office International des Epizooties* 29, 255–272.

Hockland, S. and Eng, L. (1997) *Capsicum annuum* v. *longum*: a new host record for the rice white tip nematode, *Aphelenchoides besseyi*. *International Journal of Nematology* 7, 229.

Hockland, S. and Prior, T. (2009) *Xiphinema americanum* sensu lato. PM 7/95 (1). *EPPO Bulletin* 39, 382–392.

Hockland, S., Niere, B., Grenier, E., Blok, V., Phillips, M. *et al.* (2012) An evaluation of the implications of virulence in non- European populations of *Globodera pallida* and *G. rostochiensis* for potato cultivation in Europe. *Nematology* 14, 1–13.

Hockland, S., Inserra, R.N. and Kohl, L.M. (2013) International plant health-putting legislation into practice. In: Perry, R. and Moens, M. (eds) *Plant Nematology*, 2nd edn. CAB International, Wallingford, UK, pp. 359–382.

Hodda, M. (2011) Phylum Nematoda Cobb 1932. In: Zhang, Z.-Q. (ed.) *Animal Biodiversity: An Outline of Higher-level Classification and Survey of Taxonomic Richness*. *Zootaxa* Special Issue 3148.

Hodda, M., Peters L. and Traunspurger, W. (2009) Nematode diversity in terrestrial, freshwater aquatic and marine systems. In: Wilson, M.J. and Kakouli-Duarte, T. (eds) *Nematodes as Environmental Indicators*. CABI, Wallingford, UK, pp. 45–93.

Hofmanner, R. and Menzel, R. (1914) Neue Arten freilebenden Nematoden aus der Schweiz. *Zool. Anz.* 44, 80–91.

Holgado, R. and Crump, D.H. (2003) First record on the occurrence of nematophagous fungi parasitizing cyst nematodes in Norway. *International Journal of Nematology* 13, 65–71.

Holguin, C.M., Baeza, J.A., Mueller, J.D. and Agudelo, P. (2015) High genetic diversity and geographic subdivision of three lance nematode species (*Hoplolaimus* spp.) in the United States. *Ecology and Evolution* 5(14), 2929–2944.

Holguin, C.M., Ma, X., Mueller, J.D. and Agudelo, P. (2016) Distribution of *Hoplolaimus* species in soybean fields in South Carolina and North Carolina. *Plant Disease* 100, 149–153.

Hollis, J.P. (1977) Loss in yield of rice caused by the ring nematode *Criconemoides onoensis* revealed by elimination of yellow nutsedge, *Cyperus esculentus*. *Nematologica* 23(1), 71–78.

Hollis, J.P. Jr, Keoboonrueng, S. and Nickle, W.R. (eds) (1984) Nematode parasites of rice. In: *Plant and Insect Nematodes*. Marcel Dekker, Inc., New York, pp. 95–146.

Holterman, M., Karssen, G., van den Elsen, S., van Megen, H., Bakker, J. and Helder, J. (2009) Small subunit rDNA-based phylogeny of the Tylenchida sheds light on relationships among some high-impact plant-parasitic nematodes and the evolution of plant feeding. *Phytopathology* 99(3), 227–235. doi: 10.1094/PHYTO-99-3-0227.

Holtzmann, O.V. (1968) Plant nematode associations previously unreported from Hawaii. *Plant Disease Reporter* 52(7), 515–518.

Hooks, C.R.R., Wang, K., Ploeg, A. and McSorley, R. (2010) Using marigold (*Tagetes* spp.) as a cover crop to protect crops from plant-parasitic nematodes. *Applied Soil Ecology* 46, 307–320.

Hooper, D.J. (1961) A redescription of *Longidorus elongatus* (de Man, 1876) Thorne & Swanger, 1936 (Nematoda: Dorylaimoidea), and description of five new species of *Longidorus* from Great Britain. *Nematologica* 6 (3), 237–257.

Hooper, D.J. (1972) *Ditylenchus dipsaci*. *CIH Descriptions of Plant-parasitic Nematodes*, Set 1, No. 14. Commonwealth Institute of Helminthology, St Albans, UK.

Hooper, D.J. (1973a) *Ditylenchus destructor*. *CIH Descriptions of Plant-parasitic Nematodes*, Set 2, No. 21. Commonwealth Institute of Helminthology, St Albans, UK.

Hooper, D.J. (1973b) *Longidorus elongatus*. *CIH Descriptions of Plant-parasitic Nematodes*, Set 2, No. 30. Commonwealth Institute of Helminthology, St Albans, UK.

Hooper, D.J. (1976) *Trichodorus viruliferus*. *CIH Descriptions of Plant-parasitic Nematodes*, Set 6, No. 86. Commonwealth Institute of Helminthology, St Albans, UK.

Hooper, D.J. (1991) Stem nematode (*Ditylenchus dipsaci*) a pest of faba beans (*Vicia faba*). *Aspects of Applied Biology* 27, 143–149.

Hooper, D.J. and Cowland, J.A. (1986) Fungal hosts for the chrysanthemum nematode, *Aphelenchoides ritzemabosi*. *Plant Pathology* 35(1), 128–129.

Hotchkin, P. and Giblin, R. (1984) Comparison of electrophoregrams from *Bursaphelenchus* spp. (Aphelenchoididae). *Revue de Nématologie* 7, 319–320.

Hu, K. and Yang, B. (1995) Preliminary comparison of *Bursaphelenchus xylophilus* and *B. mucronatus* utilizing enzyme electrophoresis. *Forest Research* 8, 73–77.

Hu, M., Chilton, N.B., Zhu, X.Q. and Gasser, R.B. (2002) Single-strand conformation polymorphism-based analysis of mitochondrial cytochrome c oxidase subunit 1 reveals significant substructuring in hookworm populations. *Electrophoresis* 23, 27–34.

Hu, X.Q., Yu, M., Lin, L., Wang, Y. and Yu, S. (2004) Species and distribution of rice root nematode in Yunnan Province, China. *Agricultural Sciences in China* 3(8), 598—03.

Huang, D. and Yan, G. (2017) Specific detection of the root-lesion nematode *Pratylenchus scribneri* using conventional and Real-Time PCR. *Plant Disease* 101, 359–365.

Huang, D., Yan, G., Gudmestad, N., Ye, W., Whitworth, J., Frost, K., Crow, W. and Hajihassani, A. (2019) Developing a one-step multiplex PCR assay for rapid detection of four stubby-root nematode species, *Paratrichodorus allius, P. minor, P. porosus*, and *Trichodorus obtusus. Plant Disease* 103, 404–410.

Huang, J.S. (1985) Mechanisms of resistance to root-knot nematodes. In: Barker KR, Carter CC, Sasser JN (eds) *An Advanced Treatise on Meloidogyne: Volume II.* North Carolina State University Graphics, Raleigh, NC. pp. 165–174.

Hübschen, J., Kling, L., Ipach, U., Zinkernagel, V., Brown, D. and Neilson, R. (2004) Development and validation of species-specific primers that provide a molecular diagnostic for virus-vector longidorid nematodes and related species in German viticulture. *European Journal of Plant Pathology* 110(9), 883–891. doi: 10.1007/s10658-004-4841-x

Humphreys-Pereira, D.A., Williamson, V.M., Lee, S., Coyne, D.L., Salazar, L. and Gomez-Alpizar, L. (2014) Molecular and morphological characterization of *Scutellonema bradys* from yam in Costa Rica and development of specific primers for its detection. *Nematology* 16, 137–147.

Hunt, D.J. (1977) Plant parasitic nematodes from the Windward Islands. *PANS* 23(4), 402–411.

Hunt, D.J. (1993) *Aphelenchida, Longidoridae and Trichodoridae: Their Systematics and Bionomics.* CABI, Wallingford, UK.

Hunt, D.J. (2008) A checklist of the Aphelenchoidea (Nematoda: Tylenchina). *Journal of Nematode Morphology and Systematics* 10(2), 99–135.

Hunt, D.J. (2018) Plant parasitic nematode genera and species cited. In: Sikora, R.A., Coyne, D., Hallmann, J., and Timper, P. (eds) *Plant Parasitic Nematodes in Subtropical and Tropical Agriculture*, 3rd edn. CABI Publishing, Wallingford, UK.

Hunt, D.J. and Handoo, Z.A. (2009) Taxonomy, identification and principal species. In: Perry, R.N., Moens, M. and Starr, J.L. (eds) *Root-knot Nematodes.* CABI, Wallingford, UK, pp. 55–88.

Hunt, D.J. and Rahman, L. (1991) Two new species of *Paralongidorus* (Nematoda: Dorylaimida) from deepwater rice in Bangladesh. *Afro-Asian Journal of Nematology* 1(1), 83–85.

Hunt, D.J., Bridge, J. and Machon, J.E. (1989) On *Achlysiella*, a new genus of obese Pratylenchidae (Nematoda: Tylenchoidea). *Revue de Nématologie* 12(4), 401–407.

Hunter, J.E., Ko, W.H., Kunimoto, R.K. and Higaki, T. (1974) A foliar disease of anthurium seedlings caused by *Aphelenchoides fragariae. Phytopathology* 64(2), 267–268.

I3N (2019) IABIN Invasives Information Network (I3N). Available at: http://i3n.iabin.net/index.html

Ibrahim, A.A.M., Al-Hazmi, A.S., Al-Yahya, F.A. and Alderfasi, A.A. (1999) Damage potential and reproduction of *Heterodera avenae* on wheat and barley under Saudi field conditions. *Nematology* 1, 625–630.

Ibrahim, S.K. (1991) Peroxidase isoenzymes from *Meloidogyne* spp. cultured on different hosts. *Revue de Nématologie* 14, 335–344.

Ibrahim, S.K. and Perry, R.N. (1993) Desiccation survival of rice stem nematode, *Ditylenchus angustus. Fundamental and Applied Nematology* 16, 31–38.

Ibrahim, S.K., Perry, R.N., Plowright, R.A. and Rowe, J. (1993) Hatching behaviour of the rice cyst nematodes *Heterodera sacchari* and *H. oryzicola* in relation to age of host plant. *Fundamental and Applied Nematology* 16(1), 23–29.

Ibrahim, S.K., Perry, R.N., Burrows, P.R. and Hooper, D.J. (1994a) Differentiation of species and populations of *Aphelenchoides* and of *Ditylenchus augustus* using a fragment of ribosomal DNA. *Journal of Nematology* 26, 412–421.

Ibrahim, S.K., Perry, R.N. and Hooper, D.J. (1994b) Use of esterase and protein patterns to differentiate two new species of *Aphelenchoides* on rice from other species of *Aphelenchoides* and from *Ditylenchus augustus* and *D. myceliophagus. Journal of Nematology* 40, 267–275.

Ibrahim, S.K., Perry, R.N. and Webb, R.M. (1995) Use of isoenzyme and protein phenotypes to discriminate between six *Pratylenchus* species from Great Britain. *Annals of Applied Biology* 126, 317–327.

Ibrahim, S.K., Davies, K.G. and Perry, R.N. (1996) Identification of the root-knot nematode, *Meloidogyne incognita*, using monoclonal antibodies raised to nonspecific esterases. *Physiological and Molecular Plant Pathology* 49, 79–88.

İmren, M., Waeyenberge, L., Viaene, N., Dababat, A. and Elekcioğlu, İ.H. (2012) Molecular characterization of cereal cyst nematodes from the South Anatolian Region in Turkey using ITS-rDNA sequences. *Türkiye Entomoloji Dergisi* 36(4), 491–499.

İmren, M., Waeyenberge, L., Viaene, N., Elekcioğlu, İ.H. and Dababat, A. (2015) Morphological and molecular identification of cereal cyst nematodes from the eastern Mediterranean region of Turkey. *Turkish Journal of Agriculture and Forestry* 39, 91–98. doi: 10.3906/tar-1404-56

Inácio, M.L., Nóbrega, F., Vieira, P., Bonifácio, L., Naves, P., Sousa, E. and Mota, M. (2014) First detection of *Bursaphelenchus xylophilus* associated with *Pinus nigra* in Portugal and in Europe. *Forest Pathology* 45(3), 235–238.

Inagaki, H. (1977) *Soybean nematodes.* Technical Bulletin No. 31, ASPAC Food and Fertilizer Technology Center, Taiwan.

Inoue, S. (1995) Changes of forest vegetation after pine wilting damage in Sawara forest of Kyushu University. *Bulletin of the Kyushu University Forests* 72, 21–32.

Inserra, RN., Vovlas, N. and O'Bannon, J.H. (1980) A classification of *Tylenchulus semipenetrans* biotypes. *Journal of Nematology* 12(4), 283–287.

Inserra, R.N., Vovlas, N., Griffin, G.D. and Anderson, J.L. (1983) Development of the false root-knot nematode, *Nacobbus aberrans*, on sugarbeet. *Journal of Nematology* 15(2), 288–296.

Inserra, R.N., Griffin, G.D. and Anderson, J.L. (1985) *The false root-knot nematode* Nacobbus aberrans. Research Bulletin No. 510, Utah Agricultural Experimental Station, Logan, Utah.

Inserra, RN., Vovlas, N., O'Bannon, J.H. and Esser, R.P. (1988) *Tylenchulus graminis* n. sp. and *T. palustris* n. sp. (Tylenchulidae) from native flora of Florida, with notes on *T. semipenetrans* and *T. furcus. Journal of Nematology* 20, 266–287.

Inserra, R.N., Brito, J.A., Dong, K., Handoo, Z.A., Lehman, P.S., Powers, T. and Millar, L. (2003) *Exotic nematode plant pests of agricultural and environmental significance to the United States: Heterodera latipons.* Society of Nematologists. Available at: http://nematode.unl.edu/wgroup.htm (accessed 27 February 2021)

Inserra, R.N., Stanley, J.D., O'Bannon, J.H. and Esser, R.P. (2005) Nematode quarantine and certification programmes implemented in Florida. *Nematologia Mediterranea* 33, 113–123.

IPPC (1997). *Revised International Plant Protection Convention.* Report of the 29th Session of the FAO Conference, Rome, Italy, 7–18 November 1997. Available at: www.ippc.int

IPPC Secretariat (2005) *Identification of risks and management of invasive alien species using the IPPC framework. Proceedings of the workshop on invasive alien species and the International Plant Protection Convention*, Braunschweig, Germany, 22–26 September 2003. FAO, Rome.

IPPC (2021) *International Plant Protection Convention.* Food and Agriculture Organization of United Nations. Available at: http://www.ippc.int www.ippc.int

ISEHS (1974) Plant nematology. In: *Twentieth Annual Report 1974. Rosewarne, Ellbridge and Isles of Scilly Experimental Horticulture Stations*, Camborne, UK, pp. 44–49.

Ismailov, G.M. (1976) *Ditylenchus destructor* on potato and its control in the Azerbaidzhan SSR. VIII Vsesoyuznoe soveshchanie no nematodnym boleznyam sel'skokhozyaistvennykh kul'tur. *Tezisy dokladov i soobshchenii. Izdatel'stvo "Shtiintsa"*, Kishinev, USSR, 76.

Itoh, Y., Ohshima, Y. and Ichinohe, M. (1969) A root-knot nematode, *Meloidogyne mali* n. sp. on apple-tree from Japan (Tylenchida: Heteroderidae). *Applied Entomology and Zoology* 4, 194–202.

Iwahori, H., Tsuda, K., Kanzaki, N., Izui, K. and Futai, K. (1998) PCR-RFLP and sequencing analysis of ribosomal DNA of *Bursaphelenchus* nematodes related to pine wilt disease. *Fundamental and Applied Nematology* 21, 655–666.

Jackson, G.V.H., Ruabete, T.K. and Wright, J.G. (2003) *Burrowing and lesion nematodes of banana*. Pest advisory leaflet No. 5. Secretariat of the Pacific Community, Noumea, New Caledonia.

Jacob, J., Mitreva, M., Vanholme, B. and Gheysen, G. (2008) Exploring the transcriptome of the burrowing nematode *Radopholus similis*. *Molecular Genetics and Genomics* 280(1), 1–17.

Jaehn, A., Rebel, E.K. and Lordello, L.G.E. (1980) The origin of *Meloidogyne coffeicola*. In: Lordello, L.G.E. (ed.) *Trabalhos apresentados a IV Reuniao Brasileira de Nematologia*, 16–20 de julho de 1979, Sao Paulo. Publicacao No. 4. Sociedade Brasileira de Nematologia, Piracicaba, SP, Brazil, pp. 159–161.

Jaffee, B.A. (1986) Parasitism of *Xiphinema rivesi* and *X. americanum* by zoosporic fungi. *Journal of Nematology* 18(1), 87–93.

Jaffee, B.A. and Shaffer, R.L. (1985) Evaluation of post-plant nematicides for control of dagger nematodes on peach, 1982–83. *Fungicide and Nematicide Tests, American Phytopathological Society* 40, 102–103.

Jaffee, B.A., Harrison, M.B., Shaffer, R.L. and Strang, M.B. (1987) Seasonal population fluctuation of *Xiphinema americanum* and *X. rivesi* in New York and Pennsylvania orchards. *Journal of Nematology* 19, 369–378.

Jaffee, B.A., Glenn, D.M., Forer, L.B. and Shaffer, R.L. (1988) Effect of soil texture on *Xiphinema americanum* and *X. rivesi*. *Nematologica* 34(2), 226–237.

Jain, A.K. (2009) Nematode pests of small millets – a review. *Agricultural Reviews* 30, 132–138.

Jain, R.K. and Hasan, N. (1986) Association of vesicular-arbuscular mycorrhizal (VAM) fungi and plant parasitic nematodes with forage sorghum (*Sorghum bicolor* L.). *Sorghum Newsletter* 29, 84.

Jain, R.K. and Hasan, N. (1987) Relative susceptiblity of some selected forage sorghum (*Sorghum bicolor* L.) varieties to spiral nematode, *Helicotylenchus dihystera*. *Indian Journal of Nematology* 17(2), 199–201.

Janowicz, K. (1984) Interaction of *Ditylenchus destructor* and fungi causing dry rot in stored potatoes. *Ochrona Roslin* 28(3), 10–11.

Janowicz, K. and Mazurkiewicz, K. (1982) An attempt to evaluate the effect of *Ditylenchus destructor* on infections of potato tubers with *Rhizoctonia solani*. *Ochrona Roslin* 26(5), 8–9.

Janssen, T., Karssen, G., Couvreur, M., Waeyenberge, L. and Bert, W. (2017a) The pitfalls of molecular species identification: a case study within the genus *Pratylenchus* (Nematoda: Pratylenchidae). *Nematology* 19, 1–21.

Janssen, T., Karssen, G., Topalović, O., Coyne, D. and Bert, W. (2017b) Integrative taxonomy of root-knot nematodes reveals multiple independent origins of mitotic parthenogenesis. *PLoS ONE* 12(3), e0172190. doi: 10.1371/journal.pone.0172190

Jatala, P. (1979) Review of the false root-knot nematode (*Nacobbus* spp.). Research progress. International Potato Center. In: *Report of the 2nd Planning Conference on the Developments in the Control of Nematode Pests of Potatoes*, Lima, Peru, 13–17 November 1978. International Potato Center, Lima, Peru, pp. 66–69.

Jatala, P. and Bridge, J. (1990) Nematode parasites of root and tuber crops. In: Luc, M., Sikora, R.A. and Bridge, J. (eds) *Plant Parasitic Nematodes in Subtropical and Tropical Agriculture*. CAB International, Wallingford, UK, pp. 137–180.

Jatala, P., Kaltenbach, R. and Bocangel, M. (1979) Biological control of *Meloidogyne incognita acrita* and *Globodera pallida* on potatoes. *Journal of Nematology* 11, 303.

Jaumot, M., Pochet, J. and Fernández, C. (1997) Protein analysis of root-lesion nematodes using SDS-PAGE. *Nematropica* 27, 33–39.

Javed, N., Javed, M., Ilyas, M.B., Khan, M.M. and Inam-Ul-Haq, M. (2008) Reaction of various citrus root stocks (germplasm) against citrus root nematode (*Tylenchulus semipenetrans* Cobb.). *Pakistan Journal of Botany* 40(6), 2693–2696.

Jeger, M., Bragard, C., Caffier, D., Candresse, T., Chatzivassiliou, E., Dehnen-Schumutz *et al.* (2018) Pest categorization of *Hirschmanniella* spp. *EFSA Journal* 16(6), 5297.

Jelly, N.S., Schellenbaum, P., Walter, B. and Maillot, P. (2012) Transient expression of artificial microRNAs targeting Grapevine fanleaf virus and evidence for RNA silencing in grapevine somatic embryos. *Transgenic Research* 21, 1319–1327.

Jepson, S.B. (1985) *Meloidogyne chitwoodi*. In: *CIH Descriptions of Plant-parasitic Nematodes*, Set 8, No. 106, Commonwealth Agricultural Bureaux, Farnham Royal, UK.

Jepson, S.B. (1987) *Identification of Root-knot Nematodes (*Meloidogyne *species)*. CAB International, Wallingford, UK.

Jerath, M.L. (1968) *Heterodera sacchari*, a cyst nematode pest of sugar cane new to Nigeria. *Plant Disease Reporter* 52, 237–239.

Jeronimo, J. (1988) *El nematodo enquistado del maiz Punctoder sp. y su expresion en 17 municipios del estado de Michoacan*. Thesis. Universidad Michoacana de san Nicolas de Hidalgo Uruapan, Michoacan, Mexico.

Jeszke, A., Budziszewska, M., Dobosz, R., Stachowiak, A., Protasewicz, D., Wieczorek, P. and Obrępalska-Stęplowska, A. (2013) A comparative and phylogenetic study of the *Ditylenchus dipsaci, Ditylenchus destructor* and *Ditylenchus gigas* populations occurring in Poland. *Journal of Phytopathology* 162, 61–67.

Ji, L., Wang, J.C., Yang, X.L., Huang, G.M. and Lin, M.S. (2006) PCR-RFLP patterns for differentiation of three Ditylenchus species. *Journal of Nanjing Agricultural University* 29, 39–43. [In Chinese.]

Jiang, L., Zheng, J., Chen, Z. and Wu, J. (2006) Screening and characterization of monoclonal antibody to *Bursaphelenchus xylophilus*. *Journal of Agricultural Biotechnology* 14, 412–415.

Johnson, A.W. (1968) Cactus cyst nematode found in Georgia. *Plant Disease Reporter* 52, 114.

Johnson, A.W. (1975) Resistance of sweet corn cultivars to plant-parasitic nematodes. *Plant Disease Reporter* 59(4), 373–376.

Johnson, A.W. and Chalfant, R.B. (1973) Influence of organic pesticides on nematodes and corn earworm damage on yield of sweet corn. *Journal of Nematology* 5, 177–180.

Johnson, A.W., Dowler, C.C. and Hauser, E.W. (1974) Seasonal population dynamics of selected plant parasitic nematodes on four monocultured crops. *Journal of Nematology* 6(4), 144.

Johnson, A.W., Dowler, C.C., Glaze, N.C. and Sumner, D.R. (1987) Postplant use of nemacur for control of ring nematodes in peaches, 1982–86. *Fungicide and Nematicide Test,s American Phytopathological Society* 42, Report 161.

Johnson, C.S. (1998) Tobacco. In: Barker, K.R., Pederson, G.A. and Windham, G.L. (eds) *Plant and Nematode Interactions*. Agronomy Monograph No. 36. American Society of Agronomy, Crop Science Society of America, Soil Science Society of America, Madison, Wisconsin, pp. 487–522

Johnson, C.S. and Chinheya, C. (2018) Nematode parasites of tobacco. In: Sikora, R.A., Coyne, D., Hallmann, J., and Timper, P. (eds) *Plant Parasitic Nematodes in Subtropical and Tropical Agriculture*, 3rd edn. CABI Publishing, Wallingford, UK, pp. 687–716.

Johnson, C.S., Way, J. and Barker, K.R. (2005) Nematode parasites of tobacco. In: Luc, M., Skora, R.A. and Bridge, J. (eds) *Plant Parasitic Nematodes in Subtropical and Tropical Agriculture*. CAB International, St Albans, UK, pp. 675–708.

Johnson, P.W. and Fushtey, S.G. (1966) The biology of the oat cyst nematode, *Heterodera avenae* in Canada. II. Nematode development and related anatomical changes in roots of oats and corn. *Nematologica* 12, 630–636.

Jonathan, E.I. Cannayane, I. and Kavitha, P.G. (2012) Nematode infestation in turf grasses. In: Khan, M.R. and Jairajpuri, M.S. (eds) *Nematode Infestations – Part III: Horticultural Crops*. National Academy of Sciences, Allahabad (Prayagraj), UP, India, pp. 464–483.

Jones, B.L. and Waele, D. de. (1990) Histopathology of *Ditylenchus destructor* on peanut. *Journal of Nematology* 22(3), 268–272.
Jones, F.G.W. (1961) Potato root eelworm *Heterodera rostochiensis* Woll. in India. *Current Science* 30, 187.
Jones, F.G.W. and Jones, M. (1964) *Pests of Field Crops*. Edward Arnold, London.
Jones, J.T., Haegeman, A., Danchin, E.G.J., Gaur, H.S., Helder, J., Jones, M.G.K. *et al.* (2013) Top 10 plant-parasitic nematodes in molecular plant pathology. *Molecular Plant Pathology* 14(9), 946–961.
Jones, M.G.K. and Northcote, D.H. (1972) Nematode-induced syncytium – a multinucleate transfer cell. *Journal of Cell Science* 10, 789–809.
Jones, R.K. (1978) The feeding behaviour of *Helicotylenchus* spp. on wheat roots. *Nematologica* 24(1), 88–94.
Kali Ram (1989) Studies on corn cyst nematodes (*Heterodera zea*) on maize (*Zea mays* L.). PhD thesis. Haryana Agricultural University, Hisar, India.
Kaloshian, I., Greco, N. and Saad, A.T. (1986a) Hatching of cysts and infectivity of *Heterodera ciceri* on chickpea. *Nematologia Mediterranea* 14, 129–133.
Kaloshian, I., Greco, N., Saad, A.T. and Vovlas, N. (1986b) Life cycle of *Heterodera ciceri* on chickpea. *Nematologia Mediterranea* 14, 135–145.
Kanetani, S., Gyokusen, K. and Saito, A. (1999) Relationship between the decline in the population of the endangered species, *Pinus armandii* var. Amamiana, and pine wilt disease. Sustainability of pine forests in relation to pine wilt and decline. In: *Proceedings of International Symposium*, Tokyo, Japan, 27–28 October, 1998, pp. 290–294.
Kannan, C. and Lingaraju, S. (1999) A potential biocontrol agent for pigeonpea cyst nematode (*Heterodera cajani* Koshy). *Current Science* 76(11), 1425–1427.
Kantharaju, V. and Reddy, B.M.R. (2001) Management of *Meloidogyne incognita* on potato using cultural and chemical methods. *Indian Journal of Nematology* 31, 79–98.
Kanwar, R.S. and Bajaj, H.K. (2009) Mechanism of resistance in wheat varieties Raj MR1 and AUS 15854 to cereal cyst nematode, *Heterodera avenae* Woll. *Indian Journal of Nematology* 39, 104–105.
Kanwar, R.S. and Bajaj, H.K. (2010) Cyst nematode infestation in wheat. In: Khan, M.R. and Jairajpuri, M.S. (eds) *Nematode Infestations – Part I: Food Crops*. National Academy of Sciences, Allahabad (Prayagraj), UP, India, pp. 151–170.
Kanwar, R.S. and Paruthi, I.J. (2000) Wheat yield as influenced by nematicide application in *Heterodera avenae* infested field. In: *National Nematology Symposium on Integrated Nematode Management for Sustainable Agriculture in Changing Agro-ecological and Economic Scenario in the New Millennium*. OUAT, Bhubanehswar, India, pp. 23–24.
Kanwar, R.S., Nandal, S.N., Paruthi, I. J. and Bajaj, H. K. (2007) Status of *Heterodera avenae* and losses caused by it in wheat in Haryana. *National Nematology Symposium on Nematology in 21st Century: Emerging Paradigms*. Assam Agricultural University, Jorhat, Assam, India, 22–23, p. 40.
Kaplan, D.T. and Opperman, C.H. (1997) Genome similarity implies that citrus-parasitic burrowing nematodes do not represent a unique species. *Journal of Nematology* 29(4), 430–440.
Kaplan, D.T. and Opperman, C.H. (2000) Reproductive strategies and karyotype of the burrowing nematode, *Radopholus similis*. *Journal of Nematology* 32, 126–133.
Kapur-Ghai, J., Kaur, M. and Goel, P. (2014) Development of enzyme linked immunosorbent assay (ELISA) for the detection of root-knot nematode *Meloidogyne incognita*. *Journal of Parasitic Disease* 38(3), 302–306.
Karadjova, O., Ilieva, Z., Petrova, E. and Krumov, V. (2015) *Heterodera zeae* Koshy, Swarup and Sethi, 1971 (Heteroderidae): probability of introduction on cereals in Bulgaria. *Bulgarian Journal of Agricultural Sciences* 21, 969–981.
Karnkowski, W. (1999) Quarantine pests intercepted in consignments of ornamental plants imported in to Poland in 1993/1998. *EPPO Bulletin* 29, 45–49.
Karssen, G. (1994) The use of isozyme phenotypes for the identification of root-knot nematodes (*Meloidogyne* species). *Versl. Meded. Plantenziektenk Dienst Wageningen* (Ann. Rep. Diagnostic Center) 170, 85–88.
Karssen, G. (1996) Description of *Meloidogyne fallax* n. sp. (Nematoda: Heteroderidae), a root-knot nematode from The Netherlands. *Fundamental and Applied Nematology* 19(6), 593–599.
Karssen, G. and Bolk, R.J. (2000) An additional character useful for the identification of *Pratylenchus crenatus* Loof, 1960 (Nematoda: Pratylenchidae). *Nematology* 2, 695–697.
Karssen, G. and Hoenselaar, T.V. (1998) Revision of the genus *Meloidogyne* Goeldi, 1892 (Nematoda: Heteroderidae) in Europe. *Nematologica* 44, 713–788.
Karssen, G. and Willemsen, N.M. (2010) The spiculum: an additional useful character for the identification of *Ditylenchus dipsaci* and *D. destructor* (Nematoda: Anguinidae). *EPPO Bulletin* 40, 211–212.
Karssen, G., Wesemael, W. and Moens, M. (2013) Root-knot nematodes. In: Perry, R.N. and Moens, M. (eds) *Plant Nematology*, 2nd edn. CABI Publishing, Wallingford, UK, pp. 73–105.
Kasimova, G.A. and Atakishieva, Ya.Yu. (1980) The nematode fauna of *Crocus satiyus* in Apsheron. Izvestiya Akademii Nauk Azerbaidzhanskoi SSR (Azarbajcan SSR Elmlar Akademijasynyn Habarlari), *Biologicheskie Nauki*, No. 1, 94–98.
Katagiri, S., Kaneko, N. and Nagayama, Y. (1995) Changes of species composition with vegetation recovery of damaged pine forests. *Bulletin of the Faculty of Agriculture*, Shimane University, No. 29, 7–16.
Katalan-Gateva, S.H. and Baicheva, O. (1976) The nematode fauna on two tobacco varieties, Rila 9 and Kulsko in the Blagoevgrad region. *Khelmintologiya*, Sofia, No. 1, 49–61.
Kaul, R.K. and Sethi, C.L. (1988) Effect of time of application of certain systemic nematicides on final cyst production and fecundity of *Heterodera zeae*. *Indian Journal of Nematology* 18(1), 134–135.
Kaushal, K.K., Srivastava, A.N., Pankaj, Chawla, G. and Singh, K. (2007) Cyst forming nematodes in India – a review. *Indian Journal of Nematology* 37, 1–7.
Keetch, D.P. (1982) Nematode pests of pineapple. In: Keetch, D.P. and Heyns, J. (eds) *Nematology in South Africa*. Department of Agriculture and Fisheries, Pretoria, South Africa, pp. 9–29.
Keetch, D.P. and Dalldorf, E.R. (1980) The use of grass and legume rotations for pineapple nematode control in the Eastern Cape. *Citrus and Subtropical Fruit Journal* 557, 10–12.
Keetch, D.P. and Purdon, K.T. (1979) Effect of systemic nematicides on pineapples. *Citrus Subtropical Fruit Research Institute Information Bulletin* 82, 3–4.
Kerkoud, M., Esquibet, M. and Plantard, O. (2007) Identification of *Ditylenchus* species associated with Fabaceae seeds based on a specific polymerase chain reaction of ribosomal DNA-ITS regions. *European Journal of Plant Pathology* 118, 323–332.
Kerr, E.D. (1991) Evaluation of nematicide treatments for control of nematodes on sugar beet. *Fungicide and Nematicide Tests* 46, 197.
Kerry, B., Peteira, B., Atkins, S., Franco, J. and Manzanilla-López, R. (2005) A molecular diagnostic method for detecting *Nacobbus* in soil and in potato tubers. *Nematology* 7(2), 193–202. doi: 10.1163/1568541054879539
Khan, A. and Shaukat, S.S. (1993) Effect of neem and carbofuran against population of three nematodes associated with wheat varieties. *Sarhad Journal of Agriculture* 9(2), 157–163.

Khan, A. and Shaukat, S. (2000) Effect of organic amendments and carbofuran on population density of four nematodes and growth and yield parameters of rice (*Oryza sativa* L.) var. IRRI-6. *Pakistan Journal of Zoology* 32, 145–150.

Khan, A., Khanzada, A.K., Bilqees, F.M. and Aslam, M. (1987) Influence of crop rotation on the population of some plant parasitic nematodes. *Pakistan Journal of Zoology* 19(4), 447–450.

Khan, A., Shaukat, S.S., Qamar, F., Jaffry, A.H. and Hakro, A.A. (1994) Effect of three plant extracts on nematode population and on growth parameters of wheat (var. Pirsabak-85). *Sarhad Journal of Agriculture* 10(4), 415–418.

Khan, A., Shaukat, S.S. and Islam, S. (1997) Influence of soil temperature and moisture on seasonal variation of three plant-parasitic nematodes associated with apple in Swat, Pakistan. *International Journal of Nematology* 7, 26–29.

Khan, A.A., Ramakrishnan, S., Khan, M.R., Krishnamurthy, V. and Kausar, S. (2010) Nematode infestation in tobacco. In: Khan, M.R. and Jairajpuri, M.S. (eds) *Nematode Infestations – Part II: Industrial Crops*, National Academy of Sciences, Allahabad (Prayagraj), UP, India, pp. 47–65.

Khan, A.M., Adhami, A. and Saxena, S.K. (1971) Population changes of some stylet-bearing nematodes associated with mango (*Mangifera indica* L.). *Indian Journal of Nematology* 1(2), 99–105.

Khan, B., Khan, A.A. and Khan, M.R. (2003) Pathogenic variability among isolates of *Meloidogyne javanica* on Capsicum annum. *Journal of Nematology* 35 (4), 430–432.

Khan, E. and Chawla, M.L. (1975) *CIH Description of Plant-parasitic Nematodes*, Set 5, No. 66. Commonwealth Institute of Helminthology, St Albans, UK.

Khan, E., Chawla, M.L. and Saha, M. (1978) Comments on the classification of the Longidoroidea (Nematoda) with description of three new species. *Indian Journal of Nematology* 6, 47–62.

Khan, F.A. and Misari, S.M. (1992) Plant-parasitic nematodes associated with groundnut crop in four ecological zones of Nigeria. *Journal of African Zoology* 106(3), 263–271.

Khan, M.H. (1983) Plant parasitic nematodes associated with pineapples in Rampur, Chitwan. *Journal of Institute Agriculture Animal Sciences* 4, 85–86.

Khan, M.R. (2004) Observation on the foliar nematode, *Aphelenchoides besseyi* infecting tuberose in West Bengal. *Annales of Plant Protection Science* 12, 106–109.

Khan, M.R. (2005) *Biological control of fusarial wilt and root-knot of legume crops*. Department of Biotechnology, Government of India Publication, New Delhi, India.

Khan, M.R. (2008) *Plant Nematodes Methodology, Morphology, Systematics, Biology and Ecology*. Science Publishers, New Hampshire, USA.

Khan, M.R. (2010a) Emerging nematode problems of forest trees and their management. Abstract. In: *National Conference on Innovations in Nematological Research for Agricultural Sustainibility – Challenges and a Roadmap Ahead*. TNAU, Coimbatore, pp. 23–25.

Khan, M.R. (2010b) White tip nematode infestation in rice. In: Khan, M.R. and Jairajpuri, M.S. (eds) *Nematode Infestations – Part I: Food Crops*. National Academy of Sciences, Allahabad (Prayagraj), UP, India, pp. 111–134.

Khan, M.R. (2016) Nematode biocontrol agents: diversity and effectiveness against phytonematodes in sustainable crop protection. *Indian Phytopathology* 69(4s), 453–463.

Khan, M.R. (2020) Nematode infestations, a potential threat to Indian forests. *Indian Phytopathology* 73, 397–414.

Khan, M.R. and Athar, M. (1996) Response of wheat cultivars to different inoculum levels of *Anguina tritici*. *Nematology Mediterranean* 24, 269–272.

Khan, M.R. and Athar, M. (1998) Effect of *Anguina tritici* on the growth and yield of wheat cultivars. *Tests of Agrochemicals and Cultivars* 132, 52–53.

Khan, M.R. and Haque Z. (2011) Soil application of *Pseudomonas fluorescens* and *Trichoderma harzianum* reduces root-knot nematode, *Meloidogyne incognita*, on tobacco. *Phytopathologia Mediterrenea* 50, 257−266.

Khan, M.R. and Haque, Z. (2012) Nematicides control root-knot of rice caused by *Meloidogyne graminicola*. *Phytopathologia Mediterranea* 51(2), 298–306.

Khan, M.R. and Hayat, F. (1999) Effect of microwave irradiation on the ear cockle disease and yield of wheat. *Nematology Mediterranean* 27, 225–231.

Khan, M.R. and Jairajpuri, M.S. (eds) (2010a) *Nematode Infestations – Part I: Food Crops*. National Academy of Sciences, Allahabad (Prayagraj), UP, India.

Khan, M.R. and Jairajpuri, M.S. (eds) (2010b) *Nematode Infestations – Part II: Industrial Crops*. National Academy of Sciences, Allahabad (Prayagraj), UP, India.

Khan, M.R. and Jairajpuri, M.S. (eds) (2012) *Nematode Infestations – Part III: Horticultural Crops*. National Academy of Sciences, Allahabad (Prayagraj), UP, India.

Khan, M.R. and Sharma, R.K. (2020) Fusarium-nematode wilt disease complexes, etiology and mechanism of development. *Indian Phytopathology* 73, 615–618.

Khan, M.R., Athar, M. and Khan, B. (1998) Effect of foliar application of monocrotophos on the earcockle disease of wheat. *Nematology Mediterranean* 26(2), 173–177.

Khan, M.R., Altaf, S., Mohiddin, F.A., Khan, U. and Anwer, A. (2009) Biological control of plant nematodes with phosphate solubilizing microorganisms. In: Khan, M.S. and Zaidi, A. (eds) *Phosphate Solubilizing Microbes for Crop Improvement*. Nova Science Publishers, Inc., New York, pp. 395–426.

Khan, M.R., Khan, S.M., Mohiddin, F.A. and Khan, N. (2011) A new bioprocess to produce low cost powder formulations of biocontrol bacteria and fungi. *Biological Control* 57(9), 1–11. doi: 10.1016/j.biocontrol.2011.04.007.

Khan, M.R., Reshu and Shahid, S. (2012a) Nematode infestation in solanaceous vegetables. In: Khan, M.R. and Jairajpuri, M.S. (eds) *Nematode Infestations – Part III: Horticultural Crops*. National Academy of Sciences, Allahabad (Prayagraj), UP, India, pp. 31–77.

Khan, M.R., Khan, A.A. and Khan, U. (2012b) Nematode infestation in cucurbitaceous vegetables. In: Khan, M.R. and Jairajpuri, M.S. (eds) *Nematode Infestations – Part III: Horticultural Crops*. National Academy of Sciences, Allahabad (Prayagraj), UP, India, pp. 158–181.

Khan, M.R., Haque, Z. and Kausar, N. (2014) Management of the root-knot nematode *Meloidogyne graminicola* infesting rice in the nursery and crop field by integrating seed priming and soil treatments of some pesticides. *Crop Protection* 63, 15–25.

Khan, M.R., Ansari, R.A., Rizvi, T.F. and Qasim, R.M. (2019a) Nematode problem in vegetable crops, and its biomanagement. In: Khan, M.R., Mukhopadhyay, A.N., Pandey, R.N., Thakur, M.P., Singh, D., Siddiqui, M.A. *et al.* (eds) *Bio-intensive Approaches: Application and Effectiveness in Management of Plant Nematodes, Insects and Weeds*. Today and Tomorrow's Printers and Publishers, New Delhi, India, pp. 171–198.

Khan, M.R., Haque, Z. and Sharma, R.K. (2019b) Nematode problem in pulses and their management through non-chemical practices. In: Khan, M.R., Mukhopadhyay, A.N., Pandey, R.N., Thakur, M.P., Singh, D., Siddiqui, M.A. *et al.* (eds) *Bio-intensive Approaches: Application and Effectiveness in Management of Plant Nematodes, Insects and Weeds*. Today & Tomorrow's Printers and Publishers, New Delhi, India, pp. 171–198.

Khan, M.R., Haque, Z., Ahamad, F. and Zaidi, B. (2020) Biomanagement of rice root-knot nematode *Meloidogyne graminicola* using five indigenous microbial isolates under pot and field trials. *Journal of Applied Microbiology*, 130, 424–438.

Khan, M.W. (ed.) (1993) *Nematode Interactions*. Chapman and Hall, New York.

Khan, M.W. (1997) The four major species of root-knot nematodes – current status and management approach. *Indian Phytopathology* 50(4), 445–457.

Khan, M.W. (1998) Present status of biological suppression of plant parasitic nematodes. In: Singh, S.P. and Hussaini, S.S. (eds) *Biological Suppression of Plant Diseases, Phytonematodes and Weeds*. Project Directorate of Biological Control, Bangalore, India, pp. 118–127.

Khan, M.W. and Dasgupta, M.K. (1993) The concept of interaction. In: Khan, M.W. (ed.) *Nematode Interactions*. Chapman and Hall, New York, pp. 42–54.

Khan, M.R. and Ahamad, F. (2020) Incidence of root-knot nematode (*Meloidogyne graminicola*) and resulting crop losses in paddy rice in Northern India. *Plant Disease* 104(1), 186–193. doi: 10.1094/PDIS-12-18-2154-RE

Khan, T.H., Jairajpuri, M.S. and Ahmad, W. (1989) Descriptions of some new and known species of dorylaim nematodes. *Nematologica* 35(4), 419–437.

Khan, Z., Son, S.H., Shin, H.D. and Kim, Y.H. (2008) First report of a foliar nematode *Aphelenchoides fragariae* (Aphelenchidae) on *Stachys riederi* var. *japonica*, a medicinal plant, in Korea. *Plant Pathology Journal* 24(1), 97–100.

Khun, K., Decraemer, W., Couvreur, M., Karssen, G., Steel, H. and Bert, W. (2015) Deceptive morphological variation in *Hirschmanniella mucronate* (Nematoda: Pratylenchidae) and a polytomous key to the genus. *Nematology* 17, 377–400.

Kikas, L. (1969) Potato stem nematode is a harmful pest. *Sotsialistlik Pollumajandus* 24(14), 637–639.

Kikuchi, T., Aikawa, T., Oeda, Y., Karim, N. and Kanzaki, N. (2009) A rapid and precise diagnostic method for detecting the pinewood nematode *Bursaphelenchus xylophilus* by loop-mediated isothermal amplification. *Phytopathology* 99, 1365–1369.

Kinh, D.N. (1981) Survival of *Ditylenchus angustus* in diseased stubbles. *International Rice Research Newsletter* 6, 13.

Kinloch, R.A. and Lutrick, M.C. (1975) The relative abundance of nematodes in an established field crop rotation. *Proceedings of the Soil and Crop Science Society of Florida* 34, 192–194.

Kinloch, R.A., Lutrick, M.C. and Robinson, W.K. (1971) The reproduction of *Helicotylenchus dihystera* on soybeans. *Proceedings of the Soil and Crop Science Society of Florida* 31, 262–263.

Kiontke, K. and Fitch, D.H.A. (2014) Nematodes. *Current Biology* 23(19), R862.

Kirjanova, E.S. and Terenteva, T.G. (1961) Cactus nematode – *Heterodera cacti* (Adam, 1932) Filipjev et Schuurmans-Stekhoven, 1941. *Vopr. Fitogel. Izdatel'stvo Gel'mint. Lab., Moscow*, USSR. (In Russian.)

Kirkpatrick, T.L. (1989) Report of nematode management committee – 1988. In: *Proceedings of the Beltwide Cotton Production Research Conference*, Nashville, TN. National Cotton Council of America, Memphis, Tennessee, pp. 7–10.

Kirkpatrick, T.L. and Lorenz, G. (1997) The reniform nematode, an emerging problem in Arkansas cotton. In: Oosterhuis, D.M. and Steward, J. McD. (eds) *Proceedings of the 1997 Cotton Research Meeting and 1997 Summaries of Cotton Research in Progress*. Arkansas Agricultural Experiment Station, Fayetteville, Arkansas, pp. 52–57.

Klingler, J. (1970) The reaction of *Aphelenchoides fragariae* to slit-like micro-openings and to stomatal diffusion gases. *Nematologica* 16(3), 417–422.

Klingler, J. (1975) Observations on the parasitic activity of the nematode *Macroposthonia xenoplax* on grape vine roots. *Zeitschrift fur Pflanzenkrankheiten und Pflanzenschutz* 82(11/12), 722–728.

Knight, K.W.L., Hill, C.F. and Sturhan, D. (2002) Further records of *Aphelenchoides fragariae* and *A. ritzemabosi* (Nematoda: Aphelenchida) from New Zealand. *Australasian Plant Pathology* 31, 93–94.

Koen, H. (1967) Notes on the host range, ecology and population dynamics of *Pratylenchus brachyurus*. *Nematologica* 13, 118–124.

Koenning, S.R., Kikpatrick, T.L., Starr, J.L., Wrather, J.A., Walker, N.R. and Mueller, J.D. (2004) Plant-parasitic nematodes attacking cotton in United States; old and emerging production challenges. *Plant Disease* 88, 100–113.

Kohl, L.M. (2011) *Astronauts of the nematode world: an aerial view of foliar nematode biology, epidemiology, and host range*. American Phytopathological Society (APS), St Paul, Minnesota. Available at: apsnet.org/edcenter/apsnetfeature/Pages/foliarnematodes.aspx (accessed 27 February 2021). doi: 10.1094/APSnetFeature-2011-0111.

Kolombia, Y.A., Karssen, G., Viaene, N., Kumar, P.L., Joos, L., Coyne, D.L. and Bert, W. (2017) Morphological and molecular characterization of *Scutellonema* species from yam (*Dioscorea* spp.) and a key to the species of the genus. *Nematology* 19, 751–787.

Korayem, A.M. and Koura, F.H.F. (1993) Population studies of plant parasitic nematodes associated with mango in Giza, Egypt. *Afro-Asian Journal of Nematology* 3(2), 148–151.

Kornobis, S. (1999) *Radopholus similis* and *Radopholus citrophilus* are one species. *Ochrona Roslin* 43(1), 8.

Kort, J., Ross, H., Rumpenhorst, H.J. and Stone, A.R. (1977) An international scheme for identifying and classifying pathotypes of potato cystnematodes *Globodera rostochiensis* and *G. pallida*. *Nematologica* 23, 333–339.

Korthals, G.W., Brommer, E. and Molendijk, L.P.G. (2000) *Meloidogyne chitwoodi* and *Meloidogyne fallax* a threat to potato production? In; *Proceedings of the Fourth World Potato Congress*, Amsterdam, The Netherlands, 4–6 September, pp. 207–208.

Koshy, P.K. (1967) A new species of *Heterodera* from India. *Indian Phytopathology* 20, 272–274.

Koshy, P.K. and Bridge, J. (1990) Nematode parasites of spices. In: Luc, M., Sikora, R.A. and Bridge, J. (eds) *Plant Parasitic Nematodes in Subtropical and Tropical Agriculture*. CAB International, Wallingford, UK, pp. 557–582.

Koshy, P.K. and Sosamma, V.K. (1978) Studies on the population fluctuations of *Radopholus similis* in coconut and arecanut roots. *Indian Phytopathology* 31(2), 180–185.

Koshy, P.K. and Swarup, G. (1971a) Investigations on the life history of the pigeon-pea cyst nematode, *Heterodera cajani*. *Indian Journal of Nematology* 1, 44.

Koshy, P.K. and Swarup, G. (1971b) Distribution of *Heterodera avenae*, *H. zeae*, *H.cajani*, and *Anguina tritici* in India. *Indian Journal of Nematology*, 1, 106–111.

Koshy, P.K. and Swarup, G. (1973) Susceptibility of plants to pigeon-pea cyst nematode, *Heterodera cajani*. *Indian Journal of Nematology* 2(1972), 1–6.

Koshy, P.K. and Swarup, G. (1979) Histopathology of pigeon-pea roots infested with *Heterodera cajani* Koshy, 1967. *Indian Journal of Nematology* 9(2), 178–179.

Koshy, P.K., Swarup, G. and Sethi, C.L. (1970) *Heterodera zeae* n.sp. (Nematoda: Heteroderidae), a cyst-forming nematode in *Zea mays*. *Nematologica* 16, 511–516.

Koshy, P.K., Swarup, G. and Sethi, C.L. (1971) Further notes on the pigeon-pea cyst nematode, *Heterodera cajani*. *Nematologica* 16(4), 477–482.

Koshy, P.K. Sundararaju, P., Sosamma, V.K. and Ravikumar, K. (1985) Efficacy of four systemic nematicides against *Radopholus similis* in coconut nursery. *Indian Journal of Nematology* 15: 148–151.

Kovaleva, E.S., Masler, E.P., Subbotin, S.A. and Chitwood, D.J. (2005) Molecular characterization of aldolase from *Heterodera glycine* and *Globodera rostochiensis*. *Journal of Nematology* 37(3), 292–296.

Koyama, Y., Thar, S.Y., Kizaki, C., Toyota, K., Sawada, E. and Abe, E. (2013) Development of specific primers to *Hirschmanniella* spp. causing damage to lotus and their economic threshold level in Tokushima prefecture in Japan. *Nematology* 15(7), 851–855.

Krall, E.L. (1991) Wheat and grass nematodes: *Anguina*, *Subanguina* and related genera. In Nickle, W.R. (ed.) *Manual of Agricultural Nematology*. Marcel Dekker, New York, pp. 721–760.

Krall, E. and Krall, H. (1978) Revision of the plant nematodes of the family Heteroderidae on the basis of the trophic specialisation of these parasites and their joint evolution with their host plants. In: Ryzhikov, K.M. (ed.) *Fitogel'mintologicheskie-issledovaniya*. Nauka, Moscow, pp. 39–56/

Kraus-Schmidt, H. and Lewis, S.A. (1981) Dynamics of concomitant populations of *Hoplolaimus columbus*, *Scutellonema brachyurum*, and *Meloidogyne incognita* on cotton. *Journal of Nematology* 13(1), 41–48.

Krishnaprasad, K.S. and Krishnappa, K. (1982) Reactions of a few rice varieties to the build up of stunt nematode *Tylenchorhynchus claytoni*. *Indian Journal of Nematology* 12(1), 173–174.

Krusberg, L.R. and Sardanelli, S. (1989) Survival of *Heterodera zeae* in soil in the field and in the laboratory. *Journal of Nematology* 21(3), 347–355.

Kuhn, J. (1874) Uber das Vorkommen Von Rubennematoden an den wurzeln der almfruchte. Z. Wiss. Landw-Arch. Kgl. Preuss. *Landesokon. Kolleg* 3, 47–50.

Kuhn, J. (1881) Die Ergebnisse der Versuche zur Ermittelung der Ursache der Rübenmüdigkeit und zur Erforschung der Natur der Nematoden. *Berichte Physiologischen Laboratorium und der Versuchsanstalt des Landwirtschaftlichen Instituts der Universität Halle* 3, 1–153.

Kulinich, O.A., Kruglic, I.A., Eroshenko, A.S. and Kolosova, N.V. (1994) Occurrence and distribution of the nematode *Bursaphelenchus mucronatus* in the Russian Far East. *Russian Journal of Nematology* 2, 113–119.

Kumar, A.C. (1964) A note on the occurrence of *Heterodera cacti* from Mysore. *Current Science* 33, 534.

Kumari, S. (2015) Characterization of *Pratylenchus crenatus* and *P. neglectus* (Nematoda: Pratylenchidae) associated with wheat crop. *Helminthologia* 52(3), 280–286.

Kumari, S. and Lišková, M. (2009) Molecular confirmation of *Xiphinema italiae* Meyl, 1953 (Nematoda: Longidoridae) from the Slovak Republic. *Helminthologia* 46, 2, 131–134.

Kumari, S. and Subbotin, S.A. (2012). Characterization of *Longidorus helveticus* (Nematoda: Longidoridae) from the Czech Republic. *European Journal of Plant Pathology* 133, 923–933.

Kunde, R.M., Lider, L.A. and Schmitt, R.V. (1968) A test of Vitis resistance to *Xiphinema index*. *American Journal of Enology and Viticulture* 19, 30–36.

Kuriyan, J. (1995) Nematodes problems of rice crop in India II: Rice cyst nematode and white tip nematode. In: Swarup, G., Dasgupta, D.R. and Gill, J.S. (eds.) *Nematode Pest Management – An Appraisal of Ecofriendly Approaches*. Nematological Society of India, New Delhi, India, pp. 151–155.

Kusano, T., Nakamura, K., Fujii, T., Sugawara, Y., Ichikawa, Y., Masuyama *et al.* (1999) RAPD-PCR fingerprinting patterns of six *Bursaphelenchus xylophilus* isolates and a *B. mucronatus* isolate. In: Futai, K., Togashi, K. and Ikeda, T. (eds) *Sustainability of Pine Forests in Relation to Pine Wilt and Decline. Proceedings of the Symposium*, Tokyo, Japan, 26–30 October 1998. Shokado Shoten, Kyoto, Japan, pp. 52–56.

Lakshmanan, P.L., Vadivelu, S. and Jayaraj, S. (1989) *National Seminar on Futuristic Approaches in Cotton Improvement*, Haryana Agricultural University, Hisar, India, pp. 59–60.

Lal, R. and Lal, A. (2006) Plant parasitic nematodes intercepted from seeds, soil clods and packing material under import quarantine. *Journal of New Seeds* 8, 49–60.

Lamberti, F., Robini, H.M. and Ekanayaka, K. (1983) Effect of some plant parasitic nematodes on the growth of black pepper in Sri Lanka. *Plant Protection Bulletin* 31, 163–166.

Lamberti, F., Bleve-Zacheo, T., Tuopay, D.K., Ciancio, A. and Boiboi, J.B. (1987) Relationships between *Xiphinema ifacolum* and rice in Liberia. *Nematologia Mediterranea* 15, 303–314.

Lamberti, F., Ciancio, A., Tuopay, D.K., Boiboi, J.B., Vovlas, N., Bleve-Zacheo, T. and Elia, F. (1991) Nematode threats to rice in Liberia. *Nematologia Mediterranea* 19(2), 291–303.

Lamberti, F., Ciancio, A., Boiboi, J.B., Tuopay, D.K., Bleve-Zacheo, T. and Elia, F. (1992) Pathogenicity and reproduction of two species of *Xiphinema* on selected vegetable crops in Liberia. *Nematologia Mediterranea* 20(1), 113–123.

Lamberti, F., Ekanayake, H.M.R.K. and Sasanelli, N. (1993) Effect of some plant parasitic nematodes on the growth of selected crops in Sri Lanka. *Nematologia Mediterranea* 21(1), 27–43.

Lamberti, F., Sabová, M. de L.F., Molinari, S., Agostinelli, A., Coiro, M.I. and Valocká, B. (1999) Phenotypic variations and genetic characterization of *Xiphinema* populations from Slovakia (Nematoda: Dorylaimida). *Nematologia Mediterranea* 27, 261–275.

Lamberti, F., Molinari, S., Moens, M., Taylor, C.E. and Brown, D.J.F. (2000) The *Xiphinema americanum* group. 1. Putative species, their geographical occurrence and distribution and regional polytomous identification keys for the group. *Russian Journal of Nematology* 8, 65–84.

Lamberti, F., Hockland, S., Agostinelli, A., Moens, M. and Brown, D.J.F. (2004) The *Xiphinema americanum* group. 3. Keys to species identification. *Nematologia Mediterranea* 32, 53–56.

Lamberti, F., Greco, N. and Troccoli, A. (2007) Control and management of plant-parasitic nematodes. In: Saxena, G. and Mukerji, K.G. (eds) *Management of Nematode and Insect-borne Plant Diseases*). Haworth Press, New York.

LaMondia, J.A. (1995) Hatch and reproduction of *Globodera tabacum tabacum* in response to tobacco, tomato, or black nightshade. *Journal of Nematology* 27(3), 382–386.

LaMondia, J.A. (1996) Trap crops and population management of *Globodera tabacum tabacum*. *Journal of Nematology* 28(2), 238–243.

Lawler, C. and Harmey, M.A. (1993) Immunological detection of nematode antigens on the surface of a wood section. *Fundamental and Applied Nematology* 16, 521–523.

Lawson, E.C., Carter, G.E. and Lewis, S.A. Jr (1984) Application of isoelectric focusing to the taxonomic identification of *Meloidogyne* spp. *Journal of Nematology* 16, 91–96.

Lazarova, S.S., Malloch, G., Oliveira, C.M.G., Hübschen, J. and Neilson, R. (2006) Ribosomal and mitochondrial DNA analyses of *Xiphinema americanum* group populations. *Journal of Nematology* 38, 404–410.

Le Roux, H.F., Pretorious, M.C. and Huisman, L. (2000) Citrus nematode IPM in Southern Africa. *Proceedings of the International Society of Citriculture* 2, 823–827.

Leal, I., Green, M., Allen, E., Humble, L. and Rott, M. (2005) An effective PCR-based diagnostic method for the detection of *Bursaphelenchus xylophilus* (Nematoda:Aphelenchoididae) in wood samples from lodgepole pine. *Nematology* 7, 833–842.

Leal, I., Green, M., Allen, E., Humble, L. and Rott, M. (2007) Application of a real-time PCR method for the detection of pinewood nematode, *Bursaphelenchus xylophilus*, in wood samples from lodgepole pine. *Nematology* 9, 351–362.

Leal, I., Foord, B., Allen, E., Campion, C., Rott, M. and Green, M. (2013) Development of two reverse transcription-PCR methods to detect living pinewood nematode, *Bursaphelenchus xylophilus*, in wood. *Forest Pathology* 43, 104–114.

Leal-Bertioli, S.C.M., Tenente, R.C.V. and Bertioli, D.J. (2000) ITS sequence of populations of the plant-parasitic nematode *Ditylenchus dipsaci*. *Nematologia Brasileira* 24, 83–85.

Lehman, P.S. (2004) Cost-benefits of nematode management through regulatory programs. In: Chen, Z.X., Chen, S.Y. and Dickson, D.W. (eds). *Nematology Advances and Perspectives. Volume 2*. CAB International, Wallingford, UK, pp. 1133–1177.

Lehman, P.S. and Inserra, R.N. (1990) Morphometric variation of *Rotylenchulus parvus* and *Rotylenchulus reniformis* populations in the southern United States. *Proceedings – Soil and Crop Science Society of Florida* 49, 220–226.

Lesufi, M.M., Swart, A., McDonald, A.H., Knoetze, R., Tiedt, L.R. and Truter, M. (2015) Morphological and molecular studies on *Aphelenchoides arachidis* Bos, 1977 (Tylenchina: Aphelenchoididae) from groundnuts in South Africa. *Nematology* 17 (4), 433–445.

Lewis, S.A., Smith, F.H. and Powell, W.M. (1976) Host parasite relations of *Hoplolaimus columbus* on cotton and soybean. *Journal of Nematology* 8, 264–269.

Li, G.W., Shao, G.Y., Huo, Y.L. and Xu, F.Y. (1983) Discovery of and preliminary investigations on pine wood nematodes in China. *Forest Science and Technology* 7, 25–28.

Li, H. (2008) Identification and pathogenicity of *Bursaphelenchus* species (Nematoda: Parasitaphelenchidae). PhD thesis, Ghent University, Ghent, Belgium.

Li, H., Trinh, P.Q., Waeyenberge, L. and Moens, M. (2009) Characterisation of *Bursaphelenchus* spp. isolated from packaging wood imported at Nanjing, China. *Nematology* 11(3), 375–408.

Li, W., Zonghe, Y., Nakhla, M.K. and Skantar, A.M. (2015) Real-time PCR for detection and identification of *Anguina funesta*, *A. agrostis*, *A. tritici*, and *A. pacificae*. *Plant Disease* 99, 1584–1589.

Liao, J., Zhang, L. and Feng, Z. (2001) Reliable identification of *Bursaphelenchus xylophilus* by rDNA amplification. *Nematologia Mediterranea* 29, 131–135.

Lilley, C.J., Kyndt, T. and Gheysen, G. (2011) Nematode resistant GM crops in industrialised and developing countries. In: Jones, J.T., Gheysen, G. and Fenoll, C. (eds) *Genomics and Molecular Genetics of Plant–Nematode Interactions*. Springer, Heidelberg, pp. 517–541.

Lima, E.A., Mattos, J.K., Moita, A.W., Carneiro, R.G. and Carneiro, R.M.D.G. (2009) Host status of different crops for *Meloidogyne ethiopica* control. *Tropical Plant Pathology* 34(3), 152–157.

Limber, D.P. (1973) Notes on the longevity of *Anguina tritici* (Steinbuch, 1799) Filipjev, 1936, and its ability to invade wheat seedlings after 32 years of dormancy. *Proceedings of the Helminthological Society of Washington* 40(2), 272–274.

Lin, B.R., Wang, H.H., Zhuo, K. and Liao, J.L. (2016) Loop-mediated isothermal amplification for the detection of *Tylenchulus semipenetrans* in soil. *Plant Disease* 100, 877–883.

Lin, M.S., Ding, X.F, Wang, Z.M., Zhou, F.M. and Lin, N. (2005) Description of *Aphelenchoides besseyi* from abnormal rice with small grains and erect panicles in China. *Chinese Journal Rice Sciences* 19(4), 361–365.

Lin, Y.Y., Wang, K.M. and Tsay, T.T. (1992) The occurrence of *Aphelenchoides besseyi* on Dendrobium Lady Fay, *Plant Protection Bulletin* (Taipei) 34(3), 202–215.

Linford, M.B. and Oliveira, J.M. (1940) *Rotylenchulus reniformis*, nov. gen., n. sp., a nematode parasite of roots. *Proceedings of the Helminthological Society of Washington* 7, 35–42.

Linsell, K.J, Riley, I.T., Davies, K.A., and Oldach, K.H. (2014) Characterization of resistance to *Pratylenchus thornei* (Nematoda) in wheat (*Triticum aestivum*): attraction, penetration, motility, and reproduction. *Phytopathology* 104(2), 174–187.

Litchman, E. (2010) Invisible invaders: non-pathogenic invasive microbes in aquatic and terrestrial ecosystems. *Ecology Letters* 13, 1560–1572.

Liu, G.K., Chen, J., Xiao, S., Zhang, S.S. and Pan, D.M. (2011) Development of species-specific PCR primers and sensitive detection of the *Tylenchulus semipenetrans* in China. *Agriculture Sciences, China* 10, 252–258.

Loof, P.A.A. (1960) Taxonomic studies on the genus *Pratylenchus* (Nematoda). *Tijdschr. Plantenziekten* 66, 29–90.

Loof, P.A.A. (1974) *Tylenchorhynchus claytoni*. *CIH Descriptions of Plant-parasitic Nematodes*, Set 3, No. 39. Commonwealth Institute of Helminthology St Albans, UK.

Loof, P.A.A. (1978) *The genus* Pratylenchus *Filipjev, 1936 (Nematoda: Pratylenchidae): a review of its anatomy, morphology, distribution, systematics and identification*. Swedish University of Agricultural Sciences, Research Information Centre, Uppsala Sweden.

Loof, P.A. (1991) The family Pratylenchidae Thorne, 1949. In: Nickle, W.R. (ed.). *Manual of Agricultural Nematology*. Marcel Dekker, New York, pp. 363–421.

Loof, P.A.A. and Luc, M. (1990) A revised polytomous key for the identification of species of the genus *Xiphinema* Cobb, 1913 (Nematoda: Longidoridae) with exclusion of the *X. americanum*-group. *Systematic Parasitology* 16(1), 35–66.

Loos, C.A. (1953) *Meloidogyne brevicauda*, n.sp a cause for root knot of mature tea in Ceylon. *Proceedings of the Helminthological Society of Washington* 20, 83–91.

Lopez, R. and Vilchez, H. (1991) Two new hosts of the coffee root-knot nematode, *Meloidogyne exigua*, in Costa Rica. *Agronomía Costaricense*, 15(1/2), 163–166.

Lordello, L.G.E. and Brancalion, A.M. (1986) Evaluation of the nematicide Furadan 350 FMC in the control of the nematode *Rotylenchulus reniformis* attacking cotton. *Nematologia Brasileira*, 10, 103–107.

Lordello, L.G.E. and Zamith, A.P.L. (1954) Constatacao da molestia do 'anel vermelho' do coqueiro no Estado do Rio de Janeiro. Redescricao do agente causador – *Aphelenchoides cocophilus* (Cobb, 1919) Goodey, 1933 (Nematoda, Aphelenchidae). *Anais da Escola superior de Agricultura 'Luiz de Queiroz'* 11, 125–132.

Lordello, L.G.E. and Zamith, A.P.L. (1960) *Meloidogyne coffeicola* sp. n., a pest of coffee trees in the State of Paraná, Brazil. *Review of Brasil. Biology* 20(4), 375–379.

Lordello, L.G.E., Zamith, A.P. and Boock, O.J. (1961) Two nematodes found attacking potato in Cochabamba, Bolivia. *Anais da Academia Brasileira de Ciências* 33, 209–215.

Lordello, R.R.A. and Fazuoli, L.C. (1980) *Meloidogyne decalineata* in coffee roots from Sao Tome Island (Africa). *Revista de Agricultura, Piracicaba, Brazil* 55(4), 238.

Lordello, R.R.A. and Sabino, N.P. (1985) Effect of nematode control on fibre properties of upland cotton. *Bragantia* 44, 695–699.

Lordello, R.R.A., Lordello, A.I.L., Sawazaki, E. and Aloisi-Sobrinho, J. (1985) Resistance of maize in the field to *Pratylenchus* infection. *Nematologia Brasileira* 9, 163–173.

Luambano, N.D., Kashando, B.E., Masunga, M.M., Mwenisongole, A.E., Mziray, M.F., Mbaga, J.E., Polinia, R.M. and Mgonja, D.M. (2019) Status of *Pratylenchus coffeae* in banana-growing areas of Tanzania. *Physiological and Molecular Plant Pathology* 105, 102–109.

Luc, M. (1958) Les nematodes et le fletrissement des cotonniers dans le Sudouest de Madagascar. *Cotton Fibres Trop* 13, 1–18.

Luc, M. (1961) *Xiphinema* de l'ouest africain: (Nematoda: Dorylaimidae). Deuxieme note. *Nematologica* 6, 107–122.

Luc, M. (1974). *Heterodera sacchari*. C.I.H. Descriptions of Plant-parasitic Nematodes, Set 4, No. 48: 3 pp.

Luc, M. (1986) Cyst nematodes in equatorial and hot tropical regions. In: Lamberti, F. and Taylor, C.E. (eds) *Cyst Nematodes*. Plenum Press, New York, pp. 355–372.

Luc, M. and Fortuner, R. (1975) *Hirschmanniella spinicaudata*. *CIH Descriptions of Plant-parasitic Nematodes*, Set 5, No. 68. Commonwealth Institute of Helminthology, St Albans, UK.

Luc, M. and Taylor, D.P. (1977) *Heterodera oryzae*. *CIH Descriptions of Plant-parasitic Nematodes*, Set 7, No. 91. Commonwealth Institute of Helminthology, St Albans, UK.

Luc, M. and Vilardebo, A. (1961) Les nématodes associés aux bananiers cultivés dans l'Ouest Africain. *Fruit* 16(5), 205–219; 6, 261–279.

Luc, M., Bridge, J. and Sikora, R.A. (1990) Reflections on nematology in subtropical and tropical agriculture. In: Luc, M., Sikora, R.A. and Bridges, J. (eds) *Plant Parasitic Nematodes in Subtropical and Tropical Agriculture*. CAB International, Wallingford, UK, pp. 1–10.

Luc, M., Bridge, J. and Sikora, R.A. (2005) Reflections on nematology in tropical and subtropical agriculture. In: Luc, M., Sikora, R.A. and Bridges, J. (eds) *Plant Parasitic Nematodes in Subtropical and Tropical Agriculture*. CAB International, Wallingford, U.K. pp. 1–10.

Luc, M., Taylor, D.P. and Cadet, P. (1978) Description of new tropical Heteroderidae, *Hylonemaivorensis* n. gen, n. sp. and a new outlook of the family Heteroderidae (Nematoda: Tylenchida). *Revue de Nématologie* 1, 73–86.

Lucus, L.T. (1982) Population dynamics of *Belonolaimus longicaudatus* and *Criconemella ornata* and growth response of bermudagrass and overseeded grasses on golf greens following treatment with nematicides. *Journal of Nematology* 14(3), 358–363.

Ma, F.C., Yu, H.M., Horng, F.W., Chang, N.H. and Hsui, Y.R. (2002) Hardwood natural regeneration and rehabilitation in a seriously nematode-damaged Luchu pine (*Pinus luchuensis*) plantation. *Taiwan Journal of Forest Science* 17, 269–280.

Ma, X., Agudelo, P., Mueller, J.D. and Knap, H.T. (2011) Molecular characterization and phylogenetic analysis of *Hoplolaimus stephanus*. *Journal of Nematology* 43, 25–34.

Maafi, Z.T. and Kheiri, A. (1993) Plant parasitic nematodes on banana from Hormozgan Province. *Iranian Journal of Plant Pathology* 29(1/2), 21–23.

MacGowan, J.B. (1980) Tylenchorhynchus claytoni *Steiner 1937. The tobacco stunt nematode*. Nematology Circular No. 67. Division of Plant Industry, Florida Department of Agriculture and Consumer Services, Gainesville, Florida.

MacGowan, J.B. (1983) *The stubby-root nematode,* Paratrichodorus christiei *(Allen 1957) Siddiqi 1974*. Nematology Circular No. 97. Division of Plant Industry, Florida Department of Agriculture and Consumer Services, Gainesville, Florida.

MacGowan, J.B. and Langdonm, K.R. (1989) *Hosts of the rice root-knot nematode* Meloidogyne graminicola. Nematology Circular No. 172. Division of Plant Industry, Florida Department of Agriculture and Consumer Services, Gainesville, Florida.

MacGuidwin, A.E. and Rouse, D.I. (1990) Role of *Pratylenchus penetrans* in the potato early dying disease of Russet Burbank potato. *Phytopathology* 80, 1077–1082.

Maceda, A., Silva, J.F., Dias, W.P., Santos, J.M., Soares, P.L.M., Marcondes, M.C. *et al.* (2009) Ocorrência de *Scutellonema brachyurus* Andrassy em soja no Brasil. In: *II International Congress of Tropical Nematology*, Maceio. Abstract on ONTA, No. 128.

Machado, A.C.Z., Ferraz, L.C.C.B. and Oliveira, C.M.G. (2007) Development of a species-specific reverse primer for the molecular diagnostic of *Pratylenchus brachyurus*. *Nematropica* 37, 249–257.

Machado, A.C.Z., Amaro, P.M. and Silva, S.A.D. (2019) Two novel potential pathogens for soybean. *PLoS ONE* 14(8), e0221416. doi: 10.1371/journal.pone.0221416

Machon, J.E. and Bridge, J. (1996) *Radopholus citrii* n. sp. (Tylenchida: Pratylenchidae) and its pathogenicity on citrus. *Fundamental and Applied Nematology* 19(2), 127–133.

Madani, M., Palomares-Rius J.E., Vovlas, N., Castillo, P. and Tenuta, M. (2018) *European Journal of Plant Pathology* 150, 1023–1039.

Madani, M., Vovlas, N., Castillo, P., Subbotin, S.A. and Moens, M. (2004) Molecular characterization of cyst nematode species (*Heterodera* spp.) from the Mediterranean Basin using RFLPs and sequences of ITS-rDNA. *Journal of Phytopathology* 152, 229–234.

Madani, M., Subbotin, S.A. and Moens, M. (2005) Quantitative detection of the potato cyst nematode, *Globodera pallida*, and the beet cyst nematode, *Heterodera schachtii*, using Real-Time PCR with SYBR green I dye. *Molecular and Cellular Probes* 19, 81–86.

Madani, M., Kyndt, T., Colpaet, N., Subbotin, S.A., Gheysen, G. and Moens, M. (2007) Polymorphism among sugar beet cyst nematode *Heterodera schachtii* populations as inferred from AFLP and ITS rDNA gene sequence. *Russian Journal of Nematology* 15(2), 117–128.

Madani, M., Tenuta, M., Chizhov, V.N. and Subbotin, S.A. (2015) Diagnostics of stem and bulb nematodes, *Ditylenchus weischeri* and *D. dipsaci* (Nematoda: Anguinidae), using PCR with species-specific primers. *Canadian Journal of Plant Pathology* 37(2), 212–220.

Majd-Taheri, Z., Tanha Maafi,. Z., Subbotin, S.A., Pourjam, E. and Eskandar, A. (2013) Molecular and phylogenetic studies on Pratylenchidae from Iran with additional data on *Pratylenchus delattrei*, *Pratylenchoides alkani* and two unknown species of *Hirschmanniella* and *Pratylenchus*. *Nematology* 15(6), 633–651. doi:10.1163/15685411-00002707

Malan, A.P. and Meyer, A.J. (1993) Interaction between a South African population of *Xiphinema index* and different grapevine rootstocks. *South African Journal for Enology and Viticulture* 14(1), 11–15.

Malek, R.B. (1968) The dagger nematode, *Xiphinema americanum*, associated with decline of shelterbelt trees in South Dakota. *Plant Disease Reporter* 52, 795–798.

Malek, R.B. and Appleby, J.E. (1984) Epidemiology of pine wilt in Illinois. Disease distribution. *Plant Disease* 68(3), 180–186.

Malek, R.B. and Jenkins, W.R. (1964) Aspects of the host–parasite relationship of nematodes and hairy vetch. *New Jersey Experimental Station Bulletin* 813, 1–31.

Malik, Z. and Jairajpuri, M.S. (1983) Statistical analysis of variability in *Xiphinema americanum* Cobb, 1913 sensu Siddiqi, 1959. *Indian Journal of Nematology* 13(1), 71–78.

Mamiya, Y. (1983) Pathology of the pine wilt disease caused by *Bursaphelenchus xylophilus*. *Annual Review of Phytopathology* 21, 201–220.

Mamiya, Y. (2008) Movement of the pinewood nematode *Bursaphelenchus xylopilus* through tracheids in diseased pine trees. *Japanese Journal of Nematology* 38 (1), 41–44.

Mamiya, Y. and Kiyohara, T. (1972) Description of *Bursaphelenchus lignicolus* n.sp. (Nematoda: Aphelenchoididae) from pine wood and histopathology of nematode-infested trees. *Nematologica* 18, 120–124.

Mani, A. and Dakshinamurthy, V. (1986) Citrus nematode and root-knot nematodes: distribution, crop loss, symptoms and management. In: Swarup, G. and Dasgupta, D.R. (eds) *Plant-Parasitic Nematodes of India*. IARI, New Delhi, pp. 249–260.

Mani, M.P., Rajeswari, S. and Sivakumar, C.V. (1998) Management of potato cyst nematodes, *Globodera* spp. through plant rhizosphere bacterium, *Pseudomonas fluorescens* Migula. *Journal of Biological Control* 12, 131–134.

Mankau, R. and Imbriani, J.L. (1975) The life cycle of an endoparasite in some tylenchid nematodes. *Nematologica* 21(1), 89–94.

Mansourabad, M.A, Ghaderi, R., Kashi, L. and Karegar, A. (2016) Observations on *Hoplolaimus indicus* Sher, 1963 and *Hoplolaimus seinhorsti* Luc, 1958 (Nematoda: Hoplolaimidae) from Southern Iran. *Journal of Crop Protection* 5(2), 189–201. doi: 10.18869/modares.jcp.5.2.189

Manzanilla, R., Costilla, M.A., Doucet, M., Franco, J., Inserra, R.N., Lehman, P.S., Prado, I.C.D.R.S. and Evans, K. (2002) *Nacobbus* species: systematics, distribution, biology and management. *Nematropica* 32, 149–227.

Manzanilla-López, R.H. and Marbán-Mendoza, N. (eds) (2012) *Practical Plant Nematology*. Biblioteca Básica de Agricultura, Grupo Mundi-Prensa, Mexico City.

Manzanilla-López, R.H., Franco, J., Peteira, B. and Kerry, B. (2004) *Nacobbus aberrans*: its molecular diagnosis in soil and potato tubers. *Programs and Abstracts of ONTA 34th Annual Meeting*, Puerto Vallarta, Mexico (October 4–8, 2004), p. 83.

Maqbool, M.A. (1980) Occurrence of eight cyst nematodes on some agricultural crops in Pakistan. *Journal of Science, Karachi University* 8, 103–108.

Maqbool, M.A. (1988) Present status of research on plant parasitic nematodes in cereals and food and forage legumes in Pakistan. In: Saxena, N.C., Sikora, R.A. and Srivastava, J.P. (eds) *Nematodes Parasitic to Cereals and Legumes in Temperate Semi-arid Regions*. ICARDA, Aleppo, Syria, pp. 173–180.

Maqbool, M.A. and Ghazala, P. (1988) Observation of some known species of *Hoplolaimus* von Daday, 1905 (Nemata: *Hoplolaimidae*) from Pakistan. *Pakistan Journal of Nematology* 6, 1–7.

Marban-Mendoza, N., Dicklow, M.B. and Zuckerman, B.M. (1989) Evaluation of control of *Meloidogyne incognita* and *Nacobbus aberrans* on tomato by two leguminous plants. *Revue de Nématologie* 12(4), 409–412.

Marek, M., Zouhar, M., Rysanek, P. and Havranek, P. (2005) Analysis of ITS sequences of nuclear rDNA and development of a PCR-based assay for the rapid identification of the stem nematode *Ditylenchus dipsaci* (Nematoda: Anguinidae) in plant tissues. *Helminthologia* 42, 49–56.

Maria, M., Ye, W., Yu, Q. and Gu, J. (2018) Description of *Xiphinema parachambersi* n. sp. (Nematoda: Longidoridae) from Imported Ornamental Plants in Japan with a Key to *Xiphinema* Species in Group 1. *Journal of Nematology* 50(3), 369–386. doi: 10.21307/jofnem-2018-027.

Marks, R.J. and Brodie, B.B. (1998) Introduction: Potato cyst nematodes – an international pest complex. In: Marks, R.J. and Brodie, B.B. (eds) *Potato Cyst Nematodes – Biology, Distribution and Control*. CAB International, Wallingford, UK, pp. 1–4.

Marlatt, R.B. (1970) Transmission of *Aphelenchoides besseyi* to *Ficus elastica* leaves via Sporobolus poiretii inflorescences. *Phytopathology* 60 (3), 543–544.

Marshall, J.W., Zijlstra, C. and Knight, K.W.L. (2001) First record of *Meloidogyne fallax* in New Zealand. *Australian Plant Pathology* 30(3), 283–284.

Martelli, G.P, Cohn, E. and Dalmasso, A. (1966) A Redescription of *Xiphinema Italiae* Meyl, 1953 and Its Relationship To *Xiphinema Arenarium* Luc Et Dalmasso, 1963 and *Xiphinema Conurum* Siddiqi, 1964. *Nematologica* 12(2), 183–194

Martin, E.M., Eastwood, R.F. and Ogbonnaya, F.C. (2004) Identification of microsatellite marker associated with cereal cyst nematode resistance gene Cre3 in wheat. *Australian Journal of Agriculture Research* 55, 1205–1211.

Mathews, H.J.P. (1975) *Heterodera carotae. CIH Descriptions of Plant Parasitic Nematodes*, Set 5, No. 61. Commonwealth Institute of Helminthology, St Albans, UK.

Matsunaga, K. and Togashi, K. (2004) Among-tree difference in the inhibition of systemic dispersal of *Bursaphelenchus xylophilus* (Nematoda: Aphelenchoididae) by *Pinus densiflora*. *Applied Entomology and Zoology* 39(2), 271–277.

Maung, Z.T.Z., Kyi, P.P., Myint, Y.Y., Lwin, T. and De Waele, D. (2010) Occurrence of the rice root nematode *Hirschmanniella oryzae* on monsoon rice in Myanmar. *Tropical Plant Pathology* 35, 3–10.

May, D.B., Johnson, W.A., Zuck, P.C., Chen, C.C. and Dyer, A.T. (2016) Assessment and management of root lesion nematodes in Montana wheat production. *Plant Disease* 100, 2069–2079.

McCay, A.C. and Ophel, K.M. (1993) Toxigenic *Clavibacter/Anguina* associations infecting grass seedheads. *Annual Review of Phytopathology* 31, 151–167.

McClure, M.A. (1988) First molt in *Anguina tritici*. *Journal of Nematolpgy* 20(1), 167–169.

McClure, M.A., Nischwitz, C., Skantar, A.M., Schmitt, M. E. and Subbotin, S.A. (2012) Root-knot nematodes in golf course greens of the western United States. *Plant Disease* 96, 635–647.

McCuiston, J.L., Hudson, L.C., Subbotin, S.A., Davis, E.L. and Warfield, C.Y. (2007) Conventional and PCR detection of *Aphelenchoides fragariae* in diverse ornamental host plant species. *Journal of Nematology* 39(4), 343–355.

McDonald, A.H. and Nicol, J.M. (2005) Nematode parasites of cereals. In Luc, M., Sikora, R.A. and Bridge, J. (eds) *Plant Parasitic Nematodes on Subtropical and Tropical Agriculture*, 2nd edn. CABI, Wallingford, UK, pp. 131–192.

McDonald, D., Bos, W.S. and Gumel, M.H. (1979) Effects of infestation of peanut (groundnut) seed by the testa nematode, *Aphelenchoides arachidis*, on seed infection by fungi and on seedling emergence. *Plant Disease Reporter* 63(6), 464–467.

McElroy, F.D. and Van, Gundy, S.D. (1968) Observations on the feeding processes of *Hemicycliophora arenaria*. *Phytopathology*, 58, 1558-1565.

McElroy, F.D., Sher, S.A. and Van Gundy, S.D. (1966) The sheath nematode, *Hemicycliophora arenaria* a native to California soils. *Plant Disease Reporter* 40, 581–583.

McGeachie, I. and Rahman, L. (1983) *Ufra* disease: a review and a new approach to control. *Tropical Pest Management* 29, 325–332.

McGregor, R.C. (1978) People-placed pathogens: the emigrant pests. In: Horsfall, J.G. and Cowling, E.B. (eds) *Plant Disease. An Advanced Treatise. Volume II. How Disease Develops in Populations*. Academic Press Inc., New York and London, pp. 383–396.

McSorley, R. (1978) Components of a management program for nematodes on corn. *Dissertation Abstracts International* 39B (5), 2123–2124.

McSorley, R. (1992) Nematological problems in tropical and subtropical fruit tree crops. *Nematropica* 22, 103–116.

McSorley, R. and Dickson, D.W. (1990) Vertical distribution of plant-parasitic nematodes in sandy soil under maize. *Plant and Soil* 123(1), 95–100.

McSorley, R. and Parrado, J. L. (1983) Control of nematodes on white-fleshed sweet potato with Vydate, 1982. *Fungicide and Nematicide Tests, American Phytopathological Society* 38, 1.

McSorley, R. and Parrado, J.L. (1986) *Helicotylenchus multicinctus* on bananas: an international problem. *Nematropica* 16, 73–91.

McSorley, R., Parrado, J.L. and Goldweber, S. (1981) Plant-parasitic nematodes associated with mango and relationship to tree condition. *Nematropica* 11(1), 1–9.

Meagher, J.W. (1968) The distribution of cereal cyst nematode (*Heterodera avenae*) in Victoria and its relation to soil type. *Australian Journal of Experimental Agricultural Animal Husbandry* 8, 637–640.

Meagher, J.W. (1972) Cereal cyst nematode (*Heterodera avenae* Woll.). Studies on ecology and control in Victoria. *Technical Bulletin of Department of Agriculture Victoria* 24, 50.

Meagher, J.W. (1982) The effect of environment on survival and hatching of *Heterodera avenae*. *Bulletin EPPO* 12, 325–331.

Meagher, J.W. and Rooney, D.R. (1966) The effect of crop rotations in the Victorian Wimmera on the cereal cyst nematode (*Heterodera avenae*), Nitrogen fertility and wheat yield. *Australian Journal of Experimental Agricultural Animal Husbandry* 6, 425–431.

Meagher, J.W., Brown, R.H. and Rovira, A.D. (1978) The effects of cereal cyst nematode (*Heterodera avenae*) and *Rhizoctonia solani* on the growth and yield of wheat. *Australian Journal of Agriculture Research* 29, 1127–1137.

Meena, P., Nehra, S. and Trivedi, P.C. (2009) Survey, pathogenecity and biology of *Heterodera cajani* on pigeonpea and reaction of some pigeonpea varieties against pigeonpea cyst nematode. *Asian Journal of Experimental Science* 23(1), 141–148.

Mekete, T., Reynolds, K., Lopez-Nicora, H.D., Gray, M.E. and Niblack, T.L. (2011) Distribution and diversity of root lesion nematode (*Pratylenchus* spp.) associated with *Miscanthus giganteus* and *Panicum virgatum* used for biofuels and species identification in a multiplex polymerase chain reaction. *Nematology* 13, 673–686.

Meldal, B.H.M., Debenham, N.J., De Ley, P., De Ley, I.T., Vanfleteren J.R., Vierstraete, A.R. *et al.* (2007) An improved molecular phylogeny of the Nematoda with special emphasis on marine taxa. *Molecular Phylogenetics and Evolution* 42(3), 622–636.

Mendoza-de Gives, P., Zavaleta-Mejia, E., Herrera-Rodriguez, D. and Quiróz-Romero, H. (1994) *In vitro* trapping capability of *Arthrobotrys* spp. on infective larvae of *Haemonchus contortus* and *Nacobbus aberrans*. *Journal of Helminthology* 68(3), 223–229.

Meng, Q.-P., Long, H. and Xu, J.H. (2004) PCR assays for rapid and sensitive identification of three major root-knot nematodes, *Meloidogyne incognita, M. javanica* and *M. arenaria*. *Acta Phytopathologcia Sinica* 34, 204–210.

Meng, S., Alderman, S., Fraley, C., Ludy, R., Sun, F. and Osterbauer, N. (2012) Identification of *Anguina funesta* from annual ryegrass seed lots in Oregon. *Plant Health Progress* 10, 1094. doi: 10.1094/PHP-2012-1024-01-RS.

Metge, K. and Burgermeister, W. (2006) Intraspecific variation in isolates of *Bursaphelenchus xylophilus* (Nematoda: Aphelenchoididae) revealed by ISSR and RAPD fingerprints. *Journal of Plant Diseases and Protection* 113, 275–282.

Mgonja, D.M., Temu, G.E., Ndunguru, J.C., Mziray, M.F., Lyantagaye, S.L. and Luambano, N.D. (2020) Molecular approach to confirm traditional identification of *Radopholus similis* sampled in Tanzania. *Journal of Nematology* 52, e2020-20. doi: 10.21307/jofnem-2020-020.

Mgonja, D.M., Temu, G.E., Mziraya, M.F., Kashando, B.E., Mwenisongolea, A.E., Masungaa, M.M., Lyantagaye, S.L. and Luambano, N.D. (2019) Morphological and molecular identification of *Pratylenchus goodeyi* associated with banana in Tanzania. *Tanzania Journal of Science* 45(2), 265–278.

Miah, S.A. and Bakr, M.A. (1977) Sources of resistance to *ufra* disease of rice in Bangladesh. *International Rice Research Newsletter* 2, 8.

Mian, I.H. and Latif, M.A. (1995) Ultrastructure and morphometrics of *Ditylenchus angustus* (Butler, 1913) Filipjev, 1936 (Nematoda: Anguinidae). *Japanese Journal of Nematology* 24, 14–19.

Mian, I.H. and Tsuno, K. (1988) Nematode parasite of sugarcane cultivated for production of molasses in Bangladesh. *Bulletine of Instituite Tropical Agriculture, Kyushu University*, 11, 1–13.

Micoletzky, H. (1922) Die freilebenden Erd-Nematoden mit besonderer Berucksichtigung der Steiermark und der Bukowina, zugleich mit einer Revision samtlicher nicht mariner, freilebender Nematoden in Form von Genus - Beschreibunger und Bestimmungsschlusseln. *Archiv fur Naturgeschichte*, Berlin 87 (1921).
Midha, S.K. and Swarup, G. (1972) Factors affecting development of earcockle and tundu diseases of wheat. *Indian Journal of Nematology* 2, 97–104.
Midha, S.K. and Swarup, G. (1974) Studies on the wheat seed gall nematode caused by *Anguina tritici*. *Indian Journal of Nematology* 4, 53–63.
Miller, L.I. (1986) Economic importance of cyst nematodes in North America. In: Lamberti, F. and Taylor, C.E. (eds) *Cyst Nematodes*. Plenum Press, New York, pp. 373–385.
Miller, L.I. and Gray, B.J. (1968) Horsenettle cyst nematode, *Heterodera virginiae* n. sp., a parasite of solanaceous plants. *Nematologica* 14, 535–543.
Miller, L.I. and Gray, B.J. (1972) *Heterodera solanacearum* n. sp., a parasite of solanaceous plants. *Nematologica* 18, 404–413.
Milne, D.L. (1982a) Nematode pests of citrus. In: Keetch, D.P. and Heyns, J. (eds) *Nematology in Southern Africa*. Bulletin No. 400, Department of Agriculture and Fisheries, Republic of South Africa, pp. 12–18.
Milne, D.L. (1982b) Nematode pests of litchi. In: Keetch, D.P. and Heyns, J. (eds) *Nematology in Southern Africa*. Bulletin No. 400, Department of Agriculture and Fisheries, Republic of South Africa, pp. 38–41.
Milne, D.L., de Villiers, E.A. and Holtzhausen, L.C. (1971) Litchi tree decline caused by nematodes. *Phytophylactica* 3(1), 37–44.
Milne, D.L., Appleton, M.R. and Holtzhausen, L.C. (1975) The anatomical reaction of Litchi chinensis roots attacked by three nematode species and a soil mite. *Phytophylactica* 7(1), 15–20.
Milne, D.L., de Villiers, E.A. and Smith, B.L. (1977) Response of nematode populations and pineapple yields to foliar application of systemic nematicides. *Citrus and Subtropical Fruit Journal* 525, 12–14.
Minton, N.A. and Baujard, P. (1990) Nematode parasites of peanut. In: Luc, M., Sikora, R.A. and Bridge, J. (eds) *Plant Parasitic Nematodes in Subtropical and Tropical Agriculture*. CAB International, Wallingford, UK, pp. 285–320.
Minton, N.A. and Bell, D.K. (1969) *Criconemoides ornatus* parasitic on peanuts. *Journal of Nematology* 1, 349–351.
Minz, G., Ziv, D. and Strich-Harari, D. (1960) Decline of banana plantations caused by spiral nematodes in the Jordan Valley and its control by DBCP. *Ktavim* 10, 147–157.
Mishra, S.D. and Prasad, S.K. (1990) Chemotherapeutic management of *Heterodera cajani* in infesting pigeonpea. In: *Environment Concern and Tissue Injury, Environmental Series* 4, 59–65.
Mizukubo, T. and Adachi, H. (1997) Effect of temperature on *Pratylenchus penetrans* development. *Journal of Nematology* 29, 306–314.
Mizukubo, T. and Minagawa, N. (1991) Morphometric differentiation of *Pratylenchus neglectus* (Rensch, 1924) and *P. gotohi* n. sp. (Nematoda: Pratylenchidae). *Japanese Journal of Nematology* 21, 26–42.
Moens, M., Perry, R.N. and Starr, J.L. (2009) *Meloidogyne* species – a diverse group of novel and important plant parasites. In: Perry, R.N., Moens, M. and Starr, J.L. (eds) *Root-knot Nematodes*. CAB International, Wallingford, UK, pp. 1–17.
Moens, T., Araya, M., Swennen, R. and De Waele, D. (2006) Reproduction and pathogenecity of *Helicotylenchus multicinctus*, *Meloidogyne incognita* and *Pratylenchus coffeae*, and their interaction with *Radopholus similis* on *Musa*. *Nematologica* 8(1), 45–58.
Mohammad, H.Y. (1988) The relationship between the biology of *Meloidogyne graminicola* and *Heterodera sacchari* and nematocide treatments. *Index to Theses Accepted for Higher Degrees in the Universities of Great Britain and Ireland* 37(3), 1236.
Mohammad, H.Y. and Hague, N.G.M.T.I. (1987) The effect of nematicide on root invasion by *Meloidogyne graminicola* and *Heterodera sacchari*. *Nematologica* 33, 244–247.
Mohandas, C., Ramana, K.V. and Raski, D.J. (1985) *Trophotylenchulus piperis* n. sp. parasitic on *Piper nigrum* L. in Kerala, India (Nemata: Tylenchulidae). *Revue de Nématologie* 8(2), 97–102.
Mohilal, M.M and Khan, M.R. (2012) Nematode infestation in pineapple. In: Khan, M.R. and Jairajpuri, M.S. (eds) *Nematode Infestations – Part III: Horticultural Crops*. National Academy of Sciences, Allahabad (Prayagraj), UP, India, pp. 288–321.
Mohilal, M.M. and Pramodini, M. (2012) Spiral nematode, *Helicotylenchus multicinctus* infestation in banana. In: Khan, M.R. and Jairajpuri, M.S. (eds) *Nematode Infestations – Part III: Horticultural Crops*. National Academy of Sciences, Allahabad (Prayagraj), UP, India, pp. 322–356.
Mojtahedi, H., Sturhan, D., Akhiani, A. and Barooti, S. (1980) *Xiphinema* species in Iranian vineyards. *Nematologia Mediterranea* 8(2), 165–170.
Mojumder, V. and Mishra, S.D. (2001) Seed coating of pigeonpea with neem formulations for the management of *Heterodera cajani* and other phytoparasitic nematodes. *Current Nematology* 12, 7–10.
Mokabli, A., Valette, S., Gauthier, J.P. and Rivoal, R. (2001) Influence of temperature on the hatch of *Heterodera avenae* Woll. populations from Algeria. *Nematologica* 3(2), 171–178.
Mokabli, A., Valette, S., Gauthier, J. P. and Rivoal, R. (2002) Variation in virulence of cereal cyst nematode populations from North Africa and Asia. *Nematology* 4, 521–525.
Mokrini, F., Waeyenberge, L., Viaene, N., Andaloussi, F.A. and Moens, M. (2013) Quantitative detection of the root-lesion nematode, *Pratylenchus penetrans*, using qPCR. *European Journal of Plant Pathology* 137, 403–413.
Mokrini, F., Waeyenberge, L., Viaene, N., Andaloussi, F.A. and Moens, M. (2014) The β-1, 4-endoglucanase gene is suitable for the molecular quantification of the root-lesion nematode, *Pratylenchus thornei*. *Nematology* 16, 789–796.
Mondal, A.H. and Miah, S.A. (1987a) *Ufra* problem in low lying areas of Bangladesh. *Internation Rice Research Newsletter* 12, 29–30.
Mondal, A.H. and Miah, S.A. (1987b) Control of *ufra* of rice by seedling treatment. *Bangladesh Journal of Plant Pathology* 3, 25–30.
Mondal, A.H., Rahman, L. and Miah, S.A. (1989) Yield loss due to transplanting of *ufra* infected seedlings of rice. *Bangladesh Journal of Plant Botany* 18, 67–72.
Monroy, F., van der Putten, .W.H., Yergeau, E., Mortimer, S.R., Duyts H. and Bezemer, T.M. (2012) Community patterns of soil bacteria and nematodes in relation to geographic distance. *Soil Biology and Biochemistry* 45, 1–7.
Montasser, S.A., Sweelam, M.E. and El-Mesalamy, A.F. (2008) Occurrence and distribution of *Aphelenchoides arachidis* on groundnut pods in Egypt. *Pakistan Journal of Nematology* 26, 151–152.
Montes, M.J., Andres, M.F., Sin, E., Lopez-Brana, I., Martin-Sanchez, J.A., Romero, M.D. and Delibes, A. (2008) Cereal cyst nematode resistance conferred by the Cre7 gene from *Aegilopes triuncialis* and its relationships with Cre genes from Australian wheat cultivars. *Genome* 51, 315–319.
Mor, M. and Sturhan, D. (2000) On the identity of *Heterodera latipons* Franklin. *Nematology* 2, 776–777.
Moraes, M.V. de (1977) *Systemics for Nematode Control in Coffee Nurseries*. Paper presented at II Reunion de Nematologia, Piracicaba, Brazil. Sociedade Brasileira de Nematologia, Publication no. 2, 245–248.
Mortimer, J.J., Bridge, J. and Jackson, G.V.H. (1981) *Hirschmanniella* sp. an endoparasitic nematode associated with miti-miti disease of taro corms in the Solomon Islands. *FAO Plant Protection Bulletin* 29(1/2), 9–11.

Mota, M.M. and Eisenback, J.D. (1993a) Morphology of females and cysts of *Globodera tabacum tabacum*, *G. t. virginiae*, and *G. t. solanacearum* (Nemata: Heteroderinae). *Journal of Nematology* 25, 136–147.

Mota, M.M. and Eisenback, J.D. (1993b) Morphology of second stage juveniles and males of *Globodera tabacum tabacum*, *G. t. virginiae*, and *G. t. solanacearum* (Nemata: Heteroderinae). *Journal of Nematology* 25(1), 27–33.

Mota, M.M. and Eisenback, J.D. (1993c) Morphometrics of *Globodera tabacum tabacum*, *G. virginiae* and *G. t. solanacearum* (Nemata: Heteroderinae). *Journal of Nematology* 25(2), 148–160.

Mota, M. and Vieira, P. (eds) (2004) The pinewood nematode *Bursaphelenchus xylophilus*. Proceedings of an international workshop, University of Évora, Portugal, 20–22 August 2001. Vol. 1 in: Hunt, D.J. and Perry, R.N. (series eds) *Nematological Monographs and Perspectives*. Brill, Leiden, Netherlands.

Mota, M., Braasch, H., Bravo, M.A., Penas, A.C., Burgermeister, W., Metge, K. and Sousa, E. (1999) First record of *Bursaphelenchus xylophilus* in Portugal and in Europe. *Nematology* 1, 727–734.

Moura, J.I.L., Vilela, E.F., Brasil, G.H. and Cangucu, R. (2000) Mass trapping of *Rhynchophorus palmarum* using pheromone in coconut plantation in Brazil. XXI Int. *Congrational Entomology*, Abstr. 610, Abstr. Book 1, August. Foz do Iguassu, Brazil, p. 154.

Moura, R.M., Pedrosa, E.M.R. and Prado, M.D.C. (2002) Incidence of *Pratylenchus coffeae* causing severe infection on coffee in the Northeast of Brazil. *Fitopatologia Brasileira* 27(6), 649–651

Moura, R.M. de. and Almeida, A.V. de. (1982) Preliminary studies on the occurrence of phytonematodes associated with sugar-cane in areas of low productivity in Pernambuco State. *Trabalhos apresentados a V Reuniao Brasileira de Nematologia*, 9–13 fevereiro 1981, Londrina, PR, Brasil. Publicacao No. 5, 1981, 213–220.

Mueller, J.D. and Sanders, G.B. (1987) Control of *Hoplolaimus columbus* on late planted soybean with aldicarb. *Journal of Nematology* () 19, 123–126.

Mugniery, D. (1978) Development rates of *Globodera rostochiensis* and *G. pallida*, (Nematoda: Heteroderidae) in relation to temperature. *Revue de Nematologie* 1(1), 3–12.

Mugniery, D., Bossis, M. and Pierre, J.S. (1992) Hybridization between *Globodera rostochiensis* (Wollenweber), *G. pallida* (Stone), *G. virginiae* (Miller and Gray), *G. solanacearum* (Miller and Gray) and *Globodera* "mexicana" (Campos-Vela). Description and future of the hybrids. *Fundamental and Applied Nematology* 15(4), 375–382.

Mukherjee, B. and Dasgupta, M.K. (1982) Community analyses of plant-parasitic nematodes in tea plantations of West Bengal, India. *Nematologia Mediterranea* 10(1), 1–7.

Mulk, M.M. (1976) *Meloidogyne graminicola*. *CIH Descriptions of Plant-parasitic Nematodes*, Set 6, No. 87. Commonwealth Institute of Helminthology, St Albans, UK.

Muller, J., Sturhan, D., Rumpenhorst, H.J., Braasch, H. and Unger, J.G. (1996) On the occurrence of a root-knot nematode (*Meloidogyne chitwoodi*) new to Germany. *Nachrichtenblatt des Deutschen Pflanzenschutzdienstes* 48(6), 126–131.

Mulvey, R.H. (1972) Identification of *Heterodera* cysts by terminal and cone top structures. *Canadian Journal of Zoology* 50, 1277–1292.

Mulvey, R.H. and Stone, A.R. (1976) Description of *Punctodera matadorensis* n. gen., n. sp. (Nematoda: Heteroderidae) from Saskatchewan with lists of species and generic diagnoses of Globodera (n. rank), *Heterodera*, and *Sarisodera*. *Canadian Journal of Zoology* 54(5), 772–785.

Munawar, M., Powers, T.O., Tian, Z., Harris, T., Higgins, R. and Zheng, J. (2018) Description and distribution of three Criconematid nematodes from Hangzhou, Zhejiang province, China. *Journal of Nematology* 50(2). doi: 10.21307/jofnem-2018-010.

Mundo-Ocampo, M., Baldwin, J.G. and Jeronimo, R.J. (1987) Distribution, morphological variation and host range of *Punctodera chalcoensis* (abstract). *Journal of Nematology* 19, 545–546.

Muniz, M.S., Campos, V.P., Castagnone-Sereno, P., Castro, J., Almeida, M.A. and Carneiro, R.M.D.G. (2008) Diversity of *Meloidogyne exigua* (Tylenchida: Meloidogynidae) populations from coffee and rubber tree. *Nematology* 10(6), 897–910.

Muñoz, R.C. and Sosa-Moss, C. (1983) Ciclo de vida de *Punctodera chalcoensis* Stone, Sosa Moss y Mulvey, 1976. (Nematoda: Heteroderidae). Thesis. Instituto Politecnico Nacional, Escuela Nacional de Ciencias Biologicas. Mexico.

Muraleedharan, N. (1992) Bioecology and management of tea pests in Southern India. *Journal of Plantation Crops* 20, 1–21.

Muthukrishnan, T.S., Rajendrawn, G., Ramamurthy, V.V. and Chandrashekaran, J. (1979). Pathogenecity and control of *Hirschmanniella oryzae*. *Indian Journal Nematology* 7, 8–16.

Mwamula, A.O., Kim, Y.H., Lee, H.W., Bae, E., Kim, Y.H. and Lee, D.W. (2020) Taxonomic notes on three *Tylenchorhynchus* spp. (Nematoda, Telotylenchidae) associated with turfgrass in Korea. *European Journal of Plant Pathology* 156, 1101–1117.

Nagamine, K., Hase, T. and Notomi, T. (2002) Accelerated reaction by loop-mediated isothermal amplification using loop primers. *Molecular and Cellular Probes* 16, 223–229.

Nair, K.K.R. (1979) Studies on the chemical control of banana nematodes. *Agriculture Research Journal of Kerala* 17, 232–235.

Nakasono, K. and Ichinohe, M. (1961) *Hemicriconemoides kanayaensis* n. sp. associated with tea root in Japan (Nematoda: Criconematidae). *Japanese Journal of Applied Entomology and Zoology* 5, 273–276.

Nandakumar, C. and Khera, S. (1974) *In vitro* studies of the feeding habits of *Pratylenchus mulchandi* and *Hoplolaimus indicus* on pearl millet roots. *Indian Journal of Nematology* 3, 138–142.

Nandakumar, C., Prasad, J.S., Rao, Y.S. and Rao J. (1975) Investigation on the white tip nematode (*Aphelenchoides besseyi* Christie 1942) of rice (*Oryza sativa* L.). *Indian Journal of Nematology* 5, 62–69.

Nandal, S.N., Paruthi, I.J. and Khan, M.R. (2010) Seed gall nematode infestation in wheat. In: Khan, M.R. and Jairajpuri, M.S. (eds) *Nematode Infestation – Part I: Food Crops*. National Academy of Sciences, Allahabad (Prayagraj), UP, India, pp. 135–150.

NAPPO (2019) North American Plant Protection Organization. Available at: http://www.nappo.org

Narbaev, Z.N. (1973) Nematodes of cacti in Tashkent, USSR. *Materialy Nauchnykh Konfersentii Vsesoyuznogo Obshchestva Gel'mintologov, 1971–1972* 25, 164–167.

Nath, R.C., Mukherjee, B., Dasgupta, M.K. and Siddiqi, M.R. (1997) Density, diversity and community structure of plant parasitic nematodes in pineapple plantations of Tripura, India. *International Journal of Nematology* 7, 51–61.

Nath, R.P., Akhter, S.W., Haider, M.G. and Bhide, V.S. (1975) Effect of *Hoplolaimus indicus* Sher, 1963 on sugarcane seedlings. *Indian Journal of Nematology* 4, 241–242.

NEBN-IAS, (2019) North European and Baltic Network on Invasive Alien Species. Available at http://www.nobanis.org/default.asp

Needham, T. (1743) A letter concerning certain chalky tubulous concretions called malm; with some microscopiea observations on the farina of red lily and of worms discovered in smutty corn. *Philosophical Transactions of the Royal Society of London* 42, 173, 174, 634–641.

Neher, D.A. (2010) Ecology of plant and free-living nematodes in natural and agricultural soil. *Annual Review of Phytopathology* 48, 371–394.

Nesmith, W.C., Zehr, E.I. and Dowler, W.M. (1981) Association of *Macroposthonia xenoplax* and *Scutellonema brachyurum* with the peach tree short life syndrome. *Journal of Nematology* 13(2), 220–225.

Netscher, C. (1970) Les nématodes parasites des cultures maraichFres au Sénégal. *Cahiers ORSTOM Série Biologie* 11, 209–229.

Netscher, C. and Sikora, R.A. (1990) Nematode parasites of vegetables. In: Luc, M., Sikora, R.A. and Bridge, J. (eds) *Plant Parasitic Nematodes in Subtropical and Tropical Agriculture*. CAB International, Wallingford, UK, pp. 237–283.

Nickle, W.R. (1970) A taxonomic review of the genera of the Aphelenchoidae (Fuchs, 1937) Thorne, 1949 (Nematoda: Tylenchida). *Journal of Nematology* 2(4), 375–392.

Nickle, W.R. (ed.) (1992) *Manual of Agricultural Nematology*. Marcel Dekker, New York, pp. 275–362.

Nickle, W.R., Golden, A.M., Mamiya, Y. and Wergin, W.P. (1981) On the taxonomy and morphology of the pinewood nematode, *Bursaphelenchus xylophilus* (Steiner and Buhrer, 1934) Nickle (1970). *Journal of Nematology* 13, 385–392.

Nicol, J.M. (2002) Important nematode pests. In: Anonymous (ed.) *Bread Wheat: Improvement and Production*. FAO, Rome, pp. 345–366

Nicol, J.M. and Rivol, R. (2008) Global knowledge and its application for the integrated control and management of nematodes on wheat. In: Ciancio, A. and Mukerji, K.G. (eds) *Integrated Management and Biocontrol of Vegetable and Grain Crops Nematodes*. Springer, Dordrecht, Netherlands, pp. 243–287.

Nicol, J.M., Turner, S.J., Coyne, D.L., den Nijs, L., Hockland, S. and Tahna, M.Z. (2011) Current nematode threats to world agriculture. In: Jones, J., Gheysen, G. and Fenoll, C. (eds) *Genomics and Molecular Genetics of Plant–Nematode Interactions*. Springer, Heidelberg, Germany, pp. 21–43.

Nischwitz, C., Dhiman, C., Schmitt, M. and McClure, M. (2013) Cyst nematodes in golf course greens in the western United States. *Annual Conference of the American Phytopathological Society* (Poster).

Nobbs, J.M., Ibrahim, S.K. and Rowe, J. (1992) A morphological and biochemical comparison of the four cyst nematodes species, *Heterodera elachista*, *H. oryzicola*, *H. oryzae*, and *H. sacchari* (Nematoda: Heteroderidae) known to attack rice. *Fundamental and Applied Nematology* 15, 551–562.

Nobbs, J.M., Liu, Q., Hartley, D., Handoo, Z., Williamson, V.M., Taylor, S., Walker, G. and Curran, J. (2001) First record of *Meloidogyne fallax* in Australia. *Australian Plant Pathology* 30(4), 373.

Noe, J.P. (1993) Damage functions and population-changes of *Hoplolaimus columbus* on cotton and soybean. *Journal of Nematology* 25, 440–445.

Nong, I. and Weber, G.F. (1964) *Amaryllis* diseases caused by two nematodes. *Phytopathology* 902–903.

Norton, D.C. (1967) *Xiphinema americanum* as a factor in unthriftiness of red clover. *Phytopathology* 57(12), 1390–1391.

Norton, D.C. (1984) Nematode parasites of corn. In: Nickle, W.R. (ed.) *Plant and Insect Nematodes*. Marcel Dekker, New York, pp. 61–94.

Norton, D.C. (Chmn) *et al.* (1984) *Distribution of Plant-Parasitic Nematode Species in North America*. Nematode Geographical Distribution Committee, Society of Nematologists, Mesilla Park, New Mexico.

Nose, M. and Shiraishi, S. (2007) Breeding for resistance to pine wilt disease. In: Futai, K. and Sutherland, J. (eds) *Pine Wilt Disease*. Springer, Tokyo, pp. 334–350.

Notomi, T., Okayama, H., Masubuchi, H., Yonekawa, T., Watanabe, K., Amino, N. and Hase, T. (2000) Loop-mediated isothermal amplification of DNA. *Nucleic Acids Research* 28, e63.

Nunn, G.B. (1992) Nematode Molecular Evolution. PhD thesis, University of Nottingham, Nottingham, UK.

Nwauzor, E.C. and Fawole, B. (1982) Root-knot nematodes on yams in eastern Nigeria. *Proceedings of the 3rd Research Planning Conference on root-knot nematodes, Meloidogyne* spp., Regions IV and V, 16–20 November 1981, Ibadan, Nigeria. (International *Meloidogyne* Project). International Institute of Tropical Agriculture, Ibadania, Niger, pp. 161–167.

Nyczepir, A.P. (1989) *Peach tree short life: a nematode associated disease*. Nematology Circular No. 163. Florida Department of Agriculture and Consumer Services, Gainesville, Florida.

O'Bannon, J.H. and Esser, R.P. (1970) Control of *Heterodera cacti* infecting *Zygocactus truncatus*. *Plant Disease Reporter* 54, 692–694.

Odihirin, R.A. (1977) Irrigation water as a means for the dissemination of the sugar cane cyst nematode *Heterodera sacchari* at Bacita Sugar Estate (abstract only). *Nigerian Society for Plant Protection* 2, 58.

Ohkawa, K. and Saigusa, T. (1972) Studies on the nematode damage caused by continuous cropping of *Lilium speciosum rubrum*. *Bulletin of the Kanagawa Horticultural Experiment Station* 20, 103–107.

Ohshima, Y. (1974) *Heterodera elachista* n. sp., an upland rice cyst nematode from Japan *Japanese Journal of Nematology* 4, 51–56.

Ohshima, Y, Momota, Y. and Shimizu, K. (1981) Additional information of some morphological characteristics of *Heterodera glycines* Ichinohe. *Japanese Journal of Nematology* 10 52-53

Okimoto, R., Chamberlin, H.M., Macfarlane, J.L. and Wolstenholme, D.R. (1991) Repeated sequence sets in mitochondrial DNA molecules of root knot nematodes (*Meloidogyne*): nucleotide sequences genome location and potential for host-race identification. *Nucleic Acids Research* 19, 1619–1626.

Oliveira, C.M.G., Monteiro, A.R., Antedom, O.S.R. and Inomoto, M.M. (1999) Host reaction of *Coffea* spp. to *Pratylenchus brachyurus*. *Nematropica* 29(2), 241–244.

Oliveira, C.M.G., Blok, V., Neilson, R., Mróz, T. and Roberts, D. (2017) Hydrolysis probe-based PCR for detection of *Pratylenchus crenatus*, *P. neglectus* and *P. penetrans*. *Nematology* 19, 81–91.

Olowe, T. and Corbett, D.C.M. (1976) Aspects of the biology of *Pratylenchus brachyurus* and *P. zeae*. *Nematologica* 22(2), 202–211.

Olson, M., Harris, T., Higgins, R., Mullin, P., Powers, K., Olson, S. and Powers, T.O. (2017) Species delimitation and description of *Mesocriconema nebraskense* n. sp. (Nematoda: Criconematidae), a morphologically cryptic, parthenogenetic species from North American Grasslands. *Journal of Nematology* 49, 42–66.

Olthof, T.H.A. (1986) Reaction of six *Solanum tuberosum* cultivars to *Pratylenchus penetrans*. *Journal of Nematology* 18, 54–58.

Olthof, T.H.A. and Potter, J.W. (1973) The relationship between population densities of *Pratylenchus penetrans* and crop losses in summer-maturing vegetables in Ontario. *Phytopathology* 63, 577–582.

Olthof, T.H.A. and Wolynetz, M.S. (1991) *Pratylenchus penetrans* and *P. neglectus* in tubers of potato (*Solanum tuberosum*) in Ontario. *Canadian Journal of Plant Science* 71, 1251–1256.

Olthof, T.H.A., Marks, C.F. and Elliot, J.M. (1973) Relationship between population densities of *Pratylenchus penetrans* and crop losses in flue-cured tobacco in Ontario. *Journal of Nematology* 5(2), 158–162.

Olthof, T.H.A., Potter, J.W. and Peterson, E.A. (1974) Relationship between population densities of *Heterodera schachtii* and losses in vegetable crops in Ontario. *Phytopathology* 64, 549–554.

Oostenbrink, M. (1963) Symptomen in aaltiespopulaties in 'red ring' – ziake cocos-enolie-palmen. *Mededelingen van de Land bouwhoge school en de Opzoekingsstations van de staat te Gent* 28, 663–671.

Orbin, D.P. (1973) Histopathology of soybean roots infected with *Helicotylenchus dihystera*. *Journal of Nematology* 5, 37–40.

Orbin, D.P. and Cairns, E.J. (1971) Histopathology of soyabean roots infected with *Helicotylenchus dihystera* (Cobb) Sher. *Journal of Nematology* 3(4), 322–323.

Orffer, C.J. (1977) Hot-water treatment of vine propagating material-urgently necessary in RSA. *Deciduous Fruit Grower* 27(7), 224–225, 228–231.

Orlando, V., Chitambar, J.J., Dong, K., Chizhov, V.N., Mollov, D., Bert, W. and Subbotin. S.A. (2016) Molecular and morphological characterisation of *Xiphinema americanum*-group species (Nematoda: Dorylaimida) from California, USA, and other regions, and co-evolution of bacteria from the genus *Candidatus Xiphinematobacter* with nematodes. *Nematology* 18, 1015–1043.

Orlando, V., Grove, I.G., Edwards, S.G., Prior, T., Roberts, D., Neilson, R. and Back, M. (2020) Root-lesion nematodes of potato: current status of diagnostics, pathogenicity and management. *Plant Pathology* 69(3), 405–417. doi: 10.1111/ppa.13144.

Orton Williams, K.J. (1972) *Meloidogyne javanica*. *CIH Descriptions of Plant-parasitic Nematodes*, Set 1, No. 3. Commonwealth Institute of Helminthology, St Albans, UK.

Orton Williams, K.J. (1973a) *Macroposthonia sphaerocephala*. *CIH Descriptions of Plant-parasitic Nematodes*, Set 2, No. 28. Commonwealth Institute of Helminthology, St Albans, UK.

Orton Williams, K.J. (1973b) *Hoplolaimus galeatus*. *CIH Descriptions of Plant-parasitic Nematodes*, Set 2, No. 24. Commonwealth Institute of Helminthology, St Albans, UK.

Orton Williams, K.J. (1974) *Meloidogyne hapla*. In: *CIH Descriptions of Plant-parasitic Nematodes*, Set 3, No. 31. Commonwealth Agricultural Bureaux, Farnham Royal, UK.

Orton Williams, K.J. (1975) *Meloidogyne arenaria*. In: *CIH Descriptions of Plant-parasitic Nematodes*, Set 5, No. 62. Commonwealth Agricultural Bureaux, Farnham Royal, UK.

Orton Williams, K.J. (1980) Plant parasitic nematodes of the Pacific. *Technical Report, Vol. 8, UNDP/FAO* - SPEC Survey of Agricultural Pests and Diseases in the South Pacific. CAB International, Wallingford, UK.

Orton Williams, K.J. and Siddiqi, M.R. (1973) *Radopholus similis*. *CIH Descriptions of Plant-parasitic Nematodes*, Set 2, No. 27. Commonwealth Institute of Helminthology, St Albans, UK.

Otazu, V.R., Hoopes, G., Caero. and Huayta, I. (1985) El rosario de la papa causado po *Nacobbus aberrans* (Thorne, 1935) Thorne and Allen 1944. Su efecto en el rendimiento y algunos aspectos que inciden en su propación y prevalencia en Bolivia. *Fitopatología* 20, 65–70.

Ou, S., Peng, D., Liu, X., Li, Y. and Moens, M. (2008) Identification of *Heterodera glycines* using PCR with sequence characterized amplified region (SCAR) primers. *Nematology* 10, 397–403.

Ou, S.H. (1985) *Rice Diseases*. 2nd edn. Commonwealth Mycological Institute, Kew, UK.

Overman, A.J. (1964) The effect of temperature and flooding on nematode survival in fallow sandy soil. *Proceedings Soil Crop Science Society of Florida* 24, 141–149.

Overman, A.J., Howard, C.M. and Albregts, E.E. (1987) Soil solarization for strawberries. *Proceedings of the Florida State Horticultural Society* 100, 236–239.

Ozbayrak, M. (2019) DNA barcoding of *Pratylenchus* from agroecosystems in the northern great plains of North America. *Theses, Dissertations, and Student Research in Agronomy and Horticulture* 163. Available at: https://digitalcommons.unl.edu/agronhortdiss/163.

Özturk, L., Behmand, T., Sin, B., Avci, G.G. and Elekcioğlu, I.H. (2018) Morphologic and molecular identification of *Xiphinema americanum* associated with pine trees. *International. Journal of Molecular Biology* 3(3), 94–96.doiI: 10.15406/ijmboa.2018.03.00058.

P.K.B. (2016) *Plantwise Technical Fact Sheet*: *Wheat Seed Gall Nematode*. Plantwise Knowledge Bank. Available at: http://www.plantwise.org/KnowledgeBank/Datasheet.aspx?dsid=5388

Padgham, J.L., Mazid, M.A., Duxbury, J.M., Abawi, G.S. and Hossain, M. (2004) Yield loss caused by *Meloidogyne graminicola* on lowland rainfed rice in Bangladesh. *Journal of Nematology* 36, 42–48.

Page, S.L.J. (1983) Biological studies of the cotton root nematode *Meloidogyne acronea*. PhD Thesis. London University, London.

Page, S.L.J. (1985) *Meloidogyne acronea*. In: *CIH Descriptions of Plant-parasitic Nematodes*, Set 8, No. 114. Commonwealth Agricultural Bureaux, Farnham Royal, UK.

Page, S.L.J. and Bridge, J. (1985) Observations on *Pasteuria penetrans* as a parasite of *Meloidogyne acronea*. *Nematologica* 31(2), 238–240.

Palazova, G. and Baicheva, O. (2002) Electrophoretic studies of *Ditylenchus dipsaci* (Kuhn, 1857) Filipjev, 1936 from two hosts: *Allium sativum* and *Allium cepa*. *Experimental Pathology and Parasitology* 5, 39–40.

Palmisano, A. and Ambrogioni, L. (2000) *Meloidogyne ulmi* sp. n., a root– knot nematode from elm. *Nematologia Mediterranea* 28, 279–293.

Palomares-Rius, J.E., Cantalapiedra-Navarrete, C., Archidona-Yuste, A., Subbotin, S.A. and Castillo, P. (2017) The utility of mtDNA and rDNA for barcoding and phylogeny of plant-parasitic nematodes from Longidoridae (Nematoda, Enoplea). *Scientific Reports* 7, 10905. doi: 10.1038/s41598-017-11085-4.

Palomares-Rius, J.E., Cantalapiedra-Navarrete, C., Archidona-Yuste, A., Tzortzakakis, E.A., Birmpilis, I.G., Vovlas, N., Subbotin, S.A. and Castillo, P. (2018) Prevalence and molecular diversity of reniform nematodes of the genus *Rotylenchulus* (Nematoda: Rotylenchulinae) in the Mediterranean Basin. *European Journal of Plant Pathology* 150, 439–455.

Pan, C. (1985) Studies on plant parasitic nematodes on economically important crops. III. Description of *Meloidogyne fujianensis* n. sp. (Nematoda: Meloidogynidae) infesting citrus in Nanjing county. *Acta Zoologica Sinica* 3, 263–268.

Pan, C.S., Ling, J. and Wang, S.Y. (1988) Studies on parasitic nematodes in crops of economical importance in Fujian. IV. *Meloidogyne fujianensis*. *Acta Zoologica Sinica* 34, 305–309.

Panda, M. and Seshadri, A.R. (1979) Efficacy of few nematicides in controlling the infestations of reniform nematodes in some pulse crops. Nematology symposium, Bhubhaneshwar, Orissa, March 1979. *Indian Journal of Nematology* 9(1), 84–85.

Paneque, M. and Sampedro, J. (1986) Report of the cyst nematode *Cactodera cacti* in the Holguin Province, Cuba. *Ciencia y Técnica en la Agricultura, Protección de Plantas* 9(4), 123–129.

Papova, M.B., Zelenskii, G.L and Subbotin, S.A. (1994) An assessment of resistance in cultivars of *Oryza sativa* L. to *Aphelenchoides besseyi* Christie, 1942. *Russian Journal of Nematology* 2(1), 41–44.

Parihar, A. and Siddiqi, A.U. (2010) Nematode infestation in maize. In: Khan, M.R. and Jairajpuri, M.S. (eds) *Nematode Infestation – Part I: Food Crops*. National Academy of Sciences, Allahabad (Prayagraj), UP, India, pp. 171–190.

Park, B.Y., Park, S.N., Lee, J.K. and Bae, C.H. (2009) Morphometric and genetic variability among *Tylenchulus semipenetrans* populations from citrus growing area in Korea. *Plant Pathology Journal* 25, 236–240.

Paruthi, I.J. and Bhatti, D.S. (1992) Efficacy of different practices for the control of earcockle in wheat caused by *Anguina tritici*. *Current Nematology* 3, 1–4.

Paruthi, I.J., Singh, M. and Gupta, D.C. (1987) Quantitative and qualitiative losses in wheat (*Triticum aestivum*) grains due to earcockle and 'tundu'. *Seed Research* 15, 83–86.

Paruthi, I.J., Bajaj, H.K. and Bhatti, D.S. (1990) Studies on some factors affecting the emergence of *Anguina tritici* larvae from wheat seed galls and their ascendance on to seedlings. *Indian Journal of Ecology* 17(1), 48–50.

Paruthi, I.J., Nandal, S.N. and Kumar, A. (2009) Wheat seed gall nematode (*Anguina tritici*), its status and management. In: Sharma, G.L. (ed.) *Phytonematode Management in Field Crops*. Oxford Book Co., Jaipur, India, pp. 1–20.

Parvatha, R.P., Khan, R.M. and Rao, M.S. (1991) Integrated management of the citrus nematode *Tylenchulus semipenetrans* using oil-cakes and *Paecilomyces lilacinus*. *Afro-Asian Journal of Nematology* 1, 221–222.

Patel, D.J., Desai, M.V. and Patel, D.B. (1984) Life cycle of reniform nematode, *Rotylenchulus reniformis*, Linford & Oliviera, 1940 in tobacco. *Tobacco Research* 10, 162–169.

Patel, D.J., Patel, M.B., Patel, B.A., Patel, N.B. and Patel, R.R. (1996) Nematode problems of cotton and their management in India. In: *Proceedings National Seminar on Centenary of Cotton in India*. Gujarat Agricultural University, Surat. pp. 40–51.

Pathak, D.K. (1992) Some studies on the rice stem nematode, *Ditylenchus angustus* (Butler, 1913) Filipjev, 1936 of deep water rice of Assam. AnMSc(Agri) thesis. Assam Agricultural University, Jorhat, Assam.

Paxton, J.D. (1975) Phytoalexins, phenolics, and other antibiotics in roots resistant to soil-borne fungi. In: Bruehl G.W. (ed.) *Biology and Control of Soil-borne Plant Pathogens*. American Phytopathological Society, St. Paul, Minnesota, pp. 185–192.

Payan, L.A. and Dickson, D.W. (1990) Comparison of populations of *Pratylenchus brachyurus* based on isozyme phenotypes. *Journal of Nematology* 22, 538–545.

Peacock, C.H. and Dunn, R.A. (1986) Effects of nematicide formulations on turfgrass nematodes. *Proceedings, Soil and Crop Science Society of Florida* 45, 185–188.

Pedersen, J.F. and Rodriguez-Kábana, R. (1991) Winter grass cover crop effects on nematodes and yields of double cropped soybean. *Plant and Soil*, 131(2), 287–291.

PExD, (2019) PExD – Phytosanitary Export Database Harmful Organism Reports. *Hirschmanniella*. Available at: https://pcit.aphis.usda.gov/PExD/faces/ViewPExD.jsp (accessed 21 December 2019)

Peetz, A.B. and Zasada, I.A. (2016) Species-specific diagnostics using a β-1,4-endoglucanase gene for *Pratylenchus* spp. occurring in the Pacific Northwest of North America. *Nematology* 18, 1–11.

Peñas, A.C., Metge, K., Mota, M. and Valadas, V. (2006) *Bursaphelenchus antoniae* sp. n. (Nematoda: Parasitaphelenchidae) associated with *Hylobius* sp. from *Pinus pinaster* in Portugal. *Nematology* 8, 659–669.

Peñas, A.C., Bravo, M.A., Valadas, V. and Mota, M. (2008) Detailed morphometric studies of *Bursaphelenchus xylophilus* and characterisation of other *Bursaphelenchus* species (Nematoda: Parasitaphelenchidae) associated with *Pinus pinaster* in Portugal. *Journal of Nematode Morphology and Systematics* 10(2), 137–163.

Peng, D., Gaur, H.S. and Bridge, J. (2018) Nematode parasites of rice. In: Sikora, R.A., Coyne, D., Hallmann, J., and Timper, P. (eds) *Plant Parasitic Nematodes in Subtropical and Tropical Agriculture*, 3rd edn. CABI Publishing, Wallingford, UK, pp. 120–163.

Peregrine, W.T.H. and Bridge, J. (1992) The lesion nematode *Pratylenchus goodeyi*, an important pest of Ensete in Ethiopia. *Tropical Pest Management* 38(3), 325–326.

Perez, E.E., Weingartner, D.P. and McSorley, R. (2000) Niche distribution of *Paratrichodorus minor* and *Belonolaimus longicaudatus* following fumigation on potato and cabbage. *Journal of Nematology* 32, 343–348.

Perry, R.N. (1995) Rice stem nematode, *Ditylenchus angustus* development and survival. *International Rice Research Notes* 20, 21–22.

Perry, R.N. (1998) The physiology and sensory perception of potato cyst nematodes, *Globodera* species. In: Marks, R.J. and Brodie, B.B. (eds) *Potato Cyst Nematodes – Biology, Distribution and Control*. CAB International, Wallingford, UK, pp. 27–49.

Perry, R.N. and Beane, J. (1983) *Pea Cyst Nematode Development on Field Beans*. Report for 1983, Part 1. Rothamsted Experimental Station, Rothamsted, UK.

Perry, R.N. and Moens, M. (2006) *Plant Nematology*. CABI Publishing, Wallingford, UK.

Perry, R.N. and Moens, M. (2013) *Plant Nematology*, 2nd edn. CABI Publishing, Wallingford, UK.

Perry, R.N., Moens, M. and Starr J.L. (eds) (2009) *Root-knot Nematodes*. CAB International, Wallingford, UK.

Perry, R.N., Moens, M. and Jones, J.T. (2018) *Cyst Nematodes*. CABI Publishing, Wallingford, UK.

Perry, V.G. (1958) Parasitism of two species of dagger nematodes (*Xiphinema americanum* and *X. chambersi*) to strawberry. *Phytopathology* 48, 420–423.

Perry, V.G. and Rhoades, H. (1982a) The genus *Belonolaimus*. In: Riggs, R.D. (ed.) *Nematology in the Southern Region of the United States*. Southern Cooperative Series Bulletin 276. Arkansas Agricultural Experiment Station, University of Arkansas, Fayetteville, Arkansas, pp. 144–149.

Perry, V.G. and Rhoades, H.L. (1982b) The trichodorid nematodes. In: Riggs, R.D. (ed.) *Nematology in the Southern Region of the United States*. Southern Cooperative Series Bulletin 276. Arkansas Agricultural Experiment Station, University of Arkansas, Fayetville, Arkansas, pp. 383–386.

Petersen, D.J. and Vrain, T.C. (1996) Rapid identification of *Meloidogyne chitwoodi*, *M. hapla*, and *M. fallax* using PCR primers to amplify their ribosomal intergenic spacer. *Fundamental and Applied Nematology* 19(6), 601–605.

PExD, (2019) PExD- Phytosanitary Export Database Harmful Organism Reports. *Hirschmanniella*. Available at: https://pcit.aphis.usda.gov/PExD/faces/ViewPExD.jsp (accessed 21 December 2019)

Philis, J. (1993) Distribution and ecology of *Xiphinema index* in Cyprus. *Nematologia Mediterranea* 21(2), 139–142.

Philis, J. (1994) Cultural practices for the control of *Xiphinema index* in Cyprus. *Nematologia Mediterranea* 22(2), 125–126.

Phipps, P.M. (1987) Control of northern root knot and ring nematode on peanut in Virginia, 1986. *Fungicide and Nematicide Tests, American Phytopathological Society* 42, 162.

Phipps, P.M. and Elliott, A.P. (1983) Evaluation of nematicides for control of plant parasitic nematodes on corn, 1982. *Fungicide and Nematicide Tests, American Phytopathological Society* 38, 3.

Pinochet, J. (1987) Management of plant parasitic nematodes in Central America: the Panama experience. In: Veetch, J.A. and Dickson, D.W. (eds) *Vistas on Nematology*. Society of Nematologists, Hyattsville, Maryland, pp. 105–113.

Pinochet, J. and Rowe, P.R. (1979) Progress in breeding for resistance to *Radopholus similis* in banana. *Nematropica* 9, 76–79.

Pinochet, J., Raski, D.J. and Goheen, A.C. (1976) Effects of *Pratylenchus vulnus* and *Xiphinema index* singly and in combination on growth of *Vitis vinifera* 'Thompson seedless'. *Journal of Nematology* 8(4), 300.

Pinochet, J., Rodriguez-Kábana, R., Marull, J. and McGawley, E.F. (1993) *Meloidogyne javanica* and *Pratylenchus vulnus* on pecan (*Carya illinoensis*). *Fundamental and Applied Nematology* 16(1), 73–77.

Pinochet, J., Calvet, C., Camprubi, A. and Fernández, C. (1995) Growth and nutritional response of Nemared peach rootstock infected with *Pratylenchus vulnus* and the mycorrhizal fungus *Glomus mosseae*. *Fundamental and Applied Nematology* 18(3), 205–210.

Pinochet, J., Anglis, M., Dalmau, E., Fernández, C. and Felipe, A. (1996) Prunus rootstock evaluation to root-knot and lesion nematodes in Spain. *Journal of Nematology* 28(4, supplement), 616–623.

Piotte, C., Castagnone-Sereno, P., Bongiovanni, M., Dalmasso, A. and Abad, P. (1995) Analysis of a satellite DNA from *Meloidogyne hapla* and its use as a diagnostic probe. *Phytopathology* 85, 458–462.

Pitcher, R.S. (1967) The host–parasite relations and ecology of *Trichodorus viruliferus* on apple roots, as observed from an underground laboratory. *Nematologica* 13, 547–557.

Pitcher, R.S. and McNamara, D.G. (1970) The effect of nutrition and season of year on the reproduction of *Trichodorus viruliferus*. *Nematologica* 16, 99–106.

Pitcher, R.S. and McNamara, D.G. (1973) The control of *Xiphinema diversicaudatum*, the vector of arabis mosaic virus in hops. *Annals of Applied Biology* 75(3), 468–469.

Pitcher, R.S., Siddiqi, M.R. and Brown, D.J.F. (1974) *Xiphinema diversicaudatum*. *CIH Descriptions of Plant-parasitic Nematodes*, Set 4, No. 60. Commonwealth Institute of Helminthology, St Albans, UK.

Planer, F.R. (1972) Above ground stem infection caused by *Ditylenchus destructor*. *Nematologica* 18, 417.

Platzer, E.G. (1981) Potential use of protein patterns and DNA nucleotide sequences in nematode taxonomy. In: Zuckerman, B.M. and Rohde, R.A. (eds) *Plant Parasitic Nematodes. Vol. III*. Academic Press, Inc., New York, pp. 3–21.

Ploetz, R.C., Thomas, J.E. and Slabaugh, W.R. (2003) Diseases of banana and plantain. In: Ploetz, R.D. (ed.) *Diseases of Tropical Fruit Crops*. CABI Publishing, Wallingford, UK, pp. 73–134.

Plowright, R.A. and Gill, J.R. (1994) Aspects of resistance in deep water rice to stem nematode, *Ditylenchus angustus*. *Fundamnetal and Applied Nematology* 17, 357–367.

Plowright, R.A. and Hunt, D.J. (1994) Plant parasitic nematodes of upland, hydromorphic and inland swamp rice ecosystems in Côte d'Ivoire. *Afro-Asian Journal of Nematology* 4(1), 61–67.

Pokharel, R.R., Abawi, G.S., Duxbury, J.M., Smart, C.D., Wang, X. and Brito, J.A. (2010) Variability and the recognition of two races in *Meloidogyne graminicola*. *Australasian Plant Pathology* 39(4), 326–333.

Potter, J.W. and Olthof, T.H.A. (1993) Nematode pests of vegetable crops. In: Evans, K., Trudgill, D.L. and Webster, J.M. (eds) *Plant Parasitic Nematodes in Temperate Agriculture*. CAB International, Wallingford, UK, pp. 171–207.

Powers, T.O. and Harris, T.S. (1993) A polymerase chain reaction method for identification of five major *Meloidogyne* species. *Journal of Nematology* 25, 1–6.

Powers, T.O., Szalanski, A.L., Mullin, P.G., Harris, T.S., Bertozzi, T. and Griesbach, J. A. (2001) Identification of seed gall nematodes of agronomic and regulatory concern with PCR-RFLP of ITS1. *Journal of Nematology* 33, 191–194.

Powers, T.O., Neher, D.A., Mullin, P., Esquivel, A., Giblin-Davis, R.M., Kanzaki, N., Stock, S.P., Mora, M.M. and Uribe-Lorio, L. (2009) Tropical nematode diversity: vertical stratification of nematode communities in a Costa Rican humid lowland rainforest. *Molecular Ecology* 18, 985–996.

Powers, T.O., Bernard, E.C., Harris, T., Higgins, R., Olson, M., Lodema, M., Mullin, P., Sutton, L. and Powers, K.S. (2014) COI haplotype groups in *Mesocriconema* (Nematoda: Criconematidae) and their morphospecies associations. *Zootaxa* 3827(2), 101–46.

PQ, Order-2003 (2015) Plant quarantine order (Regulation of import into India), 2003, S.O. 2496 (E) dated 15 September, 2015. Directorate of Plant Protection, Quarantine and Storage, Faridabad, Haryana, India.

Pradesh, K.S.K. and Singh, D.B. (1987) Nematode pests associated with potato and their control. In: *Annual Scientific Report, 1984*. ICAR Central Potato Research Institute, Shimla, India, pp. 122–123.

Prasad, J.S.R. and Rao, Y.S. (1984) Ectoparasitic nematode fauna of upland rice soils. *Beitrage zur Tropischen Landwirtschaft und Veterinarmedizin* 22(1), 79–82.

Prasad, J.S. and Varaprasad, K.S. (2001) Ufra nematode, *Ditylenchus angustus* is seed borne. *Crop Protection* 21(1), 75–76.

Prasad, J.S., Panwar, M.S. and Rao, Y.S. (1986) Reaction of some rice cultures to root-knot and cyst nematodes. *Oryza* 23, 59–61.

Prasad, J.S., Panwar, M.S. and Rao, Y.S. (1987) Nematode problems of rice in India. *Tropical Pest Management* 33(2), 127–136.

Prasad, J.S., Kumar, R.M. and Subbarao, L.V. (2004) Role of manures and fertilizers in the management of the root nematode *Hirschmanniella oryzae* in rice (*Oryzae sativa* L.). *Indian Journal of Nematology* 34(1), 1–4.

Prasad, K.S.K. (1992) Nematode pests of potato. In: Bhatti, D.S. and Walia, R.K. (eds) *Nematode Pests of Crops*. CBS, New Delhi, pp. 79–93.

Price, N.S. (1994) Alternate cropping in the management of *Radopholus similis* and *Cosmopolites sordidus*, two important pests of banana and plantain. *International Journal of Pest Management* 40(3), 237–244.

Price, N.S. and Bridge, J. (1995) *Pratylenchus goodeyi* (Nematoda: Pratylenchidae): a plant-parasitic nematode of the montane highlands of Africa. *Journal of African Zoology* 109(5/6), 435–442.

Price, P.C., Fisher, J.M. and Kerr, A. (1979) On *Anguina funesta* n. sp. and its association with *Corynebacterium* sp. in infecting *Lolium rigidum*. *Nematologica* 25, 76–85.

PROINPA (1991) *Annual Report*, 1990–91. Proyecto de Investigación de la Papa. Compendium, Cochabamba, Bolivia, pp. 47–56.

Prot, J.C. (1993) Monetary value estimates of nematode problems, research proposal and priorities: the rice example in South and Southeast Asia. *Fundamental and Applied Nematology* 16(5), 385–388.

Prot, J.C., Soriano, I.R.S. and Matias, D.M. (1994) Major root-parasitic nematodes associated with irrigated rice in the Philippines. *Fundamental and Applied Nematology* 17(1), 75–78.

Qiu, T.X., Yan, M.F. and Lu, Q. (1991) Study on the occurrence, regulation and control of *Aphelenchoides besseyi*. *Zhejiang Nongye Kexue* 6, 290–292.

Quasim, M. and Ghaffar, A. (1986) *Heterodera zeae* in the rhizosphere of declining almond trees. *Nematologia Mediterranea* 14(1), 159.

Raabe, F.D. and Holtzmann, O.V. (1965) A foliar nematode in hibiscus. *Phytopathology* 53(3), 478–479.

Rabie, E.C., Tustin, H.A., Martin, P.P. and Hugon, R. (1997) Pre-plant control of nematodes in pineapple. Proceedings of the Second International Pineapple Symposium, Troisllets, Martinique, 20–24 February, 1995. *Acta Horticulture* 425, 435–442.

Radewald, J.D. and Raski, D.J. (1962) A study of the life cycle of *Xiphinema index*. *Phytopathology* 52, 748.

Radewald, J.D., Maire, R.G., Shibuya, F. and Sher, S.A. (1972) Control of stunt nematode (*Tylenchorhynchus claytoni*) on azalea (*Rhododendron indicum*). *Plant Disease Reporter* 56(6), 540–543.

Radice, A.D., Halisky, P.M. and Meyers, R.F. (1985) Studies on the host range, biology, and pathogenicity of *Punctodera punctata* infecting turfgrasses. *Journal of Nematology* 17, 162–165.

Radivojevic, M., Krnjaic, D., Krnjaic, S., Bacic, J., Subbotin, S.A., Madani, M. and Moens, M. (2001) Molecular methods confirming the presence of *Globodera rostochiensis* (Wollenweber, 1923) in Yugoslavia. *Russian Journal of Nematology* 9, 139–141.

Rahman, M.F. (2010) Lance nematode infestation in paddy. In: Khan, M.R. and Jairajpuri, M.S. (eds) *Nematode Infestations – Part I: Food Crops*. National Academy of Sciences, Allahabad (Prayagraj), UP, India, pp. 74–83.

Rahman, M.L. (1993) Effect of time of nematicide application to control *ufra* disease. *Bangladesh Journal of Plant Pathology* 9, 9–12.

Rahman, M.L. and Evans, A.A.F. (1987) Studies on host parasitic relationship of rice stem nematode, *Ditylenchus angustus* (Nematoda: Tylenchida) on rice (*Oryza sativa* L.) *Nematologica* 33, 451–459.

Rahman, M.L. and Taylor, B. (1983) Nematode pests associated with deepwater rice in Bangladesh. *International Rice Research Newsletter* 8, 20–21.
Rahman, M.L., Mondal, A.H. and Bakr, M.A. (1994) Widespread *ufra* disease incidence in different rice eco-systems in Bangladesh. *International Rice Research Newsletter* 19, 27–28.
Rajan, and Lal, A. (2010) Risk of pine wilt nematode infestation in Himalayan forests. In: Khan, M.R. and Jairajpuri, M.S. (eds) *Nematode Infestations – Part II: Industrial Crops*. National Academy of Sciences, Allahabad (Prayagraj), UP, India, pp. 342–354.
Rajaram, B. (1982) Studies on the lance nematode *Hoplolaimus seinhorsti* Luc, 1958 parasitic on cotton II: vertical distribution of the nematode. *Indian Zoologist* 6(1/2), 95–97.
Rajendran, G., Naganathan, T.N. and Vadivelu, S. (1979) Studies on banana nematodes. *Indian Journal of Nematology* 9, 54.
Rama, K. and Dasgupta, M.K. (1987) Population ecology and community structure of plant parasitic nematodes associated with pineapple in West Bengal. *Indian Journal of Nematology* 17(2), 264–269.
Ramakrishnan, S. and Rajendran, G. (2002) Management of rice nematodes through *Pseudomonas fluorescens* In: *Proceedings of National Symposium on Biodiversity and Management of Nematodes in Cropping Systems for Sustainable Agriculture*, 11–13 November 2002, Jaipur, India. Nematological Society of India, IARI, New Delhi, pp. 144–147.
Ramakrishnan, S., Khan M.R., Sharma, H.K., Cannayane, I. and Shanthi A. (2013) Integrated management of nematodes infesting horticultural crops. In: Khan, M.R. and Jairajpuri, M.S. (eds) *Nematode Infestations – Part II: Industrial Crops*. National Academy of Sciences, Allahabad (Prayagraj), UP, India, pp. 548–568.
Rammah, A. and Hirschmann, H. (1988) *Meloidogyne mayaguensis* n. sp. (Meloidogynidae), a root-knot nematode from Puerto Rico. *Journal of Nematology* 20, 58–69.
Ramsdell, D.C. and Gillett, J.M. (1985) Relative susceptibility of American, French hybrid and European grape cultivars to infection by peach rosette mosaic virus. *Phytopathologia Mediterranea* 24(1/2), 41–43.
Ramsdell, D.C. and Myers, R.L. (1978) Epidemiology of peach rosette mosaic virus in a Concord grape vineyard. *Phytopathology* 68(3), 447–450.
Ramsdell, D.C., Gillett, J.M. and Bird, G.W. (1995) Susceptibility of American grapevine scion cultivars and French hybrid rootstock and scion cultivars to infection by peach rosette mosaic nepovirus. *Plant Disease* 79(2), 154–157.
Ramsdell, D.C., Bird, G.W., Warner, F.W., Davenport, J.F., Diamond, C.J. and Gillet, J.M. (1996) Field pathogenicity studies of four species of plant-pathogenic nematodes on French-American hybrid grapevine cultivars in Michigan. *Plant Disease* 80(3), 334–338.
Randhawa, R. and Chitambar J.J. (2017) *California Pest Rating for Anguina tritici*. California Department of Food and Agriculture, Sacramento, California. Available at: https://blogs.cdfa.ca.gov/Section3162/?p=3446 (accessed 28 February 2021)
Randig, O., Bongiovanni, M., Carneiro, R.M.D.G., Sarah, J.-L. and Castagnone-Sereno, P. (2002a) A species-specific satellite DNA family in the genome of the coffee root-knot nematode *Meloidogyne exigua*: application to molecular diagnostics of the parasite. *Molecular Plant Pathology* 3, 431–437.
Randig, O., Bongiovanni, M., Carneiro, R.M.D.G. and Castagnone-Sereno, P. (2002b) Genetic diversity of root-knot nematodes from Brazil and development of SCAR markers specific for the coffee-damaging species. *Genome* 45, 862–870.
Rao, G.M.V.P., Ganguly, A.K. and Ganguly, S. (1998) Variations in b-esterase profiles of two geographical populations of *Rotylenchulus reniformis*. *Indian Journal of Nematology* 28, 235–237.
Rao, J. and Prakash, A. (2002) *Rice White-tip Nematode*, AZRA Publications, Cuttack, India.
Rao, S.B., Rathi, A., Gothalwal, R., Atkinson, H., and Rao, U. (2011) A comparison of the variation in Indian populations of pigeonpea cyst nematode, *Heterodera cajani*, revealed by morphometric and AFLP analysis. *ZooKeys* 135, 1–19.
Rao, V.R. and Swarup, G. (1974) Pathogenicity of the spiral nematode, *Helicotylenchus dihystera*, to sugarcane. *Indian Journal of Nematology* 4(2), 160–166.
Rao, V.R. and Swarup, G. (1975) Factors affecting the reproduction, population development and survival of the spiral nematode, *Helicotylenchus dihystera*. *Indian Journal of Nematology* 5(2), 139–147.
Rao, Y.S. and Jayaprakash, A. (1978) *Heterodera oryzicola* n.sp. (Nematoda: Heteroderidae) a cyst nematode on rice (*Oryza sativa* L.) from Kerala State, India. *Nematologica* 24(4), 341–346.
Rao, Y.S., Prasad, J.S. and Panwar, M.S. (1986a) Stem nematode (*Ditylenchus angustus*) a potential pest of rice in Assam and West Bengal, India. *International Rice Research Newsletter* 3, 24–26.
Rao, Y.S. Prasad, J.S. and Panwar, M.S. (1986b) Nematode pests of rice: crop loss, symptomatology and management. In: Swarup, G. and Dasgupta, D.R. (eds) *Plant Parasitic Nematodes of India – Problems and Progress*. IARI, New Delhi, India, pp. 279–299.
Rashid, A. and Khan, A.M. (1974) Two new species of the genus *Helicotylenchus*, *Steiner*, 1945 from India, with a redescription of *H. solani* Rashid, 1972 (Nematoda: Hoplolaiminae). *Indian Journal of Nematology* 2(2), 123–128.
Rashid, F., Coomans, A. and Sharma, R.D. (1986) Longidoridae (Nematoda: Dorylaimida) from Bahia State, Brazil. *Nematologia Mediterranea* 14, 235–250.
Rashidifard, M. (2019) Comparative molecular and morphological identification, and reproduction potential of South African *Meloidogyne* species with emphasis on *Meloidogyne enterolobii*. PhD Thesis. North-West University, Potchefstroom, South Africa. doi: 10.13140/RG.2.2.23450.70087
Rashidifard, M., Shokoohi, E., Hoseinipour, A. and Jamali, S. (2015) Distribution, morphology, seasonal dynamics, and molecular characterization of *Tylenchulus semipenetrans* from citrus orchards in southern Iran. *Biologia*, 70/5, 1-11 (Section Zoology). doi: 10.1515/biolog-2015-0098
Rasinya, B.P. (1972) Results of Basudin and Zinophos treatments in controlling nematodes of the potato. *Kratkie doklady po voprosam zashchity rastenii*. (VIII Pribaltiiskaya konferentsiya po zashchite rastenii. Part II). Litovskii Nauchno-Issledovatel'skii Institut Zemledeliya, Kaunas, USSR, pp. 92–94.
Raski, D.J. (1957) *Trophotylenchulus* and *Trophonema* two new genera of Tylenchulinae n. fam. (Nematoda). *Nematologica* 2, 85–90.
Raski, D.J. (1958) Four new species of *Hemicycliophora* de Man, 1921, with further observations on *H. brevis* Thorne, 1955 (Nematoda: Criconematidae). *Proceedings of the Helminthological Society of Washington* 25, 125–131.
Raski, D.J., Hewitt, W.B., Goheen, A.C., Taylor, C.E. and Taylor, R.H. (1965) Survival of *Xiphinema index* and reservoirs of fanleaf virus in fallowed vineyard soil. *Nematologica* 11, 349–352.
Raski, D.J., Jones, N.O., Hafez, S.L., Kissler, J.J. and Luvisi, D.A. (1981) Systemic nematicides tested as alternatives to DBCP. *California Agriculture* 35(5/6), 10–12.
Rãssner, J. (1972) Vertickalverteilung wandernder Wurzelnematoden im Boden in AbhSngigkeit von Wassergehalt und Durchwurzelung. *Nematologica* 18, 360–372.
Rathaiah, Y. (1988) Diseases of deep water rice in Jorhat district, Assam, India. *International Rice Research Newsletter* 14, 32–33.
Rathore, Y.S. (2018) Host choice in *Rotylenchulus* species. *International Journal of Pure Applied Bioscience* 6(5), 346–354. doi: 10.18782/2320-7051.6878
Rau, G.J. (1958) A new species of sting nematode. *Proceedings of the Helminthological Society of Washington* 25, 95–98.
Rau, G.J. (1961) Amended description of *Belonolaimus gracilis* Steiner, 1949 and *B. longicaudatus* Rau, 1958 (Nematoda: Tylenchida). *Proceedings of the Helminthological Society of Washington* 28, 198–200.
Rau, J. (1976) The distribution of virus-transmitting nematodes in natural woodlands of Lower Saxony. *Zeitschrift fur Angewandte Zoologie* 63(4), 427–443.

Rawsthorne, D. and Brodie, B.B. (1986) Relationship between root growth of potato, root diffusate production, and hatching of *Globodera rostochiensis*. *Journal of Nematology* 18, 379–384.

Ray, S., Sahoo, N.K. and Mohanty, K. (1992) Plant parasitic nematodes associated with tuber crops in Orissa. *Journal of Root Crops* 18, 70–72.

Razjivin, A., Fernández, M., Ortega, J. and Quincosa, A. (1981) Nuevas especies de *Hirschmanniella* (Nematoda: Pratylenchidae), parasitas de plantas indeseables en plantaciones de arroz. *Poeyana* 216, 1–7.

Reddy, P.P. and Khan, R.M. (1988) Evaluation of *Paecilomyces lilacinus* for the biological control of *Rotylenchulus reniformis* infecting tomato as compared with carbofuran. *Nematologia Mediterranean* 16, 113–116.

Reddy, P.P., Khan, R.M. and Rao, M.S. (1991) Evaluation of chemicals for the management of the burrowing nematode, *Radopholus similis*. *Current Nematology* 2, 1–2.

Reid, A., Manzanilla-Lopez, R.H. and Hunt, D.J. (2003) *Nacobbus aberrans* (Thorne, 1935) Thorne and Allen, 1944 (Nematoda: Pratylenchidae); a nascent species complex revealed by RFLP analysis and sequencing of the ITS-rDNA region. *Nematology* 5(3), 441–451.

Reid, A., Evans, F.F., Mulholland, V., Cole, Y. and Pickup, J. (2015) High throughput diagnosis of potato cyst nematodes in soil samples. *Plant Pathology: Techniques and Protocols* 1302, 137–148.

Rhoades, H.L. (1969) Nematicide efficacy in controlling sting and stubby-root nematodes attacking onions in central Florida. *Plant Disease Reporter* 53, 728–730.

Rhoades, H.L. (1985) Comparison of fenamiphos and Arthrobotrys amerospora for controlling plant nematodes in central Florida. *Nematropica* 15(1), 1–7.

Riascos-Ortiz, D., Mosquera-Espinosa, A.T., De Agudelo, F.V., de Oliveira, C.M.G. and Muñoz-Flórez, J.E. (2019) Morpho-molecular characterization of Colombian and Brazilian populations of *Rotylenchulus* associated with *Musa* spp. *Journal of Nematology* 51, e2019-47. doi: 10.21307/jofnem-2019-047

Richardson, D.M., Pysek, P., Rejmanek, M., Barbour, M.G., Panetta, F.D. and West, C.J. (2000) Naturalization and invasion of alien plants: concepts and definitions. *Diversity and Distributions* 6, 93–107.

Richardson, P.N. and Grewal, P.S. (1993) Nematode pests of glasshouse crops and mushrooms. In: Evans, K., Trudgill, D.L. and Webster, J.M. (eds) *Plant Parasitic Nematodes in Temperate Agriculture*. CABI, Wallingford, UK, pp. 501–544.

Rickard, D.A., Barker, K.R. and Wells, J.C. (1977) Effects of *Macroposthonia ornata* and *Meloidogyne hapla* on yield and value of peanut. *Journal of Nematology* 9(4), 281.

Riedel, R.M. and Larsen, P.O. (1974) Interrelationship of *Aphelenchoides fragariae* and *Xanthomonas begoniae* on Rieger begonia. *Journal of Nematology* 6(4), 215–216.

Riggs, R.D. (1991) Resistance breaking races of plant parasitic nematodes. In: Nickle, W.R. (ed.) *Manual of Agriculture Nematology*. Marcel Dekker Inc., New York, pp. 827–851.

Riggs, R.D. and Wrather, J.A. (eds) (1992) *Biology and Management of the Soybean Cyst Nematode*. American Phytopathological Press, St Paul, Minnesota.

Riley, I.T. and McKay, A.C. (1990) Specificity of the adhesion of some plant pathogenic microorganisms to the cuticle of nematodes in the genus *Anguina* (Nematoda: Anguinidae). *Nematologica* 36, 90–103.

Riley, I.T., Reardon, T.B. and McKay, A.C. (1988) Electrophoresis resolution of species boundaries in seed-gall nematodes, *Anguina* spp. (Nematoda: Anguinidae), from some graminaceous hosts in Australia and New Zealand. *Nematologica* 34, 401–411.

Rivoal, R. and Cook, R. (1993) Nematode pests of cereals. In: Evans, K., Trudgill, D.L. and Webster, J.M. (eds) *Plant Parasitic Nematodes in Temperate Agriculture*. CAB International, Wallingford, UK, pp. 259–303

Robbins, R.T and Barker, K.R. (1974) The effects of soil type, particle size, temperature and moisture on reproduction of *Belanolaimus longicaudatus*. *Journal of Nematology* 6, 1–6.

Robbins, R.T. and Brown, D.J.F. (1991) Comments on the taxonomy, occurrence and distribution of Longidoridae (Nematoda) in North America. *Nematologica* 37(4), 395–419 (abstract/19922315350).

Roberts, P.A. and Thomason, I.J. (1981) *Sugar beet Pest Management: Nematodes*. Special Publication, 3272. Division of Agricultural Sciences, University of California, Davis, California.

Roberts, P.A. Thomason, I.J. and McKinney, H.E. (1981) Influence of nonhosts, crucifers, and fungal parasites on field populations of *Heterodera schachtii*. *Journal of Nematology* 13(2), 164–171.

Robertson G.P. and Freckman, D.W. (1995) The spatial distribution of nematode trophic groups across a cultivated ecosystem. *Ecology* 76, 1425–1432.

Robinson, A.F., Inserra, R.N., Caswell-Chen, E.P., Vovlas, N. and Troccoli, A. (1997) *Rotylenchulus* species: identification, distribution, host ranges, and crop plant resistance. *Nematropica* 27(2), 127–180.

Robinson, A.F., Cook, C.G., Westphal, A. and Bradford, J.M. (2005) *Rotylenchulus reniformis* below plow depth suppresses cotton yield and root growth. *Journal of Nematology* 37(3), 285–291.

Rodriguez, M.G., Rodriguez, I. and Sanchez, L. (1995) Species of the genera *Meloidogyne* which parasitize coffee in Cuba. Geographical distribution and symptomatology. *Revista de Proteccion Vegetal* 10, 123–128.

Rodriguez-Kabana, R. (1986) Organic and inorganic nitrogen amendments to soil as nematode suppressants. *Journal of Nematology* 18(2), 129–135.

Rodriguez-Kabana, R. and Ingram, E. (1976) Potato seed-piece treatment with the systemic nematicide phenamiphos for control of plant parasitic nematodes. *Nematropica* 6(2), 81–85.

Rodriguez-Kabana, R. and Ingram, E.G. (1977) Treatment of potato seed-pieces with oxamyl for control of plant-parasitic nematodes. *Plant Disease Reporter* 61(1), 29–31.

Rodriguez-Kabana, R. and King, P.S. (1978) Dosimetric comparison of the nematicidal activities of oxamyl and phenamiphos. *Nematropica* 8, 22.

Rodriguez-Kabana, R., Hoveland, C.S. and Haaland, R.L. (1977) Evaluation of a seed-treatment method with acetone for delivering systemic nematicides with wheat and rye. *Journal of Nematology* 9(4), 323-326

Rohde, R.A. and Jenkins, W.R. (1957) Host range of a species of *Trichodorus* and its host–parasite relationships on tomato. *Phytopathology* 47, 295–298.

Rojankovski, E. and Ciurea, A. (1986) Contributions to the study of interactions between the potato rot nematode, *Ditylenchus destructor* Thorne, and fungi in the potato disease complex. *Archiv für Phytopathologie und Pflanzenschutz* 22, 101–106.

Roman, J. and Grullon, L. (1975) Nematodes associated with sugarcane in the Dominican Republic. *Journal of Agriculture of the University of Puerto Rico* 59(2), 138–140.

Rovira, A.D. and Simon, A. (1982) Integrated control of *Heterodera avenae*. *Bulletin EPPO* 12, 517–523.

Ruehle, J.L. (1971) Nematodes parasitic on forest trees: III. Reproduction on selected hardwoods. *Journal of Nematology* 3, 170–173.

Rumpenhorst, H.J. (1985) Vergleichende elektrophoretische Untersuchungen von proteinen einiger Zystennematoden von Getreide und Gräsern. *Mitteilungen aus der Biologischen Bundesaustalt fur Land-und Fartmirtschaft*, Berlin-Dahlem, 226, 64–74.

Rybarczyk-Mydłowska, K., Mooyman, P., van Megen, H., van den Elsen, S., Vervoort, M., Veenhuizen, P. *et al.* (2012) Small subunit ribosomal DNA-based phylogenetic analysis of foliar nematodes (*Aphelenchoides* spp.) and their quantitative detection in complex DNA backgrounds. *Phytopathology* 102(12), 1153–1160.

Rybarczyk-Mydłowska, K., Dmowska, E. and Kowalewska, K. (2019) Phylogenetic studies on three *Helicotylenchus* species based on 28S rDNA and mtCOI sequence data. *Journal of Nematology* 51, 1–17. doi: 10.21307/jofnem-2019-033.

Ryss, A., Viera, P., Mota, M. and Kulinich, O. (2005) A synopsis of the genus *Bursaphelenchus* Fuchs, 1937 (Aphelenchida: Parasitaphelenchidae) with keys to species. *Nematology* 7(3), 393–458.

Sabova, M., Valocka, B., Liskova, M. and Vargova, V. (1988) The first finding of *Heterodera latipons* Franklin, 1969 on grass stands in Czechoslovakia. *Helminthologia* 25, 201–206.

Saeed, M. (1974) Studies on some stylet-bearing nematodes associated with sapodilla (*Achras zapota* L.), with special reference to *Hemicriconemoides mangiferae* Siddiqi, 1961. PhD Thesis. Department of Botany, University of Karachi,

Saeed, M. and Ghaffar, A. (1986) Seasonal population fluctuations of *Hemicriconemoides mangiferae* and other nematodes associated with sapodilla (*Achras zapota*) in Karachi. *Pakistan Journal of Nematology* 4(2), 67–74.

Saeki, Y., Kawano, E., Yamashita, C., Akao, S. and Nagatomo, Y. (2003) Detection of plant parasitic nematodes, *Meloidogyne incognita* and *Pratylenchus coffeae* by multiplex PCR using specific primers. *Soil Science and Plant Nutrition* 49, 291–295.

Sahebani, N., Kheiri, A., Rahimian, H. and Tehrani, A.S (2007) Effect of *Rathayibacter tritici* on movement of *Anguina tritici* larvae and nematode function as vector of ear rot bacterium. *Journal of Science and Tech Agriculture and Natural Resources* 10(4B), 357–365.

Salawu, E.O. (1992) Effect of *Meloidogyne incognita* and *Heterodera sacchari* singly and combined on the growth of sugarcane (abstract). *Nematologia Mediterranea* 20(2), 187–188.

Salawu, E.O. (1994) Development and life cycle of *Heterodera sacchari* Luc and Merny, 1963 on sugarcane (abstract). *Pakistan Journal of Nematology* 12(2), 131–136.

Saleh, H. and Fattah, F.A. (1990) Studies on wheat seed gall nematode. *Nematology Mediterranean* 18(1), 59–62.

Sanchez-Monge, A., Flores, L., Salazar, L., Hockland, S. and Bert, W. (2015) An updated list of the plants associated with plant-parasitic *Aphelenchoides* (Nematoda: Aphelenchoididae) and its implications for plant-parasitism within this genus. *Zootaxa* 4013(2), 207–224.

Sankranarayana C. (2010) Nematode infestation in sugarcane. In: Khan, M.R. and Jairajpuri, M.S. (eds) *Nematode Infestations – Part II: Industrial Crops*. National Academy of Sciences, Allahabad (Prayagraj), UP, India, pp. 17–46.

Santo, G.S. (1994) Biology and management of root-knot nematodes on potato in the Pacific Northwest. In: Zehner, G.W., Powelson, M.L., Jansson, R.K. and Raman, K.V. (eds) *Advances in Potato Pest Biology and Management*. APS Press, St Paul, Minnesota, pp. 193–201.

Santo, G.S. and O'Bannon, J.H. (1981) Pathogenicity of the Columbia root-knot nematode (*Meloidogyne chitwoodi*) on wheat, corn, oat and barley. *Journal of Nematology* 13(4), 548–550.

Santo, G.S., Ponti, R.P. and Wilson, J.H. (1985) Evaluation of nonfumigant nematicides for control of the ring and northern root-knot nematodes on Concord grapes, 1981–83. *Fungicide and Nematicide Tests, American Phytopathological Society* 40, 102.

Santos, M.S.N. de A. (1967) *Meloidogyne ardenesis* n. sp. (Nematoda: Heteroderidae), a new British species of root-knot nematode. *Nematologica* 13, 593–598.

Saotome, H. and Ishibashi, N. (1989) Effect of cinnamic aldehyde on nematode fauna in soil. *Proceedings of the Association for Plant Protection of Kyushu* 35, 107–109.

Sarah, J.L. (1986) Influence of *Pratylenchus brachyurus* on the growth of development of pineapple in the Ivory Coast. XVIII International symposium E.S.N., Antibes. *Revue de Nematologie* 9, 308–309.

Sarah, J.L. (1989) Banana nematodes and their control in Africa. *Nematropica* 19, 199–215.

Sarah, J.L. (1997) Knowledge of the intraspecific diversity of the nematode *Radopholus similis*. *Fruits* (Paris) 52, 261–266.

Sardanelli, S., Krusberg, L.R. and Golden, A.M. (1981) Corn cyst nematode, *Heterodera zeae*, in the United States. *Plant Disease* 65(7), 622.

Sasanelli, N. (1992) Nematicidal activity of aqueous extracts from leaves of *Ruta graveolens* on *Xiphinema index*. *Nematologia Mediterranea* 20(1), 53–55.

Sasanelli, N. and Catalano, L. (1991) In vitro nematicidal activity of aqueous extracts from pods of *Capsicum annuum* (chilli pepper) on *Xiphinema index*. *Informatore Fitopatologico* 41(10), 55–56.

Sasser, J.N. (1979) Economic importance of *Meloidogyne* in tropical countries. In: Lamberti, F. and Taylor, C.E. (eds) *Root-Knot Nematodes (*Meloidogyne *species). Systematics, Biology and Control*. Academic Press, London, pp. 359–374.

Sasser, J.N. (1989) *Plant Parasitic Nematodes: the Farmer's Hidden Enemy*. North Carolina State University Graphics, Raleigh, North Carolina.

Sasser, J.N. and Carter, C.C. (1984) Overview of the International *Meloidogyne* project, 1975–1984. In: Sasser, J.N. and Carter, C.C. (eds) *An Advanced Treatise on* Meloidogyne, *Biology and Control Volume I*. North Carolina State University Graphics, Raleigh, North Carolina, pp. 19–24.

Sato, E., Min, Y.Y., Shirakashi, T., Wada, S. and Toyota, K. (2007) Detection of the root-lesion nematode, *Pratylenchus penetrans* (Cobb), in a nematode community using real-time PCR. *Japanese Journal of Nematology* 37, 87–92.

Satyanarayana, P.J., Seshagiri, R.Y., Zaheruddeen, S.M. and Mohan, D.C. (1980) Weeds as hosts to parasitic nematodes of rice. *Current Science* 49(19), 755–756.

Saxena, G.C., Deshmukh, R. and Singh, Y.P. (1990) In vitro studies of some insecticides and fungicides on the mortality of *Helicotylenchus dihystera*. *International Nematology Network Newsletter* 7(3), 25–27.

Saxena, M.C., Greco, N. and Di Vito, M. (1992) Control of *Heterodera ciceri* by crop rotation. *Nematologia Mediterranea* 20(1), 75–78.

Saxena, S.K. (1986) Nematode infestations in vegetables. In: Swarup, G. and Dasgupta, D.R. (eds) *Plant Parasitic Nematodes of India, Problems and Progress*. Indian Agricultural Research Institute, New Delhi, India, pp. 312–327.

Sayre, R.M. (1980) Biocontrol: Bacillus penetrans and related parasites of nematodes. *Journal of Nematology* 12(4), 260–270.

Schilt, H.G. and Cohn, E. (1975) Pathogenicity and population increase of *Paratrichodorus minor* as influenced by some environmental factors. *Nematologica* 21(1), 71–80.

Schmidt, C.I. (1969) *Meloidogyne coeffeicola*, a serious root nematode problem in Brazilian coffee. *FAO Plant Protection Bulletin* 17, 56–57.

Schmitt, D.P. and Corbin, F.T. (1981) Interaction of fensulfothion and phorate with preemergence herbicides on soybean parasitic nematodes. *Journal of Nematology* 13(1), 37–41.

Schmitt, D.P. and Norton, D.C. (1972) Relationships of plant-parasitic nematodes to sites in native Iowa prairies. *Journal of Nematology* 4, 200–206.

Schmitt, D.P., Wrather, J.A. and Riggs, R.D. (eds) (2004) *Biology and Management of the Soybean Cyst Nematode*, 2nd edn. Schmitt and Associates of Marceline, Marceline, Missouri.

Schmitz, B., Burgermeister, W. and Braasch, H. (1998) Molecular genetic classification of Central European *Meloidogyne chitwoodi* and *M. fallax* populations. *Nachrichtenblatt des Deutschen Pflanzenschutzdienstes* 50, 310–317.

Schneider, S.M. and Ferris, H. (1987) Stage-specific population development and fecundity of *Paratrichodorus minor*. *Journal of Nematology* 19, 395–403.

Schuerger, A.C. and McClure, M.A. (1983) Ultrastructural changes induced by *Scutellonema brachyurum* in potato roots. *Phytopathology* 73(1), 70–81.

Sciumbato, G.L. (1991) Southern United States soybean disease loss estimate for 1990. *Proceedings of the Southern Soybean Disease Workers*, 32–38.

Scotto La Massese, C., Minot, J.C., Voisin, R., Castaing, L.R.M. and Fabre, A. (1988) Relationship between soil type, previous crop and age of plantation on the composition and the distribution of the nematofauna associated with vineyards of the south-east of France. *Acta Oecologica, Oecologia Applicata* 9(2), 137.

Seenivasan, N., Devrajan, K. and Selvaraj, N. (2007) Management of potato cyst nematodes, *Globodera* spp. through biological control. *Indian Journal of Nematology* 37, 27–29.

Seinhorst, J.W. (1977) *Pratylenchus fallax. CIH Description of Plant-parasitic Nematodes*, Set 7, No. 100. Commonwealth Institute of Helminthology, St. Albans, UK.

Senthamizh, K., Rajendran, G. and Subramanian, S. (2005) Host range, biology and races of *Heterodera cajani* Koshy, 1967 occurring in Tamil Nadu (India). *Indian Journal Nematology* 35(2), 187–191.

Seshadri, A.R. and Dasgupta, D.R. (1975) *Ditylenchus angustus. CIH Description of Plant Parasitic Nematodes*, Set 5, No. 64. Commonwealth Institute of Helminthology, St Albans, UK.

Seshadri, A.R. and Dasgupta, D.R. (1980) Nematodes In: *Hand Book of Agriculture*. ICAR, New Delhi, pp. 551–557.

Sethi, C.L. and Meher, H.C. (1991) Effect of phenamiphos on soil nematodes and yield of cowpea, pea and okra. *Indian Journal of Nematology* 19, 89–94.

Shahina, F. (1996) A diagnostic compendium of the genus *Aphelenchoides* Fischer, 1894 (Nematoda: Aphelenchida) with some new records of the group from Pakistan. *Pakistan Journal of Nematology* 14, 1–32.

Shahina, F. and Maqbool, M.A. (1989) Embryonic and postembryonic developmental stages of *Heterodera zeae* Koshy et al., 1971 on *Zea mays*. *Pakistan Journal of Nematology* 7(1), 3–16.

Shahina, F. and Maqbool, M.A. (1995) *Cyst nematodes of Pakistan (Heteroderidae) (a monograph)*. National Nematological Research Centre, University of Karachi, Karachi, Pakistan.

Sharma, N.N., Edward, J.C., Misra, S.L. and Chandrashekar, M. (1992a) Studies on phytonematodes associated with important rice growing tracts of India. *Current Nematology* 3(2), 107–109.

Sharma, R.D. and Ferraz, E.C.A. (1977) Efficiency of systemic nematicides in the control of plant parasitic nematodes associated with cacao seedlings. *Revista Theobroma* 7(1), 3–12.

Sharma, R.D. and Scolari, D.D.G. (1984) Efficiency of green manure and crop rotation in the control of nematodes under savannah conditions. *Nematologia Brasileira* 8, 193–218.

Sharma, S.B. and Nene, Y.L. (1988) Effect of *Heterodera cajani*, *Rotylenchulus reniformis* and *Hoplolaimus seinhorsti* on pigeonpea biomass. *Indian Journal of Nematology* 18(2), 273–278.

Sharma, S.B. and Swarup, G. (1984) *Cyst-forming Nematodes of India*. Cosmo Publications, New Delhi.

Sharma, S.B., Smith, D.H. and McDonald, D. (1992b) Nematode constraints of chickpea and pigeonpea production in the semiarid tropics. *Plant Disease* 76(9), 868–874.

Sharma, S.B., Pande, S., Saha, M., Kaushal, K.K., Lal, M., Singh, M., Singh, K., Pokharel, R. and Upreti, R.P. (2001) Plant parasitic nematodes associated with rice and wheat based cropping systems in Nepal. *International Journal of Nematology* 11, 35–38.

Sharma, S.N., Sain, R.S. and Singh, H. (2004) Breeding for resistance to wheat nematodes (*Heterodera avenae*). *Acta Agronomica Hungarica* 52, 189–197.

Shashi S., Satish, K., Bajaj, H.K. and Ganguly, A.K. (1998) Esterase and malate dehydrogenase patterns of races of *Heterodera cajani* and *H. zeae*. *Indian Journal of Nematology* 28, 105–109.

Shashi, S., Satish, K. and Bajaj, H.K. (1999) Electrophoretic patterns of proteins and some enzymes of some *Heterodera* species. In: *Proceedings of National Symposium on Rational Approaches in Nematode Management for Sustainable Agriculture*, 23–25 November, 1998, Anand, India. Nematological Society of India, pp. 65–70.

Shaw, J.N. and Muth, O.H. (1949) Some types of forage poisoning in Oregon cattle and sheep. *Journal of American Veterinary Medicine Association* 114(866), 315–317.

Shazad, S. and Ghaffar, A. (1986) Tomato, a natural host of *Heterodera zeae* in Pakistan. *International Nematology Network Newsletter* 3, 38.

Sheela, M.S. and Sundraraju, P. (2012) Nematode infestation in banana. In: Khan, M.R. and Jairajpuri, M.S. (eds) *Nematode Infestations – Part III: Horticultural Crops*. National Academy of Sciences, Allahabad (Prayagraj), UP, India, pp. 236–261.

Sheela, M.S., Hebsybai. and Anitha, N. (1996) Ecofriendly management strategies for burrowing nematode of banana in Kerala. In: Nayar, N.K. and George, T.E. (eds) *Proceedings Symposium on Technological Advancement in Banana/plantain Production and Processing – India – International*, 20–24 August 1996. Abstract of papers. Kerala Agricultural University, Mannuthy, India, pp. 136–137.

Shepherd, J.A. (1968) A nematode survey of tobacco soils in Rhodesia and Zambia, and the effects of grass-tobacco rotations on nematode populations. *Rhodesian Journal of Agricultural Research* 6, 19–26.

Shepherd, J.A. and Barker, K.R. (1990) Nematode parasites of tobacco. In: Luc, M. Sikora, R.A. and Bridge, J. (eds) *Plant Parasitic Nematodes in Subtropical and Tropical Agriculture,* 1st edn. CABI. Wallingford, UK, pp. 493–517.

Sher, S.A. (1961) Revision of the Hoplolaiminae (Nematoda) I. Classification of Nominal Genera and Nominal Species *Nematologica* 6, 155–169.

Sher, S.A. (1963) Revision of the Hoplolaiminae (Nematoda) II. *Hoplolaimus* Daday, 1905 and *Aorolaimus* n. gen. *Nematologica* 9, 267–295.

Sher, S.A. (1964) Revision of the Hoplolaiminae (Nematoda). III. *Scutellonema* Andrássy, 1958. *Nematologica* 9, 421–433.

Sher, S.A. (1966) Revision of the Hoplolaiminae (Nematoda) VI. *Helicotylenchus* Steiner, 1945. *Nematologica* 12, 1–56.

Sher, S.A. (1968) Revision of the genus *Hirschmanniella* Luc and Goodey, 1963 (Nematoda : Tylenchida) *Nematologica* 14, 243–275.

Sher, S.A. (1970) Revision of the genus Nacobbus Thorne & Allen, 1944 (Nematoda: Tylenchida). *Journal of Nematology* 2, 228–235.

Sher, S.A. and Allen, M.W. (1953) Revision of the genus *Pratylenchus* (Nematoda: Tylenchidae). *University of California Publications Zoology* 57, 441–470.

Shiabova, T.N. (1977) Study of gall nematodes in glasshouses in West Siberia. *Sbornik Nauchnykh Trudov Sibirskogo Nauchno-Issledovatel'skogo Instituta Khimizatsii Sel'skogo Kozyaistva* 3, 81–88.

Shiabova, T.N. (1982) Pathogenicity factors of *Heterodera avenae*. *Nachno Tekhnicheskii Byulletin VASKhNIL, Sibirskoe Otdelenie* 22, 32–36.

Shmal'ko, V.H. (1959) The cactus nematode, *Heterodera cacti* Filipjev and Schuurmans Stekhoven, 1941. Trudy Gel'mintologicheskoi Laboratorii. Akademiya Nauk SSSR [Academy of Sciences of the USSR], Moscow.

Shokoohi, E. and Duncan, L.W. (2005) Nematode parasites of citrus. In: Sikora, R.A., Coyne, D., Hallmann, J. and Timper, P. (eds) *Plant Parasitic Nematodes in Subtropical and Tropical Agriculture*, 3rd edn. CABI Publishing, Wallingford, UK, pp. 437–466.

Shokoohi, E., Marais, M., Fourie, H. and Panahi, H. (2018) Morphological and molecular characterization of three *Helicotylenchus* species from Iran. *Zootaxa* 4388(2), 207–220. doi: 10.11646/zootaxa.4388.2.4

Shreenivasa, K.R., Krishnappa, K., Reddy, B.M.R., Ravichandra, N.G., Karuna, K. and Kantharaju, V. (2005) Integrated management of nematode complex of banana. *Indian Journal of Nematology* 35, 37–40.

Siddiqi, I.A., Sher, S.A. and French, A.M. (1973) *Distribution of Plant Parasitic Nematodes in California*. State of California Department of Food and Agriculture, Division of Plant Industry, Sacramento, California.
Siddiqi, M.R. (1961) Studies on species of *Criconematoides* (Nematoda: Tylenchida). *Proceedings Helminthological Society of Washington* 28, 19–34.
Siddiqi, M.R. (1962) *Trichodorus pakistanensis* n.sp. (Nematoda: Trichodoridae) with observations on *T. porosus* Allen, 1957, *T. mirzai* Siddiqi, 1960, and *T. minor* Colbran, 1956, from India. *Nematologica* 8, 193–200.
Siddiqi, M.R. (1963a) On the classification of the Pratylenchidae (Thorne, 1949) nov. grad. (Nematoda: Tylenchida), with a description of *Zygotylenchus browni* nov. gen. et nov. sp. *Zeitschrift für Parasitenkunde* 23, 390–396.
Siddiqi, M.R. (1963b) *Trichodorus* spp. (Nematoda: Trichodoridae) from Tunisia and Nicaragua. *Nematologica* 9, 69–75.
Siddiqi, M.R. (1964a) Studies on nematode root-rot of citrus in Uttar Pradesh, India. *Proceedings of Zoological Society, Calcutta* 17, 67–75.
Siddiqi, M.R. (1964b) *Radopholus williamsi* n.sp. (Nematoda: Pratylenchidae) a parasite of sugar-cane roots at L'Étoile, Mauritius. *Indian Journal of Entomology* 26, 207–208.
Siddiqi, M.R. (1964c) *Xiphinema conurum* n.sp. and *Paralongidorus microlaimus* n.sp. with a key to the species of *Paralongidorus* (Nematoda: Longidoridae). *Proceeding of the Helminthological Society of Washington* 31, 133–137.
Siddiqi, M.R. (1972a) *Pratylenchus coffeae CIH. Descriptions of Plant-parasitic Nematodes*, Set 1, No. 6. Commonwealth Institute for Helminthology, St Albans, UK.
Siddiqi, M.R. (1972b) *Rotylenchulus reniformis. C.I.H. Descriptions of Plant-parasitic Nematodes*, Set 1, No. 5. Commonwealth Institute for Helminthology, St Albans, UK.
Siddiqi, M.R. (1972c) *Scutellonema bradys. CIH Descriptions of Plant-parasitic Nematodes*, Set 1, No. 10. Commonwealth Institute for Helminthology, St Albans, UK.
Siddiqi, M.R. (1973a) *Hirschmanniella oryzae. C.I.H. Descriptions of Plant-parasitic Nematodes*, Set 2, No. 26. Commonwealth Institute for Helminthology, St Albans, UK.
Siddiqi, M.R. (1973b) *Xiphinema americanum. C.I.H. Descriptions of Plant-parasitic Nematodes*, Set 2, No. 29. Commonwealth Institute for Helminthology, St Albans, UK.
Siddiqi, M.R. (1974a) Systematics of the genus *Trichodorus* Cobb, 1913 (Nematoda: Dorylaimida), with descriptions of three new species. *Nematologica* 19(3), 259–278.
Siddiqi, M.R. (1974b) *Tylenchulus semipentrans. CIH Descriptions of Plant-parasitic Nematodes*, Set 3, No. 34. Commonwealth Institute for Helminthology, St Albans, UK.
Siddiqi, M.R. (1974c) *Aphelenchoides ritzemabosi. CIH Descriptions of Plant-parasitic Nematodes*, Set 3, No. 32. Commonwealth Institute for Helminthology, St Albans, UK.
Siddiqi, M.R. (1974d) *Hoplolaimus pararobustus*. CIH Descriptions of Plant-parasitic Nematodes, Set 3, No. 33. Wallingford, UK: CAB International.
Siddiqi, M.R. (1975a) *Zygotylenchus guevarai. CIH Descriptions of Plant-parasitic Nematodes*, Set 5, No. 65. Commonwealth Institute for Helminthology, St Albans, UK.
Siddiqi, M.R. (1975b) *Aphelenchoides fragariae. CIH Descriptions of Plant-parasitic Nematodes*, Set 5, No. 74. Commonwealth Institute for Helminthology, St Albans, UK.
Siddiqi, M.R. (1976) *Tylenchorhynchus annulatus* (= *T. martini*). *CIH Descriptions of Plant-parasitic Nematodes*, Set 6, No. 85. Commonwealth Institute for Helminthology, St Albans, UK.
Siddiqi, M.R. (1977) *Hemicriconemoides mangiferae. CIH Descriptions of Plant-parasitic Nematodes*, Set 7, No. 99. Commonwealth Institute for Helminthology, St Albans, UK.
Siddiqi, M.R. (1986) *Tylenchida: Parasites of Plants and Insects*. CAB International, Farnham Royal, UK.
Siddiqi, M.R. (2000) *Tylenchida: Parasites of Plants and Insects*, 2nd edn. CABI, Wallingford, UK.
Siddiqui, Z.A. and Mahmood, I. (1993) Occurrence of races of *Heterodera cajani* in Uttar Pradesh, India. *Nematologia Mediterranea* 21, 185–186.
Sikora, R.A. (1988) Plant parasitic nematodes of wheat and barley in temperate and temperate semi-arid regions – a comparative analysis. In: Saxena, M.C., Sikora, R.A. and Srivastava, J.P. (eds). *Nematodes Parasitic to Cereals and Legumes in Temperate Semi-arid Regions*. International Center for Agricultural Research in the Dry Areas (ICARDA), Syria, pp. 46–68.
Sikora, R.A. and Fernandez, E. (2005) Nematode parasites of vegetables. In: Luc, M., Sikora, R.A. and Bridge, J. (eds) *Plant Parasitic Nematodes in Subtropical and Tropical Agriculture*, 2nd edn. CABI Publishing, Wallingford, UK, pp. 319–392.
Sikora, R.A. and Greco, N. (1990) Nematode parasites of food legumes. In: Luc, M., Sikora, R.A. and Bridge, J. (eds) *Plant Parasitic Nematodes in Subtropical and Tropical Agriculture*. CAB International, Wallingford, UK, pp. 181–235.
Sikora, R.A., Greco, N. and Silva, J.F.V. (2005) Nematode parasites of food legumes. In: Luc, M., Sikora, R.A. and Bridge, J. (eds) *Plant Parasitic Nematodes in Subtropical and Tropical Agriculture*, 2nd edn. CABI Publishing, Wallingford, UK, pp. 259–318.
Sikora, R.A., Coyne, D., Hallmann, J. and Timper, P. (2018) Reflection and challenges: nematology in subtropical and tropical agriculture. In: Sikora, R.A., Coyne, D., Hallmann, J. and Timper, P. (eds) *Plant Parasitic Nematodes in Subtropical and Tropical Agriculture*, 3rd edn. CABI Publishing, Wallingford, UK, pp. 1–19.
Silva, G.S. (1992) White tip and national rice production. *Informe Agropecuario Belo Horizonte*, 16(172), 57–59.
Silveira, S.G.P. da, Curi, S.M. and Leite, N. (1989) Control of the rice seed nematode *Aphelenchoides besseyi*. *Pesquisa Agropecuaria Brasileira*, 24(8), 1027–1032.
Simberloff, D. (2011) How common are invasion-induced ecosystem impacts? *Biological Invasions* 13, 1255–1268.
Simeone, A.M. (1993) Further observations on the role of root lesion nematode *Pratylenchus vulnus* on peach replant disease. Paper presented at the Symposium on Soil Sickness and Replant Diseases in Fruit Trees, Bologna, Italy,Sep. 3–4, 1990. *Acta Horticulturae* 324, 85–90,
Singh, H. (1985) Effect of crop rotation on the population of cereal cyst nematode. *Indian Journal of Nematology* 15, 263–264.
Singh, I. and Prasad, S.K. (1972) Effect of some pesticides and growth regulators on *Anguina tritici* infesting wheat. *Indian Journal of Entomology* 34(1), 77–79.
Singh, I., Sakhuja, P.K. and Sharma, S.K. (1987) Influence of some crucifers on *Heterodera avenae* – the cereal cyst nematode. *Plant Disease Research* 2, 93–94.
Singh, K. and Misra, S.R. (1976) Pathogenicity and histopathology of *Hoplolaimus indicus* on sugarcane. *Nematologica* 22, 433–436.
Singh, K.B., Vito, M. di, Greco, N. and Saxena, M.C. (1989) Reaction of wild *Cicer* spp. lines to *Heterodera ciceri*. *Nematologia Mediterranea* 17(2), 113–114.
Singh, K.B., Vito, M. di, Greco, N. and Saxena, M.C. (1996) Registration of ILWC 292, a chickpea cyst nematode-resistant germplasm of *Cicer reticulatum* Ladiz. *Crop Science* 36(5), 1421–1422.
Singh, M. K. (1995) Ph.D. Thesis: Studies on the morphology and bio-ecology of nematode fauna of Rewa. APS University, Rewa, India.
Singh, R., Benival, M.S., Karwasara, R.S. and Saharan, H.S. (2009) Effect of data on the incidence of loose smut flag smut and seedgall of wheat Bangaladesh. *Journal of Agriculture Research* 34(1), 1–4.
Singh, R.V. and Khera, S. (1979) Pathogenicity of *Rotylenchulus reniformis* on brinjal (*Solanum melongena* L.). *Indian Journal of Nematology* 9, 117–124.
Singh, R.V. and Kumar, V. (2003) Biodiversity of nematodes associated with roses, *Rosa indica* in and around Delhi Region. In: Singh, R.V., Pankaj Dhawan, S.C. and Gaur, H.S. (eds) *Proceedings of National Symposium on Biodiversity and Management of Nematodes in Cropping Systems for Sustainable Agriculture*, Jaipur, India, 11–13 November, 2002. Division of Nematology, Indian Agricultural Research Institute, New Delhi, pp. 29–31.
Singh, S., Satish, K. and Bajaj, H.K. (1998) Catalase and glutamate oxaloacetate transaminase isozymes patterns of races of *Heterodera cajani* and *H. zeae*. *Haryana Agricultural University Journal of Research* 28(4), 175–177.
Singh, S.K., Khurma, U.R. and Lockhart, P.J. (2010) Weed hosts of root-knot nematodes and their distribution in Fiji. *Weed Technology* 24, 607–612.

Singh, S.K., Hodda, M., Ash, G.J. and Banks, N.C. (2013a) Plant-parasitic nematodes as invasive species: characteristics, uncertainty and biosecurity implications. *Annals of Applied Biology* 163(3), 323–350.

Singh, S.K., Hodda, M. and Ash, G.J. (2013b) Plant-parasitic nematodes of potential phytosanitary importance, their main hosts and reported yield losses. *Bulletin OEPP/ EPPO Bulletin* 43(2), 334–374.

Širca, S., Urek, G. and Karssen, G. (2004) First report of the root-knot nematode *Meloidogyne ethiopica* on tomato in Slovenia. *Plant Disease* 88(6), 680.

Širca, S., Geric, S.B., Mavrič, P.I., Viršček, M.M., Urek, G. and Javornik, B. (2007) *Xiphinema rivesi* from Slovenia transmit Tobacco ringspot virus and Tomato ringspot virus to cucumber bait plants. *Plant Disease* 91(6), 770.

Sitaramaiah, K. and Pathak, K.N. (1993) Nematode bacterial disease interaction. In: Khan, M.W. (ed.) *Nematode Interactions*. Chapman and Hall, New York, pp. 232–250.

Sivakumar, C.V. and Khan, E. (1982) Description of *Hirschmanniella kaverii* sp. n. (Radopholidae: Nematoda) with a key for identification of *Hirschmanniella* spp. *Indian Journal of Nematology* 12, 86–90.

Sivakumar, C.V. and Sheshadri, A.R. (1979) Longevity of the reniform nematode, *Rotylenchulus reniformis* in host free soil. *Indian Journal of Nematology* 6, 138–144.

Sivakumar, M., Ambika, S. and Zareena, B.M. (2005) Biocontrol of root rot disease complex involving *Rotylenchulus reniformis*. Paper presented during National Symposium on Recent advances and Research Priorities in Indian Nematology, 9–10 December, 2005 at IARI, New Delhi.

Sivapalan, P. (1972) Nematode pests of tea. In: Webster, J.M. (ed.) *Economic Nematology*. Academic Press, New York, pp. 285–310.

Skantar, A.M. and Carta, L.K. (2004) Molecular characterization and phylogenetic evaluation of the Hsp90 gene from selected nematodes. *Journal of Nematology* 36, 466–480.

Skantar, A.M., Handoo, Z.A., Zanakis, G.N. and Tzortzakakis, E.A. (2012) Molecular and morphological characterization of the corn cyst nematode, *Heterodera zeae*, from Greece. *Journal of Nematology* 44(1), 58–66.

Skantar, A.M., Handoo, Z.A., Kantor, M.R., Hult, M.N. and Hafez, S.A. (2019) First report of the cactus cyst nematode, *Cactodera cacti*, from a cactus garden in Idaho. *Journal of Nematology* 51, e2019-44. doi: 10.21307/jofnem-2019-044 e2019-44.

Skarbilovich, T.S. (1947) Revision of the systematics of the family Anguillulinidae Baylis and Daubney, 1962. *Dokladi Akademii Nauk SSSR* 57, 307–308.

Skarbilovich, T.S. (1959) On the structure of the nematodes of the order Tylenchida Thorne, 1949. *Acta Parasitologica Polonica* 15, 117–132.

Slabaugh, W.B. (1974) The nature of resistance in tomatoes to three nematodes. *Dissertation Abstracts International* 34B(11), 5280.

Smart, G.C. and Nguyen, K.B. (1991) Sting and awl nematodes: *Belonolaimus* spp. and *Dolichodorus* spp. In: Nickle, W.R. (ed.) *Manual of Agricultural Nematology*. Marcel Dekker, New York, pp. 627–667.

Smiley, R.W., Gourlie, J.A., Yan, G. and Rhinhart, K.E.L. (2014) Resistance and tolerance of Landrace wheat in fields infested with *Pratylenchus neglectus* and *P. thornei*. *Plant Disease* 98, 797–805.

Smiley, R.W., Dababat, A.A., Iqbal, S., Jones, M.G.K., Tanha Maafi, Z., Peng, D., Subbotin, S.A. and Waeyenberge, L. (2017) Cereal cyst nematodes: a complex and destructive group of *Heterodera* species. *Plant Disease* 101, 1692–1720.

Song, Z., Cheng, J., Cheng, F., Zhang, D. and Liu, Y. (2017) Development and evaluation of loop-mediated isothermal amplification assay for rapid detection of *Tylenchulus semipenetrans* using DNA extracted from soil. *Plant Pathology Journal* 33(2), 184–192.

Sopp, E. (1994) Studies on the resistance of grapevine rootstocks to virus-transmitting nematodes with particular regard to the nematode biocoenosis in vineyard soils. Untersuchungen zur Resistenz von Unterlagsreben gegenuber virusubertragenden Nematoden unter besonderer Berucksichtigung der Nematodenzonose in Weinbergsboden Geisenheim am Rhein, Germany. *Forschungsanstalt Geisenheim*, 17, pp. 95.

Sorribas, F.J., Verdejo-Lucas, S., Forner, J.B., Alcaide, A., Pons, J. and Ornat, C. (2000) Seasonality of *Tylenchulus semipenetrans* Cobb and *Pasteuria* sp. in citrus orchards in Spain. *Journal of Nematology* 32, 622–632.

Sorribas, F.J., Verdejo-Lucas, S., Pastor, J., Ornat, C., Pons, J. and Valero, J. (2008) Population densities of *Tylenchulus semipenetrans* related to physicochemical properties of soil and yield of clementine mandarin in Spain. *Plant Disease* 92(3), 445–450.

Sosa-Moss, C. (1977) Situación taxonómica actual del nemátodo enquistado que ataca al maiz en México. *Memorias del Séptimo Congreso Nacional de Fitopatologia*. Sociedad Mexicana de Fitopatologia Mexico, D.F.

Sosa-Moss, C. (1986) Cyst nematodes in Mexico, Central and South America. *Nematologia Mediterranea* 15, 13–19.

Sosa-Moss, C. (1987) Cyst nematodes in Mexico, Central and South America. *Nematologia Mediterranea*, 15, 1–12.

Sosa-Moss, C. and Gonzales, C.P. (1973) Respuesta del maiz chalquenó fertilizado y no fertilizado a cuatro diferentes niveles de *Heterodera punctata* raza Mexicana (Nematoda: Heteroderidae). *Nematropica* 3, 13–14.

Soteros, J.J. (1985) Strawberry nematode. *New Zealand Commercial Grower* 40, 44.

Souffer, R.F., Soulen, D.M. and Smith, S.H. (1975) Spread and control of Prunus stem pitting. *Acta Horticulturae* 44, 107–112.

Sousa, A.I., Gomes, V.F. and Tenente, R.C.V. (2003) Tratamento fisico aplicado as sementes de melao (*Cucumis melo* L.), importadas da Holanda, na erradicaçao de *Ditylenchus dipsaci* (Khun, 1857) Filipjev, 1936. *Nematologia Brasileira* 27, 223–225.

Southey, J.F. (1957) Observations on *Heterodera cacti* Filipjev et Sch. *Stekhoven* and *Meloidogyne* spp. on imported cactus plants with a list of new host records. *Nematologica* 2, 1–6.

Southey, J.F. (1972) *Anguina tritici*. *CIH Descriptions of Plant Parasitic Nematodes*, Set 1, No. 13. Commonwealth Institute of Helminthology. St Albans, UK.

Southey, J.F. (1973) *Anguina agrostis*. *CIH Descriptions of Plant Parasitic Nematodes*, Set 2, No. 20. Commonwealth Institute of Helminthology, St Albans, UK.

Southey, J.F. (1974) *Anguina graminis*. *CIH Descriptions of Plant-parasitic Nematodes*, Set 4, No. 53. CAB International, Wallingford, UK.

Southey, J.F. (1986) *Laboratory Methods for Work with Plant and Soil Nematodes*. Reference Book 402. Her Majesty's Stationery Office, London.

Southey, J.F. (1993) Nematode pests of ornamental and bulb crops. In: Evans, K., Trudgill, D.L. and Webster, J. M. (eds) *Plant Parasitic Nematodes in Temperate Agriculture*. CAB International, Wallingford, UK, pp. 463–500

Southey, J.F. and Aitkenhead, P. (1972) Soil borne pests and pathogens associated with imported plants, with special reference to nematodes. *Bulletin OEPP* 7, 49–59.

Spaull, V.W. and Cadet, P. (1990) Nematode parasites of sugarcane. In: Luc, M., Sikora, R.A. and Bridge, J. (eds) *Plant Parasitic Nematodes in Subtropical and Tropical Agriculture*. CAB International, Wallingford, UK, pp. 461–491.

Speijer, P.R. and Fogain, R. (1999) Musa and Ensete nematode pest status in selected African countries. In: Frison, E.A., Gold, C.S. and Sikora, R.A. (eds) *Mobilizing IPM for Sustainable banana product in Africa*. *Proceedings of workshop on banana IPM held in Nelspruit*, South Africa, 23–28 Novemver 1998. INIBAP, Montpellier, France, pp. 98–108.

Srivastava, A.N. and Chawla, G. (2005) Maize cyst nematode *H. zeae* a key nematode pest of maize and its management. *TB-ICN* 39/2005.

Srivastava, A.N. and Sethi, C.I. (1986) Seasonal population fluctuations of *Heterodera zeae*, *Rotylenchulus reniformis* and *Tylenchorhynchus vulgaris* in maize–cowpea–wheat rotation. *Indian Journal of Nematology* 16, 16–18.

Starr, J.L. (1984) Expression of resistance in peanuts to *Pratylenchus brachyurus*: impact on screening for resistance. *Journal of Nematology* 16(4), 404–406.

Steenkamp, S., De Waele, D. and McDonald, A. (2016) Reproduction and damage potential of five geographical *Ditylenchus africanus* populations on peanut. *Journal of Nematology* 48(2), 72–78.

Steiner, G.R. and Lehew, R.R. (1933) *Hoplolaimus bradys* n. sp. (Tylenchidae, Nematoda), the cause of a disease of yam (*Dioscorea* sp.). *Zoologischer Anzeiger* 101, 260–264.

Stelter, H. and Kuhn, H. (1973) The cactus cyst nematode, *Heterodera cacti*. *Nachrichtenblatt fur den Pflanzenschutzdienst in der DDR* 27(11), 239–240.

Stirling, G.R. (2014) *Biological Control of Plant-Parasitic Nematodes*. CABI International, Wallingford, UK.

Stirling, G.R. and McCulloch, J.S. (1984) *Paralongidorus australis* n. sp. (Nematoda: Longidoridae), causing poor growth of rice in Australia. *Nematologica* 30, 387–394.

Stoeen, M. (1977) Potato tuber eelworm, *Ditylenchus destructor*: attack and damage to potato and carrot. *Gartneryrket* 67(36), 1034–1036.

Stoller, B.B. (1957) An improved test for nematodes in the soil. *Plant Disease Reporter* 41, 531–532.

Stone, A.R. (1973) *Heterodera pallida* n.sp. (Nematoda: Heteroderidae), a second species of potato cyst nematode. *Nematologica* 18, 591–606.

Stone, A.R. (1983) Three approaches to the status of a species complex, with a revision of some species of *Globodera* (Nematoda: Heteroderidae). In: Stone, A.R., Platt, H.M. and Khalil, L.F. (eds) *Concepts in Nematode Systematics*. Proceedings International Symposium, Cambridge University, 2–4 Sept. 1981. Systematics Association Special Volume, 22. Academic Press London, UK, New York, pp. 221–233.

Stone, A.R. and Rowe, J.A. (1977) *Heterodera humuli*. *CIH Descriptions of Plant-parasitic Nematodes*, Set 7, No. 105. Commonwealth Institute of Helminthology, St Albans, UK.

Stone, A.R., Sosa-Moss, C. and Mulvey, R.H. (1976) *Punctodera chalcoensis* n.sp. (Nematoda: Heteroderidae) a cyst nematode from Mexico parasitising *Zea mays*. *Nematologica* 22(4), 381–389.

Storey, G.W. and Evans, K. (1987) Interactions between *Globodera pallida* juveniles, *Verticillium dahliae* and three potato cultivars, with descriptions of associated histopathologies. *Plant Pathology* 36, 192–200.

Stouffer, R.F. and Mowery, P.D. (1980) Effect of preplant soil fumigation on the control of Prunus stem pitting and the growth of Late Sunhaven peach trees. *Acta Phytopathologica Academiae Scientiarum Hungarica* 15, 247–250.

Stoyanov, D. (1967) Additions to host records of *Meloidogyne* sp., *Helicotylenchus multicinctus* and *Rotylenchulus reniformis*. *Nematologica* 13, 173.

Stoyanov, D. (1971) Some nematological problems of citrus in Cuba. *Revista de Agricultura* (Cuba) 4(2), 65–71.

Strümpel, H. (1967) Beobachtungen zur Lebensweise von *Aphelenchoides fragariae* in Lorraine-Begonien. *Nematologica* 13(1), 67–72.

Sturhan, D. (1962) Fber neue Wirtspflanzen der BlattSlchen *Aphelenchoides fragariae* und *A. ritzemabosi*, mit Bemerkungen zu den Wirtspflanzenkreisen beider Nematodenarten. *Anzeiger für SchSdlingskunde* 35(5), 65–67.

Sturhan, D. (1963) Der pflanzenparasitische Nematode *Longidorus maximus*, seine Biologie und okologie. Z. *Angewandte Zoology* 50, 129–193.

Sturhan, D. and Brzeski, M.W. (1991) Stem and bulb nematodes, *Ditylenchus* spp. In: Nickle, W.R. (ed.) *Manual of Agricultural Nematology*. Marcel Dekker Inc., New York, pp. 423–464.

Sturz, A.V. and Kimpinski, J. (2004) End-root bacteria derived from marigolds (*Tagetes* spp.) can decrease soil population densities of root-lesion nematodes in the potato root zone. *Plant and Soil* 262, 241–249.

Subbotin, S.A. (2018) Recombinase polymerase amplification assay for rapid detection of the root-knot nematode *Meloidogyne enterolobii*. *Nematology* 21(3), 243–251. doi: 10.1163/15685411-00003210.

Subbotin, S.A. and Riley, I.T. (2012) Stem and gall nematodes. In: Manzanilla-Lopez, R H and Marbán-Mendoza, N. (eds) *Practical Plant Nematology*. Biblioteca Basica de Agricultura, Instituto Interamericano de Cooperación para la Agricultura (IICA), San José, Costa Rica, pp. 521–577.

Subbotin, S.A., Waeyenberge, L., Molokanova, I.A. and Moens, M. (1999) Identification of *Heterodera avenae* group species by morphometrics and rDNARFLPs. *Nematology* 1, 195–207.

Subbotin, S.A., Waeyenberge, L. and Moens, M. (2000) Identification of cyst forming nematodes of the genus *Heterodera* (Nematoda: Heteroderidae) based on the ribosomal DNA-RFLP. *Nematology* 2, 153–164.

Subbotin, S.A., Vierstraete, A., DeLey, P., Rowe, J., Waeyenberge, L., Moens, M. and Vanfleteren, J.R. (2001) Phylogenic relationships within the cyst-forming nematodes (Nematoda, Heteroderidae) based on analysis of sequences from the ITS regions of ribosomal DNA. *Molecular Phylogenics and Evolution* 21, 1–16.

Subbotin, S.A., Sturhan, D., Rumpenhorst, H.J. and Moens, M. (2002) Description of the Australian cereal cyst nematode *Heterodera australis* sp. n. (Tylenchida: Heteroderidae). *Russian Journal of Nematology* 10, 139–148.

Subbotin, S.A., Sturhan, D., Rumpenhorst, H.J. and Moens, M. (2003) Molecular and morphological characterisation of the *Heterodera avenae* species complex (Tylenchida: Heteroderidae). *Nematology* 5(4), 515–538.

Subbotin, S.A., Krall, E.L., Riley, I.T., Chizhov, V.N., Staelens, A., DeLoose, M. and Moens, M. (2004) Evolution of the gall-forming plant parasitic nematodes (Tylenchida: Anguinidae) and their relationships with hosts as inferred from Internal Transcribed Spacer sequences of nuclear ribosomal DNA. *Molecular Phylogenetics and Evolution* 30, 226–235.

Subbotin, S.A., Madani, M., Krall, E., Sturhan, D., and Moens, M. (2005a) Molecular diagnostics, taxonomy, and phylogeny of the stem nematode *Ditylenchus dipsaci* species complex based on the sequences of the internal transcribed spacer-rDNA. *Phytopathology* 95, 1308–1315.

Subbotin, S., Vovlas, N., Crozzoli, R., Sturhan, D., Lamberti, F., Moens, M. and Baldwin, J. (2005b) Phylogeny of Criconematina Siddiqui, 1980 (Nematoda: Tylenchida) based on morphology and D2–D3 expansion segments of the 28S-rRNA gene sequences with application of a secondary structure model. *Nematology* 7, 927–944.

Subbotin, S.A., Sturhan, D., Chizhov, V.N., Vovlas, N. and Baldwin, J.G. (2006) Phylogenetic analysis of Tylenchida Thorne, 1949 as inferred from D2 and D3 expansion fragments of the 28S rRNA gene sequences. *Nematology* 8, 455–474.

Subbotin, S.A., Sturhan, D., Vovlas, N., Castillo, P., Tanyi, T.J., Moens, M. and Baldwin, J.G. (2007) Application of the secondary structure model of rRNA for phylogeny: D2– D3 expansion segments of the LSU gene of plant-parasitic nematodes from the family Hoplolaimidae Filipjev, 1934. *Molecular Phylogenetics and Evolution* 43, 881–890.

Subbotin, S.A., Ragsdale, E.J., Mullens, T., Roberts, P.A., Mundo-Ocampo, M. and Baldwin, J.G. (2008) A phylogenetic framework for root-lesion nematodes of the genus *Pratylenchus* (Nematoda): evidence from 18S and D2–D3 expansion segments of 28S ribosomal RNA genes and morphological characters. *Molecular Phylogenetics and Evolution* 48, 491–505.

Subbotin, S.A., Mundo-Ocampo, M. and Baldwin, J.G. (2010) Systematics of cyst nematodes (Nematoda: Heteroderinae). Vol. 8A in: Hunt, D.J. and Perry, R.N. (series eds) *Nematology Monographs and Perspectives*. Brill, Leiden, Netherlands

Subbotin, S.A., Cid, D.P.V.I., Mundo-Ocampo, M. and Baldwin, J.G. (2011a) Identification, phylogeny and phylogeography of circumfenestrate cyst nematodes (Nematoda: Heteroderidae) as interfered from analysis of ITS-rDNA. *Nematology* 13, 805–824.

Subbotin, S.A., Deimi, A.M., Zheng, J. and Chizhov, V.N. (2011b) Length variation and repetitive sequences of Internal Transcribed Spacer of ribosomal RNA gene, diagnostics and relationships of populations of potato rot nematode, *Ditylenchus destructor* Thorne, 1945 (Tylenchida: Anguinidae). *Nematology* 13(7), 773–785.

Subbotin, S.A., Inserra, R.N., Marais, M., Mullin, P., Powers, T.O. *et al.* (2011c) Diversity and phylogenetic relationships within the spiral nematodes of *Helicotylenchus* Steiner, 1945 (Tylenchida: Hoplolaimidae) as inferred from analysis of the D2-D3 expansion segments of 285 rRNA gene sequences. *Nematology* 13, 333–345.

Subbotin, S.A., Waeyenberge, L. and Moens, M. (2013) Molecular systemics. In: Perry, R. and Moens, M. (eds) *Plant Nematology*, 2nd edn. CAB International, Wallingford, UK, pp. 40–72.

Subbotin, S.A., Chitambar, J.J., Chizhov, V.N., Stanley, J.D., Inserra, R.N., Doucet, M.E. *et al.* (2014) Molecular phylogeny, diagnostics, and diversity of plant-parasitic nematodes of the genus *Hemicycliophora* (Nematoda: Hemicycliophoridae). *Zoological Journal of the Linnean Society* 171, 475–506.

Subbotin, S.A., Vovlas, N., Yeates, G.W., Hallmann, J., Kiewnick, S., Chizhov, M.L.R.H., Inserra, R.N. and Castillo, P. (2015) Morphological and molecular characterisation of *Helicotylenchus pseudorobustus* (Steiner, 1914) Golden, 1956 and related species (Tylenchida: Hoplolaimidae) with a phylogeny of the genus. *Nematology* 17, 27–52.

Subbotin, S.A., Toumi, F., Elekçiog˘ Lu, I.H., Waeyenberge, L. and Tanha Maafi,.Z. (2018) DNA barcoding, phylogeny and phylogeography of the cyst nematode species of the Avenae group from the genus *Heterodera* (Tylenchida: Heteroderidae). *Nematology* 20, 671–702.

Subbotin, S.A., Cid Del Prado Vera, I., Inserra, R.N., Chizhov, V.N. and Decraemer, W. (2019a) Molecular characterisation of some stubby root nematodes (Nematoda: Trichodoridae) from the USA and other countries. *Nematology* 22(1), 39–52. doi: 10.1163/15685411-00003279.

Subbotin, S.A., Franco, J., Knoetze, R., Roubtsova, T.V., Bostock, R.M. and Cid Del Prado Vera, I. (2019b) DNA barcoding, phylogeny and phylogeography of the cyst nematode species from the genus *Globodera* (Tylenchida: Heteroderidae). *Nematology* 22(3), 269–297. doi: 10.1163/15685411-00003305.

Subramanian S., Banu G. and Umamaheswari, R. (2010) Nematode infestation in coffee. In: Khan, M.R. and Jairajpuri, M.S. (eds) *Nematode Infestations – Part II: Industrial Crops/* National Academy of Sciences, Allahabad (Prayagraj), UP, India, pp. 149–170.

Subramaniyan, S., Balasubramanian, P., Sundarababur, R., Naganathan, T.G., Lakshmanan, P.L., Rajendran, G., Sivakumar, C.V. and Vadivelu, S. (1989) Present status of the potato cyst nematodes in Nilgiris, Tamil Nadu. *Current Science* 58, 701–702.

Sud, U.C., Varaprasad, K.J., Seshadri, A.R. and Kher, K.K. (1984) Relationship between initial densities of *Rotylenchulus reniformis* and damage to cotton with fit to Seinhorst curves. *Indian Journal of Nematology* 14, 148–151.

Sudakova, I.M. (1968) Effect of temperature on the life cycle of *Aphelenchoides besseyi*. *Parasitologiya* 2, 71–74.

Sudakova, I.M. and Maslennikova, V.F. (1968) *Aphelenchoides besseyi* Christie causing white tip disease in rice. Mater. nauch. Konf. Obshch. Gel'mInt. Uzbek, September, 1968, Tashkent: *Izdat. Akadamiya Nauk. Uzbek. SSR*, pp. 103–105.

Sukul, N.C. (1970) Inhibition of nematode infestation of wheat seedlings by *Polygonium hydropiper*. *Nature* 226, 771–772.

Sullivan, M. (2013) *CPHST Pest Datasheet for Bursaphelenchus cocophilus*. USDA-APHIS-PPQ-CPHST [USDA Animal and Plant Health Inspection Service, Plant Protection and Quarantine Division, Center for Plant Health Science and Technology]

Sullivan, M., Mackesy, D. and Molet, T. (2014) CPHST Pest Datasheet for *Heterodera sacchari*. USDA-APHIS-PPQ-CPHST

Sultana, T., Kim, J., Lee, S. H., Han, H., Kim, S., Min, G. S., Nadler, S. A. and Park, J. K. (2013) Comparative analysis of complete mitochondrial genome sequences confirms independent origins of plant-parasitic nematodes. *BMC Evolutionary biology* 13, 12. doi: 10.1186/1471-2148-13-12

Sun, X.T., Hu, C.Z., Jiang, Q., Chen, Y.Y. and Cui, R.Q. (2013) Morphological identification of six rice root nematodes in Jiangxi Province. *Acta Agriculturae Universitatis Jiangxiensis* 35(6), 179–182.

Susulovska, S., Susulovsky, A. and Kornobis, F.W. (2016) Morphometrical and molecular data on plant parasitic nematodes *Longidorus attenuatus* Hooper, 1961 and *L. danuvii* Barsi et al., 2007 (Nematoda: Longidoridae) reported from Ukraine for the first time. *Helminthologia* 53, 396–400.

Swart, A., Malan, A.P. and Heyns, J. (1996) *Paralongidorus maximus* (Butschli) Siddiqi (Dorylaimida: longidoridae) from South Africa. *African Plant Protection* 2(1), 31–34.

Swarup, G. and Sosa-Moss, C. (1990) Nematode parasites of cereals. In: Luc, M., Sikora, R.A. and Bridge, J. (eds) *Plant Parasitic Nematodes in Subtropical and Tropical Agriculture*. CAB International, Wallingford, UK, pp. 109–136.

Swarup, G., Prasad, S.K. and Raski, D.J. (1964) Some Heterodera species from India. *Plant Disease Reports* 48, 236.

Swofford, D. (1999) PAUP: *Phylogenetic Analysis Using Parsimony*. Sinauer Associates, Sunderland, Massachusetts.

Szczygiel, A. (1966) Studies on the fauna and population dynamics of nematodes occurring on strawberry plantations. *Ekologia polska* 14, 651–709.

Szczygiel, A. and Hasior, H. (1971) Possibility of persistence of leaf and bud nematode (*Aphelenchoides fragariae*) on strawberry plants and in the soil. *Zesz. probl. Postep, Nauk Roln* 121, 101–106.

Szczygiel, A. and Hasior, H. (1972) Seasonal variations in population of plant parasitic nematodes in strawberry plantations. *Ekologia Polska* 20(38), 507–523.

Tacconi, R. (1972) Infestations of *Aphelenchoides fragariae*, *Aphelenchoides ritzemabosi* and *Ditylenchus dipsaci* on strawberry in some Italian provinces. *Redia* 53, 313–319.

Tacconi, R. and Ambrogioni, L. (1995) *Quarantine Nematodes*. Editrice 'Lo Scarabeo'. Bologna, Italy.

Tacconi, R. and Talame, F.M. (1995) Parasitic nematodes of stone and pome fruits and vegetable crops. *Rivista di Frutticoltura e di Ortofloricoltura* 57(3), 63–71.

Tacconi, R., Zani, A. and Pasini, P. (1982) Control trial against *Aphelenchoides fragarip* on strawberry using chemical and physical means. *Frutticoltura* 44(5), 102–106.

Tacconi, R., Santi, R. and Gironi, R. (1999) Population dynamics of *Aphelenchoides besseyi* on rice and effect of seed treatments on seed for propagation. *Nematologia Mediterranean* 27(2), 291–293.

Takagi, M., Goto, M., Wari, D., Kashima, T. and Toyota, K. (2019) Seasonal occurrence and life cycle of lotus root nematode *Hirschmanniella diversa* (Tylenchida: Pratylenchidae) in lotus roots in paddy fields. *Applied Entomology and Zoology* 54, 465–471.

Takagi, M., Goto, M., Wari, D., Saito, M., Perry, R.N. and Toyota, K. (2020) Screening of nematicides against the lotus root nematode, *Hirschmanniella diversa* Sher (Tylenchida: Pratylenchidae) and the efficacy of a selected nematicide under lotus micro-field conditions. *Agronomy*, 10, 373. doi: 10.3390/agronomy10030373

Takaji, K. (1969) Nematodes confront tea plantations in Japan. *Japanese Agricultural Research Quarterly* 4, 27–2.

Takeuchi, Y., Kanzaki, N. and Futai, K. (2005) A nested PCR-based method for detecting the pine wood nematode, *Bursaphelenchus xylophilus*, from pine wood. *Nematology* 7, 775–782.

Tandingan, D.L.I., Mundo-Ocampo, M., Yoder, M. and De Ley, P. (2007) Nematodes from vernal pools in the Santa Rosa Plateau Ecological Reserve, California I. *Hirschmanniella santarosae* sp.n. (Nematoda: Pratylenchidae), a cryptic sibling species of *H. pomponiensis* Abdel-Rahman and Maggenti, 1987. *Nematology* 9, 405–429.

Tanha Maafi, Z. and Damadzadeh, M. (2008) Incidence and control of the citrus nematode, *Tylenchulus semipenetrans* Cobb. in the north of Iran. *Nematology* 10(1), 113–122.

Tanha Maafi, Z., Subbotin, S.A. and Moens, M. (2003) Molecular identification of cyst-forming nematodes (Heteroderidae) from Iran and a phylogeny based on ITSrDNA sequences. *Nematology* 5, 99–111.

Tanha Maafi, Z., Sturham, D., Handoo, Z., Mor, M., Moens, M. and Subbotin, S.A. (2007) Morphological and molecular studies on *Heterodera sacchari*, *H. goldeni*, and *H. leuceilyma* (Nematoda: Heteroderidae). Nematology 9(4), 483–497.

Tanha Maafi, Z., Amani, M., Stanley, J.D., Inserra, R.N., Berg, E.V.D. and Subbotin, S.A. (2012) Description of *Tylenchulus musicola* sp. n. (Nematoda: Tylenchulidae) from banana in Iran with molecular phylogeny and characterisation of species of *Tylenchulus* Cobb, 1913. *Nematology* 14, 353–369.

Tarès, S., Lemontey, J.M., de Guiran, G. and Abad, P. (1993) Cloning and characterization of a highly conserved satellite DNA sequence specific for the phytoparasitic nematode *Bursaphelenchus xylophilus*. *Gene* 129, 269–273.

Tarjan, A.C. (1956) Known and suspected plant-parasitic nematodes of Rhode Island. II. *Xiphinema americanum* with notes on Tylencholaimus brevicaudatus, N. comb. *Proceedings of the Helminthological Society of Washington* 23, 88–92.

Tarjan, A.C. (1971) Migration of three pathogenic citrus nematodes through two Florida soils. *Proceedings of Soil and Crop Science Society, Florida* 31, 253–255.

Tarjan, A.C. and Frederick, J.J. (1984) Comparative nematicidal efficacies of several commercial products on bermudagrass. *Proceedings of the Florida State Horticultural Society* 97, 300–304.

Tarjan, A.C. and Weischer, B. (1965) Observations of some Pratylenchinae (Nemata), with additional data on *Pratylenchoides guevarai* Tobar Jiménez, 1963 (syn. *Zygotylenchus browni* Siddiqi, 1963 and *Mesotylus gallicus* de Guiran, 1964). *Nematologica* 11, 432–440.

Tastet, C., Val, F., Lesage, M., Renault, L., Marche, L., Bossis, M. and Mugniery, D. (2001) Application of a putative fatty-acid binding protein to discriminate serologically the two European quarantine root-knot nematodes, *Meloidogyne chitwoodi* and *M. fallax*, from other *Meloidogyne* species. *European Journal of Plant Pathology* 107, 821–832.

Taylor, A.L. (1969) Nematode parasites of rice. In: Peachey, J.E. (ed.) *Nematodes of Tropical Crops* Commonwealth Bureau of Helminthology, St Albans, UK. pp. 264–268.

Taylor, A.L. and Sasser, J.N. (1978) *Biology, Identification and Control of Root-Knot Nematodes,* Meloidogyne *spp.* North Carolina State University, Raleigh, North Carolina.

Taylor, C.E. (1967) The multiplication of *Longidorus elongatus* (de Man) on different host plants with reference to virus transmission. *Annals of Applied Biology* 59, 275–281.

Taylor, C.E. and Brown, D.J.F. (1981) Nematode–virus interactions. In: Zuckerman, B.M. and Rohde, R.A. (eds) *Plant Parasitic Nematodes, Vol. III*. Academic Press Inc., New York and London, pp. 281–301.

Taylor, C.E. and Brown. D.J.F. (1997) *Nematode Vectors of Plant Viruses*. CAB International. Wallingford, UK.

Taylor, C.E. and Gordon, S.C. (1970) A comparison of four nematicides for the control of *Longidorus elongatus* and *Xiphinema diversicaudatum* and the viruses they transmit. *Horticultural Research* 10, 133–141.

Taylor, C.E. and Murant, A.F. (1966) Nematicidal activity of aqueous extracts from raspberry canes and roots. *Nematologica* 12, 488–494.

Taylor, C.E. and Murant, A.F. (1968) Chemical control of raspberry ringspot and tomato black ring viruses in strawberry. *Plant Pathology* 17, 171–178.

Taylor, C.E. and Thomas, P.R. (1968) The association of *Xiphinema diversicaudatum* (Micoletzky) with strawberry latent ringspot and arabis mosaic viruses in a raspberry plantation. *Annals of Applied Biology* 62, 147–157.

Taylor, C.E., Alphey, T.J.W. and Brown, D.J.F. (1978) The distribution of nematode virus-vectors in Great Britain. In: Scott, P.R. and Bainbridge, A. (eds) *Plant Disease Epidemiology*. Blackwell Scientific Publications, Oxford, UK, pp. 265–273.

Taylor, C.R. and Raski, D.J. (1964) On the transmission of grape fanleaf by *Xiphinema index*. *Nematologica* 10, 489–4950.

Taylor, L.R., Sasser, J.N. and Nelson, L.A. (1982) *Relationships of climate and soil characteristics to geographical distribution of* Meloidogyne *species in agricultural soils*. Cooperative Publication, Department of Plant Pathology, North Carolina State University and US Agency for International Development, Raleigh, North Carolina.

Taylor, S.P. (2000) The root-lesion nematode, *Pratylenchus negelectus*, in field crops in Australia. PhD thesis, University of Adelaide, Australia.

Teliz, D., Grogan, R.G. and Lownsbery, B.F. (1966) Transmission of tomato ringspot, peach yellow bud mosaic, and grape yellow vein diseases by *Xiphinema americanum*. *Phytopathology* 56, 658–663.

Tenente, R.C.V. and Evans, A.A.F. (1997) Electrophoresis of proteins from several races of *Ditylenchus dipsaci*, recovered from dried infested courgette tissue. *Nematologia Brasileira* 21, 84–91.

Tenente, R.C.V. and Manso, E.S.B.G.C. (1994) Chemical and heat treatments of rice seeds infested with *Aphelenchoides besseyi*. *Nematologia Brasileira* 18 (1-2), 28–34.

Thakar, N.A., Patel, B.K., Patel, H.R. and Patel, C.C. (1987) Occurrence of white tip disease of rice in Gujarat. *Gujarat Agricultural University Research Journal* 13(1), 65.

Thakur, N.A. and Patel, C.C. (1984) Screening of few cowpea varieties against reniform nematodes *Rotylenchulus reniformis* Linford and Oliviera. *Indian Journal of Nematology* 14, 204.

Thames, W.H. and Heald, C.M. (1974) Chemical and cultural control of *Rotylenchulus reniformis* on cotton. *Plant Disease Reporter* 58, 337–341.

Thiéry, M. and Mugniéry, D. (1996) Interspecific rDNA restriction fragment length polymorphism in *Globodera* species, parasites of solanaceous plants. *Fundamental and Applied Nematology* 19, 471–479.

Thomas, P.R. (1968) *Aphelenchoides ritzemabosi* on Saintpaulia. *Plant Pathology* 17, 94–95.

Thomason, I.J. and McKenry, M. (1975) Chemical control of nematode vectors of plant viruses. In: Lamberti, F., Taylor, C.E. and Seinhorst, J.W. (eds) *Nematode Vectors of Plant Viruses*. Plenum Press. London, pp. 423–438.

Thompson, A.K., Been, B.O. and Perkins, C. (1973) Nematodes in stored yams. *Experimental Agriculture* 9, 281–286.

Thompson, J.P., Owen, K.J., Stirling, G.R. and Bell, M.J. (2008) Root-lesion nematodes (*Pratylenchus thornei* and *P. neglectus*): a review of recent progress in managing a significant pest of grain crops in northern Australia. *Australasian Plant Pathology* 37, 235–242.

Thorne, G. (1935) The sugar beet nematode and other indigenous nemic parasites of shadscale. *Journal of Agricultural Research* 51, 509–514.

Thorne, G. (1945) *Ditylenchus destructor*, n.sp., the potato rot nematode, and *Ditylenchus dipsaci* (Kühn, 1857) Filipjev, 1936, the teasel nematode (Nematoda: Tylenchidae). *Proceedings Helminthological Society of Washington* 12, 27–34.

Thorne, G. (1961) *Principles of Nematology*. McGraw Hill, New York and London.
Thorne, G. and Allen, M.W. (1944) *Nacobbus dorsalis*, nov. gen., nov. spec. (Nematoda: Tylenchidae) producing galls on roots of alfileria, *Erodium cicutarium* (L.) L'Hér. *Proceedings of the Helminthological Society of Washington* 11, 27–31.
Thorne, G. and Allen, M.W. (1950) *Paratylenchus hamatus* n.sp, and *Xiphinema index* n.sp., two nematodes associated with fig roots, with a note on *Paratylenchus* anceps Cobb. *Proceedings of the Helminthological Society of Washington* 17(1): 27–35
Tiilikkala, K., Carter, T., Heikinheimo, M. and Venalainen, A. (1995) Pest risk analysis of *Meloidogyne chitwoodi* for Finland. *Bulletin OEPP* 25(3), 419–435.
Tikhonova, L.V. (1966) Bioecology of the agent responsible for white tip disease of rice: *Aphelenchoides besseyi*. *Vest. sel'-khoz. Nauki Alma-Ata*. 2, 45–47.
Timm, R.W. (1965) *A Preliminary Study of the Plant Parasitic Nematodes of Thailand and the Philippines*. SEATO, Secretariat General, Bangkok.
Timper, P. and Brodie, B.B. (1994) Effect of *Hirsutella rhossiliensis* on infection of potato by *Pratylenchus penetrans*. *Journal of Nematology* 26, 304–307.
Tiyagi, S.A., Khan, A.V. and Alam, M.M. (2002) Biodegradable effect of oil seed cakes on plant parasitic nematodes and soil inhabiting fungi infesting. *Trigonella foenum-greacum* and *Phaseolus aureus*. *Indian Journal of Nematology* 32, 47–57.
Tobar Jiménez, A. (1963) *Pratylenchoides guevarai* n.sp., nuevo nematode Tylénchido, relacionado con el ciprés (*Cupressus sempervirens* L.). *Revista Ibérica de Parasitologia* 23, 27–36.
Tobar Jiménez, A. (1973) Nematodes of the unirrigated areas of Alhama. I. Population levels and host crops. *Revista Ibérica de Parasitologia* 33(4), 525–556.
Tobar Jiménez, A. and Martinez Sierra, C. (1966) El *Pratylenchoides guevarai* Tobar Jiménez, 1963, parásito grave de la violeta (*Viola odorata* L.) grandina. *Revista Ibérica de Parasitologia*, 26(2/3), 233–238.
Tobar Jimenez, A., Gallardo Bernal, M. and Palacios Mejia, F. (1974) Some hosts of various nematodes of the order Tylenchida. *Revista Ibérica de Parasitologia*, 34(3/4), 237–244.
Todd, E.H. and Atkins, J.G. (1958) White tip disease of rice. I. Symptoms, laboratory culture of nematodes, and pathogenicity tests. *Phytopathology*, 48, 632–637.
Tomaszewski, E.K., Khalil, M.A.M., El-Deeb, A.A., Powers, T.O. and Starr, J.L. (1994) *Meloidogyne javanica* parasitic on peanut. *Journal of Nematology* 26, 436–441.
Townshend, J.L. and Anderson, R.V. (1976) *Pratylenchus neglectus* [=*P. minyus*]. *CI. Descriptions of Plant-parasitic Nematodes*, Set 6, No. 82. Commonwealth Institute of Helminthology, St Albans, UK.
Townshend, J.L. and Davidson, T.R. (1989) Extraction, fecundity, and hosts of *Criconemella rustica*. *Nematologica* 35(1), 76–81.
Toyohara, G. and Fujihara, M. (1998) Succession of secondary forests affected by pine wilt disease in western Japan followed on vegetation maps. *Applied Vegetation Science* 1, 259–266. doi: 10.2307/1478956
Triantaphyllou, A.C. (1979) Cytogenetics of root-knot nematodes. In: Lamberti, F. and Taylor, C.E. (eds) *Root-knot Nematodes* (Meloidogyne *species*); *Systematics, Biology and Control*. Academic Press, London and New York, pp. 85–109.
Troccoli, A., Subbotin, S.A., Chitambar, J.J., Janssen, T., Waeyenberge, L., Stanley, J.D. *et al.* (2016) Characterisation of amphimictic and parthenogenetic populations of *Pratylenchus bolivianus* Corbett, 1983 (Nematoda: Pratylenchidae) and their phylogenetic relationships with closely related species. *Nematology* 18, 651–678.
Trudgill, D. (1995) Origins of root-knot nematodes (*Meloidogyne* spp., Nematoda) in relation to their cultural control. *Phytoparasitica* 23, 191–194.
Trudgill, D.L. (1967) The effect of environment on sex determination in *Heterodera rostochiensis*. *Nematologica* 13, 263–272.
Trudgill, D.L. and Alphey, T.J.W. (1976) Chemical control of the virus-vector nematode *Longidorus elongatus* and of *Pratylenchus crenatus* in raspberry plantations. *Plant Pathology* 25(1), 15–20.
Trujillo, E.E. (1968) Diseases of tuberose in Hawaii. *University of Hawaii, Cooperative Extension Service, Circular* 427, 3–13.
Tsai, B.Y. and Apt, W.J. (1979) Anhydrobiosis of the reniform nematode: survival and coiling. *Journal of Nematology* 11, 316.
Tsay, T.T. (1997) Introductive occasions of exotic plant-parasitic nematodes into Taiwan and the retarding strategies. *Plant Protection Bulletin (Taipei)* 39(1), 33–61, 105.
Tsay, T.T., Cheng, Y.H., Teng, Y.C., Lee, M.D., Wu, W.S and Lin, Y.Y. (1998) Bionomics and control of rice white tip disease nematode, *Aphelenchoides besseyi*. *Plant Protection Bulletin, Taipei* 4 (3), 277–286.
Turner, D.R. and Chapman, R.A. (1972) Infection of seedlings of alfalfa and red clover by concomitant populations of *Meloidogyne incognita* and *Pratylenchus penetrans*. *Journal of Nematology* 4, 2806.
Turner, S.J and Subbotin, S.A. (2013) Cyst nematodes. In: Perry, R.N. and Moens, M. (eds) *Plant Nematology*, 2nd edn. CAB International, Wallingford, UK, pp. 109–143.
Tzortzakakis, E.A., Cantalapiedra-Navarrete, C., Archidona-Yuste, A., Palomares-Rius, J.E. and Castillo, P. (2016) First Report of the Carolina Spiral Nematode, *Scutellonema brachyurus*, from soil of a garden in Crete, Greece. *Journal of Nematology* 48(1), 7–7.
Uehara, T., Mizukubo, T., Kushida, A. and Momota, Y. (1998a) Identification of *Pratylenchus coffeae* and *P. loosi* using specific primers for PCR amplification of ribosomal DNA. *Nematologica* 44, 357–368.
Uehara, T., Mizukubo, T., Kushida, A. and Momota, Y. (1998b) Identification of *Pratylenchus penetrans* (Cobb) by PCR using ITS-based species-specific primers. *Japanese Journal of Nematology* 28, 1–7.
Uematsu, S., Yabu, T., Yao, M., Kurihara, T. and Koga, H. (2019) Light and scanning electron microscopy of the Indian lotus roots invaded by *Hirschmanniella diversa*. *Nematological Research* 46(2), 79–82.
Umarao, G.T., Kaushal, K.K., and Sashi, V. (2008) Molecular characterization of Indian populations of maize cyst nematode, *Heterodera zeae* using PCR-RFLP of ribosomal DNA. *Indian Journal of Nematology* 38, 81–85.
USDA (2012) *Importation of* Orzya sativa *L. Seed into the United States from Countries in the Western Hemisphere and Europe. Version 2*. Animal and Plant Health Inspection Service, United States Department of Agriculture, Washington DC.
USDA (2015) *Qualitative pest risk analysis for the wheat gall nematode,* Anguina tritici, *in US*. Plant Epidemiology and Risk Analysis Laboratory, Center for Plant Health Science and Technology, Plant Protection and Quarantine, Animal and Plant Health Inspection Service, United States Department of Agriculture, Washington DC.
USDA-PCIT (2014) USDA Phytosanitary Certificate Issuance and Tracking System. Available at: https://pcit.aphis.usda.gov/pcit/ (accessed 21 August 2014)
USDA-PCIT (2017) USDA Phytosanitary Certificate Issuance and Tracking System. Available at: https://pcit.aphis.usda.gov/pcit/ (accessed 4 April 2017)
USDA-PCIT (2018) USDA Phytosanitary Certificate Issuance and Tracking System. Available at: https://pcit.aphis.usda.gov/PExD/faces/ReportHarmOrgs.jsp (accessed 4 July 2018)
USDA-PCIT (2019) Phytosanitary Certificate Issuance and Tracking System. *Trichodorus minor*. Queried February 8, 2019. Available at: https://pcit.aphis.usda.gov/pcit/
Usmanova, A.Z. (1972) *Ditylenchus destructor*, a serious parasite of potatoes. In: *Gel'minty pishchevykh produktov. Tezisy dokladov mezhrespublikanskoi nauchnoi konferentsii*, 22–25 Dekabrya 1972 goda. Samarkandskii Kooperativnyi Institut im. V.V. Kuibysheva. Samarkand, USSR, pp. 69–70.

Uzma, I., Nasira, K., Firoza, K. and Shahina, F. (2015) Review of the genus *Helicotylenchus* Steiner, 1945 (Nematoda: Hoplolaimidae) with updated diagnostic compendium. *Pakistan Journal of Nematology* 33(2), 115–160.

Vadivelu, R.S., Rajendran, G., Naganathan, T.G. and Jayaraj, S. (1987) Studies on the nematode pests of banana and their management – a review. In: *Proceedings Third Group Discussions on the Nematological Problems of Plantation Crops*. Sugarcane Breeding Institute, Coimbatore, pp. 32–33.

Vaish, S.S., Singh, U.P., Pandey, S. and Upadhyaya, V. (2016) Root knot nematode (*Meloidogyne graminicola*): an emerging concern of wheat and barley in rice growing regions. In: Vaish, S.S. (ed.) *Diseases of Wheat and their Management*. Astral International, New Delhi, India, pp. 169–196.

Van Den Berg, E. (1998) New records and notes on known specie of Hoplolaimidae (Nemata) in South Africa. *Journal of Nematode Morphology and Systematics* 1, 29–46.

Van Den Berg, E. and Heyns, J. (1973) South African Hoplolaiminae. 2. The genus *Scutellonema* Andrassy, 1958. *Phytophylactica* 5, 23–40.

Van Den Berg, E. and Heyns, J. (1975) South African Hoplolaiminae. 4. The genus *Helicotylenchus* Steiner, 1945. *Phytophylactica* 7(1), 35–52.

Van Den Berg, E. and Kirby, M.F. (1979) Some spiral nematodes from the Fiji Islands (Hoplolaimidae: Nematoda). *Phytophylactica* 11(2), 99–109.

Van Den Berg, E., Subbotin, S.A., Handoo, Z. A. and Tied, L.R. (2009) *Hirschmanniella kwazuna* sp. n. from South Africa with notes on a new record of *H. spinicaudata* (Schuurmans Stekhoven, 1944) Luc and Goodey, 1964 (Nematoda: Pratylenchidae) and on the molecular phylogeny of *Hirscmanniella* Luc and Goodey, 1964. *Nematology* 11(4), 523–540.

Van Den Berg, E., Tiedt, L.R. and Subbotin, S.A. (2012) Morphological and molecular characterisation of *Criconemoides brevistylus* Singh and Khera, 1976 and *C. obtusicaudatus* Heyns, 1962 from South Africa (Nematoda: Criconematidae) with first description of a male *C. obtusicaudatus* and proposal of new synonyms. *Nematology* 14, 961–976.

Van Den Berg, E., Tiedt, L.R., Coyne, D.L., Ploeg, A.T., Navas-Cortés, J.A., Roberts, P.A., Yeates, G.W. and Subbotin, S.A. (2013) Morphological and molecular characterisation and diagnostics of some species of *Scutellonema* Andrássy, 1958 (Tylenchida: Hoplolaimidae) with a molecular phylogeny of the genus. *Nematology* 15, 719–745.

Van Den Berg, E., Tiedt, L.R., Inserra, R.N., Stanley, J.D., Vovlas, N., Palomares-Rius, J.E., Castillo, P. and Subbotin, S.A. (2015) Characterisation of a topotype and other populations of *Hemicriconemoides strictathecatus* Esser, 1960 (Nematoda: Criconematidae) from Florida with description of *H. phoenicis* sp. n. from the USA. *Nematology* 17, 265–300.

Van Den Berg, E., Palomares-Rius, J.E., Vovlas N., Tiedt L.R., Castillo, P. and Subbotin, S.A. (2016) Morphological and molecular characterisation of one new and several known species of the reniform nematode, *Rotylenchulus* Linford and Oliveira, 1940 (Hoplolaimidae: Rotylenchulinae), and a phylogeny of the genus. *Nematology* 18, 67–107.

Van Den Berg, E., Tiedt, L.R., Stanley, J.D., Inserra, R.N. and Subbotin, S.A. (2017) Characterisation of some *Scutellonema* species (Tylenchida: Hoplolaimidae) occurring in Botswana, South Africa, Costa Rica and the USA, with description of *S. clavicaudatum* sp. n. and a molecular phylogeny of the genus. *Nematology* 19, 131–173.

Van Den Berg, E., Tiedt, L.R., Liebanas, G., Chitambar, J.J., Stanley, J.D., Inserra, R.N., Castillo, P. and Subbotin, S.A. (2018) Morphological and molecular characterization of two new *Hemicycliophora* species (Tylenchida: Hemicycliophoridae) with a revision of the taxonomic status of some known species and a phylogeny of the genus. *Nematology* 20, 319–354.

Van der Beek, J.G. and Karssen, G. (1997) Interspecific hybridization of meiotic parthenogenetic *Meloidogyne chitwoodi* and *M. fallax*. *Phytopathology* 87, 1061–1066.

Van der Vegte, F.A. and Daiber, K.C. (1983) A preliminary report on the occurrence of *Ditylenchus destructor* on the ornamental *Liatris spicata* and efforts to eradicate the former. *Proceedings of the 6th Symposium and General Meeting of the Nematological Society of Southern Africa*.

Van Gundy, S.D. (1958) The life history of the citrus nematode, *Tylenchulus semipenetrans* Cobb. *Nematologica* 3, 283–294.

Van Gundy, S.D. (1959) The life history of *Hemicycliophora arenaria* Raski (Nematoda: Criconematidae). *Proceedings of Helminthological Society, Washington* 26, 67–72.

Van Gundy, S.D. and McElroy S.D. (1969) Sheath nematode: its biology and control. *Proceeding of 1st International Citrus Symposium* 2, 985–989.

Van Gundy, S.D. and Rackham, R.L. (1961) Studies on the biology and pathogenicity of *Hemicycliophora arenaria*. *Phytopathology* 51, 393–397.

Van Megen, H., van den Elsen, S., Holterman, M., Karssen, G., Mooyman, P., Bongers, T., Holovachov, O., Bakker, J. and Helder, J. (2009) A phylogenetic tree of nematodes based on about 1200 full-length small subunit ribosomal DNA sequences. *Nematology* 11, 927–950.

Vanstone, V.A. and Nicol, J.M. (1993) Factors affecting pathogenicity and multiplication of *Pratylenchus neglectus* and *P. thornei* in inoculation experiments. In: *Proceedings of the Pratylenchus workshop in 9th Biennial APPS Conference*, Hobart, Australia, pp. 9–14.

Vanstone, V.A., Rathjen, A.J., Ware, A.H. and Wheeler, R.D. (1998) Relationship between root lesion nematodes (*Pratylenchus neglectus* and *P. thornei*) and performance of wheat cultivars. *Australian Journal of Experimental Agriculture* 38, 181–188.

Varo Alcala, J. and Tobar Jiménez, A. (1970) Lesions and reactions caused by soil nematodes on the roots of plants. *Xth International Nematology Symposium*, 8–13 September, 1970. European Society of Nematologists, Pescara, Italy.

Vazquez, J.T. (1971) Principales problemas nematológicos en México. *Nematropica* 1(1), 30.

Vega, E. (1978) Vineyard replanting and soil disinfection in Argentina. *Bulletin de l'OIV* 51(566), 250–262.

Venkitesan, T.S. (1976) Studies on the burrowing nematode *Radopholus similis* on pepper and its role in slow wilt disease. PhD thesis. University of Agricultural Science, Bangalore.

Venter, C. (1994) Strategies for the control of the peanut pod nematode on groundnuts in South Africa. In: *Proceedings of the 11th Regional Groundnut Workshop for Southern Africa*, ICRISAT, Mbabane, Swaziland, July 1994.

Venter, C., Aswegen, G., Meyer, A.J. and Waele, D. de. (1995) Histological studies of *Ditylenchus africanus* within peanut pods. *Journal of Nematology* 27(3), 284–291.

Venter, C., De Waele, D. and Meyer, A.J. (1991) Reproductive and damage potential of *Ditylenchus destructor* on peanut. Journal of Nematology 23, 12–19.

Verdejo, S. and Pinochet, J. (1991) Reproduction of the nematode *Zygotylenchus guevarai* (Nemata: Pratylenchidae) in monoxenic cultures. *Revue de Nematologie* 14(2), 291–297.

Verma, M.K. (1991) Cultivation of mustard and rape seed on the management of earcockle disease of wheat. *International Journal of tropical Plant Diseases* 9(1), 91–94.

Vieira, P., Burgermeister, W., Mota, M., Metge, K. and Silva, G. (2007) Lack of genetic variation of *Bursaphelenchus xylophilus* in Portugal revealed by RAPD-PCR analyses. *Journal of Nematology* 39, 118–126.

Vieira, P., van den Eves, A.S., Verma, R., Wantoch, S., Eisenback, J.D. and Kamo, K. (2015) The *Pratylenchuspenetrans* transcriptome as a source for the development of alternative control strategies: mining for putative genes involved in parasitism and evaluation of in planta RNAi. *PLoS ONE* 10(12), e0144674. doi: 10.1371/journal.pone.0144674

Villate, L., Esmenjaud, D., Van Helden, M., Stoeckel, S., and Plantard, O. (2010) Genetic signature of amphimixis allows for the detection and fine scale localization of sexual reproduction events in a mainly parthenogenetic nematode. *Molecular Ecology* 19(5), 856–873.

Vilsoni, F., McClure, M.A. and Butler, L.D. (1976) Occurrence, host range and histopathology of *Radopholus similis* in ginger (*Zingiber officinale*). *Plant Disease Reporter* 60(5), 417–420.

Vincent, B., Altemayer, V., Roux-Morabito, G., Naves, P., Sousa, E. and Lieutier, F. (2008) Competitive interaction between *Bursaphelenchus xylophilus* and the closely related species *Bursaphelenchus mucronatus*. *Nematology* 10, 219–230.

Visser, T. (1959) Observations on the prevalence and control of parasitic eelworms in tea. *Tea Quarterly* 30, 96–107.

Volcy, C. (2011) Past and present of the nematode *Radopholus similis* (Cobb) Thorne with emphasis on Musa: a review. *Agronomía Colombiana* 29(3), 433–440.

Volkers, E.E.L. and Gamboa, A. (1988) Nematodes on plantain. *Boletin Informativo, Manejo Integrado de Plagas* 7, 3–4.

Von Mende, N. and McNamara, D.G. (1995) Biology of the hop cyst nematode *Heterodera humuli*. *Annals of Applied Biology* 126(3), 505–516.

Vovlas, N. and Inserra, R.N. (1977) Histological alterations induced by *Pratylenchus thornei* and *Zygotylenchus guevarai* in the roots of bean. *Italia Agricola* 9, 122–125.

Vovlas, N. and Inserra, R.N. (1985) Single modified food cell induced by *Helicotylenchus pseudorobustus* in corn roots. *Journal of Nematology* 17, 371–373.

Vovlas, N. and Inserra, R.N. (2000) Root-knot nematodes as parasites of citrus. *Proceedings of the International Society of Citriculture* 2, 812–817.

Vovlas, N. and Larizza, A. (1994) Embryogenic patterns and parasitic habits of *Helicotylenchus oleae* and *H. pseudorobustus*. *Afro-Asian Journal of Nematology* 4(1), 17–21.

Vovlas, N., Inserra, R.N. and Lamberti, F. (1976) Observations on the epidemiology and pathogenicity of *Zygotylenchus guevarai*. *Nematologia Mediterranea* 4(2), 183–193.

Vovlas, N., Inserra, R.N. and Martelli, G.P. (1978) Anatomical changes induced by *Xiphinema index* and *Meloidogyne incognita* in the roots of a *Vitis vinifera* X V. rotundifolia hybrid. *Nematologia Mediterranea* 6(1), 67–75.

Vovlas, N., Greco, N. and Di Vito, M. (1985) *Heterodera ciceri* sp. n. (Nematoda: Heteroderidae) on *Cicer arietinum* L. from northern Syria. *Nematologia Mediterranea* 13, 239–252.

Vovlas, N., Frisullo, S., Santos, M.S.N. de A., Abrantes, I.M. de O. and Espirito, S.S.N. (1994) *Ceratocystis paradoxa* and *Helicotylenchus multicinctus* associated with root systems of declining bananas in the Republica Democratica de São Tomé e Principe. *Nematologia Mediterranea* 22(1), 119–121.

Vovlas, N., Nico, A.I., DeLuca, F., DeGiorgi, C. and Castillo, P. (2007) Diagnosis and molecular variability of an Argentinean population of *Nacobbus aberrans* with some observations on histopathology in tomato. *Journal of Nematology* 39, 17–26.

Vovlas, N., Subbotin, S.A., Troccoli, A., Liébanas, G. and Castillo, P. (2008) Description of *Rotylenchus montanus* sp. n. and recognition of *R. jaeni* comb. n. as a separate species with approaches to molecular phylogeny of the genus *Rotylenchus* (Nematoda, Tylenchida). *Zoologica Scripta* 37, 521–537.

Vovlas, N., Troccoli, A. and Rodrigues, C. (1995) Supplementary female morphology and male description of *Helicotylenchus varicaudatus* from banana roots. *Nematologia Mediterranea* 23(1), 93–99.

Vovlas, N., Troccoli, A., Palomares-Rius, J.E., De Luca, F., Liébanas, G., Landa, B.B., Subbotin, S.A. and Castillo, P. (2011) *Ditylenchus gigas* n.sp. parasitizing broad bean: a new stem nematode singled out from the *Ditylenchus dipsaci* species complex using a polyphasic approach with molecular phylogeny. *Plant Pathology* 60, 762–775.

Vrain, T.C. (1993) Restriction fragment length polymorphism separates species of the *Xiphinema americanum* group. *Journal of Nematology* 25, 361–364.

Vrain, T.C., Wakarchuk, A.C., Levesque, A.C. and Hamilton, R.I. (1992) Intraspecific rDNA restriction fragment length polymorphism in the *Xiphinema americanum* group. *Fundamental and Applied Nematology* 15, 563–573.

Vuong, H.H. and Rodriguez, H. (1972) Chemical control of rice nematodes in Madagascar. In: *International Symposium of Nematology* (11th), European Society of Nematologists, Reading, UK, 3–8 September, 1972. Abstracts, pp. 78–79.

Waeyenberge, L. and Moens, M. (2001) *Meloidogyne chitwoodi* and *M. fallax* in Belgium. *Nematologia Mediterranea* 29, 91–97.

Waeyenberge, L., Viaene, N. and Moens, M, (2009a) Species-specific duplex PCR for the detection of *Pratylenchus penetrans*. *Nematology* 11, 847–857.

Waeyenberge, L., Viaene, N., Subbotin, S.A. and Moens, M. (2009b) Molecular identification of *Heterodera* spp., an overview of fifteen years of research. In: Riley, I.T., Nicol, J.M. and Dababat, A.A. (eds) *Cereal Cyst Nematodes: Status, Research and Outlook*. CIMMYT, Ankara, Turkey, pp. 109–114.

Wagner, F. and Domling, M. (1974) Control of the grass seed nematode (*Anguina agrostis*) through cultivation techniques. *Gesunde Pflanzen* 26(7), 148–150.

Walia, R.K., and Bajaj, H.K. (2000) Morphological and morphometrical variations in two races of *Heterodera cajani* Koshy. *Indian Journal of Nematology* 30(2), 124–128.

Walkinshaw, C.G., Griffin, G.D. and Larson, R.H. (1961) *Trichodorus christiei* as a vector of potato corky ringspot (tobacco rattle virus). *Phytopathology* 51, 806–808.

Wallace, H.R. (1961) Chrysanthemum eelworm. In: *National Chrysanthemum Society Year Book*, pp. 106–128.

Walters, S.A. and Barker, K.R. (1994) Current distribution of five major *Meloidogyne* species in the United States. *Plant Disease* 78(8), 772–774.

Wang, L., Wang, Y., Yang, B., Hu, X. and Yu, S. (2001) Studies on relationships of *Bursaphelenchus xylophilus* and *B. mucronatus* by RAPD. *Acta Phytopathologica Sinica* 31, 225–229.

Wang, L.H. (1971) Embryology and life cycle of *Tylenchorhynchus claytoni* Steiner, 1937 (Nematoda: Tylenchoidea). *Journal of Nematology* 3(2), 101–107.

Wang, X., Bosselut, N., Castagnone, C., Voisin, R., Abad, P. and Esmenjaud, D. (2003) Multiplex polymerase chain reaction identification of single individuals of the Longidorid nematodes *Xiphinema index*, *X. diversicaudatum*, *X. vuittenezi*, and *X. italiae* using specific primers from ribosomal genes. *Phytopathology* 93, 160–166.

Wang, Y., Yu, S.F., Li, Y.G., Hu, X.Q. and Pu, W.Q. (2004) Studies on rDNA of *Tylenchulus semipenetrans* and *Pratylenchus* spp. by ITS-RFLP. *Journal of Laiyang Agricultural College*, 21, 151–153.

Wang, Z.Y., Yang-Hong, F. and Ji-MuXiang. (2006) The small grains and erect panicle of rice caused by nematode and control the disease in Jiangsu Province. *Journal of Nanjing Agriculture University* 29(1), 54–56.

Wani, A.H. (2005) Effect of cropping sequences and ploughing on plant parasitic nematodes and plant growth in field. *Indian Journal of Nematology* 35, 63–67.

Webley, D. and Lewis, S. (1977) *Punctodera punctata*. *CIH Descriptions of Plant-parasitic Nematodes*, Set 7, No. 102. Commonwealth Institute of Helminthology, St Albans, UK.

Webster, J.M. and Hooper, D.J. (1968) Serological and morphological studies on the inter- and intraspecific differences of the plant-parasitic nematodes *Heterodera* and *Ditylenchus*. *Parasitology* 58, 879–891.

Webster, J.M., Anderson, R.V., Baillie, D.L., Beckenbach, K., Curran, J. and Rutherford, T. (1990) DNA probes for differentiating isolates of the pinewood nematode species complex. *Revue de Nématologie* 13, 255–263.

Wehunt, E.J. and Edwards, D.I. (1968) *Radopholus similis* and other nematode species on banana. In: Smart, G.C. and Perry, V.G. (eds) *Tropical Nematology*. University of Florida Press, Gainesville, USA, pp. 1–19.

Weischer, B. (1993) Nematode–virus interactions. In: Khan, M.W. (ed.) *Nematode Interactions*. Chapman and Hall, New York, pp. 217–231.

Weldekidan, T., Evans, T.A., Carroll, R.B. and Mulrooney, R.P. (1992) Etiology of soybean severe stunt and some properties of the causal virus. *Plant Disease* 76(7), 747–750.

Wendt, C.D., Swart, A., Vrain, T.C. and Webster, J.M. (1995) *Ditylenchus africanus* sp. n. from South Africa; a morphological and molecular characterization. *Fundamentals of Applied Nematology* 18, 241–250.

Wendt, K.R., Vrain, T.C. and Webster, J.M. (1993) Separation of three species of *Ditylenchus* and some host races of *D. dipsaci* by restriction fragment length polymorphism. *Journal of Nematology* 25, 555–563.

Wendt, K.R., Swart, A., Vrain, T.C. and Webster, J.M. (1995) *Ditylenchus africanus* sp. n. from South Africa; a morphological and molecular characterization. *Fundamental and Applied Nematology* 18(3), 241–250.

Westcott, W., Zehr, E.I., Newall, W.C. and Cain, D.W. (1994) Suitability of Prunus Selections as Hosts for the Ring Nematode (*Criconemella xenoplax*). *Journal of the American Society for Horticultural Science* 119(5).

Wheeler, T.A. and Starr, J.L. (1987) Incidence and economic importance of plant-parasitic nematodes on peanut in Texas. *Peanut Science* 14(2), 94–96.

Whitehead, A.G. (1959) *Hoplolaimus angustulatus* n. sp. (Hoplolaiminae: Tylenchida). *Nematologica* 4, 99–105.

Whitehead, A.G. (1968) Taxonomy of *Meloidogyne* (Nematoda: Heteroderidae) with descriptions of four new species. *Transactions of the Zoological Society of London* 31, 263–401.

Whitehead, A.G. (1969) The distribution of root-knot nematodes (*Meloidogyne* spp.) in tropical Africa. *Nematologica* 15, 315–333.

Whitehead, A.G. and Hooper, D.J. (1970) Needle nematodes (*Longidorus* spp.) and stubby root nematodes (*Trichodorus* spp.) harmful to sugar beet and other field crops in England. *Annals of Applied Biology* 65, 339–350.

Willers, P. (1979) Influence of citrus nematode, *Tylenchulus semipenetrans*, on navel yield in a Sundays River Valley orchard. *Citrus and Subtropical Fruit Journal* 542, 9–10.

Willers, P. (1997) First record of *Meloidogyne mayaguensis* Rammah and Hirschmann, 1988: Heteroderidae on commercial crops in the Mpumalanga province, South Africa. *Inligtingsbulletin – Instituut vir Tropiese en Subtropiese Gewasse* 294, 19–20.

Willers, P. and Grech, N.M. (1986) Pathogenicity of the spiral nematode *Helicotylenchus dihystera* to guava. *Plant Disease* 70(4), 352.

Williams, J.R. (1960) *Studies on the nematode soil fauna of sugar cane fields in Mauritius. 4. Tylenchoidea (partim)*. Occasional Papers, Mauritius Sugarcane Industry Research Institute, Reduit, Mauritius.

Williams, T.D. (1961) The female reproductive system of *Longidorus elongatus* (de Man, 1876) Thorne and Swanger, 1936 (Nematoda: Longidorinae), with a note on the genera Longidorus (Micoletzky, 1922) Thorne and Swanger, 1936, and Xiphinema Cobb, 1913. *Canadian Journal of Zoology* 39, 413–418.

Williams, T.D. (1969) The effect of Formalin, nabam, irrigation and nitrogen on *Heterodera avenae* Woll., *Ophiobolus graminis* Sacc. and the growth of spring wheat. *Annals of Applied Biology* 64, 325–334.

Williams, T.D. and Siddiqi, M.R. (1972) *Heterodera avenae*, *CIH Descriptions of Plant-parasitic Nematodes*, Set 1, No.2. Commonwealth Institute of Helminthology, St Albans, UK.

Williamson, V.M., Caswell Chen, E.P, Westerdahl, B.B., Wu, F.F. and Caryl, G. (1997) A PCR assay to identify and distinguish single juveniles of *Meloidogyne hapla* and *M. chitwoodi*. *Journal of Nematology* 29, 9–15.

Willis, C.B. and Thompson, L.S. (1972) Birdsfoot trefoil cultivars as hosts for root-lesion nematodes and effects of nematodes on yields. *Canadian Journal of Plant Science* 52, 95–101.

Wilski, A. (1972) The control of potato rot eelworm, *Ditylenchus destructor* Thorne, in potato tubers by dipping in Nemafos. *Biuletyn Instytutu Ochrony Roslin* 54, 191–197.

Winfield, A.L. (1974) Observations on the occurrence, pathogenicity and control of *Pratylenchus vulnus*, *P. thornei* and *Xiphinema diversicaudatum* associated with glasshouse roses. *Annals of Applied Biology* 77(3), 297–307.

Winfield, A.L. and Cooke, D.A. (1975) The ecology of *Trichodorus*. In: Lamberti, F., Taylor, C.E. and Seinhorst, J.W. (eds) *Nematode Vectors of Plant Viruses*. Plenum Press, London and New York, pp. 309–340.

Wingfield, M.J. (1983) Transmission of pine wood nematode to cut timber and girdled trees. *Plant Disease* 67(1), 35–37.

Winkelheide, R. and Sturhan, D. (1993) Light-microscopical studies on development and morphology of a bacterium of the genus *Pasteuria* parasitizing *Heterodera goettingiana*. *Zentralblatt für Mikrobiologie* 148(2), 109–116.

Winslow, R.D. and Willis, R.J. (1972) Nematode diseases of potatoes. In: Webster, J.M. (ed.) *Economic Nematology*. Academic Press, New York pp.17—48.

Wiryadiputra, S. (1996) Resistance of *Robusta coffee* to coffee root lesion nematode, *Pratylenchus coffeae*. *Pelita Perkebunan* 12(3), 137–148.

Wishart, J., Phillipa, M.S. and Blok, V.C. (2002) Ribosomal intergenic spacer: a polymerase chain reaction diagnostic for *Meloidogyne chitwoodi*, *M. fallax* and *M. hapla*. *Phytopathology* 92(8), 884–892.

Wollenweber, H.W. (1923) *Krankheiten und Beschädigungen der Kartoffel*. Arbeiten des Forschungsinstitutes für Kartoffelbau, Heft 7, Paul Parey, Berlin.

Wood, T.G., Smith, R.W., Johnson, R.A. and Komolafe, P.O. (1980) Termite damage and crop loss studies in Nigeria – pre-harvest losses to yams due to termites and other soil pests. *Tropical Pest Management* 26(4), 355–370.

Wopjtowicz, M.R. and Szczygiel, A. (1990) Effect of systemic nematicides on spiral nematode and on *Asparagus sprengeri* yields. *Zeszyty Problemowe Postępów Nauk Rolniczych* 391, 123–131.

Wouts, W.M. (1985) Phylogenetic classification of the family Heteroderidae (Nematoda: Tylenchida). *Systematic Parasitology* 7, 295–328.

Wrather, J.A., Anderson, T.R., Arsyad, D.M., Tan, Y., Gai, Y.J., Ploper, L.D. *et al.* (2001) Soybean disease loss estimates for the top 10 soybean-producing countries in 1998. *Canadian Journal of Plant Pathology* 23, 115–121.

Wrather, J.A., Koenning, S.R., Anderson, T.R. (2003) Effect of diseases on soybean yields in the United States and Ontario (1999 to 2002). *Plant Health Progress*, March:0–16.

Wyss, U. (1970) Zur bestimmung von Trichodorus-Arten. *Pflanzenkrankheiten und Pflanzenschutz* 77, 506–514.

Wyss, U, (1978) Root and cell response to feeding by *Xiphinema index*. *Nematologica* 24(2), 159–166.

Xu, C.-L., Li, Y., Xie, H., Huang, X., Wu, W.-J., Yu, L. and Wang, D.-W. (2014) Morphological and karyotypic differences within and among populations of *Radopholus similis*. *ZooKeys* 444, 69–93. doi: 10.3897/zookeys.444.8186.

Xu, J., Liu, P, Meng, Q. and Long, H. (2004) Characterisation of *Meloidogyne* species from China using isozyme phenotypes and amplified mitochondrial DNA restriction fragment length polymorphism. *European Journal of Plant Protection* 110, 309–315.

Xu, Y., Ye, W., Wang, J. and Zhao, Z. (2018) Morphological and molecular characterisation of *Longidorus pinus* sp. n. (Nematoda: Longidoridae) from China and a key to known species of *Longidorus* in China. *Nematology* 20(7), 617–639.

Xu, Y.-M. and Zhao, Z.-Q. (2019) *Longidoridae and Trichodoridae (Nematoda: Dorylaimida and Triplonchida)*. Fauna of New Zealand, No. 79. Landcare Research, Lincoln, New Zealand.

Yamada, E. and Takakura, S. (1987) Ecological investigations on the strawberry nematode, *Aphelenchoides fragarip* on lilies. *Japanese Journal of Nematology* 17(12), 1–7.

Yamazaki, K., Okabe, M., Aihara, T. and Yuhara, I. (1981) Studies on injury of azalea in continuous cropping field. I. Distribution of *Tylenchorhynchus claytoni* in azalea fields in Kanagawa prefecture and its control. *Bulletin of the Kanagawa Horticultural Experiment Station* 28, 73–83

Yan, G.P. and Smiley, R.W. (2010) Distinguishing *Heterodera filipjevi* and *H. avenae* using polymerase chain reaction-restriction fragment length polymorphism and cyst morphology. *Phytopathology* 100, 216–224.

Yan, G.P., Smiley, R.W., Okubara, P.A., Skantar, A., Easley, S.A., Sheedy, J.G. and Thompson, A.L. (2008) Detection and discrimination of *Pratylenchus neglectus* and *P. thornei* in DNA extracts from soil. *Plant Disease* 92, 1480–1487.

Yan, G.P., Smiley, R.W. and Okubara, P.A. (2012) Detection and quantification of *Pratylenchus thornei* in DNA extracted from soil using real-time PCR. *Phytopathology* 102, 14–22.

Yan, G.P., Smiley, R.W., Okubara, P.A., Skantar, A.M. and Reardon, C.L. (2013) Developing a real-time PCR assay for detection and quantification of *Pratylenchus neglectus* in soil. *Plant Disease* 97, 757–764.

Yan, G.P., Plaisance, A., Huang, D., Gudmestad, N.C. and Handoo, Z.A. (2016) First report of the root-lesion nematode *Pratylenchus scribneri* infecting potato in North Dakota. *Plant Disease* 100(5), 1023.

Yang, B., Wang, Q. and Feng, R. (1988) *Meloidogyne kongi* n. sp. (Nematoda: Meloidogynidae) a root-knot nematode parasitizing citrus sp. in Guangxi, China. *Journal of Guangxi Agricultural College* 7 (3), 1–9.

Yang, B., Hu, K., Chen, H. and Zhu, W. (1990) A new species of root-knot nematode *Meloidogyne jiangyangensis* n. sp. parasitizing mandarin orange. *Acta Phytopathological Sinica* 20, 259–264.

Ye, W. (2012) Development of primetime-real-time PCR for species identification of soybean cyst nematode (*Heterodera glycines* Ichinohe, 1952) in North Carolina. *Journal of Nematology* 44, 284–290.

Ye, W. and Giblin-Davis, R.M. (2013) Molecular characterization and development of real-time PCR assay for pine-wood nematode *Bursaphelenchus xylophilus* (Nematoda: Parasitaphelenchidae). *PLoS ONE* 8(11), e78804. doi: 10.1371/journal.pone.0078804.

Yeates, G.W., Ferris, H., Moens, T. and van der Putten, W.H. (2009) The role of nematodes in ecosystems. In: M.J. Wilson and T. Kakouli-Duarte (eds) *Nematodes as Environmental Indicators*. CABI, Wallingford, UK, pp. 1–44

Yik, C.P. and Birchfield, W. (1979) Host studies and reactions of rice cultivars to *Meloidogyne graminicola*. *Phytopathology* 69(5), 497–499.

Young, H.K. and Seung, H.O. (1995) In vitro culture and factors affecting population changes of *Ditylenchus destructor* of ginseng. *Korean Journal of Plant Pathology* 11, 39–46.

Zaki, F.A. and Bhatti, D.S. (1986) Pathogenicity of pigeon-pea cystnematode, *Heterodera cajani* Koshy, 1967 on some pulse crops. *Indian Journal of Nematology* 16(1), 30–35.

Zamora, P., Rodriguez, V., Renedo, F., Sanz, A.V., Domínguez, J.C., Pérez-Escolar, G. *et al.* (2015) First report of *Bursaphelenchus xylophilus* causing pine wilt disease on *Pinus radiata* in Spain. *Disease Notes* 99(10), 1449. doi: 10.1094/PDIS-03-15-0252-PDN

Zavaleta, M.E. and Gomez, R.O. (1995) Effect of *Tagetes erecta* L.–tomato (*Lycopersicon esculentum* Mill.) intercropping on some tomato pests. *Fitopatología* 30(1), 35–46.

Zeng, Y., Ye, W., Yang, L., Huang, Y., Zhao, K., Zhang, Z., Liang, H. and Kerns, J. (2014) Morphological and molecular characterization of two isolates of *Paratrichodorus porosus* from Shenzhen, China. *Helminthologia* 51(4), 323–330.

Zepp, A.L. and Szczygiel, A. (1986) Pathogenicity of five populations of *Pratylenchus penetrans* to three fruit tree seedling rootstocks. *Zeszyty Problemowe Postepow Nauk Rolniczych* 323, 73–90.

Zhang, K.Y., Liu, H., Sun, J., Liu, J.R., Fei, K., Zhang, C.X. *et al.* (2008) Molecular phylogeny of geographical isolates of *Bursaphelenchus xylophilus*: implications on the origin and spread of this species in China and worldwide. *Journal of Nematology* 40, 127–137.

Zhang, L., Liao, J. and Feng, Z. (2001) Sequencing and PCR-SSCP analysis of ribosomal DNA of *Bursaphelenchus* nematodes. *Acta Phytopathologica Sinica* 31, 84–89.

Zhang, L., Kong, F. and Yang, B. (2002) Intra and interspecific variation in *Bursaphelenchus xylophilus* and *B. mucronatus* revealed by mtDNA polymorphism. *Forest Research* 15, 7–12.

Zhang, Q., Bai, G., Yang, W., Li, H and Xiong, H. (2006) Pathogenic cellulase assay of pine wilt disease and immunological localisation. *Bioscience Biotechnology Biochemistry* 70, 2727–2732.

Zhang, S. (1993) *Meloidogyne mingnanica* n. sp. (Meloidogynidae) parasitizing Citrus in China. *Journal of Fujian Agricultural University (Natural Sciences Edition)* 22, 69–76.

Zhang, S., Gao, R. and Weng, Z. (1990) *Meloidogyne citri* n. sp. (Meloidogynidae), a new root-knot nematode parasitizing citrus in China. *Journal of Fujian Agricultural College* 19, 305–311.

Zhang, S.L., Liub, G.K., Janssen, T., Zhang, S.S., Xiaoa, S., Li, M., Couvreur, S.T. and Bert, W. (2014) A new stem nematode associated with peanut pod rot in China: morphological and molecular characterization of *Ditylenchus arachis* n. sp. (Nematoda: Anguinidae). *Plant Pathology* 63, 1193–1206.

Zhang, X.H. and Wang, H.M. (1989) The occurrence of *Aphelenchoides fragariae* on strawberry seedlings in Tianjin port. *Plant Quarantine* 3(1), 69–70.

Zhang, X.S., Zhang, G.Z., Zhang, H.W., Peng, D.W., Du, Z.J. and Tang, C.Y. (2009) Effects of soil treatments on the quantitative dynamics of rhizospheric nematodes and their control efficacy to root rot of American ginseng caused by *Ditylenchus destructor*. *Acta Phytopathologica Sinica* 39(5), 555–560.

Zhao, B.G., Futai, K., Sutherland, J.R. and Takeuchi, Y. (2008) *Pine Wilt Disease*. Springer, Berlin.

Zhao, J., Yu, S., Yao, J., Lin, C., Ding, D. and Wang, H. (2004) PWN risk assessment in Huangshan Scenic Área II. Monitoring of nematode carried by pine borer beetles. *Forest Research* 17, 72–76.

Zhen, F., Agudelo, P. and Gerard, P. (2012) A protocol for assessing resistance to *Aphelenchoides fragariae* in hosta cultivars. *Plant Disease* 96, 1438–1444.

Zheng, J., Xu, J., Wu, Y. and Li, D. (1998) RAPD fingerprinting on inter- and infra-species of *Bursaphelenchus xylophilus* and *B. mucronatus*. *Journal of Zhejiang Agricultural University* 24, 597–601

Zheng, J., Subbotin, S.A., Waeyenberge, L. and Moens, M. (2000) Molecular characterisation of Chinese *Heterodera glycines* and *H. avenae* populations based on RFLPs and sequences of rDNA-ITS regions. *Russian Journal of Nematology* 8, 109–113.

Zheng, J., Subbotin, S.A., He, S., Gu, J. and Moens, M. (2003) Molecular characterisation of some Asian isolates of *Bursaphelenchus xylophilus* and *B. mucronatus* using PCR-RFLPs and sequences of ribosomal DNA. *Russian Journal of Nematology* 11, 17–22.

Zheng, L., Lin, M. and Zheng M. (1990) Occurrence and identification of a new disease of the citrus, Donghai root-knot nematode- *Meloidogyne donghaiensis* sp. nov. in coast soil of Fujian, in China. *Journal of Fujian Academy of Agricultural Sciences* 5 (2), 56–63.

Zhou, Q.-J., Cai, Y., Gu, J.-F., Wang, X. and Chen, J. (2017) Rapid and sensitive detection of *Meloidogyne mali* by loop-mediated isothermal amplification combined with a lateral flow dipstick. *European Journal of Plant Pathology* 148, 755–769.

Zhu, W.S., Chen, H., Lan, Y.Y. and Qian, K.M. (1992) Study on citrus nematode disease. *Acta Phytopathologica Sinica* 22, 29–33.

Zijlstra, C. (2000) Reliable identification of the quarantine root-knot nematodes *Meloidogyne chitwoodi* and *M. fallax* by PCR-based techniques. *Bulletin OEPP* 30 (3/4), 575–579.

Zijlstra, C., Donkers-Venne, D.T.H.M. and Fargette, M. (2000) Identification of *Meloidogyne incognita*, *M. javanica* and *M. arenaria* using sequence characterised amplified region (SCAR) based PCR assays. *Nematology* 2, 847–853.

Zijlstra, C., Van Hoof, R. and Donkers-Venne, D. (2004) A PCR test to detect the cereal root-knot nematode *Meloidogyne naasi*. *European Journal of Plant Pathology* 110, 855–860.

Zouhar, M., Marek, M., Licinio, J. and Ryšánek, P. (2002) Using point mutations in rDNA for differentiation of bioraces of *Ditylenchus dipsaci* from the Czech Republic. *Plant Protection Science* 38 (Special 2), 358–360.

Zouhar, M., Marek, M., Douda, O., Mazáková, J. and Ryšánek, P. (2007) Conversion of sequence characterized amplified region (SCAR) bands into high-throughput DNA markers based on RAPD technique for detection of the stem nematode *Ditylenchus dipsaci* in crucial plant hosts. *Plant Soil and Environment* 53, 97–104.

Zuckerman, B.M. and Strich-Harari, D. (1963) The life stages of *Helicotylenchus multicinctus* (Cobb) in banana roots. *Nematologica* 9, 347–353.

Zuckerman, B.M., Dicklow, M.B. and Acosta, N. (1993) A strain of *Bacillus thuringiensis* for the control of plant-parasitic nematodes. *Biocontrol Science and Technology* 3(1), 41–46.

Zunke, U. (1990a) Observation on the invasion and endoparasitic behavior of the root lesion nematode *Pratylenchus penetrans*. *Journal of Nematology* 22, 309–320.

Zunke, U. (1990b) Ectoparasitic feeding behavior of the root lesion nematode, *Pratylenchus penetrans*, on root hairs of different host plants. *Revue de Nématologie* 13, 331–337.

Zwieg, R. and Hudelson, B. (2010) *Tobacco Rattle*. University of Wisconsin garden fact sheets. Available at: https://pddc.wisc.edu/2015/08/18/tobacco-rattle/ (accessed 31 July 2010)

Zyubin, B.N. (1968) *Peronospora* infection of the poppy and plant nematodes. In: Karakeev, K.K. *et al.* (eds) *Helminths of Animals and Plants in Kirgizia*. Izdatelvo 'Ilim.', Frunze, Kyrgyzstan, pp. 141–146.

Index

Note: Page numbers in **bold** type refer to **figures**; Page numbers in *italic* type refer to *tables*

Index

Note: Page numbers in **bold** type refer to **figures**; Page numbers in *italic* type refer to *tables*